The Physics of Neutrino Interactions

This advanced text discusses the fundamental concepts of neutrinos, their properties and interactions with matter, presenting a theoretical framework for describing relativistic particles. It provides a pedagogical description of the field theory of neutrinos, necessary to understand the standard model (SM) of electroweak interactions, and neutrino scattering from leptons and nucleons. Applications of neutrino scattering processes from the nucleons and nuclei are discussed in detail. Nuclear-medium effects in quasielastic scattering, and inelastic and deep inelastic scattering are also covered in depth. A separate chapter on neutrinos in astrophysics highlights the applications of various neutrino processes in the understanding of the universe and its evolution. The text introduces the subject of neutrino oscillations and highlights the need for beyond the standard model (BSM) physics. This topical book will stimulate new ideas and avenues for research, and will form a valuable resource for advanced graduate students and academic researchers in the fields of particle physics and nuclear physics.

M. Sajjad Athar is Professor in the Department of Physics, Aligarh Muslim University, India. He is currently a member of the IUPAP-Neutrino panel and a member of the NuSTEC board. He is an active collaborator in the MINERvA experiment at Fermilab, USA. He is also a member of the proposed DUNE experiment at Fermilab.

S. K. Singh has been Professor of Physics at Aligarh Muslim University, India, and visiting scientist at the University of Mainz, Germany; the University of Valencia, Spain; and the International Centre of Theoretical Physics, Italy. His work on neutrino reactions has been used to analyze the first neutrino experiments from deuterium at Argonne National Laboratory and Brookhaven National Laboratory, USA.

The Physics of Neutrino Interactions

M. Sajjad Athar

S. K. Singh

CAMBRIDGE
UNIVERSITY PRESS

CAMBRIDGE
UNIVERSITY PRESS

University Printing House, Cambridge CB2 8BS, United Kingdom

One Liberty Plaza, 20th Floor, New York, NY 10006, USA

477 Williamstown Road, Port Melbourne, VIC 3207, Australia

314 to 321, 3rd Floor, Plot No.3, Splendor Forum, Jasola District Centre, New Delhi 110025, India

79 Anson Road, #06–04/06, Singapore 079906

Cambridge University Press is part of the University of Cambridge.

It furthers the University's mission by disseminating knowledge in the pursuit of education, learning and research at the highest international levels of excellence.

www.cambridge.org
Information on this title: www.cambridge.org/9781108489065

First published 2020

Printed in India by Nutech Print Services, New Delhi 110020

A catalogue record for this publication is available from the British Library

Library of Congress Cataloging-in-Publication Data

Names: Athar, M. Sajjad, 1970- author. | Singh, S. K., 1946- author.
Title: The physics of neutrino interactions / M. Sajjad Athar, S.K. Singh.
Description: Cambridge ; New York,NY : Cambridge University Press, 2020. |
    Includes bibliographical references and index.
Identifiers: LCCN 2020001152 (print) | LCCN 2020001153 (ebook) | ISBN
    9781108489065 (hardback) | ISBN 9781108773836 (ebook)
Subjects: LCSH: Neutrino interactions.
Classification: LCC QC793.5.N428 A84 2020 (print) | LCC QC793.5.N428
    (ebook) | DDC 539.7/215—dc23
LC record available at https://lccn.loc.gov/2020001152
LC ebook record available at https://lccn.loc.gov/2020001153

ISBN 978-1-108-48906-5 Hardback

To
Fouzia Ahmed (wife of MSA)
and
Shashi Singh (wife of SKS)

for their patience, love and support

Contents

Figures

Tables

Preface

The need for writing a self-contained comprehensive book on the physics of neutrino interactions had been in our minds for a long time, while teaching various graduate courses in high energy physics and nuclear physics and conducting research in the field of neutrino physics at the Aligarh Muslim University. We also realized the need for such a book while attending many topical workshops, conferences, and short-term schools like NuFact, NuInt, NuSTEC, etc., held in the USA, Europe, Japan, and elsewhere in the area of neutrino physics and while responding to questions asked by the young researchers in many formal and informal discussions. The aforementioned scientific events bring together research students and senior scientists working on various aspects of neutrino physics common to nuclear physics, particle physics, and astrophysics, which make the subject interdisciplinary. In recent times, the research activity in the field of neutrino physics, around the world, and its applications in the other areas of physics has attracted a large number of students to this field. It was, therefore, felt that this is an appropriate time to write a book on the physics of neutrino interactions focusing on introducing the basic mathematical and physical concepts and methods with the help of simple examples to illustrate the calculation of various neutrino processes relevant for applications in particle physics, nuclear physics, and astrophysics, for the benefit of all those interested in learning the subject.

The main aim of the book is to present a pedagogical account of the physics of neutrino interactions, with balance among its theoretical and experimental aspects, for describing various neutrino scattering processes from leptons, nucleons, and nuclei used in studying neutrino properties like its mass, charge, magnetic moment, and the newly discovered phenomenon of neutrino mixing and oscillations. The book is intended primarily for graduate students and young post-doctoral research scientists working in neutrino physics but it can also be used by advanced undergraduates who have some exposure to basic courses in special theory of relativity, quantum mechanics, nuclear physics, particle physics, and are interested in neutrino physics.

There is hardly a single book which discusses all the above-mentioned aspects of neutrino physics in one place. The reason is the diversity of the various aspects of neutrino physics in their origin, development, and technical details, which have been discussed and formulated at different times and are described in detail in different books devoted to a particular aspect of neutrino physics. For example, there are many books on V-A theory of weak interactions and neutrino, standard model of electroweak interactions and neutrino, and recently many books on the physics of neutrino mass, mixing and oscillations. Most of these books are excellent, but they mainly focus on the particle physics aspect of neutrinos and their interactions.

However, looking back at the origin of neutrino studies and the development of neutrino physics over the last 90 years, we see that neutrino physics has truly become an interdisciplinary subject involving particle physics, nuclear physics, and astrophysics. The basic reason for writing this book is to present neutrino physics as an interdisciplinary subject and describe it in a self-contained pedagogical manner with conceptually simple but technically rigorous treatment of various topics, with appropriate historical perspective, to keep the reader interested throughout the book as he/she goes through various topics like the mathematical preliminaries on relativistic field theory and local gauge field theories and then to the standard model and its applications to various neutrino scattering processes from leptons, nucleons, and nuclei, and finally the need for physics beyond the standard model. The book is based on the actual teaching material used by us in graduate courses at the Aligarh Muslim University for many years and also on the research work done by us during our collaboration with the experimental physicists working on the neutrino oscillation experiments at K2K, MiniBooNE, MINERvA, T2K, and DUNE.

After introducing various aspects of neutrino physics in Chapter 1, we describe briefly in Chapters 2–4 the mathematical preliminaries needed to follow the subject assuming no prior knowledge of relativistic quantum field theory in order to make the book self-contained. A historical approach is then followed in Chapters 5 and 6 to describe the phenomenological $V-A$ theory of weak interactions and its development starting from the Fermi theory of β-decay to the Cabibbo–Kobayashi–Maskawa (CKM) formulation of weak interactions of three-flavors of quark and lepton doublets $(u\ d')$, $(c\ s')$, $(t\ b')$ and $(\nu_e\ e)$, $(\nu_\mu\ \mu)$ and $(\nu_\tau\ \tau)$, and their antiparticles, where we also discuss the limitations of the phenomenological $V-A$ theory mediated by the massive intermediate vector bosons (IVB). The theoretical attempts for finding a convergent and renormalizable theory of weak interactions mediated by the massive intermediate vector bosons led to the concept of local gauge field theory with continuous symmetry which are spontaneously broken by the Higgs mechanism to generate mass. The concept was used by Weinberg and Salam to create a theory of leptons and their electroweak interactions which was later extended to quarks using Glashow–Iliopoulos–Maiani (GIM) mechanism to formulate the standard model (SM) for the unified theory of electromagnetic and weak interactions. All these developments are described in Chapters 7 and 8.

The standard model is then applied to perform calculations of (anti)neutrino scattering from leptons in Chapter 9 and various other scattering processes like quasielastic (QE), inelastic (IE), and deep inelastic scattering (DIS) from nucleons in Chapters 10–13, with essential details relegated to relevant appendices. Most of the (anti)neutrino scattering experiments for the cross section measurements at high energies have been done from the nuclear targets in the past while the recent experiments in the low and intermediate energy regions of few MeV to a few GeV are also being done in nuclear targets in context of neutrino oscillation experiments, where nuclear medium effects play an important role. In view of this we devote the next three chapters, 14–16, to discuss the (anti)neutrino scattering process of QE, IE, and DIS from the nuclear targets in some detail. The next two chapters, 17 and 18, are then used to discuss the various sources of neutrinos, their energy distribution, fluxes of neutrinos including the techniques used for their detection and to introduce the physics and phenomenology of neutrino mixing and oscillations and their mathematical formulation along with the present status of the progress made in understanding, theoretically as well as experimentally, the various parameters like the mass squared differences of neutrinos Δm_{ij}^2, the mixing angles $\theta_{ij}(i < j = 1-3)$, and the CP violating phase δ used for describing the phenomenology of neutrino oscillations.

Neutrinos are known to play very important role in astrophysics and cosmology, and there are excellent books written on the subject. However, we felt it useful to give an essence of the important role neutrinos play in astrophysics to motivate student's interest in this field and Chapter 19 is devoted to this topic highlighting the physics aspects without technical details. Notwithstanding the success of the standard model in describing the physics of neutrino interactions, the existence of neutrino mass, mixing and phenomenon of neutrino oscillation demonstrates that there is physics beyond the standard model (BSM). In fact there are many other phenomenon which imply the existence of physics beyond the standard model. We attempt to describe some of them in Chapter 20 and emphasize the need to make efforts to understand, theoretically as well as experimentally, BSM physics.

In summary the special features of the book are:

- It is comprehensive, self-contained, and requires only a basic knowledge of special theory of relativity and quantum mechanics, with some exposure to nuclear and particle physics at the undergraduate level, to follow the subject.

- It gives a pedagogical description of neutrino properties and neutrino interactions starting from the Fermi theory of β-decay to the phenomenological V–A theory formulation by Cabibbo–Kobayashi–Maskawa applicable to the three flavours of quark and lepton doublets.

- It describes the role of local gauge field theories in formulating the fundamental interactions and explains in detail the concept of spontaneous breaking of local gauge symmetry and generation of mass using Higgs mechanism in the formulation of standard model of electroweak interaction of quarks and leptons as done by Glashow, Salam, and Weinberg.

- It discusses in detail the applications of the standard model (SM) to neutrino scattering from leptons and hadrons. Processes like QE, IE, and DIS from nucleons and nuclei are presented, emphasizing the importance of nuclear medium effects.

- It introduces the physics of neutrino oscillations at a basic level including the matter enhancement due to the Mikheyev–Smirnov–Wolfenstein effect in two-flavor and three-flavor oscillation and discusses the present status of the subject with future prospects for the observation of CP violation, mass hierarchy, and sterile neutrinos.

- It highlights the role of neutrinos in astrophysics and also emphasizes the need for physics beyond the standard model (BSM) by describing the phenomenon of neutrinoless double beta decay (NDBD) and some other phenomenon like the existence of lepton flavor violation (LFV) and flavor changing neutral currents (FCNC).

Acknowledgments

A book of this length is not possible without the contribution of many people like the teachers, collaborators and students in our academic career. It is pleasure to express sincere thanks to our research students Farhana Zaidi, Atika Fatima, Vaniya Ansari, Prameet Gaur, Sayeed Akhter, Zubair Ahmad Dar, Faiza Akbar, and Anuj Upadhyay for the typesetting, with patience and skill; rechecking the mathematical calculations and doing the proof reading. However, mistakes and typos, that might slip through are completely owned by us. Special thanks to Atika Fatima for taking out time and helping us in the completion of several chapters. S. K. Singh would like to acknowledge with gratitude the contribution of his teacher at the Carnegie Mellon University (Late) Lincoln Wolfenstein for introducing him to the fascinating world of neutrino physics. We are also thankful to our research collaborators in the past with whom we did several collaborative papers in the area of neutrino physics and electroweak interactions like Eulogio Oset, Manuel J. Vicente Vacas and Luis Alvarez Ruso from the University of Valencia, Spain; Hartmuth Arenhovel and Dieter Drechsel, University of Mainz, Germany; Ignacio Ruiz Simo, University of Granada, Spain; Takaaki Kajita and Morihiro Honda, University of Tokyo, Japan; and Huma Haider, Rafi Alam and many others at the Aligarh Muslim University. We are thankful to Jorge Morfin, Scientist Emeritus, Fermi National Accelerator Laboratory, for being instrumental in organising NuSTEC (Neutrino Scattering Theory Experiment Collaboration) workshops and schools where the interactions with the young minds resulted in the need for writing a book on this topic.

M. Sajjad Athar is also thankful to his parents and specially to his father Professor Mohammad Ozair who has, since the beginning, motivated him to do work with dedication and honesty. He is also thankful to his teenaged sons Zain and Saud for providing a conducive atmosphere without any complaints and for their love. SKS is thankful to his daughters Asmita and Tanaya for their understanding and active cooperation enabling him to complete this work in time.

We will like to acknowledge with thanks Sudhir Jain, NPD, BARC for encouraging us to write this book and requesting Mr Gauravjeet Singh Reen then at CUP to take initiatives. We are thankful to Ms Taranpreet Kaur and Mr Reen, for their logistic support and commitment in publishing the book from CUP.

Chapter 1

Neutrino Properties and Its Interactions

1.1 Historical Introduction to Neutrinos

1.1.1 Neutrino hypothesis

Beginning with the neutrino hypothesis proposed by Pauli in 1930, the story of the neutrino has been an amazing one [1]. It all started with a letter written by Pauli to the participants of a nuclear physics conference in Tubingen, Germany, on December 4, 1930 [1], in which he proposed the existence of a new neutral weakly interacting particle of spin $\frac{1}{2}$ and called it "neutron" as a "verzweifelten Ausweg" (desperate remedy), to explain the two outstanding problems in contemporary nuclear physics which posed major difficulties with respect to the scientists' theoretical interpretations. These two problems were related with the puzzle of energy conservation in β-decays of nuclei [2, 3], discovered by Chadwick in 1914 [4], and anomalies in understanding the spin–statistics relation in the case of ^{14}N and ^{6}Li nuclei within the context of the nuclear structure model that was prevalent in the early decades of the twentieth century [5, 6] in which electrons and protons were considered to be nuclear constituents.

This proposed 'verzweifelten Ausweg' was considered so tentative by Pauli himself that he postponed its scientific publication by almost three years. Today, neutrinos, starting from being a mere theoretical idea of an undetectable particle, are known to be the most abundant particles in the universe after photons, being present almost everywhere with a number density of approximately 330/cm^3 pan universe. The history of the progress of our understanding of the physics of neutrinos is full of surprises; neutrinos continue to challenge our expectations regarding the validity of certain symmetry principles and conservation laws in particle physics. The study of neutrinos and their interaction with matter has made many important contributions to our present knowledge of physics, which are highlighted by the fact that ten Nobel Prizes have been awarded for physics discoveries in topics either directly in the field of neutrino physics or in the topics in which the role of neutrino physics has been very crucial.

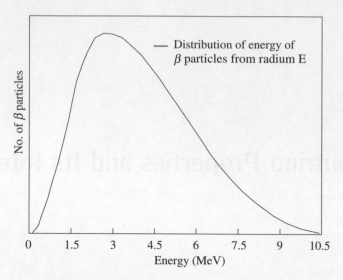

Figure 1.1 Continuous β-decay spectrum of RaE [15].

In this chapter, we will provide a historical introduction to the development of our understanding of neutrinos and their properties as they have emerged from the theoretical and experimental studies made over the last 90 years.

1.1.2 The problem of energy conservation in β-decays of nuclei

Almost three years after the discovery of nuclear radioactivity by Becquerel in 1896 [7], certain new radiation were discovered and studied by Curie, Rutherford, and others [8]. Further studies by Rutherford revealed that these radiations are of two types; Rutherford named them as α-radiation, which are readily absorbed and β-radiation which are more penetrative [9]. One more type of radiation, the γ-radiation was discovered a year later in 1900 by Villard [10]. It was realized quite early that β-radiation were electrons as identified from the study of cathode rays [11]. After the discovery of the nucleus by Rutherford in 1911 [12], it was Bohr [13] who established that β-radioactivity is a nuclear process like α- and γ-radioactivity and β-ray electrons originated from the nuclei. Further investigations by Chadwick [4] established in 1914, that the energy spectrum of the β-rays coming from nuclear β-decay was continuous. A typical β-ray spectrum of electrons from the nuclear β-decay of RaE is shown in Figure 1.1. The continuous energy spectrum of β-electrons is in complete contrast with the energy spectrum of α-decays and γ-decays of atomic nuclei which appear as discrete spectra. The discreteness of the energy spectra of α- and γ-radiation was quite consistent with the quantum description of nuclei which predicted discrete nuclear energy levels that would emit radiation of a fixed energy when de-excited to lower energy levels by the emission of nuclear radiation like α- and γ-rays. In this scenario of the nuclear energy levels being discrete, the continuous energy spectrum of β-electrons was argued by Meitner [14] to be due to the broadening of the discrete energy of primary electrons emitted in the β-decay, caused by secondary processes leading to continuous energy loss of the primary electrons as they travel through the nucleus. The other explanation given by Ellis [15] was to assume that the primary electrons emitted in the β-decay have an intrinsically continuous spectrum. This explanation of the phenomenon of primary electrons being emitted with a continuous energy spectrum posed difficulties with respect to its theoretical interpretation in the context of contemporary knowledge of the nuclear structure which, as explained earlier, described nuclear energy levels to be discrete according to quantum mechanics; the phenomenon seemed to violate the law of conservation of energy. A theoretical understanding of the nuclear β-decay depends crucially on whether the continuous energy spectrum of β-electrons is of the primary electrons or a result of the secondary processes suffered by the primary electrons, emitted in the β-decay, during their passage through the nuclear medium. This dilemma was resolved by making calorimetric measurements of the absolute heat (energy) in the absorption of β-electrons coming from the decay of RaE (^{210}Bi-nucleus) in an experiment performed by Ellis and Wooster [15]. The calorimetric measurement of the energy should result in the average energy of β-electrons if the observed spectrum was due to primary electrons according to Ellis, or to the maximum energy of the electron, if it was due to secondary processes according to Meitner.

Ellis and Wooster [15] reported the heat measurement equivalent to be 344±10% keV, confirmed by the latter measurement of 332±6% keV by Meitner and Orthmann [16] which corresponds to the average energy of the electrons and not to the maximum energy of the electrons corresponding to the spectrum as shown in Figure 1.1. This confirmed the primary origin of the continuous spectrum of β-electrons. After these experiments, it was established that the continuous energy spectrum of the electrons corresponds to the primary electrons emitted in the β-decays of nuclei. There were two very unconventional theoretical interpretations proposed to explain the continuous energy spectrum of β-decay electrons by Pauli [1] and Bohr [2], respectively. They are as follows:

1. Pauli [1] proposed that the conservation of energy holds exactly in the β-decay processes but a very penetrating neutral spin $\frac{1}{2}$ particle was emitted together with the electron.

2. Bohr [2] proposed that the conservation of energy is not exact but only statistical in interactions responsible for β-decays.

The idea of nonconservation of energy was not supported by the further developments in the study of β-decays of nuclei; therefore, Pauli's proposal was accepted by the physics community as the appropriate solution of the continuous energy problem in nuclear β-decays.

1.1.3　Anomalies in the spin–statistics relation for nuclei

In the first few decades of the twentieth century, it was generally assumed that the nuclei were made up of protons and electrons which were the only known particles at the time. In this picture of the nuclear structure, ^{14}N with charge number 7 and mass number 14 should have 14 protons and 7 electrons leading to a half integral spin and obey Fermi statistics for ^{14}N. However, Kronig [5], and Heitler and Herzberg [6] showed, using the molecular band spectra of ^{14}N, that it has spin 1 and satisfies Bose statistics.

Similar examples were found later; for example, ^{6}Li with 6 protons and 3 electrons and deuteron ^{2}H with 2 protons and 1 electron both should have a half integral spin and follow Fermi statistics according to the proton–electron model of the nucleus but were found to have spin 1 with Bose statistics. This anomalous situation in describing the nuclear structure of ^{14}N, ^{6}Li, and ^{2}H was resolved with the presence of another nuclear constituent with neutral charge and spin $\frac{1}{2}$ in Pauli's proposal. Moreover, the observation in nuclear β-decays that if the initial nucleus had integer/half integer spin then the final nucleus also had integer/half integer spin, could also be explained with the presence of two spin $\frac{1}{2}$ particles in addition to the proton in the β-decay processes, which was otherwise not possible in the proton–electron model of the nucleus.

1.1.4 Pauli's neutron/neutrino vs. Fermi's neutrino

A closer reading of Pauli's letter [1] proposing the new particle makes it clear that Pauli's neutral particle was not exactly the neutrino as we know it today. For this purpose, the original letter translated by Riesselmann has been reproduced here:

Dear Radioactive Ladies and Gentlemen[,]

As the bearer of these lines, to whom I graciously ask you to listen, will explain to you in more detail, because of the 'wrong statistics of the N and Li-6 nuclei and the continuous beta spectrum, I have hit upon a desperate remedy to save the 'exchange theorem' (1) of statistics and the law of conservation of energy. Namely, the possibility that in the nuclei there could exist electrically neutral particles, which I will call neutrons, that have spin $\frac{1}{2}$ and obey the exclusion principle and that further differ from light quanta in that they do not travel with the velocity of light. The mass of the neutrons should be of the same order of magnitude as the electron mass and in any event not larger than 0.01 proton mass. The continuous beta spectrum would then make sense with the assumption that in beta decay, in addition to the electron, a neutron is emitted such that the sum of the energies of neutron and electron is constant.

Now it is also a question of which forces act upon neutrons. For me, the most likely model for the neutron seems to be, for wave-mechanical reasons (the bearer of these lines knows more), that the neutron at rest is a magnetic dipole with a certain moment μ. The experiments seem to require that the ionizing effect of such a neutron can not be bigger than the one of a gamma-ray, and then μ is probably not allowed to be larger than e • $(10^{-13}$ cm$)$. But so far I do not dare to publish anything about this idea, and trustfully turn first to you, dear radioactive people, with the question of how likely it is to find experimental evidence for such a neutron if it would have the same or perhaps a 10 times larger ability to get through [material] than a gamma-ray. I admit that my remedy may seem almost improbable because one probably would have seen those neutrons, if they exist, for a long time. But nothing ventured, nothing gained, and the seriousness of the situation, due to the continuous structure of the beta spectrum, is illuminated by a remark of my honored predecessor, Mr Debye, who told me recently in Bruxelles: 'Oh, It's better not to think about this at all, like new taxes.' Therefore one should seriously discuss every way of rescue. Thus, dear radioactive people, scrutinize and judge. - Unfortunately, I cannot personally appear in Tübingen since I am indispensable here in Zürich because of a *ball on the night from December 6 to 7. With my best regards to you, and also to Mr Back, your humble servant*

W. Pauli

It is evident from the contents of this letter that Pauli's neutral particle had the following properties:

1. The proposed neutral spin $\frac{1}{2}$ particles are called 'neutrons' and are constituents of nuclei.
2. They do not travel with the velocity of light.
3. Their mass is similar to the electron mass but not larger than 0.01 times the proton mass.
4. The new particle (neutron) has a magnetic moment which is of the order of $e \times 10^{-13}$cm and is bound in the nucleus by magnetic forces.
5. The neutral spin $\frac{1}{2}$ particle shares the available energy with the electron leading to the continuous energy spectrum of β-electrons.

Six months later, Pauli himself first talked about the idea of the new particle in June, 1931 in the meeting of the American Physical Society in Pasadena [3, 17]; here, he abandoned the idea of the new particle being a constituent of the nuclei due to considerations of empirical masses. However, he still talked about neutrons and later in the summer of 1931, lectured in the University of Michigan, Ann Arbor about the magnetic properties of the new particle [18]. In October 1931, Pauli attended a conference on nuclear physics in Rome and participated in the discussions on deliberations where Fermi was also present. Fermi was impressed by Pauli's idea of a new particle. In fact, in the words of Pauli, '(Fermi) at once showed a lively interest in my new idea and a very positive attitude towards my new particle'. The very next year, in 1932, Chadwick [19] discovered a new neutral particle with a mass similar to the proton. This particle was named neutron, and Pauli's 'neutron' was rechristened by Fermi as 'neutrino' "little neutral one" [20]. With the discovery of the neutron by Chadwick, a clearer picture of the nuclear structure in terms of the protons and the neutrons emerged as elaborated by Heisenberg [21] and Iwanenko [22]. The theoretical interpretation of the nuclear β-decay was given by Fermi [23] and Perrin [24] in terms of the proton–neutron model of the nucleus in which neutrinos ("neutrons" as proposed by Pauli) were emitted along with the electrons. However, at this time, the interaction of neutrinos with the other material particles remained to be understood as stated by Pauli himself in the Seventh Solvay Conference in October 1933, in Brussels [25]. After the discovery of the neutron and study of its properties, it seems, in hindsight, that Pauli's 'neutron' is more like a hybrid of Chadwick's neutron and Fermi's neutrino.

1.2   Neutrino Interactions

The theory of neutrino interactions with matter has passed through many milestones before the standard model of electroweak interactions was formulated which describes the interaction of neutrinos with leptons and quarks considered to be the fundamental constituents of matter. The first attempt at a description of the nature of neutrino interactions is present in the original proposal of Pauli, where he postulated that neutrinos have a penetrating power larger than the photons implying an interaction weaker than the electromagnetic interaction and an electromagnetic interaction of neutrinos through its magnetic moment. However, in the Solvay Conference in 1933, there was no discussion on neutrino interactions except the general acceptability of the idea of the neutrino and its properties as proposed by Pauli. Soon after the Solvay Conference, Fermi [23] and Perrin [24] independently proposed the theory of β-decay, which was the first milestone in the theory of neutrino interactions with matter. The Fermi theory of β-decay, as it is known today, describes the β-decays of nuclei in which no change of angular momentum and parity of the nucleus is observed. The theory was extended by Gamow and Teller [26] and Bethe and Bacher [27] to describe the observation of nuclear β-decays with a change of one unit of angular momentum and no change in parity. A more general phenomenological theory of nuclear β-decays and other weak interaction processes was subsequently developed as more experimental data were accumulated on the weak decays of leptons, hadrons, and nuclei. In this section, we give a historical introduction of the theory of β-decay which led to the phenomenological $V - A$ (vector – axial vector) theory of weak interactions and later to the standard model of electroweak interactions.

1.2.1   Fermi theory of β-decay

The theory of β-decays by Fermi [23] and Perrin [24], in 1933, assumes that an electron–neutrino pair is created in the basic transitions of β-decay, in which neutron is converted into proton that is,

$$n \longrightarrow p + e^- + \bar{\nu}. \tag{1.1}$$

Its interaction Hamiltonian density was written in analogy with quantum electrodynamics (QED). The Hamiltonian density of the electromagnetic (EM) interactions is written as a scalar product of the electromagnetic current of the electron $j_\mu^{EM}(x)$ and the electromagnetic field $A_\mu(x)$, that is,

$$\mathcal{H}^{EM}(x) = e j_\mu^{EM}(x) A^\mu(x), \tag{1.2}$$

where $j_\mu^{EM}(x) = \bar{\psi}_e(x) \gamma_\mu \psi_e(x)$ is the electromagnetic current of the electron defined in terms of $\psi_e(x)$. In the Fermi theory of β-decay, the Hamiltonian density involving the charged fermion fields of electrons is written as:

$$\mathcal{H}(x) = G \bar{\psi}_p(x) \gamma_\mu \psi_n(x) \ \bar{\psi}_e(x) \gamma^\mu \psi_\nu(x) + \text{Hermitian conjugate (h.c.),} \tag{1.3}$$

where G is the strength of the new interaction and $\psi_p(x)$, $\psi_n(x)$, $\psi_e(x)$, and $\psi_\nu(x)$ are the spin $\frac{1}{2}$ Dirac fields of proton, neutron, electron, and neutrino, respectively and γ_μ is the Dirac gamma matrix [28].

The β-decay Hamiltonian proposed by Fermi in Eq. (1.3) represents a point interaction of four fermions (neutron, proton, electron, neutrino) and its strength G has the dimension of M^{-2}. This makes the theory nonrenormalizable and effectively a low energy theory. This is in contrast with the QED Lagrangian, which depicts the interaction of two charged fermions with the electromagnetic field; the coupling strength is $e = \sqrt{4\pi\alpha}$, where $\alpha = \frac{1}{137}$ is dimensionless, making the theory renormalizable.

The β-decay interaction is shown diagrammatically in Figure 1.2(a) to be contrasted with electromagnetic interaction shown in Figure 1.2(b). In field theoretical language, the β-decay of neutrons described by the Hamiltonian density in Eq. (1.3) occurs at a point x where a neutron field annihilates and creates a proton field, an electron field, and a neutrino field as shown in Figure 1.2(a). It should be noted that in the Fermi theory, the neutrino is created along with the electron and is not emitted being a constituent of the nuclei as proposed by Pauli in his famous letter. The electron created is not the electron of the electron–proton model of the nucleus which was discarded in favor of the nuclear model proposed by Heisenberg [21] and Iwanenko [22], after the discovery of the neutron by Chadwick [19]. The model proposed by Fermi [23] and Perrin [24] was the first successful application of quantum field theory (QFT) beyond QED processes.

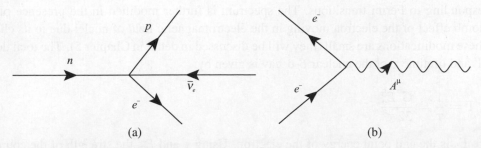

Figure 1.2 (a) Four fermions interact for β-decay; (b) Interaction of two electrons with a photon field (QED).

In nuclei, the neutrons and protons are nonrelativistic and in the nonrelativistic limit of the Dirac spinors, one may write (as shown in Appendix A):

$$\bar{\psi}_p(x)\gamma_\mu\psi_n(x) \;\rightarrow\; \chi_p{}^+\mathbb{1}\chi_n \qquad \text{for } \mu = 0$$
$$\rightarrow\; 0 \qquad \text{for } \mu = i,$$

where χ_n and χ_p are the Pauli spinors and $\mathbb{1}$ is the unit operator in nuclear coordinates and spin space. In the nonrelativistic limit,

$$[\bar{\psi}_p(x)\gamma_\mu\psi_n(x)][\bar{\psi}_e(x)\gamma^\mu\psi_\nu(x)] \propto \chi_p^\dagger\mathbb{1}\chi_n e^{-i(\vec{p}_e+\vec{p}_\nu)\cdot\vec{x}}. \tag{1.4}$$

Moreover, in the case of nuclear β-decay, the total energy available to the electron and the neutrino is a few MeV such that $|(\vec{p}_e + \vec{p}_\nu)| \cdot |\vec{x}|$ (approximated by pR, where p is the total momentum of the electron–neutrino pair and R is the nuclear radius) is of the order of 10^{-2}. Therefore, the exponential term in Eq. (1.4) can be approximated to unity such that:

$$[\bar{\psi}_p(x)\gamma_\mu\psi_n(x)][\bar{\psi}_e(x)\gamma^\mu\psi_\nu(x)] \propto \chi_p^+\mathbb{1}\chi_n. \tag{1.5}$$

Thus, the nuclear operator in the Fermi theory is a unit operator $\mathbb{1}$ in the nuclear space of coordinate and spin, and cannot induce any change in the quantum numbers of the nuclei, depending upon the space coordinates $\vec{x}$, that is, orbital angular momentum$(\vec{L})$, or spin $(\vec{S})$ and the total angular momentum $(\vec{J} = \vec{L} + \vec{S})$. Therefore, the Fermi theory describes the nuclear β-decays in which there is no change in the total angular momentum J $(\Delta J = 0)$ and parity.

The matrix element becomes independent of the electron momentum because pR is very small and is negligible compared to unity. Therefore, the decay rate depends only upon the phase space available to the electron and neutrino, and is proportional to $d\vec{p}_e d\vec{p}_\nu$ subject to energy conservation, that is, $E_e + E_\nu = \Delta = E_i - E_f$, where $(\vec{p}_e, E_e)$ and $(\vec{p}_\nu, E_\nu)$ are the momentum and energy of the electron and neutrino, respectively, in this process. After integrating over the neutrino momentum, we get the the energy spectrum of the electron, that is, $\frac{d\Gamma}{dE_e}$, given in the limit $m_e \to 0$ by:

$$\frac{d\Gamma}{dE_e} \propto p_e E_e (\Delta - E_e)^2. \tag{1.6}$$

This energy spectrum describes very well the continuous energy of nuclear β-decays corresponding to Fermi transitions. The spectrum is further modified in the presence of the Coulomb effect of the electron moving in the electromagnetic field of nuclei due to its charge but these modifications are small (they will be discussed in detail in Chapter 5). The total decay rate Γ and lifetime τ of the nuclear β-decay is given by:

$$\Gamma = \frac{1}{\tau} = \frac{G^2 E_o^5}{32\pi^3}, \tag{1.7}$$

where E_o is the end point energy of the electron. Using τ and E_o, the strength of the coupling G is found to be

$$G \approx \frac{1.0 \times 10^{-5}}{M_p^2}, \quad \text{where } M_p \text{ is the proton mass.} \tag{1.8}$$

1.2.2 Gamow–Teller theory

The Fermi Hamiltonian discussed in Section 1.2.1 is a scalar product of two vector currents in the lepton and nucleon sectors. This Hamiltonian does not describe the β-decays when the total angular momentum carried by the electron–neutrino pair is one unit (corresponding to $|\Delta J| = 1$) and there is no change in parity. Such transitions were observed and described by Gamow–Teller(GT) [26] who proposed that the Hamiltonian density responsible for these transitions is a scalar product of axial vector–axial vector current in nucleon and lepton sectors in order to conserve parity, that is,

$$\mathcal{H}^{GT}(x) = G_\beta^A [\bar{\psi}_p(x)\gamma_\mu\gamma_5\psi_n(x)][\bar{\psi}_e(x)\gamma^\mu\gamma_5\psi_\nu(x)] + h.c. \tag{1.9}$$

which in the nonrelativistic limit of the nucleon kinematics reduces to

$$\mathcal{H}_{NR}^{GT}(x) = G_\beta^A [\chi_p^\dagger \sigma_i \chi_n][\bar{\psi}_e(x)\gamma^i\gamma_5\psi_\nu(x)], \tag{1.10}$$

where σ_i $(i = 1 - 3)$ are the Pauli spin operators. The Pauli spin operators σ_i can change the total angular momentum $\vec{J}$ of the initial nucleus by one unit without changing the orbital angular momentum, implying no change in the parity. This leads to the GT transitions in β-decays corresponding to $\Delta J = 1$ transitions. The matrix element for the nuclear β-decay corresponding to the Gamow–Teller transformations is also independent of momentum as argued in Section 1.2.1 and yields the similar spectrum as predicted by Eq. (1.6). A typical electron spectrum corresponding to GT transformation is shown in Figure 1.3, which is the same as the β spectrum for Fermi transition. The total decay rate determines the coupling constant G_{β}^{A}, that is, the strength of β-decays in GT transitions. A comparison of the lifetimes of the nuclei undergoing GT transitions and Fermi transitions yields a value of $G_{\beta}^{A}(0)/G_{\beta}^{V}(0)$ ≈ 1.2. We see that the strength of β-decays in the axial vector sector is larger than that in the vector current sector. This was theoretically explained much later in the phenomenological studies of the $V - A$ currents in the weak transitions (discussed in Chapter 5).

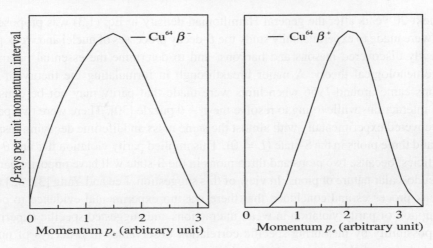

Figure 1.3 β-decay spectrum for GT transitions vs. momentum $|\vec{p}_e|$ (in arbitrary units). [29]

At the time, when the GT Hamiltonian was proposed, it was realized that following Fermi's idea of current–current interactions, the most general form of the four fermion interaction need not be limited to only two terms corresponding to the scalar product of vector–vector and axial vector–axial vector currents but could be a sum of five possible terms constructed from the scalar products of vector (V), scalar (S), pseudoscalar (P), axial vector (A), and tensor (T) currents in nucleonic and leptonic sectors assuming parity conservation [26, 27], that is,

$$\mathcal{H}_I^{\beta}(x) = \sum_{i=S,V,T,A,P} G_i \bar{\psi}_p(x) O_i \psi_n(x) \bar{\psi}_e(x) O^i \psi_{\nu}(x) + h.c. \tag{1.11}$$

with five coupling constants G_i which may be complex, implying ten real parameters to determine $\mathcal{H}_I^{\beta}(x)$. The operators O^i in Eq. (1.11) are constructed from the Dirac matrices as:

$$O^i(i = 1 - 5) = 1, \gamma^{\mu}, \sigma^{\mu\nu}, \gamma^{\mu}\gamma_5, \gamma_5.$$

With the help of these operators O_i, the bilinear covariants $\bar{\psi} O_i \psi$ are constructed in the leptonic and nucleonic sectors (discussed in Chapter 5).

The number of fundamental parameters describing the weak interaction Hamiltonian density $\mathcal{H}_I^{\beta}(x)$ seemed too many to provide a useful phenomenological theory. However, these parameters were determined using the symmetry properties of the weak interaction Hamiltonian and the numerous experimental observations made and interpreted theoretically over the next 20 years. This resulted in a successful description of most of the weak processes at low energies in terms of a very few nonvanishing parameters of $H_I^\beta(x)$. The experimental observation of many observables and their theoretical interpretation have played a decisive role in arriving at a phenomenological theory of weak interactions. We will discuss them in detail in Chapter 5, but here would like to highlight the discovery of parity violation and its interpretation in terms of the two-component theory of neutrinos.

1.2.3   Parity violation and the two-component neutrino

For the next 20 years after the general Hamiltonian density in Eq. (1.3) was proposed, many attempts were made to experimentally study the β-decay processes of nuclei and weak processes of the newly discovered mesons and baryons, and to determine the essential parameters of the phenomenological theory. A major breakthrough in formulating the theory of neutrino interactions came around 1956 when hints were made that parity may not be conserved in the weak interactions while trying to resolve the $\tau - \theta$ puzzle [30]. There were two particles τ and θ discovered experimentally with almost the same mass and lifetime decaying respectively into two and three pions in the S-state ($L = 0$). This implied parity violation if τ and θ were the same particles, because two pions and three pions in the S-state will have opposite parities due to the pseudoscalar nature of pions. In view of this suggestion, Lee and Yang [31, 32] analyzed many weak processes and concluded that there was no experimental evidence to contradict the assumption of parity violation in weak interactions and suggested specific experiments to test this possibility by measuring specific correlation observables in β-decays of nuclei and other weak decays of elementary particles which were parity violating, that is, observables that change sign under the transformation $\vec{r} \to -\vec{r}$ like $\vec{\sigma}_N \cdot \hat{p}_e$ or $\vec{\sigma}_e \cdot \hat{p}_e$, where $\hat{p}_e$ is the unit vector along the electron momentum and $\vec{\sigma}_N$ and $\vec{\sigma}_e$ are, respectively, the nucleon and electron spin operators. The first experiment to test the parity violation in β-decay was performed by Wu et al. [33] with polarized cobalt nucleus where a large asymmetry of β-electrons with respect to the spin direction of the polarized ^{60}Co was observed. Later on, many experiments on the longitudinal polarization of electrons in nuclear β^--decays and positrons in nuclear β^+-decays were conducted and the violation of parity in weak interactions was firmly established [34, 35]. Therefore, it was established by 1957, from the studies of various β-decays that parity is violated in weak interactions. The fact that neutrinos have almost zero mass, was confirmed from the study of the energy spectrum of β-decay electrons; it can especially be confirmed from the shape of the spectrum near the end point that neutrino is almost massless (see Figure 1.4 [36]) by comparing the energy spectrum with $m_\nu \neq 0$ and $m_\nu = 0$.

The evidence of the presence of parity violation led to the revival of the two-component theory for massless neutrinos by many authors [37, 38, 39]. The theory was first formulated by Weyl [40] but was not pursued further as it violated parity conservation and was not considered appropriate for application to physical processes. In a parallel development, during the time parity violation was being established experimentally, the two-component theory of neutrinos and its implications in physical processes were studied in detail. This led to the formulation of a phenomenological theory of weak interactions in terms of two-component neutrinos.

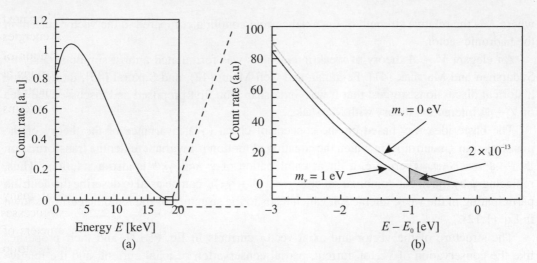

Figure 1.4 β^--decay spectrum with $m_\nu \neq 0$ and $m_\nu = 0$ [36].

1.2.4 Chiral (γ_5) invariance and $V - A$ theory of β-decays

Some major developments were made in the experimental analysis of β-decays of nuclei and many elementary particles were discovered over the next 20 years after the proposed theory by Fermi [23]. The experimental confirmation that the neutrino mass is almost negligible and the discovery of parity violation in weak interactions, along with the observation that the longitudinal polarization of the electron (positron) is $-v$ $(+v)$ and the helicity of the neutrino (antineutrino) is -1 $(+1)$ led to major developments in the phenomenological theory of weak interactions. Once parity violation is allowed, the weak interaction Hamiltonian density can have a pseudoscalar term in addition to a scalar term to accommodate the parity violating effects. The additional pseudoscalar term requires that the weak interaction Hamiltonian density should have five more coupling constants G_i' associated with the pseudoscalar term in analogy with Eq. (1.11). A vast amount of extensive work on various observables of β-decays was done to determine the various coupling constants G_i and G_i' which established that the phenomenological theory of weak interactions is of the current–current form where the currents have a $V - A$ structure in the leptonic and hadronic sectors; they do not have any other variants like scalar, tensor, or pseudoscalar as proposed in the general Hamiltonian. All the fermions, that is leptons and baryons participated in the weak interaction through their left-handed component, that is, $\psi_L = (1 - \gamma_5)\psi$, instead of ψ so that the interacting currents for the leptonic (l^μ) and the hadronic (J^μ) currents can be written as:

$$l^\mu = \bar{\psi}_e \gamma^\mu (1 - \gamma_5) \psi_\nu \tag{1.12}$$

$$J^\mu = \bar{\psi}_p \gamma^\mu (1 - \lambda \gamma_5) \psi_n, \tag{1.13}$$

where λ is the relative strength of the axial current coupling compared to the vector current in the hadronic sector.

An elegant $V - A$ theory of weak interactions was formulated almost simultaneously by Sudarshan and Marshak [41], Feynman and Gell-Mann [42], and Sakurai [43], though some historical discussions suggest that it was Sudarshan who first proposed and discussed the idea of $V - A$ interaction theory with Marshak.

The basic idea was based on the concept of chiral (γ_5) invariance of the theory of the massless spin $\frac{1}{2}$ neutrinos, in which the equation of motion is invariant under the transformation $\psi \rightarrow \psi' = \gamma_5 \psi$. Therefore, a linear combination of ψ and $\gamma_5 \psi$ is also a solution. Thus, replacing ψ by the linear combination of $\psi_L = \frac{1}{2}(1 - \gamma_5)\psi$ can be used to describe the neutrino participating in the weak interaction which leads to the structure of the leptonic current shown in Eq. (1.12).

The structure of the vector and axial vector currents in Eq. (1.13), and their properties like the conservation of vector current, partial conservation of axial current, and the relative strength of the axial vector and vector currents, that is, λ, were established later by many authors [44, 45, 46, 47, 48]. Thus, the interaction Hamiltonian H_{int} is written as:

$$H_{\text{int}} = \frac{G_F}{\sqrt{2}} l_\mu J^{\mu\dagger} + \text{h.c.} \tag{1.14}$$

where a factor $\frac{1}{\sqrt{2}}$ is introduced in the definition of H_{int} so that the constant G introduced by Fermi (in Eq. (1.3)) is consistent with G_F.

1.2.5   Intermediate vector boson (IVB)

One of the significant implications of the $V - A$ theory, in addition to explaining the parity violation and observed helicities of the leptons in a natural way in weak interactions was to give credence to the theory of weak interactions mediated by spin 1 intermediate vector bosons (IVB) in analogy with quantum electrodynamics in which the electromagnetic interaction is mediated by spin 1 photons. In the IVB theory, the basic weak interactions between the $e\nu$ pair and the np pair is mediated by a vector field W^μ and the interaction Hamiltonian is given by:

$$H_{\text{int}}^{IVB} = g \left[\bar{\psi}_e \gamma_\mu (1 - \gamma_5) \psi_\nu + \bar{\psi}_n \gamma_\mu (1 - \lambda \gamma_5) \psi_p \right] W^\mu. \tag{1.15}$$

The β-decay process $n \rightarrow p + e^- + \bar{\nu}_e$ is then a second order process as shown in Figure 1.5 and the strength g of the weak interaction of the $e\nu$ pair with the vector boson W^μ is related to the Fermi coupling constant $\frac{G_F}{\sqrt{2}} = \frac{g^2}{M_W^2}$, where M_W is the mass of the W boson, being very high compared to q^2 ($M_W^2 >> q^2$) in these processes. One of the reasons for the introduction of an IVB to mediate weak interactions was to avoid the divergences encountered in the phenomenological $V - A$ theory while extending the theory to higher energies. For example, the total cross sections for the $\nu_e - e^-$ scattering is found to increase with square of the center of mass (CM) energy (s), that is, $\sigma(\nu_e e^- \rightarrow \nu_e e^-) = \frac{G^2}{\pi} s$. The cross section would

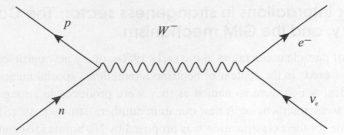

Figure 1.5 β^--decay process mediated through a W^- boson.

diverge at higher energies and violate the unitarity limit which is given by $\sigma \leq \frac{4\pi}{s}$ for $\nu_e e^- \rightarrow \nu_e e^-$ scattering as shown in Figure (1.6). In fact, the presence of such divergence problems in the Fermi theory was realized quite early by many authors. It was hoped that the $V - A$ theory mediated by a massive intermediate boson may help to remove this divergence; such a theory was theoretically proposed very early by Schwinger [49], Bludman [50], and Leite Lopes [51] in the hope that it will solve the divergence problem of Fermi theory in higher orders [52, 53, 54, 55] due to the presence of the momentum dependence of the spin-1 W^μ propagator but it does not happen. Moreover, the short range of weak interactions implying a large mass of the mediating vector boson created more problems than the theory of intermediate vector bosons was supposed to solve. Even though the intermediate vector boson of Schwinger [49], Leite Lopes [51] and others, did not solve the divergence problem, it led to other developments which contributed to the formulation of the standard model of electroweak interactions mediated by the intermediate vector bosons.

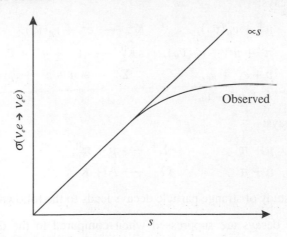

Figure 1.6 Divergence of cross section at high energies ($\sigma \propto s$) which violates unitarity, while saturation at high energies is observed experimentally.

1.2.6 Weak interactions in strangeness sector: The Cabibbo theory, and the GIM mechanism

With the advent of particle accelerators in the early 1950s, many new particles and their decay modes were discovered. In the context of neutrino interactions, special mention may be made of strange particles, which were so named as they were produced in strong interactions and decayed through weak interactions. A new quantum number, 'strangeness' (S) was assigned to them and a scheme for their classification was proposed by Nishijima [56] and Gell Mann [57] in 1956. The first strange mesons observed were K^+, K^0, K^-, and $\bar{K}^0$ and were assigned strangeness S of $+1$ for $\left(K^+ \quad K^0\right)$ and strangeness of -1 for $\left(\bar{K}^0 \quad K^-\right)$. The first strange baryons observed were named hyperons and assigned strangeness -1 for $\Lambda, \Sigma^{\pm,0}$, strangeness -2 for $\Xi^{-,0}$, and -3 for Ω^-. Accordingly, the already known mesons like pions $\left(\pi^{\pm,0}\right)$ and baryons like nucleons (p, n) were assigned strangeness quantum number $S = 0$. The strange mesons and hyperons were found to decay into lighter particles including leptons and hadrons and the decays were classified as semileptonic and nonleptonic decays, depending upon the presence or absence of leptons. Some specific examples of purely leptonic as well as semileptonic and nonleptonic decays of strange mesons and hyperons are as follows:

i) Purely leptonic decays:

$$
\begin{aligned}
\mu^{\mp} &\longrightarrow e^{\mp} + \nu_\mu(\bar{\nu}_\mu) + \bar{\nu}_e(\nu_e), \\
\tau^{\mp} &\longrightarrow e^{\mp} + \nu_\tau(\bar{\nu}_\tau) + \bar{\nu}_e(\nu_e), \\
\tau^{\mp} &\longrightarrow \mu^{\mp} + \nu_\tau(\bar{\nu}_\tau) + \bar{\nu}_\mu(\nu_\mu).
\end{aligned}
$$

ii) Semileptonic decays:

$$
\begin{aligned}
K^{\pm} &\longrightarrow \mu^{\pm} + \nu_\mu(\bar{\nu}_\mu), & K^{\pm} &\longrightarrow e^{\pm} + \nu_e(\bar{\nu}_e), \\
K^{\pm} &\longrightarrow \pi^0 + \mu^{\pm} + \nu_\mu(\bar{\nu}_\mu), & K_L^0 &\longrightarrow \pi^+ + \mu^- + \bar{\nu}_\mu, \\
\Lambda &\longrightarrow p + e^- + \bar{\nu}_e, & \Sigma^- &\longrightarrow n + e^- + \bar{\nu}_e, \\
\Xi^- &\longrightarrow \Lambda + \mu^- + \bar{\nu}_\mu.
\end{aligned}
$$

iii) Nonleptonic decays:

$$
\begin{aligned}
\Lambda &\longrightarrow p + \pi^-, & \Sigma^+ &\longrightarrow p + \pi^0, \\
\Sigma^- &\longrightarrow n + \pi^-, & \Omega^- &\longrightarrow \Lambda + K^-.
\end{aligned}
$$

A phenomenological study of strange particle decays leads to the following observations:

(i) Strange particle decays are suppressed when compared to the decays of nonstrange particles. For example, $K^+ \to \mu^+ \nu_\mu$ is suppressed by a factor of $\approx 1/5$ as compared to the $\pi^+ \to \mu^+ \nu_\mu$ transition.

(ii) Strange particle decays follow the $|\Delta S| = 1$ and $\Delta S = \Delta Q$ rule, where ΔQ and ΔS are respectively, the changes in hadronic charge and strangeness.

(iii) The strangeness changing decays in which $\Delta Q = 0$, that is, $K_L \longrightarrow \mu^+ \mu^-$ or $K^{\pm} \longrightarrow \pi^{\pm} \bar{\nu} \nu$, $K^{\pm} \longrightarrow \pi^{\pm} e^+ e^-$, $K^{\pm} \longrightarrow \pi^{\pm} \mu^+ \mu^-$, $\Sigma^+ \to p e^+ e^-$, $\Sigma^+ \to p \mu^+ \mu^-$ are highly suppressed.

(iv) The strangeness changing $|\Delta S| = 1$ currents follow the $|\Delta I| = \frac{1}{2}$ rule in contrast to the $\Delta S = 0$ currents, which follow the $\Delta I = 1$ rule, where I is the isospin of the hadrons.

In order to explain phenomenologically, the suppression of the strength of $|\Delta S| = 1$ currents as compared to $\Delta S = 0$ currents, Gell-Mann and Levy [58] and Cabibbo [59] proposed that the strength of the $|\Delta S| = 1$ weak current in the hadronic sector is suppressed as compared to the $\Delta S = 0$ currents by a factor described by a parameter to be determined experimentally from the β-decays of hyperons and strange mesons like $\Sigma^0(\Lambda) \rightarrow pe^- \bar{\nu}_e$ and $K^\pm \rightarrow \pi^0 l^\pm \nu_l(\bar{\nu}_l)$, and the leptonic decays like $K^\pm \rightarrow l^\pm \nu_l(\bar{\nu}_l)$, where $l = e, \mu$. The Gell-Mann and Levy [58] proposal was formulated in terms of the physical particles p, n, and Λ following the Sakata model [60] of elementary particles. In this model, the hadronic current is written as:

$$J_\mu^{\text{hadron}} = \frac{1}{\sqrt{1 + \epsilon^2}} \left[\bar{\psi}_p \gamma_\mu (1 - \gamma_5) \psi_n + \epsilon \bar{\psi}_p \gamma_\mu (1 - \gamma_5) \psi_\Lambda \right], \tag{1.16}$$

with the parameter ϵ describing the suppression of $|\Delta S| = 1$ currents. The Cabibbo model [59] was formulated in terms of the quark model of the hadrons which was proposed by Gell-Mann and Pais [61] and Zweig [62, 63]. In the quark model, the proton, neutron, and lambda particles are considered to be the bound states of quarks. The quark contents of proton, neutron, and lambda being *uud*, *udd* and *uds*, respectively, the weak transitions of β-decay correspond to $d \rightarrow u$ ($s \rightarrow u$) transitions in case of $\Delta S = 0$ ($|\Delta S| = 1$) transitions; the transitions are shown in Figure 1.7. Therefore, in the quark model, the weak hadronic current J_μ was written by Cabibbo as:

$$J_\mu^{\text{Cabibbo}}(x) = \cos\theta_C \bar{\psi}_u(x) \gamma_\mu (1 - \gamma_5) \psi_d(x) + \sin\theta_C \bar{\psi}_u(x) \gamma_\mu (1 - \gamma_5) \psi_s(x), \tag{1.17}$$

in which the $|\Delta S| = 1$ currents are suppressed by a factor $\tan\theta_C$. The phenomenological value of $\tan\theta_C$ was determined to be $\tan\theta_C = 0.2327$ at that time.

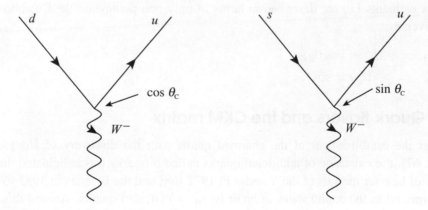

Figure 1.7 Interaction of quark with W boson field.

Equation (1.17) can be rewritten as:

$$J_\mu^{\text{Cabibbo}} = \bar{\psi}_u \gamma_\mu (1 - \gamma_5)(\cos\theta_C \psi_d + \sin\theta_C \psi_s), \tag{1.18}$$

which implies that a linear combination of d and s quarks defined as $d' = d \cos \theta_C + s \sin \theta_C$ participates in weak interactions. Thus, u and d' can be considered to form a doublet under 'weak isospin' represented by a column matrix, that is, $\begin{pmatrix} u \\ d' \end{pmatrix}$ like the isospin $\begin{pmatrix} u \\ d \end{pmatrix}$ doublet which takes part in strong interactions. This raises the question of the physical significance of the orthogonal component of d' defined by $s' = -d \sin \theta_C + s \cos \theta_C$. It was proposed by Glashow, Iliopoulos, and Maiani (GIM) [64], following the earlier suggestion of Bjorken and Glashow [65], that there exists a fourth quark 'c' named charm quark, which forms another 'weak isospin doublet' with s', that is, $\begin{pmatrix} c \\ s' \end{pmatrix}$. Its weak interaction is described by the weak quark current,

$$J_\mu^{GIM} = \bar{\psi}_c(x) \gamma_\mu (1 - \gamma_5)(- \sin \theta_C \psi_d(x) + \cos \theta_C \psi_s(x)). \tag{1.19}$$

This also implies that the neutral current now defined as:

$$\bar{u}O_\mu u + \bar{c}O_\mu c + \bar{d}'O_\mu d' + \bar{s}'O_\mu s' = \bar{u}O_\mu u + \bar{c}O_\mu c + \bar{d}O_\mu d + \bar{s}O_\mu s; \quad O_\mu = \gamma_\mu(1 - \gamma_5)$$

does not have terms like $\bar{d}s$ and $\bar{s}d$ which change strangeness. Thus, the Cabibbo theory extended by Glashow, Iliopoulos, and Maiani (GIM) explained the absence of flavor changing neutral current (FCNC) and provided the concept of quark mixing, a physical interpretation. Hence, while the d and s quarks participate in strong interactions, d' and s', that is, the mixed state of d and s participate in weak interactions. This mixing of two quarks is described, in general, by a 2×2 unitary mixing matrix U such that:

$$\begin{pmatrix} d' \\ s' \end{pmatrix} = U \begin{pmatrix} d \\ s \end{pmatrix}, \text{where} \qquad U = \begin{pmatrix} U_{ud} & U_{us} \\ U_{cd} & U_{cs} \end{pmatrix}. \tag{1.20}$$

The most general unitary 2×2 matrix is parameterized in terms of one parameter. Therefore, the matrix elements U_{ij} are described in terms of only one parameter, the Cabibbo angle θ_C and are given as:

$$\begin{aligned} U_{ud} &= U_{cs} = \cos \theta_C, \\ U_{us} &= -U_{cd} = \sin \theta_C. \end{aligned}$$

1.2.7 Quark flavors and the CKM matrix

Soon after the establishment of the charmed quark with the discovery of J/ψ particles in 1974 [66, 67], the existence of additional quarks called b-quarks was anticipated through the discovery of heavier mesons of the Υ series in 1977 [68] and the B series in 1983 [69], which were interpreted as the bound states of $b\bar{b}$ or $b\bar{q}$ ($q = u, d, s, c$) quarks. Around this time, the existence of five quarks u, d, s, c, and b was considered to be established as many baryon states with b as one of the constituents were also observed. The concept of quark–lepton symmetry was invoked to propose the existence of a sixth quark t (top quark) to form a doublet with b-quark like the doublets of (u, d) and (c, s) quarks in 4-quarks, the two-flavor doublet model. In analogy with the three-flavor doublet $(\nu_e\ e^-)$, $(\nu_\mu\ \mu^-)$, and $(\nu_\tau\ \tau^-)$ model of six leptons, the three-flavor doublet quark model of 6-quarks $(u\ d)$, $(c\ s)$, and $(t\ b)$ was considered in order to maintain symmetry between quarks and leptons. A quark mixing of three-flavor quarks, that is, $d, s,$ and b was formulated by Kobayashi and Maskawa [70] like the three-flavor mixing of neutrinos formulated earlier in 1962 by Maki et al. [71].

In a three-flavor mixing scheme, the quark flavor states participating in the weak interaction, say $q' = d', s', b'$ are assumed to be a mixture of three quark states $q = d, s, b$ participating in the strong interactions and the mixing is described by a unitary 3×3 matrix U called the CKM (Cabibbo–Kobayashi–Maskawa) matrix,

$$q_i' = \sum_{ij} U_{ij} q_j, \tag{1.21}$$

where $q_i' (= d', s', b')$ are the weak interaction eigenstates of quarks and $q_i (= d, s, b)$ are the strong interaction eigenstates. These nine matrix elements of the unitary 3×3 matrix, that is, U_{ij} are described in terms of four independent parameters. There are quite a few parameterizations of this matrix but the most popular parameterization is given by CKM in which these parameters are chosen to be three rotation angles θ_{12}, θ_{13}, and θ_{23} like Euler angles describing rotation in real three-dimensional space and a phase angle δ. Explicitly, the U matrix in this parameterization is written as:

$$U = \begin{pmatrix} c_{12}c_{13} & s_{12}c_{13} & s_{13}e^{-i\delta} \\ -s_{12}c_{23} - c_{12}s_{23}s_{13}e^{i\delta} & c_{12}c_{23} - s_{12}s_{23}s_{13}e^{i\delta} & s_{23}c_{13} \\ s_{12}s_{23} - c_{12}c_{23}s_{13}e^{i\delta} & -c_{12}s_{23} - s_{12}c_{23}s_{13}e^{i\delta} & c_{23}c_{13} \end{pmatrix}, \tag{1.22}$$

where $c_{ij} = \cos\theta_{ij}, s_{ij} = \sin\theta_{ij}$, and δ is the phase angle. For the three-flavor mixing described here, the weak interaction Lagrangian is written in terms of weak quark current as:

$$J_\mu^{CKM} = \bar{\psi}_u \gamma_\mu (1 - \gamma_5) \psi_{d'} + \begin{pmatrix} u \to c \\ d' \to s' \end{pmatrix} + \begin{pmatrix} u \to t \\ d' \to b' \end{pmatrix} + \text{h.c.} \tag{1.23}$$

The presence of the phase angle δ in the Lagrangian makes it complex which violates the time reversal invariance (T invariance). The invariance of the weak interaction under the combined CPT (charge conjugation, parity, time reversal) symmetry implies that the T violation is equivalent to CP (charge conjugation, parity) violation; therefore, the phase angle δ is used to describe the phenomenon of CP violation in weak processes.

1.2.8 Nonleptonic weak interaction and CP violation

Strange mesons and baryons (hyperons) also decay through modes which do not involve neutrinos. These decays are called nonleptonic weak decays; some examples are as follows:

$$\begin{aligned} K^\pm &\to \pi^\pm \pi^0, & K^\pm &\to \pi\pi\pi \\ K_L^0 &\to \pi^+\pi^-\pi^0,\ \pi^0\pi^0\pi^0, & K_S^0 &\to \pi^+\pi^-,\ \pi^0\pi^0 \\ \Lambda &\to p\pi^-, n\pi^0, & & \\ \Sigma^+ &\to n\pi^+, p\pi^0 & \Sigma^0 &\to p\pi^-, n\pi^0 \quad \Sigma^- \to n\pi^-. \end{aligned} \tag{1.24}$$

Many nonleptonic decays of strange mesons and hyperons were discovered in the emulsion experiments done with cosmic rays and particle beams in early accelerators where these particles were produced.

The following conclusions were drawn from the analysis of these decays.

1. Nonleptonic decays violated strangeness with $|\Delta S| = 1$ and exhibited dominance of the $\Delta I = \frac{1}{2}$ rule like the semileptonic decays.

2. Parity violation was also established in nonleptonic decays of hyperons like $\Lambda \rightarrow n\pi^0$, $\Lambda \rightarrow p\pi^-$ by a measurement of the asymmetry in the angular distribution of pions. Historically, this was one of the early observations of parity violation in particle physics which went unnoticed [72].

3. CP violation was discovered in the comparative study of nonleptonic decays of neutral kaons K_L^0 and K_S^0 in two and three pion modes. K_L^0 and K_S^0 are the neutral kaon states defined to be the eigenstates of CP corresponding to eigenvalues of -1 and $+1$. Therefore, the experimental observation of $K_L^0 \rightarrow \pi^0\pi^0$ and $K_L^0 \rightarrow \pi^+\pi^-$ would violate CP [73].

1.3   Neutrino Flavors and Universality of Neutrino Interactions

In the earlier sections, we have described the progress in the understanding of neutrinos and their interactions mainly from the study of β-decay of nuclei and nucleons. Simultaneous, with the developments in the experimental and theoretical understanding of β-decay of nuclei and nucleons, many other particles like muons, pions, kaons, and hyperons were discovered in cosmic rays and accelerator experiments at CERN, ANL, BNL, Serpukhov, etc. and their weak decays were studied, which contributed to the study of weak interactions. Neutrinos (antineutrinos) emitted in the β-decay of nucleons and nuclei were identified as electron neutrinos (antineutrinos) because they were always accompanied with positrons (electrons). Later, other heavy leptons like muons and tauons were discovered, which were found to decay weakly into lighter particles involving one or more neutrino (antineutrino). The additional neutrinos (antineutrinos) associated with these particles were later identified to be different from the neutrinos associated with electrons. Three types of neutrinos are known today; they are the electron neutrino (ν_e), muon neutrino (ν_μ), and tau neutrino (ν_τ), and they have antiparticles $\bar{\nu}_e$, $\bar{\nu}_\mu$, and $\bar{\nu}_\tau$ corresponding to the three charged leptons, e^-, μ^-, τ^- and their antiparticles e^+, μ^+, τ^+. The weak interactions of all these neutrinos have a $V - A$ structure and the same strength, leading to the universality of weak interactions. In this section, we give the historical introduction to our present understanding of neutrino flavors ν_e, ν_μ, and ν_τ and their antiparticles, and the universality of weak interactions.

1.3.1   Experimental discovery of $\bar{\nu}_e$ and $\bar{\nu}_e \neq \nu_e$

The attempts to make direct observation of neutrinos and antineutrinos possible took a very long time to succeed experimentally because of the theoretical calculations by Bethe and

Peierls [74] as well as by Fierz [55], who found the neutrino nucleus cross section to be very small, of the order of $\approx 10^{-44}$cm^2, for *MeV* neutrinos available at that time from β-decay sources. This led them to conclude that there was no possibility to observe neutrinos in the near future. Suggestions were made to observe them indirectly by measuring the recoil of the daughter nucleus in the emission of the $e\nu$ pair in nuclear β-decay [3] and experimental attempts were made early by Rodeback and Allen [75], Leipunski [54], Snell and Pleasonton [76], Jaeobsen et al. [77], Sherwin [78] and Crane and Halpern [79] with clear evidence of the existence of neutrinos. With the development of nuclear reactors where a very high flux of antineutrinos was produced from the fusion reactions of nuclei in the nuclear pile sites, it was argued by Pontecorvo [80], Alvarez [81] and Fermi [82] that direct neutrino–nucleus reactions with antineutrinos being generated as a result of fusion reactions could be observed due to the high flux of $\bar{\nu}_e$ despite small neutrino cross sections. The neutrino event rates could be made still larger by using huge targets of the order of tons of material, thus increasing the number of nucleon targets, so that the number of events could be significant enough to be observed. The attempts to observe neutrinos through the nuclear reactions induced by antineutrinos from the reactors finally succeeded when the group led by Reines and Cowan [83, 84] used a 300 L liquid scintillator target detector to observe neutrinos at the Hanford reactor in 1953 and later with a 4200 L liquid scintillator target detector at the Savannah River reactor in 1956 [85, 86]. They observed the reaction

$$\bar{\nu}_e + p \rightarrow e^+ + n \tag{1.25}$$

by making a coincidence measurement of the photons from particle annihilation $e^- + e^+ \rightarrow \gamma + \gamma$ and a neutron capture $n + {}^{108}\text{Cd} \rightarrow {}^{109}\text{Cd} + \gamma$ reaction a few microseconds later as illustrated in Figure 1.8.

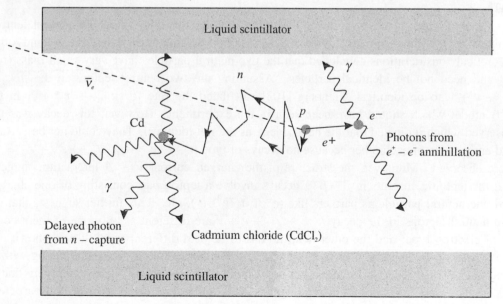

Figure 1.8 Two instant and one delayed photon in Reines and Cowan's experiment [83, 84].

They observed a cross section of

$$\bar{\sigma} = (11 \pm 2.6) \times 10^{-44} \mathrm{cm}^2/\mathrm{nucleon} \tag{1.26}$$

for the reaction in Eq. (1.25) averaged over the spectrum of $\bar{\nu}_e$ from fusion reactions, which was in good agreement with the theory.

The original proposal of Pontecorvo [80] and Alvarez [81] to use ^{37}Cl as target was followed up by Davis et al. [87, 88] who looked at the reaction $\nu_e + ^{37}$Cl $\rightarrow$ $e^- + ^{37}$Ar using the Brookhaven reactor with 4000 L of liquid CCl_4 and tried to observe the ^{37}Ar produced in the reaction. No events were observed but a limit of $\bar{\sigma}(\bar{\nu} \mid ^{37}$Cl $\rightarrow e^- + ^{37}$Ar$) < 0.9 \times 10^{-45} \mathrm{cm}^2$ was obtained while the prediction was $\approx 2.6 \times 10^{-45} \mathrm{cm}^2$. This negative result was very important as it showed that antineutrinos from reactors do not produce electrons hinting that $\bar{\nu}_e \neq \nu_e$. Moreover, the efforts made by Davis opened up a detection principle which was later used in observing solar neutrinos and played a major role in the development of neutrino physics.

1.3.2 Discovery of muons and muon neutrinos

Muons were discovered in cosmic rays by Neddermeyer and Anderson [89], Street and Stevenson [90], and Nishina et al. [91] in 1937. They were found to have mass between electrons and protons and were very penetrating with long lifetime and therefore, not to be confused with Yukawa's proposed π-mesons [92]. π-mesons were later discovered to decay into muons [93, 94, 95], that is, $\pi \rightarrow \mu + \nu$ as predicted by Tanikawa [96] and Marshak and Bethe [97]. Muons were later confirmed to decay through $\mu^- \rightarrow e^- + \nu + \nu$, that is, into an electron and two neutral leptons in cosmic ray experiments by Conversi et al. [98] as theoretically predicted earlier by Sakata and Inoue [99] and later by many others [100, 101, 102].

The experimental observation of the upper limit of the total energy of neutral leptons emitted in μ-decay, the magnitude of the Michel parameter ρ (Chapter 5), and other theoretical considerations concluded that the two neutral particles have very small mass (like ν_e) and need not be identical particles. Assuming the two neutral leptons in the $\mu^- \rightarrow e^- + \nu + \nu$ to be identical, Feinberg [103] calculated the rate for $\mu^- \rightarrow e^- + \gamma$ in the IVB model which should have been seen in the experiment. However, this decay was not observed which indicated that the neutral leptons are not identical. They could not be particle and antiparticle of each other because no decays of type $\mu^- \rightarrow e^- + \gamma$, $\mu^- \rightarrow e^- + e^- + e^+$ were observed. Moreover, the structure of the charged current $V - A$ interaction suggests that the leptonic currents in $\beta^- (\beta^+)$ decays involve a lepton pair consisting of one charged and one neutral particle as partners like $(e^-, \bar{\nu}_e)$, (e^+, ν_e), etc. This further suggests that the two neutral leptons in the decay $\mu^- \rightarrow e^- + \nu + \nu$ are different as only one of them could be of electron type; and the other neutrino has to be of a different type. It was known that in the case of the electron neutrino, ν_e and $\bar{\nu}_e$ are distinct and an e^- is emitted along with $\bar{\nu}_e$ (and not ν_e) in β^--decay and a ν_e (not $\bar{\nu}_e$) is emitted in nuclear β^+-decay, implying that in $\mu^- \rightarrow e^- + \nu + \nu$, one of the neutral leptons is $\bar{\nu}_e$. Therefore, if the other neutral particle is a neutrino and different from the electron type, it should be associated with a muon and so it was identified as ν_μ, 'the muon neutrino'. The possibility of two pairs of neutral leptons was earlier discussed theoretically by Oneda et al. [104]. Depending on the analogy of the emission of $(e^-, \bar{\nu}_e)$ and (e^+, ν_e) pairs in nuclear β-decays, it is clear that the $\pi \rightarrow \mu$ decays should proceed as $\pi^+ \rightarrow \mu^+ + \nu_\mu$ and $\pi^- \rightarrow \mu^- + \bar{\nu}_\mu$. Motivated by these developments, it was suggested by Pontecorvo [105] and Schwartz [106] to use high energy neutrino beams from pion decays to perform experiments like:

$$\nu + n \quad \longrightarrow \quad \mu^- + p \qquad\qquad \nu + n \longrightarrow e^- + p \qquad\qquad (1.27)$$

$$\bar{\nu} + n \quad \longrightarrow \quad \mu^+ + p \qquad\qquad \bar{\nu} + n \longrightarrow e^+ + p \qquad\qquad (1.28)$$

to test whether the neutrinos from pion decays produce muons or electrons. Theoretical calculations for the aforementioned processes were done by Lee and Yang [107], Cabibbo and Gatto [108], and Yamaguchi [109] using the phenomenological $V - A$ theory. The experiments performed at Brookhaven National Laboratory (BNL) by Danby et al. [110] and later by Bienlein et al. at the European Organisation for Nuclear Research (CERN) [111] observed that neutrinos from the pion decays, which were accompanied by muons, produce only muons in these reactions. This confirmed that these neutrinos are different from electron neutrinos, that is, $\nu_\mu \neq \nu_e$.

1.3.3 Lepton number conservation and $e - \mu$ universality

When it was established from the reactor experiments by Reines and Cowan [85] and Davis [87, 88], that the antineutrinos from reactors produce only positrons (and not electrons), a new quantum number called the lepton number (L_e) was proposed to phenomenologically explain this observation. Particles like electrons (e^-) and electron neutrinos (ν_e) were assigned $L_e = +1$ and their antiparticles, that is, positrons (e^+) and antineutrinos ($\bar{\nu}_e$) were assigned $L_e = -1$ while baryons and mesons like nucleons and pions were assigned $L_e = 0$. Therefore, the conservation of the lepton number would explain the observed results of the reactor experiments. Accordingly, an electron is produced along with an antineutrino in β^--decays while a positron is produced in β^+-decays. Similarly, in later experiments at Brookhaven, the ν_μ neutrinos from π^+-decays did not produce an e^- but only μ^-, demonstrating that ν_μ is different from ν_e; experiments at CERN demonstrated that ν_μ is different from $\bar{\nu}_\mu$ (as in the case of ν_e and $\bar{\nu}_e$). Therefore, a separate lepton number for muon L_μ (other than L_e) and its conservation was suggested. Separate conservation laws for L_e and L_μ explain the absence of $\mu^- \rightarrow e^- + \gamma$, $\mu^- \rightarrow e^- + e^- + e^+$, or $\mu^- +^{32}\text{S} \rightarrow e^- +^{32}\text{S}$ reactions. It is natural to extend this scheme of classification of leptons to tauons (τ^-, ν_τ) and its antiparticles which were discovered later.

During this time, various developments in the experimental and theoretical study of β-decays were taking place; other weak processes involving the muon and its decay modes, that is, $\mu^+ \rightarrow e^+ + \nu + \bar{\nu}$ and $\mu^- \rightarrow e^- + \nu + \bar{\nu}$ as well as the weak muon and electron capture processes from nuclei, that is, $\mu^- + (A, Z) \rightarrow \nu + (A, Z - 1)$ and $e^- + (A, Z) \rightarrow \nu + (A, Z - 1)$ were also discovered and studied in detail. Pontecorvo [112] compared the probability of μ^- capture with electron capture on nuclei and suggested the hypothesis of $\mu - e$ universality of weak current. According to this hypothesis, the strength of the weak interaction involving muon and nucleon is the same as that of the electron and nucleon. This was later elaborated by Puppi [113], Klein [114], and Tiomno and Wheeler [115].

The idea of the universality of weak interactions is extended to the hadronic sector where the strength of the weak interactions in the vector and axial vector sectors in the case of $n \rightarrow p$ transitions in β-decay is compared with the strength of vector and axial vector currents in μ-decays. While this universality is valid for vector currents as supported by the comparison of the strengths of weak interactions in the e and μ sectors with the strength of vector interactions in the hadronic sector, it is only approximately valid in the case of axial vector current. The analysis of the GT transitions in nuclear β-decay shows that the strength in the axial vector sector is about 20% greater than the strength in the vector sector. This was later understood in terms of the renormalization of the axial strength due to strong interactions and elaborated by Goldberger and Treiman [45] and later by Adler [47] and Weisberger [48]. The idea of the universality of weak interactions was later extended to strange particles by Gell-Mann and Levy [58] and Cabibbo [59] where a very small difference in the strength of vector interactions in the leptonic (μ and e) sector and the hadronic (n and p) sector and a large suppression in the case of the strange sector ($\Lambda \rightarrow p$ or $K \rightarrow \pi$) as compared to the strength in the $\Delta S = 0$ sector ($n \rightarrow p, \Sigma \rightarrow \Lambda$) was explained phenomenologically. Later, this universality was understood by introducing the concept of quark mixing proposed by Cabibbo [59] (discussed in Chapter 6).

In summary, the weak interaction Hamiltonian incorporating parity violation and the two-component theory of neutrino which evolved into the $V - A$ is now expressed as follows:

$$\mathcal{H}_I = \frac{G_F}{\sqrt{2}} \cos\theta_C \left(J_\mu J^{\mu^\dagger} + h.c. \right), \tag{1.29}$$

$$\text{where } J_\mu = J_\mu^l + J_\mu^h, \tag{1.30}$$

$$\text{with } J_\mu^l = \bar{\psi}_e \gamma_\mu (1 - \gamma_5) \psi_{\nu_e} + \bar{\psi}_\mu \gamma_\mu (1 - \gamma_5) \psi_{\nu_\mu} + \bar{\psi}_\tau \gamma_\mu (1 - \gamma_5) \psi_{\nu_\tau}, \tag{1.31}$$

$$J_\mu^h = \cos\theta_C \bar{\psi}_u \gamma_\mu (1 - \lambda\gamma_5) \psi_d + \sin\theta_C \bar{\psi}_u \gamma_\mu (1 - \lambda\gamma_5) \psi_s. \tag{1.32}$$

1.3.4 Discovery of tau neutrino and $e - \mu - \tau$ universality

In 1975, Perl et al. [116] discovered a heavy lepton through the scattering process $e^+ + e^- \rightarrow \tau^+ + \tau^-$ of mass around 1.776 GeV. Soon, it was established by many experimental and theoretical analysis that the heavy lepton was a spin $\frac{1}{2}$ fermion and consistent with being a point particle like electrons and muons. $\tau^\mp$ leptons undergo weak decay in all the modes of the leptonic and semileptonic decays. Like muons, it was also found to decay into leptonic modes by emitting a muon or an electron accompanied by two neutral leptons, that is,

$$\tau^- \longrightarrow \mu^- + \nu + \nu,$$
$$\tau^- \longrightarrow e^- + \nu + \nu.$$

Being a heavier lepton of mass > 1 GeV, it can also decay into two-particle and three-particle semileptonic modes like:

$$\tau^\mp \longrightarrow \pi^\mp \nu, \qquad\qquad \tau^\mp \longrightarrow K^\mp \nu,$$
$$\tau^\mp \longrightarrow \pi^\mp \pi^0 \nu, \qquad\qquad \tau^\mp \longrightarrow K^\mp \pi^0 \nu,$$

which were also observed [117]. In analogy with the muon case, it was conjectured that τ has its own neutrino ν_τ which is emitted in the τ-decay. Since there was no possibility to produce a $\nu_\tau(\bar{\nu}_\tau)$ beam in the laboratory, it was not possible to directly confirm its existence; it was observed indirectly from two body decay modes. However, it has now been observed directly in experiments with the accelerator and atmospheric neutrinos by DONUT [118], OPERA [119], and SuperK [120]. A separate leptonic number L_τ for the ν_τ was defined with the conservation of L_τ to explain phenomenologically all the leptonic and semileptonic decays of τ lepton. Thus, neutrinos are found to exist in three flavors (also known as generations) described as electron neutrinos (ν_e), muon neutrinos (ν_μ), and tau neutrinos (ν_τ). Along with their corresponding leptons, they are grouped into three doublets under the quantum number I_W called the weak isospin, that is,

$$\begin{pmatrix} \nu_e \\ e^- \end{pmatrix}, \quad \begin{pmatrix} \nu_\mu \\ \mu^- \end{pmatrix}, \quad \begin{pmatrix} \nu_\tau \\ \tau^- \end{pmatrix}$$

with similar assignments for their antiparticles. Later, it was established from $e^+ - e^-$ scattering experiments performed at very high energy that in the low energy region of $E < 46$ GeV, there are only three flavors of neutrinos. This result is also supported by analysis of the relevant cosmological data. The strength of the weak interaction of τ is found to be the same as the strength of μ particles confirming the $\mu - \tau$ universality and thus leading to $e - \mu - \tau$ universality.

1.4 Properties of Neutrinos

1.4.1 Weyl, Dirac and Majorana neutrinos

For massless neutrinos, the Dirac equation may be decoupled as discussed in Chapter 2; all neutrinos are left-handed and antineutrinos are right-handed particles in any frame of reference. This is because these neutrinos travel with the speed of light and an observer would not be able to choose a frame which can move faster than the speed of light. The operation of CPT (charge conjugation, parity, and time reversal taken together) will change a particle with left-handed helicity, say described by the spinor ψ_L^ν, to an antiparticle with right-handed helicity, described by the spinor $\bar{\psi}_R^{\bar{\nu}}$ or vice versa. In the case of Weyl neutrinos/antineutrinos, the scenario would be like that shown in Figure 1.9. Thus, according to Weyl [40], neutrinos and antineutrinos are two different particles, with opposite lepton number for a particular flavor of neutrino(say ν_e and $\bar{\nu}_e$) and would have definite helicity states, that is, a left-handed neutrino will always be left-handed and a right-handed antineutrino will always be right-handed.

However, if the neutrinos are massive, then its speed would be lesser than the speed of light and there would always be a possibility of finding a frame(say II frame) which travels faster than the I frame. For an observer in the I frame, the neutrino is left-handed described by ψ_L^ν, while for an observer in the II frame, the neutrino is right-handed ψ_R^ν, as Lorentz transformation will not change the spin of the particle. Similarly, this would be true for the antineutrino, that is, starting with $\bar{\psi}_R^{\bar{\nu}}$, there is always a possibility of finding an antiparticle with left-handed helicity which is defined by $\bar{\psi}_L^{\bar{\nu}}$. Thus, there are four spinors, viz., ψ_L^ν, ψ_R^ν, $\bar{\psi}_L^{\bar{\nu}}$, and $\bar{\psi}_R^{\bar{\nu}}$.

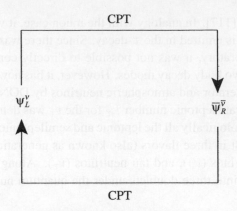

Figure 1.9 Weyl neutrinos.

Therefore, if we assume that $\psi_R^{\bar{\nu}}$ is not the same as $\bar{\psi}_R^{\bar{\nu}}$, then it implies that ψ_R^{ν} has its own CPT mirror image $\bar{\psi}_L^{\bar{\nu}}$, and there exist four states with the same mass; these four states are called Dirac neutrinos ν^D [28]. This scenario has been shown in Figure 1.10. A Dirac particle is one which is different from its antiparticle. The Dirac particle and antiparticle of the same helicity are two different objects and these two interact differently with matter. If the neutrinos happen to be Dirac neutrinos, then they will have finite magnetic and electric dipole moments.

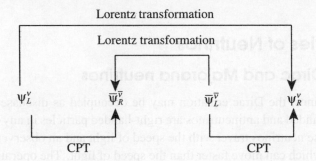

Figure 1.10 Dirac neutrinos.

The concept of a particle and its antiparticle being identical was introduced by Majorana in 1937 [121]. Consider the case of massive left-handed neutrinos (ψ_L^{ν}). If the neutrino happens to be a Majorana neutrino, then after applying Lorentz transformation to a moving reference frame, the right-handed particle ψ_R^{ν}, which is obtained is the same as the particle obtained by CPT operation as shown in Figure 1.11. Thus, unlike the case of the electron and positron which have different charges, neutrinos and antineutrinos are neutral; therefore, if we consider that ψ_R is the same as $\bar{\psi}_R$, then there are only two possible states with the same mass:

$$\psi_R \equiv \bar{\psi}_R,$$
$$\psi_L \equiv \bar{\psi}_L.$$

These sets of states are called Majorana neutrinos ν^M. If the neutrino happens to be a Majorana neutrino, then the particle is its own antiparticle, and it will have identical interactions with

CPT

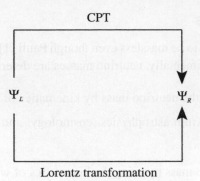

Lorentz transformation

Figure 1.11 Majorana neutrinos.

matter. A Majorana neutrino has neither electric dipole moment nor magnetic dipole moment. It has no charge radius. The lepton number is violated if neutrinos are Majorana particles.

There is a very important consequence of neutrinos being Majorana neutrinos. Some nuclei can undergo neutrinoless double β-decay, that is,

$$_Z^A X'' \rightarrow _{Z+2}^A Y'' + 2e^-,$$

which may be understood in the following way.

In general, β^- and β^+-decay are described by the following basic processes:

$$n \rightarrow p + e^- + \bar{\nu}_e$$

and

$$p \rightarrow n + e^+ + \nu_e,$$

or, in general, nuclear β-decay is given by the following reactions:

$$_Z^A X \rightarrow _{Z+1}^A Y + e^- + \bar{\nu}_e,$$
$$_Z^A X' \rightarrow _{Z-1}^A Y' + e^+ + \nu_e,$$

In a neutrinoless double β-decay, virtual neutrinos $(\nu_e = \bar{\nu}_e)$ being emitted at one vertex are

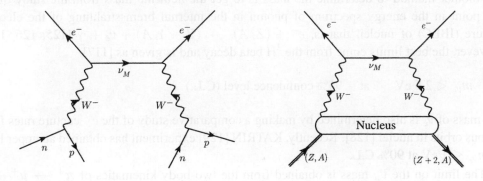

Figure 1.12 Neutrinoless double beta decay ($0\nu\beta\beta$).

absorbed at the other vertex as shown in Figure 1.12, giving electrons and positrons that violate lepton number conservation.

1.4.2 Neutrino mass

Neutrinos have been assumed to be massless even though Pauli [1] in his original letter proposed a 'tiny mass' for them. Experimentally, neutrino masses are determined by the following ways:

(i) direct determination of the neutrino mass by kinematics of weak decays,

(ii) indirect determination from astrophysics, cosmology, and neutrinoless double β-decay $(0\nu\beta\beta)$.

The determination of neutrino mass by using the kinematics of weak decay processes is model independent as it is obtained using energy and momentum conservation while the determination of neutrino mass from astrophysics, cosmology and neutrinoless double β-decay are model dependent and suffer from the uncertainties of the parameters in the theoretical analyses of the cosmological data and the double β-decay.

(i) Direct determination

For many years, neutrino mass for ν_e was determined mainly from the end point of the energy spectrum of β-electron as proposed by Fermi [23] in the β-decay of $^3\text{H} \rightarrow ^3\text{He} + e^- + \bar{\nu}_e$. In the presence of neutrino mass, the shape factor $\frac{d\Gamma}{dE_e}$ of the electron energy spectrum is given by:

$$S(E_e) \propto (Q - E_e)^2 [(Q - E_e)^2 - m_{\nu_e}^2]^{\frac{1}{2}},$$

where $Q = 18.6$ keV for β-decay of ^3H. This is subject to small atomic corrections due to the excitation of the ^3He atom [122] during the β-emission process. In modern times, the Institute of Theoretical and Experimental Physics (ITEP), Moscow group reported a nonzero neutrino mass ($m_{\nu_e} < 35\text{eV}$) [123] using β-spectrum of tritium. The ITEP group repeated the measurements and reported in 1980 [124], a neutrino mass in the range $14 \text{ eV} < m_{\nu_e} < 45 \text{ eV}$. Later, many more measurements were performed by this group.

Another method to determine the mass is to get the neutrino mass from the study of the end point in the energy spectrum of photon in the internal bremsstrahlung of the electron capture (IBEC) of nuclei, that is, $e^- + (Z, A) \rightarrow (Z - 1, A) + \nu_e + \gamma$ [125, 126, 127]. However, the best limits come from the ^3H beta decay and is given as [117]:

$$m_{\nu_e} < 2.5 \text{ eV} \qquad \text{at} \quad 95\% \text{ confidence level (C.L.)}$$

The mass of ν_e is also determined by making a comparative study of the e^- capture rates from various orbits in nuclei [128]. Recently, KATRIN [129] experiment has obtained an upper limit on $m_{\nu_e} \leq 1$ eV at 90% C.L.

The limit on the ν_μ mass is obtained from the two-body kinematics of $\pi^+ \rightarrow \mu^+ + \nu_\mu$ decay at rest (DAR). Since the pion decay is a two-body decay and takes place when the pion is at rest, the momentum and energy conservation leads to

$$m_{\nu_\mu}^2 = m_\pi^2 + m_\mu^2 - 2m_\pi \sqrt{m_\mu^2 + |\vec{p}_\mu|^2}. \tag{1.33}$$

In an experiment performed by Assamagan et al. [130], the momentum of the muon is found to be

$$|\vec{p}_\mu| = 29.792 \pm 0.00011 \text{ MeV}.$$

Using the value of $|\vec{p}_\mu|$ as quoted here and the masses of the muon and pion, the value of $m_{\nu_\mu}^2$ is found to be

$$m_{\nu_\mu}^2 = -(0.016 \pm 0.023) \text{ MeV}^2.$$

In many cases, the decaying pion is not completely at rest and may carry a small momentum before decaying. Therefore, a measurement of pion momentum and muon momentum is required to determine the ν_μ mass $m_{\nu_\mu}^2 = \left[\sqrt{|\vec{p}_\mu|^2 + m_\mu^2} - \sqrt{|\vec{p}_\pi|^2 + m_\pi^2} \right]^2 - |\vec{p}_\nu|^2$. Experiments of this kind lead to a mass limit of $\nu_\mu < 500$ keV [131]. The best limit on ν_μ mass is [117]

$$m_{\nu_\mu} < 190 \text{ keV} \quad \text{at} \quad 90\% \text{ C.L.}$$

The limit on m_{ν_τ} is obtained from the decay kinematics of $\tau^- \to 3\pi + \nu_\tau$ and $\tau^- \to 5\pi + \nu_\tau$, assuming it to be a two-body decay of τ^- at rest, that is,

$$\tau^- \longrightarrow h^- + \nu_\tau,$$

where h^- represents the hadronic system composed of three or five pions. The aforementioned decay kinematics is similar to what we have obtained in the case of m_{ν_μ} determination from the pion decay in Eq. (1.33). In the case of pion decay, we have only one particle, that is, μ^- in the final state whose momentum is measured experimentally. However, in the case of τ^-, we have a system consisting of three or five particles; therefore, the measurement of the momentum of particles is not easy and hence, E_h is determined from the invariant mass distribution of the hadronic system [132]. From energy and momentum conservation, the energy of the hadronic system is given by:

$$E_h = \frac{m_\tau^2 + m_h^2 - m_{\nu_\tau}^2}{2m_\tau}. \tag{1.34}$$

ν_τ mass is also determined from the distribution of missing energy in the three charged particle decays of τ^+ and τ^- [132, 133, 134, 135].

The best limit on m_{ν_τ} is [117]:

$$m_{\nu_\tau} < 18.2 \text{ MeV} \quad \text{at} \quad 95\% \text{ C.L.} \tag{1.35}$$

(ii) Cosmological and astrophysical observations

The limits on neutrino masses have also been obtained from cosmological observations. The presence of cosmic neutrinos, which are relics of the Big Bang, like the cosmic microwave background radiation(CMBR) is a prediction of the standard cosmological model. These cosmic neutrinos affect cosmological evolution and may be used to get information on neutrino properties, specially, the neutrino mass. The limits on neutrino masses have been obtained by measuring the anisotropy of CMBR using Planck satellite. Some cosmological observations are mainly sensitive to the sum of neutrino masses $\sum_i m_i$. In the standard model of Big Bang cosmology, the three active neutrinos of the standard model of particle physics are assumed to be massless. However, extensions of the standard cosmological model with varying neutrino masses have been used to put limits on neutrino masses. The present bounds using different combinations of current cosmological data give a range of neutrino masses, 0.14 eV$< \sum m_\nu <$0.72 eV.

In a supernova burst, almost 10^{53} neutrinos and antineutrinos of all flavors are released in a few seconds. The neutrinos released during a few seconds of supernova explosion contain mostly electron type neutrinos; the later part of the burst contains neutrino of all flavors. The observations made on the neutrinos from the supernova event (SN1987A) that took place in 1987 in the Large Magellanic Cloud just outside our Milky Way galaxy has also been used to put a limit on neutrino masses. For example, the latest analyses suggest $m_{\bar{\nu}_e} < 5.7$ eV.

(iii) Neutrinoless double β-decay

An alternative way to measure the mass of a neutrino is to look for neutrinoless double β-decay. Many experiments are being conducted to observe $0\nu\beta\beta$ decay of heavy nuclei on ^{76}Ge, ^{136}Xe, etc. The lifetime of the heavy nuclei undergoing double β-decays depends upon the mass of the neutrinos which have to be of the Majorana type ($\nu = \bar{\nu}$). In the past, the Heidelberg-Moscow experiment claims to have observed $0\nu\beta\beta$ decay, but the results are considered to be controversial and inconclusive.

The double β-decay rates depend upon the effective mass of neutrino which is defined as:

$$\langle m_\nu \rangle = \sum_{i=1}^{3} m_i U_{ei}^2, \tag{1.36}$$

where U_{ei} are the mixing matrix elements and m_i are the masses of the mass eigenstates. The electron type neutrino is represented by a mixed state, described by:

$$|\nu_e\rangle = \sum_i U_{ei} |\nu_i\rangle.$$

The present limits for the effective Majorana mass of the electron neutrino ($\langle m_{\nu_e} \rangle$), which is a coherent sum of the mass eigenvalues weighted with the square of the elements of the mixing matrix, are in the range 0.1 eV$< m_{\nu_e} <$0.4 eV.

However, these limits are model dependent due to the uncertainties in the parameters of neutrino mixing as well as the uncertainties in the knowledge of the nuclear matrix element of heavy nuclei undergoing neutrinoless double β-decay.

1.4.3   Neutrino charge and charge radius

(i) Neutrinos when proposed were assumed to be electrically neutral. However, there are attempts to measure the charge of the neutrino in β-decays by measuring the charge of the neutron Q_n and the total charge of the proton and electron, that is, $|Q_p + Q_{e^-}|$ in the decay $n \rightarrow p + e^- + \bar{\nu}_e$ [136, 137]. This gives a limit $Q_{\bar{\nu}} < (0.5 \pm 2.9) \times 10^{-21} e$. The astrophysical limit derived from the SN1987A supernova observation is [138]:

$$Q_{\bar{\nu}} < 2 \times 10^{-15} e.$$

(ii) The charge of neutrino is consistent with zero to a very high degree of precision; however, it may have a charge distribution like a neutron even though it is considered a point particle in the field theory. Attempts to determine the charge radius have been made [139] for ν_e and ν_μ from $\nu_e e$ [140], $\bar{\nu}_e e$ [141], and $\nu_\mu e$ [142] scattering. Like hadron, the mean square charge radius is deduced from a measurement of the vector form factor in the $\nu_e e$ and $\nu_\mu e$ elastic scattering using the relation

$$\langle r^2 \rangle = -6 \frac{d}{dq^2} f(q^2)|_{q^2=0}, \tag{1.37}$$

where $f(q^2)$ is the form factor corresponding to the matrix element of the vector current. In the case of neutral particles, the value of $\langle r^2 \rangle$ could be negative or positive and the following limits [143, 144, 145] are obtained in the case of ν_e and ν_μ:

$$-5.3 \times 10^{-32} < \left[\langle r^2 \rangle_{\nu_\mu}\right] < 1.3 \times 10^{-32} \text{ cm}^2,$$

$$-0.77 \times 10^{-32} < \left[\langle r^2 \rangle_{\nu_\mu}\right] < 2.5 \times 10^{-32} \text{ cm}^2,$$

$$-5.0 \times 10^{-32} < \left[\langle r^2 \rangle_{\nu_e}\right] < 10.2 \times 10^{-32} \text{ cm}^2.$$

1.4.4 Magnetic and electric dipole moments of neutrinos

In general, the electroweak properties of a spin $\frac{1}{2}$ Dirac particle are described in terms of two vector form factors called electric and magnetic form factors which in the static limit define the charge and magnetic moment, and two axial vector form factors called axial and tensor form factors which in the static limit define the axial charge and electric dipole moment. If the neutrinos are considered to be Dirac neutrinos with nonzero mass, it could have these form factors to be nonvanishing and experimental attempts can be made to study them. The electromagnetic properties like electric and magnetic dipole moments of neutrinos depend upon the nature of the neutrinos. Dirac neutrinos can have electric and magnetic moments like neutrons. It is to be noted that a magnetic moment of the order $e \times 10^{-13}$ cm^2 was proposed by Pauli [1]. However, if the neutrinos are Majorana neutrinos, their electric and magnetic moments are zero. The existence of electric dipole moment depends on the validity of CP invariance which forbids a nonzero electric dipole moment for elementary particles. Since neutrinos participate in weak interactions which violate CP invariance, they may have an electric dipole moment; on the other hand, there is no symmetry principle which forbids the existence of magnetic dipole moment for neutral particles. Theoretically, even massless particles can have magnetic dipole moment.

The standard model calculations for the magnetic moment of a neutrino with mass m_ν yield a very small magnetic moment of the order $3 \times 10^{-19} \frac{m_\nu}{eV} \mu_B$. In general, the neutrino magnetic moment need not be proportional to the neutrino mass; there are models constructed to give larger magnetic moments. Experimentally, the laboratory limits on the neutrino magnetic moments are obtained by performing elastic $\nu_e e$, $\bar{\nu}_e e$, and $\nu_\mu e$ scattering. The magnetic moment of the neutrino additionally contributes to the weak cross section due to the electromagnetic scattering of neutrinos. For a neutrino of magnetic moment μ_ν, the additional differential cross section is given by:

$$\frac{d\sigma^{EM}}{dE'_e} = \frac{\pi \alpha^2 \mu_\nu^2}{m_e^2} \left(\frac{1}{T} - \frac{1}{E_\nu} \right), \tag{1.38}$$

where T is the recoil kinetic energy of the electron. Therefore, to see the maximum effect of magnetic moment on the cross section, very low energy scattering processes are favored. Antineutrinos from reactors and neutrino beams from pions decay at rest (DAR) at the accelerators are favored. Earlier experiments were done by Reines and Cowan, but now many experiments have been performed at various reactors and accelerators around the world to determine the magnetic moment of neutrinos. A stronger limit is obtained from the astrophysical and cosmological considerations but they are model dependent. A summary of these results can be found in Refs. [143, 146, 147, 148]; the present limits are:

$$\mu_{\nu_e} < 1.8 \times 10^{-10} \mu_B.$$

In case of ν_μ, the limit on the magnetic moment μ_{ν_μ} from accelerator experiments is [117]:

$$\mu_{\nu_\mu} < 7.4 \times 10^{-10} \mu_B.$$

The limits on the magnetic moment of ν_τ are weaker and come from the study of $e^+ e^- \rightarrow \nu \bar{\nu} \gamma$ processes at accelerators [149, 150, 151] and also from $\nu_\tau e^- \rightarrow \nu_\tau e^-$ scattering from the bubble chamber experiments at CERN and BEBC [152]. The limits are [117]:

$$\mu_{\nu_\tau} \approx 5.4 \times 10^{-7} \mu_B.$$

1.4.5 Helicity of neutrino

The discovery of parity violation and the revival of the two-component theory of neutrinos in the study of the nuclear β-decays implied that the neutrinos are left-handed and antineutrino are right-handed. Indirect evidence of this property of the neutrino and antineutrino was available from observations made on the polarization of electrons and other spin momentum correlation measurements of the emitted electrons and positrons in many weak decays of elementary particles. The direct measurement of the helicity of ν_e was made in an excellent experiment performed by Goldhaber et al. [153]. They measured the polarization of photons in a weak electron capture experiment by ^{152}Eu nuclei leading to neutrino (ν_e)

and ^{152}Sm* nucleus which decays to ^{152}Sm by photon emission. From the polarization measurements of photon, it was inferred that the neutrinos emitted in such decays are left-handed. Later, experiments done on the muon capture process on ^{12}C and many other experiments confirmed that all neutrinos are left-handed and antineutrinos are right-handed. This will be discussed in some detail in Chapter 5.

1.5   New Developments

1.5.1   Standard model of electroweak interactions and neutral currents

The standard model of electroweak interactions is one of the most important milestones in our understanding of fundamental interactions. It unifies electromagnetic (EM) and weak interactions and is based on the principle of local gauge invariance of basic interactions. The Fermi theory of β-decay was formulated in analogy with electromagnetic interaction in quantum electrodynamics, which is generated from invariance under the local gauge transformations.

The Fermi theory evolved phenomenologically into the $V - A$ theory which was considered to be a low energy manifestation of a theory considered to be mediated by vector bosons W as suggested earlier by Klein [154], Bludman [50], Schwinger [49], and Leite Lopes [51]. Since weak interactions are of very short range, the mediating W bson has to be of very high mass. Some experiments were suggested by Pontecorvo [105] and Lee and Yang [107] and carried out at BNL [155] and CERN [156] laboratories to search for W bosons in the range of a few GeV without any success. The requirement for W bosons to be massive prevented the formulation of weak interactions as a local gauge theory with W bosons as gauge bosons as the principle of local gauge invariance required that gauge bosons be massless. Therefore, the theories proposed earlier to unify electromagnetic and weak interaction did not progress until Weinberg [157] and Salam [37] in 1967 used the Higgs mechanism of spontaneous breaking of symmetry to generate the masses of the gauge bosons by introducing a scalar field. The mechanism of mass generation by spontaneous breaking of the local gauge symmetry named after Higgs was developed earlier independently by Higgs [158], Englert and Brout [159], Guralnik et al. [160] and Kibble [161] in the 1960s following the works of Nambu [46], Goldstone [162], and Nambu and Jona-Lasinio [163].

The introduction of the scalar field popularly known as Higgs boson (also God particle) facilitates the generation of masses for the massless gauge bosons while preserving the local gauge invariance of the weak interaction Lagrangian. The standard model applied to the leptons reproduces the essential features of the phenomenological weak interactions like parity violation, two-component neutrino, and the massive $W^{\pm}$ bosons mediating the charged current interactions, as well as the massless photon (γ) mediating the electromagnetic interaction. In addition, it also predicts the existence of a new massive neutral gauge boson Z^0 implying neutral currents in the leptonic sector (electron and neutrino) which were discussed quite early in the history of weak interactions but were never observed. The masses of the gauge bosons $W^{\pm}$ and Z^0 and their couplings to the leptons were predicted in terms of g, the weak coupling of

$W^{\pm}$, e the electromagnetic coupling of the photon to the lepton, and a free parameter θ_W called the weak mixing angle to be determined from the experiments. The mass and the coupling of the Higgs boson remain undetermined in the model. The standard model of the leptons was later extended to the quark sector including strange quarks following the GIM mechanism and is applied, in general, to describe the electroweak interactions of leptons and hadrons.

Specifically, the standard model of the electroweak interactions is based on the local gauge invariance of the Dirac Lagrangian under the group $SU(2)_L \times U(1)_Y$ in which the left-handed leptons and hadrons are assigned to the doublets of the group $SU(2)_L$ and their right-handed partners are assigned to a singlet for each flavor as follows:

$$\begin{pmatrix} \nu_e \\ e^- \end{pmatrix}_L, \ \begin{pmatrix} \nu_\mu \\ \mu^- \end{pmatrix}_L, \ \begin{pmatrix} \nu_\tau \\ \tau^- \end{pmatrix}_L, \ \begin{pmatrix} u \\ d' \end{pmatrix}_L, \ \begin{pmatrix} c \\ s' \end{pmatrix}_L, \ \begin{pmatrix} t \\ b' \end{pmatrix}_L.$$

These left-handed doublets form the basic representation of $SU(2)$ and their right-handed partners like

$$e_R, \ \mu_R, \ \tau_R, \ u_R, \ d'_R, \ c_R, \ s'_R, \ t_R, \ b'_R$$

form the singlet representation. The requirement of the gauge invariance introduces three massless gauge fields $W^{\pm}, W^3_\mu$ corresponding to $SU(2)_L$ with a coupling g and one massless gauge field B_μ corresponding to U(1) with a coupling g'. The $W^{\pm}$ fields are charged, while W^3_μ and B_μ are neutral. The new scalar Higgs field ϕ is introduced in such a way that the linear combinations of B_μ and W^3_μ fields, that is,

$$A_\mu = \frac{1}{\sqrt{g^2 + g'^2}} \left(g' W^3_\mu + g B_\mu \right) \tag{1.39}$$

remains massless and is to be identified with the photon field; the orthogonal combination of B_μ and W^3_μ, that is,

$$Z_\mu = \frac{1}{\sqrt{g^2 + g'^2}} \left(g W^3_\mu - g' B_\mu \right) \tag{1.40}$$

along with W^+ and W^- become massive and mediate the neutral and charged current weak interactions, respectively (see Chapter 8).

The experimental observation of neutral currents (NC) was a major triumph of the standard model when they were discovered at CERN in experiments with $\nu_\mu / \bar{\nu}_\mu$ beams [164] in the reactions

$$\nu_\mu + N \longrightarrow \nu_\mu + N, \tag{1.41}$$

$$\bar{\nu}_\mu + N \longrightarrow \bar{\nu}_\mu + N. \tag{1.42}$$

The existence of NC events was later observed in many other experiments [165, 166, 167]. Purely leptonic NC events in reactions like $\nu_\mu + e^- \rightarrow \nu_\mu + e^-$ and $\bar{\nu}_\mu + e^- \rightarrow \bar{\nu}_\mu + e^-$

were also observed soon afterward [168]. Later at SLAC, NC reactions with electron beams were observed in 1978 through the observation of parity violating effects in polarized electron scattering from nucleons and nuclear targets due to the interference of photon and Z^0 exchange in $e^- p$ scattering [169].

The neutrino experiments induced by neutral currents in the leptonic as well as the hadronic sector played an important role in determining the value of the weak mixing angle θ_W.

1.5.2 Discovery of $W^\pm$, Z^0, and Higgs boson

(i) $W^\pm$ and Z^0 boson:

The weak gauge bosons $W^\pm$ and Z^0 predicted by the standard model of Glashow–Weinberg–Salam (G–W–S) are produced in hadronic collisions like the $\bar{p}p$ collisions in which one of the antiquarks in $\bar{p}$ collides with the quark in p to produce W^+ or W^- which decays through the weak processes, for example,

$$
\begin{aligned}
\bar{d} + u &\longrightarrow W^+ \longrightarrow \mu^+ + \nu_\mu \ (e^+ + \nu_e), \\
\bar{u} + d &\longrightarrow W^- \longrightarrow \mu^- + \bar{\nu}_\mu \ (e^- + \bar{\nu}_e), \\
\bar{d} + d &\longrightarrow Z^0 \longrightarrow e^- + e^+ \ (\mu^- + \mu^+), \\
\bar{u} + u &\longrightarrow Z^0 \longrightarrow e^- + e^+ \ (\mu^- + \mu^+).
\end{aligned}
\tag{1.43}
$$

$W^\pm$ and Z^0 bosons were first discovered in the $\bar{p}p$ collision experiments performed at CERN in 1983 by two groups which used the beams of protons with center of mass energy $E_{CM} = 540$ GeV [170, 171]. We see from Eq. (1.43) that the signature of $W^\pm$ bosons are single charged leptons $\mu^\pm$ with large missing transverse momenta, while the signature of Z^0 are two-charged leptons coming out at an angle θ such that

$$
M_Z^2 = 2E_+ E_- (1 - \cos\theta).
$$

After the discovery of $W^\pm$ and Z^0 at CERN, other measurements of their mass have been performed at hadron colliders at CERN and Fermilab as well as at the $e^- e^+$ colliders at LEP in CERN and SLAC [172, 173, 174]. The masses of $W^\pm$ and Z^0 bosons are measured to be [117]:

$$
\begin{aligned}
M_W &= 80.379 \pm 0.012 \text{ GeV}, \\
M_Z &= 91.1876 \pm 0.0021 \text{ GeV}.
\end{aligned}
$$

(ii) Higgs boson:

The search for Higgs boson as formulated in the standard model started quite early first at LEP in CERN and then at Tevatron in Fermilab without any success. However, the experiments at Fermilab indicated that Higgs boson, if existed, could have a mass between 115 GeV and 140 GeV [175]. In July 2012, using the large hadron collider (LHC) at CERN operating at a center of mass energy of 7 TeV in the $\bar{p}p$ scattering, the ATLAS (a toroidal LHC apparatus) and CMS (compact muon solenoid) collaborations succeeded in observing for the first time the Higgs boson through its decays in W^+W^- modes and two photon modes. A typical Higgs event observed by the CMS collaboration has been shown in Figure 1.13 [176, 177]. A combined analysis of both data at ATLAS and CMS gives [117]:

$$M_H = 125.18 \pm 0.16 \text{ GeV}.$$

Later, its mass, properties, and decay modes were measured and confirmed by the collaborations in many other experiments [178].

(a)   (b)

Figure 1.13   (a) $p\bar{p}$ collisions inside the CMS detector at CERN. This event was recorded in 2012 by the compact muon solenoid (CMS) at the large hadron collider (LHC). (b) Experimental confirmation of the Higgs boson [176, 177].

1.5.3   Neutrino mass, mixing, and oscillations

Most of the weak interaction phenomenology after the discovery of nuclear β-decays was done assuming neutrinos to be massless but the question of neutrino mass has been a subject of great interest ever since Pauli's proposal of neutrinos and the possibility of its nonzero mass [1]. A proper understanding of neutrino mass and its origin could solve many problems being faced today in neutrino physics like the nature of neutrinos, that is, Dirac or Majorana, mixing of neutrinos and neutrino oscillations, that is, three or four flavors, mass hierarchy, CP violation in lepton sector, and sterile neutrinos. We have given the experimental status of the search for neutrino masses in Section 1.4.2; a theoretical and experimental description of neutrino oscillations is deferred to Chapter 18. We present here a brief discussion on the existence of neutrino mass, mixing, and oscillations and their implications.

Pontecorvo [179] suggested that if a neutrino has nonzero mass, then it may oscillate into an antineutrino $\nu_e \rightleftarrows \bar{\nu}_e$ in analogy with the $K^0 - \bar{K}^0$ oscillation [61]. The possibility that if $\nu_e \neq \nu_\mu$; then ν_e could oscillate into ν_μ and vice versa was suggested by Maki et al. [71]. This possibility, called the flavor oscillation, was later formulated by Gribov and Pontecorvo [180] and Bahcall and Frautschi [181]. The possibility of oscillation requires neutrinos to have nonzero mass. In the case of massive neutrinos, other possibilities like the decay of a heavier neutrino into lighter neutrinos by emitting a photon or other lighter particles are also possible.

Interest in neutrino oscillation physics started to grow after the early results of the solar neutrino experiments reported by Davis et al. from the Homestake mines lab in USA [87, 88]. Through this experiment, they claimed to have observed the ν_e flux from the sun which was smaller than the flux predicted by contemporary solar models [182]. The phenomenon of neutrino oscillations in which a ν_e oscillates to other flavors offers a natural explanation for the reduction in the solar neutrino flux. However, this explanation required a large mixing of the two flavors of neutrinos and the fine tuning of neutrino oscillation parameters to explain the experimental results obtained by Davis et al. Moreover, all these conclusions were subject to the uncertainties in the parameters of the standard solar model. In view of this, many experiments to observe the neutrino oscillations using neutrino beams from other sources like nuclear reactors and accelerators were done without any positive result. However, during the 1990s the deep underground detectors at Kamiokande [183] in Japan and the IMB collaboration [184] in USA, initially planned to observe the proton decay predicted by the grand unified theories, succeeded in detecting a depletion in the ν_μ flux in atmospheric neutrinos compared to the theoretical calculations.

The Kamiokande collaboration further studied the zenith angle dependence of the atmospheric ν_μ flux and found that the up going ν_μ flux, where the distance neutrinos travel through the earth, is smaller than the down going ν_μ flux observed in detectors placed in the underground observatory. This angle dependence could be explained on the basis of neutrino oscillations. However, the statistics of the experimental data was poor and the low energy ν_μ data was not as conclusive as the higher energy data on ν_μ flux. Finally, the Super-Kamiokande experiment with a much larger detector mass, of about 15 times the original Kamiokande detector, confirmed these findings and established the phenomenon of neutrino oscillation consistent with the parameters of ν oscillation phenomenology.

After the early indications of neutrino oscillations, in the case of solar neutrinos, with the experiments of Davis and its model dependence on solar model parameters, neutrino oscillations was finally confirmed in a model independent way at the SNO Lab in Sudbury, Canada by an observation of CC (charged current) and NC (neutral current) reactions on deuterium and elastic scattering (ES) on electron targets in three independent experiments, that is,

$$\nu_e + d \longrightarrow e^- + p + p \quad \text{(CC)}, \qquad\qquad \nu_{e,\mu,\tau} + d \longrightarrow \nu_{e,\mu,\tau} + n + p \quad \text{(NC)},$$

$$\nu_e + e^- \longrightarrow \nu_e + e^- \qquad \text{(ES)}.$$

It was observed that the flux of the CC reaction [185] which is affected by the $\nu_e \to \nu_\mu$ and $\bar{\nu}_e \to \bar{\nu}_\mu$ oscillations was almost three times smaller than the NC reaction [185] which is not affected by the neutrino oscillation. These observations used the higher energy neutrinos from the decays of ^{8}B produced in the solar core. The ν_e flux of these neutrinos was smaller than ν_e flux produced in $pp \to de^+\nu_e$ reaction by a factor of 10^{-4} which are of lower energy. Experiments with these low energy ν_e sources were done at GALLEX and SAGE with the detectors using gallium with a very low Q value of 233 keV. The observed event rates showed that the ν_e flux were smaller by a factor of two compared with the predicted fluxes, thus, confirming the phenomenon of neutrino oscillations for solar neutrinos in the region of very low energy. The phenomenon of neutrino oscillations in the lower energy region was later confirmed with the reactor antineutrinos at KAMLAND, RENO, and other experiments. Finally, in many long baseline neutrino experiments performed with accelerator neutrinos at LSND, MiniBooNE, SciBooNE etc., the phenomenon of neutrino oscillations was also confirmed in the region of intermediate and high energies. A detailed description of these experiments is given in Chapter 17.

The physics of neutrino mass, mixing, and oscillations can be demonstrated by a simple example of two flavor mixing of ν_e and ν_μ in analogy with quark mixing [70]. A pure ν_e beam described by a wave function while traveling in space may develop a component of ν_μ in this beam; the mixture of the ν_μ wave function will describe the probability of finding the ν_μ component in the ν_e beam after a time t as illustrated in Figure 1.14. We assume that the flavor state ν_e and ν_μ participating in the weak interactions are mixtures of the mass eigenstates ν_1 and ν_2 and the mixing is described by a unitary mixing matrix U such that:

$$\nu_{l=e,\mu} = \sum_{i=1,2} U_{li}\nu_i. \tag{1.44}$$

The unitarity of the U matrix requires that in two-dimensional space it is described by one parameter which is generally chosen to be θ such that:

$$U = \begin{pmatrix} c_{12} & s_{12} \\ -s_{12} & c_{12} \end{pmatrix} \tag{1.45}$$

where $c_{12} = \cos\theta$ and $s_{12} = \sin\theta$. As a pure beam of ν_e at $t = 0$ propagates, the mass eigenstates $|\nu_1\rangle$ and $|\nu_2\rangle$, occurring in Eq. (1.44), would evolve according to

$$|\nu_1\rangle = \nu_1(0)e^{-iE_1 t} \tag{1.46}$$

$$|\nu_2\rangle = \nu_2(0)e^{-iE_2 t} \tag{1.47}$$

where $E_1 = \sqrt{|\vec{p}|^2 + m_1^2} \approx |\vec{p}| + \frac{m_1^2}{2|\vec{p}|}$ and $E_2 = \sqrt{|\vec{p}|^2 + m_2^2} \approx |\vec{p}| + \frac{m_2^2}{2|\vec{p}|}$, $|\vec{p}| \approx E$ being the common momentum of neutrinos with energy E_1 and E_2; m_1 and m_2 are the mass of $|\nu_1\rangle$ and $|\nu_2\rangle$ states, respectively. After a time t, the $|\nu_e(t)\rangle$ will be a different admixture of $|\nu_1\rangle$ and $|\nu_2\rangle$. The probability of finding ν_μ in the beam of ν_e at a later time t is given by (see Chapter 18)

$$P(\nu_e \to \nu_\mu) = \sin^2 2\theta \sin^2\left(\frac{\Delta m^2}{4E}L\right) = \sin^2\theta \sin^2\left(1.27\frac{\Delta m^2}{E}L\frac{[\text{eV}^2][\text{km}]}{[\text{GeV}]}\right), \tag{1.48}$$

Thus, we see that for $P(\nu_e \to \nu_\mu) \neq 0$, we need $\Delta m^2 \neq 0$ and $\theta \neq 0$, that is, we need the mass difference between the neutrino mass eigenstates to be nonzero, implying that at least one

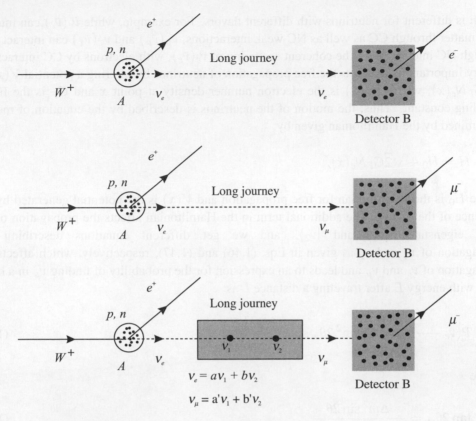

Figure 1.14 A W^+-boson interacts at the point A and produces e^+ and ν_e. The ν_e travels a long distance; it interacts with the detector B and produces a lepton e^- as shown in the top panel. If the flavor of the initial and final lepton is different, that is, $e^- \neq \mu^-$, it can be concluded that during its long journey, a neutrino which was produced initially with a flavor e gets converted into another flavor μ as shown in the central panel. This is only possible when the neutrino propagates as a linear combination of mass eigenstates ν_1 and ν_2, which are different from the flavor eigenstates ν_e and ν_μ as shown in the bottom panel.

of them is massive, and the mixing angle θ to be nonzero. Thus, if the explanation of the solar neutrino flux deficit and other deficits observed in the atmospheric, reactor, and accelerator neutrinos are explained to be due to neutrino oscillations, neutrinos should have nonzero mass and the neutrino flavors should mix.

1.5.4 Matter enhancement of neutrino oscillations and the MSW (Mikheyev–Smirnov–Wolfenstein) effect

One of the major developments in the physics of the neutrino oscillation is the inclusion of matter effects in the study of the solar neutrino problem first discussed by Wolfenstein [186] and later elaborated by Mikheyev and Smirnov [187] and Bethe [188]. Wolfenstein [186] demonstrated that solar neutrinos while propagating from the center of the sun, where they are created in nuclear reactions, to terrestrial detectors may undergo coherent scattering with matter which is different for neutrinos with different flavors. For example, while $\nu_e(\bar{\nu}_e)$ can interact with matter through CC as well as NC weak interactions, $\nu_\mu(\bar{\nu}_\mu)$ and $\nu_\tau(\bar{\nu}_\tau)$ can interact only through NC interactions. The coherent scattering of $\nu_e(\bar{\nu}_e)$, with electrons by CC interactions is very important as it affects the free propagation of neutrinos by offering a potential $V(x) = \sqrt{2}G_F N_e(x)$, where $N_e(x)$ is the electron number density at point x and G_F is the Fermi coupling constant. Thus, the motion of the neutrinos is described by the equation of motion determined by the Hamiltonian given by:

$$H = H_0 + \sqrt{2}G_F N_e(x),$$

where H_0 is the Hamiltonian for free propagation and $V(x)$ is the potential generated by the presence of the matter. The additional term in the Hamiltonian affects the propagation of the mass eigenstates $|\nu_1\rangle$ and $|\nu_2\rangle$, and we get different equations describing the propagation of ν_1 and ν_2 as given in Eqs. (1.46) and (1.47), respectively, which affects the propagation of ν_e and ν_μ and leads to an expression for the probability of finding ν_μ in a beam of ν_e with energy E after traveling a distance L as:

$$P(\nu_e \longrightarrow \nu_\mu) = \frac{1}{2}\sin^2 2\theta_m \sin^2\left(\frac{\Delta m_m^2 L}{4E}\right), \tag{1.49}$$

where

$$\tan 2\theta_m = \frac{\Delta m^2 \sin 2\theta}{\Delta m^2 \cos 2\theta - A}, \tag{1.50}$$

with

$$A = 2\sqrt{2}G_F N_e E, \tag{1.51}$$

$$\Delta m_m^2 = \sqrt{(\Delta m^2 \cos 2\theta - A)^2 + (\Delta m \sin 2\theta)^2}. \tag{1.52}$$

The mixing angle θ_m can be maximal (i.e., $\theta_m = \frac{\pi}{4}$) even for very small values of θ, if

$$\Delta m^2 \cos^2 2\theta = A = 2\sqrt{2}G_F N_e E.$$

This condition is called the resonance condition and it explained the solar neutrino problem which needed a large mixing angle. Since $N_e(x)$ is a varying function of number density ($(N_e(x) = \frac{Y_e}{m_N}\rho(x)$, where Y_e is the number of electrons per nucleon, m_N is the nucleon mass, and $\rho(x)$ is the density of the sun), there might be an energy for which at a certain point x, the condition (A) in Eq. (1.51) may be satisfied by making ν_e oscillate into ν_μ at that point x. This oscillation may be an adiabatic process or otherwise. The enhancement of neutrino oscillations is called the MSW (Mikheyev–Smirnov–Wolfenstein) effect. The details are discussed in Chapter 18.

1.5.5 Three-flavor neutrino oscillations and mass hierarchy

In the three-flavor oscillation scenario (ν_e, ν_μ, ν_τ), the mass eigenstates ν_1, ν_2, and ν_3 are related to the flavor eigenstates ν_e, ν_μ, ν_τ such that:

$$
\begin{aligned}
|\nu_e\rangle &= a_1|\nu_1\rangle + b_1|\nu_2\rangle + c_1|\nu_3\rangle \\
|\nu_\mu\rangle &= a_2|\nu_1\rangle + b_2|\nu_2\rangle + c_2|\nu_3\rangle \\
|\nu_\tau\rangle &= a_3|\nu_1\rangle + b_3|\nu_2\rangle + c_3|\nu_3\rangle
\end{aligned}
\tag{1.53}
$$

where a, b, and c are the different weight factors. If one assumes neutrinos to be Dirac particles, then in a three-flavor oscillation, the flavor and mass eigenstates are related through a unitary 3×3 unitary matrix given by:

$$
\begin{bmatrix} \nu_e \\ \nu_\mu \\ \nu_\tau \end{bmatrix} = \begin{bmatrix} U_{e1} & U_{e2} & U_{e3} \\ U_{\mu1} & U_{\mu2} & U_{\mu3} \\ U_{\tau1} & U_{\tau2} & U_{\tau3} \end{bmatrix} \begin{bmatrix} \nu_1 \\ \nu_2 \\ \nu_3 \end{bmatrix}.
\tag{1.54}
$$

The oscillation matrix is chosen to be unitary such that the total probability of oscillation (including all possible cases) is unity, that is,

$$
P(\nu_e \rightleftharpoons \nu_\mu) + P(\nu_\mu \rightleftharpoons \nu_\tau) + P(\nu_\tau \rightleftharpoons \nu_e) = 1.
\tag{1.55}
$$

The 3×3 unitary matrix is known as the PMNS (Pontecorvo, Maki, Nagakava and Sakata) matrix, whose parameters are three mixing angles θ_{12}, θ_{13}, θ_{23}, and a CP violating phase δ given by the U matrix in Eq. (1.22). Worldwide, many experiments are being performed, using neutrinos from different sources like accelerators, solar, reactor, atmospheric, etc., to determine the parameters $U_{ij}(i = e, \mu, \tau, j = 1, 2, 3, i \neq j)$ very precisely. The experiments like DUNE, T2K, NOνA, HyperK, etc. are expected to give some information about δ.

Corresponding to the three neutrino flavors, there are three mass eigenstates in which two of the neutrino mass eigenstates are nearly degenerate, that is,

$$
|\Delta m_{21}^2| \equiv \Delta m_{\text{small}}^2 \ll |\Delta m_{31}^2| \cong |\Delta m_{23}^2| \equiv \Delta m_{\text{big}}^2.
\tag{1.56}
$$

The present limits on Δm_{ij}^2 ($i \neq j = 1, 2, 3$) show that $\Delta m_{12}^2 = m_1^2 - m_2^2$ is the smallest of all the $\Delta m_{ij}^2(i \neq j)$, while Δm_{13}^2 is of the order of Δm_{23}^2. The analysis of the neutrino oscillation experiments does not give any information about the absolute masses corresponding to the three mass eigenstates. Therefore, there are two possibilities:

(i) $m_3^2 > m_1^2 \ (\approx m_2^2)$, or

(ii) $m_3^2 < m_1^2 \ (\approx m_2^2)$

as shown Figure 1.15 corresponding to the normal and inverted mass hierarchy of neutrinos. The two mass hierarchy scenarios affect many aspect of the analysis of the neutrino oscillations. The current analyses of the neutrino oscillation experiments suggest a preference for the normal mass hierarchy.

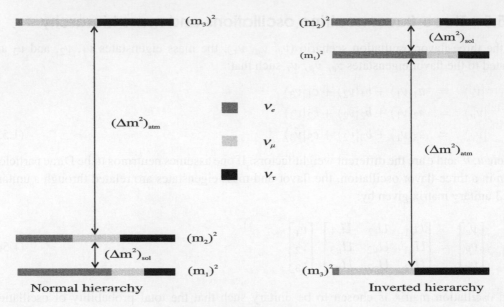

Figure 1.15 Neutrino mass hierarchy.

1.5.6 Sterile neutrinos and 3+1 flavor mixing

The phenomenology of the standard model of the electroweak interactions with three-flavor mixing is successful in describing most of the available experimental data from the various neutrino experiments done with solar, atmospheric, reactor, and accelerator (anti)neutrinos in terms of three mixing angles θ_{ij} $(i \neq j = 1,2,3)$, a phase angle δ, and the mass squared differences Δm_{ij}^2 $(i \neq j = 1,2,3)$. However, there are some 'anomalous' results from few oscillation experiments in the low energy regime from LSND, GALLEX, and reactor (anti)neutrinos which are not explained within the three-flavor scenario of the standard model and suggestions have been made for the existence of an additional flavor of neutrinos. Since there is strong experimental evidence from LEP experiments by measuring the total width of Z^0-decays into $q\bar{q}$ and $l\bar{l}$ pairs as well as from the cosmological data on the ^{4}He abundance that there are only three flavors of neutrinos in the energy region of $E < 46$ GeV, the additional neutrino must be 'sterile', that is, having no interaction with matter in this energy region even if they exist.

The simplest and most popular extension of the three-flavor mixing phenomenology within the standard model is to consider the 3+1 flavor neutrino mixing in which three active neutrinos ν_e, ν_μ, ν_τ are mixed states of the three massive neutrino states ν_1, ν_2, ν_3 and a fourth mass neutrino state ν_4. However, the fourth weak interaction state is a sterile neutrino ν_s which is mainly composed of a fourth neutrino mass state (ν_4) at the eV energy scale; the coupling of ν_s to other mass states ν_1, ν_2, ν_3 is negligible. The $3 + 1$ flavor mixing is described by a 4×4 unitary matrix which is parameterized in terms of 9 parameters, 6 of which are chosen to be the rotation angles, that is, θ_{12}, θ_{13}, θ_{14}, θ_{23}, θ_{24}, and θ_{34} and three phases, that is, δ_1, δ_2, and δ_3. Thus, there are three additional mixing angles and two additional phases making the interpretation of experimental data quite difficult and model dependent. There have been many experiments in recent years to determine these parameters. For the latest review on sterile neutrinos, please see Ref. [189, 190, 191]. Due to the presence of increased number of independent parameters, various assumptions have been made about the additional mass squared differences, mixing parameters, and phase angles for analyzing and interpreting the available data.

In the framework of 3+1 active–sterile neutrino mixing, the anomalies in the earlier experiments of LSND, GALLEX, and reactor antineutrinos are found to be sensitive to the oscillations generated by the mixing of the fourth sterile neutrino, that is, ν_4 through $\Delta m_{41}^2 \approx \Delta m_{42}^2 \approx \Delta m_{43}^2 \geq 1$ eV^2 which is much larger than the scale of Δm_{ij}^2 in solar and atmospheric neutrino mass squared differences, that is, $\Delta m_{sol}^2 = \Delta m_{21}^2 \approx 7.4 \times 10^{-5}$ eV^2 and $\Delta m_{atm}^2 = |\Delta m_{31}^2| \approx |\Delta m_{32}^2| \simeq 2.5 \times 10^{-3}$ eV^2 which generate the oscillations in the case of solar, atmospheric, and accelerator neutrinos. Moreover, in most of the analyses, the matrix elements of the 3+1 flavor mixing matrix corresponding to the mixing of the active neutrinos to the sterile neutrinos denoted by U_{e4}, $U_{\mu4}$, and $U_{\tau4}$ are considered as a perturbation on the 3×3 active neutrino mixing matrix and are taken to be very small, that is, $|U_{e4}|^2 \simeq |U_{\mu4}|^2 \approx |U_{\tau4}|^2 \ll 1$. The task of determining the parameters describing the phenomenology of 3+1 flavor mixing of neutrinos would be one of the main objectives of future experiments being done with reactor, solar, atmospheric, and accelerator neutrinos in the low and high energy region of neutrinos. There are already some attempts in this direction, but the current results reported for the values of these parameters, determined from the analyses of present experiments, are likely to change in future when experimental results with improved statistics become available.

1.6 Summary

Thus, starting from 1930, with Pauli's conjecture, neutrinos have played very important role in the understanding of weak interactions. Neutrinos are both puzzles and solution to many puzzles. With the development of precision neutrino experiments, the years to come are expected to be quite exciting as other properties associated with neutrinos are expected to be revealed. We have, in this chapter, introduced the various topics to be discussed in detail in the book. In the next three chapters, we present mathematical and quantum field theoretic preliminaries followed by the chapters on the phenomenological description of the weak interactions of leptons and hadrons leading to the standard model of electroweak interactions. These chapters are the building blocks to understand the physics of neutrino interactions. We discuss the various neutrino scattering phenomenon involving point particles as well as hadrons in separate chapters on the quasielastic, inelastic and deep inelastic scattering processes, followed by the chapters on neutrino-nucleus scattering focussing on the nuclear medium effects and their importance when scattering takes place with a bound nucleon inside a nucleus. The last few chapters deal with neutrino sources and its detection; neutrino oscillation; neutrinos in astrophysics and finally a chapter on the physics beyond the standard model.

Chapter 2

Relativistic Particles and Neutrinos

2.1 Relativistic Notation

Neutrinos are neutral particles of spin $\frac{1}{2}$; they are completely relativistic in the massless limit. In order to describe neutrinos and their interactions, we need a relativistic theory of spin $\frac{1}{2}$ particles. The appropriate framework to describe the elementary particles in general and the neutrinos in particular, is relativistic quantum mechanics and quantum field theory. In this chapter and in the next two chapters, we present the essentials of these topics required to understand the physics of the weak interactions of neutrinos and other particles of spin 0, 1, and $\frac{1}{2}$.

We shall use natural units, in which $\hbar = c = 1$, such that all the physical quantities like mass, energy, momentum, length, time, force, etc. are expressed in terms of energy. In natural units:

$$\hbar c = 0.19732697 \text{ GeV fm} = 1 \qquad \text{and} \qquad 1 \text{ fm} = 5.06773 \text{ GeV}^{-1}.$$

The original physical quantities can be retrieved by multiplying the quantities expressed in energy units by appropriate powers of the factors $\hbar, c$, and $\hbar c$. For example, mass $m = E/c^2$, momentum $p = E/c$, length $l = \hbar c / E$, and time $t = \hbar / E$, etc.

2.1.1 Metric tensor

In the relativistic framework, space and time are treated on equal footing and the equations of motion for particles are described in terms of space–time coordinates treated as four- component vectors, in a four-dimensional space called Minkowski space, defined by x^μ, where $\mu = 0, 1, 2, 3$ and $x^\mu = (x^0 = t, x^1 = x, x^2 = y, x^3 = z)$ in any inertial frame, say S. In another inertial frame, say S', which is moving with a velocity $\beta(= \frac{v}{c})$ in the positive X direction, the space–time coordinates $x^{\mu'}(x^{0'} = t', x^{1'} = x', x^{2'} = y', x^{3'} = z')$ are related to x^μ through the Lorentz transformation given by:

$$x' = \frac{x - \beta t}{\sqrt{1 - \beta^2}}, \quad y' = y, \quad z' = z, \quad \text{and} \quad t' = \frac{t - \beta x}{\sqrt{1 - \beta^2}}, \tag{2.1}$$

such that

$$x'^2 = t'^2 - \vec{x}'^2 = t^2 - \vec{x}^2 = x^2 \text{ (constant)}, \tag{2.2}$$

remains invariant under Lorentz transformations. For this reason, the quantity $\sqrt{t^2 - |\vec{x}|^2}$ is called the length of the four-component vector x^μ in analogy with the length of an ordinary vector $\vec{x}$, that is, $|\vec{x}| = \sqrt{|\vec{x}|^2}$ which is invariant under rotation in three-dimensional Euclidean space. Therefore, the Lorentz transformations shown in Eq. (2.1) are equivalent to a rotation in a four-dimensional Minkowski space in which the quantity defined as $\sqrt{t^2 - |\vec{x}|^2}$, remains invariant, that is, it transforms as a scalar quantity under the Lorentz transformation. This is similar to a rotation in the three-dimensional Euclidean space in which the length of an ordinary vector $\vec{r}$, defined as $|\vec{r}| = \sqrt{x^2 + y^2 + z^2}$ remains invariant, that is, transforms as a scalar under rotation. The scalar product of the space–time four vector x^μ with itself is defined in terms of a quantity $g_{\mu\nu}(\mu, \nu = 0, 1, 2, 3)$ called metric tensor,

$$g_{\mu\nu} = \begin{pmatrix} 1 & 0 & 0 & 0 \\ 0 & -1 & 0 & 0 \\ 0 & 0 & -1 & 0 \\ 0 & 0 & 0 & -1 \end{pmatrix} = g^{\mu\nu}, \tag{2.3}$$

such that the quantity $x^2 = x \cdot x$, defined as:

$$x \cdot x = \sum_{\mu,\nu} g_{\mu\nu} x^\mu x^\nu = t^2 - |\vec{x}|^2 = \sum g^{\mu\nu} x_\mu x_\nu,$$

transforms as a scalar and remains invariant under the Lorentz transformation. The metric tensor $g_{\mu\nu}$ allows us to define another vector x_μ as:

$$x_\mu = \sum_\nu g_{\mu\nu} x^\nu = (t, -\vec{x}), \tag{2.4}$$

such that the scalar product $x \cdot x$ can be represented as:

$$x \cdot x = \sum_\mu x^\mu x_\mu.$$

It is easy to see that $g_{\mu\nu}$ and $g^{\mu\nu}$ are symmetric tensors of second rank and satisfy the condition

$$\sum_\nu g_{\mu\nu} g^{\nu\rho} = \delta^\rho_\mu, \tag{2.5}$$

where δ_μ^ρ is the Kronecker delta defined as:

$$\begin{aligned} \delta_\mu^\rho &= 1, \quad \mu = \rho \\ &= 0, \quad \mu \neq \rho \end{aligned} \tag{2.6}$$

because

$$x^\mu = \sum_{\nu,\rho} g^{\mu\nu} g_{\nu\rho} x^\rho.$$

This is possible only if Eq. (2.5) holds, such that:

$$\sum_{\mu,\rho} g^{\mu\nu} g_{\nu\rho} x^\rho = \sum_\rho \delta_\rho^\mu x^\rho = x^\mu. \tag{2.7}$$

2.1.2 Contravariant and covariant vectors

The four vector x^μ is defined as a contravariant vector with components $x^\mu = (x^0, x^1, x^2, x^3) = (t, \vec{x})$ represented by a superscript μ and the four vector x_μ is defined as a covariant vector $x_\mu = (x_0, -x^1, -x^2, -x^3) = (t, -\vec{x})$ represented by a subscript μ. The scalar product of a contravariant and covariant vectors can be defined with the help of the metric tensor. In fact, it is the metric tensor which defines the geometry of space and transforms as a symmetric tensor of rank 2. In general, a scalar product of any two vectors A^μ and B^μ is defined as:

$$A \cdot B = A^0 B^0 - \vec{A} \cdot \vec{B} = \sum_{\mu=0}^3 A^\mu B_\mu = \sum_{\mu=0}^3 A_\mu B^\mu.$$

We follow the Einstein's summation convention, where any repeated index implies a summation over that index and write:

$$A \cdot B = g_{\mu\nu} A^\mu B^\nu = g^{\mu\nu} A_\mu B_\nu = A_\mu B^\mu = A^\mu B_\mu.$$

Another example of the four vector is the energy–momentum four vector p^μ defined as:

$$p^\mu = (E, \vec{p}) \quad \text{and} \quad p_\mu = (E, -\vec{p}) \tag{2.8}$$

and the scalar product

$$p^2 = p^\mu p_\mu = E^2 - |\vec{p}|^2 = m^2, \tag{2.9}$$

such that $p^2 = m^2$, where m is the invariant mass of the particle, identified as the rest mass, that is, the energy of the particle at rest. For a particle moving with velocity $\vec{\beta}$, the energy and momentum are given by:

$$E = \gamma m \quad \text{and} \quad \vec{p} = \gamma \vec{\beta} m, \tag{2.10}$$

where $\gamma = \frac{1}{\sqrt{1-|\vec{\beta}|^2}}$ such that

$$E^2 - |\vec{p}|^2 = \gamma^2(1 - |\vec{\beta}|^2)m^2 = m^2,$$

defines the relativistic relation between the energy E and momentum $\vec{p}$ of a real particle of mass m. It is straightforward to see that if the ordinary vector $\vec{x}$ is defined through the components ($\mu = 1, 2, 3$) of a contravariant vector x^μ, then the derivative operator $\vec{\nabla} = \frac{\partial}{\partial \vec{x}}$ is defined through the space covariant vector ∂_μ defined as:

$$\partial_\mu = \frac{\partial}{\partial x^\mu} = \left(\frac{\partial}{\partial t}, \frac{\partial}{\partial \vec{x}}\right) = \left(\frac{\partial}{\partial t}, \vec{\nabla}\right). \tag{2.11}$$

The fact that the 4-component derivative vector, that is, $\frac{\partial}{\partial x^\mu}$ transforms as a covariant vector can be seen as follows:

Consider any scalar function $\phi(x)$ for which a change $\delta\phi(x)$ in $\phi(x)$, due to a change δx^μ in x^μ, is defined through the equation:

$$\delta\phi(x) = \frac{\partial\phi}{\partial x^\mu}\delta x^\mu. \tag{2.12}$$

Since $\delta\phi$ is a scalar quantity and δx^μ transforms as a contravariant vector, the derivative operator $\frac{\partial}{\partial x^\mu} = \partial_\mu$ will transform as a covariant vector. Similarly,

$$\partial^\mu = \frac{\partial}{\partial x_\mu} = \left(\frac{\partial}{\partial t}, -\vec{\nabla}\right), \tag{2.13}$$

such that

$$\partial_\mu A^\mu = \frac{\partial A^0}{\partial t} + \vec{\nabla} \cdot \vec{A},$$

$$\text{and } \partial_\mu \partial^\mu = \frac{\partial^2}{\partial t^2} - \vec{\nabla}^2 = \square, \tag{2.14}$$

where $\square$ is known as the D'Alembertian operator.

2.2 Wave Equation for a Relativistic Particle

2.2.1 Klein–Gordon equation for spin 0 particles

In nonrelativistic quantum mechanics, the Schrödinger equation is obtained by using the nonrelativistic energy–momentum relation for a particle given by:

$$E = \frac{|\vec{p}|^2}{2m}. \tag{2.15}$$

Treating energy(E) and momentum($\vec{p}$) as operators in coordinate representation:

$$\hat{E} \rightarrow +i\frac{\partial}{\partial t}, \tag{2.16}$$

$$\hat{\vec{p}} \rightarrow -i\vec{\nabla}, \tag{2.17}$$

we obtain

$$i\frac{\partial \psi(\vec{x}, t)}{\partial t} = \left(-\frac{1}{2m}\vec{\nabla}^2\right)\psi(\vec{x}, t), \tag{2.18}$$

for wave function $\psi(\vec{x}, t)$ which describes the motion of a free particle. This equation is linear in time derivative but quadratic in space derivative. Thus, space and time dependence of the wave function are not treated on equal footing. Therefore, the Schrödinger equation is not relativistically covariant and does not retain the same form in all the inertial frames under a Lorentz transformation. To describe the motion of a relativistic free particle, we start with the relativistic energy–momentum relation, that is,

$$E^2 = |\vec{p}|^2 + m^2 \quad \text{or} \quad E^2 - |\vec{p}|^2 = p^\mu p_\mu = m^2, \tag{2.19}$$

and use the coordinate representation of the 4-component energy–momentum operators p^μ and p_μ, that is,

$$\begin{aligned}
\hat{p}^\mu &= i\partial^\mu = \left(i\frac{\partial}{\partial t}, -i\vec{\nabla}\right), \\
\hat{p}_\mu &= i\partial_\mu = \left(i\frac{\partial}{\partial t}, i\vec{\nabla}\right).
\end{aligned} \tag{2.20}$$

If $\phi(\vec{x}, t)$ is the wave function of a relativistic particle, using the operator form of p^μ and p_μ given by Eq. (2.20) in Eq. (2.19), we obtain:

$$\left((i\partial^\mu)(i\partial_\mu) - m^2\right)\phi(\vec{x}, t) = 0,$$

$$(\Box + m^2)\phi(\vec{x}, t) = 0, \tag{2.21}$$

where $\Box \equiv \partial_\mu \partial^\mu = -\vec{\nabla}^2 + \frac{\partial^2}{\partial t^2}$.

Equation (2.21) is the wave equation for a relativistic particle. Since the operator $(\Box + m^2)$ is a scalar operator, we assume $\phi(\vec{x}, t)$ to be a scalar function of $(\vec{x}, t)$, making the equation invariant under the Lorentz transformation. Moreover, being a scalar function, it describes only scalar particles, that is, particles of spin zero. This equation is known as the Klein–Gordon equation [192, 193].

We take the complex conjugate of Eq. (2.21) to obtain the relativistic wave equation for the complex conjugate wave function $\phi^*(\vec{x}, t)$, that is,

$$\phi^*(\vec{x}, t)(\Box + m^2) = 0. \tag{2.22}$$

The free particle Klein–Gordon equation given by Eq. (2.21) has a plane wave solution:

$$\phi(\vec{x}, t) = Ne^{-ip^\mu x_\mu} \equiv Ne^{-i(Et - \vec{p} \cdot \vec{x})},$$ (2.23)

where N is an appropriate normalization constant. The momentum p^μ is the eigenvalue of the momentum operator $\hat{p}^\mu$ because

$$\hat{p}^\mu \phi(\vec{x}, t) = i \frac{\partial}{\partial x_\mu} \phi(\vec{x}, t) = iN \frac{\partial}{\partial x_\mu} e^{-ip^\mu x_\mu} = p^\mu \phi(\vec{x}, t),$$

which satisfies the equation:

$$p^2 - m^2 = 0,$$

$$\text{i.e., } p_0^2 = E^2 = |\vec{p}|^2 + m^2,$$

$$\text{leading to } p_0 = \pm\sqrt{|\vec{p}|^2 + m^2} = \pm E_0.$$ (2.24)

Therefore, the Klein–Gordon equation admits both positive and negative energy solutions. The presence of negative energy solutions presents a problem toward its physical interpretation. To see this, consider Eq. (2.21) and (2.22) for $\phi(\vec{x}, t)$ and $\phi^*(\vec{x}, t)$. We multiply Eq. (2.21) by $i\phi^*$ from the left and Eq. (2.22) by $i\phi$ from the right and subtract to get:

$$i(\phi^* \Box \phi - \phi \Box \phi^*) = 0$$

leading to

$$\partial_\mu(i\phi^* \partial^\mu \phi - i\phi \partial^\mu \phi^*) = 0,$$

that is,

$$\partial_\mu J^\mu = 0 \quad \text{with}$$ (2.25)
$$J^\mu = i(\phi^* \partial^\mu \phi - \phi \partial^\mu \phi^*),$$ (2.26)

which describes the continuity equation for the probability density $J^0(= \rho)$ and the current density $(\vec{J} = \vec{j})$ given by:

$$J^0 = i\left(\phi^* \frac{\partial}{\partial t}\phi - \phi \frac{\partial}{\partial t}\phi^*\right), \quad \text{and}$$ (2.27)

$$\vec{J} = -i\left(\phi^* \vec{\nabla}\phi - \phi \vec{\nabla}\phi^*\right),$$ (2.28)

leading to the conservation of total probability P defined as $P = \int J^0 d\vec{x}$, that is,

$$\frac{dP}{dt} = \frac{\partial}{\partial t} \int J^0 d\vec{x} = -\int \vec{\nabla} \cdot \vec{J} d\vec{x} = 0, \quad \text{that is, } P \text{ is constant.}$$ (2.29)

However, in the case of the Klein–Gordon equation, we obtain the probability density ρ from the four vector J^μ as:

$$\rho = J^0 = 2p^0 |N|^2 \phi^* \phi = \pm 2E_0 |N|^2 \ (\because \phi^* \phi = 1), \tag{2.30}$$

which gives negative probability for the negative energy solutions. The main cause of the appearance of the negative energy solution in the relativistic case is the quadratic relation between energy and momentum, that is, $E^2 = p^2 + m^2$, unlike the nonrelativistic case, where $E = m + \frac{|\vec{p}|^2}{2m}$, is always positive. It was for this reason that the Klein–Gordon equation was not used for sometime as a relativistic wave equation and attempts were made to linearize the relativistic wave equation, which led to the Dirac equation [28]. The Klein–Gordon equation was revived again as a genuine relativistic equation for spin zero particle, after Pauli and Weisskopf [194] suggested that ϕ should be interpreted as a field operator in quantum field theory and not as a wave function. The field $\phi(\vec{x}, t)$ and $\phi^*(\vec{x}, t)$, with appropriate normalization constant N, could be then used to describe the particles with positive and negative charges; the probability density ρ multiplied by the charge $(+e)$, that is,

$$\rho = ie \left(\phi^* \frac{\partial \phi}{\partial t} - \phi \frac{\partial \phi^*}{\partial t} \right)$$

would describe the charge density which could be positive as well as negative.

Keeping this interpretation of probability density ρ in mind, we define the normalization constant N, such that the total number of particles in a volume V, given by $\int \rho dV$ remains unchanged. We, therefore, adopt a normalization in Eq. (2.23) for the Klein–Gordon wave function as:

$$N = \frac{1}{\sqrt{V}} .$$

With this value of N, the wave function is normalized such that there are $2E$ particles in volume V, that is, $\int \rho dV = 2E$. This is called the covariant normalization because with this normalization, the density of states, that is, the number of states per particle $\rho(E)$ is proportional to $\frac{d\vec{p}}{2E}$ which is relativistically covariant and occurs naturally in the 4-dimensional formulation of the fields in quantum field theory (this will be discussed in the next chapter).

Equivalently, if we take:

$$N = \frac{1}{\sqrt{2EV}},$$

then, there would be one particle in volume V, that is, $\int \rho dV = 1$.

2.3 Dirac Equation for Spin $\frac{1}{2}$ Particles

In order to achieve linearization of the relativistic wave equation, Dirac [28] proposed a linear relation between the energy and momentum of the type

$$E = \vec{\alpha} \cdot \vec{p} + \beta m = \sum_{i=1}^{3} \alpha_i p_i + \beta m, \tag{2.31}$$

for a free particle, where $\alpha_i (i = 1, 2, 3)$ and β are to be chosen such that the relativistic relation between energy and momentum, that is,

$$E^2 = |\vec{p}|^2 + m^2 \quad \text{, that is, } E = \pm\sqrt{|\vec{p}|^2 + m^2} = \pm E_p \tag{2.32}$$

is reproduced. This requires that α_i and β must satisfy the following conditions, assuming Eqs. (2.31) and (2.32). A summation over the repeated indices is implied.

1.

$$(\alpha_i p_i + \beta m)(\alpha_j p_j + \beta m) = E^2 = p_i p_i + m^2,$$

that is, $\alpha_i \alpha_j p_i p_j + \alpha_i p_i \beta m + \beta m \alpha_j p_j + \beta^2 m^2 = p_i p_i + m^2,$

since i and j are dummy indices, therefore, they can be interchanged and one can write:

$$\left(\frac{\alpha_i \alpha_j + \alpha_j \alpha_i}{2}\right) p_i p_j + m p_i \left(\frac{\alpha_i \beta + \beta \alpha_i}{2}\right) + m p_j \left(\frac{\alpha_j \beta + \beta \alpha_j}{2}\right) + \beta^2 m^2 = p_i p_j \delta_{ij} + m^2$$

leading to the conditions that:

$$\{\alpha_i, \alpha_j\} = 2\delta_{ij} \tag{2.33}$$

$$\{\alpha_i, \beta\} = 0 \tag{2.34}$$

$$\alpha_i^2 = \beta^2 = 1, \tag{2.35}$$

where $\{A, B\}$ is called the anticommutator of A and B operators defined as:

$$\{A, B\} = AB + BA. \tag{2.36}$$

Obviously, the quantities α_i and β are not numbers but operators satisfying an algebra given by Eqs. (2.33), (2.34) and (2.35). These operators are generally represented by matrices in n-dimensional space and are, therefore, $n \times n$ matrices with the dimension n yet to be decided.

2. If there exist α_i and β with the properties discussed earlier, the Dirac equation can be represented as:

$$H\psi(\vec{x}, t) \;=\; i\frac{\partial \psi(\vec{x}, t)}{\partial t} \tag{2.37}$$

with $\; H \;=\; \vec{\alpha}.\vec{p} + \beta m.$

Since H is a Hermitian operator, $\alpha^\dagger = \alpha$ and $\beta^\dagger = \beta$, that is, the α_i and β matrices are also Hermitian.

3. Since $\alpha_i^2 = 1$ and $\beta^2 = 1$, the eigenvalues of α_i and β are ± 1.

4. The matrices α_i and β are traceless, which can be understood as follows:

Eq. (2.33) implies that:

$$\alpha_i \alpha_j + \alpha_j \alpha_i = 0 \text{ for } i \neq j$$

that is, $\alpha_i \alpha_j = -\alpha_j a_i.$ \tag{2.38}

Multiplying by α_j on both sides and taking trace, we get

$$Tr(\alpha_i \alpha_j \alpha_j) \;=\; -Tr(\alpha_j \alpha_i \alpha_j) = -Tr(\alpha_j \alpha_j \alpha_i)$$

$$\Rightarrow Tr(\alpha_i) \;=\; -Tr(\alpha_i)$$

$$\Rightarrow Tr(\alpha_i) \;=\; 0. \tag{2.39}$$

Similarly, using Eq. (2.34)

$$\alpha_i \beta = -\beta \alpha_i$$

and multiplying by α_i on both sides and taking trace, we get

$$Tr(\beta) \;=\; 0. \tag{2.40}$$

5. Since all the four matrix operators $\alpha_i (i = 1, 2, 3)$ and β have eigenvalues ± 1 and are traceless, their dimensions must be even, that is, $n = 2, 4, 6...$. Since, there are only 3 traceless matrices in 2 dimensions, the minimum dimension of the matrices should be 4.

6. Since α_i and β are 4×4 matrices, the wave function ψ in Eq. (2.37) is a 4-component column vector.

The most popular matrix representation used for α_i and β matrices is the Pauli–Dirac parametrization, in which, the three α_i matrices are written in terms of the three Pauli matrices σ_i as:

$$\alpha_i = \begin{pmatrix} 0 & \sigma_i \\ \sigma_i & 0 \end{pmatrix}, \tag{2.41}$$

where

$$\sigma_1 = \begin{pmatrix} 0 & 1 \\ 1 & 0 \end{pmatrix} ; \quad \sigma_2 = \begin{pmatrix} 0 & -i \\ i & 0 \end{pmatrix} ; \quad \sigma_3 = \begin{pmatrix} 1 & 0 \\ 0 & -1 \end{pmatrix} \tag{2.42}$$

and

$$\beta = \begin{pmatrix} 1 & 0 \\ 0 & -1 \end{pmatrix} , \tag{2.43}$$

which satisfy the conditions given in Eqs. (2.33), (2.34) and (2.35). However, it should be emphasized that the Dirac equation is independent of any given parametrization of α_i and β matrices.

It is convenient to introduce a new set of matrices $\gamma^\mu (\mu = 0, 1, 2, 3)$ defined in terms of α_i and β as:

$$\gamma^0 \;=\; \beta \tag{2.44}$$

$$\gamma^i \;=\; \beta \alpha_i = \gamma^0 \alpha_i = \begin{pmatrix} 0 & \sigma_i \\ -\sigma_i & 0 \end{pmatrix} , \tag{2.45}$$

which satisfy the relations:

$$\{\gamma^\mu, \gamma^\nu\} \;=\; 2g^{\mu\nu} \tag{2.46}$$

$$\gamma^{0^2} \;=\; 1 ; \quad \gamma^{i^2} = -1 \tag{2.47}$$

$$\gamma^{0\dagger} \;=\; \gamma^0 ; \quad \gamma^{i\dagger} = -\gamma^i \tag{2.48}$$

$$\gamma^{\mu\dagger} \;=\; \gamma^0 \gamma^\mu \gamma^0 . \tag{2.49}$$

Multiplying the Dirac equation, Eq. (2.37), by γ^0 from the left, we get:

$$i\gamma^0 \frac{\partial \psi}{\partial t} \;=\; (\gamma^0 \alpha_i p^i + \gamma^0 \beta m)\psi$$

$$\text{or} \quad \left(i\gamma^0 \frac{\partial}{\partial t} - \gamma^i p^i - m \right) \psi(\vec{x}, t) \;=\; 0$$

$$\left(i\gamma^0 \frac{\partial}{\partial t} + i\gamma^i \partial_i - m \right) \psi(\vec{x}, t) \;=\; 0 \quad \left(\because p^i = -i\frac{\partial}{\partial x^i} = -i\partial_i \right)$$

$$(i\gamma^\mu \partial_\mu - m)\,\psi(\vec{x}, t) \;=\; 0$$

$$(i\slashed{\partial} - m)\,\psi(\vec{x}, t) \;=\; 0, \quad \text{where } \slashed{\partial} = \gamma^\mu \partial_\mu . \tag{2.50}$$

The Hermitian conjugate of $\psi(\vec{x}, t)$ satisfies:

$$\psi^\dagger(\vec{x}, t)(-i\gamma^{\mu\dagger} \overleftarrow{\partial}_\mu - m) = 0, \tag{2.51}$$

where $\overleftarrow{\partial}_\mu$ operates to the left on $\psi^\dagger(\vec{x}, t)$. Multiplying Eq. (2.51) by γ^0 from the right, we have

$$\psi^\dagger(\vec{x}, t)(-i\gamma^{\mu\dagger}\gamma^0\overleftarrow{\partial}_\mu - m\gamma^0) = 0.$$

Since $\gamma^{0^2} = 1$, we can write this equation as:

$$\psi^\dagger(\vec{x}, t)(-i\gamma^0\gamma^0\gamma^{\mu\dagger}\gamma^0\overleftarrow{\partial}_\mu - m\gamma^0) = 0$$

$$\psi^\dagger(\vec{x}, t)\gamma^0(-i\gamma^0\gamma^{\mu\dagger}\gamma^0\overleftarrow{\partial}_\mu - m) = 0.$$

Using $\psi^\dagger(\vec{x}, t)\gamma^0 = \bar{\psi}(\vec{x}, t)$ and $\gamma^0\gamma^{\mu\dagger}\gamma^0 = \gamma^\mu$, we obtain:

$$\bar{\psi}(\vec{x}, t)(-i\gamma^\mu\overleftarrow{\partial}_\mu - m) = 0,$$

$$\Rightarrow \bar{\psi}(\vec{x}, t)(i\overleftarrow{\slashed{\partial}} + m) = 0. \tag{2.52}$$

Thus, Eqs. (2.50) and (2.52) are the Dirac equations satisfied by $\psi(\vec{x}, t)$ and $\bar{\psi}(\vec{x}, t)$.

It can be shown that the following are true:

- The Dirac equation is covariant under the Lorentz transformation (discussed in Appendix A).

- There is a conserved current J^μ associated with the Dirac equation given by $J^\mu = \bar{\psi}\gamma^\mu\psi$. It can be shown that by multiplying Eq. (2.50) by $\bar{\psi}$ from the left and Eq. (2.52) by ψ from the right and then adding the modified Eq. (2.52) to the modified Eq. (2.50) we obtain:

$$i(\bar{\psi}\gamma^\mu\partial_\mu\psi + \bar{\psi}\gamma^\mu\overleftarrow{\partial}_\mu\psi) = 0,$$

$$\text{i.e., } \partial_\mu\bar{\psi}\gamma^\mu\psi = 0,$$

$$\text{or } \partial_\mu J^\mu = 0. \tag{2.53}$$

It may be noted that Eq. (2.53) is the covariant form of the continuity equation leading to the conservation of probability (Eq. (2.29)). Since $J^\mu = (J^0, J^i) = (\rho, \vec{J})$, we obtain the expressions for the probability density ρ and the probability current density $\vec{J}$ as:

$$\rho(\vec{x}, t) = \bar{\psi}(\vec{x}, t)\gamma^0\psi(\vec{x}, t) = \psi^\dagger(\vec{x}, t)\psi(\vec{x}, t), \tag{2.54}$$

$$\vec{J}(\vec{x}, t) = \bar{\psi}(\vec{x}, t)\vec{\gamma}\psi(\vec{x}, t) = \psi^\dagger(\vec{x}, t)\vec{\alpha}\psi(\vec{x}, t), \tag{2.55}$$

making the probability density $\rho(\vec{x}, t)$ always positive, and thus avoiding the situation encountered in the Klein–Gordon equation. It should be kept in mind that $E^2 = |\vec{p}|^2 + m^2$ and the Dirac equation still admits the negative energy solutions as we will see in Section 2.3.2.

2.3.1 Spin of a Dirac particle

The Hamiltonian operator of a Dirac particle is:

$$H = \vec{\alpha} \cdot \vec{p} + \beta m. \tag{2.56}$$

It is easy to see that it does not commute with the angular momentum operator $\vec{L} = \vec{r} \times \vec{p}$, that is,

$$[H, \vec{L}] = [\vec{\alpha} \cdot \vec{p} + \beta m, \vec{r} \times \vec{p}] \neq 0.$$

Therefore, $\vec{L}$ is not a constant of motion. To see this, let us consider, for simplicity, the operator L_3 and compute

$$
\begin{aligned}
[H, L_3] &= [\alpha_i p_i, (\vec{r} \times \vec{p})_3] \\
&= [\alpha_i p_i, x_1 p_2 - x_2 p_1].
\end{aligned}
\tag{2.57}
$$

Using the commutation relations for position and momentum operators in quantum mechanics

$$[p_i, x_j] = -i\delta_{ij}, \qquad [p_i, p_j] = 0,$$

and

$$[A, BC] = [A, B]C + B[A, C],$$

we obtain:

$$
\begin{aligned}
[H, L_3] &= \alpha_i [p_i, x_1] p_2 - \alpha_i [p_i, x_2] p_1 \\
&= -i(\alpha_1 p_2 - \alpha_2 p_1) \neq 0.
\end{aligned}
\tag{2.58}
$$

In the relativistic case, we show that the constant of motion is not the $\vec{L}$ operator but an operator given by:

$$\vec{J} = \vec{L} + \frac{1}{2}\vec{\Sigma}, \quad \text{where } \Sigma^i = \begin{pmatrix} \sigma^i & 0 \\ 0 & \sigma^i \end{pmatrix}, \quad i = 1 - 3 \tag{2.59}$$

is the 4-dimensional representation of Pauli matrices σ^i, that is, $\Sigma^i = \sigma^i \otimes I$, I being the unit matrix operator. To see this, define a tensor $\sigma^{\mu\nu}$ as:

$$\sigma^{\mu\nu} = \frac{i}{2}[\gamma^\mu, \gamma^\nu]. \tag{2.60}$$

Writing the space components of $\sigma^{\mu\nu}$, we have:

$$\sigma^{ij} = \frac{i}{2}[\gamma^i, \gamma^j]$$

$$= \frac{i}{2}\begin{pmatrix} -(\sigma^i\sigma^j - \sigma^j\sigma^i) & 0 \\ 0 & -(\sigma^i\sigma^j - \sigma^j\sigma^i) \end{pmatrix}$$

$$= \epsilon_{ijk}\begin{pmatrix} \sigma^k & 0 \\ 0 & \sigma^k \end{pmatrix} \quad (\because [\sigma^i, \sigma^j] = 2i\epsilon_{ijk}\sigma^k)$$

$$= \epsilon_{ijk}\Sigma^k. \tag{2.61}$$

For example,

$$\sigma^{12} = \Sigma^3 = \frac{i}{2}(\gamma^1\gamma^2 - \gamma^2\gamma^1).$$

Therefore,

$$\left[H, \frac{1}{2}\Sigma_3\right] = -\frac{i}{4}[\alpha_i p_i, (\alpha_1\alpha_2 - \alpha_2\alpha_1)]$$

$$= -i(\alpha_2 p_1 - \alpha_1 p_2) = -[H, L_3]. \quad \text{(from Eq. (2.58))} \tag{2.62}$$

Thus, we see that

$$\left[H, L_3 + \frac{1}{2}\Sigma_3\right] = 0. \tag{2.63}$$

We also see that $(\Sigma_3)^2 = 1$, implying that Σ_3 has eigenvalues ± 1. Comparing the operator $\vec{J} = \vec{L} + \frac{1}{2}\vec{\Sigma}$ with the nonrelativistic total angular momentum $\vec{J} = \vec{L} + \vec{S} = \vec{L} + \frac{1}{2}\vec{\sigma}$, we conclude that $\vec{\Sigma}$ is indeed the 4-dimensional generalization of $\vec{\sigma}$, with $\frac{1}{2}\vec{\Sigma}$ being identified with the spin s of the Dirac particle. Eigenvalues of $\frac{1}{2}\Sigma_3 = \pm\frac{1}{2}$ imply that the particle has spin $\frac{1}{2}$. We, therefore, expect that ψ would have 2 components but it has 4 components as discussed in Section 2.3. The other two components are associated with the negative energy solutions as will be discussed in the next section. It is interesting to see that while Σ_3 does not commute with H, where Σ_3 is chosen to be the component along the '3' axis in the Euclidean space (Z-axis), the component of $\vec{\Sigma}$ along the momentum direction of the particle, that is, $\vec{\Sigma} \cdot \hat{p}$ commutes with the Hamiltonian and is a constant of motion, that is,

$$\left[H, \vec{\Sigma} \cdot \hat{p}\right] = 0. \tag{2.64}$$

2.3.2   Plane wave solutions of the Dirac equation

As we have seen, the Dirac wave function $\psi(\vec{x}, t)$ satisfies the equation:

$$(i\gamma^\mu\partial_\mu - m)\psi(\vec{x}, t) = 0,$$

where $\psi(\vec{x}, t)$ is a 4-component column vector

$$\psi(\vec{x}, t) = \begin{pmatrix} \psi_1 \\ \psi_2 \\ \psi_3 \\ \psi_4 \end{pmatrix}, \tag{2.65}$$

describing spin $\frac{1}{2}$ particles. The free particle solutions of Eq. (2.65) are plane waves given by:

$$\psi(\vec{x}, t) = N u(\vec{p}) e^{-i p^\mu x_\mu}; \quad p^\mu = (E, \vec{p}). \tag{2.66}$$

Then, $u(\vec{p})$ satisfies the equation

$$(\not{p} - m) u(\vec{p}) = 0. \tag{2.67}$$

To understand the physical meaning of Eq. (2.67), let us consider the particle at rest, so that a correspondence can be made with the nonrelativistic particle of spin $\frac{1}{2}$. For $\vec{p} = 0$, the equation becomes:

$$(E\gamma^0 - m) u(\vec{p}) = 0, \quad \text{with } u(\vec{p}) = \begin{pmatrix} u_1 \\ u_2 \\ u_3 \\ u_4 \end{pmatrix}. \tag{2.68}$$

Using Eq. (2.68), one may write

$$\begin{pmatrix} E - m & 0 & 0 & 0 \\ 0 & E - m & 0 & 0 \\ 0 & 0 & -E - m & 0 \\ 0 & 0 & 0 & -E - m \end{pmatrix} \begin{pmatrix} u_1 \\ u_2 \\ u_3 \\ u_4 \end{pmatrix} = 0. \tag{2.69}$$

This has a solution given by the condition

$$\det(E\gamma^0 - mI) = 0,$$

$$\text{that is,} \quad (E - m)^2 (E + m)^2 = (E^2 - m^2)^2 = 0,$$

which has four eigenvalues, given by:

$$E_1 = E_2 = m \quad \text{and} \quad E_3 = E_4 = -m \tag{2.70}$$

with eigenvectors:

$$u_1 = \begin{pmatrix} 1 \\ 0 \\ 0 \\ 0 \end{pmatrix}; \quad u_2 = \begin{pmatrix} 0 \\ 1 \\ 0 \\ 0 \end{pmatrix}; \quad u_3 = \begin{pmatrix} 0 \\ 0 \\ 1 \\ 0 \end{pmatrix} \quad \text{and} \quad u_4 = \begin{pmatrix} 0 \\ 0 \\ 0 \\ 1 \end{pmatrix}, \tag{2.71}$$

respectively. Thus, we have two positive energy states u_1 and u_2 which are degenerate and two negative energy states u_3 and u_4, which are also degenerate. The two positive energy states obviously correspond to the two spin states of a nonrelativistic spin $\frac{1}{2}$ particle. The Dirac equation, therefore, predicts, in addition, two negative energy states which are degenerate and correspond to spin states of spin $\frac{1}{2}$.

Now, consider the general case with $\vec{p} \neq 0$ and express the 4-component spinor $u(\vec{p})$ in terms of two 2-component spinors u_a and u_b and write

$$u(\vec{p}) = \begin{pmatrix} u_a(\vec{p}) \\ u_b(\vec{p}) \end{pmatrix} \text{ with } u_a(\vec{p}) = \begin{pmatrix} u_1 \\ u_2 \end{pmatrix} \text{ and } u_b(\vec{p}) = \begin{pmatrix} u_3 \\ u_4 \end{pmatrix}. \tag{2.72}$$

The Dirac equation is then expressed as a matrix equation using the Pauli–Dirac representation for γ^μ and $p_i = -p^i$, $(i = 1 - 3)$, where p^i are ordinary vectors. Using Eq. (2.67), we obtain:

$$\begin{pmatrix} p_0 - m & -\vec{\sigma} \cdot \vec{p} \\ \vec{\sigma} \cdot \vec{p} & -p_0 - m \end{pmatrix} \begin{pmatrix} u_a(\vec{p}) \\ u_b(\vec{p}) \end{pmatrix} = 0. \tag{2.73}$$

The equation has non-trivial solutions for

$$(p_0 - m)(p_0 + m) - (\vec{\sigma} \cdot \vec{p})(\vec{\sigma} \cdot \vec{p}) = 0$$

$$\text{or } p_0{}^2 = |\vec{p}|^2 + m^2 \quad (\because (\vec{\sigma} \cdot \vec{p} \, \vec{\sigma} \cdot \vec{p}) = \vec{p} \cdot \vec{p} + i\vec{\sigma}(\vec{p} \times \vec{p}) = |\vec{p}|^2).$$

Thus, the non-trivial plane wave solutions lead to two energy eigenvalues given by:

$$p_0 = \pm E = \pm \sqrt{|\vec{p}|^2 + m^2}, \tag{2.74}$$

and both would be doubly degenerate as shown in the case of solutions obtained for the limiting case of $\vec{p} = 0$ (Eq. (2.70)). The eigen functions $u_a(\vec{p})$ and $u_b(\vec{p})$ satisfy the relations for positive energy solutions, say $p_0 = E_+ = \sqrt{|\vec{p}|^2 + m^2}$, that is,

$$(E_+ - m)u_a(\vec{p}) - (\vec{\sigma} \cdot \vec{p})u_b(\vec{p}) = 0, \tag{2.75}$$

$$(\vec{\sigma} \cdot \vec{p})u_a(\vec{p}) - (E_+ + m)u_b(\vec{p}) = 0. \tag{2.76}$$

Equation (2.76) gives:

$$u_b(\vec{p}) = \frac{\vec{\sigma} \cdot \vec{p}}{E_+ + m} u_a(\vec{p}). \tag{2.77}$$

It should be noted that Eq. (2.75) also gives the same relation between $u_a(\vec{p})$ and $u_b(\vec{p})$, that is,

$$u_a(\vec{p}) = \frac{\vec{\sigma} \cdot \vec{p}}{E_+ - m} u_b(\vec{p}), \tag{2.78}$$

Multiplying both the sides of Eq. (2.78) by $\vec{\sigma} \cdot \vec{p}$, we get:

$$(E_+ - m)(\vec{\sigma} \cdot \vec{p})\, u_a(\vec{p}) = (\vec{\sigma} \cdot \vec{p})\,(\vec{\sigma} \cdot \vec{p})\, u_b(\vec{p}).$$

Multiplying this by $(E_+ + m)$ on both the sides and using $(\vec{\sigma} \cdot \vec{p})^2 = |\vec{p}|^2$, we get:

$$u_b(\vec{p}) = \frac{E_+^2 - m^2}{|\vec{p}|^2(E_+ + m)}(\vec{\sigma} \cdot \vec{p}) u_a(\vec{p})$$

$$\text{or} \quad u_b(\vec{p}) = \frac{\vec{\sigma} \cdot \vec{p}}{E_+ + m} u_a(\vec{p}). \tag{2.79}$$

We see that the lower components $u_b(\vec{p})$ of $u(\vec{p})$ are given in terms of the upper components $u_a(\vec{p})$ through Eq. (2.79). The upper components are, therefore, independent and arbitrary and can be chosen such that they reproduce correctly the nonrelativistic limit of the solution, that is, $u_a = \chi_r(r = 1, 2)$ with $\chi_1 = \begin{pmatrix} 1 \\ 0 \end{pmatrix}$ and $\chi_2 = \begin{pmatrix} 0 \\ 1 \end{pmatrix}$ as shown in Eq. (2.71), with normalization $\chi_r^+ \chi_s = \delta_{rs}$.

The positive energy solutions are represented as:

$$u_r^+(\vec{p}) = \begin{pmatrix} 1 \\ \frac{\vec{\sigma} \cdot \vec{p}}{E_+ + M} \end{pmatrix} \chi_r = \begin{pmatrix} 1 \\ \frac{\vec{\sigma} \cdot \vec{p}}{E + M} \end{pmatrix} \chi_r. \tag{2.80}$$

Similarly, for negative energy solutions, we have:

$$p_0 = E_- = -E_+ = -\sqrt{|\vec{p}|^2 + m^2},$$

$$(E_- - m)u_a(\vec{p}) - (\vec{\sigma} \cdot \vec{p})u_b(\vec{p}) = 0,$$

$$\text{leading to} \quad u_a(\vec{p}) = \frac{\vec{\sigma} \cdot \vec{p}}{E_- - m} u_b(\vec{p}). \tag{2.81}$$

Now, choosing the independent solution for the negative energy state $u_b(\vec{p})$ as $\chi_r(r = 1, 2) = \begin{pmatrix} 1 \\ 0 \end{pmatrix}$ and $\begin{pmatrix} 0 \\ 1 \end{pmatrix}$, we obtain:

$$u_r^-(\vec{p}) = \frac{\vec{\sigma} \cdot \vec{p}}{E_- - m}\begin{pmatrix} 1 \\ 0 \end{pmatrix}, \quad \frac{\vec{\sigma} \cdot \vec{p}}{E_- - m}\begin{pmatrix} 0 \\ 1 \end{pmatrix}$$

$$= \begin{pmatrix} \frac{\vec{\sigma} \cdot \vec{p}}{E_- - m} \\ 1 \end{pmatrix} \chi_r \qquad (r = 1, 2). \tag{2.82}$$

$\because E_- = -\sqrt{|\vec{p}|^2 + m^2} = -E$, we can also write

$$u_r^-(\vec{p}) = \begin{pmatrix} -\frac{\vec{\sigma} \cdot \vec{p}}{E + m} \\ 1 \end{pmatrix} \chi_r \qquad (r = 1, 2). \tag{2.83}$$

The two degenerate states $r = 1, 2$ corresponding to the positive and negative energy states refer to the spin up and spin down states of Dirac particles. We can explicitly write the Dirac spinors as

$$u_\uparrow^+(\vec{p}) = \begin{pmatrix} 1 \\ 0 \\ \frac{\vec{\sigma}\cdot\vec{p}}{E+M} \\ 0 \end{pmatrix} ; \; u_\downarrow^+(\vec{p}) = \begin{pmatrix} 0 \\ 1 \\ 0 \\ \frac{\vec{\sigma}\cdot\vec{p}}{E+M} \end{pmatrix} ; \; u_\uparrow^-(\vec{p}) = \begin{pmatrix} -\frac{\vec{\sigma}\cdot\vec{p}}{E+M} \\ 0 \\ 1 \\ 0 \end{pmatrix} ; \; u_\downarrow^-(\vec{p}) = \begin{pmatrix} -\frac{\vec{\sigma}\cdot\vec{p}}{E+M} \\ 0 \\ 0 \\ 1 \end{pmatrix}. \quad (2.84)$$

In literature, it is customary to define the positive energy spinors as $u_r(\vec{p})$ instead of $u_r^+(\vec{p})$ and the negative energy spinors as $v_r(\vec{p})$ instead of $u_r^-(\vec{p})$ to facilitate the description of antiparticles through the relation:

$$u_r(\vec{p}) \;=\; u_r^+(\vec{p}) \qquad\qquad (2.85)$$

$$v_r(\vec{p}) \;=\; \epsilon^{rs} u_s^-(-\vec{p}), \qquad\qquad (2.86)$$

where $\epsilon_{11} = \epsilon_{22} \;=\; 0$ and $\epsilon_{12} = -\epsilon_{21} = 1$

$$\Rightarrow v_1(\vec{p}) \;=\; u_2^-(-\vec{p}) \text{ and } v_2(\vec{p}) = -u_1^-(-\vec{p}). \qquad (2.87)$$

The positive energy spinor $u_r^+(\vec{p})(= u_r(\vec{p}))$ satisfies the relation:

$$(\not{p} - m)u_r^+(\vec{p}) \;=\; 0,$$

$$\Rightarrow (\not{p} - m)u_r(\vec{p}) \;=\; 0, \qquad\qquad (2.88)$$

and $$\bar{u}_r(\vec{p})(\not{p} - m) \;=\; 0. \qquad\qquad (2.89)$$

The negative energy spinor $u_r^-(\vec{p})$ satisfies the relation:

$$(\gamma^0 p_0 - \vec{\gamma}\cdot\vec{p} - m)u_r^-(\vec{p}) \;=\; 0,$$

$$\Rightarrow (\gamma^0 E_- - \vec{\gamma}\cdot\vec{p} - m)u_r^-(\vec{p}) \;=\; 0,$$

$$\text{or } (-\gamma^0 E + \vec{\gamma}\cdot\vec{p} - m)v_r(\vec{p}) \;=\; 0,$$

$$\Rightarrow (-\not{p} - m)v_r(\vec{p}) \;=\; 0,$$

$$\Rightarrow (\not{p} + m)v_r(\vec{p}) \;=\; 0, \qquad\qquad (2.90)$$

and $$\bar{v}_r(\vec{p})(\not{p} + m) \;=\; 0. \qquad\qquad (2.91)$$

Therefore, the solution for the Dirac wave function $\psi(\vec{x}, t)$ is written as:

$$\psi(\vec{x}, t) = N \begin{cases} u_r(\vec{p})e^{-ip\cdot x} & \text{for the positive energy states} \\ v_r(\vec{p})e^{ip\cdot x} & \text{for the negative energy states} \end{cases}, \qquad (2.92)$$

where N is the normalization constant.

2.3.3 Normalization of Dirac spinors

Let us now define the normalization of the Dirac wave function in a way similar to the wave function for the Klein–Gordon equation for uniformity and write

$$\psi_r(\vec{x},t) = \frac{1}{\sqrt{V}} \begin{cases} u_r(\vec{p})e^{-ip\cdot x} \\ v_r(\vec{p})e^{ip\cdot x} \end{cases} \tag{2.93}$$

with $\quad u_r(\vec{p}) = N\begin{pmatrix} 1 \\ \frac{\vec{\sigma}\cdot\vec{p}}{E+m} \end{pmatrix}\chi_r \quad$ and $\quad v_r(\vec{p}) = N\begin{pmatrix} \frac{\vec{\sigma}\cdot\vec{p}}{E+m} \\ 1 \end{pmatrix}\chi_r. \tag{2.94}$

We use Eq. (2.83) and keep in mind that $u_r^-(-\vec{p})$ is written in terms of $v_r(\vec{p})$. Now, we determine the normalization of the spinors $u_r(\vec{p})$ and $v_r(\vec{p})$ with the normalization condition

$$\int \rho\, dV = 2E. \tag{2.95}$$

We use the expressions for ρ given in Eq. (2.54) and $u_r(\vec{p})$ and $v_r(\vec{p})$ given in Eq. (2.93) with the normalization for χ_r to obtain the condition for $u_r(\vec{p})$, that is

$$\int \rho\, dV = \int \psi^\dagger\psi\, dV = u^\dagger(\vec{p})u(\vec{p}) = 2E,$$

and $\quad u_r^\dagger(\vec{p})u_s(\vec{p}) = |N|^2\frac{2E}{E+m}\delta_{rs} = 2E\delta_{rs}, \tag{2.96}$

using $(\vec{\sigma}\cdot\vec{p})(\vec{\sigma}\cdot\vec{p}) = |\vec{p}|^2$ and $\chi_r^\dagger\chi_s = \delta_{rs}$, leading to $N = \sqrt{E+m}$. Similarly, we can obtain

$$v_r^\dagger(\vec{p})v_s(\vec{p}) = 2E\delta_{rs} \tag{2.97}$$

and $\quad u_r^\dagger(\vec{p})v_s(-\vec{p}) = 0 \tag{2.98}$

with $N=\sqrt{E+m}$.

If $\rho\, dV$ is normalized to 1, then

$$N = \sqrt{\frac{E+m}{2E}}.$$

So, for this normalization, we write

$$\psi_r(\vec{x},t) = \frac{1}{\sqrt{2EV}} \begin{cases} u_r(\vec{p})e^{-ip\cdot x} \\ v_r(\vec{p})e^{ip\cdot x}. \end{cases} \tag{2.99}$$

We, therefore, write the components $u_r(\vec{p})$ and $v_r(\vec{p})$ of the Dirac wave function $\psi(\vec{x}, t)$ as:

$$u_r(\vec{p}) = \sqrt{E+m} \begin{pmatrix} 1 \\ \frac{\vec{\sigma} \cdot \vec{p}}{E+m} \end{pmatrix} \chi_r e^{-ip \cdot x}, \qquad (2.100)$$

$$v_r(\vec{p}) = \sqrt{E+m} \begin{pmatrix} \frac{\vec{\sigma} \cdot \vec{p}}{E+m} \\ 1 \end{pmatrix} \chi_r e^{ip \cdot x}. \qquad (2.101)$$

Using these normalizations for $u_r(\vec{p})$ and $v_r(\vec{p})$, we can derive the normalization condition:

$$\bar{u}_r(\vec{p}) u_s(\vec{p}) = 2m\delta_{rs} \qquad (2.102)$$

$$\bar{v}_r(\vec{p}) v_s(\vec{p}) = -2m\delta_{rs} \qquad (2.103)$$

$$\bar{u}_r(\vec{p}) v_s(-\vec{p}) = 0, \qquad (2.104)$$

where $\bar{u} = u^\dagger \gamma^0$. Multiplying Eq. (2.88) by $\bar{u}\gamma^\mu$ from the left and Eq. (2.89) by $\gamma^\mu u$ from the right and adding them together, we get

$$\bar{u}(\vec{p})\gamma_\mu(\not{p} - m)u(\vec{p}) + \bar{u}(\vec{p})(\not{p} - m)\gamma_\mu u(\vec{p}) = \bar{u}(\vec{p})(\gamma_\mu \not{p} + \not{p}\gamma_\mu)u(\vec{p})$$
$$- 2m\bar{u}(\vec{p})\gamma_\mu u(\vec{p}) = 0, \qquad (2.105)$$

that is,

$$p^\mu \bar{u}(\vec{p}) u(\vec{p}) = m\bar{u}(\vec{p})\gamma_\mu u(\vec{p}) \qquad (\because \gamma^\mu \gamma^\nu + \gamma^\nu \gamma^\mu = 2g^{\mu\nu}).$$

Taking $\mu = 0$ component, we obtain:

$$\bar{u}_r(\vec{p}) u_s(\vec{p}) = \frac{m}{E} u_r^\dagger(\vec{p}) u_s(\vec{p}) = 2m\delta_{rs}. \qquad (2.106)$$

Similarly, using Eqs. (2.90) and (2.91) and using $p^0 = -E$ for $v_r(\vec{p})$ solutions, we obtain:

$$\bar{v}_r(\vec{p}) v_s(\vec{p}) = -2m\delta_{rs}. \qquad (2.107)$$

Since $u(\vec{p})$ is a 4-component column matrix and $\bar{u}(\vec{p})$ is a 4-component row matrix, $u(\vec{p})\bar{u}(\vec{p})$ would be a 4×4 square matrix given by:

$$\sum_{r=1}^{2} u_r(\vec{p})\bar{u}_r(\vec{p}) = \not{p} + m = p_0 \gamma^0 - p_i \gamma^i + m,$$

$$= \begin{pmatrix} E+m & -\vec{\sigma} \cdot \vec{p} \\ \vec{\sigma} \cdot \vec{p} & -E+M \end{pmatrix}. \qquad (2.108)$$

By operating Eq. (2.108), on the Dirac spinor $u_s(\vec{p})$, we get:

$$\sum_r u_r(\vec{p})\bar{u}_r(\vec{p}) u_s(\vec{p}) = (\not{p} + m) u_s(\vec{p}),$$

that is, $$\sum_r 2m\delta_{rs} u_r(\vec{p}) = (\not{p} + m) u_s(\vec{p}) = 2m u_s(\vec{p}), \qquad (2.109)$$

which is an identity. However, Eq. (2.108) can be proved by explicitly evaluating $\sum_r u_r(\vec{p})$ $\bar{u}_r(\vec{p})$ using $u_r(\vec{p})$ for $r = 1, 2$ from Eq. (2.100). Similarly, we can show that:

$$\sum_{r=1}^{2} v_r(\vec{p})\bar{v}_r(\vec{p}) = \not{p} - m, \tag{2.110}$$

such that

$$\sum_{r,s=1}^{2} \left(u_r(\vec{p})\bar{u}_s(\vec{p}) - v_r(\vec{p})\bar{v}_s(\vec{p}) \right) = 2m\delta_{rs}. \tag{2.111}$$

The relations given in Eqs. (2.108) and (2.110) are called the completeness relations for the Dirac spinors for the positive and negative energy solutions.

2.4 Negative Energy States and Hole Theory

We see that the Dirac equation admits negative energy solutions in addition to the positive energy solutions with eigenvalues of $E = \pm E_0$, where $E_0 = \sqrt{|\vec{p}|^2 + m^2}$. These solutions are described by the Dirac spinor $u_r(\vec{p})$ and $v_r(\vec{p})$ corresponding to positive and negative energy solutions, respectively, with $r(= 1, 2)$ representing the two spin states of particles. Obviously, the negative energy states lie below the positive energy states separated by an energy gap of $2E_0$ as shown in Figure 2.1. We have also seen that the negative energy solutions do not

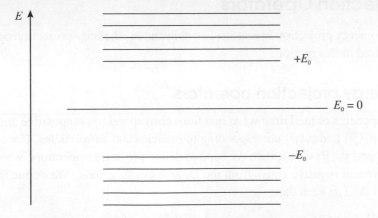

Figure 2.1 Energy spectrum of Dirac particles with energy gap $2E_0$.

present any problem in defining a positive definite probability density. It raises the question of defining the state of minimum energy, that is, vacuum state. Since $\vec{p}$ can take any value, the negative energy states have no lower bound in energy. However, in quantum mechanics, the lowest energy state, that is, ground state is considered to be the most stable state. In most of the cases, the higher states make transitions to the ground state due to interactions in physical systems. Therefore, all the positive energy states would eventually transit to negative energy states, leaving no states in the positive energy states. The situation seems quite unphysical. In this scenario, Dirac proposed that the ground state can be defined as a state in which all the negative energy states are filled. Thus, the ground state or the vacuum state in relativistic quantum mechanics is not a zero particle state but a many particle state. A given energy state can have only two spin $\frac{1}{2}$ particles in it, corresponding to the two spin states (up and down). If all the negative energy states are filled, then no other spin $\frac{1}{2}$ particles can occupy any state in the space of energy states, thus forbidding transition from positive energy states to negative energy states. This filled space of negative energy states is called the Dirac sea.

In order to make a transition to the energy states in the Dirac sea, a vacuum has to be created in the Dirac sea, that is, a negative energy state has to be excited to a positive energy state needing an energy which should be greater than $2m$, the minimum energy corresponding to $p = 0$. This will create a vacancy in the Dirac sea, called a hole, which is equivalent to the absence of a particle of negative energy and negative charge or equivalent to the presence of a particle of positive energy and positive charge. Therefore, such a transition creates an electron of positive energy state and a hole in the Dirac sea, creating an electron–hole pair. The 'hole in the Dirac sea' is interpreted as an antiparticle and the process is known as pair production, in which if sufficient energy $\geqslant 2m$ is available, then a pair of electron and its antiparticle can be created. Such an antiparticle was discovered by Anderson in 1932 and is called a positron. Since then, positron beams have been created and $e^- e^+$ scattering experiments have been performed. Positron beams have found applications in many areas of physics, chemistry, and biology.

2.5 Projection Operators

There are two energy projection operators, two spin and two helicity projection operators. They will be discussed in this section.

2.5.1 Energy projection operators

The four components of the Dirac wave functions correspond to two positive and two negative energy states $u_r(\vec{p})$ and $v_r(\vec{p})$ corresponding to particles and antiparticles. If we are interested only in $u_r(\vec{p})$ and $v_r(\vec{p})$ states, then we need to define projection operators, which can project out the positive and negative states from the Dirac wave functions. We define these operators as $\Lambda_+(p)$ and $\Lambda_-(p)$ such that:

$$\Lambda_+ u_r(\vec{p}) = u_r(\vec{p}), \qquad\qquad \Lambda_- v_r(\vec{p}) = v_r(\vec{p}),$$

$$\Lambda_+ v_r(\vec{p}) = 0, \qquad\qquad \Lambda_- u_r(\vec{p}) = 0. \tag{2.112}$$

It is easy to see that:

$$\Lambda_+ = \frac{\not{p} + m}{2m} = \Sigma_r u_r(\vec{p})\bar{u}_r(\vec{p}) \quad \text{and} \quad \Lambda_- = \frac{-\not{p} + m}{2m} = -\Sigma_r v_r(\vec{p})\bar{v}_r(\vec{p}) \tag{2.113}$$

The projection operators also satisfy the relations:

$$\Lambda_+^2 = \Lambda_+ \; ; \; \Lambda_-^2 = \Lambda_-. \tag{2.114}$$

$$\Lambda_-\Lambda_+ = \Lambda_+\Lambda_- = 0. \tag{2.115}$$

$$\Lambda_+ + \Lambda_- = 1. \tag{2.116}$$

2.5.2 Spin projection operators

We have seen that the Dirac spinor has four components, two corresponding to the particles and two corresponding to the antiparticles. The two components of the particle/antiparticle spinors correspond to the spin up and spin down state of the spin $\frac{1}{2}$ particle and are written explicitly in Eq. (2.84). We can define the spin projection operator such that a particular spin state $u_r(r = 1)$ corresponding to spin up $|\uparrow\rangle$ or $u_r(r = 2)$ corresponding to spin down $|\downarrow\rangle$ state can be chosen for describing the particle.

The spin projection operator must commute with $p\!\!\!/$ so that the four components have unique energy–momentum and spin eigenvalues. In the rest frame, the Dirac particle is described by mass m, momentum $p^\mu(= m, \vec{0})$, and unit vector $\hat{s}$ in the direction of the spin quantization axis and has only space component $\hat{s}(i = 1, 2, 3)$. In order to treat spin in a covariant way, we, therefore, define a four vector $n^\mu(= 0, \hat{s})$ in the rest system which satisfies, by construction, the conditions:

$$n^2 = n^\mu n_\mu = -1 \quad \text{and} \quad n \cdot p = n^\mu p_\mu = 0. \tag{2.117}$$

It can be shown that in an arbitrary frame

$$n^0 = \frac{\hat{s} \cdot \vec{p}}{m} \quad \text{and} \quad \hat{n} = \hat{s} + \frac{(\vec{p} \cdot \hat{s})\hat{p}}{m(E + m)}. \tag{2.118}$$

In the rest frame, it is straightforward to see that the action of the operator $\vec{\Sigma} \cdot \hat{s}$, where $\vec{\Sigma} = \begin{pmatrix} \vec{\sigma} & 0 \\ 0 & \vec{\sigma} \end{pmatrix}$ on the positive and negative energy solution $u_r(p)$ and $v_r(p)$ ($r = 1, 2$) (Eqs. (2.84), (2.85) and (2.86)) gives:

$$\vec{\Sigma} \cdot \hat{s}\, u_1(\vec{p}, \vec{s}) = u_1(\vec{p}, \vec{s}), \qquad \vec{\Sigma} \cdot \hat{s}\, v_1(\vec{p}, \vec{s}) = -v_1(\vec{p}, \vec{s}),$$

$$\vec{\Sigma} \cdot \hat{s}\, u_2(\vec{p}, \vec{s}) = -u_2(\vec{p}, \vec{s}), \qquad \vec{\Sigma} \cdot \hat{s}\, v_2(\vec{p}, \vec{s}) = v_2(\vec{p}, \vec{s}). \tag{2.119}$$

This shows that the operator $\vec{\Sigma} \cdot \hat{s}$, defined in four-dimensions as $\begin{pmatrix} \vec{\Sigma} \cdot \hat{s} & 0 \\ 0 & \vec{\Sigma} \cdot \hat{s} \end{pmatrix}$, reproduces the correct eigenvalues for spin up and spin down wave functions for the positive energy solution but not for the negative energy solutions $v_r(\vec{p})(r = 1, 2)$. However, if we define an operator as

$$\begin{pmatrix} \vec{\Sigma} \cdot \hat{s} & 0 \\ 0 & -\vec{\Sigma} \cdot \hat{s} \end{pmatrix}, \tag{2.120}$$

then the correct eigenvalues for all the four Dirac solutions corresponding to particles $u_r(\vec{p})$ and antiparticles $v_r(\vec{p})$ $(r = 1, 2)$ are reproduced. It is easy to see that the covariant generalization of this operator is $\gamma_5 \not{s}$, that is,

$$(\gamma_5 \not{s}) = \begin{pmatrix} \vec{\Sigma} \cdot \hat{s} & 0 \\ 0 & -\vec{\Sigma} \cdot \hat{s} \end{pmatrix}. \tag{2.121}$$

It can be verified that

$$[\gamma_5 \not{s}, \not{p}] = 0 \quad \text{and} \quad (\gamma_5 \not{s})^2 = 1, \tag{2.122}$$

where

$$\gamma^5 = i\gamma^0 \gamma^1 \gamma^2 \gamma^3 = \begin{pmatrix} 0 & 1 \\ 1 & 0 \end{pmatrix} = \gamma_5, \quad \gamma_5 = -i\gamma_0 \gamma_1 \gamma_2 \gamma_3 \tag{2.123}$$

demonstrating that the eigenvalues of the covariant operator $\gamma_5 \not{s}$ are ± 1 and the individual spin states $u_r(\vec{p})$ and $v_r(\vec{p})$ are also the eigenstates of the energy–momentum operator p^μ. The spin projection operators $\Lambda_+(s)$ and $\Lambda_-(s)$ are, therefore, constructed as:

$$\Lambda_{\pm}(s) = \frac{1 \pm \gamma_5 \not{s}}{2} \tag{2.124}$$

in analogy with the energy projection operators $\Lambda_{\pm}(E)$ given in Eq. (2.113). Note here, that:

$$u(\vec{p})\bar{u}(\vec{p}) = (\not{p} + m)\frac{1 + \gamma_5 \not{s}}{2}, \tag{2.125}$$

$$v(\vec{p})\bar{v}(\vec{p}) = (\not{p} + m)\frac{1 - \gamma_5 \not{s}}{2}. \tag{2.126}$$

2.5.3 Helicity and helicity projection operators

We have seen in Section 2.3.1 that the spin $\vec{s}$ of the Dirac particle does not commute with the Hamiltonian and is, therefore, not a constant of motion. However, the component of spin along the direction of the momentum, that is, $\vec{s} \cdot \hat{p} = \frac{1}{2} \frac{\vec{\sigma} \cdot \vec{p}}{|\vec{p}|}$ is a constant of motion because the operator $\frac{\vec{\sigma} \cdot \vec{p}}{|\vec{p}|}$ commutes with the Hamiltonian, that is,

$$[\vec{\sigma} \cdot \hat{p}, H] = 0 \quad \text{and} \quad (\vec{\sigma} \cdot \hat{p})^2 = 1, \tag{2.127}$$

implying that the operator $(\vec{\sigma} \cdot \hat{p})$ can be used to specify the eigenstates of the the Dirac particles with eigenvalues ± 1. The representation $\vec{\sigma} \cdot \hat{p}$ is called the helicity operator and the Dirac particle states of both positive and negative energy have two states each with eigenvalues ± 1. These are called positive and negative helicity states. In literature, sometimes, the helicity operator is defined in terms of the spin operator $\vec{S}$, where $\vec{S} = \frac{1}{2}\vec{\sigma}$, as $\vec{S} \cdot \hat{p}$. In this case, the eigenvalues of the positive and negative helicity states are $\pm\frac{1}{2}$. In four-dimensional notation, the spin $\vec{\sigma}$ is $\Sigma_i = \begin{pmatrix} \sigma_i & 0 \\ 0 & \sigma_i \end{pmatrix} = \gamma^5 \gamma^0 \gamma^i$, such that $\vec{\Sigma} \cdot \hat{p} = \gamma^5 \gamma^0 \vec{\gamma} . \hat{p}$. Moreover, the four-dimensional representation of the spin operator Σ_k can also be represented in terms of the space components of the antisymmetric tensor $(\sigma^{\mu\nu})$, as shown in Eq. (2.61). In view of the above discussions, the helicity projection operators $h_\pm$ are defined as:

$$h_\pm = \frac{1 \pm \vec{\Sigma} \cdot \vec{p}}{2}. \tag{2.128}$$

2.6 Massless Spin $\frac{1}{2}$ Particle and Weyl Equation

2.6.1 Equation of motion for massless particles

Let us consider the Dirac equation:

$$(\not{p} - m)u(\vec{p}) = 0,$$

where $u(\vec{p})$ is a 4-component spinor $u(\vec{p}) = \begin{pmatrix} u_a(\vec{p}) \\ u_b(\vec{p}) \end{pmatrix}$. Then, $u(\vec{p})$ satisfies Eqs. (2.75) and (2.76) that is,

$$p^0 u_a(\vec{p}) - (\vec{\sigma} \cdot \vec{p}) u_b(\vec{p}) = m u_a(\vec{p}), \tag{2.129}$$

$$p^0 u_b(\vec{p}) - (\vec{\sigma} \cdot \vec{p}) u_a(\vec{p}) = -m u_b(\vec{p}). \tag{2.130}$$

Adding and subtracting these equations, we obtain

$$(p^0 - \vec{\sigma} \cdot \vec{p})(u_a(\vec{p}) + u_b(\vec{p})) = m(u_a(\vec{p}) - u_b(\vec{p})), \tag{2.131}$$

$$(p^0 + \vec{\sigma} \cdot \vec{p})(u_a(\vec{p}) - u_b(\vec{p})) = m(u_a(\vec{p}) + u_b(\vec{p})). \tag{2.132}$$

Defining two linear combinations of $u_1(\vec{p})$ and $u_2(\vec{p})$ as:

$$u_L(\vec{p}) = \frac{u_a(\vec{p}) - u_b(\vec{p})}{2}, \tag{2.133}$$

$$u_R(\vec{p}) = \frac{u_a(\vec{p}) + u_b(\vec{p})}{2}, \tag{2.134}$$

which satisfy:

$$(p^0 - \vec{\sigma} \cdot \vec{p}) u_R(\vec{p}) = m u_L(\vec{p}), \tag{2.135}$$

$$(p^0 + \vec{\sigma} \cdot \vec{p}) u_L(\vec{p}) = m u_R(\vec{p}), \tag{2.136}$$

shows that the wave equation for $u_R(\vec{p})$ and $u_L(\vec{p})$ are coupled by the mass term. In the limit $m \to 0$, these equations get decoupled to give:

$$p^0 u_R(\vec{p}) = (\vec{\sigma} \cdot \vec{p}) u_R(\vec{p}), \tag{2.137}$$

$$p^0 u_L(\vec{p}) = -(\vec{\sigma} \cdot \vec{p}) u_L(\vec{p}). \tag{2.138}$$

These are known as Weyl equations; they were proposed in 1937 for massless spin $\frac{1}{2}$ particles [40].

We see that these equations are not invariant under parity transformation, which transforms $\vec{x} \to -\vec{x}$, $\vec{\sigma} \to \vec{\sigma}$, and $\vec{p} \to -\vec{p}$, such that $\vec{\sigma} \cdot \vec{p} \to -\vec{\sigma} \cdot \vec{p}$ and $p^0 \to p^0$. Since these equations violate parity, they were not used in particle physics until 1957, until parity violation was discovered in weak interactions.

The solutions of Eqs. (2.137) and (2.138) are obtained in a simple way. Multiplying both sides by p^0, we get:

$$p^{0^2} u_R(\vec{p}) = (\vec{\sigma} \cdot \vec{p}) p^0 u_R(\vec{p}) = (\vec{\sigma} \cdot \vec{p})(\vec{\sigma} \cdot \vec{p}) u_R(\vec{p}). \tag{2.139}$$

Similarly, $p^{0^2} u_L(\vec{p}) = (\vec{\sigma} \cdot \vec{p})(\vec{\sigma} \cdot \vec{p}) u_L(\vec{p}),$ $\tag{2.140}$

implying that both the wave functions $u_R(\vec{p})$ and $u_L(\vec{p})$ satisfy

$$(p^{0^2} - |\vec{p}|^2) u_{R(L)}(\vec{p}) = 0,$$
$$\Rightarrow p^{0^2} = |\vec{p}|^2 \quad \text{that is} \quad p^0 = \pm|\vec{p}|. \tag{2.141}$$

Thus, for the positive energy solutions ($p^0 = +|\vec{p}|$), using Eqs. (2.137) and (2.138), we find:

$$p^0 u_R(\vec{p}) = (\vec{\sigma} \cdot \vec{p}) u_R(\vec{p}),$$

$$p^0 u_L(\vec{p}) = -(\vec{\sigma} \cdot \vec{p}) u_L(\vec{p}),$$

$$\text{or} \quad \frac{\vec{\sigma} \cdot \hat{p} |\vec{p}|}{p^0} u_R(\vec{p}) = u_R(\vec{p}),$$

$$\text{Similarly,} \quad \frac{\vec{\sigma} \cdot \hat{p} |\vec{p}|}{p^0} u_L(\vec{p}) = -u_L(\vec{p}),$$

$$\Rightarrow (\vec{\sigma} \cdot \hat{p}) u_R(\vec{p}) = u_R(\vec{p}) \quad \text{(Using (2.141))} \tag{2.142}$$

$$\text{and} \quad (\vec{\sigma} \cdot \hat{p}) u_L(\vec{p}) = -u_L(\vec{p}). \tag{2.143}$$

$\vec{\sigma} \cdot \hat{p}$ is called the chirality operator in the case of massless particles. In the four-dimensional representation, $\vec{\sigma} \cdot \hat{p}$ is represented by $\vec{\Sigma} \cdot \hat{p}$. We have already mentioned that $\vec{\Sigma} \cdot \hat{p}$ is the helicity operator which commutes with the Dirac Hamiltonian and is used to label the spin states as helicity states. The two different Weyl equations, Eqs. (2.142) and (2.143) describe the states with opposite helicity. The negative helicity state particle has its spin aligned in a direction opposite to its momentum. If we visualize the spin of a particle arising due to a circular motion, then the motion of a particle with negative helicity would correspond to the motion of a left-handed screw. Similarly, the motion of a particle with positive helicity would correspond to the motion of a right-handed-screw. Accordingly, the particles are labeled as left-handed with helicity $h = -1$ and right-handed with helicity $h = +1$. The term 'helicity' is also referred to as chirality and is the same in the case of massless particles.

2.6.2 Equation of motion in Weyl representation

Weyl [40] showed that a massless fermion can be described by a two-component wave function, each component satisfying the fermion's equation of motion:

$$\psi = \begin{pmatrix} \phi_1 \\ \phi_2 \end{pmatrix},$$ (2.144)

where ϕ_1 and ϕ_2 are the two component spinors. Weyl's choice of gamma matrices were:

$$\gamma^5 = \begin{pmatrix} 1 & 0 \\ 0 & -1 \end{pmatrix}, \gamma^0 = \begin{pmatrix} 0 & -1 \\ -1 & 0 \end{pmatrix}, \gamma^i = \begin{pmatrix} 0 & \sigma^i \\ -\sigma^i & 0 \end{pmatrix}.$$ (2.145)

Using Eq. (2.144) we may write the Dirac equation as

$$\left(i\gamma^\mu \frac{\partial}{\partial x^\mu} - m \right) \begin{pmatrix} \phi_1 \\ \phi_2 \end{pmatrix} = 0,$$

$$\implies \left(i\gamma^0 \frac{\partial}{\partial t} + i\vec{\gamma}\cdot\vec{\nabla} - mI \right) \begin{pmatrix} \phi_1 \\ \phi_2 \end{pmatrix} = 0,$$

which results in

$$m\phi_1 = -i\frac{\partial \phi_2}{\partial t} + i\vec{\sigma}\cdot\vec{\nabla}\phi_2 \quad \text{and} \quad m\phi_2 = -i\frac{\partial \phi_1}{\partial t} - i\vec{\sigma}\cdot\vec{\nabla}\phi_1.$$ (2.146)

For a massless fermion($m = 0$), like the neutrinos:

$$i\frac{\partial \phi_2}{\partial t} = i\vec{\sigma}\cdot\vec{\nabla}\phi_2 \quad \text{and} \quad i\frac{\partial \phi_1}{\partial t} = -i\vec{\sigma}\cdot\vec{\nabla}\phi_1.$$ (2.147)

Notice that the upper and lower components are now decoupled. Using $E \to i\hbar\frac{\partial}{\partial t}, \vec{p} \to -i\hbar\vec{\nabla}$ and working in natural units ($\hbar = c = 1$), we get:

$$E\phi_2 = -\vec{\sigma}\cdot\vec{p}\phi_2 \quad \text{and} \quad E\phi_1 = \vec{\sigma}\cdot\vec{p}\phi_1.$$ (2.148)

Moreover, for a massless particle, the expectation value of energy and momentum are the same, that is, $\langle E \rangle = \langle |\vec{p}| \rangle$; this implies that for $\psi_R = \begin{pmatrix} \phi_1 \\ 0 \end{pmatrix}$, $\langle \vec{\sigma}\cdot\vec{p} \rangle = +\langle |\vec{p}| \rangle$ and for $\psi_L = \begin{pmatrix} 0 \\ \phi_2 \end{pmatrix}$, $\langle \vec{\sigma}\cdot\vec{p} \rangle = -\langle |\vec{p}| \rangle$. In this expression, ψ_R represents a right-handed neutrino (+ve helicity) and ψ_L represents the left-handed neutrino (−ve helicity).

Defining the projection operators as $\frac{1}{2}(1 \pm \gamma^5)$, leads to:

$$\frac{1}{2}(1 + \gamma^5)\psi = \begin{pmatrix} \phi_1 \\ 0 \end{pmatrix} = \psi_R,$$

$$\frac{1}{2}(1 - \gamma^5)\psi = \begin{pmatrix} 0 \\ \phi_2 \end{pmatrix} = \psi_L.$$

Thus, one is able to project out a right-handed neutrino spinor using $\frac{1}{2}(1+\gamma^5)$ and a left-handed neutrino spinor using $\frac{1}{2}(1-\gamma^5)$. The different experimental evidences agreed with the assumption that only ψ_L takes part in weak interactions.

2.6.3 Chirality and chirality projection operators

In order to elaborate the concept of chirality, let us consider the four-dimensional representation of the helicity operator

$$\vec{\Sigma} \cdot \vec{p} = \gamma^5 \gamma^0 \vec{\gamma} \cdot \vec{p}$$

and calculate the quantity

$$(\vec{\Sigma} \cdot \vec{p})u(\vec{p}) = (\gamma^5 \gamma^0 \vec{\gamma} \cdot \vec{p})u(\vec{p}).$$

Using the Dirac equation for the massless particle,

$$\not{p}u(\vec{p}) = 0 \quad \text{and} \quad p^0 = |\vec{p}|,$$

we get

$$
\begin{aligned}
(\gamma^5 \gamma^0 \vec{\gamma} \cdot \vec{p})u(\vec{p}) &= \gamma^5 \gamma^0 \gamma^0 p^0 u(\vec{p}), \\
\text{that is,} \quad (\vec{\Sigma} \cdot \vec{p})u(\vec{p}) &= \gamma^5 |\vec{p}| u(\vec{p}), \\
\Rightarrow (\vec{\Sigma} \cdot \hat{p})u(\vec{p}) &= \gamma^5 u(\vec{p}),
\end{aligned}
\tag{2.149}
$$

showing that γ^5 is the same as the helicity or chirality operator. This implies that:

$$\gamma^5 u_R(\vec{p}) = u_R(\vec{p}) \quad \text{and} \quad \gamma^5 u_L(\vec{p}) = -u_L(\vec{p}), \tag{2.150}$$

$$\Rightarrow \gamma^5(u_R(\vec{p}) + u_L(\vec{p})) = u_R(\vec{p}) - u_L(\vec{p}). \tag{2.151}$$

Writing $u(\vec{p}) = u_L(\vec{p}) + u_R(\vec{p})$, we obtain:

$$u_L(\vec{p}) = \frac{1-\gamma_5}{2} u(\vec{p}), \tag{2.152}$$

$$u_R(\vec{p}) = \frac{1+\gamma_5}{2} u(\vec{p}). \tag{2.153}$$

This has a simple interpretation. If $u(\vec{p})$ is a solution of the massless Weyl equation, that is, $\not{p}u(\vec{p}) = 0$, then $\gamma_5 u(\vec{p})$ is also a solution because $\gamma_5 \not{p}u(\vec{p}) = -\not{p}\gamma_5 u(\vec{p}) = 0$. The orthogonal linear combinations of $u(\vec{p})$ and $\gamma_5 u(\vec{p})$, that is, $\frac{1+\gamma_5}{2}u(\vec{p})$ and $\frac{1-\gamma_5}{2}u(\vec{p})$, will also be a solution, which corresponds to definite chirality. We, therefore, define chirality projection operators Λ_L and Λ_R as:

$$\Lambda_L = \frac{1-\gamma_5}{2} \quad \text{and} \quad \Lambda_R = \frac{1+\gamma_5}{2}, \tag{2.154}$$

which project out the left-handed(L) and right-handed(R) components of the massless particles. Such that:

$$\Lambda_L u_L(\vec{p}) \;=\; u_L(\vec{p}), \qquad\qquad \Lambda_R u_R(\vec{p}) = u_R(\vec{p}), \tag{2.155}$$

$$\Lambda_L u_R(\vec{p}) \;=\; 0, \qquad\qquad\quad \Lambda_R u_L(\vec{p}) = 0, \tag{2.156}$$

$$\Lambda_L{}^2 \;=\; \Lambda_L, \qquad\qquad\quad \Lambda_R{}^2 = \Lambda_R, \tag{2.157}$$

$$\Lambda_L \Lambda_R \;=\; \Lambda_R \Lambda_L = 0, \qquad \Lambda_L + \Lambda_R = 1. \tag{2.158}$$

2.7 Relativistic Spin 1 Particles

2.7.1 Massless spin 1 particles

Maxwell's equations for electromagnetic fields describe the wave equation for photons, that is, the massless spin 1 field. They are generally written in terms of the electric field ($\vec{E}(\vec{x}, t)$) and magnetic field ($\vec{B}(\vec{x}, t)$). In the covariant formulation of Maxwell's equations, they are written in terms of a 4-component vector field $A^{\mu}(\vec{x}, t)(\mu = 0, 1, 2, 3)$ (traditionally called potential). Since a real photon is transverse, it has only two nonzero components of its field $A^{\mu}(\vec{x}, t)$ along the directions perpendicular to the direction of motion. Therefore, the 4-component field $A^{\mu}(\vec{x}, t)$ is subjected to some constraints to eliminate the extra degrees of freedom. We first describe Maxwell's equations in a covariant form, that is, in terms of $A^{\mu}(\vec{x}, t)$ and discuss the constraints on $A^{\mu}(\vec{x}, t)$ to obtain a relativistic equation for massless spin 1 particle. Maxwell's equation for the fields $\vec{E}(\vec{x}, t)$ and $\vec{B}(\vec{x}, t)$ are written in natural units as:

$$\vec{\nabla}.\vec{B} \;=\; 0, \qquad\qquad\qquad \text{(Gauss's law in magnetism)} \tag{2.159}$$

$$\vec{\nabla}.\vec{E} \;=\; \rho, \qquad\qquad\qquad \text{(Gauss's law in electrostatics)} \tag{2.160}$$

$$\vec{\nabla} \times \vec{E} \;=\; -\frac{\partial \vec{B}}{\partial t}, \qquad\qquad\qquad \text{(Faraday's law)} \tag{2.161}$$

$$\vec{\nabla} \times \vec{B} \;=\; \vec{J} + \frac{\partial \vec{E}}{\partial t}, \qquad \text{(Ampere's law with Maxwell's modification),} \tag{2.162}$$

where Eqs. (2.159) and (2.161) are homogeneous differential equations and Eqs. (2.160) and (2.162) are inhomogeneous differential equations. Equation (2.159) implies that $\vec{B}$ is a divergenceless vector; therefore, it can be written as a curl of another vector, that is,

$$\vec{B} = \vec{\nabla} \times \vec{A}, \quad \text{where } \vec{A} \text{ is the vector potential.} \tag{2.163}$$

Using this definition of $\vec{B}$, Eq. (2.161) leads to

$$\vec{\nabla} \times \left(\vec{E} + \frac{\partial \vec{A}}{\partial t} \right) = 0. \tag{2.164}$$

Since curl of a gradient is zero, we may write

$$\vec{E} + \frac{\partial \vec{A}}{\partial t} = -\vec{\nabla}\phi,$$

$$\Rightarrow \qquad \vec{E} = -\frac{\partial \vec{A}}{\partial t} - \vec{\nabla}\phi, \tag{2.165}$$

where ϕ is the scalar potential. Using Eqs. (2.164) and (2.165) in Eqs. (2.160) and (2.162) lead to

$$\vec{\nabla}^2 \phi + \frac{\partial}{\partial t}(\vec{\nabla}.\vec{A}) = -\rho \tag{2.166}$$

and $\left(\dfrac{\partial^2 \vec{A}}{\partial t^2} - \vec{\nabla}^2 \vec{A} \right) + \vec{\nabla}\left(\vec{\nabla}.\vec{A} + \dfrac{\partial \phi}{\partial t} \right) = \vec{J}.$ \hfill (2.167)

We have now obtained Maxwell's equations for $\vec{E}$ and $\vec{B}$ fields in terms of the vector and scalar potentials $\vec{A}\,(\vec{x}, t)$ and $\phi\,(\vec{x}, t)$ through the coupled equations. We can decouple them using the freedom due to arbitrariness inherent in defining the vector and scalar potentials $\vec{A}\,(\vec{x}, t)$ and $\phi(\vec{x}, t)$ through Eqs. (2.163) and (2.165). Note that the vector potential defined in Eq. (2.163), is not unique as we can always add a term expressed in terms of the gradient of a scalar like $\vec{\nabla}\Lambda$, with a vanishing curl, since $\vec{\nabla} \times \vec{\nabla}\Lambda = 0$, that is

$$\vec{A} \rightarrow \vec{A}' = \vec{A} + \vec{\nabla}\Lambda, \tag{2.168}$$

without changing the magnetic field $\vec{B}$.

This change in $\vec{A}$ through Eq. (2.168) would change $\vec{E}$ defined through Eq. (2.165). However, if we make a simultaneous change in the scalar potential $\phi\,(\vec{x}, t)$ such that:

$$\phi \rightarrow \phi' = \phi - \frac{\partial \Lambda}{\partial t}, \tag{2.169}$$

where Λ is also a function of x, then

$$\begin{aligned}
\vec{E} \rightarrow \vec{E}' &= -\vec{\nabla}\phi' - \frac{\partial \vec{A}'}{\partial t} \\
&= -\vec{\nabla}\left(\phi - \frac{\partial \Lambda}{\partial t} \right) - \frac{\partial}{\partial t}(\vec{A} + \vec{\nabla}\Lambda) \\
&= -\vec{\nabla}\phi - \frac{\partial \vec{A}}{\partial t} = \vec{E} \tag{2.170}
\end{aligned}$$

remains unchanged. Therefore, $\vec{E}$ and $\vec{B}$ remain unchanged during the simultaneous change of $\phi(\vec{x}, t)$, and $\vec{A}(\vec{x}, t)$ using a scalar function $\Lambda(\vec{x}, t)$ through Eqs. (2.168) and (2.169) for scalar and vector potentials. These are called gauge transformations and the invariance of Maxwell's equations under these transformations is called gauge invariance in electrodynamics. It should be noted that these gauge transformations are dependent on the space–time coordinates and are essentially local transformations and not global transformations. Since $\Lambda(\vec{x}, t)$ is arbitrary, we are free to choose $\Lambda(\vec{x}, t)$. Historically, the two choices made, in describing the classical electrodynamics are the Lorenz gauge [195] and the Coulomb gauge. In the Lorenz gauge, $\Lambda(\vec{x}, t)$ is chosen such that:

$$\vec{\nabla}.\vec{A} + \frac{\partial \phi}{\partial t} = 0, \tag{2.171}$$

which decouples Maxwell's equations for $\phi(\vec{x}, t)$ and $\vec{A}(\vec{x}, t)$ giving us

$$\vec{\nabla}^2 \phi - \frac{\partial^2 \phi}{\partial t^2} = -\rho \quad \text{and} \tag{2.172}$$

$$\vec{\nabla}^2 \vec{A} - \frac{\partial^2 \vec{A}}{\partial t^2} = -\vec{J}, \tag{2.173}$$

implying

$$\vec{\nabla}^2 \Lambda - \frac{\partial^2 \Lambda}{\partial t^2} = 0.$$

In the Coulomb gauge, we choose

$$\vec{\nabla}.\vec{A} = 0, \tag{2.174}$$

giving us

$$\vec{\nabla}^2 \phi = -\rho \quad \text{and} \tag{2.175}$$

$$\frac{\partial^2 \vec{A}}{\partial t^2} - \vec{\nabla}^2 \vec{A} + \vec{\nabla} \left(\frac{\partial \phi}{\partial t} \right) = \vec{J}. \tag{2.176}$$

In this gauge, the solution of the equation of motion for $\phi(\vec{x}, t)$ gives

$$\phi(\vec{x}, t) = \int \rho(x') \frac{e}{|\vec{x} - \vec{x}'|} d\vec{x}', \tag{2.177}$$

which is the reason for calling it the Coulomb gauge. Moreover, because of Eq. (2.174), the Coulomb gauge is also called the transverse or radiation gauge.

In the case of free fields, $\rho = 0$, $\vec{J} = 0$, such that Eq. (2.176) becomes:

$$\Box \vec{A} = 0, \tag{2.178}$$

leading to a solution of the type

$$\vec{A}(\vec{x}, t) = \vec{\epsilon}(\vec{k}) e^{-ik \cdot x}, \tag{2.179}$$

with $\vec{k} \cdot \vec{e} = 0$, that is, the photon field $\vec{A}(x)$ is transverse in the Coulomb gauge.

In the covariant notation, the gauge transformations in Eqs. (2.168) and (2.169) are written as:

$$A^\mu \rightarrow A^{\mu\,\prime} = A^\mu - \partial^\mu \Lambda \qquad (2.180)$$

and the Lorenz condition in Eq. (2.171) is written as:

$$\partial^\mu A_\mu = 0. \qquad (2.181)$$

2.7.2 Covariant form of Maxwell's equations and gauge invariance

Maxwell's equations of electrodynamics are also expressed in the covariant form using an antisymmetric field tensor $F^{\mu\nu}(\vec{x}, t)$ defined as:

$$F^{\mu\nu}(\vec{x}, t) = \frac{\partial A^\mu}{\partial x_\nu} - \frac{\partial A^\nu}{\partial x_\mu} = -F^{\nu\mu}(\vec{x}, t), \qquad (2.182)$$

where $x^\mu = (t, \vec{x})$, $A^\mu = (\phi, \vec{A})$, and $A_\mu = g_{\mu\nu} A^\nu$. In the component form, the field tensor $F^{\mu\nu}(\vec{x}, t)$ is expressed in terms of electric and magnetic fields as:

$$F^{\mu\nu}(\vec{x}, t) = \partial^\mu A^\nu - \partial^\nu A^\mu = \begin{pmatrix} 0 & -E_1 & -E_2 & -E_3 \\ E_1 & 0 & -B_3 & B_2 \\ E_2 & B_3 & 0 & -B_1 \\ E_3 & -B_2 & B_1 & 0 \end{pmatrix}, \qquad (2.183)$$

such that, under gauge transformation:

$$
\begin{aligned}
F^{\mu\nu}(\vec{x}, t) \rightarrow F^{\prime\mu\nu}(\vec{x}, t) &= \partial^\mu A^{\prime\nu} - \partial^\nu A^{\prime\mu} = \partial^\mu A^\nu - \partial^\nu A^\mu \\
\because \quad \delta F^{\mu\nu}(\vec{x}, t) &= F^{\prime\mu\nu}(\vec{x}, t) - F^{\mu\nu}(\vec{x}, t) = \partial^\mu \delta A^\nu - \partial^\nu \delta A^\mu \\
&= -\partial^\mu \partial^\nu \Lambda(x) + \partial^\nu \partial^\mu \Lambda(x) \\
&= 0, \qquad\qquad\qquad\qquad\qquad\qquad\qquad (2.184)
\end{aligned}
$$

that is, the field tensor $F^{\mu\nu}(\vec{x}, t)$ remains invariant under the gauge transformation.

We define the four-component electromagnetic current as $J^\mu = (\rho, \vec{J})$ and can write the inhomogeneous Maxwell's equations (Eqs. (2.160) and (2.162)) as:

$$\partial_\mu F^{\mu\nu}(\vec{x}, t) = J^\nu(\vec{x}, t), \qquad (2.185)$$

and the homogeneous Maxwell's equations (Eqs. (2.159) and (2.161)) as:

$$\partial^\mu F^{\nu\lambda}(\vec{x}, t) + \partial^\nu F^{\lambda\mu}(\vec{x}, t) + \partial^\lambda F^{\mu\nu}(\vec{x}, t) = 0. \qquad (2.186)$$

It is clear from the covariant formulation of Maxwell's equations, that:

a) Maxwell's equations are gauge invariant because under the gauge transformation

$$A^\mu(\vec{x}, t) \longrightarrow A'^\mu(\vec{x}, t) = A^\mu(\vec{x}, t) - \partial^\mu \Lambda(\vec{x}, t),$$ (2.187)

$F^{\mu\nu}(\vec{x}, t)$ remains unchanged, that is,

$$F^{\mu\nu}(\vec{x}, t) \longrightarrow F'^{\mu\nu}(\vec{x}, t) = F^{\mu\nu}(\vec{x}, t).$$ (2.188)

Maxwell's equations, Eqs. (2.185) and (2.186) remain unchanged.

b) The Lorentz covariant and gauge invariant field equations satisfied by A^μ are obtained by using Eqs. (2.183) and (2.185) leading to:

$$\Box A^\nu(\vec{x}, t) - \partial^\nu(\partial_\mu A^\mu(\vec{x}, t)) = J^\nu(\vec{x}, t).$$ (2.189)

Therefore, using Eq. (2.185) it can be shown that the electromagnetic current is conserved,

$$\partial_\mu J^\mu(\vec{x}, t) = \partial_\mu \partial_\nu F^{\nu\mu}(\vec{x}, t) = 0.$$ (2.190)

This leads to the conservation of charge.

c) In the Lorenz gauge, that is,

$$\vec{\nabla}.\vec{A}(\vec{x}, t) + \frac{\partial \phi(\vec{x}, t)}{\partial t} = 0$$ (2.191)

or

$$\partial_\mu A^\mu(\vec{x}, t) = 0,$$ (2.192)

Eq. (2.189), for the free fields becomes

$$\Box A^\mu(\vec{x}, t) = 0,$$ (2.193)

implying that the electromagnetic field $A^\mu(\vec{x}, t)$ is massless.

This equation can be interpreted as a set of four wave equations for the massless scalar fields corresponding to $\mu = 0, 1, 2, 3$ components of $A^\mu(\vec{x}, t)$. However, a massive spin 1 particle has only three components of scalar fields, so that Eq. (2.193) has one additional component of scalar field which needs to be eliminated. Since, a scalar field can be constructed from $A^\mu(\vec{x}, t)$, that is, $\partial_\mu A^\mu(\vec{x}, t)$; therefore, the constraint $\partial_\mu A^\mu(\vec{x}, t) = 0$ is imposed to eliminate one of the scalar fields. This is an alternative physical explanation of the gauge condition given in Eq. (2.192). However, the Lorenz gauge condition does not completely specify the electromagnetic field $A^\mu(\vec{x}, t)$ corresponding to the real photons, which have only two components transverse to the direction of its propagation. This creates some difficulties in its quantization using covariant formulations (Chapter 3). Therefore, an appropriate basis is chosen to specify the electromagnetic field $A^\mu(\vec{x}, t)$, such that $A^\mu(\vec{x}, t)$ has only two transverse components and satisfies the condition 2.193.

2.7.3 Plane wave solution of photon

Equation (2.193) with the constraint given in Eq. (2.192) for the photon field admits a solution of the form

$$A^\mu(\vec{x}, t) = \frac{1}{\sqrt{V}} \epsilon^\mu(\vec{k}) e^{-ik \cdot x} \tag{2.194}$$

with $k^2 = 0$, that is, $k^0 = \omega_k = |\vec{k}|$

and $k_\mu \cdot \epsilon^\mu(\vec{k}) = 0,$ \hfill (2.195)

where $\epsilon^\mu(\vec{k})$ is the four-vector describing the spin or polarization states of the photon. Therefore, the four-field $A^\mu(\vec{x}, t)$ described in terms of the polarization four vector $\epsilon^\mu(\vec{k})$, can be expressed in terms of a set of four independent orthogonal polarization vectors ϵ_r^μ with ($\mu = 0, 1, 2, 3$) components just like an ordinary vector $\vec{r}$, is expressed in terms of three independent ordinary vectors $\hat{i}$, $\hat{j}$, and $\hat{k}$ along the X, Y, Z axes. Expanding the field $A^\mu(x)$ in its Fourier components in momentum space, we write $A^\mu(\vec{x}, t)$ with the normalization factor as

$$A^\mu(\vec{x}, t) = \sum_{r,k} \frac{1}{\sqrt{2V\omega_k}} \epsilon_r^\mu(\vec{k}) \left(a_r(\vec{k}) e^{-ik \cdot x} + a_r^*(\vec{k}) e^{ik \cdot x} \right). \tag{2.196}$$

A massless photon vector field of spin 1 described by $\vec{A}(\vec{x}, t)$ is a transverse field with two polarization states which are perpendicular to the direction of propagation. In contrast, the vector field $\vec{A}(\vec{x}, t)$, with mass $m \neq 0$, is described by three polarization states including the longitudinal component. In the Lorenz gauge description of $A^\mu(x)$, there are four polarization states ϵ_r^μ of with constraints in ϵ_r^μ, that is

$$\sum_r k.\epsilon_r(\vec{k}) = 0. \tag{2.197}$$

We take a photon moving in the z direction such that $k^\mu = (k^0, 0, 0, \vec{k})$ and define the polarization vector $\epsilon_r^\mu(\vec{k})$ as:

$$\epsilon_r^\mu(\vec{k}) = (0, \vec{\epsilon}_r(\vec{k})) \quad r = 1, 2, 3, \tag{2.198}$$

where $\vec{\epsilon}_1(\vec{k})$ and $\vec{\epsilon}_2(\vec{k})$ are orthogonal to each other and also to $\vec{k}$, and

$$\epsilon_3^\mu(\vec{k}) = (0, \hat{k}), \quad \text{such that} \quad \vec{\epsilon}_r(\vec{k}) \cdot \vec{\epsilon}_s(\vec{k}) = \delta_{rs}. \quad \vec{k} \cdot \vec{\epsilon}_{r(s)} = 0; \ r, s = 1, 2 \tag{2.199}$$

$\vec{\epsilon}_1(\vec{k})$ and $\vec{\epsilon}_2(\vec{k})$ are the transverse polarizations and $\vec{\epsilon}_3(\vec{k})$ is the longitudinal polarization of the photon. We also define a time-like vector, known as the scalar polarization $\epsilon_0^\mu(\vec{k})$ as:

$$\epsilon_0^\mu(\vec{k}) = n^\mu = (1, 0, 0, 0), \tag{2.200}$$

such that:

$$\epsilon_3^\mu(\vec{k}) = \frac{k^\mu - k \cdot n \, n^\mu}{\sqrt{(k \cdot n)^2 - k^2}}.$$
(2.201)

An explicit representation of $\epsilon_r^\mu(\vec{k})$ can be written as:

$$\epsilon_0^\mu(\vec{k}) = (1,0,0,0), \quad \epsilon_1^\mu(\vec{k}) = (0,1,0,0),$$

$$\epsilon_2^\mu(\vec{k}) = (0,0,1,0), \quad \epsilon_3^\mu(\vec{k}) = (0,0,0,1).$$
(2.202)

We see that the orthonormalization conditions in covariant form are written as:

$$\epsilon_{r\mu}(\vec{k}) \cdot \epsilon_{r'}^\mu(\vec{k}) = -\zeta_r \delta_{rr'},$$
(2.203)

$$\sum_r \zeta_r \epsilon_r^\mu(\vec{k}) \epsilon_r^\nu(\vec{k}) = -g^{\mu\nu},$$
(2.204)

where $\qquad \zeta_0 = -1 \quad$ and $\quad \zeta_1 = \zeta_2 = \zeta_3 = +1.$
(2.205)

2.7.4 Massive spin 1 particles

In the last section, we have developed the wave equation for the massless particle of spin 1, starting from Maxwell's equation. The general formalism of the wave equation for particle fields with spin $J \geq 1$ was first given by Dirac in 1936 [196], about 8 years after his paper on the relativistic equation for spin $\frac{1}{2}$ particles was published. The formulation of the wave equation for higher spin fields was also developed by Fierz and Pauli [197], Bargmann and Wigner [198], based on the principle that the wave equations for fields with $J \geq 1$ can be developed using the basic concepts of wave equations for spin $\frac{1}{2}$. In the specific case of spin 1 fields, it was applied by Duffin [199] and Kemmer [200] and Proca [201]. However, for the sake of simplicity and elucidation, we will rely on the analogy of the massless spin 1 photon with the equations of motion to extend it to $m \neq 0$ fields.

Massive spin 1 particles have three components of spin direction corresponding to the spin projections $S_z = \pm 1$ and 0. While a nonrelativistic spin 1 particle is described by a field $\vec{A}(\vec{x}, t)$ having three components corresponding to the spin directions, the relativistic formulation describes it in terms of a 4-vector field $A^\mu(\vec{x}, t)$ with four components subjected to subsidiary conditions to eliminate the fourth component. Then, all the components satisfy a wave equation similar to the Klein–Gordon equation for massive particles. The simplest way to visualize such an equation of motion for a free field is to replace the operator $\partial_\mu \partial^\mu$ in Eq. (2.189) with the operator $\partial_\mu \partial^\mu + M^2$, like the operator in the Klein–Gordon equation (Eq. (2.21)), and write the wave equation for the massive spin 1 field $W^\mu(x)$ as

$$(\Box + M^2) W^\mu(\vec{x}, t) - \partial^\mu \partial_\nu W^\nu(\vec{x}, t) = J^\mu(\vec{x}, t),$$
(2.206)

where $J^\mu(\vec{x}, t)$ represents a vector current coupled to $W^\mu(\vec{x}, t)$. This is called the Proca equation. In case of free fields, where $J^\mu(\vec{x}, t) = 0$, we get:

$$(\Box + M^2) W^\mu(\vec{x}, t) - \partial^\mu \partial \cdot W(\vec{x}, t) = 0.$$
(2.207)

Taking divergence of Eq. (2.207), we get:

$$(\Box + M^2)\partial_\mu W^\mu(\vec{x}, t) - \Box \partial_\nu W^\nu(\vec{x}, t) = 0, \tag{2.208}$$

which gives the condition

$$M^2 \partial_\mu W^\mu(\vec{x}, t) = 0,$$

However, $M \neq 0, \Rightarrow \partial_\mu W^\mu(\vec{x}, t) = 0, \tag{2.209}$

which is the condition to remove the fourth component as discussed in Section 2.7.1. Therefore, the equation for particle field $W^\mu(\vec{x}, t)$ of spin 1, with mass M is written as

$$(\Box + M^2)W^\mu(\vec{x}, t) \;\; = \;\; 0, \tag{2.210}$$

Equation (2.210) is the Proca equation for free fields. There is no further freedom due to gauge invariance because under a gauge transformation,

$$W^\mu(\vec{x}, t) \rightarrow W^\mu(\vec{x}, t) + \partial^\mu \chi(\vec{x}, t). \tag{2.211}$$

Thus, Eq. (2.210) becomes

$$(\Box + M^2)W^\mu(\vec{x}, t) + M^2 \partial^\mu \chi(\vec{x}, t) = 0 \;\; \text{with} \;\; \vec{\nabla}^2 \chi = 0, \tag{2.212}$$

which is not the same as Eq. (2.210) due to the mass dependent term unless $\partial^\mu \chi = 0$ in Eq. (2.211) implying no gauge transformation. Therefore, the freedom to chose a gauge such that the degree of freedom of $W^\mu(\vec{x}, t)$ is further reduced like the massless field $A^\mu(\vec{x}, t)$, is not available. Consequently, there are three degrees of freedom for $W^\mu(\vec{x}, t)$ fields. In case of the wave equations for the spin 1 field in the presence of external currents $J^\mu(x)$, the condition $\partial_\mu W^\mu(x) = 0$ is satisfied if the field is coupled to conserved currents, that is, $\partial_\mu J^\mu(x) = 0$. This can be shown after taking the divergence of the Proca equation in Eq. (2.206), such that it becomes

$$(\Box + M^2)\partial \cdot W(\vec{x}, t) - \Box \partial \cdot W(\vec{x}, t) = \partial_\mu J^\mu(\vec{x}, t). \tag{2.213}$$

Therefore, in the case of a massive vector field $W^\mu(\vec{x}, t)$, if it is coupled to a conserved current, that is, $\partial_\mu J^\mu = 0$, then $\partial_\mu W^\mu(\vec{x}, t) = 0$. In analogy with the massless spin 1 field $A^\mu(\vec{x}, t)$, the solution of Eq. (2.212) is written as

$$W^\mu(\vec{x}, t) = \frac{1}{\sqrt{V}}\epsilon^\mu e^{-ik \cdot x}, \tag{2.214}$$

where ϵ^μ is the polarization 4-vector representing the spin states of the particle of spin 1, with $k^2 = M^2$ and $\epsilon \cdot k = 0$; the 4-vector can be expressed in terms of the four independent vectors ϵ_r^μ $(r = 0, 1, 2, 3)$, as in the case of massless spin 1 particles, called polarization vectors. Because of the transversality condition $\varepsilon \cdot k = 0$, only three are independent and can be chosen to describe the polarization states of spin 1 particle, in terms of the transverse and longitudinal components of polarization. In the rest frame of the particle, the polarization states ϵ_r^μ ($r = 1, 2, 3$) can be chosen to be along the X, Y, and Z directions, that is,

$$\epsilon_1^\mu = (0, 1, 0, 0); \quad \epsilon_2^\mu = (0, 0, 1, 0); \quad \epsilon_3^\mu = (0, 0, 0, 1), \tag{2.215}$$

such that

$$\epsilon_r^\mu \epsilon_{s\mu} = -\delta_{rs} \tag{2.216}$$

and $P_{\mu\nu} = \sum_{\lambda=1}^{3} \epsilon_\mu^\lambda \epsilon_\nu^\lambda$ \hfill (2.217)

has components as

$$P_{00} = 0, \ P_{11} = P_{22} = P_{33} = 1 \ \text{and} \ P_{\mu\nu} = 0 \ \text{for} \ \mu \neq \nu. \tag{2.218}$$

In a frame, where the particle is moving with momentum $\vec{k}$ in the Z direction, we get

$$k^\mu = (k^0, 0, 0, \vec{k}),$$

$$\epsilon_1^\mu = (0, 1, 0, 0), \quad \epsilon_2^\mu = (0, 0, 1, 0), \ \text{and} \ \epsilon_3^\mu = \left(\frac{k}{m}, 0, 0, \frac{\omega}{m}\hat{k}\right). \tag{2.219}$$

Therefore, $P_{\mu\nu}$ can be derived to be

$$P_{\mu\nu} = \frac{1}{m^2} \begin{pmatrix} |\vec{k}|^2 & 0 & 0 & -k^0|\vec{k}| \\ 0 & 1 & 0 & 0 \\ 0 & 0 & 1 & 0 \\ -k^0|\vec{k}| & 0 & 0 & (k^0)^2 \end{pmatrix}, \tag{2.220}$$

either by actual substitution of the components of ϵ_r^μ given in Eq. (2.219) or by using a Lorenz transformation on P_{00}, using the transformation

$$P_{\mu\nu} = \Lambda_\mu^\rho \Lambda_\nu^\eta P_{\rho\eta}, \tag{2.221}$$

where $\Lambda_\mu^\rho(k)$ is a Lorentz transformation describing a boost along the Z-axis, described by the parameter $\gamma = \frac{\omega_k}{m}$ and $\gamma\beta = \frac{k}{m}$. It can be verified that the general form of $P_{\mu\nu}(k)$ will be given by

$$\sum_\lambda \epsilon_\mu^\lambda(k) \epsilon_\nu^\lambda(k) = -g_{\mu\nu} + \frac{k_\mu k_\nu}{m^2}. \tag{2.222}$$

2.8   Wave Equation for Particle with Spin $\frac{3}{2}$

The relativistic wave equation for spin $\frac{3}{2}$ particles can be formulated using a 3-spinor description [196, 198] or a vector–spinor description [202] or a description based on a direct higher order representation of $\left(\frac{3}{2}\,0\right) + \left(0\,\frac{3}{2}\right)$ under the Lorentz group [203, 204]. The equations of motion for the higher spin particles were also discussed by Bhabha [205] and Harish-Chandra [206].

Without going into details of these formulation, we use the vector–spinor description of spin $\frac{3}{2}$ given by Rarita and Schwinger [202] to discuss the relativistic wave equation of spin $\frac{3}{2}$ particles. In this theory, the fundamental quantity is $\psi_\alpha^\mu(\vec{x}, t)$, which has the mixed transformation properties of a Lorentz vector denoted by μ and a Dirac spinor denoted by α. The proposed equation of motion for the free spin $\frac{3}{2}$ particle is

$$(i\gamma^\nu \partial_\nu - m)\,\psi_\alpha^\mu(\vec{x}, t) = 0. \tag{2.223}$$

The main difficulty in this approach (as well as in other approaches based on multi-spinor description) is that $\psi_\alpha^\mu(\vec{x}, t)$ includes the fields corresponding to lower spin particles, that is, spin $\frac{1}{2}$ in this case. For example, in the rest frame, the spin decomposition of spin $\frac{3}{2}$ fields in the vector–spinor formalism is given by

$$\left(\frac{1}{2} \oplus \frac{1}{2}\right) + \frac{1}{2} = (1 \oplus 0) \oplus \frac{1}{2} = \frac{3}{2} \oplus \frac{1}{2} \oplus \frac{1}{2}, \tag{2.224}$$

which has two spin $\frac{1}{2}$ components in addition to the physical spin $\frac{3}{2}$ particle. Therefore, the Rarita– Schwinger (RS) spinor $\psi_\alpha^\mu(\vec{x}, t)$ is subjected to additional constraints on $\psi_\alpha^\mu(\vec{x}, t)$ to eliminate the additional degrees of freedom arising due to the presence of spin $\frac{1}{2}$ particles. Since the Dirac spinor corresponding to spin $\frac{1}{2}$ is a Lorentz scalar ψ in the Minkowski space, we have to construct the Lorentz scalar from $\psi_\alpha^\mu(\vec{x}, t)$ and constrain them to vanish as we have done in the case of spin 1 field, that is, $\partial_\mu A^\mu(\vec{x}, t) = 0$. This is the only scalar quantity we can construct from the vector fields $A^\mu(x)$ or $W^\mu(x)$. In the present case, we can construct two scalars, that is, $\gamma_\mu \psi_\alpha^\mu(\vec{x}, t)$ and $p_\mu \psi_\alpha^\mu(\vec{x}, t)$; therefore, two subsidiary conditions are imposed on the RS spinor $\psi_\mu^\alpha(\vec{x}, t)$, that is,

$$\gamma_\mu \psi_\alpha^\mu(\vec{x}, t) = 0, \tag{2.225}$$

$$p_\mu \psi_\alpha^\mu(\vec{x}, t) = 0. \tag{2.226}$$

However, the nature of the RS equation of motion is such that if Eq. (2.225) is assumed then Eq. (2.226) follows. In case of massless spin $\frac{3}{2}$ particles, the RS spinor $\psi_\alpha^\mu(\vec{x}, t)$ satisfies a gauge condition

$$\psi_\alpha^\mu(\vec{x}, t) \rightarrow \psi_{\alpha'}^\mu(\vec{x}, t) = \psi_\alpha^\mu(\vec{x}, t) + \partial^\mu \phi_\alpha(\vec{x}, t), \tag{2.227}$$

where ϕ_α is the arbitrary function, which satisfies the relation:

$$\gamma^\mu \partial_\mu \phi_\alpha(\vec{x}, t) = 0. \tag{2.228}$$

Equation (2.226) implies that in the rest frame, $\psi^0(\vec{x}, t) = 0$ and $\psi^\mu(\vec{x}, t)$ is represented by the wave function of a vector particle $\vec{\psi}(\vec{x}, t)$ in Lorentz space with 3-components ($\mu = 1, 2, 3$). Keeping in mind this simple interpretation of $\psi^\mu_\alpha(\vec{x}, t)$ at rest, we can write a solution of Eq. (2.223) in the form

$$\psi^\mu_\alpha(\vec{x}, t) = N u^\mu_\alpha(\vec{p}) e^{-ip \cdot x} \tag{2.229}$$

with $p^2 = m^2$, and an appropriate normalization constant N. The $u^\mu_\alpha(\vec{p})$ is then represented as (with α replaced by S_Δ for convenience)

$$u^\mu(\vec{p}, S_\Delta) = [\epsilon^\mu(\vec{p}, \lambda) \otimes u(\vec{p}, s)]^{\frac{3}{2}}_{S_\Delta} \tag{2.230}$$

$$= \sum_{\lambda, s} \left(\begin{array}{cc|c} 1 & \frac{1}{2} & \frac{3}{2} \\ \lambda & s & S_\Delta \end{array} \right) \epsilon^\mu(\vec{p}, \lambda) \, u(\vec{p}, s), \tag{2.231}$$

where $u(\vec{p}, s)$ and $\epsilon^\mu(\vec{p}, \lambda)$ are spin $\frac{1}{2}$ and spin 1 polarization vectors, respectively. Evaluating the Clebsch–Gordan coefficients, for example,

$$u^\mu(\vec{p}, 3/2) = \left(\begin{array}{cc|c} 1 & \frac{1}{2} & \frac{3}{2} \\ 1 & \frac{1}{2} & \frac{3}{2} \end{array} \right) \epsilon^\mu(\vec{p}, 1) \, u(\vec{p}, 1/2) = \epsilon^\mu(\vec{p}, 1) \, u(\vec{p}, 1/2), \tag{2.232}$$

$$u^\mu(\vec{p}, -3/2) = \left(\begin{array}{cc|c} 1 & \frac{1}{2} & \frac{3}{2} \\ -1 & -\frac{1}{2} & -\frac{3}{2} \end{array} \right) \epsilon^\mu(\vec{p}, -1) u(\vec{p}, -1/2) = \epsilon^\mu(\vec{p}, -1) u(\vec{p}, -1/2), \tag{2.233}$$

leads to the following explicit form of the spinors

$$u^\mu(\vec{p}, \pm 3/2) = \epsilon^\mu(\vec{p}, \pm 1) \, u(\vec{p}, \pm 1/2), \tag{2.234}$$

$$u^\mu(\vec{p}, \pm 1/2) = \sqrt{\frac{2}{3}} \, \epsilon^\mu(\vec{p}, 0) \, u(\vec{p}, \pm 1/2) + \sqrt{\frac{1}{3}} \, \epsilon^\mu(\vec{p}, \pm 1) \, u(\vec{p}, \mp 1/2). \tag{2.235}$$

The Rarita–Schwinger spinor for a spin $\frac{3}{2}$ particle is written down in the following form [202]

$$u^\mu(\vec{p}, s) = \sqrt{\frac{E_\Delta + M_\Delta}{2 M_\Delta}} \left(\begin{array}{c} \mathbf{I} \\ \frac{\vec{\sigma} \cdot \vec{p}}{E_\Delta + M_\Delta} \end{array} \right) S^\mu_{\Delta N} \chi_s, \tag{2.236}$$

where χ_s is the four-components spin states for spin $\frac{3}{2}$ particle:

$$\chi_{+\frac{3}{2}} = \begin{pmatrix} 1 \\ 0 \\ 0 \\ 0 \end{pmatrix}, \quad \chi_{+\frac{1}{2}} = \begin{pmatrix} 0 \\ 1 \\ 0 \\ 0 \end{pmatrix} \quad \chi_{-\frac{1}{2}} = \begin{pmatrix} 0 \\ 0 \\ 1 \\ 0 \end{pmatrix} \quad \chi_{-\frac{3}{2}} = \begin{pmatrix} 0 \\ 0 \\ 0 \\ 1 \end{pmatrix}$$

and $S_{\Delta N}^{\mu}$ is the four-components coupling matrices containing the Clebsch–Gordan coefficients for the coupling $1 \otimes \frac{1}{2} = \frac{3}{2}$:

$$S^0 = \frac{\vec{p}}{M_\Delta} \begin{pmatrix} 0 & \sqrt{2/3} & 0 & 0 \\ 0 & 0 & \sqrt{2/3} & 0 \end{pmatrix}, \; S^1 = \begin{pmatrix} -\sqrt{1/2} & 0 & \sqrt{1/6} & 0 \\ 0 & -\sqrt{1/6} & 0 & \sqrt{1/2} \end{pmatrix},$$

$$S^2 = i \begin{pmatrix} \sqrt{1/2} & 0 & \sqrt{1/6} & 0 \\ 0 & \sqrt{1/6} & 0 & \sqrt{1/2} \end{pmatrix}, \; S^3 = \frac{E_\Delta}{\vec{p}} S_0. \tag{2.237}$$

2.9 Discrete Symmetry: Parity, Time Reversal, and Charge Conjugation

The parity ($\hat{P}$), time reversal ($\hat{T}$) and charge conjugation ($\hat{C}$) operators, are the examples of the discrete symmetries and one can not describe them using infinitesimal transformations. The parity operator performs a reflection of the space coordinates about the origin ($\vec{r} \to -\vec{r}$) leading to a change in velocity ($\vec{v}$), momentum ($\vec{p}$) of the particle while the orbital angular momentum ($\vec{L} = \vec{r} \times \vec{p}$) and spin angular momentum ($\vec{S}$) do not change. Time reversal operator operates on the time coordinates of the system and changes velocity, momentum and angular momentum. Charge conjugation operator transforms a particle into an antiparticle in the same state such that the momentum, position etc. are unchanged, while charge, magnetic moment etc., change sign. Then we have composed symmetries like $\hat{C}\hat{P}$ and $\hat{C}\hat{P}\hat{T}$. $\hat{C}\hat{P}\hat{T}$ is conserved in all the four basic interactions and it leads to the mass, lifetime and the magnitude of the magnetic moment of particles and antiparticles to be the same. $\hat{C}$, $\hat{P}$, $\hat{T}$ and $\hat{C}\hat{P}$ are conserved in strong and electromagnetic interactions, while they are violated in the weak interaction processes. In the next subsection, we shall discuss $\hat{P}$, $\hat{T}$ and $\hat{C}$ in some detail.

2.9.1 Parity

The parity operator consists of inversion (Figure 2.2) of all three space components ($x, y, z \to -x, -y, -z$) such that

$$\hat{P}\psi(\vec{r}, t) = \eta\psi(-\vec{r}, t), \tag{2.238}$$

where $\psi(\vec{r}, t)$ is a scalar wave function and η is the phase factor, and the repetition of $\hat{P}$ operation gives back the same initial state i.e.

$$\hat{P}^2\psi(\vec{r}, t) = \eta\hat{P}\psi(-\vec{r}, t) = \eta^2\psi(\vec{r}, t) = \psi(\vec{r}, t), \tag{2.239}$$

implies that $\eta = \pm 1$, where $\eta = +1$ is known as the even parity state and $\eta = -1$ is known as the odd parity state.

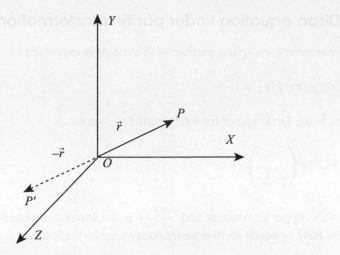

Figure 2.2 Mirror reflection resulting $\vec{r} \to -\vec{r}$ while $|\vec{r}|$ remains invariant.

Under the parity transformation the position vector $\vec{r} \xrightarrow{\hat{P}} -\vec{r}$ and velocity $\vec{v} \xrightarrow{\hat{P}} -\vec{v}$. The Maxwell's equations of motion are invariant under $\hat{P}$ as $\vec{E} \to -\vec{E}, \vec{B} \to \vec{B}$ and $\vec{\nabla} \to -\vec{\nabla}$ resulting:

$$\vec{\nabla}.\vec{B}(\vec{r},t) = 0$$

$$\vec{\nabla}.\vec{E}(\vec{r},t) = \frac{\rho}{\epsilon_0};$$

$$\vec{\nabla} \times \vec{E}(\vec{r},t) = -\frac{\partial \vec{B}(\vec{r},t)}{\partial t}$$

$$\vec{\nabla} \times \vec{B}(\vec{r},t) = \frac{\partial \vec{E}(\vec{r},t)}{\partial t} + \vec{J}(\vec{r},t).$$

When we talk about parity of a particle or a system, there could be intrinsic parity and/or orbital parity. Here we shall discuss intrinsic parity. For example, consider a particle moving with momentum $\vec{p}$, its momentum eigenfunction is given by:

$$\psi_{\vec{p}}(\vec{r},t) = e^{-i(Et-\vec{p}.\vec{r})} \tag{2.240}$$

Under parity operation

$$\psi_{\vec{p}}(\vec{r},t) \xrightarrow{\hat{P}} \eta\psi_{\vec{p}}(-\vec{r},t) = \eta\psi_{-\vec{p}}(\vec{r},t), \tag{2.241}$$

such that if the particle is at rest ($\vec{p} = 0$), $\psi_{\vec{p}}(\vec{r},t)$ is an eigenstate of $\hat{P}$ with the eigenvalue η. The parity quantum number presented by such particle is known as the intrinsic parity which is nothing but the parity presented by the particle at rest.

2.9.2   Dirac equation under parity transformation

The Dirac equation for a spin $\frac{1}{2}$ particle with mass m is expressed as

$$(i\gamma^\mu \partial_\mu - m)\psi(r) = 0, \tag{2.242}$$

where $\psi(r)$ is the Dirac spinor for a free particle given by

$$\psi(r) = N \begin{pmatrix} I \\ \frac{\vec{\sigma} \cdot \vec{p}}{E+M} \end{pmatrix} \chi e^{-ip \cdot r}, \tag{2.243}$$

where χ is the upper component and $\frac{\vec{\sigma} \cdot \vec{p}}{E+M}\chi$ is the lower component. The upper and lower components have opposite parities under parity transformation, i.e.

$$u(\vec{r}, t) \xrightarrow{\hat{P}} \eta_\psi u(-\vec{r}, t), \tag{2.244}$$

$$v(\vec{r}, t) \xrightarrow{\hat{P}} -\eta_\psi v(-\vec{r}, t), \tag{2.245}$$

where $u(\vec{r}, t)$ and $v(\vec{r}, t)$ are the upper and lower components of the Dirac field.

Therefore, while defining the parity operation for a Dirac operator, it acts on $\vec{r}$ as well as on $\psi(\vec{r}, t)$. Since $\psi(\vec{r}, t)$ is a four component spinor i.e. ψ_β, represented by a column vector; P would be a 4×4 matrix operator such that the operation of P on $\psi(\vec{r}, t)$ is written as

$$\psi(\vec{r}, t) \xrightarrow{\hat{P}} \psi^P(-\vec{r}, t) = \eta_\psi P_{\alpha\beta}\psi_\beta(-\vec{r}, t). \tag{2.246}$$

We know that operating parity twice, we get the original spinor, thus

$$\eta_\psi = \pm 1 \qquad \text{and} \qquad P^2 = I.$$

One of the possible representation of $P_{\alpha\beta}$ is γ_0.

We have discussed in Appendix-A, how the Dirac equation is invariant under Lorentz transformation resulting the identity

$$\hat{S}(\hat{\Lambda})\gamma^\nu \hat{S}^{-1}(\hat{\Lambda}) = \Lambda^\nu_\mu \gamma^\mu, \tag{2.247}$$

where Λ^ν_μ is the transformation matrix, $S(\hat{\Lambda})$ is a 4×4 matrix and γ^μ is the Dirac γ-matrix.

For the parity operation, since $t, x, y, z \rightarrow t, -x, -y, -z$, resulting the transformation matrix as

$$\Lambda^\nu_\mu = \begin{pmatrix} 1 & 0 & 0 & 0 \\ 0 & -1 & 0 & 0 \\ 0 & 0 & -1 & 0 \\ 0 & 0 & 0 & -1 \end{pmatrix} = g^{\nu\mu}, \tag{2.248}$$

and the parity operator satisfy the following constrain

$$\Lambda^\nu_\mu \gamma^\mu = \hat{P}\gamma^\nu \hat{P}^{-1} \tag{2.249}$$

$$\Lambda^\sigma_\nu \Lambda^\nu_\mu \gamma^\mu = \hat{P}\Lambda^\sigma_\nu \gamma^\nu \hat{P}^{-1}$$

$$\delta^\sigma_\mu \gamma^\mu = \hat{P}\sum_{\nu=0}^{3} g^{\sigma\nu}\gamma^\nu \hat{P}^{-1}$$

$$\hat{P}^{-1}\gamma^\sigma \hat{P} = g^{\sigma\sigma}\gamma^\sigma, \tag{2.250}$$

where $g^{\sigma\sigma}$ shows that only diagonal elements contribute. One choice of $\hat{P}$ could be $e^{i\phi}\gamma^0$, where ϕ is some unobservable arbitrary phase, such that:

$$\hat{P}^{-1} = e^{-i\phi}\gamma^0 \tag{2.251}$$

$$\hat{P} = e^{i\phi}\gamma^0$$

$$(\hat{P})^4 = 1 = (e^{i\phi})^4$$

$$\hat{P}^{-1} = e^{-i\phi}\gamma^0 = \hat{P}^\dagger, \tag{2.252}$$

i.e. $\hat{P}$ is a unitary operator. Also

$$\hat{P}^{-1} = \gamma^0 \hat{P}^\dagger \gamma^0, \tag{2.253}$$

$$\psi'(\vec{x}',t) = e^{i\phi}\gamma^0 \psi(\vec{x},t). \tag{2.254}$$

Scalar (1):

$$\overline{\psi}'(x')\psi'(x') = \psi'^\dagger(x')\gamma^0 \psi'(x')$$

$$= (P\psi(x))^\dagger \gamma^0 (P\psi(x)); \quad P = \gamma^0$$

$$= \overline{\psi}(x)\psi(x) \tag{2.255}$$

Pseudoscalar (γ_5):

$$\overline{\psi}'(x')\gamma^5 \psi'(x') = \psi'^\dagger(x')\gamma^0 \gamma^5 \psi'(x')$$

$$= -\overline{\psi}(x)\gamma^5 \psi(x) \tag{2.256}$$

Vector (γ^μ):

$$\overline{\psi}'(x')\gamma^\mu \psi'(x') = \psi'^\dagger(x')\gamma^0 \gamma^\mu \psi'(x')$$

$$= \overline{\psi}(x)\gamma_\mu \psi(x). \tag{2.257}$$

Axial vector ($\gamma^\mu \gamma_5$):

$$\overline{\psi}(\vec{r},t)\gamma^\mu \gamma_5 \psi(\vec{r},t) \xrightarrow{\hat{P}} \overline{\psi}^P(\vec{r},t)\gamma^\mu \gamma_5 \psi^P(\vec{r},t)$$

$$= |\eta_\psi|^2 \overline{\psi}(-\vec{r},t)\gamma_0 \gamma^\mu \gamma_5 \gamma_0 \psi(-\vec{r},t)$$

$$= -\overline{\psi}(-\vec{r},t)\gamma_\mu \gamma_5 \psi(-\vec{r},t). \tag{2.258}$$

Tensor ($\sigma^{\mu\nu}$): In the case of tensor interactions, we see the transformation under parity in the different components viz. σ^{0i} and σ^{ij}.

$$
\begin{aligned}
\bar{\psi}(\vec{r},t)\sigma^{0i}\psi(\vec{r},t) & \xrightarrow{\hat{P}} \bar{\psi}^P(\vec{r},t)\sigma^{0i}\psi^P(\vec{r},t) \\
& = |\eta_\psi|^2 \bar{\psi}(-\vec{r},t)\gamma_0 \frac{i}{2}\left(\gamma^0\gamma^i - \gamma^i\gamma^0\right)\gamma_0\psi(-\vec{r},t) \\
& = -\bar{\psi}(-\vec{r},t)\sigma^{0i}\psi(-\vec{r},t).
\end{aligned}
\tag{2.259}
$$

Next we see the parity transformation on σ^{ij}:

$$
\begin{aligned}
\bar{\psi}(\vec{r},t)\sigma^{ij}\psi(\vec{r},t) & \xrightarrow{\hat{P}} \bar{\psi}^P(\vec{r},t)\sigma^{ij}\psi^P(\vec{r},t) \\
& = |\eta_\psi|^2 \bar{\psi}(-\vec{r},t)\gamma_0 \frac{i}{2}\left(\gamma^i\gamma^j - \gamma^j\gamma^i\right)\gamma_0\psi(-\vec{r},t) \\
& = \frac{i}{2}\bar{\psi}(-\vec{r},t)\left(\gamma^i\gamma^j - \gamma^j\gamma^i\right)\psi(-\vec{r},t) \\
& = \bar{\psi}(-\vec{r},t)\sigma^{ij}\psi(-\vec{r},t).
\end{aligned}
\tag{2.260}
$$

Combining Eqs. (2.259) and (2.260), we find

$$
\bar{\psi}(\vec{r},t)\sigma^{\mu\nu}\psi(\vec{r},t) \xrightarrow{\hat{P}} -\bar{\psi}(-\vec{r},t)\sigma_{\mu\nu}\psi(-\vec{r},t).
\tag{2.261}
$$

Therefore, under parity transformation, scalar and vector interactions are invariant, while pseudoscalar and axial vector change sign.

2.9.3 Charge conjugation

Charge conjugation ($\hat{C}$) turns a particle into an antiparticle

$$
\hat{C}\psi(\text{particle}) \longrightarrow \psi(\text{antiparticle}) = \eta_C\psi^C(r),
\tag{2.262}
$$

in the same state leaving position, momentum and angular momentum unchanged. If a set of particles is obeying certain physical law, and the set of corresponding antiparticles, if also obey the same law then the law is said to be invariant under the charge conjugation.

Its effect on electric field $\vec{E}$ and magnetic field $\vec{B}$ is the following:

$$
\begin{aligned}
\vec{E} & = \frac{1}{4\pi\epsilon_0}\left(\frac{q}{|\vec{r}|^3}\right)\vec{r} \xrightarrow{\hat{C}} -\vec{E}, \\
\vec{B} & = \frac{\mu_0}{4\pi}\left(\frac{Id\vec{l}\times\vec{r}}{|\vec{r}|^3}\right) \xrightarrow{\hat{C}} -\vec{B}, \\
\vec{\nabla}.\vec{E} & = 4\pi\rho \xrightarrow{\hat{C}} -\vec{\nabla}.\vec{E} \ \Rightarrow\ \rho \xrightarrow{\hat{C}} -\rho, \\
\vec{\nabla}\times\vec{B} & = \frac{\partial\vec{E}}{\partial t} + \vec{J} \ \Rightarrow\ \vec{J} \xrightarrow{\hat{C}} -\vec{J}.
\end{aligned}
$$

Writing $\vec{E}$ and $\vec{B}$ in terms of potential A_μ and four current density J_μ

$$\vec{E} = -\vec{\nabla}\phi - \frac{\partial \vec{A}}{\partial t}$$

$$\vec{B} = -\vec{\nabla} \times \vec{A}, \text{ we may see that under } \hat{C}$$

$$A_\mu \overset{\hat{C}}{\to} -A_\mu$$

and $\qquad J_\mu \overset{\hat{C}}{\to} -J_\mu$

Strong and electromagnetic interactions are invariant under $\hat{C}$ operation i.e.

$$[\hat{C}, \mathcal{H}_{\text{em}}] = 0,$$
$$[\hat{C}, \mathcal{H}_{\text{strong}}] = 0,$$

while it gets violated in the case of weak interactions i.e.

$$[\hat{C}, \mathcal{H}_{\text{weak}}] \neq 0,$$

where $\mathcal{H}_{\text{em}}$, $\mathcal{H}_{\text{strong}}$ and $\mathcal{H}_{\text{weak}}$, respectively, represent the Hamiltonian of the electromagnetic, strong and weak interactions.

Dirac field

Consider a system of Dirac fermions interacting with an electromagnetic field

$$[i\gamma^\mu(\partial_\mu + ieA_\mu) - m]\psi(x) = 0. \tag{2.263}$$

Adjoint equation is

$$\bar{\psi}(x)[i\gamma^\mu(\partial_\mu - ieA_\mu) + m] = 0. \tag{2.264}$$

Taking the transpose of the adjoint equation

$$[i(\gamma^\mu)^T(\partial_\mu - ieA_\mu) + m]\bar{\psi}(x)^T = 0. \tag{2.265}$$

Recall positron is an antielectron which when interacts with an electromagnetic field would satisfy

$$[i\gamma^\mu(\partial_\mu - ieA_\mu) - m]\psi^C(x) = 0, \tag{2.266}$$

where $\psi^C(x)$ is the charge conjugation of $\psi(x)$.

How $\psi^C(x)$ is related to $\psi(x)$? Notice that Eqs. (2.265) and (2.266) are very similar. Multiply in Eq. (2.265) by C from the left, that is

$$C\left[i(\gamma^\mu)^T(\partial_\mu - ieA_\mu) + m\right]\bar{\psi}(x)^T = 0$$

$$\left[iC(\gamma^\mu)^T C^{-1}C(\partial_\mu - ieA_\mu) + CC^{-1}Cm\right]\bar{\psi}(x)^T = 0$$

$$\left[iC(\gamma^\mu)^T C^{-1}C(\partial_\mu - ieA_\mu) + Cm\right]\bar{\psi}(x)^T = 0$$

$$\left[i\gamma^\mu(\partial_\mu - ieA_\mu) - m\right]C\bar{\psi}(x)^T = 0,$$

where $C(\gamma^\mu)^T C^{-1} = -\gamma^\mu$ implies that $C^{-1}\gamma^\mu C = -(\gamma^\mu)^T$

$$\psi^C(x) = \eta_\psi C\bar{\psi}^T(x). \tag{2.267}$$

In $C^{-1}\gamma^\mu C = -(\gamma^\mu)^T$ we multiply by C from the left

$$\gamma^\mu C = -C(\gamma^\mu)^T$$

$$\gamma^\mu C = +C^T(\gamma^\mu)^T$$

$$= +(\gamma^\mu C)^T$$

$$\Rightarrow \quad C = -C^{-1} = -C^\dagger = -C^T$$

Unlike the parity transformation where we determine $P = \gamma_0$, in the case of charge conjugation, the matrix C depends on the structure of the Dirac matrices. In the Pauli-Dirac representation, C is defined as

$$C = i\gamma^2\gamma^0. \tag{2.268}$$

Using the expression of C from Eq. (2.268) in Eq. (2.267), we find

$$\psi^C(r) = i\eta_C\gamma^2\gamma^0\bar{\psi}^T(r). \tag{2.269}$$

Similarly, for the adjoint Dirac spinor, we find

$$\bar{\psi}(r) \xrightarrow{\hat{C}} \bar{\psi}^C(r) = \psi^{C\dagger}(x)\gamma_0$$

$$= -\psi^T(r)C^{-1},$$

$$= i\eta_C^*\psi^T(r)\gamma^2\gamma^0. \tag{2.270}$$

If F is a 4×4 matrix, then the bilinear covariants (Appendix-A) of the form $\bar{\psi}_b F\psi_a$ transform as (using Eqs. (2.269) and (2.270))

$$\bar{\psi}_b F\psi_a \xrightarrow{\hat{C}} \bar{\psi}_a \left(\gamma^2\gamma^0 F^T\gamma^2\gamma^0\right)\psi_b = \bar{\psi}_a \left(\gamma_2\gamma_0 F^T\gamma_2\gamma_0\right)\psi_b \tag{2.271}$$

under charge conjugation operation.

In the following we show the transformation of different bilinear covariants under C:

(i) Scalar: I

$$\bar{\psi}_b \psi_a = \bar{\psi}_a \left(\gamma_2 \gamma_0 \gamma_2 \gamma_0 \right) \psi_b = \bar{\psi}_a \psi_b. \tag{2.272}$$

(ii) Pseudoscalar: γ_5

$$\begin{aligned}
\bar{\psi}_b \gamma_5 \psi_a &= \bar{\psi}_a \left(\gamma_2 \gamma_0 \gamma_5^T \gamma_2 \gamma_0 \right) \psi_b \\
&= \bar{\psi}_a \left(\gamma_2 \gamma_0 \gamma_5 \gamma_2 \gamma_0 \right) \psi_b \\
&= \bar{\psi}_a \gamma_5 \psi_b.
\end{aligned} \tag{2.273}$$

(iii) Vector: γ^μ.

Since the transpose of γ^μ ($\mu = 0-3$) behave differently, therefore, we perform the transformation for the components $\mu = 0, 1, 2, 3$, individually. Starting with $\mu = 0$

$$\begin{aligned}
\bar{\psi}_b \gamma_0 \psi_a &= \bar{\psi}_a \left(\gamma_2 \gamma_0 \gamma_0^T \gamma_2 \gamma_0 \right) \psi_b \\
&= -\bar{\psi}_a \gamma_0 \psi_b.
\end{aligned} \tag{2.274}$$

For $\mu = 1$ and 3, $(\gamma^\mu)^T$ behaves the same way, thus,

$$\begin{aligned}
\bar{\psi}_b \gamma_1 \psi_a &= \bar{\psi}_a \left(\gamma_2 \gamma_0 \gamma_1^T \gamma_2 \gamma_0 \right) \psi_b \\
&= -\bar{\psi}_a \left(\gamma_2 \gamma_0 \gamma_1 \gamma_2 \gamma_0 \right) \psi_b \\
&= \bar{\psi}_a \gamma_0 \gamma_1 \gamma_0 \psi_b \\
&= -\bar{\psi}_a \gamma_1 \psi_b.
\end{aligned} \tag{2.275}$$

For $\mu = 2$, we find

$$\begin{aligned}
\bar{\psi}_b \gamma_2 \psi_a &= \bar{\psi}_a \left(\gamma_2 \gamma_0 \gamma_2^T \gamma_2 \gamma_0 \right) \psi_b \\
&= \bar{\psi}_a \left(\gamma_2 \gamma_0 \gamma_2 \gamma_2 \gamma_0 \right) \psi_b \\
&= -\bar{\psi}_a \gamma_2 \psi_b.
\end{aligned} \tag{2.276}$$

From Eqs. (2.274), (2.275) and (2.276), we conclude

$$\bar{\psi}_b \gamma^\mu \psi_a = -\bar{\psi}_a \gamma^\mu \psi_b. \tag{2.277}$$

(iv) Pseudoscalar: $\gamma^\mu \gamma_5$

$$
\begin{aligned}
\bar{\psi}_b \gamma^\mu \gamma_5 \psi_a &= \bar{\psi}_a \left(\gamma_2 \gamma_0 (\gamma^\mu \gamma_5)^T \gamma_2 \gamma_0 \right) \psi_b \\
&= -\bar{\psi}_a \left(\gamma_2 \gamma_0 \gamma_5 \gamma^\mu \gamma_2 \gamma_0 \right) \psi_b \\
&= \bar{\psi}_a \gamma^\mu \gamma_5 \psi_b.
\end{aligned}
\tag{2.278}
$$

(v) Tensor: $\sigma^{\mu\nu}$

$$
\begin{aligned}
\bar{\psi}_b \sigma^{\mu\nu} \psi_a &= \bar{\psi}_a \left(\gamma_2 \gamma_0 (\sigma^{\mu\nu})^T \gamma_2 \gamma_0 \right) \psi_b \\
&= \frac{i}{2} \bar{\psi}_a \left(\gamma_2 \gamma_0 (\gamma^{\nu T} \gamma^{\mu T} - \gamma^{\mu T} \gamma^{\nu T}) \gamma_2 \gamma_0 \right) \psi_b \\
&= -\frac{i}{2} \bar{\psi}_a \left(\gamma_2 \gamma^\nu \gamma^\mu \gamma_2 - \gamma_2 \gamma^\mu \gamma^\nu \gamma_2 \right) \psi_b \\
&= -\bar{\psi}_a \sigma^{\mu\nu} \psi_b.
\end{aligned}
\tag{2.279}
$$

2.9.4 Time reversal

If a physical law which is invariant when a time t changes by $-t$, we say that the law is invariant under the time reversal operation, such that

$$
\begin{aligned}
\vec{r} &\xrightarrow{\hat{T}} \vec{r} \\
t &\xrightarrow{\hat{T}} -t \\
\vec{p} &\xrightarrow{\hat{T}} -\vec{p} \\
\vec{L} &\xrightarrow{\hat{T}} -\vec{L} \\
\vec{J} &\xrightarrow{\hat{T}} -\vec{J}.
\end{aligned}
$$

Suppose we make a movie of a particle falling under the influence of gravity (Fig. 2.3), A is the initial point and B is the final point, and the motion is described by:

$$
m \frac{d^2 z}{dt^2} = -mg.
\tag{2.280}
$$

Now changing t to $-t$ is equivalent to turning the movie backwards in time i.e. now B is the initial point and A is the final point and the equation of motion is still described by Eq. (2.280). We say that the equation of motion is invariant under time reversal operation.

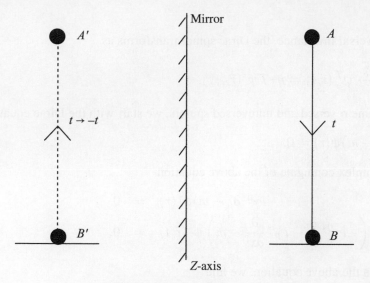

Figure 2.3 Time reversal $t \to -t$.

Under the time reversal operation,

electric field
$$\vec{E}(\vec{x},t) = \frac{1}{4\pi\epsilon_0}\frac{q}{|\vec{r}|^3}\vec{r} \xrightarrow{\hat{T}} \vec{E}(\vec{x},t)$$

magnetic field
$$\vec{B}(\vec{x},t) = \frac{\mu_0}{4\pi}\frac{Id\vec{l}\times\vec{r}}{|\vec{r}|^3} \xrightarrow{\hat{T}} -\vec{B}(\vec{x},-t)$$

charge density
$$\rho(\vec{x},t) \xrightarrow{\hat{T}} \rho^T(\vec{x},t) = \rho(\vec{x},-t)$$

current density
$$\vec{J}(\vec{x},t) \xrightarrow{\hat{T}} \vec{J}^T(\vec{x},t) = -\vec{J}(\vec{x},-t).$$

Effect of time reversal operation on the Maxwell's equations

$$\vec{\nabla}.\vec{B}(\vec{r},t) = 0$$
$$\vec{\nabla}.\vec{E}(\vec{r},t) = \frac{\rho}{\epsilon_0}$$
$$\vec{\nabla}\times\vec{E}(\vec{r},t) = -\frac{\partial\vec{B}(\vec{r},t)}{\partial t}$$
$$\vec{\nabla}\times\vec{B}(\vec{r},t) = \frac{\partial\vec{E}(\vec{r},t)}{\partial t} + \vec{J}(\vec{r},t)$$
$$\vec{E}(\vec{r},t) = -\vec{\nabla}\phi(\vec{r},t) - \frac{\partial\vec{A}(\vec{r},t)}{\partial t}$$
$$\vec{B}(\vec{r},t) = -\vec{\nabla}\times\vec{A}(\vec{r},t)$$

$$(2.281)$$

Dirac fields

Under time reversal invariance, the Dirac spinor transforms as

$$\psi(\vec{r}, t) \xrightarrow{\hat{T}} \psi^T(\vec{r}, t) = \eta_T \mathcal{T} \psi^*(\vec{r}, -t).$$

(2.282)

To relate the time reversed and unreversed spinors, we start with the Dirac equation

$$(i\gamma^\mu \partial_\mu - m)\psi(r) = 0.$$

(2.283)

Taking the complex conjugate of the above equation

$$(-i\gamma^{\mu*}\partial_\mu - m)\psi^*(r) = 0$$

$$\Rightarrow \quad \left(-i\gamma^{0*}\frac{\partial}{\partial t} - i\gamma^{i*}\frac{\partial}{\partial x^i} - m\right)\psi^*(\vec{r}, t) = 0.$$

(2.284)

Applying $\hat{T}$ on the above equation, we find

$$\left(-i\gamma^{0*}\frac{\partial}{\partial(-t)} - i\gamma^{i*}\frac{\partial}{\partial x^i} - m\right)\psi^*(\vec{r}, -t) = 0.$$

(2.285)

If $\psi^T(\vec{r}, t)$ is the solution of the Dirac equation then

$$(i\gamma^\mu \partial_\mu - m)\psi^T(r) = 0$$

$$\left(i\gamma^0 \frac{\partial}{\partial t} + i\gamma^i \frac{\partial}{\partial x^i} - m\right)\mathcal{T}\psi^*(\vec{r}, -t) = 0.$$

(2.286)

Multiplying $\mathcal{T}$ from the left in Eq. (2.285),

$$\left(i\mathcal{T}\gamma^{0*}\frac{\partial}{\partial t} - i\mathcal{T}\gamma^{i*}\frac{\partial}{\partial x^i} - \mathcal{T}m\right)\psi^*(\vec{r}, -t) = 0.$$

(2.287)

Comparing Eqs. (2.286) and (2.287)

$$\mathcal{T}\gamma^{0*}\mathcal{T}^{-1} = \gamma^0 \qquad \text{and} \qquad \mathcal{T}\gamma^{i*}\mathcal{T}^{-1} = -\gamma^i.$$

(2.288)

Similarly, we find

$$\mathcal{T}^{-1}\gamma^0\mathcal{T} = \gamma^{0*} = (\gamma^{0\dagger})^T = \gamma^{0T},$$

(2.289)

$$\mathcal{T}^{-1}\gamma^i\mathcal{T} = -\gamma^{i*} = -(\gamma^{i\dagger})^T = \gamma^{iT}.$$

(2.290)

If Eq. (2.288) holds, then the time reversed wave function $\psi^T(r)$ would satisfy the Dirac equation.

Now, we see the transformation of different γ matrices under $\mathcal{T}$:

(i) Vector: γ^μ

$$\mathcal{T}^{-1}\gamma^\mu\mathcal{T} = (\gamma^\mu)^T.$$

(2.291)

(ii) Pseudoscalar: γ^5

$$
\begin{aligned}
\mathcal{T}^{-1}\gamma^5\mathcal{T} &= -i\mathcal{T}^{-1}(\gamma^0\gamma^1\gamma^2\gamma^3)\mathcal{T} \\
&= -i(\mathcal{T}^{-1}\gamma^0\mathcal{T})(\mathcal{T}^{-1}\gamma^1\mathcal{T})(\mathcal{T}^{-1}\gamma^2\mathcal{T})(\mathcal{T}^{-1}\gamma^3)\mathcal{T}) \\
&= -i\gamma^{0T}\gamma^{1T}\gamma^{2T}\gamma^{3T} = (\gamma^5)^T.
\end{aligned}
\tag{2.292}
$$

(iii) Axial vector: $\gamma^\mu\gamma^5$

$$
\begin{aligned}
\mathcal{T}^{-1}\gamma^\mu\gamma^5\mathcal{T} &= \mathcal{T}^{-1}\gamma^\mu\mathcal{T}\mathcal{T}^{-1}\gamma^5\mathcal{T} \\
&= \gamma^{\mu T}\gamma^{5T} = -(\gamma^\mu\gamma^5)^T.
\end{aligned}
\tag{2.293}
$$

(iv) Tensor: $\sigma_{\mu\nu}$

$$
\begin{aligned}
\mathcal{T}^{-1}\sigma^{\mu\nu}\mathcal{T} &= \frac{i}{2}\mathcal{T}^{-1}(\gamma^\mu\gamma^\nu - \gamma^\nu\gamma^\mu)\mathcal{T} \\
&= \frac{i}{2}(\mathcal{T}^{-1}\gamma^\mu\mathcal{T}\mathcal{T}^{-1}\gamma^\nu\mathcal{T} - \mathcal{T}^{-1}\gamma^\nu\mathcal{T}\mathcal{T}^{-1}\gamma^\mu\mathcal{T}) \\
&= \frac{i}{2}(\gamma^\nu\gamma^\mu - \gamma^\mu\gamma^\nu)^T = -(\sigma^{\mu\nu})^T.
\end{aligned}
$$

Using Eq. (2.288) the matrix $\mathcal{T}$ is constructed. One choice of $\mathcal{T}$ is

$$
\mathcal{T} = i\gamma^1\gamma^3.
\tag{2.294}
$$

The matrix $\mathcal{T}$ satisfies the relation

$$
\mathcal{T} = -\mathcal{T}^\dagger = -\mathcal{T}^{-1} = \mathcal{T}^T,
\tag{2.295}
$$

which shows that the matrix $\mathcal{T}$ is antisymmetric.

Using Eq. (2.294) in Eq. (2.282), we find

$$
\begin{aligned}
\psi^{\mathcal{T}}(\vec{r},t) &= \eta_T\mathcal{T}\psi^*(\vec{r},-t) = i\eta_T\gamma^1\gamma^3\psi^*(\vec{r},-t) \tag{2.296} \\
&= i\eta_T\gamma^0\gamma^1\gamma^3\bar{\psi}^T(\vec{r},-t). \tag{2.297}
\end{aligned}
$$

Similarly, the adjoint Dirac spinor is expressed, under time reversal, as

$$
\begin{aligned}
\bar{\psi}^{\mathcal{T}}(\vec{r},t) &= \psi^{\mathcal{T}\dagger}\gamma^0 \\
&= -i\eta_T^*\psi^T(\vec{r},-t)\gamma^3\gamma^1\gamma^0. \tag{2.298}
\end{aligned}
$$

The bilinear term $\bar{\psi}_a F\psi_b$ transforms as

$$
\bar{\psi}_a(\vec{r},t)F\psi_b(\vec{r},t) \xrightarrow{\hat{T}} \bar{\psi}_b(\vec{r},-t)\gamma^0\gamma^1\gamma^3 F^T\gamma^3\gamma^1\gamma^0\psi_a(\vec{r},-t),
\tag{2.299}
$$

where F is a 4×4 matrix.

Now, we see the transformation of different bilinear covariants under time reversal:

Table 2.1 Transformations of Dirac spinor and adjoint spinors under various symmetry operations.

Symmetry transformation	Dirac spinor $\psi(\vec{r}, t)$	Adjoint Dirac spinor $\bar{\psi}(\vec{r}, t)$
$\hat{P}$	$\eta_P \gamma^0 \psi(-\vec{r}, t)$	$\eta_P \bar{\psi}(-\vec{r}, t) \gamma^0$
$\hat{C}$	$-i\eta_C \gamma^2 \psi^*(\vec{r}, t)$	$i\eta_C^* \psi^T(\vec{r}, t) \gamma^2 \gamma^0$
$\hat{T}$	$i\eta_T \gamma^0 \gamma^1 \gamma^3 \bar{\psi}^T(\vec{r}, -t)$	$-i\eta_T^* \psi^T(\vec{r}, -t) \gamma^3 \gamma^1 \gamma^0$
$\hat{C}\hat{P}$	$-i\eta_P \eta_C \gamma^0 \gamma^2 \psi^*(-\vec{r}, t)$	$i\eta_P \eta_C^* \bar{\psi}^T(-\vec{r}, t) \gamma^2$
$\hat{C}\hat{P}\hat{T}$	$i\eta_{CPT} \gamma^5 \psi(\vec{r}, t)$	$-i\eta_{CPT}^* \psi^T(\vec{r}, t) \gamma^5 \gamma^0$

(i) Scalar: I

$$\bar{\psi}_a(\vec{r}, t) \psi(\vec{r}, t)_b \xrightarrow{\hat{T}} \bar{\psi}_b(\vec{r}, -t) \gamma^0 \gamma^1 \gamma^3 \gamma^3 \gamma^1 \gamma^0 \psi_a(\vec{r}, -t)$$

$$= \bar{\psi}_b(\vec{r}, -t) \psi_a(\vec{r}, -t). \tag{2.300}$$

(ii) Pseudoscalar: γ_5

$$\bar{\psi}_a(\vec{r}, t) \gamma^5 \psi_b(\vec{r}, t) \xrightarrow{\hat{T}} \bar{\psi}_b(\vec{r}, -t) \gamma^0 \gamma^1 \gamma^3 \gamma^{5^T} \gamma^3 \gamma^1 \gamma^0 \psi_a(\vec{r}, -t)$$

$$= \bar{\psi}_b(\vec{r}, -t) \gamma^0 \gamma^1 \gamma^3 \gamma^5 \gamma^3 \gamma^1 \gamma^0 \psi_a(\vec{r}, -t)$$

$$= \bar{\psi}_b(\vec{r}, -t) \gamma^0 \gamma^5 \gamma^0 \psi_a(\vec{r}, -t)$$

$$= -\bar{\psi}_b(\vec{r}, -t) \gamma^5 \psi_a(\vec{r}, -t). \tag{2.301}$$

(iii) Vector: γ^μ

For γ^μ, we look for the different components. For $\mu = 0$:

$$\bar{\psi}_a(\vec{r}, t) \gamma^0 \psi_b(\vec{r}, t) \xrightarrow{\hat{T}} \bar{\psi}_b(\vec{r}, -t) \gamma^0 \gamma^1 \gamma^3 \gamma^{0^T} \gamma^3 \gamma^1 \gamma^0 \psi_a(\vec{r}, -t)$$

$$= \bar{\psi}_b(\vec{r}, -t) \gamma^0 \psi_a(\vec{r}, -t). \tag{2.302}$$

For $\mu = 1$:

$$\bar{\psi}_a(\vec{r}, t) \gamma^1 \psi_b(\vec{r}, t) \xrightarrow{\hat{T}} \bar{\psi}_b(\vec{r}, -t) \gamma^0 \gamma^1 \gamma^3 (\gamma^1)^T \gamma^3 \gamma^1 \gamma^0 \psi_a(\vec{r}, -t)$$

$$= -\bar{\psi}_b(\vec{r}, -t) \gamma^1 \psi_a(\vec{r}, -t). \tag{2.303}$$

For $\mu = 2$:

$$\bar{\psi}_a(\vec{r}, t) \gamma^2 \psi_b(\vec{r}, t) \xrightarrow{\hat{T}} \bar{\psi}_b(\vec{r}, -t) \gamma^0 \gamma^1 \gamma^3 (\gamma^2)^T \gamma^3 \gamma^1 \gamma^0 \psi_a(\vec{r}, -t)$$

$$= -\bar{\psi}_b(\vec{r}, -t) \gamma^2 \psi_a(\vec{r}, -t). \tag{2.304}$$

Table 2.2 Transformations of Dirac bilinear covariants under various symmetry operations.

Bilinear covariants → Symmetry operations ↓	$\bar{\psi}_a\psi_b$	$\bar{\psi}_a\gamma^5\psi_b$	$\bar{\psi}_a\gamma^\mu\psi_b$	$\bar{\psi}_a\gamma^\mu\gamma^5\psi_b$	$\bar{\psi}_a\sigma^{\mu\nu}\psi_b$
$\hat{P}$	$\bar{\psi}_a\psi_b$	$-\bar{\psi}_a\gamma^5\psi_b$	$\bar{\psi}_a\gamma_\mu\psi_b$	$-\bar{\psi}_a\gamma_\mu\gamma^5\psi_b$	$-\bar{\psi}_a\sigma_{\mu\nu}\psi_b$
$\hat{T}$	$\bar{\psi}_b\psi_a$	$-\bar{\psi}_b\gamma_5\psi_a$	$\bar{\psi}_b\gamma_\mu\psi_a$	$\bar{\psi}_b\gamma_\mu\gamma^5\psi_a$	$\bar{\psi}_b\sigma_{\mu\nu}\psi_a$
$\hat{C}$	$\bar{\psi}_b\psi_a$	$\bar{\psi}_b\gamma^5\psi_a$	$-\bar{\psi}_b\gamma^\mu\psi_a$	$\bar{\psi}_b\gamma^\mu\gamma^5\psi_a$	$-\bar{\psi}_b\sigma^{\mu\nu}\psi_a$
$\hat{C}\hat{P}$	$\bar{\psi}_b\psi_a$	$-\bar{\psi}_b\gamma^5\psi_a$	$-\bar{\psi}_b\gamma_\mu\psi_a$	$-\bar{\psi}_b\gamma_\mu\gamma^5\psi_a$	$\bar{\psi}_b\sigma^{\mu\nu}\psi_a$
$\hat{C}\hat{P}\hat{T}$	$\bar{\psi}_a\psi_b$	$\bar{\psi}_a\gamma^5\psi_b$	$-\bar{\psi}_a\gamma^\mu\psi_b$	$-\bar{\psi}_a\gamma^\mu\gamma^5\psi_b$	$\bar{\psi}_a\sigma^{\mu\nu}\psi_b$
$\hat{G}$	$-\bar{\psi}_b\psi_a$	$-\bar{\psi}_b\gamma^5\psi_a$	$\bar{\psi}_b\gamma^\mu\psi_a$	$-\bar{\psi}_a\gamma^\mu\gamma^5\psi_b$	$\bar{\psi}_b\sigma^{\mu\nu}\psi_a$

For $\mu = 3$:

$$
\begin{aligned}
\bar{\psi}_a(\vec{r},t)\gamma^3\psi_b(\vec{r},t) \;&\xrightarrow{\hat{T}}\; \bar{\psi}_b(\vec{r},-t)\gamma^0\gamma^1\gamma^3(\gamma^3)^T\gamma^3\gamma^1\gamma^0\psi_a(\vec{r},-t) \\
&= \bar{\psi}_b(\vec{r},-t)\gamma^0\gamma^3\gamma^0\psi_a(\vec{r},-t) \\
&= -\bar{\psi}_b(\vec{r},-t)\gamma^3\psi_a(\vec{r},-t).
\end{aligned}
\tag{2.305}
$$

Thus, we find that the zeroth component does not change sign under $\hat{T}$ operation while the ith component changes sign and we may write

$$
\bar{\psi}_a(\vec{r},t)\gamma^\mu\psi_b(\vec{r},t) \xrightarrow{\hat{T}} \bar{\psi}_b(\vec{r},-t)\gamma_\mu\psi_a(\vec{r},-t).
\tag{2.306}
$$

(iv) Axial vector: $\gamma^\mu\gamma_5$

$$
\begin{aligned}
\bar{\psi}_a(\vec{r},t)\gamma^\mu\gamma_5\psi_b(\vec{r},t) \;&\xrightarrow{\hat{T}}\; \bar{\psi}_b(\vec{r},-t)\gamma^0\gamma^1\gamma^3(\gamma^\mu\gamma_5)^T\gamma^3\gamma^1\gamma^0\psi_a(\vec{r},-t) \\
&= \bar{\psi}_b(\vec{r},-t)\gamma^0\gamma^1\gamma^3\gamma_5{}^T\gamma^{\mu T}\gamma^3\gamma^1\gamma^0\psi_a(\vec{r},-t).
\end{aligned}
\tag{2.307}
$$

For $\mu = 0$, we find

$$
\bar{\psi}_a(\vec{r},t)\gamma^0\gamma^5\psi_b(\vec{r},t) \xrightarrow{\hat{T}} \bar{\psi}_b(\vec{r},-t)\gamma^0\gamma^5\psi_a(\vec{r},-t),
\tag{2.308}
$$

following the same analogy as performed in Eq. (2.302). Similarly for $\mu = i$,

$$
\bar{\psi}_a(\vec{r},t)\gamma^i\gamma^5\psi_b(\vec{r},t) \xrightarrow{\hat{T}} -\bar{\psi}_b(\vec{r},-t)\gamma^i\gamma^5\psi_a(\vec{r},-t).
\tag{2.309}
$$

Thus

$$
\bar{\psi}_a(\vec{r},t)\gamma^\mu\gamma^5\psi_b(\vec{r},t) \xrightarrow{\hat{T}} \bar{\psi}_b(\vec{r},-t)\gamma_\mu\gamma^5\psi_a(\vec{r},-t).
\tag{2.310}
$$

(v) Tensor: $\sigma^{\mu\nu}$

$$\bar{\psi}_a(\vec{r},t)\sigma^{\mu\nu}\psi_b(\vec{r},t) \xrightarrow{\hat{T}} \bar{\psi}_b(\vec{r},-t)\gamma^0\gamma^1\gamma^3(\sigma^{\mu\nu})^T\gamma^3\gamma^1\gamma^0\psi_a(\vec{r},-t)$$

$$= \frac{i}{2}\bar{\psi}_b(\vec{r},-t)\gamma^0\gamma^1\gamma^3\left(\gamma^{\nu T}\gamma^{\mu T} - \gamma^{\mu T}\gamma^{\nu T}\right)$$

$$\gamma^3\gamma^1\gamma^0\psi_a(\vec{r},-t). \tag{2.311}$$

For $\sigma^{\mu\nu}$, we perform time reversal on the components σ^{0i} and σ^{ij}. For σ^{0i}, Eq. (2.311) becomes

$$\bar{\psi}_a(\vec{r},t)\sigma^{0i}\psi_b(\vec{r},t) \xrightarrow{\hat{T}} \frac{i}{2}\bar{\psi}_b(\vec{r},-t)\gamma^0\gamma^1\gamma^3\left(\gamma^{iT}\gamma^{0T} - \gamma^{0T}\gamma^{iT}\right)\gamma^3\gamma^1\gamma^0\psi_a(\vec{r},-t). \tag{2.312}$$

For $i = 1$, we obtain

$$\bar{\psi}_a(\vec{r},t)\sigma^{01}\psi_b(\vec{r},t) \xrightarrow{\hat{T}} -\frac{i}{2}\bar{\psi}_b(\vec{r},-t)\left[\gamma^0\gamma^1\gamma^3\gamma^1\gamma^0\gamma^3\gamma^1\gamma^0\right.$$

$$\left. - \gamma^0\gamma^1\gamma^3\gamma^0\gamma^1\gamma^3\gamma^1\gamma^0\right]\psi_a(\vec{r},-t)$$

$$= -\frac{i}{2}\bar{\psi}_b(\vec{r},-t)\left[\gamma^1\gamma^0 - \gamma^0\gamma^1\right]\psi_a(\vec{r},-t)$$

$$= \bar{\psi}_b(\vec{r},-t)\sigma^{01}\psi_a(\vec{r},-t). \tag{2.313}$$

Similarly, one can do for $i = 2$ and 3. Thus

$$\bar{\psi}_a(\vec{r},t)\sigma^{0i}\psi_b(\vec{r},t) \xrightarrow{\hat{T}} \bar{\psi}_b(\vec{r},-t)\sigma^{0i}\psi_a(\vec{r},-t). \tag{2.314}$$

For $\mu = i$ and $\nu = j$, we get

$$\bar{\psi}_a(\vec{r},t)\sigma^{ij}\psi_b(\vec{r},t) \xrightarrow{\hat{T}} \frac{i}{2}\bar{\psi}_b(\vec{r},-t)\gamma^0\gamma^1\gamma^3\left(\gamma^{jT}\gamma^{iT} - \gamma^{iT}\gamma^{jT}\right)\gamma^3\gamma^1\gamma^0\psi_a(\vec{r},-t). \tag{2.315}$$

For $i = 1$ and $j = 2$, we obtain

$$\bar{\psi}_a(\vec{r},t)\sigma^{12}\psi_b(\vec{r},t) \xrightarrow{\hat{T}} -\frac{i}{2}\bar{\psi}_b(\vec{r},-t)\left[\gamma^0\gamma^1\gamma^3\gamma^2\gamma^1\gamma^3\gamma^1\gamma^0\right.$$

$$\left. - \gamma^0\gamma^1\gamma^3\gamma^1\gamma^2\gamma^3\gamma^1\gamma^0\right]\psi_a(\vec{r},-t)$$

$$= -\bar{\psi}_b(\vec{r},-t)\sigma^{12}\psi_a(\vec{r},-t). \tag{2.316}$$

For $i = 1$ and $j = 3$, we get

$$\bar{\psi}_a(\vec{r},t)\sigma^{13}\psi_b(\vec{r},t) \xrightarrow{\hat{T}} \frac{i}{2}\bar{\psi}_b(\vec{r},-t)\left[\gamma^0\gamma^1\gamma^3\gamma^3\gamma^1\gamma^3\gamma^1\gamma^0\right.$$
$$\left. - \gamma^0\gamma^1\gamma^3\gamma^1\gamma^3\gamma^3\gamma^1\gamma^0\right]\psi_a(\vec{r},-t)$$
$$= -\bar{\psi}_b(\vec{r},-t)\sigma^{13}\psi_a(\vec{r},-t). \tag{2.317}$$

Similarly, for $i = 2$ and $j = 3$, we get

$$\bar{\psi}_a(\vec{r},t)\sigma^{23}\psi_b(\vec{r},t) \xrightarrow{\hat{T}} -\bar{\psi}_b(\vec{r},-t)\sigma^{23}\psi_a(\vec{r},-t). \tag{2.318}$$

From Eqs. (2.316), (2.317) and (2.318), it is established that

$$\bar{\psi}_a(\vec{r},t)\sigma^{ij}\psi_b(\vec{r},t) \xrightarrow{\hat{T}} -\bar{\psi}_b(\vec{r},-t)\sigma^{ij}\psi_a(\vec{r},-t). \tag{2.319}$$

Using Eqs. (2.314) and (2.319), we get

$$\bar{\psi}_a(\vec{r},t)\sigma^{\mu\nu}\psi_b(\vec{r},t) \xrightarrow{\hat{T}} \bar{\psi}_b(\vec{r},-t)\sigma_{\mu\nu}\psi_a(\vec{r},-t). \tag{2.320}$$

(vi) $\sigma^{\mu\nu}\gamma_5$

$$\bar{\psi}_a(\vec{r},t)\sigma^{\mu\nu}\gamma_5\psi_b(\vec{r},t) \xrightarrow{\hat{T}} \bar{\psi}_b(\vec{r},-t)\gamma^0\gamma^1\gamma^3(\sigma^{\mu\nu}\gamma_5)^T\gamma^3\gamma^1\gamma^0\psi_a(\vec{r},-t)$$
$$= \frac{i}{2}\bar{\psi}_b(\vec{r},-t)\gamma^0\gamma^1\gamma^3\gamma_5\left(\gamma^{\nu T}\gamma^{\mu T} - \gamma^{\mu T}\gamma^{\nu T}\right)$$
$$\gamma^3\gamma^1\gamma^0\psi_a(\vec{r},-t). \tag{2.321}$$

Using the properties of γ matrices given in Appendix-C, the components $\sigma^{0i}\gamma_5$ and $\sigma^{ij}\gamma_5$ transforms under time reversal transformation as

$$\bar{\psi}_a(\vec{r},t)\sigma^{0i}\gamma_5\psi_b(\vec{r},t) \xrightarrow{\hat{T}} -\bar{\psi}_b(\vec{r},-t)\sigma^{0i}\gamma_5\psi_a(\vec{r},-t), \tag{2.322}$$
$$\bar{\psi}_a(\vec{r},t)\sigma^{ij}\gamma_5\psi_b(\vec{r},t) \xrightarrow{\hat{T}} \bar{\psi}_b(\vec{r},-t)\sigma^{ij}\gamma_5\psi_a(\vec{r},-t). \tag{2.323}$$

Combining the two equations, we get the transformation of $\sigma^{\mu\nu}\gamma_5$ under $\hat{T}$, as

$$\bar{\psi}_a(\vec{r},t)\sigma^{\mu\nu}\gamma_5\psi_b(\vec{r},t) \xrightarrow{\hat{T}} -\bar{\psi}_b(\vec{r},-t)\sigma_{\mu\nu}\gamma_5\psi_a(\vec{r},-t). \tag{2.324}$$

In Tables 2.1 and 2.2, respectively, the transformation of Dirac spinor as well as adjoint spinor and Dirac bilinear covariants under $\hat{C}$, $\hat{P}$, $\hat{T}$, $\hat{C}\hat{P}$, $\hat{C}\hat{P}\hat{T}$ and $\hat{G}$ are shown.

Chapter 3

Quantization of Free Particle Fields

3.1   Introduction

The concept of associating particles with fields originated during the study of various physical phenomena involving electromagnetic radiation. For example, the observations and theoretical explanations of the black body radiation by Planck, the photoelectric effect by Einstein, and the scattering of a photon off an electron by Compton established that electromagnetic radiation can be described in terms of "discrete quanta of energy" called photon, identified as a massless particle of spin 1. Consequently, Maxwell's equations of classical electrodynamics, describing the time evolution of the electric $(\vec{E}(\vec{x}, t))$ and magnetic $(\vec{B}(\vec{x}, t))$ fields are interpreted to be the equations of motion of the photon, written in terms of the massless spin 1 electromagnetic field $A^{\mu}(\vec{x}, t)$. Later, the quantization of the electromagnetic field $A^{\mu}(\vec{x}, t)$ was formulated to explain the emission and absorption of radiation in terms of the creation and annihilation of photons during the interaction of the electromagnetic field with the physical systems. The concept of treating photons as quanta of the electromagnetic fields was successful in explaining the physical phenomena induced by the electromagnetic interactions; methods of field quantization were used leading to quantum electrodynamics (QED), the quantum field theory of electromagnetic interactions. The concept was later generalized by Fermi [23, 207] and Yukawa [208, 209] to formulate, respectively, the theory of weak and strong interactions in analogy with the theory of QED.

In order to describe QED, the quantum field theory of electromagnetic fields and their interaction with matter, in terms of the massless spin 1 fields $A^{\mu}(\vec{x}, t)$ corresponding to photons, the equations of motion of $A^{\mu}(\vec{x}, t)$ should be fully relativistic. This requires the reformulation of classical equations of motion for the fields to obtain the quantum equations of motion for the fields and find their solutions, in case of free fields as well as fields interacting with matter. This is generally done using perturbation theory for which a relativistically covariant perturbation theory is required.

The path of transition from a classical description of fields to a quantum description of fields, requiring the quantization of fields, their equations of motion, propagation, and interaction with matter involves understanding many new concepts and mathematical methods. For this purpose, the Lagrangian formulation for describing the dynamics of particles and their interaction with the fields is found to be suitable. In this chapter, we attempt to explain this formulation in the case of free fields; we take up the case of interacting fields in the next chapter. The mathematical treatment in this chapter and the following chapter is based on the materials contained in Refs.[210] and [211].

3.2 Lagrangian Formulation for the Dynamics of Particles and Fields

3.2.1 Equation of motion for particles

The equation of motion of a particle in classical physics is described in terms of a set of variables called generalized coordinates $q^i (i = 1 - n)$ depending upon the degrees of freedom, for the motion of the particle. For example, for a particle moving in 3-dimensions, there will be three generalized coordinates, taken to be either the cartesian (x, y, z) or the spherical polar (r, θ, ϕ) coordinates. In most of the cases, the generalized coordinates are chosen to be position coordinates $x_i (i = 1 - n)$, which evolve with time t according to the predictions made by the equation of motion. Historically, these equations of motion are derived from Newton's laws of motion but can also be derived from the "principle of least action(S)", where the action S is derived in terms of the Lagrangian L as:

$$S = \int_{t_1}^{t_2} dt\, L = \int \mathcal{L}\, d\vec{x}\, dt, \tag{3.1}$$

where $L = \int d\vec{x}\, \mathcal{L}$. $\mathcal{L}$ is known as the Lagrangian density, which is function of the generalized coordinates x_i, generalized velocity $\dot{x}_i(t) = \left(\frac{dx_i(t)}{dt}\right)$, and time t, and therefore:

$$S = \int_{A(t_1)}^{B(t_2)} dt\, d\vec{x}\, \mathcal{L}(x_i(t), \dot{x}_i(t), t). \tag{3.2}$$

The principle of least action states that out of many possibilities of the motion of particles between A and B (Figure 3.1), the classical motion of the particles is such that the action S is extremum. Any variation in $x_i(t)$ like $x_i(t) \rightarrow x_i(t) + \delta x_i(t)$ vanishes at the end points A and B, where $\delta x_i(t_1) = \delta x_i(t_2) = 0$, as shown in Figure 3.1. For any path $x_i'(t)$ other than $x_i(t)$ such that $x_i'(t) = x_i(t) + \delta x_i(t)$, the change in action δS is given by:

$$\begin{aligned}
\delta S &= \delta \int_A^B L(x_i(t), \dot{x}_i(t))dt \\
&= \int_A^B \sum_i \left[\frac{\partial L}{\partial x_i}\delta x_i + \frac{\partial L}{\partial \dot{x}_i}\delta \dot{x}_i\right] dt \qquad \left(\because \delta \dot{x}_i(t) = \frac{d}{dt}(\delta x_i)\right)
\end{aligned}$$

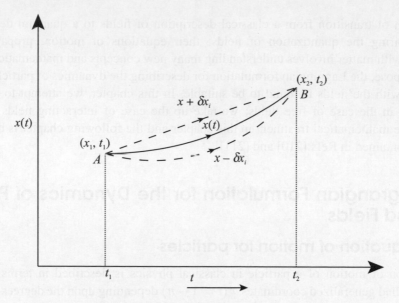

Figure 3.1 δ variation-extremum path.

$$
= \int_A^B \sum_i \left\{ \frac{\partial L}{\partial x_i} \delta x_i + \frac{\partial L}{\partial \dot{x}_i} \frac{d}{dt}(\delta x_i) \right\} dt
$$

$$
= \int_A^B \sum_i \left\{ \frac{\partial L}{\partial x_i} \delta x_i + \frac{d}{dt}\left(\frac{\partial L}{\partial \dot{x}_i}(\delta x_i) \right) - \frac{d}{dt}\left(\frac{\partial L}{\partial \dot{x}_i} \right)(\delta x_i) \right\} dt
$$

$$
= \int_A^B \sum_i \left\{ \frac{\partial L}{\partial x_i} - \frac{d}{dt}\left(\frac{\partial L}{\partial \dot{x}_i} \right) \right\} \delta x_i \, dt
$$

as $\delta x_i(t_1) = 0, \delta x_i(t_2) = 0,$ at the end points A and B.

Since the variations $\delta x_i(t)$ are arbitrary, the principle of least action, that is, $\delta S = 0$ leads to the well known Euler–Lagrange equation of motion for the particle, that is,

$$
\frac{d}{dt}\left(\frac{\partial L}{\partial \dot{x}_i} \right) - \frac{\partial L}{\partial x_i} = 0. \tag{3.3}
$$

The momentum of the particle p_i corresponding to the motion along x_i is defined as:

$$
p_i = \frac{\partial L}{\partial \dot{x}_i} \tag{3.4}
$$

and the Hamiltonian (H) of the particle is defined as:

$$
H = \sum_i p_i \dot{x}_i - L(x_i, \dot{x}_i, t) = \sum_i \frac{\partial L}{\partial \dot{x}_i} \dot{x}_i - L(x_i, \dot{x}_i, t). \tag{3.5}
$$

Let us consider the example of simple harmonic motion of a particle of mass m attached to a spring under a force $\vec{F}$, say along the x- direction, given by $F_x = -kx$, where k is the spring constant. Newton's equations of motion leads to the equations of motion for the particle being:

$$m\frac{d^2x}{dt^2} + kx = 0, \tag{3.6}$$

with the solution

$$x = x_0 \sin \omega t, \tag{3.7}$$

where $\omega = \sqrt{\dfrac{k}{m}}$.

This can be derived by considering a Lagrangian defined in terms of the position of the particle, that is, x(it has only one degree of freedom), written as:

$$L = T - V = \frac{1}{2}m\dot{x}^2 - \frac{1}{2}m\omega^2 x^2. \tag{3.8}$$

The Euler–Lagrange equation of motion (Eq. (3.3)) leads to the equation:

$$\frac{d^2x}{dt^2} + \omega^2 x = 0 \qquad \text{and} \qquad p_x = m\dot{x}. \tag{3.9}$$

The Hamiltonian H is given by:

$$H = \frac{p_x^2}{2m} + \frac{1}{2}m\omega^2 x^2 = T + V, \tag{3.10}$$

which is the total energy.

In classical physics, the equations of motion for a system derived by applying the Euler–Lagrange equation (Eq. (3.3)) are deterministic. This means that once initial conditions are given through the specification of position $\vec{x}$ and velocity $\vec{v} = \frac{\vec{p}}{m}$ at time $t = 0$, the dynamical behavior of the particle is completely determined.

In quantum physics, dynamical variables of classical physics, like the position x and the momentum p_x, are treated as operators, that is,

$$x \to \hat{x} \quad \text{and} \quad p_x \to \hat{p}_x = -i\frac{\partial}{\partial x}$$

or in general

$$\vec{x} \to \hat{\vec{x}}, \qquad \vec{p} \to \hat{\vec{p}} = -i\vec{\nabla},$$

such that $\hat{x}$ and $\hat{p}_x$ do not commute, and the commutator of $\hat{x}$ and $\hat{p}_x$ is given by:

$$[\hat{x}, \hat{p}_x] = i, \tag{3.11}$$

implying that the physical observables, like the position x_i and the momentum p_i, are the eigenvalues of the operators $\hat{x}_i$ and $\hat{p}_i$, respectively, and cannot be determined simultaneously. In the presence of many degrees of freedom, the commutation relations take the form:

$$[\hat{x}_i, \hat{p}_j] = i\delta_{ij}, \tag{3.12}$$

$$[\hat{x}_i, \hat{x}_j] = [\hat{p}_i, \hat{p}_j] = 0. \tag{3.13}$$

This procedure of obtaining quantum equations of motion is also applied to the fields.

3.2.2 Quantization of a harmonic oscillator

We start with the Hamiltonian for the harmonic oscillator, that is,

$$H = \frac{p_x^2}{2m} + \frac{1}{2}m\omega^2 x^2. \tag{3.14}$$

We first introduce the operators $\hat{q}$ and $\hat{p}$ as:

$$\hat{q} = \sqrt{\alpha}\hat{x} \quad \text{and} \quad \hat{p} = \frac{\hat{p}_x}{\sqrt{\alpha}}, \text{ where } \alpha = m\omega \tag{3.15}$$

which satisfy the commutation relations $[\hat{q}, \hat{p}] = i$.

Let us now introduce $\hat{a}^\dagger$ and $\hat{a}$ operators such that:

$$\hat{a} = \frac{1}{\sqrt{2}}(\hat{q} + i\hat{p}),$$
$$\hat{a}^\dagger = \frac{1}{\sqrt{2}}(\hat{q} - i\hat{p}), \tag{3.16}$$

then it can be shown that:

$$\left[\hat{a}, \hat{a}^\dagger\right] = \hat{a}\hat{a}^\dagger - \hat{a}^\dagger\hat{a} = 1, \tag{3.17}$$

and

$$\hat{H} = \frac{\omega}{2}(\hat{a}\hat{a}^\dagger + \hat{a}^\dagger\hat{a}), \tag{3.18}$$

$$= \omega\left(\hat{a}^\dagger\hat{a} + \frac{1}{2}\right) = \omega\left(N + \frac{1}{2}\right), \quad \text{where } N = \hat{a}^\dagger\hat{a}. \tag{3.19}$$

To see the physical interpretation of N, let us calculate the energy of a state ψ using Eq. (3.19) that is,

$$H\psi = \omega\left(N + \frac{1}{2}\right)\psi = E\psi \tag{3.20}$$

and compare it with the result in conventional quantum mechanics where the energy of the harmonic oscillator is derived to be:

$$H\psi = \left(n + \frac{1}{2}\right)\omega\psi \quad \text{with } n = 0,1,2,3... \tag{3.21}$$

Therefore, N can be identified as a number operator with integer eigenvalues 0,1,2,... such that for a state $|\psi\rangle$

$$N|\psi\rangle = n|\psi\rangle, \tag{3.22}$$

with the lowest value of $n = 0$. This can be also seen by considering the expectation value of N in any state ψ as:

$$\langle\psi|N|\psi\rangle = \langle\psi|\hat{a}^\dagger\hat{a}|\psi\rangle = \langle\hat{a}\psi|\hat{a}\psi\rangle \geq 0. \tag{3.23}$$

Therefore, the minimum value of $\langle\psi|N|\psi\rangle$ is zero corresponding to the eigenvalues of the lowest eigenstate, that is, the ground state of a harmonic oscillator. To see the physical interpretation of operators $\hat{a}$ and $\hat{a}^\dagger$, we note that they satisfy the following commutation relations

$$[N,\hat{a}^\dagger] = \hat{a}^\dagger \quad \text{and} \quad [N,\hat{a}] = -\hat{a}. \tag{3.24}$$

Operating Eq. (3.24) on a state ψ, which satisfies Eq. (3.22), we obtain:

$$[N,\hat{a}^\dagger]|\psi\rangle = \hat{a}^\dagger|\psi\rangle,$$

$$N\,\hat{a}^\dagger|\psi\rangle = (n+1)\,\hat{a}^\dagger|\psi\rangle, \tag{3.25}$$

$$\text{and similarly} \quad N\,\hat{a}|\psi\rangle = (n-1)\,\hat{a}\,|\psi\rangle. \tag{3.26}$$

Equations (3.25) and (3.26) show that $\hat{a}$ and $\hat{a}^\dagger$ lower and raise the value of n by 1. By lowering the value of n by 1 in successive operations, we can reach the ground state $|\psi_0\rangle$ corresponding to the eigenvalues zero, such that:

$$\hat{a}|\psi_0\rangle = 0, \quad \text{that is,} \quad |\psi_0\rangle = |0\rangle, \tag{3.27}$$

where $|0\rangle$ is the ground state and has energy

$$H|0\rangle = \omega(N + \frac{1}{2})|0\rangle = \frac{\omega}{2} \tag{3.28}$$

called the zero point energy of the harmonic oscillator. Since $\hat{a}$ and $\hat{a}^\dagger$ lower and raise the number by 1, respectively, we can write the action of $\hat{a}$ and $\hat{a}^\dagger$ on the normalized state $|n\rangle$, such that $\langle n|n\rangle = 1$. For $|n\rangle$ states, the following operator equations are obtained:

$$\hat{a}|n\rangle = \sqrt{n}|\,n-1\rangle \quad \text{and} \quad \hat{a}^\dagger|n\rangle = \sqrt{n+1}\,|n+1\rangle. \tag{3.29}$$

It is easy to see that a normalized state $|n\rangle$ can be generated from the ground state $|0\rangle$ by operating $\hat{a}^\dagger$, n number of terms to obtain:

$$|n\rangle = \frac{(\hat{a}^\dagger)^n}{\sqrt{n!}}|0\rangle, \quad n = 0,1,2,..., \tag{3.30}$$

$$\text{with} \quad H|n\rangle = \left(n + \frac{1}{2}\right)\omega|n\rangle. \tag{3.31}$$

3.2.3   Equation of motion for fields

The motion of a classical particle with specified degrees of freedom as a function of time t is described by its position at a given point in space. On the other hand, the interpretation of the particle as a quanta of field $\phi(\vec{x}, t)$ necessitates its description in terms of the equation of motion for the field $\phi(\vec{x}, t)$ which is spread over a region of space and varies with time. Since the field $\phi(\vec{x}, t)$ is a continuous function of space and time, this implies infinite degrees of freedom. Therefore, the equation of motion for fields are obtained by generalizing the principle of least action to a continuous system of infinite degrees of freedom. The Lagrangian for a particular field is written in terms of the fields and their derivatives using general principles of Lorentz invariance and other symmetries of the system, when in interaction. For a free field, the Lagrangian includes only the kinetic energy and mass terms and is constructed such that the equations of motion for various particles described in the last chapter are reproduced. These particles of spin 0, $\frac{1}{2}$, or 1 are now described by the fields carrying spin 0, $\frac{1}{2}$, or 1 and not by the single particle wave functions discussed in Chapter 2. These fields can then be quantized in terms of the creation and annihilation operators to describe the emission and absorption of particles corresponding to the fields.

In general, the action S in the case of fields is defined as:

$$S(\Omega) = \int L(\vec{x}, t)dt = \int_{\Omega} d^4x \mathcal{L}(\phi_i(\vec{x}, t), \phi_{i,\alpha}(\vec{x}, t), t), \tag{3.32}$$

where $L(\vec{x}, t)$ is the Lagrangian and $\mathcal{L}(\phi_i(\vec{x}, t), \phi_{i,\alpha}(\vec{x}, t))$ is the Lagrangian density for the field $\phi(\vec{x}, t)$. $d^4x(= dx^0 d\vec{x})$ is the four-dimensional element, $\phi_{i,\alpha} = \partial_\alpha \phi_i(x) = \frac{\partial \phi_i}{\partial x^\alpha}$, is the derivative of the field $\phi_i(x)$ and the integration is performed over a region of volume element Ω, in four-dimensional space–time. The principle of least action implies that under a variation of fields $\phi_i(x)$, the action S is extremum, that is, $\delta S = 0$. If we define the variation of fields $\phi_i(x)$ as $\delta \phi_i(x)$, that is,

$$\phi_i(x) \rightarrow \phi_i(x) + \delta \phi_i(x),$$

such that the variation at the surface $\mathcal{S}(\Omega)$ of volume Ω vanishes, that is, $\delta \phi_i(x) = 0$ on $\mathcal{S}(\Omega)$, then the principle of least action states that the change in the action δS is zero, that is,

$$\delta S = \int_{\Omega} \left(\frac{\partial \mathcal{L}}{\partial \phi_i(x)} \delta \phi_i(x) + \frac{\partial \mathcal{L}}{\partial \phi_{i,\alpha}(x)} \delta \phi_{i,\alpha}(x) \right) d^4x = 0 \tag{3.33}$$

as all the $\phi_i(x)$ and $\phi_{i,\alpha}(x)$ for $i = 1 - n$ are linearly independent. The fields $\phi_i(x)$ can be taken as real or complex fields. In the case of complex fields, each $\phi_i(x)$ can be treated as two independent real fields, or equivalently, two independent fields $\phi(x)$ and $\phi^*(x)$. The principle of least action in Eq. (3.33) leads to the Euler–Lagrange equation of motion for the fields $\phi_i(x)$. Consider:

$$\delta S = \int_{\Omega} d^4x \left(\frac{\partial \mathcal{L}}{\partial \phi_i(x)} \delta \phi_i(x) + \frac{\partial \mathcal{L}}{\partial (\partial_\alpha \phi_i(x))} \partial_\alpha \delta \phi_i(x) \right)$$

$$= \int_\Omega d^4x \frac{\partial \mathcal{L}}{\partial \phi_i(x)} \delta\phi_i(x) - \int_\Omega d^4x\, \partial_\alpha \left(\frac{\partial \mathcal{L}}{\partial(\partial_\alpha \phi_i(x))} \right) \delta\phi_i(x)$$

$$+ \frac{\partial \mathcal{L}}{\partial(\dot\phi_i(x))} \delta\phi_i(x) \Bigg|_{t_i}^{t_f}. \tag{3.34}$$

At the extremum, $\delta\phi_i(\vec{x}, t_i) = \delta\phi_i(\vec{x}, t_f) = 0$, which leads to:

$$\delta S = \int_\Omega d^4x \left[\frac{\partial \mathcal{L}}{\partial \phi_i(x)} - \partial_\alpha \left(\frac{\partial \mathcal{L}}{\partial(\partial_\alpha \phi_i(x))} \right) \right] \delta\phi_i(x), \tag{3.35}$$

and for an arbitrary $\delta\phi_i(x)$

$$\partial_\alpha \left(\frac{\partial \mathcal{L}}{\partial(\partial_\alpha \phi_i(x))} \right) - \frac{\partial \mathcal{L}}{\partial \phi_i(x)} = 0, \quad \text{if } \delta S = 0. \tag{3.36}$$

In order to quantize the classical field, we apply the method of canonical quantization used in nonrelativistic quantum mechanics and define the canonical momentum density $\pi_i(\vec{x}, t)$ corresponding to the field $\phi_i(\vec{x}, t)$ as:

$$\pi_i(\vec{x}, t) = \frac{\partial \mathcal{L}}{\partial \dot\phi_i(\vec{x}, t)}. \tag{3.37}$$

With the definition of the conjugate momentum(Eq. (3.37)), the fields $\phi_i(\vec{x}, t)$ and the conjugate momenta $\pi_i(\vec{x}, t)$ become operators and field quantization is implemented through the commutation relations of $\hat\phi_i(\vec{x}, t)$ and $\hat\pi_j(\vec{x}, t)$ in analogy with the quantization in particle mechanics (Eqs. (3.12) and (3.13)), that is,

$$[\hat\phi_i(\vec{x}, t), \hat\pi_j(\vec{x}', t)] = i\delta(\vec{x} - \vec{x}')\delta_{ij}, \tag{3.38}$$

$$[\hat\phi_i(\vec{x}, t), \hat\phi_j(\vec{x}, t)] = [\hat\pi_i(\vec{x}, t), \hat\pi_j(\vec{x}, t)] = 0. \tag{3.39}$$

It should be noted that these are the commutation relations of fields at equal time, that is, $t = t'$ and are called equal time commutators (ETC). Thus, the transition from the classical field theory to the quantum field theory is made by introducing the commutation relations among the field variables $\phi_i(\vec{x}, t)$ and $\pi_j(\vec{x}, t)$. The method is known as second quantization, while the commutation relations given in Eqs. (3.12) and (3.13) for $\hat{x}$ and $\hat{p}_x$, etc. are known as first quantization. The quantization of the fields leads to the description of the field in terms of the particle which is acronymed as "the field quanta". These field quanta carry energy, momentum, and charge of the field.

3.2.4 Symmetries and conservation laws: Noether's theorem

Symmetries have played an important role in the development of physics and mathematics. They manifest through the invariance of the Lagrangian under certain transformations defining the symmetry. It has been shown that the invariance of the Lagrangian describing any system under the symmetry transformations implies the existence of conserved quantities. This is known as Noether's theorem [212] and the conserved quantities are called Noether's charges or Noether's currents, so named after its discoverer. Therefore, there is a conservation law associated with every continuous symmetry. For example, translational invariance leads to the conservation of linear momentum; rotational invariance leads to the conservation of angular momentum, etc.

Let us consider infinitesimal transformations in the case of internal symmetry where the dynamical variable x is unchanged as $\phi_i(x) \to \phi_i'(x) = e^{i\alpha}\phi_i(x)$, where α is independent of x, such that:

$$\phi_i(x) \to \phi_i'(x) = \phi_i(x) + \delta\phi_i(x). \tag{3.40}$$

The variation in the Lagrangian due to the variation in the field will be

$$
\begin{aligned}
\delta L &= \sum_i \left(\frac{\partial L}{\partial \phi_i(x)} \delta\phi_i(x) + \frac{\partial L}{\partial \phi_{i,\alpha}(x)} \delta\phi_{i,\alpha}(x) \right) \\
&= \sum_i \left(\partial_\alpha \left(\frac{\partial L}{\partial \phi_{i,\alpha}(x)} \right) \delta\phi_i(x) + \frac{\partial L}{\partial \phi_{i,\alpha}(x)} \delta\phi_{i,\alpha}(x) \right) \\
&= \sum_i \partial_\alpha \left(\frac{\partial L}{\partial \phi_{i,\alpha}(x)} \delta\phi_i(x) \right)
\end{aligned}
\tag{3.41}
$$

If L is invariant under the transformation given in Eq. (3.40), that is, $\delta L = 0$, then the quantity in the parenthesis is known as the conserved current given by:

$$\partial_\alpha J_i^\alpha = 0, \tag{3.42}$$

where

$$J_i^\alpha = \frac{\partial L}{\partial(\partial_{i,\alpha}\phi(x))} \delta\phi_i(x). \tag{3.43}$$

The charge operator Q is defined as:

$$Q = \int d\vec{x}\, J^0, \tag{3.44}$$

and since the integral of a total charge vanishes, we may write

$$\frac{dQ}{dt} = \int d\vec{x}\partial_0 J^0 = -\int_V d\vec{x}\partial_i\, J^i = 0, -\int_S S_i \cdot J^i = 0 \tag{3.45}$$

which implies that Q is time independent. Thus,

$$[Q, H] = 0. \tag{3.46}$$

Let us consider the case when the symmetry transformations involve the change in space coordinates as well as the fields given as:

$$x^\mu \rightarrow x'^\mu \;\; = \;\; x^\mu + \delta x^\mu \tag{3.47}$$

$$\text{and} \;\; \phi_r(x) \rightarrow \phi'_r(x') \;\; = \;\; \phi_r(x) + \delta\phi_r(x). \tag{3.48}$$

Since

$$\delta\phi_r(x) = \phi'_r(x) - \phi_r(x), \tag{3.49}$$

we may define

$$\delta_T\phi_r(x) \;\; = \;\; \phi'_r(x') - \phi_r(x) = \phi'_r(x') - \phi_r(x') + \phi_r(x') - \phi_r(x)$$

$$= \;\; \delta\phi_r(x') + \frac{\partial\phi_r}{\partial x_\nu}\delta x_\nu. \tag{3.50}$$

For the first order, the changes will be small such that:

$$\delta\phi_r(x') \;\; \approx \;\; \delta\phi_r(x) \tag{3.51}$$

$$\Rightarrow \delta_T\phi_r(x) \;\; = \;\; \delta\phi_r(x) + \frac{\partial\phi_r}{\partial x_\nu}\delta x_\nu. \tag{3.52}$$

Invariance of the Lagrangian is required, such that:

$$\mathcal{L}(\phi'_r(x'), \phi'_{r,\mu}(x')) - \mathcal{L}(\phi_r(x)\phi_{r,\mu}(x)) = 0. \tag{3.53}$$

With the help of Eq. (3.52), we can write:

$$\delta\mathcal{L} + \frac{\partial\mathcal{L}}{\partial x^\mu}\delta x^\mu = 0. \tag{3.54}$$

Moreover,

$$\delta\mathcal{L} = \frac{\partial\mathcal{L}}{\partial\phi_r}\delta\phi_r + \frac{\partial\mathcal{L}}{\partial\phi_{r,\mu}}\delta\phi_{r,\mu}. \tag{3.55}$$

Using the Euler–Lagrange equation:

$$\delta\mathcal{L} \;\; = \;\; \frac{\partial}{\partial x^\mu}\left(\frac{\partial\mathcal{L}}{\partial\phi_{r,\mu}}\delta\phi_r\right)$$

$$= \;\; \frac{\partial}{\partial x^\mu}\left[\frac{\partial\mathcal{L}}{\partial\phi_{r,\mu}}\left(\delta_T\phi_r - \frac{\partial\phi_r}{\partial x_\nu}\delta x_\nu\right)\right]. \tag{3.56}$$

Using Eq. (3.56) in Eq. (3.54), we have:

$$\frac{\partial}{\partial x^\mu}\left[\frac{\partial \mathcal{L}}{\partial \phi_{r,\mu}}\left(\delta_T\phi_r - \frac{\partial \phi_r}{\partial x_\nu}\delta x_\nu\right)\right] + \frac{\partial \mathcal{L}}{\partial x^\mu}\delta x^\mu = 0$$

$$\Rightarrow \frac{\partial}{\partial x^\mu}\left[\frac{\partial \mathcal{L}}{\partial \phi_{r,\mu}}\delta_T\phi_r\right] - \frac{\partial T^{\mu\nu}}{\partial x^\mu}\delta x_\nu = 0, \tag{3.57}$$

where

$$T^{\mu\nu} = \frac{\partial \mathcal{L}}{\partial \phi_{r,\mu}}\frac{\partial \phi_r}{\partial x_\nu} - \mathcal{L}g^{\mu\nu}. \tag{3.58}$$

$T^{\mu\nu}$ is known as the energy–momentum tensor. It can be shown that the conserved quantity J^μ is given by:

$$J^\mu = \frac{\partial \mathcal{L}}{\delta \phi_{r,\mu}}\delta_T\phi_r - T^{\mu\nu}\delta x_\nu. \tag{3.59}$$

Taking $\mu = \nu = 0$; $\mu = 0$, $\nu = i$

$$T^{00} = \frac{\partial \mathcal{L}}{\partial \dot{\phi}}\dot{\phi} - \mathcal{L} = \mathcal{H}, T^{0i} = \frac{\partial \mathcal{L}}{\partial \dot{\phi}}\frac{\partial \phi}{\partial x_i} \tag{3.60}$$

is interpreted as the energy-momentum density of the field. The total energy and the four momentum of the field are given by:

$$H = \int T^{00}d\vec{x}, \tag{3.61}$$

$$P^\mu = \int T^{0\mu}d\vec{x}. \tag{3.62}$$

The application of the conservation of J^μ using Eq. (3.42), leads to the conservation of momentum, energy, and angular momentum when applied to the symmetries corresponding to translational, rotational, and Lorentz transformations.

3.3 Quantization of Scalar Fields: Klein–Gordon Field

3.3.1 Real scalar field: Creation and annihilation operators

We consider the real scalar field $\phi(x)$, which corresponds to a particle of spin 0 and satisfies the Klein–Gordon equation given in Chapter 2. This equation for the field $\phi(x)$ can be derived from the Lagrangian density given by:

$$\mathcal{L}(x) = \frac{1}{2}(\phi_{,\alpha}(x)\phi^{,\alpha}(x) - m^2\phi^2(x)). \tag{3.63}$$

Using the Euler–Lagrange equation, that is,

$$\frac{\partial}{\partial x^\alpha}\left(\frac{\partial \mathcal{L}}{\partial \phi_\alpha}\right) - \frac{\partial \mathcal{L}}{\partial \phi} = 0, \tag{3.64}$$

we get

$$\partial_\mu \partial^\mu \phi(x) + m^2 \phi(x) = 0,$$
$$\Rightarrow (\Box + m^2)\phi(x) = 0. \tag{3.65}$$

The conjugate momentum $\pi(x)$ is obtained by the definition:

$$\pi(\vec{x}, t) = \frac{\partial \mathcal{L}}{\partial \dot{\phi}} = \dot{\phi}(\vec{x}, t). \tag{3.66}$$

The quantization of the real field ϕ using the canonical quantization procedure of nonrelativistic quantum mechanics is achieved by treating $\phi(x)$ and $\dot{\phi}(x)$ as operators satisfying the commutation relations:

$$[\phi(\vec{x}, t), \dot{\phi}(\vec{y}, t)] = i\delta^3(\vec{x} - \vec{y}), \tag{3.67}$$

$$[\phi(\vec{x}, t), \phi(\vec{y}, t)] = [\pi(\vec{x}, t), \pi(\vec{y}, t)] = [\dot{\phi}(\vec{x}, t), \dot{\phi}(\vec{y}, t)] = 0. \tag{3.68}$$

We have seen in Chapter 2 that the Klein–Gordon equation admits solutions like:

$$\phi(\vec{x}, t) = \frac{1}{\sqrt{2V\omega_{\vec{k}}}} e^{-ik\cdot x} \quad \text{and} \quad \phi^\dagger(\vec{x}, t) = \frac{1}{\sqrt{2V\omega_{\vec{k}}}} e^{ik\cdot x},$$

where

$$e^{\pm ik\cdot x} = e^{\pm ik_\mu x^\mu} = e^{\pm ik^\mu x_\mu} = e^{\pm i(k_0 t - \vec{k}\cdot\vec{x})}.$$

Here V is the normalization volume, $\omega_{\vec{k}} = \sqrt{|\vec{k}|^2 + m^2}$ and $k^\mu = (\omega_{\vec{k}}, \vec{k})$.

We expand $\phi(\vec{x}, t)$ in terms of the complete set of solutions of the Klein–Gordon equation and write:

$$\phi(x) = \phi^+(x) + \phi^-(x) = \sum_{\vec{k}} \frac{1}{\sqrt{2V\omega_{\vec{k}}}} \left(a(\vec{k})e^{-ik\cdot x} + a^\dagger(\vec{k})e^{ik\cdot x}\right), \tag{3.69}$$

$$\text{where} \quad \phi^+(x) = \sum_{\vec{k}} \frac{1}{\sqrt{2V\omega_{\vec{k}}}} \left(a(\vec{k})e^{-ik\cdot x}\right) \text{ and} \tag{3.70}$$

$$\phi^-(x) = \sum_{\vec{k}} \frac{1}{\sqrt{2V\omega_{\vec{k}}}} \left(a^\dagger(\vec{k})e^{ik\cdot x}\right). \tag{3.71}$$

$a^\dagger(\vec{k})$ is the Hermitian conjugate of the operator $a(\vec{k})$. $\phi(x)$ is an operator in coordinate space, while $a(\vec{k})$ and $a^\dagger(\vec{k})$ are operators in momentum space. This form of $\phi(x)$ is necessitated because $\phi(x)$, being real, satisfies the relation $\phi(x) = \phi^\dagger(x)$.

The conjugate momentum field $\dot{\phi}(\vec{x}, t)$ is then derived to be:

$$\dot{\phi}(\vec{x},t) = \sum_{\vec{k}} \frac{1}{\sqrt{2V\omega_{\vec{k}}}}(-i\omega_{\vec{k}})\left(a(\vec{k})e^{-ik\cdot x} - a^\dagger(\vec{k})e^{ik\cdot x}\right). \tag{3.72}$$

Using Eqs. (3.69) and (3.72), it can be shown that:

$$a(\vec{k}) \;=\; \frac{1}{\sqrt{2V\omega_{\vec{k}}}} \int d\vec{x}\, e^{ik\cdot x}(\omega_{\vec{k}}\phi(x) + i\dot{\phi}(x)), \tag{3.73}$$

$$a^\dagger(\vec{k}) \;=\; \frac{1}{\sqrt{2V\omega_{\vec{k}}}} \int d\vec{x}\, e^{-ik\cdot x}(\omega_{\vec{k}}\phi(x) - i\dot{\phi}(x)), \tag{3.74}$$

leading to the following commutation relations for $a(\vec{k})$ and $a^\dagger(\vec{k})$.

$$\left[a(\vec{k}), a^\dagger(\vec{k}')\right] = \iint \frac{d\vec{x}d\vec{x}'}{V\sqrt{2\omega_{\vec{k}}}\sqrt{2\omega_{\vec{k}'}}} e^{-ik\cdot x}e^{ik'\cdot x'}[\omega_{\vec{k}}\phi(x) + i\dot{\phi}(x), \omega_{\vec{k}}\phi(x') - i\dot{\phi}(x')]$$

$$= \iint \frac{d\vec{x}d\vec{x}'}{V\sqrt{2\omega_{\vec{k}}}\sqrt{2\omega_{\vec{k}'}}} e^{-i(k\cdot x - k'\cdot x')}(-i\omega_{\vec{k}}[\phi(x), \dot{\phi}(x')] + i\omega_{\vec{k}'}[\dot{\phi}(x), \phi(x')])$$

$$= \frac{\omega_{\vec{k}} + \omega_{\vec{k}'}}{\sqrt{4\omega_{\vec{k}}\omega_{\vec{k}'}}} \int \frac{d\vec{x}}{V} e^{-i(k-k')\cdot x}$$

since $[\phi(x), \dot{\phi}(x')] = i\delta^3(x - x')$ the aforementioned expression results in,

$$\left[a(\vec{k}), a^\dagger(\vec{k}')\right] \;=\; \delta_{\vec{k}\vec{k}'}\;, \tag{3.75}$$

$$\left[a(\vec{k}), a(\vec{k}')\right] \;=\; \left[a^\dagger(\vec{k}), a^\dagger(\vec{k}')\right] = 0. \tag{3.76}$$

Similarly, it can be shown that these relations for $a(\vec{k})$ and $a^\dagger(\vec{k})$ are precisely the relations derived earlier in this chapter in the case of the harmonic oscillator and can be interpreted, respectively, as the annihilation and creation operators. For example, the Hamiltonian can be derived as:

$$H \;=\; \int (\pi(x)\dot{\phi}(x) - \mathcal{L}(x))d\vec{x}$$

$$=\; \int d\vec{x}\frac{1}{2}(\dot{\phi}^2(x) + (\nabla\phi(x))^2 + m^2\phi^2(x))$$

$$=\; \int d\vec{x}\frac{\omega_{\vec{k}}}{2}(a(\vec{k})a^\dagger(\vec{k}) + a^\dagger(\vec{k})a(\vec{k})) \tag{3.77}$$

$$= \sum_{\vec{k}} \omega_{\vec{k}} \left(a^\dagger(\vec{k}) a(\vec{k}) + \frac{1}{2} \right)$$

$$= \sum_{\vec{k}} \omega_{\vec{k}} \left(N(\vec{k}) + \frac{1}{2} \right). \tag{3.78}$$

Similarly, the momentum operator $\vec{P}$ is given by:

$$\vec{P} = \sum_{\vec{k}} \vec{k} \left(N(\vec{k}) + \frac{1}{2} \right), \tag{3.79}$$

where $N(\vec{k}) = a^\dagger(\vec{k}) a(\vec{k})$ is the number operator as defined in the case of the harmonic oscillator. It can be shown that:

$$[H, a(\vec{k})] = -\omega_{\vec{k}} a(\vec{k}) \quad \text{and} \quad [H, a^\dagger(\vec{k})] = \omega_{\vec{k}} a^\dagger(\vec{k}). \tag{3.80}$$

The total number operator for all the oscillators in the system may be obtained using

$$N = \int d\vec{k} \, N(\vec{k}) = \int d\vec{k} \, a^\dagger(\vec{k}) a(\vec{k}). \tag{3.81}$$

However, a major difficulty is encountered in defining the energy of the ground state, which is a vacuum state in the case of field theory. The vacuum state $|0\rangle$ is defined as a state in which there is no particle, that is,

$$N(\vec{k})|0\rangle = 0 \quad \text{or} \quad a(\vec{k})|0\rangle = 0 \quad \text{for all } \vec{k}.$$

Equivalently, in terms of the field operators:

$$\phi^+(x)|0\rangle = 0 \quad \text{and} \quad \langle 0|\phi^-(x) = 0 \text{ for all } x.$$

The energy of the vacuum state is then given by:

$$H|0\rangle = \left(\frac{1}{2} \sum_{\vec{k}} \omega_{\vec{k}} \right) |0\rangle \qquad \text{for all } \vec{k}, \tag{3.82}$$

leading to an infinite constant. However, energy of the physical states is measured with reference to the energy of the vacuum; only the energy differences are measured in physical processes. Therefore, we can reset the vacuum energy given by Eq. (3.82); energies are measured with reference to this point. Thus, the occurrence of this constant albeit being infinite is absorbed in the definition of vacuum state energy. This is achieved by considering the "normal ordering" of the creation(annihilation) operators $a^\dagger(\vec{k})(a(\vec{k}))$ or the field operators $\phi(x)(\phi^\dagger(x))$, which assures that a normal ordered product of the field operators gives zero when operated on a vacuum state. It should be noted that the factor of $\frac{1}{2} \sum_{\vec{k}} \omega_{\vec{k}}$ appears when we rewrite $a(\vec{k})a^\dagger(\vec{k})$ in terms of $a^\dagger(\vec{k})a(\vec{k})$ in Eq. (3.77) to bring $a(\vec{k})$, the annihilation operator to the right of the creation operator. As long as all the annihilation operators are kept to the right and physical measurements are made treating vacuum as reference point, we can ignore the problem of infinite constant. This is called the "normal ordering" of the product of operators. It is needed only when both the creation and annihilation operators appear together in the product and not in the case between two annihilation or two creation operators like $a^\dagger(\vec{k})a^\dagger(\vec{k})$ or $a(\vec{k})a(\vec{k})$, as they commute. For example:

$$
\begin{aligned}
N[a^\dagger(\vec{k}_1)a(\vec{k}_2)a(\vec{k}_3)a^\dagger(\vec{k}_4)] &= N[a^\dagger(\vec{k}_1)a^\dagger(\vec{k}_4)a(\vec{k}_2)a(\vec{k}_3)] \\
\text{or } N[\phi(x)\phi(y)] &= N\left[(\phi^+(x)+\phi^-(x))(\phi^+(y)+\phi^-(y))\right] \\
&= N\left[\phi^+(x)\phi^+(y)+\phi^+(x)\phi^-(y)+\phi^-(x)\phi^+(y)\right. \\
&\quad \left. +\phi^-(x)\phi^-(y)\right] \\
&= \phi^+(x)\phi^+(y)+\phi^-(y)\phi^+(x)+\phi^-(x)\phi^+(y) \\
&\quad +\phi^-(x)\phi^-(y). \quad\quad (3.83)
\end{aligned}
$$

Thus, keeping the "normal ordering" in mind and the removal of the infinite constant appearing in the energy of the vacuum, the energy and momentum operators are given by:

$$
\begin{aligned}
H &= \sum_{\vec{k}} \omega_{\vec{k}}\, a^\dagger(\vec{k})a(\vec{k}), \\
\vec{P} &= \sum_{\vec{k}} \vec{k}\, a^\dagger(\vec{k})a(\vec{k}).
\end{aligned}
\quad\quad (3.84)
$$

The energy of a state $|\phi\rangle$ is given by:

$$
\begin{aligned}
\langle\phi|H|\phi\rangle &= \langle\phi|\sum_{\vec{k}}\omega_{\vec{k}}a^\dagger(\vec{k})a(\vec{k})|\phi\rangle \\
&= \sum_{\vec{k}}\omega_{\vec{k}}\langle\phi|a^\dagger(\vec{k})a(\vec{k})|\phi\rangle \\
&= \sum_{\vec{k}}\omega_{\vec{k}}|a(\vec{k})|\phi\rangle|^2. \quad\quad (3.85)
\end{aligned}
$$

which is always positive. It also shows that the vacuum expectation value of any normal order product of the fields is always zero since $a(\vec{k})|0\rangle = 0$ or $\phi^+(x)|0\rangle = 0$.

3.3.2 Fock space

Fock space in quantum field theory is the space in which the states are defined by the occupation number of particles in that state. It is characterized by the eigenvalues of the number operator N defined as:

$$
N = \sum_{\vec{k}} a^\dagger(\vec{k})a(\vec{k}), \quad\quad (3.86)
$$

which satisfies the commutation relations:

$$
[N, a^\dagger(\vec{k})] = a^\dagger(\vec{k}), \quad [N, a(\vec{k})] = -a(\vec{k}) \quad\quad (3.87)
$$

and shows that $a(\vec{k})$ and $a^\dagger(\vec{k})$, respectively, annihilates and creates one particle in a state, in the Fock space. Now operating Eq. (3.87) on a state $|n(\vec{k})\rangle$ such that $N|n(\vec{k})\rangle = n|n(\vec{k})\rangle$ with n particles, we find:

$$[N, a^\dagger(\vec{k})]|n(\vec{k})\rangle = a^\dagger(\vec{k})|n(\vec{k})\rangle \quad \text{and} \quad [N, a(\vec{k})]|n(\vec{k})\rangle = -a|n(\vec{k})\rangle$$

that is, $Na^\dagger(\vec{k})|n(\vec{k})\rangle = (n+1)a^\dagger(\vec{k})|n(\vec{k})\rangle \quad \text{and} \quad Na(\vec{k})|n(\vec{k})\rangle = (n-1)a(\vec{k})|n(\vec{k})\rangle.$

$$(3.88)$$

This shows that $a^\dagger(\vec{k})$ creates one particle while $a(\vec{k})$ annihilates one particle. For example, the vacuum is a zero particle state. The one particle state, two particle states, and multi- particle states are created by the repeated operation of the creation operator $a^\dagger(\vec{k})$. For example, a single particle state $|\vec{k}\rangle$ is created as:

$$|\vec{k}\rangle = a^\dagger(\vec{k})|0\rangle \qquad (3.89)$$

which satisfies $N|\vec{k}\rangle = |\vec{k}\rangle$, $H|\vec{k}\rangle = \omega_{\vec{k}}|\vec{k}\rangle$, $P|\vec{k}\rangle = \vec{k}|\vec{k}\rangle$, etc.

Similarly, a two particle state $|\vec{k}_1, \vec{k}_2\rangle$ is created by two operations of the creation operator, that is,

$$|\vec{k}_1, \vec{k}_2\rangle = a^\dagger(\vec{k}_1)a^\dagger(\vec{k}_2)|0\rangle. \qquad (3.90)$$

For bosons, these states are even under particle interchange, that is, $|\vec{k}_1, \vec{k}_2\rangle = |\vec{k}_2, \vec{k}_1\rangle$, and satisfy

$$|\vec{k}_1, \vec{k}_2\rangle = |\vec{k}_2, \vec{k}_1\rangle = a^\dagger(\vec{k}_1)a^\dagger(\vec{k}_2)|0\rangle. \qquad (3.91)$$

Moreover, they satisfy

$$\begin{aligned} H|\vec{k}_1, \vec{k}_2\rangle &= \left(\omega_{\vec{k}_1} + \omega_{\vec{k}_2}\right)|\vec{k}_1, \vec{k}_2\rangle, \\ \vec{P}|\vec{k}_1, \vec{k}_2\rangle &= \left(\vec{k}_1 + \vec{k}_2\right)|\vec{k}_1, \vec{k}_2\rangle. \end{aligned} \qquad (3.92)$$

A zero particle state defines the vacuum, that is, $|0\rangle$, and satisfies

$$\begin{aligned} \langle 0|0\rangle &= 0, \\ H|0\rangle &= 0, \\ \vec{P}|0\rangle &= 0. \end{aligned} \qquad (3.93)$$

A multi-particle state $|\vec{k}_1, \vec{k}_2, ..., \vec{k}_n\rangle$ is written as:

$$|\vec{k}_1, \vec{k}_2, ..., \vec{k}_n\rangle = a^\dagger(\vec{k}_1)a^\dagger(\vec{k}_2).....a^\dagger(\vec{k}_n)|0\rangle \qquad (3.94)$$

which has to be properly normalized. It can be shown that a properly normalized state with n particles of momentum is written as:

$$|k(n)\rangle = \frac{1}{\sqrt{n!}}(a^\dagger(\vec{k}))^n|0\rangle. \qquad (3.95)$$

3.4 Complex Scalar Field

3.4.1 Creation and annihilation operators

We now consider the case of complex scalar fields and apply all the results derived in the earlier section. The Lagrangian density for the complex scalar fields is written as:

$$\mathcal{L} = N(\phi^{\dagger}_{,\alpha}(x)\phi'^{\alpha}(x) - m^2\phi(x)^{\dagger}\phi(x)), \tag{3.96}$$

where N denotes the normal ordering of the fields $\phi(x)$ and $\phi^{\dagger}(x)$. These fields are treated as independent fields and are complex. The normal ordering is explicitly written here but is generally understood in the context of all the field operators when they appear in product form. The fields $\phi(x)(\phi^{\dagger}(x))$ can always be expressed as a sum (difference) of two real independent fields ($\phi_1(x)$ and $\phi_2(x)$) like:

$$\phi(x) = \frac{\phi_1(x) + i\phi_2(x)}{\sqrt{2}}, \tag{3.97}$$

$$\phi^{\dagger}(x) = \frac{\phi_1(x) - i\phi_2(x)}{\sqrt{2}}. \tag{3.98}$$

The Lagrangian density can also be written in terms of the two real fields $\phi_1(x)$ and $\phi_2(x)$:

$$\mathcal{L} = \frac{N}{2} \sum_{i=1,2} (\phi_{i,\mu}(x)\phi^{i,\mu}(x) - m^2\phi_i^2). \tag{3.99}$$

We will, however, work with the complex fields $\phi(x)$ and $\phi^{\dagger}(x)$ in the following. Using the Euler–Lagrange equation, the Lagrangian density in Eq. (3.96) leads to the Klein–Gordon equation:

$$(\Box + m^2)\phi(x) = 0 \quad \text{or} \quad (\Box + m^2)\phi^{\dagger}(x) = 0. \tag{3.100}$$

The conjugate fields, that is, $\frac{\partial \mathcal{L}}{\partial \phi}$ and $\frac{\partial \mathcal{L}}{\partial \phi^{\dagger}}$ to ϕ and $\phi^{\dagger}$ are given by:

$$\pi(x) = \frac{\partial \mathcal{L}}{\partial \dot{\phi}(x)} = \dot{\phi}^{\dagger}(x) \quad \text{and} \quad \pi^{\dagger}(x) = \frac{\partial \mathcal{L}}{\partial \dot{\phi}^{\dagger}(x)} = \dot{\phi}(x), \tag{3.101}$$

leading to the equal time commutation relations for the purpose of quantization being:

$$[\phi(\vec{x},t), \dot{\phi}^{\dagger}(\vec{y},t)] = i\delta^3(\vec{x} - \vec{y}), \tag{3.102}$$

$$\left[\phi^{\dagger}(\vec{x},t), \dot{\phi}(\vec{y},t)\right] = i\delta^3(\vec{x} - \vec{y}), \tag{3.103}$$

$$[\phi(\vec{x},t), \phi(\vec{y},t)] = [\dot{\phi}^{\dagger}(\vec{x},t), \dot{\phi}(\vec{y},t)] = [\phi^{\dagger}(\vec{x},t), \phi^{\dagger}(\vec{y},t)] =$$

$$[\dot{\phi}(\vec{x},t), \dot{\phi}(\vec{y},t)] = \left[\phi(\vec{x},t), \phi^{\dagger}(\vec{y},t)\right] = [\dot{\phi}(\vec{x},t), \dot{\phi}^{\dagger}(\vec{y},t)] = 0. \tag{3.104}$$

In analogy with the real field, we expand the fields $\phi(\vec{x}, t)$ and $\phi^\dagger(\vec{x}, t)$ in terms of the complete set of solutions, that is,

$$\phi(\vec{x}, t) = \phi^+ + \phi^- = \sum_{\vec{k}} \frac{1}{\sqrt{2V\omega_{\vec{k}}}} \left[a(\vec{k}) e^{-ik \cdot x} + b^\dagger(\vec{k}) e^{ik \cdot x} \right], \tag{3.105}$$

$$\phi^\dagger(\vec{x}, t) = (\phi^+)^\dagger + (\phi^-)^\dagger = \sum_{\vec{k}} \frac{1}{\sqrt{2V\omega_{\vec{k}}}} \left[b(\vec{k}) e^{-ik \cdot x} + a^\dagger(\vec{k}) e^{ik \cdot x} \right]. \tag{3.106}$$

It should be noted that in this case, $\phi(\vec{x}, t) \neq \phi^\dagger(\vec{x}, t)$. The fields, $\phi^+(\vec{x}, t)$, and $\phi^-(\vec{x}, t)$ are the positive and negative energy parts of $\phi(x)$. Using Eqs. (3.104), (3.105) and (3.106), we can derive the following commutation relation for $a(\vec{k}), a^\dagger(\vec{k}), b(\vec{k}), b^\dagger(\vec{k})$ i.e.

$$\left[a(\vec{k}), a^\dagger(\vec{k}') \right] = \left[b(\vec{k}), b^\dagger(\vec{k}') \right] = \delta_{\vec{k}\,\vec{k}'}, \tag{3.107}$$

$$\left[a(\vec{k}), a(\vec{k}') \right] = \left[b(\vec{k}), b(\vec{k}') \right] = \left[a^\dagger(\vec{k}), a^\dagger(\vec{k}') \right]$$

$$= \left[b^\dagger(\vec{k}), b^\dagger(\vec{k}') \right] = \left[a(\vec{k}), b^\dagger(\vec{k}') \right] = 0. \tag{3.108}$$

We have already explained the interpretation of $a(\vec{k})$ and $a^\dagger(\vec{k})$ as the annihilation and creation operators of field $\phi(x)$. In an analogous way, $b(\vec{k})$ and $b^\dagger(\vec{k})$ are interpreted as annihilation and creation of particles corresponding to field $\phi^\dagger(x)$. It is clear that if we continue with the Lagrangian density and the fields in terms of $\phi_1(x)$ and $\phi_2(x)$, then using Eq. (3.99) we would obtain:

$$\pi_1(x) = \frac{\partial \mathcal{L}}{\partial \dot{\phi}_1} = \dot{\phi}_1(x), \tag{3.109}$$

$$\pi_2(x) = \frac{\partial \mathcal{L}}{\partial \dot{\phi}_2} = \dot{\phi}_2(x). \tag{3.110}$$

From Eq. (3.97), we have:

$$\phi(\vec{x}, t) = \frac{1}{\sqrt{2}} (\phi_1(\vec{x}, t) + i\phi_2(\vec{x}, t))$$

$$= \frac{1}{\sqrt{2}} \sum_{\vec{k}} \frac{1}{\sqrt{2V\omega_{\vec{k}}}} \left[e^{-ik \cdot x} (a_1(\vec{k}) + ia_2(\vec{k})) + e^{ik \cdot x} (a_1^\dagger(\vec{k}) + ia_2^\dagger(\vec{k})) \right], \tag{3.111}$$

where $a_1^\dagger(\vec{k})$ and $a_2^\dagger(\vec{k})$ will create quanta of fields $\phi_1(x)$ and $\phi_2(x)$, while $a_1(x)$ and $a_2(x)$ will annihilate quanta of $\phi_1(x)$ and $\phi_2(x)$. Similarly,

$$\phi^\dagger(\vec{x}, t) = \frac{1}{\sqrt{2}} \sum_{\vec{k}} \frac{1}{2V\omega_{\vec{k}}} \left[e^{-ik \cdot x} (a_1(\vec{k}) - ia_2(\vec{k})) + e^{ik \cdot x} (a_1^\dagger(\vec{k}) - ia_2^\dagger(\vec{k})) \right]. \tag{3.112}$$

Equations (3.105) and (3.106) can be reproduced from Eqs. (3.111) and (3.112), using the following relations:

$$a(\vec{k}) = \frac{1}{\sqrt{2}} (a_1(\vec{k}) + ia_2(\vec{k})), \tag{3.113}$$

$$b(\vec{k}) = \frac{1}{\sqrt{2}}(a_1(\vec{k}) - ia_2(\vec{k})), \tag{3.114}$$

$$a^\dagger(\vec{k}) = \frac{1}{\sqrt{2}}(a_1^\dagger(\vec{k}) - ia_2^\dagger(\vec{k})), \tag{3.115}$$

$$b^\dagger(\vec{k}) = \frac{1}{\sqrt{2}}(a_1^\dagger(\vec{k}) + ia_2^\dagger(\vec{k})). \tag{3.116}$$

However, there is a physical reason to consider the formulation of quantized fields of complex fields in term of $\phi(x)$ and $\phi^\dagger(x)$ instead of $\phi_1(x)$ and $\phi_2(x)$.

Using Eqs. (3.113)–(3.116) and Eqs. (3.58) and (3.62), the energy–momentum operator for the complex Klein–Gordon field can be derived to be:

$$P^\mu = (H, \vec{P}) = \frac{1}{2}\sum_{\vec{k}} k^\mu \left(a^\dagger(\vec{k})a(\vec{k}) + a(\vec{k})a^\dagger(\vec{k}) + b^\dagger(\vec{k})b(\vec{k}) + b(\vec{k})b^\dagger(\vec{k})\right). \tag{3.117}$$

Using the commutation relations in Eq. (3.107) and the concept of normal ordering discussed in Section 3.3.1, the operator P^μ is written as:

$$P^\mu = \sum_{\vec{k}} k^\mu \left(a^\dagger(\vec{k})a(\vec{k}) + b^\dagger(\vec{k})b(\vec{k})\right). \tag{3.118}$$

We can, therefore, define number operators for both types of particles created by the creation operators $a^\dagger(\vec{k})$ and $b^\dagger(\vec{k})$:

$$N_a = \sum_{\vec{k}} a^\dagger(\vec{k})a(\vec{k}), \tag{3.119}$$

$$N_b = \sum_{\vec{k}} b^\dagger(\vec{k})b(\vec{k}). \tag{3.120}$$

The vacuum state $|0\rangle$ is defined as the state of minimum energy (redefined to zero energy) through the operation of $a(\vec{k})$ and $b(\vec{k})$ as:

$$a(\vec{k})|0\rangle = 0 \qquad \text{and} \qquad b(\vec{k})|0\rangle = 0 \tag{3.121}$$

or equivalently $\phi^+(\vec{x})|0\rangle = 0$ and $(\phi^-)^\dagger(\vec{x})|0\rangle = 0$.

The one particle states in the Fock space are created by the action of $a^\dagger(\vec{k})|0\rangle$ and $b^\dagger(\vec{k})|0\rangle$ as:

$$|\vec{k}\rangle = a^\dagger(\vec{k})|0\rangle \quad \text{and} \quad |\vec{k}\rangle = b^\dagger(\vec{k})|0\rangle. \tag{3.122}$$

We already know the type of particles created by $a^\dagger(\vec{k})$. To understand the type of particles created by $b^\dagger(\vec{k})$, let us go to the next section.

3.4.2 Charge of the complex scalar field: Particles and antiparticles

The Lagrangian given in Eq. (3.96) is invariant under the global phase transformation

$$\phi(x) \to e^{-iq\theta}\phi(x), \qquad \phi^\dagger(x) \to \phi^\dagger(x)e^{iq\theta}, \tag{3.123}$$

where θ is independent of space–time and q is an arbitrary parameter. For infinitesimal θ, we have:

$$\delta\phi = -iq\theta\phi, \qquad \delta\phi^\dagger = iq\theta\phi^\dagger. \tag{3.124}$$

According to Noether's theorem, the conserved current J^μ associated with this global transformation is given by:

$$J^\mu = iq((\partial^\mu\phi)\phi^\dagger - (\partial^\mu\phi^\dagger)\phi), \tag{3.125}$$

such that $\partial_\mu J^\mu = 0$, corresponding to the conservation of the quantity $Q = \int d\vec{x}\, J^0(x)$, that is,

$$
\begin{aligned}
Q &= q\sum_{\vec{k}} \left[a^\dagger(\vec{k})a(\vec{k}) - b^\dagger(\vec{k})b(\vec{k}) \right] \\
&= q\sum_{\vec{k}} \left(N_a(\vec{k}) - N_b(\vec{k}) \right).
\end{aligned} \tag{3.126}
$$

Q is also known as the normal ordered charge operator. Q when operating on a vacuum state gives 0, that is,

$$Q|0\rangle = q\sum_{\vec{k}} \left[a^\dagger(\vec{k})a(\vec{k}) - b^\dagger(\vec{k})b(\vec{k}) \right]|0\rangle = 0. \tag{3.127}$$

This can be seen as follows:
Q operating on the first of the two one particle states $|\vec{k}\rangle$ gives:

$$
\begin{aligned}
qN_a|\vec{k}\rangle &= q\sum_{\vec{k}'} \left[a^\dagger(\vec{k}')a(\vec{k}') \right] a^\dagger(\vec{k})|0\rangle \\
&= q\sum_{\vec{k}'} \left[a^\dagger(\vec{k}')a(\vec{k}')a^\dagger(\vec{k}) \right]|0\rangle \\
&= q\sum_{\vec{k}'} a^\dagger(\vec{k}')\delta_{\vec{k}\vec{k}'}|0\rangle \\
&= q\, a^\dagger(\vec{k})|0\rangle \\
&= q\,|\vec{k}\rangle. \tag{3.128}
\end{aligned}
$$

Similarly, $\qquad qN_b|\vec{k}\rangle = +q\,|\vec{k}\rangle,$ $\qquad\qquad\qquad$ (3.129) such that

$$q(N_a - N_b)|0\rangle = 0|0\rangle. \tag{3.130}$$

Using Eq. (3.118), the Hamiltonian H is defined as:

$$H = \sum_{\vec{k}} \omega_{\vec{k}} \left[a^\dagger(\vec{k})a(\vec{k}) + b^\dagger(\vec{k})b(\vec{k}) \right], \tag{3.131}$$

and it can be shown that:

$$[H, Q] = 0 \tag{3.132}$$

so that the energy states could be simultaneously characterized by the eigenvalues of the charge operator Q. If the charge q is identified to be the charge of the particle created by $a^\dagger(\vec{k})$, then the particle created by $b^\dagger(\vec{k})$ has a charge $-q$. Since the energy E(or equivalent mass as the energy at rest) and charge are the only two quantum numbers associated with a charged scalar particle, they represent the particles and antiparticles of spin zero. The physical example of such particles are the charged pions π^+ and π^-. On the other hand, for a real particle field, the charge operator Q defined in Eq. (3.128) or Eq. (3.129) will be zero because $\phi(x) = \phi^\dagger(x)$ corresponding to $a = b$ and $a^\dagger = b^\dagger$.

If we had persisted with the description of the complex fields in terms of two real fields like $\phi_1(x)$ and $\phi_2(x)$, then both fields would separately correspond to neutral particles of zero charge and would not be adequate to describe the charged scalar particles. This is the reason why we describe the field quantization in this case in terms of $\phi(x)$ and $\phi^+(x)$ fields.

3.4.3 Covariant commutation relation

In this section, we will describe the concept of covariant commutation relations and its relation to the field propagators in the case of real scalar fields which would be generalized to the complex scalar fields. The commutation relations for the two fields $\phi(x_1)$ and $\phi(x_2)$ were defined for equal times in Eqs. (3.102)–(3.104) in context of the canonical quantization procedure of the fields. The covariant commutation relation (CCR) is defined for the two fields $\phi(x_1)$ and $\phi(x_2)$ at the two arbitrary space–time points x_1 and x_2 in terms of $\Delta(x_1 - x_2)$ as:

$$\Delta(x_1 - x_2) = -i[\phi(x_1), \phi(x_2)], \tag{3.133}$$

where

$$\phi(x_1) = \phi^+(x_1) + \phi^-(x_1),$$

$$\phi(x_2) = \phi^+(x_2) + \phi^-(x_2),$$

and

$$\begin{aligned}
[\phi(x_1), \phi(x_2)] &= \left[\phi^+(x_1) + \phi^-(x_1), \phi^-(x_2) + \phi^+(x_2)\right] \\
&= [\phi^+(x_1), \phi^-(x_2)] + [\phi^-(x_1), \phi^+(x_2)]. \tag{3.134}
\end{aligned}$$

Since $[\phi^+(x_1), \phi^+(x_2)] = [\phi^-(x_1), \phi^-(x_2)] = 0$, they contain only annihilation or only creation operators which commute. Consider the first term, that is,

$$[\phi^+(x_1), \phi^-(x_2)] = \frac{1}{2V} \sum_{\vec{k}} \sum_{\vec{k}'} \frac{1}{\sqrt{\omega_{\vec{k}} \omega_{\vec{k}'}}} [a(\vec{k}), a^\dagger(\vec{k}')](e^{-ikx_1 + ik'x_2}),$$

$$= \frac{1}{2V} \sum_{\vec{k}} \frac{1}{\omega_{\vec{k}}} e^{-ik(x_1 - x_2)}. \tag{3.135}$$

Since $\vec{k} = \vec{k}'$, $\omega_{\vec{k}} = \sqrt{|\vec{k}|^2 + m^2} = \sqrt{|\vec{k}'|^2 + m^2} = \omega_{\vec{k}'}$. Taking the limit $V \to \infty$ and replacing the summation over $\vec{k}$ with an integral over $\vec{k}'$, we write:

$$[\phi^+(x_1), \phi^-(x_2)] = \frac{1}{2(2\pi)^3} \int \frac{d\vec{k}'}{\omega_{\vec{k}}} e^{-ik(x_1 - x_2)}. \tag{3.136}$$

We introduce the function $\Delta^+(x)$ as:

$$\Delta^+(x) = \frac{-i}{2(2\pi)^3} \int \frac{d\vec{k}}{\omega_{\vec{k}}} e^{-ik \cdot x}$$

such that $\quad [\phi^+(x_1), \phi^-(x_2)] = i\Delta^+(x_1 - x_2). \tag{3.137}$

Similarly, we define the second term in the Eq. (3.134) in terms of $\Delta^-(x)$:

$$\Delta^-(x) = \frac{+i}{2(2\pi)^3} \int \frac{d\vec{k}}{\omega_{\vec{k}}} e^{+ik \cdot x} \tag{3.138}$$

such that $\quad [\phi^-(x_1), \phi^+(x_2)] = i\Delta^-(x_1 - x_2). \tag{3.139}$

It is easy to see that:

$$[\phi^-(x_1), \phi^+(x_2)] = i\Delta^-(x_1 - x_2) = -i\Delta^+(x_2 - x_1). \tag{3.140}$$

Therefore,

$$\Delta(x_1 - x_2) = \Delta^+(x_1 - x_2) + \Delta^-(x_1 - x_2)$$

$$= -\frac{1}{(2\pi)^3} \int \frac{d\vec{k}}{\omega_{\vec{k}}} \sin(k(x_1 - x_2)). \tag{3.141}$$

Thus, $\Delta(x_1 - x_2)$ satisfies the following properties:

(i) it is a real function of $(x_1 - x_2)$,

(ii) it is an odd function of $(x_1 - x_2)$,

(iii) $\Delta(x_1 - x_2)|_{(x_1)_0 = (x_2)_0} = 0$,

(iv) the equation $(\Box + m^2)\Delta(x_1 - x_2) = 0$ is satisfied.

This $\Delta(x)$ defined in Eq. (3.141) can be written in covariant form as:

$$\Delta(x) = -\frac{i}{(2\pi)^3} \int d^4k \, \delta(k^2 - m^2) \epsilon(k_0) e^{-ik \cdot x},$$

where $\epsilon(k_0) = +1$ for $k_0 > 0$

$$= -1 \quad \text{for } k_0 < 0 \tag{3.142}$$

and $d^4k = dk_0 \, d\vec{k}$. This can be verified by using the property of the δ function:

$$\delta(k^2 - m^2) = \delta(k_0^2 - (|\vec{k}|^2 + m^2)) = \delta(k_0^2 - \omega_{\vec{k}}^2) = \delta\left[(k_0 - \omega_{\vec{k}})(k_0 + \omega_{\vec{k}})\right]$$

$$= \frac{1}{2\omega_{\vec{k}}}[\delta(k_0 - \omega_{\vec{k}}) + \delta(k_0 + \omega_{\vec{k}})]. \tag{3.143}$$

(v) The equal time condition $(x_1)_0 - (x_2)_0 = 0$ implies that $(x_1 - x_2)^2 = ((x_1)_0 - (x_2)_0)^2 - (\vec{x}_1 - \vec{x}_2)^2 < 0$, that is, it corresponds to a space- like separation for which $[\phi(x_1), \phi(x_2)] = 0$. The invariance of $\Delta(x_1 - x_2)$, therefore, implies that it is zero for all space-like separation. This means that the physical observables depending on the fields at the two points separated by space- like separation do not interfere with each other. This is consistent with the special theory of relativity in which two events separated by space-like intervals cannot be connected by the Lorentz transformation. This is called the principle of microcausality.

(vi) A simple representation of the functions $\Delta^{\pm}(x_1 - x_2)$ is written in terms of the contour integral in the complex k_0 plane as:

$$\Delta^{\pm}(x_1 - x_2) = -\frac{1}{(2\pi)^4} \int_{C^{\pm}} \frac{d^4k \, e^{-ik \cdot (x_1 - x_2)}}{k^2 - m^2}, \tag{3.144}$$

where the contours $C^{\pm}$ are shown in Figure 3.2. The poles of the integrand in Eq. (3.144) are at $k_0 = \pm \omega_{\vec{k}}$ with $\omega_{\vec{k}} = \sqrt{\vec{k}^2 + m^2}$ and the integration over $C^+(C^-)$ gives $\Delta^+(\Delta^-)$.

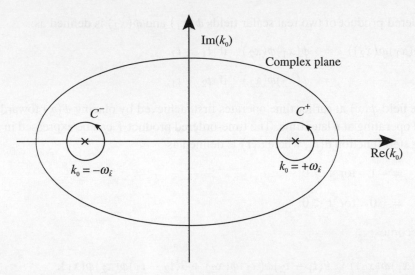

Figure 3.2 Contours for Eq. (3.144). C^+ is for Δ_F^+ and C^- is for Δ_F^-. The poles of the integral given in Eq. (3.144) are at $k_0 = \omega_{\vec{k}}$ for C^+ and $k_0 = -\omega_{\vec{k}}$ for C^-.

3.5 Time-ordered Product and Propagators for Scalar Fields

The covariant commutators $\Delta^{\pm}(x_1 - x_2)$ can be considered as the vacuum expectation values of the product of fields and can, therefore, be treated as the matrix element for creating and annihilating particles from vacuum as will be discussed in this section. Consider the product of two real scalar fields $\phi(x_1)\phi(x_2)$, that is,

$$
\begin{aligned}
\phi(x_1)\phi(x_2) &= (\phi^+(x_1) + \phi^-(x_1))(\phi^+(x_2) + \phi^-(x_2)) \qquad (3.145)\\
&= \phi^+(x_1)\phi^+(x_2) + \phi^+(x_1)\phi^-(x_2) + \phi^-(x_1)\phi^+(x_2) + \phi^-(x_1)\phi^-(x_2)\\
&= \phi^+(x_1)\phi^+(x_2) + \phi^+(x_1)\phi^-(x_2) + \phi^-(x_1)\phi^+(x_2)\\
&\quad + \phi^-(x_1)\phi^-(x_2) + \phi^-(x_2)\phi^+(x_1) - \phi^-(x_2)\phi^+(x_1)\\
&= \phi^+(x_1)\phi^+(x_2) + \phi^-(x_2)\phi^+(x_1) + \phi^-(x_1)\phi^-(x_2)\\
&\quad + \phi^-(x_1)\phi^+(x_2) + [\phi^+(x_1), \phi^-(x_2)] \qquad (3.146)\\
&= N(\phi(x_1)\phi(x_2)) + [\phi^+(x_1), \phi^-(x_2)]. \qquad (3.147)
\end{aligned}
$$

Taking the vacuum expectation value on both sides of Eq. (3.147)

$$
\langle 0|\phi(x_1)\phi(x_2)|0\rangle = \langle 0|N(\phi(x_1)\phi(x_2))|0\rangle + \langle 0|[\phi^+(x_1), \phi^-(x_2)]|0\rangle. \qquad (3.148)
$$

Since the vacuum expectation value of a normal-ordered product vanishes, we get

$$
\begin{aligned}
\langle 0|\phi(x_1)\phi(x_2)|0\rangle &= \langle 0|[\phi^+(x_1), \phi^-(x_2)]|0\rangle.\\
&= i\Delta^+(x_1 - x_2). \qquad (3.149)
\end{aligned}
$$

A time-ordered product of two real scalar fields $\phi(x_1)$ and $\phi(x_2)$ is defined as:

$$
\begin{aligned}
T\left(\phi(x_1)\phi(x_2)\right) &= \phi(x_1)\phi(x_2) \quad \text{if } t_1 > t_2 \\
&= \phi(x_2)\phi(x_1) \quad \text{if } t_2 > t_1,
\end{aligned}
\tag{3.150}
$$

That is, the field $\phi(x)$ at earlier time operates first, achieved by placing $\phi(x)$ towards the right of the field operating at a later time. The time-ordered product T can be expressed in a compact form using step function $\theta(t)$, where $\theta(t)$ is defined as:

$$
\begin{aligned}
\theta(t) &= 1 \quad \text{for } t > 0 \\
&= 0 \quad \text{for } t < 0.
\end{aligned}
\tag{3.151}
$$

Then, T becomes

$$
T(\phi(x_1)\phi(x_2)) = \theta(t_1 - t_2)\phi(x_1)\phi(x_2) + \theta(t_2 - t_1)\phi(x_2)\phi(x_1).
\tag{3.152}
$$

Another function called the Feynman Δ-function, $\Delta_F(x_1 - x_2)$, is defined as:

$$
i\Delta_F(x_1 - x_2) = \langle 0|T\{\phi(x_1)\phi(x_2)\}|0\rangle.
\tag{3.153}
$$

That is,

$$
\Delta_F(x) = \theta(t)\Delta^+(x) - \theta(-t)\Delta^-(x),
\tag{3.154}
$$

which includes the vacuum expectation value of the products of the fields for both cases of time-ordering. Explicitly,

$$
\Delta_F(x) = \pm\Delta^\pm(x) \quad \text{for } t \gtrless 0.
\tag{3.155}
$$

These Δ-functions are physically interpreted to describe the propagation of particles between x_1 and x_2 depending upon $t_1 > t_2$ or $t_2 > t_1$. Let us consider $\Delta_F(x_1 - x_2)$ for $t_1 > t_2$. In this case:

$$
\begin{aligned}
\Delta_F(x_1 - x_2) &= -i\langle 0|\phi(x_1)\phi(x_2)|0\rangle \quad t_1 > t_2 \\
&= -i\langle 0|\phi(x_2)\phi(x_1)|0\rangle \quad t_1 < t_2,
\end{aligned}
\tag{3.156}
$$

in which the particle is created at x_2 and gets annihilated at x_1. In case $t_1 < t_2$, the particle is created at x_1 and gets annihilated at x_2. The physical processes are shown in Figure 3.3, where for a fixed t_1, the particle's creation and annihilation are depicted depending upon either $t_1 > t_2$ or $t_2 > t_1$.

In the case of Feynman Δ-function, $\Delta_F(x_1 - x_2)$, both time-ordered diagrams are included. In actual calculations, all values of x_1 and x_2 are integrated. Therefore, the Feynman Δ-function $\Delta_F(x_1 - x_2)$ is the appropriate description of the propagation of particles between its creation followed by its annihilation. This is therefore called the particle propagator. An integral

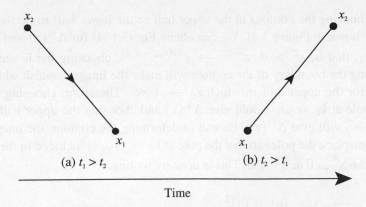

(a) $t_1 > t_2$ (b) $t_2 > t_1$

Time

Figure 3.3 Propagation of bosons between x_1 and x_2 for (a) $t_1 > t_2$ and (b) $t_2 > t_1$.

representation of $\Delta_F(x_1 - x_2)$ which combines both $\Delta^+(x)$ and $\Delta^-(x)$ functions in Eq. (3.154) is given by:

$$\Delta_F(x) = \frac{1}{(2\pi)^4} \int_{C_F} \frac{d^4 k \, e^{-ik \cdot x}}{k^2 - m^2}, \tag{3.157}$$

where C_F is the contour specified to reproduce the results given in Eq. (3.154). This is achieved by performing the k_0 integration in the complex k_0 plane in Eq. (3.157) using Cauchy's integral formula. The denominator of the integrand is written as:

$$k^2 - m^2 = (k_0 - \omega_{\vec{k}})(k_0 - \omega_{\vec{k}}), \tag{3.158}$$

where $\omega_{\vec{k}} = \sqrt{|\vec{k}|^2 + m^2}$, so the integral gets contribution from the poles at $k_0 = \pm \omega_{\vec{k}}$ provided the integral vanishes at the boundaries of the contour C_F which is assured by

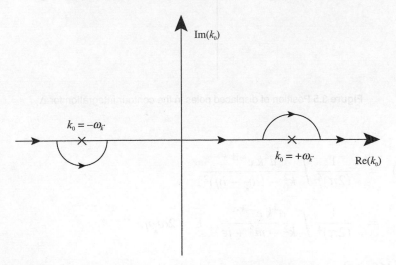

Figure 3.4 Contour for the boson propagator Δ_F.

appropriately choosing the contour in the upper half or the lower half to perform the contour integration (as shown in Figure 3.4). We can obtain Eq. (3.144) for $\Delta^+(x)$ and $\Delta^-(x)$. In the case of $t_1 > t_2$, that is, $x^0 > 0$, $e^{-ik\cdot x} \to e^{-i(k_0 x^0 - \vec{k}\cdot\vec{x})}$, choosing the lower half in which $k_0 \to -i\infty$ along the boundary of the contour will make the integral vanish while for $x^0 < 0$, it will happen for the upper half in which $k_0 \to +i\infty$. Therefore, choosing the lower half including the pole at $k_0 = \omega_{\vec{k}}$ would give $\Delta^+(x)$ and choosing the upper half including the pole at $k_0 = -\omega_{\vec{k}}$ will give $\Delta^-(x)$. Instead of deforming the contour, the integrand can also be modified to displace the poles so that the pole at $k_0 = \pm\omega_{\vec{k}}$ is included in the lower (upper) half as we choose $x^0 > 0$ or $x^0 < 0$. This is done by writing:

$$k_0^2 - \omega_{\vec{k}}^2 \quad \longrightarrow \quad k_0^2 - (\omega_{\vec{k}} - i\eta)^2 \tag{3.159}$$

such that the poles are now at $k_0 = \pm(\omega_{\vec{k}} - i\eta)$ as shown in Figure 3.5. The integral representation of $\Delta_F(x)$ now becomes:

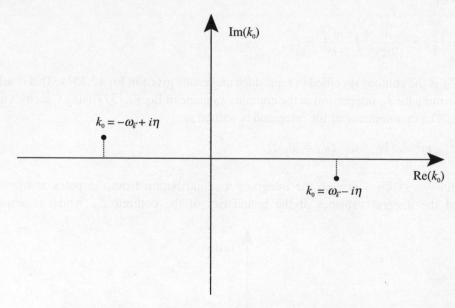

Figure 3.5 Position of displaced poles in the contour integration for Δ_F.

$$
\begin{aligned}
\Delta_F(x) &= \frac{1}{(2\pi)^4} \int \frac{d^4k\, e^{-ik\cdot x}}{k_0^2 - (\omega_{\vec{k}} - i\eta)^2} \\
&= \frac{1}{(2\pi)^4} \int \frac{d^4k\, e^{-ik\cdot x}}{k^2 - m^2 + i\epsilon}, \quad E = 2\omega_{\vec{k}}\eta.
\end{aligned} \tag{3.160}
$$

Equation (3.160) is the integral representation of the Feynman propagator for scalar particles.

3.6 Quantization of Spin $\frac{1}{2}$ Fields

The Dirac equation for spin $\frac{1}{2}$ particles was discussed in Chapter 2. The equation for the Dirac field can be derived from the following Lagrangian:

$$\mathcal{L} = \overline{\psi}(x)\left(i\gamma^{\mu}\frac{\partial}{\partial x^{\mu}} - m\right)\psi(x), \tag{3.161}$$

to give:

$$\left(i\gamma^{\mu}\frac{\partial}{\partial x^{\mu}} - m\right)\psi(x) = 0, \tag{3.162}$$

and

$$\overline{\psi}(x)\left(i\gamma^{\mu}\frac{\overleftarrow{\partial}}{\partial x^{\mu}} + m\right) = 0. \tag{3.163}$$

The conjugate fields are given by:

$$\pi(x) = \frac{\partial \mathcal{L}}{\partial \dot{\psi}} = i\psi^{\dagger}, \tag{3.164}$$

$$\overline{\pi}(x) = \frac{\partial \mathcal{L}}{\partial \dot{\overline{\psi}}} = 0. \tag{3.165}$$

We have already seen from the solution of the Dirac equation that $\psi(x)$ is a 4-component column vector, so we denote it by $\psi_r(x)$ which can be represented by two 2-component column vectors $\psi_1(x)$ and $\psi_2(x)$ which are expressed in terms of 2-component spinors $u_r(p)$ and $v_r(p)(r = 1,2)$. In view of this, we rewrite the expression of the Dirac field in terms of the complete set of solutions as:

$$\psi(x) = \sum_{\vec{k}}\sum_{r=1,2}\frac{1}{\sqrt{2\omega_{\vec{k}}}}\left[a_r(\vec{k})u_r(\vec{k})e^{-ik\cdot x} + b_r^{\dagger}(\vec{k})v_r(\vec{k})e^{ik\cdot x}\right] \tag{3.166}$$

and

$$\psi^{\dagger}(x) = \sum_{\vec{k}}\sum_{r=1,2}\frac{1}{\sqrt{2\omega_{\vec{k}}}}\left[a_r^{\dagger}(\vec{k})u_r^{\dagger}(\vec{k})e^{-ik\cdot x} + b_r(\vec{k})v_r^{\dagger}(\vec{k})e^{ik\cdot x}\right], \tag{3.167}$$

where $\omega_{\vec{k}} = \sqrt{|\vec{k}|^2 + m^2}$.

The Hamiltonian density can be written as:

$$\begin{aligned}
\mathcal{H}(x) &= \pi(x)\dot{\psi}(x) - \mathcal{L} = i\psi^{\dagger}(x)\dot{\psi}(x) - \mathcal{L} \\
&= i\psi^{\dagger}(x)\dot{\psi}(x) - i\psi^{\dagger}(x)\dot{\psi}(x) - i\psi^{\dagger}\gamma_0\gamma_i\frac{\partial}{\partial x^i}\psi + m\overline{\psi}\psi, \\
&= \psi^{\dagger}(x)\left(-i\gamma^0\vec{\gamma}\cdot\vec{\nabla} + \gamma^0 m\right)\psi(x) \\
&= \overline{\psi}(x)\left(-i\vec{\gamma}\cdot\vec{\nabla} + m\right)\psi(x), \tag{3.168}
\end{aligned}$$

leading to the Hamiltonian:

$$
\begin{aligned}
H &= \int d\vec{x}\, \mathcal{H} \\
  &= \int d\vec{x}\, \overline{\psi}(-i\vec{\gamma}\cdot\vec{\nabla}+m)\psi(x)
\end{aligned}
\tag{3.169}
$$

and the momentum operator

$$
\vec{P} = -i\int d\vec{x}\,\overline{\psi}(x)\vec{\nabla}\psi(x),
\tag{3.170}
$$

Using $J^{\mu}=\overline{\psi}\gamma^{\mu}\psi$, we obtain:

$$
Q = q\int d\vec{x}\, J^{0} = q\int d\vec{x}\, \psi^{\dagger}(x)\psi(x).
\tag{3.171}
$$

Using Eqs. (3.166) and (3.167), respectively, for $\psi(x)$ and $\psi^{\dagger}(x)$, we find:

$$
H = \sum_{\vec{k},r}\omega_{\vec{k}}[a_r^{\dagger}(\vec{k})a_r(\vec{k})-b_r(\vec{k})b_r^{\dagger}(\vec{k})],
\tag{3.172}
$$

$$
\vec{P} = \sum_{\vec{k},r}\vec{k}[a_r^{\dagger}(\vec{k})a_r(\vec{k})-b_r(\vec{k})b_r^{\dagger}(\vec{k})].
\tag{3.173}
$$

If we perform the quantization procedure using the creation and annihilation operators according to the commutation relation discussed in the case of complex scalar fields, given in Eqs. (3.107) and (3.108), we encounter the following difficulties.

i) Using the commutation relation for $b(\vec{k})$ operators, we obtain

$$
b_r(\vec{k})b^{\dagger}(\vec{k}) = N_b(\vec{k})+1,
\tag{3.174}
$$

$$
\text{where}\quad N_b(\vec{k}) = \sum_r b_r^{\dagger}(\vec{k})b_r(\vec{k})
\tag{3.175}
$$

leading to

$$
H = \sum_{\vec{k},r}\omega_{\vec{k}}[N_a(\vec{k})-N_b(\vec{k})-1],
\tag{3.176}
$$

where $N_a(\vec{k})=a_r^{\dagger}(\vec{k})a_r(\vec{k})$.

Therefore, $N_b(\vec{k})$ is the occupation number corresponding to the particle created by $b^{\dagger}(\vec{k})$. It can take any integer value making the contribution of the second term in H negative. This leads to energy eigenvalues which are unbounded from below. It will not allow the determination of a state of minimum energy, that is, the vacuum cannot be defined in this case.

ii) The number operator $N_a(\vec{k})$ and $N_b(\vec{k})$ can take eigenvalues $0, 1, 2...$, that is, any integral number of particles occupying the states in Fock space. However, in the case of fermions, a state can have only one particle according to Pauli's exclusion principle. Therefore, these commutation rules are not appropriate for describing the quantization of fermions.

Instead, we postulate the following commutation (called anticommutation) rules for the creation and annihilation operators. We can then describe the quantization of "fermions" and apply for quantization of $a_r(\vec{k})$ and $b_r(\vec{k})$ operators as:

$$\{a_r(\vec{k}), a_s^\dagger(\vec{k}')\} = \{b_r(\vec{k}), b_s^\dagger(\vec{k}')\} = \delta_{rs}\delta_{\vec{k}\vec{k}'}, \tag{3.177}$$

$$\{a_r(\vec{k}), a_s(\vec{k})\} = \{a_r^\dagger(\vec{k}), a_s^\dagger(\vec{k})\} = \{b_r(\vec{k}), b_s(\vec{k})\} = \{b_r^\dagger(\vec{k}), b_s^\dagger(\vec{k})\} = 0. \tag{3.178}$$

In this case, we see that:

$$a_r(\vec{k})a_r(\vec{k}) = b_r(\vec{k})b_r(\vec{k}) = 0 \tag{3.179}$$

$$\text{and} \quad N_a^2 = \sum_{r,k}(a_r^\dagger(\vec{k})a_r(\vec{k})a_r^\dagger(\vec{k})a_r(\vec{k}))$$

$$= \sum_r(a_r^\dagger(\vec{k})(1 - a_r^\dagger(\vec{k})a_r(\vec{k}))a_r(\vec{k}))$$

$$= N_a - \sum_r a_r^\dagger(\vec{k})a_r^\dagger(\vec{k})a_r(\vec{k})a_r(\vec{k})$$

$$= N_a. \tag{3.180}$$

That is, $N_a(N_a - 1) = 0$ implying $N_a = 0, 1$ consistent with the exclusion principle. Similarly for N_b, we obtain $N_b(N_b - 1) = 0$. Therefore, using the anticommutation relation in Eq. (3.177), we may write:

$$b_r^\dagger(\vec{k})b_r(\vec{k}) - 1 = -b_r(\vec{k})b_r^\dagger(\vec{k}) \tag{3.181}$$

and the expressions for H and P given in Eqs. (3.172) and (3.173) as:

$$H = \sum_{k,r}\omega_{\vec{k}}(a_r^\dagger(\vec{k})a_r(\vec{k}) + b_r^\dagger(\vec{k})b_r(\vec{k}) - 1), \tag{3.182}$$

$$P = \sum_{k,r}\vec{k}\,(a_r^\dagger(\vec{k})a_r(\vec{k}) + b_r^\dagger(\vec{k})b_r(\vec{k}) - 1). \tag{3.183}$$

Therefore, using the normal-ordering for the operators and redefining the vacuum state as is done in the case of scalar fields, we obtain:

$$H = \sum_{k,r}\omega_{\vec{k}}[N_r(\vec{k}) + \bar{N}_r(\vec{k})], \tag{3.184}$$

$$\vec{P} = \sum\vec{k}[N_r(\vec{k}) + \bar{N}_r(\vec{k})]. \tag{3.185}$$

Similarly, the charge operator Q is given by:

$$Q = q \sum_{k,r} [N_r(\vec{k}) - \bar{N}_r(\vec{k})], \tag{3.186}$$

for an electron $q = -e$. Equation (3.166) can be inverted to obtain $a_r(\vec{k})$ and $b_r^\dagger(\vec{k})$ in terms of ψ as:

$$a_r(\vec{k}) = \int d\vec{x} \frac{1}{\sqrt{2\omega_{\vec{k}} V}} e^{ik \cdot x} u_r^\dagger(\vec{k}) \psi(\vec{x}, t), \tag{3.187}$$

$$b_r^\dagger(\vec{k}) = \int d\vec{x} \frac{1}{\sqrt{2\omega_{\vec{k}} V}} e^{ik \cdot x} v_r^\dagger(\vec{k}) \psi(\vec{x}, t).. \tag{3.188}$$

These equations can be used to derive the commutation relations for the fields $\psi(\vec{x}, t)$ and $\psi^\dagger(\vec{x}, t)$. However, we can apply the rules of canonical quantization to postulate the anticommutation rules (in place of commutation rules) between the fields $\psi(\vec{x}, t)$ and $\psi^\dagger(\vec{x}, t)$ and their canonically conjugate fields $\pi_\psi(\vec{x}, t)$ and $\pi_{\psi^\dagger}(\vec{x}, t)$. Since $\pi_\psi(\vec{x}, t) = i\psi^\dagger(\vec{x}, t)$ and $\pi_{\psi^\dagger}(\vec{x}, t) = 0$, we obtain:

$$\{\psi_\alpha(\vec{x}, t), \psi_\beta(\vec{y}, t)\} = 0 \tag{3.189}$$

$$\{\psi_\alpha(\vec{x}, t), \psi_\beta^\dagger(\vec{y}, t)\} = \delta(\vec{x} - \vec{y}) \delta_{\alpha\beta} \tag{3.190}$$

$$\{\psi_\alpha^\dagger(\vec{x}, t), \psi_\beta^\dagger(\vec{y}, t)\} = 0. \tag{3.191}$$

These anticommutation rules can be obtained by using Eqs. (3.177) and (3.178). Conventionally, we can use Eqs. (3.166) and (3.167) for the anticommutation relations for the fields $\psi(\vec{x}, t)$ and $\psi^\dagger(\vec{x}, t)$ to obtain the anticommutation relations for the creation and annihilation operators as done in the case of scalar fields in Section 3.3.

With the help of the expansion of $\psi(\vec{x}, t)$ and $\psi^\dagger(\vec{x}, t)$ fields in terms of the creation and annihilation operators and their anticommutation relation, we can demonstrate the following:

i) These anticommutation rules for the Dirac field operators correctly reproduce the Dirac equation for $\psi(\vec{x}, t)$ and $\psi^\dagger(\vec{x}, t)$, that is,

$$i\dot{\psi}(\vec{x}, t) = [\psi(\vec{x}, t), H] \tag{3.192}$$

leading to $(i\gamma^\mu \partial_\mu - m)\psi(\vec{x}, t) = 0$ \tag{3.193}

and $\overline{\psi}(\vec{x}, t)(i\gamma^\mu \partial_\mu + m) = 0.$ \tag{3.194}

ii) The vacuum state $|0\rangle$ is defined through the action of the annihilation operators:

$$a_r(\vec{k})|0\rangle = b_r(\vec{k})|0\rangle = 0 \quad \text{for each } r \text{ and } \vec{k} \tag{3.195}$$

$$N|0\rangle = 0. \tag{3.196}$$

iii) The relation $[\psi(\vec{x},t),Q] = \psi(\vec{x},t)$ shows that if $Q|q\rangle = |q\rangle$, then

$$\langle q'|[\psi(\vec{x},t),Q]|q\rangle = \langle q'|\psi(\vec{x},t)|q\rangle, \tag{3.197}$$

$$\text{that is,} \quad (q - q')\langle q'|\psi(\vec{x},t)|q\rangle = \langle q'|\psi(\vec{x},t)|q\rangle, \tag{3.198}$$

implying that $q' = q - 1$, that is, the action of the field $\psi(\vec{x},t)$ is to reduce charge by one unit.

iv) $[Q,H] = 0$, implying that the charge is conserved.

v) The particles annihilated by $\psi^+(\vec{x},t)$ and the particles created by $\psi^-(\vec{x},t)$ have opposite charge with the same mass and spin. Therefore, they can be treated as particles and antiparticles.

vi) The operator $a^\dagger(\vec{k})$ and $b^\dagger(\vec{k})$ created particles and antiparticles in states characterized by spin s and momentum $\vec{k}$. A multiparticle state in the Fock space may therefore be written as:

$$|n\rangle = |n_1 n_2 ... a_n\rangle = (a_{s_1}^\dagger(\vec{k}_1))^{n_1}(a_{s_2}^\dagger(\vec{k}_2))^{n_2}...(b_{r_1}^\dagger(\vec{k}_1))^{n_1}(a_{r_2}^\dagger(\vec{k}_2))^{n_2}...|0\rangle, \tag{3.199}$$

where $n_1, n_2, ..., n_n$ can be either 0 or 1 for each of the $(2s_i + 1)$ or $(2r_i + 1)$ spin states.

vii)

$$Q|\vec{k},s\rangle = |\vec{k},s\rangle, \tag{3.200}$$

$$Q|\vec{k},\bar{s}\rangle = -|\vec{k},\bar{s}\rangle, \tag{3.201}$$

$$\text{where} \quad |\vec{k},s\rangle = a_s^\dagger(\vec{k})|0\rangle \quad \text{and} \quad |\vec{k},\bar{s}\rangle = b_s^\dagger(\vec{k})|0\rangle. \tag{3.202}$$

3.7 Covariant Anticommutators and Propagators for Spin $\frac{1}{2}$ Fields

In analogy with the Klein–Gordon fields, we will introduce the covariant anticommutators for Dirac fields using the expansion for $\psi(\vec{x},t)$ and $\psi^\dagger(\vec{x},t)$ to obtain:

$$\{\psi(\vec{x},t),\psi(\vec{y},t)\} = \{\bar{\psi}(\vec{x},t),\bar{\psi}(\vec{y},t)\} = 0 \tag{3.203}$$

$$\{\psi(\vec{x},t),\bar{\psi}(\vec{y},t)\} = \{\psi^+(\vec{x},t) + \psi^-(\vec{x},t),\bar{\psi}^+(\vec{y},t) + \bar{\psi}^-(\vec{y},t)\}$$

$$= \{\psi^+(\vec{x},t),\bar{\psi}^-(\vec{y},t)\} + \{\psi^-(\vec{x},t),\bar{\psi}^+(\vec{y},t)\}. \tag{3.204}$$

Using the commutation relation for $a_r(\vec{k})$ and $b_r(\vec{k})$ or $a_r^\dagger(\vec{k})$ and $b_r^\dagger(\vec{k})$ given in Eqs. (3.177) and (3.178),

$$\{\psi_r^+(\vec{x},t), \bar{\psi}_s^-(\vec{y},t)\} = \frac{1}{2V} \sum_{r,\vec{k}} \sum_{s,\vec{k}'} \frac{1}{\sqrt{\omega_{\vec{k}} \omega_{\vec{k}'}}} \{a_r(\vec{k}), a_s^\dagger(\vec{k}')\} e^{-ik \cdot x + ik' y} u_r(\vec{k}) \bar{u}_s(\vec{k}') \qquad (3.205)$$

$$= \frac{1}{2V} \sum_{r,\vec{k}} \frac{1}{\omega_{\vec{k}}} e^{-ik(x-y)} u_r(\vec{k}) \bar{u}_r(\vec{k}) \qquad (3.206)$$

$$= \int \frac{d\vec{k}}{(2\pi)^3 2\omega_{\vec{k}}} (\not{k} + m) e^{-ik \cdot (x-y)}, \qquad (3.207)$$

where we also used $\sum_r u_r(\vec{k}) \bar{u}_r(\vec{k}) = (\not{k} + m)$ from Chapter 2.

$$\{\psi_r^+(\vec{x},t), \bar{\psi}_s^-(\vec{x},t)\} = iS_{rs}^+(\vec{x} - \vec{y}), \qquad (3.208)$$

where

$$S_{rs}^+ = -\frac{i}{(2\pi)^3} \int d\vec{k} \frac{(\not{k} + m)_{rs}}{2\omega_{\vec{k}}} e^{-ik \cdot x} \qquad (3.209)$$

$$= \left(i \frac{\partial}{\partial x^\mu} + m\right)_{rs} \Delta^+(x - y). \qquad (3.210)$$

We can now use these results for the integral representation for $\Delta^\pm(x)$ to obtain the integral representation for $S^\pm(x)$, that is,

$$S^\pm = -(i\not{\partial} + m) \int_{C^\pm} \frac{d^4k}{(2\pi)^4} \frac{e^{-ik \cdot x}}{k^2 - m^2} \qquad (3.211)$$

$$= -\frac{1}{(2\pi)^4} \int_{C^\pm} d^4k \frac{1}{\not{k} - m} e^{-ik \cdot x}, \qquad (3.212)$$

where $C^\pm$ are the contours in the complex (k_0) plane shown in Figure 3.2.

3.8 Time-ordered Products and Feynman Propagators

We define the time-ordered product or T-product for fermion fields $\psi(x)$ as:

$$T(\psi(x_1)\bar{\psi}(x_2)) = \psi(x_1)\bar{\psi}(x_2) \qquad \text{(for } t_1 > t_2) \qquad (3.213)$$

$$= -\bar{\psi}(x_2)\psi(x_1) \qquad \text{(for } t_2 > t_1) \qquad (3.214)$$

$$\text{or} \quad T(\psi(x), \bar{\psi}(x_2)) = \theta(t_1 - t_2)\psi(x_1)\bar{\psi}(x_2) - \theta(t_2 - t_1)\bar{\psi}(x_2)\psi(x_1). \qquad (3.215)$$

This definition of time-ordered product for the fermion fields or T-product is different from the definition of T-product for boson fields in the sign of the second term. This also occurs in the definition of normal product.

In the case of fermion fields:

$$
\begin{aligned}
\psi(x_1)\bar{\psi}(x_2) &= (\psi^+(x_1)+\psi^-(x_1))(\bar{\psi}^+(x_2)+\bar{\psi}^-(x_2)) \\
&= \psi^+(x_1)\bar{\psi}^+(x_2)+\psi^+(x_1)\bar{\psi}^-(x_2)+\psi^-(x_1)\bar{\psi}^+(x_2)+\psi^-(x_1)\bar{\psi}^-(x_2) \\
&= \bar{\psi}^-(x_2)\psi^+(x_1)-\bar{\psi}^-(x_2)\psi^+(x_1)+\psi^+(x_1)\bar{\psi}^+(x_2) \\
&\quad +\psi^+(x_1)\bar{\psi}^-(x_2)+\psi^-(x_1)\bar{\psi}^+(x_2)+\psi^-(x_1)\bar{\psi}^-(x_2) \\
&= N(\psi(x_1)\bar{\psi}(x_2))+\{\psi^+(x_1),\bar{\psi}^-(x_2)\},
\end{aligned}
\tag{3.216}
$$

where

$$
N(\psi(x_1)\bar{\psi}(x_2)) = \psi^-(x_1)\bar{\psi}^+(x_2)+\psi^-(x_1)\bar{\psi}^-(x_2)-\bar{\psi}^-(x_2)\psi^+(x_1)+\psi^+(x_1)\bar{\psi}^+(x_2).
$$

Taking the vacuum expectation value of the T-order product, we get:

$$
\begin{aligned}
\langle 0|T\{\psi(x_1)\bar{\psi}(x_2)\}|0\rangle &= \langle 0|\theta(t_1-t_2)\psi(x_1)\bar{\psi}(x_2) \\
&\quad -\theta(t_1-t_2)\bar{\psi}(x_2)\psi(x_1) \\
&= \langle 0|\{\psi^+(x_1)\bar{\psi}^-(x_2)\}|0\rangle \\
&= iS^+(x_1-x_2) \quad \text{for } t_1 > t_2
\end{aligned}
\tag{3.217}
$$

and similarly,
$$
\begin{aligned}
\langle 0|T\{\bar{\psi}(x_2)\psi(x_1)\}|0\rangle &= \langle 0|\{\bar{\psi}^-(x_2),\psi^+(x_1)\}|0\rangle \\
&= iS^-(x_1-x_2) \quad \text{for } t_2 > t_1.
\end{aligned}
\tag{3.218}
$$

The Feynman propagator S_F is defined as:

$$
\langle 0|T\{\psi(x_1)\bar{\psi}(x_2)\}|0\rangle = iS_F(x_1-x_2),
\tag{3.219}
$$

where
$$
\begin{aligned}
S_F(x_1-x_2) &= \theta(t_1-t_2)S^+(x_1-x_2)-\theta(t_2-t_1)S^-(x_1-x_2) \\
&= \left(i\gamma^\mu\frac{\partial}{\partial x^\mu}+m\right)\Delta_F(x_1-x_2).
\end{aligned}
\tag{3.220}
$$

Here, $\Delta_F(x_1-x_2)$ is given in Eq. (3.160).

We obtain the integral representation of $S_F(x)$ as:

$$
S_F(x) = \frac{1}{(2\pi)^4}\int_{C_F}\frac{d^4k\,e^{-ik\cdot x}(\not{k}+m)}{k^2-m^2+i\epsilon}.
\tag{3.221}
$$

Using the integral representation of $\Delta_F(x)$ and $S_F(x)$ given in Eqs. (3.160) and (3.220), respectively, the physical interpretation of $S^+(x_1-x_2)$ and $S^-(x_1-x_2)$ are given in Figure 3.6. In case of $t_1 > t_2$, the Dirac particle is created at x_2 and travels to x_1, where it is annihilated, while for $t_2 > t_1$, the Dirac antiparticle is created at x_1 (through $b^\dagger(\vec{k})$) and travels to x_2, where it is annihilated (through $b(\vec{k})$).

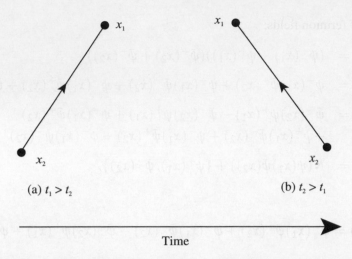

Figure 3.6 Propagation of a fermion between x_1 and x_2 for (a) $t_1 > t_2$ and (b) $t_2 > t_1$.

3.9   Quantization of Massless Electromagnetic Fields: Photons

The equation of motion for a massless spin 1 field $A^\mu(x)$ corresponding to photons has been discussed in Chapter 2, using Maxwell's equations of classical electrodynamics for electric field $\vec{E}(x)$ and magnetic field $\vec{B}(x)$. Defining an antisymmetric tensor of second rank $F^{\mu\nu}$ as:

$$F^{\mu\nu} = \partial^\nu A^\mu(x) - \partial^\mu A^\nu(x),\tag{3.222}$$

Maxwell's equations of motion can be written as:

$$\partial_\nu F^{\mu\nu}(x) = J^\mu(x)\tag{3.223}$$

$$\partial^\lambda F^{\mu\nu} + \partial^\mu F^{\nu\lambda} + \partial^\nu F^{\lambda\mu} = 0,\tag{3.224}$$

or in terms of $A^\mu(x)$ as:

$$\Box A^\mu - \partial^\mu(\partial_\nu A^\nu(x)) = J^\mu(x).\tag{3.225}$$

These equations of motion for $A^\mu(x)$ can be obtained by choosing a Lagrangian $\mathcal{L}(x)$ for the free field as:

$$\mathcal{L}(x) = -\frac{1}{4}F_{\mu\nu}(x)F^{\mu\nu}(x)\tag{3.226}$$

and applying the Euler–Lagrange equation. However, this Lagrangian is not suitable for the quantization procedure as the conjugate field $\pi^\mu(x)$ corresponding to $A^\mu(x)$ is given by:

$$\pi^\mu(x) = \frac{\partial \mathcal{L}}{\partial \dot{A}_\mu(x)} = -F^{\mu 0}(x),\tag{3.227}$$

implying that $\pi^0 = -F^{00}(x) = 0$, that is, one of the canonical fields is zero. This is not compatible with the canonical quantization conditions that need to be satisfied for all the four components of the $A^\mu(x)$ field and the conjugate fields $\Pi^\mu(x)$.

A Lagrangian which is suitable for the quantization and reproduces the equations of motion was first proposed by Fermi for the free field:

$$\mathcal{L} = -\frac{1}{2}(\partial_\nu A_\mu(x))(\partial^\nu A^\mu(x)). \tag{3.228}$$

Differentiating Eq. (3.228) with respect to $A^\mu_{,\alpha}$, we get:

$$\frac{\partial \mathcal{L}}{\partial A^\mu_{,\alpha}} = -\partial_\alpha A^\mu, \quad \frac{\partial \mathcal{L}}{\partial \dot{A}_\mu} = -\dot{A}^\mu. \tag{3.229}$$

Using the Euler–Lagrange equation of motion, we obtain the equation of motion for the $A^\mu(x)$ field as:

$$\Box A^\mu(x) = 0, \tag{3.230}$$

and the conjugate field $\pi^\mu(x)$

$$\pi^\mu(x) = -\dot{A}^\mu, \tag{3.231}$$

which facilitates the process of canonical quantization.

It can be shown that the Lagrangian density given in Eqs. (3.226) and (3.228) lead to same action(S) in the Lorenz gauge giving the same equations of motion after applying the principle of least action.

The solutions of the equations of motion given in Eq. (3.230) can be written in analogy with the scalar field for each component of $A^\mu(x)$, that is,

$$A^\mu(x) = \sum_{r,\vec{k}} \sqrt{\frac{1}{2V\omega_{\vec{k}}}} \left[\epsilon^\mu_r(\vec{k}) a_r(\vec{k}) e^{-ik \cdot x} + \epsilon^{*\mu}_r(\vec{k}) a^\dagger_r(\vec{k}) e^{ik \cdot x} \right], \tag{3.232}$$

$$= A^+_\mu(x) + A^-_\mu(x), \tag{3.233}$$

where k satisfies the relation $k^2 = k_0^2 - \vec{k}^2 = 0$, that is, $k_0 = |\vec{k}|$ and $r = 0, 1, 2, 3$ labels the four polarization states for the four-component vector field $A^\mu(x)$, represented by $\epsilon^\mu_r(\vec{k})$. As noted earlier in Chapter 2, all the four components are not independent but subject to the Lorenz condition; the choice of gauge can eliminate two of the independent components for real photons. The polarization vectors $\epsilon^\mu_r (r = 0 - 3)$ are chosen such that ϵ^μ_0 is time-like and $\epsilon^\mu_r (r = 1, 2, 3)$ are space-like. The operators $a_r(\vec{k})$ and $a^\dagger_r(\vec{k})$ are the annihilation and creation operators for the quanta of the $A^\mu(x)$ field, that is, the photon, in the state of polarization $r(= 0, 1, 2, 3)$ corresponding to the scalar $(r = 0)$, longitudinal $(r = 3)$ or transverse $(r = 1, 2)$ photons, assuming the Z-direction to be the direction of propagation. The normalization of the polarization vectors as discussed in Chapter 2 are given as:

$$\epsilon_r(\vec{k})\epsilon_s(\vec{k}) = \epsilon_{r\mu}(\vec{k})\epsilon_s^\mu(\vec{k}) = -\rho_r\delta_{rs}, \tag{3.234}$$

where $\rho_0 = -1, \rho_1 = \rho_2 = \rho_3 = 1$, and the completeness relation is

$$\sum_r \rho_r\epsilon_r^\mu(\vec{k})\epsilon_s^\nu(\vec{k}) = -g^{\mu\nu}. \tag{3.235}$$

3.10   Commutation Relations and Quantization of $A^\mu(x)$

The canonical quantization of the electromagnetic field A^μ describing the photons is achieved by postulating the equal time commutation relations (ETCR) for $A^\mu(x)$ and the conjugate field $\pi^\mu(x)(= -\dot{A}^\mu(x))$ as:

$$\left[A^\mu(\vec{x},t), A^\nu(\vec{x}',t)\right] = \left[\dot{A}^\mu(\vec{x},t), \dot{A}^\nu(\vec{x}',t)\right] = 0, \tag{3.236}$$

$$\left[A^\mu(\vec{x},t), \dot{A}^\nu(\vec{x}',t)\right] = -ig^{\mu\nu}\delta(\vec{x}-\vec{x}'). \tag{3.237}$$

It can be seen that these equations are like the ETCR for the scalar Klein–Gordon field defining the quantization, except for the factor $-g^{\mu\nu}$ which takes care of the four independent components of $A^\mu(x)$. However, there is still some problem with the quantization, that is, the commutation relation in Eq. (3.237) does not satisfy the condition $\partial_\mu A^\mu = 0$. In order to implement this condition, we first quantize in the general case and impose this condition after quantization using the theory of Gupta [213] and Bleuler [214] as discussed later in Section 3.11.

The canonical commutation relations for the fields given in Eqs. (3.236) and (3.237) lead to the commutation relations for the creation and annihilation operators as:

$$\left[a_r(\vec{k}), a_s^\dagger(\vec{k}')\right] = \rho_r\delta_{rs}\delta_{\vec{k}\vec{k}'}, \tag{3.238}$$

$$\left[a_r(\vec{k}), a_s(\vec{k}')\right] = \left[a_r^\dagger(\vec{k}), a_s^\dagger(\vec{k}')\right] = 0. \tag{3.239}$$

These commutation relations are similar to the commutation relations for the scalar fields except in the case of $r = 0$, that is, the creation and annihilation operators corresponding to the scalar photons with polarization $\vec{\epsilon}_0$. This creates a problem as it allows the creation of scalar photon states with negative normalization, which can be seen as follows.

The vacuum state $|0\rangle$ is defined through the action of the annihilation operator as:

$$a_r(\vec{k})|0\rangle = 0 \quad \text{and} \quad \langle 0|a_r^\dagger(\vec{k}) = 0, \tag{3.240}$$

while the one photon state $|\vec{k},r\rangle$ with momentum k^μ and polarization ϵ_r^μ is defined through the action of the creation operator $a_r^\dagger(\vec{k})$ as:

$$a_r^\dagger(\vec{k})|0\rangle = |\vec{k},r\rangle, \quad \langle 0|a_r(\vec{k}) = \langle \vec{k},r|$$

$$\text{and} \quad \langle \vec{k}',0|\vec{k},0\rangle = \langle 0|a_0(\vec{k}')a_0^\dagger(\vec{k})|0\rangle$$

$$= \langle 0|[a_0(\vec{k}'), a_0^\dagger(\vec{k})]|0\rangle$$

$$= \langle 0|0\rangle \rho_0 \, \delta_{\vec{k}\vec{k}'} = -\delta_{\vec{k}\vec{k}'}. \tag{3.241}$$

The Hamiltonian is calculated in terms of the creation and annihilation operators from the general expression of $\mathcal{H}$ as:

$$H = \int d\vec{x} \left[\pi^\mu(x)\dot{A}_\mu(x) - \mathcal{L}(x) \right], \tag{3.242}$$

with $\pi^\mu = -\dot{A}^\mu$ to give

$$H = \sum_{r,\vec{k}} \rho_r \omega_{\vec{k}} a_r^\dagger(\vec{k}) a_r(\vec{k})$$

$$= \sum_{r,\vec{k}} \omega_{\vec{k}} N_r(\vec{k}), \tag{3.243}$$

with $N_r(\vec{k}) = \rho_r a_r^\dagger(\vec{k}) a_r(\vec{k})$. We see that the number operator for scalar photons $r = 0$ is negative. This is related to the negative normalization of these photons. However, it does not create any problems in calculating the energy.

The energy of a single particle state $|\vec{k}, r\rangle$ is, therefore, given by E_{kr} and is obtained from the eigenvalue equation

$$H|\vec{k}, r\rangle = \omega_{\vec{k}}|\vec{k}, r\rangle = \omega_{\vec{k}} a_r^\dagger(\vec{k})|0\rangle, \tag{3.244}$$

since,

$$H|\vec{k}, r\rangle = \sum_{\vec{k}', r'} \omega'_{\vec{k}} \rho'_r a_{r'}^\dagger(\vec{k}') a_{r'}(\vec{k}') a_r^\dagger(\vec{k})|0\rangle$$

$$= \sum_{\vec{k}', r'} \omega'_{\vec{k}} \rho'_r a_{r'}^\dagger(\vec{k}') [a_{r'}(\vec{k}'), a_r^\dagger(\vec{k})]|0\rangle$$

$$= \sum_{\vec{k}', r'} \omega'_{\vec{k}} \rho'_r a_{r'}^\dagger(\vec{k}') \rho_r \delta_{rr'} \delta_{kk'}|0\rangle$$

$$= \rho_r^2 \omega_{\vec{k}} a_r^\dagger(\vec{k})|0\rangle = \omega_{\vec{k}}|\vec{k}, r\rangle. \tag{3.245}$$

Therefore, a number operator $N_r(\vec{k})$ can still be defined as:

$$N_r(\vec{k}) = \rho_r a_r^\dagger(\vec{k}) a_r(\vec{k}), \tag{3.246}$$

ignoring the problem of the negative normalization of the scalar photon at present, which is solved by choosing the appropriate gauge through the mechanism of gauge fixing.

3.11   Lorenz Condition and Gupta–Bleuler Formalism

The covariant formalism of the quantization of massless photons presented in the previous section in analogy with the covariant formalism for the quantization of the scalar field is beset with the following problems:

i) It can be seen that the Lorenz conditions which are needed in this formalism to reproduce Maxwell's equations are not implemented in the commutation relations postulated in Eqs. (3.236) and (3.237), that is,

$$[\partial_\mu A^\mu(\vec{x},t), \dot{A}^\nu(\vec{x}',t)] = -i\partial_\mu g^{\mu\nu}\delta(\vec{x}-\vec{x}') \neq 0. \tag{3.247}$$

ii) The commutation relations for the annihilation and creation operators corresponding to the $r=0$ component of polarization vectors, that is, ϵ_0^μ, are not consistent with the canonical commutation relations for the scalar fields.

Both the aforementioned problems are manifestation of the non-implementation of the Lorenz condition. Therefore, the covariant theory is not equivalent to the massless photon corresponding to Maxwell's theory. The problem was resolved by Gupta [213] and Bleuler [214] independently, by imposing a weaker condition, that is,

$$\partial_\mu A^{\mu+}|\psi\rangle = 0, \quad \langle\psi|\partial_\mu A^{\mu-} = 0, \tag{3.248}$$

involving absorption operators only, implying that at the level of matrix element:

$$\langle\psi|\partial_\mu A^\mu|\psi\rangle = \langle\psi|\partial_\mu A^{\mu+} + \partial_\mu A^{\mu-}|\psi\rangle = 0. \tag{3.249}$$

The Lorenz condition [195] is realized weakly at the expectation values level and not at the operator identity level. Since the physical observables are theoretically obtained through the calculation of expectation values of the operators, this implementation of the Lorenz condition should be considered satisfactory.

The equation implies that:

$$\langle\psi|\partial_\mu A^{\mu+}|\psi\rangle = -i\langle\psi|\sum_{\vec{k}}\sum_{r=0}^{3}\frac{1}{\sqrt{2V\omega_{\vec{k}}}}k_\mu\epsilon_r^\mu(\vec{k})a_r(\vec{k})e^{-ik\cdot x}|\psi\rangle = 0. \tag{3.250}$$

Let us assume that the photon is moving along the z-axis, that is, $\vec{k} = |\vec{k}|\hat{z}$ such that:

$$k^\mu = (k_0, 0, 0, |\vec{k}|) \text{ and } k_\mu = (k_0, 0, 0, -|\vec{k}|). \tag{3.251}$$

Since $k_0 = |\vec{k}| = \omega_{\vec{k}}$, with the choice of ϵ_r^μ for $r = 0, 1, 2, 3$ given in Eq. (2.202), we obtain:

$$k_\mu \sum_{r=0}^{3} a_r \epsilon_r^\mu = \omega_{\vec{k}}\left(a_0(\vec{k}) - a_3(\vec{k})\right), \tag{3.252}$$

which gives

$$\langle\psi|\partial_\mu A^{\mu+}|\psi\rangle = -i\langle\psi|\sum_{\vec{k}}\frac{1}{\sqrt{2V\omega_{\vec{k}}}}\omega_{\vec{k}}(a_0(\vec{k}) - a_3(\vec{k}))e^{-ik\cdot x}|\psi\rangle$$

$$= 0, \tag{3.253}$$

that is, $a_0(\vec{k})|\psi\rangle - a_3(\vec{k})|\psi\rangle = 0 \quad \text{for all } \vec{k} \tag{3.254}$

and $\langle\psi|a_0^\dagger(\vec{k}) = \langle\psi|a_3^\dagger(\vec{k}). \tag{3.255}$

This implies that the contribution of the scalar photons and longitudinal photons cancel each other. Since the physical operators in Fock space will involve summation over all the photon polarizations in a covariant formulation, only transverse photons will contribute. The theory then becomes equivalent to Maxwell's theory. This can be seen in case of the Hamiltonian operator:

$$\langle\psi|H|\psi\rangle = \langle\psi|\sum_{\vec{k}}\sum_{r=0}^{3}\rho_r\omega_{\vec{k}}a_r^\dagger(\vec{k})a_r(\vec{k})|\psi\rangle$$

$$= \langle\psi|\omega_{\vec{k}}(a_1^\dagger(\vec{k})a_1(\vec{k}) + a_2^\dagger(\vec{k})a_2(\vec{k}) + a_3^\dagger(\vec{k})a_3(\vec{k}) - a_0^\dagger(\vec{k})a_0(\vec{k}))|\psi\rangle$$

$$= \langle\psi|\sum_{\vec{k}}\sum_{r=1}^{2}\omega_{\vec{k}}a_r^\dagger(\vec{k})a_r(\vec{k})|\psi\rangle. \tag{3.256}$$

Therefore, in the case of free fields of real photons, only transverse components will contribute. It is important to realize that the Lorenz condition eliminates both the additional photon degrees of freedom arising due to the scalar and longitudinal components included due to covariant formulation. However, in the case of interacting fields where virtual photons can be produced in the intermediate states, the situation is not so simple. In such cases, the contribution of scalar and longitudinal photons cannot be ignored. In any physical process involving real photons, the initial and final states will be transverse photons which are described by free fields, being asymptotic states, while the intermediate states will be virtual states, needing a covariant description of photons in terms of all the polarization components included in the photon propagator.

3.12 Time-ordered Product and Propagators for Spin 1 Fields

In Section 3.3, we defined the covariant commutators and their vacuum expectation values and related them to the propagators $\Delta^\pm(x)$ and $\Delta(x)$ for the scalar field $\phi(x)$. We also defined the T-order product of scalar fields and their expectation values and related them to the propagators $\Delta^\pm(x)$ and $\Delta_F(x)$. We can extend the formalism to apply them directly in case of the four-component vector field $A^\mu(x)$ in the following way:

We define the covariant commutator of the fields $A^\mu(x)$ in analogy with Eq. (3.133) to write:

$$[A^\mu(x_1), A^\nu(x_2)] = iD^{\mu\nu}(x_1 - x_2),\tag{3.257}$$

where

$$D^{\mu\nu}(x_1 - x_2) = \lim_{m \to 0}(-g^{\mu\nu}\Delta(x_1 - x_2)).\tag{3.258}$$

Here $\Delta(x_1 - x_2)$ is the invariant Δ-function given as:

$$\Delta(x_1 - x_2) = -\frac{1}{(2\pi)^3}\int \frac{d\vec{k}}{\omega_{\vec{k}}} \sin[k(x_1 - x_2)].\tag{3.259}$$

Similarly, the vacuum expectation value of the time-ordered product(T-product) is defined to give the Feynman propagator $D_F^{\mu\nu}(x)$ as:

$$\langle 0|T[A^\mu(x_1), A^\nu(x_2)]|0\rangle = iD_F^{\mu\nu}(x_1 - x_2),\tag{3.260}$$

where

$$\begin{aligned}
D_F^{\mu\nu}(x_1 - x_2) &= -\lim_{m \to 0}(g^{\mu\nu}\Delta_F(x_1 - x_2))\\[4pt]
&= -\frac{g^{\mu\nu}}{(2\pi)^4}\int_{CF} \frac{d^4k\, e^{-ik\cdot(x_1 - x_2)}}{k^2 + i\epsilon}\tag{3.261}\\[4pt]
&= \frac{1}{(2\pi)^4}\int_{CF} d^4k\, D_F^{\mu\nu}(\vec{k})\, e^{-ik\cdot(x_1 - x_2)}.\tag{3.262}
\end{aligned}$$

Here

$$D_F^{\mu\nu}(\vec{k}) = \frac{-g^{\mu\nu}}{k^2 + i\epsilon},\tag{3.263}$$

is the propagator in momentum space. In terms of the explicit polarization states $\epsilon_r(\vec{k})$, $D_F^{\mu\nu}(\vec{k})$ can be written using all the polarization states, that is,

$$\begin{aligned}
D_F^{\mu\nu}(\vec{k}) &= \frac{1}{k^2 + i\epsilon}\left[\sum_{k,r}\rho_r \epsilon_r^\mu(\vec{k})\epsilon_r^\nu(\vec{k})\right]\\[4pt]
&= \frac{1}{k^2 + i\epsilon}\left[\sum_{k,r=1}^{2}\epsilon_r^\mu(\vec{k})\epsilon_r^\nu(\vec{k}) + \frac{(k^\mu - (k.n)n^\mu)(k^\nu - (k.n)n^\nu)}{(k.n)^2 - k^2} - n^\mu n^\nu\right].\tag{3.264}
\end{aligned}$$

Using the explicit values of the polarization vectors,

$$\epsilon_0^\mu(\vec{k}) = n^\mu = (1,0,0,0) \quad \text{and} \quad \epsilon_3^\mu(\vec{k}) = \left[\frac{k^\mu - (k.n)n^\mu}{\sqrt{(k.n)^2 - k^2}}\right].\tag{3.265}$$

The three terms in Eq. (3.264) correspond to the propagation of the transverse photons, longitudinal photons, and the scalar photons. However we express $D_F^{\mu\nu}(\vec{k})$ as:

$$D_F^{\mu\nu}(\vec{k}) = D_{F,T}^{\mu\nu}(\vec{k}) + D_{F,C}^{\mu\nu}(\vec{k}) + D_{F,R}^{\mu\nu}(\vec{k}), \tag{3.266}$$

where

$$D_{F,T}^{\mu\nu}(\vec{k}) = \frac{1}{k^2 + i\epsilon} \left[\sum_{\vec{k},r=1}^{2} \epsilon_r^\mu(\vec{k}) \epsilon_r^\nu(\vec{k}) \right] \tag{3.267}$$

$$D_{F,C}^{\mu\nu}(\vec{k}) = \frac{n^\mu n^\nu}{(k.n)^2 - k^2} \tag{3.268}$$

$$D_{F,R}^{\mu\nu}(\vec{k}) = \frac{1}{k^2 + i\epsilon} \left[\frac{k^\mu k^\nu - (k.n)(k^\mu n^\nu + k^\nu n^\mu)}{(k.n)^2 - k^2} \right]. \tag{3.269}$$

To understand the physical interpretation of the longitudinal and scalar terms, let us consider the term proportional to $n^\mu n^\nu$, that is,

$$D_{F,C}^{\mu\nu}(\vec{k}) = \frac{n^\mu n^\nu}{(k.n)^2 - k^2}$$

and the rest, $D_{F,R}^{\mu\nu}(\vec{k})$ as:

$$D_{F,R}^{\mu\nu}(\vec{k}) = \frac{1}{k^2 + i\epsilon} \frac{k^\mu k^\nu - (k.n)(k^\mu n^\nu + k^\nu n^\mu)}{(k.n)^2 - k^2}. \tag{3.270}$$

The physical interpretation of $D_{F,C}^{\mu\nu}(\vec{k})$ can be seen by calculating it in coordinate space in which

$$D_{F,C}^{\mu\nu}(x) = \frac{1}{(2\pi)^4} \int \frac{n^\mu n^\nu}{(k.n)^2 - k^2} e^{-ik \cdot x} d^4x. \tag{3.271}$$

Using n^μ from Eq. (3.265), we obtain the only non-vanishing component of $D_{F,C}^{\mu\nu}(x)$ as:

$$\begin{aligned} D_{F,C}^{00}(x) &= \frac{1}{(2\pi)^3} \int \frac{d\vec{k} e^{-i\vec{k} \cdot \vec{x}}}{|\vec{k}|^2} \delta(x_0) \\ &= \frac{1}{4\pi |\vec{x}|} \delta(x_0) = \phi(x), \end{aligned} \tag{3.272}$$

that is, instantaneous Coulomb potential between the two static charges. On the other hand, the contribution of the $D_{F,R}^{\mu\nu}(\vec{k})$ term to any physical process involving real photons specified by the polarization vector $\epsilon^\mu(\vec{k})$ and the external electromagnetic current j^μ of a charged particle described by:

$$j^\mu(x) = \frac{1}{(2\pi)^4} \int d^4k \, j^\mu(\vec{k}) e^{-ikx}, \tag{3.273}$$

which is conserved, vanishes. This is because any matrix element corresponding to a physical process involving electromagnetic interaction, shown in Figure 3.7 between the two fermion currents $j^\mu(x_1)$ and $j^\mu(x_2)$ through the photon propagator gets contribution from $D_{F,R}^{\mu\nu}$ as:

$$\int d^4x_1 \, d^4x_2 \, j_\mu(x_1) \, D_{F,R}^{\mu\nu}(x_1 - x_2) \, j_\nu(x_2). \tag{3.274}$$

Since the currents $j^\mu(x_1)$ and $j^\mu(x_2)$ are conserved, that is, $k \cdot j_\mu^{(1)} = k \cdot j_\mu^{(2)} = 0$, the contribution of $D_{F,R}^{\mu\nu}$ to the physical process vanishes. Thus, the covariant description of the photon field is equivalent to the description of photon fields in Coulomb gauge where it is described by $A^\mu(= \phi, \vec{A})$ as discussed in Chapter 2.

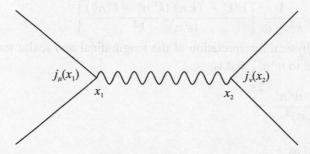

Figure 3.7 Interaction of two electromagnetic currents through photon exchange.

Chapter 4

Interacting Fields and Relativistic Perturbation Theory

4.1 Introduction

In the previous chapter, we have described free particles and fields in terms of Lagrangian, equations of motion, their solutions and quantization. However, in the physical world, particles and fields are visualized by their interactions with other particles and fields or among themselves. For example, the simple processes of Compton scattering, photoelectric effect, and Coulomb scattering involving photons and electrons, are well known. All the known elementary particles interact with each other through the four fundamental interactions, that is, electromagnetic, weak, strong, and gravitational, through the exchange of gauge fields. Since the particles themselves can be described in terms of fields, all the physical processes governed by the four fundamental interactions are examples of various types of fields in interaction with each other including self interaction, allowed by the general principles of physics. These interactions are described by an interaction Lagrangian $L_{int}(x)$, to be included along with the free Lagrangian $L_{free}(x)$, described in Chapter 2 for a quantum description of the evolution of physical systems. The interaction Lagrangians can be obtained by using the symmetry properties of the physical system defined by certain transformations called local gauge transformations and imposing the invariance of the free Lagrangian under these transformations. These will be discussed in some detail in Chapter 8, in the case of electromagnetic, weak, and strong interactions of scalar, vector, and spin $\frac{1}{2}$ particles.

In this chapter, we give some simple examples of interaction Lagrangians involving spin 0, spin $\frac{1}{2}$, and spin 1 particles and illustrate the general principles to write them. We use the example of electromagnetic interaction to demonstrate the general method of the relativistic perturbation theory to find out the solution of the equations of motion of fields in the presence of the interaction Lagrangian $L_{int}(x)$. It is assumed that the strength of the interaction Lagrangian can be quantified by a parameter which is small, so that perturbation theory can be applied. This is normally the situation in the case of electromagnetic and weak interactions; in a limited range of kinetic variables, it is also true in the case of strong interactions. The relativistic perturbation theory has been very useful in describing physical processes.

4.2 Simple Forms of Interaction Lagrangians of Fields

The form of the interaction Lagrangian of fields depends upon the type of field, that is, spin 0, spin $\frac{1}{2}$ and spin 1. However, we have seen in Chapter 2 that quadratic terms for various fields and their derivatives which form a scalar like ϕ^2, $\phi^*\phi$, $\bar{\psi}\psi$ or $W_\mu W^\mu$, give rise to the mass term, while the lowest order scalars formed from derivatives like $\partial_\mu \phi^* \partial^\mu \phi$, $\bar{\psi}\gamma^\mu \partial_\mu \psi$ or $(\partial_\mu A^\nu)(\partial^\mu A_\nu)$ give rise to the kinetic energy term in the Lagrangian. Therefore, any addition of such quadratic terms would redefine the mass or the kinetic energy term. The terms used to describe the interaction of fields should avoid such terms. Therefore, the interaction Lagrangian for one field (i.e., self interaction) is written in terms of fields having powers higher than the quadratic terms. On the other hand, the interaction Lagrangian for more than one type of fields can be written involving all the fields using the general principle of symmetries of interactions like Lorentz invariance, parity and T-invariance.

In the following sections, we give examples of simple forms of interaction Lagrangian for various interactions.

4.2.1 Electromagnetic interactions

These interactions involve the interaction of charged particles with the electromagnetic field. Since the electromagnetic field is described by a vector field $A^\mu(x)$, we should construct a vector current for particles of spin 0, spin $\frac{1}{2}$, or higher spin to construct a scalar quantity because electromagnetic interactions conserve parity. The simple vector currents which carry charge can be constructed for complex scalar fields ϕ or spin $\frac{1}{2}$ fields like ψ as:

$$j_\mu^\phi(x) = \phi^*(x)\partial_\mu \phi(x) - \phi \partial_\mu \phi^*(x), \tag{4.1}$$

$$j_\mu^\psi(x) = \bar{\psi}(x)\gamma_\mu \psi(x). \tag{4.2}$$

Hence, the simplest interaction Lagrangian conserving parity can be written as:

$$\mathcal{L}_{\text{int}} = g j_\mu^\phi(x) A^\mu(x) \qquad \text{for spin 0,} \tag{4.3}$$

$$\text{and} \quad \mathcal{L}_{\text{int}} = g j_\mu^\psi(x) A^\mu(x) \qquad \text{for spin } \frac{1}{2}, \tag{4.4}$$

where g is the strength of the interaction. In fact, such an interaction can be generated by the principle of minimal coupling or the local gauge invariance (see Chapter 7), which predicts $g = e$.

4.2.2 Weak interactions

In the case of weak interactions, parity is violated; therefore, the interaction Lagrangian would have scalar and pseudoscalar terms simultaneously to give parity violating effects through their interference. Since weak interactions are four fermion interactions as proposed by Fermi (Chapter 1), we construct scalars and pseudoscalars from the bilinear convariants formed from fermion fields. We have seen in Chapter 2 that in the case of spinors, five bilinear covariants can be constructed which transform as scalar, vector, tensor, axial vector or pseudoscalar. Therefore, a scalar can be constructed using any of the five covariants. Consequently, the Lagrangian can be constructed as:

$$\mathcal{L}_{\text{int}}^{\text{spinors}}(x) = \sum_{i=S,V,T,A,P} g_i \bar{\psi}(x) O^i \psi(x) \bar{\psi}(x) O_i \psi(x), \tag{4.5}$$

where $O^i (i = S, V, T, A, P)$ are the operators involving Dirac matrices such that the bilinear covariants $\bar{\psi} O^i \psi$ transform as S, V, T, A, P. There would be additional terms in the Lagrangian transforming as pseudoscalar such that the Lagrangian is written as:

$$\mathcal{L}_{\text{int}}^{\text{weak}}(x) = \sum_i \bar{\psi}(x) O^i \psi(x) \bar{\psi}(x) (C_i + C_i' \gamma_5) \psi(x), \tag{4.6}$$

where C_i and C_i' are, respectively, the strength of the scalar and pseudoscalar couplings.

4.2.3 Strong interactions

The simplest interaction Lagrangian between nucleon fields $\psi(x)$ and scalar fields $\sigma(x)$ or pseudoscalar fields $\phi(x)$ was first given by Yukawa [208] and is written as:

$$\mathcal{L}_{\text{int}}^{\text{strong}}(x) = g_s \bar{\psi}(x) \psi(x) \sigma(x) \qquad \text{(for scalar fields)} \tag{4.7}$$

and

$$\mathcal{L}_{\text{int}}^{\text{strong}}(x) = g_{ps} \bar{\psi}(x) \gamma_5 \psi(x) \phi(x) \text{ or } g_{pv} \bar{\psi}(x) \gamma_\mu \gamma_5 \psi(x) \partial_\mu \phi(x) \quad \text{(for pseudoscalar fields),} \tag{4.8}$$

where g_s is the strength of the scalar coupling of field $\sigma(x)$ and g_{ps} and g_{pv} are respectively the strengths of the pseudoscalar and pseudovector coupling of the pseudoscalar fields $\phi(x)$.

4.2.4 Self interaction of various fields

In the earlier sections, we have given examples of simple interaction Lagrangians in which two different fields interact. There may exist self interactions of spin 0, spin $\frac{1}{2}$, or spin 1 fields. In case of real scalar fields, the interaction Lagrangian could have terms which are higher than the quadratic term in fields $\phi(x)$, that is,

$$\mathcal{L}_{\text{int}}^{\text{real scalar}}(x) = \sum_{n \geq 3} g_n \phi^n(x), \tag{4.9}$$

where g_n is the strength of the interaction in the nth term. It should be noted that the dimension of g_n in the case of real fields should be $(4-n)$ as $\int \mathcal{L} d\vec{x}$ has dimensions of energy. In the case of complex scalar fields, the interaction Lagrangian can be written as:

$$\mathcal{L}_{\text{int}}^{\text{complex scalar}}(x) = \sum_{n \geq 2} g_n (\phi^*(x)\phi(x))^n. \tag{4.10}$$

Similarly, the self interaction Lagrangian can be written for higher spin fields like the Yang–Mills field or gluon fields and their form is predicted by the local gauge field theories as will be discussed in Chapter 7.

4.3   Evolution of Physical Systems and the S-matrix

In quantum mechanics, the physical observables correspond to the expectation value of certain operators between the states describing the system and are generally dependent on time. If they are independent of time, they are conserved. Energy, momentum, and angular momentum are simple examples of such observables, which are the expectation values of the Hamiltonian, momentum, and angular momentum operators, respectively, in a given state $|\phi(t)\rangle$, in Hilbert space. We use a generic notation $\phi(t)$ to represent a state which could be specified later as $\chi(t)$, $\phi(t)$, or $\psi(t)$ corresponding to scalar, pseudoscalar, spin $\frac{1}{2}$, or any other particle of higher spin. Thus, the expectation value of any operator $\langle O(t) \rangle$ between these states is defined as:

$$\langle O(t) \rangle = \frac{\langle \phi(t)|O(t)|\phi(t) \rangle}{\langle \phi(t)|\phi(t) \rangle}. \tag{4.11}$$

In Eq. (4.11), the ket $|\phi(t)\rangle$, the bra $\langle \phi(t)|$, and the operator $O(t)$ are all time dependent. We can always perform unitary transformation to transfer the time dependence of the state $|\phi(t)\rangle$ to the operator $O(t)$ and vice versa, without changing the expectation value $\langle O(t) \rangle$. Therefore, there is some ambiguity in specifying the time dependence of $|\phi(t)\rangle$ and/or $O(t)$. This ambiguity gives rise to various ways of describing the evolution of the physical system, described by the state $|\phi(t)\rangle$ and operator $O(t)$, known as the Schrödinger picture (S), Heisenberg picture (H), and interaction picture (I). The interaction picture is most suitable for formulating the relativistic perturbation theory. In the following sections, we will introduce these pictures before describing the relativistic perturbation theory.

4.3.1   Schrödinger, Heisenberg, and Interaction picture

We are familiar with the Schrödinger picture in which the equation of motion is given by:

$$i\frac{\partial \psi^S(t)}{\partial t} = H\psi^S(t), \tag{4.12}$$

where we have put a superscript S over state $\psi(t)$ and operator H to refer to the Scrödinger picture. In this case, the time dependence is carried by the state $\psi^S(t)$; the Hamiltonian operator, H, is independent of time in most cases, leading to the conservation of energy. Equation (4.12) can be solved to get:

$$\psi^S(t) = e^{-iH(t-t_0)}\psi^S(t_0),\tag{4.13}$$

where $\psi^S(t_0)$ is the initial state at $t = t_0$, specified by the initial conditions. For simplicity of notation, let us set $t_0 = 0$. Then, Eq. (4.13) can be written as a transformation equation:

$$\psi^S(t) = U^S(t)\psi^S(0).\tag{4.14}$$

Here, H is a hermitian operator and $U^S(t) = e^{-iHt}$ is a unitary operator. In the Heisenberg picture, the time dependence is carried by the operator and the state is taken to be independent of time, defined at $t = t_0(= 0)$ by $\psi^H(0)$. Therefore,

$$\psi^H(t) = \psi^H(0)(= \psi^S(0))$$

and is related to $\psi^S(t)$ by a unitary transformation

$$\psi^H = U^{S^\dagger}(t)\psi^S(t). \qquad (\because \psi^S(0) = U^{S^\dagger}\psi^S(t)).\tag{4.15}$$

Therefore, in the Heisenberg picture, any operator O^S in the Schrödinger picture is given by:

$$O^H(t) = U^{S^\dagger}(t)O^S U^S(t),\tag{4.16}$$

making $\langle\psi^S(t)|O^S|\psi^S(t)\rangle = \langle\psi^H|O^H(t)|\psi^H\rangle$ invariant. Differentiating Eq. (4.16), we can derive the well-known Heisenberg equation of motion as:

$$
\begin{aligned}
\frac{d}{dt}O^H(t) &= \frac{\partial}{\partial t}(U^{S^\dagger}(t)O^S U^S(t)) \\[2mm]
&= \frac{\partial U^{S^\dagger}(t)}{\partial t}O^S U^S(t) + U^{S^\dagger}(t)O^S\frac{\partial U^S(t)}{\partial t} + U^{S^\dagger}(t)\frac{\partial O^S}{\partial t}U^S(t) \\[2mm]
&= iU^{S^\dagger}(t)HO^S U^S(t) - iU^{S^\dagger}(t)O^S HU^S(t) \qquad \left(\because \frac{\partial O^S}{\partial t} = 0\right) \\[2mm]
&= iHO^H(t) - iO^H(t)H \\[2mm]
&= -i[O^H(t), H] \\[2mm]
\Rightarrow \quad i\frac{d}{dt}O^H(t) &= [O^H(t), H].
\end{aligned}
\tag{4.17}
$$

These equations are well-suited to derive the equations of motion for free particles and find their solutions. In the case of Interaction picture, the Hamiltonian is given in terms of the free Hamiltonian (H_0) and an interaction Hamiltonian (H_I), that is,

$$H = H_0 + H_I, \tag{4.18}$$

$$\text{where} \quad H_0 = \int (\pi(x)\dot{\phi}(x) - \mathcal{L}_{\text{free}})d\vec{x}, \tag{4.19}$$

$$\text{and} \quad H_I = -\int \mathcal{L}_{\text{int}}d\vec{x}, \tag{4.20}$$

where $\pi(x)$ is the momentum conjugate to $\phi(x)$.

In the presence of the interaction Lagrangian $\mathcal{L}_{\text{int}}(x)$, the equations of motion cannot be solved exactly as in the case of free fields. Therefore, the time dependence is defined by the state $\psi^I(t)$ in the interaction picture which is evolved from the Schrödinger picture $\psi^S(t)$ using H_0 instead of H, that is, it corresponds to the solution of the free Hamiltonian H_0, defined as:

$$\psi^I(t) = U_0^\dagger(t)\psi^S(t), \tag{4.21}$$

$$\text{where} \quad U_0(t) = e^{-iH_0t} \tag{4.22}$$

$$\text{such that} \quad O^I(t) = U_0^\dagger(t)O^S U_0(t). \tag{4.23}$$

It should be noted that the free Hamiltonian is the same in all pictures, that is,

$$H_0^I = H_0^S = H_0. \tag{4.24}$$

Differentiating Eq. (4.23), we obtain the equation of motion in the interaction picture as:

$$i\frac{d}{dt}O^I(t) = [O^I(t), H_0]. \tag{4.25}$$

Putting Eq. (4.21) in Eq. (4.12), we obtain

$$i\frac{d}{dt}\psi^I(t) = H_I^I(t)\psi^I(t), \tag{4.26}$$

$$\text{where} \quad H_I^I(t) = e^{iH_0t}H_I^S e^{-iH_0t}. \tag{4.27}$$

Equation (4.26) determines the time development of $\psi^I(t)$ in the interaction picture, which could be solved to determine the evolution of the physical system in the presence of the interaction Hamiltonian calculated in this picture. It can be further shown that:

$$\psi^I(t) = U_0^\dagger(t)\psi^S(t) = U_0^\dagger(t)U^S(t)\psi^H(0) = U(t)\psi^H(0), \tag{4.28}$$

$$\text{with} \quad U(t) = U_0^\dagger(t)U^S(t) = e^{iH_0t}e^{-iHt}, \tag{4.29}$$

$$\text{and} \quad \psi^I(t) = U(t)\psi^H(0)$$

$$= U(t)U^\dagger(t')\psi^I(t')$$

$$= U(t, t')\psi^I(t'),$$ (4.30)

$$\text{with } U(t, t') = U(t)U^\dagger(t')$$ (4.31)

connecting $\psi^I(t)$ at two time points t and t'.

4.3.2 S-Matrix and relativistic perturbation theory

Consider the equation of motion of $\psi^I(t)$ in the interaction picture

$$\frac{d}{dt}\psi^I(t) = -iH_I(t)\psi^I(t).$$ (4.32)

When $H_I(t) = 0$, that is, there is no interaction, $\psi^I(t)$ is a constant of motion. As soon as the interaction is switched on, it develops a time dependence and $\psi^I(t)$ evolves with time, depending upon the interaction Hamiltonian H_I. In a scattering process, the physical system is in an initial state $|i\rangle$, where the particles are not interacting; then, interaction (i.e., collision or scattering) takes place and depending upon the nature of the interaction, it leads to a set of final states of the system. After a considerable lapse of time, particles in the set of final state of the system are again noninteracting. We introduce the concept of asymptotic states to define the initial state at $t = -\infty$, that is, $\psi(-\infty) = |i\rangle$, where all the particle states are specified and the final states at $t = +\infty$, that is, $\psi(+\infty)$, which consists of particles in states to be specified depending upon the action of the interaction.

S-matrix is defined as an operator which changes $\psi(-\infty)$ to $\psi(\infty)$ as:

$$\psi^I(\infty) = S\psi^I(-\infty) = S|i\rangle$$ (4.33)

and describes the action of H_I on the system. The probability amplitude that a system is in state $|f\rangle$, after the collision is given by:

$$\langle f|\psi^I(\infty)\rangle = \langle f|S|i\rangle = S_{fi},$$ (4.34)

which describes the probability amplitude for the state $|i\rangle$ to evolve into $|f\rangle$. Obviously, the normalization of $\psi^I(\infty)$, that is, $\langle\psi^I(\infty)|\psi^I(\infty)\rangle = 1$, ensures the conservation of probability, that is,

$$\sum_f |\langle f|S|i\rangle|^2 = \sum_f |S_{fi}|^2 = 1.$$ (4.35)

In order to calculate the S-matrix, we need to solve Eq. (4.32) to obtain $\psi^I(\infty)$. A formal solution of Eq. (4.32) is written as:

$$\psi^I(t) = -i\int H_I(t)\psi^I(t)dt,$$ (4.36)

$$\text{that is, } \psi^I(t) = |i\rangle + (-i)\int_{-\infty}^{t} dt_1 H_I(t_1)\psi^I(t_1),$$ (4.37)

using the initial condition of $\psi^I(-\infty) = |i\rangle$.

Equation (4.37) can be solved iteratively, that is,

$$\psi^I(t) = |i\rangle + (-i) \int_{-\infty}^{t} dt_1 H_I(t_1)|i\rangle + (-i)^2 \int_{-\infty}^{t} dt_1 \int_{-\infty}^{t_1} dt_2 H_I(t_1) H_I(t_2)\psi^I|t_2\rangle. \quad (4.38)$$

This would lead to an infinite series involving one additional multiplicative factor $H_I(t)$ in each order. Such a series will converge only if the interaction Hamiltonian H_I is weak and is specified by some parameter (say λ) which is less than 1. Using the expansion in Eq. (4.38), the S-matrix is defined as a perturbative series given by:

$$S = \sum_{n=0}^{\infty} (-i)^n \int_{-\infty}^{\infty} dt_1 \int_{-\infty}^{t_1} dt_2 ..... \int_{-\infty}^{t_{n-1}} dt_n\, H_I(t_1) H_I(t_2)....H_I(t_n) \quad (4.39)$$

$$= \sum_{n=0}^{\infty} S^{(n)} \quad (4.40)$$

and the matrix element is given by

$$S_{fi} = \langle f|S|i\rangle. \quad (4.41)$$

This is known as the S-matrix expansion in the relativistic perturbation theory.

4.4 Dyson Expansion and Wick's Theorem

4.4.1 Dyson expansion

In order to calculate the expansion series in Eq. (4.40), Dyson [215] proposed a method which simplifies the calculation. To illustrate the Dyson expansion, let us consider the second term $S^{(2)}$ in the expansion integrated between t_0 and t, that is,

$$S^{(2)} = (-i)^2 \int_{t_0}^{t} dt_1 \int_{t_0}^{t_1} dt_2\, H_I(t_1) H_I(t_2). \quad (4.42)$$

The limits of t_2 integration are from t_0 to t_1, that is, $t_2 \leq t_1$ and then t_1 is integrated between t_0 to t. Let us consider the (t_1, t_2) plane for integration as shown in Figure 4.1. The straight line shows the $t_1 = t_2$ line in the (t_1, t_2) plane. The integrand lying in the upper half shaded region ADB has $t_2 > t_1$ while the lower half region ABC, shaded by vertical lines, has $t_2 < t_1$. If $H_I(t_1)$ and $H_I(t_2)$ were commuting operators, then the integration over t_1 and t_2 in Eq. (4.42), can be interchanged giving equal results; the total result could be written as half of the integral over the region ADBC, that is,

$$\int_{t_0}^{t} dt_1 \int_{t_0}^{t_1} dt_2 H_I(t_1) H_I(t_2) = \frac{1}{2} \int_{t_0}^{t} dt_1 \int_{t_0}^{t} dt_2\, H_I(t_1)\, H_I(t_2). \quad (4.43)$$

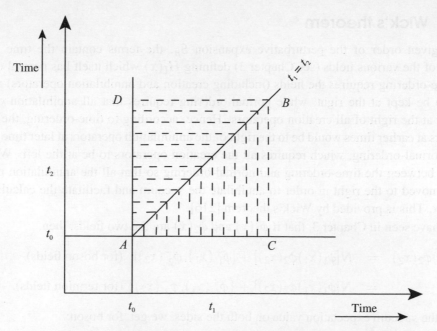

Figure 4.1 (t_1, t_2) plane for integration.

This can be done even in the case of non-commuting operators $H_I(t_1)H_I(t_2)$ by defining the time–ordered products of $H_I(t_1)H_I(t_2)$ such that: when $t_2 > t_1$ the integration is done along ADB but when $t_2 < t_1$ the integration is done along ACB i.e.

$$(-i)^2 \int_{t_0}^{t} dt_1 \int_{t_0}^{t} dt_2 \, H_I(t_1)H_I(t_2) = \frac{(-i)^2}{2} \int_{t_0}^{t} dt_1 \int_{t_0}^{t} dt_2 \, T[H_I(t_1)H_I(t_2)]. \quad (4.44)$$

Using this relation, Eq. (4.39) can be generalized to

$$S^{(n)} = \frac{(-i)^n}{n!} \int_{-\infty}^{\infty} dt_1 \int_{-\infty}^{\infty} dt_2 ... \int_{-\infty}^{\infty} dt_n \, T\left[H_I(t_1)H_I(t_2)...H_I(t_n)\right]. \quad (4.45)$$

Writing Eq. (4.45) in terms of the Hamiltonian density $\mathcal{H}(x)$, we obtain the following results for $S^{(n)}$:

$$S^{(n)} = \frac{(-i)^n}{n!} \int ... \int d^4x_1 \, d^4x_2 ... d^4x_n \, T\left[\mathcal{H}_I(x_1)\mathcal{H}_I(x_2)...\mathcal{H}_I(x_n)\right]. \quad (4.46)$$

The calculation of the matrix element $S_{fi} = \langle f|S|i \rangle = \sum_n \langle f|S_n|i \rangle$ can be performed in each order of perturbation expansion using the initial states $|i\rangle$ and $|f\rangle$, which are the eigenstates of the unperturbed free field Hamiltonian. These are assumed to be the states in Hilbert space which describe physical particles even though they are treated to be asymptotic states in the S-matrix theory considered here. This assumption needs further justification which can be found in standard textbooks on quantum field theory, for example, the text by Bjorken and Drell [211] and that of Mandl and Shaw [210].

4.4.2 Wick's theorem

In any given order of the perturbative expansion S_n, the terms contain the time ordered-product of the various fields (see Chapter 3) defining $H_I(x)$ which itself has normal ordering. The time-ordering requires the fields (including creation and annihilation operators) at earlier times to be kept at the right, while normal-ordering requires that all annihilation operators are kept at the right of all creation operators. Hence, according to time-ordering, the creation operators at earlier times would be to the right of the annihilation operators at later time contrary to the normal-ordering, which requires all the creation operators to be at the left. We need a relation between the time-ordering and normal-ordering so that all the annihilation operators can be moved to the right in order to annihilate the vacuum and facilitate the calculations of S-matrix. This is provided by Wick's theorem [216].

We have seen in Chapter 3, that if $\phi_1(x)$ and $\phi_2(x)$ are the two fields, then:

$$\phi_1(x_1)\phi_2(x_2) = N[\phi_1(x_1)\phi_2(x_2)] + [\phi_1^+(x_1), \phi_2^-(x_2)] \quad \text{(for boson fields)} \tag{4.47}$$

$$= N[\phi_1(x_1)\phi_2(x_2)] + \{\phi_1^+(x_1), \phi_2^-(x_2)\} \text{ (for fermion fields)}. \tag{4.48}$$

Taking the vacuum expectation value on both the sides, we get, for boson:

$$\langle 0|\phi_1(x_1)\phi_2(x_2)|0\rangle = \langle 0|N(\phi_1(x_1)\phi_2(x_2))|0\rangle + \langle 0|[\phi_1^+(x_1), \phi_2^-(x_2)]|0\rangle \tag{4.49}$$

and a similar equation for the fermions involving the vacuum expectation value of anticommutators $\langle 0|\{\phi_1(x_1), \phi_2(x_2)\}|0\rangle$ in place of commutators.

Since the vacuum expectation value of a normal product vanishes by definition, we get:

$$\langle 0|\phi_1(x_1)\phi_2(x_2)|0\rangle = \langle 0|[\phi_1^+(x_1), \phi_2^-(x_2)]|0\rangle \tag{4.50}$$

$$= [\phi_1^+(x_1), \phi_2^-(x_2)]. \tag{4.51}$$

Equation (4.51) follows because the commutator (or anticommutator in the case of fermions) is a c-number not an operator. Therefore, using Eqs. (4.49) and (4.51), we can write Eqs. (4.47) and (4.48) as:

$$\phi_1(x_1)\phi_2(x_2) = N(\phi_1(x_1)\phi_2(x_2)) + \langle 0|\phi_1(x_1)\phi_2(x_2)|0\rangle \tag{4.52}$$

with

$$N(\phi_1(x_1)\phi_2(x_2)) = +N(\phi_2(x_2)\phi_1(x_1)) \quad \text{(for bosons)} \tag{4.53}$$

$$= -N(\phi_2(x_2)\phi_1(x_1)) \quad \text{(for fermions)}, \tag{4.54}$$

which requires all the creation operators to be at the left. Taking the time-ordering of Eq. (4.52), we get:

$$T(\phi_1(x_1)\phi_2(x_2)) = \theta(x_1^0 - x_2^0)N(\phi_1(x_1)\phi_2(x_2)) + \theta(x_2^0 - x_1^0)N(\phi_2(x_2)\phi_1(x_1))$$

$$+ \langle 0|T(\phi_1(x_1)\phi_2(x_2))|0\rangle. \tag{4.55}$$

For boson fields $\phi_1(x_1)\phi_2(x_2) = \phi_2(x_2)\phi_1(x_1)$ and $\theta(x_1^0 - x_2^0) + \theta(x_2^0 - x_1^0) = 1$, the equation can be written as:

$$T(\phi_1(x_1)\phi_2(x_2)) = N(\phi_1(x_1)\phi_2(x_2)) + \langle 0|T(\phi_1(x_1)\phi_2(x_2))|0\rangle. \tag{4.56}$$

This above equation is also valid in the case of fermions because a -1 appears due to the interchange of two fields $\phi_1(x_1)\phi_2(x_2)$ in anticommutators and another factor of -1 appears in the definition of the time-ordered product for the fermion fields.

A special notation is used for the vacuum expectation value of the time-ordered product of commutators as:

$$\underline{\phi_1(x_1)\phi_2(x_2)} = \langle 0|T(\phi_1(x_1)\phi_2(x_2))|0\rangle \tag{4.57}$$

called 'contraction' of $\phi_1(x_1)$ and $\phi_2(x_2)$. Using this notation, the time-ordered product of the two field operators is written as:

$$T(\phi_1(x_1)\phi_2(x_2)) = N(\phi_1(x_1)\phi_2(x_2)) + \underline{\phi_1(x_1)\phi_2(x_2)}. \tag{4.58}$$

A nonzero value of the 'contraction' of the product of the two fields taken at $x_1^0 \neq x_2^0$ will appear only when a product state created by the operator field $\phi_2(x_2)$ is annihilated by the operator fields of $\phi_1(x_1)$; otherwise, it will vanish. The non-vanishing 'contraction' in Eq. (4.57) are Feynman propagators which have been discussed in Chapter 3. For example, in the case of scalar, spinor, and vector fields, they are written as:

$$\underline{\phi(x_1)\phi(x_2)} = i\Delta_F(x_1 - x_2). \tag{4.59}$$

$$\underline{\phi(x_1)\phi^\dagger(x_2)} = \phi^\dagger(x_2)\phi(x_1) = i\Delta_F(x_1 - x_2). \tag{4.60}$$

$$\underline{\psi_\alpha(x_1)\bar{\psi}_\beta(x_2)} = -\bar{\psi}_\beta(x_2)\psi_\alpha(x_1) = iS_{F\alpha\beta}(x_1 - x_2). \tag{4.61}$$

$$\underline{A^\mu(x_1)A^\nu(x_2)} = iD_F^{\mu\nu}(x_1 - x_2). \tag{4.62}$$

The definition of the time-ordered product is generalized to the product of n fields which is proved by following the method of induction. The generalization to n fields is written as:

$$T((\phi(x_1)...\phi(x_n))) = N(\phi(x_1)\phi(x_2)...\phi(x_n)) + \text{all possible contractions.} \tag{4.63}$$

Since the contraction is done by considering the product of two fields in pairs, all possible contractions of n fields would consist of various terms, where the contraction is done with 1 pair of two fields, 2 pairs of two fields each, 3 pairs of 2 fields each, and so on, until all the fields are exhausted. In the case of n being even, all the fields will be contracted in pairs, while for the odd n, one field will be left out. For example, in the case of n=4, using $\phi(x_\alpha) = \phi_\alpha$,

$$T(\phi_1\phi_2\phi_3\phi_4) = N(\phi_1\phi_2\phi_3\phi_4 + \underbrace{\phi_1\phi_2}\,\phi_3\phi_4 + \underbrace{\phi_1\phi_2\phi_3}\,\phi_4$$

$$+ \underbrace{\phi_1\phi_2\phi_3}\phi_4 + \phi_1\underbrace{\phi_2\phi_3}\,\phi_4 + \phi_1\underbrace{\phi_2\phi_3\phi_4}$$

$$+ \underbrace{\phi_1\phi_2}\,\underbrace{\phi_3\phi_4} + \underbrace{\phi_1\phi_2}\,\underbrace{\phi_3\phi_4} + \phi_1\,\overbrace{\phi_2\phi_3}\,\phi_4 + \phi_1\,\underbrace{\phi_2}\,\underbrace{\phi_3\,\phi_4}). \tag{4.64}$$

In the case of n fields, we, therefore write:

$$T(\phi_1\phi_2...\phi_n) = N(\phi_1\phi_2...\phi_n) + N(\underbrace{\phi_1\phi_2}\,\phi_3...\phi_n)$$

$$+ N(\underbrace{\phi_1\phi_2}\,\underbrace{\phi_3\phi_4}\,...\phi_n)... + (\phi_1\phi_2...\,\underbrace{\phi_{n-3}\phi_{n-2}}\,\underbrace{\phi_{n-1}\phi_n})$$

$$+ ... + N(\underbrace{\phi_1\phi_2}\,\underbrace{\phi_3\phi_4}\,...\,\underbrace{\phi_{n-1}\phi_n}) \quad \text{for even } n. \tag{4.65}$$

In case we consider the fermion field $\psi(x)$, in place of $\phi(x)$, then the normal product of n fields is given as:

$$N(\psi_1, \psi_2, ..., \psi_n) = (-1)^P(\psi_1', \psi_2', ..., \psi_n'), \tag{4.66}$$

where $\psi_\alpha'(\alpha = 1...n)$ are the rearrangement of the fields $\psi_\alpha(\alpha = 1...n)$ and P is the number of interchanges of neighboring fermion field operators as they anticommute. There is no additional factor of -1 when boson fields are interchanged.

Equation (4.64) contains 0, 1, and 2 pairs of contractions to be continued until the last term, where all the fields (except one) are contracted in case n is even (odd).

When the interaction Hamiltonian density $H_I(x)$ contains a mixed time-ordered product like $T(H_I(x_1)H_I(x_2)...)$, then

$$T[H_I(x_1)H_I(x_2)...H_I(x_n)] = T[N(\phi_1(x_1)\phi_2(x_1)...)N(\phi_1(x_2)\phi_2(x_2)...)...$$
$$N(\phi_1(x_n)\phi_2(x_n))]. \tag{4.67}$$

It should be noted that the contraction terms from the right-hand side will contribute only when the two fields are evaluated at different times, that is, $t_1 \neq t_2$. Therefore, contraction of field products like $\phi_1(x_1)\phi_2(x_1)$ will not contribute because corresponding to the same group, they would have equal time commutators which would vanish [217].

This is known as Wick's theorem [216]. We see that the contracted terms give the Feynman propagators which describe the propagation of virtual particles. The non-contracted fields in a normal product contain a set of creation and annihilation operators. Therefore, when the S-matrix element like $\langle f|S|i \rangle$ is calculated, the action of the set of annihilation and creation operators in S acting on the initial state $|i\rangle$ must produce the particles present in the final state $|f\rangle$ for a non-vanishing matrix element. Such a matrix element may include propagation of virtual particles if the normal product contains contracted terms like the terms in Eqs. (4.64) and (4.65). Thus, using Wick's theorem for the expansion of the S-matrix, the transition matrix elements of $\langle f|S|i \rangle$ are calculated in any order of relativistic perturbation theory.

4.5 S-matrix and Feynman Diagrams

The calculation of individual terms $S^{(n)}$ given in Eq. (4.46), in the expansion of S-matrix and its matrix element $S_{fi}^{(n)} = \langle f|S^{(n)}|i\rangle$ between the initial state $|i\rangle$ and the final state $|f\rangle$ is facilitated with the help of Feynman diagrams. This will be described in this section. For the purpose of illustration, let us consider the interaction Lagrangian $\mathcal{L}_I(x)$ for the interaction of the electron with the electromagnetic field given in Eq. (4.4), that is,

$$\mathcal{L}_I^{em} = e\bar{\psi}(x)\gamma^\mu\psi(x)A_\mu = e\bar{\psi}(x)\slashed{A}\psi(x) \tag{4.68}$$

such that

$$H_I^{em}(x) = -e\bar{\psi}(x)\slashed{A}(x)\psi(x), \tag{4.69}$$

where $\psi(x)$ and $A_\mu(x)$ are the fields for the electron and photon, respectively. In the case of quantized fields, the normal-ordering of fields in $H_I(x)$ is to be understood such that:

$$S^{(1)} = +ie\int d^4x\, T(\bar{\psi}(x)\slashed{A}\psi(x))$$
$$= +ie\int d^4x\, N[(\bar{\psi}^+(x)+\bar{\psi}^-(x))(\slashed{A}^+(x)+\slashed{A}^-(x))(\psi^+(x)+\psi^-(x))]. \tag{4.70}$$

There are eight terms in $S^{(1)}$ and they contribute to eight basic processes taking place at a point x in which four of them are associated with the absorption of photons (due to the $\phi^+(x)$ term) and the other four with the creation of photons. These four diagrams are associated with $\bar{\psi}^+(x)\psi^+(x), \bar{\psi}^+(x)\psi^-(x), \bar{\psi}^-(x)\psi^+(x),$ and $\bar{\psi}^-(x)\psi^-(x)$ and correspond to the annihilation of e^-e^+ pair, emission and absorption of e^+ and e^-, and creation of e^-e^+ pair as shown in Figure 4.2. However, all these processes are not possible for real physical electrons and photons due to the energy momentum conservation, so $S_{fi}^{(1)} = \langle f|S^{(1)}|i\rangle = 0$ for all $|i\rangle$

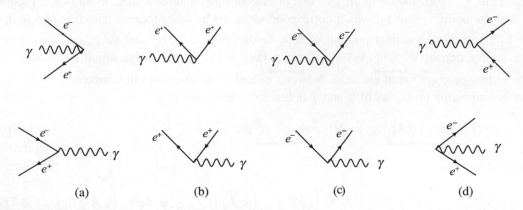

Figure 4.2 Feynman diagrams for eight basic processes in QED. (a) e^+e^- annihilation, (b) e^+ scattering, (c) e^- scattering, and (d) e^+e^- creation, accompanied by the absorption (upper row) and emission (lower row) of a photon.

and $|f\rangle$, as shown in the next section. That is, all the processes possible from the mathematical structure of the S-matrix may not be physical. This statement is valid in each order of the S–matrix expansion, that is, $S^{(n)}$. These diagrams depicting the physical as well as unphysical processes appearing in the S-matrix expansion are called Feynman diagrams. In order to obtain the real processes from $H_I(x)$ of Eq. (4.46), we go to the next order in perturbation theory in which we can also demonstrate the physical interpretation of the contracted terms in context of the real processes. Let us now consider the second order term $S^{(2)}$:

$$S^{(2)} = \sum_{i=0}^{m} S_i^{(2)}, \tag{4.71}$$

where m is the maximum number of contractions for $x_1^0 \neq x_2^0$, given by:

$$S_0^{(2)} = -\frac{e^2}{2!} \int d^4x_1 \int d^4x_2 \, N\Big((\bar{\psi} A \psi)_{x_1} (\bar{\psi} A \psi)_{x_2} \Big), \tag{4.72}$$

$$S_1^{(2)} = -\frac{e^2}{2!} \int d^4x_1 \int d^4x_2 \, N\Big((\bar{\psi} A \underline{\psi})_{x_1} (\underline{\bar{\psi}} A \psi)_{x_2} + (\underline{\bar{\psi}} A \psi)_{x_1} (\bar{\psi} A \underline{\psi})_{x_2}$$
$$+ (\bar{\psi} A \psi)_{x_1} (\bar{\psi} A \psi)_{x_2} \Big), \tag{4.73}$$

$$S_2^{(2)} = -\frac{e^2}{2!} \int d^4x_1 \int d^4x_2 \, N\Big((\bar{\psi} A \psi)_{x_1} (\bar{\psi} A \psi)_{x_2} + (\bar{\psi} A \psi)_{x_1} (\bar{\psi} A \psi)_{x_2}$$
$$+ (\bar{\psi} A \psi)_{x_1} (\bar{\psi} A \psi)_{x_2} \Big), \tag{4.74}$$

$$S_3^{(2)} = -\frac{e^2}{2!} \int d^4x_1 \int d^4x_2 \, N\Big(\bar{\psi} A \psi)_{x_1} (\bar{\psi} A \psi)_{x_2} \Big). \tag{4.75}$$

The term $S_0^{(2)}$ corresponds to the two sets of disconnected diagrams each, as shown in Figure 4.2 at the point x_1 and x_2, which correspond to an unphysical process. The first two terms in $S_1^{(2)}$ include the normal product of four fields $(\bar{\psi} A)_{x_1} (A \psi)_{x_2}$ and $(A \psi)_{x_1} (\bar{\psi} A)_{x_2}$ with the contraction of $\psi(x_1)\bar{\psi}(x_2)$ and $\bar{\psi}(x_1)\psi(x_2)$ fields, respectively, which correspond to the virtual propagation of electrons between x_1 and x_2 as discussed in Chapter 3. Using the anticommutation properties of ψ and $\bar{\psi}$ fields, it can be shown that:

$$N\big((\bar{\psi} A \psi)_{x_1} (\bar{\psi} A \psi)_{x_2} \big) = N\big((\bar{\psi} A \psi)_{x_2} (\bar{\psi} A \psi)_{x_1} \big), \tag{4.76}$$

leading to:

$$S_1^{(2)} = -e^2 \int d^4x_1 d^4x_2 \, N\left[(\bar{\psi} A \psi)_{x_1} (\bar{\psi} A \psi)_{x_2} + \frac{1}{2} (\bar{\psi} A \psi)_{x_1} (\bar{\psi} A \psi)_{x_2} \right] \tag{4.77}$$

$$= S_{1A}^{(2)} + S_{1B}^{(2)}, \tag{4.78}$$

where

$$S_{1A}^{(2)} = -ie^2 \int d^4x_1 d^4x_2 \, N\left[(\bar{\psi}A)_{x_1}(A\psi)_{x_2}S_F(x_1 - x_2)\right] \tag{4.79}$$

$$S_{1B}^{(2)} = -\frac{ie^2}{2} \int d^4x_1 d^4x_2 \, N\left[(\bar{\psi}\psi)_{x_1}(\bar{\psi}\psi)_{x_2}D_F(x_1 - x_2)\right], \tag{4.80}$$

Here, $S_F(x)$ and $D_F(x)$ are the Feynman propagators for the electron and photon, respectively. The term $S_{1A}^{(2)}$ contains two uncontracted field operators for electrons and two uncontracted operators for photons which create or absorb real particles at x_1 and x_2. They are represented by the external lines in the Feynman diagram and connected through the exchange of virtual electrons described by the Feynman propagator $S_F(x_1 - x_2)$.

There are many processes described by this term but real physical processes will correspond to those processes in which there are two particles in the initial state $|i\rangle$ and two particles in the final state $|f\rangle$ satisfying the energy–momentum conservation and the combination of creation and annihilation operators such that $\langle f|S|i \rangle$ is non-vanishing. The various real processes which are described by this term are, therefore,

$$(i) \; \gamma + e^- \to \gamma + e^- \quad (ii) \; \gamma + e^+ \to \gamma + e^+$$
$$(iii) \; \gamma + \gamma \to e^+ + e^- \quad (iv) \; e^+ + e^- \to \gamma + \gamma.$$

known as the Compton scattering of (i) electrons and (ii) positrons; (iii) e^+e^- pair creation and (iv) e^+e^- annihilation.

The Feynman diagrams corresponding to these processes are shown in Figure 4.3. On the other hand, the term $S_{1B}^{(2)}$ has four uncontracted fermion operators which can describe the

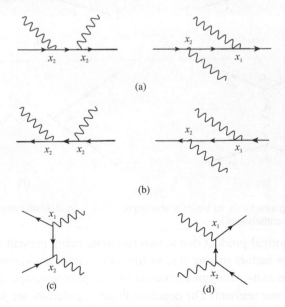

Figure 4.3 Feynman diagrams for (a) Compton scattering for e^-, (b) Compton scattering for e^+, (c) pair annihilation, and (d) pair creation.

absorption and creation of electrons or positrons at point x_1 and x_2 connected by the photon propagator $D_F^{\mu\nu}(x)$; the processes would, therefore, consist of the absorption or creation of $e^-(e^+)$ as external lines and photons as propagators and can be described as follows.

$$(i)\ e^- + e^- \to e^- + e^-\quad (ii)\ e^+ + e^+ \to e^+ + e^+\quad (iii)\ e^- + e^+ \to e^- + e^+.$$

These processes are known as Møller scattering of (i) electron and (ii) positron; and (iii) Bhabha scattering of electron and positron. The Feynman diagrams corresponding to these processes are given in Figure 4.4. However, in these cases, there are additional constraints

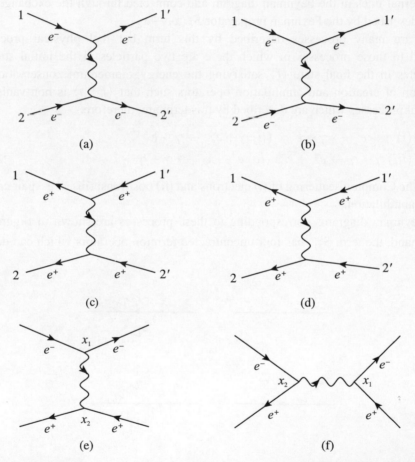

Figure 4.4   Feynman diagrams for (a, b) Møller scattering of $e^- - e^-$, (c, d) Møller scattering of $e^+ - e^+$, and (e, f) Bhabha scattering of $e^- - e^+$.

while evaluating the normal product due to two fermions being present in the initial and final states. The appropriate factors arising due to the antisymmetrization of two electrons or an electron and a positron state, or symmetrization of two photon states, in the initial and final states need to be taken into account. The details of these calculations are to be found in standard texts on quantum field theory (QFT) [210].

The next term with two contraction $S_2^{(2)}$ can be written as:

$$S_2^{(2)} = S_{2A}^{(2)} + S_{2B}^{(2)}, \text{ where} \tag{4.81}$$

$$S_{2A}^{(2)} = -\frac{e^2}{2!} \int d^4x_1 d^4x_2 (N(\bar{\psi}\slashed{A}\psi)_{x_1}(\bar{\psi}\slashed{A}\psi)_{x_2} + (\bar{\psi}\slashed{A}\psi)_{x_1}(\bar{\psi}\slashed{A}\psi)_{x_2}) \tag{4.82}$$

$$S_{2B}^{(2)} = -\frac{e^2}{2!} \int d^4x_1 d^4x_2 N(\bar{\psi}\slashed{A}\psi)_{x_1}(\bar{\psi}\slashed{A}\psi)_{x_2}. \tag{4.83}$$

The $S_{2A}^{(2)}$ term contains two uncontracted fermion fields $N(\bar{\psi}(x_1)\psi(x_2))$ and $N(\psi(x_1)\bar{\psi}(x_2))$ with two contractions corresponding to the electron and photon propagator $S_F(x_1 - x_2)$ and $D_F^{\mu\nu}(x_1 - x_2)$, respectively. The uncontracted fermion field will describe two physical processes according to the electrons (positrons) being presented in the initial and final states at x_1 and x_2 connected with electron and photon propagators between x_1 and x_2. The Feynman diagram corresponding to these processes are shown in Figure 4.5.

Figure 4.5 Electron (left) and positron (right) self energy.

These are called self energy diagrams for the electrons (positrons) and correspond to the modification of the properties of electrons (positron) converting a bare (mathematical) electron (positron) to the physical electron (positron). In actual calculations, the contribution of this diagram diverges but the divergent contribution is absorbed in the definition of physical properties of electrons through the concept of renormalization. It should be noted that in the higher order of perturbation theory, that is, $S^{(4)}, S^{(6)}$, etc, there will more number of such loop diagrams.

Similarly, the term $S_{2B}^{(2)}$ has two uncontracted electromagnetic fields $A^\mu(x_1), A^\mu(x_2)$ along with two fermion propagators $S_F(x_1 - x_2)$ and $S_F(x_2 - x_1)$. This describes a physical process in which there is a photon in the initial as well as in the final states connected by two fermion propagators for the electrons and the positrons corresponding to the Feynman diagram shown in Figure 4.6.

Figure 4.6 Photon self energy.

This diagram is called the photon self energy diagram in which an interacting photon can create a virtual $e^- e^+$ pair which recombines to give back the photon state. This can happen when a photon is passing through an external electromagnetic field or changing the distribution of virtual e^-e^+ pairs, like the polarization in dielectrics. These diagrams are, therefore, called vacuum polarization diagrams and give divergent contributions. Such divergences are removed through the renormalization procedure in QED [210].

Finally, the last term $S_3^{(2)}$ with three contractions does not have any free fields either in the initial or in the final state corresponding to three propagators between x_1 and x_2 as shown in Figure 4.7.

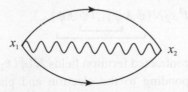

Figure 4.7 The vacuum diagram.

Such diagrams do not correspond to any physical process. Thus, the first term and the last term in $S^{(2)}$, which correspond to disconnected and fully connected Feynman diagrams, respectively, do not correspond to any physical process and can be ignored in the applications of the relativistic perturbation theory to physical processes.

4.6   Invariant Matrix Elements and Feynman Diagrams

4.6.1   Matrix elements in first order perturbation theory

We have seen in the last section that the various terms in the S-matrix expansion generate transition which may lead to physical processes involving the initial state $|i\rangle$ and the final state $|f\rangle$, subject to the conservation of energy and momentum. The uncontracted field operators appearing in the S-matrix expansion are combinations of fields such as $\psi(x)$, $\bar{\psi}(x)$, and $A^\mu(x)$, which are expressed in terms of creation and annihilation operators leading to the emission and absorption of real particles represented by the external lines in Feynman diagrams. For example, using the expression of $\psi(x)$, $\bar{\psi}(x)$, and $A^\mu(x)$, we can write

$$\bar{\psi}^-(x)|0\rangle = \sum_{r,\vec{p}} \sqrt{\frac{1}{2E_{\vec{p}}V}} \bar{u}_r(\vec{p})e^{ip\cdot x}|e_r^-(\vec{p})\rangle, \tag{4.84}$$

$$\psi^-(x)|0\rangle = \sum_{r,\vec{p}} \sqrt{\frac{1}{2E_{\vec{p}}V}} v_r(\vec{p})e^{ip\cdot x}|e_r^+(\vec{p})\rangle, \tag{4.85}$$

$$A^{\mu-}(x)|0\rangle = \sum_{r,\vec{k}} \sqrt{\frac{1}{2E_{\vec{k}}V}} \epsilon_r^\mu(\vec{k})e^{ik\cdot x}|\gamma(\vec{k})\rangle. \tag{4.86}$$

Similarly, the action of $\psi(x)$, $\bar{\psi}(x)$, and $A^\mu(x)$ on the initial state $|i\rangle$, which consists of particle states of $e^-(\vec{p})$, $e^+(\vec{p})$, and $\gamma(\vec{k})$, will also lead to vacuum states, that is,

$$\psi^+(x)|e^-(\vec{p})\rangle = \sqrt{\frac{1}{2E_{\vec{p}}V}}u(\vec{p})e^{-ip\cdot x}|0\rangle, \tag{4.87}$$

$$\bar{\psi}^+|e^+(\vec{p})\rangle = \sqrt{\frac{1}{2E_{\vec{p}}V}}\bar{v}(\vec{p})e^{-ip\cdot x}|0\rangle, \tag{4.88}$$

and

$$A^{\mu+}(x)|\gamma(\vec{k})\rangle = \sqrt{\frac{1}{2E_{\vec{k}}V}}\epsilon^\mu(\vec{k})e^{-ik\cdot x}|0\rangle \tag{4.89}$$

due to the action of the respective creation and annihilation operators.

Let us apply the first order relativistic perturbation theory to calculate a process such as $e^-(\vec{p}) \to e^-(\vec{p}\,') + \gamma(\vec{k}')$, which is one of the processes given in Figure 4.2 and described by the term $S^{(1)}$, given by:

$$S^{(1)} = ie \int d^4x N\left(\bar{\psi}(x)A(x)\psi(x)\right). \tag{4.90}$$

In this case, the initial state $|i\rangle$ and final state $|f\rangle$ are written as:

$$|i\rangle = |e^-(\vec{p})\rangle = a_e^\dagger(\vec{p})|0\rangle, \tag{4.91}$$

$$|f\rangle = |e^-(\vec{p}\,')\gamma(\vec{k}\,')\rangle = a_e^\dagger(\vec{p}\,')a_\gamma^\dagger(\vec{k}\,')|0\rangle, \tag{4.92}$$

where we have used a subscript $e(\gamma)$ on creation operators $a^\dagger$ to show that they create an electron(e) and a photon (γ). The S-matrix element $\langle f|S^{(1)}|i\rangle$ is calculated using Eq. (4.90) to give:

$$\langle f|S^{(1)}|i\rangle = ie\langle e^-(\vec{p}\,')\gamma(\vec{k}\,')|\int d^4x N\left(\bar{\psi}(x)A(x)\psi(x)\right)|e^-(\vec{p})\rangle. \tag{4.93}$$

Only the $\bar{\psi}^- A^- \psi^+(x)$ term in $N\left(\bar{\psi}(x)A(x)\psi(x)\right)$ will give a nonzero contribution to $\langle f|S^{(1)}|i\rangle$ for the process $e^- \to e^- + \gamma$.

$$\langle f|S^{(1)}|i\rangle = ie \int d^4x \sqrt{\frac{1}{2E_{\vec{p}}V}}\sqrt{\frac{1}{2E_{\vec{p}\,'}V}}\sqrt{\frac{1}{2VE_{\vec{k}'}}}\bar{u}(\vec{p}\,')\epsilon u(\vec{p})e^{i(p'+k'-p)\cdot x}$$

$$= (2\pi)^4\delta^4(p'+k'-p)\sqrt{\frac{1}{2E_{\vec{p}}V}}\sqrt{\frac{1}{2E_{\vec{p}\,'}V}}\sqrt{\frac{1}{2VE_{\vec{k}'}}}\mathcal{M}_{fi}, \tag{4.94}$$

where $\mathcal{M}_{fi} = ie\bar{u}(\vec{p}\,')\epsilon_\mu\gamma^\mu(\vec{k}')u(\vec{p})$. Equation (4.94) is the final result for $\langle f|S^{(1)}|i\rangle$. The matrix element $\mathcal{M}_{fi}$ is called the invariant matrix element or Feynman matrix element for the process $e^-(p) \to e^-(p') + \gamma(k')$. The factor $(2\pi)^4\delta^4(p'+k'-p)$ represents the conservation of momentum and energy. The Feynman diagram in momentum space corresponding to this process is given in Figure 4.8, where the lines labeled $e^-(p)$, $e^-(p')$ are the external lines describing the real electron in the initial and final states and the line $\gamma(k')$ is the external line describing the real photon. The vertex labeled $ie\gamma^\mu$ shows the strength and structure of the vertex corresponding to this transition. However, in this case, the energy momentum conservation condition, that is, $p' + k' = p$ is not compatible with the real particles satisfying $p^2 = p'^2 = m^2$, $k'^2 = 0$ and this process does not take place. Similarly, all the first order processes shown in Figures 4.2(a)–4.2(d) are not physical processes.

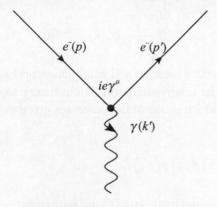

Figure 4.8 The process $e^- \to e^- + \gamma$.

4.6.2 Matrix elements in second order perturbation theory

Let us consider the term $S_1^{(2)}$ (Section 4.5), the second order of perturbation theory, to illustrate the use of Feynman propagators in momentum space for calculating the invariant matrix element $\mathcal{M}$. For definiteness, we take the physical process of Compton scattering, that is, $\gamma(k) + e^-(p) \to \gamma(k') + e^-(p')$. In this case,

$$|i\rangle = a_e^\dagger(\vec{p})a_\gamma^\dagger(\vec{k})|0\rangle \; ; \quad |f\rangle = a_e^\dagger(\vec{p}\,')a_\gamma^\dagger(\vec{k}\,')|0\rangle \tag{4.95}$$

and the relevant diagrams in $S_1^{(2)}$ will come from the $S_{1A}^{(2)}$ term from Eq. (4.79), where the initial photon is absorbed at x_1 and emitted at x_2 or absorbed at x_2 and emitted at x_1, corresponding to the two diagrams shown in Figure 4.3(a). These diagrams are reproduced in Figure 4.9 with appropriate labeling of the momentum of electrons and photons, to illustrate the difference between the two diagrams. The non-vanishing matrix element of $\langle f|S_{1A}^{(2)}|i\rangle$, corresponding to Figure 4.9, would be written as:

$$\langle f|S_{1A,a}^{(2)}|i\rangle = -e^2 \int d^4x_1\, d^4x_2 \frac{1}{\sqrt{2VE_{\vec{p}}}} \frac{1}{\sqrt{2VE_{\vec{p}\,'}}} \frac{1}{\sqrt{2VE_{\vec{k}}}} \frac{1}{\sqrt{2VE_{\vec{k}\,'}}}$$

$$\bar{u}(p')e^{ip'\cdot x_1}\slashed{\epsilon}(\vec{k}\,')e^{ik'\cdot x_1}$$

$$\times \frac{1}{(2\pi)^4}\int d^4q\; iS_F(q)e^{-iq\cdot(x_1-x_2)}\slashed{\epsilon}(\vec{k})e^{-ik\cdot x_2}u(\vec{p})e^{-ip\cdot x_2}. \tag{4.96}$$

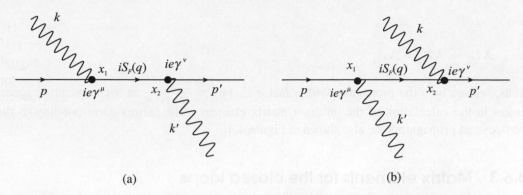

(a)   (b)

Figure 4.9 Compton scattering of electrons via (a) s-channel and (b) u-channel Feynman diagrams.

Performing the d^4x_1 and d^4x_2 integration,

$$= -ie^2 \frac{(2\pi)^4(2\pi)^4}{(2\pi)^4} \int d^4q \delta^4(p'+k'-q)\delta^4(q-p-k)\bar{u}(\vec{p}')\slashed{\epsilon}(\vec{k}')S_F(q)\slashed{\epsilon}(\vec{k})u(\vec{p})$$

$$\times \frac{1}{\sqrt{2VE_{\vec{p}}}}\frac{1}{\sqrt{2VE_{\vec{p}'}}}\frac{1}{\sqrt{2VE_{\vec{k}}}}\frac{1}{\sqrt{2VE_{\vec{k}'}}}. \tag{4.97}$$

Performing the d^4q integration, we get

$$\langle f|S^{(2)}_{1A,a}|i\rangle = -ie^2(2\pi)^4\delta^4(p'+k'-p-k)\bar{u}(\vec{p}')\slashed{\epsilon}(\vec{k}')S_F(q=p+k)\slashed{\epsilon}(\vec{k})u(\vec{p})$$

$$= (2\pi)^4\delta^4(p'+k'-p-k)\mathcal{M}^a_{fi}$$

$$\times \frac{1}{\sqrt{2VE_{\vec{p}}}}\frac{1}{\sqrt{2VE_{\vec{p}'}}}\frac{1}{\sqrt{2VE_{\vec{k}}}}\frac{1}{\sqrt{2VE_{\vec{k}'}}}, \tag{4.98}$$

where

$$\mathcal{M}^a_{fi} = -ie^2\bar{u}(\vec{p}')\epsilon_\nu(\vec{k}')\gamma^\nu S_F(q=p+k)\epsilon_\mu(\vec{k})\gamma^\mu u(\vec{p}). \tag{4.99}$$

Similarly, for the second diagram, as shown in Figure 4.9, we have:

$$\langle f|S^{(2)}_{1A,b}|i\rangle = (2\pi)^4\delta^4(p'+k'-p-k)\mathcal{M}^b_{fi}$$

$$\times \frac{1}{\sqrt{2VE_{\vec{p}}}}\frac{1}{\sqrt{2VE_{\vec{p}'}}}\frac{1}{\sqrt{2VE_{\vec{k}}}}\frac{1}{\sqrt{2VE_{\vec{k}'}}}, \tag{4.100}$$

where

$$\mathcal{M}^b_{fi} = -ie^2\bar{u}(\vec{p}')\epsilon_\nu(\vec{k})\gamma^\nu S_F(q=p-k')\epsilon_\mu\gamma^\mu(\vec{k}')u(\vec{p}) \tag{4.101}$$

and

$$\mathcal{M}_{fi} = \mathcal{M}_{fi}^a + \mathcal{M}_{fi}^b.$$

Thus, we see how the propagator with a factor $iS_F(p) = \frac{i}{\not{p}-m+i\epsilon}$ in the momentum space, enters in the calculation of the invariant matrix element. The factors corresponding to the vertices and propagators are also shown in Figure 4.9.

4.6.3 Matrix elements for the closed loops

Some new feature appears in addition to the external lines and propagators when the invariant matrix element corresponding to the closed loops, which correspond to the second order terms in perturbation theory like $S_2^{(2)}$ in Eq. (4.74) are calculated. Consider a typical Feynman diagram corresponding to the self energy of the electron as shown in Figure 4.10, where the external and internal lines are properly labeled with corresponding momenta, in which

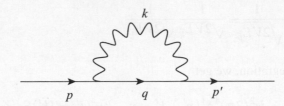

Figure 4.10 Electron self energy as an example of closed loop diagram.

an electron of momentum p emits a photon of momentum k, which is again absorbed by the electron such that $p = k + q = p'$. Thus, for a given momentum of photon k, the virtual electron has momentum $q = p - k$. Since, the photon momentum k can take any value, which fixes the electron momentum in virtual state to be $q = p - k$ and $k^2 \neq 0$, the contribution of all values of k and q should be included. This means that we should integrate over all values of k and q subject to the condition that $k + q = p$, that is, effectively integrating over only one of them, either k or q. This becomes evident if we perform the momentum integration in the S-matrix element corresponding to the term $S_{2A}^{(2)}$ in Eq. (4.82) with $|i\rangle = a_e^\dagger(\vec{p})|0\rangle$ and $|f\rangle = a_e^\dagger(\vec{p}\,')|0\rangle$. The non-vanishing term corresponding to $S_{2A}^{(2)}$ after integrating over d^4x_1 and d^4x_2 is given by:

$$\langle f|S_{2A}^{(2)}|i\rangle = -e^2 \frac{1}{\sqrt{2VE_{\vec{p}}}} \frac{1}{\sqrt{2VE_{\vec{p}\,'}}} \int d^4q \, d^4k \, \delta^4(p'-k-q) \times$$

$$(2\pi)^4 \delta^4(k+q-p) i D_F^{\mu\nu}(k)\bar{u}(\vec{p}\,')\gamma_\mu iS_F(q)\gamma_\nu u(\vec{p}), \qquad (4.102)$$

where $D_F^{\mu\nu}(k)$ is the photon propagator and $S_F(q)$ is the electron propagator. Performing the q integration, we obtain:

$$\langle f|S_{2A}^{(2)}|i\rangle = \frac{1}{\sqrt{2VE_{\vec{p}}}}\frac{1}{\sqrt{2VE_{\vec{p}}{}'}}(2\pi)^4\delta^4(p'-p)\mathcal{M}_{fi}, \tag{4.103}$$

$$\text{where } \mathcal{M}_{fi} = -\frac{e^2}{(2\pi)^4}\int d^4k\, i\, D_F^{\mu\nu}(k)\bar{u}(\vec{p})\gamma_\mu\, iS_F(p-k)\gamma_\nu u(\vec{p}). \tag{4.104}$$

Thus, the calculation of $\mathcal{M}_{fi}$ involves an integration over the loop momenta k, corresponding to the internal line of the photon. Such integration over the loop momenta also appears in the self energy of photon, as shown in Figure 4.6. It is this momentum integration over the loop in second order and higher orders, which gives rise to divergences, necessitating the need for the renormalization in QED.

4.6.4 Feynman rules: A summary

We use the standard formula for the S-matrix element $\langle f|S|i\rangle$ in terms of the invariant matrix element as

$$\langle f|S|i\rangle = \delta_{fi} + (2\pi)^4\delta^4(P_f - P_i)\prod_{i=\text{initial particles}}\frac{1}{\sqrt{2VE_i}}\prod_{f=\text{final particles}}\frac{1}{\sqrt{2VE_f}}\mathcal{M}_{fi}, \tag{4.105}$$

where P_f and P_i are the sum of 4-momenta of all the particles in final and initial state, respectively, and

$$\mathcal{M}_{fi} = \sum_n \mathcal{M}_{fi}^{(n)}, \tag{4.106}$$

where $M_{fi}^{(n)}$ is the invariant matrix element calculated for all the connected diagrams in the nth order term in the S-matrix $S^{(n)}$. The calculation of $M_{fi}^{(n)}$ in momentum space is done using the following rules in QED:

(1) External Lines   (i) Spin 0:

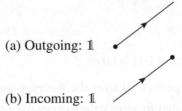

(a) Outgoing: $\mathbb{1}$

(b) Incoming: $\mathbb{1}$

(ii) Spin $\frac{1}{2}$:

(a) Incoming particle: $u(\vec{p}, s)$

(b) Incoming antiparticle: $\bar{v}(\vec{p}, s)$

(c) Outgoing particle: $\bar{u}(\vec{p}, s)$

(d) Outgoing antiparticle: $v(\vec{p}, s)$

(iii) Spin 1:

(a) Incoming: $\epsilon_\mu(\vec{p}, \lambda)$

(b) Outgoing: $\epsilon_\mu^*(\vec{p}, \lambda)$

(2) Vertex factors: Electromagnetic interaction: $ie\gamma^\mu$

(3) Propagators: (i) Spin 0: $\dfrac{i}{q^2 - m^2}$

(ii) Spin $\frac{1}{2}$: $\dfrac{i(\slashed{q} + m)}{q^2 - m^2}$ Fermion propagator

(iii) Spin 1:
          (a) Massless: $\dfrac{-ig^{\mu\nu}}{q^2}$

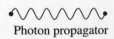

Photon propagator

(4) For a closed Fermion loop, a factor of (-1) is taken.

(5) For a closed loop, integration is performed over the 4-momentum of the loop, that is, $\int \dfrac{d^4q}{(2\pi)^4}$. The factor $\dfrac{1}{(2\pi)^4}$ is inserted to satisfy the energy–momentum condition given consistently by a factor $(2\pi)^4 \delta(P_f - P_i)$.

(b) Massive: $\dfrac{-i\left(g^{\mu\nu}-\frac{q^\mu q^\nu}{M^2}\right)}{q^2-M^2}$

W, Z propagator

(6) For the vertex, a factor depending upon the strength and nature of interaction Lagrangian is added. In case of QED, the factor is $+ie\gamma^\mu$. It should be noted that an additional factor of (-1) is needed when we go from the Lagrangian to the Hamiltonian required to calculate $S^{(n)}$.

(7) Proper care of the factors of (-1) and $(+1)$ should be taken when interchanging the fermion fields in the normal-ordered product of field operators.

(8) A definite order should be followed, that is, from right to left in writing the invariant matrix element $\mathcal{M}$, from the Feynman diagrams in momentum space.

4.7   Scattering Cross Sections and Particle Decay Rates

4.7.1   Scattering cross sections

The physical concept of cross sections in a scattering process is very useful in making quantitative studies of the elastic, inelastic, and deep inelastic scattering processes in the respective kinematical regions.

The name 'cross section' is derived from the study of collision processes in classical mechanics, in which a particle collides with a fixed spherical target with radius 'r', lying in the interaction region of total volume V and area A. The probability P that the incoming particle collides with the sphere of radius 'r' is given by:

$$P = \frac{\pi r^2}{A},\tag{4.107}$$

such that, πr^2 is the quantity representing the cross section of area relevant for the collision process and is generally represented by σ. It can be given by:

$$\sigma = PA.\tag{4.108}$$

Thus, the cross section 'σ' is effectively an area in which the incident and target particle interact for the scattering to take place. This concept of cross section σ is generalized to any collision process in which a beam of particles scatter with a fixed target or two beams of particles from opposite directions collide with each other. Consider a beam of particles of density ρ, that is, number of particles per unit volume $(= \frac{n}{V})$ moving with velocity v. In time t, this beam, passing through an effective area of interaction A, will have $n = \rho v t A$ particles, that is,

$$A = \frac{n}{\rho v t} = \frac{1}{\rho v t} \quad \text{(if } n \text{ is normalized to 1)} \tag{4.109}$$

such that:

$$\sigma = \frac{nP}{\rho v t} = \frac{P/t}{(v/V)} = \frac{P/t}{1/At}. \tag{4.110}$$

In Eq. (4.110), the numerator is the probability per unit time while the denominator is the flux of the incident particle. Therefore, the cross section is defined as:

$$\sigma = \frac{\text{Probability per unit time}}{\text{Flux of the incident particle}} = \frac{\text{Probability/time}}{\text{Incident flux}}. \tag{4.111}$$

Since the probability in the present case is $|\langle f|S|i\rangle|^2$, the cross section is given by:

$$\sigma = \frac{|\langle f|S|i\rangle|^2 \, V}{T v}. \tag{4.112}$$

Using the expression for $\langle f|S|i\rangle$ given in Eq. (4.105), we obtain for $f \neq i$,

$$\sigma = \left((2\pi)^4 \delta^4(P_f - P_i) \right)^2 \frac{V}{T v} \prod_{i=1,2} \prod_{f=1-n} \frac{1}{2E_f V} \frac{1}{2E_i V} |\mathcal{M}_{fi}|^2. \tag{4.113}$$

In Eq. (4.113), we obtain square of the $\delta^4(P_f - P_i)$ function, an operation which may not be normally well defined but can be evaluated in the limit of $V \to \infty$ and $T \to \infty$ using the integral representation of $\delta^4(P)$ function. For a rigorous proof, please see Ref.[218]. Here, we shall, demonstrate its evaluation in a heuristic way. For any function $f(P)$, the quantity

$$\delta^4(P)f(P) = \delta^4(P)f(P=0), \tag{4.114}$$

assuming that the left-hand side always occurs under an integration over P. Therefore,

$$\delta^4(P)\delta^4(P) = \delta^4(P)\delta^4(P=0). \tag{4.115}$$

Since the integral representation of $\delta^4(P)$ is given by:

$$\delta^4(P) = \frac{1}{(2\pi)^4} \int_{VT} e^{ip \cdot x} d^4 x, \tag{4.116}$$

we can formally write,

$$(2\pi)^4 \delta^4(0) = VT \quad \text{(for } P \to 0\text{) leading to} \tag{4.117}$$

$$\sigma = (2\pi)^4 \delta^4(P_f - P_i) \frac{V^2}{v} \prod_i \left(\frac{1}{2E_i V} \right) \prod_f \left(\frac{1}{2E_f V} \right) |\mathcal{M}_{fi}|^2. \tag{4.118}$$

The factor V^2 in Eq. (4.118) is compensated by $\frac{1}{V^2}$ occurring due to the normalization of two particle states in the initial state which are involved in the collision, making Eq. (4.118) independent of V, except for the powers of V, that is, V^{n_f}, occurring in the normalization of the particle states in the final state, where there are n_f particles which cancels with the V^{n_f} coming from the density of state of final particles as explained below.

In quantum mechanics, the momentum is not fixed, but always lies in between a range of $\vec{p}$ and $\vec{p} + d\vec{p}$. Therefore, we have to multiply Eq. (4.118), for the cross section, by the density of states $\rho(p)dp$, that is, the number of states lying between $\vec{p}$ and $\vec{p} + d\vec{p}$. This is given by the density of states in phase space

$$\rho(p)dp = \frac{V d\vec{p}}{(2\pi)^3}$$

for each particle in the final state, in the standard case of plane–wave normalization of states. Therefore, the final expression for the cross section is given as:

$$\sigma = \frac{1}{4E_1 E_2 \, v} \int \prod_f \frac{d\vec{p}_f}{(2\pi)^3 (2E_f)} (2\pi)^4 \delta^4(P_f - P_i) |\mathcal{M}_{fi}|^2. \qquad (4.119)$$

For a fixed target, v is the velocity of the incident particles; otherwise, it would be relative velocity, that is, $|v| = |\vec{v}_{rel}|$, for which

$$v_{rel} = \left| \frac{\vec{p}_1}{E_1} - \frac{\vec{p}_2}{E_2} \right|. \qquad (4.120)$$

In such cases, $E_1 E_2 \, v = |\vec{p}_1 E_2 - \vec{p}_2 E_1| = \sqrt{(p_1 \cdot p_2)^2 - m_1^2 m_2^2}$.

Therefore, we get the covariant expression for the scattering cross section σ as:

$$\sigma = \frac{1}{4\sqrt{(p_1 \cdot p_2)^2 - m_1^2 m_2^2}} \int \prod_f \frac{d\vec{p}_f}{(2\pi)^3 2E_f} (2\pi)^4 \delta^4(P_f - P_i) |\mathcal{M}_{fi}|^2. \qquad (4.121)$$

4.7.2 Particle decay rates

In case of a particle of momentum $P^\mu (= E, \vec{P})$ decaying at rest or in flight in two or more particles, that is, $H(P) \to 1(p_1) + 2(p_2) + ...$, the S-matrix for the transition is given by:

$$\langle f|S|i \rangle = (2\pi)^4 \delta^4 \left(P - \sum_f p_f \right) \frac{1}{\sqrt{2E_i V}} \prod_{f=1}^{n} \sqrt{\frac{1}{2VE_f}} |\mathcal{M}_{fi}| \qquad (4.122)$$

such that the probability for transition per unit time, that is, $\frac{|\langle f|S|i \rangle|^2}{T}$ is given by:

$$\frac{|\langle f|S|i \rangle|^2}{T} = (2\pi)^4 \delta^4 \left(P - \sum_f p_f \right) \cdot \frac{VT}{T} \frac{1}{2E_i V} \prod_{f=1}^{n} \frac{1}{2VE_f} |\mathcal{M}_{fi}|^2. \qquad (4.123)$$

Integrating over the density of final states $\rho(p)$, we obtain the decay rate Γ as:

$$\Gamma = \frac{1}{2E_i} \int \prod_f \frac{d\vec{p}_f}{(2\pi)^3 2E_f} (2\pi)^4 \delta^4(P - \sum_f p_f)|\mathcal{M}_{fi}|^2. \tag{4.124}$$

The life time τ of the particles is inverse of the decay rate, that is, $\tau = \frac{1}{\Gamma}$.

It should be noted that in both cases of scattering cross section and particle decay rates in which more than one identical particles are produced in the final state, due consideration should be made of the (anti)symmetrization of states depending upon (fermions) bosons in calculating the density of state in phase space.

Chapter 5

Phenomenological Theory I: Nuclear β-decays and Weak Interaction of Leptons

5.1 Introduction

The history of the phenomenological theory of neutrino interactions and weak interactions in general, begins with the attempts to understand the physics of nuclear radiation known as β-rays. These highly penetrating and ionizing component of the radiation discovered by Becquerel [9] in 1896 were subsequently established to be electrons by doing many experiments in which their properties like charge, mass, and energy were studied [8]. Since the energy of these β-ray electrons was found to be in the range of a few MeV, they were believed to be of nuclear origin in the light of the basic structure of the nucleus known at that time [4]. It was assumed that the electrons are emitted in a nuclear process called β-decay in which a nucleus in the initial state goes to a final state by emitting an electron. The energy distribution of the β-ray electrons was found to be continuous lying between m_e, the mass of the electron, and a maximum energy $E_{\max}$ corresponding to the available energy in the nuclear β-decay, that is, $E_{\max} = E_i - E_f$, where E_i and E_f are the energies of the initial and final nuclear states. A typical continuous energy distribution for the electrons from the β-decay of RaE is shown in Figure 1.1 of Chapter 1. It was first thought that the electrons in the β-decay process were emitted with a fixed energy $E_{\max}$ and suffered random losses in their energy due to secondary interactions with nuclear constituents as they traveled through the nucleus before being observed leading to a continuous energy distribution. However, the calorimetric heat measurements performed by Ellis et al. [15] and confirmed later by Meitner et al. [16] in the β-decays of RaE, established that the electrons emitted in the process of the nuclear β-decay have an intrinsically continuous energy distribution. The continuous energy distribution of the electrons from the β-decay posed a difficult problem toward its theoretical interpretation in the context of the contemporary model of the nuclear structure and seemed to violate the law of conservation of energy. In order to save the law of energy conservation, Pauli proposed in

1930 [1], the existence of a new, neutral, almost massless, spin $\frac{1}{2}$ particle which was assumed to be emitted along with the electron in the β-decay and share the available energy with the electron leading to the continuous energy distribution of the electron. The new particle was later named neutrino by Fermi [219] and a theory of the nuclear β-decay was given for the first time by Fermi [219] and independently by Perrin [24] in 1933. The theory was extended by Gamow and Teller [26] and Bethe and Bacher [27] in order to understand other types of nuclear β-decays, which were observed later. It was followed by a large amount of experimental and theoretical works on β-decays and other weak interaction processes of elementary particles and nuclei before a phenomenological theory of weak interactions was formulated. The experimental determination of the various parameters of the phenomenological theory took almost 20 years leading to the $V - A$ theory of weak interactions [41, 42, 43]. During this period, many landmark discoveries in physics like parity violation in weak interactions, two-component neutrinos, neutrino helicities, lepton number and its conservation, and the existence of more than one neutrino flavor were made. The $V - A$ theory was later extended to the strangeness sector by Cabibbo [59] in 1963, to the charm sector by Glashow –Iliopoulos–Maiani (GIM) [64] in 1970, and to the heavy quark flavor sector by Kobayashi and Maskawa [70] in 1973. The phenomenological theory was quite successful in explaining the weak interaction processes at low energies but faced many difficulties when extended to higher energies due to the divergences in higher orders and non-renormalizability of the theory. However, the $V - A$ theory implied the existence of massive vector bosons as mediators of weak interactions; these bosons played an important role in the formulation of the standard model of the unified theory of the electromagnetic and weak interactions [220, 157, 37].

In the following sections, we describe the experimental and theoretical developments in the study of nuclear β-decays and weak decays of leptons leading to the $V - A$ theory of weak interactions and its application to weak processes [221, 222]. We also include a separate section at the end of this chapter which describes the limitations of the $V - A$ theory.

5.2 Development of Phenomenological Theory

5.2.1 Fermi and Gamow–Teller theories of β-decays

Fermi [219] proposed the theory of the nuclear β-decay in analogy with quantum electrodynamics (QED), the theory of electromagnetic interactions in which a photon is created in the interaction between a charged particle and the electromagnetic field according to the principles of creation and annihilation of particles in quantum field theory(QFT) as shown in Figure 5.1.

The interaction Hamiltonian density $\mathcal{H}_{int}^{em}(x)$ for the electromagnetic interaction of a charged particle, say electron (with charge e) with the electromagnetic field $A^{\mu}(x)$ is written in QED as a scalar product of the four-component electromagnetic current $j_{\mu}^{em}(x)$ and the four-component electromagnetic field $A^{\mu}(x)$, that is,

$$\mathcal{H}_{int}^{em}(x) \;=\; e j_{\mu}(x) A^{\mu}(x), \tag{5.1}$$

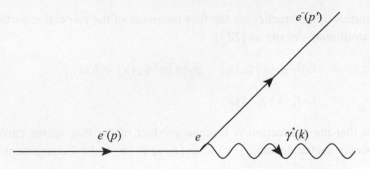

Figure 5.1 Interaction of a charged particle with the electromagnetic field.

where

$$j_\mu(x) = \bar{\psi}_e(x)\gamma_\mu\psi_e(x), \tag{5.2}$$

and $e = \sqrt{4\pi\alpha}$, with $\alpha(= \frac{1}{137})$ being the fine structure constant. The function $\psi_e(x)$ is the spin $\frac{1}{2}$ field of the electron which satisfies the Dirac equation, that is,

$$\left(i\gamma^\mu\frac{\partial}{\partial x^\mu} - m\right)\psi_e(x) = 0 \tag{5.3}$$

and is expressed in terms of the Dirac spinor, and the four momenta of the particle $(E, \vec{p})$, as:

$$\psi_e(x) = Nu_e(\vec{p})e^{-i(Et-\vec{p}\cdot\vec{x})}, \tag{5.4}$$

where N is the normalization factor and u_e is a four-component Dirac spinor given by:

$$u_e = \begin{pmatrix} 1 \\ \frac{\vec{\sigma}\cdot\vec{p}}{E+M} \end{pmatrix}\chi_e^s. \tag{5.5}$$

Here, $\vec{\sigma}$ is the Pauli spin operator and χ_e^s is a two-component spin wave function of the electron corresponding to the spin up $(S = \frac{1}{2}, S_z = +\frac{1}{2})$ and spin down $(S = \frac{1}{2}, S_z = -\frac{1}{2})$ states of the electron.

The interaction described by the Hamiltonian density given in Eq. (5.1) is diagrammatically represented by the Feynman diagram shown in Figure 5.1, where the direction of the arrow shows the direction of the electromagnetic current $j_\mu(x)$ of the electron with initial momentum p^μ and final momentum p'^μ interacting with the electromagnetic field $A^\mu(x)$ at space–time x creating a photon of momentum k^μ such that $p^\mu = p'^\mu + k^\mu$, with strength e.

In analogy with QED, Fermi assumed that, in the process of β-decay, a neutron in the nucleus is converted into a proton by creating a pair of electron and neutrino (established later as an antineutrino) from the physical vacuum through a weak interaction process:

$$n(p) \rightarrow p(p') + e^-(k') + \bar{\nu}_e(k), \tag{5.6}$$

where the quantities in the bracket are the four momenta of the respective particles. He wrote the β-decay Hamiltonian density as [223]:

$$\mathcal{H}_{\text{int}}^{\text{Fermi}}(x) = G\bar{\psi}_e\gamma_\mu(x)\psi_\nu(x) \times \bar{\psi}_p(x)\gamma^\mu\psi_n(x) + h.c.$$

$$= G\, j_\mu^e(x)\, j_h^\mu(x), \qquad (5.7)$$

which assumes that the interaction is a scalar product of the two vector currents, that is, a lepton 4-component vector current, $j_\mu^e(x) = \bar{\psi}_e(x)\gamma_\mu\psi_\nu(x)$ and a 4-component vector nucleon current $j_h^\mu(x) = \bar{\psi}_p(x)\gamma^\mu\psi_n(x)$. For this reason, it is called a current–current interaction of β-decay. It is also called a four Fermi point interaction because there are four spin $\frac{1}{2}$ fermion fields of proton, neutron, electron, and antineutrino described by the function $\psi_i(x)(i = n, p, e, \bar{\nu}_e)$ given in Eq. (5.4) interacting at the space–time point x. The strength of the interaction is described by the coupling constant G which has the dimension of $(\text{mass})^{-2}$ in contrast to the electromagnetic interaction, where strength of the coupling $e(= \sqrt{4\pi\alpha})$ is dimensionless. Diagrammatically, the interaction described by Eq. (5.7), is represented by the Feynman diagram shown in Figure 5.2. Here a nucleon current $j_\mu^h(x)$, in which a neutron of momentum p^μ is converted into a proton of momentum p'^μ, interacts with a lepton current $j_e^\nu(x)$ with strength G, in which an electron of momentum k'^ν and antineutrino of momentum k^ν is created (or equivalently, a neutrino of momentum k^ν is annihilated) such that:

$$p^\nu = p'^\nu + k^\nu + k'^\nu = p'^\nu + q^\nu \text{ with } q^\nu = k^\nu + k'^\nu, \qquad (5.8)$$

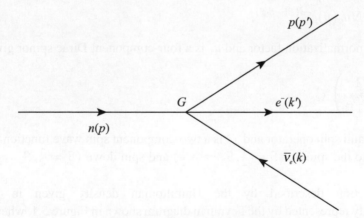

Figure 5.2 Four point Fermi interaction.

at the space–time point x. It should be noted that in the weak interaction process of β-decay, both currents, that is, the nucleon and the lepton currents carry charge of one unit unlike the electromagnetic current, that is, $j_\mu^{\text{em}}(x)$ which carries no charge. The weak currents in the β-decay are, therefore, known as charge changing (CC) weak currents.

In the case of nuclear β-decays, a neutron inside a nucleus at the initial state $|i\rangle$ decays into a proton defined by a final state $|f\rangle$, and a pair of electron and antineutrino is created in the interaction. The S matrix for the process, in the Fermi theory, is given by:

$$S = -i \langle f| \int dt H^{\text{Fermi}}_{\text{int}}(t) |i\rangle = i \langle f| \int d^4x \mathcal{L}^{\text{Fermi}}_{\text{int}}(x) |i\rangle$$

$$= iG \langle f| \int d^4x \left\{ \bar{\psi}_e(x)\gamma_\mu\psi_\nu(x)\bar{\psi}_p(x)\gamma^\mu\psi_n + h.c. \right\} |i\rangle, \tag{5.9}$$

where $\psi_n(x)$, $\psi_p(x)$, $\psi_e(x)$, and $\psi_\nu(x)$ are the wave functions of neutrons, protons, electrons, and neutrinos given by:

$$\psi_n(x) = \frac{1}{\sqrt{2E_n V}}\psi_n(\vec{x})e^{-iE_n t}, \tag{5.10}$$

$$\psi_p(x) = \frac{1}{\sqrt{2E_p V}}\psi_p(\vec{x})e^{-iE_p t}, \tag{5.11}$$

$$\psi_e(x) = \frac{1}{\sqrt{2E_e V}}\psi_e(\vec{x})e^{-iE_e t}, \tag{5.12}$$

$$\psi_\nu(x) = \frac{1}{\sqrt{2E_\nu V}}\psi_\nu(\vec{x})e^{-iE_\nu t}. \tag{5.13}$$

In nuclear β-decay, the neutrons and protons are bound in the nucleus and the electron moves in the Coulomb field of the final nucleus after being created in the process of β-decay. The electron–neutrino pair created in the process carries an energy in the region of a few MeV resulting in a very small, almost negligible recoil energy of the nucleus $\left(\simeq \frac{(|\vec{k}+\vec{k'}|)^2}{2AM} \approx \right.$ a few keV $\left. \right)$. Therefore, the final nucleus along with the initial nucleus can be considered to be at rest. Moreover, the nucleons bound in the nucleus move with their Fermi momentum corresponding to a kinetic energy of about 8–10 MeV and can be treated non-relativistically. This allows us to take the non-realtivistic limit of the hadronic current $j^\mu_h(x)$ using the following form of the wave functions $\psi_n(x)$ and $\psi_p(x)$, for the neutron and proton respectively, apart from a normalization factor, that is,

$$\psi_{p,n}(\vec{x}) = \begin{pmatrix} 1 \\ \frac{\vec{\sigma}\cdot\vec{p}}{E+M} \end{pmatrix} \phi_{p,n}(\vec{x}). \tag{5.14}$$

In the limit $\vec{p} \to 0$ (Appendix A),

$$\bar{\psi}_p(\vec{x})\gamma^\mu\psi_n(\vec{x}) \to \phi^\dagger_p(\vec{x})\mathbb{1}\phi_n(\vec{x}), \quad \text{when } \mu = 0$$
$$\to 0. \quad\quad\quad \text{when } \mu = 1,2,3 \tag{5.15}$$

Therefore, using Eqs. (5.9) and (5.15), we can write the S matrix as:

$$S = i2\pi\delta(\Delta_{fi} - E_e - E_\nu)G \int d\vec{x}\, e^{-i\vec{q}\cdot\vec{x}}\phi^\dagger_p(\vec{x})\mathbb{1}\phi_n(\vec{x})\bar{u}_e\gamma^0 u_\nu, \tag{5.16}$$

where $\Delta_{fi} = E^n_i - E^p_f$ is the available energy for the decay and $\vec{q} = \vec{k} + \vec{k'}$. Since $|\vec{q}|$ is of the order of a few MeV and $|\vec{x}|$ is of the order of the nuclear radius, that is, a few Fermi, the product $|\vec{q} \cdot \vec{x}|$ is of the order of 10^{-2} ($\because \ \ \hbar c = 197$ MeV-fm). Thus, the exponent in Eq. (5.16) can be expanded in powers of $\vec{q} \cdot \vec{x}$, that is,

$$e^{-i\vec{q}\cdot\vec{x}} = 1 + (-i\vec{q}\cdot\vec{x}) + \frac{(-i\vec{q}\cdot\vec{x})^2}{2!} + ... \tag{5.17}$$

Writing $\phi_p^{\dagger}\mathbb{1}\phi_n = \phi^{\dagger}\tau^{+}\phi$, where τ^{+} is the isospin raising operator, that is, $\tau^{+} = \frac{\tau_x + i\tau_y}{2}$, and $\phi = \begin{pmatrix} \phi_p \\ \phi_n \end{pmatrix}$, is a two-component isospinor, the S matrix is written as:

$$
\begin{aligned}
S &= \pm i2\pi\delta(\Delta_{fi} - E_e - E_v)G \\
&\times \int d\vec{x} \left(1 + (-i\vec{q}\cdot\vec{x}) + \frac{(-i\vec{q}\cdot\vec{x})^2}{2!} + .. \right) \phi^{\dagger}(x)\mathbb{1}\tau^{+}\phi(x)\overline{u}_e\gamma^0 u_v.
\end{aligned}
\tag{5.18}
$$

Taking the matrix element of the S matrix in the initial and final states and retaining only the first term in the expansion, we get

$$\langle f|S|i\rangle = S_{fi} = i2\pi\delta(\Lambda_{if} - E_e - E_v)G\mathcal{M}_{fi} \tag{5.19}$$

with

$$\mathcal{M}_{fi} = \overline{u}_e\gamma^0 u_v \mathcal{M}_F, \text{ where } \mathcal{M}_F = \langle f|\mathbb{1}\tau^{+}|i\rangle.$$

It can be seen that in the Fermi theory, the transition operator in the nuclear space is $\mathbb{1}\tau^{+}$, that is, it becomes independent of $\vec{x}$ and does not induce any change in the spin ($\vec{S}$) or the angular momentum ($\vec{L}$) implying no change in parity. Therefore, the operator describes the nuclear β-decays known as the allowed Fermi transitions corresponding to $\Delta J = 0$ with no change in the parity. It should be noted that transition operators which include the higher order terms in $\vec{q} \cdot \vec{x}$ in the expansion (Eq. (5.17)) are called forbidden transitions with decay rates smaller by many orders of magnitude due to $|\vec{q} \cdot \vec{x}| \approx 10^{-2}$.

The decay probability per unit time $d\Gamma$ for a process like:

$$^{14}O(0^{+}) \rightarrow^{14} N^{*}(0^{+}) + e^{+} + v_e, \tag{5.20}$$

that undergoes Fermi transition, is then derived using the aforementioned S matrix elements S_{fi} with the general expression of the decay probability per unit time $d\Gamma = \frac{|S_{fi}|^2}{T}$ integrated over the available phase space (see Section 5.4 for details), that is,

$$d\Gamma = \frac{1}{T}\overline{\sum}_i \sum_f |S_{fi}|^2 \frac{d\vec{k}}{(2\pi)^3(2E_e)} \frac{d\vec{k}'}{(2\pi)^3 2E_v} \tag{5.21}$$

$$= G^2 \frac{\delta(\Delta_{fi} - E_e - E_v)}{(2\pi)^5 2E_e 2E_v} \overline{\sum}_i \sum_f |\mathcal{M}_{fi}|^2 d\vec{k}\, d\vec{k}', \tag{5.22}$$

where

$$\mathcal{M}_{fi} = \overline{u}_e\gamma^0 u_v \mathcal{M}_F \text{ with } \mathcal{M}_F = \langle f|\mathbb{1}\tau^{+}|i\rangle, \tag{5.23}$$

leading to

$$\sum\sum|\mathcal{M}_{fi}|^2 = 4|\mathcal{M}_F|^2 E_e E_\nu(1 + \beta_e \cos\theta_{e\nu}), \quad \beta_e = \frac{|\vec{k}'|}{E_e}, \cos\theta_{e\nu} = \hat{k}' \cdot \hat{k} \ . \tag{5.24}$$

Performing the integration over all the neutrino variables assuming neutrinos to be massless, we find:

$$d\Gamma = \frac{G^2}{8\pi^4}|\mathcal{M}_F|^2 E_e|\vec{k}'|(\Delta_{fi} - E_e)^2 dE_e d\Omega_e(1 + \beta_e \cos\theta_{e\nu})$$

and integrating over the electron angle, we get

$$\frac{d\Gamma}{dE_e} = \frac{G^2}{2\pi^3}|\mathcal{M}_F|^2 E_e|\vec{k}'|(\Delta_{fi} - E_e)^2. \tag{5.25}$$

Integrating over the electron's energy and using the limit $m_e \to 0$, we obtain:

$$\Gamma = \frac{G^2\Delta_{fi}^5}{60\pi^3}|\mathcal{M}_F|^2. \tag{5.26}$$

In deriving the aforementioned equations, we have assumed the electron and the neutrino to be free particles and represented them by plane waves. While this is true for neutrinos, it is not true for electrons as they move in the Coulomb field of the nucleus. Therefore, instead of a plane wave, the electron wave function should be obtained by solving a relativistic Dirac equation for the continuum states of electrons moving in the Coulomb field of the nucleus $A(Z, N)$ undergoing β-decay. This is done by treating the nucleus $A(Z, N)$ as a point nucleus with charge Ze. It results in multiplying the decay probability/ time, $d\Gamma$ by a factor known as the Fermi function [224]:

$$F(Z, E_e) = |\psi_{e,Z}(x)|^2/|\psi_{e,Z}(0)|^2 \tag{5.27}$$

given by:

$$|\psi_{e,Z}(x)|^2/|\psi_{e,Z}(0)|^2 = 2(1+\gamma)(2|\vec{k}'||x|)^{-2(1-\gamma)}\exp(\pi Z\alpha E/|\vec{k}'|)\frac{|\Gamma(\gamma \pm iZ\alpha E_e/|\vec{k}'|)|^2}{|\Gamma(2\gamma+1)|^2}, \tag{5.28}$$

where $\gamma = (1 - Z^2\alpha^2)^{\frac{1}{2}}, \eta = Z\alpha$ and Γ is the gamma function. The aforementioned expression diverges at $x = 0$ and is therefore evaluated at $x = R$, where R is the nuclear radius. However, in a realistic situation, the nucleus is taken to have a charge distribution rather than a point nucleus and the Dirac equation is solved numerically. Following this method, the Coulomb correction is obtained and the Fermi function is tabulated to be used in the study of nuclear β-decays.

Performing integration over electron angles, we obtain the expression for the energy distribution of the electron as:

$$\frac{d\Gamma}{dE_e} = \frac{G^2}{2\pi^3}\zeta|\vec{k}'|E_e(\Delta_{fi} - E_e)^2, \quad \text{where} \ \zeta = |\mathcal{M}_F|^2. \tag{5.29}$$

Integrating over electron's energy, the total decay rate Γ is calculated as:

$$\Gamma = \frac{G^2}{2\pi^3}\zeta f_0(Z, \Delta_{fi}),\tag{5.30}$$

where

$$f_0(Z, \Delta_{fi}) = \int_{E_e=m_e}^{\Delta_{fi}} F(Z, E_e)|\vec{k}'|E_e(\Delta_{fi} - E_e)^2 dE_e,\tag{5.31}$$

where $f_0(Z, \Delta)$ is called the Fermi integral. The mean life $\tau = \frac{1}{\Gamma}$ is then defined as:

$$\tau = \frac{1}{\Gamma} = \frac{2\pi^3}{G^2}\frac{1}{\zeta f_0}\tag{5.32}$$

$$\text{and the half life} \quad t_{\frac{1}{2}} = \tau \ln 2.\tag{5.33}$$

The nuclear β-decays are generally characterized by their half life $t_{\frac{1}{2}}$ represented by a number called ft value. The value is defined as:

$$ft = \frac{2\pi^3 \ln 2}{G^2 \zeta}, \quad \text{where} \quad ft = f_0 \times t_{\frac{1}{2}}.\tag{5.34}$$

The ft values for some of the super-allowed Fermi decays are listed in Table 5.1. When the effect of the Coulomb interaction of the electron with the final nucleus is neglected, while the mass of the electron is retained, we obtain the following expression for the total decay rate Γ:

$$\Gamma = \frac{|\mathcal{M}_F|^2 G^2}{2\pi^3} f_0,$$

$$\text{where} \quad f_0 = \frac{d^3}{30}(\Delta_{fi}^2 + 4m_e^2) - \frac{1}{4}d\Delta_{fi}^2 m_e^2 + \frac{1}{4}m_e^4\Delta_{fi}\ln\left(\frac{\Delta_{fi}+d}{m_e}\right)\tag{5.35}$$

and $d = \sqrt{\Delta_{fi}^2 - m_e^2}$. We see that in the limit of $m_e \to 0$, the equation reduces to Eq. (5.26).

In general, a calculation of the nuclear matrix $\mathcal{M}_F$ is complicated involving knowledge of the nuclear wave functions. However, in the case of the allowed Fermi transitions corresponding to $0^+ \to 0^+$ transitions (β^+-decays), for example, $^{10}C \to^{10} B$, $^{14}O \to^{14} N^*$, $^{34}Cl \to^{34} S$, $^{42}Sc \to^{42} Ca$, $^{46}V \to^{46} Ti$, $^{54}Co \to^{54} Fe$, etc., the calculations of the matrix elements are simple. These transitions are between the two members of the isospin triplet states, that is, $|T = 1 \ T_3 = \pm 1\rangle$ and $|T = 1 \ T_3 = 0\rangle$ giving the matrix element

$$\langle T = 1 \ T_3 = 0|T_-|T = 1 \ T_3 = 1\rangle = \sqrt{2},$$

such that:

$$\mathcal{M}_F = \sqrt{2}.$$

Table 5.1 ft values from super-allowed Fermi decays [225, 226].

Parent	ft (s)
^{10}C	3078.0 ± 4.5
^{14}O	3071.4 ± 3.2
^{22}Mg	3077.9 ± 7.3
^{26m}Al	3072.9 ± 1.0
^{34}Cl	3070.7 ± 1.8
^{34}Ar	3065.6 ± 8.4
^{38m}K	3071.6 ± 2.0
^{38}Ca	3076.4 ± 7.2
^{42}Sc	3072.4 ± 2.3
^{46}V	3074.1 ± 2.0
^{50}Mn	3071.2 ± 2.1
^{54}Co	3069.8 ± 2.6
^{62}Ga	3071.5 ± 6.7
^{74}Rb	3076.0 ± 11.0

Using this value for $\mathcal{M}_F$, the experimental values of nuclear energy difference Δ_{fi}, and the ft values given in Table 5.1, the coupling constant G is evaluated. Using this information, we find $G = 1.166 \times 10^{-5}$ GeV^{-2}.

In addition to the allowed Fermi transitions, nuclear β-decays were also observed where the angular momentum J of the initial nucleus is changed by one unit, that is, $\Delta J = 1$ without any change in parity. There are many examples of such decays, some of which are listed in Table 5.2. The Hamiltonian density to describe such decays must have $\vec{J}(= \vec{L} + \vec{S})$ dependent operators in the non-relativistic limit to allow for the change of $\vec{J}$ of the initial nucleus without any change in parity, that is, the spin ($\vec{S}$) dependent operators. Therefore, Gamow and Teller [26] proposed the Hamiltonian density to be a scalar product of axial vector currents in the hadronic and leptonic sectors to explain such decays.

$$\mathcal{H}_{\text{int}}^{\text{GT}}(x) = G_A \bar{\psi}_p(x) \gamma^\mu \gamma_5 \psi_n(x) \bar{\psi}_e(x) \gamma_\mu \gamma_5 \psi_\nu(x) + h.c., \tag{5.36}$$

Performing a non-relativistic reduction of the nucleonic current $J_h^\mu(x) = \bar{\psi}_p(x) \gamma^\mu \gamma_5 \psi_n(x)$, using Eq. (5.4), we obtain:

$$\bar{\psi}_p(\vec{x}) \gamma^\mu \gamma_5 \psi_n(\vec{x}) \quad \rightarrow \quad 0 \qquad\qquad i = 0 \tag{5.37}$$

$$\rightarrow \quad \phi^\dagger(\vec{x}) \sigma^i \phi(\vec{x}) \quad i = 1, 2, 3$$

$$= \quad \phi^\dagger(\vec{x}) \vec{\sigma} \tau^+ \phi, \tag{5.38}$$

leading to the matrix element $\mathcal{M}_{fi}$ corresponding to the Gamow–Teller transitions as

$$\mathcal{M}_{fi}^{GT} = \langle f | \vec{\sigma} \tau^+ | i \rangle \bar{u}_e(k') \gamma^i \gamma_5 u_\nu(k) = \mathcal{M}_{GT} \bar{u}(k') \gamma^i \gamma_5 u(k), \tag{5.39}$$

Table 5.2 The columns list the parent nucleus in the transition, the initial, and final spins, and the transition type. F stands for a pure Fermi transition and GT stands for a Gamow–Teller transition [225, 226].

Parent	J_i	J_f	Type	Reference
^{6}He	0	1	GT/β^-	[227]
^{32}Ar	0	0	F/β^+	[228]
^{38m}K	0	0	F/β^+	[229]
^{60}Co	5	4	GT/β^-	[230]
^{67}Cu	3/2	5/2	GT/β^-	[231]
^{114}In	1	0	GT/β^-	[232]
^{14}O/^{10}C			F-GT/β^+	[233]
^{26}Al/^{30}P			F-GT/β^+	[234]

where

$$\mathcal{M}_{GT} = \langle f|\vec{\sigma}\tau^+|i\rangle. \tag{5.40}$$

The presence of the Pauli spin operator $\vec{\sigma}$ in $\mathcal{M}_{GT}$ describes the change in spin of the nucleus by one unit that does not affect the orbital angular momentum state implying no change in parity, that is, $|\Delta J| = 0, 1$, with no $0^+ \rightarrow 0^+$ transition. These transitions are called allowed Gamow–Teller (GT) β-transitions. A calculation of $\overline{\sum}\sum|\mathcal{M}_{fi}^{GT}|^2$, from Eq. (5.39) gives:

$$\overline{\sum}\sum|\mathcal{M}_{fi}^{GT}|^2 = 4|\mathcal{M}_{GT}|^2 E_e E_\nu (1 - \frac{1}{3}\beta_e \cos\theta),$$

where $\mathcal{M}_{GT}$ is given by Eq. (5.40) and $|\mathcal{M}_{GT}|^2 = \langle\sigma\rangle^2$.

After integrating over the neutrino variables, the expressions for the energy and angular distribution of the electrons emitted in Gamow–Teller β-decays are obtained as:

$$\frac{d\Gamma}{dE_e d\Omega_e} = \frac{G_A^2}{8\pi^4}|\mathcal{M}_{GT}|^2 |\vec{k}'|E_e(\Delta_{fi} - E_e)^2(1 - \frac{1}{3}\beta_e\cos\theta_{ev}),$$

$$\frac{d\Gamma}{dE_e} = \frac{G_A^2}{2\pi^3}|\mathcal{M}_{GT}|^2 |\vec{k}'|E_e(\Delta_{fi} - E_e)^2, \tag{5.41}$$

and the decay rate Γ is given by:

$$\Gamma = \frac{G_A^2}{2\pi^3}f_0\xi, \tag{5.42}$$

with $\xi = |\mathcal{M}_{GT}|^2$ and f_0 given in Eq. (5.31). A comparison with the experimental results on the decay rate gives a value of $G_A = 1.2723(23)G$ [235]. We see that the energy distribution of the electron and the total decay rate are the same in the case of allowed Fermi and allowed GT transitions. Moreover, the energy distribution of electrons in both types of transitions are consistent with the observation that $m_{\nu_e} = 0$ (Fig. 1.4).

However, many of the nuclear β-decays are of mixed type, that is, they have both the allowed Fermi and GT transitions. The most studied decays of such type are the β-decays of the neutron and ^3_1H, that is,

$$n \rightarrow p + e^- + \bar{\nu}_e, \tag{5.43}$$

$$^3_1H \rightarrow ^3_2He + e^- + \bar{\nu}_e, \tag{5.44}$$

which involve $(\frac{1}{2})^+ \rightarrow (\frac{1}{2})^+$ transitions resulting in $\Delta J = 0$, $|\Delta J| = 1$ with no change in parity corresponding to the allowed Fermi and GT transitions. Some examples of such types of the nuclear transitions are listed in Table 5.2. Before these decays are discussed we describe the general form of the β-decay Hamiltonian density in the next section.

5.2.2 General form of the Hamiltonian and the parity violation in β-decays

We have seen in the earlier sections that the $H^{\text{Fermi}}_{\text{int}}(x)$ for the Fermi theory is a scalar product of two vector currents in the nucleon and electron sector (VV) while the $H^{GT}_{\text{int}}(x)$ for the GT theory is a scalar product of the two axial vector currents (AA). It was proposed by Gamow and Teller [26], and Bethe and Bacher [27] that, in general, the interaction Hamiltonian could be a sum of a scalar product of all the bilinear covariants, that is, scalar (S), vector (V), pseudoscalar (P), axial vector (A), and tensor (T) which can be formed from the nucleon and lepton fields. Therefore, the general form of the Hamiltonian density for the β-decay interaction can be written as:

$$\mathcal{H}^{\beta\ decay}_{\text{int}}(x) = G \sum_{i=S,V,T,A,P} C_i \bar{\psi}_p(x) O^i \psi_n(x) \, \bar{\psi}_e(x) O_i \psi_\nu(x) + h.c., \tag{5.45}$$

where $O^i = 1$, γ^μ, $\gamma^\mu \gamma_5$, γ_5, $\sigma^{\mu\nu}$ for the scalar (S), vector (V), axial vector (A), pseudoscalar (P), and tensor (T) interactions, respectively and C_is are the coupling strengths of these interactions, which could be complex quantities. There would be, thus, five complex or ten real parameters needed to describe the β-decay interaction which should then be determined on the basis of various experimental results obtained in the study of β-decays of nuclei and other elementary particles. The transition operators for describing the nuclear β-decays corresponding to each bilinear term in the interaction Hamiltonian density given in Eq. (5.45), is obtained by taking the non-relativistic limit of the bilinear covariants for which the results are given in Table 5.3 (Appendix A). We observe from Table 5.3 that nuclear β-decays do not provide any information about pseudoscalar interactions while the scalar and vector interactions both describe the Fermi transitions. The axial vector and the tensor interactions describe the Gamow–Teller transitions. The knowledge of the nuclear structure (wave functions), therefore, does not give any information to distinguish between scalar and vector interactions in the case of Fermi transitions or between the axial vector and the tensor interactions in the case of GT transitions. It is only the structure of the leptonic part which would lead to different values for various observables in the nuclear β-decays due to scalar and vector interactions in the case of Fermi transitions and due to the axial vector and tensor interaction in the case of GT transitions.

Table 5.3 Nucleon operators for the decay $n \rightarrow p\, e^- \, \bar{\nu}_e$ in the relativistic and non-relativistic limits, where $\sigma_{\mu\nu} = \frac{i}{2}[\gamma_\mu, \gamma_\nu]$, and ψs are the nuclear wave functions.

Coupling	Relativistic limit	Non-relativistic limit
Pseudoscalar (P)	$\bar{\psi}(x)\gamma_5\psi(x)$	0
Scalar (S)	$\bar{\psi}(x)\psi(x)$	$\mathbb{1}\,\tau^+$
Vector (V)	$\bar{\psi}(x)\gamma_\mu\psi(x)$	$\begin{cases} \mathbb{1}; & \text{for } \mu = 0, \\ 0; & \text{for } \mu = i \end{cases}$
Axial vector (A)	$\bar{\psi}(x)\gamma_\mu\gamma_5\psi(x)$	$\begin{cases} 0; & \text{for } \mu = 0, \\ \sigma_i\tau^+; & \text{for } \mu = i \end{cases}$
Tensor (T)	$\bar{\psi}(x)\sigma_{\mu\nu}\psi(x)$	$\begin{cases} 0; & \text{for } \mu = 0,\ \nu = 0, \\ \epsilon_{ijk}\sigma^k & \text{for } \mu = i,\ \nu = j \end{cases}$

The various experiments performed to determine the coupling constants C_V, C_A, C_S, and C_T (Eq. (5.45)) often led to conflicting conclusions in the beginning. The early experiments seemed to prefer the S, T combination, even though there were many experiments favoring the V, A combination.

It took more than 20 years before a clear picture started to emerge for a phenomenological theory of β-decays and weak interactions. The real breakthrough came in 1957 after the discovery of the parity violation in weak processes [33], which was implied by the resolution of the $\tau - \theta$ puzzle encountered in the study of weak decays of the K mesons in the two and three pion modes. The solution of the puzzle indicated that parity may not be conserved in weak interactions.

There were two particles τ and θ discovered experimentally with almost the same mass and lifetime decaying respectively into two and three pions in the S-wave state [39]. This implied parity violation if τ and θ were the same particles, because two pions and three pions in an S-wave will have opposite parities due to the pseudoscalar nature of pions. In view of this suggestion, Lee and Yang [31, 32] analyzed many weak processes and concluded that there was no experimental evidence to contradict the assumption of parity violation in weak interactions. They suggested specific experiments to test this possibility by measuring specific correlation observables in β-decays of nuclei and other weak decays of elementary particles which were parity violating (i.e., changing sign under the transformation $\vec{r} \rightarrow -\vec{r}$) like $\vec{\sigma}_N \cdot \hat{p}_e$ or $\vec{\sigma}_e \cdot \hat{p}_e$, where $\hat{p}_e$ is the unit vector along the electron momentum and $\vec{\sigma}_N$ and $\vec{\sigma}_e$ are, respectively, the nucleon and electron spin operators. The first experiment to test the parity violation in β-decay was performed by Wu et al. [33] with a polarized cobalt nucleus, where a large asymmetry of β-electrons with respect to the spin direction of the polarized ^{60}Co was observed. In this experiment, a ^{60}Co ($J = 5$) nucleus undergoes β-decay to ^{60}Ni*($J = 4$) along with an electron and an antineutrino (see Figure 5.3(a)):

$$^{60}\text{Co}(J = 5) \longrightarrow {}^{60}\text{Ni}^*(J = 4) + e^-(J = \tfrac{1}{2}) + \bar{\nu}_e(J = \tfrac{1}{2}). \tag{5.46}$$

An observation of the nonzero parity violating spin correlation like $\langle \vec{J}_N \cdot \vec{p}_e \rangle$, where $\vec{J}_N$ is the total angular momentum of the nucleus and $\vec{p}_e$ is the momentum of the emitted electron in

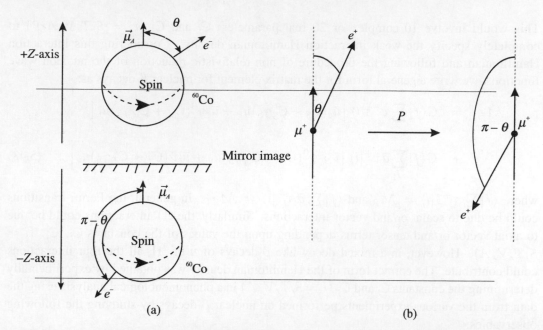

Figure 5.3 (a) Effect of parity transformation on the process $^{60}\text{Co} \longrightarrow ^{60}\text{Ni}^* + e^- + \bar{\nu}_e$. The electrons coming at $\pi - \theta$ direction with respect to the nuclear spin were not observed leading to front–back asymmetry; hence, indication of parity violation in weak interactions. (b) Parity transformation of a muon decaying into a positron.

β-decay demonstrated that parity is violated in weak interactions. The direction of the nuclear angular momentum was determined by putting the nucleus in a magnetic field which aligns the angular momentum along the direction of the magnetic field. The direction of the magnetic field was reversed to change the alignment of the angular momentum. The number of electrons were measured in both cases and a nonzero asymmetry was measured.

Around the same time, the observation of a large asymmetry in the emission of positron in the decay of polarized muon confirmed the phenomenon of parity violation in weak processes other than nuclear β-decays [236, 237] as illustrated in Figure 5.3(b). Later on, many experiments on the observable $\langle \vec{\sigma}_e \cdot \hat{p}_e \rangle$, that is, the longitudinal polarization of the electrons in the nuclear β⁻-decays and the positrons in the nuclear β⁺-decays were conducted and the violation of parity in weak interactions was firmly established [34, 35]. The intervening years between 1934 and 1957 were filled with experimental and theoretical activities in the study of weak interaction processes. All the activities made valuable contributions toward the understanding of the physics of weak interactions.

In the presence of parity violation, the interaction Hamiltonian $\mathcal{H}_{\text{int}}^{\beta \, \text{decay}}(x)$ would be a mixture of scalar and pseudoscalar products of the two weak currents, that is, the leptonic and the hadronic currents and can be written, without loss of generality, in the form

$$\mathcal{H}_{\text{int}}^{\beta \, \text{decay}}(x) = \sum_{i=S,T,V,A,P} G \bar{\psi}_p(x) O^i \psi_n(x) \bar{\psi}_e(x) O_i (C_i + C_i' \gamma_5) \psi_\nu(x) + \text{h.c.} \qquad (5.47)$$

This would involve 10 complex or 20 real parameters C_i and C_i' $(i = S, T, V, A, P)$ to completely specify the weak interaction Hamiltonian density. Considering this interaction Hamiltonian and following the procedure of non-relativistic reduction of the nucleon wave functions, we write a general form for the matrix element for nuclear β-decays as:

$$
\mathcal{M}_{fi} = G\langle f| \sum_i \tau^+ \mathbb{1} |i\rangle \left[\bar{u}_e (C_S + C_S' \gamma_5) u_\nu + \bar{u}_e \gamma^0 (C_V + C_V' \gamma_5) u_\nu \right]
$$

$$
+ \; G\langle f| \sum_i \vec{\sigma} \tau^+ |i\rangle \left[\bar{u}_e \vec{\sigma} \gamma^0 (C_A + C_A' \gamma_5) u_\nu + \bar{u}_e \vec{\sigma} (C_T + C_T' \gamma_5) u_\nu \right], \quad (5.48)
$$

where $\langle f| \sum_i \tau_i^+ \mathbb{1} |i\rangle = \mathcal{M}_F$ and $\langle f| \sum_i \vec{\sigma}_i \tau_i^+ |i\rangle = \mathcal{M}_{GT}$. In general, the Fermi transitions could be due to scalar or/and vector interactions. Similarly, the GT interactions could be due to axial vector or/and tensor terms depending upon the values of the constants C_i, C_i', $(i = S, T, V, A)$. However, in a mixed decay like β-decays of n or ^{3}H, all the four interactions could contribute. The correct form of the Hamiltonian density was obtained by experimentally determining the constants C_i and C_i' $(i = S, T, V, A)$ in a phenomenological analysis using the data from the various experiments performed on nuclear β-decays by studying the following observables:

- The energy and angular distributions of electrons (positrons) in the β-decay of unpolarized nuclei.

- The longitudinal polarization of electrons (positrons) from β-decay of the unpolarized nuclei.

- Helicity of the neutrino.

- The spin–momentum correlation like $\langle \vec{J} \cdot \vec{p}_e \rangle$ and $\langle \vec{J} \cdot \vec{p}_\nu \rangle$, where $\vec{J}$ is the spin of the polarized nucleus, $\vec{p}_e$ is the momentum of the electron (positron), and $\vec{p}_\nu$ is the momentum of the antineutrino (neutrino) in the $\beta^- (\beta^+)$-decays of polarized nuclei.

In the following sections, we discuss briefly the conclusions drawn from the study of these observables from various experiments in nuclear β-decays.

5.2.3   The energy and angular distribution of electrons (positrons) for the β^+ (β^-)-decay of unpolarized nuclei

A general calculation of the electron (positron) spectrum following the methods outlined in Section 5.2.1 using the matrix elements given in Eq. (5.48) yields the following result for the energy and angular distributions of electrons (positrons) for the β-decay of unpolarized nuclei $\beta^- (\beta^+)$ decay as [238, 239, 240, 241]:

$$
\frac{d\Gamma}{dE_e d\Omega_e} = G^2 F(Z, E_e) \frac{p_e E_e (\Delta_{fi} - E_e)^2}{8\pi^4} \cdot K \left(1 + a\beta_e \cos\theta_{e\nu} \pm b \frac{m_e}{E_e} \right), \quad (5.49)
$$

where $\quad K = |\mathcal{M}_F|^2 (D_{SS} + D_{VV}) + |\mathcal{M}_{GT}|^2 (D_{TT} + D_{AA}),$

$$Ka = |\mathcal{M}_F|^2(-D_{SS} + D_{VV}) + \frac{1}{3}|\mathcal{M}_{GT}|^2(D_{TT} - D_{AA}),$$

$$Kb = |\mathcal{M}_F|^2 2Re(D_{SV}) + |\mathcal{M}_{GT}|^2 2Re(D_{TA}),$$

and $\qquad D_{ij} = C_i C_j^* + C_i' C_j'^* \qquad$ for $i, j = S, V, T, A.$ $\qquad$ (5.50)

Integrating over the solid angle, the energy distribution is obtained as:

$$\frac{d\Gamma}{dE_e} = G^2 F(Z, E_e) \frac{p_e E_e (\Delta - E_e)^2}{2\pi^3} K \left(1 \pm b \frac{m_e}{E_e}\right), \qquad (5.51)$$

and integrating over the electron's energy, the total decay rate may be obtained as:

$$\Gamma = \int \frac{d\Gamma}{dE_e} dE_e.$$

A comparison with the experimental results on energy (Figure 5.4) and the angular distribution leads to the following conclusions:

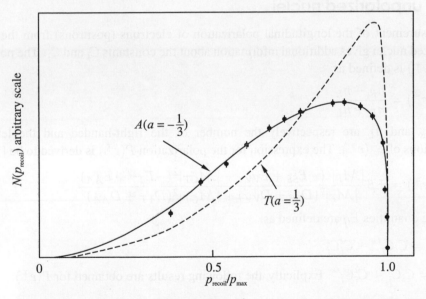

Figure 5.4 Recoil momentum spectrum for the decay $^6\text{He} \rightarrow {}^6\text{Li} + e^- + \bar{\nu}_e$ together with the predictions for pure A and pure T couplings [242].

1. The energy spectrum can be fitted with a value of $b = -0.02 \pm 0.09$ for the Fermi transition and $b = 0.0008 \pm 0.0020$ for the Gamow–Teller transition [243, 244], showing that there is no $\frac{1}{E}$ dependency in the energy spectrum. This implies that $D_{SV} = 0$ for the Fermi transition and $D_{TA} = 0$ for the GT transitions. Assuming the coupling constants C_is and C_i's to be real (due to T-invariance), this means that the Fermi transitions are either S or V type while GT transitions are either A or T type and there is no interference between them.

2. The endpoint energy of the electron in the energy spectrum shown in Figure 1.4 is $E_{max} = 20$ keV. Therefore, it is concluded that mass of the electron neutrino, that is, m_{ν_e} is consistent with zero. In fact, in modern times, the very high precision data obtained from the β-decay of ^{3}H is being used to determine the mass of m_{ν_e} with the results that $m_{\nu_e} < 0.5$ eV (see Section 1.4).

3. The experimental results on the angular distribution of $e^-(e^+)$ in the $\beta^-(\beta^+)$ decay of nuclei are consistent with the value $a = 0.97 \pm 0.14$ [75] for 4 Fermi transitions and -0.3343 ± 0.003 [227] for GT transitions. These values are close to 1 and $-\frac{1}{3}$ predicted for the vector interactions in Fermi transitions and the axial vector interactions in GT transitions, respectively (Fig. 5.4).

Therefore, the study of the energy spectrum of electrons (positrons) in nuclear $\beta^-(\beta^+)$- decays established that Fermi transitions are due to vector interactions while GT transitions are due to axial vector interactions.

5.2.4   The longitudinal polarization of $e^-(e^+)$ from $\beta^-(\beta^+)$-decays of unpolarized nuclei

The measurement of the longitudinal polarization of electrons (positrons) from the decay of unpolarized nuclei gives additional information about the constants C_i and C_i'. The polarization P of $e^-(e^+)$ is defined as:

$$P(e^\mp) = \frac{n_R - n_L}{n_R + n_L},\tag{5.52}$$

where n_R and n_L are respectively the number of the right-handed and the left-handed polarizations of $e^-(e^+)$. The expression for the polarization $P(e^\mp)$ is derived to be [238]:

$$P(e^\mp) = \pm\beta \frac{|\mathcal{M}_F|^2(-E_{SS} + E_{VV}) + |\mathcal{M}_{GT}|^2(-E_{TT} + E_{AA})}{|\mathcal{M}_F|^2(D_{SS} + D_{VV}) + |\mathcal{M}_{GT}|^2(D_{TT} + D_{AA})},\tag{5.53}$$

where the quantities E_{ij} are defined as:

$$E_{ij} = C_i C_j'^* + C_i' C_j^*\tag{5.54}$$

and $D_{ij} = C_i C_j^* + C_i' C_j'^*$. Explicitly, the following results are obtained for $P(e^\mp)$:

$$
\begin{aligned}
P(e^\mp) &= \mp\beta\frac{E_{SS}}{D_{SS}} = \frac{C_S C_S^{*\prime} + C_S' C_S^*}{C_S C_S^* + C_S' C_S'^*} \quad \text{for scalar interaction}\\[2mm]
&= \pm\beta\frac{E_{VV}}{D_{VV}} = \frac{C_V C_V^{*\prime} + C_V' C_V^*}{C_V C_V^* + C_V' C_V'^*} \quad \text{for vector interaction}\\[2mm]
&= \mp\beta\frac{E_{TT}}{D_{TT}} = \frac{C_T C_T^{*\prime} + C_T' C_T^*}{C_T C_T^* + C_T' C_T'^*} \quad \text{for tensor interaction}\\[2mm]
&= \pm\beta\frac{E_{AA}}{D_{AA}} = \frac{C_A C_A^{*\prime} + C_A' C_A^*}{C_A C_A^* + C_A' C_A'^*} \quad \text{for axial vector interaction.}
\end{aligned}\tag{5.55}
$$

A comparison with the experimental results from the measurements of the polarization of $e^-(e^+)$ shows that $P(e^-) = -(0.99 \pm 0.009)\beta$ for the GT decay of ^{32}P by Brosi et al. [245] which is consistent with $P(e^\mp) = \mp\beta$[246, 247].

This implies that in the case of pure Fermi transitions for which $P(e^-) = -1$, one has

$$\frac{E_{SS}}{D_{SS}} = 1 \text{ i.e., } |C_S - C_{S'}|^2 = 0 \implies C_S = C'_S \text{ for scalar interaction,} \tag{5.56}$$

$$\frac{E_{VV}}{D_{VV}} = -1 \text{ i.e., } |C_V + C_{V'}|^2 = 0 \implies C_V = -C'_V \text{ for vector interaction,} \tag{5.57}$$

and in the case of pure GT transition,

$$\frac{E_{TT}}{D_{TT}} = 1 \text{ i.e., } |C_T - C_{T'}|^2 = 0 \implies C_T = C'_T \text{ for the tensor interaction,} \tag{5.58}$$

$$\frac{E_{AA}}{D_{AA}} = -1 \text{ i.e., } |C_A + C_{A'}|^2 = 0 \implies C_A = -C'_A \text{ for axial vector interaction.} \tag{5.59}$$

From these results, it may be inferred that if the Fermi interaction is a vector interaction then $C_V = -C'_V$; and if the GT transition is an axial vector interaction, then $C_A = -C'_A$ with the consequence that for V and A interactions, the $\mathcal{H}_{\text{int}}(x)$ in Eq. (5.47) can be written as:

$$\mathcal{H}_{\text{int}}(x) = \frac{G_F}{\sqrt{2}} \overline{\psi}_p (C_V \gamma^\mu + C_A \gamma^\mu \gamma^5) \psi_n \overline{\psi}_e \gamma_\mu (1 - \gamma_5) \psi_\nu, \quad G_F = \sqrt{2}G. \tag{5.60}$$

Therefore, the leptonic part in $\mathcal{H}_{\text{int}}(x)$ is written as:

$$l_\mu = \overline{\psi}_e \gamma_\mu (1 - \gamma_5) \psi_\nu \tag{5.61}$$

implying that ψ_ν can be replaced by $(1 - \gamma_5)\psi_\nu$ in writing the weak interaction Hamiltonian density. Therefore, the Dirac spinor for neutrino u_ν enters as $(1 - \gamma_5)u_\nu$, that is, ν is left-handed. On the other hand, for the S and T interactions, it enters as $(1 + \gamma_5)u_\nu$. This leads to the important conclusion that the helicity of the neutrino (antineutrino) is −1(+1) as discussed in Chapter 2. This was experimentally confirmed by Goldhaber et al. [153] in nuclear β-decays and by Garwin et al. [236] in the case of pion decays.

5.2.5 Helicity of the neutrino

The discovery of parity violation and the experimental evidence that neutrino mass is zero, that is, $m_{\nu_e} = 0$ revived the two-component theory of neutrino. The theory implied that the neutrino (antineutrinos) are left(right)-handed particles or vice versa, as can be seen from the formulation of the two-component theory of neutrino (discussed in Chapter 2). The evidence that neutrinos (antineutrinos) is left (right)-handed was available from the observations made on the longitudinal polarization of the electrons and other spin momentum correlation measurements of the emitted electrons and positrons in many weak decays of nuclei and elementary particles. The helicity of ν_e was directly measured in an excellent experiment

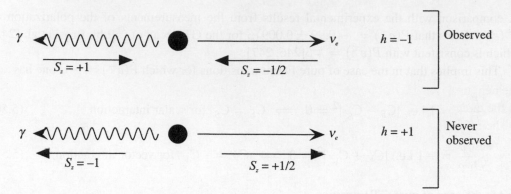

Figure 5.5 Helicity of neutrino indirectly determined in Goldhaber experiment.

performed by Goldhaber et al. [153]. They measured the polarization of photons in an electron capture experiment using ^{152}Eu nuclei leading to ^{152}Sm* and its radiative decay to ^{152}Sm.

$$^{152}\text{Eu} + e^- \longrightarrow \nu_e + ^{152}\text{Sm}^*,$$
$$^{152}\text{Sm}^* \longrightarrow ^{152}\text{Sm} + \gamma,$$

resulting in the final reaction:

$$^{152}\text{Eu} + e^- \longrightarrow ^{152}\text{Sm}^* + \nu_e + \gamma. \tag{5.62}$$

The electron from the K-shell is captured at rest from the ^{152}Eu nucleus and the momentum conservation requires that $\vec{p}_{Sm^*} = -\vec{p}_\nu$. The emission of photons from ^{152}Sm* to the ground state of Sm in the forward direction stops the ^{152}Sm implying that $\vec{p}_\gamma = -\vec{p}_{\nu_e}$ in Eq. (5.62), that is, neutrinos and photons are emitted in opposite direction to each other. Hence, if we choose the neutrino direction to be along the Z-axis, then the photon is emitted in the negative Z-direction. The spin considerations (the angular momentum J of ^{152}Eu (J_i) and ^{152}Sm (J_f) in the initial and the final states being zero, that is, $J_i = J_f = 0$) of initial and final particles in Eq. (5.62) show that the polarization state of the emitted photons determine the helicity of the neutrino. The electron in the K-shell has $J = \frac{1}{2}$ ($L = 0, S = \frac{1}{2}$) while ^{152}Eu has $J = 0$ leading to ^{152}Sm* which has $J = 1$ and decays into ^{152}Sm with $J = 0$. Therefore, the total angular momentum in the initial state is $\frac{1}{2}$ due to the electron spin and can have $J_Z = \pm\frac{1}{2}$. The photon is transverse so it can have only $S_Z = \pm 1$. Therefore, if $J_Z^e = +\frac{1}{2}$, it can come from $J_Z^\gamma = +1$, $S_Z^\nu = -\frac{1}{2}$, and if $J_Z^e = -\frac{1}{2}$, it can come only from $J_Z = -1$, $S_Z^\nu = +\frac{1}{2}$ (Figure 5.5). Hence, if the neutrino has negative helicity, that is, $S_Z^\nu = -\frac{1}{2}$, then only $S_Z^\gamma = +1$ is possible. Thus, the polarization state of the photon determines the helicity of the neutrino. If the photon has left-handed polarization, the helicity of the neutrino is -1 and if the photon has right-handed polarization, the neutrino has helicity $+1$. In an elegant experiment, Goldhaber et al. [153] measured the circular polarization of photons emitted by Compton scattering in a magnetized iron block and confirmed that the emitted photons were left-handed, thus determining the helicity of ν_e conclusively to be -1.

In later experiments done on the muon capture process using ^{12}C, that is,

$$\mu^- +^{12}\mathrm{C}(0) \to^{12} \mathrm{B}(1^+) + \nu_\mu, \tag{5.63}$$

the helicity of ν_μ was determined by measuring the recoil polarization of the final nucleus ^{12}B in the direction of its recoil momentum. A typical measurement by Roesch et al. [248] found the helicity of ν_μ to be -1.00 ± 0.11 consistent with negative helicity as predicted theoretically.

Another experiment was performed by Garwin et al. [236], who observed pions decay at rest. Consider the charged pions $\pi^\pm$ decaying to $\mu^\pm + \nu_\mu(\bar{\nu}_\mu)$. The momentum of muons

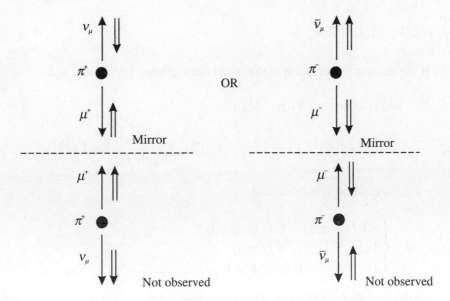

Figure 5.6 Pion decay at rest.

will be opposite in direction to the momentum of neutrinos. To conserve the total angular momentum, their spins should also be oppositely directed (Figure 5.6). We can define the helicity as follows.

$$h = \frac{\vec{\sigma} \cdot \vec{p}}{|\vec{p}|} = +1 \text{ for } \vec{\sigma} \uparrow\uparrow \vec{p} \text{ and } -1 \text{ for } \vec{\sigma} \uparrow\downarrow \vec{p}.$$

The top part of Figure 5.6 would therefore imply $h = -1(+1)$ for both μ^+ and $\nu_\mu(\mu^-$ and $\bar{\nu}_\mu)$. The mirror reflection of this process would be equivalent to parity inversion. In this case, $h = +1(-1)$ for both μ^+ and $\nu_\mu(\mu^-$ and $\bar{\nu}_\mu)$, that is, if parity is conserved, both helicities should be equally probable. However, it was observed from π^+ decay that the polarization of μ^+ is negative and that for π^- decay, the polarization of μ^- is positive. Therefore, it was concluded that parity is not conserved in weak interactions. This experiment also provided the evidence that the helicity of ν_μ $(\bar{\nu}_\mu)$ coming from $\pi^+(\pi^-)$ decay is negative (positive), that is, $h = -1(+1)$, corresponding to upper diagrams in Figure (5.6).

5.2.6 Spin–momentum correlations in the β-decay of polarized nuclei

In the case of nuclear β-decays of polarized nuclei, the direction of the polarization can be used as a reference direction to measure the angular distribution of electrons (positrons) as was done in the famous experiment of Wu et al. [33]. The asymmetry of the electron (positron) angular distribution is sensitive to the relative sign between C_i and C_i' just like the longitudinal polarization of electrons (positrons) in the case of β-decays of unpolarized nuclei. The angular distribution of electrons from the β-decays of a nucleus corresponding to the transition $J_i \rightarrow J_f$ is given by [238, 239, 240, 241]:

$$\frac{d\Gamma}{d\Omega} \propto \xi(1 + A\langle \hat{J}_i \rangle \cdot \hat{p}_e), \tag{5.64}$$

where $\langle \hat{J}_i \rangle$ is the nuclear polarization of the initial nuclear state with spin $\vec{J}$ and

$$\xi = |\mathcal{M}_F|^2 D_{VV} + |\mathcal{M}_{GT}|^2 D_{AA},$$

$$A\xi = \pm \Lambda_{J_i J_f} |\mathcal{M}_{GT}|^2 E_{AA} - 2\delta_{J_i J_f} |\mathcal{M}_F||\mathcal{M}_{GT}| \sqrt{\frac{J_i}{J_i + 1}} Re(E_{VA}), \tag{5.65}$$

where $+(-)$ is for electron (positron) and

$$\Lambda_{J_i J_f} = \begin{cases} -J_i/(J_i + 1) & \text{if } J_f = J_i + 1, \\ 1/(J_i + 1) & \text{if } J_f = J_i \\ 1 & \text{if } J_f = J_i - 1. \end{cases} \tag{5.66}$$

In the case of $^{60}\text{Co} \rightarrow \, ^{60}\text{Ni} \, e^- \bar{\nu}_e$ decay, which is a GT transition corresponding to $J_i = 5 \rightarrow J_f = 4$, $\Lambda_{J_i J_f} = 1$ giving $A = \frac{E_{AA}}{D_{AA}}$ which was found to be -1 in the case of Wu's experiment. From Eq. (5.59), this implies that $C_A = -C_A'$.

5.2.7 Mixed β-transitions and sign of $\frac{C_A}{C_V}$

We see from Eq. (5.65) that a mixed transition having Fermi as well as Gamow–Teller transitions will give information about the interference term $E_{VA} = C_V C_A^* + C_V^* C_A$ from an observation of the spin correlation A. For simplicity, let us consider the case of the polarized free neutron decays which correspond to the β-decay of $n(\frac{1}{2}^+) \rightarrow p(\frac{1}{2}^+) + e^- + \bar{\nu}_e$. In this case, the transition rate for the polarized neutron is expressed as [226, 241]:

$$d\Gamma(\hat{\sigma}_n, \vec{p}_e, \vec{p}_\nu) \propto F(E_e) d\Omega_e d\Omega_\nu \left[1 + a\frac{\vec{p}_e \cdot \vec{p}_\nu}{E_e E_\nu} + b\frac{m_e}{E_e} \right.$$

$$+ \left. \langle \vec{\sigma}_n \rangle \cdot \left(A\frac{\vec{p}_e}{E_e} + B\frac{\vec{p}_\nu}{E_\nu} + D\frac{\vec{p}_e \times \vec{p}_\nu}{E_e E_\nu} + R\frac{\vec{\sigma}_e \times \vec{p}_e}{E_e} + ... \right) \right], \tag{5.67}$$

Table 5.4 Values of the coefficients A and B appearing in Eq. (5.67).

Coefficient	Value	Year	$\langle m_e / E_e \rangle$	Reference
A	$-0.1146(19)$	1986	0.581	[249]
	$-0.1160(9)(12)$	1997	0.582	[250]
	$-0.1135(14)$	1997	0.558	[251]
	$-0.11926(31)(42)$	2013	0.559	[252]
	$-0.12015(34)(63)$	2018	0.586	[253]
	$-0.11869(99)$		0.569	Average (S=2.6)
B	$0.9894(83)$	1995	0.554	[254]
	$0.9801(46)$	1998	0.594	[255]
	$0.9670(120)$	2005	0.600	[256]
	$0.9802(50)$	2007	0.598	[257]
	$0.9805(30)$		0.591	Average

where

$$a = \frac{1 - |\lambda|^2}{1 + 3|\lambda|^2}, \qquad A = -2\frac{|\lambda|^2 + \mathrm{Re}\lambda}{1 + 3|\lambda|^2},$$

$$B = 2\frac{|\lambda|^2 - \mathrm{Re}\lambda}{1 + 3|\lambda|^2}, \qquad D = 2\frac{\mathrm{Im}\lambda}{1 + 3|\lambda|^2},$$

with $\lambda = \frac{C_A}{C_V}$.

The term D represents the measure of T-violation. Assuming T-invariance, we find that $\lambda = \frac{C_A}{C_V}$ is real and can be obtained by a measurement of the spin correlation of the polarized neutron with momentum of electron ($\vec{p}_e$) or neutrino ($\vec{p}_\nu$), through the measurement of coefficients A or B. If both can be measured, we determine $\mathrm{Re}\lambda$ directly because $A + B = -\frac{4\mathrm{Re}\lambda}{1 + 3\lambda^2}$. The values of A and B are given in Table 5.4.

It should be noted that in the case of neutron decay, the decay rate Γ is given by

$$ft = \frac{2\pi^3 \ln 2}{G^2 C_V^2 (1 + 3\lambda^2)}. \tag{5.68}$$

Therefore, $|\lambda|^2$ is determined from the ft value as well as from the coefficient $a(= \frac{1-\lambda^2}{1+3\lambda^2})$ from the angular distribution. Information regarding the spin–momentum correlation measurements have come from the extensive experiments that have been done over the last 20 years at ILL Grenoble [226] using the electron spectrometer PERKEO and also in other experiments. The current value of λ determined from these experiments is given by [117]

$$\lambda = \frac{C_A}{C_V} = -1.2732 \pm 0.0023. \tag{5.69}$$

5.3   Two-component Neutrino and the $V - A$ Theory

The study of various observables in the nuclear β-decay like the decay rates, energy and angular distributions of electrons (positrons), their polarizations and spin–momentum correlations with the polarization of the initial nucleus established the following:

1. Neutrinos are almost massless spin $\frac{1}{2}$ particles.

2. The parity is maximally violated in β-decays.

3. Neutrinos (antineutrinos) have negative (positive) helicity.

4. Fermi transitions are mainly due to vector interactions while Gamow–Teller transitions are mainly due to axial vector transitions.

5. The relative strength of the axial vector and the vector interactions is ≈ 1.26, that is, $\left|\frac{C_A}{C_V}\right| = 1.26$.

6. The relative sign between the vector and the axial vector interactions is negative, that is, $\frac{C_A}{C_V} = -1.26$.

These observations imply that

i) neutrinos, which can be considered to be massless, can then be described by a two-component theory of massless spin $\frac{1}{2}$ particles (Chapter 3).

ii) only the left-handed neutrinos (and the right-handed antineutrinos) couple through lepton currents with vector and axial vector current of hadrons in the β -decay processes through a combination $V - A$.

In view of these observations, the two-component theory of massless spin $\frac{1}{2}$ fermions proposed earlier by Weyl was revived to formulate a theory for weak interactions by Marshak and Sudarshan [41], Feynman and Gell-Mann [42], and Sakurai [43] almost simultaneously, which is known as the $V - A$ theory of weak interactions. This theory is explained here.

We write the wave function ψ_ν of neutrinos as:

$$\psi_\nu = \psi_\nu^L + \psi_\nu^R, \tag{5.70}$$

where $\psi_\nu^L = \frac{1-\gamma_5}{2}\psi_\nu$ and $\psi_\nu^R = \frac{1+\gamma_5}{2}\psi_\nu$ are respectively the left-handed and the right-handed components of the neutrino wave functions. Similarly, we write the electron wave function $\psi_e = \psi_e^L + \psi_e^R$, where ψ_e^L is the left-handed electron and ψ_e^R is the right-handed electron. It is straightforward to see that ψ_ν^L couples to ψ_e^L only through vector (V) and axial vector (A) interactions and not through scalar (S), or tensor (T) interactions, that is,

$$\begin{aligned}
C_S \bar{\psi}_e^L \psi_\nu^L &= 0, & \text{(Scalar)} \\
C_T \bar{\psi}_e^L \sigma_{\mu\nu} \psi_\nu^L &= 0, & \text{(Tensor)} \\
C_V \bar{\psi}_e^L \gamma_\mu \psi_\nu^L &= C_V \bar{\psi}_e \gamma_\mu \frac{(1-\gamma_5)}{2} \psi_\nu, & \text{(Vector)} \\
C_A \bar{\psi}_e^L \gamma_\mu \gamma_5 \psi_\nu^L &= C_A \bar{\psi}_e \gamma_\mu \gamma_5 \frac{(1-\gamma_5)}{2} \psi_\nu. & \text{(Axial vector)}
\end{aligned} \tag{5.71}$$

Therefore, assuming that only the left-handed components of the lepton fields and the hadron fields couple in the theory of β-decay, we arrive at the $V - A$ theory of weak interactions.

5.4 Weak Interaction of Muon

Muon was discovered in cosmic rays in 1937 [89]; its properties were determined from studies of its weak interaction through its decays, scattering, and capture from nucleons and nuclei. It was established to be a heavy lepton of mass $m_\mu = 105.658$ MeV with spin $\frac{1}{2}$ and its own neutrino ν_μ and lepton number L_μ. The properties of electrons, muons, and their neutrinos are listed in Table 5.5. The interaction of muons μ and their neutrinos ν_μ with matter is described through the following processes:

Table 5.5 Electron and muon families with their respective lepton numbers [117].

Particle	L_e	L_μ	Mass (MeV)	Life time (s)
ν_e	+1	0	$< 2 \times 10^{-6}$	-
e^-	+1	0	0.51099	-
ν_μ	0	+1	< 0.19	-
μ^-	0	+1	105.65837	2.197×10^{-6}

1. Weak decays of muon

$$\mu^- \to e^- + \bar{\nu}_e + \nu_\mu. \tag{5.72}$$

2. Inverse muon-decay and ν_μ scattering

$$\nu_\mu + e^- \to \nu_e + \mu^- \quad \text{and} \quad \nu_e + \mu^- \to \nu_\mu + e^-. \tag{5.73}$$

3. Muon capture from nucleons and nuclei

$$\mu^- + p \to n + \nu_\mu \quad \text{and} \quad \mu^- + A(Z,N) \to \nu_\mu + A'(Z-1, N+1). \tag{5.74}$$

4. Neutrino scattering from nucleons and nuclei

$$\nu_\mu + n \to \mu^- + p, \qquad \bar{\nu}_\mu + p \to \mu^+ + n,$$

$$\nu_\mu + A(Z,N) \to \mu^- + A'(Z+1, N-1),$$
$$\bar{\nu}_\mu + A(Z,N) \to \mu^+ + A'(Z-1, N+1). \tag{5.75}$$

In this section, we describe the weak interactions of muons focusing mainly on muon-decay and inverse muon-decay processes for which the Hamiltonian density is described by:

$$\mathcal{H}_{\text{int}}(x) = \frac{G_\mu}{\sqrt{2}} \, \overline{\psi}_{\nu_\mu} \gamma_\mu (1 - \gamma_5) \psi_\mu \, \overline{\psi}_{\nu_e} \gamma^\mu (1 - \gamma_5) \psi_e + h.c. \tag{5.76}$$

involving the point interaction of electron and muon currents as shown in Figure 5.7.

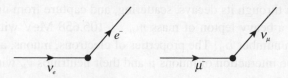

Figure 5.7 Lepton current connects ν_e to e^- and ν_μ to μ^-.

5.4.1 Weak decay of muons

The weak decay of the muon

$$\mu^-(p') \rightarrow e^-(p) + \bar{\nu}_e(k) + \nu_\mu(k') \tag{5.77}$$

is diagrammatically shown in Figure 5.8, where p' and k' are the momenta of the muon (μ^-) and the muon neutrino(ν_μ) and p and k are the momenta of the electron (e^-) and the electron antineutrino($\bar{\nu}_e$). All the particles in this decay are point particles and therefore, the vertex has no structure (Figure 5.8). The coupling G_μ is taken to be constant and is determined using the life time of the muon. Since the emission of the electron in the weak interaction processes

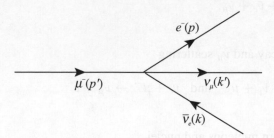

Figure 5.8 Muon-decay at rest.

is known to be accompanied by an electron type antineutrino ($\bar{\nu}_e$), the other neutral particle was identified as a neutrino associated with a muon, that is, muon neutrino(ν_μ) and the decay probability per unit time for the initial muon is given as:

$$d\Gamma = \frac{1}{(2\pi)^5 2m_\mu} \frac{d\vec{p}}{2E_e} \frac{d\vec{k}}{2k^0} \frac{d\vec{k'}}{2k^{0'}} \delta^4(p' + p + k - k') \sum_{s_{\nu_e}, s_{\nu_\mu}} |\mathcal{M}|^2, \tag{5.78}$$

where s_{ν_e} and s_{ν_μ} are neutrino spins.

The transition matrix element $\mathcal{M}$ is given by:

$$\mathcal{M} = \frac{G_\mu}{\sqrt{2}} \left[\bar{u}_{\nu_\mu}(k', s_{\nu_\mu}) \gamma^\rho (1 - \gamma_5) u_\mu(p', s') \right] \left[\bar{u}_e(p, s) \gamma_\rho (1 - \gamma_5) u_{\nu_e}(k, s_{\nu_\mu}) \right].$$

We define

$$\sum_{\text{initial spins}} \sum_{\text{final spins}} |\mathcal{M}|^2 = \frac{G_\mu^2}{2} \mathcal{M}^{\rho\lambda}(\mu) \mathcal{M}_{\rho\lambda}(e),$$

where

$$\mathcal{M}^{\rho\lambda}(\mu) = \mathcal{M}^\rho(\mu) \mathcal{M}^{\lambda\dagger}(\mu) = \sum \left[\bar{u}_{\nu_\mu} \gamma^\rho (1 - \gamma_5) u_\mu \right] \left[\bar{u}_{\nu_\mu} \gamma^\lambda (1 - \gamma_5) u_\mu \right]^\dagger,$$

$$\mathcal{M}_{\rho\lambda}(e) = \mathcal{M}_\rho(e) \mathcal{M}_\lambda^\dagger(e) = \sum \left[\bar{u}_{\nu_e} \gamma_\rho (1 - \gamma_5) u_{\nu_e} \right] \left[\bar{u}_{\nu_e} \gamma_\lambda (1 - \gamma_5) u_{\nu_e} \right]^\dagger.$$

Considering that the electron and muon are both polarized with spins s and s', respectively, we can write:

$$u_\mu \bar{u}_\mu = (\not{p}' + m_\mu) \left(\frac{1 + \gamma_5 \not{s}'}{2} \right), \tag{5.79}$$

$$u_e \bar{u}_e = (\not{p} + m_e) \left(\frac{1 + \gamma_5 \not{s}}{2} \right), \tag{5.80}$$

$$
\begin{aligned}
\mathcal{M}^{\rho\lambda}(\mu) &= \frac{1}{2} Tr \left[(\not{p}' + m_\mu)(1 + \gamma_5 \not{s}') \gamma^\lambda (1 - \gamma_5) \not{k}' \gamma^\rho (1 - \gamma_5) \right] \\
&= 4 \left\{ (p'^\lambda - m_\mu s'^\lambda) k'^\rho - (p' - m_\mu s') . k' g^{\rho\lambda} \right. \\
&\quad \left. + (p'^\rho - m_\mu s'^\rho) k'^\lambda + i \epsilon^{\alpha\lambda\beta\rho} (p'_\alpha - m_\mu s'_\alpha) k'_\beta \right\},
\end{aligned}
\tag{5.81}
$$

with similar results for $\mathcal{M}_{\rho\lambda}(e)$, with $p' \to p$, $s' \to s$, and $m_\mu \to m_e$, $k' \to k$ leading to:

$$\sum_{s_{\nu_e}, s_{\nu_\mu}} |\mathcal{M}|^2 = \frac{G_\mu^2}{2} 64 (p' - m_\mu s).k(p - m_e s).k', \tag{5.82}$$

where the spin vectors s^μ and $s^{\mu'}$ are given in the rest frame of muons such that:

$$s'^\mu = (0, \hat{s}'), \qquad s^\mu = \left(\frac{\vec{p}_e . \hat{s}}{m_e}, \hat{s} + \frac{(\vec{p}_e . \hat{s}) \, \vec{p}_e}{m_e (E + m_e)} \right). \tag{5.83}$$

Putting this value of $|\mathcal{M}|^2$ in Eq. (5.78) we obtain the decay probability of $d\Gamma$ as:

$$
\begin{aligned}
d\Gamma &= \frac{64\,G_\mu^2}{2}\,\frac{d\vec{p}}{(2\pi)^5}\frac{1}{2E_e 2m_\mu}\int\frac{d\vec{k}}{2E_\nu}\frac{d\vec{k}'}{2E_\nu'}\,\delta^4(p+k+k'-p') \\
&\quad (p'-m_\mu s').k\,(p-m_e s)\cdot k' \\[4pt]
&= \frac{2\,G_\mu^2}{(2\pi)^5}\,\frac{d\vec{p}}{m_\mu E_e}\int\frac{d\vec{k}\,d\vec{k}'}{E_\nu E_\nu'}\delta^4(p+k+k'-p')\,(p'-m_\mu s').k\,(p-m_e s)\cdot k' \\[4pt]
&= \frac{2\,G_\mu^2}{(2\pi)^5}\,\frac{d\vec{p}}{m_\mu E_e}\,(p'^{\alpha}-m_\mu s'^{\alpha})\,(p^{\beta}-m_e s^{\beta})\,I_{\alpha\beta},
\end{aligned}
\tag{5.84}
$$

where

$$
I_{\alpha\beta} = \int\frac{d\vec{k}\,d\vec{k}'}{E_\nu E_\nu'}\,\delta^4(k+k'-q)\,k_\alpha'\,k_\beta.
\tag{5.85}
$$

Here, $q = k - k' = p' - p$ and $I_{\alpha\beta}$ is a symmetric tensor of rank 2 depending only on momentum q_α. Therefore, we can write, in the most general form,

$$
I_{\alpha\beta} = Aq^2 g_{\alpha\beta} + Bq_\alpha q_\beta.
\tag{5.86}
$$

It can be shown using the expression for $I_{\alpha\beta}$ in Eq. (5.85) that $A = \frac{\pi}{6}$, $B = \frac{\pi}{3}$ such that $I_{\alpha\beta} = \frac{\pi}{6}(q^2 g_{\alpha\beta} + 2q_\alpha q_\beta)$ giving

$$
d\Gamma = \frac{\pi}{3}\,\frac{G_\mu^2}{(2\pi)^5}\,\frac{d\vec{p}}{m_\mu E_e}\,(p'-m_\mu s')^{\alpha}\,(p-m_e s)^{\beta}\,(q^2 g_{\alpha\beta}+2q_\alpha q_\beta).
\tag{5.87}
$$

Evaluating in the rest frame of muon, where

$$
\begin{aligned}
p^{\alpha'} &= (m_\mu, 0)\,;\ s^{\alpha'} = (0, \vec{s}\,'), \\
p^{\beta} &= (E_e, \vec{p}_e)\,;\ s^{\beta} = \left(\frac{\hat{s}\cdot\vec{p}_e}{m_e}, \hat{s}+\frac{(\vec{p}_e\cdot\hat{s})\vec{p}_e}{(m_e+E_e)m_e}\right), \\
q^{\alpha} &= (m_\mu - E_e, -\vec{p}_e),
\end{aligned}
\tag{5.88}
$$

we obtain:

$$
\begin{aligned}
d\Gamma &= \frac{\pi G_\mu^2}{3(2\pi)^5 m_\mu}d\Omega_e|\vec{p}_e|dE_e\left[\left(m_\mu^2 + E_e^2 - 2m_\mu E_e - |\vec{p}_e|^2\right)\right. \\[4pt]
&\quad \times \left[(E_e - \vec{p}_e\cdot\hat{s})\,m_\mu + m_\mu\left(\vec{p}_e - m_e\hat{s} - \frac{\vec{p}_e\cdot\hat{s}}{E_e+m_e}\vec{p}_e\right).\hat{s}'\right] \\[4pt]
&\quad +\ 2\left((E_e - \vec{p}_e\cdot\hat{s})\,(m_\mu - E_e) + \left(\vec{p}_e - m_e\hat{s} - \frac{\vec{p}_e\cdot\hat{s}}{E_e+m_e}\vec{p}_e\right)\cdot\vec{p}_e\right) \\[4pt]
&\quad \left.\times\ \left(m_\mu^2 - m_\mu E_e - m_\mu\hat{s}'\cdot\vec{p}_e\right)\right].
\end{aligned}
\tag{5.89}
$$

Since $m_e << m_\mu$, we may evaluate the decay rate in the extreme relativistic limit for the electron, that is, $m_e \to 0$, $|\vec{p}_e| \to E_e$ and $\vec{p}_e \to \hat{n} E_e$, and obtain:

$$d\Gamma = \frac{\pi G_\mu^2}{3(2\pi)^5 m_\mu} d\Omega_e \, m_\mu^3 \, E_e^2 \, dE_e (1 - \hat{n} \cdot \hat{s}) \left[\left(3 - \frac{4E_e}{m_\mu} \right) + \left(1 - \frac{4E_e}{m_\mu} \right) \hat{n} \cdot \hat{s}' \right]. \qquad (5.90)$$

It can be shown that

$$E_e^{\text{max}} = \frac{m_\mu^2 + m_e^2}{2m_\mu} \simeq \frac{m_\mu}{2} \qquad (5.91)$$

and $\quad |\vec{p}_e^{\text{max}}| = \dfrac{m_\mu^2 - m_e^2}{2m_\mu} \simeq \dfrac{m_\mu}{2}. \qquad (5.92)$

We define a dimensionless variable

$$\epsilon = \frac{E_e}{E_{\text{max}}} = \frac{2E_e}{m_\mu}, \qquad (5.93)$$

and write

$$d\Gamma = \frac{G_\mu^2 m_\mu^5}{192\pi^3} 2\epsilon^2 (3 - 2\epsilon) \left(\frac{1 - \hat{n} \cdot \hat{s}}{2} \right) \left[1 + \left(\frac{1 - 2\epsilon}{3 - 2\epsilon} \right) \hat{n} \cdot \hat{s}' \right] \frac{d\Omega}{4\pi} d\epsilon. \qquad (5.94)$$

This equation gives the angular and energy distributions of the electron, information about the helicity of the emitted electron, as well as the asymmetry in the electron direction with respect to the spin of the muon in case of the decay of the polarized muon. For the spin sum over the final electron and spin average over the muon spin, let us define the spin direction such that $\hat{n} \cdot \hat{s}' = P \cos\theta$, where P is the degree of polarization of the muon and write

$$d\Gamma = \frac{G_\mu^2 m_\mu^5}{192\pi^3} \left(\frac{1 - \hat{n} \cdot \vec{s}}{2} \right) A [1 + B(P \cos\theta)] \frac{d\Omega}{4\pi} d\epsilon, \qquad (5.95)$$

where

$$A = 2\epsilon^2 (3 - 2\epsilon) \text{ and } B = \left(\frac{1 - 2\epsilon}{3 - 2\epsilon} \right).$$

In the case of the unpolarized muon, we average over the spin directions of the muon by integrating over the angle θ and obtain:

$$d\Gamma = \frac{G_\mu^2 m_\mu^5}{192\pi^3} \left(\frac{1 - \hat{n} \cdot \hat{s}}{2} \right) 2\epsilon^2 [3 - 2\epsilon] d\epsilon. \qquad (5.96)$$

(i) Decay rate

The decay rate for the right-handed electron ($\hat{n} \cdot \hat{s} = +1$) is zero. Taking $\hat{n} \cdot \hat{s} = -1$, we obtain:

$$dT = \frac{G_\mu^2 m_\mu^5}{192\pi^3} 2\epsilon^2 [3 - 2\epsilon] d\epsilon \tag{5.97}$$

leading to the decay rate

$$\Gamma = \frac{G_\mu^2 m_\mu^5}{192\pi^3}. \tag{5.98}$$

Taking into account the electron mass terms in the decay rate which we have neglected, we obtain $\Gamma = f(m_e)\Gamma(m_e = 0)$, where

$$
\begin{aligned}
f(m_e) &= 1 - 8\frac{m_e^2}{m_\mu^2} + 8\frac{m_e^4}{m_\mu^4} - \frac{m_e^8}{m_\mu^8} - 12\frac{m_e^4}{m_\mu^4} ln\frac{m_e^2}{m_\mu^2} \\
&= 1 - 1.87 \times 10^{-4}.
\end{aligned}
\tag{5.99}
$$

This result does not include the radiative corrections to muon-decay. The calculation of these corrections leads to additional correction f^{rad} and the decay rate is given by:

$$\Gamma = \Gamma_0(m_e = 0) f^{\text{rad}}, \tag{5.100}$$

and f^{rad} is given as:

$$f^{\text{rad}} = 1 - \frac{\alpha}{2\pi}\left(\pi^2 - \frac{25}{4}\right) \simeq 0.9958 \simeq 1 - 4.3 \times 10^{-3}. \tag{5.101}$$

Comparing the theoretical value of the decay rate,

$$\Gamma = \frac{G_\mu^2 \, m_\mu^5}{192\pi^3} f_e(m_e) \, f_{\text{rad}} \Rightarrow \tau = \frac{1}{\Gamma} = \frac{192\pi^3}{G_\mu^2 m_\mu^5 f_e f^{\text{rad}}} \tag{5.102}$$

with the experimental value of $\tau = (2.19703 \pm 0.00004)\mu$s and $m_\mu = (105.658387 \pm 0.000034)$ MeV, we obtain $G_\mu = (1.166367 \pm 0.000004) \times 10^{-5}$ GeV^{-2} which is very close to the value of $G_F (= (1.136 \pm 0.003) \times 10^{-5}GeV^{-2})$ obtained from nuclear β-decay.

(ii) The energy spectrum of the electron
After integrating over the angle (θ, ϕ), we obtain for the energy distribution of the electron,

$$\frac{dT}{d\epsilon} = \frac{G^2 m_\mu^5}{192\pi^3} 2\epsilon^2 [3 - 2\epsilon], \tag{5.103}$$

where $\epsilon = \frac{2E_e}{m_\mu}$. The experimental results are in a very good agreement with the $V - A$ theory of weak interactions. For an example, see the discussion in Ref.[238, 239].

(iii) Electron helicity and parity violation
We see from Eq. (5.95) that the decay rate dT is nonzero only for $\hat{n} \cdot \hat{s} = -1$ in the limit $m_e \to 0$, implying that the electron coming out of the muon- decay is left-handed. This result

is independent of energy, that is, the coefficient of $\hat{n} \cdot \hat{s}$ is unity. It is also independent of energy in the limit of vanishing m_e. Thus, parity is maximally violated in muon-decay. This can be diagrammatically understood in a simple way considering the case of maximum energy, that is, $\epsilon \simeq 1$ corresponding to the case when $\bar{\nu}_e$ and ν_μ are emitted in the same direction (see Figure 5.9). Since $\bar{\nu}_e$ and ν_μ have opposite helicities and move in the same direction, the total spin projection carried by the $(\bar{\nu}_e, \nu_\mu)$ pair is zero. The spin carried by the electron is in the same direction as the muon. Since the direction of the emission of the electron is in the –Z-direction, its helicity is –1 if the muon's spin is quantized in the +Z-direction. Therefore, the electrons emitted in the muon-decay are left-handed as in the case of β-decay.

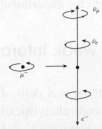

Figure 5.9 A schematic representation of muon decay in case of maximum electron energy ($\epsilon = 1$) . Neutrinos have equal momenta and opposite spins. Hence, electron must be emitted in the opposite direction and has the spin in the same direction as that of the muon.

(iv) Electron asymmetry in the decay of polarized muons

We can write the angular distribution of the electrons from the decay of polarized muons as:

$$\frac{d\Gamma}{dE_e d\cos\theta} = K[A(E_e) + B(E_e)\cos\theta]\Theta(E_e^{\max} - E_e), \tag{5.104}$$

where (keeping $m_e \neq 0$)

$$K = \frac{G^2 m_\mu |\vec{p}_e| E_e}{12\pi^3}, A = 3E_e^{\max} - 2E_e - 2\frac{m_e^2}{E_e}, B = \frac{|\vec{p}_e|}{E_e}\left(E_e^{\max} - 2E_e + \frac{2m_e^2}{m_\mu}\right)$$

and $E_e^{\max}$ is the maximum energy of the electron.

Under the parity transformation $\theta \to \pi - \theta$, where θ is the angle between the muon's spin and the electron's momentum (as shown in Figure 5.10), that is, under this transformation,

$$\sin\theta \xrightarrow{\theta \to \pi-\theta} +\sin\theta \ , \cos\theta \xrightarrow{\theta \to \pi-\theta} -\cos\theta,$$

$$\frac{d\Gamma(\pi - \theta)}{dE_e \, d\cos\theta} \quad \to \quad K[A(E_e) - B(E_e)\cos\theta].$$

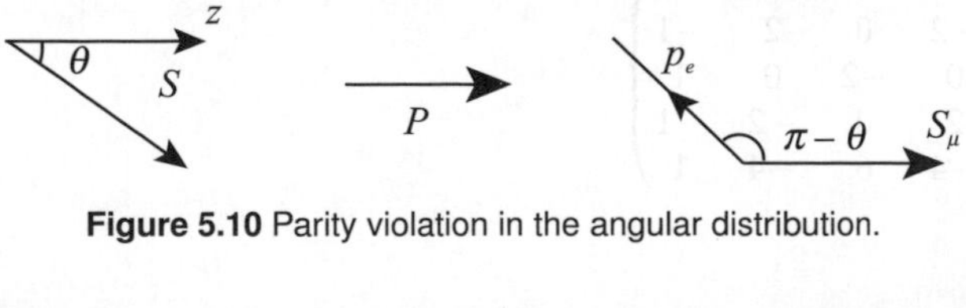

Figure 5.10 Parity violation in the angular distribution.

Therefore, $d\Gamma(\theta)$ and $d\Gamma(\pi - \theta)$ are not the same, implying parity violation. The asymmetry parameter B is a measure of the parity violation.

After integrating over the electron's energy, the angular distribution $d\Gamma/d\cos\theta$ is given by:

$$\frac{d\Gamma}{d\cos\theta} \sim K[1 - P\cos\theta], \tag{5.105}$$

that is, the electrons are emitted dominantly in the direction opposite to the spin of the muon, that is, $\theta = \pi$. Note that the asymmetry is $\frac{P}{3}$, where P is the magnitude of polarization of the initial muon. For a completely polarized muon along the $\hat{z}$ direction ($P = +1$), the asymmetry is $A = \frac{1}{3}$, which is to be compared with the experimental value of $A = +0.325 \pm 0.005$ for e^+ in the case of μ^+-decay.

5.4.2 General structure of weak interaction in muon-decay and Michel parameters

The $V - A$ structure of weak interactions was derived from the study of nuclear β-decays involving nucleons which were treated non-relativistically. Nevertheless, the process of muon-decay is also described very successfully using the $V - A$ theory. Attempts were made to arrive at a theory of muon-decay using a fully relativistic treatment of the most general structure of the weak interaction Hamiltonian. We therefore consider once again the general structure of $H_I^{\text{Weak}}(x)$ as:

$$\mathcal{H}_I^{\text{Weak}}(x) = \frac{G_\mu}{\sqrt{2}} \sum_i [\bar{u}_{\nu_\mu}(x)O_i u_\mu(x)] \, [\bar{u}_e(x)O^i(A_i - A_i'\gamma_5)u_{\nu_e}(x)], \tag{5.106}$$

where $O_i = S, V, T, A, P$ operators are defined in terms of the Dirac gamma matrices in Eq. (5.106) and coefficients A_i and A_i' are the strength of various terms. The properties of the Dirac gamma matrices and the permutation symmetry between μ^-, e^-, $\bar{\nu}_e$, ν_μ allows us to use the Fierz transformation [258] and write H_I^{Weak} in Eq. (5.106) as:

$$\mathcal{H}_I^{\text{Weak}}(x) = \frac{G_\mu}{\sqrt{2}} \sum_i [\bar{u}_e(x)O_i u_\mu(x)] \, [\bar{u}_{\nu_\mu}(x)O^i(C_i - C_i'\gamma_5)u_{\nu_e}(x)], \tag{5.107}$$

where the coefficients C_i and C_i' in Eq. (5.107) are given in terms of the coefficients A_i and A_i' used in Eq. (5.106) as:

$$C_i = \sum_j \Lambda_{ij} A_j \;\;, C_i' = \sum_j \Lambda_{ij} A_j' \tag{5.108}$$

with [259]

$$\Lambda_{ij} = \frac{1}{4} \begin{pmatrix} 1 & 4 & 6 & 4 & 1 \\ 1 & -2 & 0 & 2 & -1 \\ 1 & 0 & -2 & 0 & 1 \\ 1 & 2 & 0 & -2 & -1 \\ 1 & -4 & 6 & -4 & 1 \end{pmatrix}.$$

One can check that the $V - A$ structure remains invariant (from an overall sign) under the aforementioned permutation, that is,

$$C_V = \frac{1}{4}(A_S - 2A_V + 2A_A - A_P).$$

$$C_A = \frac{1}{4}(A_S + 2A_V - 2A_A - A_P).$$

$$C_V - C_A = \frac{1}{4}(-4A_V + 4A_A) = -(A_V - A_A). \tag{5.109}$$

Therefore, let us start with the most general Lorentz invariant, four Fermi point interaction matrix element for a muon-decay $\mu^-(p_\mu) \to e^-(p_e) + \bar{v}_e(k_{v_e}) + v_\mu(k_{v_\mu})$ using Eq. (5.107)

$$\mathcal{M} = \frac{G}{\sqrt{2}} \sum_{i=S,V,A,T,P} \left[\bar{u}_e(\vec{p}_e, s_e)\Gamma^i u_\mu(\vec{p}_\mu, s_\mu) \right] \left[\bar{u}_{v_\mu}(\vec{k}_{v_\mu}, s_{v_\mu})\Gamma_i \left(C_i - C'_i \gamma_5\right) u_{v_e}(\vec{k}_{v_e}, s_{v_e}) \right], \tag{5.110}$$

Table 5.6 Michel parameters in $V - A$ theory and their values determined in the experiments.

Michel parameter	$V - A$ values	Exp. values [117]
ρ	$\frac{3}{4}$	0.74979 ± 0.00017
η	0	0.057 ± 0.034
ζ	$+1$	$1.009^{+0.0016}_{-0.0007}$
δ	$\frac{3}{4}$	0.75047 ± 0.0034
h	$+1$	1.00 ± 0.04

where Γ^i stand for the five bilinear covariants, viz., scalar (S), vector (V), axial vector (A), tensor (T), and pseudoscalar (P) terms having 1,4,4,6, and 1 components, respectively. To calculate decay rate, $|\mathcal{M}|^2$ is required; $|\mathcal{M}|^2$ should be proportional to the product of leptonic tensor($L_{e\mu}^{\alpha\beta}$) and the neutrino tensor ($L_{\alpha\beta}^{v_\mu v_e}$), that is,

$$|\mathcal{M}|^2 = L_{e\mu}^{\alpha\beta} L_{\alpha\beta}^{v_\mu v_e}, \text{where} \tag{5.111}$$

$$L_{\alpha\beta}^{v_\mu v_e} = \sum_{i,j} \sum_{s_{v_\mu} s_{v_e}} \left[\bar{u}_{v_\mu}(\vec{k}_{v_\mu}, s_{v_\mu})\Gamma_\alpha^i \left(C_i - C'_i \gamma_5\right) u_{v_e}(\vec{k}_{v_e}, s_{v_e}) \right] \left[\bar{u}_{v_e}(\vec{k}_{v_e}, s_{v_e}) \left(C_j^* + C_j'^* \gamma_5\right) \right.$$

$$\left. \Gamma_\beta^j u_{v_\mu}(\vec{k}_{v_\mu}, s_{v_\mu}) \right]$$

$$= \sum_{i,j} Tr \left[\left(A_{ij} \pm B_{ij}\gamma_5\right) \Gamma_\alpha^i \not{k}_{v_e} \Gamma_\beta^j \not{k}_{v_\mu} \right], \tag{5.112}$$

$$A_{ij} = C_i C_j^* + C'_i C_j'^*, \quad B_{ij} = C_i C_j'^* + C'_i C_j^*$$

'+' sign is for S, P, and T terms and '−' sign is for V and A terms.

$$L_{e\mu}^{\alpha\beta} = \sum_{i,j}\sum_{s_e} \left[\bar{u}_e(\vec{p}_e,s_e)\Gamma_i^\alpha u_\mu(\vec{p}_\mu,s_\mu)\right]\left[\bar{u}_\mu(\vec{p}_\mu,s_\mu)\Gamma_j^\beta u_e(\vec{p}_e,s_e)\right]$$

$$= \sum_{i,j} Tr\left[\Gamma_i^\alpha\left(\not{p}_\mu + m_\mu\right)\frac{1+\gamma_5\not{s}_\mu}{2}\Gamma_j^\beta\not{p}_e\right].$$

Assuming that the muons are at rest and neglecting radiative corrections, the charged lepton energy spectrum is given by [259]:

$$\frac{d\Gamma(\mu)}{d\Omega dx} = \frac{G_\mu^2 m_\mu^5}{192\pi^4}x^2\left(\frac{1}{1+4\eta\frac{m_e}{m_\mu}}\left[4(x-1)+\frac{2}{3}\rho(4x-3)+6\frac{m_e}{m_\mu}\frac{1-x}{x}\eta\right]\right.$$

$$\left. - \xi\cos\theta_{\hat{s}\cdot\hat{p}_e}\left[(1-x)+\frac{2}{3}\delta(4x-3)\right]\right), \tag{5.113}$$

$$x = \frac{E_e}{E_{max}}, \quad E_{max} = \frac{m_\mu}{2}(1+\frac{m_e^2}{m_\mu^2}), \quad x_0 = \frac{m_e}{E_{max}}, \tag{5.114}$$

where $\theta_{\hat{s}\cdot\hat{p}_e}$ is the angle between the muon spin and the electron momentum. ρ, η, ξ, and δ of the aforementioned expression are known as Michel parameters [218] and are given in terms of the strengths of the various couplings $C_i(C_i')$ (i = S,V,T,A,P) as [259]:

$$\rho = \frac{1}{16}\left(3a_V+6a_T+3a_A\right), \qquad \xi = -\frac{1}{16}\left(4b'+3a'-14c'\right),$$

$$\eta = \frac{1}{16}\left(a_S-2a_V+2a_A-a_P\right), \qquad \delta = -\frac{1}{16\xi}\left(3b'-6c'\right),$$

where $a_i = |C_i|^2+|C_i'|^2$, $i=S,V,A,T,P$.

$$a' = 2Re\left(C_SC_P'^*+C_S'C_P^*\right), \quad b' = 2Re\left(C_VC_A'^*+C_V'C_A^*\right),$$

$$c' = 2Re\left(C_TC_T'^*\right).$$

The constants a_is are constrained as $a_S+4a_V+4a_A+6a_T+a_P=16$, corresponding to the 16 components of the bilinear spinors. In case of a $V-A$ interaction, $C_S=C_P=C_T=C_S'=C_P'=C_T'=0$ and $C_V=C_V'=-C_A=-C_A'=1$. On the other hand, for SPT interaction, $C_V=C_V'=C_A=C_{A'}=0$.

For the $V-A$ interaction, Table 5.6 shows the predicted values of the Michel parameters ρ, η, ξ, and δ along with the experimental values, where a good agreement between the theoretical values predicted by the $V-A$ theory and the experimental values can be seen.

5.4.3 Radiative corrections to μ-decays

Muons are charged particles which decay by emitting charged electrons and $\bar{\nu}_e$ and ν_μ. Therefore, muon-decay is subject to the radiative corrections in which a photon is emitted by the charged particles and is reabsorbed. Diagrammatically, they are shown in Figure 5.11(a),(b) and (c); which contribute to the radiative corrections in the lowest order in perturbation theory. The figures correspond to the loop diagrams discussed in Chapter 4. The radiative corrections are categorized as the (i) vertex correction (Figure 5.11(a)) and the (ii) self-energy diagram (Figure 5.11(b) and 5.11(c)). The diagrams corresponding to the charged particle Bremsstrahlung in which a photon is emitted from a charged particle as shown in Figures 5.11(d) and 5.11(e) are also considered in addition to the loop diagrams because the very low energy real photons, which are emitted below the threshold of experimental detection, are indistinguishable from those which are emitted and re-absorbed. The early calculations of

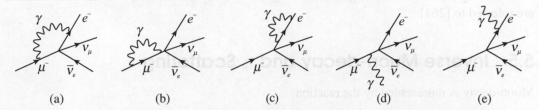

    (a)   (b)   (c)   (d)   (e)

Figure 5.11 The higher order corrections to the muon-decay: (a) Vertex correction, (b, c) Self energy correction and (d, e) Bremsstrahlung contributions.

these diagrams using QED have been done by Behrends et al. [260], Knoshita and Sirlin [261], Berman and Sirlin [262], and Berman [263]. For a detailed review, please see Ref.[264]. The Bremsstrahlung diagrams, Figure 5.11(d) and 5.11(e), help to remove the infrared divergences due to the other diagrams. Ultraviolet divergences are removed by using the renormalization procedure to regularize the divergent integrals in an appropriate gauge using Ward's identities. In the case of $V - A$ theory, one finds the following correction to the momentum (energy) distribution [239]:

$$\frac{d\Gamma^{\text{corrected}}}{dx} = \frac{d\Gamma^{\text{uncorrected}}}{dx}(1 + f(x)); \quad x = \frac{|\vec{p}_e|}{|\vec{p}_e^{\text{max}}|}, \tag{5.115}$$

where

$$f(x) = \frac{1}{3 - 2x}\frac{\alpha}{2\pi}\left[(6 - 4x)R(x) + (6 - 6x)\ln(x) + \frac{1 - x}{3x^2}\right.$$

$$\left.\left\{(5 + 17x - 34x^2)(\omega + \ln(x))(-22x + 34x^2)\right\}.\right]$$

$$R(x) = 2\sum_{n=1}^{\infty}\frac{x^n}{n^2} - \frac{\pi^2}{3} - 2 + \omega\left(\frac{3}{2} + 2\ln\frac{1 - x}{x}\right)$$

$$- \ln(x)(2\ln(x) - 1) + (3\ln(x) - 1) - \frac{1}{x}\ln(1 - x).$$

$$\omega = \ln\frac{m_\mu}{m_e},$$

leading to the following modification in the decay rate (in the limit $m_e \to 0$), resulting in:

$$\Gamma_{\text{corrected}} = \frac{G_\mu^2 m_\mu^5}{192\pi^3}\left[1 - \frac{\alpha}{2\pi}\left(\pi^2 - \frac{25}{4}\right)\right] \approx 0.9958, \tag{5.116}$$

where $\frac{G_\mu^2 m_\mu^5}{192\pi^3}$ is the uncorrected value of the decay rate Γ; numerically, the corrections are very small $\approx 0.4\%$. Using Eq. (5.116), one obtains a precise value of G_μ by taking the measured lifetime $\tau = \frac{1}{\Gamma}$ and mass of the muons as the inputs. The calculation of the radiative corrections shown in Figure 5.11 and the higher order diagrams are beyond the scope of this book; readers are referred to [264].

5.5 Inverse Muon-decay and ν_μ Scattering

Muon-decay is represented by the reaction

$$\mu^- \to e^- + \bar{\nu}_e + \nu_\mu$$

and the inverse muon-decay process would be a scattering process like:

$$\nu_e + \mu^- \to e^- + \nu_\mu \qquad \text{or} \qquad \nu_\mu + e^- \to \mu^- + \nu_e$$

depending upon the incident beam and the target. Since preparing a muon target is difficult due to the muon being an unstable particle, the reaction $\nu_\mu + e^- \to \mu^- + \nu_e$ with ν_μ beams and an electron target is more amenable to experiments. Such experiments have been done at CERN SPS [265], Fermilab Tevatron [266], and CHARM II at CERN [267]. In this section, we show the calculation of the cross section for this process in $V - A$ theory, in the first order of perturbation theory. The Feynman diagram for these processes is shown in Figure 5.12. The matrix element $\mathcal{M}$ corresponding to the Feynman diagram shown in Figure 5.12 is

$$\mathcal{M} = \frac{G_\mu}{\sqrt{2}}[\bar{u}_{\nu_e}(k',s'_{\nu_e})\gamma_\mu(1-\gamma_5)u_e(p,s_e)][\bar{u}_\mu(p',s'_e)\gamma^\mu(1-\gamma_5)u_{\nu_\mu}(k,s_{\nu_\mu})], \tag{5.117}$$

where (k,s_{ν_μ}), (k',s'_{ν_e}) are the momenta and spin of initial and final neutrinos, ν_μ and ν_e and (p,s_e), (p',s'_e) are the momenta and spin of the initial and final charged leptons e and μ.

Following standard methods for calculating the differential cross section, we write:

$$
\begin{aligned}
d\sigma &= \frac{(2\pi)^4\delta^4(p+k-p'-k')}{4((k.p)^2 - m_\nu^2 m_e^2)^{\frac{1}{2}}}\prod\frac{d\vec{p}'}{(2\pi)^3 p'_0}\frac{d\vec{k}'}{(2\pi)^3 k'_0}\sum_{spins}|\mathcal{M}|^2 \\
&= \frac{1}{(2\pi)^2}\frac{\delta^4(p+k-p'-k')}{4k.p}\frac{d\vec{p}'}{p'_0}\frac{d\vec{k}'}{k'_0}\overline{\sum}\sum|\mathcal{M}|^2. \tag{5.118}
\end{aligned}
$$

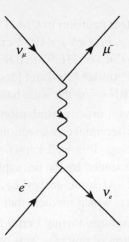

Figure 5.12 Inverse muon decay: $\nu_\mu + e^- \rightarrow \mu^- + \nu_e$.

The spin sums $\overline{\sum}\sum |\mathcal{M}|^2$ is calculated in a way similar to the calculations done in muon- decay with the result:

$$\sum_i \sum_f |\mathcal{M}|^2 = 64 G_\mu^2 (k.p)(k'.p').$$

Since $k'.p' = \frac{1}{2}(s - m_\mu^2)$, where $s = 4E_\nu^2$ is the CM (center of mass) energy of the neutrino. We get:

$$\frac{d\sigma}{d\Omega} = \frac{G_\mu^2}{2\pi^2}(s - m_\mu^2) \int \frac{d\vec{p}'}{p_0'} k_0' dk_0' \delta^4(p + k - p' - k').$$

Performing the momentum integration over p', we obtain:

$$\frac{d\sigma}{d\Omega} = \frac{G_\mu^2}{2\pi^2}(s - m_\mu^2)\frac{s - m_\mu^2}{2s}$$

$$= \frac{G_\mu^2}{4\pi^2}s\left(1 - \frac{m_\mu^2}{s}\right)^2, \tag{5.119}$$

where we have used

$$\int \int \frac{d\vec{p}'}{p_0'} k_0' dk_0' \delta^4(p + k - p' - k') = \frac{s - m_\mu^2}{2s}, \tag{5.120}$$

which, in the high energy limit $s \gg m_\mu^2$ gives:

$$\frac{d\sigma}{d\Omega} = \frac{G_\mu^2 s}{4\pi^2} \quad \text{and} \quad \sigma = \frac{G_\mu^2 s}{\pi} = \frac{4G_\mu^2}{\pi} E_{\nu}^2, \tag{5.121}$$

where E_ν is the energy of the incident neutrino in CM. The standard model prediction for the inverse muon-decay is $\sigma = 17.23 \times E_\nu(\text{GeV}) \times 10^{-42} \text{cm}^2/\text{GeV}$. This is in good agreement with the experimentally observed values of $(16.93 \pm 0.85 \pm 0.52) \times E_\nu(\text{GeV}) \times 10^{-42} \text{ cm}^2/\text{GeV}$ by CCFR collaboration at the Fermilab [266] and $(16.51 \pm 0.93) \times E_\nu(\text{GeV}) \times 10^{-42} \text{cm}^2/\text{GeV}$ by Bergsma et al. [267] at CERN using the wide band beam.

The calculation done in the lowest order perturbation theory shows that the cross section increases with energy as E_ν^2. The differential cross section $\frac{d\sigma}{d\Omega}$ given in Eq. (5.119) is isotropic showing that it is an s-wave scattering. It is well-known that the s-wave scattering amplitude and the scattering cross section is bounded by the principle of unitarity and cannot exceed the unitarity limit as energy increases. At some energy E_ν, the cross section calculated earlier will violate the principle of unitarity, which predicts that $\frac{d\sigma}{d\Omega} \leq \frac{1}{4E_\nu^2}$. The fact that the higher order calculations in perturbation expansion using Fermi theory and also the $V - A$ theory do not solve this problem of the violation of unitarity, was realized long back by Fierz [258] and Heisenberg [268]. This is one of the limitations of $V - A$ theory and will be discussed in some detail in Section 5.7.

5.6 Muon Capture and $\mu - e$ Universality

Historically, the leptonic decay of muon has been the most studied process following its discovery. The evidence that muons also interact weakly with nucleons was soon reported by Conversi et al.[269] by observing the process of the weak muon capture in heavy nuclei through the process $\mu^- + p \rightarrow n + \nu_\mu$. With improved experimental techniques, the process of μ capture was also observed later in light nuclei like ^3He and ^{12}C, though the observation of muon capture in protons could only be made in 1962 [270]. In this process, the muon is absorbed by the proton in nuclei before decaying into electrons and only one neutrino is produced. The process competes with the decay process $\mu^- \rightarrow e^- + \bar{\nu}_e + \nu_\mu$ in the nucleus and dominates for $Z \geq 6$. Since its first observation, the process of nuclear muon capture has been studied in numerous experiments involving complex nuclei. Theoretically, it was discussed in the early days by Pontecorvo [112], Fermi, Teller and Weisskopf [271]. The weak process of ordinary K-capture of electrons from nuclei, that is, $e^- + p \rightarrow n + \nu_e$ was already predicted by Yukawa and Sakata [272] and observed by Alvarez [273]. Later, a comparative study of muon capture in nuclei, electron capture in nuclei, muon decay, and nuclear β-decay was made by Tiomno and Wheeler [115], Lee, Rosenbluth and Yang [274] and Puppi [113] and Klein [114]. It was shown that the strength of the coupling in μ decay, e^- capture, μ^- capture, and β-decay in nuclei are close to each other as far as Fermi interactions are concerned. This suggested the principle of universal Fermi interaction(UFI) for the first time and implies that the strength of the Fermi interaction, that is, the vector current interaction is the same for the electrons and muons in their interaction with the nucleons and nuclei and $G_F \approx G_\mu$. This is known as $\mu - e$ universality.

In the case of μ^- capture at rest, the energy transferred to the nuclear system is of the order of $m_\mu(100 \text{ MeV})$ and momentum transfer $q^2(\approx m_\mu^2)$ is large as compared to the energies involved in a nuclear β-decay. Therefore, the structure of the matrix elements of the $V - A$

currents, taken between the nucleon states, involves more terms in addition to the C_V and C_A terms which depend upon momentum. The additional term in the matrix element of the vector current term is proportional to $\bar{u}_p \sigma_{\mu\nu} q^\nu u_n(p)$ and is known as the weak magnetism term; and in the axial term is proportional to $\bar{u}_p(p') q^\mu \gamma_5 u_n(p)$ and is known as the pseudoscalar term. The process of muon capture in nuclei plays a very important role in determining the strength of coupling of these terms, specially in the case of the pseudoscalar term. This is because the pseudoscalar term can be shown to be proportional to the mass of the lepton and is therefore negligible in all weak processes involving electrons. The reactions with muons in the initial or final state are the only reactions where the pseudoscalar term is determined specially at very low energies. In the region of low energy weak interactions involving nuclei, the muon capture plays a major role, in determining the pseudoscalar coupling as emphasized in the work of Goldberger-Trieman [275] , Leite Lopes [276], and Wolfenstein [277].

5.7   Limitations of the Phenomenological Theory

5.7.1   High energy behavior of $\nu_l - l^-$ scattering and unitarity

We consider the reactions $\nu_l(k) + l^-(p) \to l^-(k') + \nu_l(p')$,$(l = e, \mu)$ with $k(k')$ and $p(p')$ being the momenta of the initial (final) neutrino and lepton respectively; they are calculated in the theory using the $(J^l)^\dagger_\mu (J^l)^\mu$ piece of interaction Hamiltonian $H_{int}(x)$ given in Eq. (5.106). We obtain the cross section $\frac{d\sigma}{d\Omega}$ and the total scattering cross section σ, in the CM frame, as:

$$\frac{d\sigma}{d\Omega} = \frac{G_F^2 s}{4\pi^2}\left(1 - \frac{m_l^2}{s}\right)^2, \tag{5.122}$$

$$\sigma = \frac{G_F^2 s}{\pi}\left(1 - \frac{m_l^2}{s}\right)^2, \tag{5.123}$$

where s is the center of mass (CM) energy of the ν_l and l^-, that is, $s = (p + k)^2$. For the present purpose of taking high energy limit, we can neglect the second term as $m_l^2 << s$ and obtain:

$$\sigma = \frac{G_F^2}{\pi} s. \tag{5.124}$$

We note that

i) The angular distribution is isotropic.

ii) The total cross section increases with energy.

In general, the two-particle scattering amplitude $f(\theta)$ can be expanded in terms of the partial wave amplitudes as:

$$f(\theta) = \frac{1}{|\vec{p}_{CM}|}\sum_{J=0}^{\infty}\left(J + \frac{1}{2}\right)\mathcal{M}_J P_J(\cos\theta), \tag{5.125}$$

where $\vec{p}_{CM}$ is the momentum of the incident particle in the CM frame and $\mathcal{M}_J$ is the amplitude in the Jth partial wave. In the CM frame, $\vec{k} + \vec{p} = 0$ and $s = (k + p)^2 = (k_0 + p_0)^2_{CM}$,

$$
\begin{aligned}
s &= \left(\sqrt{m_e^2 + |\vec{k}|^2} + \sqrt{m_\nu^2 + |\vec{p}|^2} \right)^2 = m_e^2 + |\vec{k}|^2 + m_\nu^2 + |\vec{p}|^2 \\
&\quad + 2\sqrt{m_e^2 + |\vec{k}|^2}\sqrt{m_\nu^2 + |\vec{p}|^2} \\
&\simeq 4|\vec{p}_{CM}|^2, \text{ in the limit } m_e, m_\nu \rightarrow 0.
\end{aligned} \tag{5.126}
$$

We find the differential scattering cross section in the CM frame as:

$$
\frac{d\sigma}{d\Omega} = |f(\theta)|^2 = \frac{1}{4|\vec{p}_{CM}|^2} |\sum_{J=0}^{\infty} (2J+1)\, \mathcal{M}_J P_J(\cos\theta)|^2
$$

leading to

$$
\frac{d\sigma}{d\Omega} = \frac{1}{4|\vec{p}_{CM}|^2} |\mathcal{M}_0|^2
$$

as only $J = 0$ contributes, because the $\nu_e e^-$ interaction is local in $V - A$ theory. This leads to:

$$
\sigma = \frac{\pi}{|\vec{p}_{CM}|^2} |\mathcal{M}_0|^2 = \frac{4\pi}{s} |\mathcal{M}_0|^2. \tag{5.127}
$$

The unitarity of the S matrix, that is, $S^\dagger S = 1$ implies that the scattering amplitude $\mathcal{M}_J$ in the Jth partial wave can be written as:

$$
\mathcal{M}_J = e^{i\delta_J} \sin\delta_J,
$$

where δ_J is the phase shift. This implies an upper bound for $\mathcal{M}_J$, that is, $|\mathcal{M}_J| \leq 1$ for each J.

We, therefore, find a unitarity upper limit for $\sigma(\nu_l e^- \rightarrow l\nu_e) = \frac{4\pi}{s}$. Since the phenomenological theory predicts that the cross section increases with s, that is, $\frac{G^2 s}{\pi}$, if $\frac{G^2 s}{\pi} \geq \frac{4\pi}{s}$, then the unitarity limit is violated. This implies that for the CM energy of $s \geqslant \frac{2\pi}{G}$, the unitarity limit is violated. This corresponds to the lab energy of

$$
E_\nu^{lab} = \frac{s}{2m_e} = \frac{\pi}{m_e G} \approx 6 \times 10^8 \text{ GeV}.
$$

This is too high an energy to be tested in the laboratory but in principle, the theory violates unitarity at very high energies of $E \approx 10^8$ GeV when the cross sections are calculated in the lowest order of perturbation theory.

5.7.2 Divergence and renormalization

The problem of the violation of the unitarity described in the previous section arises in the first order calculations of the neutrino–electron scattering cross section using perturbation theory. The higher order contributions to $\nu_l e^-$ scattering should also be considered. The contribution in the higher order of perturbation theory comes due to the diagrams shown in Figure 5.13. The

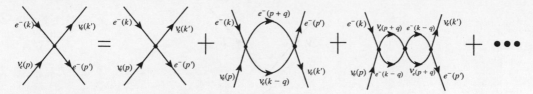

Figure 5.13 First order, second order, and higher order diagrams in $V - A$ theory for $\nu_e - e^-$ scattering.

matrix element $\mathcal{M}(2)$ for the second order diagram is

$$\mathcal{M}(2) \quad \propto \quad G_F^2 \int \frac{d^4q}{(2\pi)^4} \left[\bar{u}_e(p',s')\gamma_\mu(1-\gamma_5)\frac{1}{\slashed{p}+\slashed{q}-m_e}\gamma_\nu(1-\gamma_5)u_e(p,s) \right]$$
$$\times \left[\bar{u}_{\nu_e}(k',t')\gamma^\mu(1-\gamma_5)\frac{1}{\slashed{k}-\slashed{q}}\gamma^\nu(1-\gamma_5)u_{\nu_e}(k,t) \right] + \dots. \tag{5.128}$$

The high energy behavior of this integral can be understood by counting the powers of the momentum $\vec{q}$ in the numerator and the denominator of the integrand, which behave like

$$\mathcal{M}(2) \quad \propto \quad \int_0^\infty \frac{d^4q}{q^2} \simeq \int_0^\infty \frac{q^3\,dq}{q^2} \simeq \int_0^\infty q\,dq \tag{5.129}$$

and therefore diverges. The higher order diagrams give even more divergent integrals. This was realized quite early by Heisenberg [268] and Fierz [258]. Such divergences also occur in QED but they are absorbed to all orders, in terms of the renormalization of the charge, mass, and the vertex function. In this case, divergence of such integrals appear in all orders requiring a new renormalization constant each time. The Fermi theory and the phenomenological $V - A$ theory is, therefore, not renormalizable. This is one of the major difficulties faced by the $V - A$ theory.

In order to solve this problem, it was suggested quite early in the development of the phenomenological theory that if the charged current weak interactions are mediated by massive vector bosons like the electromagnetic interactions are mediated by vector A_μ fields, then the additional q^2 dependence of the massive vector boson propagator may help to solve the problem of divergence. However, the IVB theory did not succeed in removing the divergence.

5.7.3 Intermediate vector boson (IVB) theory

In this theory, the weak interaction of leptons/nucleons is mediated by the vector bosons $W_\mu^\pm$ with the interaction Hamiltonian given by:

$$\mathcal{H} = g_W J_\mu^\dagger W^\mu + \text{h.c.}, \tag{5.130}$$

where g_W is the weak coupling strength.

The $\nu_e - e^-$ scattering is then described in the lowest order by the Feynman diagram shown in Figure 5.14.

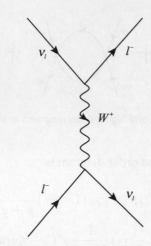

Figure 5.14 $v_l - l^-$ scattering mediated by W^+ boson.

The matrix element $\mathcal{M}$ is written as:

$$\mathcal{M} = g_W^2 \bar{u}_e \gamma_\mu (1 - \gamma_5) u_{v_e} D^{\mu\nu} \bar{u}_{v_e} \gamma_\nu (1 - \gamma_5) u_e, \tag{5.131}$$

where $D^{\mu\nu} = -i \dfrac{\left(g^{\mu\nu} - \dfrac{q^\mu q^\nu}{M_W^2} \right)}{q^2 - M_W^2}$

is the $W^\pm$ propagator with M_W being the mass of $W^\pm$ bosons. For small values of q^2 implying $q^2 << M_W^2$, Eq. (5.131) reduces to the expression obtained using $V - A$ theory provided

$$\frac{g_W^2}{M_W^2} \approx \frac{G_F}{\sqrt{2}}.$$

Carrying out the computation for the cross section in the CM frame using the expression of the amplitude from Eq. (5.131), we obtain:

$$\frac{d\sigma}{d\Omega} = \frac{1}{64\pi^2 s} |\mathcal{M}|^2 \tag{5.132}$$

$$\Rightarrow \frac{d\sigma}{d\Omega} = \frac{2g_W^4 |\vec{k}|^2}{\pi^2 \left(q^2 - M_W^2 \right)^2}, \tag{5.133}$$

where $|\vec{k}|$ is the momentum of the initial particle in the CM frame. This leads to:

$$\sigma = \frac{4G_F^2 |\vec{k}|^2}{\pi} \left(1 + \frac{4|\vec{k}|^2}{M_W^2} \right)^{-1}, \tag{5.134}$$

in which $k^2 << M_W^2$ limit gives the $V - A$ result, that is,

$$\sigma = \frac{G_F^2}{\pi} s$$

and in the limit $|\vec{k}| \to \infty$,

$$\sigma = \frac{G_F^2 M_W^2}{\pi} = \text{constant.}$$

thus preventing the violation of the unitarity limit. However, the partial wave unitarity in the S-wave is still violated. It can be shown that the partial wave amplitude in the S-wave, f_0 corresponding to the transition amplitude calculated in the W exchange is given by [278]:

$$f_0 = \frac{G_F M_W^2}{\sqrt{2}\pi} \log \left(1 + \frac{4|\vec{k}^2|}{M_W^2} \right)$$

which still violates the unitarity limit, that is,

$$f_0 > 1, \quad \text{for } k \approx \frac{M_W}{2} \exp \left(\frac{\pi}{\sqrt{2}G_F M_W^2} \right).$$

In the second order of the perturbation theory in the IVB theory, there would be two IVB propagators bringing additional q^2 dependence proportional to $(q^2)^2$ in the denominator, making the loop integral in the diagram behave as:

$$M^{(2)} \sim \int \frac{d^4 q}{q^2} \frac{1}{(q^2)^2} \approx \int \frac{dq}{q^3},$$

which is no longer divergent as $q \to \infty$. It should be noted that the term proportional to $q^\mu q^\nu$ in the numerator of the W propagator does not contribute when contracted with the leptonic part in $q^\mu l_\mu$ as $q^\nu l_\nu \sim 0$ in the limit of $m_e \to 0$. Thus, the processes in which the IVB appear in virtual state are not divergent in the IVB theory.

However, the situation is quite different in the processes, where real IVB's are produced that is, $\nu_\mu + \bar{\nu}_\mu \to W^+ + W^-$ as shown in Figure 5.15.

The matrix element $\mathcal{M}$ for this process is written as:

$$\mathcal{M} \propto g_W^2 \epsilon_\mu^{-*}(k_2, \lambda_2) \epsilon_\nu^{+*}(k_1, \lambda_1)$$
$$\times \bar{v}(p_2) \gamma^\mu (1 - \gamma_5) \frac{1}{(\not{p}_1 + \not{p}_2) - m_l} \gamma^\nu (1 - \gamma_5) u(p_1), \qquad (5.135)$$

where ϵ_μ and ϵ_ν are the polarization vectors of the W^μ bosons, that is, $\epsilon_\mu^{-*}(k_2, \lambda_2)$ is associated with W^- and $\epsilon_\nu^{+*}(k_1, \lambda_1)$ with W^+; k_1 and k_2 are four momenta; and λ_1 and λ_2 are the polarization states of W^+ and W^- respectively. In order to calculate $\sum_{\text{spins}} |\mathcal{M}|^2$, we have to

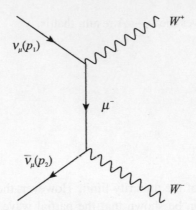

Figure 5.15 Feynman diagram depicting $\nu_\mu \bar{\nu}_\mu \to W^+ W^-$ via μ^- exchange.

sum over the polarization states of W_μ^+ and W_μ^- using:

$$\sum_{\lambda=0,\pm1} \epsilon_\mu(k,\lambda)\epsilon_\nu^*(k,\lambda) = -g_{\mu\nu} + \frac{k_\mu k_\nu}{M_W^2}. \tag{5.136}$$

Since we are interested in the high energy behavior of the cross section, we consider the contribution of the $\frac{k_\mu k_\nu}{M_W^2}$ term in Eq. (5.136). It can be shown that this momentum dependent term comes from the longitudinal component of the polarization vector $\epsilon_\mu(k,\lambda)$ of the W bosons (Chapter 2). We, therefore, calculate the $\sum_{\text{spin}} |\mathcal{M}_{00}|^2$ term corresponding to the production of the longitudinally polarized W bosons. The following result is obtained for $\sum |\mathcal{M}_{00}|^2$, using Eqs. (5.135) and (5.136) [278, 279]

$$\sum |\mathcal{M}_{00}|^2 \sim \frac{g_W^4}{M_W^4}(p_1.k_1)\,(p_2.k_2) \sim \frac{g_W^4}{M_W^4}E^4(1-\cos^2\theta), \tag{5.137}$$

where E is the CM energy and θ is the CM scattering angle. Using this equation, we find

$$\frac{d\sigma}{d\Omega}(\nu_\mu\bar{\nu}_\mu \to W^+W^-) = G_F^2\frac{E^2\sin^2\theta}{8\pi^2} \tag{5.138}$$

the energy dependence of the differential cross section similar to the result for the differential cross section for the process $\nu e^- \to \nu e^-$, leading to the violation of unitarity. Therefore, the IVB model has the same problems of bad high energy behavior of the cross section as the local $V - A$ theory. This is the main reason for the lack of renormalization in the phenomenological theory, even in the presence of intermediate vector bosons.

5.7.4 Radiative corrections

Another difficulty of considerable importance was faced by the Fermi type theory as well as the IVB type theory, which relates to the bad high energy behavior in the calculation of radiative correction to weak processes. One finds that in the leptonic sector, the radiative corrections to the μ-decay are finite, but the corrections to $v_e + e^- \rightarrow v_e + e^-$ diverge in the order α^2. In the case of the semileptonic sector, like $n \rightarrow p + e^- + \bar{v}_e$, the corrections diverge in order α even in the absence of strong interactions. The presence of form factors due to the structure of hadrons, which are expected to provide damping of the amplitude with increasing E and q^2, do not help to cure the divergences. The divergences appear both in the vector as well as in the axial vector contributions and do not cancel with each other. The problem essentially arises due to the bad high energy behavior of the transition amplitudes when calculated in the higher orders of perturbation because of the non-renormalizability of the theory. Numerous attempts were made to explain this difficulty without success [280]. The problem of radiative corrections could be solved only after the formulation of the unified theory of electroweak interactions, which came in the early 1970s.

5.8 τ Lepton and Its Weak Decays and $e - \mu - \tau$ Universality

5.8.1 τ lepton and its properties

The tau(τ) lepton first discussed theoretically by Tsai [281] and Thacker and Sakurai [282] in 1971, was discovered in 1975 by Perl et al. [116] at the SPEAR storage ring in the SLAC-LBL laboratory and later by Burmester et al. [283] at DESY in 1977. For a detailed review, please see Ref. [284]. In a storage ring, the electron and positron beams circulate in opposite directions and are forced to collide in a region where detectors are placed for observing the final particles produced in the collision, like electrons, muons, and hadrons within a large solid angle. In the e^+e^- collision experiments, a number of anomalous events with electrons and muons like $e^-\mu^+$ or $e^+\mu^-$ pairs, accompanied by some neutral particles, that is in the reaction, $e^- + e^+ \rightarrow e^- + \mu^+$ and at least two unobserved particles, were reported. Since the lepton number is conserved, $e^{\pm}\mu^{\mp}$ cannot be produced through a direct interaction but only as decay products of some intermediate particles produced in the collision process of e^- and e^+, consistent with the conservation of both lepton numbers L_e and L_μ.

These intermediate particles could be heavy hadrons or leptons. The threshold energy for the production of these $e^{\pm}\mu^{\mp}$ events was in the region of $3.6 - 4.0$ GeV. A characteristic feature of these $e^{\pm}\mu^{\mp}$ events was that with increase in energy, the electrons and muons were collinearly emitted in opposite directions indicating that they are most likely the decay products of a particle–antiparticle pair produced in the collision.

The threshold energy of $3.6 - 4.0$ GeV implies that these particles $\tau^{-(+)}$ would have a mass $m \geq 1.7$ GeV. This expectation is consistent with the earlier results from CERN [284] that there are no heavy leptons with mass $m \leq 1.5$ GeV. These particles could be heavy leptons, undergoing three body decays like muons

$$\tau^{-(+)} \rightarrow e^-(e^+) + v_\tau(\bar{v}_\tau) + \bar{v}_e(v_e), \tau^-(\tau^+) \rightarrow \mu^-(\mu^+) + v_\tau(\bar{v}_\tau) + \bar{v}_\mu(\bar{v}_\mu). \tag{5.139}$$

The other possibilities are that the $e^- \mu^+$ pairs come from the intermediate state where two photons are produced through the reaction $e^+ e^- \rightarrow 2\gamma \rightarrow e^+ e^-, \mu^+ \mu^-$ or through the sequential decays of heavy baryons or bosons produced in the $e^- e^+$ collision. All these processes would yield an $e^- \mu^+$ energy spectrum which carries the characteristic signatures of the parent intermediate particle produced in the $e^- e^+$ collisions. A detailed analysis of the energy (momentum) distribution of $e^- (e^+)$ and $\mu^+ (\mu^-)$ emitted in these processes ruled out the possibility of the intermediate particle being heavy bosons, photons, or heavy hadrons. The signature of the intermediate particles being heavy leptons was the most likely, and these particles were called τ leptons.

The properties of τ leptons like their mass, lifetime, spin, and their structure were determined by studying the observable $R_{e\mu}(s)$ and $R_{eh}(s)$ ($\sqrt{s}$ is the CM energy) defined as

$$R_{e\mu} = \frac{\sigma(e^+ e^- \rightarrow \tau^+ \tau^- \rightarrow e\mu)}{\sigma(e^+ e^- \rightarrow \mu^+ \mu^-)}, \tag{5.140}$$

$$R_{eh} = \frac{\sigma(e^+ e^- \rightarrow \tau^+ \tau^- \rightarrow eh)}{\sigma(e^+ e^- \rightarrow \mu^+ \mu^-)}, \tag{5.141}$$

where h is hadron.

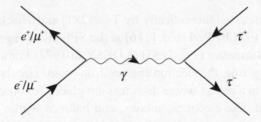

Figure 5.16 $e^- e^+ (\mu^- \mu^+) \rightarrow \tau^- \tau^+$ process mediated by a photon exchange.

Using the standard procedure for calculating cross sections for $e^- e^+$ collisions and assuming photon exchange corresponding to Figure 5.16, the cross section $\sigma_{\tau\tau}$ for the process $e^+ e^- \rightarrow \tau^+ \tau^-$ is given by:

$$\sigma_{\tau\tau} = \sigma_{\mu\mu} |F_\tau(s)|^2 F(\beta), \tag{5.142}$$

where

$$\sigma_{\mu\mu} = \frac{4\pi\alpha^2}{3s} \tag{5.143}$$

is the cross section for $e^+ e^- \rightarrow \mu^+ \mu^-$. $F(\beta)$ is a function of the velocity $\beta(= \frac{|\vec{p}|}{E})$ of τ^+ given by:

$$F(\beta) = \frac{1}{4}\beta^3 \qquad \qquad \text{for spin } 0, \tag{5.144}$$

$$= \frac{1}{2}\beta(3 - \beta^2) \qquad \text{for spin } \frac{1}{2}, \tag{5.145}$$

$$= \beta^3 \left[\left(\frac{s}{4m_\tau^2} \right)^2 + 5\frac{s}{4m_\tau^2} + \frac{3}{4} \right] \quad \text{for spin 1,} \tag{5.146}$$

and $F_\tau(s)$ is the form factor at the $\gamma \tau^+ \tau^-$ vertex (Figure 5.16) and taken to be

$$F_\tau(s) = 1 \pm \frac{s}{\Lambda_\pm^2}, \tag{5.147}$$

where Λ is the cut-off parameter. In Figure 5.17, we show the behavior of R_{eh} and $R_{e\mu}$ for

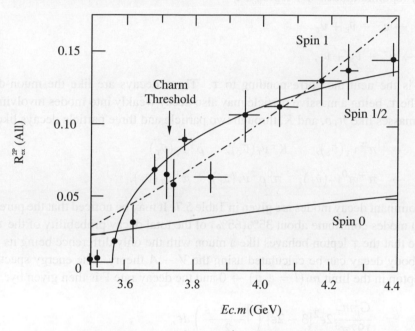

Figure 5.17 The production cross section ratio $R_{eX}^{2p} = \frac{\sigma(e^+e^- \to eX)}{\sigma(e^+e^- \to \mu^+\mu^-)}$ vs. the CM energy $\sqrt{s}$ for all eX events with no detected photons [285]. The three fitted curves indicate the threshold behavior for different spins of τ.

various values of spin $J = 0, \frac{1}{2}, 1$. It was inferred from Figure 5.17 and Eqs. (5.144)–(5.146) that

i) τ has a mass $m_\tau = 1.784^{+0.0027}_{-0.0036}$ GeV.

ii) τ has spin $\frac{1}{2}$.

iii) $\Lambda > 50$ GeV implying that $F(s) \approx 1.0014$, that is, τ is a point particle.

Therefore, it is concluded that the τ particles called tauons are heavy leptons of spin $\frac{1}{2}$, like muons, and have their own ν_τ associated with them, since:

$$\tau^- \nrightarrow e^- + \gamma \text{ or } \tau^- \nrightarrow \mu^- + \gamma$$

decays have never been observed. The non-observation of these decays leads to a separate lepton number L_τ for τ and ν_τ leptons and their conservation law. It also establishes that there are three generations (flavors) of neutrinos, that is, ν_e, ν_μ, and ν_τ. The name 'heavy lepton' given to τ leptons is a misnomer because the word 'lepton' is taken from Greek which describes something which is light.

5.8.2   Weak decays of τ leptons and $e - \mu - \tau$- universality

The purely leptonic modes of τ lepton are

$$\tau^- \to e^- + \bar{\nu}_e + \nu_\tau \tag{5.148}$$

$$\tau^- \to \mu^- + \bar{\nu}_\mu + \nu_\tau, \tag{5.149}$$

where ν_τ is the neutrino corresponding to τ. These decays are like the muon-decays. In addition, the τ, being a massive particle may also decay weakly into modes involving particles of higher masses like π, ρ, and K, through two particle and three particle decays like:

$$\tau^\mp \quad \to \quad \pi^\mp \nu_\tau(\bar{\nu}_\tau), \quad K^\mp \nu_\tau(\bar{\nu}_\tau), \quad \rho^\mp \nu_\tau(\bar{\nu}_\tau) \tag{5.150}$$

$$\tau^\mp \quad \to \quad \pi^\mp \pi^0 \nu_\tau(\bar{\nu}_\tau), \quad \pi^0 \rho^\mp \nu_\tau(\bar{\nu}_\tau), \quad \pi^\mp \nu_\tau(\bar{\nu}_\tau). \tag{5.151}$$

A list of dominant decay modes are given in Table 5.7. It may be noticed that the purely leptonic (hadronic) modes contribute about 35%(65%) of the total decay probability of the τ lepton. If we assume that the τ lepton behaves like a muon with the only difference being its mass, then the three-body decay can be calculated using the $V - A$ theory. The energy spectrum of the emitted lepton in the limit $m_l(l = e, \mu) \to 0$ and the decay rate Γ is then given by:

$$\frac{d\Gamma}{dE_e} = \frac{G_\mu^2 m_\tau^5}{192\pi^3} 2\epsilon^2 [3 - 2\epsilon] \left(\frac{1 - \hat{n} \cdot \hat{s}}{2} \right) d\epsilon, \tag{5.152}$$

Table 5.7 Decay modes of tau lepton [117].

Mode	BR(%)
$e^- \bar{\nu}_e \nu_\tau$	17.82 ± 0.04
$\mu^- \bar{\nu}_\mu \nu_\tau$	17.39 ± 0.04
$\pi^- \nu_\tau$	10.82 ± 0.05
$K^- \nu_\tau$	$(6.96 \pm 0.1) \times 10^{-3}$
$\rho^- \nu_\tau$	(21.8 ± 2.0)
$\pi^- \rho^0 \nu_\tau$	(5.4 ± 1.7)
$\pi^- \pi^0 \nu_\tau$	25.49 ± 0.09
$e^- \bar{\nu}_e \nu_\tau \gamma$	1.83 ± 0.05
$\pi^- 2\pi^0 \nu_\tau$	9.26 ± 0.1
$K^- 2\pi^0 \nu_\tau$	$(6.5 \pm 2.2) \times 10^{-4}$
$\pi^- \pi^- \pi^+ \nu_\tau$	9.31 ± 0.05
$\pi^- \pi^- \pi^+ \pi^0 \nu_\tau$	4.62 ± 0.05

where $\epsilon = \frac{2E_l}{m_\tau}$, E_l is the energy of the outgoing lepton and $\hat{s}$ represents the spin of the outgoing lepton. Assuming the outgoing lepton to be left-handed, that is, $\hat{n} \cdot \hat{s} = -1$, we obtain:

$$\Gamma = \frac{G_\mu^2 m_\tau^5}{192\pi^3}. \tag{5.153}$$

This expression is obtained in the limit $m_l = 0$, where m_l represents the mass of the outgoing lepton (e^- or μ^-). Taking into account the lepton mass, the expression of the decay rate becomes

$$\Gamma = \frac{G_\mu^2 m_\tau^5}{192\pi^3}\left(1 - 8\epsilon^2 - 24\epsilon^4 \ln(\epsilon) + 8\epsilon^6 - \epsilon^8\right), \quad \text{where } \epsilon = \frac{m_l}{m_\tau}$$

The electron energy spectrum given in Eq. (5.152) is shown in Figure 5.18, where it is compared with the experimental results [133]. The agreement between the theory and the experiment is very good.

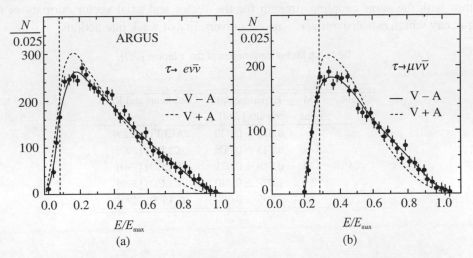

Figure 5.18 Normalized (left) electron and (right) muon energy distributions for tau decays compared with V–A (solid line) and V+A (dashed line) spectra[286].

Using this result in $V - A$ theory, we can also obtain:

$$\Gamma_{\tau^- \to e^- \nu_e \nu_\tau} = \Gamma_{\mu^- \to e^- \bar{\nu}_e \nu_\mu}\left[\frac{m_\tau^5}{m_\mu^5}\left(1 - \frac{8m_e^2}{m_\tau^2}\right)\right]$$

$$\Gamma_{\tau^- \to \mu^- \nu_\mu \nu_\tau} = \Gamma_{\mu^- \to e^- \bar{\nu}_e \nu_\mu}\left[\frac{m_\tau^5}{m_\mu^5}\left(1 - \frac{8m_\mu^2}{m_\tau^2}\right)\right]$$

$$\Rightarrow \quad R = \frac{\Gamma_{\tau \to \mu}}{\Gamma_{\tau \to e}} = \frac{\left(1 - \frac{8m_\mu^2}{m_\tau^2}\right)}{\left(1 - \frac{8m_e^2}{m_\tau^2}\right)}. \tag{5.154}$$

Using the values of m_e, m_μ, and m_τ, we obtain:

$$\Gamma_{\tau \to e} = 0.620 \times 10^{12}/s$$

$$\Gamma_{\tau \to \mu} = 0.603 \times 10^{12}/s$$

$$\text{and } R = \frac{\Gamma_{\tau \to \mu}}{\Gamma_{\tau \to e}} = 0.972, \tag{5.155}$$

which is in agreement with the experimental values of $\frac{\Gamma_{\tau \to \mu}}{\Gamma_{\tau \to e}} = 0.9 \pm 0.1$. Of course, the radiative corrections are different in all these cases as they are mass dependent and are large for $\tau^\pm$ decays.

The Michel parameters ρ, η, ζ, and δ determined experimentally from the analysis of the three particle weak decays of τ leptons are shown in Table 5.8 along with theoretical predictions of $V - A$ theory. The agreement between the theoretical and experimental values of the Michel parameters is very good. This shows that the weak interaction of τ leptons have a $V - A$ structure with the same coupling strength for the vector and axial vector currents as in the muon-decay which establishes the $e - \mu - \tau$ universality of weak interactions.

Table 5.8 Michel parameters of the τ lepton [287].

Name	SM value	Experimental results [288]	Comments and Ref.
η	0	0.013 ± 0.020	(ALEPH) [289]
ρ	3/4	0.745 ± 0.008	(CLEO) [290]
$\zeta\delta$	3/4	0.746 ± 0.021	(CLEO) [290]
ζ	1	1.007 ± 0.040	(CLEO) [290]
ζh	1	0.995 ± 0.007	(CLEO) [290]

Chapter 6

Phenomenological Theory II: Weak Decays of Hadrons

6.1 Introduction

Hadrons are strongly interacting particles which also participate in weak and electromagnetic interactions. They are not elementary particles and are composed of quarks, which come in six flavors; and each flavor comes in three colors (R, B, and G) as shown in Table 6.1. Hadrons are specified by their quantum numbers like spin, parity, isospin, strangeness, baryon number, etc., and are classified as mesons and baryons on the basis of their quark content. Mesons are the bound states of a quark and an antiquark pair ($q\bar{q}$), while baryons are the bound states of three quarks (qqq). They are bound in such a way that the physical states of mesons and baryons

Table 6.1 Quarks are spin $\frac{1}{2}$ fermions with baryon number $\frac{1}{3}$; they have been assigned positive parity. $Q(|e|)$ is the quark's charge in the units of electronic charge, I is the isospin and its 3rd component is I_3. S, C, B, and T stand for the strangeness, charm, bottom, and top quantum numbers, respectively [117].

Quark flavors	$Q(|e|)$	I	I_3	S	C	B	T
u	+2/3	1/2	+1/2	0	0	0	0
d	−1/3	1/2	−1/2	0	0	0	0
s	−1/3	0	0	−1	0	0	0
c	+2/3	0	0	0	+1	0	0
b	−1/3	0	0	0	0	−1	0
t	+2/3	0	0	0	0	0	+1

are color singlets. The strong forces which bind the quarks together, in the case of baryons, or quarks and antiquarks together, in the case of mesons, are provided by the exchange of massless vector fields between them called the gluons. The dynamics of the strong forces, that is, the binding of the quarks and antiquarks and their interactions is described by the theory of strong interactions known as quantum chromodynamics (QCD), in a way similar to QED

which describes the interactions among charged particles. In Tables 6.2 and 6.3, some of the
low lying mesons and baryons, which are considered in this chapter while discussing their weak
interactions, are listed along with their quark contents and other quantum numbers.

Table 6.2 Mesons ($J^P = 0^-$) and their properties: quark content, J represents angular momentum, P the parity,
I the isospin, I_3 the 3rd component of I, S the strangeness, M the mass and τ the lifetime of a given
meson [117].

Particles (quark content)	I	I_3	S	M (MeV)	τ(s)
$\pi^+(u\bar{d})$	1	+1	0	139.57018	26×10^{-9}
$\pi^0(u\bar{u} \text{ or } d\bar{d})$	1	0	0	134.9766	8.4×10^{-17}
$\pi^-(d\bar{u})$	1	-1	0	139.57018	26×10^{-9}
$K^+(u\bar{s})$	1/2	+1/2	+1	493.677 ± 0.013	$(12.380 \pm 0.021) \times 10^{-9}$
$K^-(s\bar{u})$	1/2	$-1/2$	-1	493.677 ± 0.013	$(12.380 \pm 0.021) \times 10^{-9}$
$K^0(d\bar{s})$	1/2	+1/2	+1	497.614 ± 0.024	-
$\bar{K}^0(\bar{d}s)$	1/2	$-1/2$	-1	497.614 ± 0.024	-
$K_s(\frac{d\bar{s}-s\bar{d}}{\sqrt{2}})$	1/2	-	-	-	$(89.54 \pm 0.04) \times 10^{-12}$
$K_l(\frac{d\bar{s}+s\bar{d}}{\sqrt{2}})$	1/2	-	-	-	$(51.16 \pm 0.21) \times 10^{-9}$

Table 6.3 Baryons ($J^P = \frac{1}{2}^+$) and their properties: quark content, isospin(I) and its 3rd component (I_3), total
angular momentum (J) and parity (P), strangeness (S), mass (M), and lifetime(τ) [117].

Particles (quark content)	I	I_3	S	M (MeV)	τ(s)
$p(uud)$	$\frac{1}{2}$	$+\frac{1}{2}$	0	938.28	stable
$n(udd)$	$\frac{1}{2}$	$-\frac{1}{2}$	0	939.57	8.815×10^2
$\Lambda(uds)$	0	0	-1	1115.683	2.6×10^{-3}
$\Sigma^+(uus)$	1	+1	-1	1189.37	$(8.018 \pm 0.026) \times 10^{-11}$
$\Sigma^0(uds)$	1	0	-1	1192.642(24)	$7.4 \pm 0.7 \times 10^{-20}$
$\Sigma^-(dds)$	1	-1	-1	1197.449(30)	$1.479 \pm 0.011 \times 10^{-10}$
$\Xi^0(uss)$	1/2	+1/2	-2	1314.83(20)	$(2.90 \pm 0.09) \times 10^{-10}$
$\Xi^-(dss)$	1/2	$-1/2$	-2	1321.31(13)	$1.639 \pm 0.015 \times 10^{-10}$

The theory of the weak interaction of hadrons is not as simple as the weak interaction of
leptons, which has been described in the previous chapter. We have seen that the interaction
Hamiltonian of the weak interaction, in the case of leptons, is described by

$$H_{\text{int}} = \frac{G}{\sqrt{2}} l_\mu l^{\mu\dagger} + \text{h.c.,} \tag{6.1}$$

where

$$l_\mu = \sum_{l=e,\mu,\tau} \bar{\psi}_{\nu_l} \gamma_\mu (1 - \gamma_5) \psi_l \tag{6.2}$$

The currents here are of the $V - A$ type. These currents l_μ are constructed from the lepton fields $\psi_l(x)$ described by the Dirac equation for a point spin $\frac{1}{2}$ particle. In the case of hadrons, the interaction is also of the current×current type with $V - A$ currents at the level of quarks which are considered as point particles. However, the hadrons are not point particles. Therefore, the general structure of the matrix element of the vector and axial vector currents has to be determined from the quantum fields describing the hadrons using the principles of Lorentz covariance, parity, time reversal, etc. For example, the nuclear β-decay in which a neutron decays to give a proton, electron, and neutrino, that is, $n \rightarrow p + e^- + \bar{\nu}_e$, is a weak interaction process involving hadrons where the currents are of $V - A$ type but with the difference that there would be more terms contributing to the matrix element of the hadronic current, in addition to that appearing in the case of leptons.

The weak interactions of the hadrons are classified into three types of processes:

   i) Semileptonic weak processes,

   ii) Nonleptonic weak processes,

   iii) Radiative weak processes,

which are described briefly in the following sections.

6.2 Semileptonic Weak Decays of Hadrons without Strangeness

Semileptonic processes are defined as those processes in which both the leptons and hadrons are involved either in a decay process like $n \rightarrow p + e^- + \bar{\nu}_e$ or in a scattering process like $\nu_\mu + n \rightarrow \mu^- + p$. We have discussed the semileptonic decays of spin $\frac{1}{2}$ particles like the neutron decay, that is, $n \rightarrow p + e^- + \bar{\nu}_e$ in some detail in the last chapter and will take up neutrino scattering in Chapter 9. There are simpler processes in which a hadron of spin zero like a pion or a kaon decays into two leptons, that is, $\pi^\pm \rightarrow e^\pm \nu_e(\bar{\nu}_e)$, $\pi^\pm \rightarrow \mu^\pm \nu_\mu(\bar{\nu}_\mu)$, $K^\pm \rightarrow e^\pm \nu_e(\bar{\nu}_e)$, and $K^\pm \rightarrow \mu^\pm \nu_\mu(\bar{\nu}_\mu)$. There are also decays like $\pi^\pm \rightarrow \pi^0 e^\pm \nu_e \ (\bar{\nu}_e)$, $K^\pm \rightarrow \pi^0 e^\pm \nu_e(\bar{\nu}_e)$, and $K^\pm \rightarrow \pi^+ \pi^- e^\pm \nu_e \ (\bar{\nu}_e)$ in which three or four particles are involved. In Tables 6.4 and 6.5, we have shown the decay modes of some low lying mesons and baryons, respectively. In this section, we describe the semileptonic weak decays of spin zero particles with strangeness $S = 0$ like pions.

6.2.1 Two-body decay of pions: πl_2 decays

Pions are pseudoscalar particles (0^-) with spin zero and negative parity, first predicted by Yukawa [272] in 1935 as the carrier of the strong nuclear force. They were first discovered by Powell and collaborators [291] in 1947 in cosmic ray experiments. Pions exist in three charged states π^+, π^-, and π^0 which form an isospin triplet. Their properties are listed in Table 6.2. The two-body weak decays of π^+ and π^- take place through electron and muon channels,

$$\pi^- \longrightarrow l^- + \bar{\nu}_l, \tag{6.3}$$

$$\pi^+ \longrightarrow l^+ + \nu_l, \quad l = e, \mu. \tag{6.4}$$

Table 6.4 Decay modes with the corresponding branching ratios of charged as well as neutral pions and kaons [117].

Particle	Decay mode	Branching ratio (%)	Particle	Decay mode	Branching ratio (%)
π^+	$\mu^+\nu_\mu$	99.98770 ± 0.00004	K_s	$\pi^0\pi^0$	30.69 ± 0.05
	$\mu^+\nu_\mu\gamma$	$(2.00\pm0.25)\times10^{-4}$		$\pi^+\pi^-$	69.20 ± 0.05
	$e^+\nu_e$	$(1.230\pm0.004)\times10^{-4}$		$\pi^\pm e^\mp \bar{\nu}_e(\nu_e)$	$(7.04\pm 0.08)\times 10^{-4}$
				$\pi^\pm\mu^\mp\bar{\nu}_\mu(\nu_\mu)$	$(4.69\pm 0.05)\times 10^{-4}$
π^0	2γ	98.823 ± 0.034			
	$e^+e^-\gamma$	1.174 ± 0.035			
$K^\pm$	$\mu^+\nu_\mu(\mu^-\bar{\nu}_\mu)$	63.56 ± 0.11	K_l	$\pi^\pm e^\mp\bar{\nu}_e(\nu_e)$	40.55 ± 0.11
	$e^+\nu_e(e^-\bar{\nu}_e)$	$(1.582\pm 0.007)\times 10^{-5}$		$\pi^\pm\mu^\mp\bar{\nu}_\mu(\nu_\mu)$	27.04 ± 0.07
	$\pi^0\, e^\pm\, \nu_e(\bar{\nu}_e)$	5.07 ± 0.04		$\pi^0\pi^0\pi^0$	19.52 ± 0.12
	$\pi^0\mu^\pm\nu_\mu\,(\bar{\nu}_\mu)$	3.352 ± 0.033		$\pi^+\pi^-\pi^0$	12.54 ± 0.05
	$\pi^\pm\pi^0$	20.67 ± 0.08		$\pi^+\pi^-$	$(1.97\pm 0.01)\times 10^{-3}$
	$\pi^0\pi^\pm\pi^0$	1.760 ± 0.023		$\pi^0\pi^0$	$(8.64\pm 0.06)\times 10^{-4}$
	$\pi^\pm\pi^+\pi^-$	5.583 ± 0.024			

Table 6.5 Decay modes of hyperons with branching ratios [117]. Neutron decays to $pe^-\bar{\nu}_e$ with 100% branching ratio.

Particle	Mode	Branching ratio (%)	Particle	Mode	Branching ratio (%)
Λ	$p\pi^-$	63.9 ± 0.5	Ξ^0	$\Lambda\pi^0$	99.524 ± 0.012
	$n\pi^0$	35.8 ± 0.5		$\Sigma^+e^-\bar{\nu}_e$	$(2.52\pm0.08)\times10^{-4}$
	$n\gamma$	$(1.75\pm 0.15)\times 10^{-3}$		$\Lambda\gamma$	$(1.17\pm 0.07)\times 10^{-3}$
	$pe^-\bar{\nu}_e$	$(8.32\pm 0.14)\times 10^{-4}$			
	$p\mu^-\bar{\nu}_\mu$	$(1.57\pm 0.35)\times 10^{-4}$			
Σ^+	$p\pi^0$	51.57 ± 0.3	Ξ^-	$\Lambda\pi^-$	99.887 ± 0.035
	$p\gamma$	1.23×10^{-3}		$\Lambda e^-\bar{\nu}_e$	$(5.63\pm 0.31)\times 10^{-4}$
	$n\pi^+$	48.31 ± 0.3			
	$\Lambda e^+\nu_e$	$(2.0\pm0.5)\times10^{-5}$			
Σ^0	$\Lambda\gamma$	100			
	Λe^+e^-	5×10^{-3}			
Σ^-	$n\pi^-$	99.848	Ω^-	ΛK^-	67.8 ± 0.7
	$ne^-\bar{\nu}_e$	1.02×10^{-3}		$\Xi^0\pi^-$	23.6 ± 0.7
	$n\mu^-\bar{\nu}_\mu$	$(4.5\pm 0.4)\times 10^{-4}$		$\Xi^-\pi^0$	8.6 ± 0.4
				$\Xi^0e^-\bar{\nu}_e$	$(5.6\pm 0.28)\times 10^{-3}$

In the $V-A$ theory, the S matrix element for the processes $\pi^-(p) \rightarrow l^-(k') + \bar{\nu}_l(k)$ corresponding to the Feynman diagram shown in Figure 6.1, is written as

$$S = i\frac{G_F}{\sqrt{2}}\int d^4x\langle l^-\bar{\nu}_l|J_h^\mu(x)J_\mu^l(x)|\pi^-\rangle \tag{6.5}$$

$$= i\frac{G_F}{\sqrt{2}}(2\pi)^4\delta^4(p-k-k')\langle 0|V^\mu - A^\mu|\pi^-(p)\rangle\bar{u}(k)\gamma_\mu(1-\gamma_5)v(k') \tag{6.6}$$

and the decay rate is given by

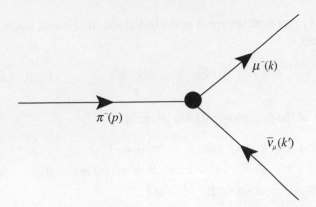

Figure 6.1 Feynman diagram for the pion decay at rest, $\pi^-(p) \longrightarrow l^-(k) + \bar{\nu}_l(k')$ (for $l = \mu$). The quantities in the brackets are the respective momenta of the particles.

$$d\Gamma = \frac{1}{2m_\pi}(2\pi)^4\delta^4(p - k - k')\frac{d\vec{k'}}{2E_l(2\pi)^3}\frac{d\vec{k}}{2E_\nu(2\pi)^3}|\mathcal{M}|^2, \tag{6.7}$$

where

$$\mathcal{M} = \frac{G_F}{\sqrt{2}}\langle 0|V^\mu - A^\mu|\pi^-(p)\rangle\bar{u}(k)\gamma_\mu(1 - \gamma_5)v(k'). \tag{6.8}$$

The hadronic matrix element is expressed as:

$$\langle 0|V^\mu - A^\mu|\pi^-(p)\rangle = \langle 0|V^\mu|\pi^-(p)\rangle - \langle 0|A^\mu|\pi^-(p)\rangle. \tag{6.9}$$

In order to write the matrix element in Eq. (6.9), it should be noted that the matrix element of the vector current between $|\pi\rangle$ and $|0\rangle$ state vanishes as the Lorentz structure of the matrix element $\langle 0|V^\mu|\pi^-(p)\rangle$ is axial vector in nature: V^μ is a vector and $|\pi^-\rangle$ is pseudoscalar and no axial vector can be constructed with the only momentum available, that is, the momentum of the pion p_μ. Using the same argument, the Lorentz structure of the matrix element $\langle 0|A^\mu|\pi^-(p)\rangle$ is a vector and the matrix element is constructed using the momentum of the pion p_μ. Therefore,

$$\langle 0|V^\mu|\pi^-(p)\rangle = 0,$$

$$\langle 0|A^\mu|\pi^-(p)\rangle = if_\pi p^\mu, \tag{6.10}$$

where f_π is the pion decay constant.

Using the expression for the matrix element given in Eq. (6.8) leads to:

$$\overline{\sum_{s_f}}\sum_{s_i}|\mathcal{M}|^2 = \left(\frac{G_F^2}{2}\right)f_\pi^2 p_\mu p_\nu \sum_{s_i}|\bar{u}(k)\gamma^\mu(1 - \gamma^5)v(k')|^2, \tag{6.11}$$

$$\sum_{s_i}|\bar{u}(k)\gamma^\mu(1 - \gamma^5)v(k')|^2 = \sum_{s_i}\{\bar{u}(k)\gamma^\mu(1 - \gamma^5)v(k')\}\{\bar{u}(k)\gamma^\nu(1 - \gamma^5)v(k')\}^\dagger,$$

$$= 8(k'^\mu k^\nu - k\cdot k'g^{\mu\nu} + k'^\nu k^\mu) - 8ik'_\alpha k_\beta\epsilon^{\alpha\mu\beta\nu}. \tag{6.12}$$

where $s_i (= 0)$ and s_f represent the spins of the initial and final states, respectively. Contracting it with $p_\mu p_\nu$ will lead to

$$\overline{\sum_{s_f} \sum_{s_i}} |\mathcal{M}|^2 = 4G_F^2 f_\pi^2 (2p \cdot k p \cdot k' - p^2 k \cdot k') \qquad (\because p = k + k'). \qquad (6.13)$$

The scalar products of the four momenta are given by:

$$\begin{aligned} p \cdot k &= (k + k') \cdot k = k^2 + k \cdot k' = m_l^2 + k \cdot k' \\ p \cdot k' &= (k + k') \cdot k' = k \cdot k' + k'^2 = k \cdot k' ; \quad (\because m_\nu = 0) \\ p^2 &= (k + k')^2 = m_l^2 + 2k \cdot k' = m_\pi^2. \end{aligned}$$

Equation (6.13) gives

$$\begin{aligned} \overline{\sum_{s_f} \sum_{s_i}} |\mathcal{M}|^2 &= 4G_F^2 f_\pi^2 \left(2[m_l^2 + k \cdot k']k \cdot k' - m_\pi^2 k \cdot k'\right) \\ &= 4G_F^2 f_\pi^2 \left(2\left[m_l^2 + \frac{(m_\pi^2 - m_l^2)}{2}\right]\left[\frac{m_\pi^2 - m_l^2}{2}\right] - m_\pi^2 \left[\frac{m_\pi^2 - m_l^2}{2}\right]\right) \\ &= 2G_F^2 f_\pi^2 (m_\pi^2 - m_l^2) m_l^2. \end{aligned}$$

Using $\overline{\sum\sum}|\mathcal{M}|^2$ as obtained here and performing the momentum integration over $d\vec{k}$ in Eq. (6.7), we obtain

$$\Gamma = \frac{G_F^2 f_\pi^2}{8\pi m_\pi} \frac{m_l^2}{m_\pi^2} (m_\pi^2 - m_l^2)^2.$$

Using the numerical values of m_π, m_μ and $\Gamma = 38.46 \ \mu s^{-1}$, we find $f_\pi = 131$ MeV.

We note that

i)

$$R_\pi = \frac{\Gamma(\pi^- \longrightarrow e^- \bar{\nu}_e)}{\Gamma(\pi^- \longrightarrow \mu^- \bar{\nu}_\mu)} = \frac{m_e^2}{m_\mu^2} \frac{(1 - \frac{m_e^2}{m_\pi^2})^2}{(1 - \frac{m_\mu^2}{m_\pi^2})^2} = 1.2834 \times 10^{-4}.$$

This is in good agreement with the experimental value of $R_\pi = 1.23 \pm 0.02 \times 10^{-4}$ [292]. Including the radiative corrections, the value of R_π becomes $R_\pi = 1.233 \times 10^{-4}$.

ii) In the limit $m_e \to 0$, $\Gamma(\pi^- \to e^- \bar{\nu}_e) \to 0$, and $\Gamma(\pi^+ \to e^+ \nu_e) \to 0$, πe_2 decay modes are forbidden. This really follows if the neutrinos are left-handed and the weak interactions have $V - A$ structure, contributing through the axial vector part.

iii) There were indications in the pre $V - A$ era that only axial vector interactions could explain these decays in the calculations made by Ruderman and Finkelstein [293], in view of the experimental limit of $R < 5 \times 10^{-3}$ available at that time for these decays.

6.2.2 Three-body decays of pions: πl_3 decays

Pions also decay into three particles:

$$\pi^+/\pi^-(p_1) \longrightarrow \pi^0(p_2) + e^+/e^-(k) + \nu_e/\bar{\nu}_e(k') \tag{6.14}$$

and are called pion β-decays in analogy with the neutron β-decay, that is, $n \to p + e^- + \bar{\nu}_e$. The S matrix element in the $V - A$ theory, for the process given in Eq. (6.14) is written as

$$S = i\frac{G_F}{\sqrt{2}}(2\pi)^4\delta^4(p_1 - p_2 - k - k')\langle\pi^0(p_2)|V^\mu - A^\mu|\pi^-(p_1)\rangle\bar{v}_{\bar{\nu}_e}\gamma_\mu(1 - \gamma_5)u_e(k).$$

Using the general expression for $d\Gamma$, the decay rate for three particles in the final state, we write:

$$d\Gamma = (2\pi)^4\delta^4(p_1 - p_2 - k - k')\frac{1}{2E_\pi}\frac{d\vec{k}}{2E_e(2\pi)^3}\frac{d\vec{k'}}{2E_\nu(2\pi)^3}\frac{d\vec{p_2}}{2E_\pi(2\pi)^3}\overline{\sum}\sum|\mathcal{M}|^2, \tag{6.15}$$

where

$$\mathcal{M} = \frac{G_F}{\sqrt{2}}\langle\pi^0(p_2)|V^\mu - A^\mu|\pi^-(p_1)\rangle\bar{v}_{\bar{\nu}_e}(k')\gamma_\mu(1 - \gamma_5)u_e(k). \tag{6.16}$$

The hadronic matrix element $\langle\pi^0(p_2)|V^\mu - A^\mu|\pi^-(p_1)\rangle$ is written as:

$$\langle\pi^0(p_2)|V^\mu - A^\mu|\pi^-(p_1)\rangle = \langle\pi^0(p_2)|V^\mu|\pi^-(p_1)\rangle - \langle\pi^0(p_2)|A^\mu|\pi^-(p_1)\rangle. \tag{6.17}$$

Again the Lorentz structure of the matrix element of the axial vector current demands that $\langle\pi^0|A^\mu|\pi^-\rangle$ vanishes because no covariant axial vector term can be constructed from the momenta p_1 and p_2, and therefore,

$$\langle\pi^0|A^\mu|\pi^\pm\rangle = 0. \tag{6.18}$$

Similar arguments show that $\langle\pi^0|V^\mu|\pi^\pm\rangle$ is a vector which is written as

$$\langle\pi^0|V^\mu|\pi^\pm\rangle = f_1(q^2)p_1^\mu + f_2(q^2)p_2^\mu, \tag{6.19}$$

where $f_1(q^2)$ and $f_2(q^2)$ are the form factors. These form factors depend upon the scalar quantities that can be constructed from p_1^μ and p_2^μ, that is, p_1^2, p_2^2, $p_1 \cdot p_2$. Since $p_1^2 = m_{\pi^+}^2$ and $p_2^2 = m_{\pi^0}^2$, only $p_1 \cdot p_2$ is the independent variable and is generally expressed in terms of $q^2 = m_{\pi^+}^2 + m_{\pi^0}^2 - 2p_1 \cdot p_2$. Defining the aforementioned matrix element in terms of $q = p_1 - p_2$ and $P = p_1 + p_2$, we get

$$\langle\pi^0|V^\mu|\pi^+\rangle = f_+(q^2)P^\mu + f_-(q^2)q^\mu, \tag{6.20}$$

where the form factors $f_+(q^2)$ and $f_-(q^2)$ are defined as

$$f_+(q^2) = \frac{f_1(q^2) + f_2(q^2)}{2} \quad \text{and} \quad f_-(q^2) = \frac{f_1(q^2) - f_2(q^2)}{2}. \tag{6.21}$$

The matrix element in Eq. (6.16) is then given by

$$\mathcal{M} = \frac{G_F}{\sqrt{2}} \bar{v}_{\bar{v}_e}(k') \left(f_+(q^2)\slashed{P} + f_-(q^2)\slashed{q} \right)(1 - \gamma_5)u_e(k). \tag{6.22}$$

Since $\bar{v}_{\bar{v}_e}(k')\slashed{q}u_e(k) = \bar{v}_{\bar{v}_e}(k')(\slashed{k} + \slashed{k}')u_e(k) = m_e\bar{v}_{\bar{v}_e}(k')u_e(k)$, the contribution of the $f_-(q^2)$ term is very small and can be neglected. In fact, the contribution of this term vanishes if one uses the hypothesis of the conserved vector current (CVC) to be discussed later in this chapter.

Using Eq. (6.22), the matrix element for the three-body decay of the pion, is written as:

$$\mathcal{M} = \frac{G_F}{\sqrt{2}} f_+(q^2) \left[\bar{v}_{\bar{v}_e}(k')\slashed{P}(1 - \gamma_5)u_e(k) \right].$$

Using $P^\mu = p_1^\mu + p_2^\mu = 2p_1^\mu - k^\mu - k'^\mu$ and the Dirac equation for u_e and $\bar{v}_{\bar{v}_e}$, we get:

$$\mathcal{M} = \sqrt{2}\, G_F f_+(q^2)\, m_\pi [\bar{v}_{\bar{v}_e}(k')\gamma_0(1 - \gamma_5)u_e(k)]. \tag{6.23}$$

Evaluating $\overline{\sum}\sum |\mathcal{M}|^2$ in the rest frame of the pion leads to the decay rate Γ as:

$$\begin{aligned}
\Gamma &= \frac{16G_F^2 \cos^2\theta_c f_+^2(q^2)m_\pi^2}{(2\pi)^5 2m_\pi} \int \frac{E_e E_{v_e} d\vec{p}_2 d\vec{k} d\vec{k}' \delta^4(p_1 - p_2 - k - k')}{8m_\pi E_e E_{v_e}} \\
&= \frac{G_F^2 \cos^2\theta_c f_+^2(q^2)}{(2\pi)^5} \int d\vec{k} d\vec{k}' \delta(\Delta - E_e - E_{v_e}), \quad \text{where } \Delta = m_{\pi^\pm} - m_{\pi^0} \\
&= \frac{4\pi G_F^2 \cos^2\theta_c f_+^2(q^2)}{(2\pi)^5} \int_0^\Delta 4\pi E_e^2 dE_e (\Delta - E_e)^2 \\
&= \frac{G_F^2 \cos^2\theta_c f_+^2(q^2)\Delta^5}{60\pi^3}.
\end{aligned} \tag{6.24}$$

Experimentally using the value of the πe_3 decay rate, $\Gamma = 0.39855$ s^{-1} and $\Delta = 4.594$ MeV, and the other numerical constants in Eq. (6.24), $f_+(0)$ is obtained as [292]:

$$|f_+(0)| = 1.37 \pm 0.02.$$

The significance of this value of $|f_+(0)|$ is discussed further in Section 6.3.5.

6.3 Symmetry Properties of the Weak Hadronic Current

The weak hadronic current $J_\mu(x)$ has the vector $V_\mu(x)$ and the axial vector $A_\mu(x)$ terms which under Lorentz transformation are constructed as bilinear covariants from the nucleon fields. These bilinear covariants have certain definite properties under discrete transformation like C, P and T as well as internal symmetries like the isospin and unitary symmetry. These symmetry properties are exploited in writing the matrix elements of these currents between the initial and final states of spin $\frac{1}{2}$ or spin zero particles. In this section, we will discuss these symmetry properties and their role in writing the general structure of the matrix elements.

6.3.1 Lorentz transformation properties and matrix elements

The weak hadronic current $J_\mu(x)$ is written as

$$J_\mu(x) = V_\mu(x) - A_\mu(x) \tag{6.25}$$

and the general structure of the matrix elements of the vector and the axial vector currents between the neutron and proton states of momentum p and p' is written as

$$\langle \dot{u}_p(\vec{p}')|V_\mu|u_n(\vec{p})\rangle = \sum_i \bar{u}_p(\vec{p}')O^i_\mu u_n(\vec{p}),$$

$$\langle u_p(\vec{p}')|A_\mu|u_n(\vec{p})\rangle = \sum_i \bar{u}_p(\vec{p}')O^i_\mu \gamma_5 u_n(\vec{p}),$$

where $u_n(\vec{p})$ and $u_p(\vec{p}')$ are the Dirac spinors for the neutrons and protons with four momenta p and p', respectively. O^i_μ are the operators constructed from the four-momenta of the initial and final particles and four Dirac matrices like γ_μ, p_μ, p'_μ, $\sigma_{\mu\nu}p^\mu$ and $\sigma_{\mu\nu}p'^\mu$ such that the quantity $\bar{u}_p(p')O^i u_n(p)$ transforms as a vector. In general, there are five terms but they are not all independent; they may be related to each other using the Dirac equation for $u_n(\vec{p})$ and $u_p(\vec{p}')$. It can be shown that there are only three independent operators which are generally chosen to be γ_μ, q_μ, and $\sigma_{\mu\nu}q^\nu$. Similarly, the matrix elements of the axial vector currents are also written in terms of the three independent operators. These are explained in detail in Chapter 10 and we use some of those results here for introducing the subject. The three independent operators imply that using the Lorentz invariance, the matrix elements of the vector and axial vector currents are defined in terms of the six form factors, that is, $f_i(q^2)$ and $g_i(q^2)$ with $i = 1, 2, 3$. The matrix elements for the transition $n \to p$ is written as [294]:

$$\langle u_p(p')|V_\mu|u_n(p)\rangle = \bar{u}_p(p') \left[f_1(q^2)\gamma_\mu + i\frac{\sigma_{\mu\nu}q^\nu}{2M}f_2(q^2) + f_3(q^2)\frac{q^\mu}{M} \right] u_n(p), \tag{6.26}$$

$$\langle u_p(p')|A_\mu|u_n(p)\rangle = \bar{u}_p(p') \left[g_1(q^2)\gamma_\mu + i\frac{\sigma_{\mu\nu}q^\nu}{2M}g_2(q^2) + g_3(q^2)\frac{q^\mu}{M} \right] \gamma_5 u_n(p), \tag{6.27}$$

where we have taken $M_p = M_n = M$, with M_p and M_n being the masses of the proton and neutron, respectively. $f_i(q^2)$ and $g_i(q^2)$, $(i = 1 - 3)$ are the vector and axial vector form factors. This is similar to the matrix elements written for the electromagnetic current for the proton($\mathcal{M}^p_{em}$) and the neutron($\mathcal{M}^n_{em}$) (see Chapter 10), that is,

$$\mathcal{M}^p_{em} = \bar{u}_p(p') \left[F^p_1(q^2)\gamma_\mu + i\frac{\sigma_{\mu\nu}q^\nu}{2M}F^p_2(q^2) \right] u_p(p), \tag{6.28}$$

$$\mathcal{M}^n_{em} = \bar{u}_n(p') \left[F^n_1(q^2)\gamma_\mu + i\frac{\sigma_{\mu\nu}q^\nu}{2M}F^n_2(q^2) \right] u_n(p), \tag{6.29}$$

where $F^p_{1,2}(q^2)$ are the electromagnetic form factors for the transition $p \to p$ and $F^n_{1,2}(q^2)$ are the electromagnetic form factors for the transition $n \to n$. In the limit $q^2 \to 0$, these form factors are normalized as $F^p_1(0) = 1$, $F^p_2(0) = \mu_p(= 1.7928\mu_N)$, $F^n_1(0) = 0$, $F^n_2(0) = \mu_n(= -1.91\mu_N)$ and the q^2 dependence is given by a dipole form.

6.3.2   Isospin properties of the weak hadronic current

The weak hadronic currents between the neutron and proton states involve a change of charge $\Delta Q = \pm 1$ in the case of $n \to p + e^- + \bar{\nu}_e$ and $p \to n + e^+ + \nu_e$. Since $Q = I_3 + \frac{B}{2}$, $\Delta Q = \pm 1$ implies $\Delta I_3 = \pm 1$ using baryon number conservation. Since protons and neutrons are assigned to a doublet under isospin representation corresponding to $I = \frac{1}{2}$, $I_3 = +\frac{1}{2}$ and $I_3 = -\frac{1}{2}$, respectively, they can be written as a two-component isospinor under the group of isospin transformation, given by

$$u = \begin{pmatrix} u_p \\ u_n \end{pmatrix}. \tag{6.30}$$

This isospin group of transformations is generated by the three 2×2 Pauli matrices τ_i, defined as $\tau_1 = \begin{pmatrix} 0 & 1 \\ 1 & 0 \end{pmatrix}$, $\tau_2 = \begin{pmatrix} 0 & -i \\ i & 0 \end{pmatrix}$, and $\tau_3 = \begin{pmatrix} 1 & 0 \\ 0 & -1 \end{pmatrix}$ with the algebra described by $[\tau_i, \tau_j] = i\epsilon_{ijk}\tau_k$, where ϵ_{ijk} is the antisymmetric Levi–Civita tensor, which is $+1(-1)$ for cyclic (anticyclic) permutations and zero for any two repeated indices, that is, $\epsilon_{ijk} = -\epsilon_{jik}$ and $\epsilon_{iik} = 0$. By defining the isospin raising and lowering operators $\tau^{\pm} = \frac{\tau_1 \pm i\tau_2}{2}$, we can write

$$\bar{u}_p O_W^\mu u_n = \bar{u} O_W^\mu \tau^+ u,$$

$$\bar{u}_n O_W^\mu u_p = \bar{u} O_W^\mu \tau^- u, \tag{6.31}$$

where u is the isospinor in the isospin space $u = \begin{pmatrix} u_p \\ u_n \end{pmatrix}$. Similarly, in the case of the electromagnetic interaction,

$$\bar{u}_p O_{em}^\mu u_p = \bar{u} O_{em}^\mu \frac{1 + \tau_3}{2} u,$$

$$\bar{u}_n O_{em}^\mu u_n = \bar{u} O_{em}^\mu \frac{1 - \tau_3}{2} u, \tag{6.32}$$

implying that the isoscalar and isovector current matrix elements are expressed as

$$\bar{u}\, \mathbb{1} O_{em}^\mu u = \bar{u}_p O_{em}^\mu u_p + \bar{u}_n O_{em}^\mu u_n,$$

$$\bar{u}\tau_3 O_{em}^\mu u = \bar{u}_p O_{em}^\mu u_p - \bar{u}_n O_{em}^\mu u_n. \tag{6.33}$$

If we parameterize the matrix element of the isoscalar and isovector components as

$$\bar{u}\, \mathbb{1} O_{em}^\mu u = \bar{u} \left[F_1^S(q^2)\gamma_\mu + iF_2^S(q^2)\frac{\sigma_{\mu\nu}q^\nu}{2M} \right] u,$$

$$\bar{u}\tau_3 O_{em}^\mu u = \bar{u} \left[F_1^V(q^2)\gamma_\mu + i\frac{\sigma_{\mu\nu}q^\nu}{2M}F_2^V(q^2) \right] \tau_3 u, \tag{6.34}$$

and using the electromagnetic matrix element of protons and neutrons as given in Eqs. (6.28) and (6.29), we can write:

$$F_{1,2}^S = F_{1,2}^p(q^2) + F_{1,2}^n(q^2), \tag{6.35}$$

$$F_{1,2}^V = F_{1,2}^p(q^2) - F_{1,2}^n(q^2). \tag{6.36}$$

We see that while the weak currents transform as τ^+ and τ^- components of an isovector current, the electromagnetic current transforms as the sum of an isoscalar and isovector current.

6.3.3 T invariance

It may be recalled from Chapter-3 that the time reversal invariance holds if

$$\mathcal{M}' = \mathcal{M}^*, \tag{6.37}$$

where $\mathcal{M}'$ represents the time reversed matrix element and $\mathcal{M}^*$ represents the Hermitian conjugate of the unreversed matrix element. Under time-reversal invariance, the initial and final state particles are interchanged as well as the initial and final state particle's spin and angular momenta are reversed.

Now, we examine what happens, when time reversal invariance [295] is applied on the weak vector and axial vector matrix elements defined in Eqs. (6.26) and (6.27). The transformation of the form factors under T invariance is defined as:

$$\bar{u}_p(p')f(q^2)u_n(p) \longrightarrow \bar{u}_n(p)\gamma_0\gamma_1\gamma_3\tilde{f}(q^2)\gamma_3\gamma_1\gamma_0 u_p(p), \tag{6.38}$$

where f and $\tilde{f}$ represent the unreversed and time reversed form factors, and p and n represent the unreversed initial and the final state particles. Taking all the bilinear covariants used with the form factors in the vector and the axial vector current individually, we obtain the transformation of the vector and axial vector form factors under T invariance as:

$$\bar{u}_p u_n \xrightarrow{\hat{T}} \bar{u}_n u_p, \qquad\qquad \bar{u}_p \gamma_5 u_n \xrightarrow{\hat{T}} -\bar{u}_n \gamma_5 u_p,$$

$$\bar{u}_p \gamma^\mu u_n \xrightarrow{\hat{T}} \bar{u}_n \gamma_\mu u_p, \qquad\qquad \bar{u}_p \gamma^\mu \gamma_5 u_n \xrightarrow{\hat{T}} \bar{u}_n \gamma_\mu \gamma_5 u_p,$$

$$\bar{u}_p \sigma^{\mu\nu} u_n \xrightarrow{\hat{T}} \bar{u}_n \sigma_{\mu\nu} u_p, \qquad\qquad \bar{u}_p \sigma^{\mu\nu} \gamma_5 u_n \xrightarrow{\hat{T}} -\bar{u}_n \sigma_{\mu\nu} \gamma_5 u_p.$$

The hadronic current J_μ is defined in Eq. (6.25) with V_μ and A_μ defined in Eqs. (6.26) and (6.27), respectively. The time reversed current J'_μ is obtained as

$$
\begin{aligned}
J'_\mu = & \; \bar{u}_n \left[f_1(q^2)\gamma_\mu + (-i)(\sigma_{\mu\nu})\frac{q^\nu}{2M}f_2(q^2) + \frac{q_\mu}{M}f_3(q^2) - g_1(q^2)\gamma_\mu\gamma_5 \right. \\
& \left. - (-i)(-\sigma_{\mu\nu}\gamma_5)\frac{q^\nu}{2M}g_2(q^2) - \frac{q_\mu}{M}(-\gamma_5)g_3(q^2) \right] u_p.
\end{aligned}
\tag{6.39}
$$

Hermitian conjugate of Eq. (6.25) is written as

$$
\begin{aligned}
\tilde{J}_\mu = & \; \bar{u}_p \left[f_1^*(q^2)\gamma_\mu + (-i)\sigma_{\mu\nu}\frac{q^\nu}{2M}f_2^*(q^2) + \frac{q_\mu}{M}f_3^*(q^2) - g_1^*(q^2)\gamma_\mu\gamma_5 \right. \\
& \left. - (-i)(-\sigma_{\mu\nu}\gamma_5)\frac{q^\nu}{2M}g_2^*(q^2) - \frac{q_\mu}{M}(-\gamma_5)g_3^*(q^2) \right] u_n.
\end{aligned}
\tag{6.40}
$$

Comparing Eqs. (6.39) and (6.40), we find that $f_i = f_i^*$ and $g_i = g_i^*$ which implies that if time reversal invariance holds, the form factors must be real.

6.3.4   Conserved vector current hypothesis

The hypothesis of the conserved vector current was proposed by Gershtein and Zeldovich [44] and Feynman and Gell-Mann [42]. They made an important observation in the study of the nuclear β-decays in Fermi transitions ($\Delta J = 0$) driven by vector currents with no change in parity. They observed that the strength of the weak vector coupling (weak charge) for the muon and neutron decays are the same, just like in the case of the electromagnetic interactions where the strength of the electromagnetic coupling, that is, electric charge (e) remains the same for electrons and protons. Since the equality of the charge coupling, also known as the universality of the electromagnetic interactions follows from the conservation of the electromagnetic current, it was suggested that the weak vector current is also conserved, that is, $\partial_\mu V^\mu(x) = 0$, which leads to the equality of the weak vector coupling for leptons and hadrons.

In fact, they proposed a stronger hypothesis of the isotriplet of the vector currents which goes beyond the hypothesis of CVC and predicts the form factors $f_{1,2}(q^2)$ describing the matrix elements of the weak vector current in terms of the electromagnetic form factors of hadrons. According to this hypothesis, the weak currents V_μ^+, V_μ^-, and the isovector part of the electromagnetic current V_μ^{em} are assumed to form an isotriplet under the isospin symmetry such that f_1 and f_2 are given in terms of the isovector electromagnetic form factors,

$$f_1(q^2) = F_1^p(q^2) - F_1^n(q^2),$$
$$f_2(q^2) = F_2^p(q^2) - F_2^n(q^2).$$

The conservation of the vector current hypothesis, that is, $\partial_\mu V^\mu(x) = 0$ then follows from the conservation of electromagnetic current and implies, using the Dirac equation for $u(p)$ and $\bar{u}(p')$ in Eq. (6.26), that

$$f_3(q^2) = 0.$$

It should be noted that while the isotriplet current hypothesis implies CVC due to the isospin symmetry, the vice versa is not true. In the literature, the term CVC is mostly used to refer to both the isotriplet hypothesis of weak vector currents V_μ^+ and V_μ^- and the conservation of the vector current.

6.3.5   Implications of the CVC hypothesis

(i) **Pion β-decay:** πl_3 **decays**
   The matrix element for the pion β-decays $\pi^\pm \to \pi^0 e^\pm \nu_e(\bar{\nu}_e)$ was derived in Section 6.2.2. This matrix is rewritten using the isospin notation in which $V_\mu^- = V_\mu I^-$, where I^- is the isospin lowering operator in three-dimensional representation, as:

$$\mathcal{M} = \langle \pi^0 | V_\mu^- | \pi^+ \rangle = (f_+(q^2) P_\mu + f_-(q^2) q_\mu) \langle \pi^0 | I^- | \pi^+ \rangle. \tag{6.41}$$

   The conservation of V_μ implies that $q^\mu \langle \pi^0 | V_\mu^- | \pi^+ \rangle = 0$, that is,

$$f_+(q^2) q.P + f_-(q^2) q^2 = 0. \tag{6.42}$$

Since $q.P = m^2_{\pi^+} - m^2_{\pi^0} \approx 0$, Eq. (6.42) implies that $f_-(q^2) = 0$. Moreover, the non-vanishing form factor $f_+(q^2)$ is given in terms of the electromagnetic form factor appearing in the hadronic matrix element for the process $e^- \pi^+ \to e^- \pi^+$, that is,

$$\langle \pi^+ | V^{em}_\mu | \pi^+ \rangle = f^{em}(q^2) P_\mu \langle \chi^+_\pi | \chi^+_\pi \rangle,$$

where $f^{em}(q^2)$ is the charge form factor of the pion with $f^{em}(0) = 1$. Since $\langle \pi^0 | I^- | \pi^+ \rangle = \sqrt{2} \langle \pi^+ | \pi^+ \rangle = \sqrt{2}$, we get $f_+(q^2) = \sqrt{2} f^{em}(q^2)$. In the limit $q^2 \to 0$, $f_+(0) = \sqrt{2} f^{em}(0) = \sqrt{2}$. Therefore, the hypothesis of CVC predicts $f_+(0) = \sqrt{2}$ which is in fair agreement with the experimental value $|f_+(0)| = 1.37 \pm 0.02$, obtained from the decay of $\pi^\mp \to \pi^0 + e^\mp + \bar{\nu}_e(\nu_e)$.

(ii) **Weak magnetism**

The conserved vector current hypothesis predicts that in the low $q^2(\to 0)$ limit,

$$f_1(0) = F^p_1(0) - F^n_1(0) = 1,$$
$$f_2(0) = F^p_2(0) - F^n_2(0) = \mu_p - \mu_n = 3.7 \mu_N. \tag{6.43}$$

This was first verified in the $\beta^\mp$-decays of the $A = 12$ triplet nuclei ^{12}B, ^{12}N, and ^{12}C*, all having $J^P = 1^+$. These states decay by β^-, β^+, and γ decays of magnetic dipole transitions:

$$^{12}\text{B} \longrightarrow {}^{12}\text{C} \, e^- \, \nu_e, \qquad {}^{12}\text{N} \longrightarrow {}^{12}\text{C} \, e^+ \, \nu_e,$$

$$^{12}\text{C}^* \longrightarrow {}^{12}\text{C} \, \gamma,$$

as shown in Figure 6.2. The energy spectrum of the $\beta^-(\beta^+)$-decays are given by [239, 296]:

$$d\Gamma \simeq F(Z, E_e) p_e E_e (\Delta - E_e)^2 \left(1 \pm \frac{8}{3} a E_e \right) dE_e, \tag{6.44}$$

where

$$a \approx \frac{\mu_p - \mu_n}{2M} \left| \frac{f_2(0)}{g_1(0)} \right|, \text{using the conserved vector current hypothesis.} \tag{6.45}$$

We see from the non-relativistic limit of the vector and axial vector currents (Appendix A) that in the limit of small q^2, if the term of the order of $\frac{|\vec{q}|}{2M}$ is retained, then there is an additional contribution to the matrix element of $\beta^\mp$-decays due to the vector terms containing the f_2 term. The correction term to the $\beta^\mp$ spectrum, that is, $\pm\frac{8}{3} aE$ comes due to the interference of the vector and the axial vector terms if the CVC value of $f_2(0) = \mu_p - \mu_n$ is assumed. Several research groups have measured the $\beta^\mp$ spectrum of ^{12}B and ^{12}N and studied the correction factor $\frac{8}{3} aE$. The shape of the correction factor is found to be in agreement with the prediction of CVC given in Eq. (6.44). This term is

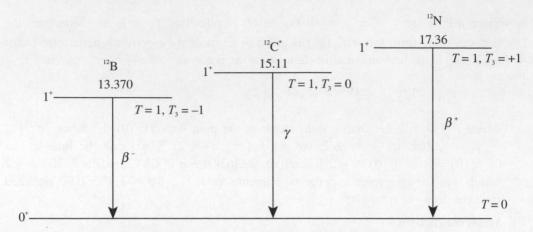

Figure 6.2 β-decay of $A = 12$ nuclei, viz. ^{12}B, ^{12}N, and ^{12}C* [239].

called the weak magnetism because of its structural similarity to the F_2 term in the matrix element of the electromagnetic current (Eq. (6.28)), which describes the electromagnetic interactions due to the magnetic moments of the charged particles. This is one of the major confirmation of the hypothesis of CVC from nuclear β-decays. The hypothesis of CVC has also been confirmed in the case of β-decays of other nuclei and the process of muon capture in nuclei.

6.3.6 Partial conservation of axial vector current (PCAC)

In contrast to the vector current which is conserved, as shown in the previous section, the axial vector current is not conserved. To see this explicitly, we define the matrix element of the axial vector current between one pion state and vacuum which enters in the πl_2 decay of pion (see Section 6.2.1),

$$< 0|A^\mu(x)|\pi^-(q) >= i f_\pi q^\mu e^{-iq.x}, \tag{6.46}$$

where q is the four-momentum of the pion. Taking the divergence of Eq. (6.46),

$$< 0|\partial_\mu A^\mu(x)|\pi^-(q) > = (-i)if_\pi q_\mu q^\mu e^{-iq.x} = f_\pi m_\pi^2 e^{-iq.x}, \tag{6.47}$$

since $q^2 = m_\pi^2$. If the axial vector current A^μ is divergenceless, then either $f_\pi = 0$ or $m_\pi = 0$, implying that the pion is massless or that it does not decay. Since $m_\pi \neq 0$, conservation of the axial vector current implies $f_\pi = 0$, which is not true. Therefore, the axial vector current is not conserved. However, since the pion is the lightest hadron, we can work in the limit of $m_\pi \to 0$, and say that the axial vector current is conserved,

$$\lim_{m_\pi \longrightarrow 0} \partial_\mu A^\mu(x) = 0,$$

which is termed as the partial conservation of axial vector current (PCAC). The hypothesis of PCAC has been very useful in calculating many processes in weak interaction physics and deriving relations between various processes in the limit $m_\pi \to 0$. However, the real predictive power of PCAC lies in making further assumptions about the divergence of the axial vector field $\partial_\mu A^\mu(x)$ that establishes a connection between weak and strong interaction physics. The success of PCAC in various applications where physical processes are calculated is based on the following assumptions:

(i) The divergence of the axial vector field is a pseudoscalar; and the pion is also described by a pseudoscalar field. If it is assumed that both are the same, then the physical pion field is described by the divergence of the axial vector, that is

$$\partial_\mu A^\mu(x) \propto \phi_\pi(x), \tag{6.48}$$

such that

$$\partial_\mu A^\mu(x) = C_\pi \phi_\pi(x), \tag{6.49}$$

and

$$\langle 0|\partial_\mu A^\mu|\pi^-\rangle = f_\pi q^2 e^{-iq\cdot x} = C_\pi \langle 0|\phi_\pi(x)|\pi^-(q)\rangle \tag{6.50}$$

This assumption makes it possible to relate the weak interaction processes induced by A^μ to the pion physics in the strong interaction processes through the matrix element of its derivative, that is, $\partial_\mu A^\mu$.

(ii) Taking the limit $m_\pi \to 0$ (corresponding to the conserved axial vector current) in the processes involving pions and nucleons, makes it easier to evaluate the transition amplitude in many weak processes. If further assumption is made that these amplitudes vary smoothly with q^2 and do not change much over the range of q^2 involved in the processes, then the amplitudes evaluated at $q^2 = 0$ can be extrapolated to the physical limit of $q^2 = m_\pi^2$. This is called the soft pion limit widely used in weak interaction physics. However, there remains an ambiguity whether to take the limit as $m_\pi^2 \to 0$ or $m_\pi \to 0$.

6.3.7 Implications of PCAC

There are important implications of PCAC in the evaluation of the matrix elements of the axial vector current between the hadronic states. Some of the most celebrated applications of PCAC are:

(i) Goldberger–Treiman [297] relation and pseudoscalar coupling,

(ii) Adler–Weisberger [47, 48] relation for axial vector coupling,

(iii) Soft pion theorems and their applications.

In this section, we describe each one of them, very briefly.

(i) Goldberger-Treiman relation and the pseudoscalar coupling

The Goldberger–Treiman relation gives a relation between the axial vector coupling in the weak interaction, $g_1(q^2)$, and the pion–nucleon coupling in the strong interaction, $g_{\pi NN}$ using the hypothesis of PCAC. Consider the matrix element of the axial vector current between $n \to p$ states, that is,

$$< p|A_\mu|n > = \bar{u}(p') \left(g_1(q^2)\gamma_\mu + q_\mu \frac{g_3(q^2)}{M} \right) \gamma_5 u(p). \tag{6.51}$$

The divergence of the current given in this equation in the momentum space is given by:

$$\begin{aligned} q^\mu < p|A_\mu|n > \ &= \ \bar{u}(p') \left(g_1(q^2)\slashed{q} + q^2 \frac{g_3(q^2)}{M} \right) \gamma_5 u(p), \\ &= \ \bar{u}(p') \left(g_1(q^2)(\slashed{p}' - \slashed{p}) + q^2 \frac{g_3(q^2)}{M} \right) \gamma_5 u(p), \\ &= \ \left(2M g_1(q^2) + q^2 \frac{g_3(q^2)}{M} \right) \bar{u}(p')\gamma_5 u(p). \end{aligned} \tag{6.52}$$

Assuming that the matrix element of A_μ between the nucleon states is dominated by one pion pole, as shown in Figure 6.3, we can write:

$$< p|A_\mu|n > \ = \ -\sqrt{2}g_{\pi NN}\bar{u}_p(p')\gamma_5 u_n(p) \frac{1}{q^2 - m_\pi^2} f_\pi q_\mu. \tag{6.53}$$

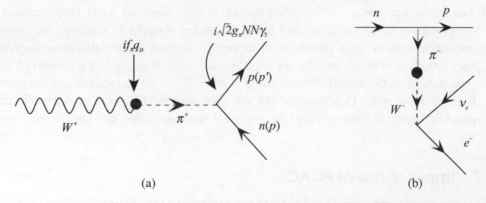

Figure 6.3 Coupling of the axial vector current to a particle of zero mass.

Calculating the divergence of this equation, we obtain

$$q^\mu < p|A_\mu|n > = -\sqrt{2}g_{\pi NN}\bar{u}_p(p')\gamma_5 u_n(p) \frac{1}{q^2 - m_\pi^2} f_\pi q^\mu q_\mu, \tag{6.54}$$

which results in

$$q^\mu < p|A_\mu|n > = -\sqrt{2}g_{\pi NN}\bar{u}_p(p')\gamma_5 u_n(p) \frac{1}{q^2 - m_{\pi^2}} f_\pi m_\pi^2. \tag{6.55}$$

In the pion pole dominance, the pseudoscalar form factor $g_3(q^2)$ given in Eq. (6.51) is related to the matrix element given in Eq. (6.53). Comparing the second term of Eq. (6.52) with Eq. (6.55), we obtain

$$-\sqrt{2}g_{\pi NN}\bar{u}(p')\gamma_5 u(p)\frac{1}{q^2-m_\pi^2}f_\pi m_\pi^2 = q^2\frac{g_3(q^2)}{M}m_\pi\bar{u}(p')\gamma_5 u(p), \qquad (6.56)$$

which gives

$$\frac{g_3(q^2)}{M} = -\frac{\sqrt{2}g_{\pi NN}f_\pi}{q^2-m_\pi^2}. \qquad (6.57)$$

Using this value of $g_3(q^2)$ in Eq. (6.52), we get

$$q^\mu\langle p|A_\mu|n\rangle = \left(2Mg_1(q^2)+q^2\left(-\frac{\sqrt{2}g_{\pi NN}f_\pi}{q^2-m_\pi^2}\right)\right)\bar{u}(p')\gamma_5 u(p). \qquad (6.58)$$

In the chiral limit ($m_\pi \to 0$), the axial vector current is conserved. that is

$$\left(2Mg_1(q^2)-\sqrt{2}g_{\pi NN}f_\pi\right)\bar{u}(p')\gamma_5 u(p) = 0. \qquad (6.59)$$

The above exprssion yields

$$\sqrt{2}Mg_1(q^2) = g_{\pi NN}f_\pi, \qquad (6.60)$$

with $g_1(q^2) = \dfrac{g_1(0)}{\left(1-\dfrac{q^2}{M_A^2}\right)}$ and $g_1(0) = 1.267$. Equation (6.60) is known as the Goldberger–

Treiman relation. This relation was derived by Goldberger and Treiman [297], Wolfenstein [277], and Leite Lopes [276] around the same time (in 1958) using the dispersion relation and perturbation theory.

(ii) Renormalization of axial vector coupling g_1

Adler [47] and Weisberger [48] used the hypothesis of PCAC and methods of current algebra to express the axial vector coupling $g_1(0)$ in terms of the off shell ($m_\pi^2 \to 0$) pion–nucleon scattering cross sections. Defining quantities called chirality $\chi^i(t)$ in terms of the axial vector current $A_i^\mu(x)$ as

$$\chi_i(t) = \int d\vec{x} A_i^0(x) \qquad (i=1,2,3) \qquad (6.61)$$

and using the PCAC relation in Eq. (6.49), one can write

$$\frac{d}{dt}\chi_i(t) = C_\pi \int d\vec{x}\,\phi_i^\pi(x), \qquad (6.62)$$

relating the axial vector current with the pion fields through its derivative. Assuming that the weak current satisfies a current algebra corresponding to $SU(2)\times SU(2)$ symmetry, we obtain the commutation relation for $A_i(x)$ as

$$[A_i^0(x_1), A_j^0(x_2)]_{x_0=y_0} = \delta(\vec{x} - \vec{y})i\epsilon_{ijk}V_k^0, \tag{6.63}$$

leading to the following commutation relation for chiralities $\chi_i(t)$.

$$[\chi_+(t), \chi_-(t)] = 2I_3, \tag{6.64}$$

where $+$, $-$, and 3 are the raising, lowering, and third components of the isospin. Taking the matrix element of Eq. (6.64) on both sides between the proton states, we get

$$\langle p|[\chi_+(t), \chi_-(t)]|p\rangle = 2\langle p|I_3|p\rangle, \tag{6.65}$$

in the limit of $q^2 \to 0$, the R.H.S. gives 1 while the L.H.S. is evaluated by inserting a complete set of intermediate states. The one-particle nucleon state in the intermediate state gives g_1. The contribution from other intermediate states like Δ resonance and higher resonances are expressed in terms of the cross section for these processes in the pion-nucleon scattering. The following result is obtained

$$1 - \frac{1}{g_A^2} = \frac{4M^2}{g_{\pi NN}^2}\frac{1}{\pi}\int_{M+m_\pi}^{\infty}\frac{WdW}{W^2 - M^2}[\sigma_0^+(W) - \sigma_0^-(W)], \tag{6.66}$$

where W is the center of mass energy and $\sigma_0^\pm(W)$ are the cross sections for zero mass pion ($m_\pi^2 \to 0$) for $\pi^\pm$ scattering on proton. Using the experimental values of the cross section at $q^2 = m_\pi^2$ and extrapolating it to $m_\pi^2 \to 0$ limit, the integral is performed numerically to evaluate g_A, which is found to be 1.24, in good agreement with the experiments. It demonstrated that the renormalisation of g_A from its quark value of 1 is due to the presence of strong interactions through the πN scattering.

(iii) Soft pion theorems and pion physics

The hypothesis of PCAC and current algebra have been used to obtain various predictions on the production of pions induced by photons, electrons, or pions in the very low energy region, where the extrapolation of the results for $m_\pi^2 \to 0$, that is, the soft pion limit, to the physical pion could be valid. Some of the most important results are as follows [298]:

(a) S wave amplitude in the threshold photoproduction of pions.

(b) Determination of $g_1(q^2)$ in terms of the amplitudes of the threshold electroproduction of pions.

(c) Relations between the threshold one-pion production to the two-pion production induced by photons, electrons, and neutrinos.

(d) Adler's consistency conditions in neutrino nucleon scattering in the $q^2 \to 0$ region.

(e) Various consistency conditions for the pion nucleon scattering amplitudes in the very low energy of pions.

6.3.8 G-parity and second class currents

G-parity is a multiplicative quantum number first used to classify the multipion states in pp and πp collisions [299] and later used by Weinberg [300] to classify weak hadronic currents. It is defined as the product of C, the charge conjugation, and a rotation by $180°$ about the Y-axis in the isotopic spin (isospin) space,

$$G = Ce^{i\pi I_Y}. \tag{6.67}$$

Since strong interactions are invariant under C and isospin, they are also invariant under G-parity. The G-parity is a very useful concept in the study of pion production in N$\bar{\text{N}}$ collisions. Since weak currents involve bilinear covariants formed out of nucleon fields $\bar{\psi}(p')$ and $\psi(p)$, their transformations can be well defined under G-parity. The weak vector and axial vector currents between a neutron and a proton are defined in Eqs. (6.26) and (6.27). The vector and axial vector terms corresponding to f_1 and g_1 enter in the case of leptons and contain no q^2 dependence because they are point particles and $f_1 = g_1 = 1$. However, in the case of hadrons, which have a composite structure and are strongly interacting particles, there are additional terms in the definition of the matrix elements of the vector and the axial vector currents as defined in Eqs. (6.26) and (6.27). Consequently, all the form factors acquire a q^2 dependence, that is, $f_i(q^2)$ and $g_i(q^2)$ and deviate from unity to get renormalized even at $q^2 = 0$ (except in the case of $f_1(q^2)$, where $f_1(0) = 1$ due to CVC). Thus, the additional terms like $f_2(q^2)$, $f_3(q^2)$, $g_2(q^2)$, and $g_3(q^2)$ which are induced due to the presence of the strong interactions should respect the symmetries of the strong interactions and should have the same G-parity as the original $f_1(q^2)$ and $g_1(q^2)$, respectively, in the vector and axial vector sector, that is, $GV^\mu G^{-1} = V^\mu$ and $GA^\mu G^{-1} = -A^\mu$. Since the currents belong to the triplet representation of the isospin, all the terms have similar transformation properties under the rotation $e^{i\pi I_Y}$. It is their transformation properties under C-parity which defines their relative transformation properties under G-parity. Under G-parity, the bilinear terms in Eqs. (6.26) and (6.27) transform as:

$$\bar{u}_p u_n \xrightarrow{G} -\bar{u}_p u_n \qquad \text{(associated with } f_3\text{)} \tag{6.68}$$

$$\bar{u}_p \gamma_5 u_n \xrightarrow{G} -\bar{u}_p \gamma_5 u_n \qquad \text{(associated with } g_3\text{)} \tag{6.69}$$

$$\bar{u}_p \gamma_\mu \gamma_5 u_n \xrightarrow{G} -\bar{u}_p \gamma_\mu \gamma_5 u_n \qquad \text{(associated with } g_1\text{)} \tag{6.70}$$

while

$$\bar{u}_p \gamma^\mu u_n \xrightarrow{G} \bar{u}_p \gamma^\mu u_n \qquad \text{(associated with } f_1\text{)} \tag{6.71}$$

$$\bar{u}_p \sigma^{\mu\nu} u_n \xrightarrow{G} \bar{u}_p \sigma^{\mu\nu} u_n \qquad \text{(associated with } f_2\text{)} \tag{6.72}$$

$$\bar{u}_p \sigma^{\mu\nu} \gamma_5 u_n \xrightarrow{G} \bar{u}_p \sigma^{\mu\nu} \gamma_5 u_n. \qquad \text{(associated with } g_2\text{)} \tag{6.73}$$

What is observed from Eqs. (6.68)–(6.73) is that the bilinear terms associated with f_2 transforms the same way as f_1 does, while f_3 transforms in the opposite way. Similarly, g_3 transforms the same way as g_1 does while g_2 transforms in a different way. It was Weinberg [300] who first used the G-properties to classify weak currents. He called the currents associated with f_1, f_2, g_1, and g_3, which are invariant under G-parity, as first class currents, and the currents associated with f_3 and g_2, which violate G-parity, as second class currents. Consequently, if the G invariance is valid in weak interactions, only currents with form factors $f_1(q^2)$, $f_2(q^2)$, $g_1(q^2)$, and $g_3(q^2)$ should exist and $f_3(q^2) = g_2(q^2) = 0$. It should be noted that $f_3(q^2) = 0$ is also predicted as a consequence of CVC hypothesis.

6.4 Semileptonic Weak Decays of Hadrons with Strangeness

Around the time pions were discovered, a new class of particles named strange particles were also discovered in the cosmic ray experiments; these particles belonged to both categories of bosons and fermions. Strange bosons are known as K-mesons and strange fermions are known as hyperons. They are copiously produced in strong interactions and decay through weak and electromagnetic processes; they have a long lifetime. It was the study of the production of K-mesons and their decay into pions that gave rise to the $\tau - \theta$ puzzle leading to the discovery of parity violation in weak interactions. With the availability of electron and neutrino beams having high energy and high intensity, these strange particles can also be produced through the electromagnetic and weak interactions in the processes shown in Table 6.6; however, the production cross sections are small. In electromagnetic interactions, the cross sections are of the order of $\sim 10^{-31}$ cm^2 while in weak interactions, the cross sections are extremely small, that is, of the order of $\sim 10^{-41}$ cm^2 as compared to the strong interaction production cross section ($\sim 10^{-27}$ cm^2).

The properties and the quark content of strange mesons and hyperons are given in Tables 6.2 and 6.3 and the dominant decay modes of K mesons and hyperons are given in Tables 6.4 and 6.5, respectively. Strange particles decay through semileptonic, nonleptonic, and radiative

Table 6.6 Production modes of kaons.

Strong interaction	Electromagnetic interaction	Weak interaction
$\pi^- + p \longrightarrow \Lambda + K^0$	$l^- + p \longrightarrow l^- + \Lambda + K^+$	$\nu_l + p \longrightarrow l^- + p + K^+$
$\longrightarrow \Sigma^0 + K^0$	$\longrightarrow l^- + \Sigma^0 + K^+$	$\nu_l + n \longrightarrow l^- + n + K^+$
$\longrightarrow \Sigma^- + K^+$	$\longrightarrow l^- + \Sigma^+ + K^0$	$\longrightarrow l^- + p + K^0$
$\pi^+ + p \longrightarrow \Sigma^+ + K^+$	$l^- + n \longrightarrow l^- + \Lambda + K^0$	
$\longrightarrow K^+ + \bar{K}^0 + p$	$\longrightarrow l^- + \Sigma^0 + K^0$	
$p + \bar{p} \longrightarrow \pi^+ + K^- + K^0$	$l = e, \mu, \tau$	
$\longrightarrow \pi^- + K^+ + \bar{K}^0$		

decay modes. Semileptonic and nonleptonic as well as weak radiative decays of mesons and hyperons have been the focus of experimental studies for a long time. The following observations can be made from the study of weak decays of strange mesons and hyperons:

i) The strangeness changing semileptonic decays follow the $|\Delta S| = 1$ and $\Delta S = \Delta Q$ rule, that is, $\Delta Q = +1$, $\Delta S = +1$, or $\Delta Q = -1$, $\Delta S = -1$ like the processes

$$\Lambda\,(\Sigma^0)\ \longrightarrow\ p + l^- + \bar{\nu}_l, \qquad K^{\pm} \longrightarrow \pi^0 + l^{\pm} + \nu_l(\bar{\nu}_l),$$
$$\Sigma^- \ \longrightarrow\ n + l^- + \bar{\nu}_l, \qquad K^0_L \longrightarrow \pi^{\pm} + l^{\mp} + \bar{\nu}_l(\nu_l),$$

are allowed, but $\Delta Q = -1$, $\Delta S = +1$ like the processes

$$\Sigma^+ \ \nrightarrow\ n + e^+ + \nu_e$$

are not allowed

The selection rule $\Delta Q = \Delta S$ implies the isospin properties of these currents to be $\Delta I = \frac{1}{2}$.

ii) The strangeness changing semileptonic decays with $\Delta S = 1$ are suppressed as compared to the strangeness conserving $\Delta S = 0$ decay by a factor of $20 - 25$. For example:

$$\frac{\Gamma(\Lambda \longrightarrow p + e^- + \bar{\nu}_e)}{\Gamma(n \longrightarrow p + e^- + \bar{\nu}_e)} \approx\ 0.05 \qquad \text{and} \qquad \frac{\Gamma(K^- \longrightarrow l^- + \bar{\nu}_l)}{\Gamma(\pi^- \longrightarrow l^- + \bar{\nu}_l)} \approx 0.05.$$

iii) The strangeness changing semileptonic decays with $\Delta S = 1$ and $\Delta Q = 0$ are highly suppressed by a factor of $10^{-8} - 10^{-9}$. This is known as the absence of the flavor changing neutral current (FCNC). For example,

$$\frac{\Gamma(K_L \longrightarrow \mu^+ \mu^-)}{\Gamma(K^+ \longrightarrow \mu^+ \nu)} \approx\ 10^{-9}.$$

iv) The suppression of the strangeness changing currents in the strange mesons and hyperons are of the same order, suggesting a universal suppression of the strength of the coupling of $\Delta S = 1$ current.

v) The nonleptonic weak decays of the strange mesons and hyperons obey the $\Delta I = \frac{1}{2}$ rule. This means that in decays like $K^{\pm} \rightarrow \pi^{\pm}\pi^0, K^0 \rightarrow \pi^{\pm}\pi^{\mp}$, or $Y \rightarrow N\pi$, the transition amplitude corresponding to $\Delta I = \frac{1}{2}$ transitions dominate over the amplitudes corresponding to $\Delta I \geq \frac{3}{2}$ transitions.

vi) The nonleptonic weak decays of neutral kaons, that is, K^0_L, violate CP invariance; this phenomenon has been observed in the decays $K^0_L \rightarrow \pi^+ \pi^-$ and $K^0_L \rightarrow \pi^0 \pi^0$.

vii) The hypothesis of CVC is also valid in the case of the strangeness changing decays but the hypothesis of PCAC (partially conserved axial current) and its predictions are not established at the same level of validity as in the case of $\Delta S = 0$ currents.

Based on these observations, Cabibbo [59] in 1963 proposed an extension of the $V - A$ theory which describes the weak interaction of strange particles based on the SU(3) symmetry of weak currents. The theory established the universality of weak interactions involving both $\Delta S = 0$ and $\Delta S = 1$ currents and also led to the concept of quark mixing in weak interactions. In the following sections, we will describe Cabibbo's extension of the phenomenological theory of $V - A$ interaction to the strangeness sector [59].

6.4.1 The Cabibbo theory and the universality of weak interactions

The Cabibbo theory was proposed against the backdrop of two outstanding problems of that time. One was the apparent suppression of the $\Delta S = 1$ currents with respect to the $\Delta S = 0$ current couplings. The other problem was the small discrepancy in the coupling strengths of the vector currents in the case of weak decays of muons and neutrons. The coupling strengths G_μ in the muon decay($\mu^- \to e^- + \bar{\nu}_e + \nu_\mu$) and G_β in the neutron decay($n \to p + e^- + \bar{\nu}_e$) were found to be $G_\mu = 1.1663787 \times 10^{-5}$ GeV^{-2} and $G_\beta - G_F \approx 1.136 \pm 0.003010^{-5}$ GeV^{-2}, respectively.

Motivated by the approximate equality of G_μ and G_β, the hypothesis of the conservation of vector current (CVC) was proposed [44, 42] with the hope that the small discrepancy may be explained by the different contributions of the radiative corrections in the case of μ decays and neutron β decays. However, detailed calculations by Berman [263], Kinoshita, and Sirlin [261] showed that the corrections enhanced the discrepancy instead of reducing it, casting doubts about the validity of the CVC hypothesis. The difference in the numerical values of G_β and G_μ seemed to have a deeper origin in the physics of weak interactions and needed to be understood. In this background, Gell-Mann and Levy [58] suggested a mechanism based on the Sakata model [60]. In the Sakata model, the proton, neutron, and lambda hyperon are elementary particles; all the other particles are made up of these three particles. In the Gell-Mann and Levy model, the $\Delta S = 1$ currents appear in the interaction Lagrangian with a smaller strength as compared to the $\Delta S = 0$ currents. The hadronic current $J_h^\mu(x)$ is written as

$$J_h^\mu = \frac{1}{\sqrt{1+\epsilon^2}}\overline{\psi}_p(x)\gamma^\mu(1-\gamma_5)[\psi_n(x) + \epsilon\psi_\Lambda(x)]. \tag{6.74}$$

With $\epsilon \approx 20\%$, one is able to explain the suppression of the strength of the $\Delta S = 1$ currents. Moreover with $\epsilon = 20\%$, the factor $(1+\epsilon^2)^{-\frac{1}{2}} = 0.97$, which is of the right order of magnitude to make $\frac{G_\mu}{\sqrt{1+\epsilon^2}}$ consistent with G_β. This model could explain the suppression of the strength of the $\Delta S = 1$ current coupling as compared to the $\Delta S = 0$ coupling and reduce the strength of G_β to restore the relation $G_\beta \approx G_\mu$. However, the model could not explain the enhancement of nonleptonic weak decays like $\Lambda \to n\pi^0(p\pi^-)$ compared to the strengths of semileptonic weak decays with $\Delta S = 1$ and the $\Delta I = \frac{1}{2}$ rule in the nonleptonic decays. Moreover, the model was based on the Sakata model which was later found to be inadequate for explaining the structure of hadrons. Therefore, the mechanism proposed by Gell-Mann and Levy was not pursued any further.

Cabibbo considered hadronic currents in the context of SU(3) symmetry in which $\Delta S = 0$ and $\Delta S = 1$ currents would appear as members of the octet current. The implementation of the universality between leptonic and hadronic currents would require that the $\Delta S = 0$ and $\Delta S = 1$ hadronic current, which appear in the weak interaction Hamiltonian, may have different strengths allowing the suppression of $\Delta S = 1$ current with respect to $\Delta S = 0$ current; however, the sum of $\Delta S = 0$ and $\Delta S = 1$ currents have be normalized to the normalization of the lepton currents, that is, e and μ currents. This will restore the $e - \mu$ universality as well as the CVC in the case of weak vector currents. Cabibbo, therefore, proposed the hadronic weak vector current entering the weak interaction Hamiltonian to be

$$V_\mu^h = aV_\mu(\Delta S = 0) + bV_\mu(\Delta S = 1), \tag{6.75}$$

such that $|a|^2 + |b|^2 = 1$. A convenient parameterization for a and b, that is, $a = \cos\theta_C$ and $b = \sin\theta_C$ was taken by Cabibbo, where θ_C is known after him as the Cabibbo angle. Cabibbo also extended this analogy to axial vector currents to write the weak hadronic current as

$$J_\mu^h = \cos\theta_C J_\mu(\Delta S = 0) + \sin\theta_C J_\mu(\Delta S = 1), \tag{6.76}$$

with $J_\mu = V_\mu - A_\mu$. This provided a natural suppression of the $\Delta S = 1$ weak currents as compared to the $\Delta S = 0$ weak currents by a factor of $\tan\theta_C$. It also provided a suppression of the $\Delta S = 0$ transition in the hadronic vector current by a factor of $\cos\theta_C$ to bring about universality of weak vector currents making $G_\mu \cos\theta_C = G_F \cos\theta_C \approx G_\beta$. An experimental analysis of the $\Delta S = 1$ β-decays of hyperons and kaons can provide the value of θ_C.

In an earlier work [59], Cabibbo determined the value of θ_C from the ratios of the leptonic and semileptonic decays of kaons and pions; he found $\tan\theta_C = 0.2623$ from leptonic decays and $\tan\theta_C = 0.266$ from semileptonic decays. However, the analysis of the available data on semileptonic decays of Λ and other hyperons yields $\tan\theta_C = 0.2327$, which is the current global value [117]. In formulating the theory, Cabibbo ignored the enhancement of $\Delta I = \frac{1}{2}$ currents in nonleptonic decays and believed it to have a different origin than the physics of the Cabibbo angle and the suppression of the semileptonic $\Delta S = 1$ decays. In fact, the theoretical origin of the Cabibbo angle is still not understood; however, phenomenologically, the theory has been very successful in explaining the experimental data on $\Delta S = 1$ decays of strange mesons and hyperons and has provided important inputs toward formulating the standard model of electroweak interactions. The theory also succeeded in solving the two problems posed at the beginning of the section.

6.4.2 The Cabibbo theory in the quark model and quark mixing

Interestingly, around the time Cabibbo proposed the idea of an SU(3) structure for weak currents, the quark model of the hadrons proposed by Gell-Mann [61] and Zweig [62, 63] with SU(3) symmetry became very successful in explaining the structure of hadrons as well as their decay properties. The Cabibbo theory of $V - A$ currents was then formulated in terms of quarks. The Feynman diagrams for some of the processes at the quark level have been shown in Figure 6.4; the diagrams show that the mesonic and baryonic decays in terms of the quark–quark transitions, are described as:

$$n \longrightarrow p + e^- + \bar{\nu}_e \implies d \longrightarrow u + e^- + \bar{\nu}_e$$
$$\Lambda \longrightarrow p + e^- + \bar{\nu}_e \implies s \longrightarrow u + e^- + \bar{\nu}_e$$
$$K^+ \longrightarrow \mu^+ + \nu_\mu \implies \bar{s}u \longrightarrow |0\rangle \implies u \to s$$
$$K^- \longrightarrow \mu^- + \bar{\nu}_\mu \implies \bar{u}s \longrightarrow |0\rangle \implies s \to u$$
$$\pi^+ \longrightarrow \mu^+ + \nu_\mu \implies \bar{d}u \longrightarrow |0\rangle \implies u \to d$$

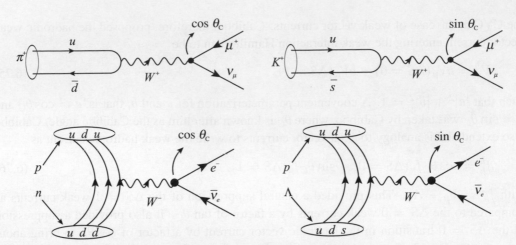

Figure 6.4 Top panel: (left) pion and (right) kaon decay. Bottom panel: (left) neutron and (right) lambda decay.

In the pre-Cabibbo $V - A$ theory, the hadronic vector and axial vector currents in the quark model are written in term of the quark fields $q(= u, d, s)$ as:

$$J_\mu^h \sim g \left[\bar{u}_u(p')\gamma_\mu(1 - \gamma_5)u_d(p) + \bar{u}_u(p')\gamma_\mu(1 - \gamma_5)u_s(p) \right]. \tag{6.77}$$

In the Cabibbo theory, they are written as

$$J_\mu^h \sim g \left[\bar{u}_u(p')\gamma_\mu(1 - \gamma_5)(d\cos\theta_C + s\sin\theta_C) \right]. \tag{6.78}$$

If we define the linear combination of the d and s quark in Eq. (6.78) as d', that is,

$$d' = d\cos\theta_C + s\sin\theta_C,$$

then the hadronic current J_μ^h can be rewritten as

$$J_\mu^h \sim g \left[\bar{u}_u(p')\gamma_\mu(1 - \gamma_5)d' \right], \tag{6.79}$$

meaning that d' takes part in the weak interaction and not the d quark. The following observations may be made from the aforementioned discussions:

(i) The $\Delta S = 0$ current in the nuclear β-decays occurs with a reduced strength of $G_F \cos\theta_C$ and G_F becomes comparable to G_μ, the coupling strength in the case of pure leptonic decays like $\mu \longrightarrow e^- + \nu_\mu + \bar{\nu}_e$. This makes the universality of the weak current in the leptonic and hadronic sectors more apparent, that is, $G_\mu = G_F$ in the case of vector currents.

(ii) The Cabibbo theory explains satisfactorily the suppression of $\Delta S = 1$ decay rate as compared to the $\Delta S = 0$ decay rate by a factor of $\tan^2\theta_C$, an observation for which the theory was formulated.

(iii) In this model, $\Delta S = 0$ currents are due to $d(\bar{d}) \rightleftharpoons u(\bar{u})$ transitions corresponding to $\Delta I = 1$ isovector currents, while $\Delta S = 1$ transitions are due to $s(\bar{s}) \rightarrow u(\bar{u})$ transitions corresponding to $\Delta I = \frac{1}{2}$ both with $\Delta Q = \Delta S$ rules. This provides a natural explanation for the absence of $\Delta Q = -\Delta S$ currents.

6.4.3 Applications of Cabibbo theory: K decays

(i) Kl_2 decays

We see from Table 6.4 that only $K^\pm$ decays into two leptons through charged current interactions like the decay $K^\pm \rightarrow l^\pm \nu_l(\bar{\nu}_l)$, which is suppressed as compared to πl_2 decays (Section 6.2.1); due to a suppression factor of $\tan^2 \theta_C$ in the decay rate. The two-body decays of neutral kaons like $K_{L,S} \rightarrow \mu^+\mu^-$, e^+e^- are highly suppressed ($B.R. < 10^{-8}$). The matrix element for the Kl_2 mode of the charged kaon decay is obtained by following the same method as that used for writing the matrix elements for πl_2 decays, that is,

$$\mathcal{M} = \frac{G_F}{\sqrt{2}} \langle 0|V^\mu - A^\mu|K(p)\rangle \left[\bar{u}_l \gamma_\mu (1 - \gamma_5) u_\nu(k)\right]. \tag{6.80}$$

We write the matrix element of $V - A$ current in analogy with the πl_2 decays (Section 6.2.1) as:

$$\langle 0|A^\mu|K^\pm(p)\rangle = i\sin\theta_C f_K p_\mu, \qquad \text{and} \qquad \langle 0|V^\mu|K^\pm(p)\rangle = 0,$$

leading to the decay rate

$$\Gamma(K^\pm \longrightarrow l_\nu^\pm + \nu_l(\bar{\nu}_l)) = \frac{G_F^2 f_K^2 \sin^2\theta_C}{8\pi} m_l^2 m_K \left(1 - \frac{m_l^2}{m_K^2}\right)^2. \tag{6.81}$$

The experimentally observed value of $\tau = \frac{1}{\Gamma} = 1.237 \pm 0.003 \times 10^{-8}$ s [117], may be obtained with $\sin\theta_C = 0.221$, $f_K = 160$ MeV.

Moreover, a comparison with the decay rate of electron/muon decay modes yields

$$R^{e\mu} = \frac{\Gamma(K^- \longrightarrow e^- \bar{\nu})}{\Gamma(K^- \longrightarrow \mu^- \bar{\nu})} = \frac{m_e^2}{m_\mu^2} \left(\frac{m_K^2 - m_e^2}{m_K^2 - m_\mu^2}\right)^2 = 2.58 \times 10^{-5}, \tag{6.82}$$

which is in agreement with the experimentally observed value for this ratio, $R^{\text{exp}} = (2.425 \pm 0.012) \times 10^{-5}$. The branching ratio (BR) is [117]

$$\frac{\Gamma(K^- \longrightarrow \mu^- \bar{\nu}_\mu)}{\Gamma(K^- \longrightarrow all)} = 0.6356. \tag{6.83}$$

Using the experimental value of $\Gamma(K^- \rightarrow \mu^- \bar{\nu}_\mu)$ as $5.75 \times 10^7 \text{s}^{-1}$ and the present value of $\sin\theta_C$, the ratio $\frac{f_K}{f_\pi}$ is obtained as 1.1928 [117]. Theoretically, the value of f_K/f_π is calculated using lattice QCD (quantum chromodynamics) and is found to be very close to this value [301].

(ii) Kl_3 **decays**

The three-particle semileptonic decay of charged and neutral mesons are

$$
\begin{aligned}
K^+ &\longrightarrow \pi^0 \mu^+ \nu_\mu, & K_l &\longrightarrow \pi^- \mu^+ \nu_\mu, & (6.84)\\
&\longrightarrow \pi^0 e^+ \nu_e, & &\longrightarrow \pi^+ \mu^- \bar{\nu}_\mu, & (6.85)\\
K^- &\longrightarrow \pi^0 \mu^- \bar{\nu}_\mu, & &\longrightarrow \pi^- e^+ \nu_e, & (6.86)\\
&\longrightarrow \pi^0 e^- \bar{\nu}_e, & &\longrightarrow \pi^+ e^- \bar{\nu}_e. & (6.87)
\end{aligned}
$$

The amplitude in the Cabibbo $V - A$ theory for the process $K^+(p_K) \to \pi^0(p_\pi)\mu^+(p_l)$ $\nu_\mu(p_{\nu_l})$ is written in analogy with the πl_3 decays and has the general form

$$
\mathcal{M} = \frac{G_F}{\sqrt{2}} \sin \theta_C [f_+(q^2) P^\mu + f_-(q^2) q^\mu] \bar{u}_l \gamma_\mu (1 - \gamma_5) u_{\nu_l}, \text{where } l = e, \mu
$$

with
$$
P^\mu = p_K^\mu + p_\pi^\mu, q^\mu = p_K^\mu - p_\pi^\mu = p_l^\mu + p_{\nu_l}^\mu. \tag{6.88}
$$

The condition of the conservation of the weak vector current, that is, $q_\mu \langle p_K | V_\mu | p_\pi \rangle = 0$ does not lead to $f_-(q^2) = 0$ because $q.P = m_K^2 - m_\pi^2 \neq 0$. However, we can see that the contribution of $f_-(q^2)$ will be small because

$$
f_-(q^2) q^\mu \bar{u}_l \gamma_\mu (1 - \gamma_5) u_{\nu_l} = f_-(q^2) m_l \bar{u}_l (1 - \gamma_5) u_{\nu_l} \tag{6.89}
$$

and is proportional to the lepton mass. Therefore, Ke_3 decays are sensitive only to $f_+(q^2)$ while $K\mu_3$ decays are sensitive to both $f_+(q^2)$ and $f_-(q^2)$. Defining $\xi = \frac{f_-}{f_+}$, we get

$$
\begin{aligned}
\mathcal{M} &= \frac{G_F}{\sqrt{2}} \sin \theta_C [f_+(q^2)(p_K^\mu + p_\pi^\mu) \bar{u}_l \gamma_\mu (1 - \gamma_5) u_{\nu_l} + \xi f_+(q^2) m_l \bar{u}_l (1 - \gamma_5) u_{\nu_l}] \\
&= \frac{G_F}{\sqrt{2}} \sin \theta_C f_+(q^2) [2 p_K^\mu \bar{u}_l \gamma_\mu (1 - \gamma_5) u_{\nu_l} + (\xi - 1) m_l \bar{u}_l (1 - \gamma_5) u_{\nu_l}].
\end{aligned}
$$

The form factors are the functions of q^2 and are generally parameterized as

$$
f_\pm(q^2) = \left[1 + \lambda_\pm \left(\frac{q^2}{m_\pi^2} \right) \right] f_\pm(0). \tag{6.90}
$$

The analysis of $K\mu_3$ decays is done in terms of $f_+(q^2)$ and $f_0(q^2)$, where $f_0(q^2)$ is defined in terms of $f_-(q^2)$ and $f_+(q^2)$ as

$$
f_0(q^2) = f_+(q^2) + \frac{q^2}{m_K^2 - m_\pi^2} f_-(q^2) = f_0(0) \left(1 + \lambda_0 \frac{q^2}{m_\pi^2} \right). \tag{6.91}
$$

Here $f_0(q^2)$ is called a scalar form factor. If $f_-(q^2)$ is well behaved near $q^2 \approx 0$, that is, it does not have a pole, then $f_0(0) = f_+(0)$ follows from Eq. (6.91) and $K\mu_3$ decays are described in terms of the parameters $f_+(0)$, λ_+, and λ_0.

The energy distribution is given by [63, 238, 239, 302, 303]:

$$
\frac{d\Gamma}{dE_\pi dE_\mu} \propto G_F^2 \sin^2 \theta_C f_+^2(q^2) [A + B\xi(q^2) + C\xi^2(q^2)], \tag{6.92}
$$

where $\quad \zeta = \dfrac{f_-(q^2)}{f_+(q^2)}, \quad$ and

$$A = m_K \left[2(E_l E_\nu - m_K E'_\pi) + m_l^2 \left(\dfrac{E'_\pi}{4} - E_\nu \right) \right],$$

$$B = m_l^2 \left(E_\nu - \dfrac{E'_\pi}{2} \right),$$

$$C = \dfrac{m_l^2 E'_\pi}{4}, \quad E'_\pi = \dfrac{m_K^2 + m_\pi^2 - m_l^2}{2m_K} - E_\pi.$$

The three parameters $f_+(q^2), \lambda_+$, and λ_0 are determined from the experimental results of $K\mu3$ and $Ke3$ decay of charged and neutral kaons. The parameters are found by analyzing the total decay rate, energy distribution of the charged particles, branching ratio for $K\mu3$ $(Ke3)$ decay, and the polarization of muons (electrons) in the final state. It can be shown that in the case of $Ke3$ decays $(m_e \to 0)$, the decay rate is given by:

$$d\Gamma = \dfrac{G_F^2 \sin^2 \theta_C}{12\pi^3} m_K f_+^2(q^2) |\vec{p}_\pi|^3 dE_\pi, \tag{6.93}$$

which determines $f_+(q^2)$.

The following are the main features of the aforementioned study:

i) The $Ke3$ decay determines the value of $f_+(q^2)$ while $K\mu3$ decays give the values of $f_+(0), \lambda_+$, and λ_0. An average value of $\sin \theta_C |f_+(0)|$ from these experiments is found to be 0.2163 within 0.1% experimental uncertainties [301], from which $f_+(0)$ is found to be 0.961 ± 0.008 [304].

ii) Using the expression given in Eq. (6.92), the function $\rho(E_\pi, E_\mu)$ given by

$$\rho(E_\pi, E_\mu) = \sin^2 \theta_C f_+^2(q^2)[A + B\zeta + C\zeta^2] \tag{6.94}$$

is studied and the values of f_+ and ζ are determined which give λ_+ and λ_0.

iii) The branching ratio of $\dfrac{\Gamma(K\mu3)}{\Gamma(Ke3)}$ is determined theoretically using the values of $f_+(q^2), \lambda_+$, and λ_0 and is compared with the experimentally observed values.

iv) These parameters are also sensitive to the muon polarization $(\mathcal{P})$, which is given as $\mathcal{P} = \dfrac{\vec{A}}{|\vec{A}|}$, where

$$\vec{A} = a_1(\zeta)\vec{p}_\mu - a_2(\zeta)\left[\dfrac{\vec{p}_\mu}{m_\mu}[m_K - E_\pi + \dfrac{\vec{p}_\pi \cdot \vec{p}_\mu}{|\vec{p}_\mu|^2}(E_\mu - m_\mu)] + \vec{p}_\pi \right]$$

$$+ m_K \operatorname{Im}\zeta(\vec{p}_\pi \times \vec{p}_\mu) \tag{6.95}$$

with $\quad a_1(\zeta) = \left(\dfrac{2m_K^2}{m_\mu} \right) \left[E_\nu + E'_\pi \operatorname{Re} b(q^2) \right],$

$$a_2(\xi) = m_K^2 + 2\,\mathrm{Re}\,b(q^2)m_K E_\mu + m_\mu^2|b(q^2)|^2,$$

$$b(\xi) = \frac{1}{2}[\xi(q^2) - 1].$$

v) Any component of the polarization lying in a direction perpendicular to the decay plane is due to $\mathrm{Im}\xi \neq 0$ (see the last term in Eq. (6.95)). The presence of any non-zero value of the imaginary component of ξ implies T violation. This result has been used to test time-reversal invariance in the charged kaon sector using $K\mu_3$ decay.

In addition to the dominant decay modes discussed here, kaons also decay into two pions and two leptons known as Kl_4 decays and radiative decays in which a photon is emitted in addition to leptons and hadrons. These decays have branching ratios in the range $10^{-4} - 10^{-5}$ (except $K^\pm \to \mu^\pm \nu_\mu(\bar{\nu}_\mu)\gamma$) and are not discussed here.

6.4.4 Semileptonic decays of hyperons (Y)

The study of semileptonic decays has played a very important role in the theory of weak interactions. We have already seen in the earlier sections that the Cabibbo theory explains the weak decays in the case of the strange sector. In this section, we now use the Cabibbo theory of $V - A$ interactions to describe the semileptonic decays of hyperons. The properties and decay modes of hyperons are given in Tables 6.3 and 6.5. These decays obey the $\Delta S = 1$, $\Delta Q = \Delta S$ rule (except for $\Sigma^- \to \Lambda e^- \bar{\nu}_e$, which is a $\Delta S = 0$ decay like $n \to p + e^- + \bar{\nu}_e$). These decays are described in terms of the $N - Y$ transition form factors occurring in the definition of the matrix elements of the vector and axial vector currents in analogy with the nucleon form factors defined in Eqs. (6.26) and (6.27) but with a coupling strength of $\frac{G_F}{\sqrt{2}}\sin\theta_C$. Hyperon beams are also produced with significant polarization leading to the study of these decays with polarized hyperons. The polarization as well as the spin correlation of the final leptons have also been studied in the decays of unpolarized and polarized hyperons. High energy polarized hyperon beams were used to carry out precision measurements of heavy hyperon decays which helped to establish the Cabibbo theory.

The transition matrix element for the generic hyperon decay $Y(p) \to B(p') + l^-(k) + \bar{\nu}_l(k')$, where Y and B are the initial and final baryon states, is written as:

$$\mathcal{M} = \frac{G_s}{\sqrt{2}}\bar{u}_B(p')[O_\mu^V - O_\mu^A]u_Y(p)\bar{u}_l\gamma^\mu(1 - \gamma_5)\nu_l, \tag{6.96}$$

where $G_s = G_F\cos\theta_C$ for $\Delta S = 0$ transitions, $G_s = G_F\sin\theta_C$ for $|\Delta S| = 1$ transitions, and [294]

$$O_\mu^V = f_1(q^2)\gamma^\mu + i\frac{f_2(q^2)}{M_B + M_Y}\sigma_{\mu\nu}q^\nu + \frac{2f_3(q^2)}{M_B + M_Y}q_\mu, \tag{6.97}$$

$$O_\mu^A = g_1(q^2)\gamma^\mu\gamma_5 + i\frac{g_2(q^2)}{M_B + M_Y}\sigma_{\mu\nu}q^\nu\gamma_5 + \frac{2g_3(q^2)}{M_B + M_Y}q_\mu\gamma_5 \tag{6.98}$$

are the operators defining the matrix elements of the vector and the axial vector currents, respectively. M_Y and M_B, are, respectively, the masses of the initial hyperon and the final baryon. There are other parameterizations in the literature which are equivalent to each other in the calculation of the physical observables except in the definition of $f_2(q^2)$ [59, 303, 305].

In the Cabibbo $V - A$ theory with quarks, the vector and axial vector currents V_μ^i and A_μ^i ($i = 1 - 8$) are the members of two octets defined as (Appendix B)

$$V_\mu^i = \bar{q}\frac{\lambda^i}{2}\gamma_\mu q, \qquad A_\mu^i = \bar{q}\frac{\lambda^i}{2}\gamma_\mu\gamma_5 q, \tag{6.99}$$

where $\frac{\lambda^i}{2}$ are the generators of SU(3). In this notation, the weak charged current in the $\Delta S = 0$ sector are given by $J_\mu^{1\pm i2} = V_\mu^{1\pm i2} - A_\mu^{1\pm i2}$ with $V_\mu^{1\pm i2} = \bar{q}\frac{\lambda_1\pm i\lambda_2}{2}\gamma_\mu q$, $A_\mu^{1\pm i2} = \bar{q}\frac{\lambda_1\pm i\lambda_2}{2}\gamma_\mu\gamma_5 q$. The electromagnetic current J_μ^{em} is given by

$$J_\mu^{em} = V_\mu^3 + \frac{1}{\sqrt{3}}V_\mu^8, \tag{6.100}$$

while the $\Delta S = 1$ current corresponding to the $s \rightarrow u$ transition is given by

$$J_\mu^{4+i5} = V_\mu^{4+i5} - A_\mu^{4+i5}. \tag{6.101}$$

The weak hadronic current(CC) is written as

$$J_\mu^{\pm cc} = V_{ud}(V_\mu^{1\pm i2} - A_\mu^{1\pm i2}) + V_{us}(V_\mu^{4\pm i5} - A_\mu^{4\pm i5}),$$
$$\text{where} \quad V_{ud} = \cos\theta_C, \ V_{us} = \sin\theta_C \tag{6.102}$$

The Cabibbo theory, assuming SU(3) symmetry, describes the relation among the various form factors of the weak transition, that is, $f_i(q^2)$ and $g_i(q^2)$ ($i = 1, 2, 3$). It relates the vector form factors $f_1(q^2)$ and $f_2(q^2)$ to the electromagnetic form factors of the proton and the neutron; it also relates the axial vector form factor $g_1(q^2)$ to the axial vector form factor in the nucleon sector. This is because the matrix elements of an octet operator O_μ^i; $i = V, A$, taken between two octets of baryons Y and B can be written as:

$$\langle Y_i|O_j^{V,A}|B_k\rangle = f_{ijk}F^{V,A} + d_{ijk}D^{V,A}, \tag{6.103}$$

where D and F correspond to symmetric and antisymmetric coupling of the two octets to an octet (i.e., octet of B_k and O_j coupling to the octet Y_i) in the decomposition $8 \otimes 8 = 1 \oplus 8_S \oplus 8_A \oplus 10 \oplus \bar{10} \oplus 27$. The superscripts V, A in Eq. (6.103) correspond to the vector and axial vector currents and the coefficient f_{ijk} and d_{ijk} are the structure constants of SU(3). Therefore, all the vector form factors corresponding to the vector current operator O^V for all the decays shown in Table 6.7 are expressed in terms of F^V and D^V which are determined in terms of the electromagnetic form factors $F_{i=1,2}^{p,n}(q^2)$. Similarly, all the axial vector form factors are determined in terms of F^A and D^A which are in turn given in terms of the axial vector form factor for the process $n \rightarrow p + e^- + \bar{\nu}_e(g_1 = F^A + D^A)$ and one unknown axial vector parameter chosen to be D^A/F^A. This unknown parameter is experimentally determined from the analysis of semileptonic hyperon decays. Following the methods shown in detail in Appendix B, we obtain the form factors for various hyperon decays as shown in Table 6.7, where the superscript A on D and F are dropped for brevity.

Table 6.7 Cabibbo model predictions for the β-decays of baryons [306]. Here, for simplicity, we have dropped the superscript A from F, D in $g_1(0)$ and g_1/f_1. The variables used in the table are defined as: $V_{ud} = \cos\theta_C$, $V_{us} = \sin\theta_C$, $\mu_p = 1.7928\mu_N$, and $\mu_n = -1.9130\mu_N$.

Decay	Scale	$f_1(0)$	$g_1(0)$	g_1/f_1	f_2/f_1
$n \to pe^-\bar{\nu}$	V_{ud}	1	$D+F$	$F+D$	$\mu_p - \mu_n$
$\Xi^- \to \Xi^0 e^-\bar{\nu}$	V_{ud}	-1	$D-F$	$F-D$	$\mu_p + 2\mu_n$
$\Sigma^\pm \to \Lambda e^\pm \nu$	V_{ud}	0^a	$\sqrt{\frac{2}{3}}D$	$\sqrt{\frac{2}{3}}D$	$-\sqrt{\frac{3}{2}}\mu_n$
$\Sigma^- \to \Sigma^0 e^-\bar{\nu}$	V_{ud}	$\sqrt{2}$	$\sqrt{2}F$	F	$\frac{(2\mu_p+\mu_n)}{2}$
$\Sigma^0 \to \Sigma^+ e^-\bar{\nu}$	V_{ud}	$\sqrt{2}$	$-\sqrt{2}F$	$-F$	$\frac{(2\mu_p+\mu_n)}{2}$
$\Xi^0 \to \Sigma^+ e^-\bar{\nu}$	V_{us}	1	$D+F$	$F+D$	$\mu_p - \mu_n$
$\Xi^- \to \Sigma^0 e^-\bar{\nu}$	V_{us}	$\frac{1}{\sqrt{2}}$	$\frac{1}{\sqrt{2}}(D+F)$	$F+D$	$\mu_p - \mu_n$
$\Sigma^- \to ne^-\bar{\nu}$	V_{us}	-1	$D-F$	$F-D$	$\mu_p + 2\mu_n$
$\Sigma^0 \to pe^-\bar{\nu}$	V_{us}	$\frac{-1}{\sqrt{2}}$	$\frac{1}{\sqrt{2}}(D-F)$	$F-D$	$\mu_p + 2\mu_n$
$\Lambda \to pe^-\bar{\nu}$	V_{us}	$-\sqrt{\frac{3}{2}}$	$-\frac{1}{\sqrt{6}}(D+3F)$	$F+\frac{D}{3}$	μ_p
$\Xi^- \to \Lambda e^-\bar{\nu}$	V_{us}	$\sqrt{\frac{3}{2}}$	$-\frac{1}{\sqrt{6}}(D-3F)$	$F-\frac{D}{3}$	$-(\mu_p+\mu_n)$

^{a}Since $f_1(0) = 0$ for $\Sigma^\pm \to \Lambda e^\pm \nu$, predictions are given for f_2 and g_1 rather than f_2/f_1 and g_1/f_1.

In the vector sector, the scalar form factor $f_3(q^2)$ vanishes due to the hypothesis of the conserved vector current as well as the G invariance; in the axial vector sector, the weak electric form factor $g_2(q^2)$ violates the G-parity in the limit of exact SU(3) symmetry and is assumed to vanish, that is, $g_2(q^2) = 0$. The G-parity and the absence of the second class currents are well supported in the $\Delta S = 0$ current of the nucleon sector as the isospin symmetry is a very good symmetry. However, the validity of G-parity in the case of $|\Delta S| = 1$ processes is only as good as the validity of SU(3) symmetry and is expected to be at the level of 15–20%. Nevertheless, G invariance is generally assumed in analyzing semileptonic decays of hyperons.

6.4.5 Physical observables in semileptonic hyperon decays

Experimental studies of semileptonic hyperon decays have been made possible by the availability of high intensity polarized beams of hyperons at CERN, BNL, and FNAL. The following precision measurements on the following observables were analyzed using the Cabibbo theory: (*i*) decay rate, (*ii*) the lepton–neutrino angular correlations, (*iii*) the asymmetry coefficients for the decay of polarized hyperons, and (*iv*) the polarization of the final state baryons in the case of unpolarized hyperon decays. The analysis was made using the following general assumptions:

i) SU(3) symmetry is present but some symmetry breaking effects are included by retaining the different masses of nucleons and hyperons, that is, $\delta = M_Y - M_B$.

ii) A dipole parameterization is used for the $N - Y$ transition form factors $f_i(q^2)$ and $g_i(q^2)$, ($i = 1 - 3$).

iii) T invariance is assumed which implies that all the vector and axial vector form factors are real.

iv) The CVC hypothesis requires that $f_3(q^2) = 0$, while the assumption of G invariance, that is, the absence of second class currents, requires $f_3(q^2) = 0$ and $g_2(q^2) = 0$.

v) There are contributions proportional to $\frac{m_e}{M}$ and $\frac{|\vec{q}|}{M}$ appearing in the analysis of the experiments which are expected to be small. While the terms dependent upon $\frac{m_e}{M}$ are neglected, the terms up to second order $O(\frac{q^2}{M^2})$ are included. The muon decay modes are generally analyzed separately by taking the term $O(\frac{m_\mu}{M})$ also into account.

With these assumptions, an effective Hamiltonian for the decay $Y \to Be\bar{\nu}$ is defined as:

$$\mathcal{M} = <Be\bar{\nu}|H_{\text{eff}}|Y>$$

with [306],

$$H_{\text{eff}} = \sqrt{2}G_S \frac{1 - \vec{\sigma}_e.\hat{p}_e}{2} \left[G_V + G_A\vec{\sigma}_e.\vec{\sigma}_B + G_P^e\vec{\sigma}_B.\hat{p}_e + G_P^v\vec{\sigma}_B.\hat{p}_v \right] \frac{1 - \vec{\sigma}_e.\hat{p}_v}{2}, \qquad (6.104)$$

where the effective couplings $G_i(i = V, A, P)$ are derived in terms of $f_i(q^2)$ and $g_i(q^2)$ using the Dirac spinors for the baryons $|Y>$ and $|B>$; $\hat{p}_e$, $\hat{p}_v$ are the unit vectors along the direction of the electron and antineutrino momenta; and $\vec{\sigma}_e$ and $\vec{\sigma}_B$ are the spins of the electron and the final baryon operating on the two-component spin wave function of the electron and final baryon, respectively. In the rest frame of the decaying hyperon Y, the couplings are given (neglecting g_3 as being proportional to the lepton mass, its contribution is negligible) as:

$$G_V = f_1 - \delta f_2 - \frac{E_v + E_e}{2M_Y} f_1', \qquad G_A = -g_1 + \delta g_2 + \frac{E_v - E_e}{2M_Y} f_1'$$

$$G_P^e = \frac{E_e}{2M_Y}[-f_1' + g_1 + \Delta g_2], \qquad G_P^v = \frac{E_v}{2M_Y}[f_1' + g_1 + \Delta g_2],$$

with $f_1' = f_1 + \Delta f_2$, $\delta = \frac{M_Y - M_B}{M_Y}$, $\Delta = \frac{M_Y + M_B}{2M_Y}$. In the rest frame of the initial hyperon with polarization $\vec{\mathcal{P}}_Y$, the decay distribution is given as:

$$d\Gamma = \frac{E_B + M_B}{2M_Y} \frac{E_e^2 E_v^3}{E_e^{\text{max}} - E_e} \frac{|\mathcal{M}|^2}{(2\pi)^5} dE_e d\Omega_e d\Omega_v, \qquad (6.105)$$

where

$$|\mathcal{M}|^2 = G_s^2\xi[1 + a\hat{p}_e.\hat{p}_v + A\vec{\mathcal{P}}_Y \cdot \hat{p}_e + B\vec{\mathcal{P}}_Y \cdot \hat{p}_v + A'(\vec{\mathcal{P}}_Y \cdot \hat{p}_e)(\hat{p}_e.\hat{p}_v)$$

$$+B'(\vec{\mathcal{P}}_Y \cdot \hat{p}_v)(\hat{p}_e.\hat{p}_v) + D\vec{\mathcal{P}}_Y \cdot (\hat{p}_e \times p_v)], \qquad (6.106)$$

and the coefficients ξ, a, A, B, A', B', and D are given in terms of the couplings G_i ($i = V, A, P$). Similar expressions can be derived for the decay of the polarized hyperon in which the polarization of the final baryon is observed. The complete expressions for the decay distribution of leptons in the case of the decays of polarized hyperons and for the polarization of the final baryon in the case of the decay of unpolarized hyperons are not reproduced here and can be found in Ref. [306].

Since the vector form factors are given in terms of the electromagnetic form factors, which are already determined from electron scattering, the analysis of semileptonic hyperon decays are done in terms of the Cabibbo angle θ_C and the axial vector form factors. The axial vector form factors are given in terms of the couplings F^A and D^A. The q^2 dependence of the D^A and F^A are taken to be the same as the q^2 dependence of the axial vector form factor $g_1(q^2)$ in neutron decay or in quasielastic scattering processes like $\nu_\mu + n \rightarrow \mu^- + p$ because $g_1(q^2) = D^A(q^2) + F^A(q^2)$. Therefore, the following parameterizations are used for the vector and axial vector form factors:

$$\frac{f_1(q^2)}{f_1(0)} = \frac{f_2(q^2)}{f_2(0)} = \frac{1}{\left(1 - \frac{q^2}{M_V^2}\right)^2} \quad \text{and} \quad g_1(q^2) = \frac{g_1(0)}{\left(1 - \frac{q^2}{M_A^2}\right)^2} \quad (6.107)$$

with $M_V = 0.84 \pm 0.04$ GeV, $M_A = 1.08 \pm 0.08$ GeV; $g_1(0) = D^A + F^A = 1.26$ is taken from the neutron β-decay and the quasielastic (anti)neutrino–nucleon scattering. Many experiments have been done to measure the following observables:

  i) The rate of various $\Delta S = 1$ and $\Delta S = 0$ decays.

 ii) The electron–neutrino correlations corresponding to the term a in Eq. (6.106).

iii) Electron correlations with respect to the initial polarization of the hyperons corresponding to term A in Eq. (6.106).

iv) The polarization of the final baryons in the decays of unpolarized hyperons.

An analysis of these results gives [306]:

$$F^A + D^A = 1.2670 \pm 0.0030, \qquad \sin\theta_C = 0.2250 \pm 0.0027$$
$$F^A - D^A = -0.314 \pm 0.016.$$

These analyses have also been used to test the hypothesis of the G invariance and the presence of second class currents. A non-zero value of the form factor g_2 corresponding to the second class current has been reported in these decays but the results are not conclusive. Table 6.8 presents the values of $\sin\theta_C$ obtained from the semileptonic decays of hyperons. It should be noted that all semileptonic decays of strange mesons and hyperons satisfy the $\Delta I = \frac{1}{2}$ rule experimentally, which is predicted in a natural way in the Cabibbo model through Eq. (6.78) in which an s-quark changes to a u-quark.

Table 6.8 Results from $\sin \theta_C$ analysis using measured g_1/f_1 values. The table has been taken from Cabibbo et al. [306].

Decay process	Rate (μs^{-1})	g_1/f_1	$\sin \theta_C$
$\Lambda \to p e^- \bar{\nu}$	3.161(58)	0.718(15)	0.2224 ± 0.0034
$\Sigma^- \to n e^- \bar{\nu}$	6.88(24)	$-0.340(17)$	0.2282 ± 0.0049
$\Xi^- \to \Lambda e^- \bar{\nu}$	3.44(19)	0.25(5)	0.2367 ± 0.0099
$\Xi^0 \to \Sigma^+ e^- \bar{\nu}$	0.876(71)	$1.32(+.22/-.18)$	0.209 ± 0.027
Combined	—	—	0.2250 ± 0.0027

6.5 Nonleptonic Decays of Strange Particles

6.5.1 Nonleptonic decays of K-mesons

Nonleptonic decays are those decays in which a heavier meson decays into lighter mesons involving no leptons. Charged kaons like K^+ and K^- as well as neutral kaons like K_L^0 and K_S^0 are both observed to decay by these modes. The dominant nonleptonic decay modes of $K^\pm$ and $K_{L,S}^0$ mesons are listed in Table 6.4. Since $K_L^0 \longrightarrow \pi^+ \pi^-$ and $K_L^0 \longrightarrow \pi^0 \pi^0$ are CP violating decays, we describe them separately in the next section. In this section, we will discuss two and three particle decay modes assuming CP invariance.

(i) $K\pi_2$ decays

The two pions in the decays of K^+, K^-, and K_S^0 are in zero angular momentum state, which is always symmetric in space. Therefore, the symmetry of the final state is described by their isospin contents. Since the pions are isovector, the pionic states can have total isospin $I = 0, 1, 2$. The pion in $I = 0$ or $I = 2$ states are symmetric; whereas, $I = 1$ state is antisymmetric. Since K^+, K^0, K^-, and $\bar{K}^0$ states belong to isospin $I = \frac{1}{2}$, the interaction for $K\pi_2$ decays should have $\Delta I = \frac{1}{2}, \frac{3}{2}, \frac{5}{2}$. If we assume that $\Delta I = \frac{1}{2}$ amplitude dominates, then only $I = 0$, $I_3 = 0$ state would be possible in the final state for K^0 decays. Writing

$$|I = 0, I_3 = 0\rangle = \frac{1}{\sqrt{3}}|\pi^+(1)\pi^-(2) + \pi^-(1)\pi^+(2) - \pi^0(1)\pi^0(2)\rangle, \qquad (6.108)$$

it is predicted that

$$\frac{\Gamma(K_S^0 \longrightarrow \pi^+ \pi^-)}{\Gamma(K_S^0 \longrightarrow \pi^0 \pi^0)} = 2, \qquad (6.109)$$

apart from the phase factor. In the case of $K^+ \longrightarrow \pi^+ \pi^0$, the final state has $I_3 = 1$; therefore, $I = 0$ is excluded. Hence in the final state I should be 2 implying that only $\Delta I = \frac{3}{2}$ and $\Delta I = \frac{5}{2}$ amplitudes will contribute. With the assumption of $\Delta I = \frac{1}{2}$ dominance, the rate of $K^\pm \longrightarrow \pi^\pm \pi^0$ should be very small (forbidden in exact limit) as compared to $K_S^0 \longrightarrow \pi^+ \pi^- (\pi^0 \pi^0)$.

Denoting the amplitudes in $\Delta I = \frac{1}{2}, \frac{3}{2}$, and $\frac{5}{2}$ channels as a_1, a_3, and a_5, it can be shown that [239]:

$$\frac{\Gamma(K^+ \longrightarrow \pi^+ \pi^0)}{\Gamma(K_s \longrightarrow \pi^+ \pi^-) + \Gamma(K_s \longrightarrow \pi^0 \pi^0)} = \frac{\frac{3}{4}|a_3 - \frac{2}{5}a_5|^2}{A_0^2 + (\mathrm{Re}A_2)^2} \simeq \frac{\frac{3}{4}|a_3 - \frac{2}{5}a_5|^2}{|a_1|^2}, \qquad (6.110)$$

where

$$A_0 = \langle \pi\pi \rangle_{I=0}^S |H_{wk}|K^0\rangle, \qquad\qquad A_2 = \langle (\pi\pi)_{I=2}^S |H_{wk}|K^0\rangle,$$

$$a_1 = A_0 e^{i\delta_0}, \qquad\qquad\qquad\qquad a_3 + a_5 = A_2 e^{i\delta_2}.$$

Here δ_0 and δ_2 are the pion–pion scattering phase shifts in $I = 0$ and $I = 1$ channels. A detailed analysis and comparison with the experimental values of $K\pi_2$ decay rates gives [239]:

$$\left|\frac{a_3}{a_1}\right| = 0.045 \pm 0.005 \qquad \text{and} \qquad \left|\frac{a_5}{a_1}\right| = 0.001 \pm 0.003. \qquad (6.111)$$

This supports $\Delta I = \frac{1}{2}$ dominance in $K\pi_2$ decays.

(ii) $K\pi_3$ decays

Three pion decays of charged and neutral kaons are given in Table 6.4. The final state has three pions, so they can have $I = 0, 1, 2, 3$ with

$$1 \oplus 1 \oplus 1 = (0 \oplus 1 \oplus 2) \oplus 1 = \overbrace{1} \oplus \overbrace{0 \oplus 1 \oplus 2} \oplus \overbrace{1 \oplus 2 \oplus 3}. \qquad (6.112)$$

This has multiplicities of 1 for $I = 0$; 3 for $I = 1$; 2 for $I = 2$; and 1 for $I = 3$ states. Decays, where two pions are either $\pi^+\pi^+$ or $\pi^-\pi^-$ or $\pi^0\pi^0$, will be in symmetric states under the exchange of the first two particles. Hence, let us consider

$$K^+ \longrightarrow \pi^+\pi^+\pi^-, \qquad\qquad K^- \longrightarrow \pi^-\pi^-\pi^+,$$

$$K_L^0 \longrightarrow \pi^+\pi^-\pi^0, \qquad\qquad K_L^0 \longrightarrow \pi^0\pi^0\pi^0.$$

$I = 0$ state is completely antisymmetric; $I = 3$ state is completely symmetric; while $I = 1$ and $I = 2$ will be of mixed symmetric states. The symmetry considerations in this case are not very simple so let us first make the assumption that $\Delta I = \frac{1}{2}$ rule is followed and compare its predictions with the experimental results. Assuming $\Delta I = \frac{1}{2}$ rule, the final state could be $I = 0$ or $I = 1$. Since $I = 0$ is completely antisymmetric, the dominant contribution will come from the $I = 1$ state, while the $I = 2$ state will contribute only if there is a contribution from the $\Delta I = \frac{3}{2}$ amplitudes. The three pion states can be expressed in terms of the isospin states as:

$$|\pi^+\pi^+\pi^-\rangle = \frac{2}{\sqrt{5}}|I = 1, I_3 = 1\rangle + \frac{1}{\sqrt{5}}|I = 3, I_3 = 1\rangle,$$

$$|\pi^+\pi^0\pi^0\rangle = \frac{1}{\sqrt{5}}|I = 1, I_3 = 1\rangle + \frac{2}{\sqrt{5}}|I = 3, I_3 = 1\rangle,$$

$$|\pi^+\pi^-\pi^0\rangle = \sqrt{\frac{2}{5}}|I=1, I_3=0\rangle + \sqrt{\frac{3}{5}}|I=3, I_3=0\rangle,$$

$$|\pi^0\pi^0\pi^0\rangle = -\sqrt{\frac{3}{5}}|I=1, I_3=0\rangle + \sqrt{\frac{2}{5}}|I=3, I_3=0\rangle.$$

If we assume that only $I = 1$ state contributes due to the $\Delta I = \frac{1}{2}$ rule, then assuming CP invariance, we get the following relations:

$$\frac{\Gamma(K^+ \longrightarrow \pi^+\pi^+\pi^-)}{\Gamma(K^+ \longrightarrow \pi^+\pi^0\pi^0)} = 4,$$

$$\frac{\Gamma(K_L^0 \longrightarrow \pi^+\pi^-\pi^0)}{\Gamma(K_L^0 \longrightarrow \pi^0\pi^0\pi^0)} = \frac{2}{3},$$

$$\frac{\Gamma(K_L^0 \longrightarrow \pi^+\pi^-\pi^0)}{\Gamma(K^+ \longrightarrow \pi^+\pi^0\pi^0)} = 2,$$

$$\frac{\Gamma(K_L^0 \longrightarrow \pi^0\pi^0\pi^0)}{\Gamma(K^+ \longrightarrow \pi^+\pi^+\pi^-) - \Gamma(K_L^0 \longrightarrow \pi^+\pi^0\pi^0)} = 1.$$

Detailed analysis of $K\pi_3$ decays and comparison with experimental data shows that [239]:

$$\left|\frac{a_3}{a_1}\right| = 0.06 \pm 0.01, \tag{6.113}$$

supporting the dominance of $\Delta I = \frac{1}{2}$ rule.

There is no theoretical explanation of the $\Delta I = \frac{1}{2}$ dominance in the $V - A$ theory.

6.5.2 Nonleptonic decays of hyperons

Nonleptonic decays of hyperons are of the type $Y \to B + M$, where the hyperon (Y) decays into a baryon (B) which can be a nucleon or a hyperon and a meson (M). The list of hyperon two-body decays is given in Table 6.5. All the decays have $\Delta S = 1$; no decay with $\Delta S > 1$ has been observed. In fact, the decay $\Lambda \to p\pi^-$ was historically the first decay to be observed which indicates the existence of parity violation; this violation was not noticed at that time [72].

In the phenomenological theory, these decays as well as the nonleptonic mesonic decays are supposed to be explained by the term of the type $\frac{G}{\sqrt{2}}(J_\mu)^h(J^\mu)^{h\dagger}$, that is,

$$\frac{G}{\sqrt{2}}\left[\bar{s}\gamma^\mu(1-\gamma_5)u + \text{h.c.}\right]\left[\bar{u}\gamma^\mu(1-\gamma_5)d + \text{h.c.}\right]^\dagger.$$

Since the first term has $\Delta I = \frac{1}{2}$ and the second term has $\Delta I = 1$, these decays in the phenomenological theory could have $\Delta I = \frac{1}{2}, \frac{3}{2}$ in the isospin space. Therefore, the $V - A$ theory does not predict the dominance of $\Delta I = \frac{1}{2}$ transitions.

We write the general and simplest term of the matrix element for $Y(\frac{1}{2}^+) \to B(\frac{1}{2}^+) + M(0^-)$ transitions allowing for parity violation, as

$$\mathcal{M} = G_F m_M^2 \bar{u}_Y (a - b\gamma_5) u_B. \tag{6.114}$$

Evaluating the matrix element in the rest frame of the initial baryon, using

$$u_i = \begin{pmatrix} \chi_i \\ 0 \end{pmatrix}, \quad \bar{u}_i u_i = 1$$

$$u_f = \sqrt{\frac{E_B + M_B}{2M_B}} \begin{pmatrix} \chi_f \\ \frac{\vec{\sigma}.\vec{p}}{E_B + M_B} \chi_f \end{pmatrix}, \quad \bar{u}_f u_f = 1$$

$$\Rightarrow \quad \mathcal{M} = G_F m_M^2 \sqrt{\frac{E_B + M_B}{2M_B}} \chi_f^\dagger \left(a + b \frac{\vec{\sigma}.\vec{p}}{E_B + M_B} \right) \chi_i$$

$$= G_F m_M^2 \sqrt{\frac{E_B + M_B}{2M_B}} \chi_f^\dagger (S + P\vec{\sigma}.\hat{n}) \chi_i,$$

where $S = a$, $P = \frac{b|\vec{p}|}{E_B + M_B}$, $\hat{n} = \frac{\vec{p}}{|\vec{p}|}$, and M_B is the mass of the final baryon. If $\Delta I = \frac{1}{2}$ dominance is assumed, then

$$\frac{\Gamma(\Lambda \longrightarrow p\pi^-)}{\Gamma(\Lambda \longrightarrow n\pi^0)} \simeq 2. \tag{6.115}$$

Moreover, a relation between the amplitudes of the various decay modes of hyperons is obtained as

$$A(\Sigma^+ \longrightarrow n\pi^+) - A(\Sigma^- \longrightarrow n\pi^-) = -\sqrt{2} A(\Sigma^+ \longrightarrow p\pi^0), \tag{6.116}$$

which seems to be satisfied quite well experimentally, once the appropriate corrections are made for the final state interactions in the respective pion nucleon final states.

A careful analysis of nonleptonic hyperons decays gives [239]:

$$\Lambda \longrightarrow N\pi : \quad \left| \frac{a_3}{a_1} \right| = 0.03^{\pm 0.01 \text{ (for s-wave)}}_{\pm 0.03 \text{ (for p-wave)}} \quad \text{and} \quad \Sigma \longrightarrow N\pi : \quad \left| \frac{a_{I \geq \frac{3}{2}}}{a_1} \right| = 0.07 \pm 0.03.$$

It is to be noted that a_1 and a_3 amplitudes are expected to be of the same order of magnitude as that obtained in the processes satisfying $\Delta Q = \Delta S$, $\Delta I = \frac{1}{2}$ rules in the semileptonic decays of strange particles. However, the amplitude a_1 is higher than a_3 by a factor of $15 - 30$. In view of this, Cabibbo formulated a model for the semileptonic weak interactions assuming that the physics of the weak interaction processes in the semileptonic decays are completely different from the nonleptonic $\Delta S = 1$ decays.

In the case of a hyperon decay like $\Lambda \to N\pi, \Sigma \to N\pi$ etc., using the matrix element given in Eq. (6.114), the decay rate is obtained as [239],

$$d\Gamma \propto 1 + \gamma \hat{s}.\hat{s}' + (1 - \gamma)\vec{s}.\hat{n}\, \vec{s}'.\hat{n} + \alpha(\vec{s}.\hat{n} + \vec{s}'.\hat{n}) + \beta(\vec{s} \times \vec{s}').\hat{n} \tag{6.117}$$

where $\vec{s}$ and $\vec{s}'$ are the spins of the initial and final baryons.

$$\alpha = 2\mathrm{Re}\frac{SP^*}{|S|^2 + |P|^2}, \qquad \beta = 2\mathrm{Im}\frac{SP^*}{|S|^2 + |P|^2},$$

$$\gamma = \frac{|S|^2 - |P|^2}{|S|^2 + |P|^2},$$

with $\alpha^2 + \beta^2 + \gamma^2 = 1$. For unpolarized hyperon decays, summing over all the directions of $\vec{s}$, we find:

$$d\Gamma \propto 1 + \alpha\vec{s}'.\hat{n}, \tag{6.118}$$

where α gives the polarization of the final baryon. For polarized hyperon decays, summing over the directions of $\vec{s}'$, we find:

$$d\Gamma \propto 1 + \alpha\vec{s}.\hat{n}. \tag{6.119}$$

The parameter α gives the asymmetry of the final baryon with respect to the spin polarization of the initial hyperon. α is used to determine the polarization of the initial hyperon. In the presence of T invariance, the amplitudes S and P are real and imply $\beta = 0$. Therefore, a non-zero value of β gives a measure of T non-invariance in these decays. However, an analysis of the final state interactions (FSI) in the πN system need to be done to study the presence of T non-invariance, if any. Table 6.9 lists the decay rates and asymmetry observed in these decays.

Table 6.9 Nonleptonic decays of hyperons [117].

Decay	Asymmetries	Branching ratios	Decay rate ($10^3\ \mu s^{-1}$)
$\Lambda \to p\pi^-$	0.750 ± 0.010	$(63.9 \pm 0.5)\%$	2.428
$\Lambda \to n\pi^0$	0.692 ± 0.017	$(35.8 \pm 0.5)\%$	1.36
$\Sigma^+ \to p\pi^0$	-0.980 ± 0.015	$(51.57 \pm 0.3)\%$	6.432
$\Sigma^+ \to n\pi^+$	0.068 ± 0.013	$(48.31 \pm 0.3)\%$	6.025
$\Sigma^- \to n\pi^-$	-0.068 ± 0.008	$(99.848 \pm 0.005)\%$	4.571
$\Xi^0 \to \Lambda\pi^0$	-0.347 ± 0.010	$(99.524 \pm 0.012)\%$	3.432
$\Xi^- \to \Lambda\pi^-$	-0.392 ± 0.008	$(99.887 \pm 0.035)\%$	6.094

6.5.3 Radiative weak decays

Radiative weak decays are those in which a photon is emitted in a weak decay process. There are two types of radiative decays:

(i) Decays in which a photon is emitted in an allowed weak decay at the quark level, for example, in the decays like:

$$\Lambda \longrightarrow p\pi^-\gamma; \qquad \pi^\pm \longrightarrow \mu^\pm \nu(\bar{\nu})\gamma$$

$$\Sigma^\pm \longrightarrow n\pi^\pm\gamma; \qquad K^\pm \longrightarrow \mu^\pm \nu(\bar{\nu})\gamma.$$

In these decays, there are two types of processes which contribute as shown in Figure 6.5. The decays shown in Figures 6.5(a) and 6.5(b) are the internal Bremsstrahlung (IB) contributions in which the charged particle radiates after being emitted; the decay shown in Figure 6.5(c) is called a structure dependent (SD) or direct emission (DE) process in which the photon is emitted from the intermediate states in the transition. There

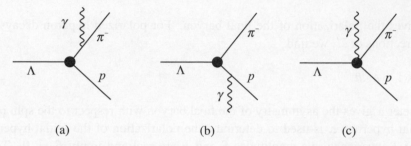

(a)   (b)   (c)

Figure 6.5 Internal Bremsstrahlung and direct emission radiative decays.

could also be contribution from the interference between the two processes. While the weak coupling shown in Figures 6.5(a) and 6.5(b) are known from $V - A$ theory, the form factors and couplings for the decay shown in Figure 6.5(c) are not known and are determined from the requirement of the gauge invariance of the total amplitude in the absence of specific knowledge of the intermediate states.

(ii) Weak radiative decays in which a photon is emitted corresponding to the weak forbidden decays at the quark level like $s \to d\gamma$ or $c \to u\gamma$: Examples of such decays are the radiative decays of hyperons like $\Sigma^+ \to p\gamma$ and $\Lambda \to n\gamma$:

These processes get contributions from three types of decays shown in Figure 6.6(a), (b), and (c) corresponding to one quark, two quark, and three quark transitions. The calculation of the one quark transitions with two spectator quarks is model dependent due to the unknown structure of the direct emission diagram in which many states of intermediate particles can also be excited before radiating. A theoretical model is needed for evaluating all such diagrams. In the simplest model, as shown in Figure 6.6, the photon is either radiated from the virtual W boson or the virtual quark. The two quark transition corresponds to the W exchange between the two quarks in which one of the participating quark radiates with one spectator quark and three quarks transition with no spectator quark in which the third quark not participating in the W exchange radiates. All these contributions are subject to QCD corrections due to the gluon exchange.

There are two types of experimental measurements made on these decays: decay rate as well as the asymmetry parameter of the photon. The amplitude for transitions like

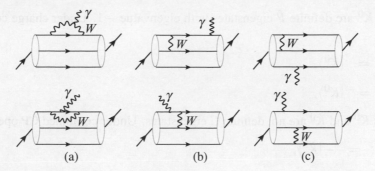

Figure 6.6 Radiative decays corresponding to (a) one quark, (b) two quark, and (c) three quark, transitions.

$B(p) \rightarrow B'(p') + \gamma(k)$ is similar to the structure of nonleptonic decays $B(p) \rightarrow B'(p') + \pi(k)$ and can be written as:

$$\mathcal{M}(B \longrightarrow B'\gamma) = ieG_F\bar{u}(p')\sigma_{\mu\nu}[A_\gamma + B_\gamma\gamma_5]\epsilon^\mu k^\nu, \tag{6.120}$$

where $A_\gamma(B_\gamma)$ are the parity conserving (parity violating) amplitudes. The decay rate (Γ) and the asymmetry (α) is expressed as:

$$\Gamma \propto |A|_\gamma^2 + |B|_\gamma^2, \qquad \alpha \propto \frac{2Re(A_\gamma^* B_\gamma)}{|A|_\gamma^2 + |B|_\gamma^2}. \tag{6.121}$$

The theory can be easily extended to weak radiative decays involving heavy flavors, that is, $c \rightarrow u\gamma$ or $b \rightarrow s\gamma$, to describe the radiative decays of charmed and other particles with heavier flavors. The experimental results for the decay rates have been explained satisfactorily in many models by using heavy quarks as shown in Figure 6.6. However, there is no model which describes the decay rate as well as the asymmetry, simultaneously. This remains a problem yet to be understood in the context of the Cabibbo theory.

6.6 CP Violation in the Neutral Kaon Sector

6.6.1 Neutral kaons, CP eigenstates, and $K^0 - \bar{K}^0$ oscillations

Pure K^0 and $\bar{K}^0$ beams are states of definite strangeness and parity, that is, under parity(P) operation,

$$\hat{P}|K^0\rangle = -|K^0\rangle,$$
$$\hat{P}|\bar{K}^0\rangle = -|\bar{K}^0\rangle. \tag{6.122}$$

Thus, K^0 and $\bar{K}^0$ are definite $\hat{P}$ eigenstates with eigenvalue -1. Under charge conjugation(C) operation,

$$\hat{C}|K^0\rangle = |\bar{K}^0\rangle,$$
$$\hat{C}|\bar{K}^0\rangle = |K^0\rangle, \qquad (6.123)$$

which implies K^0 and $\bar{K}^0$ are not definite $\hat{C}$ eigenstates. Under combined CP operation,

$$\hat{C}\hat{P}|K^0\rangle = -|\bar{K}^0\rangle,$$
$$\hat{C}\hat{P}|\bar{K}^0\rangle = -|K^0\rangle. \qquad (6.124)$$

Thus, K^0 and $\bar{K}^0$ are not definite eigenstates of combined CP operation. However, their linear combinations are eigenstates of CP. We can define

$$|K_1^0\rangle = \frac{1}{\sqrt{2}}\left\{|K^0\rangle + |\bar{K}^0\rangle\right\}$$
$$|K_2^0\rangle = \frac{1}{\sqrt{2}}\left\{|K^0\rangle - |\bar{K}^0\rangle\right\} \qquad (6.125)$$

such that

$$\hat{C}\hat{P}|K_1^0\rangle = -\frac{1}{\sqrt{2}}\left\{|\bar{K}^0\rangle + |K^0\rangle\right\} = -|K_1^0\rangle,$$
$$\hat{C}\hat{P}|K_2^0\rangle = -\frac{1}{\sqrt{2}}\left\{|\bar{K}^0\rangle - |K^0\rangle\right\} = +|K_2^0\rangle. \qquad (6.126)$$

Notice that the $|K_1^0\rangle$ and $|K_2^0\rangle$ states are definite CP eigenstates with eigenvalue -1 and $+1$, respectively. Since $|K_1^0\rangle$ and $|K_2^0\rangle$ are not particle and antiparticle states, their masses as well as lifetimes may differ. K-mesons decay either into two or three pions. Since the pions have negative intrinsic parity and the two pion and three pion states are in S-states for kaons decaying at rest, the 2π states have $(CP)_{2\pi} = +1$ and the three pion states have $(CP)_{3\pi} = -1$. Therefore, CP conservation implies $K_1^0 \to 3\pi$ and $K_2^0 \to 2\pi$. The 2π decay is faster because of the availability of more phase space than the 3π decay; thus, K_2^0 has a shorter lifetime of $\tau_S = 0.892 \times 10^{-10}$s as compared to K_1^0 which has a longer lifetime of $\tau_L = 5.18 \times 10^{-8}$s. For this reason, K_1^0 is called K_L^0 and K_2^0 is called K_S^0. The strangeness eigenstates $|K^0\rangle$ and $|\bar{K}^0\rangle$ are expressed in terms of $|K_S^0\rangle$ and $|K_L^0\rangle$ as

$$|K^0\rangle = \frac{1}{\sqrt{2}}\left[|K_S^0\rangle + |K_L^0\rangle\right], \qquad (6.127)$$

$$|\bar{K}^0\rangle = \frac{1}{\sqrt{2}}\left[|K_L^0\rangle - |K_S^0\rangle\right]. \qquad (6.128)$$

Since strangeness is conserved in the strong interactions, K^0 and $\bar{K}^0$ can be produced in processes like $\pi^- p \to K\Lambda$ but they decay into 2π and 3π modes through weak interactions which violate strangeness. However, if we assume T invariance or CP invariance (due to the CPT invariance) of the physical laws, then only K_1^0 would decay into pion states with CP$=-1$;

K_2^0 would decay into pion states with CP = +1. Since strangeness is not conserved in weak interactions, and both K^0 to $\bar{K}^0$ are neutral particles, it was suggested that K^0 may get converted into $\bar{K}^0$ while propagating and vice versa. However, this transformation of K^0 to $\bar{K}^0$ or $\bar{K}^0$ to K^0 is not possible in the first order at the tree level because of the $\Delta S = 1$ rule. It is possible only in the second order with the exchange of two W bosons as shown in Figure 6.7. This simply means that $|K^0\rangle$ and $|\bar{K}^0\rangle$ are just production states, and they propagate in space as a linear combination of some other states, say $|K_1^0\rangle$ and $|K_2^0\rangle$.

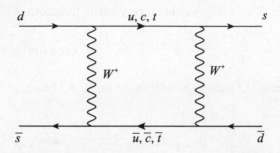

Figure 6.7 $K^0 \longrightarrow \bar{K}^0$ oscillation via two W bosons exchange.

Assuming CP invariance, we describe the phenomena of the evolution of a kaon beam and kaon oscillation ($K^0 \leftrightarrow \bar{K}^0$). We start with a K^0 beam produced at, say, time $t = 0$, represented by a state $|\psi\rangle$ as

$$|\psi(t = 0)\rangle = |K^0\rangle = \frac{1}{\sqrt{2}} \left[|K_S^0\rangle + |K_L^0\rangle \right]. \tag{6.129}$$

This state evolves with time such that at a later time t, the time-dependent wave function of $|K^0\rangle$, using the expression for the time evolution of a state is:

$$|\psi(t)\rangle = |\psi(t = 0)\rangle e^{-iEt} = |\psi(t = 0)\rangle e^{-i\left(m - \frac{i\Gamma}{2}\right)t}, \tag{6.130}$$

where Γ^{-1} is the lifetime. Equation (6.130) may be rewritten as

$$\begin{aligned}
|\psi(t)\rangle &= \frac{1}{\sqrt{2}} \left[e^{-im_S t} e^{-\frac{\Gamma_S t}{2}} |K_S^0\rangle + e^{-im_L t} e^{-\frac{\Gamma_L t}{2}} |K_L^0\rangle \right] \\
&= \frac{e^{-im_S t}}{\sqrt{2}} \left[e^{-\frac{\Gamma_S t}{2}} |K_S^0\rangle + e^{-i(m_L - m_S)t} e^{-\frac{\Gamma_L t}{2}} |K_L^0\rangle \right] \\
&= \frac{e^{-im_S t}}{\sqrt{2}} \left[e^{-\frac{\Gamma_S t}{2}} |K_S^0\rangle + e^{-i\Delta m t} e^{-\frac{\Gamma_L t}{2}} |K_L^0\rangle \right],
\end{aligned} \tag{6.131}$$

where $\Delta m = m_L - m_S$, with m_S and m_L being the masses of shorter and longer lived kaons, respectively. Using Eq. (6.125), we find:

$$|\psi(t)\rangle = \frac{e^{-im_S t}}{2} \left[e^{-\frac{\Gamma_S t}{2}} \left\{ |K^0\rangle - |\bar{K}^0\rangle \right\} + e^{-i\Delta m t} e^{-\frac{\Gamma_L t}{2}} \left\{ |K^0\rangle + |\bar{K}^0\rangle \right\} \right]. \tag{6.132}$$

The probability of observing $|K^0\rangle$ after a time t is

$$
\begin{aligned}
P(K^0 \longrightarrow K^0) &= |\langle K^0|\psi(t)\rangle|^2 \\
&= \langle K^0|\psi(t)\rangle \, \langle K^0|\psi(t)\rangle^* \\
&= \frac{1}{4}\left[e^{-\Gamma_S t} + e^{-\Gamma_L t} + e^{-\left(\frac{\Gamma_S + \Gamma_L}{2}\right)t} e^{i\Delta mt} + e^{-\left(\frac{\Gamma_S + \Gamma_L}{2}\right)t} e^{-i\Delta mt} \right]
\end{aligned}
$$

$\therefore \frac{\tau_S}{\tau_L} \sim \frac{1}{570} \Rightarrow \Gamma_L \sim 570\Gamma_S \Rightarrow t\Gamma_S$ would be very small, therefore, $e^{-t\Gamma_S} \approx 1$,

$$
P(K^0 \longrightarrow K^0) = \frac{1}{4}\left[e^{-\Gamma_S t} + e^{-\Gamma_L t} + 2e^{-\left(\frac{\Gamma_L}{2}\right)t}(\cos \Delta mt) \right]. \tag{6.133}
$$

The probability of finding $|\bar{K}^0\rangle$ after a time t, in the initial K^0 beam, is

$$
\begin{aligned}
P(K^0 \longrightarrow \bar{K}^0) &= |\langle \bar{K}^0|\psi(t)\rangle|^2 \\
&= \frac{1}{4}\left[e^{-\Gamma_S t} - e^{-\Gamma_L t} - e^{-\left(\frac{\Gamma_S + \Gamma_L}{2}\right)t} e^{i\Delta mt} - e^{-\left(\frac{\Gamma_S + \Gamma_L}{2}\right)t} e^{-i\Delta mt} \right] \\
&= \frac{1}{4}\left[e^{-\Gamma_S t} + e^{-\Gamma_L t} - 2e^{-\left(\frac{\Gamma_L}{2}\right)t}(\cos \Delta mt) \right]. \tag{6.134}
\end{aligned}
$$

It can be seen from Eqs. (6.133) and (6.134) that:

$$
\begin{aligned}
P(K^0 \longrightarrow K^0) &= P(\bar{K}^0 \longrightarrow \bar{K}^0) \quad \text{and} \\
P(K^0 \longrightarrow \bar{K}^0) &= P(\bar{K}^0 \longrightarrow K^0). \tag{6.135}
\end{aligned}
$$

Figure 6.8 plots the probability of the neutral kaons oscillation. The curve (a) represents the surviving probability of K^0, that is, $P(K^0 \to K^0)$ while the curve (b) represents the oscillating

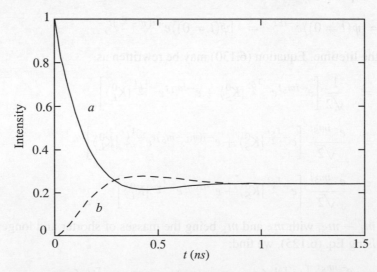

Figure 6.8 Probability of neutral kaon oscillation. (a) represents the survival probability of K^0, that is, $P(K^0 \to K^0)$, while (b) represents the oscillation probability, $P(K^0 \to \bar{K}^0)$, for a fixed $\Delta m \neq 0$.

probability of K^0 to $\bar{K}^0$, that is, $P(K^0 \rightarrow \bar{K}^0)$, for a fixed value of Δm. The intensity of $\bar{K}^0$ rises from zero and shows oscillatory behavior with the same frequency. Using the observed data for oscillations ($K^0 \leftrightarrow \bar{K}^0$), the mass difference $m_L - m_S$ is found to be 3.52×10^{-12} MeV. It has been observed experimentally that the mass difference between K_1^0 and K_2^0 is $3.484 \pm 0.006 \times 10^{-12}$ MeV. $\Delta m = m_L - m_S$ is a positive quantity; this implies that the shorter lived kaon is less massive than the longer lived kaon.

6.6.2   CP violation in the neutral kaon decays

In 1964, Christenson et al. [73] observed that one in a thousand events $|K_L^0\rangle$ also decays to a two pion mode which was a clear evidence of CP violation. The degree of CP violation is generally presented by the ratio of the amplitudes for the processes

$$K_L^0 \longrightarrow \pi^+ \pi^- \qquad \text{and} \qquad K_S^0 \longrightarrow \pi^+ \pi^-,$$

$$K_L^0 \longrightarrow \pi^0 \pi^0 \qquad \text{and} \qquad K_S^0 \longrightarrow \pi^0 \pi^0. \tag{6.136}$$

In the presence of CP violation, K_L^0 and K_S^0 would not be the pure CP eigen states as K_1^0 and K_2^0, but would involve a small admixture of the state with opposite CP. We, therefore define [239]:

$$|K_S^0\rangle = \frac{1}{\sqrt{2(1+|\epsilon_1|^2)}} \left[(1+\epsilon_1)|K_0\rangle - (1-\epsilon_1)|\bar{K}^0\rangle \right], \tag{6.137}$$

$$|K_L^0\rangle = \frac{1}{\sqrt{2(1+|\epsilon_2|^2)}} \left[(1+\epsilon_2)|K_0\rangle + (1-\epsilon_2)|\bar{K}^0\rangle \right], \tag{6.138}$$

where ϵ_1, ϵ_2 are complex numbers. If CP violation is accompanied by T violation such that CPT is conserved, then $\epsilon_1 = \epsilon_2 = \epsilon$, while if T is conserved and CPT is violated, then $\epsilon_1 \neq \epsilon_2$. We calculate the decay amplitudes of K_S^0 and K_L^0 into two pions as given in Eq. (6.136), assuming CPT invariance. Since pions are bosons, the wavefunction of the final state should be symmetric. The total isospin of the final state should be $I = 0$ or $I = 2$ with $I_3 = 0$. There are four amplitudes which may correspond to these decays:

$$\langle \pi\pi, I = 0|H_{wk}|K_S^0\rangle, \qquad\qquad \langle \pi\pi, I = 2|H_{wk}|K_S^0\rangle,$$

$$\langle \pi\pi, I = 0|H_{wk}|K_L^0\rangle, \qquad\qquad \langle \pi\pi, I = 2|H_{wk}|K_L^0\rangle. \tag{6.139}$$

Using the Clebsch–Gordan coefficients, we may write

$$|\pi^+ \pi^-\rangle = \sqrt{\frac{1}{3}}|\pi\pi, I = 2\rangle + \sqrt{\frac{2}{3}}|\pi\pi, I = 0\rangle, \tag{6.140}$$

$$|\pi^0 \pi^0\rangle = \sqrt{\frac{2}{3}}|\pi\pi, I = 2\rangle - \sqrt{\frac{1}{3}}|\pi\pi, I = 0\rangle, \tag{6.141}$$

where the charged pion state is defined as $|\pi^+ \pi^-\rangle = \frac{1}{\sqrt{2}}(|\pi^{+(1)}\pi^{-(2)}\rangle + |\pi^{+(2)}\pi^{-(1)}\rangle)$. The pions in the final state undergo final state interactions described by the phase shifts δ_0 and

δ_2, respectively, for $I = 0$ and $I = 2$ final states. The decay amplitudes for the decay of K^0 in $I = 0$ and $I = 2$ states are defined as:

$$\langle \pi\pi, I = 0 | H_{wk} | K^0 \rangle = A_0 e^{i\delta_0}, \tag{6.142}$$

$$\langle \pi\pi, I = 2 | H_{wk} | K^0 \rangle = A_2 e^{i\delta_2}. \tag{6.143}$$

The amplitudes for $\bar{K}^0$ are obtained, using the CPT invariance, as:

$$\langle \pi\pi, I = 0 | \longrightarrow |\pi\pi, I = 0\rangle; \qquad \langle \pi\pi, I = 2 | \longrightarrow |\pi\pi, I = 2\rangle; \qquad |K^0\rangle \longrightarrow -\langle \bar{K}^0 |,$$

which leads to the decay amplitudes for $\bar{K}^0$ becoming

$$\langle \pi\pi, I = 0 | H_{wk} | \bar{K}^0 \rangle = -A_0^* e^{i\delta_0}, \tag{6.144}$$

$$\langle \pi\pi, I = 2 | H_{wk} | \bar{K}^0 \rangle = -A_2^* e^{i\delta_2}. \tag{6.145}$$

Taking A_0 to be real, which eliminates the arbitrary phase, the amplitudes for the decays given in Eq. (6.136) are obtained using Eqs. (6.137)–(6.145) assuming $\epsilon_1 = \epsilon_2$ as

$$\langle \pi^+\pi^- | H_{wk} | K_L^0 \rangle$$

$$= \frac{1}{\sqrt{3}} \langle \pi\pi, I = 2 | H_{wk} | K_L^0 \rangle + \sqrt{\frac{2}{3}} \langle \pi\pi, I = 0 | H_{wk} | K_L^0 \rangle,$$

$$= \frac{1}{\sqrt{6(1 + |\epsilon|^2)}} \left[(1 + \epsilon) \langle \pi\pi, I = 2 | H_{wk} | K^0 \rangle + (1 - \epsilon) \langle \pi\pi, I = 2 | H_{wk} | \bar{K}^0 \rangle \right]$$

$$+ \frac{1}{\sqrt{3(1 + |\epsilon|^2)}} \left[(1 + \epsilon) \langle \pi\pi, I = 0 | H_{wk} | K^0 \rangle + (1 - \epsilon) \langle \pi\pi, I = 0 | H_{wk} | \bar{K}^0 \rangle \right],$$

$$= \frac{1}{\sqrt{6(1 + |\epsilon|^2)}} \left[(A_2 - A_2^*) e^{i\delta_2} + \epsilon (A_2 + A_2^*) e^{i\delta_2} + 2\sqrt{2}\epsilon A_0 e^{i\delta_0} \right]. \tag{6.146}$$

Similarly, other amplitudes are obtained as

$$\langle \pi^+\pi^- | H_{wk} | K_S^0 \rangle = \frac{1}{\sqrt{6(1 + |\epsilon|^2)}} \left[(A_2 + A_2^*) e^{i\delta_2} + (A_2 - A_2^*) \epsilon e^{i\delta_2} + 2\sqrt{2} A_0 e^{i\delta_0} \right], \tag{6.147}$$

$$\langle \pi^0\pi^0 | H_{wk} | K_L^0 \rangle = \frac{1}{\sqrt{3(1 + |\epsilon|^2)}} \left[(A_2 - A_2^*) e^{i\delta_2} + (A_2 + A_2^*) \epsilon e^{i\delta_2} - 2\sqrt{2}\epsilon A_0 e^{i\delta_0} \right], \tag{6.148}$$

$$\langle \pi^0\pi^0 | H_{wk} | K_S^0 \rangle = \frac{1}{\sqrt{3(1 + |\epsilon|^2)}} \left[(A_2 + A_2^*) e^{i\delta_2} + (A_2 - A_2^*) \epsilon e^{i\delta_2} - 2\sqrt{2} A_0 e^{i\delta_0} \right]. \tag{6.149}$$

The experimental observations are made on the ratios of the amplitude defined as:

$$\eta_{+-} = \frac{\langle \pi^+\pi^- | H_{wk} | K_L^0 \rangle}{\langle \pi^+\pi^- | H_{wk} | K_S^0 \rangle}, \qquad \eta_{00} = \frac{\langle \pi^0\pi^0 | H_{wk} | K_L^0 \rangle}{\langle \pi^0\pi^0 | H_{wk} | K_S^0 \rangle}. \tag{6.150}$$

Since the values ϵ and $|A_2|$ are small with the experimental limit $|\frac{A_2}{A_0}| \sim \frac{1}{20}$, neglecting second order terms in ϵ and $|A_2|$, we obtain

$$\eta_{+-} \simeq \frac{(A_2 - A_2^*)e^{i\delta_2} + \epsilon(A_2 + A_2^*)e^{i\delta_2} + 2\sqrt{2}\epsilon A_0 e^{i\delta_0}}{(A_2 + A_2^*)e^{i\delta_2} + 2\sqrt{2}A_0 e^{i\delta_0}}$$

$$= \epsilon + \frac{2i\, \mathrm{Im}\,(A_2)e^{i\delta_2}}{2\, \mathrm{Re}(A_2)e^{i\delta_2} + 2\sqrt{2}A_0 e^{i\delta_0}}.$$

Since $A_2 \ll A_0$, we get

$$\eta_{+-} \simeq \epsilon + \frac{1}{\sqrt{2}}\, \mathrm{Im}\left(\frac{A_2}{A_0}\right) e^{i(\pi/2 + \delta_2 - \delta_0)},$$

$$= \epsilon + \epsilon', \tag{6.151}$$

with $\epsilon' = \frac{1}{\sqrt{2}}\, \mathrm{Im}\left(\frac{A_2}{A_0}\right) e^{i(\pi/2 + \delta_2 - \delta_0)}$. Similarly, one may obtain

$$\eta_{00} \simeq \epsilon - 2\epsilon'. \tag{6.152}$$

The present limit of these parameters are [117]:

$$|\eta_{+-}| = (2.22 \pm 0.011) \times 10^{-3},$$

$$|\eta_{00}| = (2.232 \pm 0.011) \times 10^{-3}.$$

6.7 Flavour Changing Neutral Currents (FCNC) and GIM Mechanism

The Cabibbo theory was able to explain several decay rates including the decays of strange as well as non-strange particles. However, the theory allowed decays like $K_L^0 \to \mu^+\mu^-$, $K^\pm \to \pi^\pm \nu\bar{\nu}$ which were experimentally found to be highly suppressed. This is because in the Cabibbo model, the hadronic current $J_\mu^\pm$ is given by:

$$J_\mu^\pm = g\bar{q}\gamma_\mu\tau^\pm q_L, \qquad \text{where} \quad q_L = \frac{(1 - \gamma_5)}{2}q. \tag{6.153}$$

With $q = \begin{pmatrix} u \\ d' \end{pmatrix}$, we get

$$J_\mu^+ \simeq g\bar{u}\gamma_\mu(1 - \gamma_5)(d\cos\theta_C + s\sin\theta_C),$$

$$J_\mu^- \simeq g(\bar{d}\cos\theta_C + \bar{s}\sin\theta_C)\gamma_\mu(1 - \gamma_5)u. \tag{6.154}$$

If u and d' belong to a doublet representation of SU(2) in weak isospin space, then the symmetry group implies the existence of the third component of the current

$$
\begin{aligned}
J_\mu^0 &\simeq 2g\bar{q}\gamma_\mu(1-\gamma_5)\tau_3 q \qquad\qquad (\because [\tau^+,\tau^-]=2\tau_3) \\
&= g\Big(\bar{u}\gamma_\mu(1-\gamma_5)u - \bar{d}'\gamma_\mu(1-\gamma_5)d'\Big) \\
&= g\Big(\bar{u}\gamma_\mu(1-\gamma_5)u - \bar{d}\gamma_\mu(1-\gamma_5)d\cos^2\theta_C - \bar{s}\gamma_\mu(1-\gamma_5)s\sin^2\theta_C \\
&\quad - \Big[\bar{d}\gamma_\mu(1-\gamma_5)s + \bar{s}\gamma_\mu(1-\gamma_5)d\Big]\sin\theta_C\cos\theta_C\Big).
\end{aligned}
\tag{6.155}
$$

The last term predicts the flavor changing neutral current (FCNC) decays with $|\Delta S|=1$ and $\Delta Q=0$ with a strength of $g\sin\theta_C\cos\theta_C$, which is suppressed by a factor $\sin\theta_C$ comparable to the $\Delta S=0$ decays with strength $g\cos\theta_C$ in contrast to the experimental results showing a suppression of the order of $10^{-8} - 10^{-9}$.

Glashow, Iliopoulos, and Maiani [64] proposed a mechanism to suppress these decays by postulating the existence of a fourth quark c (known now as the charm quark) following the earlier proposal of Bjorken and Glashow [65] which forms a weak doublet with the quark s', the orthogonal combination of d' proposed earlier by Cabibbo

$$
s' = -d\sin\theta_C + s\cos\theta_C
\tag{6.156}
$$

such that we now have another doublet of quarks, that is, $q' = \begin{pmatrix} c \\ s' \end{pmatrix}$ in addition to $q = \begin{pmatrix} u \\ d' \end{pmatrix}$. Adding this doublet to the quark picture, we get additional weak currents as

$$
\begin{aligned}
J_\mu^{\pm\prime} &= g\bar{q}'\gamma_\mu(1-\gamma_5)\tau^\pm q', \\
J_\mu^{0\prime} &= 2g\bar{q}'\gamma_\mu(1-\gamma_5)\tau_3 q',
\end{aligned}
$$

giving

$$
\begin{aligned}
J_\mu^{+\prime} &= g\bar{c}\gamma_\mu(1-\gamma_5)[-d\sin\theta_C + s\cos\theta_C], \tag{6.157}\\
J_\mu^{-\prime} &= g[-\bar{d}\sin\theta_C + \bar{s}\cos\theta_C]\gamma_\mu(1-\gamma_5)c, \tag{6.158}\\
J_\mu^{0\prime} &= g\Big(\bar{c}\gamma_\mu(1-\gamma_5)c + \bar{d}\gamma_\mu(1-\gamma_5)d\sin^2\theta_C + \bar{s}\gamma_\mu(1-\gamma_5)s\cos^2\theta_C \\
&\quad - [\bar{d}\gamma_\mu(1-\gamma_5)s + \bar{s}\gamma_\mu(1-\gamma_5)d]\sin\theta_C\cos\theta_C\Big). \tag{6.159}
\end{aligned}
$$

In addition to predicting the new weak charged current coupling of c quarks, with d and s quarks, in Eqs. (6.157) and (6.158), the total neutral current now becomes

$$
\begin{aligned}
J_\mu^{\mathrm{NC}} &= J_\mu^0 + J_\mu^{0\prime} \\
&= g\Big(\bar{u}\gamma_\mu(1-\gamma_5)u + \bar{c}\gamma_\mu(1-\gamma_5)c - \bar{d}\gamma_\mu(1-\gamma_5)d - \bar{s}\gamma_\mu(1-\gamma_5)s\Big).
\end{aligned}
\tag{6.160}
$$

Thus, there are no FCNC in the first order. This is called the Glashow–Iliopoulos–Maiani (GIM) mechanism. In this model, the FCNC could occur in the second order as shown in Figures 6.9(a) and 6.9(b).

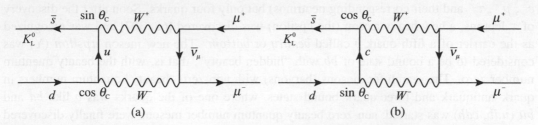

Figure 6.9 Feynman diagrams for the K^0 decay into $\mu^+\mu^-$ with a (a) u-quark exchange and (b) c-quark exchange.

Without the GIM mechanism, the amplitude will be proportional to $\simeq f(m_u)g^4 \sin\theta_C$ $\cos\theta_C$, where $f(m_u)$ is a factor which depends upon the mass m_u obtained from the loop diagram; this mass would give too large a value for the decay rate for FCNC decays. With the GIM mechanism, there will be another term proportional to $\simeq g^4 f(m_c) \sin\theta_C \cos\theta_C$, which will cancel the contribution from the first term. There would be an almost complete cancelation if $m_c \simeq m_u$, and a limit on m_c/m_u can be obtained using the experimental limit of FCNC decays. A calculation by Gaillard and Lee [305] put a limit of $m_c \simeq 1 - 3$ GeV. This extension of the Cabibbo model of quark mixing to four quarks solved the problem of the FCNC and provided a way to estimate the mass of the new quark.

GIM mechanism re-established the symmetry between the leptons and quarks. Therefore, instead of u and d, it is now u and d' which are the counterparts of e^- and ν_e in the leptonic sector; c and s' are the counterparts of μ^- and ν_μ. Figure 6.10 shows the quark mixing between the two generations of quarks with their couplings to W bosons. In 1973, Kobayashi and Maskawa [70] extended the idea to six flavors of quarks and provided a quark mixing model which explains CP violation.

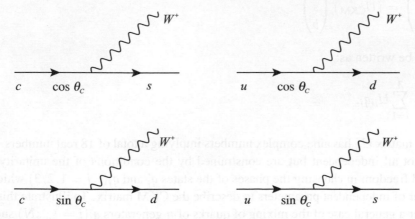

Figure 6.10 Quark mixing.

6.7.1 Six quark mixing and CKM matrix

In 1975, a new lepton called τ [116] was discovered, which spoiled the symmetry between the leptons and quarks proposed by Bjorken and Glashow [65]. There were now six leptons (i.e., e^-, μ^-, τ^- and their corresponding neutrinos) but only four quarks. Soon after the discovery of τ leptons, a new heavy meson (the upsilon) was discovered [68, 307] and was recognized as the carrier of a fifth quark b called *beauty* or *bottom*. The new meson *upsilon* (Υ) was considered to be a bound state of $b\bar{b}$ with "hidden beauty", that is, with the beauty quantum number zero. The search for mesons (baryons) with non-zero "beauty" quantum numbers in quark–antiquark and three quark bound states, where one of the quarks was b like $b\bar{d}$ and $b\bar{u}$ (udb, cdb) was started; non-zero beauty quantum number mesons were finally discovered in the 1980s [69, 308, 309]. Such mesons are called B-mesons and baryons like Λ_b, Σ_b, etc. Therefore, the extension of the lepton–quark symmetry to six leptons necessitated the existence of a sixth quark t called "top", to make the total number of quarks equal to six. They are called u, d, s, c, b, t in the order they appear in literature. The top quark was also discovered in 1995 at Fermilab [310, 311], with a mass $m_t = 173.0 \pm 0.4$ GeV.

In order to describe the weak interaction of all these quarks and leptons, the Cabibbo– GIM model of the quark mixing was extended to six quarks by Kobayashi and Maskawa [70]. For this purpose, these quarks are classified according to the weak isospin and weak hypercharge as in the case of the Cabibbo theory. In this scheme, the left-handed components of the quarks are assigned to a doublet under SU(2), that is, $\begin{bmatrix} u \\ d \end{bmatrix}_L$, $\begin{bmatrix} c \\ s \end{bmatrix}_L$, $\begin{bmatrix} t \\ b \end{bmatrix}_L$; the right-handed components are assigned to a singlet, that is, u_R, d_R, c_R, s_R, t_R, b_R.

The quark mixing theory of Cabibbo was extended to six quarks in which the d, s, b quarks mix to give the states d', s', b'; these states participate in the weak interactions. The states (d', s', b') are written in terms of (d, s, b) states by a 3×3 unitary matrix known as the Cabibbo–Kobayashi–Maskawa (CKM) matrix, that is,

$$\begin{pmatrix} d' \\ s' \\ b' \end{pmatrix} = (U_{\text{CKM}}) \begin{pmatrix} d \\ s \\ b \end{pmatrix} \qquad (6.161)$$

which can be written as:

$$q'_i = \sum_{i=1}^{3} U_{ij} q_j. \qquad (6.162)$$

The CKM matrix U_{ij} has nine complex numbers implying a total of 18 real numbers. However, they are not all independent but are constrained by the conditions of the unitarity of the U matrix and freedom in choosing the phases of the states q'_i and $q_j (i, j = 1, 2, 3)$ which reduces the number of independent parameters to describe the CKM matrix. To illustrate this point, let us consider a general case of the mixing of quarks of n generators $q_i (i = 1, ...N)$ such that:

$$q'_i = \sum U_{ij} q_j, \qquad i, j = 1, n. \qquad (6.163)$$

The matrix U_{ij} is defined by n^2 complex numbers, that is, $2n^2$ real numbers with the following conditions:

i) The unitarity condition in n dimension, that is, $U^\dagger U = 1$ leads to n^2 conditions reducing the independent parameters to $2n^2 - n^2 = n^2$.

ii) There are $2n$ quark states (n states q_i and n rotated states q_i'), with $2n$ phases which are not physical. Leaving out one overall phase, there would be $2n - 1$ phases which can be eliminated by redefining the states leaving $n^2 - (2n - 1) = (n - 1)^2$ independent parameters.

iii) In order to choose $(n - 1)^2$ independent parameters to define an $n \times n$ matrix in n dimensions, an orthogonal matrix in n dimension can be chosen which has $n(n - 1)/2$ real parameters leaving $(n - 1)^2 - \frac{n(n-1)}{2} = \frac{(n-1)(n-2)}{2}$ independent parameters, chosen to be phases inserted in any of the matrix elements U_{ij}.

Therefore, in n dimensions, an $n \times n$ unitary matrix is specified by $\frac{n(n-1)}{2}$ real angles and $\frac{(n-1)(n-2)}{2}$ phases. We see that in two dimensions, we have one angle, that is, the Cabibbo angle θ_C. In three dimensions, there would be three real angles and one phase. Kobayashi and Maskawa chose the three Euler angles $\theta_{12}, \theta_{23}, \theta_{13}$ and a phase $\delta_{13}(= \delta)$ associated with the third mixing angle θ_{13}. The matrix is explained in Chapter 1.

It can be shown that one may choose all the angles θ_{12}, θ_{23}, and θ_{13} to lie in the range $0 < \theta_{ij} < \frac{\pi}{2}$ so that $c_{ij} \geq 0$ and $s_{ij} \geq 0$ and the phase δ is in the range $0 \leq \delta \leq 2\pi$. The phase δ appearing in the U matrix makes the interaction Hamiltonian, written in terms of q_i', non-real, which violates CP invariance. Therefore, the phase δ can be used to describe the CP violation in hadron physics. This was emphasized by Kobayashi and Maskawa in their original work. The current values of the CKM matrix elements have been determined in the experiments using semileptonic decays; they are given in Eq. (6.165). For example, $|U_{ud}|$ is determined while comparing $n \to p + e^- + \bar{\nu}_e$ and $\mu^- \to e^- + \nu_\mu + \bar{\nu}_e$ decay rates, while the comparison of the decay rates of $K^- \to \pi^0 + e^- + \bar{\nu}_e$ and $\mu^- \to e^- + \nu_\mu + \bar{\nu}_e$ is used to determine U_{us}. $|U_{cs}|$ is obtained by comparing the decay rates of $\bar{D}^0 \to K^- + e^+ + \bar{\nu}_e$ with $\mu^- \to e^- + \nu_\mu + \bar{\nu}_e$. Similarly, the other parameters have been determined by comparing the decay rates of other particles involving strange, charm, bottom, and top quarks.

However, the strength of the transition involving the third generation of quarks is quite small except the diagonal element U_{tb}. This can be seen in a transparent way by using Wolfenstein's parameterization of the CKM matrix; Wolfenstein emphasized the hierarchical character of the mixing between the generations of quarks. Wolfenstein's parameterization of the CKM matrix is written as [312]:

$$U_{CKM} = \begin{pmatrix} 1 - \lambda^2/2 & \lambda & \rho\lambda^3 e^{i\phi} \\ -\lambda & 1 - \lambda^2/2 & \lambda^2 \\ \lambda^3(1 - \rho e^{i\phi}) & -\lambda^2 & 1 \end{pmatrix}, \tag{6.164}$$

where $\lambda \approx \sin\theta_C \approx 0.22$, $\rho < 1$, and ϕ is the CP-violating phase factor. The structure of matrix elements in Equation (6.164) shows that $b \to u$, $t \to d$, $t \to d$, $b \to c$ transitions are quite small. The experimentally determined values of the CKM matrix at present are given as follows [117]:

$$U_{CKM} = \begin{pmatrix} 0.97446 \pm 0.00010 & 0.22452 \pm 0.00044 & 0.00365 \pm 0.00012 \\ 0.22438 \pm 0.00044 & 0.97354 \pm 0.00010 & 0.04214 \pm 0.00076 \\ 0.00896 \pm 0.00024 & 0.04133 \pm 0.00074 & 0.999105 \pm 0.000032 \end{pmatrix}. \quad (6.165)$$

6.8 Weak Interaction of Hadrons with Charm and Heavy Flavors

6.8.1 Discovery of charm and heavy flavors

One of the reasons that the fourth quark called charm and denoted by c was proposed was to restore the lepton–quark symmetry [65]. There were four leptons, that is, the electron, the muon, and their respective neutrinos but only three quarks, that is, u, d, and s. The existence of a fourth quark would bring a complete symmetry between the quarks and the leptons. The idea of the fourth quark was the basic ingredient to formulate the GIM mechanism which resolved the problem of FCNC. The existence of a fourth quark will extend the flavor symmetry of the strong interactions from $SU(3)_f$ to $SU(4)_f$. The $SU(3)_f$ symmetry is broken due to the large mass of s quark compared to the mass of the u, d quarks. $SU(4)_f$ is also expected to be broken due to the large mass of c quark as compared to mass of the u, d, and s quarks. However, the following multiplets of mesonic $Q\bar{q}$ or $q\bar{Q}$ and baryonic Qqq, qQq, QQq, or QQQ states, where $Q(=c)$ is a heavy quark and $q(=u,d,s)$ is a light quark, are predicted according to the group structure of $SU(4)_f$.

$$4 \otimes \bar{4} = 1 \oplus 15 \tag{6.166}$$

$$\text{and } 4 \otimes 4 \otimes 4 = 20_S + 20_M + 20_M + \bar{4}_A. \tag{6.167}$$

The mesons are predicted to occur in the multiplet of a singlet and 15-plet. This implies that in addition to 9 well-known mesons corresponding to $3 \otimes \bar{3} = 1 + 8$ decomposition under $SU(3)_f$, we would have seven more states corresponding to one singlet states and 6 states belonging to the 15-plet under $SU(4)$. They are identified as $c\bar{c}$, $c\bar{u}$, $c\bar{d}$, $c\bar{s}$, $\bar{c}u$, $\bar{c}d$, $\bar{c}s$ states. The $c\bar{c}$ state is called a charmonium state with total charm quantum number zero; the state has mesons with hidden charm content. This state is like a positronium system which is a bound state of e^-e^+ with lepton number zero. However, the potentials responsible for the binding of the e^-e^+ system and $q\bar{q}$ system are entirely different. The other $q\bar{Q}$ and $Q\bar{q}$ states are mesonic states with charm quantum number $C = \pm 1$ like kaon states with strangeness quantum number $S = \pm 1$.

The discovery of the charmonium states was made as the result of a narrow resonance J/ψ observed simultaneously at BNL and SLAC [313, 314]. Later, many states of charmonium were observed which provided information about the strong interaction force between c and $\bar{c}$ quarks. A list of some $c\bar{c}$ states is given in Table 6.10. The direct evidence of the existence of the charmed particles with $C = +1$ like $c\bar{u}$, $c\bar{d}$, $c\bar{s}$ called D^0, D^+, and D_s^+ and with $C = -1$ like $\bar{c}u$, $\bar{c}d$, $\bar{c}s$ called $\bar{D}^0$, D^-, and D_s^-, respectively, were observed soon after the discovery of the charmonium state. These mesons decay by weak interactions. The particles with heavier quark content decay into particles with lighter quark content. Soon after the observation of charmed mesons, some baryonic states with a charm content of cqq with $q = u, d, s$ were also observed [315], which decay through weak interactions.

The discovery of the τ lepton in 1975, at the Stanford Positron-Electron Asymmetric Ring (SPEAR) [116] and its decay into other lighter leptons like e, μ, and hadrons like π, K, through weak interactions accompanied by its own neutrino ν_τ due to the conservation of the τ lepton number(L_τ), makes the number of leptons six. The lepton–quark symmetry then requires the existence of two more quarks in the list of (u, d), (c, s); the two quarks were named (t, b), the top and bottom. Soon after the discovery of τ leptons, resonances similar to the $b\bar{b}$ states were observed in 1977. A spectrum of $b\bar{b}$ states was observed; they are known as η_b (9.4 GeV), χ_{b_1} (9.9 GeV), η_b (10.0 GeV) (see Table 6.10). Mesonic states like $q\bar{Q}$ or $\bar{q}Q$, where one of the quarks Q is a heavy quark b, were also observed and are called B mesons like the D mesons in the case of the charmed quark and K mesons in the case of the s quark.

Table 6.10 Predicted $c\bar{c}$ and $b\bar{b}$ states with principal quantum numbers n=1 and 2, and radial quantum number $n_r = n - L$, compared with experimentally observed states. Masses are given in MeV/c^2.

$^{2S+1}L_J$	n	n_r	J^{PC}	$c\bar{c}$ state	$b\bar{b}$ state
$^1S_0$	1	1	0^{-+}	$\eta_c(2984)$	$\eta_b(9398)^{(*)}$
$^3S_1$	1	1	1^{--}	$J/\psi(3097)$	$\Upsilon(9460)$
$^3P_0$	2	1	0^{++}	$\chi_{c0}(3415)$	$\chi_{b0}(9859)$
$^3P_1$	2	1	1^{++}	$\chi_{c1}(3511)$	$\chi_{b1}(9893)$
$^3P_2$	2	1	2^{++}	$\chi_{c2}(3556)$	$\chi_{b2}(9912)$
$^1P_1$	2	1	1^{+-}	$h_c(3525)$	$h_b(9899)$
$^1S_0$	2	2	0^{-+}	$\eta_c(3639)$	$\eta_b(9999)$
$^3S_1$	2	2	1^{--}	$\psi(3686)$	$\Upsilon(10023)$

The discovery of the top quark, t, took a much longer time than the bottom quark due to the very high energies involved in producing $t\bar{t}$ pairs or mesonic states like $t\bar{q}$ and baryonic states like tqq, etc. The top quark was discovered in 1995 in the hadron collider at Fermilab. It is produced in pairs through the process $q\bar{q} \to t\bar{t}$ and $gg \to t\bar{t}$. The expected decay modes are:

$$t\bar{t} \longrightarrow W^+ b W^- \bar{b}.$$

The decays of $W^+(W^-)$ in the hadronic and leptonic modes along with the production of jets due to $b\bar{b}$ are the signatures of $t\bar{t}$ production in the high energy hadronic collision. Some recent results combined with the old ones for a few detectors at CDF [310], DØ [311] and Tevatron [316] have measured cross sections in the range of 7.55 to 7.65 pb:

$$\sigma_{t\bar{t}} = 7.63 \pm 0.50 \ pb \ \text{(CDF)}, \tag{6.168}$$

$$\sigma_{t\bar{t}} = 7.56 \pm 0.59 \ pb \ \text{(DØ)}, \tag{6.169}$$

$$\sigma_{t\bar{t}} = 7.60 \pm 0.41 \ pb \ \text{(Tevatron)}, \tag{6.170}$$

which are in agreement with the standard model expectation of $7.35^{+0.28}_{-0.33}$ pb in the perturbative QCD for $m_t = 172.5$ GeV/c^2. Therefore, the mass of the top quark is determined to be 172.5 GeV/c^2. It is a very heavy quark and can decay weakly into b quarks and the intermediate vector bosons. Its lifetime is 5×10^{-25} s. At LHC, 90% production of the top quarks is through gg collisions at the energy $\sqrt{s} = 14$ TeV and 80% at energy $\sqrt{s} = 7$ TeV.

6.8.2   Weak decays of particles with charm and heavy flavors

The weak semileptonic and nonleptonic decays of particles with charm and heavy flavors are understood in terms of the quark model in which a quark current is involved in the weak decay through its coupling with the lepton current via the vector meson exchange of W^+ or W^-. This model is subject to QCD corrections involving the gluon exchange, thus, involving more than one quark. However, we will not go into the details of these corrections and describe the single quark model, that is, spectator model in which other quarks do not participate in the weak transition. For simplicity, we first describe the weak decays of particles with charm quark using the four-quark Cabibbo–GIM model of quark mixing and extend it to particles with heavier flavors using the six quark mixing of CKM.

6.8.3   Weak decays of particles with charm

Semileptonic decays of charmed mesons

(i)   Two-body decay modes

The two-body semileptonic decay of mesons like $P(q\bar{q} \to l\nu_l)$, where P could be π^+, K^+, D^+, D_s^+ are depicted in Figure 6.11. The decay rates for the two-body decays of a pseudoscalar

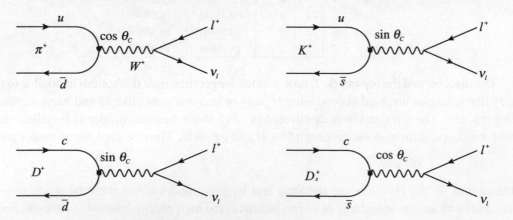

Figure 6.11 Leptonic weak decays of mesons.

meson P are proportional to $U_{ij}^2 f_P^2 m_l^2 \left(1 - \frac{m_l^2}{m_P^2}\right)^2 m_P$ (see Eq. (6.81)), where $U_{ij} = \cos\theta_C$ ($\sin\theta_C$) for the $\Delta S = 0$ ($\Delta S = 1$) transitions, m_P and f_P are, respectively, the mass and decay constant of meson P. Therefore, the predictions for charmed particle decays are:

$$\frac{\Gamma(D^+ \longrightarrow \mu^+ \nu_\mu)}{\Gamma(K^+ \longrightarrow \mu^+ \nu_\mu)} \approx \frac{m_{D^+}}{m_{K^+}}, \qquad \frac{\Gamma(D_s^+ \longrightarrow \mu^+ \nu_\mu)}{\Gamma(K^+ \longrightarrow \mu^+ \nu_\mu)} = \cot^2 \theta_C \frac{m_{D_s^+}}{m_{K^+}},$$

which are in agreement with the experimentally observed rates.

(ii) Three body decay modes

In the spectator model, the semileptonic decays of mesons with a charm quark is depicted in Figure 6.12. Similar diagrams for the semileptonic decays of charmed baryons are depicted in Figure 6.13.

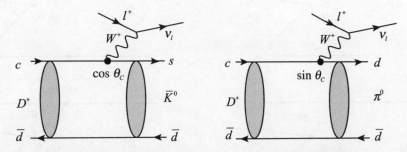

Figure 6.12 Semileptonic weak decays of charmed mesons.

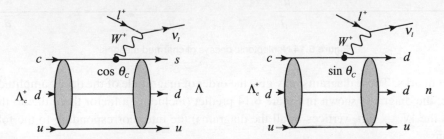

Figure 6.13 Semileptonic weak decays of charmed baryons.

In the case of the charmed mesons, the decays with $\Delta C = \Delta S = \Delta Q$ are Cabibbo allowed with $\Delta I = 0$, while the decays with $\Delta C = \Delta Q = \pm 1$ with $\Delta S = 0$ are Cabibbo suppressed with $\Delta I = 1/2$. Comparing the strength of the $c \leftrightarrows s$ and $c \leftrightarrows d$ transitions. It can be shown that:

$$\frac{\Gamma(D^+ \longrightarrow \bar{K}^0 l^+ \nu)}{\Gamma(K^0 \longrightarrow \pi^- l^+ \nu)} = \left(\frac{M_{D^+}}{M_K}\right)^5 \cot^2 \theta_C \frac{f(m_K/m_D)}{f(m_\pi/m_K)},$$

where $f(x)$ is a phase factor given by:

$$f(x) = 1 - 8x^2 + 8x^6 - x^8 - 24x^4 \ln x.$$

In this case, the agreement of the theoretical calculations in the spectator model with the experimental results is not very good as the QCD corrections due to gluon exchange could be large.

(iii) Nonleptonic decays of charmed mesons

In the spectator model, the nonleptonic decays of the charmed mesons are diagrammatically represented in Figure 6.14. Similar diagrams may be depicted for baryon decays with two

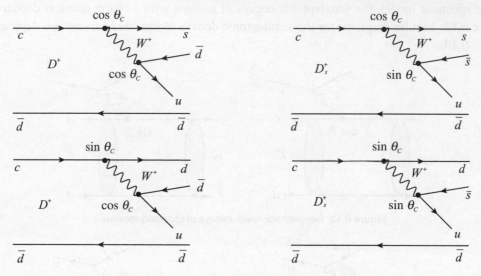

Figure 6.14 Nonleptonic decays of charmed mesons.

spectator quarks. These diagrams may give the order of magnitude of the decay amplitudes. For example, the diagrams shown in Figure 6.14 predict (including a factor three due to the color factor at the $W \rightarrow u_i d_j$ vertices in all the diagrams) the rates corresponding to the following transitions

$$c \longrightarrow s\,u\,\bar{d} \propto 3\cos^4\theta_C, \qquad c \longrightarrow s\,u\,\bar{s} \propto 3\cos^2\theta_C \sin^2\theta_C,$$
$$c \longrightarrow d\,u\,\bar{s} \propto 3\sin^4\theta_C, \qquad c \longrightarrow d\,\bar{d}\,u \propto 3\sin^2\theta_C \cos^2\theta_C.$$

These predictions compare very poorly with the experimentally observed values. This is because the QCD effects of hadrons with charm and the heavy flavors due to the quark–quark interactions arising due to kinematical and/or dynamical effects become more prominent as compared to the case of the nonleptonic decays of strange mesons and hyperons. These effects have been shown to play an important role in theoretically explaining the dominance of $\Delta I = \frac{1}{2}$ transition amplitudes in the case of nonleptonic decays of strange mesons and hyperons. Some of these corrections may also arise due to the weak interaction effects between two quarks inside the hadron as a result of the W exchange and W annihilation (shown in Figure 6.15). They are further affected by the final state interactions. However, the major correction arises due to the QCD effects involving gluon exchanges as shown in Figure 6.16. All these corrections improve the agreement with the experimental data [278].

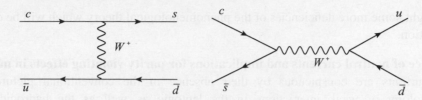

Figure 6.15 W boson exchanges.

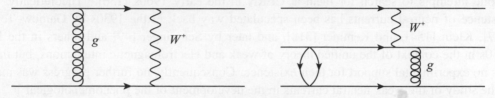

Figure 6.16 QCD corrections.

6.8.4 Weak decays of particles with heavy flavors

In order to describe the weak decays of particles with heavy flavors like b and t quarks, we consider the six quark mixing model of Cabibbo–Kobayashi–Maskawa using the CKM matrix. In this model of weak interactions, the hadronic current J_μ^h is written as:

$$J_\mu^h = q\gamma_\mu(1-\gamma_5)q' = \begin{pmatrix} \bar{u} & \bar{c} & \bar{t} \end{pmatrix} \gamma_\mu(1-\gamma_5) \begin{pmatrix} U_{ud} & U_{us} & U_{ub} \\ U_{cd} & U_{cs} & U_{cb} \\ U_{td} & U_{ts} & U_{tb} \end{pmatrix} \begin{pmatrix} d \\ s \\ b \end{pmatrix}, \qquad (6.171)$$

where, $U_{ij}(i = u, c, t ; j = d, s, b)$ are the matrix elements of the CKM matrix, described in terms of the Euler angles θ_1, θ_2, θ_3 and the phase δ as given in Section 6.7.1. In the case of four quarks, the mixing matrix is described in terms of one parameter, the Cabibbo angle θ_C which is determined from the weak decays of strange particles (Section 6.4.1). The other terms in the matrix elements of the matrix U_{ij} are determined from the decays of heavier particles involving $b \to c$, $b \to u$, and $t \to b$ decays. These numerical values are used to test the unitarity relations between these matrix elements, that is,

$$|U_{ud}|^2 + |U_{us}|^2 + |U_{ub}|^2 = 1,$$
$$|U_{cd}|^2 + |U_{cs}|^2 + |U_{cb}|^2 = 1, \qquad (6.172)$$
$$|U_{td}|^2 + |U_{ts}|^2 + |U_{tb}|^2 = 1.$$

6.9 Limitations of the Phenomenological Theory from the Hadron Sector

We have discussed the limitations of the phenomenological theory by studying the weak processes in the leptonic sector. The study of the various weak processes in the hadronic sector bring to light some more deficiencies of the phenomenological theory which will be discussed in this section.

(i) Absence of neutral currents and implications for parity violating effects in nuclei

Neutral currents are conspicuous by their absence in the conventional picture of the phenomenology of weak interactions in the leptonic as well as the hadronic sectors. Experimentally, there was no evidence for the existence of such currents until 1973 [164], despite attempts to search for them at CERN in the early 1960s [156]. Theoretically, the existence of neutral currents has been speculated way back in the 1930s by Gamow–Teller [317], Klein [154], and Kemmer [318], and later by Schwinger [49] and others in the late 1950s in the context of the unified theory of weak and electromagnetic interactions, but there was no experimental support for their existence. Consequently, no further progress was made in the study of the weak neutral currents in the development of the phenomenological $V - A$ theory of weak interactions.

It is to be noted that in the conventional picture of the weak interaction theory, the interaction between two nucleons is described in terms of the charged currents mediated by the exchange of charged W bosons between the nucleons as shown in Figure 6.17, leading to a parity violating (PV) pseudoscalar nucleon–nucleon potential $V^{PV}(r)$. Such a mechanism also leads to a parity violating πNN coupling constant $f_{\pi NN}$ giving rise to a PV potential in the one boson exchange model similar to the parity conserving $N - N$ potential as shown in Figure 6.18 [319]. This parity violating potential $V^{PV}(r)$ gives rise to an admixture $\delta\psi_i$ of the opposite parity states in a given nuclear state ψ_i (characterized by definite parity and isospin) and is calculated in the usual way by using the first order perturbation to be

$$\delta\psi_i = \sum_{j \neq i} \frac{\langle \psi_j | V^{PV} | \psi_i \rangle}{E_i - E_j}, \tag{6.173}$$

where ψ_i and ψ_j are the eigenstates of the parity conserving nuclear interaction Hamiltonian. The presence of such an admixture in the nuclear states is responsible for the parity violating effects in some nuclear processes of scattering and decays involving nucleons and nuclei driven by strong and electromagnetic interactions. Some examples of such processes with experimentally observed parity violating observables are given in Table 6.11.

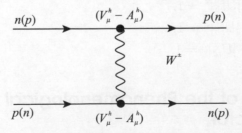

Figure 6.17 Nucleon–nucleon scattering through a W exchange.

The theoretical attempts to calculate $V^{PV}(r)$ using the phenomenological $V - A$ theory with charged currents only and explain the parity violating effects are not satisfactory. Even in the context of the Cabibbo theory with GIM mechanism, in which weak neutral currents are present in the $\Delta S = 0$ sector, the conventional theory is not able to explain these PV effects in a consistent manner. Inspite of the theoretical uncertainties inherent in the calculation of $V^{PV}(r)$ and the experimental uncertainties reported in the parity violating observables shown in Table 6.11, the phenomenological $V - A$ theory is found to be inadequate.

Table 6.11 Some of the observed parity violating effects in nucleons and nuclei [319, 320]. A_L represents asymmetry and P_L represents the polarization of the hadron(photon) in the final state.

	Process	Value of A_L and P_L	Experiment
Scattering	$\vec{p}p \to \vec{p}p$	$A_L = (-0.93 \pm 0.20 \pm 0.05) \times 10^{-7}$	Bonn
		$(-1.7 \pm 0.8) \times 10^{-7}$	LANL
		$(-1.57 \pm 0.23) \times 10^{-7}$	PSI
		$(0.84 \pm 0.34) \times 10^{-7}$	TRIUMP
	$\vec{p}\,^4He \to \vec{p}\,^4He$	$A_L = -(3.3 \pm 0.9) \times 10^{-9}$	PSI
Decay	$^{18}F(0^-) \to{}^{18}F(1^+) + \gamma$	$P_\gamma = (-7 \pm 20) \times 10^{-4}$	Caltech/Seattle
		$(-10 \pm 18) \times 10^{-4}$	Mainz
		$(3 \pm 6) \times 10^{-4}$	Florence
		$(2 \pm 6) \times 10^{-4}$	Queens
	$^{19}F(\frac{1}{2}^-) \to{}^{19}F(\frac{1}{2}^+) + \gamma$	$A_\gamma = (-8.5 \pm 2.6) \times 10^{-5}$	Seattle
		$(-6.8 \pm 1.8) \times 10^{-5}$	Mainz
Capture	$np \to d\vec{\gamma}$	$P_\gamma = (1.8 \pm 1.8) \times 10^{-7}$	LNPI
	$\vec{n}p \to d\gamma$	$A_\gamma = (0.6 \pm 2.1) \times 10^{-7}$	Grenoble
		$(-1.2 \pm 1.9 \pm 0.2) \times 10^{-7}$	LANL

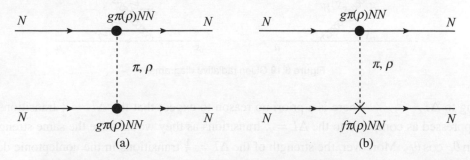

Figure 6.18 Nucleon–nucleon potential in the the meson exchange model. (a) Parity-conserving (PC) potential, (b) PV potential. $g\pi(\rho)NN$ and $f\pi(\rho)NN$ are the parity conserving and parity violating couplings.

(ii) $\Delta I = \frac{1}{2}$ rule in hadronic decays

We have seen that the $\Delta I = \frac{1}{2}$ rule in the semileptonic decays of strange particles like K-mesons and hyperons follows from the structure of the hadronic current appearing in the interaction Lagrangian, that is,

$$\mathcal{L}_{int}^{\Delta S=1} = \frac{G_F}{\sqrt{2}} \sin\theta_C [\bar{u}\gamma_\mu(1 - \gamma_5)s \, l^\mu + h.c.].$$

Since (u, d) belong to the isospin doublet $I = \frac{1}{2}$ and s is an isosinglet $I = 0$, the $\Delta I = \frac{1}{2}$ rule for the $\Delta S = 1$ semileptonic decays follows. However, in the case of nonleptonic decays like $K^{\pm} \to \pi^{+}\pi^{-}$, $K_s \to \pi^0 \pi^0$, $\Lambda \to p\pi^-$, $\Sigma^+ \to p\pi^0$, $\Sigma^+ \to n\pi^+$, etc., the $\mathcal{L}_{\text{int}}^{\text{nonleptonic}}(x)$ is written as:

$$\mathcal{L}_{\text{int}}^{\text{nonleptonic}} = \frac{G_F}{\sqrt{2}} J_\mu^{h\dagger} J^{\mu h} + \text{h.c.}$$

in which, the $\Delta S = 1$ decays are described by the Lagrangian:

$$\mathcal{L}_{\text{int}}^{\Delta S=1}(x) = \frac{G_F}{\sqrt{2}} \sin \theta_C \cos \theta_C \left\{ [\bar{u}\gamma_\mu(1-\gamma_5)d] \right.$$

$$\left. [\bar{u}\gamma^\mu(1-\gamma_5)s]^\dagger \right\},$$

a product of two currents with $\Delta I = 1$ in the $\Delta S = 0$ sector and $\Delta I = \frac{1}{2}$ in the $\Delta S = 1$ sector

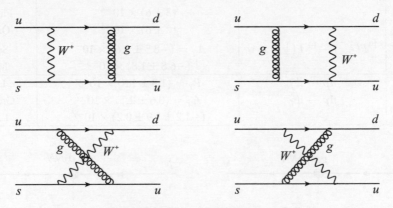

Figure 6.19 Gluon radiative diagrams.

leading to $\Delta I = \frac{1}{2}, \frac{3}{2}$. There is a priori no reason to expect that the $\Delta I = \frac{3}{2}$ transitions will be suppressed as compared to the $\Delta I = \frac{1}{2}$ transitions as they would have the same strength as $\frac{G_F}{\sqrt{2}} \sin \theta_C \cos \theta_C$. Moreover, the strength of the $\Delta I = \frac{1}{2}$ transitions in the nonleptonic decays was experimentally found to be much larger than the strength of the $\Delta I = \frac{1}{2}$ transitions in semileptonic decays, while the $V - A$ theory predicts it to be only slightly smaller by a factor $\cos \theta_C$. A satisfactory explanation of the strength of the nonleptonic weak decays pose quite severe problems to be explained by the $V - A$ theory when applied to the strangeness sector.

It is now believed that the suppression of $\Delta I \geq \frac{3}{2}$ transitions in the nonleptonic decays of mesons and hyperons and the enhancement of $\Delta I = \frac{1}{2}$ transitions in the nonleptonic decays as compared to the semileptonic decays are dynamical effects arising due to the renormalization of the weak couplings in quantum chromodynamics when the effects of gluon exchanges are included. There are two types of gluon exchange diagrams that contribute to the renormalization of the weak couplings. These are shown in Figures 6.19 and 6.20 and are called the gluon

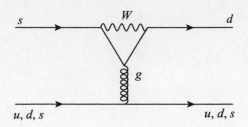

Figure 6.20 Penguin diagrams.

radiative and penguin diagrams, respectively. The details of these calculations are beyond the scope of this book and the readers are referred to textbooks on QCD [321]. In this context, the intuition of Cabibbo that the origin of the physics of the $\Delta I = \frac{1}{2}$ rule in nonleptonic decays is beyond the scope of the phenomenological weak interaction and lies elsewhere was indeed correct.

Chapter **7**

Gauge Field Theories and Fundamental
Interactions

7.1 Introduction

Our present understanding of the physical phenomena in nature and the laws governing them is based on the assumption that quarks and leptons are the basic constituents of matter which interact with each other through strong (quarks only), weak, electromagnetic, and gravitational interactions. One of the major aims of scientists in the physics community has been to formulate a unified theory of all these fundamental interactions to describe the natural phenomenon. The first and earliest step in this direction was to unify electromagnetic and gravitational interactions, both of them being long range interactions, that is, proportional to $\frac{1}{r}$; many attempts were made to unify them in the early twentieth century. It was then believed that these were the only two fundamental interactions and the interactions could be described by field theories based on the principle of invariance under certain transformations called local gauge transformations because of their explicit dependence on space–time coordinates. In this type of field theories, the electromagnetic interaction between two charged particles is described by the exchange of a massless vector field $A_\mu(x)$ as proposed by Weyl [40], while the gravitational interaction between the two objects is described by the exchange of a tensor field $g_{\mu\nu}(x)$ as proposed by Weyl [322] and Einstein [323]. Later, after the discovery of the atomic nucleus and the experimental studies of the structure of nuclei and the phenomenon of nuclear radioactivity, two more fundamental interactions, viz., strong and weak interactions were revealed. The existence of strong interaction is responsible for binding neutrons and protons together and the weak interaction enables them to decay inside the nucleus. Both the interactions were found to be of short range. The need was felt to formulate a unified theory of all the four fundamental interactions, viz., the electromagnetic, strong, weak, and gravitational interactions. In analogy with the theory of electromagnetic interactions being mediated by a massless neutral vector field $A_\mu(x)$ called the photon, Yukawa [92] proposed that the newly discovered short range strong interactions are mediated by complex scalar field $U(x)$ (in fact, the scalar component of a four-vector field like ϕ in the case of electromagnetic interactions and $\bar{\Psi}_n \gamma_0 \Psi_n = \Psi^\dagger \Psi$ in the case of weak interactions) corresponding to a "heavy particle" called meson with a mass 'm' appropriate to the short range 'r' of the strong interactions through the relation $m \propto \frac{1}{r}$. He further proposed that these meson fields $U(x)$ coupled strongly to the nucleons through strong interactions to provide nuclear binding, and coupled weakly to the leptons and nucleons through weak interactions describing the processes of β-decay, that is, $n \to p + e^- + \bar{\nu}_e$, and $p \to n + e^+ + \nu_e$ in nuclei. This idea led to the generalization of the concept of mediating fields which was used later to formulate the theoretical models of fundamental interactions. Such theoretical models of fundamental interactions were proposed earlier by Klein [324] and Kemmer [318] in the 1930s; in hindsight, these models can be considered as extensions of Weyl's theory of electromagnetism to describe strong and weak interactions.

In the theory of electromagnetism formulated by Weyl [40], the vector field $A_\mu(x)$ mediating the electromagnetic interaction was a neutral massless gauge field arising due to the requirement of invariance of the free electron Lagrangian under a local gauge transformation corresponding to U(1) symmetry. Kemmer [318] extended the formalism of the gauge field theory to describe nuclear interactions based on the SU(2) symmetry in isospin space which was an experimentally discovered symmetry in nuclear interactions. This model introduced a triplet of massless vector gauge fields corresponding to the three generators of SU(2). The two charge gauge fields of the triplet could describe the strong interactions between neutrons and protons or the weak interactions between them. In the case of weak interactions, this would imply the existence of neutral currents corresponding to the third component of the triplet for which there were no experimental evidence, even though their existence was speculated earlier by Wentzel [325] and Gamow and Teller [317].

Klein [192] also suggested the existence of a triplet of vector fields in the context of the Kaluza–Klein theory in the five dimensional space–time, deducing from the compactification of the fifth dimension. This triplet of vector fields is similar to the massless gauge fields corresponding to the SU(2) symmetry suggested by Kemmer. However, in this model, while the two charged fields of the triplet are used to describe the strong or weak charged interactions, the third neutral field was assigned to describe the electromagnetic interactions. Assigning charged and neutral fields to the same triplet implied that the strong (weak), and electromagnetic interactions had the same strength of coupling, a prediction not supported by experiments. Therefore, the study of the models proposed by Kemmer and Klein to explain fundamental interactions in analogy with the gauge field theoretical model of electromagnetic interactions given by Weyl [40] was not pursued any further for a long time.

It took more than 20 years before interest in the study of gauge field theoretical models of the fundamental interactions and their role in obtaining a unified theory of fundamental interactions was revived in the mid-1950s. The revival of interest was due to two major developments in our understanding of weak and strong interactions of elementary particles.

One was in the field of strong interactions, where an elegant extension of Weyl's theory to non-abelian gauge field theories was formulated to describe strong interactions [326, 327, 328]; the other was in the field of weak interactions, where a very successful phenomenological theory of the weak interaction was evolved from extensive experimental as well as theoretical studies of various weak interaction processes for almost 25 years after the Fermi's theory [329], leading to the $V - A$ theory of weak interactions. In the formulation of the gauge field theory of strong interactions[326, 327, 328], the strong interaction was assumed to be invariant under local gauge transformation corresponding to SU(2) symmetry in isospin space. The requirement of invariance under local SU(2) symmetry introduced the existence of a triplet of massless vector gauge fields called Yang-Mills fields, which were self interacting. While the masslessness and the self interaction of the gauge field assured that the theory could be renormalized, it was not appropriate to describe phenomenologically the strong or weak interactions, as the masslessness of the gauge fields implied long range. Introducing a mass term in the interaction or introducing more particles to generate mass through their interaction with the gauge fields would destroy the renormalizability of the theory. The attempts in this direction to formulate a gauge theory of strong and weak interactions were not successful.

The $V - A$ theory was very successful in describing the weak processes at low energies but was found inadequate at higher energies due to the violations of unitarity and the presence of divergences in the higher orders of perturbation theory. The theory was essentially not renormalizable. However, it was considered to be a low energy manifestation of an appropriate high energy theory of weak interactions presumably mediated by charged intermediate vector bosons (IVB) that are massive with a mass M corresponding to the range r of weak interactions; the bosons interact with the weak charged vector and axial vector ($V - A$) currents of nucleons and leptons. The introduction of such naive IVB helped to partially solve the divergence problems present in the $V - A$ theory, but the theory still remained non-renormalizable(see Chapter 5). However, the possibility that the weak interactions are mediated by vector bosons which if identified with massless gauge fields corresponding to a local SU(2) symmetry in weak isospin space could lead to a renormalizable theory of weak and electromagnetic interactions. This was an attractive idea proposed by many during the 1950s [50, 51, 49], but it encountered many difficulties similar to those faced by the models proposed by Kemmer and Klein. In these models, weak interactions are mediated by the triplet of massless isovector vector fields, like the gauge fields of the Yang–Mills theory [326], which interact weakly with the vector, and axial vector currents of nucleons and leptons. While the charged components of the isovector vector fields interact with the charged weak currents, the third and the neutral component of the isovector fields is assumed to interact either with the electromagnetic current or a possible neutral weak current. Such models faced the following problems in achieving a satisfactory description of the weak interactions and its unification with the electromagnetic interactions.

a. The two charged gauge fields proposed to be mediating the weak interactions were predicted to be massless. The masslessness of the gauge fields rendered the theory renormalizable, but implied the range of weak interactions to be infinite which was phenomenologically incorrect.

b. If the third neutral component of the isovector vector gauge field is proposed to be weakly interacting like in the models of Kemmer [318] and Bludman [50], then the model predicts the existence of neutral currents, a result not supported by experiments at that time.

c. If the third neutral component of the isovector vector gauge field is identified with the electromagnetic field $A_\mu(x)$ putting the two charged weak fields and the electromagnetic field in the same triplet like in the models of Klein [324], Schwinger [49], and Leite Lopes [51], it leads to the following difficulties.

i) It was hard to explain the simultaneous presence of different types of coupling of the three components of an isovector field to the nucleonic and leptonic weak currents. While the third neutral component of the isovector field would have a parity conserving coupling to the electromagnetic currents of charged leptons and nucleons, the charged components of the same isovector field would have parity conserving as well as parity violating couplings to the charged weak currents of nucleons and leptons. This is not allowed by the SU(2) symmetry if the three fields belong to the same triplet.

ii) The same effective coupling strength for weak and electromagnetic interactions, that is, $e^2 = g^2/M_W^2$, where e and g are respectively, the strength of the coupling of the electromagnetic field (A_μ) and the weak fields ($W^\pm$) to the electromagnetic and weak currents implying a mass of vector field $M_W \sim \sqrt{\frac{g^2}{e^2}} \sim 30$ GeV. This is not compatible with the vector gauge fields being massless.

Thus, faced with the problem of massless vector bosons, a new idea of spontaneous symmetry breaking (SSB) inspired by the Bardeen–Cooper–Schrieffer (BCS) theory of superconductivity in solid state physics was brought to particle physics by Nambu [330] and Goldstone [162] to generate masses for the vector gauge fields in the theory. The initial application of such models of SSB to gauge field theories in the interaction Lagrangian for massless fermions and bosons with exact continuous symmetry like U(1) or SU(2) led to the appearance of massless spin 0 bosons known as the Nambu–Goldstone bosons. The theory succeeded in dynamically generating the mass of fermions and bosons, whereas experimentally there was no evidence for the existence of the massless and spinless, Nambu–Goldstone bosons, predicted by this type of theories. At first, it seemed to be a model-dependent result but soon, it was proved to be quite a general result in all the Lorentz invariant field theories with exact symmetry under the continuous symmetry transformation that are broken spontaneously [331, 332]. Therefore, the gauge field theories which are invariant under exact continuous local symmetry and are broken spontaneously by the Nambu–Goldstone mechanism had two undesirable features which prevented them from being a suitable theory of weak interactions: one was the appearance of the massless vector gauge bosons and the other was the appearance of the massless scalar Nambu–Goldstone bosons. Thus, the theories based on SSB were inapplicable to strong or weak interactions.

However, following the suggestion of Schwinger [333] that the invariance under the exact gauge symmetry need not always imply the existence of massless vector bosons if their coupling with the matter fields is strong, Anderson explicitly showed that [334] in a system with high density of electrons which are interacting via electromagnetic interactions, the massless photon acquires a mass through its interaction with the free electron gas. This was similar to the situation in superconductivity, where the photon acquires a mass responsible for explaining the Meissner effect through the spontaneous breaking of gauge symmetry of electromagnetic interactions. In both the cases of high density electrons and superconductivity, the physical systems were essentially non-relativistic. However, in a series of papers published independently by Englert and Brout [159], Higgs [158] and Guralnik [335, 336], Guralnik and Hagen [337], and Kibble [161], it was shown to be the case also in the Lorentz covariant relativistic field theories if the exact continuous symmetry is local and not global. In the case of a local gauge invariant field theory, which is spontaneously broken, the massless spin 0 field becomes the longitudinal component of the massless spin-1 gauge field, which appears due to the requirement of the local gauge invariance, making thc field massive. This is known as the Higgs mechanism.

The Higgs mechanism applied to spontaneously broken local gauge field theories provided a way to generate masses for gauge fields. With an appropriate choice of the symmetry group to be $SU(2) \times U(1)$. Weinberg [157] and Salam [37] independently proposed a model for the unified theory of weak and electromagnetic interactions of leptons, which was later extended to quarks and hadrons using the GIM mechanism [64] for describing the weak interactions of hadrons. The model correctly reproduces all the aspects of the $V - A$ theory at low energies and predicts the existence of neutral currents which were observed in 1973.

Salam [37] and Weinberg [157] both speculated that spontaneously broken local gauge field theory was also renormalizable like the exact symmetric gauge field theories but gave no proof. It was t'Hooft and Veltman [338] and Lee and Zinn-Justin [339] who proved the renormalizability of spontaneously broken local gauge field theories using the formalism based on Feynman's path integrals. Thus, with the renormalizability of the model established firmly, the Weinberg–Salam–Glashow model provided a successful unified theory of weak and electromagnetic interactions.

In summary, the main concepts leading to a unified theory of electromagnetic and weak interactions are as follows:

- Gauge invariance of field theory.
- Spontaneously broken symmetries of gauge field theories.
- Higgs mechanism for the generation of mass of gauge fields.
- Choice of the gauge symmetry group and its fundamental representation.

In this chapter, we shall develop an understanding of the principle of gauge invariance and its consequences leading to the introduction of massless gauge fields as carriers of fundamental interactions and apply the concept of spontaneous symmetry breaking of local gauge field theories to generate the mass of gauge fields using Higgs mechanism.

7.2 Gauge Invariance in Field Theory

The idea of gauge invariance as a dynamical principle to describe electromagnetic interactions and its nomenclature was conceptualized by Weyl during his efforts to find a unified theory of electromagnetic and gravitational interactions [40]. Since then, efforts have been made to use this concept as the symmetry principle to formulate the theory of weak, strong, and electromagnetic interactions. In fact, the principle of gauge invariance is equivalent to the principle of phase invariance in quantum mechanics and field theory in the context of wave functions and fields; however, the terminology of gauge invariance is used even now to describe the physical concepts and underlying principles of phase invariance. As a necessary prerequisite to understand these efforts, we first describe the principle of gauge invariance in classical electrodynamics and then elaborate the role of gauge invariance as a dynamical principle in describing electromagnetic interactions in field theory.

In Chapter 2, we have already introduced the concept of gauge invariance in field theory, where we discussed the gauge invariance in classical electrodynamics and reviewed Maxwell's theory of electrodynamics. In this section, we will first obtain Maxwell's equations from a variational principle.

7.2.1 Maxwell's equations from a variational principle

It is instructive to note that Maxwell's equation for the electromagnetic field $A_\mu(x_\mu)$ can be derived from a suitable Lagrangian $\mathcal{L}(x)$ for the electromagnetic field in the presence of sources densities $J^\mu(x)$, that is,

$$\mathcal{L}(x) = -\frac{1}{4}F_{\mu\nu}(x)F^{\mu\nu}(x) - J_\mu(x)A^\mu(x) \tag{7.1}$$

or from the Lagrangian of free electromagnetic field, $\mathcal{L}_{\text{free}}$ given by:

$$\begin{aligned}
\mathcal{L}_{\text{free}}(x) &= -\frac{1}{4}F_{\mu\nu}(x)F^{\mu\nu}(x) \\
&= -\frac{1}{4}(\partial_\mu A_\nu - \partial_\nu A_\mu)(\partial^\mu A^\nu - \partial^\nu A^\mu) \\
&= -\frac{1}{2}\partial_\mu A_\nu(\partial^\mu A^\nu - \partial^\nu A^\mu) \tag{7.2}
\end{aligned}$$

$$\Rightarrow \quad \frac{\partial \mathcal{L}_{\text{free}}}{\partial A_\nu} = 0, \quad \text{and} \quad \frac{\partial \mathcal{L}_{\text{free}}}{\partial(\partial_\mu A_\nu)} = -F^{\mu\nu}. \tag{7.3}$$

Putting this in the Euler Lagrange equation:

$$\frac{\partial \mathcal{L}_{\text{free}}}{\partial A_\nu} - \partial_\mu \frac{\partial \mathcal{L}_{\text{free}}}{\partial(\partial_\mu A_\nu)} = 0, \tag{7.4}$$

which results in

$$\partial_\mu F^{\mu\nu} = 0. \tag{7.5}$$

We may write the action using the Lagrangian in Eq. (7.1)

$$S = \int \mathcal{L}d^4x = \int \left[-\frac{1}{4}g_{\mu\lambda}g_{\nu\rho}F^{\lambda\rho}F^{\mu\nu} - J^\mu A_\mu \right] d^4x.$$

Any variation in S leads to:

$$\delta S = \int \left[-\frac{1}{2} g_{\mu\lambda} g_{\nu\rho} F^{\lambda\rho} \delta F^{\mu\nu} - J^\mu \delta A_\mu \right] d^4x.$$

Using the property $F^{\lambda\rho} = -F^{\rho\lambda}$, we may write:

$$\delta S = \int \left[-F^{\lambda\rho} \partial_\lambda \delta A_\rho - J^\mu \delta A_\mu \right] d^4x.$$

Using the method of integration by parts in the first part of the right-hand side (R.H.S), we obtain:

$$\delta S = \int \left[\partial_\lambda F^{\lambda\rho} - J^\rho \right] \delta A_\rho d^4x$$

which vanishes ($\delta S = 0$) for arbitrary δA_ρ, such that:

$$\partial_\mu F^{\mu\nu} = J^\nu \tag{7.6}$$

which is nothing but the inhomogeneous Maxwell's equation in the medium.

It is also seen from Eqs. (7.1) and (7.2), that the presence of a mass term like $m^2 A_\mu A^\mu$ is not allowed in the Lagrangian since $m^2 A_\mu A^\mu$ transforms under the local gauge transformation as:

$$A^\mu(x) A_\mu(x) \rightarrow A'^\mu(x) A'_\mu(x) = (A^\mu(x) - \partial^\mu \Lambda)(A_\mu(x) - \partial_\mu \Lambda) \neq A^\mu(x) A_\mu(x) \tag{7.7}$$

and violates the local gauge invariance. Thus, the invariance under local gauge invariance requires the electromagnetic field to be massless, a property inherent in Maxwell's equations of electrodynamics.

7.2.2 Gauge invariance in field theory and phase invariance

In field theory, the free particles of spin 0, 1/2, 1 are described by the scalar, spinor, and vector fields represented by ϕ, ψ, A^μ, which are solutions of the equations of motion for these fields derived from the given Lagrangian using Euler–Lagrange equations. For example, the Lagrangian for free scalar and spinor fields, ϕ and ψ, respectively, are written as:

$$\mathcal{L}^\phi_{\text{free}} = \frac{1}{2}(\partial_\mu \phi(x))^* \partial^\mu \phi(x) - \frac{1}{2} m^2 \phi^*(x) \phi(x), \qquad \phi = \phi_1 + i\phi_2,$$

$$\mathcal{L}^\psi_{\text{free}} = \overline{\psi}(x)(i\gamma_\mu \partial^\mu - m)\psi(x). \tag{7.8}$$

$\mathcal{L}^A_{\text{free}}$ for the electromagnetic field A is given in Eq. (7.2).

A phase transformation on the fields $\phi(x)$ or $\psi(x)$ is defined as the transformation in which

$$\phi(x) \rightarrow \phi'(x) = e^{i\alpha} \phi(x); \qquad \phi^*(x) \rightarrow \phi'^*(x) = \phi^*(x) e^{-i\alpha} \tag{7.9}$$

$$\text{and} \quad \psi(x) \rightarrow \psi'(x) = e^{i\alpha} \psi(x); \qquad \psi^*(x) \rightarrow \psi'^*(x) = \psi^*(x) e^{-i\alpha}, \tag{7.10}$$

where α is a parameter describing the phase transformations. If α is a constant, independent of the space–time coordinates $(\vec{x}, t)$, the transformation is called the global phase transformation (Figure 7.1), that is, the symmetry transformation is carried out by the same amount at each point in space and time. If the parameter depends on space and time, that is, $\alpha = \alpha(\vec{x}, t)$,

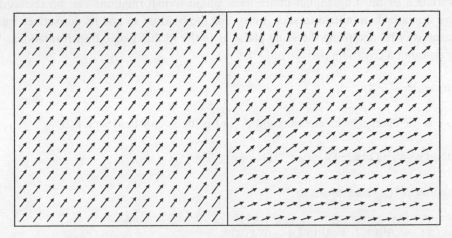

Figure 7.1 (Left) In a global transformation, α has a fixed value; therefore, $e^{i\alpha}$ has a constant value throughout space–time. (Right) In a local transformation, $e^{i\alpha(x)}$, where $x = x_\mu$ is a point in space–time, has a different value at each space–time point.

the transformations are called local phase transformations. For example, in case of rotation, the angles of rotation can change as one moves at different locations and times. It is clear that under the global phase transformation, the fields $\phi(\psi)$ and their derivatives $\partial^\mu \phi(\partial^\mu \psi)$ transform in the same way, that is,

$$\partial^\mu \phi(x) \to \partial^\mu \phi'(x) = e^{i\alpha} \partial^\mu \phi(x),$$

$$\partial^\mu \psi(x) \to \partial^\mu \psi'(x) = e^{i\alpha} \partial^\mu \psi(x), \tag{7.11}$$

with corresponding transformations for the complex conjugate of the field derivatives $\partial^\mu \phi^*(x)$ $[\partial^\mu \psi^*(x)]$. These transformations leave the Lagrangian for scalar (spinor fields) in Eq. (7.8) invariant under global phase transformations.

However, the situation is not so in the case of local phase transformations. Let us illustrate this for the case of spinor field ψ. Under the local phase transformation, prescribed by the parameter $\alpha(x)$, the field ψ transforms as:

$$\psi(x) \to \psi'(x) = e^{i\alpha(x)} \psi(x), \tag{7.12}$$

The field derivative transforms as:

$$\partial^\mu \psi(x) \to \partial^\mu \psi'(x) = e^{i\alpha(x)} [\partial^\mu \psi(x) + i(\partial^\mu \alpha(x)) \psi(x)]. \tag{7.13}$$

The presence of the second term shows that the field derivative undergoes more changes in addition to the change in the phase of the initial field derivative, spoiling the invariance of the

Lagrangian given in Eq. (7.8), which now acquires an extra term $\mathcal{L}_{add}$ given by:

$$\mathcal{L}_{add} = -\bar{\psi}(x)\gamma_\mu\partial^\mu\alpha(x)\psi(x). \tag{7.14}$$

The invariance of the Lagrangian under local phase transformations can be restored if the derivative operator is modified in such a way that the field $\psi(x)$ and its derivatives transform in the same way by acquiring the same phase as in the global gauge transformations. This is done by defining a covariant derivative D^μ as:

$$D^\mu = \partial^\mu - ieA^\mu(x), \tag{7.15}$$

where $-e$ is the charge on an electron and $A^\mu(x)$ is the vector field such that:

$$D^\mu\psi(x) \to (D^\mu\psi(x))' = e^{i\alpha(x)}D^\mu\psi(x)$$

requiring that $A^\mu(x)$ transforms as:

$$A^\mu(x) \to A^\mu(x') = A^\mu(x) + \frac{1}{e}\partial^\mu\alpha(x) \tag{7.16}$$

and making the term like $\bar{\psi}\gamma^\mu D_\mu\psi$ invariant under the local phase transformation. Thus, substituting the ordinary derivative ∂^μ everywhere in the Lagrangian by the covariant derivative D^μ, the Lagrangian for the scalar, spinor, or the vector fields can be made locally phase invariant. If the vector field $A^\mu(x)$ is identified with the electromagnetic field and e with the charge, then the transformation given in Eq. (7.16) as a requirement of the local phase invariance is same as the condition of the local gauge invariance in Maxwell's electrodynamics with $\Lambda(x) = -\frac{1}{e}\alpha(x)$(see Chapter 3). Therefore, the terms "local gauge invariance" and "local phase invariance" are equivalent in the context of quantum mechanics and field theory to describe the interactions of spin 0, $\frac{1}{2}$, and spin 1 particles. Earlier, transformations with the constant α were called gauge transformations of first kind and transformations with $\alpha(x)$ being space–time dependent were called gauge transformations of the second kind.

7.3 Local Gauge Symmetries and Fundamental Interactions

7.3.1 Introduction

Since the early days of twentieth century, symmetry principles have played an important role in formulating the theories of fundamental interactions like gravitation and electromagnetism. In general, symmetry principles are stated in terms of the invariance of the Lagrangian (describing a classical or quantum system of particles and fields) under certain transformations, which could be independent or dependent on the space–time coordinates, respectively. Accordingly, the corresponding symmetries are called global or local symmetries. The principle of invariance of classical systems under local transformations was used by Einstein and Weyl to formulate theories of gravitation and electromagnetism around the period when the special theory of relativity and quantum mechanics was proposed. It was Weyl who introduced the word "gauge" (*eich* in German) transformations to specify the way in which the transformations were to be implemented. He also formulated a field theory of electrodynamics based on the local gauge transformations in 1929 [40]. Einstein's theory of gravitation is also considered by many to be a gauge field theory based on the invariance under symmetry transformations that depend on space–time coordinates. Attempts to formulate the gauge field theories of nuclear interactions, like weak and strong interactions, started quite early in the 1930s; however, it took more than 30 years for these attempts to succeed. The aforementioned developments underscore the importance of symmetries in modern physics.

The invariance of the Lagrangian under gauge symmetry transformations leads, in general, to the Euler–Lagrange equations of motion and conserved current associated with the symmetry in terms of the fields describing the Lagrangian. In the case of local symmetry transformations, the invariance requirements necessitate the introduction of new vector fields in the Lagrangian known as gauge fields, which are massless and coupled to the conserved current. The nature and form of the coupling describe the interaction of the new gauge fields with the fields describing the Lagrangian.

In this way, the discovery of the principle of local gauge symmetry, which determines the form of the electromagnetic, weak and strong interactions, can be considered to be a major triumph of twentieth century physics, comparable to the special theory of relativity and quantum mechanics. While the special theory of relativity and quantum mechanics correctly describe the kinematics and dynamics of relativistic particles by defining equations of motion, the principle of local gauge symmetry provides the form and strength of the interactions (force) and it also specifies the interaction Lagrangian depending upon the underlying symmetry.

The theory of electromagnetic interaction of charged particles like electrons, which is mediated by photons, is formulated as a local gauge field theory based on abelian U(1) symmetry, with massless gauge fields as photons. It was Weyl who first demonstrated that the interaction Lagrangian for charged particles in an electromagnetic field can be derived from the free Lagrangian for the charged particles by imposing the invariance of free electron Lagrangian under the local gauge transformation corresponding to an abelian U(1) symmetry. The idea was applied by Yang and Mills [326], Shaw [328] and Utiyama [327] to extend the U(1) symmetry to the non-abelian local symmetries SU(2) in isospin space in order to formulate the theory of strong interactions and by many others to formulate the theory of weak interactions [37, 157, 220, 340, 341, 342]. The role of the local gauge symmetries in formulating a unified theory of fundamental interactions was indeed very important as evident from the early papers of Salam and Ward [341, 342] who stated in 1964 that: "It should be possible to generate the strong, weak and electromagnetic interaction terms with all their correct symmetry properties (as well as with clues regarding their relative strengths) by making local gauge transformations in the kinetic energy terms of the free Lagrangian for all particles."

In the following subsections, we attempt to demonstrate the use of the principle of local gauge symmetry in formulating the electromagnetic interaction Lagrangian of spin 0 and spin $\frac{1}{2}$ particles and extend it for deriving the interaction Lagrangian of non-abelian Yang–Mills fields and gluons toward formulating a theory of strong interactions based on SU(2) and SU(3) symmetry.

7.3.2 $U(1)$ gauge symmetry and the electromagnetic interactions of spin 0 particles

We start with the Lagrangian for the complex scalar field:

$$\mathcal{L} = \frac{1}{2}(\partial_\mu\phi(x))^*\partial^\mu\phi(x) - \frac{1}{2}m^2\phi^*(x)\phi(x). \tag{7.17}$$

The Euler–Lagrange equation gives the equation of motion for the scalar field as:

$$\left.\begin{aligned}(\partial_\mu\partial^\mu + m^2)\phi(x) &= 0 \\ \text{and} \quad (\partial_\mu\partial^\mu + m^2)\phi^*(x) &= 0.\end{aligned}\right\} \tag{7.18}$$

This Lagrangian is invariant under the global U(1) transformation:

$$\left.\begin{aligned}\phi(x) &\xrightarrow{U(1)} \phi'(x) = e^{i\alpha}\phi(x) \\ \text{and} \quad \phi^*(x) &\xrightarrow{U(1)} \phi'^*(x) = e^{-i\alpha}\phi^*(x).\end{aligned}\right\}, \tag{7.19}$$

where α is a constant parameter. For an infinitesimal variation:

$$\left.\begin{aligned}\delta\phi &= i\alpha\phi, & \delta\phi^* &= -i\alpha\phi^*, \\ \delta(\partial_\mu\phi) &= i\alpha\partial_\mu\phi, & \delta(\partial_\mu\phi^*) &= -i\alpha\partial_\mu\phi^*.\end{aligned}\right\} \tag{7.20}$$

Now the invariance of $\mathcal{L}$ under U(1) transformation means $\delta\mathcal{L} = 0$, where

$$\delta\mathcal{L} = \frac{\partial\mathcal{L}}{\partial\phi}\delta\phi + \frac{\partial\mathcal{L}}{\partial(\partial_\mu\phi)}\delta(\partial_\mu\phi) + \frac{\partial\mathcal{L}}{\partial\phi^*}\delta\phi^* + \frac{\partial\mathcal{L}}{\partial(\partial_\mu\phi^*)}\delta(\partial_\mu\phi^*). \tag{7.21}$$

Using the Euler–Lagrange equation $\frac{\partial\mathcal{L}}{\partial\phi} = \partial_\mu\left(\frac{\partial\mathcal{L}}{\partial(\partial_\mu\phi)}\right)$, we may rewrite Eq. (7.21) as:

$$\delta\mathcal{L} = \partial_\mu\left(\frac{\partial\mathcal{L}}{\partial(\partial_\mu\phi)}\right)\delta\phi + \frac{\partial\mathcal{L}}{\partial(\partial_\mu\phi)}\delta(\partial_\mu\phi) + \partial_\mu\left(\frac{\partial\mathcal{L}}{\partial(\partial_\mu\phi^*)}\right)\delta\phi^* + \frac{\partial\mathcal{L}}{\partial(\partial_\mu\phi^*)}\delta(\partial_\mu\phi^*). \tag{7.22}$$

Using Eq. (7.20), Eq. (7.22) becomes:

$$\delta\mathcal{L} = i\alpha\partial_\mu\left[\frac{\partial\mathcal{L}}{\partial(\partial_\mu\phi)}\phi - \frac{\partial\mathcal{L}}{\partial(\partial_\mu\phi^*)}\phi^*\right] = 0. \tag{7.23}$$

Using $\mathcal{L}$ from Eq. (7.17) in Eq. (7.23), we obtain:

$$\Rightarrow \delta\mathcal{L} = \frac{i\alpha}{2}\left[\partial_\mu(\partial^\mu\phi^*)\phi - \partial_\mu(\partial^\mu\phi)\phi^*\right]$$
$$= \alpha\partial_\mu j^\mu,$$

where $j^\mu = -\frac{i}{2}\left[\phi^*(\partial^\mu\phi) - (\partial^\mu\phi^*)\phi\right].$ \hfill (7.24)

Thus, if the Lagrangian is invariant under the global transformation defined by Eq. (7.19), then the current j^μ defined by Eq. (7.24) is conserved. Current conservation implies charge conservation.

Instead of the transformation given by Eq. (7.19), if we apply a local gauge transformation to the Lagrangian given by Eq. (7.17),

$$\phi(x) \xrightarrow{U(1)} \phi'(x) = e^{i\alpha(x)}\phi(x) \text{ and}$$
$$\phi^*(x) \xrightarrow{U(1)} \phi'^*(x) = e^{-i\alpha(x)}\phi^*(x), \hfill (7.25)$$

it leads to:

$$\partial_\mu\phi(x) \xrightarrow{U(1)} \partial_\mu\phi'(x) = e^{i\alpha(x)}\left[\partial_\mu\phi(x) + i(\partial_\mu\alpha(x))\phi(x)\right],$$
$$\partial_\mu\phi^*(x) \xrightarrow{U(1)} \partial_\mu\phi'^*(x) = e^{-i\alpha(x)}\left[\partial_\mu\phi^*(x) - i(\partial_\mu\alpha(x))\phi^*(x)\right], \hfill (7.26)$$

which destroys the invariance of the Lagrangian in Eq. (7.23) under the local gauge transformation. In order to preserve the invariance, a covariant derivative D_μ is defined as $D_\mu = \partial_\mu - ieA_\mu(x)$. The new vector field $A_\mu(x)$ is constrained to transform as:

$$A_\mu(x) \rightarrow A'_\mu(x) = A_\mu(x) + \frac{1}{e}\partial_\mu\alpha(x), \hfill (7.27)$$

as the covariant derivative field $D_\mu\phi(x)$ transforms as:

$$D_\mu\phi(x) \xrightarrow{U(1)} e^{i\alpha(x)}D_\mu\phi(x),$$

such that the Lagrangian remains invariant. The new Lagrangian written in terms of the covariant derivative $D_\mu\phi(x)$ is given by:

$$\mathcal{L} = \frac{1}{2}\left(D_\mu\phi(x)\right)^*\left(D^\mu\phi(x)\right) - \frac{1}{2}m^2\phi^*(x)\phi(x),$$

which, in terms of the ordinary derivative ∂_μ, may be written as:

$$\mathcal{L} = \frac{1}{2}\left[(\partial_\mu + ieA_\mu(x))\phi^*(x)\right]\left[(\partial^\mu - ieA^\mu(x))\phi(x)\right] - \frac{1}{2}m^2\phi^*(x)\phi(x)$$
$$= \frac{1}{2}\partial_\mu\phi^*(x)\partial^\mu\phi(x) + \frac{ie}{2}A_\mu(x)\phi^*(x)(\partial^\mu\phi(x)) - \frac{ie}{2}(\partial_\mu\phi(x))^*A^\mu(x)\phi(x)$$
$$+ \frac{e^2}{2}A_\mu(x)A^\mu(x)\phi^*(x)\phi(x) - \frac{1}{2}m^2\phi^*(x)\phi(x)$$

$$
\begin{aligned}
= & \ \frac{1}{2}\partial_\mu\phi^*(x)\partial^\mu\phi(x) - \frac{1}{2}m^2\phi^*(x)\phi(x) + \frac{ie}{2}A_\mu(x)\left[\phi^*(x)(\partial^\mu\phi(x))\right. \\
& \left. -\phi(x)(\partial^\mu\phi^*(x))\right] \\
+ & \ \frac{e^2}{2}A_\mu(x)A^\mu(x)\phi^*(x)\phi(x).
\end{aligned}
\tag{7.28}
$$

Comparing Eqs. (7.17) and (7.28), we observe that there are additional terms which are given by:

$$
\begin{aligned}
\mathcal{L}_{\text{add}} &= \frac{ie}{2}A_\mu(x)\left[\phi^*(x)(\partial^\mu\phi(x)) - \phi(x)(\partial^\mu\phi(x))^*\right] + \frac{e^2}{2}A_\mu(x)A^\mu(x)\phi^*(x)\phi(x) \\
&= -A_\mu(x)j^\mu(x) + \frac{e^2}{2}A_\mu(x)A^\mu(x)\phi^*(x)\phi(x),
\end{aligned}
\tag{7.29}
$$

where

$$
j^\mu(x) = -\frac{ie}{2}\left[\phi^*(x)(\partial^\mu\phi(x)) - \phi(x)(\partial^\mu\phi^*(x))\right].
\tag{7.30}
$$

The two additional terms in the new Lagrangian, given by Eq. (7.29) give the interaction of the new field A_μ with the conserved current generated by ϕ and a quadratic interaction term of the field ϕ with the new field. If we identify the new vector field as an electromagnetic field, then the first term describes the interaction of the electromagnetic field with the electromagnetic current of the charged particle with the coupling strength e. To this, we add the kinetic energy term for the A_μ field in a gauge invariant way by introducing the electromagnetic field tensor $F^{\mu\nu}$. Therefore, the full Lagrangian for the scalar field consistent with the invariance under local U(1) gauge transformation becomes:

$$
\begin{aligned}
\mathcal{L} &= \frac{1}{2}\left(\partial^\mu\phi^*(x)\partial_\mu\phi(x) - m^2\phi^*(x)\phi(x)\right) - j_\mu(x)A^\mu(x) \\
&\quad + \frac{e^2}{2}A_\mu(x)A^\mu(x)\phi^*(x)\phi(x) - \frac{1}{4}F_{\mu\nu}F^{\mu\nu}.
\end{aligned}
\tag{7.31}
$$

It should be noted that the new field A_μ is a vector field. It need not necessarily represent a spin 1 field, as in the case of electrodynamics, but could be written as $A_\mu(x) = \partial_\mu\lambda(x)$ involving derivatives, where $\lambda(x)$ is a scalar field, satisfying the transformation property:

$$
\lambda'(x) \to \lambda(x) + \frac{1}{e}\alpha(x).
\tag{7.32}
$$

This will, however, involve additional momenta due to derivative coupling and would present problems in the renormalizability of the theory.

7.3.3 $U(1)$ **gauge symmetry and the electromagnetic interactions of spin $\frac{1}{2}$ particles**

Let us consider the Lagrangian density for a free spin $\frac{1}{2}$ Dirac particle:

$$\mathcal{L}_D = \bar{\psi}(x)(i\gamma_\mu \partial^\mu - m)\psi(x), \tag{7.33}$$

where $\psi(x)$ is a four-component complex spinor function and $\bar{\psi}(x) = \psi^\dagger(x)\gamma^0$. Consider the symmetry properties of Eq. (7.33) under a global phase transformation:

$$\left.\begin{array}{l} \psi(x) \to \psi'(x) = e^{i\alpha}\psi(x), \\ \bar{\psi}(x) \to \bar{\psi}'(x) = e^{-i\alpha}\bar{\psi}(x), \end{array}\right\} \tag{7.34}$$

where α is a real constant parameter. For infinitesimal transformations of $\psi(x)$:

$$\left.\begin{array}{l} \psi(x) \to \psi'(x) = (1 + i\alpha)\psi(x) \Rightarrow \delta\psi(x) = \psi'(x) - \psi(x) = i\alpha\psi(x), \\ \bar{\psi}(x) \to \bar{\psi}'(x) = (1 - i\alpha)\bar{\psi}(x) \Rightarrow \delta\bar{\psi}(x) = \bar{\psi}'(x) - \bar{\psi}(x) = -i\alpha\bar{\psi}(x). \end{array}\right\} \tag{7.35}$$

The Lagrangian density remains invariant as:

$$\begin{aligned} \Rightarrow \delta\mathcal{L} &= i\delta\bar{\psi}(x)\gamma_\mu\partial^\mu\psi(x) + i\bar{\psi}(x)\gamma_\mu\partial^\mu(\delta\psi(x)) - m\delta\bar{\psi}(x)\psi(x) - m\bar{\psi}(x)\delta\psi(x) \\ &= i[-i\alpha\bar{\psi}(x)]\gamma_\mu\partial^\mu\psi(x) + i\bar{\psi}(x)\gamma_\mu\partial^\mu[i\alpha\psi(x)] \\ &\quad -m[-i\alpha\bar{\psi}(x)]\psi(x) - m\bar{\psi}(x)[i\alpha\psi(x)] \\ &= 0. \end{aligned} \tag{7.36}$$

This invariance of the Lagrangian density shows that $\psi(x)$ as well as $e^{i\alpha}\psi(x)$ have the same physical predictions, which in turn implies that α cannot be measured explicitly, that is, the phase of ψ remains arbitrary. Such transformations are known as global gauge transformations. α is, therefore, a parameter characterizing the transformation generated by a 1×1 unitary matrix which forms the group U(1). The group multiplication is commutative, that is, $U(\alpha_1)U(\alpha_2) = U(\alpha_2)U(\alpha_1)$, and, therefore, the sets of unitary transformations form a unitary abelian group.

The invariance of the Lagrangian, shown in Eq. (7.36) also leads to:

$$\delta\mathcal{L} = \frac{\partial\mathcal{L}}{\partial\psi}\delta\psi + \frac{\partial\mathcal{L}}{\partial(\partial_\mu\psi)}\delta(\partial_\mu\psi) + \frac{\partial\mathcal{L}}{\partial\bar{\psi}}\delta\bar{\psi} + \frac{\partial\mathcal{L}}{\partial(\partial_\mu\bar{\psi})}\delta(\partial_\mu\bar{\psi}) = 0. \tag{7.37}$$

Using infinitesimal transformations of $\psi(x)$ as discussed in Eq. (7.35), Eq. (7.37) results in:

$$\begin{aligned} \delta\mathcal{L} &= i\alpha\left[\frac{\partial\mathcal{L}}{\partial\psi}\psi + \frac{\partial\mathcal{L}}{\partial(\partial_\mu\psi)}\partial_\mu\psi\right] - i\alpha\left[\frac{\partial\mathcal{L}}{\partial\bar{\psi}}\bar{\psi} + \frac{\partial\mathcal{L}}{\partial(\partial_\mu\bar{\psi})}\partial_\mu\bar{\psi}\right] = 0 \\ &\Rightarrow i\alpha\left[\partial_\mu\left(\frac{\partial\mathcal{L}}{\partial(\partial_\mu\psi)}\right)\psi + \frac{\partial\mathcal{L}}{\partial(\partial_\mu\psi)}\partial_\mu\psi\right] \\ &\qquad -i\alpha\left[\partial_\mu\left(\frac{\partial\mathcal{L}}{\partial(\partial_\mu\bar{\psi})}\right)\bar{\psi} + \frac{\partial\mathcal{L}}{\partial(\partial_\mu\bar{\psi})}\partial_\mu\bar{\psi}\right] = 0 \end{aligned}$$

$$\Rightarrow \quad i\alpha\partial_\mu\left[\frac{\partial\mathcal{L}}{\partial(\partial_\mu\psi)}\psi\right] - i\alpha\partial_\mu\left[\frac{\partial\mathcal{L}}{\partial(\partial_\mu\bar{\psi})}\bar{\psi}\right] = 0$$

$$\Rightarrow \quad i\alpha\partial_\mu\left[\frac{\partial\mathcal{L}}{\partial(\partial_\mu\psi)}\psi - \frac{\partial\mathcal{L}}{\partial(\partial_\mu\bar{\psi})}\bar{\psi}\right] = 0. \tag{7.38}$$

Using the definition of current j_μ as:

$$\begin{aligned} j^\mu &= i\alpha\left[\frac{\partial\mathcal{L}}{\partial(\partial_\mu\psi)}\psi - \bar{\psi}\frac{\partial\mathcal{L}}{\partial(\partial_\mu\bar{\psi})}\right] \\ &= -\alpha\bar{\psi}\gamma_\mu\psi, \end{aligned} \tag{7.39}$$

and making use of Eq. (7.38), we get an equation for conserved current:

$$\partial_\mu j^\mu = 0. \tag{7.40}$$

Equation (7.40) also ensures that the charge $Q = \int j_0 d\vec{x}$ is conserved under U(1) global gauge transformation. Charge conservation is, therefore, a consequence of the invariance of the Dirac Lagrangian under global U(1) transformation. If the parameter α is made dependent on x, that is, $\alpha(x)$, then the transformation becomes local, as $\alpha(x)$ is different at each x and the group U(1) becomes the local U(1) group. If the parameter α is space–time dependent and the field $\psi(x)$ transforms as:

$$\begin{aligned} \psi(x) \rightarrow \psi'(x) &= e^{i\alpha(x)}\psi(x) \text{ , and} \tag{7.41} \\ \bar{\psi}(x) \rightarrow \bar{\psi}'(x) &= e^{-i\alpha(x)}\bar{\psi}(x) , \tag{7.42} \\ \partial_\mu\psi(x) \rightarrow \partial_\mu\psi'(x) &= (\partial_\mu\psi(x))e^{i\alpha(x)} + i(\partial_\mu\alpha(x))\psi(x)e^{i\alpha(x)}, \tag{7.43} \end{aligned}$$

then the mass term of the Lagrangian given in Eq. (7.33) will remain invariant as:

$$m\bar{\psi}(x)\psi(x) \rightarrow m\bar{\psi}'(x)\psi'(x) = m\bar{\psi}(x)e^{-i\alpha(x)}e^{i\alpha(x)}\psi(x) = m\bar{\psi}(x)\psi(x).$$

However, the kinetic energy term will not remain invariant as:

$$\begin{aligned} i\bar{\psi}(x)\gamma^\mu\partial_\mu\psi(x) &\rightarrow i\bar{\psi}'(x)\gamma^\mu\partial_\mu\psi'(x) \\ &= i\bar{\psi}(x)\gamma^\mu\partial_\mu\psi(x) - \bar{\psi}(x)(\partial_\mu\alpha(x))\psi(x) \\ &\neq i\bar{\psi}(x)\gamma^\mu\partial_\mu\psi(x). \end{aligned} \tag{7.44}$$

Therefore, unlike in the case of global gauge transformation, here ψ and $\partial_\mu\psi$ do not have the same transformation properties, making $\mathcal{L}' \neq \mathcal{L}$ under U(1). In order to impose invariance of $\mathcal{L}$ under local U(1) gauge transformation, a new field A_μ is introduced and the field transforms as:

$$A_\mu \rightarrow A'_\mu = A_\mu + \frac{1}{e}\partial_\mu\alpha(x) \tag{7.45}$$

and the derivative is modified as $\partial_\mu \to D_\mu = \partial_\mu - ieA_\mu$, such that ψ and $D_\mu\psi$ have the same transformation properties under U(1), that is,

$$
\begin{aligned}
\psi(x) \to \psi'(x) &= e^{i\alpha(x)}\psi(x) \text{ and} \\
D_\mu\psi(x) \to (D_\mu\psi(x))' &= e^{i\alpha(x)}D_\mu\psi(x)
\end{aligned}
\tag{7.46}
$$

and similarly, for the conjugate field.

Therefore, under local U(1) gauge transformation, $\mathcal{L}'_D = i\bar{\psi}(\gamma_\mu D^\mu - m)\psi$ is invariant and

$$
\begin{aligned}
\mathcal{L}_D \to \mathcal{L}'_D &= i\bar{\psi}'(x)\gamma^\mu D_\mu\psi'(x) - m\bar{\psi}'(x)\psi'(x) \\
&= i\bar{\psi}(x)e^{-i\alpha(x)}\gamma^\mu \left[\partial_\mu - ieA_\mu - i\partial_\mu\alpha(x)\right] e^{i\alpha(x)}\psi(x) \\
&\quad - m\bar{\psi}(x)e^{-i\alpha(x)}e^{i\alpha(x)}\psi(x) \\
&= i\bar{\psi}(x)\gamma^\mu(\partial_\mu - ieA_\mu)\psi(x) - m\bar{\psi}(x)\psi(x) \\
&= \bar{\psi}(x)(i\gamma^\mu\partial_\mu - m)\psi(x) + e\bar{\psi}(x)\gamma^\mu\psi(x)A_\mu \\
&= \mathcal{L}_D + e\bar{\psi}(x)\gamma^\mu\psi(x)A_\mu.
\end{aligned}
\tag{7.47}
$$

Thus, the demand of local gauge invariance is fulfilled by introducing a covariant derivative D_μ. The price we have paid is the introduction of a new field $A_\mu(x)$ that also transforms as shown in Eq. (7.45) to make the Lagrangian locally invariant. Local gauge invariance has got great physical significance, for example, lepton number, charge, etc. are locally conserved.

Thus, the QED Lagrangian is obtained as:

$$
\mathcal{L} = \bar{\psi}(x)i\gamma^\mu\partial_\mu\psi(x) - m\bar{\psi}(x)\psi(x) - \frac{1}{4}F_{\mu\nu}F^{\mu\nu} - j^\mu A_\mu.
\tag{7.48}
$$

Equation (7.48) demonstrates that the principle of local gauge invariance generates the interaction Lagrangian for the interaction of the electromagnetic field A_μ known as the gauge field which couples to the Dirac particle (charge $-e$), with the current density $j^\mu(= -e\bar{\psi}(x)\gamma^\mu \psi(x)A_\mu)$. The gauge field A_μ is massless.

7.3.4 $SU(2)$ **gauge symmetry and Yang–Mills fields**

Suppose there are two non-interacting spin $\frac{1}{2}$ particles. The Lagrangian for the free particle is then given by:

$$
\mathcal{L} = i\bar{\psi}_1(x)\gamma^\mu\partial_\mu\psi_1(x) - m_1\bar{\psi}_1(x)\psi_1(x) + i\bar{\psi}_2(x)\gamma^\mu\partial_\mu\psi_2(x) - m_2\bar{\psi}_2(x)\psi_2(x).
\tag{7.49}
$$

If the particles belong to a doublet representation under SU(2) group like the protons and neutrons which belong to a doublet representation of $SU(2)$ in isospin space, then

$$
\psi(x) = \begin{pmatrix} \psi_1(x) \\ \psi_2(x) \end{pmatrix},
\tag{7.50}
$$

and the Lagrangian $\mathcal{L}$ is written as:

$$\mathcal{L} = i\bar{\psi}(x)\gamma^\mu \partial_\mu \psi(x) - M\bar{\psi}(x)\psi(x) , \qquad (7.51)$$

where $\psi(x) = \begin{pmatrix} \psi_1(x) \\ \psi_2(x) \end{pmatrix}$ is a two-component column vector and $M = \begin{pmatrix} m_1 & 0 \\ 0 & m_2 \end{pmatrix}$.

In the limit of SU(2) symmetry, where $m_1 = m_2 = m$,

$$\mathcal{L} = i\bar{\psi}(x)\gamma^\mu \partial_\mu \psi(x) - m\bar{\psi}(x)\psi(x). \qquad (7.52)$$

If the global SU(2) transformation is applied to the Lagrangian given in Eq. (7.51), the transformation is given by:

$$\psi(x) \to \psi'(x) \;\; = \;\; U\psi(x) \qquad \text{and} \qquad \bar{\psi}(x) \to \bar{\psi}'(x) = \bar{\psi}(x)U^\dagger, \qquad (7.53)$$

where U is a 2×2 unitary matrix and given by:

$$U = e^{\frac{i}{2} \sum\limits_{i=1}^{3} \tau_i . a_i} , \qquad (7.54)$$

where τ_1, τ_2, τ_3 are the three components of Pauli matrices and a_1, a_2, a_3 are real numbers. $e^{\frac{i}{2}\vec{\tau}.\vec{a}}$ is a 2×2 matrix with determinant unity. Under such transformations:

$$\psi(x) \to \psi'(x) = U\psi(x) = e^{\frac{i}{2} \sum\limits_{i=1}^{3} \tau_i . a_i} \psi(x). \qquad (7.55)$$

It may be verified that this Lagrangian remains invariant under the aforementioned transformation.

In 1954, Yang and Mills [326] extended the idea of global SU(2) gauge transformation to the local SU(2) gauge transformation. The transformation for an isodoublet Dirac field

$$\psi(x) = \begin{pmatrix} \psi_1(x) \\ \psi_2(x) \end{pmatrix} \qquad (7.56)$$

is

$$\psi(x) \xrightarrow{SU(2)} \psi'(x) = U\psi(x) = e^{i \sum\limits_{i=1}^{3} \alpha_i(x)T_i} \psi(x), \qquad (7.57)$$

where

$$\alpha_i(x)T_i = \alpha_1(x)T_1 + \alpha_2(x)T_2 + \alpha_3(x)T_3, \qquad (7.58)$$

α_{1-3}s are real parameters which depend on x_μ and $T_i = \frac{1}{2}\tau_i$, where $\tau_i's$ are the 2×2 isospin equivalent of Pauli matrices and

$$[T_i, T_j] = i\epsilon_{ijk}T_k. \qquad (7.59)$$

Now consider the Dirac Lagrangian:

$$\mathcal{L} = \bar{\psi}(x)(i\gamma^\mu \partial_\mu - m)\psi(x), \tag{7.60}$$

which for an infinitesimal transformation,

$$\delta\psi(x) = i\alpha_i(x)T_i\psi(x) \text{ and } \delta\bar{\psi}(x) = -i\bar{\psi}(x)\alpha_i(x)T_i, \tag{7.61}$$

changes to say $\mathcal{L}'$. The change is given by:

$$\delta\mathcal{L} = \left[\frac{\partial\mathcal{L}}{\partial\psi(x)}\delta\psi(x) + \frac{\partial\mathcal{L}}{\partial(\partial_\mu\psi(x))}\delta(\partial_\mu\psi(x))\right]$$
$$+ \left[\frac{\partial\mathcal{L}}{\partial\bar{\psi}(x)}\delta\bar{\psi}(x) + \frac{\partial\mathcal{L}}{\partial(\partial_\mu\bar{\psi}(x))}\delta(\partial_\mu\bar{\psi}(x))\right]. \tag{7.62}$$

Let us consider the change in Lagrangian density $(\mathcal{L})$ resulting from a transformation $\delta\psi = i\alpha_i(x)T_i\psi$. Then:

$$\delta\mathcal{L} = \left[\frac{\partial\mathcal{L}}{\partial\psi(x)}\delta\psi(x) + \frac{\partial\mathcal{L}}{\partial(\partial_\mu\psi(x))}\delta(\partial_\mu\psi(x))\right]$$
$$= \partial_\mu\left[\frac{\partial\mathcal{L}}{\partial(\partial_\mu\psi(x))}i\alpha_i(x)T_i\psi(x)\right]. \tag{7.63}$$

Using the definition of current as:

$$j^\mu = i\alpha_i(x)\frac{\partial\mathcal{L}}{\partial(\partial_\mu\psi(x))}T_i\psi(x) \tag{7.64}$$

and making use of the Dirac Lagrangian, Eq. (7.60), we obtain:

$$j^\mu = -\alpha_i(x)\bar{\psi}(x)\gamma^\mu\vec{T}\psi(x). \tag{7.65}$$

For constant α_is, the Lagrangian will remain invariant, that is, $\delta\mathcal{L} = 0$ if $\partial_\mu j^\mu = 0$.

Now we consider the invariance under local gauge transformation, and define a covariant derivative D_μ as:

$$D_\mu = \partial_\mu - igW_\mu(x), \tag{7.66}$$

where

$$W_\mu(x) = W_\mu^i(x)T_i. \tag{7.67}$$

Here, W_μ^is are the triplet $W_\mu^1, W_\mu^2, W_\mu^3$ of gauge fields, corresponding to each component of the isospin matrix T_i, such that:

$$(D_\mu\psi(x))' = U(D_\mu\psi(x))$$
$$\Rightarrow \quad (\partial_\mu - igW_\mu')U\psi(x) = U(\partial_\mu - igW_\mu)\psi(x)$$

$$\Rightarrow \quad (\partial_\mu U)\psi(x) + U\partial_\mu\psi(x) - igW'_\mu U\psi(x) \ = \ U\partial_\mu\psi(x) - igUW_\mu\psi(x)$$

$$\Rightarrow \quad\quad\quad\quad (\partial_\mu U - igW'_\mu U)\psi(x) \ = \ -igUW_\mu\psi(x)$$

$$\Rightarrow \quad\quad\quad\quad\quad\quad \partial_\mu U - igW'_\mu U \ = \ -igUW_\mu.$$

Multiplying the right-hand of the equation by U^{-1}, we get:

$$(\partial_\mu U)U^{-1} - igW'_\mu UU^{-1} \ = \ -igUW_\mu U^{-1}$$

$$\Rightarrow W'_\mu \ = \ UW_\mu U^{-1} - \frac{i}{g}(\partial_\mu U)U^{-1} \tag{7.68}$$

$$\Rightarrow W'^i_\mu T_i \ = \ UW^i_\mu T_i U^{-1} - \frac{i}{g}(\partial_\mu U)U^{-1}.$$

For an infinitesimal transformation, that is, $U = 1 + i\alpha_i T_i$ and $U^{-1} = 1 - i\alpha_i T_i$, we may write:

$$\Rightarrow W'^i_\mu T_i \ = \ (1 + i\alpha_j T_j)W^i_\mu T_i(1 - i\alpha_j T_j) - \frac{i}{g}(i(\partial_\mu\alpha_j)T_j)(1 - i\alpha_j T_j)$$

$$= \ W^i_\mu T_i - i\alpha_j W^i_\mu.i\epsilon_{ijk}T_k + \frac{1}{g}(\partial_\mu\alpha_j)T_j$$

$$= \ \left(W^i_\mu - \epsilon_{ijk}\alpha_j W^k_\mu + \frac{1}{g}(\partial_\mu\alpha_i)\right)T_i$$

$$\Rightarrow W'^i_\mu \ = \ W^i_\mu - \epsilon_{ijk}\alpha_j W^k_\mu + \frac{1}{g}(\partial_\mu\alpha_i)$$

$$\Rightarrow \vec{W}'_\mu \ = \ \vec{W}_\mu - \vec{\alpha} \times \vec{W}_\mu + \frac{1}{g}(\partial_\mu\vec{\alpha}). \tag{7.69}$$

We add three new fields W^1_μ, W^2_μ, W^3_μ and to make the field dynamic, the kinetic energy term $\frac{1}{4}G^i_{\mu\nu}G^{\mu\nu}_i$, where $G^i_{\mu\nu} = (\partial_\mu W^i_\nu - \partial_\nu W^i_\mu)$, is to be added to the Lagrangian. Defining a gauge invariant $G_{\mu\nu}$ as:

$$G_{\mu\nu} \ = \ D_\mu W_\nu - D_\nu W_\mu$$

$$= \ \partial_\mu W_\nu - \partial_\nu W_\mu - ig[W_\mu, W_\nu]$$

$$\Rightarrow G^i_{\mu\nu}T^i \ = \ (\partial_\mu W^i_\nu - \partial_\nu W^i_\mu)T^i - igW^i_\mu W^j_\nu[T_i, T_j]$$

$$= \ (\partial_\mu W^i_\nu - \partial_\nu W^i_\mu)T^i - igW^i_\mu W^j_\nu i\epsilon^{ijk}T^k$$

$$= \ (\partial_\mu W^i_\nu - \partial_\nu W^i_\mu)T^i + g\epsilon_{ijk}W^j_\mu W^k_\nu T^i.$$

For an arbitrary T^i,

$$G^i_{\mu\nu} \ = \ \partial_\mu W^i_\nu - \partial_\nu W^i_\mu + g\epsilon_{ijk}W^j_\mu W^k_\nu = -G^i_{\nu\mu}.$$

The transformation of $G_{\mu\nu}$ under the SU(2) would give:

$$
\begin{aligned}
G_{\mu\nu} \rightarrow G'_{\mu\nu} &= \partial_\mu W'_\nu - \partial_\nu W'_\mu \\
&= \partial_\mu \left[UW_\nu U^{-1} - \frac{i}{g}(\partial_\nu U)U^{-1} \right] - \partial_\nu \left[UW_\mu U^{-1} - \frac{i}{g}(\partial_\mu U)U^{-1} \right] \\
&= \partial_\mu (UW_\nu U^{-1}) - \frac{i}{g}\partial_\mu \left[(\partial_\nu U)U^{-1} \right] - \partial_\nu \left[UW_\mu U^{-1} \right] \\
&\quad + \frac{i}{g}\partial_\nu \left[(\partial_\mu U)U^{-1} \right] \\
&= \frac{i}{g} \left[(\partial_\mu U)(\partial_\nu U^{-1}) - (\partial_\nu U)(\partial_\mu U^{-1}) \right] + (\partial_\mu U)W_\nu U^{-1} \\
&\quad + UW_\nu(\partial_\mu U^{-1}) \\
&\quad - (\partial_\nu U)W_\mu U^{-1} - UW_\mu(\partial_\nu U^{-1}) + U(\partial_\mu W_\nu - \partial_\nu W_\mu)U^{-1}. \quad (7.70)
\end{aligned}
$$

Therefore, $G_{\mu\nu}$ is not invariant under gauge transformation. We redefine $G^i_{\mu\nu} = (\partial_\mu W^i_\nu - \partial_\nu W^i_\mu) - ig[W_\mu, W_\nu]$, where $[W_\mu, W_\nu]$ transforms as:

$$
\begin{aligned}
[W_\mu, W_\nu] \rightarrow [W'_\mu, W'_\nu] &= (W'_\mu W'_\nu - W'_\nu W'_\mu) \\
&= \left(UW_\mu U^{-1} - \frac{i}{g}(\partial_\mu U)U^{-1} \right)\left(UW_\nu U^{-1} - \frac{i}{g}(\partial_\nu U)U^{-1} \right) \\
&\quad - \left(UW_\nu U^{-1} - \frac{i}{g}(\partial_\nu U)U^{-1} \right)\left(UW_\mu U^{-1} - \frac{i}{g}(\partial_\mu U)U^{-1} \right) \\
&= U[W_\mu, W_\nu]U^{-1} + \frac{1}{g^2}(\partial_\mu U \partial_\nu U^{-1} - \partial_\nu U \partial_\mu U^{-1}) \\
&\quad + \frac{i}{g} \left[\partial_\mu U W_\nu U^{-1} + UW_\nu \partial_\mu U^{-1} - \partial_\nu U W_\mu U^{-1} \right. \\
&\quad \left. - UW_\mu \partial_\nu U^{-1} \right]. \quad (7.71)
\end{aligned}
$$

Here, we have used the relation: $U^{-1}(\partial_\mu U) = -(\partial_\mu U^{-1})U$, inspired by $UU^{-1} = 1$.

When the transformations performed using Eqs. (7.70) and (7.71), are considered together, then the Lagrangian is found to be invariant under the local gauge transformation.

Thus, the complete Lagrangian invariant under SU(2) gauge transformation is given by:

$$
\mathcal{L} = \bar{\psi}(i\gamma^\mu \partial_\mu - m)\psi + g\bar{\psi}\gamma^\mu \psi W_\mu - \frac{1}{4}G^i_{\mu\nu}G^{\mu\nu}_i, \quad (7.72)
$$

where the first term represents the free Lagrangian, the second term represents the interaction of the current with the gauge boson while the third term describes the self interaction of the gauge bosons. This is shown in Figure 7.2. Thus, it may be observed from Eqs. (7.60)–(7.71), that for the local gauge invariance of the Lagrangian under SU(2), the following aspects should be kept in mind:

Figure 7.2

- ∂_μ should be replaced by a covariant derivative $D_\mu = \partial_\mu - igW_\mu$, where $W_\mu = W^i_\mu T^i$. W^i_μ stands for the three fields W^1_μ, W^2_μ, W^3_μ.

- W_μ should transform as:
  $$W'_\mu = UW_\mu U^{-1} - \frac{i}{g}(\partial_\mu U)U^{-1}$$

- The kinetic energy term $\left(-\frac{1}{4}G^i_{\mu\nu}G^{\mu\nu}_i\right)$ is added to the field dynamics, where
  $$G^i_{\mu\nu} = (\partial_\mu W^i_\nu - \partial_\nu W^i_\mu) - ig[W^i_\mu, W^i_\nu]$$
  which transforms as :
  $$G^i_{\mu\nu} \xrightarrow{SU(2)} G^{i'}_{\mu\nu} = UG^i_{\mu\nu}U^{-1}.$$

- The Yang–Mills Lagrangian:
  $$\mathcal{L}_{YM} = \bar{\psi}(i\gamma^\mu D_\mu - m)\psi - \frac{1}{4}G^i_{\mu\nu}G^{\mu\nu}_i$$
  is invariant under the local SU(2) gauge transformation.

- The gauge fields $W^i_\mu(x)$ are massless.

7.3.5 $SU(3)_c$ **gauge symmetry and QCD**

The Lagrangian for free quark fields is defined as:

$$\mathcal{L} = \sum_{j=1}^{3} \bar{q}_j(i\gamma^\mu\partial_\mu - m_j)q_j, \tag{7.73}$$

where q_js are the color quark fields. It is assumed that the free Lagrangian is invariant under SU(3)$_C$ symmetry transformation in color space. The SU(3)$_C$ symmetry is generated by the 3×3 Gell–Mann matrices $\frac{1}{2}\lambda_i$, which are $n^2 - 1(3^2 - 1 = 8)$ in number; the matrices are also traceless and hermitian. The matrices λ_i are chosen in analogy with σ_i matrices of SU(2), extended to three dimensions and are given in Appendix B. These matrices λ_is, satisfy the following commutation and anticommutation relations:

$$\left[\frac{\lambda_i}{2}, \frac{\lambda_j}{2}\right] = if_{ijk}\left(\frac{\lambda_k}{2}\right), \tag{7.74}$$

$$\left\{\frac{\lambda_i}{2}, \frac{\lambda_j}{2}\right\} = \frac{1}{3}\delta_{ij} + d_{ijk}\left(\frac{\lambda_k}{2}\right), \tag{7.75}$$

where f_{ijk} are known as the structure constants.

$f_{ijk}(d_{ijk})$ are anti-symmetric (symmetric) under the interchange of any pair of indices (i, j, k) and are listed in Appendix B.

Under the global SU(3) gauge transformation, the quark field $q(x)$ transforms as:

$$q(x) \rightarrow q'(x) = Uq(x) = e^{i\alpha_i T_i} q(x) \Big\}$$
$$\bar{q}(x) \rightarrow \bar{q}'(x) = U^\dagger \bar{q}(x) = e^{-i\alpha_i T_i} \bar{q}(x) \Big\}, \tag{7.76}$$

where U is an arbitrary 3×3 unitary matrix, $T_i = \frac{\lambda_i}{2}$, i=1–8 are the generators of the SU(3) group and α_i are the group parameters. For an infinitesimal transformation, we may write:

$$q(x) \rightarrow q'(x) = (1 + i\alpha_i T_i) q(x), \Big\}$$
$$\partial_\mu q(x) \rightarrow \partial_\mu q'(x) = (1 + i\alpha_i T_i) \partial_\mu q(x). \Big\} \tag{7.77}$$

The Lagrangian in Eq. (7.73) is invariant under the transformation defined in Eqs. (7.76) and (7.77).

Proceeding the same way as discussed in the case of U(1) and SU(2) gauge transformations, the required Lagrangian should also be invariant under local SU(3)$_C$ transformation, where the field transforms as:

$$q(x) \rightarrow q'(x) = Uq(x) = e^{i\alpha_i(x) T_i} q(x). \tag{7.78}$$

For an infinitesimal phase transformation:

$$q(x) \rightarrow q'(x) = [1 + i\alpha_i(x) T_i] q(x) \Big\}$$
$$\partial_\mu q(x) \rightarrow \partial_\mu q'(x) = [\partial_\mu + i(\partial_\mu \alpha_i(x)) T_i + i\alpha_i(x) T_i \partial_\mu] q(x) \Big\}. \tag{7.79}$$

To fulfill the requirement of local gauge invariance, we must replace the ordinary derivative ∂_μ by the covariant derivative $D_\mu (= \partial_\mu + ig T_i G^i_\mu)$ in Eq. (7.73), where G^i_μ are the eight gauge fields transforming as:

$$G^i_\mu(x) \rightarrow G'^i_\mu(x) = G^i_\mu(x) - \frac{1}{g} \partial_\mu \alpha_i(x), \tag{7.80}$$

following the discussion given in Section 7.3.4.

Introducing a kinetic energy term for the propagating field of gluons, the complete Lagrangian invariant under SU(3)$_c$ gauge transformation is given by:

$$\mathcal{L} = \sum_j \bar{q}_j (i\gamma^\mu \partial_\mu - m) q_j - g G^i_\mu \sum_j \bar{q}_j \gamma^\mu T_i q_j - \frac{1}{4} G^i_{\mu\nu} G^{i\mu\nu}. \tag{7.81}$$

For Eq. (7.81) to be locally gauge invariant under SU(3), the field strength can be defined as:

$$G^i_{\mu\nu} = \partial_\mu G^i_\nu - \partial_\nu G^i_\mu - g f_{ijk} G^j_\mu G^k_\nu. \tag{7.82}$$

In Eq. (7.81), the first term is the free Lagrangian, the second term describes the interaction of the quark current with the gauge boson of the theory, that is, gluons G_μ^i, $[i = 1, 2, ..., 8]$(like the electromagnetic field in the case of U(1) symmetry). The last term involves three type of terms, that is,

$$
\begin{aligned}
G_{\mu\nu}^i G^{i\mu\nu} &= (\partial_\mu G_\nu^i - \partial_\nu G_\mu^i)(\partial^\mu G_i^\nu - \partial^\nu G_i^\mu) - g f_{ijk}(G_\mu^j G_\nu^k(\partial^\mu G_i^\nu - \partial^\nu G_i^\mu)) \\
&\quad - g f_{ilm}(G_l^\mu G_m^\nu(\partial_\mu G_\nu^i - \partial_\nu G_\mu^i)) + g^2 f_{ijk} f_{ilm} G_\mu^j G_\nu^k G_l^\mu G_m^\nu, \qquad (7.83)
\end{aligned}
$$

which describes free gluons, and self interaction of gluons which is trilinear as well as quartic in the gluon field G_μ^i. Diagrammatically, this is shown in Figure 7.3. The Lagrangian given in

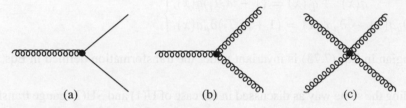

(a)   (b)   (c)

Figure 7.3 (a) Quark–gluon vertex, (b) triple gluon vertex, and (c) four gluon vertex.

Eq. (7.81) can be quantized and Feynman diagrams can be derived for doing calculations with gluon exchange. It should be noted that there are no mass terms like $G_\mu^i G^{i\mu}$ showing that all the gauge fields are massless. The Lagrangian describes the interaction of quarks with gluons and is the most reliable theory of strong interactions known as QCD.

7.3.6 Exact, broken, and spontaneously broken symmetries

i) **Exact symmetries:** In field theory, exact symmetry is characterized by two conditions. The first condition is that the interaction Lagrangian density $\mathcal{L}$ is invariant under symmetry transformations, that is, under a symmetry transformation U, the interaction Lagrangian of the system transforms as:

$$
\mathcal{L} \xrightarrow{U} \mathcal{L}' = \mathcal{L}, \quad \text{such that} \qquad \delta\mathcal{L} = 0. \qquad (7.84)
$$

This leads to the Euler–Lagrange equations of motion which also satisfy the symmetry of the Lagrangian. The Euler–Lagrange equations of motion further lead to a current J^μ, which is conserved, that is $\partial_\mu J^\mu = 0$, leading to conservation of charge Q, defined as:

$$
Q = \int J^0(\vec{x}, t) d\vec{x}. \qquad (7.85)
$$

The second condition is that the ground state of the physical system or the physical vacuum in case of field theory, is also invariant under the symmetry transformation and is, therefore, unique. These two conditions imply that there exist degenerate multiplets of states having the same energy as a requirement of the exact symmetry (as shown in the

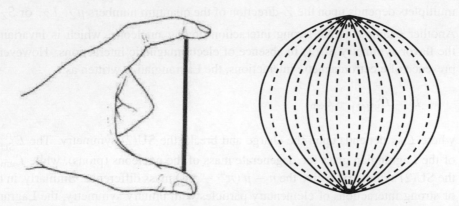

Figure 7.4 Unbroken symmetry (left): The plastic strip is in a rotationally invariant state. Spontaneously broken symmetry (right): If you squeeze the ends of a thin plastic strip together, it will bend in any direction, that is, there are infinite possibilities. As soon as it opts a particular direction, rotational invariance is lost.

left panel of Figure 7.4). Such degeneracy of states has been observed in many physical systems in the study of atoms, nuclei, and particles.

ii) **Broken or approximate symmetries:** However, many natural processes observed in the physical world do not exhibit exact symmetry, but only an approximate symmetry. These processes are described by separating the Lagrangian into two terms and writing it in the form $\mathcal{L} = \mathcal{L}_0 + \epsilon \mathcal{L}_1$, where $\mathcal{L}_0$ is invariant under the symmetry transformation and $\mathcal{L}_1$ is the part which breaks the symmetry explicitly; $\mathcal{L}_1$ is generally small with the smallness parameterized by a parameter ϵ. In many cases, it so happens that $\mathcal{L}_1$ corresponds to a lower symmetry as compared to the symmetry of $\mathcal{L}_0$; its transformation properties under the lower symmetry are used to lift the degeneracy of the multiplets of the states implied by the symmetry of $\mathcal{L}_0$. This form of treating approximate symmetries has been very useful in the study of atomic, nuclear, and particle spectroscopy. For example, in atomic or nuclear spectroscopy, the $\mathcal{L}_0$ Lagrangian is written as $\mathcal{L}_0 = \frac{p^2}{2M} - V(r)$, where $\vec{p}$ is the momentum operator $\vec{p}(= -i\vec{\nabla})$ of the electrons or nucleons and the potential $V(r)$ is either the Coulomb potential $\left(V(r) = -\frac{Ze^2}{|\vec{r}|} \right)$ in the case of atomic systems or a central potential of the Wood–Saxen type $\left(V(r) = \frac{V_0}{1 + e^{\frac{r-R}{a}}} \right)$ in the case of nuclear systems. These potentials depend upon the radial distance $|\vec{r}|$ only, leaving $\mathcal{L}_0$ invariant under rotation; this leads to degenerate multiplets. In the presence of the magnetic field $\vec{B}$ or the spin orbit interaction, the symmetry breaking Lagrangian $\mathcal{L}_1$ is either $-\vec{\mu}.\vec{B}$ or $\vec{L}.\vec{S}$, where $\vec{\mu}$, $\vec{L}$, $\vec{S}$ are the magnetic moments, angular momenta, and the spin of the electron or nucleon, respectively. In the presence of $\mathcal{L}_1$, the rotational symmetry is reduced to only the rotational symmetry about the direction of the magnetic field $\vec{B}$ or the quantization axis of $\vec{L}(\text{or } \vec{S})$ which can be chosen to be along the Z-axis. The degeneracy of the states in the multiplets is, therefore, lifted and the energy of the states forming the multiplets depends upon the Z-direction of the quantum numbers μ_Z, L_Z, or S_Z.

Another example is the strong interaction among nucleons, which is invariant under the flavor symmetry SU(2) in absence of electromagnetic interactions. However, in the presence of electromagnetic interactions, the Lagrangian is written as:

$$\mathcal{L} = \mathcal{L}^0_{SU(2)} + \mathcal{L}_{em}, \tag{7.86}$$

where $\mathcal{L}_{em}$ depends upon the charge and breaks the SU(2) symmetry. The $\mathcal{L}^0_{SU(2)}$ part of the Lagrangian gives the degenerate mass of the nucleons (pions), while $\mathcal{L}_{em}$ breaks the SU(2) symmetry to give the $n - p \, (\pi^+ - \pi^-)$ mass difference. Similarly, in the case of strong interactions of elementary particles with unitary symmetry, the Lagrangian is written as:

$$\mathcal{L} = \mathcal{L}^0_{SU(3)} + \mathcal{L}_{int}. \tag{7.87}$$

In this case, the $\mathcal{L}^{SU(3)}$ part of the Lagrangian is invariant under SU(3) and gives the degenerate mass of the octet and decouplet states of SU(3). The $\mathcal{L}_{int}$ part of the Lagrangian breaks the SU(3) symmetry to give the splitting between the different isospin multiplets constituting the octet and decouplet and lifts the degeneracy.

iii) **Spontaneously broken symmetries:** There is another type of symmetry in which the Lagrangian is invariant under a symmetry transformation leading to the equations of motion, which are also invariant; however, the ground state of the physical system or the physical vacuum in the case of field theory is not invariant under the symmetry transformation leading to a vacuum state, which is not unique. There could be many states of lowest energy, that is, ground state, which are degenerate (as shown in the right panel of Figure 7.4). This means that under the symmetry transformation, while $\delta\mathcal{L} = 0$, leading to some local conservation laws as a consequence of properties of exact symmetry, the non-uniqueness of the vacuum destroys the mass (energy) degeneracy of the states in the multiplets, showing that the symmetry is no longer exact but is broken. This type of symmetry breaking is referred to as the dynamical or the spontaneous breaking of symmetry and the system is described to have a "hidden symmetry". It is not unusual to have a physical system in which the symmetry is broken by both mechanisms explicitly as well as spontaneously. A well-known example of such physical systems is a material exhibiting ferromagnetism near the Curie temperature T_c. In the case of $T > T_c$, if the system is placed in a magnetic field, then the spin dipoles are aligned in the direction of the magnetic field. The rotational symmetry is explicitly broken down to the rotational symmetry about the axis of rotation in the direction of the magnetic field $\vec{B}$. For $T > T_c$, in the paramagnetic phase, all the spin dipoles are randomly oriented (Figure 7.5). The system displays rotational symmetry and the ground state is rotationally invariant. For $T < T_c$, in the ferromagnetic phase, all the spin dipoles are aligned in a parallel direction (spontaneous magnetization), which is arbitrary; the rotational

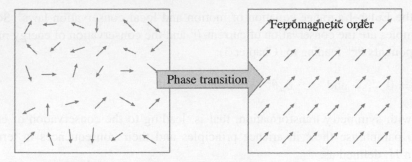

Figure 7.5 Phase transition in ferromagnetic materials. For $T > T_c$ (left), all the spin dipoles are randomly oriented. For $T < T_c$ (right), in the ferromagnetic phase, all the spin dipoles are aligned in a parallel direction (spontaneous magnetization).

symmetry is broken to a lower level to a symmetry around that arbitrary direction. This is the ground state corresponding to the lowest energy but it could correspond to any direction or orientation of spin and is, therefore, infinitely degenerate. This is an example of spontaneous breaking of symmetry(Figures 7.5 and 7.6). In the following sections,

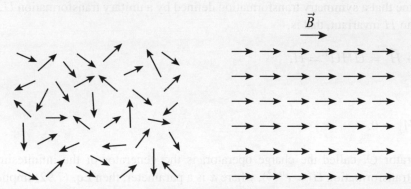

Figure 7.6 Paramagnetic material in the absence of external magnetic field (left); in the presence of external magnetic field (right), symmetry is lost.

we first describe some formal results regarding spontaneous symmetry breaking (SSB) in field theory and then discuss simple models to demonstrate them in some cases of discrete and continuous abelian and non-abelian field theories.

7.3.7 Spontaneous symmetry breaking: Some formal results

We have seen in earlier sections that the presence of an exact symmetry implies the invariance of the Lagrangian $\mathcal{L}$ of the system under some unitary transformation U, such that

$$\mathcal{L} \to \mathcal{L}' = U\mathcal{L}U^\dagger = \mathcal{L}, \tag{7.88}$$

leading to the Euler–Lagrange equation of motion and local conservation laws. Some well-known examples are the conservation of current J^μ and the conservation of energy momentum tensor components $\theta^{\mu\nu}$ leading to (Chapter 3):

$$\partial_\mu J^\mu = 0 \qquad \text{and} \qquad \partial_\mu \theta^{\mu\nu} = 0, \tag{7.89}$$

associated with symmetry transformation, that is, leading to the conservation of charge and energy. To paraphrase these invariance principles and their consequences in terms of the Hamiltonian H, defined as:

$$H = \int \mathcal{H} d\vec{x}, \tag{7.90}$$

where $\mathcal{H}$ is the Hamiltonian density,

$$\mathcal{H} = \sum_i \frac{\partial \mathcal{L}}{\partial \dot{\phi}_i} \dot{\phi}_i - \mathcal{L}(\phi_i, \dot{\phi}_i, t), \tag{7.91}$$

let us assume that a symmetry transformation defined by a unitary transformation U, leaves the Hamiltonian H invariant, that is,

$$H \rightarrow H' = UHU^\dagger = H. \tag{7.92}$$

Then,

$$[U, H] = 0. \tag{7.93}$$

If the operator Q, called the charge operator, is the generator of the infinitesimal unitary symmetry transformation $U(= e^{i\alpha Q})$, where α is a parameter, then Eq. (7.92) implies that:

$$(1 + i\alpha Q)H(1 - i\alpha Q) = H, \qquad \text{that is,} \qquad [Q, H] = 0. \tag{7.94}$$

Using Heisenberg's equation of motion, this leads to the equation of motion for Q, that is, $\frac{dQ}{dt} = [Q, H] = 0$ implying Q is a constant of motion.

This means that the quantity corresponding to the infinitesimal generator of the unitary transformation (U) is a constant of motion if U is an exact symmetry. The symmetry of H manifests in the degeneracies of the energy eigenstates corresponding to the irreducible representation of the unitary symmetry U. To demonstrate this, let us assume that $|A\rangle$ and $|B\rangle$ are the states corresponding to a given representation of U. Then they are connected by U as:

$$U|A\rangle = |B\rangle. \tag{7.95}$$

The energy E_B of the state $|B\rangle$ is defined as:

$$
\begin{aligned}
E_B &= \langle B|H|B\rangle \\
&= \langle AU^\dagger|H|UA\rangle \\
&= \langle A|H|A\rangle = E_A,
\end{aligned}
\tag{7.96}
$$

leading to $E_B = E_A$; this shows that the states $|A\rangle$ and $|B\rangle$ are degenerate. Such degeneracies in the physical states have been observed experimentally in the atomic, nuclear, and hadron spectroscopy. However, the results in Eqs. (7.95) and (7.96) are not straightforward and depend upon the invariance of the ground state (the physical vacuum state in the case of field theory) implicitly implied in the aforementioned derivation. This can be seen by the simple argument: assume that $\phi^\dagger$ is the field operator which creates a single particle state $|A\rangle$ and $|B\rangle$ from the vacuum state $|0\rangle$, then:

$$
|A\rangle = \phi_A^\dagger|0\rangle, \qquad |B\rangle = \phi_B^\dagger|0\rangle.
\tag{7.97}
$$

The operators $\phi_A^\dagger$ and $\phi_B^\dagger$ are related by the unitary transformation U as:

$$
\phi_B^\dagger = U\phi_A^\dagger U^\dagger \qquad \Rightarrow \qquad \phi_B^\dagger U = U\phi_A^\dagger
$$

giving us

$$
\begin{aligned}
\phi_B^\dagger U|0\rangle &= U\phi_A^\dagger|0\rangle = U|A\rangle = |B\rangle, \\
\phi_B^\dagger U|0\rangle &= |B\rangle = \phi_B^\dagger|0\rangle.
\end{aligned}
\tag{7.98}
$$

Equation (7.98) is possible only if $U|0\rangle = |0\rangle$, that is, the physical vacuum state $|0\rangle$ is invariant under U. Since $U = 1 + i\alpha Q$, this leads to $Q|0\rangle = 0$, which implies that Q, the generator of the infinitesimal transformation, annihilates the physical vacuum state. In this case, the vacuum state $|0\rangle$ is unique because any symmetry operation U on $|0\rangle$ will again give $|0\rangle$.

However, there exists another possibility, where $U|0\rangle \neq |0\rangle \Rightarrow Q|0\rangle \neq 0$ [343]. In this case, if $U|0\rangle$ exists, then it is not unique and, therefore, it is degenerate. Moreover, it is infinitely degenerate in the case of field theory, where the field variable $\phi(x)$ can take any value depending upon x.

This is demonstrated by the Fabri–Picasso theorem [343]. Consider the following vacuum expectation value of the operator $Q^2 \equiv QQ$,

$$
\langle 0|QQ|0\rangle = \langle 0|\int d\vec{x} J^0(x) Q|0\rangle = \int d\vec{x} \langle 0|J^0(x) Q|0\rangle.
\tag{7.99}
$$

Since Q is derivable from the current operator $J^\mu(x)$ which is conserved under the continuous symmetry operation U (Eq. (7.40), Section 7.3.3), using the translational invariance, we can write:

$$
J^\mu(x) = e^{-iP.x} J^\mu(0) e^{iP.x},
\tag{7.100}
$$

and

$$\langle 0|J^0(x)Q\rangle = \langle 0|e^{-iP.x}J^0(0)e^{iP.x}Q|0\rangle, \tag{7.101}$$

where P^μ is the momentum operator, the generator of the translation symmetry, and Q is the symmetry operator in internal space, such that

$$P^\mu|0\rangle = 0, \qquad [P^\mu, Q] = 0, \tag{7.102}$$

which gives

$$\langle 0|J^0(x)Q|0\rangle = \langle 0|J^0(0)Q|0\rangle \tag{7.103}$$

implying that $\langle 0|J^0(0)Q|0\rangle$ is independent of x and, therefore,

$$\langle 0|QQ|0\rangle = \int d\vec{x}\langle 0|j^0(x)Q|0\rangle = \int \langle 0|J^0(0)Q|0\rangle d\vec{x}. \tag{7.104}$$

This diverges because $\langle 0|j^0(0)Q|0\rangle$ is independent of x, and x extends to infinity in field theory. It implies that either $Q|0\rangle = 0$, making the norm finite or $Q|0\rangle$ has infinite norm. In the case of $Q|0\rangle = 0$, the physical vacuum is unique. This is because the two vacuum states $|A\rangle$ and $|B\rangle$ are related by the unitary transformation, that is, $|0\rangle_A$ and $|0\rangle_B$.

$$|0\rangle_B = e^{i\alpha Q}|0\rangle_A, \tag{7.105}$$

where α is a parameter and the infinitesimal generator of U.

$$|0\rangle_B = (1 + i\alpha Q + ...)|0\rangle_A, \tag{7.106}$$

implies that if $Q|0\rangle_A = 0$, then $|0\rangle_B = |0\rangle_A$.

In the other case, $Q|0\rangle$ has infinite norm. Therefore, either $Q|0\rangle$ does not exist, or if it exists, the states are infinitely degenerate leading to the infinite norm. In this case, the vacuum is not unique and the condition leading to the degeneracy of the eigenstate of the Hamiltonian is not satisfied; thus, destroying the degeneracy of the spectra of the states. This is the manifestation of a symmetry that is broken spontaneously. The realization is manifested through non-degenerate states, even though the Lagrangian and equations of motion are invariant(covariant) under symmetry transformations. Spontaneous breaking is also known as dynamical breaking of the symmetry. It happens in the case of ferromagnets for $T < T_C$, when all the spin dipoles are aligned in one direction which is arbitrary and, therefore, leads to infinite degeneracy.

7.3.8 Spontaneous symmetry breaking (SSB) and Goldstone's theorem

The phenomenon of spontaneous symmetry breaking in field theory was first studied by Nambu [330] and Goldstone [162] based on the analogy with the theory of superconductivity and applied to particle physics. Nambu's models were based on the BCS theory of superconductivity. In Nambu's model, the spontaneous breaking of chiral symmetry was achieved by quasi-particle-like excitations arising due to the interaction of fermion fields; an example of such an interaction is the creation of Cooper pairs of electrons in the BCS theory of superconductivity [344]. In Goldstone's model, the spontaneous breaking of gauge symmetry was achieved by introducing self interacting scalar fields and giving them non-zero vacuum expectation values leading to the scalar excitations of the fields, in analogy with the Ginzberg–Landau theory [345]. In both the approaches, while masses of fermions or vector bosons were generated, massless scalar or pseudoscalar bosons also appeared as a result of the SSB of an exact continuous symmetry; these massless particles were called Nambu–Goldstone bosons. In Nambu's model of the SSB of chiral symmetry, the appearance of massless pseudoscalar Nambu–Goldstone bosons may be interpreted as pions in the limit of $m_\pi \to 0$. However, in the case of other gauge symmetries, massless scalar bosons were generated. In the case of SSB of such gauge symmetries, the predictions of the existence of massless scalar Nambu–Goldstone bosons were not supported by any experimental evidence. The appearance of massless Nambu–Goldstone boson fields was thus an obstacle for these type of theories to be of any use in the phenomenology of weak interactions. At first, these were thought to be model-dependent results which could be improved by using a sophisticated field theoretic model but soon a general theorem by Goldstone, Salam, and Weinberg [331] proved the existence of massless boson fields to be an essential feature of any Lorentz invariant relativistic field theory. This put an end to such theories in which "the Goldstone bosons sit like a snake hiding in the grass ready to strike". The theorem states that "In a manifestly Lorentz invariant quantum field theory, if there is continuous symmetry under which the Lagrangian is invariant, then either the vacuum state is also invariant or there must exist spinless (scalar) particles of zero mass." This is the most quoted statement of the Goldstone theorem.

In Goldstone's approach, the spontaneous breaking of symmetry was achieved by introducing self interacting scalar fields and giving nonzero vacuum expectation values (VEV) of some scalar fields in the theory. This is possible if the physical vacuum is not invariant under symmetry transformations. Let us suppose that there exists a symmetry of the Lagrangian leading to a conserved current J_a^μ given by Noether's theorem such that:

$$\partial_\mu J_a^\mu(x) = 0, \qquad \text{[for some } a]$$

leading to a charge operator Q_a, which is defined as:

$$Q_a = \int J_a^0(x)d\vec{x}, \quad \text{such that} \tag{7.107}$$

Q_a is conserved, reflecting the symmetry of the Lagrangian under which the physical vacuum is not invariant, that is,

$$[Q_a, H] = 0, \text{ but}$$
$$Q_a|0\rangle \neq 0. \tag{7.108}$$

The fields ϕ_α in a given representation of the symmetry group will transform as:

$$[Q_a, \phi_\alpha(x)] = T^a_{\alpha\beta}\phi_\beta(x),$$

where $T^a_{\alpha\beta}$ is the matrix representation of Q_a. Taking the VEV on both sides gives:

$$T^a_{\alpha\beta}\langle 0|\phi_\beta(x)|0\rangle = \langle 0|[Q_a, \phi_a(x)]|0\rangle. \tag{7.109}$$

This shows that $\langle 0|\phi_\beta(x)|0\rangle \neq 0$ if $Q_a|0\rangle \neq 0$ and the symmetry is broken spontaneously. It means that if a component of a field in a given representation is assigned a non-zero VEV, then all the operators which do not commute with that field component break the symmetry spontaneously. For example, in the case of isotriplet of field $\phi_\alpha(x)(\alpha = 1, 2, 3)$, under SU(2) if $\langle 0|\phi_3(x)|0\rangle \neq 0$, then Q_1 and Q_2 generators do not annihilate the vacuum and break the SU(2) symmetry spontaneously.

To prove Goldstone's theorem [331], let us consider the Fourier transform of the VEV of the commutator $[J^\mu(x), \phi_\alpha(0)]$, that is,

$$M^\mu_\alpha(k) = \int d^4x e^{ik.x}\langle 0|[J^\mu(x), \phi_\alpha(0)]|0\rangle, \tag{7.110}$$

where k^μ is an arbitrary vector. For simplicity and definiteness, we drop the indices a as well as α, as it is valid for any a and α.

Let us saturate the commutator in Eq. (7.110) with a complete set of states $|n\rangle$, that is, $\sum_n |n\rangle\langle n| = 1$ satisfying the relation:

$$P^\mu|n\rangle = p^\mu_n|n\rangle,$$

where P^μ is the momentum operator and $|n\rangle$ is an eigenstate of momentum P^μ corresponding to the eigenvalue p^μ_n. The Goldstone theorem then states that one of the states $|n\rangle$ is necessarily a spin 0 massless state corresponding to $p^2_n = 0$ state. We write, after inserting complete set of $|n\rangle$ states,

$$M^\mu_\alpha(k) = \sum_n \int d^4x \left[\langle 0|J^\mu(x)|n\rangle\langle n|\phi_\alpha(0)|0\rangle - \langle 0|\phi_\alpha(0)|n\rangle\langle n|J^\mu(x)|0\rangle\right] e^{ik.x}. \tag{7.111}$$

Using translational invariance, we express $J^\mu(x)$ as:

$$J^\mu(x) = e^{-iP.x}J^\mu(0)e^{iP.x} \tag{7.112}$$

and use it in Eq. (7.111) to give:

$$
\begin{aligned}
M_\alpha^\mu(k) &= \sum_n \int d^4x \left[e^{i(p_n+k)\cdot x} \langle 0|J^\mu(0)|n\rangle\langle n|\phi_\alpha(0)|0\rangle \right. \\
&\qquad\qquad\qquad \left. - e^{-i(p_n-k)\cdot x} \langle 0|\phi_\alpha(0)|n\rangle\langle n|J^\mu(0)|0\rangle \right] \\
&= \sum_n (2\pi)^4 \left[\delta^4(p_n+k)\langle 0|J^\mu(0)|n\rangle\langle n|\phi_\alpha(0)|0\rangle \right. \\
&\qquad\qquad\qquad \left. - \delta^4(p_n-k)\langle 0|\phi_\alpha(0)|n\rangle\langle n|J^\mu(0)|0\rangle \right].
\end{aligned}
$$

In order to have spontaneous breaking of symmetry, $M_\alpha^\mu(k) \neq 0$ (see Eq. (7.101) and (7.111)), which implies that

1. $\langle n|\phi_\alpha(0)|0\rangle \neq 0$, that is, $|n\rangle$ is a single particle state created from the physical vacuum by the action of the field ϕ_α and is, therefore, a spinless boson due to $\phi_\alpha(0)$ being a scalar field.

2. The matrix element $\langle 0|J^\mu(0)|n\rangle$ should be proportional to a four vector as J^μ is a four vector implying that

$$
\langle 0|J^\mu(0)|n\rangle = a(p_n^2)p_n^\mu, \tag{7.113}
$$

with $a(p_n^2) \neq 0$ from the requirement of the Lorentz covariance. It should be noted that this is not the case in the non-relativistic theories [332], where additional terms could be present.

We also define

$$
\langle n|J^\mu(0)|0\rangle = b(p_n^2)p_n^\mu, \tag{7.114}
$$

where $b(p_n^2) \neq 0$, is another scalar which may be related to $a(p_n^2)$ depending upon the properties of $J^\mu(0)$. Using the properties of $\delta^4(p_n+k)$ and $\delta^4(p_n-k)$ and the requirement of the states $|n\rangle$ to be physical, that is, $p_n^0 = E_n > 0$, which leads to $\delta(E_n+k_0) = 0$ if $k_0 > 0$, and $\delta(E_n-k_0) = 0$ if $k_0 < 0$. Therefore, we will use the step function Θ in the expression of $M_\alpha^\mu(k)$ as:

$$
M_\alpha^\mu(k) = (2\pi)^4[-a(k^2)k_\mu\langle n|\phi_\alpha(0)|0\rangle\Theta(-k_0) - b(k^2)k_\mu\langle 0|\phi_\alpha(0)|n\rangle\Theta(k_0)]. \tag{7.115}
$$

Now we replace the step function Θ by a function $\epsilon(k_0)$ as:

$$
\epsilon(k_0) = \begin{cases} -1, & \text{if } k_0 < 0 \\ +1, & \text{if } k_0 \geq 0 \end{cases}. \tag{7.116}
$$

We write:

$$M_\alpha^\mu(k) = k^\mu[\rho_{1\alpha}(k^2)\epsilon(k_0) + \rho_{2\alpha}(k^2)], \tag{7.117}$$

where

$$\rho_{1\alpha}(k^2) = \frac{1}{2}(2\pi)^4[a(k^2)\langle n|\phi_\alpha(0)|0\rangle - b(k^2)\langle 0|\phi_\alpha(0)|n\rangle]$$

$$\rho_{2\alpha}(k^2) = \frac{1}{2}(2\pi)^4[-a(k^2)\langle n|\phi_\alpha(0)|0\rangle - b(k^2)\langle 0|\phi_\alpha(0)|n\rangle]. \tag{7.118}$$

3. The condition of the current being divergenceless, that is, $\partial_\mu J^\mu(x) = 0$ gives the corresponding condition on $M^\mu(k)$ as

$$\partial_\mu M_\alpha^\mu(k) = 0 \quad \Rightarrow \quad k^2[\epsilon(k_0)\rho_{1\alpha}(k^2) + \rho_{2\alpha}(k^2)] = 0$$

$$\Rightarrow \quad k^2[-\rho_{1\alpha}(k^2) + \rho_{2\alpha}(k^2)] = 0 \qquad \text{for} \qquad k_0 < 0$$

$$\text{and} \quad k^2[\rho_{1\alpha}(k^2) + \rho_{2\alpha}(k^2)] = 0 \qquad \text{for} \qquad k_0 > 0$$

which implies that $\rho_{1\alpha}(k^2)$ and $\rho_{2\alpha}(k^2)$ can be written in terms of arbitrary parameters $c_{1\alpha}$ and $c_{2\alpha}$ as:

$$\rho_{1\alpha}(k^2) = c_{1\alpha}\delta(k^2) \qquad \text{and} \qquad \rho_{2\alpha}(k^2) = c_{2\alpha}\delta(k^2). \tag{7.119}$$

Therefore,

$$M_\alpha^\mu(k) = k^\mu[c_{1\alpha}\epsilon(k_0) + c_{2\alpha}]\delta(k^2). \tag{7.120}$$

If $c_{1\alpha}$ and $c_{2\alpha}$ are both zero, $M_\alpha^\mu(k) = 0$, implying that there is no spontaneous symmetry breaking. If either of them is non-zero, spontaneous breaking would be present. To evaluate $c_{1\alpha}$ and $c_{2\alpha}$, we evaluate $M_\alpha^0(\vec{k} = 0, k^0)$ in two ways.

First, using Eq. (7.110), we write:

$$M_\alpha^0(\vec{k} = 0, k^0) = \int d^4x \langle 0|[J^0(x), \phi_\alpha(0)]|0\rangle e^{ik_0x_0}$$

$$= \int d\vec{x}\, 2\pi\delta(k_0) < 0|[J^0(x), \phi_\alpha(0)]|0>, \text{where}$$

$$\int dx_0 e^{ik_0x_0} = 2\pi\delta(k_0)$$

$$= 2\pi\delta(k_0)\langle 0|[Q(t), \phi_\alpha(0)]|0\rangle. \tag{7.121}$$

Since the equations of motion respect the symmetry of the Lagrangian, the operator $Q(t)$ is independent of time (Eq. (7.107)). Therefore,

$$M_\alpha^0(\vec{k} = 0, k^0) = 2\pi\delta(k_0)\eta, \qquad \text{where} \quad \eta = \langle 0|[Q, \phi_\alpha(0)]|0\rangle \tag{7.122}$$

such that

$$\int M_\alpha^0(\vec{k}=0,k^0)dk^0 = \int_{-\infty}^{+\infty} 2\pi\delta(k_0)\eta dk^0 = 2\pi\eta. \tag{7.123}$$

This integral can also be evaluated using the covariant form of $M_\alpha^\mu(k)$ in Eq. (7.120), that is,

$$
\begin{aligned}
\int M_\alpha^0(\vec{k}=0,k^0)dk^0 &= \int_{-\infty}^{+\infty} dk_0[c_{1\alpha}\epsilon(k_0)+c_{2\alpha}]k_0\delta(k_0^2-\vec{k}^2) \\
&= \int_{-\infty}^{+\infty} dk_0[c_{1\alpha}\epsilon(k_0)+c_{2\alpha}]k_0 \\
&\quad \left(\frac{\delta(k_0-|\vec{k}|)+\delta(k_0+|\vec{k}|)}{2|\vec{k}|}\right) \\
&= \int_{-\infty}^{0} dk_0[-c_{1\alpha}k_0+c_{2\alpha}k_0]\left(\frac{\delta(k_0+|\vec{k}|)}{2|\vec{k}|}\right) \\
&\quad + \int_{0}^{+\infty} dk_0[c_{1\alpha}k_0+c_{2\alpha}k_0]\left(\frac{\delta(k_0-|\vec{k}|)}{2|\vec{k}|}\right) \\
&= \frac{|\vec{k}|}{2|\vec{k}|}(c_{1\alpha}-c_{2\alpha}) + \frac{|\vec{k}|}{2|\vec{k}|}(c_{1\alpha}+c_{2\alpha}) \\
&= c_{1\alpha}. \tag{7.124}
\end{aligned}
$$

Comparing the two evaluations, Eqs. (7.123) and (7.124) we get: $c_{1\alpha} = 2\pi\eta$. Therefore, if $\eta \neq 0$, then $M_\alpha^\mu(k) \neq 0$. Using Eqs. (7.113), and (7.114), we can write:

$$
\begin{aligned}
M_\alpha^\mu(k) &= (2\pi)^4 \sum_n \left[\delta(k+p_n)a(p_n^2)\langle n|\phi_\alpha(0)|0\rangle - \delta(k-p_n)b(p_n^2)\langle 0|\phi_\alpha(0)|n\rangle\right]p_n^\mu \\
&= 2\pi\eta k^\mu \delta(k^2)\epsilon(k). \tag{7.125}
\end{aligned}
$$

If $\eta \neq 0$, only $p_n = \pm k$ states in Eq. (7.125) contribute. both have $p_n^2 = k^2 = 0$, that is, massless bosons, in order that $M_\alpha^\mu(k) \neq 0$. These are called Goldstone bosons. Therefore, there is at least one Goldstone boson corresponding to the spontaneous breaking of symmetry. There could be more than one. This is the Goldstone theorem.

7.3.9 Spontaneously broken discrete symmetry: A simple model

Let us take the Lagrangian for a self interacting real scalar field ϕ as:

$$\mathcal{L} = T - V = \frac{1}{2}(\partial_\mu\phi(x))(\partial^\mu\phi(x)) - \frac{1}{2}\mu^2\phi^2(x) - \frac{1}{4}\xi\phi^4(x). \tag{7.126}$$

ϕ^4 represents self interaction of the field with coupling strength ζ ($\zeta > 0$); the potential is symmetric about ϕ, that is, $V(\phi) = V(-\phi)$. The lowest energy corresponding to the state of minimum energy pertaining to the minimum of the Hamiltonian

$$H = T + V = -\frac{1}{2}\dot{\phi}^2 + (\nabla\phi)^2 + V(\phi), \quad \text{where } \dot{\phi} = \frac{\partial\phi}{\partial t} \tag{7.127}$$

is obtained by equating

$$\frac{\partial V}{\partial \phi} = \mu^2\phi + \zeta\phi^3 = (\mu^2 + \zeta\phi^2)\phi = 0. \tag{7.128}$$

Notice from Eq. (7.128), that for the minimum, there are two possibilities that is: either $\phi = 0$ or $\mu^2 + \zeta\phi^2 = 0$

1. if the ground state is at $\phi = 0$ and $\mu^2 > 0$, then the potential has a unique minima, which corresponds to the vacuum state(Figure 7.7).

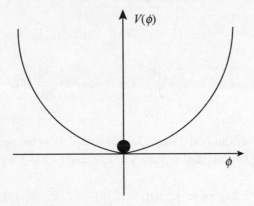

Figure 7.7 The potential $V(\phi) = \frac{1}{2}\mu^2\phi^2(x) + \frac{1}{4}\zeta\phi^4(x)$ for $\mu^2 > 0$.

2. if ϕ satisfies the equation $\mu^2 + \zeta\phi^2 = 0$,

$$\Rightarrow \qquad \phi = \pm\sqrt{\frac{-\mu^2}{\zeta}} = \pm\lambda \text{ (say)}, \tag{7.129}$$

which corresponds to two degenerate lowest energy states as shown in Figure 7.8(a) for $\mu^2 < 0$. Therefore, the real minima is not $\phi = 0$ but $\phi = +\lambda$ or $\phi = -\lambda$, and $\phi = 0$ is an unstable point. Choosing any one of them (either Figure 7.8(b) or Figure 7.8(c)) will break the symmetry.

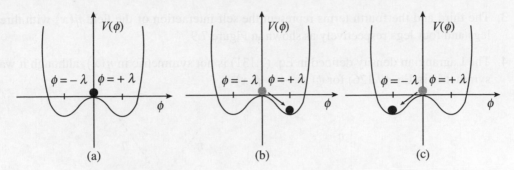

Figure 7.8 The potential $V(\phi) = \frac{1}{2}\mu^2\phi^2(x) + \frac{1}{4}\xi\phi^4(x)$ for $\mu^2 < 0$. There is an equal probability of the ball going to the left as to the right; both are the ground states for the system. The moment the ball takes a position, the left–right symmetry is broken as the system has preferred a particular place. This is an example of discrete symmetry, with just two ground states.

Let us take the case when $\phi = +\lambda$, and rescale the field $\phi(x) = \lambda + \eta(x)$, where $\eta(x)$ represents the fluctuations of ϕ around $\phi_0 = \lambda$ in terms of the new field $\eta(x)$. The Lagrangian given in Eq. (7.126) is now written as:

$$\mathcal{L} = \frac{1}{2}(\partial_\mu\eta(x))(\partial^\mu\eta(x)) - \frac{1}{2}\mu^2(\eta^2(x) + 2\eta(x)\lambda + \lambda^2)$$
$$- \frac{1}{4}\xi\left(\eta^4(x) + 4\eta^3(x)\lambda + 6\eta^2(x)\lambda^2 + 4\eta(x)\lambda^3 + \lambda^4\right). \tag{7.130}$$

Using $\mu^2 = -\xi\lambda^2$ from Eq. (7.129), and dropping constant terms like $\xi\lambda^4$ as they would not contribute to the field equations for the system, the Lagrangian is obtained as

$$\mathcal{L} = \frac{1}{2}(\partial_\mu\eta(x))(\partial^\mu\eta(x)) - \xi\lambda^2\eta^2(x) - \xi\lambda\eta^3(x) - \frac{1}{4}\xi\eta^4(x). \tag{7.131}$$

Recall the expression for the Lagrangian density for a free real scalar field that gives the Klein–Gordon equation with mass m, that is,

$$\mathcal{L} = \frac{1}{2}(\partial_\mu\phi(x))(\partial^\mu\phi(x)) - \frac{1}{2}m^2\phi^2(x). \tag{7.132}$$

When we compare it with Eq. (7.131), we obtain

$$\xi = \frac{m^2}{2\lambda^2} \quad \text{or equivalently} \quad m = \sqrt{2\lambda^2\xi} = \sqrt{-2\mu^2}, \quad \text{using Eq. (7.129)}$$

The physical significance of the different terms of the Lagrangian density given in Eq. (7.131) is as follows.

1. The first term represents the kinetic energy of the scalar field $\eta(x)$.

2. The second term represents the mass term which has been identified by expanding the powers of the new field $\eta(x)$, with $m_\eta = \sqrt{-2\mu^2}$.

3. The third and the fourth terms represent the self interaction of the field $\eta(x)$ with three legs and four legs respectively as shown in Figure 7.9

4. The Lagrangian density defined in Eq. (7.131) is not symmetric in $\eta(x)$, although it was symmetric in Eq. (7.126) for $\phi(x) \rightarrow -\phi(x)$.

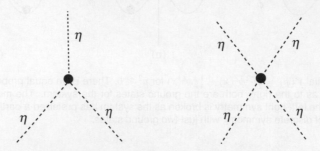

Figure 7.9 Self interaction of the $\eta(x)$ field with three legs and four legs.

The symmetry is, therefore, broken in the revised Lagrangian. This is an example of the spontaneous breaking of the discrete symmetry when one of the minima is chosen to be the physical ground state which generates mass of the $\eta(x)$ field.

7.3.10   Spontaneously broken continuous global symmetry under $U(1)$ gauge transformation and Goldstone boson

Let us consider a more realistic situation, a complex scalar field ϕ (for spin 0 particle) such that $\phi = \frac{1}{\sqrt{2}}(\phi_1 + i\phi_2)$, where ϕ_1 and ϕ_2 are the two real scalar fields and writing the Lagrangian density as:

$$\mathcal{L} = (\partial_\mu \phi(x))^*(\partial^\mu \phi(x)) - \mu^2 \phi^*(x)\phi(x) - \xi(\phi^*(x)\phi(x))^2 \tag{7.133}$$

$$= \frac{1}{2}(\partial_\mu \phi_1(x))(\partial^\mu \phi_1(x)) + \frac{1}{2}(\partial_\mu \phi_2(x))(\partial^\mu \phi_2(x)) - \frac{\mu^2}{2}(\phi_1^2(x) + \phi_2^2(x))$$

$$- \frac{\xi}{4}\left(\phi_1^4(x) + \phi_2^4(x) + 2\phi_1^2(x)\phi_2^2(x)\right), \tag{7.134}$$

where the first two terms are the kinetic energy terms and the last two terms are the potential energy terms. The symmetry of the Lagrangian may be described by rotation in (ϕ_1, ϕ_2) space such that for any rotation angle θ:

$$\begin{pmatrix} \phi_1' \\ \phi_2' \end{pmatrix} = \begin{pmatrix} \cos\theta & \sin\theta \\ -\sin\theta & \cos\theta \end{pmatrix} \begin{pmatrix} \phi_1 \\ \phi_2 \end{pmatrix}, \tag{7.135}$$

which gives $\phi_1'^2 + \phi_2'^2 = \phi_1^2 + \phi_2^2$ i.e. under the group SO(2) of rotation in the plane, where the Lagrangian is invariant.

If $\mu^2 > 0$, then we have the vacuum state defined by:

$$\langle \phi \rangle = \frac{1}{\sqrt{2}}\begin{pmatrix} \phi_1 \\ \phi_2 \end{pmatrix} = \begin{pmatrix} 0 \\ 0 \end{pmatrix}$$

and for small fluctuations of the field i.e. neglecting quartic terms of ϕ_1 and ϕ_2, the Lagrangian density given in Eq. 7.134, simply becomes the Lagrangian for a two particle non-interacting system (with the same mass μ) i.e.

$$\mathcal{L} = \frac{1}{2}(\partial_\mu \phi_1(x))^2 + \frac{1}{2}(\partial_\mu \phi_2(x))^2 - \frac{\mu^2}{2}(\phi_1^2(x) + \phi_2^2(x)). \tag{7.136}$$

For $\mu^2 < 0$, the minimum of the potential lie on the circle of radius $\frac{\mu^2}{\zeta}$ as shown in Figure 7.10, we rewrite the Lagrangian in Eq. (7.134), with a reversed sign ($\mu^2 \to -\mu^2$) for the mass term, that is

$$\begin{aligned} \mathcal{L} &= \frac{1}{2}(\partial_\mu \phi_1(x))(\partial^\mu \phi_1(x)) + \frac{1}{2}(\partial_\mu \phi_2(x))(\partial^\mu \phi_2(x)) + \frac{\mu^2}{2}(\phi_1^2(x) + \phi_2^2(x)) \\ &\quad - \frac{\zeta}{4}\left(\phi_1^4(x) + \phi_2^4(x) + 2\phi_1^2(x)\phi_2^2(x)\right). \end{aligned} \tag{7.137}$$

It should be noted that the mass term appears in the Lagrangian with a 'wrong' sign, which has been done to ensure that $\mu^2 < 0$ and we get a real value of vacuum expectation, that is:

$$\Rightarrow (\phi_1)^2_{\min} + (\phi_2)^2_{\min} = \frac{\mu^2}{\zeta} \tag{7.138}$$

and for a particular ground state corresponding to the vacuum, the choice can be:

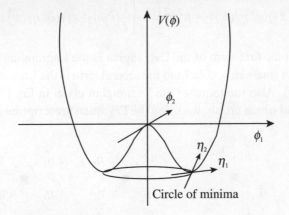

Figure 7.10 The potential $V(\phi)$ for $\mu^2 < 0$ and $\zeta > 0$.

$$(\phi_1)_{\min} = \frac{\mu}{\sqrt{\zeta}} = \lambda_1(say), \qquad (\phi_2)_{\min} = 0 = \lambda_2(say)$$

i.e. $$\langle \phi \rangle_0 = \frac{1}{\sqrt{2}}\begin{pmatrix} \lambda_1 \\ 0 \end{pmatrix} \tag{7.139}$$

which has been shown in Figure 7.10.

This time for the fluctuation about the local minima we need two fields say $\eta_1(x)$ and $\eta_2(x)$ such that:

$$\phi = \frac{1}{\sqrt{2}}\left(\frac{\mu}{\sqrt{\zeta}} + \eta_1(x) + i\eta_2(x)\right)$$

which results:

$$\phi_1(x) = \frac{\mu}{\sqrt{\zeta}} + \eta_1(x),$$

$$\phi_2(x) = \eta_2(x)$$

$$\phi_1^2(x) + \phi_2^2(x) = \eta_1^2(x) + 2\frac{\mu}{\sqrt{\zeta}}\eta_1(x) + \frac{\mu^2}{\zeta} + \eta_2^2(x). \tag{7.140}$$

Using the above expression in Eq. 7.137, we get:

$$\begin{aligned}
\mathcal{L} = &\frac{1}{2}(\partial_\mu\eta_1(x))(\partial^\mu\eta_1(x)) + \frac{1}{2}(\partial_\mu\eta_2(x))(\partial^\mu\eta_2(x)) + \frac{\mu^2}{2}\left(\frac{\mu^2}{\zeta} + \eta_1^2 + 2\frac{\mu}{\sqrt{\zeta}}\eta_1 + \eta_2^2\right)\\
&- \frac{\zeta}{4}\left(\eta_1^4 + \frac{\mu^4}{\zeta^2} + 6\frac{\mu^2}{\zeta}\eta_1^2 + 4\frac{\mu^3}{\zeta\sqrt{\zeta}}\eta_1 + 4\frac{\mu}{\sqrt{\zeta}}\eta_1^3 + \eta_2^4\right)\\
&- \frac{1}{2}\zeta\left(\frac{\mu^2}{\zeta}\eta_2^2 + \eta_1^2\eta_2^2 + 2\frac{\mu}{\sqrt{\zeta}}\eta_1\eta_2^2\right).
\end{aligned}\tag{7.141}$$

Looking for the mass terms in the Lagrangian, and writing only the kinetic energy term and the term proportional to η_1^2 or η_2^2 we find the Lagrangian as:

$$\mathcal{L} = \left[\frac{1}{2}(\partial_\mu\eta_1(x))(\partial^\mu\eta_1(x)) - \mu^2\eta_1^2\right] + \left[\frac{1}{2}(\partial_\mu\eta_2(x))(\partial^\mu\eta_2(x))\right]. \tag{7.142}$$

It may be noticed that the first term of the Lagrangian is the Lagrangian for the Klein-Gordon field for free particle of (mass m = $\sqrt{2}\mu$) and the second term is the Lagrangian for a field which is massless (i.e. m=0). Also the terms of the Lagrangian given in Eq. (7.141) define different interactions of the field which are shown using the Feynman prescription in Figure 7.11. It may

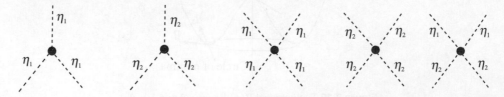

Figure 7.11 Different interaction terms for η_1 and η_2 fields.

be noticed that the Lagrangian which was symmetric in Eq. (7.133), in its new form given in Eq. (7.141) does not look symmetric and that means symmetry has been broken by making a choice of a particular vacuum state. One of the fields, defined by η_2 is massless and is a consequence of Goldstone's theorem discussed in Section 7.3.8, which says that spontaneous breaking of a continuous global symmetry is always accompanied by at least one massless scalar boson known as the Goldstone boson.

7.3.11 Spontaneously broken continuous local $U(1)$ symmetry and Higgs mechanism

Let us start by considering locally gauge invariant Lagrangian that gives rise to spontaneously broken symmetries

$$\mathcal{L} = (D_\mu \phi(x))^* (D^\mu \phi(x)) - \mu^2 \phi^*(x)\phi(x) - \xi(\phi^*(x)\phi(x))^2 - \frac{1}{4}F_{\mu\nu}F^{\mu\nu}, \tag{7.143}$$

where $\phi(x) = \frac{1}{\sqrt{2}}(\phi_1(x) + i\phi_2(x))$. The covariant derivative $D_\mu = \partial_\mu - ieA_\mu$ is required for the local invariance of the Lagrangian and $F_{\mu\nu} = \partial_\mu A_\nu - \partial_\nu A_\mu$.

For a local gauge transformation under $U(1)$

$$\phi(x) \rightarrow \phi'(x) = e^{i\alpha(x)}\phi(x) \qquad \text{and} \qquad \phi'^*(x) = e^{-i\alpha(x)}\phi^*(x)$$

and

$$A^\mu \rightarrow A^{\mu\prime} = A^\mu + \frac{1}{e}\partial^\mu \alpha(x).$$

These transformations lead to:

$$\mathcal{L} = ((\partial_\mu - ieA_\mu)\phi(x))^* ((\partial^\mu - ieA^\mu)\phi(x)) - \mu^2 \phi^*(x)\phi(x)$$
$$- \xi(\phi^*(x)\phi(x))^2 - \frac{1}{4}F_{\mu\nu}F^{\mu\nu}. \tag{7.144}$$

For $\mu^2 > 0$ the potential has a minimum identified at $\phi = 0$ and the symmetry of the Lagrangian is preserved. Whereas for $\mu^2 < 0$, we follow the same analogy as that has been used in the previous section and we expand $\phi_1(x)$ and $\phi_2(x)$ about their minimum values:

$$\phi_1(x) = \eta_1(x) + \frac{\mu}{\sqrt{\xi}}, \qquad \phi_2(x) = \eta_2(x), \tag{7.145}$$

the Lagrangian density becomes:

$$\mathcal{L} = \left[\frac{1}{2}(\partial_\mu \eta_1(x))(\partial^\mu \eta_1(x)) - \mu^2 \eta_1^2(x)\right] + \left[\frac{1}{2}(\partial_\mu \eta_2(x))(\partial^\mu \eta_2(x))\right]$$
$$+ \left[-\frac{1}{4}F_{\mu\nu}F^{\mu\nu} + \frac{1}{2}\frac{e^2\mu^2}{\xi}A_\mu A^\mu\right]$$
$$- \left([e(\eta_1(x)\partial_\mu \eta_2(x) - \eta_2(x)\partial_\mu \eta_1(x))]A^\mu + \frac{\mu}{\sqrt{\xi}}e^2\eta_1(x)A_\mu A^\mu\right.$$
$$+ \frac{e^2}{2}(\eta_1^2(x) + \eta_2^2(x))A_\mu A^\mu - \mu\sqrt{\xi}(\eta_1^3(x) + \eta_1\eta_2^2(x))$$
$$\left. - \frac{1}{4}\xi(\eta_1^2(x) + \eta_2^2(x))^2\right) - \frac{\mu e}{\sqrt{\xi}}(\partial_\mu \eta_2(x))A^\mu + \left(\frac{\mu^2}{2\sqrt{\xi}}\right)^2. \tag{7.146}$$

It may be noticed that the first two terms of the Lagrangian are the same as Eq. 7.142, a scalar field $\eta_1(x)$ with mass μ, another field $\eta_2(x)$ which is massless corresponding to the Goldstone boson. Interestingly now there is an additional mass term appearing with the gauge field A^μ, which is the result of spontaneous symmetry breaking. When this is compared with the Lagrangian for a massive vector field with mass M

$$\mathcal{L}_{Proca} = -\frac{1}{4}F_{\mu\nu}F^{\mu\nu} + \frac{1}{2}M^2 A_\mu A^\mu,$$

i.e. the Lagrangian that leads to Proca equation

$$\partial_\nu(\partial^\nu A^\mu - \partial^\mu A^\nu) + M^2 A^\mu = 0,$$

gives $M = \frac{\mu e}{\sqrt{\zeta}}$. Other terms of the Lagrangian in Eq. 7.146 lead to various couplings of the

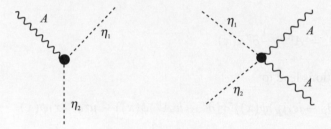

Figure 7.12 New interaction.

fields like shown in Figure 7.11, as well as some new ones due to the presence of vector field A^μ for example as shown in Figure 7.12. The Lagrangian density given in Eq. (7.146) describes

Figure 7.13 Interaction of massless field η_2 with A^μ.

the interaction of a massive vector field $A^\mu(x)$ and two scalar fields, one massive ($\eta_1(x)$) and one massless ($\eta_2(x)$) field (which is the Goldstone field), as shown in Figure 7.13. This means that the transformed Lagrangian has now fields with five degrees of freedom in contrast with the original Lagrangian which had fields with four degrees of freedom i.e. two for massless gauge field $A^\mu(x)$ and two for scalar fields $\phi_1(x)$ and $\phi_2(x)$. Therefore, the transformation has created a spurious degree of freedom for the fields which is not physical. We, therefore make a transformation using the gauge freedom such that this spurious degree of freedom disappears. We would see that this also removes the mixing of the gauge field A^μ with the Goldstone field which appears through the term $\frac{\mu e}{\sqrt{\xi}}(\partial_\mu \eta_2(x))A^\mu$ in Eq. (7.146). For this let us consider:

$$\phi = \frac{1}{\sqrt{2}}\left(\eta_1(x) + \frac{\mu}{\sqrt{\xi}} + i\eta_2(x)\right)$$

$$\simeq \frac{1}{\sqrt{2}}\left(\eta_1(x) + \frac{\mu}{\sqrt{\xi}}\right)e^{i\eta_2(x)\sqrt{\xi}/\mu} \tag{7.147}$$

to the lowest order in $\eta_2(x)$. If we substitute a new transformed set of real fields as in Eq. (7.147) (for convenience still calling them $\eta_1(x)$ and $\eta_2(x)$) and make the following transformation on the gauge field A^μ:

$$A^\mu \to A^\mu + \frac{\sqrt{\xi}}{e\mu}\partial^\mu \eta_2(x) \tag{7.148}$$

in our original Lagrangian given in Eq.7.144, the expression for the Lagrangian density is obtained as:

$$\mathcal{L} = \left[\frac{1}{2}(\partial_\mu \eta_1(x))(\partial^\mu \eta_1(x)) - \mu^2 \eta_1^2(x)\right] + \left[-\frac{1}{4}F_{\mu\nu}F^{\mu\nu} + \frac{1}{2}\frac{e^2\mu^2}{\xi}A_\mu A^\mu\right]$$

$$+ \left(\frac{\mu}{\sqrt{\xi}}e^2\eta_1(x)A_\mu A^\mu + \frac{e^2}{2}\eta_1^2(x)A_\mu A^\mu - \mu\sqrt{\xi}\eta_1^3(x) - \frac{1}{4}\xi\eta_1^4(x)\right) + \left(\frac{\mu^2}{2\sqrt{\xi}}\right)^2. \tag{7.149}$$

From Eq. 7.149, if we look at the mass terms in the Lagrangian it may be observed that there are two terms, one associated with single scalar field $\eta_1(x)$ (Higgs) and another term associated with the massive vector field A^μ, while the field associated with the Goldstone boson has disappeared.

7.3.12 Spontaneously broken local $SU(2)$ gauge symmetry and Higgs mechanism

We consider $SU(2)$ doublet of complex scalar field ϕ, consists of ϕ_α and ϕ_β with each component $\phi_1 + i\phi_2$ and $\phi_3 + i\phi_4$ respectively, such that:

$$\phi = \begin{pmatrix} \phi_\alpha \\ \phi_\beta \end{pmatrix} = \frac{1}{\sqrt{2}}\begin{pmatrix} \phi_1 + i\phi_2 \\ \phi_3 + i\phi_4 \end{pmatrix} ; \phi_{1-4} \text{ are real scalar fields, and} \tag{7.150}$$

$$\phi^\dagger \phi = \frac{1}{2}\left(\phi_1^2 + \phi_2^2 + \phi_3^2 + \phi_4^2\right). \tag{7.151}$$

Let us write the Lagrangian density in terms of complex scalar field:

$$\mathcal{L} = (\partial_\mu \phi(x))^\dagger (\partial^\mu \phi(x)) - \mu^2 \phi^\dagger(x)\phi(x) - \lambda(\phi^\dagger(x)\phi(x))^2, \tag{7.152}$$

where μ is the mass of the particle associated with complex scalar field and λ is a real parameter. The first term in Eq. 7.152 corresponds to the kinetic energy term and the second term, for $\mu^2 > 0$ corresponds to the mass term and the third term represents ϕ^4 self interaction. If λ is equal to zero, then the above Lagrangian is nothing but the Lagrangian for the Klein-Gordon field with $\mu = m$.

The above Lagrangian is invariant under $SU(2)$ global gauge transformation of the field

$$\phi(x) \rightarrow \phi'(x) = e^{i\alpha_i T_i}\phi(x). \tag{7.153}$$

In the above equation α is a parameter independent of space and time and $T_i = \frac{\tau_i}{2}$, where τ_i are the components of the Pauli matrices.

We are interested in the invariance under local $SU(2)$ gauge transformation:

$$\phi(x) \rightarrow \phi'(x) = e^{i\alpha_i(x)T_i}\phi(x). \tag{7.154}$$

Notice α_i is now a function of x, showing the space-time dependence. To ensure the invariance, we follow the algebra described in Section 7.3.4 and replace ∂_μ by the covariant derivative D_μ as:

$$D_\mu = \partial_\mu + igT_iW^i_\mu, \quad \text{where } i = 1 - 3$$

and W^i_μ are the three gauge fields. For an infinitesimal gauge transformation under $SU(2)$

$$\phi(x) \rightarrow \phi'(x) = (1 + i\alpha_i(x)T_i)\phi(x)$$
$$\Rightarrow \qquad \delta\phi(x) = i\alpha_i(x)T_i\phi(x).$$

Following the prescription of Section 7.3.4 the three gauge fields transform as:

$$W^i_\mu \rightarrow W^{i'}_\mu = W^i_\mu - \epsilon_{ijk}\alpha_j W^k_\mu - \frac{1}{g}(\partial_\mu\alpha_i(x)). \tag{7.155}$$

The gauge invariant Lagrangian is then given by:

$$\mathcal{L} = (\partial_\mu\phi + igT_iW^i_\mu\phi)^\dagger(\partial^\mu\phi + igT_iW^{\mu i}\phi) - V(\phi) - \frac{1}{4}G^i_{\mu\nu}G^{\mu\nu}_i, \text{ where}\tag{7.156}$$
$$G^i_{\mu\nu} = \partial_\mu W^i_\nu - \partial_\nu W^i_\mu + g(W_\mu \times W_\nu)^i$$
$$V(\phi) = \mu^2\phi^\dagger\phi + \lambda(\phi^\dagger\phi)^2. \tag{7.157}$$

For $\mu^2>0$, the Lagrangian in Eq. 7.156 describes a system of four scalar particles, each of mass μ, interacting with the three massless gauge bosons W^i_μ. For $\mu^2 < 0$, the potential $V(\phi)$ of Eq. 7.157 has a minimum at a finite value of $|\phi|$ i.e.

$$\mu^2 + 2\lambda\phi^\dagger\phi = 0 \tag{7.158}$$
$$\Rightarrow \qquad \phi^\dagger\phi = \frac{1}{2}(\phi_1^2 + \phi_2^2 + \phi_3^2 + \phi_4^2) = -\frac{\mu^2}{2\lambda} = \zeta^2 > 0. \tag{7.159}$$

Various choices of ϕ satisfying Eq. 7.159 are possible i.e. $\phi(x)$ may be expanded about some local minima and that choice is not unique. For example, one choice could be:

$$\phi_1 = \phi_2 = \phi_4 = 0 \quad \text{and} \quad \phi_3^2 = -\frac{\mu^2}{\lambda} = v^2. \tag{7.160}$$

Applying the above constrain is equivalent to applying spontaneous symmetry breaking of the $SU(2)$ symmetry.

By expanding $\phi(x)$, about this vacuum

$$\phi(0) = \begin{pmatrix} 0 \\ \frac{v}{\sqrt{2}} \end{pmatrix} \tag{7.161}$$

and due to gauge invariance, the field $\phi(x)$ may be expanded as:

$$\phi(x) = \begin{pmatrix} 0 \\ \frac{v+h(x)}{\sqrt{2}} \end{pmatrix}, \quad \text{where } h(x) \text{ is Higgs field} \tag{7.162}$$

and use it in Eq. 7.156. Thus, it appears that we had started with four scalar fields, and are left finally with the Higgs field $h(x)$. This can be understood as follows:

Suppose we parameterize the fluctuations from the vacuum $\phi(0)$ in terms of four real fields viz. $\theta_1(x)$, $\theta_2(x)$, $\theta_3(x)$ and $h(x)$ using:

$$\begin{aligned} \phi(x) &= e^{i\vec{\tau}.\vec{\theta}(x)/v} \begin{pmatrix} 0 \\ \frac{v+h(x)}{\sqrt{2}} \end{pmatrix}, \text{ which for small perturbations} \\ &= \left(1 + \frac{i}{v}[\tau_1\theta_1 + \tau_2\theta_2 + \tau_3\theta_3]\right) \begin{pmatrix} 0 \\ \frac{v+h(x)}{\sqrt{2}} \end{pmatrix} \\ &= \begin{pmatrix} \frac{\theta_2+i\theta_1}{\sqrt{2}} \\ \frac{v+h(x)-i\theta_3}{\sqrt{2}} \end{pmatrix}. \end{aligned}$$

It may be noticed that the four fields are independent and are completely able to parameterize the deviations from the vacuum $\phi(0)$. Since the Lagrangian is locally invariant under $SU(2)$, therefore, the three fields defined by $\theta(x)$ corresponding to the massless Goldstone bosons can be gauged. The masses generated from the gauge bosons W_μ^i can be determined by substituting $\phi(0)$ from Eq. 7.161 into the Lagrangian, and the relevant term in the Lagrangian of our interest is

$$\left(igT_i W_\mu^i \phi(x)\right)^\dagger \left(igT_i W^{i\mu} \phi(x)\right) = \frac{g^2 v^2}{8}\left\{(W_\mu^1)^2 + (W_\mu^2)^2 + (W_\mu^3)^2\right\}. \tag{7.163}$$

Comparing it with mass term associated with a boson field say $\frac{1}{2}M^2 A_\mu^2$, we conclude

$$M = \frac{1}{2}gv. \tag{7.164}$$

Thus the Lagrangian describes the three massive gauge fields and one massive scalar field h(x). The gauge fields have eaten up the Goldstone bosons and in turn have become massive. Therefore, with the help of Higgs mechanism, we get rid of massless particles. However, this is not the end of the story, as we need the theory to be renormalizable, the proof of which was provided by 't 'Hooft as discussed in the next chapter, where we shall also discuss Glashow Weinberg-Salam theory of electroweak interaction.

Chapter 8

Unified Theory of Electroweak Interactions

8.1 Introduction

The unified theory of weak and electromagnetic interactions was formulated independently by Weinberg [157] and Salam [37] for leptons. It was later extended to the quark sector using the GIM mechanism of quark mixing proposed by Glashow, Iliopoulos, and Maiani (GIM) [64]. The theory is popularly known as the "standard model of electroweak interactions". S. L. Glashow, A. Salam, and S. Weinberg were awarded the Nobel Prize in Physics in 1979 for the formulation of this theory. In this chapter, we describe the formulation of the standard model for leptons and quarks and its applications to other interactions. The predictions of the model and their experimental confirmations are also presented. The limitations of the model and some processes suggesting the need for the physics beyond the standard model are discussed later in Chapter 20.

In this section, the essential features of the electromagnetic interactions and the phenomenological $V - A$ theory of electroweak interactions, well-known from the experimental and theoretical studies of the various weak processes are summarized. They are used as inputs in formulating the standard model. They are as follows:

1. Both electromagnetic and weak interactions involve, in general, all the elementary particles, that is, leptons and quarks, unlike the strong interaction, which affect only the quarks and the hadrons built from these quarks.

2. Both electromagnetic and weak interactions are mediated by vector fields. Electromagnetic interaction is known to be mediated by photons described by the electromagnetic field A_μ, which is massless. The observed weak interactions are presumed to be mediated by charged intermediate vector bosons (IVB), $W_\mu^\pm$, which are massive with mass M_W.

3. Weak interactions involve a pair of leptons, like $(\nu_e,\ e^-)$ or $(\overline{\nu}_e,\ e^+)$, in which the charged vector bosons, $W_\mu^\pm$, interact with the charged lepton currents $l_\mu^\pm$, where only the left-handed leptons participate, that is, $l_\mu^- = \overline{\psi}_{e_L}\gamma_\mu\psi_{\nu_L}$, with $\psi_{e_L(\nu_L)} = \frac{1-\gamma_5}{2}\psi_{e(\nu)}$. The right-handed leptons, that is, $\psi_{e_R(\nu_R)} = \frac{1+\gamma_5}{2}\psi_{e(\nu)}$ are not involved in the interaction. The interaction Lagrangian is written as:

$$\mathcal{L}^W = g l_\mu^\mp W^{\pm\mu} + \text{h.c.}$$

4. Electromagnetic interactions involve the electromagnetic field A_μ, which interacts with the electromagnetic current $l_\mu^{\text{em}} = e\overline{\psi}_e\gamma_\mu\psi_e$, implying that both the left- and the right-handed components of the electron participate, because $\overline{\psi}_e\gamma_\mu\psi_e = \overline{\psi}_{e_L}\gamma_\mu\psi_{e_L} + \overline{\psi}_{e_R}\gamma_\mu\psi_{e_R}$. The interaction Lagrangian is written as:

$$\mathcal{L}^{\text{em}} = e j_\mu^{\text{em}} A^\mu(x).$$

5. Both electromagnetic and weak interactions are universal interactions, that is, they have the same coupling strength for leptons and quarks. However, the coupling strengths of the electromagnetic and weak interactions to the quarks (leptons) are different from each other.

6. The weak mediating fields $W_\mu^\pm$ couple to $l_\mu^\pm$, involving vector (V) and axial vector (A) leptonic currents, because $\overline{\psi}_{e_L}\gamma_\mu\psi_{\nu_L} \approx \overline{\psi}_e\gamma_\mu(1-\gamma_5)\psi_\nu = V_\mu - A_\mu$, implying parity violation through their interference, as well as parity conserving coupling to $W_\mu^\pm$, while the electromagnetic field A_μ couples to the electromagnetic current l_μ^{em} only with the parity conserving coupling.

7. The theory of electromagnetic interactions is renormalizable, being a gauge field theory, with massless gauge field, while the weak interaction theory is not renormalizable due to the mediating fields $W_\mu^\pm$ being massive; this gives divergent results in the higher orders of the perturbation theory.

The major obstacles in formulating a unified theory of electromagnetic and weak interactions arise due to the following reasons:

i) The different masses of the mediating fields, A_μ and $W_\mu^\pm$ in the two interactions.

ii) The difference in the nature and strength of the couplings of the mediating fields to lepton and quark currents.

iii) The differences in the renormalizability properties of the two theories, that is, the quantum electrodynamical theory of electromagnetic interaction (QED) is renormalizable, while the $V - A$ theory of the weak interaction with IVB is not renormalizable.

These obstacles are removed in the theory of weak and electromagnetic (electroweak) interactions proposed by Weinberg [157] and Salam [37], in which the following assumptions are made:

1. The left-handed and the right-handed components of leptons, (e_L, ν_L) and (e_R, ν_R), are used to write the Lagrangian in such a way that the invariance under local gauge symmetry generates the interaction Lagrangian for the interaction of leptons with the massless gauge fields corresponding to the local symmetry. In this Lagrangian, only the left-handed components (e_L, ν_L) participate in the weak interaction; while both the components of the electrons participate in the electromagnetic interaction and ν_R has no interaction with matter.

2. The concept of a spontaneously broken local gauge field theory, introducing Higgs mechanism, is used to generate masses for the gauge fields corresponding to the IVB mediating the weak interactions while keeping the gauge field corresponding to photons mediating the electromagnetic interaction, massless.

3. The renormalizability of the spontaneously broken gauge theories was only speculated in the model; it was proved later by t' Hooft and Veltman [338] and Lee and Zinn- Justin [339].

We will discuss the model in detail in the following sections.

8.2 Description of the Weinberg–Salam Model for Leptons

The model uses the local gauge field theory based on $SU(2)_I \times U(1)_Y$ symmetry, where I is the weak isospin and Y is the weak hypercharge. It is, in fact, a minimal extension of the local gauge field theory (LGFT) based on $SU(2)_I$ symmetry, which was found to be inadequate to unify weak and electromagnetic interactions. The model was similar to the earlier works of Glashow [340], Salam, and Ward [341] in which the limitations of the local gauge field theory based on $SU(2)_I$ for achieving a unified theory were discussed and need for a higher symmetry was emphasized. We begin by listing the fermions and other particles considered in the model and limit ourselves, in this section, to leptons only (shown in Table 8.1). We also show the weak isospin (I and I_3) and weak hypercharge assignments of leptons and charge ($Q = I_3 + \frac{Y}{2}$) defined in analogy with the Gell–Mann–Nishijima relation in strong interactions. In weak

Table 8.1 Weak isospin (I), its third component (I_3), charge ($Q(|e|)$) and hypercharge ($Y = 2(Q - I_3)$) of leptons and scalar mesons in the W–S model with $l = e, \mu, \tau$.

Name	I	I_3	$Q(|e|)$	Y
(Leptons) ν_L	$\frac{1}{2}$	$+\frac{1}{2}$	0	-1
l_L	$\frac{1}{2}$	$-\frac{1}{2}$	-1	-1
ν_R	0	0	0	0
l_R	0	0	-1	-2
(Scalar ϕ^+	$\frac{1}{2}$	$+\frac{1}{2}$	1	1
mesons) ϕ^0	$\frac{1}{2}$	$-\frac{1}{2}$	0	1

interactions, electrons(e) and neutrinos (ν_e) always interact in pairs through their left-handed components, that is, (ν_L, e_L) or $(\bar{\nu}_L, e_L^+)$; the right-handed components $\nu_R, e_R, \bar{\nu}_R, \bar{e}_R$ are not involved in weak interactions. Therefore, the left-handed components are assigned to a weak isospin doublet ψ_L, that is,

$$\psi_L = \begin{pmatrix} \nu_L \\ e_L \end{pmatrix}$$

and the right-handed components are assigned to a singlet representation, that is,

$$e_R \quad \text{and} \quad \nu_R.$$

The weak hypercharge Y is assigned accordingly to reproduce the correct charge Q of leptons, through the relation:

$$Q = I_3 + \frac{Y}{2} \tag{8.1}$$

as shown in Table 8.1.

The Weinberg–Salam (W–S) Lagrangian for free leptons is written considering only the electron type leptons, that is, e^- and ν_e, for the simplicity of presentation. It can be extended to other flavors of leptons like μ^- and ν_μ, and τ^- and ν_τ in a straightforward manner. Using the concepts developed in Chapter 7, the W–S Lagrangian for interacting leptons is derived in the following four steps.

i) A Lagrangian for the free electrons and neutrinos for the left-handed doublet ψ_L and the right-handed singlets e_R and ν_R is written, which is invariant under the global symmetry group SU(2)$_I \times$ U(1)$_Y$, as:

$$\mathcal{L} = i\bar{\psi}_L \partial\!\!\!/ \psi_L + i\bar{e}_R \partial\!\!\!/ e_R + i\bar{\nu}_R \partial\!\!\!/ \nu_R. \tag{8.2}$$

There is no mass term like $m^2\bar{\psi}_e\psi_e$ as it violates the invariance of the Lagrangian under SU(2)$_I \times$ U(1)$_Y$, because $\bar{\psi}_e\psi_e = \bar{e}_L e_R + \bar{e}_R e_L$, where e_L is a member of an isospin doublet and e_R is a singlet.

ii) To make the Lagrangian invariant under the local symmetry group SU(2)$_I \times$ U(1)$_Y$, defined by the unitary transformation $U = U_1 U_2$, where $U_1 = e^{\frac{ig}{2}\vec{\tau}.\vec{a}(x)}$ and $U_2 = e^{i\frac{g'}{2}Y.B}$, $\vec{\tau}$ are the Pauli matrices and Y is a unit operator, the ordinary derivatives $\partial^\mu = \frac{\partial}{\partial x_\mu}$ are replaced by the covariant derivatives D^μ defined as:

$$D^\mu = \partial^\mu + i\frac{g}{2}\vec{\tau}.\vec{W}^\mu + i\frac{g'}{2}YB^\mu, \tag{8.3}$$

where

$$\vec{\tau}.\vec{W}^\mu = \sum_{i=1}^{3} \tau^i W^{i\mu}$$

$$= \tau^1 W^{1\mu} + \tau^2 W^{2\mu} + \tau^3 W^{3\mu}$$

$$= \begin{pmatrix} W^{3\mu} & W^{1\mu} - iW^{2\mu} \\ W^{1\mu} + iW^{2\mu} & -W^{3\mu} \end{pmatrix}. \tag{8.4}$$

We have introduced a triplet of gauge fields W^i_μ corresponding to the generators of $SU(2)_I$ and B_μ, the gauge fields corresponding to the generator Y of $U(1)_Y$. The constants g and g' are the coupling strengths of the $SU(2)$ gauge fields W^i_μ and $U(1)$ gauge field B_μ, respectively. A factor of $\frac{1}{2}$ is introduced with the B_μ field for later convenience and in analogy with the coupling of W^i_μ fields, in the expression for D^μ in Eq. (8.3). The Lagrangian is thus written as:

$$\mathcal{L} = i\overline{\psi}_L \slashed{D}\psi_L + i\overline{e}_R \slashed{D}e_R + i\overline{\nu}_R \slashed{D}\nu_R - \frac{1}{4}B_{\mu\nu}B^{\mu\nu} - \frac{1}{4}G^i_{\mu\nu}G_i^{\mu\nu}, \tag{8.5}$$

where

$$D^\mu\psi_L = \left(\partial^\mu + \frac{ig}{2}\vec{\tau}\cdot\vec{W}^\mu - \frac{ig'}{2}B^\mu\right)\psi_L,$$

$$D^\mu e_R = (\partial^\mu - ig'B^\mu)e_R,$$

$$D^\mu\nu_R = \partial^\mu\nu_R, \tag{8.6}$$

using the values of Y for ψ_L, e_R, and ν_R from Table 8.1 and

$$B_{\mu\nu} = \partial_\mu B_\nu - \partial_\nu B_\mu \tag{8.7}$$

is the kinetic energy of the massless gauge field B^μ.

$$G^i_{\mu\nu} = (\partial_\mu W^i_\nu - \partial_\nu W^i_\mu) + g\epsilon_{ijk}W^j_\mu W^k_\nu, \tag{8.8}$$

which describes the kinetic energy and the self coupling of the massless $W^{\mu i}$ fields. Hence, the Lagrangian in Eq. (8.5) can be written as:

$$\mathcal{L} = i\overline{\psi}_L\slashed{\partial}\psi_L - \frac{g}{2}\overline{\psi}_L\gamma_\mu \begin{pmatrix} W^{3\mu} & W^{1\mu} - iW^{2\mu} \\ W^{1\mu} + iW^{2\mu} & -W^{3\mu} \end{pmatrix}\psi_L + \frac{g'}{2}\overline{\psi}_L\gamma_\mu B^\mu\psi_L$$

$$+ i\overline{e}_R\slashed{\partial}e_R + i\overline{\nu}_R\slashed{\partial}\nu_R + g'\overline{e}_R\gamma_\mu B^\mu e_R$$

$$= i\overline{\psi}_L\slashed{\partial}\psi_L + i\overline{e}_R\slashed{\partial}e_R + i\overline{\nu}_R\slashed{\partial}\nu_R$$

$$- \frac{g}{\sqrt{2}}(\overline{\nu}_L\gamma_\mu W^{\mu+}e_L + \overline{e}_L\gamma_\mu W^{\mu-}\nu_L)$$

$$- \frac{1}{2}\overline{\nu}_L\gamma_\mu(gW^{3\mu} - g'B^\mu)\nu_L + \frac{1}{2}\overline{e}_L\gamma_\mu(gW^{3\mu} + g'B^\mu)e_L + g'\overline{e}_R\gamma_\mu B^\mu e_R. \tag{8.9}$$

iii) The terms in the first line of Eq. (8.9) are the kinetic energy terms of the fields ψ_L, e_R, and ν_R and the terms in the second and third lines describe the interaction of these fields with the gauge fields W_μ^i and B_μ, which are massless. In order to generate the masses of some of the gauge fields corresponding to the vector bosons mediating the weak interactions, a set of scalar fields are introduced in the model. This is done by introducing a doublet of self interacting complex scalar fields $\phi(x)$ given by:

$$\phi(x) = \begin{pmatrix} \phi^+(x) \\ \phi^0(x) \end{pmatrix} = \frac{1}{\sqrt{2}} \begin{pmatrix} \phi_1(x) + i\phi_2(x) \\ \phi_3(x) + i\phi_4(x) \end{pmatrix}, \tag{8.10}$$

where $\phi_i(x)$ $(i = 1 - 4)$ are real fields. The weak isospin I and hypercharge Y quantum numbers of the $\phi(x)$ fields are shown in Table 8.1, for which the interaction Lagrangian is written as:

$$\mathcal{L}_{\text{scalar}}(\phi) = D_\mu \phi^* D^\mu \phi - V(\phi), \tag{8.11}$$

where
$$D^\mu \phi(x) = \left(\partial^\mu + \frac{ig}{2} \vec{\tau} \cdot \vec{W}^\mu + \frac{ig'}{2} B^\mu \right) \phi(x) \tag{8.12}$$

and
$$V(\phi) = \mu^2 \phi^* \phi + \lambda (\phi^* \phi)^2, \ \mu^2 < 0. \tag{8.13}$$

The minimum of the potential $V(\phi)$ occurs at $\langle \phi^* \phi \rangle_0 = -\frac{\mu^2}{2\lambda}$ corresponding to the physical ground state of the scalar fields $\phi(x)$, which is infinitely degenerate for $\mu^2 < 0$. The $SU(2)_I \times U(1)_Y$ symmetry is spontaneously broken through the Higgs mechanism [346] by choosing a particular ground state in such a way that one of the four generators, that is, $\frac{1}{2}\tau_i$ and Y of the $SU(2)_I \times U(1)_Y$ symmetry, corresponding to the electromagnetic gauge fields A_μ, that is, the charge operator $Q = \frac{1}{2}\tau_3 + \frac{Y}{2}$ leaves the ground state invariant, keeping the field A_μ massless. Thus, the $SU(2)_I \times U(1)_Y$ symmetry is broken to a lower symmetry $U(1)_Q$. The other generators like $\frac{1}{2}\tau_1$, $\frac{1}{2}\tau_2$ and $\frac{1}{2}\tau_3 - \frac{Y}{2}$ break the symmetry spontaneously, generating masses for the corresponding gauge fields. This is done by choosing a particular ground state from the infinitely degenerate states to be:

$$\phi_1 = \phi_2 = \phi_4 = 0 \text{ and } \phi_3 \neq 0$$

such that

$$\langle \phi^* \phi \rangle_0 = \frac{1}{2} \langle 0 | (\phi_1^2 + \phi_2^2 + \phi_3^2 + \phi_4^2) | 0 \rangle = -\frac{\mu^2}{2\lambda}, \tag{8.14}$$

implying

$$\langle 0 | \phi(x) | 0 \rangle = \frac{1}{\sqrt{2}} \langle 0 | \phi_3(x) | 0 \rangle = \sqrt{\frac{-\mu^2}{2\lambda}}. \tag{8.15}$$

Using the doublet notation, we can write

$$\langle 0 | \phi(x) | 0 \rangle = \begin{pmatrix} 0 \\ \frac{v}{\sqrt{2}} \end{pmatrix}, \tag{8.16}$$

where $v = \sqrt{\frac{\mu^2}{\lambda}}, \mu^2 > 0$.

Using this choice of the physical ground state, it can be shown that while τ_1, τ_2, τ_3 break the invariance of the vacuum, the operator $Q = T_3 + \frac{Y}{2} = \frac{1}{2}(\tau_3 + Y)$ leaves the vacuum invariant, that is, $Q\langle\phi\rangle_0 = \frac{1}{2}(\tau_3 + Y)\langle\phi\rangle_0 = 0$, by calculating:

$$\tau_1\langle\phi\rangle_0 = \begin{pmatrix} 0 & 1 \\ 1 & 0 \end{pmatrix}\begin{pmatrix} 0 \\ \frac{v}{\sqrt{2}} \end{pmatrix} = \begin{pmatrix} \frac{v}{\sqrt{2}} \\ 0 \end{pmatrix} \neq 0, \tag{8.17}$$

$$\tau_2\langle\phi\rangle_0 = \begin{pmatrix} 0 & -i \\ i & 0 \end{pmatrix}\begin{pmatrix} 0 \\ \frac{v}{\sqrt{2}} \end{pmatrix} = \begin{pmatrix} -\frac{iv}{\sqrt{2}} \\ 0 \end{pmatrix} \neq 0, \tag{8.18}$$

$$\tau_3\langle\phi\rangle_0 = \begin{pmatrix} 1 & 0 \\ 0 & -1 \end{pmatrix}\begin{pmatrix} 0 \\ \frac{v}{\sqrt{2}} \end{pmatrix} = \begin{pmatrix} 0 \\ -\frac{v}{\sqrt{2}} \end{pmatrix} \neq 0, \tag{8.19}$$

$$Y\langle\phi\rangle_0 = +1\langle\phi_0\rangle \neq 0, \tag{8.20}$$

and

$$\frac{1}{2}(\tau_3 + Y)\langle\phi\rangle_0 = 0, \tag{8.21}$$

$$\frac{1}{2}(\tau_3 - Y)\langle\phi\rangle_0 = \begin{pmatrix} 0 & 0 \\ 0 & -1 \end{pmatrix}\begin{pmatrix} 0 \\ \frac{v}{\sqrt{2}} \end{pmatrix} = \begin{pmatrix} 0 \\ -\frac{v}{\sqrt{2}} \end{pmatrix} \neq 0. \tag{8.22}$$

Therefore, by giving $\langle\phi\rangle_0$ a non-zero expectation value, only the generator $Q = T_3 + \frac{Y}{2}$ leaves the ground state invariant. This leads to the breaking of the symmetry $SU(2)_I \times U(1)_Y$, leaving the gauge field corresponding to the generator of the symmetry $U(1)_Q$, that is, the electromagnetic field A_μ massless and giving mass to the other three gauge fields, corresponding to the generators τ_1, τ_2, and $\frac{1}{2}(\tau_3 - Y)$.

iv) Now, we write the interaction Lagrangian by expanding the scalar field $\phi(x)$ around the minimum (v) of the Higgs potential by writing

$$\phi(x) = \frac{1}{\sqrt{2}}\begin{pmatrix} 0 \\ v + H(x) \end{pmatrix}, \tag{8.23}$$

where $H(x)$ is called the Higgs field. In fact, the expression for the scalar field $\phi(x)$ given in Eq. (8.10) can be shown to be equivalent to the field $\phi(x)$ given in Eq. (8.23) using gauge invariance and working in the unitary gauge. In order to show this, using gauge freedom, let us consider:

$$\phi(x) = e^{\frac{i\vec{\alpha}(x)\cdot\vec{\tau}}{2}}\begin{pmatrix} 0 \\ v + H(x) \end{pmatrix} \tag{8.24}$$

and choose $\vec{\alpha}(x) = \frac{2}{v}\vec{\theta}(x)$. For $\theta(x) \leq v$ and $H(x) \leq v$, corresponding to the small perturbation around vacuum results in

$$
\begin{aligned}
\phi(x) &\approx \frac{1}{\sqrt{2}} \begin{pmatrix} 1 + \frac{i\theta_3}{v} & \frac{i(\theta_1 - i\theta_2)}{v} \\ \frac{i(\theta_1 + i\theta_2)}{v} & 1 - \frac{i\theta_3}{v} \end{pmatrix} \begin{pmatrix} 0 \\ v + H(x) \end{pmatrix} \\
&\simeq \frac{1}{\sqrt{2}} \begin{pmatrix} \theta_2 + i\theta_1 \\ v + H(x) - i\theta_3 \end{pmatrix},
\end{aligned} \tag{8.25}
$$

which is the same as Eq. (8.10) for $\theta_1 = \phi_2$, $\theta_2 = \phi_1$, $v + H(x) = \phi_3$, and $\theta_3 = -\phi_4$. Hence, using the gauge freedom, we can choose a gauge transformation U, such that, under this gauge:

$$
\phi(x) \rightarrow \phi'(x) = U\phi(x) = \frac{1}{\sqrt{2}} \begin{pmatrix} 0 \\ v + H(x) \end{pmatrix}. \tag{8.26}
$$

This gauge is called the unitary gauge in which the fields $\theta_i(x)(i = 1, 2, 3)$ are gauged away, except the Higgs field. Strictly speaking, in order to gauge away the gauge fields corresponding to the generators which break the symmetry, we should use τ_1, τ_2, and $\tau_3' = \frac{1}{2}(\tau_3 - Y)$, which is orthogonal to Q, instead of τ_3. However, since τ_3' and τ_3 both break the symmetry, the effect will be the same. Moreover, the situation of the gauge transformation U and the relation between the fields $\vec{\alpha}(x)$ and $\vec{\theta}(x)$ would be different.

v) The interaction Lagrangian is, therefore, written in terms of the scalar field $\phi'(x)$ and the lepton fields ψ_L', e_R', and ν_R' calculated in the unitary gauge, given by:

$$
\phi \quad \rightarrow \quad \phi' = U\phi, \qquad\qquad \psi_L \rightarrow \psi_L' = U\psi_L, \tag{8.27}
$$

$$
\nu_R \quad \rightarrow \quad \nu_R' = U\nu_R, \qquad\qquad e_R \rightarrow e_R' = Ue_R, \tag{8.28}
$$

$$
W_\mu \quad \rightarrow \quad W_\mu' = UW_\mu U^{-1} + \frac{1}{ig}(\partial_\mu U)U^{-1}, \tag{8.29}
$$

$$
B_\mu \quad \rightarrow \quad B_\mu' = B_\mu - \frac{1}{g}\partial_\mu\alpha_3(x). \tag{8.30}
$$

However, we use the earlier notation ϕ, ψ, B_μ etc. for the new fields ϕ', ψ', B_μ' and write the Lagrangian as

$$
\mathcal{L}_{\text{int}}^{\text{WS}} \quad = \quad D^\mu\phi^* D_\mu\phi - \mu^2\phi^*\phi - \lambda(\phi^*\phi)^2 + i\bar{\psi}_L \slashed{D}\psi_L + i\bar{e}_R \slashed{D}e_R. \tag{8.31}
$$

This Lagrangian describes the interaction of the scalar field ϕ and the lepton fields ψ_L, e_R, and ν_R with the gauge fields. We add to this the interaction of the lepton fields ψ_L and e_R with the scalar field $\phi(x)$ assuming Yukawa interaction, that is,

$$
\mathcal{L}_{\text{leptons}} = -f_e(\bar{e}_R\phi^\dagger\psi_L + \bar{\psi}_L\phi e_R), \tag{8.32}
$$

since ν_R is non-interacting. The full W–S Lagrangian is thus written as:

$$\mathcal{L}^{\text{WS}} = i\bar{\psi}_L \slashed{D} \psi_L + i\bar{e}_R \slashed{D} e_R - \frac{1}{4} B_{\mu\nu} B^{\mu\nu} - \frac{1}{4} G^{\mu\nu} G_{\mu\nu}$$

$$+ D^\mu \phi^* D_\mu \phi - \mu^2 \phi^* \phi - \lambda (\phi^* \phi)^2$$

$$- f_e \left(\bar{e}_L \phi e_R + \bar{e}_R \phi^\dagger e_L \right). \tag{8.33}$$

8.3 Predictions of the Weinberg–Salam Model

8.3.1 Masses of gauge bosons

Consider the term $(D^\mu \phi)^* D_\mu \phi$ of the Lagrangian given in Eq. (8.31). Using Eq. (8.12) for $D^\mu(\phi)$ and Eqs. (8.4) and (8.23), we write:

$$D_\mu \phi = \begin{pmatrix} \partial_\mu + \frac{igW_\mu^3 + ig'B_\mu}{2} & ig\frac{W_\mu^1 - iW_\mu^2}{2} \\ ig\frac{W_\mu^1 + iW_\mu^2}{2} & \partial_\mu - \left(\frac{igW_\mu^3 - ig'B_\mu}{2} \right) \end{pmatrix} \begin{pmatrix} 0 \\ \frac{v + H(x)}{\sqrt{2}} \end{pmatrix}$$

$$= \begin{pmatrix} \partial_\mu & 0 \\ 0 & \partial_\mu \end{pmatrix} \begin{pmatrix} 0 \\ \frac{v + H(x)}{\sqrt{2}} \end{pmatrix} + \frac{1}{2} \begin{pmatrix} igW_\mu^3 + ig'B_\mu & ig(W_\mu^1 - iW_\mu^2) \\ ig(W_\mu^1 + iW_\mu^2) & -(igW_\mu^3 - ig'B_\mu) \end{pmatrix} \begin{pmatrix} 0 \\ \frac{v + H(x)}{\sqrt{2}} \end{pmatrix}$$

$$= \frac{1}{\sqrt{2}} \begin{pmatrix} 0 \\ \partial_\mu H(x) \end{pmatrix} + \frac{i}{2} \begin{pmatrix} g(W_\mu^1 - iW_\mu^2)(\frac{v + H(x)}{\sqrt{2}}) \\ (g'B_\mu - gW_\mu^3)(\frac{v + H(x)}{\sqrt{2}}) \end{pmatrix} \tag{8.34}$$

and

$$(D^\mu \phi)^* = \frac{1}{\sqrt{2}} \begin{pmatrix} 0 & \partial^\mu H(x) \end{pmatrix} - \frac{i}{2} \begin{pmatrix} g(W^{1\mu} + iW^{2\mu})(\frac{v + H(x)}{\sqrt{2}}) & (g'B^\mu - gW^{3\mu})\left(\frac{v + H(x)}{\sqrt{2}}\right) \end{pmatrix}. \tag{8.35}$$

The kinetic energy term, $(D^\mu \phi)^* (D_\mu \phi)$ is therefore given by:

$$(D^\mu \phi)^* (D_\mu \phi) = \frac{1}{2} \partial^\mu H(x) \partial_\mu H(x) + \frac{v^2 g^2}{8} (W_\mu^1 W^{1\mu} + W_\mu^2 W^{2\mu})$$

$$+ \frac{v^2}{8} (gW^{3\mu} - g'B^\mu)(gW_\mu^3 - g'B_\mu) + \frac{g^2}{8} (H^2 + 2Hv)$$

$$(W_\mu^1 W^{1\mu} + W_\mu^2 W^{2\mu})$$

$$+ \frac{g^2}{8} W_\mu^3 W^{3\mu} (H^2 + 2Hv) - \frac{gg'}{8} (B_\mu W^{3\mu})(H^2 + 2Hv)$$

$$- \frac{gg'}{8} (B^\mu W_\mu^3)(H^2 + 2Hv) + \frac{g'^2}{8} B_\mu B^\mu (H^2 + 2Hv). \tag{8.36}$$

We define,

$$W_\mu^\pm = \frac{W_\mu^1 \mp iW_\mu^2}{\sqrt{2}} \tag{8.37}$$

such that

$$W_\mu^1 W^{\mu 1} + W_\mu^2 W^{\mu 2} = |W_\mu^+|^2 + |W_\mu^-|^2 \quad (\because (W_\mu^+)^\dagger = (W_\mu^-))$$

and Eq. (8.36) may be rewritten as:

$$
\begin{aligned}
(D^\mu \phi)^*(D_\mu \phi) = & \frac{1}{2}\partial^\mu H(x)\partial_\mu H(x) + \frac{v^2 g^2}{8}(|W_\mu^+|^2 + |W_\mu^-|^2) \\
& + \frac{v^2}{8}(gW^{3\mu} - g'B^\mu)(gW_\mu^3 - g'B_\mu) + \frac{g^2}{8}(H^2 + 2Hv) \\
& \quad (|W_\mu^+|^2 + |W_\mu^-|^2) \\
& + \frac{g^2}{8}W_\mu^3 W^{3\mu}(H^2 + 2Hv) - \frac{gg'}{8}(B_\mu W^{3\mu})(H^2 + 2Hv) \\
& - \frac{gg'}{8}(B^\mu W_\mu^3)(H^2 + 2Hv) + \frac{g'^2}{8}B_\mu B^\mu(H^2 + 2Hv). \tag{8.38}
\end{aligned}
$$

We see that W^+ and W^- have acquired a mass $M_{W^\pm} = \frac{vg}{2}$. Defining the normalized orthogonal combinations of W_μ^3 and B_μ as:

$$Z_\mu = \frac{gW_\mu^3 - g'B_\mu}{\sqrt{g^2 + g'^2}}, \tag{8.39}$$

$$A_\mu = \frac{g'W_\mu^3 + gB_\mu}{\sqrt{g^2 + g'^2}}, \tag{8.40}$$

we rewrite Eq. (8.38) as:

$$
\begin{aligned}
(D^\mu \phi)^*(D_\mu \phi) = & \frac{1}{2}\partial^\mu H(x)\partial_\mu H(x) + \frac{v^2 g^2}{8}(|W_\mu^+|^2 + |W_\mu^-|^2) \\
& + \frac{g^2}{8}(H^2 + 2Hv)(|W_\mu^+|^2 + |W_\mu^-|^2) + \left(\frac{g^2 + g'^2}{4}\right) \\
& \left(\frac{H^2 + 2Hv + v^2}{2}\right)Z_\mu Z^\mu \\
& + \left[\frac{g^2 g'^2}{4(g^2 + g'^2)}(H^2 + 2Hv) - \frac{g^2 g'^2}{4(g^2 + g'^2)}(H^2 + 2Hv)\right]A_\mu A^\mu. \tag{8.41}
\end{aligned}
$$

We see from Eq. (8.41) that the field A_μ remains massless, which is identified with the electromagnetic field of the photon, while Z_μ acquires a mass $M_Z = \frac{\sqrt{g^2 + g'^2}}{2}v$, such that $\frac{M_Z}{M_W} = \sqrt{1 + \frac{g'^2}{g^2}} \geq 1$. The model predicts the relative magnitude of M_W and M_Z, but to obtain their absolute values, one needs to know the values of g and g' and the vacuum expectation value (VEV) of the scalar field ($\langle \phi \rangle_0 = \langle v \rangle$) from the phenomenology of the weak and electromagnetic processes.

8.3.2 Charged current weak interactions and electromagnetic interactions

In order to determine the couplings g and g' of the gauge bosons W_μ^i ($i = 1, 2, 3$) and B_μ, we need to derive the electromagnetic and weak interactions of these fields with weak and electromagnetic lepton currents. For this purpose, consider the last two terms of the Lagrangian given in Eq. (8.31), that is,

$$\mathcal{L} = i\bar{\psi}_L \gamma^\mu (\partial_\mu + \frac{ig}{2}\vec{\tau}.\vec{W}_\mu - \frac{ig'}{2}B_\mu)\psi_L + i\bar{e}_R\gamma^\mu(\partial_\mu - ig'B_\mu)e_R + i\bar{\nu}_R\gamma_\mu\partial^\mu\nu_R.$$

The terms describing the interaction of leptons with the gauge fields are:

$$
\begin{aligned}
\mathcal{L}_{\text{int}} &= \bar{\psi}_L\gamma^\mu\frac{1}{2}(g'B_\mu - g\vec{\tau}.\vec{W}_\mu)\psi_L + g'\bar{e}_R\gamma^\mu B_\mu e_R \\
&= (\bar{\nu}_L \quad \bar{e}_L)\,\gamma_\mu\frac{1}{2}\begin{pmatrix} g'B_\mu - gW_\mu^3 & g(-W_\mu^1 + iW_\mu^2) \\ -g(W_\mu^1 + iW_\mu^2) & g'B_\mu + gW_\mu^3 \end{pmatrix}\begin{pmatrix} \nu_L \\ e_L \end{pmatrix} + g'\bar{e}_R\gamma_\mu B_\mu e_R \\
&= -\frac{g}{\sqrt{2}}\bar{\nu}_L\gamma_\mu W_\mu^+ e_L + \frac{1}{2}\bar{\nu}_L\gamma^\mu(g'B_\mu - gW_\mu^3)\nu_L - \frac{g}{\sqrt{2}}\bar{e}_L\gamma_\mu W_\mu^- \nu_L \\
&\quad + \frac{1}{2}\bar{e}_L\gamma^\mu(g'B_\mu + gW_\mu^3)e_L + g'\bar{e}_R\gamma_\mu B_\mu e_R \\
&= -\frac{g}{\sqrt{2}}\bar{\nu}_L\gamma_\mu W_\mu^+ e_L - \frac{g}{\sqrt{2}}\bar{e}_L\gamma_\mu W_\mu^- \nu_L + \frac{g'}{2}(\bar{\nu}_L\gamma_\mu B_\mu\nu_L + \bar{e}_L\gamma_\mu B_\mu e_L + 2\bar{e}_R\gamma_\mu B_\mu e_R) \\
&\quad + \frac{g}{2}(\bar{e}_L\gamma_\mu W_\mu^3 e_L - \bar{\nu}_L\gamma_\mu W_\mu^3\nu_L).
\end{aligned}
$$

Since,

$$W_\mu^3 = \frac{gZ_\mu + g'A_\mu}{\sqrt{g^2 + g'^2}},$$

$$B_\mu = \frac{gA_\mu - g'Z_\mu}{\sqrt{g^2 + g'^2}}. \tag{8.42}$$

We obtain:

$$
\begin{aligned}
\mathcal{L}_{\text{int}} &= -\frac{g}{2\sqrt{2}}\left(\bar{\nu}_e\gamma^\mu(1 - \gamma_5)eW_\mu^+ + \bar{e}\gamma^\mu(1 - \gamma_5)\nu_e W_\mu^-\right) - \frac{\sqrt{g^2 + g'^2}}{2}\bar{\nu}_L\gamma^\mu\nu_L Z_\mu \\
&\quad + \frac{gg'}{\sqrt{g^2 + g'^2}}\bar{e}\gamma^\mu e A_\mu + \frac{Z_\mu}{\sqrt{g^2 + g'^2}}[-g'^2\bar{e}_R\gamma^\mu e_R + \frac{g^2 - g'^2}{2}\bar{e}_L\gamma^\mu e_L]. \tag{8.43}
\end{aligned}
$$

From Eq. (8.43), we see that the charged current weak interactions in which the charged raising (lowering) lepton current couples to the W_μ^+ and W_μ^- fields, is given by the Lagrangian:

$$\mathcal{L}_{\text{int}}^{\text{CC}} = -\frac{g}{2\sqrt{2}}\bar{\nu}_e\gamma^\mu(1 - \gamma_5)eW_\mu^+ + \text{h.c.}, \tag{8.44}$$

where

$$\left(\frac{g}{2\sqrt{2}}\right)^2 \frac{1}{M_W^2} = \frac{G_F}{\sqrt{2}} \qquad \text{i.e.} \quad g^2 = 4\sqrt{2}M_W^2 G_F = \sqrt{2}v^2 g^2 G_F. \tag{8.45}$$

This leads to the determination of the vacuum expectation value v in terms of the weak Fermi coupling constant G_F, that is, $v = (\sqrt{2}G_F)^{-\frac{1}{2}} \simeq 246$ GeV using $G_F \approx 10^{-5}$ GeV^{-2}. The electromagnetic current $\bar{e}\gamma^\mu e$ couples with the electromagnetic field A_μ through the interaction Lagrangian given by:

$$\mathcal{L}_{\text{int}} = \frac{gg'}{\sqrt{g^2 + g'^2}}\bar{e}\gamma^\mu e A_\mu \tag{8.46}$$

implying that $\dfrac{gg'}{\sqrt{g^2+g'^2}} = e$ (electronic charge), leading to a relation between $e, g,$ and g', that is,

$$\frac{1}{e^2} = \frac{1}{g^2} + \frac{1}{g'^2}. \tag{8.47}$$

8.3.3   Neutral current interaction and weak mixing angle

We see from the Lagrangian given by Eq (8.43), that the W–S model predicts the existence of a neutral heavy boson Z^μ which interacts via the neutral current carried by neutrinos and electrons. The neutral current interaction Lagrangians for neutrinos and electrons are given by:

$$\mathcal{L}_{\text{NC}}^\nu = -\frac{\sqrt{g^2+g'^2}}{2}\bar{\nu}_L\gamma^\mu\nu_L Z_\mu, \qquad\qquad \text{(for neutrinos)} \tag{8.48}$$

$$\mathcal{L}_{\text{NC}}^e = \frac{Z_\mu}{\sqrt{g^2+g'^2}}\left[-g'^2\bar{e}_R\gamma^\mu e_R + \frac{g^2-g'^2}{2}\bar{e}_L\gamma^\mu e_L\right], \quad \text{(for electrons)} \tag{8.49}$$

in which the strength of the coupling Z_μ is different for $\nu_L, e_L,$ and e_R. Obviously, there is no interaction of ν_R with any of the gauge fields or electrons.

It is convenient to parameterize the orthogonal and normalized mixing of the neutral gauge bosons in terms of an angle θ_W, known as the weak mixing angle, such that:

$$Z_\mu = \cos\theta_W W_{3\mu} - \sin\theta_W B_\mu \quad \text{and} \tag{8.50}$$

$$A_\mu = \sin\theta_W W_{3\mu} + \cos\theta_W B_\mu, \tag{8.51}$$

where

$$\cos\theta_W = \frac{g}{\sqrt{g^2+g'^2}}, \tag{8.52}$$

$$\sin\theta_W = \frac{g'}{\sqrt{g^2+g'^2}}. \tag{8.53}$$

Since $e = \dfrac{gg'}{\sqrt{g^2+g'^2}} = g\sin\theta_W = g'\cos\theta_W$, we obtain:

$$g = \frac{e}{\sin\theta_W}, \qquad g' = \frac{e}{\cos\theta_W}, \qquad \text{and} \qquad M_W = M_Z\cos\theta_W, \tag{8.54}$$

implying that $g > e$ and $g' > e$. This implies that the gauge field couplings of the physical $W^{\pm}$ and Z bosons to the lepton currents are $\frac{g}{2\sqrt{2}}$ and $\frac{g}{2\cos\theta_W}$, respectively. Therefore, the intrinsic weak couplings of gauge bosons are not small as compared to the electromagnetic coupling; however, the effective couplings are small because of the large mass of W and Z bosons. Moreover, the masses of the W and Z bosons can be predicted in terms of the weak mixing angle using Eqs. (8.45) and (8.54),

$$M_W^2 = \frac{g^2}{4\sqrt{2}G_F} = \frac{e^2}{4\sqrt{2}G_F \sin^2\theta_W} = \frac{\pi\alpha}{\sqrt{2}G_F \sin^2\theta_W}$$

$$\Rightarrow \qquad M_W = \frac{37.3}{\sin\theta_W} \text{ GeV,} \tag{8.55}$$

$$\text{and} \qquad M_Z = \frac{M_W}{\cos\theta_W} = \frac{37.3}{\sin\theta_W \cos\theta_W} \text{ GeV.} \tag{8.56}$$

It is convenient to express the W–S Lagrangian in terms of the electromagnetic and weak currents and their couplings to the vector boson fields $W_\mu^{\pm}$, Z_μ, and A_μ, that is, e, g, and θ_W as:

$$
\begin{aligned}
\mathcal{L}^{\text{WS}} &= -\frac{g}{\sqrt{2}} \left(\bar{\nu}_L \gamma^\mu e_L W_\mu^+ + \bar{e}_L \gamma^\mu \nu_L W_\mu^- \right) + e\, \bar{e} \gamma^\mu e A_\mu - \frac{g}{2\cos\theta_W} \bar{\nu}_L \gamma^\mu \nu_L Z_\mu \\
&\quad - \frac{g}{2\cos\theta_W} \left(2\sin^2\theta_W \bar{e}_R \gamma^\mu e_R + (2\sin^2\theta_W - 1)\bar{e}_L \gamma^\mu e_L \right) Z_\mu, \\
&= -\frac{g}{2\sqrt{2}} \left[\bar{\nu}_e \gamma^\mu (1 - \gamma_5) e W_\mu^+ + \bar{e}\gamma^\mu (1 - \gamma_5)\nu_e W_\mu^- \right] - \frac{g}{4\cos\theta_W} \bar{\nu}_e \gamma^\mu \\
&\quad (1 - \gamma_5)\nu_e Z_\mu \\
&\quad + e\, \bar{e}\gamma^\mu e A_\mu - \frac{g}{4\cos\theta_W} \left[2\sin^2\theta_W \bar{e}\gamma^\mu (1 + \gamma_5)e + (2\sin^2\theta_W - 1) \right. \\
&\quad \left. \bar{e}\gamma^\mu (1 - \gamma_5)e \right] Z_\mu.
\end{aligned}
$$

Therefore, in the W–S model

$$\mathcal{L}^{\text{WS}} = \mathcal{L}^{\text{em}} + \mathcal{L}^{\text{weak}},$$
$$\text{where } \mathcal{L}^{\text{weak}} = \mathcal{L}^{\text{CC}} + \mathcal{L}^{\text{NC}}$$

and

$$\mathcal{L}^{\text{em}} = e\, \bar{e}\gamma^\mu e A_\mu, \tag{8.57}$$

$$\mathcal{L}^{\text{CC}} = -\frac{g}{2\sqrt{2}} \left[\bar{\nu}_e \gamma^\mu (1 - \gamma_5) e W_\mu^+ + \text{h.c.} \right], \tag{8.58}$$

$$\mathcal{L}^{\text{NC}} = -\frac{g}{4\cos\theta_W} [\bar{\nu}_e \gamma^\mu (1 - \gamma_5)\nu_e + \bar{e}\gamma^\mu (g_V^e - g_A^e \gamma_5)e] Z_\mu, \tag{8.59}$$

with

$$g_V^e = 4\sin^2\theta_W - 1, \qquad g_A^e = -1, \tag{8.60}$$

$$\frac{g}{2\sqrt{2}} = \left(\frac{G_F M_W^2}{\sqrt{2}}\right)^{\frac{1}{2}}, \tag{8.61}$$

$$\frac{g}{4\cos\theta_W} = \frac{1}{\sqrt{2}}\left(\frac{G_F M_Z^2}{\sqrt{2}}\right)^{\frac{1}{2}}. \tag{8.62}$$

8.3.4 Mass of the electron

The mass of the electron is generated through the Yukawa coupling of the electron with the scalar field (ϕ) by introducing an interaction Lagrangian, with a coupling strength f_e, given by:

$$
\begin{aligned}
\mathcal{L}_{\text{int}} &= -f_e \bar{\psi}_L \phi \psi_R + \text{h.c.} \\
&= -f_e \left(\bar{\nu}_L \ \bar{e}_L\right) \begin{pmatrix} 0, \\ \frac{v+H(x)}{\sqrt{2}} \end{pmatrix} e_R + \text{h.c.}, \\
&= -f_e \frac{v+H(x)}{\sqrt{2}} \bar{e}_L e_R + \text{h.c.}, \\
&= -\frac{f_e v}{\sqrt{2}}(\bar{e}_L e_R + \bar{e}_R e_L) - \frac{f_e H(x)}{\sqrt{2}}(\bar{e}_L e_R + \bar{e}_R e_L), \\
&= -\frac{f_e v}{\sqrt{2}}\bar{\psi}_e \psi_e - \frac{f_e H(x)}{\sqrt{2}}\bar{\psi}_e \psi_e.
\end{aligned}
\tag{8.63}
$$
$$\tag{8.64}$$

The mass of the electron is predicted to be:

$$m_e = \frac{f_e v}{\sqrt{2}} = \frac{f_e}{\sqrt{2}(G_F\sqrt{2})^{\frac{1}{2}}}. \tag{8.65}$$

The strength of the coupling f_e can be experimentally determined from the coupling of the Higgs field($H(x)$) to $e^- e^+$ pairs, that is, Higgs $\rightarrow e^- e^+$. Theoretically, f_e can be determined from the charged weak decays of $W^\pm$, that is,

$$f_e = \frac{\sqrt{2} m_e}{v} = 2^{\frac{3}{4}} m_e (G_F)^{\frac{1}{2}} \simeq 3 \times 10^{-6}. \tag{8.66}$$

8.4 Extension to the Leptons of Other Flavors

In preceding sections, the essentials of the unified gauge theory in the W–S model are described for electrons and neutrinos. This can be extended to other flavors of the leptons. In this section, we consider the extension of the W–S model to other flavors of the leptons in a formalism which can then be applied to the quark sector in a straightforward way.

We write the Lagrangian for the interaction of leptons $(e^-\ \nu_e)$, $(\mu^-\ \nu_\mu)$, and $(\tau^-\ \nu_\tau)$ with the gauge fields in terms of the covariant derivatives D^μ, making the Lagrangian invariant under the local $SU(2)_I \times U(1)_Y$ gauge symmetry as:

$$\mathcal{L}_{\text{leptons}}^{\text{int}} = \sum_{f=e,\mu,\tau} \bar{l}_{f_L} i \not{D} l_{f_L} + \sum_{f=e,\mu,\tau} \bar{l}_{f_R} \not{D} l_{f_R}, \tag{8.67}$$

where l_{f_L} is the left-handed SU(2) doublet of leptons of flavor f, that is,

$$l_{f_L} = \begin{pmatrix} \nu_{f_L} \\ f_L \end{pmatrix} = \begin{pmatrix} \nu_e \\ e^- \end{pmatrix}_L, \begin{pmatrix} \nu_\mu \\ \mu^- \end{pmatrix}_L \text{ and } \begin{pmatrix} \nu_\tau \\ \tau^- \end{pmatrix}_L \tag{8.68}$$

and l_{f_R} is the right-handed SU(2) singlet of leptons of flavor f, that is,

$$l_{f_R} = \nu_{e_R},\ e_R^-,\ \nu_{\mu_R},\ \mu_R^-,\ \nu_{\tau_R},\ \tau_R^-. \tag{8.69}$$

The $\sum_{f=e,\mu,\tau}$ implies the sum over all the doublets and singlets corresponding to the flavor $f(= e,\mu,\tau)$. The weak isospin, weak hypercharge, and electric charge of all the leptons discussed in this section are given in Table 8.1. The covariant derivatives D^μ for the doublet and singlet fields l_{f_L} and l_{f_R} are defined as:

$$D^\mu = \left(\partial^\mu + ig \frac{\vec{\tau}.\vec{W}^\mu}{2} + ig' \frac{Y_L}{2} B^\mu \right) \tag{8.70}$$

for all the SU(2) doublets L of flavor f and

$$D^\mu = \partial^\mu + ig' \frac{Y_R}{2} B^\mu \tag{8.71}$$

for all the SU(2) singlets R of flavor f, where $\vec{W}^\mu$ and B^μ are the isovector and isoscalar gauge fields corresponding to $SU(2)_I$ and $U(1)_Y$, respectively. From Eqs. (8.67), (8.70), and (8.71), we obtain the following Lagrangian for the interaction of all the lepton flavors f with the gauge fields W^μ and B^μ as:

$$\mathcal{L}_{\text{leptons}}^{\text{int}} = - \sum_{f=e,\mu,\tau} \left[\bar{l}_{f_L} \left(\frac{g}{2} \vec{\tau}.\vec{W}^\mu + g' \frac{Y^L}{2} B^\mu \right) \gamma_\mu f_{f_L} + \bar{l}_{f_R} g' \frac{Y^R}{2} B^\mu \gamma_\mu l_{f_R} \right]. \tag{8.72}$$

Using

$$\frac{\vec{\tau}.\vec{W}^\mu}{2} = \frac{1}{\sqrt{2}} \left[\tau_+ \frac{W_1^\mu - iW_2^\mu}{\sqrt{2}} + \tau_- \frac{W_1^\mu + iW_2^\mu}{\sqrt{2}} \right] + \frac{\tau_3}{2} W_3^\mu, \tag{8.73}$$

where $\tau_+ = \frac{\tau_1 + i\tau_2}{2}$ and $\tau_- = \frac{\tau_1 - i\tau_2}{2}$ are the isospin raising and lowering operators and $\frac{W_1^\mu \mp iW_2^\mu}{\sqrt{2}}$ are the vector field components which create $W^\pm$ vector bosons.

The charge current weak interaction Lagrangian using Eq. (8.4) is given by:

$$\mathcal{L}_{\text{cc}}^{\text{int}} = - \frac{g}{\sqrt{2}} \sum_{f=e,\mu,\tau} \left[\bar{l}_{f_L} \gamma^\mu \tau^+ l_{f_L} W_\mu^+ + \bar{l}_{f_L} \gamma^\mu \tau^- l_{f_L} W_\mu^- \right]. \tag{8.74}$$

Using $f_L = \frac{1-\gamma_5}{2}f$ and $\overline{f}_L = \frac{1+\gamma_5}{2}\overline{f}$, we find

$$\mathcal{L}_{cc}^{int} = -\frac{g}{2\sqrt{2}} \sum_{f=e,\mu,\tau} \left[\overline{\nu}_f \gamma^\mu (1-\gamma_5) f W_\mu^+ + \overline{f}\gamma^\mu(1-\gamma_5)\nu_f W_\mu^- \right] \tag{8.75}$$

$$= -\frac{g}{2\sqrt{2}} \sum_{f=e,\mu,\tau} \left[j_f^\mu W_\mu^+ + \text{h.c.} \right], \tag{8.76}$$

where $j_\mu^f = \overline{\nu}_f \gamma^\mu (1-\gamma_5)f$ and $\frac{g}{2\sqrt{2}}$ is the magnitude of the coupling of the vector and the axial vector charged lepton currents for each flavor e, ν, τ, giving in a natural way, the lepton universality. It also reproduces the relative sign between the vector and the axial vector couplings. The strength of the coupling g is determined in terms of the universal Fermi constant G_F as:

$$\frac{g^2}{8M_W^2} = \frac{G_F}{\sqrt{2}}, \quad \text{that is,} \quad g = 2\sqrt{G_F(M_W)^2\sqrt{2}}. \tag{8.77}$$

The lepton universality of the charged current vector interaction is therefore inbuilt in the model.

Similarly from Eq. (8.4), the neutral current interaction Lagrangian is given by:

$$\mathcal{L}_{NC}^{leptons} = - \sum_{f=e,\mu,\tau} \left[\overline{l}_{f_L} \left(g\frac{\tau_3}{2}W^3 + g'\frac{Y^L}{2}B \right) l_{f_L} + \overline{l}_{f_R} g'\frac{Y^R}{2}B l_{f_R} \right] \tag{8.78}$$

Since $\tau_3 l_{f_R} = 0$ and τ_3 acts only on the doublet L, we can write:

$$\mathcal{L}_{NC}^{leptons} = - \sum_{f=e,\mu,\tau} \sum_{i=L,R} \overline{l}_{f_i} \left[g\frac{\tau_3^{if}}{2}W^3 + g'\frac{Y^{if}}{2}B \right] l_{f_i}.$$

We have

$$Q^{if} = \frac{1}{2}\tau_3^{if} + \frac{Y^{if}}{2}. \tag{8.79}$$

Here,

$$Q^{if} = Q^{fi} \text{ for example, if } f = e, \mu, \tau, \; i = L, R; \; Q^{fi} = -1$$
$$\text{and } Y^{if} = Y^{fi} \text{ (see Table 8.1).} \tag{8.80}$$

We now express $W^{3\mu}$ and B^μ in terms of the neutral bosons Z^μ and A^μ using Eqs. (8.42), (8.52) and (8.53) and write:

$$\mathcal{L}_{NC}^{leptons} = -\sum_{f,i} \overline{l}_{f_i}\gamma_\mu \left[g\frac{\tau_3^{if}}{2}(\cos\theta_W Z^\mu + \sin\theta_W A^\mu) + g'(Q^{if} - \frac{\tau_3^{if}}{2}) \right.$$
$$\left. (-\sin\theta_W Z^\mu + \cos\theta_W A^\mu) \right] l_{f_i}$$

$$= -\sum_{f,i} \overline{l}_{f_i}\gamma_\mu \left[A^\mu (g\sin\theta_W \frac{\tau_3^{if}}{2} - g'\cos\theta_W \frac{\tau_3^{if}}{2} + g'Q^{if}\cos\theta_W) \right.$$

$$+ Z^\mu (g \cos\theta_W \frac{\tau_3^{if}}{2} - g' \sin\theta_W (Q^{if} - \frac{\tau_3^{if}}{2})) \Big] l_{f_i}. \tag{8.81}$$

Since $g \sin\theta_W = g' \cos\theta_W = e$, we obtain:

$$\mathcal{L}_{\text{NC}}^{\text{leptons}} = -\sum_{f,i} \bar{l}_{f_i} \gamma_\mu \Big[A^\mu g' \cos\theta_W Q^{if} + Z^\mu$$

$$\Big(g \cos\theta_W \frac{\tau_3^{if}}{2} - g' \sin\theta_W Q^{if} + g' \sin\theta_W \frac{\tau_3^{if}}{2} \Big) \Big] l_{f_i}.$$

The first term on the right-hand side is the interaction of the electromagnetic field A^μ with a lepton,

$$\mathcal{L}_{\text{em}}^{\text{leptons}} = -e j_\mu A^\mu,$$

$$\text{where} \qquad j_\mu = \sum_{f,i} \bar{l}_{f_i} \gamma_\mu Q^{if} l_{f_i} = -\sum_f \bar{l}_f \gamma_\mu l_f \tag{8.82}$$

for $Q^i = Q = -1$, with $g' \cos\theta_W = e = g \sin\theta_W$. The second term in Eq. (8.81) given by:

$$\mathcal{L}_{\text{NC}}^{\text{leptons}} = -\sum_{f,i} Z^\mu \bar{l}_{f_i} \gamma_\mu \left((g \cos\theta_W + g' \sin\theta_W) \frac{\tau_3^{if}}{2} - Q^{if} g' \sin\theta_W \right) l_{f_i}, \tag{8.83}$$

can be simplified using $g \sin\theta_W = g' \cos\theta_W = e$, to give:

$$\mathcal{L}_{\text{NC}}^{\text{leptons}} = -\frac{e}{2 \sin\theta_W \cos\theta_W} \sum_{f,i} Z^\mu \bar{l}_{f_i} \gamma^\mu \left(\tau_3^{if} - 2Q^{if} \sin^2\theta_W \right) l_{f_i}$$

$$= -\frac{e}{2 \sin\theta_W \cos\theta_W} j_\mu^Z Z^\mu, \tag{8.84}$$

where

$$j_\mu^Z = \sum_{f,i} \bar{l}_{f_i} \gamma_\mu (\tau_3^{if} - 2Q^{if} \sin^2\theta_W) l_{f_i}$$

$$= \sum_f \bar{l}_f (\tau_3^f - 2Q^f \sin^2\theta_W) \gamma_\mu \frac{1 - \gamma_5}{2} l_f + \bar{l}_f (-2Q^f \sin^2\theta_W) \gamma_\mu \frac{1 + \gamma_5}{2} l_f$$

$$= \sum_f \bar{l}_f \gamma_\mu \left[\left(\frac{1}{2} \tau_3^f - 2Q^f \sin^2\theta_W \right) - \frac{1}{2} \tau_3^f \gamma_5 \right] l_f$$

$$= \sum_f \bar{l}_f \gamma_\mu (g_V^f - g_A^f \gamma_5) l_f. \tag{8.85}$$

Using

$$g_V^f = \frac{1}{2} \tau_3^f - 2Q^f \sin^2\theta_W, \tag{8.86}$$

$$g_A^f = \frac{1}{2} \tau_3^f, \tag{8.87}$$

we may write Eq. (8.85) as

$$j_\mu^Z = \sum_f \bar{l}_f \gamma_\mu \left[\frac{1}{2}(1-\gamma_5)\tau_3^f - 2Q^f \sin^2\theta_W \right] l_f,$$

$$= j_\mu^3 - 2\sin^2\theta_W j_\mu^{em}. \qquad (8.88)$$

Now writing explicitly in terms of the fermions, for neutrinos with $Q = 0$, $\tau_3 = +1$ and for the charged leptons e^-, μ^-, and τ^- with $Q = -1$, $\tau_3 = -1$, we get:

$$j_\mu^Z(\nu_f) = \sum_{f=e,\,\mu,\,\tau} \bar{\nu}_f \gamma_\mu \frac{(1-\gamma_5)}{2} \nu_f$$

$$j_\mu^Z(\text{charged leptons}) = \sum_{f=e,\,\mu,\,\tau} \bar{f} \gamma_\mu \left[C_L^f \frac{(1-\gamma_5)}{2} + C_R^f \frac{(1+\gamma_5)}{2} \right] f,$$

$$\text{where } C_L^f = -1 + 2\sin^2\theta_W,$$

$$C_R^f = +2\sin^2\theta_W.$$

Equivalently, we can write:

$$j_\mu^Z(\text{charged leptons}) = \sum_{f=e,\,\mu,\,\tau} \frac{1}{2}\bar{f}\gamma_\mu \left[(-1 + 4\sin^2\theta_W) + \gamma_5 \right] f$$

$$= \sum_{f=e,\,\mu,\,\tau} \bar{f} \left[\gamma_\mu(g_V^f - g_A^f \gamma_5) \right] f,$$

where $g_V^f = \frac{-1+4\sin^2\theta_W}{2}$ and $g_A^f = -\frac{1}{2}$. The values of g_V and g_A for the charged leptons and neutrinos are given in Table 8.2.

Again, the lepton universality of neutral current weak interaction is inbuilt in the model.

Table 8.2 Couplings of the leptons to Z_μ field.

States	g_V	g_A
ν_l	$1/2$	$1/2$
l	$-\frac{1}{2} + 2\sin^2\theta_W$	$-1/2$

8.4.1 Extension to the quark sector using GIM mechanism

We have seen in Chapter 6 that the weak interaction of (u, d) quarks is described in terms of the quark doublet with the Cabibbo rotated quark states, that is, (u, d'), with $d' = \cos\theta_C d + \sin\theta_C s$, where θ_C is the Cabibbo angle and s is the strange quark. The weak charged currents are, therefore, written as:

$$j^\mu(u, d, s) = \frac{1}{2}\bar{u}\gamma^\mu(1-\gamma_5)d'$$

$$= \frac{1}{2} \left[\cos\theta_C \bar{u}\gamma^\mu(1-\gamma_5)d + \sin\theta_C \bar{u}\gamma^\mu(1-\gamma_5)s \right],$$

which describes the weak charged currents in the $\Delta S = 0$ sector as well as in the $\Delta S \neq 0$ sector satisfying $|\Delta S| = 1$ and the $\Delta S = \Delta Q$ rule. It also predicts the existence of neutral current processes like $K_L^0 \to l^+l^-$ and $K^\pm \to \pi^\pm \nu \bar{\nu}$. In order to address the problem of stringent limits on the non-existence of neutral currents in the strangeness sector and maintain the quark–lepton symmetry proposed earlier by Bjorken and Glashow [65], Glashow, Ilioupolis, and Maiani [64] proposed the existence of a fourth quark c, as discussed in Chapter 6. Accordingly, the weak interaction of quarks can be described in terms of three quark doublets with left-handed quarks:

$$\begin{pmatrix} u \\ d' \end{pmatrix}_L, \quad \begin{pmatrix} c \\ s' \end{pmatrix}_L, \quad \begin{pmatrix} t \\ b' \end{pmatrix}_L \tag{8.89}$$

with

$$\begin{pmatrix} d' \\ s' \\ b' \end{pmatrix} = \begin{pmatrix} U_{ud} & U_{us} & U_{ub} \\ U_{cd} & U_{cs} & U_{cb} \\ U_{td} & U_{ts} & U_{tb} \end{pmatrix} \begin{pmatrix} d \\ s \\ b \end{pmatrix}. \tag{8.90}$$

In this section, we apply the formalism developed in the last section to the quark sector. The weak isospin, weak hypercharge, and electric charge of all the quarks discussed in this section are given in Table 8.3.

Table 8.3 Isospin (I), its third component (I_3), charge ($Q(|e|)$), and hypercharge ($Y = 2(Q - I_3)$) of the first and second generation of quarks, that is, (u, d) and (c, s) quarks.

Name	I	I_3	$Q(|e|)$	Y
$u_L(c_L)$	1/2	+1/2	+2/3	1/3
$d_L(s_L)$	1/2	−1/2	−1/3	1/3
$u_R(c_R)$	0	0	+2/3	4/3
$d_R(s_R)$	0	0	−1/3	−2/3

We first write the weak charged current interaction Lagrangian, in the four quark sector, in analogy with the Lagrangian for the weak interaction for leptons (Eq. (8.74)),

$$\mathcal{L}_{cc}^{int}(\text{quarks}) = -\frac{g}{\sqrt{2}} \sum_q \left(\bar{q}_L \gamma^\mu \tau^+ q_L W_\mu^+ + \bar{q}_L \gamma^\mu \tau^- q_L W_\mu^- \right) \tag{8.91}$$

$$= -\frac{g}{2\sqrt{2}} \left[\left(\bar{u} \gamma^\mu (1 - \gamma^5) d' + \bar{c} \gamma^\mu (1 - \gamma^5) s' \right) W_\mu^+ + \text{h.c.} \right], \tag{8.92}$$

The neutral current weak interaction Lagrangian for the quarks is written in analogy with the neutral weak interaction Lagrangian for the charged leptons (Eq. (8.84))as:

$$\mathcal{L}_{int}^{NC} = -j_\mu^{NC}(\text{quark}) Z^\mu, \tag{8.93}$$

with $\quad j_\mu^{NC}(\text{quark}) = \dfrac{e}{2 \sin \theta_W \cos \theta_W} \sum_q \bar{q} \gamma^\mu (g_V^q - g_A^q \gamma_5) q, \tag{8.94}$

where $\quad g_V^q = \dfrac{1}{2} \tau_3^q - 2 \sin^2 \theta_W Q_q \quad$ and $\tag{8.95}$

$$g_A^q = \frac{1}{2} \tau_3^q. \tag{8.96}$$

For example, the values of g_V and g_A for u and d quarks are given in Table 8.4.

Table 8.4 Couplings of the quarks (u, d) to Z_μ field

States	g_V	g_A
u	$\frac{1}{2} - \frac{4}{3}\sin^2\theta_W$	$1/2$
d	$-\frac{1}{2} + \frac{2}{3}\sin^2\theta_W$	$-1/2$

Explicitly for the case of two flavors, to demonstrate the application of the GIM mechanism and to explain the absence of neutral currents in the $\Delta S = 1$ sector, let us consider the neutral current in the case of two flavors,

$$
J_{\text{NC}}^\mu(u, c, d', s') = \sum_{q=u,c,d',s'} \bar{q} O^q q; \text{ where } O^q = \gamma^\mu(g_V^q - g_A^q \gamma_5),
$$

$$
= \bar{u} O^u u + \bar{c} O^c c + \bar{d} O^{d'} d + \bar{s} O^{s'} s. \tag{8.97}
$$

Since τ_3^q and Q^q are the same for d' and s', both corresponding to $\tau_3 = 2I_3 = -1$, $Q = -\frac{1}{3}$, we can write $O^{d'} = O^{s'}$. Therefore, $\bar{d} O^{d'} d + \bar{s} O^{s'} s = \bar{d} O^d d + \bar{s} O^s s$ and the neutral current

$$
J_{\text{NC}}^\mu(u, c, d', s') = \bar{u} O^u u + \bar{c} O^c c + \bar{d} O^d d + \bar{s} O^s s \tag{8.98}
$$

is diagonal in the quark mass states; no terms like $\bar{d}s$ or $\bar{s}d$ corresponding to $\Delta S = 1$ neutral currents appear explaining the absence of FCNC(flavor changing neutral current). This can be extended to six quark mixing using the CKM (Cabibbo–Kobayashi–Maskawa) matrix given in Eq. (8.90).

8.4.2 Triumphs of the Weinberg–Salam–Glashow Model

The standard model of electroweak interactions formulated by Weinberg and Salam is generally known as the Weinberg–Salam–Glashow (WSG) model after its extension to the quark sector. The model makes definite predictions for neutral currents in the neutrino and electron sectors and has implications in all areas of physics like particle physics, nuclear physics, atomic physics, astrophysics, and geophysics. While the neutral current of neutrinos is independent of the weak mixing angle, the neutral weak current carried by the charged leptons and the quarks has explicit dependence on the weak mixing angle θ_W. Therefore, the scattering processes like $\nu_l l^- \to \nu_l l^-$ and $\nu_l q \to \nu_l q$, involving leptons or quarks induced by the neutral current will give the value of the weak mixing angle θ_W. This in turn will give information about the masses and couplings of $W^\pm$, Z, and the Higgs boson. With a meticulous choice of physical processes involving neutrinos (antineutrinos) and the charged leptons, nucleons and nuclei, the masses of W^+, W^-, Z, and Higgs (and their decay modes) have been determined and compared with the predictions of the standard model. The history of neutrino physics over the last 50 years is a story of the quest for a unified theory of electroweak interactions leading to the standard model, its formulations, and its triumphs. In the following section, we will describe the major triumphs of the standard model during this period.

8.4.3 Discovery of the neutral currents in ν-scattering

Historically, weak neutral currents were postulated in the mid-1930s, but experimentally, their search started in the mid-1960s at CERN; these early searches were inconclusive. The first evidences for neutral currents were reported from the experiments done at CERN [168] on neutrino scattering in 1973 and at SLAC, on the electron scattering in 1978 [347], which provided the limits on the value of the weak mixing angle θ_W, confirming the standard model. Since then, many experiments have been done on neutrino scattering using leptons, nucleons, and nuclear targets; polarized electron scattering from the nucleons and nuclear targets; as well as parity violating experiments in atomic physics, in which the effects of the weak neutral currents have been observed.

i) **Neutrino scattering from leptons**
   The existence of neutral currents in the lepton sector implies the following scattering processes with ν_e and ν_μ beam on electrons.

$$\nu_\mu e^- \;\rightarrow\; \nu_\mu e^-, \qquad\qquad \bar{\nu}_\mu e^- \rightarrow \bar{\nu}_\mu e^-,$$
$$\nu_e e^- \;\rightarrow\; \nu_e e^-, \qquad\qquad \bar{\nu}_e e^- \rightarrow \bar{\nu}_e e^-.$$

It should be noted that $\nu_\mu(\bar{\nu}_\mu)e^- \rightarrow \nu_\mu(\bar{\nu}_\mu)e^-$ scattering is possible only due to the weak neutral currents through the Z exchange, while the $\nu_e e^-$ process is also possible with the weak neutral, as shown in Figure 8.1 (See Chapter 9 for details). The experiments with

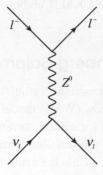

Figure 8.1 $\nu_l - l^-$ neutral current scattering.

ν_μ and $\bar{\nu}_\mu$ have been done at CERN [168] as well as at BNL [348, 349] and FNAL [350]. The statistically most significant data from CHARM II collaboration in CERN [351] gives:

$$\sin^2 \theta_W = 0.2324 \pm 0.0058(\text{stat}) \pm 0.0059(\text{syst}).$$

Scattering experiments with ν_e beams on electron targets have been done at Los Alamos while experiments with the $\bar{\nu}_e$ beam have been done at various nuclear reactors around the world. The value of the mixing angle has been determined from these experiments.

ii) **Neutrino experiments with nucleons and nuclei**

The elastic ν scattering processes like

$$\bar{\nu}_l p \;\rightarrow\; \bar{\nu}_l p, \qquad\qquad \nu_l n \rightarrow \nu_l n,$$
$$\nu_l p \;\rightarrow\; \nu_l p, \qquad\qquad \bar{\nu}_l n \rightarrow \bar{\nu}_l n,$$

are possible in the W–S model through the interaction Lagrangian $\mathcal{L}_{\mathrm{NC}}$ given in Eqs. (8.59) and (8.93). These processes have been studied earlier at BNL [352] and recently at MiniBooNE [353]. The quoted values from the BNL experiments are $\sin^2 \theta_W = 0.220 \pm 0.016(\mathrm{stat})^{+0.023}_{-0.031}(\mathrm{syst})$ [352].

The inelastic antineutrino reactions in the very low energy region have been done at various nuclear reactors [354] with reactor antineutrinos and at the Sudbury Neutrino Observatory (SNO) with solar neutrinos [355] for the neutral current induced processes on the deuterium target, that is,

$$\bar{\nu}_l + D \;\rightarrow\; \bar{\nu}_l + n + p \,,$$
$$\nu_l + D \;\rightarrow\; \nu_l + n + p.$$

These processes at very low energies correspond to a $D(^3S, \; I = 0) \rightarrow np(^1S, \; I = 1)$ transition and are possible only through the axial vector current as they involve change in spin. Therefore, they are independent of the weak mixing angle in the standard model. However, these experiments confirmed the existence of neutral currents which led towards the confirmation of the W–S model. At higher energies, inelastic processes of pion (π^0, π^+, π^-) production from nucleon targets like:

$$\nu_l + p \;\rightarrow\; \nu_l + p + \pi^0, \qquad\qquad \nu_l + n \rightarrow \nu_l + n + \pi^0,$$
$$\nu_l + p \;\rightarrow\; \nu_l + n + \pi^+, \qquad\qquad \nu_l + n \rightarrow \nu_l + p + \pi^-,$$

and coherent and incoherent production of π^0 from nuclear targets like:

$$\nu_l + A \rightarrow \nu_l + A + \pi^0, \qquad\qquad \nu_l + A \rightarrow \nu_l + X + \pi^i,$$

where $i = \pm 1, 0$, have been observed at various laboratories. The first experiments were done at ANL and BNL during the 1980s, using hydrogen and deuterium targets but in recent times, experiments have been done at FNAL and JPARC by many collaborative groups like SciBooNE, K2K, MiniBooNE, T2K, etc. to study the neutrino (antineutrino) reactions on the nuclear targets induced by neutral currents. These experiments are analyzed using the value of $\sin^2 \theta_W$ obtained from purely leptonic processes. The main emphasis here is to do an isospin analysis and understand the nuclear effects in the context of modeling the $\nu(\bar{\nu})$ nucleus cross section for analyzing the neutrino oscillation experiments as most of these experiments are done on nuclear targets [356].

iii) **Deep inelastic neutrino scattering experiments**
The deep inelastic scattering (DIS) of neutrinos on the nucleon target induced by neutral currents, that is,

$$\nu_\mu + p \rightarrow \nu_\mu + X, \qquad \bar{\nu}_\mu + p \rightarrow \bar{\nu}_\mu + X,$$
$$\nu_\mu + n \rightarrow \nu_\mu + X, \qquad \bar{\nu}_\mu + n \rightarrow \bar{\nu}_\mu + X,$$

where X is a jet of hadrons, is characterized by the absence of a charged lepton in the final state, unlike the charged current processes in which a $\mu^\pm$ is present,

$$\nu_\mu + p \rightarrow \mu^- + X, \qquad \bar{\nu}_\mu \rightarrow \mu^+ + X,$$
$$\nu_\mu + n \rightarrow \mu^- + X, \qquad \bar{\nu}_\mu \rightarrow \mu^+ + X.$$

The first evidence of the presence of neutral current induced DIS processes induced by neutrinos was reported at CERN by the Gargemelle collaboration in 1973 [168]. Since then, many experiments have been done using the DIS process induced by ν_μ and $\bar{\nu}_\mu$ on nucleon and nuclear targets at CERN, FNAL, and BNL by collaborations like CCFR [357], BEBC [358], and others. Most DIS experiments with neutrino (antineutrinos) induced by neutral currents are analyzed by studying the following ratios of the cross section:

$$R^*_{\nu p} = \frac{\sigma(\nu_\mu p \rightarrow \nu_\mu X)}{\sigma(\nu_\mu p \rightarrow \mu^- X)} \quad ; \quad R_{\bar{\nu} p} = \frac{\sigma(\bar{\nu}_\mu p \rightarrow \bar{\nu}_\mu X)}{\sigma(\bar{\nu}_\mu p \rightarrow \mu^+ X)};$$

$$R^*_{\nu n} = \frac{\sigma(\nu_\mu n \rightarrow \nu_\mu X)}{\sigma(\nu_\mu n \rightarrow \mu^- X)} \quad ; \quad R_{\bar{\nu} n} = \frac{\sigma(\bar{\nu}_\mu n \rightarrow \bar{\nu}_\mu X)}{\sigma(\bar{\nu}_\mu n \rightarrow \mu^+ X)};$$

on nucleons as well as on the isoscalar targets like deuterium and ^{12}C. For nonisoscalar targets like ^{56}Fe and ^{208}Pb, non-isoscalarity corrections are applied. The data on nucleon and nuclear targets are then used to determine the neutral current couplings of the u and d quarks. Neglecting the contribution of strange and heavier quarks, the Paschos–Wolfenstein relation [359] obtained as

$$R^{PW} = \frac{\sigma(\nu_\mu N \rightarrow \nu_\mu X) - \sigma(\bar{\nu}_\mu N \rightarrow \bar{\nu}_\mu X)}{\sigma(\nu_\mu N \rightarrow \mu^- X) - \sigma(\bar{\nu}_\mu N \rightarrow \mu^+ X)} = \left(\frac{1}{2} - \sin^2 \theta_W \right), \qquad (8.99)$$

is used to obtained the value of $\sin^2 \theta_W$, where θ_W is the Weinberg angle. The values of the weak mixing angle $\sin^2 \theta_W$, obtained from various experiments are given in Table 8.5.

With the availability of more data on cross sections with improved precision, the contribution of sea quarks like $\bar{u}, \bar{d}, s, \bar{s}, c, \bar{c}$ are also determined.

Table 8.5 Values of $\sin^2 \theta_W$ calculated from various data sets from different experiments [117].

Data	$\sin^2 \theta_W$
All data	0.22332(7)
M_H, M_Z, Γ_Z, m_t	0.22351(13)
LHC	0.22332(12)
Tevatron+M_Z	0.22295(30)
LEP	0.22343(47)
SLD+ M_Z, Γ_Z, m_t	0.22228(54)
$\mathcal{A}_{FB}, M_Z, \Gamma_Z, m_t$	0.22503(69)
$\mathcal{A}_{LR}$	0.2220(5)
$\nu_\mu(\bar{\nu}_\mu)p \to \nu_\mu(\bar{\nu}_\mu)p$	0.203(32)
$\nu_\mu(\bar{\nu}_\mu)e \to \nu_\mu(\bar{\nu}_\mu)e$	0.221(8)
Atomic parity violation	0.220(3)

8.5 Discovery of Neutral Currents in Electron Scattering

The W–S model predicts the neutral current coupling of electrons to various quarks through the Z exchange. Therefore, the electron nucleon scattering which is mediated by the photon exchange gets extra contribution from the Z exchange as shown in Figure 8.2. Since the Z exchange diagram involves both vector and axial vector currents due to the structure of the weak neutral current in the lepton and quark sectors, the interference between the photon and Z exchange diagrams would lead to parity violating effects. This can be observed as an asymmetry in the differential cross section $\frac{d\sigma}{dq^2}$ in the elastic scattering of polarized electrons from nucleons and nuclear targets. In principle, one observes the asymmetry A defined by:

$$A(q^2) \quad = \quad \frac{\sigma_R(q^2) - \sigma_L(q^2)}{\sigma_L(q^2) + \sigma_R(q^2)}, \tag{8.100}$$

where $\sigma_{L,R}(q^2)$ are the differential scattering cross sections $\left(\frac{d\sigma}{dq^2}\right)_{L,R}$ for the left-handed (L) and the right-handed (R) polarized electrons. The weak neutral currents in the electron sector were first discovered in the deep inelastic scattering of polarized electrons from deuterons at SLAC in 1978 [347]. Since then, experiments, which observe parity violating asymmetry in the elastic, inelastic, and deep inelastic scattering of polarized electrons from nucleon and nuclear targets like hydrogen(^{1}H), deuterium(^{2}D), helium(^{4}He), carbon(^{12}C), and lead(^{208}Pb), have been done at various electron accelerators like MIT-BATES [360, 361], MAINZ [362, 363], and JLab [364, 365, 366], in the region of low, medium, and high energies [367]. A special feature of the physics of the scattering of polarized electrons from nuclei and asymmetry measurements is that the neutral current coupling of an electron to the neutron is much stronger than its coupling to the proton. This is used to determine, independently, the neutron density of nuclei in polarized electron scattering from heavy nuclei like ^{208}Pb [368]. Thus, weak neutral currents provide a new tool to study the neutron distribution in nuclei. Moreover, experiments on electron annihilation from positrons also provide evidence for neutral currents in the $\mu^- \mu^+$ and $\tau^- \tau^+$ sectors through the various asymmetry measurements in processes like $e^+e^- \rightarrow$

$\mu^+\mu^-$ and $e^+e^- \rightarrow \tau^+\tau^-$. These experiments have decisively confirmed the effect of the weak neutral currents in the charged lepton sector as predicted by the W–S model. In the following section, we briefly present these effects.

Elastic scattering of polarized electrons

The parity violating asymmetries in the elastic scattering of polarized electrons arise due to the interference between the contributions coming from the photon (γ) and the Z exchange diagrams shown in Figure 8.2, for which the matrix elements are written as:

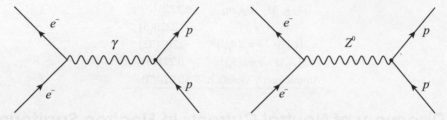

Figure 8.2 Electrons (point particles) getting scattered from a nucleon through electromagnetic (left) and weak (right) interactions. The blob depicts the fact that the nucleon has got a structure.

$$M_\gamma = -\frac{e^2}{q^2}\bar{u}_e(k')\gamma^\mu u_e(k)\bar{u}(p')J_\mu^\gamma u(p), \tag{8.101}$$

$$M_Z = -\frac{G_F}{\sqrt{2}}\bar{u}_e(k')\gamma^\mu(g_V^e - \gamma_5 g_A^e)\bar{u}(p')J_\mu^Z u(p), \tag{8.102}$$

where $g_V^e = -\frac{1}{2} + 2\sin^2\theta_W$, $g_A^e = -\frac{1}{2}$ (as given in Table 8.2), and J_μ^γ and J_μ^Z are given as:

$$\langle p'|J_\mu^\gamma|p\rangle = \bar{u}(p')\left[F_1^\gamma(q^2)\gamma_\mu + i\sigma_{\mu\nu}\frac{q^\nu}{2M}F_2^\gamma(q^2)\right]u(p), \tag{8.103}$$

$$\langle p'|J_\mu^Z|p\rangle = \bar{u}(p')\left[F_1^Z(q^2)\gamma_\mu + i\sigma_{\mu\nu}\frac{q^\nu}{2M}F_2^Z(q^2) + G_A^Z(q^2)\gamma_\mu\gamma_5\right]u(p). \tag{8.104}$$

In writing the aforementioned matrix elements, CVC and G-invariance have been assumed and the pseudoscalar form factor term is neglected as its contribution is negligible. The form factors $F_{1,2}^{\gamma,Z}(q^2)$ are the Pauli–Dirac electromagnetic (γ) and weak (Z) neutral current form factors discussed in detail in Chapter 10. They are expressed in terms of the Sachs electric and magnetic form factors, $G_{E,M}^{\gamma,Z}(q^2)$:

$$G_E^{\gamma,Z} = F_1^{\gamma,Z} + \frac{q^2}{4M^2}F_2^{\gamma,Z}(q^2), \tag{8.105}$$

$$G_M^{\gamma,Z} = F_1^{\gamma,Z}(q^2) + F_2^{\gamma,Z}(q^2). \tag{8.106}$$

$G_A^Z(q^2)$ is the weak neutral axial vector form factor. Using the isotriplet hypothesis of weak current (Chapter 5), we obtain $F_{1,2}^{\gamma,Z}$ and G_A^Z of the nucleons in terms of the electromagnetic and weak charged current form factors of the nucleons, that is,

$$G_{E,M}^{Z(p,n)} = (1 - 4\sin^2\theta_W)G_E^{\gamma(p,n)}(q^2) - G_{E,M}^{\gamma(n,p)}(q^2), \tag{8.107}$$

$$G_A^{Z,p}(q^2) = -G_A^{Z,n}(q^2) = \frac{1}{2}G_A(q^2). \tag{8.108}$$

The parity violating asymmetry $\mathcal{A}_{LR}$ is then proportional to

$$
\begin{aligned}
\mathcal{A}_{LR} &\propto \frac{\mathrm{Re}|M^\gamma M^{Z*}|}{|M^\gamma|^2 + |M^Z|^2} \simeq \frac{\mathrm{Re}|M^\gamma M^{Z*}|}{|M^\gamma|^2} \\
&= \frac{G_F q^2}{4\sqrt{2}\pi\alpha}(A_E^p(q^2) + A_M^p(q^2) + A_A^p(q^2)),
\end{aligned} \tag{8.109}
$$

where $A_E^p(q^2)$ and $A_M^p(q^2)$ are due to the interference of the electron axial vector and nucleon vector currents. The term $A_A^p(q^2)$ is due to the interference of the electron vector and nucleon axial vector currents and is small, being proportional to $(1 - 4\sin^2\theta_W)$. The expression for the asymmetry becomes very simple in the case of ^{4}He, which is a spin zero isoscalar target, so that only spin independent isoscalar hadronic currents, which exist only for the vector current in the G–W–S model contribute through the $F_1^{I=0}(q^2)$ form factor, resulting in

$$A_{LR}^{^4He} = -\frac{G_F q^2}{\pi\sqrt{2}\alpha}\sin^2\theta_W. \tag{8.110}$$

The experimental results are summarized in Table 8.6; these results are consistent with the predictions of the W–S–G model. In 2005, an experiment for fixed target Møller scattering, that is, $e^-e^- \to e^-e^-$ measured $\mathcal{A} = -0.131 \pm 13\%$ ppm [369]. The values of asymmetries obtained from different experiments are listed in Table 8.6.

Table 8.6 Results from selected parity violating (PV) experiments. Asymmetries are given in parts per million (ppm).

Experiment	$\mathcal{A}_{LR}$ (Elastic)
Mainz [362]	$-9.4 \pm 20\%$
Mainz–A4 [363]	$-17.2 \pm 5\%$
MIT–Bates [360]	$1.62 \pm 24\%$
SAMPLE [361]	$-5.61 \pm 20\%$
HAPPEX [364]	$-15.05 \pm 7.5\%$
GØ [365]	$-2 \pm 13\%$
HAPPEX–He [366]	$6.40 \pm 4.1\%$

Inelastic electron scattering

The presence of neutral currents have been observed in the inelastic scattering of electrons from the proton target in the reaction $\vec{e} + p \to \vec{e} + \Delta^+$ at JLab [370]. In this case, the matrix elements corresponding to γ and Z exchange, shown in Figure 8.3, will be similar to those given in Eqs. (8.101) and (8.102), respectively, with the matrix elements of hadronic currents $J_\mu^{\gamma,Z}$ defined as $\langle\Delta(p')|J_\mu^\gamma|p(p)\rangle$ and $\langle\Delta(p')|J_\mu^Z|p(p)\rangle$, which are parameterized in terms of four vector and four axial vector form factors $C_i^V(q^2)$ and $C_i^A(q^2)$, $(i = 3 - 6)$ defined and discussed in detail in Chapter 12.

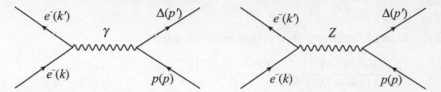

Figure 8.3 Inelastic electron–proton scattering ($e^- p \longrightarrow e^- \Delta$) through the electromagnetic(left) and weak(right) interactions.

Since the transition $p \to \Delta^+$ can take place only through $\Delta I = 1$ transitions, only the isovector parts of the currents J_μ^γ and J_μ^Z contribute to the Δ excitations for which the isovector form factors are summarized as (Chapter 11):

i) The form factors $C_i^V(q^2)$ $(i = 3, 4, 5)$ are given in terms of $C_i^\gamma(q^2)$, assuming the isotriplet properties of weak currents.

ii) $C_6^\gamma(q^2)$ and $C_6^Z(q^2)$ vanish due to CVC.

iii) $C_6^A(q^2)$ is negligible being proportional to the electron mass.

iv) Using M1 dominance of $\Delta \to p\gamma$ decay and $ep \to e\Delta$ scattering, it may be shown that $C_5^\gamma(0) = 0$, $C_4^\gamma(0) = -\frac{M}{M+M_\Delta}C_3^\gamma(0)$.

v) Using the data available in the weak excitation of Δ in ν-scattering, we obtain, $C_3^A(0) = 0$, $C_4^A(0) = -0.35$, $C_5^A(0) = 1.20$, $C_3^\gamma(0) = 1.85$, and $C_4^\gamma(0) = -0.89$.

The asymmetry is calculated to be:

$$A_{\text{inel}} = \frac{G_F q^2}{4\pi\alpha\sqrt{2}}[\Delta^V(q^2) + \Delta^A(q^2)], \tag{8.111}$$

where $\Delta^V(q^2)$ includes the contribution arising due to the interference of the electron axial vector (g_A^e) and hadron vector terms and $\Delta^A(q^2)$ is the contribution arising from the interference of the electron vector (g_V^e) and hadron axial vector terms. Neglecting the contribution of the non-resonant terms, which are found to be small ($\approx 1.5\%$), $\Delta^V(q^2)$ becomes independent of the hadronic structure.

The $\Delta^A(q^2)$ term, on the other hand, which involves the hadronic form factors $C_i^V(q^2)$ corresponding to the photon exchange and $C_i^A(q^2)$ corresponding to the Z exchange, is complicated and its discussion is given in Ref. [371, 372, 373]. Fortunately, the contribution of $\Delta^A(q^2)$ is small ($\approx 5\%$) because it is proportional to $(1 - 4\sin^2\theta_W)$. Therefore, the resonance contribution to $\Delta^V(q^2)$ is dominant, leading to $\mathcal{A} = -32.2 \times 10^{-6}$ using $\sin^2\theta_W = 0.2353$. The total contribution of $\Delta^V(q^2)$ and $\Delta^A(q^2)$ (including the non-resonant contributions) gives a theoretical value of $\mathcal{A} = (34.6 \pm 1.0) \times 10^{-6}$.

The experimental analysis of the JLab experiment [370] reports a value of the asymmetry as:

$$\mathcal{A} = (-33.4 \pm 5.3 \pm 5.1) \times 10^{-6} \quad \text{for the proton target,}$$
$$\mathcal{A} = (-43.6 \pm 14.6 \pm 6.2) \times 10^{-6} \quad \text{for the deuteron target,}$$

which is in reasonable agreement with the predictions of the W–S–G model.

Deep inelastic scattering

In the case of the deep inelastic scattering of the polarized electrons from the nucleons, the scattering takes place from the point-like constituents of the nucleons, that is, quarks as shown in Figure 8.4, through the interaction Lagrangian described in Eqs. (8.84) and (8.93) for the electromagnetic and weak neutral current.

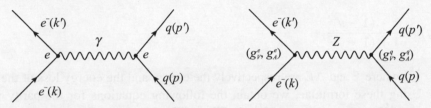

Figure 8.4 Deep inelastic scattering of an electron from a quark inside a nucleon through electromagnetic interaction (left) with a coupling strength e and weak interaction (right) with vector and axial vector coupling strengths g_V^e and g_A^e, respectively.

The strength of the vertex is e for the $ee\gamma$ and $qq\gamma$ interaction, $g_V^e(g_A^e)$ for the eeZ, and $g_V^q(g_A^q)$ for the qqZ interactions with the vector (axial vector) couplings, where:

$$g_V^i = \frac{1}{2}\tau_3^i - 2Q^i \sin^2\theta_W, \qquad g_A^i = \frac{1}{2}\tau_3^i,$$

and Q^i is the charge (in units of $|e|$), for the ith fermion. The parity violating part of the interaction Lagrangian, coming from the interference of the vector and axial vector currents in the Z exchange term, that is, VA and AV terms is given by:

$$\mathcal{L}^{PV} = \frac{G_F}{\sqrt{2}} \left[\bar{e}\gamma^\mu\gamma_5 e(C_{1u}\bar{u}\gamma_\mu u + C_{1d}\bar{d}\gamma_\mu d) + \bar{e}\gamma^\mu e(C_{2u}\bar{u}\gamma_\mu\gamma_5 u + C_{2d}\bar{d}\gamma_\mu\gamma_5 d) \right], \quad (8.112)$$

where

$$C_{1u} = -\frac{1}{2} + \frac{4}{3}\sin^2\theta_W,$$
$$C_{1d} = \frac{1}{2} - \frac{2}{3}\sin^2\theta_W,$$
$$C_{2u} = -\frac{1}{2} + 2\sin^2\theta_W,$$

$$C_{2d} = \frac{1}{2} - 2\sin^2\theta_W = -C_{2u}. \tag{8.113}$$

The asymmetry $(\mathcal{A}_{LR})$ for the deep inelastic scattering of polarized electrons with right-handed and left-handed electrons from the quarks is then given by:

$$\mathcal{A}_{\text{DIS}} \simeq \frac{G_F q^2}{4\sqrt{2}\pi\alpha}(a_1 + f(y)a_3)$$

$$\text{with } a_1 = \frac{2\sum_q e_q C_{1q}(q(x) + \bar{q}(x))}{\sum_q e_q^2(q(x) + \bar{q}(x))},$$

$$a_3 = \frac{2\sum_q e_q C_{2q}(q(x) - \bar{q}(x))}{\sum_q e_q^2(q(x) + \bar{q}(x))}, \tag{8.114}$$

where $q(x)$ and $\bar{q}(x)$ are the parton distribution functions (PDFs) for quarks and $f(y)$ is the kinematic factor given by:

$$f(y) = \frac{1 - (1-y)^2}{1 + (1-y)^2}, \tag{8.115}$$

with $y = \frac{\Delta E}{E}$, where E and ΔE are respectively the energy and the energy loss of the incident electron. Using these formulae, we obtain the following equations for the parity violating asymmetry in $\mathcal{A}_{\text{DIS}}^p$ in $\vec{e}p \to \vec{e}p$ and $\mathcal{A}_{\text{DIS}}^D$ in $\vec{e}D \to \vec{e}D$ scatterings:

$$\mathcal{A}_{\text{DIS}}^p = \frac{3G_F q^2}{4\sqrt{2}\pi\alpha}\left[\frac{C_{1u} - \frac{1}{2}C_{1d} \cdot \frac{d(x)}{u(x)}}{1 + \frac{1}{4}\frac{d(x)}{u(x)}} + \frac{C_{2u} - \frac{1}{2}C_{2d} \cdot \frac{d(x)}{u(x)}}{1 + \frac{1}{4}\frac{d(x)}{u(x)}} \cdot f(y)\right],$$

$$\mathcal{A}_{\text{DIS}}^D = \frac{3G_F q^2}{5\sqrt{2}\pi\alpha}\left[(C_{1u} - \frac{1}{2}C_{1d}) + (C_{2u} - \frac{1}{2}C_{2d}) \cdot f(y)\right]. \tag{8.116}$$

We see that while $\mathcal{A}_{\text{DIS}}^p$ depends upon the PDFs through the ratio $\frac{d(x)}{u(x)}$ for the d and u quarks, $\mathcal{A}_{\text{DIS}}^D$ is independent of the hadronic structure and depends only upon the parameters $C_{1q,2q}(q = u, d)$ of the G-W-S model which depend on $\sin^2\theta_W$. It should be noted that, in general, C_{2q} is small as compared to C_{1q} for $q = u, d$ in the valence quark model. Therefore, an experimental determination of C_{2q} would be important to test the validity of the valence quark description and to study the sea quark contributions. The first experiment in the deep inelastic scattering region was done at SLAC in 1978 [347] and then at JLab in 2014 [374]. The experimental results are given in Table 8.7.

Table 8.7 Results from selected PV experiments. Asymmetries are given in ppm.

Experiment	$-\mathcal{A}_{LR}(DIS) \times 10^{-6}$ ppm
SLAC–E122 [347]	$-120 \pm 8\%$
JLab–Hall A [374]	$-160 \pm 4.4\%$

In most recent times, experiments have been done in the region extending from the higher resonance region, that is, beyond the Δ resonance to the DIS region, in a wide range of W, the centre of mass energy. The results are shown in Figure 8.5, where the theoretical and experimental results are taken from Ref. [375, 376, 377, 378]

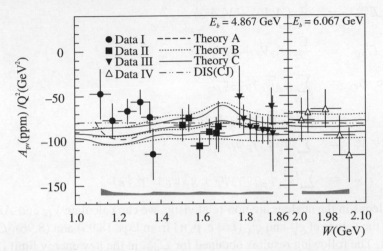

Figure 8.5 A_{pv} vs. W studied in $\vec{e}-^2$H scattering experiment. The physics asymmetry results, A_{PV}^{phys}, for the four kinematics I, II, III, and IV (solid circles, solid squares, solid triangles, and open triangles, respectively), in ppm, are scaled by $1/Q^2$ and compared with calculations from Ref. [375] (theory A, dashed lines), Ref. [376] (theory B, dotted lines), Ref. [377] (theory C, solid lines), and the DIS estimation (dash-double-dotted lines) with the extrapolated CJ PDF [378].

Weak neutral current in atomic physics

The presence of the weak neutral boson Z and its interaction with electrons and nucleons through eeZ and NNZ couplings having a $V - A$ structure as shown in Figure 8.6 gives rise to the parity violating term in the interaction Hamiltonian for $e - e$, $e - N$, and $N - N$ systems, arising through its interference with the photon exchange diagrams. The parity violating interactions between $e - N$ and $e - e$ systems are relevant to atomic physics.

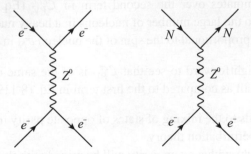

Figure 8.6 Z exchange in electron–electron (left) and electron–nucleon(right) scattering.

The parity violating part of the interaction Lagrangian for an orbital electron with other electrons and nucleons in the nucleus is written as (suppressing the $\vec{x}$ dependence):

$$\mathcal{L}_{PV}^{int} = \mathcal{L}_{PV}^{eN} + \mathcal{L}_{PV}^{ee}, \tag{8.117}$$

$$\text{where} \quad \mathcal{L}_{PV}^{eN} = \frac{G}{\sqrt{2}}(A_e V_N + V_e A_N) \tag{8.118}$$

$$\mathcal{L}_{PV}^{ee} = \frac{G}{\sqrt{2}}(A_e V_e + V_e A_e), \tag{8.119}$$

$$\text{where} \quad A_e = \sum_{i=N} g_A^e \overline{\psi}_e \gamma_\mu \gamma_5 \psi_e,$$

$$V_N = \sum (g_V^p \overline{\psi}_p \gamma^\mu \psi_p + g_V^n \overline{\psi}_n \gamma^\mu \psi_n),$$

$$V_e = \sum_{i=N} g_V^e \overline{\psi}_e \gamma_\mu \psi_e,$$

$$A_N = \sum (g_A^p \overline{\psi}_p \gamma^\mu \gamma_5 \psi_p + g_A^n \overline{\psi}_n \gamma^\mu \gamma_5 \psi_n).$$

Since the nucleons in the nucleus are non-relativistic, we can calculate V_N and A_N in the limit $\vec{p} \to 0$; using the values of g_V^i and g_A^i ($i = e, p, n$) from Eqs. (8.95) and (8.96), $\mathcal{L}_{PV}^{eN}$ and $\mathcal{L}_{PV}^{ee}$ are calculated. The following result is obtained for $\mathcal{L}_{PV}^{eN}$, in the low energy limit [239]

$$\mathcal{L}_{PV}^{eN} = \frac{G_F}{4\sqrt{2}m_e} \Big[Q_{W\rho_N}(x)(\vec{\sigma} \cdot \vec{p}\, \delta^3(\vec{x}) + \delta^3(\vec{x})\, \vec{\sigma} \cdot \vec{p})$$
$$+ (1 - 4\sin^2\theta_W) g_A\, \vec{\sigma}_N \cdot \vec{\sigma} \left(\vec{\sigma} \cdot \vec{p}\, \delta^3(\vec{x}) + \delta^3(\vec{x})\, \vec{\sigma} \cdot \vec{p} \right) \Big],$$

where $Q_{W\rho_N} = (1 - 4\sin^2\theta_W)Z - N$ is called the weak charge of the nucleus. It should be noted that:

i) In view of the smallness of $1 - 4\sin^2\theta_W$ in the W–S model, the weak charge is mainly determined by the number of neutrons, a fact also emphasized in the scattering of polarized electrons from the heavy nuclei.

ii) The first term dominates over the second term in $\mathcal{L}_{PV}^{eN}$ (Eq. (8.118)) as there is no enhancement due to the large number of nucleons in a heavy nucleus in the second term; this is because it is proportional to the spin of the nucleon $\vec{\sigma}_N$ in the non-relativistic limit.

iii) Similarly, it is straightforward to see that $\mathcal{L}_{PV}^{ee}$ is of the same order as the second term and is therefore small as compared to the first term in Eq. (8.119).

The presence of $\mathcal{L}_{PV}^{eN}$ leads to the mixing of states of opposite parity in the atomic states which can be calculated in the perturbation theory.

In general, any state $|\psi\rangle$ with a given parity will be mixed with the states of opposite parity $|\chi_n\rangle$ given by:

$$|\psi\rangle = |\psi\rangle + \sum_n |\chi_n\rangle \frac{\langle \chi_n | H_{PV} | \psi \rangle}{E(\psi) - E(\chi_n)}, \quad H_{PV} = -\mathcal{L}_{PV} \tag{8.120}$$

It is found that the mixing depends upon Z^3, where Z is the nuclear charge; therefore, the effect is enhanced for heavier atoms [379].

The atomic parity violation experiments involve observation of magnetic dipole ($M1$) transitions between the initial and final states which are now accompanied by a small amplitude of electric dipole ($E1$) transitions due to the admixture of the opposite parity states in the initial and final states. If $M1$ transition between the states $|\psi_i\rangle$ and $|\psi_f\rangle$ is written as

$$M1 = \langle \psi_f | M1 | \psi_i \rangle,$$

then

$$E1 = \sum_n \frac{\langle \psi_f | E1 | \chi_n \rangle \langle \chi_n | H^{PV} | \psi_i \rangle}{E_{\psi_i} - E_{\chi_n}} + \frac{\langle \psi_f | H^{PV} | \chi_n \rangle \langle \chi_n | E1 | \psi_i \rangle}{E_{\psi_f} - E_{\chi_n}},$$

such that initially, a pure $M1$ amplitude is then given by $M1'$ where

$$M1' = M1 \pm E1$$

leading to the parity violating effects proportional to the interference term between $M1$ and $E1$ terms, that is,

$$\frac{2\,\mathrm{Im}(M1E1^*)}{|M1 + E1|^2} \simeq \frac{2\mathrm{Im}(M_1 E_1^*)}{|M_1|^2}. \tag{8.121}$$

Therefore, in the atomic transitions where the strength of $M1$ is small, the parity violating effects could be large and may lead to observable effects. The transitions studied are therefore $6\,^2S_{\frac{1}{2}} \to 7\,^2S_{\frac{1}{2}}$ in Cs with $Z = 55$ and $6\,^2P_{\frac{1}{2}} \to 7\,^2P_{\frac{1}{2}}$ in Tl with $Z = 81$ which are single valence electron atoms with minimum theoretical uncertainties in describing the atomic wave functions.

The first, experiments were performed on the circular polarization of photons in atomic transitions in ^{133}Cs [380] and later on the optical rotation of photons in ^{209}Bi [381]. The interpretation of experimental results are subject to the theoretical uncertainties of the atomic structure calculations of heavy atoms but the results for the weak mixing angles are consistent with the results obtained from other parity violating experiments as shown in Table 8.5. Some recent reviews of parity violating effects in atoms and molecules are given in [382, 383, 384].

Weak neutral currents in $e^- e^+$ annihilation experiments

Weak neutral currents in the lepton sector have also been observed in $e^- e^+$ colliders at LEP and SLAC, where $e^- e^+ \to f\bar{f}$, $f = \mu$, τ, u, d, etc., are produced. In addition to the photon exchange, there is an additional contribution due to the Z exchange, in the $e^- e^+ \to f\bar{f}$ processes, as shown in Figure 8.7. The presence of the Z exchange leads to the forward–backward asymmetry $\mathcal{A}$ in the production of fermion pairs as well as their polarization P due to the interference between the two diagrams shown in Figure 8.7. The differential scattering cross section in the center of mass (CM) system is given by [279]

$$\frac{d\sigma}{d\cos\theta} = \frac{\pi\alpha^2}{2s} \left[(1 + \cos^2\theta)A + (\cos\theta)B \right], \tag{8.122}$$

where s is the CM energy of the electron and the $\cos\theta$ dependent term gives the forward–backward asymmetry for $0 \le \cos\theta \le 1$ and $-1 \le \cos\theta \le 0$ regions. The coefficients A and

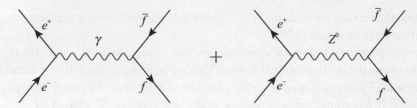

Figure 8.7 Electron–positron annihilation to give $f\bar{f}$ through electromagnetic (left) and weak (right) interactions. The net contribution is a combination of both.

B depend upon the vector and axial vector couplings of the electrons and the final fermions to the Z bosons, that is, g_V^e, g_A^e, g_V^f, and g_A^f and are given as:

$$
\begin{aligned}
A &= 1 + 2g_V^e g_V^f \text{Re}(F(s)) + [(g_V^e)^2 + (g_A^e)^2][(g_V^f)^2 + (g_A^f)^2]|F(s)|^2, \\
B &= 4g_A^e g_A^f \text{Re}(F(s)) + 8g_A^e g_V^e g_A^f g_V^f |F(s)|^2,
\end{aligned}
$$

where $\quad F(s) = \dfrac{s}{s - M_Z^2 + is\Gamma_Z/M_Z}.$ (8.123)

The cross section in Eq. (8.122) yields a forward–backward asymmetry $\mathcal{A}_{FB} = \frac{N_F - N_B}{N_F + N_B}$, where N_F is the number scattered in the region $0 \le \cos\theta \le 1$ and N_B is the number scattered in the region $-1 \le \cos\theta \le 0$. It is easy to see that $\mathcal{A}_{FB} = \frac{3B}{8A}$. Since, the second term in B involves $g_V^e = 1 - 4\sin^2\theta_W$, which is very small for $\sin^2\theta_W \approx 0.23$, it is the first term which dominates the asymmetry.

Sizeable asymmetries of the order of $\approx 10^{-2}$ are predicted by the W–S model and have been observed in many experiments done at LEP and SLAC. The asymmetries could be large at energies E corresponding to the Z peak, that is, $s \approx M_Z^2$. These observations confirm the presence of weak neutral currents in the lepton sector, involving muons and tauons in addition to electrons. It should be noted that a non-zero forward–backward asymmetry also appears in purely electromagnetic processes due to the interference between the one photon and two photon exchange diagrams; however, this effect is very small.

8.6 Discovery of $W^\pm$ and Z Bosons

8.6.1 Properties of $W^\pm$ and Z boson

The W–S–G model predicts the vector bosons $W^\pm$ and Z with spin 1 and

$$
\begin{aligned}
M_W^2 &= \frac{\pi\alpha}{\sqrt{2}G_F \sin^2\theta_W}, \\
M_Z^2 &= \frac{M_W^2}{\cos^2\theta_W}.
\end{aligned}
$$

Their interaction Lagrangian with leptons $(l = \nu, e)$ and quarks $(q = u, d)$ are given by:

$$\mathcal{L}_{\text{int}} = -\left(\frac{G_F M_W^2}{\sqrt{2}}\right)^{\frac{1}{2}} \left[\bar{\nu}_e \gamma^\mu (1 - \gamma_5) e\, W_\mu^+ + \bar{e} \gamma^\mu (1 - \gamma_5) \nu_e W_\mu^- \right.$$

$$\left. + \bar{q} \gamma^\mu (1 - \gamma_5) q W_\mu^+ + \text{h.c.}\right]$$

$$- \frac{1}{\sqrt{2}} \left(\frac{G_F M_Z^2}{\sqrt{2}}\right)^{\frac{1}{2}} \left[\bar{f} \gamma^\mu (g_V^f - g_A^f \gamma_5) f\right] Z_\mu, \qquad (8.124)$$

where $f = e, \nu$ for leptons and $f = q$ for quarks with:

$$g_V^f = \frac{\tau_3^f}{2} - 2Q^f \sin^2 \theta_W, \qquad \text{and} \qquad g_A^f = \frac{\tau_3^f}{2}.$$

Diagrammatically, the couplings are shown in Figure 8.8.

We see from Eq. (8.124) and Figure 8.8 that the properties and the phenomenology of weak interactions of $W^\pm$ and Z bosons are defined in terms of M_W, M_Z, $\sin^2 \theta_W$, and G_F, which can be determined in their interactions with leptons and quarks through weak processes involving their production and decay. They can be produced in $e^+ e^-$ collisions or in $p\bar{p}$ collisions through processes like $e^+ e^- \to W^+ W^-, ZZ$ or $q\bar{q} \to W^+ W^-, ZZ$ and can be observed through their decay processes like $W^\pm \to e^\pm \nu_e(\bar{\nu}_e), Z \to \nu\bar{\nu}, l^+ l^-$ $(l = e, \mu)$.

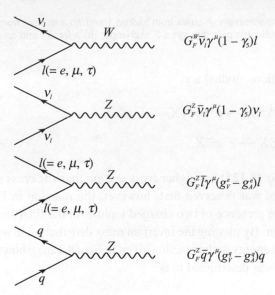

Figure 8.8 Coupling of W, Z bosons with leptons and quarks, where $G_F^W = \left(\frac{G_F M_W^2}{\sqrt{2}}\right)^{\frac{1}{2}}$ and $G_F^Z = \frac{1}{\sqrt{2}}\left(\frac{G_F M_Z^2}{\sqrt{2}}\right)^{\frac{1}{2}}$.

The first attempts to search for intermediate vector bosons like $W^\pm$ were made at CERN in the mid-1960s [385] in high energy neutrino experiments motivated by the theoretical speculations that weak interactions are mediated by IVB. However, with the success of the

W–S model, which made definite predictions for $W^\pm$, with mass M_W and also the existence of a neutral boson Z with mass M_Z in terms of G_F and $\sin^2\theta_W$, new efforts were made to search for $W^\pm$ and Z bosons, first at CERN and later at other accelerators.

The weak IVB bosons $W^\pm$ and Z were first discovered experimentally in 1983 in the hadron collider at CERN [170, 171, 386, 387]. Later, at the high energy electron–positron collider machines like SLC at SLAC and LEP at CERN, the direct production of $W^\pm$ and Z through $e^+e^- \rightarrow W^+W^-$ and $e^+e^- \rightarrow ZZ$ reactions, were studied to make a precise measurement of the masses of $W^\pm$ and Z bosons. The important features of the properties of the $W^\pm$ and Z bosons are summarized here:

i) **Masses of $W^\pm$ and Z bosons**

In hadron colliders with $p\bar{p}$ collision, the production of $W^\pm, Z$ bosons is thought to be due to the Drell–Yan mechanism, in which an antiquark from $\bar{p}$ collides with a quark in p to produce lepton pairs through $W^\pm$ and Z productions, as shown in Figure 8.9.

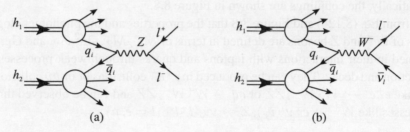

(a) (b)

Figure 8.9 The Drell–Yan mechanism. A quark from hadron 1 and an antiquark from hadron 2 combine to give (a) a lepton–antilepton pair, through a Z exchange; (b) a lepton and an antineutrino through a W^- exchange.

The specific reactions studied are:

$$\bar{p}p \rightarrow W^\pm X \rightarrow e^\pm \nu(\bar{\nu})X, \tag{8.125}$$

$$\bar{p}p \rightarrow ZX \rightarrow e^+e^- X. \tag{8.126}$$

The reaction in Eq. (8.125) has higher cross section than the cross section for the reaction in Eq. (8.126) and was observed first; however, the reaction in Eq. (8.126) is easier to analyze due to the presence of two charged leptons in the final state, almost in a back to back configuration. By plotting the invariant mass distribution in which both the electrons have well-defined energy in the reaction given in Eq. (8.126) (shown in Figure 8.10), mass M_Z of Z bosons was determined to be:

$$M_Z = 93.0 \pm 1.4(\text{stat}) \pm 3.2(\text{syst})\text{ GeV} \quad \text{UA1 [386]}, \tag{8.127}$$

$$= 92.5 \pm 1.3(\text{stat}) \pm 1.5(\text{syst})\text{ GeV} \quad \text{UA2 [387]}. \tag{8.128}$$

In the case of $e^+e^- \rightarrow$ hadrons, the invariant mass distribution looks like Figure 8.11 ([388]), corresponding to $M_Z = 91.19$ GeV. The determination of the mass of the W

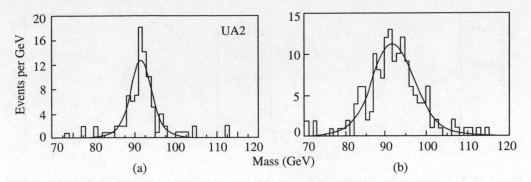

Figure 8.10 Invariant mass spectra for two $Z \to e^+ e^-$ event samples, as measured by UA2 [387]. The curves are best fit to the data, using m_Z as a free parameter.

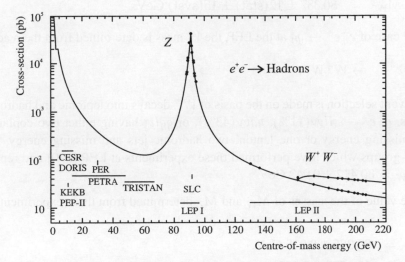

Figure 8.11 Evidence for the existence of Z boson at LEP-I [388]. Energy regions of other experiments have also been shown.

boson is not as straightforward as that of the Z boson. Since $W \to l\nu$ and the neutrino is not observed, the invariant mass distribution of $l\nu$ pair cannot be constructed. Therefore, the W mass determination requires an indirect measurement of the mass using variables which are related to M_W, like the momenta of electron and neutrino. The kinematical variable which is used in the analysis is the transverse mass:

$$M_T^2 = (E_{e_T} + E_{\nu_T})^2 - (\vec{P}_{e_T} + \vec{P}_{\nu_T})^2 \tag{8.129}$$

$$\approx 2 P_{e_T} P_{\nu_T} (1 - \cos\theta_{ev}), \tag{8.130}$$

where θ_{ev} is the angle between the momenta of the electron and neutrino, that is, $\vec{P}_{e_T}$ and $\vec{P}_{\nu_T}$ in the transverse plane. A simulation procedure is used to generate the M_T distribution corresponding to the different values of M_W. The observed M_T distribution from the CDF II experiment [389], is shown in Figure 8.12, from which a value of M_W is obtained to be:

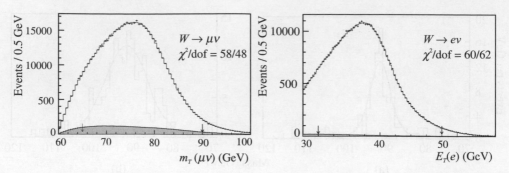

Figure 8.12 Transverse mass distribution for $W \to \mu\nu$(left) and $W \to e\nu$(right) from the CDF II experiment [389].

$$M_W = 80.387 \pm 12(\text{stat}) \pm 15(\text{syst}) \text{ GeV.} \tag{8.131}$$

In the case of $e^+e^- \to q\bar{q}$ at the LEP, the W mass is determined from the reaction

$$e^+e^- \to W^+W^-. \tag{8.132}$$

The event selection is made on the basis of $W^\pm$ decays into leptonic and hadronic modes, that is, $e^+e^- \to ll\nu\nu(11\%)$, $q\bar{q}\nu\nu(43\%)$, or $qql\nu$, having either two coplanar leptons and missing energy or one lepton, two hadronic jets and missing energy. There are many groups which have performed these experiments at LEP [390] and report a value of $M_W = 80.412 \pm 0.042$ GeV.

The average value of the masses of M_W and M_Z determined from these experiments is quoted as:

$$M_W = 80.385 \pm 0.015 \text{ GeV,}$$
$$M_Z = 91.1875 \pm 0.0021 \text{ GeV.}$$

Since $M_W = M_Z \cos\theta_W$, it leads to a value of $\sin^2\theta_W = 0.23122 \pm 0.00003$ [117].

ii) **Decay width of $W^\pm$ and Z bosons and neutrino flavors**
The decay width of $W^\pm$ and Z bosons can be calculated at the tree level using the Lagrangians for the charged current and neutral current weak interactions of $W^\pm(Z)$ given in Eqs. (8.58), (8.59), (8.91), and (8.93). The partial width for W^- to the leptonic and hadronic decays are obtained as:

$$\Gamma(W^- \to l^-\bar{\nu}_l) = \frac{G_F M_W^3}{6\pi\sqrt{2}}, \tag{8.133}$$

$$\Gamma(W^- \to \bar{u}_i d_j) = N_c V_{ij}^2 \frac{G_F M_W^3}{6\pi\sqrt{2}}. \tag{8.134}$$

We see that all the leptonic modes have the same width Γ, in the limit of the lepton mass $m_l \to 0$. The width for the quark modes depend upon N_c, the number of color and U_{ij}, the matrix element of the quark mixing matrix.

The decay width for Z decays are given as:

$$\Gamma(Z \to f\bar{f}) = N_f \frac{G_F M_Z^3}{6\pi\sqrt{2}} (|g_V^f|^2 + |g_A^f|^2), \tag{8.135}$$

where g_V^f and g_A^f are the strengths of the vector and axial vector couplings that depend upon f, the type of fermion and therefore, are different for each fermion; $N_f = 1$ for $f = l$ and $N_f = 3$ for $f = q$. Substituting the numerical values, in the appropriate equations we obtain the following:

$$\Gamma(W^- \to e^- \bar{\nu}_e) = 205 \text{ MeV}, \tag{8.136}$$

$$\Gamma(Z \to \nu\bar{\nu}) = 85 \text{ MeV}, \tag{8.137}$$

using $\sin^2 \theta_W = 0.2312$.

Assuming the lepton universality as predicted by the W–S model and neglecting the quark masses and the decay of the top quark as it is not allowed due to its high mass, we get the following results:

$$\frac{Br(W^- \to l^- \nu_l)}{(W^- \to \text{all})} = \frac{1}{3 + 2N_c} = 11.1\% \tag{8.138}$$

$$\text{and } \Gamma_N^{\text{total}} = 1845 \text{ MeV}. \tag{8.139}$$

Experimentally, we find the decay widths as shown in Table 8.8 ([391, 392, 393, 394]), which is in good agreement with theoretical predictions and provide further support for the $e - \mu - \tau$ universality of the weak interaction. Moreover, if Γ_{inv} is an Z decay into invisible modes, that is, $\nu\bar{\nu}$, and N_ν is the number of the neutrino flavor, then:

$$\frac{\Gamma_{\text{inv}}}{\Gamma_l} = \frac{N_\nu \Gamma(\nu\bar{\nu})}{\Gamma_l} = N_\nu \frac{|g_V^\nu|^2 + |g_A^\nu|^2}{|g_V^l|^2 + |g_A^l|^2} = \frac{2N_\nu}{(1 - 4\sin^2 \theta_W)^2 + 1}. \tag{8.140}$$

The experimental value for this quantity is 5.943 ± 0.016 and provides strong support for $N_\nu = 3$.

8.7 Higgs Boson

The most significant aspect of introducing the phenomenon of spontaneous symmetry breaking through the Higgs mechanism, by introducing a doublet of complex scalar fields (or four real

Table 8.8 Experimental determinations of the ratios g_l/g_v [391, 392, 393, 394].

	e	μ	τ	l
$Br(W^- \to \bar{\nu}_l l^-)$ (%)	10.75 ± 0.13	10.57 ± 0.15	11.25 ± 0.20	10.80 ± 0.09
$\Gamma(Z \to l^+ l^-)$ (MeV)	83.91 ± 0.12	83.99 ± 0.18	84.08 ± 0.22	83.984 ± 0.086

fields) in the Weinberg–Salam–Glashow theory was to generate the masses for the three gauge bosons $W_\mu^\pm$ and Z. Three of the real scalar fields are absorbed by the massless gauge fields corresponding to $SU(2)_L \times U(1)_I$ generators, to make the gauge fields massive. However, one of the four scalar fields ϕ, acquires mass and is called the Higgs boson, with spin zero and mass:

$$M_H = \sqrt{2}\mu = \sqrt{2}\lambda v. \tag{8.141}$$

μ and λ parameterize the Higgs potential and v is defined by the vacuum expectation value of the scalar field ϕ (see Eq. (8.16)), which is determined by the phenomenology of the charged weak current to be:

$$v = (\sqrt{2}G_F)^{-\frac{1}{2}}. \tag{8.142}$$

$$\lambda = \frac{M_H^2}{2v^2} = \frac{M_H^2 G_F}{\sqrt{2}}. \tag{8.143}$$

Therefore, the mass of the Higgs boson is undetermined unless λ (or μ) is determined or estimated by some other theoretical or experimental physics considerations.

The early experiments at LEP collaborations ALEPH, DELPHI, L3, and OPAL [395], to search for M_H at the center of mass energies between 189 and 209 GeV, indicated that the Higgs boson is of very high mass, most likely to be $M_H > 114.4$ GeV at the 95% confidence level. A very high mass for the Higgs boson would imply very high values of λ, the strength of the quartic self coupling of Higgs which may create problems with the unitarity and renormalizability of the theory. It is, therefore, important to theoretically study the implications of λ being very high and obtain physical constraints on λ so that the theory remains perturbative, making it relevant and applicable to the physical phenomenon. We have discussed earlier that the production of the longitudinal components of the vector boson W_L is responsible for the increase in cross section with energy, leading to violation of unitarity. In a similar way, a systematic analysis of the production of the longitudinal components of W in e^+e^- and $p\bar{p}$ collisions, that is, production of the vector bosons $W_L^+W_L^-, Z_L Z_L$, leads to a condition on the parameter λ, which implies that [396]

$$M_H < \left(\frac{8\pi\sqrt{2}}{3G_F}\right)^{\frac{1}{2}} \approx 1 \text{ TeV}. \tag{8.144}$$

The value of λ in terms of the vacuum expectation value of the field v and M_H determined in Eq. (8.143) is obtained in classical field theory. In renormalizable quantum field theory, this value is subject to the quantum field theoretic one loop and higher order loop corrections, which make λ energy dependent, given by [279]

$$\lambda(E) = \frac{\lambda(v)}{1 - \frac{3\lambda(v)}{8\pi^2}\ln\frac{E}{v}}, \tag{8.145}$$

which implies that λ increases with energy. In fact, the theory becomes non-perturbative at the energy scale of $E > \Lambda$, with $\Lambda \approx v \, \exp\left(\frac{8\pi^2}{3\lambda(v)}\right)$, which implies that:

$$
M_H < v \left[\frac{4\pi^2}{3\ln(\frac{\Lambda}{v})}\right]^{\frac{1}{2}}, \tag{8.146}
$$

for $\Lambda = 10^{16}$ GeV, $M_H \lesssim 160$ GeV. Since $\lambda \propto M_H^2$, a higher M_H will decrease the value of Λ, setting the non-perturbative region much earlier in the energy scale than 10^{16} GeV. These theoretical considerations provide a limit on the mass of Higgs boson so that the weak interaction remains weak and perturbative. In view of the high mass of Higgs, $M_H \approx 160$ GeV, by theoretical considerations and the LEP limit of $M_H \geq 114.4$ GeV, very high energy accelerators are required to produce them in laboratories. It is possible to study the Higgs boson and its properties through its decays in which leptons, hadrons, and gauge bosons $W^{\pm}$ and Z are produced. The interaction Lagrangian, $\mathcal{L}_S$ involving the scalar Higgs boson can be derived using Eq. (8.33),

$$
\mathcal{L}_S = \frac{1}{4}\lambda v^4 + \mathcal{L}_H + \mathcal{L}_{HG^2} + \mathcal{L}_{fH}, \tag{8.147}
$$

where

$$
\mathcal{L}_{fH} = -\frac{f_e}{\sqrt{2}}\bar{e}eH = -\frac{m_e}{v}\bar{e}eH \tag{8.148}
$$

$$
\mathcal{L}_H = \frac{1}{2}\partial_\mu H \partial^\mu H - \frac{1}{2}M_H^2 H^2 - \frac{M_H^2}{2v}H^3 - \frac{M_H^2}{8v^2}H^4, \tag{8.149}
$$

$$
\mathcal{L}_{HG^2} = M_W^2 W_\mu^\dagger W^\mu \left\{1 + \frac{2}{v}H + \frac{H^2}{v^2}\right\} + \frac{1}{2}M_Z^2 Z_\mu Z^\mu \left\{1 + \frac{2}{v}H + \frac{H^2}{v^2}\right\}, \tag{8.150}
$$

using the values of v, λ and μ^2 in terms of M_H, M_W, M_Z and m_e. All Higgs coupling are, therefore, determined in terms of the masses M_H, M_W, M_Z, m_e and the vacuum expectation value, $v = (\sqrt{2}G_F)^{\frac{-1}{2}}$. The couplings are shown in Figure 8.13. Since all the coupling parameters except M_H are known from experiments, the coupling of H with physical particles through their decay and production can be determined to give its mass.

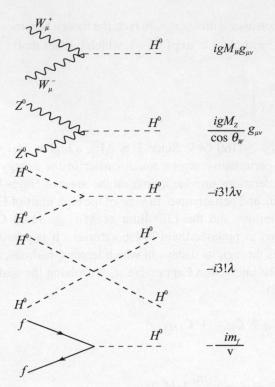

Figure 8.13 The couplings of the Higgs boson to the vector bosons, fermions, and self coupling.

8.7.1 Discovery of Higgs boson

The Higgs boson can be produced at very high energy in the $p\bar{p}$ and $e^{-}e^{+}$ collisions. In $e^{-}e^{+}$ collisions, it can be produced through scattering and annihilation processes like

$$e^{+}e^{-} \longrightarrow e^{+}e^{-}H,$$
$$e^{+}e^{-} \longrightarrow ZH,$$
$$e^{+}e^{-} \longrightarrow \nu\bar{\nu}\,W^{+}W^{-} \to \nu\bar{\nu}H.$$

In the hadronic collisions like pp scattering, the Higgs bosons are produced in the quark–quark interaction or the gluon–gluon interactions through processes like (Figure 8.14):

$$gg \longrightarrow HX, \qquad gg \longrightarrow HWX,$$
$$qq \longrightarrow qqH, \qquad qq \longrightarrow HX, \qquad qq \longrightarrow HWX.$$

The dominant production is through the gluon fusion reaction, that is, $gg \to HX$ in which the production cross section could be as large as $\sigma \approx 10^{2}\ pb$ at the CM energy of 40 GeV for $M_{H} \sim 130 - 150$ GeV and is an order of magnitude larger than other processes.

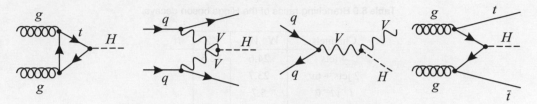

Figure 8.14 Higgs production mechanisms: ggF, VBF, VH, and $t\bar{t}H$.

Once produced, the Higgs boson will decay through processes like (Figure 8.15 and 8.16):

$$H \longrightarrow f\bar{f}, \qquad H \longrightarrow f\bar{f}\gamma, \qquad H \longrightarrow gg,$$
$$H \longrightarrow VV, \qquad H \longrightarrow W^+W^-, \qquad H \longrightarrow ZZ,$$

with f = leptons, quarks and $V = \gamma, W^{\pm}, Z$ bosons.

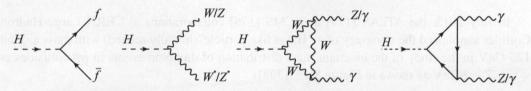

Figure 8.15 Feynman diagrams for dominant Higgs decay processes.

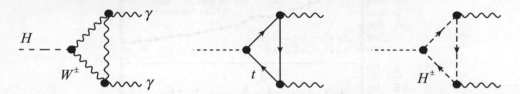

Figure 8.16 One-loop contributions to $H \to \gamma\gamma$. The third diagram shows a possible non-SM contribution from a charged scalar.

The branching ratios of the various decay modes are shown in Table 8.9 where the decay rates are given by [397]

$$\Gamma(H \to f\bar{f}) = N_f \frac{G_F m_f^2 M_H}{4\sqrt{2}} \left(1 - \frac{4m_f^2}{M_H^2}\right)^{\frac{3}{2}},$$

with $N_f = 3$ for the quarks and $N_f = 1$ for the leptons,

$$\Gamma(H \longrightarrow W^+W^-) = \frac{G_F^2 M_H^3}{8\pi\sqrt{2}} \left(1 - \frac{4M_W^2}{M_H^2} + 12\frac{M_W^4}{M_H^4}\right)^{\frac{3}{2}}$$

$$\Gamma(H \longrightarrow ZZ) = \frac{1}{2}\Gamma(H \to W^+W^-)_{M_W = M_Z}.$$

Table 8.9 Branching ratios of the Higgs boson decays.

Channels	W^+W^-	ZZ	$t\bar{t}$
4 jets	24.6	12.6	
2 jets + 6ν	23.7		
$l^+\nu l^-\bar{\nu}$	5.7		
2 jets + l^-l^+		3.6	
2 jets + $\nu\bar{\nu}$		7.3	
$l^-l^+l^-l^+$		0.3	
$l^-l^+\nu\bar{\nu}$		1.1	
$\nu\bar{\nu}\nu\bar{\nu}$		1.1	
6 jets			9.1
4 jets + $l\nu$			8.1
2 jets + $l^-\bar{\nu}l^+\nu$			2.1

In July 2012, the ATLAS [177] and CMS [176] collaborations at CERN Large Hadron Collider announced the discovery of a "Higgs like particle"(initially named) with mass around 125 GeV in the study of the invariant mass distribution of diphoton events in $p\bar{p}$ collisions at $\sqrt{s} = 7 - 8$ TeV as shown in Figure 8.17 ([398]).

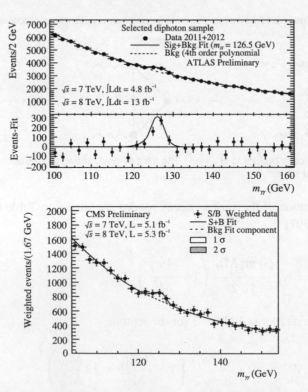

Figure 8.17 The invariant mass distribution of diphoton candidates, with each event weighted by the signal-to-background ratio in each event category, observed by ATLAS [398] at Run 2.

Later with better statistics, the discovery of Higgs boson was confirmed.

The masses measured by the two experiments are in good agreement, giving the average value [399]:

$$M_H = (125.09 \pm 0.21 \pm 0.11) \text{ GeV} = (125.09 \pm 0.24) \text{ GeV}. \tag{8.151}$$

Chapter 9

Neutrino and Electron Scattering from Point Particles

9.1 Introduction

We have seen in Chapter 8, that the standard model for the leptons developed by Salam [37] and Weinberg [157], which was later extended to the quark sector by Glashow [64], unifies weak and electromagnetic interactions. It predicts, in a unique way, the interaction Lagrangian for charge changing (CC) weak interactions of leptons and neutrinos of all flavors with charged gauge vector bosons, $W^{\pm}$ and the electromagnetic interactions of charged leptons with photons. It also predicts the existence of neutral current(NC) weak interactions of charged leptons and neutrinos of all flavors with the neutral gauge boson, Z^0. The strength of the interaction of the charged, neutral, and electromagnetic currents with the $W^{\pm}$, Z^0, and A gauge bosons are described in terms of the weak coupling constants g, electromagnetic coupling constant e, and a free parameter θ_W called the weak mixing angle. Specifically, the interaction Lagrangian discussed in Chapter 8 is written here again as:

$$L_I = -e \left[\frac{1}{2\sqrt{2}\sin\theta_W} \left(j_\mu^{CC} W^{\mu+} + \text{h.c.} \right) + \frac{1}{2\sin\theta_W \cos\theta_W} j_\mu^{NC} Z^\mu + j_\mu^{EM} A^\mu \right], \quad (9.1)$$

where $W_\mu^{\pm}$, Z_μ, and A_μ are the charged, neutral and electromagnetic gauge fields and

$$j_\mu^{CC} = \sum_{l=e,\mu,\tau} \overline{\psi}_l \gamma_\mu (1 - \gamma^5) \psi_{\nu_l}, \qquad (9.2)$$

$$j_\mu^{NC} = \sum_{l=e,\mu,\tau} \left[\overline{\psi}_l \gamma_\mu (g_V^l - g_A^l \gamma^5) \psi_l + \overline{\psi}_{\nu_l} \gamma_\mu (g_V^{\nu_l} - g_A^{\nu_l} \gamma^5) \psi_{\nu_l} \right], \qquad (9.3)$$

$$j_\mu^{EM} = \sum_{l=e,\mu,\tau} \overline{\psi}_l \gamma_\mu \psi_l, \qquad (9.4)$$

with $\sin \theta_W = \frac{e}{g}$, $g_V^l = -\frac{1}{2} + 2\sin^2 \theta_W$, $g_A^l = -\frac{1}{2}$, $g_V^{\nu_l} = \frac{1}{2}$ and $g_A^{\nu_l} = \frac{1}{2}$, where $e = \sqrt{4\pi\alpha}$, α is the fine structure constant. In the following sections, we use the interaction Lagrangian in Eq. (9.1) to calculate the cross sections for some weak and electromagnetic processes using point particles, that is, charged leptons and neutrinos.

9.2 $e^- + \mu^- \longrightarrow e^- + \mu^-$ scattering

This scattering process can take place through an electromagnetic process mediated by a virtual photon as well as by the weak neutral current mediated by a Z^0 boson in the standard model [400].

When an electron interacts with a photon field (Figure 9.1(a)), the interaction Lagrangian is given by:

$$\mathcal{L}_I^{\text{em}} = -e\overline{\psi}_e(\vec{k}')\gamma^\mu \psi_e(\vec{k}) A_\mu \tag{9.5}$$

and when it interacts with the Z^0 boson field (Figure 9.2(a)), the Lagrangian is given by:

$$\mathcal{L}_I^{\text{NC}} = \frac{-g}{2\cos\theta_W} \overline{\psi}(\vec{k}')\gamma_\mu(g_V^e - g_A^e\gamma^5)\psi(\vec{k})Z^\mu. \tag{9.6}$$

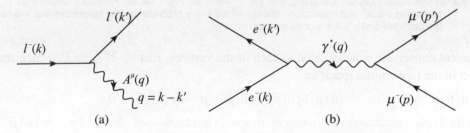

Figure 9.1 (a) Left panel: Interaction of a charged lepton with a photon field. (b) Right panel: Feynman diagram for $e^- - \mu^-$ scattering. The quantities in the brackets are the respective four momenta of the particles and q is the four momentum transfer.

Using the aforementioned Lagrangians corresponding to Figure 9.1(b) and following the Feynman rules the transition amplitude $\mathcal{M}$ for the process

$$e^-(\vec{k}, E_e) + \mu^-(\vec{p}, E_\mu) \longrightarrow e^-(\vec{k}', E_e') + \mu^-(\vec{p}', E_\mu') \tag{9.7}$$

mediated through virtual photon exchange of momentum $q = k - k'$, in the lowest order, may be written as:

$$-i\mathcal{M}_\gamma = \bar{u}(\vec{k}')ie\gamma^\mu u(\vec{k})\left(\frac{-ig_{\mu\nu}}{q^2}\right)\bar{u}(\vec{p}')ie\gamma^\nu u(\vec{p}) \tag{9.8}$$

and for the process mediated through the virtual Z^0 exchange, shown by the Feynman diagram in Figure 9.2(b) as:

$$-i\mathcal{M}_{Z^0} = \left[\bar{u}(\vec{k}')\left(\frac{-ig}{2\cos\theta_W}\right)\gamma_\mu(g_V^e - g_A^e\gamma_5)u(\vec{k})\right]\left(\frac{-ig^{\mu\nu}}{M_Z^2}\right)\left[\bar{u}(\vec{p}')\left(\frac{-ig}{2\cos\theta_W}\right)\right.$$

$$\times \; \gamma_\nu (g_V^\mu - g_A^\mu \gamma_5) u(\vec{p}) \Big],$$

$$\Rightarrow \mathcal{M}_{Z^0} = \left(\frac{-g^2}{4 M_Z^2 \cos^2 \theta_W} \right) \Big[\bar{u}(\vec{k}') \gamma_\mu (g_V^e - g_A^e \gamma_5) u(\vec{k}) \Big] \Big[\bar{u}(\vec{p}') \gamma^\mu (g_V^e - g_A^e \gamma_5) u(\vec{p}) \Big],$$

(9.9)

where $g_V^e = g_V^\mu = -\frac{1}{2} + 2 \sin^2 \theta_W$ and $g_A^e = g_A^\mu = -\frac{1}{2}$. The process proceeding through Z^0 is highly suppressed as compared to the photon exchange ($\sigma_\gamma \approx 10^5 \times \sigma_{Z^0}$), therefore, in the present case, we present the cross section for the process given in Eq. (9.7) mediating through γ-exchange.

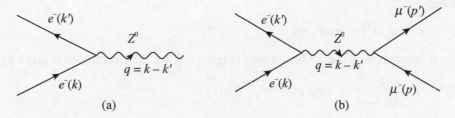

Figure 9.2 (a) Left panel: Lepton interacting with the Z^0 field. (b) Right panel: Feynman diagram for $e^- - \mu^-$ scattering via Z^0 field interaction. The quantities in the brackets are the respective four momenta of the particles and q is the four momentum transfer.

Current conservation holds good at each of the vertices, that is, $\partial^\mu J_\mu = 0$, which may be verified in the momentum space as:

$$\bar{u}(\vec{p}') q^\mu \gamma_\mu u(\vec{p}) = \bar{u}(\vec{p}') q u(\vec{p}) = \bar{u}(\vec{p}')(p' - p) u(\vec{p}) = 0,$$

where the Dirac equation (in the momentum space) has been used, that is, $(p - m) u(\vec{p}) = 0$. In the case of unpolarized e^- and μ^-, we average over the initial spins and sum over the final spins. Using the standard projection operators for spin $\frac{1}{2}$ spinors (Appendix D), we may write:

$$\overline{\sum} \sum |\mathcal{M}_\gamma|^2 = \frac{1}{2} \cdot \frac{1}{2} \cdot \frac{e^4}{q^4} \cdot Tr \Big[(k' + m_e) $$
$$\gamma^\mu (k + m_e) \gamma^\nu \Big] \cdot Tr \Big[(p' + m_\mu) \gamma_\mu (p + m_\mu) \gamma_\nu \Big]$$

$$= \frac{e^4}{4 q^4} L_e^{\mu\nu} L_{\mu\nu}^{\text{muon}},$$

(9.10)

where the symbol $\overline{\sum}$ represents the average taken over initial spins which gives a factor of $\frac{1}{2}$ each for e^- and μ^- and $\sum$ represents the sum over the final spin states. m_e and m_μ are the masses of electron and muon, respectively. The electronic tensor $L_e^{\mu\nu}$ and the muonic tensor $L_{\mu\nu}^{\text{muon}}$ are given as (Appendix D):

$$L_e^{\mu\nu} = Tr \Big[(k' + m_e) \gamma^\mu (k + m_e) \gamma^\nu \Big] = 4 [k^\mu k'^\nu + k'^\mu k^\nu - (k \cdot k' - m_e^2) g^{\mu\nu}], \quad (9.11)$$

$$L_{\mu\nu}^{\text{muon}} = Tr \Big[(p' + m_\mu) \gamma_\mu (p + m_\mu) \gamma_\nu \Big] = 4 [p_\mu p'_\nu + p'_\mu p_\nu - (p \cdot p' - m_\mu^2) g_{\mu\nu}]. \quad (9.12)$$

Contraction of $L_e^{\mu\nu}$ and $L_{\mu\nu}^{\text{muon}}$ gives us the following expression for the matrix element squared, when averaged over the initial spin states and summed over the final spin states leading to

$$\Rightarrow \ \overline{\sum}\sum |\mathcal{M}_\gamma|^2 \ = \ \frac{4e^4}{q^4}\left\{2(k\cdot p)(k'\cdot p') + 2(k\cdot p')(k'\cdot p) - 2(p\cdot p' - m_\mu^2)(k\cdot k')\right.$$

$$\left. -2(k\cdot k' - m_e^2)(p\cdot p') + 4(k\cdot k' - m_e^2)(p\cdot p' - m_\mu^2)\right\}. \quad (9.13)$$

Defining the four momentum transfer $q = k - k' = p' - p$, and $q^2 = -2k.k' = -2EE'(1 - \cos\theta)$, in the limit of $m_e \to 0$, the expression reduces to

$$\overline{\sum}\sum |\mathcal{M}_\gamma|^2 = \left[\frac{8e^4}{q^4}\left\{2k'\cdot p\,k\cdot p + (-q^2/2)k\cdot p + (q^2/2)k'\cdot p - m_\mu^2 k\cdot k'\right\}\right].$$

In the Lab frame, the target particle is at rest(Figure 9.3), that is, $\vec{p} = 0$, $E_\mu = m_\mu$, which leads to $k\cdot p = m_\mu E_e$ and $k'\cdot p = m_\mu E_e'$.

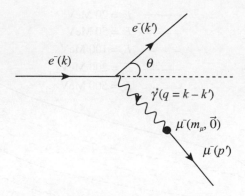

Figure 9.3 $e^- - \mu^-$ scattering in the laboratory frame. q is the four momentum transfer, that is, $q = k - k' = p' - p$. θ is the Lab scattering angle.

In this frame, $\overline{\sum}\sum |\mathcal{M}_\gamma|^2$ is derived to be (Appendix D):

$$\overline{\sum}\sum |\mathcal{M}_\gamma|^2 \ = \ \left[\frac{8e^4}{q^4}\left\{2m_\mu^2 E_e E_e' - \frac{q^2}{2}m_\mu(E_e - E_e') + \frac{m_\mu^2 q^2}{2}\right\}\right]$$

$$= \ \left\{\frac{8e^4}{q^4}2m_\mu^2 E_e E_e'\left[\cos^2\frac{\theta}{2} - \frac{q^2}{2m_\mu^2}\sin^2\frac{\theta}{2}\right]\right\}. \quad (9.14)$$

If the outgoing electron is to be observed, then the integration over momenta of the outgoing muon is to be performed and vice versa.

In the Lab frame, the differential scattering cross section is given by (Appendix E):

$$\left.\frac{d\sigma}{d\Omega_e}\right|_{\text{Lab}} \ = \ \frac{1}{64\pi^2 m_\mu E_e}\overline{\sum}\sum |\mathcal{M}_\gamma|^2 \frac{|\vec{k}'|^3}{((E_e + m_\mu)|\vec{k}'|^2 - \vec{k}\cdot\vec{k}'E_e')} \quad (9.15)$$

which leads to

$$\frac{d\sigma}{d\Omega_e}\bigg|_{\text{Lab}} = \frac{\alpha^2}{4E_e^2 \sin^4\left(\frac{\theta}{2}\right)} \frac{E_e'}{E_e} \left\{ \cos^2\frac{\theta}{2} - \frac{q^2}{2m_\mu^2} \sin^2\frac{\theta}{2} \right\} \tag{9.16}$$

and the total scattering cross section σ is given by:

$$\sigma|_{\text{Lab}} = 2\pi \int_{-1}^{+1} d(\cos\theta) \frac{\alpha^2}{4E_e^2 \sin^4\left(\frac{\theta}{2}\right)} \frac{E_e'}{E_e} \left\{ \cos^2\frac{\theta}{2} - \frac{q^2}{2m_\mu^2} \sin^2\frac{\theta}{2} \right\}. \tag{9.17}$$

Figure 9.4 presents the results for $\frac{d\sigma}{d\Omega}\big|_{\text{Lab}}$ vs. $\cos\theta$ at various values of incident electron energies, viz., $E_e = 20, 50, 100, 200$ and 500 MeV. It may be observed that the differential scattering cross section is forward peaked. When the expression given in Eq. (9.16) is compared

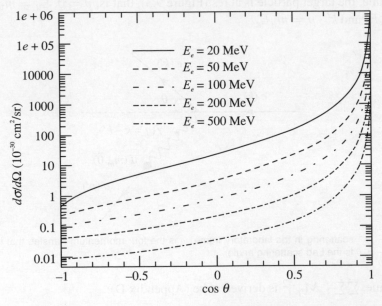

Figure 9.4 $\frac{d\sigma}{d\Omega}$ vs. $\cos\theta$ at various values of incident electron energies viz. $E_e = 20, 50, 100, 200$ and 500 MeV evaluated in the Lab frame.

with the results of $d\sigma/d\Omega$ for the scattering of e^- with a spinless ($J = 0$) point target (i.e., without any structure) known as Mott scattering cross section, which will be discussed in Chapter 10, one obtains:

$$\frac{d\sigma}{d\Omega}\bigg|_{\text{spinless}} = \frac{\alpha^2}{4E_e^2 \sin^4\theta/2} \cdot \frac{E_e'}{E_e} \cos^2\frac{\theta}{2}$$

$$\Rightarrow \qquad \frac{\frac{d\sigma}{d\Omega}\big|_{e^-\mu^- \to e^-\mu^-}}{\frac{d\sigma}{d\Omega}\big|_{\text{spinless}}} = 1 + \frac{Q^2}{2m_\mu^2} \tan^2\frac{\theta}{2}, \tag{9.18}$$

where $Q^2 = -q^2 \geq 0$. The extra factor in Eq. (9.18) (i.e., $\frac{Q^2}{2m_\mu^2} \sin^2 \frac{\theta}{2}$) is due to the presence of the spin content of the muon and is considered to be the contribution of the muon magnetic moment to the scattering cross section. One may also express the cross section in the center of

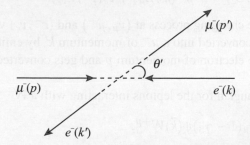

Figure 9.5 $e^- - \mu^-$ scattering in the center of mass (CM) frame. θ' is the CM scattering angle.

mass (CM) frame (Figure 9.5), in which

$$k \cdot p \;=\; k' \cdot p' = \frac{s - m_e^2 - m_\mu^2}{2}; \qquad\qquad k \cdot k' = -\frac{t - 2m_e^2}{2};$$

$$k \cdot p' \;=\; k' \cdot p = -\frac{u - m_e^2 - m_\mu^2}{2}; \qquad\qquad p \cdot p' = -\frac{t - 2m_\mu^2}{2}, \tag{9.19}$$

where s, t, u are the Mandelstam variables defined as:

$$s = (k + p)^2 = (k' + p')^2; \;\; t = (k - k')^2 = (p' - p)^2 \;\text{ and }\; u = (k - p')^2 = (k' - p)^2. \tag{9.20}$$

The general expression for the differential scattering cross section in CM frame is given as (Appendix E):

$$\left.\frac{d\sigma}{d\Omega}\right|_{CM} = \frac{1}{64\pi^2 s} \sum\sum |\mathcal{M}_\gamma|^2. \tag{9.21}$$

Using the values of $s, t,$ and u from Eq. (9.20) in Eq. (9.21), the differential cross section in the CM frame becomes,

$$\left.\frac{d\sigma}{d\Omega}\right|_{CM} = \frac{1}{16\pi^2 s}\frac{e^4}{t^2}\left[\frac{(s - m_e^2 - m_\mu^2)^2}{2} + \frac{(u - m_e^2 - m_\mu^2)^2}{2} - \left(\frac{t}{2}\right)(t - 2m_e^2)\right.$$

$$\left. - \left(\frac{t}{2}\right)(t - 2m_\mu^2) + t^2\right]. \tag{9.22}$$

In the relativistic limit, that is, $E_e >> m_e$ and $E_\mu >> m_\mu$, we get

$$\left.\frac{d\sigma}{d\Omega}\right|_{CM} = \frac{1}{32\pi^2 s}e^4\left(\frac{s^2 + u^2}{t^2}\right). \tag{9.23}$$

9.3 $\nu_\mu + e^- \to \mu^- + \nu_e$ **scattering**

This process is also known as inverse muon decay and is expressed as:

$$\nu_\mu(\vec{k}, E_{\nu_\mu}) + e^-(\vec{p}, E_e) \longrightarrow \mu^-(\vec{k}', E_\mu) + \nu_e(\vec{p}\,', E_{\nu_e}). \tag{9.24}$$

The scattering is a charge changing process at (ν_μ, μ^-) and (e^-, ν_e) vertices. In this process, a ν_μ of momentum k gets converted into a μ^- of momentum k' by emitting a virtual W^+ boson which is absorbed by the electron of momentum p and gets converted into a ν_e of momentum p'.

The interaction Lagrangian for the leptons interacting with a W^+ field is given by:

$$\mathcal{L}_I = \frac{-g}{2\sqrt{2}} \bar{\psi}(\vec{k}') \gamma_\mu (1 - \gamma_5) \psi(\vec{k}) W^{+\mu}. \tag{9.25}$$

Using this Lagrangian, we obtain the transition amplitude using Feynman rules, in the limit of $q^2 << M_W^2$, for the Feynman diagram depicted in Figure 9.6 as:

$$-i\mathcal{M}_{CC} = \bar{u}(\vec{k}') \left[\frac{-ig}{2\sqrt{2}} \gamma^\mu (1 - \gamma_5) \right] u(\vec{k}) \left(\frac{-ig_{\mu\nu}}{M_W^2} \right) \bar{u}(\vec{p}\,') \left[\frac{-ig}{2\sqrt{2}} \gamma^\nu (1 - \gamma_5) \right] u(\vec{p})$$

$$\implies \quad \mathcal{M}_{CC} = \frac{-g^2}{8 M_W^2} [\bar{u}(\vec{k}') \gamma^\mu (1 - \gamma_5) u(\vec{k})][\bar{u}(\vec{p}\,') \gamma_\mu (1 - \gamma_5) u(\vec{p})] \tag{9.26}$$

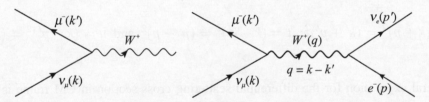

Figure 9.6 (a) Left panel: ν_μ interaction with a W field. (b) Right panel: Charged current interaction $\nu_\mu e^- \to \mu^- \nu_e$.

and

$$\overline{\sum_i} \sum_f |\mathcal{M}|^2 = \frac{1}{2} \frac{G_F^2}{2} L_{\text{muon}}^{\mu\nu} L_{\mu\nu}^{\text{electron}}, $$

where $\overline{\sum}_i$ represents the average taken over the spin of initial particles which gives a factor of $\frac{1}{2}$ due to the electron, as ν_μ has only one component being left handed. In this expression,

$$L_{\text{muon}}^{\mu\nu} = \sum_{\mu,\nu=0}^{3} Tr\left[(\not{k}' + m_\mu) \gamma^\mu (1 - \gamma_5)(\not{k}) \gamma^\nu (1 - \gamma_5) \right], \tag{9.27}$$

which leads to (Appendix D):

$$L_{\text{muon}}^{\mu\nu} = 8 \left[k^\mu k'^\nu + k'^\mu k^\nu - g^{\mu\nu} k \cdot k' + i\epsilon^{\mu\nu\alpha\beta} k_\alpha k'_\beta \right] \tag{9.28}$$

and

$$L_{\mu\nu}^{\text{electron}} = \sum_{\mu,\nu=0}^{3} Tr\left[\not{p}'\gamma_\mu(1-\gamma_5)(\not{p}+m_e)\gamma_\nu(1-\gamma_5)\right], \tag{9.29}$$

which leads to (Appendix D):

$$L_{\mu\nu}^{\text{electron}} = 8\left[p_\mu p'_\nu + p'_\mu p_\nu - g_{\mu\nu}p\cdot p' - i\epsilon_{\mu\nu\sigma\rho}p^\sigma p'^\rho\right]. \tag{9.30}$$

Contraction of $L_{\text{muon}}^{\mu\nu}$ with $L_{\mu\nu}^{\text{electron}}$ is similar to what has been discussed in the case of $e^-\mu^-$ scattering, but besides the symmetric terms, there will be an additional term from the antisymmetric part because of the following property of the tensors

$$\epsilon_{\mu\nu\alpha\beta}\epsilon^{\mu\nu\sigma\rho} = 2(\delta_\alpha^\sigma\delta_\beta^\rho - \delta_\alpha^\rho\delta_\beta^\sigma). \tag{9.31}$$

Incorporating this property, the matrix element squared becomes

$$\overline{\sum_i}\sum_f |\mathcal{M}|^2 = 64G_F^2(k\cdot p)(k'\cdot p'). \tag{9.32}$$

In the Lab frame for the initial electron at rest, we have

$$k\cdot p = m_e E_{\nu_\mu}; \ k\cdot k' = E_{\nu_\mu}E_\mu - E_{\nu_\mu}|\vec{k}'|\cos\theta \ \text{and} \ k'\cdot p' = k'\cdot(k-k'+p),$$

which results in

$$\overline{\sum_i}\sum_f |\mathcal{M}|^2 = 64G_F^2(m_e E_{\nu_\mu})\left[E_{\nu_\mu}E_\mu - E_{\nu_\mu}|\vec{k}'|\cos\theta + m_e E_\mu - m_\mu^2\right]. \tag{9.33}$$

The expression for the differential cross section in the Lab frame is then given by:

$$\frac{d\sigma}{d\Omega}\bigg|_{\text{Lab}} = \frac{1}{\pi^2 m_e}G_F^2\frac{E_\mu^2}{E_{\nu_\mu}}\left[E_{\nu_\mu}E_\mu - E_{\nu_\mu}|\vec{k}'|\cos\theta + m_e E_\mu - m_\mu^2\right] \tag{9.34}$$

and the total scattering cross section σ, in the relativistic limit, that is, $|\vec{p}| = E$, is given by:

$$\sigma = 2\pi\int_{-1}^{+1} d(\cos\theta)\frac{1}{\pi^2 m_e}G_F^2\frac{E_\mu^2}{E_{\nu_\mu}}\left[E_{\nu_\mu}E_\mu - E_{\nu_\mu}|\vec{k}'|\cos\theta + m_e E_\mu - m_\mu^2\right]. \tag{9.35}$$

The expression for the differential cross section in the CM frame is given as:

$$\frac{d\sigma}{d\Omega}\bigg|_{\text{CM}} = \frac{G_F^2}{4\pi^2 s}\left[(s-m_e^2)(s-m_\mu^2)\right], \tag{9.36}$$

where $s = (k+p)^2 = E_{\text{CM}}^2$.

In 1980, the CHARM collaboration studied the inverse muon decay using wide band neutrino and antineutrino beams from CERN SPS; a total number of 171 ± 29 events for $\nu_\mu + e^- \rightarrow \mu^- + \nu_e$ were observed [265]. The observed rates were in agreement with the prediction derived for $V - A$ theory and for left-handed two-component neutrinos. Later, in 1990, CHARM-II collaboration performed the $\nu_\mu + e^- \rightarrow \mu^- + \nu_e$ scattering experiment and found the value of the cross section to be $(18.16 \pm 1.36) \times 10^{-42} \text{cm}^2 \text{GeV}^{-1}$ [401]. A similar experiment was performed at the Tevatron, Fermilab, where the cross section was measured and found to be $\sigma(\nu_\mu e^- \longrightarrow \mu^- \nu_e) = [16.93 \pm 0.85(stat.) \pm 0.52(syst.)] \times E \times 10^{-42} \text{ cm}^2$ [266], which is consistent with the theoretically predicted value obtained using the standard model.

9.4 $\nu_\mu + e^- \rightarrow \nu_\mu + e^-$ scattering

The process

$$\nu_\mu(\vec{k}, E_{\nu_\mu}) + e^-(\vec{p}, E_e) \rightarrow \nu_\mu(\vec{k}', E'_{\nu_\mu}) + e^-(\vec{p}', E'_e) \tag{9.37}$$

is a neutral current induced process, mediated by a Z^0 boson, in the standard model. In this process, a ν_μ of momentum k emits a Z^0 boson that gets absorbed by an electron of momentum p.

The interaction Lagrangian for $\nu\nu Z$ interaction, shown in Figure 9.7(a), is given as:

$$\mathcal{L}_I^{\text{neutrino}} = \frac{-g}{2\cos\theta_W}\bar{\psi}(\vec{k}')\gamma_\mu(g_V^\nu - g_A^\nu\gamma^5)\psi(\vec{k})Z^\mu. \tag{9.38}$$

Similarly, the interaction Lagrangian at the electron vertex shown in Figure 9.7(b) is given by:

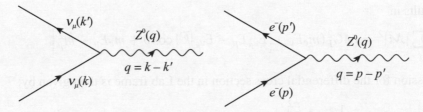

Figure 9.7 (a) Left panel: Neutrino interacting with the Z^0 field. (b) Right panel: Electron interacting with the Z^0 field.

$$\mathcal{L}_I^{\text{electron}} = \frac{-g}{2cos\theta_W}\bar{\psi}(\vec{p}')\gamma_\mu(g_V^e - g_A^e\gamma^5)\psi(\vec{p})Z^\mu. \tag{9.39}$$

Using the interaction Lagrangians given in Eqs. (9.38) and (9.39), the transition matrix element, corresponding to the Feynman diagram shown in Figure 9.8 is given by:

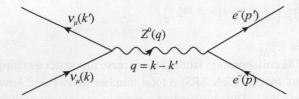

Figure 9.8 Neutral current reaction for the process $\nu_\mu + e^- \longrightarrow \nu_\mu + e^-$.

$$-i\mathcal{M}_{NC} = \left[\bar{u}(\vec{k}')\left(\frac{-ig}{2\cos\theta_W}\right)\gamma_\mu(g_V^\nu - g_A^\nu\gamma_5)u(\vec{k})\right]\left(\frac{-ig^{\mu\nu}}{M_Z^2}\right)\left[\bar{u}(\vec{p}')\left(\frac{-ig}{2\cos\theta_W}\right)\right.$$

$$\left.\times\ \gamma_\nu(g_V^e - g_A^e\gamma_5)u(\vec{p})\right],$$

with $g_V^\nu = \frac{1}{2}, g_A^\nu = \frac{1}{2}, g_V^e = -\frac{1}{2} + 2\sin^2\theta_W$, and $g_A^e = -\frac{1}{2}$. The expression for square of the invariant matrix element for the reaction given in Eq. (9.37) is obtained as:

$$\sum_i\sum_f|\mathcal{M}|^2 = \frac{1}{2}\left(\frac{4G_F^2}{2}\right)L_{\mu\nu}^{\text{neutrino}}L_{\text{electron}}^{\mu\nu}$$

where $\frac{1}{2}$ is due to the averaging over the initial electron spin, $\frac{G_F}{\sqrt{2}} = \frac{g^2}{8M_Z^2\cos^2\theta_W}$ and the leptonic tensors are obtained as:

$$L_{\mu\nu}^{\text{neutrino}} = 2\left[k_\mu k'_\nu + k'_\mu k_\nu - g_{\mu\nu}k\cdot k' - i\epsilon_{\sigma\mu\rho\nu}k^\rho k'^\sigma\right],$$

$$L_{\text{electron}}^{\mu\nu} = 4[\{(g_V^e)^2 + (g_A^e)^2\}(p'^\mu p^\nu - p'\cdot pg_{\mu\nu} + p'^\nu p^\mu) + 2p'_\lambda p_\theta i\epsilon^{\lambda\mu\theta\nu}g_V^e g_A^e$$

$$+m_e^2 g^{\mu\nu}\{(g_V^e)^2 - (g_A^e)^2\}].$$

The contraction of the leptonic tensors ($L_{\mu\nu}^{\text{neutrino}}$ and $L_{\text{electron}}^{\mu\nu}$), gives:

$$\sum_i\sum_f|\mathcal{M}|^2 = 16\,G_F^2\left[(g_V^e + g_A^e)^2(k'\cdot p')(k\cdot p) + (g_V^e - g_A^e)^2(k'\cdot p)(k\cdot p')\right.$$

$$\left. -\ m_e^2\{(g_V^e)^2 - (g_A^e)^2\}(k\cdot k')\right].$$

In the Lab frame, the initial electron is at rest and we have

$$k\cdot p = m_e E_\nu,\ k\cdot p' = E_\nu E'_e - E_\nu|\vec{p}'|\cos\theta,\quad \theta \text{ is the Lab scattering angle}$$

$$k\cdot k' = E_\nu m_e - E_\nu E'_e + E_\nu|\vec{p}'|\cos\theta,\ k'\cdot p' = E_\nu E'_e - E_\nu|\vec{p}'|\cos\theta + m_e E'_e - m_e^2$$

and $k'\cdot p = (k+p-p')\cdot p = m_e E_\nu + m_e^2 - m_e E'_e.$ \hfill (9.40)

Therefore,

$$\sum_i\sum_f|\mathcal{M}|^2 = 16\,G_F^2 m_e E_\nu\left[(g_V^e + g_A^e)^2(E_\nu E'_e - E_\nu|\vec{p}'|\cos\theta + m_e E'_e - m_e^2)\right.$$

$$+\ (g_V^e - g_A^e)^2(E_\nu + m_e - E'_e)(E'_e - |\vec{p}'|\cos\theta)$$

$$\left. -\ m_e\{(g_V^e)^2 - (g_A^e)^2\}(m_e - E'_e + |\vec{p}'|\cos\theta)\right], \hfill (9.41)$$

and the expression for the differential cross section in the Lab frame is obtained as:

$$\frac{d\sigma}{d\Omega}\bigg|_{Lab} = \frac{1}{4\pi^2 m_e}G_F^2\left(\frac{E_e'^2}{E_\nu}\right)\Big[(g_V^e + g_A^e)^2(E_\nu E_e' - E_\nu|\vec{p}\,'|\cos\theta + m_e E_e' - m_e^2)$$

$$+ (g_V^e - g_A^e)^2(E_\nu + m_e - E_e')(E_e' - |\vec{p}\,'|\cos\theta)$$

$$- m_e\{(g_V^e)^2 - (g_A^e)^2\}(m_e - E_e' + |\vec{p}\,'|\cos\theta)\Big]. \tag{9.42}$$

The total scattering cross section is then given as:

$$\sigma = 2\pi\int_{-1}^{+1}d(\cos\theta)\frac{1}{4\pi^2 m_e}G_F^2\left(\frac{E_e'^2}{E_\nu}\right)\Big[(g_V^e + g_A^e)^2(E_\nu E_e' - E_\nu|\vec{p}\,'|\cos\theta + m_e E_e' - m_e^2)$$

$$+ (g_V^e - g_A^e)^2(E_\nu + m_e - E_e')(E_e' - |\vec{p}\,'|\cos\theta) - m_e\{(g_V^e)^2 - (g_A^e)^2\}$$

$$(m_e - E_e' + |\vec{p}\,'|\cos\theta)\Big].$$

The expression for the differential cross section in CM frame is obtained as:

$$\frac{d\sigma}{d\Omega}\bigg|_{CM} = \frac{1}{4\pi^2 s}G_F^2\Big[(g_V^e + g_A^e)^2\Big(\frac{s - m_e^2}{2}\Big)^2 + (g_V^e - g_A^e)^2\Big(\frac{u - m_e^2}{2}\Big)^2$$

$$+ \frac{m_e^2}{2}\{(g_V^e)^2 - (g_A^e)^2\}t\Big], \tag{9.43}$$

where $s = (k + p)^2 = E_{CM}^2$. A similar expression may be obtained for $\bar{\nu}_\mu + e^- \to \bar{\nu}_\mu + e^-$ differential scattering cross section, using the matrix element squared, given in Eq. (9.50) and Table 9.1.

In 1973, at CERN, the Gargamelle bubble chamber experiment [402] was used to observe for the first time $\nu_\mu e^-$ and $\bar{\nu}_\mu e^-$ events; cross section measurements were performed in the energy region of ≈ 1 GeV. The experiment also measured the weak $\sin^2\theta_W$ and put some limit on it. The CHARM collaboration [351] in 1977, performed the $\nu_\mu e^-$ cross section measurements using the CERN narrow band beam (NBB) as well as the wide band neutrino beam (WBB). In total, 83 ± 16 $\nu_\mu e^-$ and 116 ± 21 $\bar{\nu}_\mu e^-$ events were reported; the cross sections were measured and have been tabulated in Table 9.2. At the Brookhaven National Laboratory (BNL), the E734 experiment [145] measured the $\nu_\mu e^-$ and $\bar{\nu}_\mu e^-$ scattering cross sections using an alternating gradient synchrotron source (AGS) with an average energy of $\nu_\mu(\bar{\nu}_\mu) \approx 1.3$ GeV. A total number of $160 \pm 17 \pm 4$ and $97 \pm 13 \pm 5$ events were reported for $\nu_\mu e^-$ and $\bar{\nu}_\mu e^-$, respectively. Moreover, CHARM-II collaboration [403] at CERN in 1987 used CERN-SPS WBB and measured the $\nu_\mu e^-$ cross section; they also performed a high precision measurement of $\sin^2\theta_W$. The results of the experiment have been tabulated in Table 9.2. $\nu_\mu e^-$ scattering cross sections were also measured at the Fermi National Accelerator Laboratory (FNAL). The experiments were performed using wide band ν_μ beam with an average energy of 20 GeV and a maximum energy of 100 GeV. The reported cross section is [166]

$$\sigma_{expt}(\nu_\mu e^- \to \nu_\mu e^-) = (1.40 \pm 0.30) \times 10^{-40} E_{\nu_\mu} cm^2/GeV.$$

These results are consistent with the standard model if one takes $\sin^2\theta_W = 0.23$.

Table 9.1 Values of α, β, and γ for $\nu_\mu e^-$, $\overline{\nu}_\mu e^-$, $\nu_e e^-$, and $\overline{\nu}_e e^-$ scattering.

Process	α	β	γ
$\nu_\mu e^- \rightarrow \nu_\mu e^-$	$(g_V^e + g_A^e)^2$	$(g_V^e - g_A^e)^2$	$(g_A^e)^2 - (g_V^e)^2$
$\overline{\nu}_\mu e^- \rightarrow \overline{\nu}_\mu e^-$	$(g_V^e - g_A^e)^2$	$(g_V^e + g_A^e)^2$	$(g_A^e)^2 - (g_V^e)^2$
$\nu_e e^- \rightarrow \nu_e e^-$	$(g_V' + g_A')^2$	$(g_V' - g_A')^2$	$g_A'^2 - g_V'^2$
$\overline{\nu}_e e^- \rightarrow \overline{\nu}_e e^-$	$(g_V' - g_A')^2$	$(g_V' + g_A')^2$	$g_A'^2 - g_V'^2$

9.5 $\nu_e + e^- \rightarrow \nu_e + e^-$ scattering

The process

$$\nu_e(\vec{k}, E_{\nu_e}) + e^-(\vec{p}, E_e) \rightarrow \nu_e(\vec{k}', E'_{\nu_e}) + e^-(\vec{p}\,'E'_e) \tag{9.44}$$

induced by the Z^0 boson at the $(\nu_e\nu_e)$ and the (e^-e^-) vertices as well as by W^+ boson at $(\nu_e e^-)$ vertices are mediated via both the neutral and the charged current interactions in the standard model, respectively. The Lagrangian for the charged current W^+ boson exchange is given by (Figure 9.9 (a)):

$$\mathcal{L}_{\nu_e e^- - W^+} = \frac{-g}{2\sqrt{2}}\overline{\psi}(\vec{p}\,')\gamma_\mu(1 - \gamma_5)\psi(\vec{k})W^{+\mu} \tag{9.45}$$

The Lagrangians for the neutral current Z^0 exchange between the two electrons and two neutrinos is given as (Figure 9.9 (b) and (c)):

$$L^{\nu\nu Z} = \frac{-g}{2\cos\theta_W}\overline{\psi}(\vec{k}')\gamma_\mu(g_V^\nu - g_A^\nu\gamma^5)\psi(\vec{k})Z^\mu, \tag{9.46}$$

$$L^{eeZ} = \frac{-g}{2\cos\theta_W}\overline{\psi}(\vec{p}\,')\gamma_\mu(g_V^e - g_A^e\gamma^5)\psi(\vec{p})Z^\mu, \tag{9.47}$$

with $g_V^\nu = \frac{1}{2}, g_A^\nu = \frac{1}{2}, g_V^e = -\frac{1}{2} + 2\sin^2\theta_W$ and $g_A^e = -\frac{1}{2}$.

Therefore, the invariant matrix element for the Feynman diagram shown in Figure 9.10(a) for the charged current reaction is written as:

$$-i\mathcal{M}^{CC} = \left[\overline{u}(\vec{p}\,')\frac{-ig}{2\sqrt{2}}\gamma_\mu(1 - \gamma_5)u(\vec{k})\right]\left(-\frac{ig^{\mu\nu}}{M_W^2}\right)\left[\overline{u}(\vec{k}')\frac{-ig}{2\sqrt{2}}\gamma_\nu(1 - \gamma_5)u(\vec{p})\right],$$

$$\Rightarrow \mathcal{M}^{CC} = \frac{G_F}{\sqrt{2}}\left[\overline{u}(\vec{p}\,')\gamma_\mu(1 - \gamma_5)u(\vec{k})\right] \cdot \left[\overline{u}(\vec{k}')\gamma^\mu(1 - \gamma_5)u(\vec{p})\right], \tag{9.48}$$

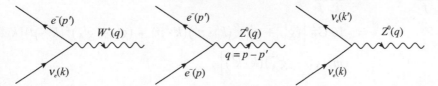

Figure 9.9 (a) Left panel: Neutrinos interacting with the W^+ field. (b) Middle panel: Electrons interacting with the Z^0 field. (c) Right panel: Neutrinos interacting with the Z^0 field.

where

$$\frac{G_F}{\sqrt{2}} = \frac{g^2}{8M_W^2}.$$

Similarly, the matrix element for the Feynman diagram shown in Figure 9.10(b) for the neutral current reaction is given as:

$$
\begin{aligned}
-i\mathcal{M}_{NC} &= \left[\bar{u}(\vec{k}')\frac{-ig}{2\,cos\theta_W}\gamma_\mu(g_V^\nu - g_A^\nu\gamma_5)u(\vec{k}) \right] \times \left(-\frac{ig^{\mu\nu}}{M_Z^2} \right) \\
&\quad \times \left[\bar{u}(\vec{p}')\frac{-ig}{2\,cos\theta_W}\gamma_\nu(g_V^e - g_A^e\gamma_5)u(\vec{p}) \right], \\
\Rightarrow \mathcal{M}_{NC} &= \frac{G_F}{\sqrt{2}}\left[\bar{u}(\vec{k}')\gamma_\mu(1 - \gamma_5)u(\vec{k}) \right] \cdot \left[\bar{u}(\vec{p}')\gamma^\mu(g_V^e - g_A^e\gamma_5)u(\vec{p}) \right],
\end{aligned}
$$

where $\frac{G_F}{\sqrt{2}} = \frac{g^2}{8M_Z^2 cos^2\theta_W}$.

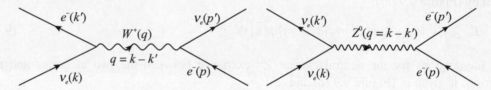

Figure 9.10 (a) Left panel: Charged current. (b) Right panel: Neutral current reactions.

Hence, the total contribution for this process from both the channels is given as:

$$
\begin{aligned}
\mathcal{M}_{CC} + \mathcal{M}_{NC} &= \frac{G_F}{\sqrt{2}}\Big[\left[\bar{u}(\vec{p}')\gamma_\mu(1 - \gamma_5)u(\vec{k}) \right] \cdot \left[\bar{u}(\vec{k}')\gamma^\mu(1 - \gamma_5)u(\vec{p}) \right] \\
&\quad + \left[\bar{u}(\vec{k}')\gamma_\mu(1 - \gamma_5)u(\vec{k}) \right] \cdot \left[\bar{u}(\vec{p}')\gamma^\mu(g_V^e - g_A^e\gamma_5)u(\vec{p}) \right] \Big].
\end{aligned}
$$

Making use of the Fierz transformation [258], this equation can be rewritten as:

$$\mathcal{M}_{CC} + \mathcal{M}_{NC} = \frac{G_F}{\sqrt{2}}\left[\bar{u}(\vec{k}')\gamma_\mu(1 - \gamma_5)u(\vec{k}) \right] \cdot \left[\bar{u}(\vec{p}')\gamma^\mu(g_V' - g_A'\gamma_5)u(\vec{p}) \right] \quad (9.49)$$

with $g_V' = g_V^e + 1, g_A' = g_A^e + 1$. The expression for the matrix element square is given by:

$$
\begin{aligned}
\overline{\sum_i}\sum_f |\mathcal{M}|^2 &= \overline{\sum_i}\sum_f \left(|\mathcal{M}_{CC}|^2 + \mathcal{M}_{CC}\mathcal{M}_{NC}^* + \mathcal{M}_{NC}\mathcal{M}_{CC}^* + |\mathcal{M}_{NC}|^2 \right) \\
&= 16\,G_F^2\Big[(g_V' + g_A')^2(k' \cdot p')(k \cdot p) + (g_V' - g_A')^2(k' \cdot p)(k \cdot p') \\
&\quad - m_e^2(g_V'^2 - g_A'^2)(k \cdot k') \Big],
\end{aligned}
$$

where the interference term is obtained as:

$$\mathcal{M}_{CC}\mathcal{M}_{NC}^* + \mathcal{M}_{NC}\mathcal{M}_{CC}^* = 16\left[m_e^2(g_A - g_V)\,(k \cdot k') + 2(g_A + g_V)\,(k \cdot p)\,(k' \cdot p') \right].$$

This expression may be written, in general, for other possible reactions like $\nu_\mu e^- \to \nu_\mu e^-$, $\bar{\nu}_\mu e^- \to \bar{\nu}_\mu e^-$ and $\bar{\nu}_e e^- \to \bar{\nu}_e e^-$, which may be obtained by considering both the charged and the neutral current interactions for $\nu_e e^- \to \nu_e e^-$, and only for the neutral current interactions for $\nu_\mu e^- \to \nu_\mu e^-$, $\bar{\nu}_\mu e^- \to \bar{\nu}_\mu e^-$ and $\bar{\nu}_\mu e^- \to \bar{\nu}_\mu e^-$ processes. For the antineutrino induced processes $g_A \to -g_A$, while for the ν_μ induced processes $g_V' \to g_V, g_A' \to g_A$, such that:

$$\overline{\sum_i} \sum_f |\mathcal{M}|^2 = 16 \, G_F^2 \left[\alpha \, (k' \cdot p')(k \cdot p) + \beta \, (k' \cdot p)(k \cdot p') - \gamma \, m_e^2 (k \cdot k') \right], \quad (9.50)$$

where the values of α, β, and γ for the various neutrino and antineutrino induced scattering off the leptons are given in Table 9.1. In the Lab frame, for the $\nu_e e^- \to \nu_e e^-$ process, the initial electron is at rest, that is, $\vec{p} = 0$. Therefore, we obtain

Table 9.2 Measured values of the total cross section and $\sin^2 \theta_W$ from different experiments for $\nu_\mu e^-$ and $\bar{\nu}_\mu e^-$ scattering at 90% confidence level.

Experiment	$\sigma(\nu_\mu e^-) (\times 10^{-42} \frac{cm^2}{GeV})$	$\sigma(\bar{\nu}_\mu e^-)$	$\sin^2 \theta_W$
Gargamelle (PS) [404]	< 1.4	$1.0^{+2.1}_{-0.9}$	$0.1 < x < 0.4$
Aachen-Padova (PS) [167]	1.1 ± 0.6	2.2 ± 1.0	0.35 ± 0.08
Gargamelle (SPS) [404]	$2.4^{+1.2}_{-0.9}$	< 2.7	$0.12^{+0.11}_{-0.07}$
VMWOF(FNAL) [166]	$1.4 \pm 0.3 \pm 0.4$		$0.25^{+0.07}_{-0.05} \pm 0.8$
BNL-COL (AGS) [165]	1.67 ± 0.44		$0.20^{+0.06}_{-0.05}$
BEBC-TST (SPS) [405]		< 3.4	< 0.45
15-feet. BC (FNAL) [406]		< 2.1	< 0.37
CHARM (SPS) [144]	$2.2 \pm 0.4 \pm 0.4$	$1.6 \pm 0.3 \pm 0.3$	$0.211 \pm 0.035 \pm 0.011$
BNL E734 (AGS) [145]	$1.8 \pm 0.2 \pm 0.25$	$1.17 \pm 0.16 \pm 0.13$	$0.195 \pm 0.018 \pm 0.013$
CHARM-II (SPS) [267]	$1.53 \pm 0.04 \pm 0.12$	$1.39 \pm 0.04 \pm 0.10$	$0.237 \pm 0.007 \pm 0.007$

$$\overline{\sum_i} \sum_f |\mathcal{M}|^2 = 16 \, G_F^2 m_e E_\nu \left[(g_V' + g_A')^2 (E_\nu E_e' - E_\nu |\vec{p}'| \cos \theta + m_e E_e' - m_e^2) \right.$$

$$+ (g_V' - g_A')^2 (E_\nu + m_e - E_e')(E_e' - |\vec{p}'| \cos \theta)$$

$$\left. - m_e (g_V'^2 - g_A'^2)(m_e - E_e' + |\vec{p}'| \cos \theta) \right]. \quad (9.51)$$

The expression for the differential cross section in the Lab frame becomes

$$\left. \frac{d\sigma}{d\Omega} \right|_{Lab} = \frac{1}{4\pi^2 m_e} G_F^2 \left(\frac{E_e'^2}{E_\nu} \right) \left[(g_V' + g_A')^2 (E_\nu E_e' - E_\nu |\vec{p}'| \cos \theta + m_e E_e' - m_e^2) \right.$$

$$+ (g_V' - g_A')^2 (E_\nu + m_e - E_e')(E_e' - |\vec{p}'| \cos \theta)$$

$$\left. - m_e (g_V'^2 - g_A'^2)(m_e - E_e' + |\vec{p}'| \cos \theta) \right]. \quad (9.52)$$

The total scattering cross section is given as:

$$
\begin{aligned}
\sigma|_{\text{Lab}} \;=\; & 2\pi \int_{-1}^{+1} d(\cos\theta)\frac{1}{4\pi^2 m_e}G_F^2\left(\frac{E_e'^2}{E_\nu}\right)\Big[(g_V'+g_A')^2(E_\nu E_e' - E_\nu|\vec{p}'|\cos\theta \\
& + m_e E_e' - m_e^2) + (g_V'-g_A')^2(E_\nu + m_e - E_e')(E_e' - |\vec{p}'|\cos\theta) \\
& - m_e(g_V'^2 - g_A'^2)(m_e - E_e' + |\vec{p}'|\cos\theta)\Big],
\end{aligned}
\tag{9.53}
$$

where $\vec{p}\,' = \vec{k} - \vec{k}' = \vec{q}$.

The expression for the differential cross section in the CM frame is obtained as:

$$
\begin{aligned}
\frac{d\sigma}{d\Omega}\Big|_{\text{CM}} \;=\; & \frac{1}{4\pi^2 s}G_F^2\Big[(g_V'+g_A')^2\left(\frac{s-m_e^2}{2}\right)^2 + (g_V'-g_A')^2\left(\frac{u-m_e^2}{2}\right)^2 \\
& + \frac{m_e^2}{2}\left\{(g_V')^2 - (g_A')^2\right\}t\Big],
\end{aligned}
\tag{9.54}
$$

where $s = (k+p)^2 = E_{\text{CM}}^2$.

During the late 1980s, the E225 experiment [407] was conducted at the Los Alamos Meson Physics Facility (LAMPF) to measure $\nu_e e$ scattering cross section; a total of 236 ± 35 events were reported. The collaboration reported the measurement of an interference term arising due to the contribution of charged current (CC) and neutral current (NC) amplitudes. The value was

$$
I = -1.07 \pm 0.17 \pm 0.11,
$$

which was found to be in good agreement with the standard model prediction of -1.07 with $\sin^2\theta_W = 0.233$. The collaboration also reported a lower limit on the magnetic moment of the neutrino. LAMPF used neutrino beams from the muon decay at rest with an average energy of $\langle E_\nu \rangle = 31.7$ MeV. Their reported result of the cross section is

$$
\sigma(\nu_e e^-) = (3.18 \pm 0.56) \times 10^{-43}\text{cm}^2,
$$

which is found to be consistent with the prediction of the standard model.

Table 9.3 Cross section of neutrino–electron $\nu_e e$ ($\bar{\nu}_e e$) scattering processes from different experiments. [a] Region in visible energy [1.5–3.0] MeV. [b] Region in visible energy [3.0–4.5] MeV.

Experiment	$\frac{\sigma(\nu_e e)}{E_{\nu_e}}(\times 10^{-42}\frac{cm^2}{GeV})$	$\sigma(\bar{\nu}_e e)\times 10^{-46}cm^2$	$\sin^2\theta_W$
Savannah River [354, 408]		7.6 ± 2.2^a	0.25 ± 0.05
(Reactor)		1.86 ± 0.48^b	
Kurchatov (Reactor) [409]		6.8 ± 4.5	0.29 ± 0.10
LAMPF E225 (LAMPF) [407]	$10.0 \pm 1.5 \pm 0.9$		0.249 ± 0.063
LSND [140]	$10.1 \pm 1.1 \pm 1.0$		

Moreover, $\nu_e e^-$ cross section measurements have been performed by the liquid scintillator neutrino detector (LSND) at the Los Alamos Neutron Science Center and 191 ± 22 events were reported. The measured value of the cross section [140], given in Table 9.3, as well as the magnetic moment of the neutrino, given in Table 9.4, are consistent both with the E225 results

Table 9.4 Magnetic moment of neutrinos as measured by the different experiments [117].

Experiment	$\mu_{\nu_e}(\mu_B)$	$\mu_{\nu_\mu}(\mu_B)$
E225 (LAMPF)	$< 0.6 \times 10^{-9}$	
MUNU	$< 9.0 \times 10^{-11}$	
GEMMA	$< 2.9 \times 10^{-11}$	
TEXONO	$\leq 2.2 \times 10^{-10}$	
BOREXINO	$\leq 3.1 \times 10^{-11}$	
LSND	1.1×10^{-9}	6.8×10^{-10}
DONUT		3.9×10^{-7}
E734 (AGS)		$< 0.85 \times 10^{-9}$

as well as with the predictions of the standard model. Moreover, the cross section for the reaction $\bar{\nu}_e + e^- \rightarrow \bar{\nu}_e + e^-$ has also been determined with reactor antineutrinos at Savannah river and Kurchatov experiments.

9.5.1 Determination of the magnetic moment of neutrinos

In the standard model of electroweak interactions, neutrinos are massless and neutral. Therefore, these neutrinos can neither have a Dirac magnetic moment due to the absence of vector coupling of neutrinos with the photon field, nor can they have anomalous magnetic moment as it arises from chirality-flipping interactions and there are no right-chiral neutrinos in the standard model.

With the observation of neutrino oscillation phenomena, it is now well established that neutrinos have non-zero masses, while the absolute mass is yet to be determined for a particular flavor of neutrino. If the mass of the neutrino is taken into account, then one is discussing a physics beyond the standard model. In the minimal extension of the standard model, it is possible to include a finite neutrino mass and their interactions with the magnetic field. There are models which discuss the Dirac neutrinos with a finite magnetic dipole moment $\mu_\nu = \frac{3eG_F}{8\sqrt{2}\pi^2}m_{\nu_i} = 3.2 \times 10^{-19}\left(\frac{m_{\nu_i}}{eV}\right)\mu_B$ [117, 410].

In the standard model, $\nu_e(k) + e^-(p) \rightarrow \nu_e(k') + e^-(p')$ scattering takes place via the exchange of W^+ and Z^0 bosons and the corresponding diagrams are shown in Figure 9.10. However, if neutrinos have a magnetic moment, then there is an additional diagram which is possible through photon exchange as shown in Figure 9.11. The amplitude arising due to W^+ and Z^0 exchange, that is, $\mathcal{M}_W + \mathcal{M}_Z$ will not coherently add up to the amplitude arising from the photon exchange. This is because of the helicity of the outgoing neutrino which gets flipped due to magnetic moment coupling and will not be the same as obtained in the case of W^+ and Z^0 exchange.

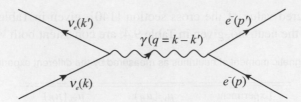

Figure 9.11 Neutral current reaction mediated by a photon.

The transition matrix element for the $\nu_e e^-$ scattering mediating through a γ-exchange is written as:

$$\mathcal{M}_\gamma = \mu_T \bar{u}(\vec{k}') i\sigma_{\mu\nu} q^\nu u(\vec{k}) \frac{e}{q^2} \bar{u}(\vec{p}') \gamma^\mu u(\vec{p}), \tag{9.55}$$

where $\mu_T = \frac{e\mu_\nu}{2m_e}$ with μ_ν being the magnetic moment of the neutrinos.

Using the trace properties of the γ-matrices, the transition matrix element squared is obtained as:

$$|\mathcal{M}_\gamma|^2 = \frac{4\pi^2 \alpha^2 \mu_\nu^2}{m_e^2 Q^2} \left[32 \left\{ k \cdot p'(2m_e^2 - p \cdot p') + k \cdot p(4k \cdot p' + p \cdot p' - 2m_e^2) \right\} \right]. \tag{9.56}$$

Using $|\mathcal{M}_\gamma|^2$ from Eq. (9.56), the differential scattering cross section in the Lab frame is obtained as:

$$\begin{aligned}
\left. \frac{d\sigma}{d\Omega} \right|_\gamma &= \frac{\alpha^2 \mu_\nu^2}{m_e^2 E_\nu} \frac{E_e'}{m_e^2 Q^2} \left\{ E_e'(1 - \cos\theta)(2m_e^2 - E_e' m_e) + m_e \left(4E_\nu E_e'(1 - \cos\theta) \right. \right. \\
&\quad + \left. \left. E_e' m_e - 2m_e^2 \right) \right\}.
\end{aligned} \tag{9.57}$$

Therefore, the differential scattering cross section for the $\nu_e e^-$ scattering, where the magnetic moment of the neutrino has been taken into account, may be written as:

$$\left. \frac{d\sigma}{d\Omega} \right|_{\nu_e e^-} = \left. \frac{d\sigma}{d\Omega} \right|_{W+Z} + \left. \frac{d\sigma}{d\Omega} \right|_\gamma. \tag{9.58}$$

The expressions for $\left. \frac{d\sigma}{d\Omega} \right|_{W+Z}$ and $\left. \frac{d\sigma}{d\Omega} \right|_\gamma$ are given in Eqs. (9.52) and (9.57), respectively.

The upper limits on the magnetic moment of neutrinos have been determined by many experiments using neutrino–electron scattering. For example, MUNU, GEMMA, TEXONO, and DONUT experiments have determined an upper limit which are listed here in Table 9.4 [117].

9.6 $e^- + e^+ \longrightarrow \mu^- + \mu^+$

The process

$$e^-(E_e, \vec{k}) + e^+(E_e', \vec{k}') \longrightarrow \mu^-(E_\mu, \vec{p}) + \mu^+(E_\mu', \vec{p}') \tag{9.59}$$

is induced by the exchange of a photon, a Z^0 boson as well as the Higgs boson in the standard model as shown in Figure 9.13. The quantities in the brackets are the four momenta of the incoming and outgoing particles, respectively. The Lagrangians for this process mediated by these bosons are, respectively, given by:

(i) photon exchange

$$L_\gamma = \sum_{l=e,\mu,\tau} \bar{\psi}_l(-\vec{p}')(-e\gamma^\mu)\psi_l(\vec{p})A_\mu. \tag{9.60}$$

(ii) Z^0 boson,

$$L_{Z^0} = \sum_{l=e,\mu,\tau} \bar{\psi}_l(-\vec{p}') \left(\frac{-g}{2\cos\theta_W}\right) \gamma^\mu (g_V^l - g_A^l \gamma^5)\psi_l(\vec{p})Z_\mu. \tag{9.61}$$

(iii) Higgs boson,

$$L_{\text{Higgs}} = \sum_{l=e,\mu,\tau} \bar{\psi}_l(-\vec{p}') \left(\frac{-m_l}{v}\right) \psi_l(\vec{p})\phi^*. \tag{9.62}$$

These interactions have also been represented by Feynman diagrams for photon, Z^0, and Higgs boson exchange, respectively as shown in Figure 9.12. Here, A^μ, Z^μ, and H^μ represent the photon, Z boson, and Higgs field, respectively and

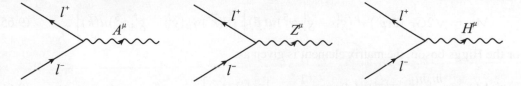

Figure 9.12 Interaction of charged leptons with photon, Z and Higgs fields.

$$g_V^l = -\frac{1}{2} + 2\sin^2\theta_W,$$

$$g_A^l = -\frac{1}{2},$$

where $l = e^-, \mu^-$. $\left(\frac{-m_l}{v}\right)$ represents the strength when leptons interact through Higgs boson, with

$$v = \frac{1}{\sqrt{\sqrt{2}G_F}} \approx 246 \text{ GeV}. \tag{9.63}$$

Using the Feynman rules, we can write the matrix element for the aforementioned diagrams. For a photon exchange, the matrix element is given as:

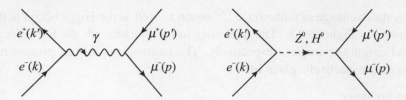

Figure 9.13 Feynman diagram for the $e^- e^+ \to \mu^- \mu^+$ via γ (left), Z^0, and H^0 (right).

$$-i\mathcal{M}_\gamma = \left[i e \bar{v}(\vec{p}') \gamma_\mu u(\vec{p}) \right] \left(\frac{-i g^{\mu\nu}}{q^2} \right) \left[i e \bar{v}(\vec{k}') \gamma_\nu u(\vec{k}) \right],$$

where the four momentum transfer q is defined as $q = k + k'$.

$$\mathcal{M}_\gamma = -\frac{e^2}{q^2} \left[\bar{v}(\vec{p}') \gamma_\mu u(\vec{p}) \right] \left[\bar{v}(\vec{k}') \gamma^\mu u(\vec{k}) \right]. \tag{9.64}$$

For the Z^0 boson, the matrix element is given as:

$$\begin{aligned}
-i\mathcal{M}_Z &= \left[\bar{v}(\vec{p}') i \left(\frac{-g}{2 \cos \theta_W} \right) \gamma^\mu (g_V^l - g_A^l \gamma^5) u(\vec{p}) \right] \left(-i \left(\frac{g_{\mu\nu} - \frac{q_\mu q_\nu}{M_Z^2}}{q^2 - M_Z^2} \right) \right) \\
&\times \left[\bar{v}(\vec{k}') i \left(\frac{-g}{2 \cos \theta_W} \right) \gamma^\nu (g_V^l - g_A^l \gamma^5) u(\vec{k}) \right].
\end{aligned}$$

In the limit $M_Z^2 >> q^2$, the transition matrix element is obtained as:

$$\mathcal{M}_Z = \sqrt{2} G_F \left[\bar{v}(\vec{p}') \gamma^\mu (g_V^l - g_A^l \gamma^5) u(\vec{p}) \right] \left[\bar{v}(\vec{k}') \gamma_\mu (g_V^l - g_A^l \gamma^5) u(\vec{k}) \right]. \tag{9.65}$$

For the Higgs boson, the matrix element is given as:

$$\mathcal{M}_H = \frac{m_e m_\mu}{v^2} \left[\bar{v}(\vec{p}') u(\vec{p}) \right] \frac{1}{q^2 - M_H^2} \left[\bar{v}(\vec{k}') u(\vec{k}) \right]. \tag{9.66}$$

The total Feynman amplitude is obtained by taking the contribution from all the three channels of the aforementioned process , that is, via γ, Z^0 and Higgs, that is,

$$\mathcal{M} = \mathcal{M}_\gamma + \mathcal{M}_Z + \mathcal{M}_H. \tag{9.67}$$

In order to obtain the differential and hence, the total scattering cross section, we have to calculate $|\mathcal{M}|^2$ which is obtained as:

$$\begin{aligned}
\overline{\sum_i} \sum_f |\mathcal{M}|^2 &= \overline{\sum_i} \sum_f [(\mathcal{M}_\gamma + \mathcal{M}_Z + \mathcal{M}_H)^* (\mathcal{M}_\gamma + \mathcal{M}_Z + \mathcal{M}_H)] \\
&= \overline{\sum_i} \sum_f \left[|\mathcal{M}_\gamma|^2 + |\mathcal{M}_Z|^2 + |\mathcal{M}_H|^2 + \mathcal{M}_\gamma^* \mathcal{M}_Z + \mathcal{M}_Z^* \mathcal{M}_\gamma \right. \\
&\quad + \left. \mathcal{M}_Z^* \mathcal{M}_H + \mathcal{M}_H^* \mathcal{M}_Z + \mathcal{M}_H^* \mathcal{M}_\gamma + \mathcal{M}_\gamma^* \mathcal{M}_H \right], \tag{9.68}
\end{aligned}$$

where

$$\overline{\sum_i}\sum_f |\mathcal{M}_\gamma|^2 = \frac{e^4}{q^4}\left(8\left(m_e^2\left(2m_\mu^2 + p\cdot p'\right) + m_e^2 k\cdot k' + k\cdot p'k'\cdot p + k\cdot pk'\cdot p'\right)\right),$$
(9.69)

$$\overline{\sum_i}\sum_f |\mathcal{M}_Z|^2 = 2G_F^2\left(8\left(2m_e^2 m_\mu^2(g_A^{l2} - g_V^{l2})^2 - m_e^2(g_A^{l4} - g_V^{l4})p\cdot p'\right.\right.$$

$$- m_\mu^2(g_A^{l4} - g_V^{l4})k\cdot k' + k\cdot p'k'\cdot p\left(g_A^{l4} + g_V^{l4} - 2g_A^{l2}g_V^{l2}\right)$$

$$\left.\left. + k\cdot pk'\cdot p'\left(g_A^{l4} + g_V^{l4} + 6g_A^{l2}g_V^{l2}\right)\right)\right),$$
(9.70)

$$\overline{\sum_i}\sum_f |\mathcal{M}_H|^2 = \frac{m_e^2 m_\mu^2}{v^4}\frac{1}{(q^2 - M_H^2)^2}\left(4\left(m_e^2 - k\cdot k'\right)\left(m_\mu^2 - p\cdot p'\right)\right),$$
(9.71)

$$\overline{\sum_i}\sum_f \mathcal{M}_\gamma^* \mathcal{M}_Z = \overline{\sum_i}\sum_f \mathcal{M}_\gamma \mathcal{M}_Z^* = \frac{e^2}{q^2}\sqrt{2}G_F\left(8\left(2m_e^2 m_\mu^2 g_V^{l2} + m_e^2 g_V^{l2}p\cdot p' + m_\mu^2 g_V^{l2}k\cdot k'\right.\right.$$

$$\left.\left. + k\cdot p'k'\cdot p(g_V^{l2} - g_A^{l2}) + k\cdot pk'\cdot p'(g_A^{l2} + g_V^{l2})\right)\right),$$
(9.72)

$$\overline{\sum_i}\sum_f \mathcal{M}_\gamma^* \mathcal{M}_H = \overline{\sum_i}\sum_f \mathcal{M}_\gamma \mathcal{M}_H^* = \frac{e^2}{q^2}\frac{m_e m_\mu}{v^2}\frac{1}{q^2 - M_H^2}\left(4m_e m_\mu(k\cdot p - k\cdot p' - k'\cdot p\right.$$

$$\left. + k'\cdot p')\right),$$
(9.73)

$$\overline{\sum_i}\sum_f \mathcal{M}_H^* \mathcal{M}_Z = \overline{\sum_i}\sum_f \mathcal{M}_H \mathcal{M}_Z^* = \sqrt{2}G_F\frac{m_e m_\mu}{v^2}\frac{1}{q^2 - M_H^2}\left(4m_e m_\mu g_V^{l2}(k\cdot p - k\cdot p'\right.$$

$$\left. - k'\cdot p + k'\cdot p')\right).$$
(9.74)

In the Lab frame, $\vec{k}' = 0$ and $\vec{p}' = \vec{k} + \vec{k}' = \vec{q}$, and the scalar products become

$$k\cdot k' = m_e E_e, \quad k'\cdot p = m_e E_\mu, \quad k\cdot p = E_e E_\mu(1 - \cos\theta),$$

$$k\cdot p' = m_e^2 + m_e E_e - E_e E_\mu(1 - \cos\theta),$$

$$p\cdot p' = E_e E_\mu(1 - \cos\theta) + m_e E_\mu - m_\mu^2,$$

$$k'\cdot p' = m_e E_e + m_e^2 - m_e E_\mu.$$

The differential cross section in the Lab frame is evaluated in terms of the scalar products and is given as:

$$\left.\frac{d\sigma}{d\Omega}\right|_{\text{Lab}} = \frac{1}{32m_e^2\pi^2}\frac{E_\mu^2}{E_e^2}\left(-\frac{2m_e m_\mu^2 e^2\left(m_e^2 + m_\mu^2 - 2E_e E_\mu(1 - \cos\theta)\right)}{v^2(m_e + E_e)\left(2m_e^2 + 2m_e E_e - M_H^2\right)}\right.$$

$$\left. - \frac{4\sqrt{2}m_e^2 m_\mu^2 G_F g_V^{l2}\left(m_e^2 + m_\mu^2 - 2E_e E_\mu(1 - \cos\theta)\right)}{v^2\left(2m_e^2 + 2m_e E_e - M_H^2\right)}\right.$$

$$+ \quad \frac{2m_e^2 m_\mu^2 \left(-m_e^2 + m_e(E_\mu - E_e) + m_\mu^2 \right) \left(m_e^2 - E_e E_\mu (1 - \cos\theta) \right)}{v^4 \left(-2m_e^2 - 2m_e E_e + M_H^2 \right)^2}$$

$$+ \quad 8G_F^2 \left(-m_e^3 (g_A^{l4} - g_V^{l4})(m_e + E_e - E_\mu) + 2m_e^2 m_\mu^2 (g_A^{l2} - g_V^{l2})^2 \right.$$

$$+ \quad m_e E_\mu \left(g_A^{l2} - g_V^{l2} \right)^2 \times \left(m_e^2 + m_e E_e - E_e E_\mu (1 - \cos\theta) \right)$$

$$+ \quad m_e E_e \left(g_A^{l4} + 6g_A^{l2} g_V^{l2} + g_V^{l4} \right) \left(E_\mu (m_e + E_e(1 - \cos\theta)) \right.$$

$$- \quad m_\mu^2 \left. \right) + m_\mu^2 E_e E_\mu (1 - \cos\theta) \left(g_V^{l4} - g_A^{l4} \right) \left. \right) + \frac{e^4}{m_e^2 (m_e + E_e)^2} \left(m_e^4 + m_e^3 E_e \right.$$

$$+ \quad 2m_e^2 \left(m_\mu^2 + E_e E_\mu \right) - m_e E_e \left(m_\mu^2 + E_\mu (E_\mu - E_e)(1 - \cos\theta) \right) + m_\mu^2 E_e E_\mu (1 - \cos\theta) \left. \right)$$

$$+ \quad \frac{1}{m_e(m_e + E_e)} 4\sqrt{2} e^2 G_F \left(m_e^4 g_V^{l2} + m_e^3 \left(E_e g_V^{l2} - E_\mu g_A^{l2} \right) + 2m_e^2 g_V^{l2} \left(m_\mu^2 + E_e E_\mu \right) \right.$$

$$+ \quad m_e E_e \left(E_\mu \left(E_e(1 - \cos\theta) \left(g_A^{l2} + g_V^{l2} \right) + E_\mu \left(g_A^{l2} - g_V^{l2} \right) \right) - m_\mu^2 \left(g_A^{l2} + g_V^{l2} \right) \right)$$

$$+ \quad m_\mu^2 E_e E_\mu (1 - \cos\theta) g_V^{l2} \left. \right). \tag{9.75}$$

In the center of the mass frame, $\vec{k} = -\vec{k}'$, $E_e = E'_e = E_l$ and $E_\mu = E'_\mu = E_l$ and the scalar products become

$$k \cdot k' = 2E_l^2 - m_e^2,$$

$$k \cdot p = E_l^2 + \sqrt{(E_l^2 - m_e^2)(E_l^2 - m_\mu^2)} \cos\theta',$$

$$k \cdot p' = E_l^2 - \sqrt{(E_l^2 - m_e^2)(E_l^2 - m_\mu^2)} \cos\theta',$$

$$k' \cdot p = E_l^2 - \sqrt{(E_l^2 - m_e^2)(E_l^2 - m_\mu^2)} \cos\theta',$$

$$k' \cdot p' = E_l^2 + \sqrt{(E_l^2 - m_e^2)(E_l^2 - m_\mu^2)} \cos',$$

$$p \cdot p' = 2E_l^2 - m_e^2.$$

Using these expressions, the differential scattering cross section in the center of mass frame is obtained as:

$$\frac{d\sigma}{d\Omega}\bigg|_{CM} = \frac{1}{128 E_l^2 \pi^2} \sqrt{\frac{m_\mu^2 - E_l^2}{m_e^2 - E_l^2}} \left[\frac{1}{E_l^2} 4\sqrt{2} e^2 G_F \left(g_V^{l2} \cos^2\theta' \left(m_e^2 - E_l^2 \right) \left(m_\mu^2 - E_l^2 \right) \right. \right.$$

$$+ \quad 2E_l^2 \cos\theta' g_A^{l2} \sqrt{\left(m_e^2 - E_l^2 \right) \left(m_\mu^2 - E_l^2 \right)} + E_l^2 g_V^{l2} \left(m_e^2 + m_\mu^2 + E_l^2 \right) \right)$$

$$+ \quad \frac{e^4}{2E_l^4} \left(\left(m_e^2 \left(m_\mu^2 \cos^2\theta' - \left(\cos^2\theta' - 1 \right) E_l^2 \right) + E_l^2 \left(\left(\cos^2\theta' + 1 \right) E_l^2 \right. \right. \right.$$

$$
\begin{aligned}
- \quad & m_\mu^2 \left(\cos^2 \theta' - 1 \right) \bigr) \bigr) \bigr) \bigr) + 16 G_F^2 \left(E_l^2 \left(8 \cos \theta' g_A^{l2} g_V^{l2} \sqrt{\left(m_e^2 - E_l^2 \right) \left(m_\mu^2 - E_l^2 \right)} \right. \right. \\
- \quad & m_\mu^2 \left(g_A^{l2} + g_V^{l2} \right) \left(\left(1 + \cos^2 \theta' \right) g_A^{l2} \left(\cos^2 \theta' - 1 \right) g_V^{l2} \right) \\
+ \quad & \left(1 + \cos^2 \theta' \right) E_l^2 \left(g_A^{l2} + g_V^{l2} \right)^2 \right) + m_e^2 \left(m_\mu^2 \left(\left(2 + \cos^2 \theta' \right) g_A^{l4} \right. \right. \\
+ \quad & 2 \left(\cos^2 \theta' - 1 \right) g_A^{l2} g_V^{l2} + \cos^2 \theta' g_V^{l4} \right) - E_l^2 \left(g_A^{l2} + g_V^{l2} \right) \left(\left(\cos^2 \theta' + 1 \right) g_A^{l2} \right. \\
+ \quad & \left. \left. \left. \left(\cos^2 \theta' - 1 \right) g_V^{l2} \right) \right) \right) + \frac{4 m_e^2 m_\mu^2 e^2 \cos \theta' \sqrt{\left(m_e^2 - E_l^2 \right) \left(m_\mu^2 - E_l^2 \right)}}{E_l^2 v^2 \left(4 E_l^2 - M_H^2 \right)} \\
+ \quad & \frac{16 \sqrt{2} m_e^2 m_\mu^2 \cos \theta' G_F g_V^{l2} \sqrt{\left(m_e^2 - E_l^2 \right) \left(m_\mu^2 - E_l^2 \right)}}{v^2 \left(4 E_l^2 - M_H^2 \right)} \\
+ \quad & \frac{8 m_e^2 m_\mu^2 \left(m_e^2 - E_l^2 \right) \left(m_\mu^2 - E_l^2 \right)}{v^4 \left(M_H^2 - 4 E_l^2 \right)^2} \Biggr] .
\end{aligned}
\tag{9.76}
$$

Figure 9.14 presents the results for the angular distribution in the CM frame, that is, $\left. \frac{d\sigma}{d\Omega} \right|_{CM}$ for $e^- e^+ \longrightarrow \mu^- \mu^+$ scattering process at various CM energies.

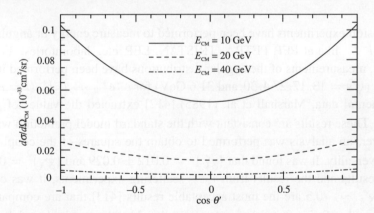

Figure 9.14 $d\sigma / d\Omega|_{CM}$ vs. $\cos \theta'$ at different values of center of mass energies, viz., $E_{CM} = 10, 20$, and 40 GeV for the process $e^- + e^+ \longrightarrow \mu^- + \mu^+$.

In Figure 9.15, the results for the contribution to the cross section from each term, that is, photon exchange, Z^0 exchange, and Higgs exchange for the CM energy of 10, 30, 50, and 90 GeV are presented. At low CM energies ($E_{CM} = \sqrt{s} \leq 10$ GeV), the effect of Z^0 is almost negligible and with the increase in CM energy, the contribution of Z^0 increases. It may be observed that for the CM energy close to the mass of Z^0 boson (i.e., at Z^0 pole), the cross section for $e^- e^+ \longrightarrow \mu^- \mu^+$ mediated by Z^0 becomes large.

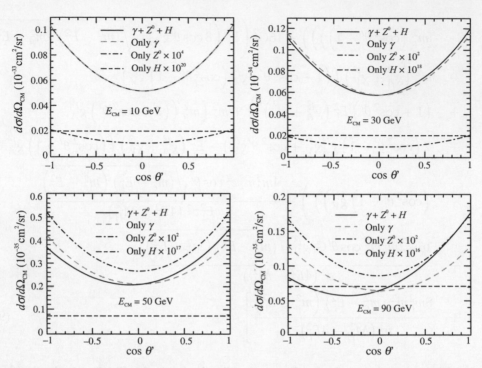

Figure 9.15 Comparison of the differential cross section with different propagators at $E_{CM} = 10$ GeV (upper left panel), 30 GeV (upper right panel), 50 GeV (lower left panel), and 90 GeV (lower right panel) for the process $e^- + e^+ \longrightarrow \mu^- + \mu^+$.

$e^- e^+$ collision experiments have been performed to measure energy or angular distribution for $e^- e^+ \to l^- l^+$ like at PEP, PETRA, TRISTAN, LEP, etc. laboratories. For example, at PETRA [411], measurements of the angular distributions have been performed in the different energy ranges ($\sqrt{s} = 13, 17, 27.4, 30$, and 31.6 GeV) for $e^- e^+ \longrightarrow l^- l^+$ ($l = e, \mu, \tau$). Using these experimental data, Marshall et al. (1985) [412] extracted the values of $g_V^e g_V^\mu$, $g_V^e g_V^\tau$, $g_A^e g_A^\mu$, $g_A^e g_A^\tau$. These results are consistent with the standard model prediction with $\sin^2 \theta_W = 0.23$. Moreover, an analysis was performed to obtain the squares of the couplings assuming $e - \mu - \tau$ universality. It was found that $\left(g_V^e\right)^2 = 0.012 \pm 0.029$ and $\left(g_A^e\right)^2 = 0.268 \pm 0.024$. Using these results and the results from $\nu_e e^-$ scattering experiments, it was concluded that $g_V^e \sim 0$ and $g_A^e \sim -0.5$ are the most acceptable results [413] that are comparable with the value of $g_V^e = -\frac{1}{2} + 2 \sin^2 \theta_W$ and $g_A^e = -\frac{1}{2}$ predicted by the standard model.

Chapter 10

Neutrino scattering Cross Sections from Hadrons: Quasielastic Scattering

10.1 Introduction

In the last chapter, we have discussed how to calculate the cross sections for the scattering of two point-like particles. Now the question arises, what happens when an electron interacts with a charge which is distributed in space like the one shown in Figure 10.1. The standard technique to measure the charge distribution and get information about the structure of the hadron is to measure the differential/total scattering cross sections of electron with a hadron and compare it with the cross section of electron scattering with a spinless ($J = 0$) point target (known as Mott scattering cross section). The ratio of these two is generally expressed as

$$\frac{\left(\frac{d\sigma}{d\Omega}\right)_{\text{extended}}}{\left(\frac{d\sigma}{d\Omega}\right)_{\text{Mott}}} = \left|F(q^2)\right|^2, \tag{10.1}$$

where $F(q^2)$, in literature, is known as the form factor. This accounts for the spatial extent of the scatterer. $F(q^2)$ not only tells about the distribution of the charge in space but using it, one can estimate the size of the target particle as well as its charge distribution and density of magnetization. Thus, for an extended charge distribution, the probability amplitude for a point-like scatterer is modified by a form factor.

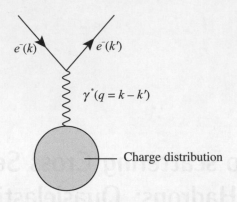

Figure 10.1 Interaction of an electron with a charge distribution.

10.2   Physical Significance of the Form Factor

Consider the elastic scattering of a "spinless" electron from a static "spinless" point object having charge $Z|e|$. In the Born approximation, where the perturbation is assumed to be weak, the scattering amplitude is written as

$$f_B = \int \psi_{\vec{k}'}^*(\vec{r}) V(\vec{r}) \psi_{\vec{k}}(\vec{r}) d\vec{r}, \quad V(\vec{r}) = -\frac{Ze^2}{4\pi r} \tag{10.2}$$

where $\psi_{\vec{k}}$ and $\psi_{\vec{k}'}$ are the wave functions of the initial and final electron with momentum $\vec{k}$ and $\vec{k}'$, respectively. These waves are assumed to be plane waves such that

$$f_B = \int e^{i\vec{q}\cdot\vec{r}} V(\vec{r}) d\vec{r}, \quad \text{where} \quad \vec{q} = \vec{k} - \vec{k}'. \tag{10.3}$$

Instead of a point charge distribution, if we assume an extended charge distribution $Z|e|\rho(\vec{r})$ with normalization $\int \rho(\vec{r}) d\vec{r} = 1$, then the potential felt by the electron located at $\vec{r}$ is given by

$$V(\vec{r}) = -\frac{Z|e^2|}{4\pi} \int \frac{\rho(\vec{r}')}{|\vec{r} - \vec{r}'|} d\vec{r}', \tag{10.4}$$

where $\vec{r}'$ is the maximum range of the charge distribution. The scattering amplitude modifies to

$$f_B = -\frac{Z|e^2|}{4\pi} \int e^{i\vec{q}\cdot\vec{r}} \int \frac{\rho(\vec{r}')}{|\vec{r} - \vec{r}'|} d\vec{r} \, d\vec{r}'. \tag{10.5}$$

Assuming $\vec{R} = \vec{r} - \vec{r}'$, which leads to $d\vec{R} = d\vec{r}$,

$$f_B = -\frac{Z|e^2|}{4\pi} \int \frac{e^{i\vec{q}\cdot\vec{R}}}{|\vec{R}|} d\vec{R} \left[\int e^{i\vec{q}\cdot\vec{r}'} \rho(\vec{r}') d\vec{r}' \right]. \tag{10.6}$$

The term in the square brackets on the right-hand side of Eq. (10.6) is known as the form factor, which is nothing but the Fourier transform of the charge density distribution, given as

$$F(\vec{q}) = \int e^{i\vec{q}\cdot\vec{r}'} \rho(\vec{r}') d\vec{r}'. \tag{10.7}$$

In field theory, if we consider the scattering of a spin $\frac{1}{2}$ electron from an external electromagnetic field (shown in Figure 10.2), the electromagnetic field in the momentum space is written as

$$A_{em}^\mu(\vec{q}) = \frac{1}{(2\pi)^3} \int e^{i\vec{q}\cdot\vec{r}} A_{em}^\mu(\vec{r})d\vec{r}, \tag{10.8}$$

where $\vec{q} = \vec{k}' - \vec{k}$ is the momentum transferred from the field source (represented by a cross in Figure 10.2) to the electron and $A_{em}^\mu(\vec{q})$ is the electromagnetic field in the momentum space. The transition amplitude, following the Feynman rules, may be written as

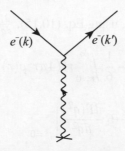

$e^-(k)$ $e^-(k')$

Figure 10.2 Feynman diagram in momentum space for electron scattering from an external source of electromagnetic field.

$$\mathcal{M} = e\bar{u}(\vec{k}')\gamma_\mu u(\vec{k}) A_{em}^\mu(\vec{q}), \tag{10.9}$$

where $u(\vec{k})$ and $\bar{u}(\vec{k}')$ are, respectively, the Dirac spinors for the incoming electron and the adjoint Dirac spinor for the outgoing electron. Square of this amplitude is given as

$$|\mathcal{M}|^2 = e^2|\bar{u}(\vec{k}')\rlap{A}{/}_{em}u(\vec{k})|^2. \tag{10.10}$$

Using Coulomb gauge for the electromagnetic field created by a point charge, that is,

$$A_{em}^\mu(\vec{x}) = \left(\frac{Ze}{4\pi|\vec{x}|}, 0, 0, 0\right), \quad (\text{in natural units}) \tag{10.11}$$

written in the momentum space by using Fourier transform as

$$A_{em}^\mu(\vec{q}) = \left(\frac{Ze}{|\vec{q}|^2}, 0, 0, 0\right), \tag{10.12}$$

we obtain the expression for the differential scattering cross section for a static external source in the massless limit of the electron, using the general expression of $\frac{d\sigma}{d\Omega}$ (given in Appendix E) as

$$\frac{d\sigma}{d\Omega}\bigg|_{Mott} = \frac{\alpha^2}{4E^2\sin^4\frac{\theta}{2}}\cos^2\frac{\theta}{2}. \tag{10.13}$$

Instead of a point charge distribution, if we assume the nucleus to have a spherical charge distribution $Ze\rho(r)$ with $\int \rho(r)d\vec{r} = 1$, then the expression for the differential cross section is obtained as

$$\frac{d\sigma}{d\Omega} = \left(\frac{d\sigma}{d\Omega}\right)_{Mott}|F(\vec{q})|^2, \tag{10.14}$$

where $F(q^2)$ is the Fourier transform of the charge density distribution and we may write

$$
\begin{aligned}
F(q^2) &= \int_{r=0}^{\infty} \int_{\cos\theta=-1}^{+1} \int_{\phi=0}^{2\pi} r^2 dr\, d\cos\theta\, d\phi\, \rho(r)\, e^{i|\vec{q}||\vec{r}|\cos\theta} \\
&= \frac{4\pi}{q^2} \int_{r=0}^{\infty} (qr) \sin(qr)\rho(r)dr; \quad |\vec{q}| = q \\
&= \frac{4\pi}{q^2} \int_{r=0}^{\infty} \left(q^2 r^2 - \frac{q^4 r^4}{6} + \ldots \right) \rho(r)dr.
\end{aligned} \tag{10.15}
$$

In the limit of small angle scattering, using Eq. (10.15), $\frac{dF}{dq^2}$ may be written as

$$
\frac{dF(q^2)}{dq^2}\bigg|_{q^2=0} = -\frac{1}{6} \int_{r=0}^{\infty} r^2 (4\pi r^2 \rho(r))dr = -\frac{1}{6}\langle r^2 \rangle
$$

$$
\Rightarrow \qquad \langle r^2 \rangle = -6 \frac{dF(q^2)}{dq^2}\bigg|_{q^2=0}, \tag{10.16}
$$

where $\langle r^2 \rangle$ is the mean charge radius squared of the scatterer, resulting in $r_{\rm rms} = \sqrt{\langle r^2 \rangle}$, which gives the rms charge radius of the scatterer.

If the charge distribution has an exponential form like $\rho(r) = \rho(0)e^{-m_v r}$, then the form factor $F(q^2)$ turns out to be:

$$
\begin{aligned}
F(q^2) &= \frac{4\pi}{q^2} \int_{r=0}^{\infty} (qr) \sin(qr)\rho(0)e^{-m_v r}\, dr \\
&= \frac{4\pi}{q^2} \int_{r=0}^{\infty} (qr) \frac{e^{-(m_v-iq)r} - e^{-(m_v+iq)r}}{2i}\rho(0)dr \\
&= \frac{8\pi\rho_0}{m_v^3} \frac{1}{\left(1 + \frac{q^2}{m_v^2}\right)^2}
\end{aligned}
$$

known as the dipole form factor with m_v as the vector dipole mass.

For a Yukawa type charge distribution $\rho(r) = \rho(0)\frac{e^{-mr}}{r}$, the form factor $F(q^2)$ is of a monopole form:

$$
F(q^2) \propto \frac{1}{\left(1 - \frac{q^2}{m^2}\right)},
$$

where m^2 is in the units of q^2.

In the next section, we will see how, in the theoretical formalism, the scattering cross sections are expressed in terms of the form factors. For simplicity, this has been demonstrated by taking a pion (spin 0) target, which is described by a single form factor $F_\pi(q^2)$. Experimentally, however, one measures the differential or the total scattering cross sections and from there, information about the form factors is obtained. The best description of the experimental data by a given set of form factor(s) using least square fitting method determines the parameterization of the form factor. With these form factors, other information like the charge density distribution, size of the scatterer, etc. are obtained.

10.3 $e^- - \pi^{\pm}$ Elastic Scattering

Elastic electron–pion scattering takes place via the exchange of a virtual photon in the reaction

$$e^-(k) + \pi^{\pm}(p) \longrightarrow e^-(k') + \pi^{\pm}(p'),$$

where k, k' and p, p' are the incoming and outgoing momenta of the electrons and pions, respectively.

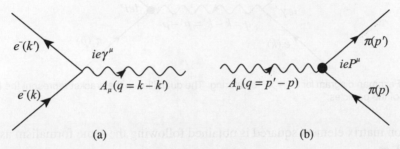

Figure 10.3 (a) Leptonic and (b) hadronic vertices for the $e^- - \pi^{\pm}$ elastic scattering.

When an electron interacts with a photon field (Figure 10.3(a)), the interaction Lagrangian is given by

$$\mathcal{L}_I^e = -\bar{\psi}(\vec{k}')e\gamma^{\mu}\psi(\vec{k})A_{\mu}, \tag{10.17}$$

and when a pion interacts with a photon field (Figure 10.3(b)), the interaction Lagrangian is given by

$$\mathcal{L}_I^{\pi} = -j_{\mu}^{\pi}A^{\mu} \qquad \text{with} \tag{10.18}$$

$$j_{\mu}^{\pi} = f_1(p^2, p'^2, p \cdot p')P_{\mu} + f_2(p^2, p'^2, p \cdot p')q_{\mu}, \tag{10.19}$$

where $P^{\mu} = p^{\mu} + p'^{\mu}$ and $f_1(p^2, p'^2, p \cdot p')$ and $f_2(p^2, p'^2, p \cdot p')$ are unknown functions which depend upon the scalar quantities constructed from the four momenta of the incoming and the outgoing pions p_{μ} and p'_{μ}, respectively, that is, p^2, p'^2 and $p \cdot p'$. Since $p^2 = p'^2 = m_{\pi}^2$ and $p \cdot p' = m_{\pi}^2 - \frac{q^2}{2}$, only q^2 is a variable quantity. Therefore, in the case of elastic scattering, the form factors are the function of q^2 only.

Conservation of the vector current at the hadronic vertex, that is, $\partial^{\mu}j_{\mu}^{\pi} = 0$ leads to $f_2(q^2) = 0$, and one is left only with one scalar function $f_1(q^2)$. For simplicity, we rename this form factor $f_1(q^2)$ as $F_{\pi}(q^2)$. Using the Lagrangians defined in Eqs. (10.17) and (10.18), the transition amplitude for the process $e^- \pi^+ \to e^- \pi^+$, as depicted in Figure 10.4, is written as

$$-i\mathcal{M} = \underbrace{\bar{u}(\vec{k}')ie\gamma^\mu u(\vec{k})}_{\text{leptonic current}} \underbrace{\left(\frac{-ig_{\mu\nu}}{q^2}\right)}_{\text{propagator}} \underbrace{ieP_\mu F_\pi(q^2)}_{\text{hadronic current}}$$

$$\Rightarrow \quad \mathcal{M} = -\frac{e^2}{q^2}\left[\bar{u}(\vec{k}')\gamma^\mu u(\vec{k})\right] P_\mu F_\pi(q^2). \tag{10.20}$$

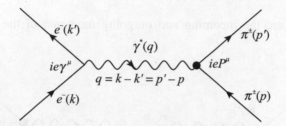

Figure 10.4 Feynman diagram for $e^- - \pi^\pm$ scattering. The quantities in the bracket represent the four momenta of the particles.

The transition matrix element squared is obtained following the same formalism as mentioned in Chapter 9, as

$$\sum\sum|\mathcal{M}|^2 = \frac{e^4}{q^4}\frac{1}{2}\overline{\sum}\overline{\sum}|\left[\bar{u}(\vec{k}')\gamma^\mu u(\vec{k})\right]|^2 P_\mu P_\nu F_\pi^2(q^2) \tag{10.21}$$

where $P = p + p' = 2p + q$. In the massless limit of the electron, Eq. (10.21) reduces to

$$\sum\sum|\mathcal{M}|^2 = \frac{4e^4}{q^4}(4(k\cdot p)(k'\cdot p) + q^2 m_\pi^2)F_\pi^2(q^2), \tag{10.22}$$

where pion is on the mass shell, that is, $p^2 = m_\pi^2$. Using the expression of the differential scattering cross section in the Lab frame (Appendix E), where $p = (m_\pi, \vec{0}), k\cdot p = m_\pi E$, $k'\cdot p = m_\pi E'$, $\frac{d\sigma}{d\Omega}\big|_{\text{Lab}}$ is obtained as

$$\frac{d\sigma}{d\Omega}\bigg|_{\text{Lab}} = \frac{1}{64\pi^2 m_\pi E}\sum\sum|\mathcal{M}|^2\frac{|\vec{k}'|^3}{((E+m_\pi)|\vec{k}'|^2 - \vec{k}\cdot\vec{k}'E')} \tag{10.23}$$

which may be further solved to give

$$\frac{d\sigma}{d\Omega}\bigg|_{\text{Lab}} = \frac{\alpha^2}{4E^2\sin^4\frac{\theta}{2}}\frac{E'}{E}F_\pi^2(q^2)\cos^2\frac{\theta}{2}. \tag{10.24}$$

In the experimental measurements of $e^- - \pi^\pm$ scattering as well as through some other inelastic processes like $e^- p \to e^- \pi^+ n$, it has been observed that with a form of

$$F_\pi(q^2) = \frac{1}{1 - \langle r_\pi^2\rangle\frac{q^2}{6}},$$

the experimental data may be explained using $r_\pi^{\text{rms}} = \sqrt{\langle r_\pi^2 \rangle} = 0.657 \pm 0.012$ fm.

Thus, for the elastic $e^- \pi^+$ scattering, the cross section is described in terms of a form factor. When we compare this cross section (given in Eq. (10.24)) with the Mott scattering cross section given in Eq. (10.13), in the case of $e^- \pi^+$ scattering, the modifications are the presence of $F_\pi^2(q^2)$, due to the structure of the pion and $\frac{E'}{E}$, due to the recoil of the pion. Moreover, when the cross section given in Eq. (10.24) is compared with the cross section obtained for the $e^- \mu^-$ scattering (discussed in Chapter 9), we realize that an additional contribution, in the case of $e^- \mu^-$ scattering, arises due to the spin of the muon. In the next section, we will see how the form factors play an important role in $e^- p$ scattering where the proton is a spin $\frac{1}{2}$ particle similar to that of a muon but the proton will also have a structure.

10.4 Electromagnetic Scattering of Electrons with Nucleons

10.4.1 Matrix element and form factors

First, let us consider the electromagnetic scattering of electrons with protons expressed as

$$e^-(k) + p(p) \longrightarrow e^-(k') + p(p'),$$

where k, k' and p, p' are the incoming and outgoing momenta of the electron and proton, respectively. The Lagrangian at the leptonic and hadronic vertices for the electron and the proton (presented in Figure 10.5) interacting with a photon field is given as

$$
\begin{aligned}
\mathcal{L}_I^l &= -e\overline{\psi}(\vec{k}')\gamma^\mu A_\mu \psi(\vec{k}), \\
\mathcal{L}_I^h &= -e\overline{\psi}(\vec{p}')\Gamma^\mu A_\mu \psi(\vec{p}),
\end{aligned}
$$

where γ^μ is a Dirac four vector and Γ^μ shows our lack of knowledge about the hadronic vertex represented by a blob in Figure 10.5(b). The transition amplitude $\mathcal{M}$ has to be a Lorentz scalar. Since the leptonic vertex is a four vector, the hadronic vertex also has to be a four vector (Γ_μ). To construct Γ_μ at the hadronic vertex, the two independent four vectors are p_μ and p'_μ. Since the proton is a spin $\frac{1}{2}$ Dirac particle and the interaction is electromagnetic in nature where parity is conserved, the most general form of the hadronic current with possible bilinear

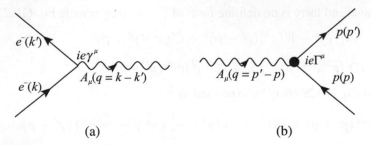

Figure 10.5 (a) Leptonic and (b) hadronic vertices for the elastic $e^- p$ scattering.

covariants (see Appendix A) transforming as a vector can be constructed from p_μ, p'_μ, γ_μ and $\sigma_{\mu\nu}(=\frac{i}{2}[\gamma_\mu, \gamma_\nu])$ operators and written as

$$\Gamma^\mu = A(p^2, p'^2, p \cdot p')\gamma^\mu + B(p^2, p'^2, p \cdot p')p'^\mu + C(p^2, p'^2, p \cdot p')p^\mu$$

$$+ iD(p^2, p'^2, p \cdot p')\sigma^{\mu\nu}p'_\nu + iE(p^2, p'^2, p \cdot p')\sigma^{\mu\nu}p_\nu. \tag{10.25}$$

Here, A, B, C, D, and E are undetermined scalar functions depending upon q^2 in the case of the elastic scattering.

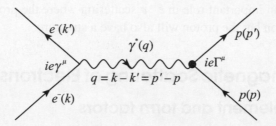

Figure 10.6 Feynman diagram for $e^- p$ scattering. The quantities in the bracket represent the four momenta of the particles.

Invariant amplitude for the elastic $e^- p$ scattering, depicted in Figure 10.6, is written as

$$-i\mathcal{M} = \underbrace{(i\,j^{e(\mu)})}_{\text{leptonic current}} \underbrace{\left(-i\frac{g_{\mu\nu}}{q^2}\right)}_{\text{propagator}} \underbrace{(i\,J^{p(\mu)})}_{\text{hadronic current}},$$

where

$$j^{e(\mu)} = \bar{u}(\vec{k}')\, e\, \gamma^\mu u(\vec{k}), \tag{10.26}$$

$$J^{p(\mu)} = \bar{u}(\vec{p}')\, e\, \Gamma^\mu u(\vec{p}). \tag{10.27}$$

Using the currents defined in Eqs. (10.26) and (10.27), the transition amplitude is obtained as

$$\mathcal{M} = -\frac{e^2}{q^2}\left[\bar{u}(\vec{k}')\gamma^\mu u(\vec{k})\right]\left[\bar{u}(\vec{p}')\Gamma_\mu u(\vec{p})\right], \tag{10.28}$$

with Γ^μ as given in Eq. (10.25).

Since Γ^μ is constructed from the four vectors available at the hadronic vertex using the bilinear covariants and there is no definite form of Γ^μ, we may rewrite Eq. (10.25) as

$$\Gamma^\mu = A(q^2)\gamma^\mu + B'(q^2)(p' - p)^\mu + C'(q^2)(p' + p)^\mu$$

$$+ iD'(q^2)\sigma^{\mu\nu}(p' - p)_\nu + iE'(q^2)\sigma^{\mu\nu}(p' + p)_\nu. \tag{10.29}$$

The last term of Eq. (10.29) may be expressed as

$$\bar{u}(\vec{p}')[i\sigma^{\mu\nu}(p' + p)_\nu]u(\vec{p}) = \bar{u}(\vec{p}')[-\frac{1}{2}(\gamma^\mu\gamma^\nu - \gamma^\nu\gamma^\mu)(p' + p)_\nu]u(\vec{p})$$

$$= -\bar{u}(\vec{p}')[(p' - p)^\mu]u(\vec{p}), \tag{10.30}$$

where we have used Dirac equations, that is, $(\not{p} - M)u(\vec{p}) = 0$ and $\bar{u}(\vec{p}')(\not{p}' - M) = 0$. From Eq. (10.30), it may be noticed that the last term in Eq. (10.29) can be written in terms of the second term in Eq. (10.29). Similarly, we obtain

$$\bar{u}(\vec{p}')[i\sigma^{\mu\nu}(p'-p)_\nu]u(\vec{p}) = \bar{u}(\vec{p}')[2M\gamma^\mu - (p'+p)^\mu]u(\vec{p}), \tag{10.31}$$

which is called Gordon decomposition.

Therefore, Eq. (10.31) can be rewritten as

$$\bar{u}(\vec{p}')(p'+p)^\mu u(\vec{p}) = \bar{u}(\vec{p}')[2M\gamma^\mu - i\sigma^{\mu\nu}(p'-p)_\nu]u(\vec{p}). \tag{10.32}$$

Thus, it may be deduced that out of the five terms from which Γ^μ is constructed, only three are linearly independent. Using Eqs. (10.30) and (10.32) in Eq. (10.29), we obtain

$$\Gamma^\mu = \gamma^\mu(A(q^2) + 2MC'(q^2)) + i\sigma^{\mu\nu}q_\nu(D'(q^2) - C'(q^2)) + q^\mu(B'(q^2) - E'(q^2)). \tag{10.33}$$

Redefining the form factors, one can obtain Γ^μ as

$$\Gamma^\mu = \gamma^\mu F_1(q^2) + i\sigma^{\mu\nu}q_\nu \frac{F_2(q^2)}{2M} + q^\mu \frac{F_3(q^2)}{M}. \tag{10.34}$$

Applying current conservation in momentum space, that is, $q_\mu J^{p(\mu)} = 0$, where $J^{p(\mu)}$ is defined in Eq. (10.27) with Γ^μ given in Eq. (10.34):

$$q_\mu J^{p(\mu)} = q_\mu \bar{u}(\vec{p}')e\left[\gamma^\mu F_1(q^2) + i\sigma^{\mu\nu}q_\nu \frac{F_2(q^2)}{2M} + q^\mu \frac{F_3(q^2)}{M}\right]u(\vec{p}) = 0$$

$$\Rightarrow \bar{u}(\vec{p}')e\left[F_1(q^2)(\not{p}' - \not{p}) - \frac{1}{2}(\gamma^\mu\gamma^\nu - \gamma^\nu\gamma^\mu)q_\mu q_\nu \frac{F_2(q^2)}{2M} + q^2 \frac{F_3(q^2)}{M}\right]u(\vec{p}) = 0. \tag{10.35}$$

Therefore, from current conservation, the first term on the left-hand side is zero when Dirac equation is applied; the second term will give $q^2 - q^2 = 0$; and from the last term, $F_3(q^2) = 0$ as $q^2 \neq 0$. Hence, the hadronic current becomes

$$\bar{u}(\vec{p}')e\Gamma_\mu u(\vec{p}) = \bar{u}(\vec{p}')e\left[\gamma_\mu F_1(q^2) + \frac{i\sigma_{\mu\nu}q^\nu}{2M}F_2(q^2)\right]u(\vec{p}). \tag{10.36}$$

Using Gordon decomposition in Eq. (10.36),

$$\bar{u}(\vec{p}')e\Gamma_\mu u(\vec{p}) = \bar{u}(\vec{p}')e\left[\gamma_\mu\left\{F_1(q^2) + F_2(q^2)\right\} - \frac{1}{2M}P_\mu F_2(q^2)\right]u(\vec{p}), \tag{10.37}$$

where we have used Eq. (10.31).

$F_1(q^2)$ and $F_2(q^2)$ are, respectively, known as the Dirac and Pauli form factors. In this section, we will discuss the scattering of electrons with free protons; therefore, in the rest of the text, we call these form factors as $F_1^p(q^2)$ and $F_2^p(q^2)$. Moreover, if the scattering of electrons takes place with free neutrons, then the only change in Eq. (10.37) is in the form factors; for the e^-n scattering, $F_1^n(q^2)$ and $F_2^n(q^2)$ are the Dirac and Pauli neutron form factors.

Besides, hermiticity of electromagnetic current demands that

$$\langle p|j_\mu^\dagger|p'\rangle = \langle p'|j_\mu|p\rangle^* = \langle p|j_\mu|p'\rangle. \tag{10.38}$$

From Eq. (10.37), we have

$$\Rightarrow \{\bar{u}(\vec{p}')e\Gamma^\mu u(\vec{p})\}^* = e\left[u^\dagger(\vec{p}')\left\{\gamma^0\gamma^\mu\left\{F_1(q^2) + F_2(q^2)\right\} - \gamma^0\frac{1}{2M}P^\mu F_2(q^2)\right\}u(\vec{p})\right]^*$$

$$= u^\dagger(\vec{p})e\left[\gamma^{\mu\dagger}\gamma^{0\dagger}\left\{F_1^*(q^2) + F_2^*(q^2)\right\} - \gamma^{0\dagger}\frac{1}{2M}P^\mu F_2^*(q^2)\right]u(\vec{p}')$$

$$= \bar{u}(\vec{p})e\left[\gamma^\mu\left\{F_1^*(q^2) + F_2^*(q^2)\right\} - \frac{1}{2M}P^\mu F_2^*(q^2)\right]u(\vec{p}'), \quad (10.39)$$

where we have used $(\gamma^0\gamma^\mu)^\dagger = \gamma^0\gamma^\mu$. Requirement of Eq. (10.38) demands that when Eq. (10.37) and Eq. (10.39) are compared, we must have $F_1(q^2) = F_1^*(q^2)$; $F_2(q^2) = F_2^*(q^2)$. This implies that the form factors $F_1(q^2)$ and $F_2(q^2)$ are real.

10.4.2 Physical interpretation of the form factors

In order to give a physical interpretation of the electromagnetic form factors $F_1(q^2)$ and $F_2(q^2)$, it is demonstrative to evaluate the matrix elements of the electromagnetic currents in the Breit frame.

The Breit frame (or the brickwall frame), depicted in Figure 10.7 is defined as the frame in which the target of mass M is assumed to be very massive in comparison to the mass of the projectile m such that there is no transfer of energy to the target, that is, $E_p = E_{p'}$ and the only change is in the direction of the projectile, that is, $\vec{p} + \vec{p}' = 0 \Rightarrow \vec{p}' = -\vec{p}$. In this frame, the four momentum transfer q becomes ($q_0 = 0, \vec{q} = -2\vec{p}$).

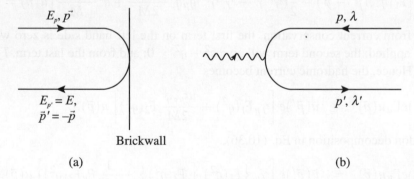

Figure 10.7 The brickwall frame.

We recall that the expression for the hadronic current J^μ is given by

$$J^\mu = \bar{u}(\vec{p}')e\left[\gamma^\mu(F_1(q^2) + F_2(q^2)) - \frac{P^\mu}{2M}F_2(q^2)\right]u(\vec{p}).$$

In the Breit frame, the zeroth component of the current four vector J^μ becomes

$$J^0 = \bar{u}(\vec{p}')e\left[\gamma^0(F_1(q^2) + F_2(q^2)) - \frac{P^0}{2M}F_2(q^2)\right]u(\vec{p}), \quad (10.40)$$

where $P^0 = E_p + E_{p'} = 2E_p$, resulting in

$$\bar{u}(-\vec{p})\gamma^0 u(\vec{p}) = u^\dagger(-\vec{p})u(\vec{p}) = 2M, \tag{10.41}$$

$$\bar{u}(-\vec{p})u(\vec{p}) = u^\dagger(-\vec{p})\gamma^0 u(\vec{p}) = 2E_p. \tag{10.42}$$

Using Eqs. (10.41) and (10.42) in Eq. (10.40), the zeroth component of the hadronic current becomes

$$J^0 = \rho = 2Me\left(F_1(q^2) + \frac{q^2}{4M^2}F_2(q^2)\right).$$

The quantity in the bracket is identified as the Sachs electric form factor $G_E(q^2)$ as

$$\rho = 2Me\, G_E(q^2). \tag{10.43}$$

The ith component of the hadronic current becomes

$$\vec{J} = \bar{u}(-\vec{p})e\left[\gamma^i(F_1(q^2) + F_2(q^2)) - \frac{\vec{P}}{2M}F_2(q^2)\right]u(\vec{p}). \tag{10.44}$$

Since $\vec{P} = \vec{p}' + \vec{p} = 0$, Eq. (10.44) becomes

$$\vec{J} = \bar{u}(-\vec{p})e\left[\gamma^i(F_1(q^2) + F_2(q^2))\right]u(\vec{p})$$

$$= -e\,\chi^{s'\dagger}(\vec{\sigma} \times \vec{q})\chi^s G_M(q^2). \tag{10.45}$$

$F_1(q^2) + F_2(q^2)$ is known as the Sachs magnetic form factor $G_M(q^2)$.

Therefore, from Eqs. (10.43) and (10.45), it may be deduced that in the Breit frame, $J^0 (= \rho)$ and $\vec{J}$ are related to the electric and magnetic Sachs form factors $G_E(q^2)$ and $G_M(q^2)$, respectively. This means that in the Breit frame, $G_E(q^2)$ and $G_M(q^2)$ are, respectively, the Fourier transforms of the charge density and the density of magnetization.

These Sachs form factors of the nucleon are expressed as

$$G_E^N(q^2) = F_1^N(q^2) + \frac{q^2}{4M^2}F_2^N(q^2), \tag{10.46}$$

$$G_M^N(q^2) = F_1^N(q^2) + F_2^N(q^2). \tag{10.47}$$

Equivalently, $F_1^N(q^2)$, $F_2^N(q^2)$ and their combination $F_M^N(q^2)(= F_1^N(q^2) + F_2^N(q^2))$ may be redefined in terms of $G_E^N(q^2)$ and $G_M^N(q^2)$ as

$$F_1^N(q^2) = \frac{G_E^N(q^2) + \tau G_M^N(q^2)}{1 + \tau}, \tag{10.48}$$

$$F_2^N(q^2) = \frac{G_M^N(q^2) - G_E^N(q^2)}{1 + \tau}, \tag{10.49}$$

$$F_M^N(q^2) = F_1^N(q^2) + F_2^N(q^2) = G_M^N(q^2), \qquad \text{where } \tau = -\frac{q^2}{4M^2}. \tag{10.50}$$

10.4.3 Cross sections and the Rosenbluth separation

With Γ_μ defined in Eq. (10.36), the transition matrix element squared may be obtained as

$$\overline{\sum}\sum |\mathcal{M}|^2 = \frac{e^4}{q^4}\overline{\sum}\sum | \left[\bar{u}(\vec{k}')\gamma^\mu u(\vec{k})\right]|^2 | \left[\bar{u}(\vec{p}')\Gamma_\mu u(\vec{p})\right]|^2,$$

where the sum over the spin of the final state particles and the average over the spin of the initial state particles have been taken into account, resulting in

$$\overline{\sum}\sum |\mathcal{M}|^2 = \frac{e^4}{q^4} \cdot \frac{1}{2} \cdot \frac{1}{2} \cdot Tr\left[(\slashed{k}' + m_e)\gamma^\mu(\slashed{k} + m_e)\gamma^\nu\right]$$

$$Tr\left((\slashed{p}' + M)\left[\gamma_\mu\left\{F_1^p(q^2) + F_2^p(q^2)\right\} - \frac{1}{2M}P_\mu F_2^p(q^2)\right]\right.$$

$$\left.(\slashed{p} + M)\left[\gamma_\nu\left\{F_1^p(q^2) + F_2^p(q^2)\right\} - \frac{1}{2M}P_\nu F_2^p(q^2)\right]\right).$$

Using $\left\{F_1^p(q^2) + F_2^p(q^2)\right\} = F_M^p(q^2)$, $\overline{\sum}\sum |\mathcal{M}|^2$ may be rewritten as

$$\overline{\sum}\sum |\mathcal{M}|^2 = \frac{4e^4}{q^4} \cdot [k^\mu k'^\nu + k'^\mu k^\nu - (k \cdot k' - m_e^2)g^{\mu\nu}]$$

$$\times \left(F_M^{p\,2}(q^2)\left[p_\mu p'_\nu + p'_\mu p_\nu - (p \cdot p' - M^2)g_{\mu\nu}\right]\right.$$

$$- \frac{F_M^p(q^2)F_2^p(q^2)}{2M}M\left(p'_\mu P_\nu + P_\mu p'_\nu + p_\mu P_\nu + P_\mu p_\nu\right)$$

$$\left.+ \frac{F_2^{p2}(q^2)}{4M^2}\left(p \cdot p' + M^2\right)P_\mu P_\nu\right). \tag{10.51}$$

The expression for $\overline{\sum}\sum |\mathcal{M}|^2$ becomes

$$\overline{\sum}\sum |\mathcal{M}|^2 = \frac{8e^4}{q^4}\left\{F_M^{p\,2}(q^2)\left[k \cdot pk' \cdot p' + k \cdot p'k' \cdot p - M^2k \cdot k' - m_e^2 p \cdot p' + 2m_e^2 M^2\right]\right.$$

$$- F_M^p(q^2)\frac{F_2^p(q^2)}{2M}M\left[2P \cdot kP \cdot k' - (k \cdot k' - m_e^2)P^2\right]$$

$$\left.+ \frac{F_2^{p2}(q^2)}{4M^2}\frac{1}{2}P^2\left[P \cdot kP \cdot k' - \frac{(k \cdot k' - m_e^2)}{2}P^2\right]\right\}.$$

Defining the four momentum transfer, $q = k - k' = p' - p$, and the sum of the momenta, $P = p + p'$ and applying the momentum conservation $k + p = k' + p'$, which gives $p' = k + p - k'$, the transition matrix element squared may be obtained as

$$\overline{\sum\sum} |\mathcal{M}|^2 = \frac{8e^4}{q^4} \left\{ F_M^{p\,2}(q^2) \left[-2k \cdot p \left(m_e^2 - k' \cdot p \right) + 2m_e^2 k' \cdot p + m_e^2 M^2 \right. \right.$$

$$\left. - k \cdot k' \left(-k \cdot p + k' \cdot p + M^2 \right) \right] - M F_M^p(q^2) \frac{F_2^p(q^2)}{2M} \left[4 \left(m_e^2 M^2 - M^2 k \cdot k' \right. \right.$$

$$\left. + 2k \cdot pk' \cdot p) \right] + \frac{F_2^{p\,2}(q^2)}{4M^2} \left[2 \left(k \cdot p - k' \cdot p + 2M^2 \right) \left(m_e^2 M^2 - M^2 k \cdot k' \right. \right.$$

$$\left. \left. + 2k \cdot pk' \cdot p) \right] \right\}. \tag{10.52}$$

Evaluating the kinematics in the Lab frame $k = (E, \vec{k})$, $p = (M, \vec{0})$, $k' = (E', \vec{k}')$, in the limit $m_e \to 0$ leads to the following relations

$$k \cdot p = ME, \qquad k' \cdot p = ME',$$

$$k \cdot k' = EE'(1 - \cos\theta),$$

$$q^2 = -2k \cdot k' = -2EE'(1 - \cos\theta).$$

Thus, Eq. (10.52) in the Lab frame is obtained as

$$\overline{\sum\sum} |\mathcal{M}|^2 = \frac{8e^4}{q^4} \left\{ F_M^{p\,2}(q^2) \left[ME \left\{ ME' - \frac{q^2}{2} \right\} + ME' \left\{ ME + \frac{q^2}{2} \right\} + M^2 \frac{q^2}{2} \right] \right.$$

$$- M F_M^p(q^2) \frac{F_2^p(q^2)}{2M} \left[2M^2 q^2 + 8M^2 EE' \right]$$

$$\left. + \frac{F_2^{p\,2}(q^2)}{4M^2} \left[\left\{ 2M^2 + ME - ME' \right\} \left\{ M^2 q^2 + 4M^2 EE' \right\} \right] \right\}, \tag{10.53}$$

and the differential scattering cross section is given by

$$\frac{d\sigma}{d\Omega} = \frac{\alpha^2}{4 E^2 \sin^4 \frac{\theta}{2}} \frac{E'}{E} \left\{ \frac{(F_1^p(q^2) + \frac{q^2}{4M^2} F_2^p(q^2))^2 + \tau F_M^{p\,2}(q^2)}{1 + \tau} \cos^2 \frac{\theta}{2} \right.$$

$$\left. + 2\tau F_M^{p\,2}(q^2) \sin^2 \frac{\theta}{2} \right\}. \tag{10.54}$$

Using $F_1^N(q^2)$, $F_2^N(q^2)$ and $F_M^N(q^2)$ from Eqs. (10.48), (10.49) and (10.50), we obtain

$$\frac{d\sigma}{d\Omega} = \frac{\alpha^2}{4 E^2 \sin^4 \frac{\theta}{2}} \frac{E'}{E} \left\{ \frac{G_E^{p\,2}(q^2) + \tau G_M^{p\,2}(q^2)}{1 + \tau} \cos^2 \frac{\theta}{2} + 2\tau G_M^{p\,2}(q^2) \sin^2 \frac{\theta}{2} \right\}, \tag{10.55}$$

Equation (10.55) is known as the Rosenbluth separation of the cross section for e^-p scattering. The equation is used to experimentally determine the electric and magnetic form factors $G_E(q^2)$ and $G_M(q^2)$ as described in the next section.

Figure 10.8 presents the results for the differential scattering cross section $\frac{d\sigma}{d\Omega}$ vs. $\cos\theta$ for the elastic e^-p scattering at different energies of the incoming electron viz. $E_e = 250, 500, 750,$ and 1000 MeV. It may be observed that the differential scattering cross section is forward peaked and is similar to what we obtain in the case of $e^-\mu^-$ scattering.

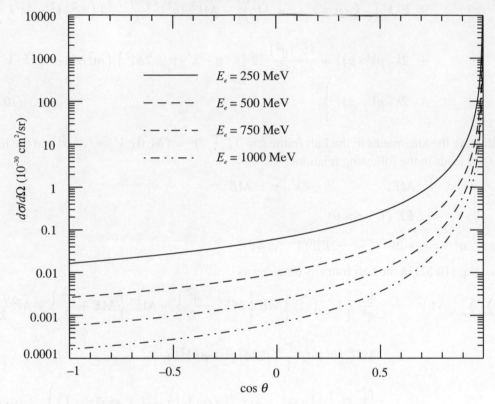

Figure 10.8 $\frac{d\sigma}{d\Omega}$ vs. $\cos\theta$ for the elastic e^-p scattering at different incoming electron energies, viz., $E_e = 250, 500, 750,$ and 1000 MeV.

10.4.4   Experimental determination of the form factors

Comparing Eqs. (10.55) and (10.13) (in the limit $E' = E$), one obtains

$$\frac{d\sigma}{d\Omega}\bigg|_{ep} = \frac{d\sigma}{d\Omega}\bigg|_{Mott}\left\{\frac{G_E^{p2}(q^2) + \tau G_M^{p\,2}(q^2)}{1+\tau} + 2\tau G_M^{p\,2}(q^2)\tan^2\frac{\theta}{2}\right\}, \qquad (10.56)$$

where the quantity in the parentheses represents the structure of the nucleon.

Equation (10.56) may be rewritten as

$$\frac{\frac{d\sigma}{d\Omega}\big|_{ep}}{\frac{d\sigma}{d\Omega}\big|_{Mott}} = \left\{ \Phi(q^2) + \Psi(q^2) \tan^2\frac{\theta}{2} \right\}, \tag{10.57}$$

which is the equation of a straight line for a fixed q^2 with

$$\Phi(q^2) = \frac{G_E^{p2}(q^2) + \tau G_M^{p\,2}(q^2)}{1+\tau}, \qquad \Psi(q^2) = 2\tau G_M^{p\,2}(q^2),$$

as the intercept and slope, respectively, as shown in Figure 10.9. The experimental measurements for $e^- p$ elastic scattering have been performed and found to be consistent with the theoretical predictions.

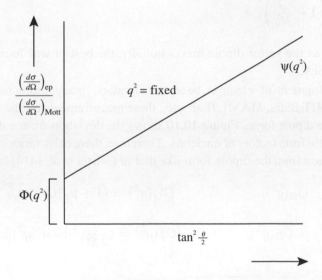

Figure 10.9 $\frac{\frac{d\sigma}{d\Omega}|_{ep}}{\frac{d\sigma}{d\Omega}|_{Mott}}$ vs. $\tan^2\frac{\theta}{2}$ for a fixed q^2.

The experimental values for the proton and neutron Sachs electric and magnetic form factors $G_E^{p,n}(q^2)$ and $G_M^{p,n}(q^2)$ at $q^2 = 0$ are given as

$$G_E^p(0) = 1, \qquad\qquad G_E^n(0) = 0,$$
$$G_M^p(0) = 2.7928\ \mu_N, \qquad G_M^n(0) = -1.913\ \mu_N,$$

where μ_N is the nucleon magnetic moment. $G_E^{p,n}(0)$ gives the value of the electric charge for the proton and neutron. Similarly, $G_M^{p,n}(0)$ gives the magnetic moments of the proton and neutron. If the nucleons are assumed to be point particles, then the magnetic moments for the proton and neutron should be $G_M^p(0) = \mu_N$ and $G_M^n(0) = 0$. Therefore, the difference from μ_N in the case of the proton and from 0 in the case of the neutron represents the anomalous magnetic moments of the nucleons which provides evidence that nucleons are not point particles and have got structure.

10.4.5 Numerical parameterization of the electromagnetic form factors

Initially, it was observed that the experimental data for the electromagnetic $e^- p$ scattering may be explained if one assumes the form factors to have a dipole form, that is,

$$G_E^p(q^2) = \frac{G_M^p(q^2)}{\mu_p} = \frac{G_M^n(q^2)}{\mu_n} = G_D(q^2), \tag{10.58}$$

where $G_D(q^2)$ is given as

$$G_D(q^2) = \frac{1}{\left(1 - \frac{q^2}{M_V^2}\right)^2}, \tag{10.59}$$

and M_V is known as the vector dipole mass. Initially, the best fit was found to be consistent with $M_V = 0.84$ GeV.

With the development of electron beam accelerators, many more measurements were performed like at MIT-Bates, MAMI, JLab, etc.; these measurements for the form factors found deviations from the dipole form. Figure 10.10 shows the deviation from a dipole form for the electric and magnetic form factors of nucleons. Therefore, there exist various parameterizations which show deviation from the dipole form like that of Galster et al. [414], which is given as

$$G_E^p(q^2) = G_D(q^2), \qquad\qquad G_M^p(q^2) = (1 + \mu_p)G_D(q^2),$$

$$G_M^n(q^2) = \mu_n G_D(q^2), \qquad\qquad G_E^n(q^2) = (\frac{q^2}{4M^2})\mu_n G_D(q^2)\xi_n,$$

$$\xi_n = \frac{1}{\left(1 - \lambda_n \frac{q^2}{4M^2}\right)},$$

with $\mu_p = 1.7927\mu_N$, $\mu_n = -1.913\mu_N$, $M_V = 0.84$ GeV, and $\lambda_n = 5.6$.

Recently, several new parameterizations [415, 416, 417, 418] for the electromagnetic isovector form factors have been presented. In the following, we present the parameterizations given by Bradford et al. [416] known as BBBA-05 as well as the parameterization given by Kelly [417] and the modifications made by Punjabi et al. [418] in Kelly's parameterization for the electric and magnetic form factors of nucleons.

(i) **BBBA-05**

The expression for electric and magnetic Sachs form factor given by Bradford et al. [416] (BBBA-05) is

$$G_E^p(q^2) = \frac{1 - 0.0578\tau}{1 + 11.1\tau + 13.6\tau^2 + 33.0\tau^3},$$

$$\frac{G_M^p(q^2)}{\mu_p} = \frac{1+0.15\tau}{1+11.1\tau+19.6\tau^2+7.54\tau^3},$$

$$G_E^n(q^2) = \frac{1.25\tau+1.30\tau^2}{1-9.86\tau+305\tau^2-758\tau^3+802\tau^4},$$

$$\frac{G_M^n(q^2)}{\mu_n} = \frac{1+1.81\tau}{1+14.1\tau+20.7\tau^2+68.7\tau^3}, \qquad \tau = -\frac{q^2}{4M^2}. \qquad (10.60)$$

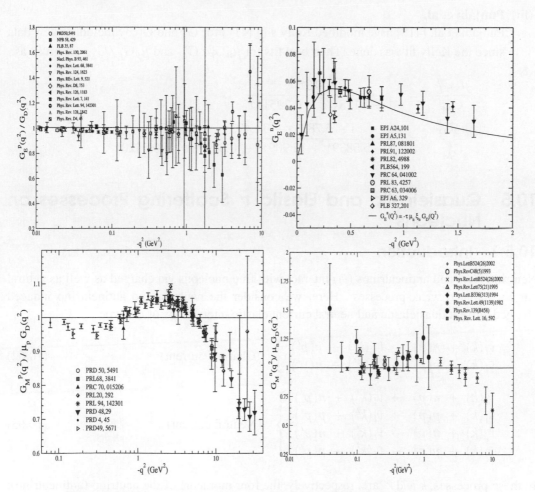

Figure 10.10 Experimental data for the electric and magnetic form factors for the proton.

(ii) **Kelly**

The parameterization for $G_E^{p,n}(q^2)$ and $G_M^{p,n}(q^2)$ given by Kelly [417] is

$$G_E^p(q^2) = \frac{1-0.24\tau}{1+10.98\tau+12.82\tau^2+21.97\tau^3},$$

$$\frac{G_M^p(q^2)}{\mu_p} = \frac{1 + 0.12\tau}{1 + 10.97\tau + 18.86\tau^2 + 6.55\tau^3},$$

$$G_E^n(q^2) = \frac{1.7\tau}{1 + 3.3\tau}\frac{1}{(1 - q^2/(0.84)^2)^2},$$

$$\frac{G_M^n(q^2)}{\mu_n} = \frac{1 + 2.33\tau}{1 + 14.72\tau + 24.20\tau^2 + 84.1\tau^3}. \tag{10.61}$$

(iii) **Punjabi et al.**

Punjabi et al. [418] have modified Kelly's fit [417] for G_E^n and G_E^p by including new data since the Kelly fit was done. Their best fits for $\mu_n G_E^n / G_M^n$ and $\mu_p G_E^p / G_M^p$ are given as:

$$\frac{\mu_n G_E^n}{G_M^n} = \frac{2.6316\tau}{1 + 4.118\sqrt{\tau} + 0.29516\tau},$$

$$\frac{\mu_p G_E^p}{G_M^p} = \frac{1 - 5.7891\tau + 14.493\tau^2 - 3.5032\tau^3}{1 - 5.5839\tau + 12.909\tau^2 + 0.88996\tau^3 + 0.5420\tau^4}.$$

10.5 Quasielastic and Elastic ν Scattering Processes on Nucleons

10.5.1 Introduction

Neutrinos (ν_l) and antineutrinos ($\bar{\nu}_l$) interact with free nucleons via charged as well as neutral current induced weak processes. Here, we consider the neutrino and antineutrino induced charged current quasielastic and neutral current elastic interactions of the type

$$\left.\begin{array}{l} \nu_l(k) + n(p) \rightarrow l^-(k') + p(p') \\ \bar{\nu}_l(k) + p(p) \rightarrow l^+(k') + n(p') \end{array}\right\} \quad \text{(Charged current)} \tag{10.62}$$

$$\left.\begin{array}{l} \nu_l(k) + n(p) \rightarrow \nu_l(k') + n(p') \\ \nu_l(k) + p(p) \rightarrow \nu_l(k') + p(p') \\ \bar{\nu}_l(k) + n(p) \rightarrow \bar{\nu}_l(k') + n(p') \\ \bar{\nu}_l(k) + p(p) \rightarrow \bar{\nu}_l(k') + p(p') \end{array}\right\} \quad \text{(Neutral current)} \tag{10.63}$$

In these processes, k and k' are, respectively, the four momenta of the neutrino (antineutrino); the corresponding charged lepton and p and p' are the four momenta of the incoming and outgoing nucleons. Feynman diagrams corresponding to reactions given in Eqs. (10.62) and (10.63) are shown in Figure 10.11.

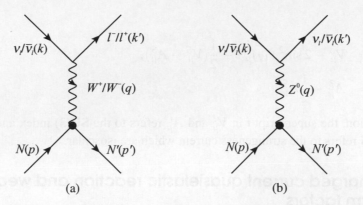

Figure 10.11 (a) Quasielastic and (b) Elastic ν scattering processes on nucleon ($N = n, p$ and $N' = p, n$) target.

10.5.2   Interaction Lagrangian

The interaction Lagrangian for the charged as well as the neutral current induced neutrino and antineutrino reactions on the free nucleon target in the standard model is written as (Chapter 8 for details)

$$\mathcal{L}_{\text{int}} = -\frac{g}{2\sqrt{2}} J_\mu^{CC} W^{+\mu} + \text{h.c} - \frac{g}{2\cos\theta_W} J_\mu^{NC} Z^\mu,$$

where the expression for the charged current is given by:

$$J_\mu^{CC} = \bar{\psi}_q' \gamma_\mu (1 - \gamma_5) U_{\text{CKM}} \psi_q + \bar{\nu}_l \gamma_\mu (1 - \gamma_5) l + \text{h.c.}$$

with $q' = (u, c, t)$, $q = (d, s, b)$. U_{CKM} is the CKM mixing matrix defined in Chapter 6.

The neutral current (NC) is given by:

$$J_\mu^{NC} = \Sigma_i \bar{\psi}_i \gamma_\mu (I_{3i} - Q_i \sin^2 \theta_W) \psi_i,$$

where i runs over all the fermions of the standard model, that is, $\begin{pmatrix} \nu_l \\ l^- \end{pmatrix}_L$, ν_{lR}, l_R, $l = e, \mu, \tau$.
I_3 is a weak isospin and Q is the charge of the ith fermion as well as for all the quark flavors.

At the nucleon level in the limit M_W^2, $M_Z^2 >> q^2$, the Lagrangian for the charged current (CC) reaction is expressed as:

$$\mathcal{L}_{\text{int}} = -\frac{G_F}{\sqrt{2}} a J_\mu^h l^\mu + \text{h.c.}$$

with

$$
\begin{array}{llll}
J_\mu^h & = & V_\mu^{1+i2} - A_\mu^{1+i2} & \quad \text{and} \quad a = \cos\theta_c & \quad \text{for } \Delta S = 0 \text{ CC reactions,} \\
J_\mu^h & = & V_\mu^{4+i5} - A_\mu^{4+i5} & \quad \text{and} \quad a = \sin\theta_c & \quad \text{for } \Delta S = 1 \text{ CC reactions, and} \\
J_\mu^h & = & V_\mu^{NC} - A_\mu^{NC} & \quad \text{and} \quad a = 1 & \quad \text{for the } \Delta S = 0 \text{ NC reactions}
\end{array}
$$

where

$$V_\mu^{NC} = V_\mu^3 - 2\sin^2\theta_W J_\mu^{em} - \frac{1}{2}(V_\mu^S - A_\mu^S),$$

$$A_\mu^{NC} = A_\mu^3 - \frac{1}{2}A_\mu^S.$$

In this expression, the superscript i in V_μ^i and A_μ^i refers to the SU(3) index and are discussed in Chapter 8. S refers to the strangeness current which are isoscalar.

10.5.3 Charged current quasielastic reaction and weak nucleon form factors

In a charged current reaction, the basic reaction (Figure 10.11(a)) for the quasielastic scattering process is a neutrino (antineutrino) interacting with a free neutron (proton). It is given by

$$\nu_l(k) + n(p) \longrightarrow l^-(k') + p(p'), \tag{10.64}$$

$$\bar{\nu}_l(k) + p(p) \longrightarrow l^+(k') + n(p'), \qquad l = e, \mu, \tau. \tag{10.65}$$

The invariant matrix element for the charged current quasielastic reaction of a neutrino and an antineutrino with a nucleon, given by Eqs. (10.64) and (10.65), is written as

$$\mathcal{M} = \frac{G_F}{\sqrt{2}}\cos\theta_C \, l_\mu \, J^\mu, \tag{10.66}$$

where G_F is the Fermi coupling constant, θ_C is the Cabibbo angle, and the leptonic weak current is given by

$$l_\mu = \bar{u}(\vec{k}')\gamma_\mu(1 \mp \gamma_5)u(\vec{k}). \tag{10.67}$$

$-(+)$ shows that it is a neutrino (antineutrino) induced quasielastic scattering process. The hadronic current J^μ is given by

$$J_\mu = \bar{u}(\vec{p}')\Gamma_\mu u(\vec{p}) \tag{10.68}$$

with

$$\Gamma_\mu = V_\mu - A_\mu. \tag{10.69}$$

The matrix elements of the vector (V_μ) and the axial vector (A_μ) currents are given by:

$$\langle N'(\vec{p}')|V_\mu|N(\vec{p})\rangle = \bar{u}(\vec{p}')\left[\gamma_\mu f_1(q^2) + i\sigma_{\mu\nu}\frac{q^\nu}{(M+M')}f_2(q^2)\right.$$

$$\left. + \frac{2q_\mu}{(M+M')}f_3(q^2)\right]u(\vec{p}), \tag{10.70}$$

$$\langle N'(\vec{p}')|A_\mu|N(\vec{p})\rangle = \bar{u}(\vec{p}')\left[\gamma_\mu\gamma_5 g_1(q^2) + i\sigma_{\mu\nu}\frac{q^\nu}{(M+M')}\gamma_5 g_2(q^2)\right.$$

$$\left. + \frac{2q_\mu}{(M+M')}\gamma_5 g_3(q^2)\right]u(\vec{p}), \tag{10.71}$$

where $N, N' = n, p$, $q^2 = (k - k')^2$ is the four momentum transfer squared. M and M' are the masses of the initial and the final nucleon, respectively. $f_1(q^2)$, $f_2(q^2)$, and $f_3(q^2)$ are the vector, weak magnetic, and induced scalar form factors and $g_1(q^2)$, $g_2(q^2)$, and $g_3(q^2)$ are the axial vector, induced tensor (or weak electric), and induced pseudoscalar form factors, respectively. The details about these form factors are discussed in Chapter 6.

Using the leptonic and hadronic currents given in Eqs. (10.67) and (10.68) in Eq. (10.66), the matrix element squared is obtained as

$$|\mathcal{M}|^2 = \frac{G_F^2}{2}\cos^2\theta_C\, L^{\mu\nu}J_{\mu\nu}. \tag{10.72}$$

The leptonic tensor $L^{\mu\nu}$ is calculated to be (Appendix D)

$$L^{\mu\nu} = 8\left[k^\mu k'^\nu + k'^\mu k^\nu - g^{\mu\nu}\,k\cdot k' \pm i\epsilon^{\mu\nu\alpha\beta}\,k'_\alpha k_\beta\right]. \tag{10.73}$$

$+(-)$ shows that it is a neutrino (antineutrino) induced process.

The hadronic tensor $J_{\mu\nu}$ given in Eq. (10.72), is obtained using Eq. (10.68) averaged over the initial spin state of the nucleon and summed over the final spin state as:

$$J_{\mu\nu} = \overline{\sum}\sum J_\mu^\dagger J_\nu$$

$$= \frac{1}{2}\text{Tr}\left[(\not{p}' + M)\Gamma_\mu(\not{p} + M)\tilde{\Gamma}_\nu\right], \tag{10.74}$$

where

$$\Gamma_\mu = \left[f_1(q^2)\gamma_\mu + i\sigma_{\mu\nu}\frac{q^\nu}{(M+M')}f_2(q^2) + \frac{2q_\mu}{(M+M')}f_3(q^2) - g_1(q^2)\gamma_\mu\gamma_5\right.$$

$$\left. - i\sigma_{\mu\nu}\frac{q^\nu}{(M+M')}\gamma_5 g_2(q^2) - \frac{2q_\mu}{(M+M')}\gamma_5 g_3(q^2)\right], \tag{10.75}$$

and $\tilde{\Gamma}_\nu = \gamma^0\,\Gamma_\nu^\dagger\,\gamma^0$. Using Eq. (10.75) in Eq. (10.74), we obtain $J_{\mu\nu}$ as

$$J_{\mu\nu} = \frac{1}{2}\left[4f_1^2(q^2)\left(p'_\mu p_\nu + p'_\nu p_\mu - (p\cdot p' - MM')g_{\mu\nu}\right)\right.$$

$$+ 4\frac{f_2^2(q^2)}{(M+M')^2}\left(MM'q^2 g_{\mu\nu} + q_\mu\left(-q_\nu(MM' + p\cdot p') + p'_\nu p\cdot q + p_\nu p'\cdot q\right)\right.$$

$$\left. - 2g_{\mu\nu}p\cdot q\, p'\cdot q + q^2 g_{\mu\nu}p\cdot p' - q^2 p_\mu p'_\nu + p_\mu q_\nu p'\cdot q + p'_\mu\left(q_\nu p\cdot q - q^2 p_\nu\right)\right)$$

$$+ \frac{16f_3^2(q^2)}{(M+M')^2}\left(q_\mu q_\nu(MM' + p\cdot p')\right) + 4g_1^2(q^2)\left((p'_\mu p_\nu + p_\mu p'_\nu) - (p\cdot p' + MM')g_{\mu\nu}\right)$$

$$+ \frac{4g_2^2(q^2)}{(M+M')^2}\left(-MM'q^2 g_{\mu\nu} + q_\mu\left(q_\nu(MM' - p\cdot p') + p'_\nu p\cdot q + p_\nu p'\cdot q\right)\right.$$

$$
\begin{aligned}
- \quad & 2g_{\mu\nu} p \cdot q p' \cdot q + q^2 g_{\mu\nu} p \cdot p' - q^2 p_\mu p'_\nu + p_\mu q_\nu p' \cdot q + p'_\mu \left(q_\nu p \cdot q - q^2 p_\nu \right) \Big) \\[4pt]
+ \quad & \frac{16 g_3^2(q^2)}{(M+M')^2} \left(q_\mu q_\nu (p' \cdot p - MM') \right) \\[4pt]
+ \quad & \frac{4 f_1(q^2) f_2(q^2)}{(M+M')} \left(q_\mu \left(M' p_\nu - M p'_\nu \right) + 2M g_{\mu\nu} p' \cdot q - M p'_\mu q_\nu - 2M' g_{\mu\nu} p \cdot q + M' p_\mu q_\nu \right) \\[4pt]
+ \quad & \frac{8 f_1(q^2) f_3(q^2)}{(M+M')} \left(q_\mu \left(M p'_\nu + M' p_\nu \right) + q_\nu \left(M p'_\mu + M' p_\mu \right) \right) + 8i f_1(q^2) g_1(q^2) \left(\epsilon_{\mu\nu\alpha\beta} \, p^\alpha p'^\beta \right) \\[4pt]
+ \quad & \frac{8i f_1(q^2) g_2(q^2)}{(M+M')} \left(M' \epsilon_{\mu\nu\alpha\beta} p^\alpha q^\beta - M \epsilon_{\mu\nu\alpha\beta} p'^\alpha q^\beta \right) + \frac{8 f_2(q^2) f_3(q^2)}{(M+M')^2} \left(q_\nu \left(p_\mu p' \cdot q - p'_\mu p \cdot q \right) \right. \\[4pt]
+ \quad & q_\mu \left(p_\nu p' \cdot q - p'_\nu p \cdot q \right) \Big) + 8i \left(\frac{f_2(q^2) g_1(q^2)}{(M+M')} \right) \left(M \epsilon_{\mu\nu\alpha\beta} p'^\alpha q^\beta + M' \epsilon_{\mu\nu\alpha\beta} p^\alpha q^\beta \right) \\[4pt]
+ \quad & \frac{8i f_2(q^2) g_2(q^2)}{(M+M')^2} \left(q_\mu \epsilon_{\nu\alpha\beta\delta} p^\alpha p'^\beta q^\delta - q_\nu \epsilon_{\mu\alpha\beta\delta} p^\alpha p'^\beta q^\delta + q^2 \epsilon^{\mu\nu\alpha\beta} p^\alpha p'^\beta + 2p \cdot q \epsilon^{\mu\nu\alpha\beta} p'^\alpha q^\beta \right) \\[4pt]
+ \quad & \frac{8i f_2(q^2) g_3(q^2)}{(M+M')^2} \left(q_\mu \epsilon_{\nu\alpha\beta\delta} p^\alpha p'^\beta q^\delta - q_\nu \epsilon_{\mu\alpha\beta\delta} p^\alpha p'^\beta q^\delta \right) \\[4pt]
+ \quad & \frac{8i f_3(q^2) g_2(q^2)}{(M+M')^2} \left(q_\mu \epsilon_{\nu\alpha\beta\delta} p^\alpha p'^\beta q^\delta - q_\nu \epsilon_{\mu\alpha\beta\delta} p^\alpha p'^\beta q^\delta \right) \\[4pt]
+ \quad & \frac{4 g_1(q^2) g_2(q^2)}{(M+M')} \left(q_\mu \left(M p'_\nu + M' p_\nu \right) - 2M g_{\mu\nu} p' \cdot q + M p'_\mu q_\nu - 2M' g_{\mu\nu} p \cdot q + M' p_\mu q_\nu \right) \\[4pt]
+ \quad & \frac{8 g_1(q^2) g_3(q^2)}{(M+M')} \left(q_\mu \left(M' p_\nu - M p'_\nu \right) + q_\nu \left(M' p_\mu - M p'_\mu \right) \right) \\[4pt]
+ \quad & \frac{8 g_2(q^2) g_3(q^2)}{(M+M')^2} \left(q_\nu \left(p_\mu p' \cdot q - p'_\mu p \cdot q \right) + q_\mu \left(p_\nu p' \cdot q - p'_\nu p \cdot q \right) \right) \Bigg].
\end{aligned}
\tag{10.76}
$$

Contraction of the various terms of the hadronic tensor $J_{\mu\nu}$ with the leptonic tensor $L^{\mu\nu}$ yields

$$
|\mathcal{M}|^2 = \frac{G_F^2}{2} \cos^2 \theta_c N(q^2),
\tag{10.77}
$$

where $N(q^2) = L^{\mu\nu} J_{\mu\nu}$ and is given in Appendix F. Using the general expression for the differential scattering cross section in the Lab frame (Figure 10.12), we obtain

$$
\frac{d\sigma}{dq^2} = \frac{G_F^2 \cos^2 \theta_c}{8\pi M^2 E_{\nu(\bar\nu)}^2} N(q^2),
\tag{10.78}
$$

where $E_{\nu(\bar\nu)}$ is the energy of the incoming (anti)neutrino.

The differential scattering cross section may also be expressed in terms of the Mandelstam variables s, t, u as:

$$
\frac{d\sigma}{dq^2} = \frac{G_F^2 M^2 \cos^2 \theta_C}{8\pi E_\nu^2} \left[A(q^2) \mp B(q^2) \frac{(s-u)}{M^2} + C(q^2) \frac{(s-u)^2}{M^4} \right].
\tag{10.79}
$$

The negative (positive) sign before the $B(q^2)$ term refers to neutrino (antineutrino) scattering. The following assumptions have been taken into account while obtaining the expression in

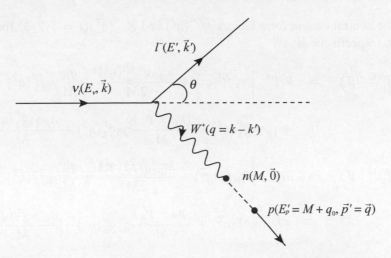

Figure 10.12 Kinematics in the Lab frame showing $\nu_l + n \rightarrow l^- + p$ scattering process.

Eq. (10.79): $M' = M$, $s - u = 4ME_\nu + q^2$ and there are no second class currents, that is, $f_3(q^2) = g_2(q^2) = 0$.

The factors $A(q^2)$, $B(q^2)$, and $C(q^2)$ are given as:

$$A(q^2) = \frac{m^2 - q^2}{4M^2} \left[\left(4 - \frac{q^2}{M^2} \right) g_1^2(q^2) - \left(4 + \frac{q^2}{M^2} \right) f_1^2(q^2) - \frac{q^2}{M^2} \left(1 + \frac{q^2}{4M^2} \right) f_2^2(q^2) \right.$$
$$- \frac{4q^2}{M^2} f_1(q^2) f_2(q^2) - \frac{m_l^2}{M^2} \left(\left(f_1(q^2) + f_2(q^2) \right)^2 + \left(g_1(q^2) + 2g_3(q^2) \right)^2 \right.$$
$$\left. \left. + \left(\frac{q^2}{M^2} - 4 \right) g_3^2(q^2) \right) \right],$$

(10.80)

$$B(q^2) = -\frac{q^2}{M^2} g_1^2(q^2) \left[f_1(q^2) + f_2(q^2) \right],$$

(10.81)

$$C(q^2) = \frac{1}{4} \left[g_1^2(q^2) + (f_1(q^2))^2 - \frac{q^2}{4M^2} (f_2(q^2))^2 \right].$$

(10.82)

The parameterizations for $f_i(q^2)$ and $g_i(q^2)$ are given in Section 10.5.6.

10.5.4 Neutral current elastic reactions and weak nucleon form factors

We first define the matrix elements for the neutral current processes on the proton and the neutron targets, that is,

$$\nu_l(k) + p(p) \longrightarrow \nu_l(k') + p(p'),$$
$$\nu_l(k) + n(p) \longrightarrow \nu_l(k') + n(p'),$$

in terms of the neutral current form factors $\tilde{f}_i^{p,n}(q^2)$ and $\tilde{g}_i^{p,n}(q^2)(i=1,2,3)$ for the protons and neutrons, respectively, as

$$
\left\langle \vec{p}' \mid J_\mu^{NC} \mid \vec{p} \right\rangle_p = \bar{u}(\vec{p}') \left[\gamma_\mu \tilde{f}_1^p(q^2) + \frac{i\sigma_{\mu\nu}q^\nu \tilde{f}_2^p(q^2)}{2M} + \frac{q_\mu}{M}\gamma_5 \tilde{f}_3^p(q^2) \right.
$$

$$
\left. + \gamma_\mu \gamma_5 \tilde{g}_1^p(q^2) + \frac{(p_\mu + p_\mu')}{M}\gamma_5 \tilde{g}_2^p(q^2) + \frac{q_\mu \gamma_5 \tilde{g}_3^p(q^2)}{M} \right] u(\vec{p}),
$$

$$
\left\langle \vec{p}' \mid J_\mu^{NC} \mid \vec{p} \right\rangle_n = \bar{u}(\vec{p}') \left[\gamma_\mu \tilde{f}_1^n(q^2) + \frac{i\sigma_{\mu\nu}q^\nu \tilde{f}_2^n(q^2)}{2M} + \frac{q_\mu}{M}\gamma_5 \tilde{f}_3^n(q^2) \right.
$$

$$
\left. + \gamma_\mu \gamma_5 \tilde{g}_1^n(q^2) + \frac{(p_\mu + p_\mu')}{M}\gamma_5 \tilde{g}_2^n(q^2) + \frac{q_\mu \gamma_5 \tilde{g}_3^n(q^2)}{M} \right] u(\vec{p}),
$$

where $\tilde{f}_1(q^2)$, $\tilde{f}_2(q^2)$, $\tilde{g}_1(q^2)$, and $\tilde{g}_3(q^2)$ are known as the first class neutral current form factors while $\tilde{f}_3(q^2)$ and $\tilde{g}_2(q^2)$ are known as the second class form factors. The parameterizations for $\tilde{f}_i(q^2)$ and $\tilde{g}_i(q^2)$ are given in Section 10.5.6.

10.5.5 Symmetry properties of weak hadronic currents and form factors

The weak form factors are constrained by the following symmetry properties of the weak hadronic currents:

(a) T invariance implies that all the form factors $f_{1-3}(q^2)$ and $g_{1-3}(q^2)$ are real.

(b) The assumption that the charged weak vector current and its conjugate along with the isovector part of the electromagnetic current form an isotriplet implies that the charged weak vector form factors $f_1(q^2)$ and $f_2(q^2)$ are related to the isovector electromagnetic form factors of the nucleon.

(c) The principle of conserved vector current (CVC) of weak vector currents implies that $f_3(q^2) = 0$.

(d) The principle of G-invariance implies that $f_3(q^2) = 0$ and $g_2(q^2) = 0$.

(e) The hypothesis of the partially conserved axial vector current (PCAC) relates $g_3(q^2)$ to $g_1(q^2)$ through the Goldberger–Treiman (GT) relation.

The implications of these symmetry properties have already been discussed in Chapter 6.

10.5.6 Parameterization of the weak form factors

(i) Vector form factors

In the case of charged current interactions, the hadronic current contains two isovector form factors $f_{1,2}(q^2)$ of the nucleons, which can be related to the isovector combination of the Dirac-Pauli form factors $F_{1,2}^p(q^2)$ and $F_{1,2}^n(q^2)$ of the proton and the neutron using the relation

$$f_{1,2}(q^2) = F_{1,2}^p(q^2) - F_{1,2}^n(q^2). \tag{10.83}$$

The Dirac and Pauli form factors are, in turn, expressed in terms of the experimentally determined Sachs electric and magnetic form factors as discussed in Section 10.4 (Eqs. (10.46) and (10.47)).

The vector form factors for the neutral current induced processes are obtained as

$$\tilde{f}_{1,2}^p(q^2) = \left(\frac{1}{2} - 2\sin^2\theta_W\right) F_{1,2}^p(q^2) - \frac{1}{2}F_{1,2}^n(q^2) - \frac{1}{2}F_{1,2}^s(q^2), \tag{10.84}$$

$$\tilde{f}_{1,2}^n(q^2) = \left(\frac{1}{2} - 2\sin^2\theta_W\right) F_{1,2}^n(q^2) - \frac{1}{2}F_{1,2}^p(q^2) - \frac{1}{2}F_{1,2}^s(q^2), \tag{10.85}$$

where θ_W is the Weinberg angle, and $F_1^s(q^2)$ and $F_2^s(q^2)$ are the strangeness vector form factors, which are discussed in this Section under the heading Strangeness form factors..

(ii) Axial vector form factor

The isovector axial vector form factor is parameterized as

$$g_1(q^2) = g_A(0) \left[1 - \frac{q^2}{M_A^2}\right]^{-2}, \tag{10.86}$$

where $g_A(0)(= C_A)$ is determined experimentally from the β-decay of neutrons as discussed in Chapter 5. M_A is known as the axial dipole mass and is obtained from the quasielastic neutrino and antineutrino scattering as well as from pion electroproduction data (Figure 10.13). The dipole parameterization is extensively used in the analysis of various experiments in quasielastic scattering. However, recently a new parameterization based on Z-expansion has been proposed in literature [419, 420]. Theoretically, $g_1(q^2)$ is calculated in various models of lattice gauge theory [419, 421, 422, 423, 424] and quark models [425].

The numerical value of M_A to be used in the calculations of neutrino–nucleon cross section is a subject of intense discussion in the neutrino physics community and a wide range of M_A have been recently discussed in the literature [356, 426, 427]. The old data available on (anti)neutrino scattering on hydrogen and deuterium targets [428, 429, 430] reanalyzed by Bodek et al. [431] gives a value of $M_A = 1.014 \pm 0.014$ GeV, while a recent analysis of the same data by Meyer et al. [420] gives a value in the range of 1.02–1.17 GeV depending upon which, data of ANL [428], BNL [429], and FNAL [430] experiments are considered. In 2002, Bernard et al. have reanalyzed the data of the neutrino and antineutrino scattering on hydrogen and deuterium targets as well as the electroproduction data and got the best χ^2 fit for M_A as:

$$M_A = 1.026 \pm 0.021 \text{ GeV}.$$

In recent years, high statistics data on quasielastic neutrino–nucleus scattering have been obtained and analyzed from neutrino and antineutrino scattering on nuclear targets both at low and intermediate energies. The data from NOMAD [432], MINERνA [433] favor a lower value

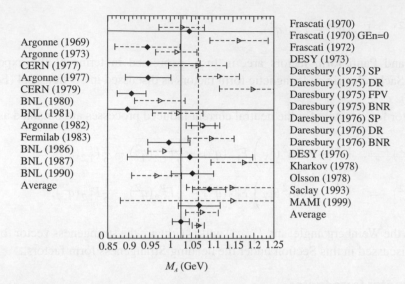

Figure 10.13 Axial mass M_A extractions from (quasi)elastic neutrino and antineutrino scattering experiments on hydrogen and deuterium targets (diamond) and from charged pion electroproduction experiments (triangle). The weighted average from the neutrino/antineutrino experiments is $M_A = 1.026 \pm 0.021$ GeV and from the electroproduction experiments is $M_A = 1.069 \pm 0.016$ GeV.

of M_A, around 1.03 GeV, while the data from MiniBooNE [434, 353, 435], MINOS [436, 437], K2K [438], T2K [439], and SciBooNE [440, 441] favor a higher value of M_A which lies in the range of 1.2–1.35 GeV. The recently suggested values of M_A from these experiments have been tabulated in Table 10.1. We shall discuss these experiments and the values of M_A determined from them alongwith the role of nuclear medium effects in Chapter 14.

In the case of neutral current induced reactions, the axial vector form factor for the nucleon is given by:

$$\tilde{g}^1_{p,n}(q^2) = \pm \frac{1}{2} g_1(q^2) - \frac{1}{2} F^s_A(q^2), \tag{10.87}$$

where $g_1(q^2)$ is given in Eq. (10.86) with $M_A = 1.026$ GeV. $F^s_A(q^2)$ is the strangeness axial vector form factor.

Table 10.1 Recent measurements of the axial dipole mass (M_A).

Experiment	M_A (GeV)	Experiment	M_A (GeV)
MINERνA [433, 442]	0.99	SciBooNE [440]	1.21±0.22
NOMAD [432]	1.05±0.02±0.06	K2K-SciBar [438]	1.144±0.077
MiniBooNE [434, 353, 435]	1.23±0.20	K2K-SciFi [438]	1.20±0.12
MINOS [436, 437]	1.19($Q^2 > 0$)	World Average	1.026± 0.021 [443]
	1.26($Q^2 > 0.3$ GeV^2)		1.014±0.014 [431]

(iii) Pseudoscalar form factor

In the charged current sector, the pseudoscalar form factor $g_3(q^2)$ is dominated by the pion pole. It is given in terms of the Goldberger–Treiman relation near $q^2 \approx 0$ if PCAC is assumed and is related to the axial vector form factor $g_1(q^2)$ as

$$g_3(q^2) = \frac{2M^2 g_1(q^2)}{m_\pi^2 - q^2}. \tag{10.88}$$

However, in the literature, there are other versions of the pseudoscalar form factor like [444]:

$$g_3(Q^2) = -\frac{M}{q^2} \left[\left(\frac{2m_\pi^2 f_\pi}{m_\pi^2 - q^2} \right) \left(\frac{M g_A(0)}{f_\pi} - \frac{g_{\pi NN}(0) \Delta q^2}{m_\pi^2} \right) + 2M g_1(q^2) \right], \tag{10.89}$$

where $g_{\pi NN}(0) = 13.21$, $f_\pi = 92.42$ MeV, and $\Delta = 1 + \frac{M g_A(0)}{f_\pi g_{\pi NN}(0)}$.

ahe pseudoscalar form factor using chiral perturbation theory (ChPT) is given by [444, 445]

$$g_3(0) = \frac{2M g_{\pi NN}(0) f_\pi}{m_\pi^2 - q^2} + \frac{g_A(0) M^2 r_A^2}{3}, \tag{10.90}$$

where axial radius $r_A = \frac{2\sqrt{3}}{M_A}$. There are many more.

The contribution from the pseudoscalar form factor is proportional to the mass of the lepton and hence, in the case of neutral current reactions, it vanishes.

(iv) Second class current form factors

In the $\Delta S = 0$ sector, the violation of G-parity due to difference in mass of u and d quarks or the intrinsic charge symmetry violation of the strong interaction is very small. The form factors $f_3(q^2)$ and $g_2(q^2)$ are expected to be very small too. Moreover, in the vector sector, the charged weak vector currents V_μ along with the isovector part of the electromagnetic current (J_μ^{em}) is assumed to form an isotriplet, which leads to the hypothesis of CVC and predicts $f_3(q^2) = 0$. However, in the axial vector sector, there is no such constraint on the form factor $g_2(q^2)$ and it could be nonvanishing albeit small. It is because of this reason that most of the experiments in the $\Delta S = 0$ sector are analyzed for the search of the second class current (SCC) assuming $f_3(q^2) = 0$ with a nonvanishing $g_2(q^2)$ which is found to be small. Generally, the form factor $g_2(q^2)$, in analogy with $g_1(q^2)$, is expressed as

$$g_2(q^2) = g_2(0) \left[1 - \frac{q^2}{M_2^2} \right]^{-2}, \tag{10.91}$$

where for simplicity, $M_2 = M_A$.

This form factor $g_2(q^2)$ may also give information about the time reversal invariance (TRI). If TRI is assumed, then all the form factors must be real; in the absence of TRI, the form factor

$g_2(q^2)$ can be taken as imaginary. We have explored the possibility of both real and imaginary $g_2(q^2)$, which are represented later in the text as $g_2^R(q^2)$ and $g_2^I(q^2)$, respectively.

(v) Strangeness form factors

The strangeness vector form factors $F_1^s(q^2)$ and $F_2^s(q^2)$ may be redefined in terms of the strangeness Sachs electric and magnetic form factors as:

$$G_E^s(q^2) \;=\; F_1^s(q^2) - \tau F_2^s(q^2), \tag{10.92}$$

$$G_M^s(q^2) \;=\; F_1^s(q^2) + F_2^s(q^2). \tag{10.93}$$

At $q^2 = 0$, the Sachs electric form factor gives the net strangeness of the nucleon, that is, $G_E^s(0) = 0$. At low momentum transfer, the electric form factor is expressed in terms of ρ^s, that is,

$$\rho^s = \frac{dG_E^s(q^2)}{dq^2} = -\frac{1}{6}\langle r_s^2 \rangle, \tag{10.94}$$

where $\langle r_s^2 \rangle$ is the strangeness radius. Similarly, at $q^2 = 0$, $G_M^s(q^2) = \mu^s$, the strangeness magnetic moment. Therefore, these two parameters ρ^s and μ^s determine the neutral current form factors $F_1^s(q^2)$ and $F_2^s(q^2)$. The q^2 dependence of $G_E^s(q^2)$ and $G_M^s(q^2)$ are obtained as:

$$G_E^s = \frac{\rho^s \tau}{\left(1 - \frac{q^2}{\Lambda_E^2}\right)}, \qquad G_M^s = \frac{\mu^s}{\left(1 - \frac{q^2}{\Lambda_M^2}\right)}, \tag{10.95}$$

where one of the fits [446] for ρ^s and μ^s assuming $\Lambda_{E,M}^s$ to be very large, gives

$$\rho^s = 0.13 \pm 0.21 \qquad \text{and} \qquad \mu^s = 0.035 \pm 0.053.$$

The strangeness axial vector form factor $F_A^s(q^2)$ is taken to be of dipole form:

$$F_A^s(q^2) = \frac{\Delta s}{\left(1 - \frac{q^2}{M_A^2}\right)^2}, \tag{10.96}$$

where Δs is the strange quark contribution to the nucleon spin and is determined from the neutral current neutrino scattering and scattering of polarized electrons from nucleons [447, 353].

10.5.7   Cross sections for charged current processes

Figure 10.14 presents the results for the differential scattering cross section $\frac{d\sigma}{dq^2}$ vs. $-q^2$, for the neutrino as well as antineutrino induced quasielastic processes, that is, $\nu_\mu + n \to \mu^- + p$ and $\bar{\nu}_\mu + p \to \mu^+ + n$, at the two values of incoming (anti)neutrino energies, viz., $E_{\nu_\mu(\bar{\nu}_\mu)} = 500$ MeV and 1 GeV. Moreover, we have presented the results for two different values of the axial dipole mass, viz., $M_A = 1.026$ and 1.2 GeV. It may be noticed that with the increase in the value of M_A, the differential scattering cross section increases. In the case of the neutrino induced process, 20% increase in the value of M_A, increases the cross section by about 20%; in the case of the antineutrino induced process, 20% increase in the value of M_A results in an increment in the cross section by about 5%.

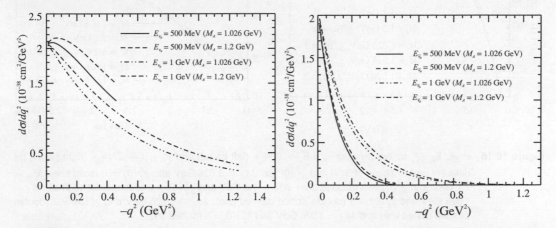

Figure 10.14 $\frac{d\sigma}{dq^2}$ vs. $-q^2$ for the $\nu_\mu + n \rightarrow \mu^- + p$ (left panel) and $\bar{\nu}_\mu + p \rightarrow \mu^+ + n$ (right panel) at the different (anti)neutrino energies, viz., $E_{\nu_\mu(\bar{\nu}_\mu)} = 500$ MeV and 1 GeV and at $M_A = 1.026$ and 1.2 GeV.

In Figure 10.15, the results for the total scattering cross section as a function of incoming (anti)neutrino energy are presented for the processes $\nu_l + n \rightarrow l^- + p$ and $\bar{\nu}_l + p \rightarrow l^+ + n$ with $l = e, \mu$ at $M_A = 1.026$ and 1.2 GeV. It may be observed that at low (anti)neutrino energies $< (0.3)0.8$ GeV, the cross section for ν_e ($\bar{\nu}_e$) is larger than the cross section for ν_μ ($\bar{\nu}_\mu$), which is basically a threshold effect. However, with increase in energy, that is, $E_{\nu_l(\bar{\nu}_l)} > 0.8(0.3)$ GeV, the cross sections for ν_e ($\bar{\nu}_e$) and ν_μ ($\bar{\nu}_\mu$) are almost comparable.

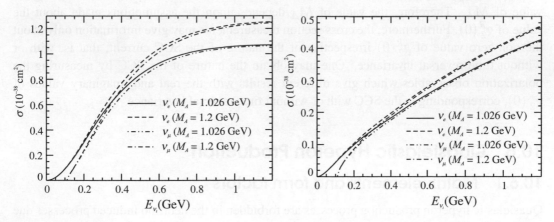

Figure 10.15 σ vs. $E_{\nu_l(\bar{\nu}_l)}$ for the $\nu_l + n \rightarrow l^- + p$ (left panel) and $\bar{\nu}_l + p \rightarrow l^+ + n$ (right panel) with $l = e, \mu$ at $M_A = 1.026$ and 1.2 GeV.

In Figure 10.16, we show the dependence of the total scattering cross section on M_A with or without the presence of $g_2^R(0)$. It may be observed from the figure that, in the case of

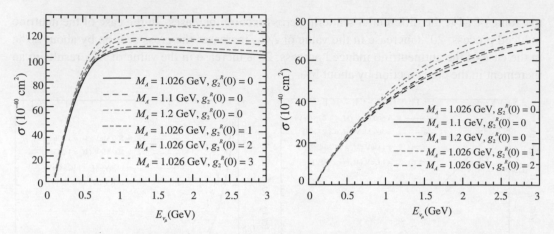

Figure 10.16 σ vs. $E_{\nu_\mu(\bar{\nu}_\mu)}$ for the process $\nu_\mu + n \rightarrow \mu^- + p$ (left panel) and $\bar{\nu}_\mu + p \rightarrow \mu^+ + n$ (right panel) for different combinations of M_A, and $g_2^R(0)$ viz. $M_A = 1.026$ GeV and $g_2^R(0) = 0$ (solid line), $M_A = 1.1$ GeV and $g_2^R(0) = 0$ (dashed line), $M_A = 1.2$ GeV and $g_2^R(0) = 0$ (dashed dotted line), $M_A = 1.026$ GeV and $g_2^R(0) = 1$ (double dotted dashed line), $M_A = 1.026$ GeV and $g_2^R(0) = 2$ (double dashed dotted line) and $M_A = 1.026$ GeV and $g_2^R(0) = 3$ (dotted line).

neutrino induced processes, that is, $\nu_\mu + n \rightarrow \mu^- + p$, the results obtained by taking $M_A = 1.1$ GeV and $g_2^R(0) = 0$ are comparable to the results obtained with $M_A = 1.026$ GeV and $g_2^R(0) = 2$; whereas, the results obtained by taking $M_A = 1.2$ GeV and $g_2^R(0) = 0$ are comparable to the results obtained using $M_A = 1.026$ GeV and $g_2^R(0) = 3$. In the case of an antineutrino induced process, that is, $\bar{\nu}_\mu + p \rightarrow \mu^+ + n$, the results obtained by taking $M_A = 1.1$ GeV and $g_2^R(0) = 0$ are comparable to the results obtained with $M_A = 1.026$ GeV and $g_2^R(0) = 1$; whereas, the results obtained by taking $M_A = 1.2$ GeV and $g_2^R(0) = 0$ are slightly lower than the results obtained using $M_A = 1.026$ GeV and $g_2^R(0) = 2$. Thus, a higher value of $\sigma(E_{\bar{\nu}_\mu})$ may be obtained by either taking a non-zero value of $g_2^R(0)$ or increasing the value of M_A. Therefore, the value of M_A depends upon the assumptions made about the value of $g_2^R(0)$. Furthermore, the cross section measurements may give information only about the non-zero value of $g_2^R(0)$ irrespective of the nature of the SCC current, that is, with or without time reversal invariance. One may obtain the nature of the SCC by measuring the polarization observables which give different results with the real and imaginary values of $g_2^R(0)$, corresponding to the SCC with or without time reversal invariance.

10.6 Quasielastic Hyperon Production

10.6.1 Matrix elements and form factors

Quasielastic hyperon production processes are forbidden in the neutrino induced processes due to the $\Delta S \neq \Delta Q$ rule, but are possible with antineutrinos as discussed in Chapter 6. The following processes are induced when an antineutrino interacts with a nucleon to produce a hyperon and an antilepton (Figure 10.17):

$$\bar{\nu}_l(k) + p(p) \rightarrow l^+(k') + \Lambda(p'), \tag{10.97}$$

$$\bar{\nu}_l(k) + p(p) \quad \rightarrow \quad l^+(k') + \Sigma^0(p'), \tag{10.98}$$

$$\bar{\nu}_l(k) + n(p) \quad \rightarrow \quad l^+(k') + \Sigma^-(p'), \qquad l = e, \mu, \tau, \tag{10.99}$$

where the quantities in the brackets represent the four momenta of the particles. The transition matrix element for the processes presented in Eqs. (10.97)–(10.99) is written as

$$\mathcal{M} = \frac{G_F}{\sqrt{2}} \sin\theta_c \, l^\mu J_\mu. \tag{10.100}$$

The leptonic current (l^μ) is given in Eq. (10.67). The hadronic current (J_μ) for the quasielastic hyperon production can be written in analogy with the antineutrino–nucleon scattering except that the mass of the final nucleon is replaced by the mass of the hyperon and the electroweak form factors of the nucleons are replaced by the $N - Y$ transition form factors. In general, J_μ is given in Eq. (10.68) with Γ_μ defined in Eq. (10.69). The matrix elements of the vector (V_μ) and the axial vector (A_μ) currents between a hyperon $Y(= \Lambda, \Sigma^0,$ and Σ^-) and a nucleon $N = n, p$ are written as:

$$\langle Y(\vec{p}')|V_\mu|N(\vec{p})\rangle = \bar{u}(\vec{p}')\left[\gamma_\mu f_1^{NY}(q^2) + i\sigma_{\mu\nu}\frac{q^\nu}{M+M'}f_2^{NY}(q^2)\right.$$

$$\left. + \frac{2\,q_\mu}{M+M'}f_3^{NY}(q^2)\right]u(\vec{p}), \tag{10.101}$$

$$\langle Y(\vec{p}')|A_\mu|N(\vec{p})\rangle = \bar{u}(\vec{p}')\left[\gamma_\mu\gamma_5 g_1^{NY}(q^2) + i\sigma_{\mu\nu}\frac{q^\nu}{M+M'}\gamma_5 g_2^{NY}(q^2)\right.$$

$$\left. + \frac{2\,q_\mu}{M+M'}\gamma_5 g_3^{NY}(q^2)\right]u(\vec{p}), \tag{10.102}$$

where M and M' are the masses of the nucleon and hyperon, respectively. $f_1^{NY}(q^2)$, $f_2^{NY}(q^2)$ and $f_3^{NY}(q^2)$ are the vector, weak magnetic, and induced scalar $N - Y$ transition form factors. $g_1^{NY}(q^2)$, $g_2^{NY}(q^2)$ and $g_3^{NY}(q^2)$ are the axial vector, induced tensor (or weak electric), and induced pseudoscalar form factors, respectively.

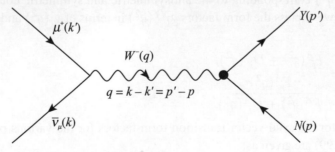

Figure 10.17 Feynman diagram for the process $\bar{\nu}_l(k) + N(p) \rightarrow l^+(k') + Y(p')$, where $N(= p, n)$ and $Y(= \Lambda, \Sigma^0, \Sigma^-)$ represents the initial nucleon and the final hyperon, respectively.

The transition matrix element squared is obtained as

$$\overline{\sum}\sum|\mathcal{M}|^2 = \frac{G_F^2 \sin^2\theta_c}{2}J^{\mu\nu}L_{\mu\nu}, \tag{10.103}$$

where $J^{\mu\nu}$ and $L_{\mu\nu}$ are obtained in a similar way, as mentioned in Section 10.5.3.

The differential scattering cross section $d\sigma/dq^2$ for the processes given in Eq. (10.97)–(10.99) in the Lab frame is then obtained as

$$\frac{d\sigma}{dq^2} = \frac{G_F^2 \sin^2 \theta_c}{8\pi M^2 E_{\bar{\nu}_l}^2} N(q^2),$$

(10.104)

where $N(q^2) = J^{\mu\nu} L_{\mu\nu}$. It is given in Appendix F.

The weak transition form factors $f_i(q^2)$ and $g_i(q^2)$, $i = 1 - 3$, are determined using the Cabibbo theory of $V - A$ interaction extended to the strange sector. The details are given in Chapter 6 and Appendix B.

10.6.2 Vector form factors

The expressions for the vector form factors in terms of the electromagnetic form factors $F_{1,2}^p(q^2)$ and $F_{1,2}^n(q^2)$ for the various processes given in Eqs. (10.97)–(10.99), are obtained using SU(3) symmetry in the Cabibbo model. The details are given in Appendix B. These form factors are as follows:

$$f_{1,2}^{p\Lambda}(q^2) = -\sqrt{\frac{3}{2}} \, F_{1,2}^p(q^2),$$

(10.105)

$$f_{1,2}^{n\Sigma^-}(q^2) = -\left[F_{1,2}^p(q^2) + 2F_{1,2}^n(q^2) \right],$$

(10.106)

$$f_{1,2}^{p\Sigma^0}(q^2) = -\frac{1}{\sqrt{2}} \left[F_{1,2}^p(q^2) + 2F_{1,2}^n(q^2) \right].$$

(10.107)

10.6.3 Axial vector form factors

The axial vector form factors $g_i^{NY}(q^2)(i = 1, 2, 3)$ are expressed in terms of the two functions $F_i^A(q^2)$ and $D_i^A(q^2)$ corresponding to the antisymmetric and symmetric couplings of the two octets. However, we express the form factors $g_i^{NY}(q^2)$ in terms of $g_i(q^2)$ and $x_i(q^2)$ which are defined as

$$g_i(q^2) = F_i^A(q^2) + D_i^A(q^2),$$

(10.108)

$$x_i(q^2) = \frac{F_i^A(q^2)}{F_i^A(q^2) + D_i^A(q^2)}, \qquad i = 1 - 3.$$

(10.109)

The expressions for the axial vector transition form factors for the various processes given in Eqs. (10.97)–(10.99) are given as:

$$g_{1,2}^{p\Lambda}(q^2) = -\frac{1}{\sqrt{6}}(1 + 2x_{1,2})g_{1,2}(q^2),$$

(10.110)

$$g_{1,2}^{n\Sigma^-}(q^2) = (1 - 2x_{1,2})g_{1,2}(q^2),$$

(10.111)

$$g_{1,2}^{p\Sigma^0}(q^2) = \frac{1}{\sqrt{2}}(1 - 2x_{1,2})g_{1,2}(q^2).$$

(10.112)

In the following, we describe the explicit forms of the axial vector form factors used for calculating the numerical results.

(a) Axial vector form factor $g_1^{NY}(q^2)$

We note from Eq. (10.108) that $g_1(q^2)$ is the axial vector form factor for $n \rightarrow p$ transition. It is defined in Eq. (10.86). The parameter $x_1(q^2)$ occurring in Eqs. (10.110)–(10.112) for $g_1^{NY}(q^2)$ ($Y = \Lambda, \Sigma^0, \Sigma^-$) is determined at low Q^2 from the analysis of semileptonic hyperon decay (SHD) and is found to be $x_1(Q^2 \approx 0) = 0.364$. There is no experimental information about the Q^2 dependence of $x_1(q^2)$; therefore, we assume it to be constant, that is, $x_1(q^2) \approx x_1(0) = 0.364$ for convenience.

(b) Second class current form factor $g_2^{NY}(q^2)$

The expression for $g_2^{NY}(q^2)$ for the hyperons $\Lambda, \Sigma^-, \Sigma^0$ are given in Eqs. (10.110)–(10.112) in terms of $g_2(q^2)$ and $x_2(q^2)$, where $g_2(q^2)$ is parameterized in Eq. (10.91). There is some information about $g_2(q^2)$ from the neutrino and antineutrino scattering off the nucleons. It is known that the determination of the experimental value of $g_2(0)$ is correlated with the value of $M_2(= M_A)$ (Eq. (10.91)) used in the analysis. There exist theoretical calculations for $g_2^{R(np)}(0)$ and $g_2^{R(NY)}(0)$ for $Y = \Lambda, \Sigma^-, \Sigma^0$. In the literature, various values of $g_2^I(0)$ for nucleons and hyperons have been used, which are in the range 1–10 [448, 449, 450]. However, there is no information about $x_2(q^2)$. To see the dependence of $g_2^R(0)$ and $g_2^I(0)$ on the differential and the total scattering cross section, we have varied $g_2^R(0)$ and $g_2^I(0)$ in the range $0 - 3$ and used $M_2 = M_A$. For the q^2 dependence of the form factor, that is, $g_2^{NY}(q^2)$, we use the SU(3) symmetric expressions for $g_2(q^2)$ taken to be of dipole form given in Eq. (10.91) for the various transitions given in Eqs. (10.110)–(10.112), treating $x_2(q^2)$ to be constant. Let us take $x_2 = x_1$ for simplicity.

(c) The induced pseudoscalar form factor $g_3^{NY}(q^2)$

In general, the contribution of $g_3(q^2)$ to the (anti)neutrino scattering cross sections is proportional to m_l^2, where m_l is the mass of the corresponding charged lepton, and is small in e^- and μ^- induced processes. It is significant in the process involving τ^- leptons. For $g_3^{NY}(q^2)$, Nambu [46] has given a generalized parameterization using PCAC and generalized GT relation for the $\Delta S = 1$ currents, that is,

$$g_3^{NY}(q^2) = \frac{(M + M_Y)^2}{2(m_K^2 - q^2)} g_1^{NY}(q^2), \tag{10.113}$$

where m_K is the mass of kaon and $g_1^{NY}(q^2)$ is given in Eqs. (10.110)–(10.112) for $Y = \Lambda, \Sigma^-, \Sigma^0$.

10.6.4 Cross sections: Experimental results

Figure 10.18 presents the results for $\bar{\nu}_\mu$ induced Λ production from the free proton in the energy region of $E_{\bar{\nu}_\mu} < 10\,\text{GeV}$ and compares them with the experimental results from the Gargamelle bubble chamber at CERN [451, 452, 453] using propane with a small admixture of freon and from Serpukhov SKAT bubble chambers [454] using freon and from the BNL experiment using a hydrogen target [455]. We have presented the theoretical results by taking the different values of $g_2(0)$, viz., $g_2^{R,I}(0) = 0$, $g_2^R(0) = 3$ and $g_2^I(0) = 3$. It may be observed that the non-zero value of $g_2(0)$ (whether real or imaginary) increases the cross section; in the case $g_2^R(0) = 3$ or $g_2^I(0) = 3$, the cross section increases by about 20%.

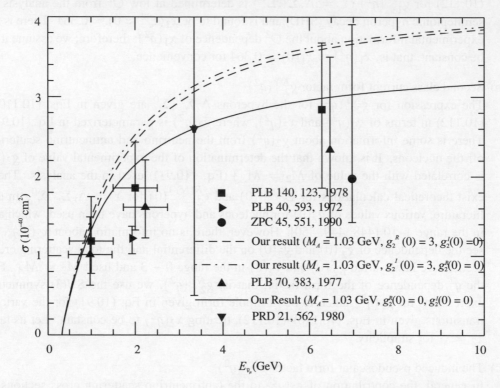

Figure 10.18 σ vs. $E_{\bar{\nu}_\mu}$, for $\bar{\nu}_\mu + p \to \mu^+ + \Lambda$ process. Experimental results (triangle right [451], square [452], triangle up [453], circle [454]), triangle down($\sigma = 2.6^{+5.9}_{-2.1} \times 10^{-40} \text{cm}^2$) [455]) are shown with error bars. The solid line represents the results obtained with $M_A = 1.03$ GeV and $g_2 = 0$ while the dashed (double dotted dashed line) represents the results obtained with $M_A = 1.03$ GeV and $g_2^R(0) = 3$ ($g_2^I(0) = 3$).

10.7 Polarization of Final Hadrons and Leptons

10.7.1 Introduction

In elastic $e^- p$ scattering, the experiments with polarized electron beam and the polarized proton target play an important role in determining the vector form factors. In the weak sector, the vector form factors are expressed in terms of the electromagnetic form factors of the nucleons. In the axial vector sector, information about the form factors is obtained from the semileptonic decays of nucleons and hyperons at low q^2. One may also obtain information about these form factors by measuring the polarization of the final hadron. Experimentally, it is difficult to study polarization of the final nucleon in quasielastic ($\Delta S = 0$) scattering as one requires a double scattering measurement. However, it is easier to study polarization observables in quasielastic hyperon production as the produced hyperons decay into pions which gives information about the polarization of the hyperon.

10.7.2 Polarization of the final hadron

The polarization four-vector ξ^τ of the hadron produced in the final state in the reactions shown in Eqs. (10.97)–(10.99) is written as (Appendix C and Ref. [456] for details):

$$\xi^\tau = \frac{\text{Tr}[\gamma^\tau \gamma_5 \, \rho_f(p')]}{\text{Tr}[\rho_f(p')]}, \tag{10.114}$$

where the spin density matrix $\rho_f(p')$ corresponding to the final hadron of momentum p' is given by

$$\rho_f(p') = L^{\alpha\beta} \, \text{Tr}[\Lambda(p')\Gamma_\alpha \Lambda(p)\tilde{\Gamma}_\beta \Lambda(p')]. \tag{10.115}$$

In this expression, Γ_α is given in Eq. (10.69). Using the following relations,

$$\Lambda(p')\gamma^\tau \gamma_5 \Lambda(p') = 2M' \left(g^{\tau\sigma} - \frac{p'^\tau p'^\sigma}{M'^2} \right) \Lambda(p')\gamma_\sigma \gamma_5, \tag{10.116}$$

$$\Lambda(p')\Lambda(p') = 2M'\Lambda(p'), \tag{10.117}$$

where M' corresponds to the mass of the final hadron, ξ^τ defined in Eq. (10.114) may be rewritten as (Appendix C):

$$\xi^\tau = \left(g^{\tau\sigma} - \frac{p'^\tau p'^\sigma}{M'^2} \right) \frac{L^{\alpha\beta} \text{Tr}\left[\gamma_\sigma \gamma_5 \Lambda(p')\Gamma_\alpha \Lambda(p)\tilde{\Gamma}_\beta\right]}{L^{\alpha\beta} \text{Tr}\left[\Lambda(p')\Gamma_\alpha \Lambda(p)\tilde{\Gamma}_\beta\right]}. \tag{10.118}$$

Note that in Eq. (10.118), ξ^τ is manifestly orthogonal to p'^τ, that is, $p' \cdot \xi = 0$. Moreover, the denominator is directly related to the differential scattering cross section given in Eq. (10.104).

With $J^{\alpha\beta}$ and $L_{\alpha\beta}$ given in Eqs. (10.74) and (10.73), respectively, an expression for ξ^τ is obtained in terms of the four-momenta of the particles. Here, we have considered two cases:

Case I: When time reversal invariance is assumed.
We evaluate the polarization vector ξ^τ defined in Eq. (10.118) in the lab frame, that is, when the initial nucleon is at rest, $\vec{p} = 0$. Moreover, from the momentum conservation, $\vec{k}'$ is resolved in terms of $\vec{k}$, $\vec{p}$, and $\vec{p}'$. If the time reversal invariance is assumed, then all the form factors defined in Eq. (10.69) are real and the polarization vector can be expressed as

$$\vec{\xi} = \frac{\left[A^h(q^2)\vec{k} + B^h(q^2)\vec{p}' \right]}{N(q^2)}, \tag{10.119}$$

where the expressions of $A^h(q^2)$, $B^h(q^2)$, and $N(q^2)$ are given in Appendix F. In Eq. (10.119), $A^h(q^2)$, $B^h(q^2)$, and $N(q^2)$ given in Appendix F are taken in the limit $f_3(q^2) = 0$ and $g_2(0) = g_2^R(0)$ to ensure the time reversal invariance.

From Eq. (10.119), it follows that the polarization lies in the plane of reaction and there is no component of polarization in a direction perpendicular to the reaction plane. This is a consequence of time reversal invariance which makes the transverse polarization in the direction perpendicular to the reaction plane vanish. We now expand the polarization vector $\vec{\zeta}$ along the orthogonal directions, $\hat{e}_L^h$, $\hat{e}_P^h$, and $\hat{e}_T^h$ in the reaction plane corresponding to the longitudinal, perpendicular, and transverse directions, defined as

$$\hat{e}_L^h = \frac{\vec{p}\,'}{|\vec{p}\,'|}, \qquad \hat{e}_P^h = \hat{e}_L^h \times \hat{e}_T^h, \qquad \text{where} \qquad \hat{e}_T^h = \frac{\vec{p}\,' \times \vec{k}}{|\vec{p}\,' \times \vec{k}|}, \tag{10.120}$$

and depicted in Figure 10.19(b). We then write $\vec{\zeta}$ as:

$$\vec{\zeta} = \zeta_L \hat{e}_L^h + \zeta_P \hat{e}_P^h, \tag{10.121}$$

such that the longitudinal and perpendicular components of the polarization vector ($\vec{\zeta}$) in the laboratory frame are given by

$$\zeta_L(q^2) = \vec{\zeta} \cdot \hat{e}_L^h, \qquad \zeta_P(q^2) = \vec{\zeta} \cdot \hat{e}_P^h. \tag{10.122}$$

From Eq. (10.122), the longitudinal $P_L^h(q^2)$ and perpendicular $P_P^h(q^2)$ components of the polarization vector defined in the rest frame of the final hadron are given by

$$P_L^h(q^2) = \frac{M'}{E_{p'}} \zeta_L(q^2), \qquad P_P^h(q^2) = \zeta_P(q^2), \tag{10.123}$$

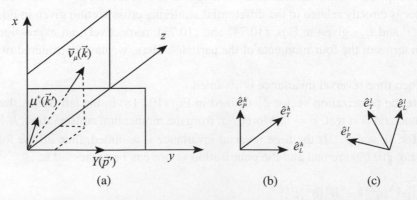

(a)   (b)   (c)

Figure 10.19 (a) Momentum and polarization directions of the final baryon and lepton. $\hat{e}_L^{h,l}$, $\hat{e}_P^{h,l}$, and $\hat{e}_T^{h,l}$ represent the orthogonal unit vectors corresponding to the longitudinal, perpendicular, and transverse directions with respect to the momentum of the final hadron in (b) and the final lepton in (c).

where $\frac{M'}{E_{p'}}$ is the Lorentz boost factor along $\vec{p}'$. With the help of Eqs. (10.119), (10.120), (10.122), and (10.123), the longitudinal $P_L^h(q^2)$ and perpendicular $P_P^h(q^2)$ components are calculated to be:

$$P_L^h(q^2) = \frac{M'}{E_{p'}} \frac{A^h(q^2)\vec{k} \cdot \vec{p}' + B^h(q^2)|\vec{p}'|^2}{N(q^2)\,|\vec{p}'|}, \tag{10.124}$$

$$P_P^h(q^2) = \frac{A^h(q^2)[(\vec{k} \cdot \vec{p}')^2 - |\vec{k}|^2|\vec{p}'|^2]}{N(q^2)\,|\vec{p}'|\,|\vec{p}' \times \vec{k}|}. \tag{10.125}$$

To see the dependence of $g_2^R(0)$ on the polarization observables, the results for $P_L^\Lambda(Q^2)$ and $P_P^\Lambda(Q^2)$ are presented as a function of Q^2 in Figure 10.20 for the process $\bar{\nu}_\mu + p \to \mu^+ + \Lambda$ using $g_2^R(0) = 0, \pm 1$, and ± 3 at $E_{\bar{\nu}_\mu} = 1$ GeV. It may be observed that $P_L^\Lambda(Q^2)$ shows large variations as we change $|g_2^R(0)|$ from 0 to 3. For example, in the peak region of Q^2, the difference is about 50% as $|g_2^R(0)|$ is changed from 0 to 3. In the case of $P_P^\Lambda(Q^2)$ also, the Q^2 dependence is quite strong and similar to $P_L^\Lambda(Q^2)$.

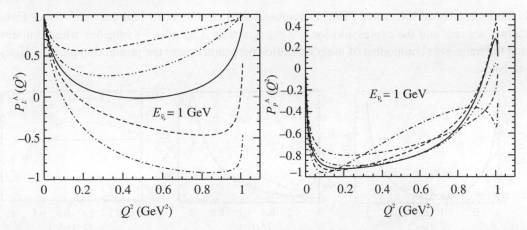

Figure 10.20 $P_L^\Lambda(Q^2)$ vs. Q^2 (left panel) and $P_P^\Lambda(Q^2)$ vs. Q^2 (right panel) for the process $\bar{\nu}_\mu + p \to \mu^+ + \Lambda$ at the incoming antineutrino energy, $E_{\bar{\nu}_\mu} = 1$ GeV for the polarized Λ in the final state, at different values of $g_2^R(0)$, viz., $g_2^R(0) = 0$ (solid line), 1 (dashed line), 3 (dashed dotted line), -1 (double dotted dashed line) and -3 (double dashed dotted line).

Case II: When time reversal violation is assumed.

In the absence of time reversal invariance, the polarization vector $\vec{\xi}$ is calculated as

$$\vec{\xi} = \frac{A^h(q^2)\vec{k} + B^h(q^2)\vec{p}' + C^h(q^2)\epsilon^{\alpha\beta\gamma\delta}k_\beta p_\gamma p_\delta'}{N(q^2)}$$

$$= \frac{A^h(q^2)\vec{k} + B^h(q^2)\vec{p}' + C^h(q^2)[\epsilon^{ijk}k_j p_k' M]}{N(q^2)}$$

$$= \frac{A^h(q^2)\vec{k} + B^h(q^2)\vec{p}' + C^h(q^2)M(\vec{k} \times \vec{p}')}{N(q^2)}, \tag{10.126}$$

where the expressions of $C^h(q^2)$ is given in Appendix F.

The polarization vector $\vec{\xi}$ may be written in terms of the longitudinal, perpendicular, and transverse components as

$$\vec{\xi} = \xi_L \hat{e}_L^h + \xi_P \hat{e}_P^h + \xi_T \hat{e}_T^h, \tag{10.127}$$

where the unit vectors are defined in Eq. (10.120). The longitudinal and perpendicular components are given in Eqs. (10.124) and (10.125), respectively. The transverse component of polarization in the rest frame of the final hadron is given as

$$P_T(q^2) = \xi_T(q^2) = \vec{\xi} . \hat{e}_T. \tag{10.128}$$

Using Eqs. (10.120) and (10.127) in Eq. (10.128), we obtain

$$P_T^h(q^2) = \frac{C^h(q^2) M [(\vec{k} \cdot \vec{p}')^2 - |\vec{k}|^2 |\vec{p}'|^2]}{N(q^2) \, |\vec{p}' \times \vec{k}|}. \tag{10.129}$$

If the time reversal invariance (TRI) is assumed, then all the vector and the axial vector form factors are real and the expression for $C^h(q^2)$ (given in Appendix F) vanishes which implies that the transverse component of the polarization perpendicular to the production plane, $P_T^h(q^2)$ vanishes.

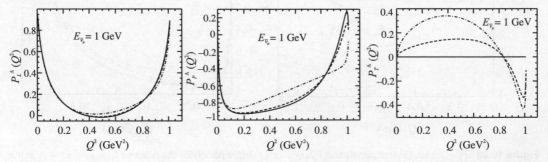

Figure 10.21 $P_L^\Lambda(Q^2)$ vs. Q^2 (left panel), $P_P^\Lambda(Q^2)$ vs. Q^2 (middle panel) and $P_T^\Lambda(Q^2)$ vs. Q^2 (right panel) for the process $\bar{\nu}_\mu + p \rightarrow \mu^+ + \Lambda$ at the incoming antineutrino energy, $E_{\bar{\nu}_\mu} = 1$ GeV for the polarized Λ in the final state, at different values of $g_2^I(0)$, viz., $g_2^I(0) = 0$ (solid line), 1 (dashed line), and 3 (dashed dotted line).

To see the dependence of $g_2^I(0)$ on the polarization observables, the results are presented in Figure 10.21 for $P_L^\Lambda(Q^2)$, $P_P^\Lambda(Q^2)$, and $P_T^\Lambda(Q^2)$ as a function of Q^2 using $g_2^I(0) = 0$, 1, and 3 at $E_{\bar{\nu}_\mu} = 1$ GeV. It may be deduced that while $P_L^\Lambda(Q^2)$ is less sensitive to $g_2^I(0)$ at low antineutrino energies, $P_P^\Lambda(Q^2)$ is sensitive to $g_2^I(0)$ at $E_{\bar{\nu}_\mu} = 1$ GeV. Moreover, $P_T^\Lambda(Q^2)$ shows 40% variations at $Q^2 = 0.4$ GeV^2, $E_{\bar{\nu}_\mu} = 1$ GeV, when $g_2^I(0)$ is varied from 0 to 3.

10.7.3 Polarization of the final lepton

Instead of the final hadron polarization, if one assumes the final lepton to be polarized, then the polarization four-vector(ζ^τ) in reactions, Eqs. (10.97)–(10.99) is written as

$$\zeta^\tau = \frac{\text{Tr}[\gamma^\tau \gamma_5 \rho_f(k')]}{\text{Tr}[\rho_f(k')]}, \tag{10.130}$$

and the spin density matrix for the final lepton $\rho_f(k')$ is given by

$$\rho_f(k') = J^{\alpha\beta} \, \text{Tr}[\Lambda(k')\gamma_\alpha(1+\gamma_5)\Lambda(k)\tilde{\gamma}_\beta(1+\tilde{\gamma}_5)\Lambda(k')], \tag{10.131}$$

with $\tilde{\gamma}_\alpha = \gamma^0 \gamma_\alpha^\dagger \gamma^0$ and $\tilde{\gamma}_5 = \gamma^0 \gamma_5^\dagger \gamma^0$.

Using Eqs. (10.116) and (10.117), ζ^τ defined in Eq. (10.130) may also be rewritten as

$$\zeta^\tau = \left(g^{\tau\sigma} - \frac{k'^\tau k'^\sigma}{m_l^2}\right) \frac{J^{\alpha\beta}\text{Tr}\left[\gamma_\sigma \gamma_5 \Lambda(k')\gamma_\alpha(1+\gamma_5)\Lambda(k)\tilde{\gamma}_\beta(1+\tilde{\gamma}_5)\right]}{J^{\alpha\beta}\text{Tr}\left[\Lambda(k')\gamma_\alpha(1+\gamma_5)\Lambda(k)\tilde{\gamma}_\beta(1+\tilde{\gamma}_5)\right]}, \tag{10.132}$$

where m_l is the charged lepton mass. In Eq. (10.132), the denominator is directly related to the differential scattering cross section given in Eq. (10.104).

With $J^{\alpha\beta}$ and $L_{\alpha\beta}$ given in Eqs. (10.74) and (10.73), respectively, an expression for ζ^τ is obtained. In the laboratory frame, where the initial nucleon is at rest, the polarization vector $\vec{\zeta}$ is calculated to be a function of three-momenta of incoming antineutrino $(\vec{k})$ and outgoing lepton $(\vec{k}')$, and is given as

$$\vec{\zeta} = \frac{\left[A^l(q^2)\vec{k} + B^l(q^2)\vec{k}' + C^l(q^2)M(\vec{k} \times \vec{k}')\right]}{N(q^2)}, \tag{10.133}$$

where the expressions of $A^l(q^2)$, $B^l(q^2)$, and $C^l(q^2)$ are given in Appendix F.

One may expand the polarization vector $\vec{\zeta}$ along the orthogonal directions, $\hat{e}_L^l$, $\hat{e}_P^l$, and $\hat{e}_T^l$ in the reaction plane corresponding to the longitudinal, perpendicular, and transverse directions, defined as

$$\hat{e}_L^l = \frac{\vec{k}'}{|\vec{k}'|}, \qquad \hat{e}_P^l = \hat{e}_L^l \times \hat{e}_T^l, \qquad \text{where} \qquad \hat{e}_T^l = \frac{\vec{k} \times \vec{k}'}{|\vec{k} \times \vec{k}'|}, \tag{10.134}$$

and depicted in Figure 10.19(c). We then write $\vec{\zeta}$ as:

$$\vec{\zeta} = \zeta_L \hat{e}_L^l + \zeta_P \hat{e}_P^l + \zeta_T \hat{e}_T^l, \tag{10.135}$$

such that the longitudinal, perpendicular, and transverse components of the $\vec{\zeta}$ in the laboratory frame are given by

$$\zeta_L(q^2) = \vec{\zeta} \cdot \hat{e}^l_L, \qquad \zeta_P(q^2) = \vec{\zeta} \cdot \hat{e}^l_P, \qquad \zeta_T(q^2) = \vec{\zeta} \cdot \hat{e}^l_T. \tag{10.136}$$

From Eq. (10.136), the longitudinal $P^l_L(q^2)$, perpendicular $P^l_P(q^2)$, and transverse $P^l_T(q^2)$ components of the polarization vector defined in the rest frame of the final lepton are given by

$$P^l_L(q^2) = \frac{m_l}{E_{k'}} \zeta_L(q^2), \qquad P^l_P(q^2) = \zeta_P(q^2), \qquad P^l_T(q^2) = \zeta_T(q^2), \tag{10.137}$$

where $\frac{m_l}{E_{k'}}$ is the Lorentz boost factor along $\vec{k}'$. Using Eqs. (10.133), (10.134), and (10.136) in Eq. (10.137), the longitudinal $P^l_L(q^2)$, perpendicular $P^l_P(q^2)$, and transverse $P^l_T(q^2)$ components are calculated to be

$$P^l_L(q^2) = \frac{m_\mu}{E_{k'}} \frac{A^l(q^2)\vec{k} \cdot \vec{k}' + B^l(q^2)|\vec{k}'|^2}{N(q^2)\,|\vec{k}'|}, \tag{10.138}$$

$$P^l_P(q^2) = \frac{A^l(q^2)[|\vec{k}|^2|\vec{k}'|^2 - (\vec{k} \cdot \vec{k}')^2]}{N(q^2)\,|\vec{k}'|\,|\vec{k} \times \vec{k}'|}, \tag{10.139}$$

$$P^l_T(q^2) = \frac{C^l(q^2)M[(\vec{k} \cdot \vec{k}')^2 - |\vec{k}|^2|\vec{k}'|^2]}{N(q^2)\,|\vec{k} \times \vec{k}'|}. \tag{10.140}$$

The lepton polarization observables do not give any significant information about T-noninvariance as they are very small and difficult to measure. In the case of τ-lepton, the polarization of τ will be substantial because of its mass but the experiments are very difficult to perform at the present time.

Chapter 11

Neutrino Scattering from Hadrons:
Inelastic Scattering (I)

11.1 Introduction

The inelastic scattering processes of (anti)neutrinos from nucleons are relevant in the region starting from the neutrino energy corresponding to the threshold production of a single pion. For neutral current (NC) induced 1π production, this starts at $E^{th}_{\nu(\bar{\nu})} = 144.7$ MeV for ν_l reactions. In the case of charged current (CC) induced 1π production, the threshold energy is higher because of the massive leptons produced in the final state; it corresponds to $E_{\nu(\bar{\nu})} \geq$ 150.5 MeV (277.4 MeV) for ν_e (ν_μ) reactions. As the neutrino energy increases, inelastic processes of multiple pion production, viz., 2π, 3π, etc., and the production of strange mesons (K) and hyperons (Y) start; both of which are the most relevant inelastic processes in the region of a few GeV. These inelastic processes have been studied very extensively, both theoretically and experimentally, in various reactions induced by photons and electrons which probe the interaction of the electromagnetic vector currents with hadrons in the presence of other strongly interacting particles like mesons and hyperons. In weak processes induced by neutrinos and antineutrinos, the inelastic processes provide a unique opportunity to study the interaction of the weak vector as well as the axial vector currents with hadrons in the presence of strongly interacting particles like mesons and hyperons. Moreover, a study of these weak processes from nucleons and nuclei is of immense topical importance in the context of the present neutrino oscillation experiments being done with the accelerator and atmospheric neutrinos in the energy region of a few GeV. The specific reactions to be studied in the inelastic channels are the various processes induced by the charged and neutral weak currents of neutrinos and antineutrinos, given in Table 11.1.

The first four reactions in Table 11.1 are strangeness conserving ($\Delta S = 0$) reactions and the last one is a strangeness changing ($\Delta S = 1$) reaction. The generic Feynman diagrams describing these reactions are shown in Figures 11.1(a) and 11.1(b), where $\nu_l(\bar{\nu}_l)$ and $l^-(l^+)$

Table 11.1 Charged and neutral current induced inelastic processes. N, N' represents protons and neutrons, $Y = \Lambda, \Sigma$ represents hyperons, $K = K^+, K^0$ represents kaons, $\bar{K} = K^-, \bar{K}^0$ represents antikaons, and $l = e, \mu, \tau$ represents leptons.

S. No.	CC induced $\nu(\bar{\nu})$ reactions	NC induced $\nu(\bar{\nu})$ reactions
1.	$\nu_l(\bar{\nu}_l) + N \longrightarrow l^-(l^+) + N' + \pi$	$\nu_l(\bar{\nu}_l) + N \longrightarrow \nu_l(\bar{\nu}_l) + N' + \pi$
2.	$\nu_l(\bar{\nu}_l) + N \longrightarrow l^-(l^+) + N' + n\pi$	$\nu_l(\bar{\nu}_l) + N \longrightarrow \nu_l(\bar{\nu}_l) + N' + n\pi$
3.	$\nu_l(\bar{\nu}_l) + N \longrightarrow l^-(l^+) + N' + \eta$	$\nu_l(\bar{\nu}_l) + N \longrightarrow \nu_l(\bar{\nu}_l) + N' + \eta$
4.	$\nu_l(\bar{\nu}_l) + N \longrightarrow l^-(l^+) + Y + K$	$\nu_l(\bar{\nu}_l) + N \longrightarrow \nu_l(\bar{\nu}_l) + Y + K$
5.	$\nu_l(\bar{\nu}_l) + N \longrightarrow l^-(l^+) + N' + K(\bar{K})$	$\nu_l(\bar{\nu}_l) + N \longrightarrow \nu_l(\bar{\nu}_l) + N' + K(\bar{K})$

are leptons interacting through the $W^{\pm}(Z)$ exchanges with the nucleon (N) producing the final nucleon (N') and hyperons (Y) and mesons like pions (π) and kaons (K). In Figure 11.1, the first vertex is the weak vertex (W) describing the weak interactions of leptons with $W^{\pm}(Z)$ bosons in the standard model (SM), while the second vertex is a mixed vertex (M) describing the weak interaction of nucleons in the SM and the strong interactions of the meson–baryon system described by a phenomenological Lagrangian consistent with the symmetries of strong interaction or effective Lagrangian motivated by the symmetries of quantum chromodynamics (QCD) like the chiral symmetry.

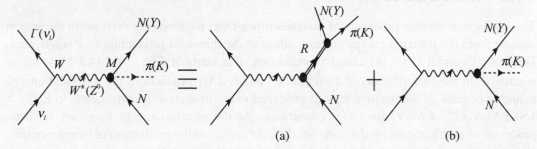

Figure 11.1 Generic Feynman diagrams representing the charged and neutral current induced inelastic processes given in Table 11.1.

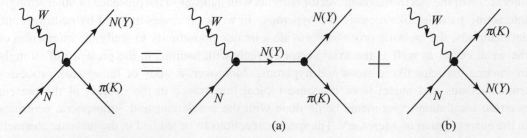

Figure 11.2 Generic Feynman diagrams representing the non-resonant background terms contributing to the inelastic processes.

It has been observed from the study of these inelastic reactions induced by photons and electrons in the case of electromagnetic interactions that the production processes are dominated by the excitation of nucleon resonances in the entire energy region as shown in Figure 11.1(a) except the region of very low energies corresponding to the threshold production of mesons where the non-resonant production of mesons shown in Figure 11.1(b) also makes important contribution. In Figure 11.1(a), R represents the excited resonance which decays into a nucleon and a pion or a hyperon and a strange meson in the case of $\Delta S = 0$ reactions; and to a nucleon and a strange meson in the case of $\Delta S = 1$ reactions. The diagrams showing the non-resonant direct production of mesons in Figure 11.1(b) receive contribution from two types of diagrams, that is, Born diagrams and contact diagrams, shown in Figure 11.2(a) and 11.2(b). The Born diagrams that are shown in Figure 11.2(a) contribute through the s, t, and u channel Feynman diagrams; while Figure 11.2(b) shows the contact diagram which appears in certain models of the effective Lagrangian for the meson–hadron interactions motivated by the chiral symmetry of QCD, and is responsible for the gauge invariance in the vector part of the hadronic current. These points are elaborated later in the text.

The nucleon resonances which are excited in the inelastic reactions are characterized by their mass, parity, spin, and isospin. They are represented by the symbol $R_{IJ}(M_R)$, where R is the name of the resonance given on the basis of the nucleon's orbital angular momentum, that is, $L = 0, 1, 2$. The resonances are named S, P, D, etc. M_R is the mass, while I and J specify their isospin and total angular momentum quantum numbers. The first resonance which is excited is called the Δ resonance, with positive parity, mass of 1232 MeV, isospin $3/2$, and spin $3/2$; it is therefore represented by the symbol $P_{33}(1232)$. The next excited resonance is called the Roper resonance which has negative parity, mass 1440 MeV, with spin $1/2$ and isospin $1/2$. The generic name given to a resonance with $I = 1/2$ is N^* resonance and to a resonance with $I = 3/2$ is Δ^* resonance. A list of resonances below the mass $M_R \leq 2.1$ GeV with spin $1/2$ and $3/2$ are given in Tables 11.2 and 11.3 along with the branching ratios of their dominant decays into various meson, nucleon, and hyperon channels.

Table 11.2 Properties of the spin $1/2$ resonances available in the Particle Data Group (PDG) [117] Breit–Wigner mass M_R, the total decay width Γ, isospin I, parity P, and the branching ratio of different meson–baryon systems like $N\pi$, $N\eta$, $K\Lambda$, and $K\Sigma$.

Resonance	Status	M_R	Γ	I	P	Branching ratios			
		(GeV)	(GeV)			$N\pi$	$N\eta$	$K\Lambda$	$K\Sigma$
$P_{11}(1440)$	****	1.370 ± 0.01	0.175 ± 0.015	$1/2$	$+$	65%	< 1%	-	-
$S_{11}(1535)$	****	1.510 ± 0.01	0.130 ± 0.020	$1/2$	$-$	42%	42%	-	-
$S_{31}(1620)$	****	1.600 ± 0.01	0.120 ± 0.020	$3/2$	$-$	30%	-	-	-
$S_{11}(1650)$	****	1.655 ± 0.015	0.135 ± 0.035	$1/2$	$+$	60%	25%	10%	-
$P_{11}(1710)$	****	1.700 ± 0.02	0.120 ± 0.040	$1/2$	$+$	10%	30%	15%	< 1%
$P_{11}(1880)$	***	1.860 ± 0.04	0.230 ± 0.050	$1/2$	$+$	6%	30%	20%	17%
$S_{11}(1895)$	****	1.910 ± 0.02	0.110 ± 0.030	$1/2$	$-$	10%	25%	18%	13%
$S_{31}(1900)$	****	1.865 ± 0.035	0.240 ± 0.060	$3/2$	$-$	8%	-	-	-
$P_{31}(1910)$	****	1.860 ± 0.03	0.300 ± 0.100	$3/2$	$+$	20%	-	-	9%
$S_{31}(2150)$	*	2.140 ± 0.08	0.200 ± 0.080	$3/2$	$-$	8%	-	-	-

Table 11.3 Properties of the spin $3/2$ resonances available in the PDG [117] with Breit–Wigner mass M_R, the total decay width Γ, isospin I, parity P, and the branching ratio of different meson–baryon systems like $N\pi$, $N\eta$, $K\Lambda$, and $K\Sigma$.

Resonance	Status	M_R (GeV)	Γ (GeV)	I	P	Branching ratios			
						$N\pi$	$N\eta$	$K\Lambda$	$K\Sigma$
$P_{33}(1232)$	$****$	1.210 ± 0.001	0.100 ± 0.002	$3/2$	$+$	99.4%	-	-	-
$D_{13}(1520)$	$****$	1.510 ± 0.005	$0.110^{+0.010}_{0.005}$	$1/2$	$-$	60%	0.08%	-	-
$P_{33}(1600)$	$****$	1.510 ± 0.05	0.270 ± 0.07	$3/2$	$+$	16%	-	-	-
$D_{13}(1700)$	$***$	1.700 ± 0.05	0.200 ± 0.100	$1/2$	$-$	12%	12%	-	-
$D_{33}(1700)$	$****$	1.665 ± 0.025	0.250 ± 0.05	$3/2$	$-$	15%	-	-	-
$P_{13}(1720)$	$****$	1.675 ± 0.015	$0.250^{+0.100}_{0.150}$	$1/2$	$+$	11%	3%	4.5%	-
$D_{13}(1875)$	$***$	1.900 ± 0.05	0.160 ± 0.06	$1/2$	$-$	7%	$< 1\%$	0.2%	0.7%
$P_{13}(1900)$	$****$	1.920 ± 0.02	0.150 ± 0.05	$1/2$	$+$	10%	8%	11%	5%
$P_{33}(1920)$	$***$	1.900 ± 0.05	0.300 ± 0.1	$3/2$	$+$	12%	-	-	4%
$D_{33}(1940)$	$**$	1.950 ± 0.1	0.350 ± 0.15	$3/2$	$-$	4%	-	-	-
$D_{13}(2120)$	$***$	2.100 ± 0.05	0.280 ± 0.08	$1/2$	$-$	10%	-	-	-

In addition to nucleon resonances, there are meson and hyperon resonances which also contribute to the inelastic production of mesons and hyperons through t and u channels. Their contributions are traditionally included along with the contribution of the non-resonant part coming from the Born diagrams in s, t, and u channels. A list of low-lying meson and hyperon resonances are given in Tables 11.4 and 11.5; these resonances are generally included in many calculations of inelastic processes. They are calculated using an effective Lagrangian for the weak interaction vertex and an effective Lagrangian for the meson–baryon vertex. Since most of the calculations were done earlier for the single meson production dominated by the pion production at intermediate energies, an effective phenomenological Lagrangian for the pion nucleon interaction with pseudoscalar (ps) or pseudovector (pv) coupling is used. After the

Table 11.4 Properties of the meson (pion and kaon) resonances available in PDG [117] with Breit–Wigner mass M_R, the total decay width Γ, isospin I, spin J, and parity P.

Resonance	M_R (MeV)	Γ (MeV)	I	J	P
$\rho(770)$	775.26 ± 0.25	147.8 ± 0.9	1	1	$-$
$\omega(782)$	782.65 ± 0.12	8.49 ± 0.08	0	1	$-$
$K^*(892)$	891.66 ± 0.26	50.8 ± 0.9	$1/2$	1	$-$
$f_0(980)$	990 ± 20	$10 - 100$	0	0	$+$
$\phi(1020)$	1019.461 ± 0.016	4.249 ± 0.013	0	1	$-$
$h_1(1170)$	1170 ± 20	360 ± 40	0	1	$+$
$b_1(1235)$	1229.5 ± 3.2	142 ± 9	1	1	$+$
$K_1(1270)$	1272 ± 7	90 ± 20	$1/2$	1	$+$
$K_1(1400)$	1669 ± 7	64^{+10}_{14}	1	$3/2$	$-$

Table 11.5 Properties of the spin $1/2$ and $3/2$ hyperon resonances available in PDG [117] with Breit–Wigner mass M_R, the total decay width Γ, isospin I, spin J, and parity P.

Resonance	Status	M_R (MeV)	Γ (MeV)	I	J	P
$\Lambda(1405)$	$****$	$1405.1\pm^{1.3}_{1.0}$	50.5 ± 2	0	1/2	−
$\Lambda(1520)$	$****$	1517 ± 4	$15\pm^{10}_{8}$	0	3/2	−
$\Lambda(1600)$	$***$	1544 ± 3	$112\pm^{12}_{2}$	0	1/2	+
$\Lambda(1690)$	$****$	1697 ± 6	65 ± 14	0	3/2	−
$\Lambda(1710)$	$*$	1713 ± 13	180 ± 40	0	1/2	+
$\Lambda(1800)$	$***$	$1800\pm^{50}_{80}$	300 ± 100	0	1/2	−
$\Lambda(2050)$	$*$	2056 ± 22	493 ± 60	0	3/2	−
$\Sigma(1385)$	$****$	1383.7 ± 1	36 ± 5	1	3/2	+
$\Sigma(1670)$	$****$	1669 ± 7	$64\pm^{10}_{14}$	1	3/2	−
$\Sigma(1730)$	$*$	1727 ± 27	276 ± 90	1	3/2	+
$\Sigma(1750)$	$***$	$1750\pm^{50}_{20}$	$90\pm^{70}_{30}$	1	1/2	−
$\Sigma(1940)$	$***$	$1940\pm^{10}_{40}$	$220\pm^{80}_{70}$	1	3/2	−

importance of the role of chiral symmetry in the study of strong interactions was realized, effective Lagrangians inspired by chiral symmetry have been used to describe meson–nucleon interactions. For this purpose, an effective Lagrangian based on chiral SU(2) symmetry has been used to describe the pion–nucleon interaction which is extended to chiral SU(3) symmetry to describe the kaon–nucleon interaction.

Early inelastic (anti)neutrino experiments were done on nuclear targets at CERN and later at ANL and BNL, the experiments were done on nucleons using hydrogen and deuterium targets [457, 458, 459]. In order to analyze the experimental results of the neutrino and antineutrino inelastic reactions on the nuclear targets, it is important to have a good understanding of the basic inelastic processes on nucleon targets. It is in this context that we give in this chapter, a theoretical formulation of the inelastic neutrino and antineutrino scattering processes on hadrons involving single meson production. There have been many approaches to calculate the single meson production from nucleons and nuclear targets induced by neutrinos and antineutrinos. Most of these methods have evolved from the treatment of photo- and electro- production of mesons from nucleons and nuclei which are induced by vector currents. In the case of weak production, there is an additional contribution from the axial vector current; the interference between the vector and axial vector contributions leads to parity violating observables in the single meson production.

Historically, weak pion production has been studied for a long time starting from 1962 [460, 461, 462, 463]. The early calculations were based on

(i) dynamical models with dispersion theory,

(ii) quark models with higher symmetry like SU(6),

(iii) phenomenological Lagrangians for the interaction of mesons with nucleons and higher resonances.

These calculations have been comprehensively summarized by Adler [464] and Llewellyn Smith [294]; extensive calculations have also been done by Adler [464] in the dynamical model and in the quark model by Albright [465, 466] and Kim [467]. The calculations have been compared with the early experimental results from CERN [457]. After the results of the hydrogen and deuterium bubble chamber experiments from ANL [458] and BNL [459] and later experiments from CERN [468, 469, 470] were obtained, many new calculations were made using the phenomenological Lagrangians [471, 472, 473, 474, 475, 476, 477, 478, 479, 480] or the Lagrangians based on the chiral symmetry [481, 482, 483, 484, 485, 486, 487, 488, 489, 490, 491] or quark models [492].

As emphasized earlier in this section, the contributions from the resonance excitations and their subsequent decays play a more important role than the non-resonant diagrams. In the following sections, we outline a general formalism to calculate the resonance and non-resonant contribution to the single meson produced in the $\nu(\bar{\nu})$ reactions induced by the $\Delta S = 0$ and $\Delta S = 1$ weak charge current (CC) and also the $\Delta S = 0$ neutral current (NC).

11.2   Inelastic Scattering through CC Excitation of Resonances

Tables 11.2 and 11.3 presented the properties of spin $1/2$ and $3/2$ resonances. There also exist, in nature, many baryon resonances with higher spins like spin $5/2, 7/2, 9/2$, etc. For example, $D_{15}(1675)$ is spin $5/2$ resonance with a $****$ rating, which means that experimentally $D_{15}(1675)$ is a well-studied resonance but theoretically, the physics of such higher spin resonances like $D_{15}(1675)$ is very complicated and beyond the scope of this book. In Sections 11.2.1 and 11.2.2, we present the formalism to study charged current induced resonance excitations with spin $1/2$ and $3/2$.

11.2.1   CC excitation of spin $1/2$ resonances

Excitation of spin $1/2$ and isospin $1/2$ resonances
The basic (anti)neutrino induced spin $1/2$ resonance excitations on the nucleon target are the following:

$$\nu_l/\bar{\nu}_l\,(k) + N(p) \longrightarrow l^-/l^+\,(k') + R_{1/2}(p_R). \tag{11.1}$$

The invariant matrix element for the process, given in Eq. (11.1) and depicted in Figure 11.3, is written as

$$\mathcal{M} = \frac{G_F}{\sqrt{2}}\cos\theta_C\,l_\mu\,J^{\mu CC}_{\frac{1}{2}}, \tag{11.2}$$

where G_F is the Fermi coupling constant and θ_C is the Cabibbo mixing angle. The leptonic current l^μ is the same as in the case of quasielastic scattering; however, for the sake of completeness, the expression is given as:

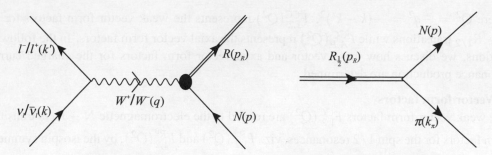

Figure 11.3 Feynman diagram for the charged current induced spin 1/2 and 3/2 resonance production (left panel) and for the decay of the resonance into a nucleon and a pion (right panel). The quantities in the parentheses represent the momenta of the corresponding particles.

$$l_\mu = \bar{u}(k')\gamma_\mu(1 \mp \gamma_5)u(k). \tag{11.3}$$

$(+)-$ represents the (anti)neutrino induced processes. The hadronic current for the spin $\frac{1}{2}$ resonance is given by

$$J_{\frac{1}{2}}^{\mu CC} = \bar{u}(p_R)\Gamma_{\frac{1}{2}}^\mu u(p), \tag{11.4}$$

where $u(p)$ and $\bar{u}(p_R)$, respectively, are the Dirac spinor and adjoint Dirac spinor for the initial nucleon and the final spin 1/2 resonance. $\Gamma_{\frac{1}{2}}^\mu$ is the vertex function. For a positive parity resonance, $\Gamma_{\frac{1}{2}}^\mu$ is given by

$$\Gamma_{\frac{1}{2}+}^\mu = V_{\frac{1}{2}}^\mu - A_{\frac{1}{2}}^\mu, \tag{11.5}$$

while for a negative parity resonance, $\Gamma_{\frac{1}{2}}^\mu$ is given by

$$\Gamma_{\frac{1}{2}-}^\mu = \left[V_{\frac{1}{2}}^\mu - A_{\frac{1}{2}}^\mu\right]\gamma_5, \tag{11.6}$$

where $V_{\frac{1}{2}}^\mu$ represents the vector current and $A_{\frac{1}{2}}^\mu$ represents the axial vector current.

The vector and axial vector currents are parameterized in terms of the vector and axial vector $N - R_{1/2}$ transition form factors, respectively, as,

$$V_{\frac{1}{2}}^\mu = \frac{F_1^{CC}(Q^2)}{(2M)^2}\left(Q^2\gamma^\mu + \slashed{q}q^\mu\right) + \frac{F_2^{CC}(Q^2)}{2M}i\sigma^{\mu\alpha}q_\alpha \tag{11.7}$$

$$A_{\frac{1}{2}}^\mu = \left[F_A^{CC}(Q^2)\gamma^\mu + \frac{F_P^{CC}(Q^2)}{M}q^\mu\right]\gamma_5, \tag{11.8}$$

where $Q^2 = -q^2 = -(k - k')^2$. $F_{1,2}^{CC}(Q^2)$ represents the weak vector form factors for the $N - R_{1/2}$ transitions while $F_{A,P}^{CC}(Q^2)$ represents the axial vector form factors. In the following sections, we discuss how these vector and axial vector form factors for the charged current resonance production are determined.

(i) Vector form factors

The weak vector form factors $F_{1,2}^{CC}(Q^2)$ are related to the electromagnetic $N - R_{1/2}$ transition form factors for the spin $1/2$ resonances, viz., $F_{1,2}^{R^+}(Q^2)$ and $F_{1,2}^{R^0}(Q^2)$, by the isospin symmetry and are expressed as:

$$F_i^{CC}(Q^2) = F_i^{R^+}(Q^2) - F_i^{R^0}(Q^2), \qquad i = 1, 2. \tag{11.9}$$

The hypothesis of the conserved vector current (CVC) states that the components of the weak charged vector currents and isovector electromagnetic current form an isotriplet in the isospin space, that is, the charged current vector form factors are related to the electromagnetic form factors by isospin symmetry. The isovector part of the weak current $\mathcal{J}_\mu^i$ can be written as

$$\mathcal{J}_\mu^i = (\mathcal{J}_\mu^1, \mathcal{J}_\mu^2, \mathcal{J}_\mu^3) = \mathcal{V}_\mu^i - \mathcal{A}_\mu^i, \qquad i = 1 - 3 \tag{11.10}$$

where

$$\mathcal{V}_\mu^i = (\mathcal{V}_\mu^1, \mathcal{V}_\mu^2, \mathcal{V}_\mu^3) = V_\mu \frac{\tau^i}{2}, \tag{11.11}$$

$$\mathcal{A}_\mu^i = (\mathcal{A}_\mu^1, \mathcal{A}_\mu^2, \mathcal{A}_\mu^3) = A_\mu \frac{\tau^i}{2}, \tag{11.12}$$

with V_μ and A_μ as defined in Eqs. (11.7) and (11.8), respectively. τ_is are the Pauli spin matrices.

Let us start with the electromagnetic current which is given by the following expression

$$J_\mu^{EM} = \frac{1}{2} \mathcal{V}_\mu^I + \mathcal{V}_\mu^3, \tag{11.13}$$

where $\mathcal{V}_\mu^3$ represents the third component of the isovector vector current ($|I, I_3\rangle = |1, 0\rangle$) and $\mathcal{V}_\mu^I$ is the isoscalar current ($|I, I_3\rangle = |0, 0\rangle$).

In the case of the isospin $1/2$ resonances, both isoscalar and isovector components of the electromagnetic current given in Eq. (11.13) contribute. The isoscalar part of the electromagnetic current, $\mathcal{V}_\mu^I$ is defined as

$$\mathcal{V}_\mu^I = V_\mu^I I_{2\times 2}, \tag{11.14}$$

where $I_{2\times 2}$ represents the 2×2 identity matrix.

In terms of the form factors, the matrix element for the electromagnetic current for the transition $N - R_{1/2}$ is given as:

$$\langle R_{1/2}^+| J_\mu^{EM} |p\rangle = \bar{u}(p_R)\left[\frac{F_1^{R_{1/2}^+}(Q^2)}{(2M)^2}\left(Q^2\gamma^\mu + q^\mu \slashed{q}\right) + \frac{F_2^{R_{1/2}^+}(Q^2)}{2M}i\sigma^{\mu\alpha}q_\alpha\right]\Gamma u(p) = V_\mu^{R_{1/2}^+},$$

(11.15)

$$\langle R_{1/2}^0| J_\mu^{EM} |n\rangle = \bar{u}(p_R)\left[\frac{F_1^{R_{1/2}^0}(Q^2)}{(2M)^2}\left(Q^2\gamma^\mu + q^\mu \slashed{q}\right) + \frac{F_2^{R_{1/2}^0}(Q^2)}{2M}i\sigma^{\mu\alpha}q_\alpha\right]\Gamma u(p) = V_\mu^{R_{1/2}^0},$$

(11.16)

where $\Gamma = 1$ (γ_5) for the positive (negative) parity resonances.

The transition matrix element for the process $\gamma N \to R_{1/2}$ is written as

$$\langle R_{1/2}| \mathcal{J}_\mu^{EM} |N\rangle = \langle R_{1/2}| \mathcal{V}_\mu^3 + \frac{1}{2}\mathcal{V}_\mu^I |N\rangle,$$

(11.17)

where $|N\rangle = \begin{pmatrix} p \\ n \end{pmatrix}$ represents the nucleon field and $|R_{1/2}\rangle = \begin{pmatrix} R_{1/2}^+ \\ R_{1/2}^0 \end{pmatrix}$ represents the spin $1/2$ resonance field. Using Eqs. (11.11) and (11.14) in Eq.(11.17), we obtain

$$
\begin{aligned}
\langle R_{1/2}| \mathcal{J}_\mu^{EM} |N\rangle &= \langle R_{1/2}| V_\mu\frac{\tau_3}{2} + V_\mu^I\frac{I_{2\times 2}}{2} |N\rangle, \\
&= \begin{pmatrix} \bar{R}_{1/2}^+ & \bar{R}_{1/2}^0 \end{pmatrix}\begin{pmatrix} \frac{V_\mu+V_\mu^I}{2} & 0 \\ 0 & \frac{-V_\mu+V_\mu^I}{2} \end{pmatrix}\begin{pmatrix} p \\ n \end{pmatrix}, \\
&= \bar{R}_{1/2}^+\left(\frac{V_\mu + V_\mu^I}{2}\right)p + \bar{R}_{1/2}^0\left(\frac{-V_\mu + V_\mu^I}{2}\right)n.
\end{aligned}
$$

(11.18)

Comparing Eqs. (11.15) and (11.16) with the first and second terms of Eq. (11.18), we obtain

$$V_\mu^{R_{1/2}^+} = (V_\mu + V_\mu^I)/2,$$

(11.19)

$$V_\mu^{R_{1/2}^0} = (-V_\mu + V_\mu^I)/2.$$

(11.20)

From Eqs. (11.19) and (11.20), V_μ and V_μ^I are obtained as

$$V_\mu = V_\mu^{R_{1/2}^+} - V_\mu^{R_{1/2}^0},$$

(11.21)

$$V_\mu^I = V_\mu^{R_{1/2}^+} + V_\mu^{R_{1/2}^0}.$$

(11.22)

The vector current for the charged current process (given in Eq. (11.7)) is written as

$$V_\mu^{CC} = \mathcal{V}_\mu^1 + i\mathcal{V}_\mu^2,$$

(11.23)

where $\mathcal{V}_\mu^1$ and $\mathcal{V}_\mu^2$ are the components of the isovector currents.

Now, writing the transition matrix element for the vector part of the charged current process, $nW^+ \to R_{1/2}^+$, we have

$$
\begin{aligned}
\langle R_{1/2}^+ | V_\mu^{CC} | n \rangle &= \langle R_{1/2}^+ | \mathcal{V}_\mu^1 + i\mathcal{V}_\mu^2 | n \rangle, \\
&= \langle R_{1/2}^+ | V_\mu \tau_+ | n \rangle, \\
&= V_\mu, \\
&= V_\mu^{R_+^{\frac{1}{2}}} - V_\mu^{R_0^{\frac{1}{2}}}
\end{aligned}
\tag{11.24}
$$

where $\tau_+ = \frac{1}{2}(\tau_1 + i\tau_2)$. Thus, Eqs. (11.21), (11.22), and (11.24) relates the electromagnetic and weak vector form factors with each other and the relation is given in Eq. (11.9).

It must be noted that the relation between the electromagnetic and weak vector form factors, as obtained here, depends only on the isospin of the resonance considered. Therefore, the weak vector form factors for the spin 3/2 and isospin 1/2 resonances are related to the electromagnetic form factors in a similar manner as obtained in Eq. (11.9) for the spin 1/2 and isospin 1/2 resonances.

(ii) Electromagnetic form factors and the helicity amplitude

The electromagnetic form factors $F_i^{R^+, R^0}(Q^2)$, $(i = 1, 2)$ are derived from the helicity amplitudes extracted from real and/or virtual photon scattering experiments. In order to determine the helicity amplitudes $A_{1/2}$ and $S_{1/2}$, assume the interaction of a nucleon with a virtual/real photon to produce a spin 1/2 resonance. The helicity amplitudes for the process $\gamma N \to R_{1/2}$ are expressed in terms of the polarization of the photon and the spins of the incoming nucleon and the outgoing spin 1/2 resonance. They are depicted in Figure 11.4, where we have fixed the spin of the resonance in the positive Z direction, that is, $J_z^R = +1/2$. It must be noted that it is our choice to fix $J_z^R = +1/2$; one may obtain the expressions for the helicity amplitudes by fixing $J_z^R = -1/2$. The expressions for $A_{1/2}$ and $S_{1/2}$ are defined as [483]:

$$
A_{1/2}^N = \sqrt{\frac{2\pi\alpha}{K_R}} \langle R, J_z^R = +1/2 | \epsilon_\mu^+ \Gamma^\mu | N, J_z^N = -1/2 \rangle \zeta,
\tag{11.25}
$$

$$
S_{1/2}^N = -\sqrt{\frac{2\pi\alpha}{K_R}} \frac{|\vec{q}|}{\sqrt{Q^2}} \langle R, J_z^R = +1/2 | \epsilon_\mu^0 \Gamma^\mu | N, J_z^N = +1/2 \rangle \zeta,
\tag{11.26}
$$

where $K_R = (M_R^2 - M^2)/2M_R$ is the momentum of the real photon measured in the resonance rest frame and $|\vec{q}|$ is the momentum of the virtual photon measured in the laboratory frame given as

$$
|\vec{q}| = \sqrt{\frac{(M_R^2 - M^2 - Q^2)^2}{(2M_R)^2} + Q^2}.
\tag{11.27}
$$

Γ^μ is the electromagnetic transition current for the positive and negative parity resonances, defined as $\Gamma^\mu = V^\mu$. The expressions for V^μ is given in Eq. (11.7). $\zeta = e^{i\phi}$, where ϕ is

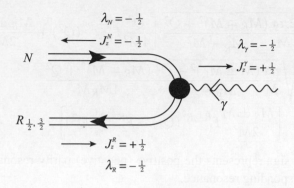

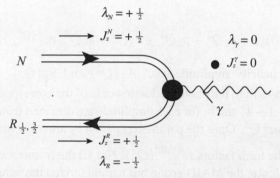

Figure 11.4 Diagrammatic representation of the helicity amplitudes, $A_{1/2}$ (top) and $S_{1/2}$ (bottom) for spin 1/2 and 3/2 resonances.

the phase factor which relates how the amplitude of the resonances and the nucleons are added. Generally, ϕ is taken to be 0 but this is not a rule of thumb. ϵ_{μ} represents the photon polarization vector. The transverse polarized photon vector, $\epsilon_{\mu}^{\pm}$ is given as

$$\epsilon_{\mu}^{\pm} = \mp \frac{1}{\sqrt{2}}(0, 1, \pm i, 0). \tag{11.28}$$

For the longitudinal polarization of the photon, ϵ_{μ}^{0} is given as

$$\epsilon_{\mu}^{0} = \frac{1}{\sqrt{Q^2}}(|\vec{q}|, 0, 0, q^0), q^0 = |\vec{q}|. \tag{11.29}$$

From Eqs. (11.25), (11.26), (11.28), and (11.29), one may observe that for the spin 1/2 resonances, $A_{1/2}$ represents the interaction of the transverse polarized photons with the $NR_{1/2}$ vertex. $S_{1/2}$ represents the interaction of the longitudinally polarized photons with the $NR_{1/2}$ vertex.

Using Eqs. (11.25)–(11.29), one can calculate the explicit relations between the form factors $F_i^{R^+, R^0}(Q^2)$ and the helicity amplitudes $A_{\frac{1}{2}}^{p,n}(Q^2)$ and $S_{\frac{1}{2}}^{p,n}(Q^2)$. These are given by [483]:

$$A_{\frac{1}{2}}^{p,n} = \sqrt{\frac{2\pi\alpha}{M} \frac{(M_R \mp M)^2 + Q^2}{M_R^2 - M^2}} \left[\frac{Q^2}{4M^2} F_1^{R^+,R^0}(Q^2) + \frac{M_R \pm M}{2M} F_2^{R^+,R^0}(Q^2) \right],$$

$$S_{\frac{1}{2}}^{p,n} = \mp \sqrt{\frac{\pi\alpha}{M} \frac{(M \pm M_R)^2 + Q^2}{M_R^2 - M^2}} \frac{(M_R \mp M)^2 + Q^2}{4M_R M}$$

$$\left[\frac{M_R \pm M}{2M} F_1^{R^+,R^0}(Q^2) - F_2^{R^+,R^0}(Q^2) \right], \tag{11.30}$$

where upper (lower) sign represents the positive (negative) parity resonance state and M_R is the mass of the corresponding resonance.

The Q^2 dependence of the helicity amplitudes is parameterized by the MAID group as [493]

$$\mathcal{A}_\alpha(Q^2) = \mathcal{A}_\alpha(0)(1 + a_1 Q^2 + a_2 Q^4 + a_3 Q^6 + a_4 Q^8) e^{-b_1 Q^2}, \tag{11.31}$$

where $\mathcal{A}_\alpha(Q^2)$ is the helicity amplitude, viz., $A_{\frac{1}{2}}(Q^2)$ and $S_{\frac{1}{2}}(Q^2)$. Parameters $\mathcal{A}_\alpha(0)$ are generally determined by a fit to the photoproduction data of the corresponding resonance, while the parameters a_i ($i = 1 - 4$) and b_1 for each amplitude are obtained from the electroproduction data available at different Q^2. Once the parameters a_i and b_1 are fixed for $A_{\frac{1}{2}}(Q^2)$ and $S_{\frac{1}{2}}(Q^2)$ amplitudes, one gets the form factors $F_{1,2}^{R^+,R^0}(Q^2)$. Not all the resonances quoted in Table 11.2 are well understood. Hence, the MAID group has parameterized the values of these parameters for the resonances which are experimentally studied in the photo- and electro- productions. The values of these parameters are presented in Tables 11.6 and 11.7 for proton and neutron targets, respectively. It must be noted that for the isospin $3/2$ resonances, irrespective of the spin of the resonance, the different parameters for $A_{1/2}$ and $S_{1/2}$ given in Eq. (11.31) have the same values for the proton and neutron targets.

Table 11.6 MAID parameterization [493] of the transition form factors for the spin $1/2$ resonance on a proton target. $\mathcal{A}_\alpha(0)$ is given in units of $10^{-3}\,\mathrm{GeV}^{-\frac{1}{2}}$ and the coefficients a_1, a_2, a_4, b_1 in units of GeV^{-2}, GeV^{-4}, GeV^{-8}, GeV^{-2}, respectively. For all fits, $a_3 = 0$. For $S_{11}(1535)$, $S_{31}(1620)$, and $S_{11}(1650)$, resonance a_2 and a_4 are taken to be 0.

N^*	Amplitude	$\mathcal{A}_\alpha(0)$	a_1	a_2	a_4	b_1
$P_{11}(1440)$	$A_{\frac{1}{2}}$	-61.4	0.871	-3.516	-0.158	1.36
	$S_{\frac{1}{2}}$	4.2	40.0	0	1.50	1.75
$S_{11}(1535)$	$A_{\frac{1}{2}}$	66.4	1.608	0	0	0.70
	$S_{\frac{1}{2}}$	-2.0	23.9	0	0	0.81
$S_{31}(1620)$	$A_{\frac{1}{2}}$	65.6	1.86	0	0	2.50
	$S_{\frac{1}{2}}$	16.2	2.83	0	0	2.00
$S_{11}(1650)$	$A_{\frac{1}{2}}$	33.3	1.45	0	0	0.62
	$S_{\frac{1}{2}}$	-3.5	2.88	0	0	0.76

Table 11.7 MAID parameterization [493] for a neutron target ($a_{2,3,4} = 0$) for spin 1/2 resonances.

N^*	Amplitude	$\mathcal{A}_\alpha(0)$	a_1	b_1
$P_{11}(1440)$	$A_{\frac{1}{2}}$	54.1	0.95	1.77
	$S_{\frac{1}{2}}$	−41.5	2.98	1.55
$S_{11}(1535)$	$A_{\frac{1}{2}}$	−50.7	4.75	1.69
	$S_{\frac{1}{2}}$	28.5	0.36	1.55
$S_{11}(1650)$	$A_{\frac{1}{2}}$	9.3	0.13	1.55
	$S_{\frac{1}{2}}$	10.	−0.5	1.55

(iii) Axial vector form factors

The axial vector current consists of two form factors, viz., $F_A^{CC}(Q^2)$ and $F_P^{CC}(Q^2)$. Experimentally, information regarding the axial vector form factors is scarce. As in the case of quasielastic scattering, the pseudoscalar form factor $F_P^{CC}(Q^2)$ is related to $F_A^{CC}(Q^2)$ by the assumption of partially conserved axial vector current (PCAC) and the pion pole dominance. The value of $F_A^{CC}(0)$ is determined by the off-diagonal Goldberger–Treiman relation. We have seen earlier in Chapters 6 and 10 that the axial vector current is conserved only in the chiral limit, that is, $m_\pi \to 0$.

The divergence of the axial vector current in Eq. (11.8) yields

$$
\begin{aligned}
\partial_\mu A^\mu_{1/2\pm} &= -iq_\mu A^\mu_{1/2\pm} = i\bar{u}(p_R)\left[F_A^{CC}\slashed{q}\gamma_5 + \frac{F_P^{CC}}{M}q^2\gamma_5\right]\Gamma u(p), \\
&= i\bar{u}(p_R)\left[F_A^{CC}(M_R \pm M)\gamma_5 + \frac{F_P^{CC}}{M}q^2\gamma_5\right]\Gamma u(p),
\end{aligned}
\tag{11.32}
$$

where $\Gamma = 1$ (γ_5) for positive (negative) parity resonances. According to the PCAC hypothesis, this expression must be proportional to the square of the pion mass. The second term in Eq. (11.32) with the pseudoscalar form factor has a pion pole.

The pion pole contribution can be obtained by applying the same procedure as we have done for obtaining the Goldberger–Treiman relation in the case of nucleons in Chapter 6. Consider that a nucleon transforms to a spin 1/2 resonance by emitting a pion and this pion then decays to a $l^-\bar{\nu}_l$ pair as shown in Figure 11.5. The matrix element for the axial vector current is written in terms of

(i) the strong $NR_{1/2}\pi$ vertex,

(ii) the pion propagator, and

(iii) the current at the leptonic vertex ($\pi \to l^-\bar{\nu}_l$),

as:

$$
A^\mu = (A^{N\to R_{1/2}\pi}) \times \left(\frac{i}{k_\pi^2 - m_\pi^2}\right) \times (-i\sqrt{2}f_\pi k_\pi^\mu),
\tag{11.33}
$$

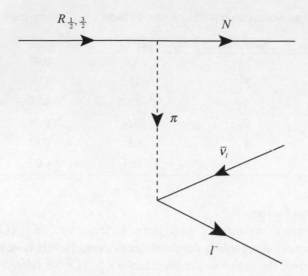

Figure 11.5 Feynman diagram of the process $R_{\frac{1}{2},\frac{3}{2}} \to Nl^-\bar{\nu}_l$ through pion decay.

where k_π is the momentum of the pion and f_π is the pion decay constant. The current at the $NR_{1/2}\pi$ vertex is determined by the pseudovector Lagrangian $\mathcal{L}_{NR_{1/2}\pi}$, given as

$$\mathcal{L}_{NR_{1/2}\pi} = C_{\text{iso}}\frac{f_{NR_{1/2}\pi}}{m_\pi}\bar{\psi}_R \Gamma^\mu \partial_\mu \vec{\phi} \cdot \vec{\tau}\,\psi, \tag{11.34}$$

where C_{iso} is the isospin factor. For $N \longrightarrow R_{1/2}^+$ transition, the value of C_{iso} is

$$C_{\text{iso}} = \sqrt{2}. \tag{11.35}$$

$f_{NR_{1/2}\pi}$ is the coupling of the $NR_{1/2}\pi$ vertex which is obtained using the partial decay width of $R_{1/2} \to N\pi$. This coupling is later discussed in the text. $\Gamma^\mu = \gamma^\mu\gamma_5\,(\gamma^\mu)$ stands for the positive (negative) parity resonances and ϕ represents the triplet of the pion field. Using the Lagrangian given in Eq. (11.34), the matrix element for the axial vector current for the $N(p) \to R_{1/2}(p_R) + \pi(k_\pi)$ vertex is written as

$$A^{N\to R_{1/2}\pi} = -i\bar{u}(p_R)\left(C_{\text{iso}}\frac{f}{m_\pi}\Gamma_\mu\partial^\mu\phi\right)u(p). \tag{11.36}$$

Using Eq. (11.36) in Eq. (11.33), the matrix element for the pion pole contribution is obtained as

$$\begin{aligned}
A^\mu &= -iC_{\text{iso}}\,\bar{u}(p_R)\frac{f_{NR_{1/2}\pi}}{m_\pi}\Gamma_\alpha\partial^\alpha\phi u(p)\,\frac{\sqrt{2}f_\pi k_\pi^\mu}{k_\pi^2 - m_\pi^2}\\
&= -iC_{\text{iso}}\,\bar{u}(p_R)\frac{f_{NR_{1/2}\pi}}{m_\pi}\gamma_\alpha(-ik_\pi^\alpha)\Gamma u(p)\,\frac{\sqrt{2}f_\pi k_\pi^\mu}{k_\pi^2 - m_\pi^2}
\end{aligned}$$

$$= -C_{\text{iso}} \, \bar{u}(p_R) \frac{f_{NR_{1/2}\pi}}{m_\pi} (M_R \pm M) \Gamma u(p) \frac{\sqrt{2} f_\pi k_\pi^\mu}{k_\pi^2 - m_\pi^2}. \tag{11.37}$$

From the definitions of q and k_π, it must be observed that the two represent the same quantity, that is, $q^\mu = p_R^\mu - p^\mu = k_\pi^\mu$. Comparing Eq. (11.37) with the second part of Eq. (11.8), the pseudoscalar form factor F_P^{CC} is obtained as

$$F_P^{CC} = -C_{\text{iso}} \, \sqrt{2} f_\pi \frac{f_{NR_{1/2}\pi}}{m_\pi} \frac{(M_R \pm M)M}{Q^2 + m_\pi^2}. \tag{11.38}$$

Using the expression of F_P^{CC} as obtained here, Eq. (11.32) may be rewritten as

$$\partial_\mu A_{1/2\pm}^\mu = i\bar{u}(p_R) \left[F_A^{CC}(M_R \pm M)\gamma_5 - C_{\text{iso}} \, \sqrt{2} f_\pi \frac{f_{NR_{1/2}\pi}}{m_\pi} \frac{(M_R \pm M)}{Q^2 + m_\pi^2} q^2 \gamma_5 \right] \Gamma u(p). \tag{11.39}$$

In the chiral limit, $m_\pi \to 0$, this expression reduces to

$$\partial_\mu A_{1/2\pm}^\mu = i\bar{u}(p_R) \left[F_A^{CC}(M_R \pm M)\gamma_5 + C_{\text{iso}} \, \sqrt{2} f_\pi \frac{f_{NR_{1/2}\pi}}{m_\pi} (M_R \pm M)\gamma_5 \right] \Gamma u(p). \tag{11.40}$$

The chiral limit demands $\partial_\mu A_{1/2\pm}^\mu = 0$, which is possible only if

$$F_A^{CC}(M_R \pm M) = -C_{\text{iso}} \, \sqrt{2} f_\pi \frac{f_{NR_{1/2}\pi}}{m_\pi} (M_R \pm M) \tag{11.41}$$

which yields the generalized Goldberger–Treiman relation as

$$F_A^{CC} = -C_{\text{iso}} \, \sqrt{2} f_\pi \frac{f_{NR_{1/2}\pi}}{m_\pi}. \tag{11.42}$$

This expression gives the values of $F_A(0)$ for the isospin $I = 1/2$ resonances using the value of C_{iso} given in Eq. (11.35) as:

$$F_A^{CC}(0) \Big|_{I=1/2} = -2 f_\pi \frac{f_{NR_{1/2}\pi}}{m_\pi}. \tag{11.43}$$

There is not much information about the Q^2 dependence of the axial vector form factor F_A^{CC} (Q^2), therefore, a dipole parameterization of $F_A^{CC}(Q^2)$, in analogy with the quasielastic scattering, is assumed:

$$F_A^{CC}(Q^2) = F_A^{CC}(0) \left(1 + \frac{Q^2}{M_A^2} \right)^{-2} \tag{11.44}$$

with $M_A \sim 1$ GeV.

Using Eq. (11.42) in Eq. (11.38), the pseudoscalar form factor $F_P^{CC}(Q^2)$ is obtained in terms of $F_A^{CC}(Q^2)$ as

$$F_P^{CC}(Q^2) = \frac{(MM_R \pm M^2)}{m_\pi^2 + Q^2} F_A^{CC}(Q^2), \tag{11.45}$$

where the $+(-)$ sign is for positive (negative) parity resonances. The Q^2 dependence of the form factors, thus, obtained are shown in Figures 11.6 and 11.7, respectively for $S_{11}(1650)$ and $P_{11}(1440)$ resonances.

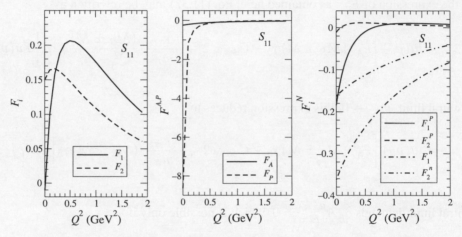

Figure 11.6 Q^2 dependence of different form factors of $S_{11}(1650)$ resonance. In the left panel, the form of isovector form factors F_1^{CC} and F_2^{CC} are shown in the form in which they are given in Eq. (11.9). In the middle panel, we have shown the axial form factors $F_{A,P}$ where for F_A, we took the dipole form and F_P is obtained from Eq. (11.45). In the right panel, we have shown the explicit dependence of $F_i^{R^+, R^0}$, $i = 1, 2$ which may be obtained from the helicity amplitudes as given in Eq. (11.30).

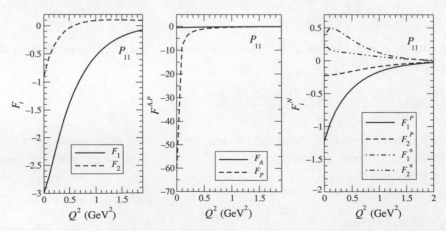

Figure 11.7 Q^2 dependence of different form factors of $P_{11}(1440)$ resonance. The lines have the same meaning as in Figure 11.6.

Excitation of spin $1/2$ and isospin $3/2$ R_{13} resonances

(i) Vector form factors

In this section, we discuss how the weak vector form factors are related to electromagnetic ones when the transition of N to isospin $3/2$ and spin $1/2$ resonances is taken into account. The vector and axial vector currents, which constitute the electroweak current given in Eq. (11.10), for the isospin $1/2 \to 3/2$ transition are defined using transition operator $T^{\dagger}$ as

$$\mathcal{V}_{\mu} = -\sqrt{\frac{3}{2}} V_{\mu} T^{\dagger}, \tag{11.46}$$

$$\mathcal{A}_{\mu} = -\sqrt{\frac{3}{2}} A_{\mu} T^{\dagger}. \tag{11.47}$$

Let us start with the isospin $1/2 \to 3/2$ transition operator $T^{\dagger}$, which is a collection of three 4×2 matrices that connects the nucleon field with the corresponding isospin $3/2$ resonance field. Each 4×2 matrix is basically a Clebsch–Gordan array. This transition operator $T^{\dagger}$ is defined in terms of the matrix elements of the components of T_{λ}, where λ is defined using a spherical basis. The normalization of the transition operator $T_{\lambda}^{\dagger}$ is given as:

$$\langle \frac{3}{2}, M | T_{\lambda}^{\dagger} | \frac{1}{2}, m \rangle = \left(1, \frac{1}{2}, \frac{3}{2} | \lambda, m, M \right). \tag{11.48}$$

The right-hand side of Eq. (11.48) represents the Clebsch–Gordan coefficients for the process $N(|I, I_3 = 1/2, \pm 1/2\rangle) + \gamma(|I, I_3 = 1, 0\rangle) \to R_{1/2}(|I, I_3 = 3/2, \pm 3/2 \pm 1/2\rangle)$. The three 4×2 matrices for the $\langle \frac{3}{2}, M | T_{\lambda}^{\dagger} | \frac{1}{2}, m \rangle$, where $\lambda = -1, 0, +1$, are given as

$$T_{-1}^{\dagger} = \begin{pmatrix} (1,\frac{1}{2},\frac{3}{2}|-1,+\frac{1}{2},+\frac{3}{2}) & (1,\frac{1}{2},\frac{3}{2}|-1,-\frac{1}{2},+\frac{3}{2}) \\ (1,\frac{1}{2},\frac{3}{2}|-1,+\frac{1}{2},+\frac{1}{2}) & (1,\frac{1}{2},\frac{3}{2}|-1,-\frac{1}{2},+\frac{1}{2}) \\ (1,\frac{1}{2},\frac{3}{2}|-1,+\frac{1}{2},-\frac{1}{2}) & (1,\frac{1}{2},\frac{3}{2}|-1,-\frac{1}{2},-\frac{1}{2}) \\ (1,\frac{1}{2},\frac{3}{2}|-1,+\frac{1}{2},-\frac{3}{2}) & (1,\frac{1}{2},\frac{3}{2}|-1,-\frac{1}{2},-\frac{3}{2}) \end{pmatrix},$$

$$T_{0}^{\dagger} = \begin{pmatrix} (1,\frac{1}{2},\frac{3}{2}|0,+\frac{1}{2},+\frac{3}{2}) & (1,\frac{1}{2},\frac{3}{2}|0,\frac{-1}{2},+\frac{3}{2}) \\ (1,\frac{1}{2},\frac{3}{2}|0,+\frac{1}{2},+\frac{1}{2}) & (1,\frac{1}{2},\frac{3}{2}|0,\frac{-1}{2},+\frac{1}{2}) \\ (1,\frac{1}{2},\frac{3}{2}|0,+\frac{1}{2},\frac{-1}{2}) & (1,\frac{1}{2},\frac{3}{2}|0,\frac{-1}{2},\frac{-1}{2}) \\ (1,\frac{1}{2},\frac{3}{2}|0,+\frac{1}{2},\frac{-3}{2}) & (1,\frac{1}{2},\frac{3}{2}|0,\frac{-1}{2},\frac{-3}{2}) \end{pmatrix}, \tag{11.49}$$

$$T_{+1}^{\dagger} = \begin{pmatrix} (1,\frac{1}{2},\frac{3}{2}|+1,+\frac{1}{2},+\frac{3}{2}) & (1,\frac{1}{2},\frac{3}{2}|+1,-\frac{1}{2},+\frac{3}{2}) \\ (1,\frac{1}{2},\frac{3}{2}|+1,+\frac{1}{2},+\frac{1}{2}) & (1,\frac{1}{2},\frac{3}{2}|+1,-\frac{1}{2},+\frac{1}{2}) \\ (1,\frac{1}{2},\frac{3}{2}|+1,+\frac{1}{2},-\frac{1}{2}) & (1,\frac{1}{2},\frac{3}{2}|+1,-\frac{1}{2},-\frac{1}{2}) \\ (1,\frac{1}{2},\frac{3}{2}|+1,+\frac{1}{2},-\frac{3}{2}) & (1,\frac{1}{2},\frac{3}{2}|+1,-\frac{1}{2},-\frac{3}{2}) \end{pmatrix}.$$

The relations between the spherical basis and the isospin space for $T^{\dagger}$ are defined as:

$$T_{\pm 1}^{\dagger} = \mp \frac{T_1^{\dagger} \pm T_2^{\dagger}}{\sqrt{2}}, \qquad T_0^{\dagger} = T_3^{\dagger}. \tag{11.50}$$

For the transition $I = 1/2$ to $I = 3/2$, the electromagnetic current J_μ^{EM} given in Eq. (11.13) is purely isovector in nature; hence, it may be rewritten as

$$J_\mu^{EM} = \mathcal{V}_\mu^3 = -\sqrt{\frac{3}{2}} V_\mu T_0^\dagger, \tag{11.51}$$

where we have used Eq. (11.46). The transition matrix element for the electromagnetic process $\gamma N \rightarrow R_{1/2}^{I=3/2}$, where $R_{1/2}^{I=3/2}$ represents the spin $1/2$ and isospin $3/2$ resonances, is obtained as

$$
\begin{aligned}
\langle R_{1/2}^{I=3/2} | J_\mu^{EM} | N \rangle &= -\sqrt{\frac{3}{2}} \langle R_{1/2}^{I=3/2} | V_\mu T_0^\dagger | N \rangle \\
&= -\sqrt{\frac{3}{2}} \begin{pmatrix} \bar{R}_{1/2}^{++} & \bar{R}_{1/2}^{+} & \bar{R}_{1/2}^{0} & \bar{R}_{1/2}^{-} \end{pmatrix} V_\mu \begin{pmatrix} 0 & 0 \\ \sqrt{\frac{2}{3}} & 0 \\ 0 & \sqrt{\frac{2}{3}} \\ 0 & 0 \end{pmatrix} \begin{pmatrix} p \\ n \end{pmatrix} \\
&= -(\bar{R}_{1/2}^{+} V_\mu p + \bar{R}_{1/2}^{0} V_\mu n), \tag{11.52}
\end{aligned}
$$

where $|R_{1/2}^{I=3/2}\rangle = \begin{pmatrix} R_{1/2}^{++} \\ R_{1/2}^{+} \\ R_{1/2}^{0} \\ R_{1/2}^{-} \end{pmatrix}$ represents the spin $1/2$ and isospin $3/2$ resonance field. From this expression, it may be noticed that the electromagnetic matrix elements and hence, the electromagnetic form factors for the proton and neutron induced isospin $3/2$ resonance excitations are the same, that is,

$$F_i^{R_{1/2}^{+}}(Q^2) = F_i^{R_{1/2}^{0}}(Q^2). \tag{11.53}$$

Using Eq. (11.46) in Eq. (11.23), the vector current for the weak charged current process is obtained as

$$V_\mu^{CC} = -\sqrt{\frac{3}{2}}(V_\mu T_1^\dagger + iV_\mu T_2^\dagger) = \sqrt{3} V_\mu T_{+1}^\dagger. \tag{11.54}$$

The transition matrix element for the weak charged current induced isospin $3/2$ resonance production on the nucleon target is written as

$$
\begin{aligned}
\langle R_{1/2}^{I=3/2} | V_\mu^{CC} | N \rangle &= \langle R_{1/2}^{I=3/2} | \sqrt{3} V_\mu T_{+1}^\dagger | N \rangle, \\
&= \sqrt{3} \begin{pmatrix} \bar{R}_{1/2}^{++} & \bar{R}_{1/2}^{+} & \bar{R}_{1/2}^{0} & \bar{R}_{1/2}^{-} \end{pmatrix} V_\mu \begin{pmatrix} 1 & 0 \\ 0 & \frac{1}{\sqrt{3}} \\ 0 & 0 \\ 0 & 0 \end{pmatrix} \begin{pmatrix} p \\ n \end{pmatrix} \\
&= (\sqrt{3} \bar{R}_{1/2}^{++} V_\mu p + \bar{R}_{1/2}^{+} V_\mu n). \tag{11.55}
\end{aligned}
$$

From Eqs. (11.52) and (11.55), it may be concluded that the electromagnetic and weak currents are expressed in terms of V_μ. Therefore, for the isospin $3/2$ and spin $1/2$ resonances, the weak vector form factors for the transition $\langle R_{1/2}^+| V_\mu^{CC} |n\rangle$ are related to the electromagnetic form factors by the following relation:

$$F_i^{CC}(Q^2) = -F_i^{R_{1/2}}(Q^2),$$ (11.56)

while for the transition $\langle R_{1/2}^{++}| V_\mu^{CC} |p\rangle$, the vector form factors defined in this expression should be multiplied by $\sqrt{3}$ as shown in the first term of Eq. (11.55).

(ii) Axial vector form factors

Next, we discuss the determination of the axial vector form factors for the transition $N - R_{31}$ by the assumption of PCAC and the pion pole dominance. We proceed in a similar manner as we have done in the last section for the transition $N - R_{1/2}$. The only change is in the definition of the Lagrangian, which now becomes

$$\mathcal{L}_{NR_{1/2}\pi} = C_{\text{iso}} \frac{f_{NR_{1/2}\pi}}{m_\pi} \bar{\psi}_R \Gamma^\mu \partial_\mu \phi \cdot T^\dagger \psi,$$ (11.57)

where $C_{\text{iso}} = -\frac{1}{\sqrt{3}}$ is the isospin factor for $I = 3/2$ resonances and $T^\dagger$ is the isospin $1/2$ to $3/2$ transition operator. Using the value of C_{iso} in Eq. (11.42), the value of $F_A^{CC}(0)$ for the isospin $I = 3/2$ resonances is obtained as

$$F_A(0)|_{I=3/2} = \sqrt{\frac{2}{3}} f_\pi \frac{f_{NR_{1/2}\pi}}{m_\pi}.$$ (11.58)

The Q^2 dependence of the axial vector form factor $F_A^{CC}(Q^2)$ is given in Eq. (11.44) and the pseudoscalar form factor $F_P^{CC}(Q^2)$ is determined in terms of $F_A^{CC}(Q^2)$ as given in Eq. (11.45).

(iii) Determination of the strong coupling constants $f_{NR_{1/2}\pi}$

The couplings of the $NR_{1/2}\pi$ vertex for the different resonances is determined by the partial decay width of the resonance in $N\pi$. In order to determine the partial decay width of the resonance, we start with the Lagrangian given in Eq. (11.34). From the expression of the Lagrangian, it may be observed that different isospins ($I = 1/2, 3/2$) of resonance yield different expressions of decay width but the spin of the resonance does not play any role in the Lagrangian or in the decay width. In the following, we present in detail the evaluation of the decay width of the spin $1/2$ resonances with isospin $1/2$.

The scalar product of the derivative pion field with the Pauli matrices yields

$$\partial_\mu \vec{\phi} \cdot \vec{\tau} = \partial_\mu \phi_1 \tau_1 + \partial_\mu \phi_2 \tau_2 + \partial_\mu \phi_3 \tau_3 = \sqrt{2}(\partial_\mu \phi_+ \tau_- + \partial_\mu \phi_- \tau_+) + \partial_\mu \phi_0 \tau_0,$$ (11.59)

where the last expression in Eq. (11.59) represents the scalar product in spherical basis. $\tau_\pm$ and $\phi_\pm$ are defined as

$$\tau_\pm = \frac{1}{2}(\tau_1 \pm i\tau_2), \qquad \tau_0 = \tau_3,$$ (11.60)

$$\phi_{\pm} = \frac{1}{\sqrt{2}}(\phi_1 \pm i\phi_2), \qquad \phi_0 = \phi_3. \tag{11.61}$$

The complex pion field $\phi(x)$ represents particles with charge $|e|$ and antiparticles with charge $-|e|$. The field operator $\hat{\phi}(x)$ is given as:

$$\hat{\phi}(x) = \int \frac{d^3k}{(2\pi)^{3/2}\sqrt{2\omega_k}} \left[\hat{a}(\vec{k})e^{-ik\cdot x} + \hat{b}^\dagger(\vec{k})e^{ik\cdot x}\right], \tag{11.62}$$

and the adjoint of the field operator $\hat{\phi}^\dagger(x)$ is given as

$$\hat{\phi}^\dagger(x) = \int \frac{d^3k}{(2\pi)^{3/2}\sqrt{2\omega_k}} \left[\hat{a}^\dagger(\vec{k})e^{ik\cdot x} + \hat{b}(\vec{k})e^{-ik\cdot x}\right]. \tag{11.63}$$

The operator $\hat{a}(\vec{k})$ annihilates a particle with momentum $\vec{k}$; $\hat{b}^\dagger(\vec{k})$ creates a particle; $\hat{a}^\dagger(\vec{k})$ annihilates an antiparticle; and $\hat{b}(\vec{k})$ creates an antiparticle. ϕ_- given in Eq. (11.61) creates a π^- or annihilates a π^+; ϕ_+ creates a π^+ or annihilates a ϕ_-; and ϕ_0 creates or annihilates a π^0, which has been discussed in detail in Chapter 2.

Using Eq. (11.59) in Eq. (11.34), the Lagrangian for the $NR_{1/2}\pi$ vertex becomes

$$
\begin{aligned}
\mathcal{L}_{NR_{1/2}\pi} &= \frac{f_{NR_{1/2}\pi}}{m_\pi} \bar{\psi}_R \Gamma^\mu (\sqrt{2}(\partial_\mu\phi_+\tau_- + \partial_\mu\phi_-\tau_+) + \partial_\mu\phi_3\tau_3)\psi, \\
&= \frac{f_{NR_{1/2}\pi}}{m_\pi} \langle R_{1/2}| \Gamma^\mu \begin{pmatrix} \partial_\mu\phi_0 & \sqrt{2}\partial_\mu\phi_- \\ \sqrt{2}\partial_\mu\phi_+ & -\partial_\mu\phi_0 \end{pmatrix} |N\rangle, \\
&= \frac{f_{NR_{1/2}\pi}}{m_\pi} [\bar{R}_{1/2}^+ \Gamma^\mu \partial_\mu\phi_0 p + \sqrt{2}\bar{R}_{1/2}^+ \Gamma^\mu \partial_\mu\phi_+ n + \bar{R}_{1/2}^0 \Gamma^\mu \partial_\mu\phi_- p \\
&\quad - \bar{R}_{1/2}^0 \Gamma^\mu \partial_\mu\phi_0 n].
\end{aligned}
\tag{11.64}
$$

In Eq. (11.64), the first term on the right-hand side represents the absorption of a π^0 by the proton to produce a spin $1/2$ resonance with positive charge. Similarly, one may identity the different transitions represented by the different terms in Eq. (11.64). Now, focusing on the first term of the Lagrangian given in Eq. (11.64), that is,

$$\mathcal{L}_{NR_{1/2}\pi} = \frac{f_{NR_{1/2}\pi}}{m_\pi} \bar{R}_{1/2}^+ \Gamma^\mu \partial_\mu\phi_0 p, \tag{11.65}$$

we can evaluate the decay width of the resonance. The Hermitian conjugate of the Lagrangian can be written as

$$L_{NR_{\frac{1}{2}}\pi}^{h.c.} = \frac{f_{NR_{1/2}\pi}}{m_\pi} \bar{p}\Gamma^\mu \partial_\mu\phi_0^\dagger R_{1/2}^+. \tag{11.66}$$

$\Gamma^\mu = \gamma^\mu\gamma_5 \, (\gamma^\mu)$ for the positive (negative) parity resonances. In the following, we present the evaluation of the decay width considering the negative parity resonances. However, one may

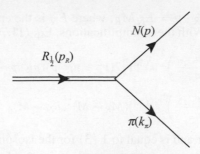

Figure 11.8 Feynman diagram for the decay of a spin 1/2 resonance to a nucleon and a pion. The quantities in the parentheses represent the four momenta of the corresponding particles.

obtain similar results while considering positive parity resonances and the expression for the decay width is given in terms of both positive and negative parity resonances.

Using Eqs. (11.65) and (11.66) and applying Feynman rules, the transition matrix element for the process $R_{1/2}(p_R) \rightarrow N(p) + \pi(k_\pi)$ as shown in Figure 11.8, is written as

$$\mathcal{M} = \frac{if_{NR_{1/2}\pi}}{m_\pi} \bar{u}(p)\gamma_\mu k_\pi^\mu u(p_R). \tag{11.67}$$

The spin averaged transition matrix element squared is obtained as

$$\overline{\sum}\sum|\mathcal{M}|^2 = \frac{1}{2}\left(\frac{f_{NR_{1/2}\pi}}{m_\pi}\right)^2 k_\pi^\mu k_\pi^\nu |\bar{u}(p)\gamma_\mu u(p_R)|^2$$

$$= \frac{1}{2}\left(\frac{f_{NR_{1/2}\pi}}{m_\pi}\right)^2 k_\pi^\mu k_\pi^\nu [\text{Tr}(\gamma_\mu(\not{p}_R + M_R)\gamma_\nu(\not{p} + M))]$$

$$= 2\left(\frac{f_{NR_{1/2}\pi}}{m_\pi}\right)^2 (2p_R \cdot k_\pi p \cdot k_\pi - m_\pi^2(p_R \cdot p - MM_R)). \tag{11.68}$$

Using the expression of the two-body decay Γ given in Appendix E, we get

$$\frac{d\Gamma}{d\Omega_\pi} = \frac{|\vec{k}_\pi|}{32\pi^2 M_R^2} \overline{\sum}\sum|\mathcal{M}|^2. \tag{11.69}$$

In this expression of Γ, we have to perform the integration over the solid angle of the pion, but the transition matrix element squared given in Eq. (11.68), when evaluated in the resonance rest frame ($\vec{p}_R = 0$), does not contain any angular dependence. Thus, the following formula for the width is obtained:

$$\Gamma = \frac{|\vec{k}_\pi^{CM}|}{8\pi M_R^2}\left(2\left(\frac{f_{NR_{1/2}\pi}}{m_\pi}\right)^2 (2p_R \cdot k_\pi p \cdot k_\pi - m_\pi^2(p_R \cdot p - MM_R))\right), \tag{11.70}$$

$|\vec{k}_\pi^{CM}|$ represents the momentum of the outgoing pion in the resonance rest frame. Substituting $k_\pi = p_R - p$ in these expressions yields the decay width in terms of p_R and p. The scalar product of p_R with p gives $p_R \cdot p = E_N M_R$, where E_N is the energy of the outgoing nucleon in the resonance rest frame. With these simplifications, Eq. (11.70) can be rewritten as

$$
\begin{aligned}
\Gamma_{R_{1/2} \to N\pi} &= \frac{1}{4\pi M_R^2} \left(\frac{f_{NR_{1/2}\pi}}{m_\pi} \right)^2 |\vec{k}_\pi^{\mathrm{CM}}| \left[2(p_R^2 - p_R \cdot p)(p_R \cdot p - p^2) - m_\pi^2 M_R E_N + m_\pi^2 M M_R) \right] \\
&= \frac{I_R}{4\pi M_R} \left(\frac{f_{NR_{1/2}\pi}}{m_\pi} \right)^2 |\vec{k}_\pi^{\mathrm{CM}}| (M_R - M)^2 (E_N + M),
\end{aligned}
\tag{11.71}
$$

where I_R is the isospin factor and is equal to 1 (3) for the isospin 3/2 (1/2) resonances.

Now, for positive parity resonances, the expression for the decay width is given as

$$
\Gamma_{R_{1/2} \to N\pi} = \frac{I_R}{4\pi M_R} \left(\frac{f_{NR_{1/2}\pi}}{m_\pi} \right)^2 |\vec{k}_\pi^{\mathrm{CM}}| (M_R + M)^2 (E_N - M),
\tag{11.72}
$$

where

$$
E_N = \frac{W^2 + M^2 - m_\pi^2}{2M_R},
\tag{11.73}
$$

$$
|\vec{k}_\pi^{\mathrm{CM}}| = \frac{\sqrt{(W^2 - m_\pi^2 - M^2)^2 - 4m_\pi^2 M^2}}{2M_R}.
\tag{11.74}
$$

W is the total center of mass energy carried by the resonance.

11.2.2   CC excitation of spin 3/2 resonances

Excitation of spin 3/2 and isospin 1/2 R_{13} resonances

The basic reactions for the (anti)neutrino induced spin 3/2 resonance excitations on the nucleon target, shown in the left-hand side of Figure 11.9, are written as:

$$
\nu_l (\bar{\nu}_l) \ (k) + N(p) \longrightarrow l^- (l^+) \ (k') + R_{3/2}(p_R)
\tag{11.75}
$$

for which the transition amplitude is written as

$$
\mathcal{M} = \frac{G_F}{\sqrt{2}} \cos\theta_C \ l_\mu \ J_{\frac{3}{2}}^{\mu CC},
\tag{11.76}
$$

where the leptonic current is the same as given in Eq. (11.3) and the general structure for the hadronic current for spin 3/2 resonance excitation is written as [294]:

$$
J_{\frac{3}{2}}^{\mu CC} = \bar{\psi}_\nu(p') \Gamma_{\frac{3}{2}}^{\nu\mu} u(p).
\tag{11.77}
$$

Here $\psi^\mu(p)$ is the Rarita–Schwinger spinor for the spin 3/2 resonances and $\Gamma_{\nu\mu}^{\frac{3}{2}}$ has the following general structure for the positive and negative parity resonance states:

$$
\Gamma_{\nu\mu}^{\frac{3}{2}^+} = \left[V_{\nu\mu}^{\frac{3}{2}} - A_{\nu\mu}^{\frac{3}{2}} \right] \gamma_5,
\tag{11.78}
$$

$$
\Gamma_{\nu\mu}^{\frac{3}{2}^-} = V_{\nu\mu}^{\frac{3}{2}} - A_{\nu\mu}^{\frac{3}{2}},
\tag{11.79}
$$

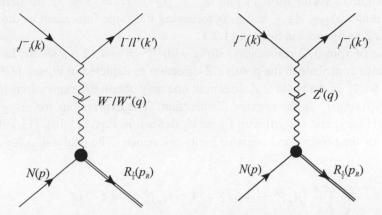

Figure 11.9 Feynman diagram for the charged current (left) and neutral current (right) induced spin 3/2 resonance production. The quantities in the parentheses represent the four momenta of the corresponding particles.

where $V_{\frac{3}{2}}\,(A_{\frac{3}{2}})$ is the vector (axial vector) current for spin 3/2 resonances and are given by

$$V_{\nu\mu}^{\frac{3}{2}} = \left[\frac{C_3^V}{M}(g_{\mu\nu} - q_\nu\gamma_\mu) + \frac{C_4^V}{M^2}(g_{\mu\nu}q\cdot p' - q_\nu p'_\mu) + \frac{C_5^V}{M^2}(g_{\mu\nu}q\cdot p - q_\nu p_\mu) + g_{\mu\nu}C_6^V \right],$$
(11.80)

$$A_{\nu\mu}^{\frac{3}{2}} = -\left[\frac{C_3^A}{M}(g_{\mu\nu} - q_\nu\gamma_\mu) + \frac{C_4^A}{M^2}(g_{\mu\nu}q\cdot p' - q_\nu p'_\mu) + C_5^A g_{\mu\nu} + \frac{C_6^A}{M^2}q_\nu q_\mu \right]\gamma_5.$$
(11.81)

In this expression, C_i^V and C_i^A are the vector and axial vector charged current transition form factors which are functions of Q^2.

(i) Vector form factors

The vector current for the R_{13} resonances is given in Eq. (11.80) with corresponding form factors C_i^V defined for each resonances. The hypothesis of CVC leads to $C_6^V(Q^2) = 0$. The determination of the weak vector form factors in terms of the electromagnetic form factors depends upon the isospin structure of the current. For isospin $1/2$ resonances, the relation between the weak vector form factor and the electromagnetic form factors is given by Eq. (11.9). Similarly, for the isospin $3/2$ resonances, the relation is given by Eq. (11.56). The isovector $C_i^V(Q^2), i = 3, 4, 5$ form factors for any $J = \frac{3}{2}$, $I = \frac{1}{2}$ resonance like $D_{13}(1520)$, $P_{13}(1720)$, etc., are written as [482],

$$C_i^V(Q^2) = C_i^{R^+}(Q^2) - C_i^{R^0}(Q^2), \qquad i = 3, 4, 5.$$
(11.82)

The electromagnetic vector form factors $C_i^{R^+,R^0}(Q^2)$, $(i = 3,4,5)$ are derived from the helicity amplitudes $A_{1/2}$, $A_{3/2}$, and $S_{1/2}$ following the same formalism as discussed earlier for the spin $1/2$ resonances in Section 11.2.1.

In the case of spin $3/2$ resonances, along with $J_z^R = +1/2$ (as shown in Figure 11.4), $J_z^R = +3/2$ also contribute in the positive Z-direction as depicted in Figure 11.10. Again it is our choice to fix J_z^R in the positive Z-direction; one may obtain the expressions for the helicity amplitudes by fixing J_z^R in the negative Z-direction. The expressions for $A_{1/2}$ and $S_{1/2}$ are given in Eqs. (11.25) and (11.26) with $\Gamma_\mu \simeq V_\mu$ defined in Eqs. (11.78), (11.79) and (11.80), respectively, for the positive and negative parity resonance. The expression for $A_{3/2}$ is given as:

$$A_{3/2}^N = \sqrt{\frac{2\pi\alpha}{K_R}} \langle R, J_z^R = +3/2|\, \epsilon_\mu^+ \Gamma^\mu \,|N, J_z^N = +1/2\rangle \, \zeta, \qquad (11.83)$$

where the variables in Eq. (11.83) are already defined in Section 11.2.1.

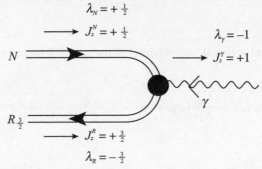

Figure 11.10 Diagrammatic representation of the helicity amplitude, $A_{3/2}$ for spin $3/2$ resonances.

The explicit relations between the form factors $C_i^{R^+,R^0}(Q^2)$ and the helicity amplitudes $A_{1/2,3/2}^{p,n}(Q^2)$ and $S_{\frac{1}{2}}^{p,n}(Q^2)$ are given by [483]

$$A_{\frac{3}{2}}^{p,n}(Q^2) = \sqrt{\frac{\pi\alpha}{M} \frac{(M_R \mp M)^2 + Q^2}{M_R^2 - M^2}} \left[\frac{C_3^{R^+,R^0}(Q^2)}{M}(M \pm M_R) \right.$$
$$\left. \pm \frac{C_4^{R^+,R^0}(Q^2)}{M^2} \frac{M_R^2 - M^2 - Q^2}{2} \pm \frac{C_5^{R^+,R^0}(Q^2)}{M^2} \frac{M_R^2 - M^2 + Q^2}{2} \right], \qquad (11.84)$$

$$A_{\frac{1}{2}}^{p,n}(Q^2) = \sqrt{\frac{\pi\alpha}{3M} \frac{(M_R \mp M)^2 + Q^2}{M_R^2 - M^2}} \left[\frac{C_3^{R^+,R^0}(Q^2)}{M} \frac{M^2 + MM_R + Q^2}{M_R} - \frac{C_4^{R^+,R^0}(Q^2)}{M^2} \right.$$
$$\left. \times \frac{M_R^2 - M^2 - Q^2}{2} - \frac{C_5^{R^+,R^0}(Q^2)}{M^2} \frac{M_R^2 - M^2 + Q^2}{2} \right], \qquad (11.85)$$

$$S_{\frac{1}{2}}^{p,n}(Q^2) = \pm\sqrt{\frac{\pi\alpha}{6M} \frac{(M_R \mp M)^2 + Q^2}{M_R^2 - M^2}} \frac{\sqrt{Q^4 + 2Q^2(M_R^2 + M^2) + (M_R^2 - M^2)^2}}{M_R^2}$$
$$\times \left[\frac{C_3^{R^+,R^0}(Q^2)}{M} M_R + \frac{C_4^{R^+,R^0}(Q^2)}{M^2} M_R^2 + \frac{C_5^{R^+,R^0}(Q^2)}{M^2} \frac{M_R^2 + M^2 + Q^2}{2} \right], \qquad (11.86)$$

where the upper (lower) sign represents the positive (negative) parity resonance state, M_R is the mass of the corresponding resonance, $A_{\frac{3}{2},\frac{1}{2}}(Q^2)$ and $S_{\frac{1}{2}}(Q^2)$ are the amplitudes corresponding to the transverse and longitudinal polarizations of the photon, respectively, and are parameterized at different Q^2 using Eq. (11.31).

The Q^2 dependence of the helicity amplitudes is parameterized by MAID (Eq. (11.31)). The values of the parameters appearing in Eq. (11.31) for the spin $3/2$ resonances are presented in Tables 11.8 and 11.9 for the proton and neutron targets, respectively. As already mentioned, for the isospin $3/2$ resonances, the different parameters for $A_{1/2,3/2}$ and $S_{1/2}$ given in Eq. (11.31) have the same values for the proton and neutron targets.

Table 11.8 MAID parameterization [493] of the transition form factors for the spin $3/2$ resonance on a proton target. Different components of this table have the same meaning as in Table 11.6.

N^*	Amplitude	$\mathcal{A}_\alpha(0)$	a_1	a_2	a_4	b_1
$D_{13}(1520)$	$A_{\frac{1}{2}}$	-27.4	8.580	-0.252	0.357	1.20
	$A_{\frac{3}{2}}$	160.6	-0.820	0.541	-0.016	1.06
	$S_{\frac{1}{2}}$	-63.5	4.19	0	0	3.40
$D_{33}(1700)$	$A_{\frac{1}{2}}$	226.0	1.91	0	0	1.77
	$A_{\frac{3}{2}}$	210.0	0.88	1.71	0	2.02
	$S_{\frac{1}{2}}$	2.1	0	0	0	2.0
$P_{13}(1720)$	$A_{\frac{1}{2}}$	73.0	1.89	0	0	1.55
	$A_{\frac{3}{2}}$	-11.5	10.83	-0.66	0	0.43
	$S_{\frac{1}{2}}$	-53.0	2.46	0	0	1.55

Table 11.9 MAID [493] parameterization for neutron target ($a_{2,3,4} = 0$) for spin $3/2$ resonances.

N^*	Amplitude	$\mathcal{A}_\alpha(0)$	a_1	b_1
$D_{13}(1520)$	$A_{\frac{1}{2}}$	-76.5	-0.53	1.55
	$A_{\frac{3}{2}}$	-154.0	0.58	1.75
	$S_{\frac{1}{2}}$	13.6	15.7	1.57
$P_{13}(1720)$	$A_{\frac{1}{2}}$	-2.9	12.7	1.55
	$A_{\frac{3}{2}}$	-31.0	5.00	1.55
	$S_{\frac{1}{2}}$	0	0	0

(ii) Axial vector form factors

The form factors $C_i^A(Q^2)$, $(i = 3, 4, 5, 6)$ corresponding to the axial vector current have not been studied in the case of higher resonances except for $P_{33}(1232)$ resonance. However, in the literature, PCAC and the Goldberger–Treiman relation are used to determine $C_5^A(Q^2)$ and $C_6^A(Q^2)$; the other form factors are taken to be zero. The divergence of the axial vector current given in Eq. (11.81) yields

$$\partial_\mu A^\mu_{3/2\pm} = -iq_\mu A^\mu_{3/2\pm}$$

$$= -i\bar{u}_\nu(p_R)\left[C_5^A q^\nu + \frac{C_6^A}{M^2}q^2 q^\nu\right]\Gamma u(p),$$

$$= -i\bar{u}_\nu(p_R)q^\nu \left[C_5^A + \frac{C_6^A}{M^2}q^2\right]\Gamma u(p), \tag{11.87}$$

where $\Gamma = 1\ (\gamma_5)$ for positive (negative) parity resonances. Now, we evaluate the pion pole contribution of the axial vector current and relate it with the second term in Eq. (11.87), which is identified as the induced pseudoscalar form factor.

We have discussed in detail the contribution of the pion pole in the case of the spin $1/2$ resonances in Section 11.2.1. Following the same analogy, the matrix element for the process $N \to R_{3/2}l\nu$ (depicted in Figure 11.5) in the pion pole dominance is written as

$$A_\mu = (A^{N\to R_{3/2}\pi}) \times \left(\frac{i}{k_\pi^2 - m_\pi^2}\right) \times (-i\sqrt{2}f_\pi k_\pi^\mu). \tag{11.88}$$

The current $A^{N\to R_{3/2}\pi}$ at the $NR_{3/2}\pi$ vertex is determined by the following Lagrangian

$$\mathcal{L}_{NR_{3/2}\pi} = C_{\text{iso}}\frac{f_{NR_{3/2}\pi}}{m_\pi}\bar{\psi}_R^\mu\,\Gamma\,\partial_\mu\vec{\phi}\cdot\vec{\tau}\,\psi \tag{11.89}$$

where $C_{\text{iso}} = \sqrt{2}$ is the isospin factor. $f_{NR_{3/2}\pi}$ is the coupling of the $NR_{3/2}\pi$ vertex which is obtained using the partial decay width of $R_{3/2} \to N\pi$ and is later discussed in the text. $\Gamma = 1(\gamma_5)$ stands for the positive (negative) parity resonances and τ represents the Pauli spin matrices. Using the Lagrangian given in Eq. (11.89), the axial vector current for the $N(p) \to R_{3/2}(p_R) + \pi(k_\pi)$ vertex is written as

$$A^{N\to R_{3/2}\pi} = -i\bar{u}_\mu(p_R)\left(C_{\text{iso}}\frac{f_{NR_{3/2}\pi}}{m_\pi}\Gamma\partial^\mu\phi\right)u(p). \tag{11.90}$$

Using Eq. (11.90) in Eq. (11.88), the transition matrix element for the process $N \to R_{3/2}l\nu$ in the pion pole dominance is obtained as

$$A^\mu = -iC_{\text{iso}}\,\bar{u}_\alpha(p_R)\frac{f_{NR_{3/2}\pi}}{m_\pi}\Gamma\partial^\alpha\phi u(p)\frac{\sqrt{2}f_\pi k_\pi^\mu}{k_\pi^2 - m_\pi^2}$$

$$= -C_{\text{iso}}\frac{f_{NR_{3/2}\pi}}{m_\pi}\,\bar{u}_\alpha(p_R)\Gamma k_\pi^\alpha u(p)\frac{\sqrt{2}f_\pi k_\pi^\mu}{k_\pi^2 - m_\pi^2}, \tag{11.91}$$

with $q^\mu = p_R^\mu - p^\mu = k_\pi^\mu$. Comparing Eq. (11.91) with the last term (pseudoscalar form factor C_6^A) of Eq. (11.81), we obtain

$$C_6^A = -C_{\text{iso}}\sqrt{2}f_\pi\frac{f_{NR_{3/2}\pi}}{m_\pi}\frac{M^2}{Q^2 + m_\pi^2}. \tag{11.92}$$

Using the value of C_6^A as obtained here in Eq. (11.87), we get

$$\partial_\mu A_{3/2\pm}^\mu = i\bar{u}_\nu(p_R)q^\nu \left[C_5^A - C_{\text{iso}}\sqrt{2}f_\pi \frac{f_{NR_{3/2}\pi}}{m_\pi}\frac{1}{Q^2+m_\pi^2}q^2\right]\Gamma u(p). \quad (11.93)$$

In the chiral limit ($Q^2 + m_\pi^2 \sim Q^2$), this expression reduces to

$$\partial_\mu A_{3/2\pm}^\mu = i\bar{u}_\nu(p_R)q^\nu \left[C_5^A + C_{\text{iso}}\sqrt{2}f_\pi \frac{f_{NR_{3/2}\pi}}{m_\pi}\right]\Gamma u(p). \quad (11.94)$$

The chiral limit demands that $\partial_\mu A_{1/2\pm}^\mu = 0$, and one may relate C_5^A with $f_{NR_{1/2}\pi}$ as

$$C_5^A = -C_{\text{iso}}\sqrt{2}f_\pi \frac{f_{NR_{3/2}\pi}}{m_\pi}, \quad (11.95)$$

which is nothing but the Goldberger–Treiman relation for the spin $3/2$ resonances.

Using the value of C_{iso} as given here, $C_5^A(0)$ for the isospin $1/2$ resonances are obtained as:

$$C_5^A(0)\Big|_{I=1/2} = -2f_\pi \frac{f_{NR_{3/2}\pi}}{m_\pi}. \quad (11.96)$$

In the case of higher resonances, the information about the Q^2 dependence of the axial vector form factor $C_5^A(Q^2)$ is scarce; therefore, a dipole parameterization, is generally assumed:

$$C_5^A(Q^2) = C_5^A(0)\left(1 + \frac{Q^2}{M_{AR}^2}\right)^{-2} \quad (11.97)$$

with $M_{AR} = 1$ GeV. However, deviations from the dipole have also been discussed by many authors [484].

Using Eq. (11.95) in Eq. (11.92), the pseudoscalar form factor $C_6(Q^2)$ is obtained in terms of $C_5(Q^2)$ as

$$C_6^A(Q^2) = \frac{M^2}{Q^2 + m_\pi^2}C_5^A(Q^2). \quad (11.98)$$

$C_3^A(Q^2)$ and $C_4^A(Q^2)$ are taken to be 0. The Q^2 dependence of the form factors, thus, obtained are shown in Figures 11.11 and 11.12, respectively, for $D_{13}(1520)$ and $P_{13}(1720)$ resonances.

Excitation of spin $3/2$ and isospin $3/2$ resonances

(i) Vector form factors for Δ resonance

In this section, we discuss the vector and axial vector form factors for the Δ resonance, which is the most studied resonance theoretically as well as experimentally. In the next sections, we discuss the parameterization of the vector and axial vector form factors for the other spin $3/2$ resonances. From the CVC hypothesis, one takes $C_6^V(Q^2) = 0$. The three vector form factors $C_i^V(Q^2), i = 3,4,5$, for the $\Delta(1232)$ resonance are given in terms of the isovector

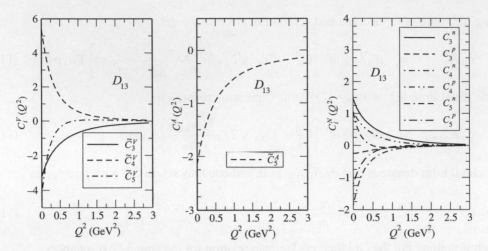

Figure 11.11 Q^2 dependence of different form factors of D_{13} resonance. From left to right panel: C_3^V, C_4^V, and C_5^V as mentioned in Eq. (11.82); C_5^A as mentioned in Eq. (11.97); and $C_i^{n,p}$, $i = 3, 4, 5$ as mentioned in Eqs. (11.84)–(11.86).

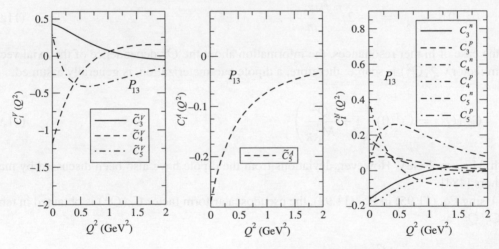

Figure 11.12 Q^2 dependence of different form factors of P_{13} resonance. From left to right panel: C_3^V, C_4^V, and C_5^V as mentioned in Eq. (11.82); C_5^A as mentioned in Eq. (11.97); and $C_i^{n,p}$, $i = 3, 4, 5$ as mentioned in Eqs. (11.84)–(11.86).

electromagnetic form factors for $p \rightarrow \Delta^+$ transition and the parameterization which are taken from the Ref. [484],

$$C_3^V(Q^2) = \frac{2.13}{\left(1 + \frac{Q^2}{M_V^2}\right)^2} \times \frac{1}{1 + \frac{Q^2}{4M_V^2}},$$

$$C_4^V(Q^2) = \frac{-1.51}{\left(1 + \frac{Q^2}{M_V^2}\right)^2} \times \frac{1}{1 + \frac{Q^2}{4M_V^2}},$$

$$C_5^V(Q^2) = \frac{0.48}{\left(1 + \frac{Q^2}{M_V^2}\right)^2} \times \frac{1}{1 + \frac{Q^2}{0.776M_V^2}}, \tag{11.99}$$

with the vector dipole mass taken as $M_V = 0.84$ GeV. From the isospin analysis, it can be shown easily that

$$(C_i^V)_{\Delta^{++}} = \sqrt{3}(C_i^V)_{\Delta^+}. \tag{11.100}$$

(ii) Axial vector form factors for Δ resonance

The axial vector form factors $C_i^A(Q^2)$, $(i = 3, 4, 5)$ are generally determined by using the hypothesis of PCAC with pion pole dominance through the off diagonal Goldberger–Treiman relation or obtained in quark model calculations [494, 495]. The earliest model used to determine the axial vector form factors was the Adler's model [464], and was applied by Schreiner and von Hippel [471]; the model was consistent in a way with the hypothesis of PCAC and generalized Goldberger– Treiman relation. These considerations give $C_6^A(Q^2)$ in terms of $C_5^A(Q^2)$ using generalized Goldberger-Treiman relation [471]:

$$C_6^A(Q^2) = C_5^A(Q^2) \frac{M^2}{Q^2 + m_\pi^2}, \tag{11.101}$$

$$C_5^A(0) = f_\pi \frac{f_{\Delta N\pi}}{2\sqrt{3}M}, \tag{11.102}$$

where $f_{\Delta N\pi}$ is the $\Delta N\pi$ coupling strength for $\Delta \to N\pi$ decay.

The Q^2 dependence of $C_3^A(Q^2)$ and $C_4^A(Q^2)$ obtained in Adler's model are given as [464, 471]:

$$C_4^A(Q^2) = -\frac{1}{4}C_5^A(Q^2); \qquad C_3^A(Q^2) = 0. \tag{11.103}$$

The Q^2 dependence of C_5^A has been parameterized by Schreiner and von Hippel [471] in Adler's model [464] and is given by

$$C_5^A(Q^2) = \frac{C_5^A(0)\left(1 + \frac{aQ^2}{b + Q^2}\right)}{\left(1 + Q^2/M_{A\Delta}^2\right)^2}, \tag{11.104}$$

where a and b are unknown parameters which are determined from experiments and found to be $a = -1.21$ and $b = 2$ GeV^2 [458, 459]. $M_{A\Delta}$ is the axial dipole mass.

Most of the recent theoretical calculations [481, 482, 484, 485] use a simpler modification to the dipole form, viz.,

$$C_5^A(Q^2) = \frac{C_5^A(0)}{\left(1 + Q^2/M_{A\Delta}^2\right)^2} \frac{1}{1 + Q^2/(3M_{A\Delta}^2)}. \tag{11.105}$$

$M_{A\Delta}$ is generally chosen to be 1.026 GeV corresponding to the world average value obtained from the experimental analysis of quasielastic scattering events [443].

(iii) Vector form factors

The vector current for the spin $3/2$ and isospin $3/2$ resonances is given in Eq. (11.80) with corresponding form factors C_i^V defined for each resonances. The hypothesis of CVC leads to $C_6^V(Q^2) = 0$. The weak vector form factors $C_i^V(Q^2)$ are related to the electromagnetic ones by the following relations

$$C_i^V(Q^2) = -C_i^R(Q^2), \qquad i = 3, 4, 5, \tag{11.106}$$

for $R_{3/2}^+$ production. For the $R_{3/2}^{++}$ production, the form factors given in Eq. (11.106) should be multiplied by $\sqrt{3}$.

(iv) Axial vector form factors

Next, we discuss the determination of the axial vector form factors for the transition of $N - R_{33}$. We proceed in a similar manner as done in the last section for spin $3/2$ and isospin $1/2$ resonances; the only change is in the definition of the Lagrangian, which now becomes

$$\mathcal{L}_{NR_{3/2}\pi} = C_{\text{iso}} \frac{f_{NR_{3/2}\pi}}{m_\pi} \bar{\psi}_R^\mu \Gamma \partial_\mu \vec{\phi} \cdot \vec{T}^\dagger \psi \tag{11.107}$$

where $C_{\text{iso}} = -\frac{1}{\sqrt{3}}$ is the isospin factor for the $I = 3/2$ resonances; $T^\dagger$ is the isospin $1/2$ to $3/2$ transition operator. Using the value of C_{iso} in Eq. (11.42), the value of $C_5^A(0)$ for R_{33} resonances is obtained as

$$C_5^A(0)\Big|_{I=3/2} = \sqrt{\frac{2}{3}} f_\pi \frac{f_{NR_{3/2}\pi}}{m_\pi}. \tag{11.108}$$

The Q^2 dependence of the axial vector form factor $C_5^A(Q^2)$ is given in Eq. (11.97) and the pseudoscalar form factor $C_6^A(Q^2)$ is determined in terms of $C_5^A(Q^2)$ as given in Eq. (11.98).

(v) Determination of the coupling constants

The couplings $f_{NR_{3/2}\pi}$ of the $NR_{3/2}\pi$ vertex are determined in terms of the partial decay width of the resonance in $N\pi$. We follow the same procedure as we have done for the determination of $f_{NR_{1/2}\pi}$ in Section 11.2.1. The Lagrangian for the $NR_{3/2}\pi$ vertex given in Eq. (11.89) contains the isospin factor $t = \tau\ (T^\dagger)$ for $I = 1/2\ (3/2)$ resonances but no explicit dependence of spin on the Lagrangian is observed. In the following, we discuss in detail the evaluation of the decay width of the spin $3/2$ resonances with isospin $3/2$, where $t = T^\dagger$.

The scalar product of the derivative pion field with the isospin $1/2 \rightarrow 3/2$ transition operator $T^\dagger$, yields

$$\begin{aligned}
\partial_\mu \vec{\phi} \cdot \vec{T}^\dagger &= \partial_\mu \phi_1 T_1^\dagger + \partial_\mu \phi_2 T_2^\dagger + \partial_\mu \phi_3 T_3^\dagger, \\
&= \partial_\mu \phi_+ T_{-1}^\dagger - \partial_\mu \phi_- T_{+1}^\dagger + \partial_\mu \phi_0 T_0^\dagger,
\end{aligned} \tag{11.109}$$

where the last expression in the equation represents the scalar product in spherical basis. $\phi_{\pm,0}$ and $T^\dagger_{\pm,0}$ are defined in Eqs. (11.61) and (11.50), respectively.

Using Eq. (11.109) in Eq. (11.89), the Lagrangian for the $NR_{3/2}\pi$ vertex becomes (Figure 11.13)

$$\mathcal{L}_{NR_{3/2}\pi} = C_{\text{iso}} \frac{f_{NR_{3/2}\pi}}{m_\pi} \langle R^\mu_{3/2} | \Gamma(\partial_\mu \phi_+ T^\dagger_{-1} - \partial_\mu \phi_- T^\dagger_{+1} + \partial_\mu \phi_0 T^\dagger_0) |N\rangle, \quad (11.110)$$

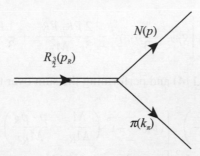

$N(p)$

$R_{\frac{3}{2}}(p_R)$

$\pi(k_\pi)$

Figure 11.13 Feynman diagram for the decay of a spin 3/2 resonance in a nucleon and a pion. The quantities in the parentheses represent the four momenta of the corresponding particles.

where the matrices $T^\dagger_{\pm 1,0}$ are defined in Eq. (11.49). Let us focus on the second term of the Lagrangian given in Eq. (11.110), that is,

$$\mathcal{L}_{NR_{3/2}\pi} = -\frac{f_{NR_{3/2}\pi}}{m_\pi} \langle R^\mu_{3/2} | \Gamma \partial_\mu \phi_- T^\dagger_{+1} |N\rangle$$

$$= -\frac{f_{NR_{3/2}\pi}}{m_\pi} \Gamma \partial_\mu \phi_- \begin{pmatrix} \bar{R}^{\mu++}_{3/2} & \bar{R}^{\mu+}_{3/2} & \bar{R}^{\mu 0}_{3/2} & \bar{R}^{\mu-}_{3/2} \end{pmatrix} \begin{pmatrix} 1 & 0 \\ 0 & \frac{1}{\sqrt{3}} \\ 0 & 0 \\ 0 & 0 \end{pmatrix} \begin{pmatrix} p \\ n \end{pmatrix}$$

$$= -\frac{f_{NR_{3/2}\pi}}{m_\pi} \Gamma \partial_\mu \phi_- (\bar{R}^{\mu++}_{3/2} p + \frac{1}{\sqrt{3}} \bar{R}^{\mu+}_{3/2} n). \quad (11.111)$$

The Hermitian conjugate of this Lagrangian can be written as

$$L^{h.c.}_{NR_{\frac{3}{2}}\pi} = -\frac{f_{NR_{3/2}\pi}}{m_\pi} \Gamma \partial_\mu \phi^\dagger_- (\bar{p} R^{\mu++}_{3/2} + \frac{1}{\sqrt{3}} \bar{n} R^{\mu+}_{3/2}). \quad (11.112)$$

$\Gamma = 1$ (γ_5) represents the positive (negative) parity resonances. In the following, the evaluation of the decay width considering the positive parity resonances is discussed. Using Eqs. (11.111) and (11.112) and applying Feynman rules, the transition matrix element for the decay $R^{++}_{3/2}(p_R)$ $\rightarrow p(p) + \pi^+(k_\pi)$ as shown in Figure 11.13, is written as

$$\mathcal{M} = \frac{i f_{NR_{3/2}\pi}}{m_\pi} \bar{u}(p) k^\mu_\pi u_\mu(p_R). \quad (11.113)$$

The square of the transition matrix element is obtained as

$$|\mathcal{M}|^2 = \left(\frac{f_{NR_{1/2}\pi}}{m_\pi}\right)^2 k_\pi^\mu k_\pi^\nu |\bar{u}(p)u_\mu(p_R)|^2,$$

$$= \left(\frac{f_{NR_{1/2}\pi}}{m_\pi}\right)^2 k_\pi^\mu k_\pi^\nu [\text{Tr}(\mathcal{P}_{\mu\nu}(p_R)(\slashed{p}+M)), \tag{11.114}$$

where $\mathcal{P}_{\mu\nu}(p_R)$ is the projection operator for the spin $3/2$ resonances and is expressed as:

$$\mathcal{P}_{\mu\nu}(p_R) = -(\slashed{p}_R + M_R)\left[g_{\mu\nu} - \frac{1}{3}\gamma_\mu\gamma_\nu - \frac{2}{3}\frac{p_{R\mu}p_{R\nu}}{M_R^2} + \frac{1}{3}\frac{p_{R\mu}\gamma_\nu - p_{R\nu}\gamma_\mu}{M_R}\right]. \tag{11.115}$$

Using Eq. (11.115) in Eq. (11.114) and performing the sum over the initial spin, we obtain

$$\overline{\sum}\sum|\mathcal{M}|^2 = \frac{2}{3}\left(\frac{f_{NR_{3/2}\pi}}{m_\pi}\right)^2\left[(p_R \cdot k_\pi)^2\left(\frac{M}{M_R} + \frac{p \cdot p_R}{M_{R^2}}\right) - m_\pi^2(p \cdot p_R + MM_R)\right],$$

$$= \frac{2}{3}\left(\frac{f_{NR_{3/2}\pi}}{m_\pi}\right)^2\left[\frac{(p_R \cdot k_\pi)^2 - m_\pi^2 M_R^2}{M_R^2}\left(M_R^2 - p_R \cdot k_\pi + MM_R\right)\right], \tag{11.116}$$

where we have used the momentum conservation, $p_R = p + k_\pi$. $\overline{\sum}\sum|\mathcal{M}|^2$ is evaluated in the resonance rest frame.

Using the expression of the two-body decay Γ given in Appendix E in the resonance rest frame, we obtain

$$\Gamma = \frac{|\vec{k}_\pi^{CM}|}{12\pi M_R^2}\left(\frac{f_{NR_{3/2}\pi}}{m_\pi}\right)^2\left[\frac{(p_R \cdot k_\pi)^2 - m_\pi^2 M_R^2}{M_R^2}\left(M_R^2 - p_R \cdot k_\pi + MM_R\right)\right], \tag{11.117}$$

$|\vec{k}_\pi^{CM}|$ represents the momentum of the outgoing pion in the resonance rest frame. Substituting $k_\pi = p_R - p$ in this expression yields the following expression for the decay width

$$\begin{aligned}
\Gamma_{R_{3/2}\to N\pi} &= \frac{I_R}{12\pi M_R^2}|\vec{k}_\pi^{CM}|\left(\frac{f_{NR_{3/2}\pi}}{m_\pi}\right)^2\left[\frac{(M_R^2 - M_R E_N)^2 - m_\pi^2 M_R^2}{M_R^2}\right. \\
&\quad \times \left.\left(M_R^2 - (M_R^2 - M_R E_N) + MM_R\right)\right], \\
&= \frac{I_R}{12\pi M_R}\left(\frac{f_{NR_{3/2}\pi}}{m_\pi}\right)^2|\vec{k}_\pi^{CM}|^3(E_N + M), \tag{11.118}
\end{aligned}$$

where I_R is the isospin factor and is equal to 1 (3) for the isospin $3/2$ $(1/2)$ resonances. E_N represents the energy of the outgoing nucleon in the resonance rest frame. Similarly, the expression for the decay width of the negative parity resonances is given as

$$\Gamma_{R_{3/2} \to N\pi} = \frac{I_R}{12\pi M_R} \left(\frac{f_{NR_{3/2}\pi}}{m_\pi} \right)^2 |\vec{k}_\pi^{CM}|^3 (E_N - M).$$ (11.119)

The expressions for E_N and $|\vec{k}_\pi^{CM}|$ are given in Eqs. (11.73) and (11.74), respectively.

11.3 Neutral Current Reactions

11.3.1 Excitation of spin 1/2 resonances

Excitation of spin 1/2 and isospin 1/2 resonances

The (anti)neutrino induced neutral current (NC) production of the spin 1/2 resonance on the nucleon target is the following:

$$\nu_l / \bar{\nu}_l (k) + N(p) \longrightarrow \nu_l / \bar{\nu}_l(k') + R_{1/2}(p_R).$$ (11.120)

The invariant matrix element for the neutral current induced process, given in Eq. (11.120) and depicted in the right panel of Figure 11.3, is written as

$$\mathcal{M} = \frac{G_F}{\sqrt{2}} \, l_\mu \, J_{\frac{1}{2}}^{\mu NC},$$ (11.121)

where the leptonic current l^μ is given in Eq. (11.3). The hadronic current $J_{\frac{1}{2}}^{\mu NC}$ is defined as

$$J_{\frac{1}{2}}^{\mu NC} = \bar{u}(p_R)\Gamma_{\frac{1}{2}}^{\mu NC} u(p).$$ (11.122)

$\Gamma_{\frac{1}{2}}^{\mu NC}$ is the vertex function and for a positive parity resonance, $\Gamma_{\frac{1}{2}}^{\mu NC}$ is given by

$$\Gamma_{\frac{1}{2}+}^{\mu NC} = V_{\frac{1}{2}}^{\mu NC} - A_{\frac{1}{2}}^{\mu NC},$$ (11.123)

while for a negative parity resonance, $\Gamma_{\frac{1}{2}}^{\mu NC}$ is given by

$$\Gamma_{\frac{1}{2}-}^{\mu NC} = \left[V_{\frac{1}{2}}^{\mu NC} - A_{\frac{1}{2}}^{\mu NC} \right] \gamma_5,$$ (11.124)

where $V_{\frac{1}{2}}^{\mu NC}$ ($A_{\frac{1}{2}}^{\mu NC}$) represents the neutral current vector (axial vector) currents and are parameterized in terms of neutral current vector and axial vector $N - R_{1/2}$ transition form factors, respectively, as,

$$V_{\frac{1}{2}}^\mu = \frac{F_1^{NC}(Q^2)}{(2M)^2} \left(Q^2\gamma^\mu + q^\mu \slashed{q} \right) + \frac{F_2^{NC}(Q^2)}{2M} i\sigma^{\mu\alpha} q_\alpha,$$ (11.125)

$$A_{\frac{1}{2}}^\mu = \left[F_A^{NC}(Q^2)\gamma^\mu + \frac{F_P^{NC}(Q^2)}{M} q^\mu \right] \gamma_5,$$ (11.126)

where $F_i^{NC}(Q^2)$, $i = 1, 2$ are the NC vector form factors; these form factors are, in turn, expressed in terms of the electromagnetic form factors $F_i^{R^+, R^0}(Q^2)$. $F_A^{NC}(Q^2)$ and $F_P^{NC}(Q^2)$ are the axial vector form factors for the NC induced processes. Since the contribution of the pseudoscalar form factor $F_P^{NC}(Q^2)$ is proportional to the lepton mass therefore, for NC induced processes, $F_P^{NC}(Q^2)$ vanishes. In the following sections, we determine the vector and axial vector form factors for the neutral current induced resonance production.

(i) Vector form factors

This section deals with the relation of the weak NC vector form factors $F_{1,2}^{NC}(Q^2)$ with the electromagnetic form factors $F_{1,2}^{R^+}(Q^2)$ and $F_{1,2}^{R^0}(Q^2)$ by isospin symmetry. Electromagnetic current is given in Eq. (11.13) and the vector current for the neutral current induced processes (given in Eq. (11.125)) is written as

$$V_\mu^{NC} = (1 - 2\sin^2\theta_W)\mathcal{V}_\mu^3 - \sin^2\theta_W \mathcal{V}_\mu^I. \tag{11.127}$$

In analogy with the charged current reactions, the transition matrix element for the $ZN \to R_{1/2}$ process is written, using Eq. (11.127), as

$$\langle R_{1/2} | V_\mu^{NC} | N \rangle = \langle R_{1/2} | (1 - 2\sin^2\theta_W)\mathcal{V}_\mu^3 - \sin^2\theta_W \mathcal{V}_\mu^I | N \rangle. \tag{11.128}$$

Using Eqs. (11.11) and (11.14) in Eq.(11.127), we obtain

$$
\begin{aligned}
\langle R_{1/2} | V_\mu^{NC} | N \rangle &= \langle R_{1/2} | (1 - 2\sin^2\theta_W) V_\mu \frac{\tau_3}{2} - \sin^2\theta_W V_\mu^I I_{2\times 2} | N \rangle \\
&= \begin{pmatrix} \bar{R}_{1/2}^+ & \bar{R}_{1/2}^0 \end{pmatrix} \left[\begin{pmatrix} (1/2 - \sin^2\theta_W)V_\mu & 0 \\ 0 & -(1/2 - \sin^2\theta_W)V_\mu \end{pmatrix} \right. \\
&\quad \left. - \begin{pmatrix} \sin^2\theta_W V_\mu^I & 0 \\ 0 & \sin^2\theta_W V_\mu^I \end{pmatrix} \right] \begin{pmatrix} p \\ n \end{pmatrix} \\
&= \bar{R}_{1/2}^+ \left[\left(\frac{1}{2} - \sin^2\theta_W \right) V_\mu - \sin^2\theta_W V_\mu^I \right] p \\
&\quad + \bar{R}_{1/2}^0 \left[-\left(\frac{1}{2} - \sin^2\theta_W \right) V_\mu - \sin^2\theta_W V_\mu^I \right] n \tag{11.129} \\
&= V_\mu^{NC(R_{1/2}^+)} + V_\mu^{NC(R_{1/2}^0)}. \tag{11.130}
\end{aligned}
$$

From this expression, it may be observed that the NC current consists of the components of the electromagnetic current, viz., V_μ and V_μ^I. Using Eqs. (11.21) and (11.22) in Eq. (11.129), we obtain the relation between the NC form factors $F_{1,2}^{NC}(Q^2)$ and the charged current form factors $F_{1,2}^{R^+, R^0}(Q^2)$ as

$$
\begin{aligned}
F_i^{NC(R_{1/2}^+)}(Q^2) &= \left(\frac{1}{2} - \sin^2\theta_W \right) (F_i^{R^+}(Q^2) - F_i^{R^0}(Q^2)) \\
&\quad - \sin^2\theta_W (F_i^{R^+}(Q^2) + F_i^{R^0}(Q^2)),
\end{aligned}
$$

$$= F_i^{R^+}(Q^2)\left(\frac{1}{2} - 2\sin^2\theta_W\right) - \frac{1}{2}F_i^{R^0}(Q^2), \tag{11.131}$$

$$F_i^{NC(R_{1/2}^0)}(Q^2) = F_i^{R^0}(Q^2)\left(\frac{1}{2} - 2\sin^2\theta_W\right) - \frac{1}{2}F_i^{R^+}(Q^2). \tag{11.132}$$

(ii) Axial vector form factors

As we have seen in charged current induced processes (Eq. (11.32)), the axial vector form factor $F_A^{CC}(Q^2)$ is determined in the pion pole dominance by the application of the PCAC hypothesis. The pseudoscalar form factor $F_P^{CC}(Q^2)$ contains the pion pole. Since in neutral current induced processes, the pseudoscalar form factor $F_P^{NC}(Q^2)$ is absent, the axial vector form factor $F_A^{NC}(Q^2)$ cannot be determined by the PCAC hypothesis. In the neutral current sector, $F_A^{NC}(Q^2)$ is determined in terms of $F_A^{CC}(Q^2)$ using the isospin symmetry, where it is assumed that the axial vector current has the same structure as that of the vector current (see Eqs. (11.11) and (11.12)). With this assumption, the axial vector current for charged and neutral current induced processes is given as:

$$A_\mu^{CC} = \mathcal{A}_\mu^1 + i\mathcal{A}_\mu^2, \tag{11.133}$$

$$A_\mu^{NC} = \mathcal{A}_\mu^3. \tag{11.134}$$

Furthermore, it is assumed that $\mathcal{A}_\mu^1$, $\mathcal{A}_\mu^2$, and $\mathcal{A}_\mu^3$ form an isotriplet in the isospin space just like the vector currents. Starting with the charged current induced processes, the transition matrix element for the process $W^+ n \to R_{1/2}^+$ in the case of axial vector current using Eq. (11.133) is written as

$$\langle R_{1/2}^+| A_\mu^{CC} |n\rangle = \langle R_{1/2}^+| \mathcal{A}_\mu^1 + i\mathcal{A}_\mu^2 |n\rangle,$$

$$= \langle R_{1/2}^+| A_\mu \tau_+ |n\rangle,$$

$$= A_\mu. \tag{11.135}$$

Similarly, the transition matrix element for the process $Zp \to R_{1/2}^+$ using Eq. (11.134) is written as

$$\langle R_{1/2}| A_\mu^{NC} |N\rangle = \langle R_{1/2}| \mathcal{A}_\mu^3 |N\rangle,$$

$$= \langle R_{1/2}| A_\mu \frac{\tau_3}{2} |N\rangle,$$

$$= \bar{R}_{1/2}^+ \frac{A_\mu}{2} p + \bar{R}_{1/2}^0 \frac{-A_\mu}{2} n. \tag{11.136}$$

From this expression, it may be observed that A_μ connects the NC axial vector current with the CC axial vector current. From Eqs. (11.135) and (11.136), the neutral current axial vector form factor for the proton and neutron targets is given as

$$F_A^{NC(R_{1/2}^+)}(Q^2) = \frac{1}{2}F_A^{CC}(Q^2), \tag{11.137}$$

$$F_A^{NC(R_{1/2}^0)}(Q^2) = -\frac{1}{2}F_A^{CC}(Q^2). \tag{11.138}$$

Excitation of spin $1/2$ and isospin $3/2$ resonances

(i) Vector form factors

Now we shall discuss the relation between the weak NC vector form factors and the electromagnetic form factors for the transition $N - R_{31}$, where the currents are purely isovector in nature. The electromagnetic current is given in Eq. (11.51) and the vector current for the weak neutral current induced processes is given as:

$$V_\mu^{NC} = (1 - 2\sin^2\theta_W)\mathcal{V}_\mu^3 = -(1 - 2\sin^2\theta_W)\sqrt{\frac{3}{2}}V_\mu T_0^\dagger. \tag{11.139}$$

Using Eq. (11.49), the transition matrix element for the neutral current induced process $ZN \rightarrow R_{1/2}^{I=3/2}$ ($R_{1/2}^{I=3/2}$ represents the spin $1/2$ and isospin $3/2$ resonances), is obtained as

$$
\begin{aligned}
\langle R_{1/2}^{I=3/2}| V_\mu^{NC} |N\rangle &= -(1 - 2\sin^2\theta_W)\sqrt{\frac{3}{2}}\langle R_{1/2}^{I=3/2}| V_\mu T_0^\dagger |N\rangle, \\
&= -\sqrt{\frac{3}{2}}(1 - 2\sin^2\theta_W)\begin{pmatrix} \bar{R}_{1/2}^{++} & \bar{R}_{1/2}^{+} & \bar{R}_{1/2}^{0} & \bar{R}_{1/2}^{-} \end{pmatrix} \\
&\quad V_\mu \begin{pmatrix} 0 & 0 \\ \sqrt{\frac{2}{3}} & 0 \\ 0 & \sqrt{\frac{2}{3}} \\ 0 & 0 \end{pmatrix}\begin{pmatrix} p \\ n \end{pmatrix} \\
&= -(1 - 2\sin^2\theta_W)(\bar{R}_{1/2}^{+}V_\mu p + \bar{R}_{1/2}^{0}V_\mu n). \tag{11.140}
\end{aligned}
$$

From this expression, one may notice that the NC matrix elements and hence, the NC form factors for the proton and neutron induced isospin $3/2$ resonances are the same, that is,

$$F_i^{R_{1/2}^+}(Q^2) = F_i^{R_{1/2}^0}(Q^2). \tag{11.141}$$

Comparing Eqs. (11.52) and (11.140), it is found that the weak NC vector form factors $F_i^{NC(R_{1/2}^+)}(Q^2)$ and $F_i^{NC(R_{1/2}^0)}(Q^2)$ are related to the electromagnetic form factors $F_i^{R_{1/2}^+}(Q^2)$ and $F_i^{R_{1/2}^0}(Q^2)$, respectively, through the relation:

$$F_i^{NC(R_{1/2}^{+,0})}(Q^2) = (1 - 2\sin^2\theta_W)F_i^{R_{1/2}^{+,0}}(Q^2). \tag{11.142}$$

(ii) Axial vector form factors

Here, we discuss the relation between the neutral and charged current axial vector form factors for the transition $N - R_{31}$. The structure of the axial vector current is given in Eq. (11.47). Since in the isospin $1/2$ to $3/2$ transitions only the isovector current contributes, the axial vector current for the neutral current induced processes becomes

$$A_\mu^{NC} = \mathcal{A}_\mu^3 = -\sqrt{\frac{3}{2}}A_\mu T_0^\dagger, \tag{11.143}$$

and the axial vector current for the charged current induced processes is given in Eq. (11.133). The transition matrix element for the charged current induced process $W^+ N \to R_{3/2}$ is written as

$$
\langle R_{3/2}| \, \mathcal{A}_\mu^{CC} \, |N\rangle \;=\; \langle R_{3/2}| \, \mathcal{A}_\mu^1 + i\mathcal{A}_\mu^2 \, |N\rangle = \sqrt{3}\, \langle R_{3/2}| \, A_\mu T_{+1}^\dagger \, |N\rangle
$$

$$
= \; \sqrt{3} \left(\bar{R}_{3/2}^{++} \quad \bar{R}_{3/2}^+ \quad \bar{R}_{3/2}^0 \quad \bar{R}_{3/2}^- \right) A_\mu \begin{pmatrix} 1 & 0 \\ 0 & \frac{1}{\sqrt{3}} \\ 0 & 0 \\ 0 & 0 \end{pmatrix} \begin{pmatrix} p \\ n \end{pmatrix}
$$

$$
= \; (\sqrt{3}\bar{R}_{3/2}^{++} A_\mu p + \bar{R}_{3/2}^+ A_\mu n). \tag{11.144}
$$

This expression implies that for the $R_{3/2}^{++}$ production, the form factors should be multiplied by $\sqrt{3}$.

Similarly, for the neutral current induced processes $ZN \to R_{3/2}$, the transition matrix element is expressed as

$$
\langle R_{3/2}| \, \mathcal{A}_\mu^{NC} \, |N\rangle \;=\; \langle R_{3/2}| \, \mathcal{A}_\mu^3 \, |N\rangle = -\sqrt{\frac{3}{2}}\, \langle R_{3/2}| \, A_\mu T_0^\dagger \, |N\rangle
$$

$$
= \; -\sqrt{\frac{3}{2}} \left(\bar{R}_{3/2}^{++} \quad \bar{R}_{3/2}^+ \quad \bar{R}_{3/2}^0 \quad \bar{R}_{3/2}^- \right) A_\mu \begin{pmatrix} 0 & 0 \\ \sqrt{\frac{2}{3}} & 0 \\ 0 & \sqrt{\frac{2}{3}} \\ 0 & 0 \end{pmatrix} \begin{pmatrix} p \\ n \end{pmatrix}
$$

$$
= \; -(\bar{R}_{3/2}^+ A_\mu p + \bar{R}_{3/2}^0 A_\mu n). \tag{11.145}
$$

From this equation, it can be observed that the axial vector form factors for the R_{31} resonances are the same as that in the case of proton and neutron targets, that is,

$$
F_A^{NC(R_{1/2}^+)} = F_A^{NC(R_{1/2}^0)}. \tag{11.146}
$$

Equations (11.144) and (11.145) relate the charged and neutral current axial vector form factors for R_{31} resonances by the following relation:

$$
F_A^{NC}(Q^2) = -F_A^{CC}(Q^2). \tag{11.147}
$$

11.3.2 Excitation of spin $3/2$ resonances

Excitation of spin $3/2$ and isospin $1/2$ resonances

The reaction for the spin $3/2$ resonance excitations induced by the neutral current on the nucleon target is written as:

$$
\nu_l(\bar{\nu}_l)\,(k) + N(p) \longrightarrow \nu_l(\bar{\nu}_l)\,(k') + R_{3/2}(p_R). \tag{11.148}
$$

The invariant transition amplitude for the process given in Eq. (11.148), is written as

$$
\mathcal{M} = \frac{G_F}{\sqrt{2}}\, l_\mu \, J_{\frac{3}{2}}^{\mu NC}. \tag{11.149}
$$

The leptonic current l_μ is given in Eq. (11.3) and the hadronic current $J_{\frac{3}{2}}^{\mu NC}$ for the spin 3/2 resonance excitation in the neutral current sector is given as

$$J_{\frac{3}{2}}^{\mu NC} = \bar{\psi}_\nu(p')\Gamma_{\frac{3}{2}}^{\nu\mu NC} u(p), \tag{11.150}$$

where $\Gamma_{\nu\mu NC}^{\frac{3}{2}}$ has the following general structure for the positive and negative parity resonance states:

$$\Gamma_{\nu\mu NC}^{\frac{3}{2}^+} = \left[V_{\nu\mu NC}^{\frac{3}{2}} - A_{\nu\mu NC}^{\frac{3}{2}} \right] \gamma_5 \tag{11.151}$$

$$\Gamma_{\nu\mu NC}^{\frac{3}{2}^-} = V_{\nu\mu NC}^{\frac{3}{2}} - A_{\nu\mu NC}^{\frac{3}{2}}. \tag{11.152}$$

In these expressions, $V_{\frac{3}{2}}^{NC}$ ($A_{\frac{3}{2}}^{NC}$) is the vector (axial vector) current for spin 3/2 resonances for the neutral current induced processes and are given by

$$V_{\nu\mu NC}^{\frac{3}{2}} = \left[\frac{C_3^{V(NC)}}{M}(g_{\mu\nu} - q_\nu\gamma_\mu) + \frac{C_4^{V(NC)}}{M^2}(g_{\mu\nu}q\cdot p' - q_\nu p_\mu') + \frac{C_5^{V(NC)}}{M^2} \right.$$
$$\left. (g_{\mu\nu}q\cdot p - q_\nu p_\mu) + g_{\mu\nu}C_6^{V(NC)} \right], \tag{11.153}$$

$$A_{\nu\mu NC}^{\frac{3}{2}} = -\left[\frac{C_3^{A(NC)}}{M}(g_{\mu\nu} - q_\nu\gamma_\mu) + \frac{C_4^{A(NC)}}{M^2}(g_{\mu\nu}q\cdot p' - q_\nu p_\mu') + C_5^{A(NC)}g_{\mu\nu} \right.$$
$$\left. + \frac{C_6^{A(NC)}}{M^2}q_\nu q_\mu \right] \gamma_5, \tag{11.154}$$

where $C_i^{V(NC)}$ and $C_i^{A(NC)}$ are the vector and axial vector neutral current transition form factors. These neutral current form factors $C_i^{V(NC)}$ and $C_i^{A(NC)}$ are determined in terms of the electromagnetic form factors $C_i^{R^+,R^0}(Q^2)$ and the charged current form factors C_i^V and C_i^A using the isospin symmetry. They are discussed in the next sections.

(i) Vector form factors
In the case of the neutral current induced spin 3/2 resonance production, the vector current is given in Eq. (11.153) with the corresponding form factors $C_i^{V(NC)}$, $i = 3, 4, 5, 6$. $C_6^{V(NC)}(Q^2)$ = 0 as required by the CVC hypothesis. As we have already discussed in earlier sections, the determination of the weak vector form factors in terms of the electromagnetic form factors depends upon the isospin of the resonance irrespective of the spin of the resonance; therefore, we directly use the expressions obtained in Subsection 11.3.1 to relate the electromagnetic and

weak vector form factors. For the isospin $1/2$ and spin $3/2$ resonances, using Eqs. (11.131) and (11.132), the neutral current form factors are expressed as

$$C_i^{NC(R^+)}(Q^2) = C_i^{R^+}(Q^2)\left(\frac{1}{2} - 2\sin^2\theta_W\right) - \frac{1}{2}C_i^{R^0}(Q^2), \tag{11.155}$$

$$C_i^{NC(R^0)}(Q^2) = C_i^{R^0}(Q^2)\left(\frac{1}{2} - 2\sin^2\theta_W\right) - \frac{1}{2}C_i^{R^+}(Q^2), \tag{11.156}$$

$i = 3, 4, 5$. $C_i^{R^+}(Q^2)$ and $C_i^{R^0}(Q^2)$ are the electromagnetic form factors for the resonances R^+ and R^0, which are parameterized in terms of the helicity amplitudes $A_{1/2}$, $A_{3/2}$ and $S_{1/2}$, given in Eq. (11.84)–(11.86).

(ii) Axial vector form factors

In neutral current induced processes, the pseudoscalar form factor $C_6^{A(NC)}$ is zero due to the negligible mass of the neutrino. The axial vector form factor $C_5^{A(NC)}(Q^2)$ is determined in terms of the charged current axial vector form factor $C_5^A(Q^2)$ using the isospin symmetry which has been discussed in Subsection 11.3.1. Using Eqs. (11.137) and (11.138), the axial vector form factor for the transition $N - R_{13}$ is obtained as

$$C_5^{A(NC\ R_{1/2}^+)}(Q^2) = \frac{1}{2}C_5^A(Q^2), \tag{11.157}$$

$$C_5^{A(NC\ R_{1/2}^0)}(Q^2) = -\frac{1}{2}C_5^A(Q^2). \tag{11.158}$$

$C_5^A(Q^2)$ is already defined in Eq. (11.97).

1.3.2.2 Excitation of spin $3/2$ and isospin $3/2$ resonances

(i) Vector form factors

For the transition $N - R_{33}$, the vector neutral current form factors $C_i^{V(NC)}(Q^2)$, $i = 3, 4, 5$ are obtained in a similar manner as we have done for the transition $N - R_{31}$ using Eq. (11.142), and are expressed as

$$C_i^{V(NC)}(Q^2) = (1 - 2\sin^2\theta_W)C_i^V(Q^2), \qquad i = 3, 4, 5. \tag{11.159}$$

(ii) Axial vector form factors

The neutral current axial vector form factor $C_5^{A(NC)}$ for the transition $N - R_{33}$, in analogy with the neutral current axial vector form factor for the transition $N - R_{31}$, is expressed as

$$C_5^{A(NC)}(Q^2) = -C_5^A(Q^2). \tag{11.160}$$

11.4 Non-resonant Contributions

The non-resonant diagrams, shown in Figure 11.2, give the essential contribution to the single meson production through the Born diagrams in s, t, and u channels as shown in Figure 11.14.

While the s channel diagram consist of direct nucleon poles, the t channel has meson poles and the u channel has the exchange baryon pole. Some phenomenological Lagrangians based on the pseudovector coupling or effective Lagrangians based on the chiral symmetry also include, in addition to the Born terms, the contact diagrams as shown in Figure 11.2(b).

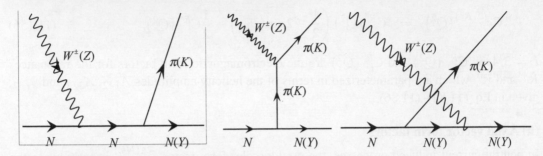

Figure 11.14 Generic Feynman diagrams for the s, t, and u channel Born terms.

In quite a general approach, the matrix element of the hadronic current is written formally as

$$\langle B(p')M(p_M)|J^\mu|N(p)\rangle = \langle B(p')M(p_M)|V^\mu + A^\mu|N(p)\rangle \tag{11.161}$$
$$= M^\mu + N^\mu, \tag{11.162}$$

where $N(p), B(p')$, and $M(p_M)$ are, respectively, the initial nucleon N of momentum p and the final baryon B of momentum p' which could be a nucleon or a hyperon and a meson π (K) of momentum p_M in case of $\Delta S = 0$ CC and NC reactions. In the case of $|\Delta S| = 1$ CC reactions, the final state has a nucleon N and a strange meson K. These particles satisfy the momentum conservation law, that is,

$$q + p = p' + p_M, \tag{11.163}$$

where $q = k - k'$ is the momentum transfer given to the hadron system, k and k' being, respectively, the momenta of the incoming neutrino and the outgoing lepton. The matrix elements M^μ and N^μ for the vector and the axial vector currents are written as

$$M^\mu = \sum_{i=1}^{8} A_i(E, E', E_\pi)\bar{u}(p')O_i^\mu(V)u(p), \tag{11.164}$$

$$N^\mu = \sum_{i=1}^{8} C_i(E, E', E_\pi)\bar{u}(p')O_i^\mu(A)u(p), \tag{11.165}$$

where $O_i^\mu(V)$ and $O_i^\mu(A)$ are eight independent covariant operators constructed from the operators γ^μ, q^μ, p^μ and k^μ, and discussed explicitly by Adler [464] and Marshak et al. [223]. The quantities $A_i(E, E', E_\pi)$ and $C_i(E, E', E_\pi)$ are the energy dependent strength of these amplitudes corresponding to the operators $O_i^\mu(V)$ and $O_i^\mu(A)$.

The gauge invariance imposed on the vector amplitudes M^μ, that is,

$$k_\mu M^\mu = 0, \tag{11.166}$$

reduces then to six independent covariant amplitudes. The expressions for $O_i^\mu(V), i = 1 - 6$ and $O_i^\mu(A), i = 1 - 8$ are given in Table 11.10. The contribution of all the Born diagrams corresponding to the non-resonant part are explicitly calculated using a phenomenological Lagrangian for πNN interaction. The contribution of the higher resonances are calculated using the dispersion relations [460, 461, 464, 223]. Another method based on a dynamical model starting from the effective Lagrangian with bare pion–nucleon couplings obtained in the quark model, is used to construct a T matrix. Thereafter, a Lippmann–Schwinger equation is formulated and solved using coupled channel equations for pion production. In this way, it combines the effective Lagrangian with dynamical models [473, 475, 476]. In the case of effective Lagrangian approaches, the explicit contribution from individual non-resonant Born diagrams and the higher resonances are explicitly calculated in terms of the parameters describing the effective Lagrangian.

Table 11.10 Expressions for $O_i^\mu(V), i = 1 - 6$ and $O_i^\mu(A), i = 1 - 8$, where $P = p + p'$ [464].

$O_i^\mu(V)$		$O_i^\mu(A)$	
$O_1^\mu(V)$	$\frac{1}{2}i\gamma_5[\gamma^\mu \not q - \not q \gamma^\mu]$	$O_1^\mu(A)$	$\frac{i}{2}[\not p_M \gamma^\mu - \gamma^\mu \not p_M]$
$O_2^\mu(V)$	$2i\gamma_5[P^\mu p_M \cdot q - P \cdot q p_M^\mu]$	$O_2^\mu(A)$	$2iP^\mu$
$O_3^\mu(V)$	$\gamma_5[\gamma^\mu p_M \cdot q - \not q p_M^\mu]$	$O_3^\mu(A)$	ip_M^μ
$O_4^\mu(V)$	$2\gamma_5[\gamma^\mu P \cdot q - \not q P^\mu - \frac{1}{2}iM(\gamma^\mu \not q - \not q \gamma^\mu)]$	$O_4^\mu(A)$	$-M\gamma^\mu$
$O_5^\mu(V)$	$i\gamma_5[q^\mu p_M \cdot q - q^2 p_M^\mu]$	$O_5^\mu(A)$	$-2\not q P^\mu$
$O_6^\mu(V)$	$\gamma_5[q^\mu \not q - q^2 \gamma^\mu]$	$O_6^\mu(A)$	$-\not q p_M^\mu$
		$O_7^\mu(A)$	iq^μ
		$O_8^\mu(A)$	$-\not q q^\mu$

Recently, effective Lagrangians based on the chiral symmetry have been used by many authors to calculate the inelastic reactions specifically the one pion production. One class of the models [496] uses Lagrangians containing nucleon, pion, σ, ω and ρ fields consistent with chiral symmetry while another class of models is based on the non-linear sigma model incorporating chiral symmetry [481, 497, 498, 499].

In the following section, we outline the formalism to write an effective Lagrangian based on the chiral symmetry which has been used to illustrate the contributions from the non-resonant background terms given in the next chapter.

11.4.1 Chiral symmetry

The Lagrangian for QCD can be written as

$$\mathcal{L}_{QCD} = \bar{q}(i\not D - m_q)q - \frac{1}{4}G_{\mu\nu}^\alpha G^{\alpha\mu\nu} \tag{11.167}$$

where $q = \begin{pmatrix} u \\ d \\ s \end{pmatrix}$ denotes the quark field and $G_{\mu\nu}^{\alpha}$ is the gluon field strength tensor with α as a color index. D_{μ} is defined as

$$D_{\mu} = \partial_{\mu} + ig\frac{\lambda^{\alpha}}{2}G_{\mu\alpha}, \tag{11.168}$$

where g is the quark–gluon coupling strength and $G_{\mu\alpha}$ is the vector gluon field. The Lagrangian written in Eq. (11.167) does not preserve chiral symmetry in its present form; however, in the limit when quark masses are assumed to be zero, the QCD Lagrangian preserves chiral symmetry. Today, it is well established that all the quarks have non-zero mass; although the current quark masses for u, d, s are small as compared to the nucleon mass. Thus, in the case of strong interactions, chiral symmetry is preserved in the limit $m_u, m_d, m_s \to 0$. The consequence of the symmetries of the Lagrangian leads to conserved currents. Vector current is conserved in nature due to isospin symmetry (Chapter 6). Similarly, the axial vector current is conserved in the presence of chiral symmetry. If the chiral symmetry is broken spontaneously, it leads to the existence of massless Goldstone bosons which are identified as pions in the limit $m_\pi \to 0$.

11.4.2 Transformation of mesons under chiral transformation

We have seen in Chapter 6 that the axial vector current is partially conserved and its consequences lead to the Goldberger–Treiman relation which relates the strong and weak couplings. The spectrum of mesons does not respect chiral symmetry. Now, we show the transformation of pion and rho mesons under the vector (Λ_V) and axial vector (Λ_A) transformations which are defined as

$$\Lambda_V \psi = e^{-i\frac{\vec{\tau} \cdot \vec{\Theta}}{2}} \psi \simeq \left(1 - i\frac{\vec{\tau} \cdot \vec{\Theta}}{2} \right) \psi, \tag{11.169}$$

$$\Lambda_A \psi = e^{-i\gamma_5 \frac{\vec{\tau} \cdot \vec{\Theta}}{2}} \psi \simeq \left(1 - i\gamma_5 \frac{\vec{\tau} \cdot \vec{\Theta}}{2} \right) \psi, \tag{11.170}$$

where $\psi = \begin{pmatrix} u \\ d \end{pmatrix}$ represents the quark doublet, Θ is the rotation angle, $\vec{\tau}$ represents the Pauli matrices. The pion and rho mesons can be expressed as

$$\vec{\pi} = i\bar{\psi}\vec{\tau}\gamma_5\psi, \qquad \vec{\rho}_{\mu} = \bar{\psi}\vec{\tau}\gamma_{\mu}\psi, \tag{11.171}$$

where $\vec{\pi}, \vec{\rho}$ represents the isovector pion and rho meson states, respectively. The subscript μ represents the vector mesons.

The vector transformation (Eq. (11.169)) when applied on the pion state yields

$$\pi_i = i\bar{\psi}\tau_i\gamma_5\psi \rightarrow \Lambda_V^\dagger \bar{\psi}\tau_i\gamma_5\Lambda_V\psi,$$

$$= i\bar{\psi}\left(1 + \frac{i\tau_j\Theta^j}{2}\right)\tau_i\gamma_5\left(1 - \frac{i\tau_j\Theta^j}{2}\right)\psi,$$

$$= i\bar{\psi}\tau_i\gamma_5\psi + i\epsilon_{ijk}\Theta^j\bar{\psi}\tau_k\gamma_5\psi,$$

$$\vec{\pi} \rightarrow \vec{\pi} + \vec{\Theta} \times \vec{\pi}. \tag{11.172}$$

This expression represents the rotation of the pion state through the isospin direction by the angle Θ. Similarly, the vector transformation of the ρ mesons gives

$$\vec{\rho}_\mu \rightarrow \vec{\rho}_\mu + \vec{\Theta} \times \vec{\rho}_\mu. \tag{11.173}$$

From Eqs. (11.172) and (11.173), it can be concluded that the vector transformation of mesons leads to rotation along the isospin direction, which means that the conservation of the vector current is associated with the isospin symmetry.

Next, we see the axial vector transformation of these meson states, starting with the pion state. For this, we start with Eq. (11.170) and obtain

$$\pi_i = i\bar{\psi}\tau_i\gamma_5\psi \simeq \Lambda_A^\dagger \bar{\psi}\tau_i\gamma_5\Lambda_A\psi,$$

$$= i\bar{\psi}\left(1 - \frac{i\gamma_5\tau_j\Theta^j}{2}\right)\tau_i\gamma_5\left(1 - \frac{i\gamma_5\tau_j\Theta^j}{2}\right)\psi,$$

$$= i\bar{\psi}\tau_i\gamma_5\psi + \Theta^j\bar{\psi}(\delta_{ij})\psi,$$

$$\Rightarrow \quad \vec{\pi} \simeq \vec{\pi} + \vec{\Theta}\bar{\psi}\psi, \tag{11.174}$$

$$= \vec{\pi} + \vec{\Theta}\sigma, \tag{11.175}$$

if σ is identified with the scalar particle associated with $\bar{\psi}\psi$. Eq. (11.174) represents the rotation of the pion into a linear combination of π and sigma meson when the axial vector transformation is applied on the pion state. Similarly, under axial vector transformation, a scalar meson $\sigma\ (= \bar{\psi}\psi)$ transforms as:

$$\sigma = \bar{\psi}\psi \simeq \Lambda_A\bar{\psi}\Lambda_A\psi,$$

$$= \bar{\psi}\left(1 - i\gamma_5\frac{\tau_j\Theta^j}{2}\right)\left(1 - i\gamma_5\frac{\tau_j\Theta^j}{2}\right)\psi,$$

$$\Rightarrow \quad \sigma \simeq \sigma - \vec{\Theta}\cdot\vec{\pi}. \tag{11.176}$$

From Eqs. (11.174) and (11.176), it is inferred that the pion and sigma mesons under axial vector transformation, are rotated into each other.

Similarly, the transformation of the axial vector current on the ρ mesons gives

$$\rho_{\mu_i} \simeq \bar{\psi}\left(1 - i\gamma_5\frac{\tau_j\Theta^j}{2}\right)\tau_i\gamma_\mu\left(1 - i\gamma_5\frac{\tau_j\Theta^j}{2}\right)\psi,$$

$$\Rightarrow \quad \vec{\rho}_\mu \simeq \vec{\rho}_\mu + \vec{\Theta} \times \vec{a}_{1_\mu}, \tag{11.177}$$

where $\vec{a}_{1_\mu} = \bar{\psi}\vec{\tau}\gamma_\mu\gamma_5\psi$ represents a vector meson a_1 with spin 1. The axial vector transformation of rho mesons shows the existence of a_1 mesons. Moreover, the two, that is, rho and a_1 mesons are rotated into one another by the axial vector transformation.

Therefore, if the chiral symmetry is good, then (π, σ) and (ρ, a_1) should be degenerate, which is not true experimentally. This is because we know that σ is not observed experimentally. In the case of ρ and a_1 meson states, the mass of ρ is $m_\rho = 0.77$ GeV while the mass of a_1 is $m_{a_1} = 1.23 \pm 0.04$ GeV. Since there is a large mass difference between the masses of ρ and a_1, the chiral symmetry is broken in nature at the nucleon level. However, if the chiral symmetry is broken spontaneously, then the degeneracy of states is not a required consequence (Chapter 7). Moreover, in this case, massless Goldstone bosons appear which are identified as pions. A small mass of pions can be generated by assigning a non-zero but very small mass to the fermions in the theory which leads to an axial vector current consistent with PCAC [500, 163]. Thus the degeneracy of mass spectrum is not present in the case of spontaneous breaking of the symmetry, which generates pion mass and leads to PCAC.

11.4.3   Linear sigma model

The linear sigma model is an effective chiral model introduced by Gell-Mann and Levy in 1960 to study the chiral symmetry in the pion–nucleon system before the formulation of QCD. Spontaneous symmetry breaking and PCAC are the natural consequences of this model. The structure of the Lagrangian is Lorentz scalar; it is also invariant under vector (Λ_V) and axial vector (Λ_A) transformations. We have studied in the earlier sections that the pion as well as sigma fields are not invariant under axial vector transformations. Our task is to first construct a field variable which is invariant under both Λ_V and Λ_A and then write the Lagrangian using it.

We have discussed in Section 11.4.2 that the vector transformation is nothing but the isospin rotation; thus, the squares of these fields are also invariant under vector transformation:

$$\pi^2 \xrightarrow{\Lambda_V} \pi^2 \qquad\qquad \sigma^2 \xrightarrow{\Lambda_V} \sigma^2, \tag{11.178}$$

while under the axial vector transformation, even the square of the meson fields are not invariant and yields the following expressions in the limit of small $\vec{\Theta}$:

$$\pi^2 \xrightarrow{\Lambda_A} \pi^2 + 2\sigma\vec{\Theta}\cdot\vec{\pi} \qquad\qquad \sigma^2 \xrightarrow{\Lambda_A} \sigma^2 - 2\sigma\vec{\Theta}\cdot\vec{\pi}. \tag{11.179}$$

Furthermore, from Eqs. (11.178) and (11.179), one may notice that the combination $\sigma^2 + \pi^2$ is invariant under both vector and axial vector transformations. This combination is also Lorentz invariant; hence, the Lagrangian for the linear sigma model can be constructed around $\sigma^2 + \pi^2$.

The most general Lagrangian of the linear sigma model for the pion–nucleon interaction is written as [501, 502]:

$$\begin{aligned}
\mathcal{L}_{LSM} &= i\bar{\psi}\partial_\mu\gamma^\mu\psi + \frac{1}{2}\partial_\mu\pi\partial^\mu\pi + \frac{1}{2}\partial_\mu\sigma\partial^\mu\sigma \\
&\quad - g_\pi(i\bar{\psi}\gamma_5\vec{\tau}\psi\vec{\pi} + \bar{\psi}\psi\sigma) - \frac{\lambda}{4}\left((\pi^2 + \sigma^2) - f_\pi^2\right)^2,
\end{aligned} \tag{11.180}$$

where ψ, π, and σ represent the nucleon, pion, and sigma fields, respectively, g_π is the pion–nucleon coupling, and f_π represents the pion decay constant. The first term in Eq. (11.180) represents the kinetic energy of the nucleon, which is the Lagrangian of the massless nucleons. The second and third terms represent the kinetic energy of the pion and sigma mesons. The fourth term represents the pion–nucleon interaction term which is generally expressed by the term $g_\pi(i\bar{\psi}\gamma_5\vec{\tau}\psi)\vec{\pi}$ and transforms like π^2 under Λ_V and Λ_A transformations, while π^2 is not invariant under Λ_A transformation and one requires a term which transforms like σ^2 to make the potential chiral invariant. The simplest choice for a Lagrangian which transforms like σ^2 is $g_\pi\bar{\psi}\psi\sigma$; thus, sigma is incorporated in the pion–nucleon potential to make it chiral invariant. The last term in Eq. (11.180) represents the pion sigma potential. The vacuum expectation value of σ is generated by this potential; thus, the chiral invariance requires that the potential must be a function of $\pi^2 + \sigma^2$. The simplest form of this potential is given by the last term in the aforementioned Lagrangian, where f_π represents the minimum of this potential.

In the Lagrangian given in Eq. (11.180), all the interaction terms between pion, nucleon, and sigma are present except the mass terms. The mass term for the nucleon is generated without breaking the chiral symmetry, by its interaction with the sigma field which is given by the potential $g_\pi(\bar{\psi}\psi)\sigma$. This is achieved by giving a finite vacuum expectation value to the sigma fields

$$< \sigma >= \sigma_0, \tag{11.181}$$

which describes the spontaneous breaking of the chiral symmetry. By this mechanism, the nucleon mass is generated. Due to the spontaneously broken chiral symmetry, the pion remains massless and sigma obtains a mass term through its coupling with the vacuum expectation value of the sigma field from the last terms in Eq. (11.180). Thus, in the linear sigma model, the pion is massless but sigma is massive.

11.4.4 Explicitly broken chiral symmetry

The chiral symmetry is a good symmetry in the limit of vanishing quark masses. In the presence of the quark mass terms in the Lagrangian, although being very small for the lowest lying u, d, s quarks, the chiral symmetry is broken explicitly.

One can visualize the effect of the explicit symmetry breaking as shown in Figure 11.15. In the case of explicit symmetry breaking, the Hamiltonian and in turn, the potential is not symmetric under rotation. The ground state of the potential is shifted but the shift is so small that the rotation along the pion axis and the radial excitation along the σ-axis in the case of spontaneous breaking of the chiral symmetry remains almost undisturbed. Spontaneous and explicit symmetry breaking generate, respectively, nucleon and pion masses. Thus, in the limit of the small explicit breaking, the effect of the spontaneous breaking of the chiral symmetry still dominates the effect of the explicit breaking. This means that the chiral symmetry is good even in the limit of the small explicit symmetry breaking effect. The effect of the explicit symmetry breaking on the linear sigma model makes the pions massive. However, the problem lies with the massive σ field as the σ meson has not been observed experimentally. Therefore, a non-linear sigma model was proposed which is discussed in the next section [501].

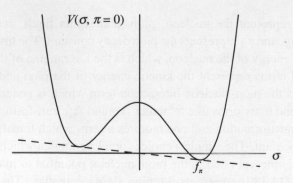

Figure 11.15 Effect of explicit symmetry breaking.

11.4.5 Non-linear sigma model

In the non-linear sigma model, this massive σ field is removed by taking an infinitely large coupling λ which results in an infinite mass of the σ meson; the potential gets infinitely steep in the sigma direction as depicted in Figure 11.16. The minimum of this potential defines a circle (known in the literature as the chiral circle but, in principle, it is a sphere not a circle) described by

$$\pi^2 + \sigma^2 = f_\pi^2. \tag{11.182}$$

The dynamics of the system is confined to the rotation along this circle. Thus, the pion and sigma meson fields can be expressed in terms of the pion fields $\vec{\Phi}(x)$ and the radius of the circle f_π, as

$$\sigma(x) \;=\; f_\pi \cos\left(\frac{\Phi(x)}{f_\pi}\right), \tag{11.183}$$

$$\vec{\pi}(x) \;=\; f_\pi \hat{\Phi} \sin\left(\frac{\Phi(x)}{f_\pi}\right), \tag{11.184}$$

where $\Phi = \sqrt{\vec{\Phi} \cdot \vec{\Phi}}$ and $\hat{\Phi} = \frac{\vec{\Phi}}{\Phi}$.

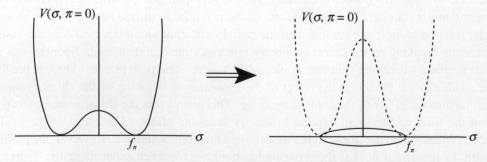

Figure 11.16 Figure depicting the infinitely steep potential in the σ direction.

The pion and sigma fields can be expressed in the complex form as

$$U(x) = e^{\frac{i\vec{\tau}\cdot\vec{\Phi}(x)}{f_\pi}} = \cos\left(\frac{\Phi(x)}{f_\pi}\right) + i\vec{\tau}\hat{\Phi}\sin\left(\frac{\Phi(x)}{f_\pi}\right) = \frac{1}{f_\pi}(\sigma + i\vec{\tau}\cdot\vec{\pi}), \qquad (11.185)$$

where $U(x)$ is the unitary 2×2 matrix. In terms of the complex field, the chiral circle is expressed as

$$\frac{1}{2}\text{Tr}(U^\dagger U) = \frac{1}{f_\pi}(\sigma^2 + \pi^2) = 1. \qquad (11.186)$$

As in the case of the vector current, isospin symmetry corresponds to the rotational symmetry; analogously, the chiral symmetry corresponds to the rotational symmetry along the chiral circle.

The Lagrangian for the linear sigma model given in Eq. (11.180) is now expressed in terms of Φ or the complex representation $U(x)$.

Writing the kinetic energy terms of the mesons in terms of $U(x)$, we get

$$\frac{1}{2}(\partial_\mu\sigma\partial^\mu\sigma + \partial_\mu\vec{\pi}\partial^\mu\vec{\pi}) = \frac{f_\pi^2}{4}(\partial_\mu U^\dagger\partial^\mu U). \qquad (11.187)$$

Similarly, the nucleon–meson coupling term modifies as

$$\begin{aligned}
-g_\pi(\bar{\psi}\psi\sigma + i\bar{\psi}\gamma_5\vec{\tau}\psi\vec{\pi}) &= -g_\pi\bar{\psi}\left(f_\pi\cos\left(\frac{\Phi}{f_\pi}\right) + i\gamma_5\vec{\tau}f_\pi\hat{\Phi}\sin\left(\frac{\Phi}{f_\pi}\right)\right)\psi, \\
&= g_\pi f_\pi\bar{\psi}e^{\frac{i\gamma_5\vec{\tau}\cdot\hat{\Phi}}{f_\pi}}\psi, \\
&= g_\pi f_\pi\bar{\psi}\Lambda\Lambda\psi, \qquad (11.188)
\end{aligned}$$

with $\Lambda = e^{\frac{i\gamma_5\vec{\tau}\cdot\hat{\Phi}}{2f_\pi}}$.

Redefining the nucleon field ψ in terms of ψ_W as

$$\psi_W = \Lambda\psi, \qquad (11.189)$$

the nucleon–meson interaction becomes

$$-g_\pi(\bar{\psi}\psi\sigma + i\bar{\psi}\gamma_5\vec{\tau}\psi\vec{\pi}) = -g_\pi f_\pi\bar{\psi}_W\psi_W = -M\bar{\psi}_W\psi_W, \qquad (11.190)$$

where the Goldberger–Treiman relation ($g_\pi = \frac{g_{\pi NN}}{\sqrt{2}}$; $g_\pi f_\pi = M$) is used. Thus, in the non-linear sigma model, the nucleon–meson interaction term reduces to the nucleon mass term. In terms of the redefined nucleon field ψ_W, the nucleon kinetic energy term is modified as:

$$\begin{aligned}
\bar{\psi}(i\partial\!\!\!/)\psi &= \bar{\psi}_W\Lambda^\dagger(i\partial^\mu\gamma_\mu)\Lambda^\dagger\psi_W, \\
&= \bar{\psi}_W(i\partial\!\!\!/ + \gamma_\mu V^\mu + \gamma_\mu\gamma_5 A^\mu)\psi_W, \qquad (11.191)
\end{aligned}$$

where V^μ and A^μ are defined in terms of the unitary matrix u as

$$V^\mu = \frac{1}{2}[u^\dagger \partial^\mu u + u \partial^\mu u^\dagger], \tag{11.192}$$

$$A^\mu = \frac{i}{2}[u^\dagger \partial^\mu u - u \partial^\mu u^\dagger], \tag{11.193}$$

and

$$u = e^{\frac{i\vec{\tau}\cdot\vec{\Phi}}{2f\pi}}; \quad U = u^2. \tag{11.194}$$

The last term of the Lagrangian given in Eq. (11.180) vanishes as the potential between the pion and sigma fields vanishes in the chiral limit ($\pi^2 + \sigma^2 = f_\pi^2$). Thus, the Lagrangian for the non-linear sigma model in the chiral limit becomes

$$\mathcal{L}_{NLSM} = \bar{\psi}_W (i\slashed{\partial} + \gamma_\mu V^\mu + \gamma_\mu \gamma_5 A^\mu - M)\psi_W + \frac{f_\pi^2}{4}(\partial_\mu U^\dagger \partial^\mu U), \tag{11.195}$$

where the first term of this Lagrangian represents the interaction between the nucleons and pions (in general, between baryons and mesons) and the second term represents the interaction between pions (in general, mesons). The Lagrangian in Eq. (11.195) can be expanded in terms of the single variable, $\vec{\Phi}(x)$, which represents the pion field. Noticeably, the sigma field which was present in the linear sigma model disappeared in the Lagrangian for the non-linear sigma model. An important point to keep in mind regarding the Lagrangian given in Eq. (11.195) is that it represents only the interaction between pions and nucleons but not the interaction of these particles with the gauge bosons.

In reality, all the interactions proceed via the exchange of gauge bosons. Thus, our next task is to incorporate the gauge bosons in the meson–meson and meson–baryon Lagrangians which can be discussed in the next section. It is also worth mentioning that the unitary matrix U is expressed in terms of the Pauli matrices $\vec{\tau}$ representing the SU(2) isospin symmetry; however, one may extend the Lagrangian given in Eq. (11.195) to the SU(3) isospin symmetry, that is, through the inclusion of the octet of mesons and baryons.

11.4.6   Lagrangian for the meson–meson and meson–gauge boson interactions

The Lagrangian for the meson–meson interaction is given in Eq. (11.195) (second term), where in the unitary matrix $U = e^{\frac{i\vec{\tau}\cdot\vec{\Phi}(x)}{f\pi}}$, $\vec{\tau}$ stands for the SU(2) isospin symmetry. In order to extend the meson–meson Lagrangian for the SU(3) symmetry (where the three massless quarks; u, d, s are considered), the Lagrangian remains the same; the only change is the modification of U, which in the SU(3) symmetry becomes

$$U(x) = e^{\frac{i\vec{\lambda}\cdot\vec{\Phi}(x)}{f\pi}} = e^{\frac{i\lambda_i \Phi_i(x)}{f\pi}}; \qquad i = 1 - 8, \tag{11.196}$$

where λ represents the Gell-Mann matrices which are given in Appendix B and Φ_i represents the octet of the meson fields, expressed as

$$\vec{\lambda} \cdot \vec{\Phi} = \sum_{i=1}^{8} \lambda_i \Phi_i = \begin{pmatrix} \Phi_3 + \frac{1}{\sqrt{3}}\Phi_8 & \Phi_1 - i\Phi_2 & \Phi_4 - i\Phi_5 \\ \Phi_1 + i\Phi_2 & -\Phi_3 + \frac{1}{\sqrt{3}}\Phi_8 & \Phi_6 - i\Phi_7 \\ \Phi_4 + i\Phi_5 & \Phi_6 + i\Phi_7 & -\frac{2}{\sqrt{3}}\Phi_8 \end{pmatrix}$$

$$= \begin{pmatrix} \pi^0 + \frac{1}{\sqrt{3}}\eta & \sqrt{2}\pi^+ & \sqrt{2}K^+ \\ \sqrt{2}\pi^- & -\pi^0 + \frac{1}{\sqrt{3}}\eta & \sqrt{2}K^0 \\ \sqrt{2}K^- & \sqrt{2}\bar{K}^0 & -\frac{2}{\sqrt{3}}\eta \end{pmatrix}. \tag{11.197}$$

With these modifications, the second term of the Lagrangian given in Eq. (11.195) gives the interaction among the octets of the pseudoscalar mesons. Using Eqs. (11.196) and (11.197) in the second term of the Lagrangian given in Eq. (11.195), one obtains the Lagrangians for different meson–meson interactions. Here, for the sake of completeness, we write down the Lagrangians for the $\pi^+\pi^+ \to \pi^+\pi^+$ and $\pi^+\pi^- \to K^+K^-$ interactions,

$$\mathcal{L}_{\pi\pi\pi\pi} = \frac{\pi^+ \partial_\mu \pi^- \pi^+ \partial^\mu \pi^-}{2f_\pi^2}, \tag{11.198}$$

$$\mathcal{L}_{\pi\pi KK} = \frac{\pi^+ \pi^- \partial_\mu K^- \partial^\mu K^+}{2f_\pi^2}. \tag{11.199}$$

However, in the real world (as we will see in the next chapter), we need the interaction of these mesons with the gauge bosons. The gauge boson fields are incorporated in the meson–meson Lagrangian by replacing the partial derivative ∂_μ with the covariant derivative D_μ, that is,

$$\mathcal{L}_{MM} = \frac{f_\pi^2}{4}(D_\mu U^\dagger D^\mu U), \tag{11.200}$$

where $D_\mu U$ and $D_\mu U^\dagger$ is written as

$$D_\mu U = \partial_\mu U - ir_\mu U + iUl_\mu, \tag{11.201}$$

$$D_\mu U^\dagger = \partial_\mu U^\dagger + iU^\dagger r_\mu - il_\mu U^\dagger. \tag{11.202}$$

r_μ and l_μ, respectively, represents the right- and left-handed currents, defined in terms of the vector (v_μ) and axial vector (a_μ) fields as

$$l_\mu = \frac{1}{2}(v_\mu - a_\mu), \qquad r_\mu = \frac{1}{2}(v_\mu + a_\mu). \tag{11.203}$$

The vector and axial vector fields are different for the interaction of the different gauge bosons with the meson fields. In the following, we explicitly discuss the interaction of the photon, $W^\pm$ and Z^0 with the meson–meson Lagrangian for the electromagnetic, charged current and neutral current induced processes, respectively.

Interaction with the photon field

Since we know that the electromagnetic interactions are purely vector in nature, the axial vector current does not contribute; thus, the left- and right-handed currents are identical and are expressed as

$$l_\mu = r_\mu = -e\hat{Q}A_\mu, \tag{11.204}$$

where e is the electric charge, A_μ represents the photon field, and $\hat{Q} = \begin{pmatrix} 2/3 & 0 & 0 \\ 0 & -1/3 & 0 \\ 0 & 0 & -1/3 \end{pmatrix}$

represents the charge of u, d, s quarks.

Using Eqs. (11.204), (11.201), and (11.202) in Eq. (11.200), one obtains the Lagrangian for the mesons interacting with the photons. For example, here we write the Lagrangians for the $\gamma\pi^+ \rightarrow \pi^+$ and $\pi^+\pi^- \rightarrow \gamma\gamma$ processes, as

$$\mathcal{L}_{\gamma\pi\pi} = -ie\pi^+\partial_\mu\pi^- \Lambda^\mu, \tag{11.205}$$

$$\mathcal{L}_{\gamma\gamma\pi\pi} = -e^2\pi^+\pi^- A_\mu A^\mu. \tag{11.206}$$

Lagrangians for the different interactions are obtained in a similar manner.

Interaction with $W^\pm$ field

In the case of weak charged and neutral current induced processes, both vector and axial vector fields contribute; thus, the left- and right-handed currents are expressed as

$$l_\mu = -\frac{g}{2}(W_\mu^+ T_+ + W_\mu^- T_-), \qquad r_\mu = 0, \tag{11.207}$$

where $g = \frac{e}{\sin\theta_W}$, θ_W is the Weinberg angle, $W_\mu^\pm$ represents the W boson field and $T_\pm$ is defined as

$$T_+ = \begin{pmatrix} 0 & V_{ud} & V_{us} \\ 0 & 0 & 0 \\ 0 & 0 & 0 \end{pmatrix}, \qquad \text{and} \qquad T_- = \begin{pmatrix} 0 & 0 & 0 \\ V_{ud} & 0 & 0 \\ V_{us} & 0 & 0 \end{pmatrix}, \tag{11.208}$$

with $V_{ud} = \cos\theta_C$ and $V_{us} = \sin\theta_C$ being the elements of the Cabibbo–Kobayashi–Maskawa matrix and θ_C being the Cabibbo angle.

Using Eqs. (11.207), (11.201), and (11.202) in Eq. (11.200), the Lagrangian for the interaction of mesons with W bosons is written. The Lagrangians for $W^+\pi^0 \rightarrow \pi^+$ and $W \rightarrow \pi$ processes are written as:

$$\mathcal{L}_{W\pi\pi} = \frac{ig}{2}V_{ud}\pi^-\partial_\mu\pi^0 W^{+\mu}, \tag{11.209}$$

$$\mathcal{L}_{W\pi} = -\frac{g}{2}f_\pi V_{ud}\partial_\mu\pi^- W^{+\mu}. \tag{11.210}$$

Interaction with Z^0 field

The left- and right-handed currents for the neutral current induced processes are expressed as

$$l_\mu = \left(-\frac{g}{\cos\theta_W} + e\tan\theta_W\right) Z_\mu \frac{\lambda_3}{2}, \qquad r_\mu = g\tan\theta_W \sin\theta_W Z_\mu \frac{\lambda_3}{2}, \quad (11.211)$$

where Z_μ represents the Z boson field and λ_3 is the third component of the Gell-Mann matrices.

Using Eqs. (11.211), (11.201), and (11.202) in Eq. (11.200), the Lagrangian for the interactions of the Z boson with the meson field is obtained. The Lagrangians for the $Z\pi \to \pi$ and $Z \to \pi$ processes are written as:

$$\mathcal{L}_{Z\pi\pi} = \frac{ig}{2\cos\theta_W}(1 - 2\sin^2\theta_W)\pi^+\partial_\mu\pi^- Z^\mu, \qquad (11.212)$$

$$\mathcal{L}_{Z\pi} = -\frac{g}{2\cos\theta_W}f_\pi\partial_\mu\pi^0 Z^\mu. \qquad (11.213)$$

11.4.7 Lagrangian for the meson–baryon–gauge boson interaction

The Lagrangian for the meson–baryon interaction is given as the first term of Eq. (11.195), where V_μ and A_μ are defined in terms of a unitary matrix u given in Eq. (11.194) for the SU(2) symmetry. The modification of u by the following expression:

$$u(x) = e^{\frac{i\vec{\Phi}\cdot\vec{\lambda}}{2f_\pi}}, \qquad (11.214)$$

leads to the meson–baryon Lagrangian in the SU(3) symmetry. The Lagrangian for the meson–baryon interaction can be rewritten as

$$\mathcal{L}_{MB} = \text{Tr}[\bar{B}(i\slashed{D} - M)B] - \frac{D}{2}\text{Tr}[\bar{B}\gamma_\mu\gamma_5\{u^\mu, B\}] - \frac{F}{2}\text{Tr}[\bar{B}\gamma_\mu\gamma_5[u^\mu, B]], \qquad (11.215)$$

where M represents the baryon mass. D and F are the axial vector coupling constants for the baryon octet obtained from the analysis of the semileptonic decays of neutron and hyperons. B represents the baryon field, defined as

$$B(x) = \sum_{i=1}^{8} \frac{1}{\sqrt{2}}b_i\lambda_i = \frac{1}{\sqrt{2}}\begin{pmatrix} b_3 + \frac{1}{\sqrt{3}}b_8 & b_1 - ib_2 & b_4 - ib_5 \\ b_1 + ib_2 & -b_3 + \frac{1}{\sqrt{3}}b_8 & b_6 - ib_7 \\ b_4 + ib_5 & b_6 + ib_7 & -\frac{2}{\sqrt{3}}b_8 \end{pmatrix}$$

$$= \begin{pmatrix} \frac{1}{\sqrt{2}}\Sigma^0 + \frac{1}{\sqrt{6}}\Lambda & \Sigma^+ & p \\ \Sigma^- & -\frac{1}{\sqrt{2}}\Sigma^0 + \frac{1}{\sqrt{6}}\Lambda & n \\ \Xi^- & \Xi^0 & -\frac{2}{\sqrt{6}}\Lambda \end{pmatrix}, \qquad (11.216)$$

with b_i being the component of the baryon field. The quantities $D_\mu B$ and u^μ are defined as

$$D_\mu B = \partial_\mu B + [\Gamma_\mu, B], \qquad (11.217)$$

$$u^\mu = i[u^\dagger(\partial^\mu - ir^\mu)u - u(\partial^\mu - il^\mu)u^\dagger], \qquad (11.218)$$

with

$$\Gamma_\mu = \frac{1}{2}[u^\dagger(\partial^\mu - ir^\mu)u - u(\partial^\mu - il^\mu)u^\dagger].$$ (11.219)

u is defined in Eq. (11.194). r_μ and l_μ, respectively, represents the right- and left-handed currents, defined in terms of the vector (v_μ) and axial vector (a_μ) fields and are given in Eq. (11.203). These currents represent the interaction of the gauge bosons with the meson–baryon Lagrangian. Thus, the vector and axial vector fields are different for the interaction of the different gauge bosons. In the following sections, we present the Lagrangians for the interaction of the photon, $W^\pm$, and Z^0 bosons with the mesons and baryons for the electromagnetic, charged current, and neutral current induced processes, respectively.

Interaction with the photon field

The left- and right-handed currents are given in Eq. (11.204) using Eqs. (11.204) and (11.216)–(11.219) in Eq. (11.215). The Lagrangian for the interaction of mesons and baryons among themselves and with the photon (γ) fields is obtained. For example, here we write the Lagrangians for the $\gamma p \to p$, $p \to p\pi^0$ and $\gamma p \to n\pi^+$ processes, as

$$\mathcal{L}_{\gamma pp} = -e\,\bar{p}\gamma_\mu p\,A^\mu,$$ (11.220)

$$\mathcal{L}_{pp\pi^0} = -\frac{(D+F)}{2f_\pi}\bar{p}\gamma_\mu\gamma_5 p\,\partial^\mu\pi^0,$$ (11.221)

$$\mathcal{L}_{\gamma pn\pi^+} = \frac{ie}{\sqrt{2}f_\pi}(D+F)\,\bar{n}\gamma_\mu\gamma_5 p\,A^\mu\pi^-.$$ (11.222)

In a similar manner, one may obtain the Lagrangians for the different possible interactions of the meson–baryon system with the photon field.

Interaction with $W^\pm$ field

In the case of weak charged current induced processes, the left- and right-handed currents are expressed as in Eq. (11.207). Using Eqs. (11.207) and (11.216)–(11.219) in Eq. (11.215), the Lagrangian for the interaction of the W bosons with the meson–baryon system is obtained. The Lagrangians for the processes $W^+n \to p$ and $W^+n \to p\pi^0$ are written as:

$$\mathcal{L}_{Wnp} = -\frac{g}{2\sqrt{2}}V_{ud}\,\bar{p}[\gamma_\mu - (D+F)\gamma_\mu\gamma_5]n\,W^{+\mu},$$ (11.223)

$$\mathcal{L}_{Wnp\pi^0} = -\frac{ig}{2\sqrt{2}f_\pi}V_{ud}\,\bar{p}[\gamma_\mu - (D+F)\gamma_\mu\gamma_5]n\,W^{+\mu}.$$ (11.224)

Interaction with Z^0 field

The expressions for the left- and right-handed currents in the case of neutral current induced processes are given in Eq. (11.211). The Lagrangian for the interactions of the Z-boson with the meson–baryon system is obtained using Eqs. (11.211) and (11.216)–(11.219) in Eq. (11.215).

The Lagrangians for the interactions $Zp \to p$ and $Zp \to n\pi^+$ processes are written as:

$$\mathcal{L}_{Zpp} = -\frac{g}{4\cos\theta_W}\, \bar{p}[\gamma_\mu(1 - 2\sin^2\theta_W) - (D + F)\gamma_\mu\gamma_5]p\, Z^\mu, \qquad (11.225)$$

$$\mathcal{L}_{Zpn\pi^+} = \frac{ig}{2\sqrt{2}f_\pi\cos\theta_W}\, \bar{p}[\gamma_\mu - (1 - 2\sin^2\theta_W)(D + F)\gamma_\mu\gamma_5]n\, \pi^+ Z^\mu. \tag{11.226}$$

The Lagrangians which we have discussed in this chapter are used to write the matrix element for specific processes. The matrix element for the different processes and the differential scattering cross sections are obtained in the next chapter.

Chapter **12**

Neutrino Scattering from Hadrons: Inelastic Scattering (II)

12.1 Introduction

The study of the various inelastic processes induced by photons, electrons, pions, and (anti) neutrinos from nucleons is very important as it provides information about the excitation mechanism of nucleons. This enables us to investigate the structure of nucleons as a composite of quarks and the role of gluons in the quark–quark forces using quantum chromodynamics (QCD). The inelastic processes induced by (anti)neutrinos are a unique source to determine the axial vector aspects of the nucleon structure and relate it to the pion physics. Moreover, in recent years, inelastic reactions induced by (anti)neutrinos leading to the production of mesons like pions, kaons, and ηs have become more relevant in the search for proton decay and for neutrino oscillations.

Within the standard model, proton stability is associated with baryon number conservation, as the proton being the lightest baryon, cannot decay into any other baryon when it is in the free state. It is believed that the baryon number conservation law is not a fundamental law; and there are models of grand unified theory (GUT) [503], which predict the lifetime of the proton to be of the order of 10^{31} years. To estimate a proton's lifetime, some calculations have also been performed using supersymmetry (SUSY) GUTs like SUSY SU(5) and SUSY SO(10). Generally, SUSY GUT models favor kaons (K^0 or K^+) in the final state if a proton decays. However, some of these models also predict decay modes where an eta meson is produced in the final state [504]. In the minimal SU(5) GUT, the predicted proton lifetime and decay to $e^+\pi^0$ is $10^{31\pm1}$ years, which has been ruled out by IMB [505], Kamiokande [506, 507], and Super-Kamiokande [508]. Although these experiments have not observed any event for the decay $p \rightarrow e^+\pi^0$, they have provided the limit for the proton's lifetime to be $10^{33\pm1}$ years. The best limit on the proton's lifetime comes from Super-Kamiokande [509] for the channel $p \rightarrow \nu K^+ > 5.9 \times 10^{33}$ years. Soudan-2 [510] looked for eta in the final state in the proton decay searches and found the best limit for the channels $p \rightarrow \mu^+ \eta$ and $p \rightarrow e^+ \eta$ to be 8.9×10^{31} years and 8.1×10^{31} years, respectively.

The search for proton decay had been going on for a long time in the deep underground experiment at the Kolar Gold Field (KGF) mines in India and simultaneously in the Homestake mines in South Dakota, which were started to study atmospheric neutrinos. No evidence has ever been found for proton decay but the best limits given by KGF and Homestake mines are 9×10^{30} years and $\sim 10^{31}$ years, respectively. In all these experiments of proton decay searches, atmospheric neutrino induced meson production is a significant part of the background. Therefore, it is important to estimate the cross sections for the neutrino interactions off the nucleon target in the few GeV energy region beyond the center of mass (CM) energy, $W > 1.2$ GeV, where inelastic processes are significant.

In the case of neutrino oscillation experiments, the inelastic production of pions is a major source of background through the decay of pions in charged leptons and photons. In the analysis of these experiments, the signature of oscillations depends on the yield of charged leptons like $\mu^{\mp}$, which are identified by their tracks and $e^{\mp}$ which are identified by the Bremsstrahlung photons produced by $e^{\mp}$. In the presence of inelastic channels, the produced $\pi^{\pm}$s decay into $\mu^{\pm}$ and the π^0s decay into two photons, which add to the background of genuine charged lepton production in the analysis of neutrino oscillation experiments. A similar situation arises at higher energies, that is, $E_{\nu/\bar{\nu}} \geq 1$ GeV where kaons and η mesons are also produced. In view of these considerations, the inelastic production of pions, kaons, and η mesons have become very important. In the following sections, we will study the production of single pions, kaons, and eta mesons and the associated production of kaons.

12.2 Single Pion Production

The various possible reactions which contribute to the single pion production induced by (anti)neutrinos on a nucleon target through charged current(CC) induced processes are the following:

$$\nu_l(k) + p(p) \rightarrow l^-(k') + p(p') + \pi^+(k_\pi) \qquad \bar{\nu}_l(k) + n(p) \rightarrow l^+(k') + n(p') + \pi^-(k_\pi)$$
$$\nu_l(k) + n(p) \rightarrow l^-(k') + n(p') + \pi^+(k_\pi) \qquad \bar{\nu}_l(k) + p(p) \rightarrow l^+(k') + p(p') + \pi^-(k_\pi)$$
$$\nu_l(k) + n(p) \rightarrow l^-(k') + p(p') + \pi^0(k_\pi) \qquad \bar{\nu}_l(k) + p(p) \rightarrow l^+(k') + n(p') + \pi^0(k_\pi).$$

$$(12.1)$$

The neutral current(NC) induced processes are:

$$\nu_l(k) + p(p) \rightarrow \nu_l(k') + n(p') + \pi^+(k_\pi) \qquad \bar{\nu}_l(k) + p(p) \rightarrow \bar{\nu}_l(k') + p(p') + \pi^0(k_\pi)$$
$$\nu_l(k) + p(p) \rightarrow \nu_l(k') + p(p') + \pi^0(k_\pi) \qquad \bar{\nu}_l(k) + p(p) \rightarrow \bar{\nu}_l(k') + n(p') + \pi^+(k_\pi)$$
$$\nu_l(k) + n(p) \rightarrow \nu_l(k') + n(p') + \pi^0(k_\pi) \qquad \bar{\nu}_l(k) + n(p) \rightarrow \bar{\nu}_l(k') + n(p') + \pi^0(k_\pi)$$
$$\nu_l(k) + n(p) \rightarrow \nu_l(k') + p(p') + \pi^-(k_\pi) \qquad \bar{\nu}_l(k) + n(p) \rightarrow \bar{\nu}_l(k') + p(p') + \pi^-(k_\pi),$$

$$(12.2)$$

where $l = e, \mu, \tau$. The quantities in the parentheses represent the four momenta of the corresponding particles.

Single pion production induced by neutrinos and antineutrinos was the first inelastic process to be studied in detail using the methods for analyzing photo- and electro- production of pions induced by electromagnetic currents which are vector currents. The weak vector current contribution to (anti)neutrino induced pion production is obtained in terms of the amplitudes of the electroproduction of pions assuming the hypothesis of conserved vector current (CVC). In the case of axial vector current, the hypothesis of the partially conserved axial vector current (PCAC) and other symmetry properties of the axial vector current are used to calculate its contribution to the inelastic pion production.

Various methods have been used in the last 50 years to calculate the single pion production such as the dynamical model with dispersion relations [462, 463, 464, 294, 465, 466, 511, 512], quark models with higher symmetry [471, 513, 492, 514, 515, 516], phenomenological Lagrangians [473, 517, 518], and effective Lagrangians based on the chiral symmetry [481, 482, 483, 485, 484, 487]. In the following, we illustrate the salient features of the method based on the effective Lagrangian with chiral symmetry which has been outlined in Chapter 11, to calculate the processes shown in Eqs. (12.1) and (12.2) and extend it to SU(3) to describe the production of kaons and η particles.

In the following sections, charged and neutral current induced single pion production on a free nucleon target is discussed. The contribution of the non-resonant background terms and contributions from the spin $1/2$ and $3/2$ resonances excitations are discussed separately.

12.2.1 Charged current

The differential scattering cross section for the charged current induced process mentioned in Eq. (12.1), may be written as

$$d\sigma = \frac{1}{4(2\pi)^5 k \cdot p} \frac{d\vec{k}'}{(2E')} \frac{d\vec{p}'}{(2E_p')} \frac{d\vec{p}_m}{(2E_m)} \delta^4(k + p - k' - p' - k_\pi) \overline{\sum}\sum |\mathcal{M}|^2, \quad (12.3)$$

where E', E_p' represent the energy of the outgoing lepton and nucleon, respectively and M is the nucleon mass. In the case of pion production, $E_m = E_\pi$ and $\vec{p}_m = \vec{p}_\pi$, where E_π and $\vec{p}_\pi$, respectively, are the energy and three-momentum of the outgoing pion. $\overline{\sum}\sum |\mathcal{M}|^2$ is the square of the transition amplitude averaged (summed) over the spins of the initial (final) state. The transition amplitude is expressed as

$$\mathcal{M} = \frac{G_F}{\sqrt{2}} \cos\theta_C \, l_\mu j^\mu, \quad (12.4)$$

where θ_C is the Cabibbo angle, l_μ and j^μ are the leptonic and hadronic currents, respectively, and G_F is the Fermi coupling constant. The weak leptonic current is written as

$$l_\mu = \bar{u}(k')\gamma_\mu(1 \mp \gamma_5)u(k), \quad (12.5)$$

where negative (positive) sign is for the (anti)neutrino induced processes. j^μ describes the total hadronic matrix element for the $W^\pm + N \rightarrow N' + \pi$ interaction for which both the non-resonant background terms and the resonance excitations followed by their decay to $N\pi$ contribute; they are obtained using an effective Lagrangian for $W^i + N \rightarrow N' + \pi$ interactions. The Feynman diagrams which may contribute to the matrix element of the hadronic current are shown in Figure 12.1. The non-resonant background terms include five diagrams viz., direct nucleon pole (NP), cross nucleon pole (CP), contact term (CT), pion pole (PP), and pion in flight (PF) terms. For $P_{33}(1232)$ resonance, we have included both direct (s channel) and cross (u channel) diagrams. Apart from the $P_{33}(1232)$ resonance, we have also included contributions from $P_{11}(1440)$, $S_{11}(1535)$ and, $S_{11}(1650)$ spin half resonances and $D_{13}(1520)$ and $P_{13}(1720)$ spin 3/2 resonances and considered both s channel and u channel contributions. In the following sections, we will write the matrix elements for the non-resonant diagrams using the chiral Lagrangian given in Eqs. (11.220)–(11.226) and for the resonance excitations and their subsequent decay using the formalism given in Chapter 11.

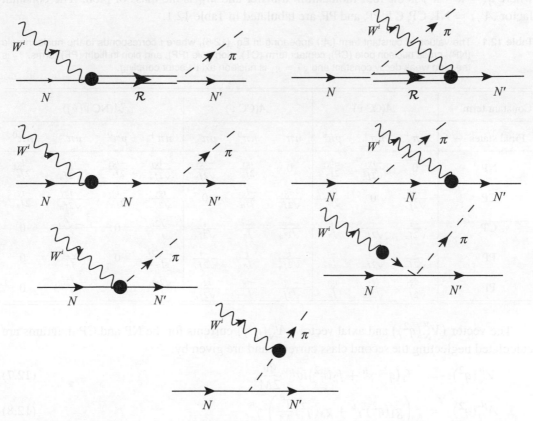

Figure 12.1 Feynman diagrams contributing for the process $W^i N \rightarrow N' \pi^{\pm,0}$, where $(W^i \equiv W^\pm ; i = \pm)$ for the charged current processes and $(W^i \equiv Z^0 ; i = 0)$ for the neutral current processes with $N, N' = p$ or n. The first row represents the direct and cross diagrams for the resonance production where R stands for the different resonances. The second row represents the nucleon and cross nucleon terms. The contact and pion pole terms are shown in the third row and the last row represents the pion in flight terms.

Non-resonant background contribution

The contribution from the non-resonant background terms may be obtained using the non-linear sigma model discussed in Chapter 11. They are expressed as [481]

$$
\begin{aligned}
j^\mu\big|_{NP} &= \mathcal{A}^{NP}\bar{u}(p')\,\slashed{k}_\pi\gamma_5\frac{\slashed{p}+\slashed{q}+M}{(p+q)^2-M^2+i\epsilon}\left[V_N^\mu(q^2)-A_N^\mu(q^2)\right]u(p), \\
j^\mu\big|_{CP} &= \mathcal{A}^{CP}\bar{u}(p')\left[V_N^\mu(q^2)-A_N^\mu(q^2)\right]\frac{\slashed{p}'-\slashed{q}+M}{(p'-q)^2-M^2+i\epsilon}\,\slashed{k}_\pi\gamma_5 u(p), \\
j^\mu\big|_{CT} &= \mathcal{A}^{CT}\bar{u}(p')\gamma^\mu\left(g_1 f_{CT}^V(q^2)\gamma_5 - f_\rho\left((q-k_\pi)^2\right)\right)u(p), \\
j^\mu\big|_{PP} &= \mathcal{A}^{PP}f_\rho\left((q-k_\pi)^2\right)\frac{q^\mu}{m_\pi^2-q^2}\bar{u}(p')\,\slashed{q}\,u(p), \\
j^\mu\big|_{PF} &= \mathcal{A}^{PF}f_{PF}(q^2)\frac{(2k_\pi-q)^\mu}{(k_\pi-q)^2-m_\pi^2}2M\bar{u}(p')\gamma_5 u(p),
\end{aligned}
\tag{12.6}
$$

where $q(=k-k')$ is the four-momentum transfer and m_π is the mass of pion. The constant factor $\mathcal{A}^i$, $i =$ NP, CP, CT, PP, and PF, are tabulated in Table 12.1.

Table 12.1   The values of constant term ($\mathcal{A}^i$) appearing in Eq. (12.6), where i corresponds to the nucleon pole (NP), cross nucleon pole (CP), contact term (CT), pion pole (PP), and pion in flight (PF) terms. f_π is the pion weak decay constant and $g_A = g_1$ is nucleon axial vector coupling.

Constant term →	$\mathcal{A}(CC\ \nu)$			$\mathcal{A}(CC\ \bar{\nu})$			$\mathcal{A}(NC\ \nu(\bar{\nu}))$			
Final states →	$p\pi^+$	$n\pi^+$	$p\pi^0$	$n\pi^-$	$n\pi^0$	$p\pi^-$	$n\pi^+$	$p\pi^0$	$p\pi^-$	$n\pi^0$
NP	0	$\frac{-ig_1}{\sqrt{2}f_\pi}$	$\frac{-ig_1}{2f_\pi}$	0	$\frac{ig_1}{2f_\pi}$	$\frac{-ig_1}{\sqrt{2}f_\pi}$	$\frac{-ig_1}{\sqrt{2}f_\pi}$	$\frac{-ig_1}{2f_\pi}$	$\frac{ig_1}{\sqrt{2}f_\pi}$	$\frac{-ig_1}{2f_\pi}$
CP	$\frac{-ig_1}{\sqrt{2}f_\pi}$	0	$\frac{ig_1}{2f_\pi}$	$\frac{-ig_1}{\sqrt{2}f_\pi}$	$\frac{-ig_1}{2f_\pi}$	0	$\frac{ig_1}{\sqrt{2}f_\pi}$	$\frac{-ig_1}{2f_\pi}$	$\frac{-ig_1}{\sqrt{2}f_\pi}$	$\frac{-ig_1}{2f_\pi}$
CT	$\frac{-i}{\sqrt{2}f_\pi}$	$\frac{i}{\sqrt{2}f_\pi}$	$\frac{i}{f_\pi}$	$\frac{-i}{\sqrt{2}f_\pi}$	$\frac{-i}{f_\pi}$	$\frac{i}{\sqrt{2}f_\pi}$	$\frac{\sqrt{2}i}{f_\pi}$	0	$\frac{-\sqrt{2}i}{f_\pi}$	0
PP	$\frac{i}{\sqrt{2}f_\pi}$	$\frac{-i}{\sqrt{2}f_\pi}$	$\frac{-i}{f_\pi}$	$\frac{i}{\sqrt{2}f_\pi}$	$\frac{i}{f_\pi}$	$\frac{-i}{\sqrt{2}f_\pi}$	$\frac{-\sqrt{2}i}{f_\pi}$	0	$\frac{\sqrt{2}i}{f_\pi}$	0
PF	$\frac{-ig_1}{\sqrt{2}f_\pi}$	$\frac{ig_1}{\sqrt{2}f_\pi}$	$\frac{ig_1}{f_\pi}$	$\frac{-ig_1}{\sqrt{2}f_\pi}$	$\frac{-ig_1}{f_\pi}$	$\frac{ig_1}{\sqrt{2}f_\pi}$	$\frac{\sqrt{2}ig_1}{f_\pi}$	0	$\frac{-\sqrt{2}ig_1}{f_\pi}$	0

The vector $(V_N^\mu(q^2))$ and axial vector $(A_N^\mu(q^2))$ currents for the NP and CP diagrams are calculated neglecting the second class currents and are given by:

$$
V_N^\mu(q^2) = f_1(q^2)\gamma^\mu + f_2(q^2)i\sigma^{\mu\nu}\frac{q_\nu}{2M},
\tag{12.7}
$$

$$
A_N^\mu(q^2) = \left(g_1(q^2)\gamma^\mu + g_3(q^2)\frac{q^\mu}{M}\right)\gamma^5,
\tag{12.8}
$$

where $f_{1,2}(q^2)$ and $g_{1,3}(q^2)$ are the vector and axial vector form factors for the nucleons. The isovector form factors viz. $f_{1,2}(q^2)$ are expressed as:

$$
f_{1,2}(q^2) = F_{1,2}^p(q^2) - F_{1,2}^n(q^2),
\tag{12.9}
$$

where $F_i^{p,n}(q^2)$; $i = 1, 2$ are the Dirac-Pauli form factors of nucleons. These form factors are, in turn, expressed in terms of the experimentally determined Sachs electric $G_E^{p,n}(q^2)$ and magnetic $G_M^{p,n}(q^2)$ form factors [414], which have been discussed in Chapter 10.

The axial vector form factor $(g_1(q^2))$ is generally taken to be of dipole form and is given by

$$g_1(q^2) = g_A(0) \left[1 - \frac{q^2}{M_A^2}\right]^{-2}, \tag{12.10}$$

where the values of $g_A(0)$ and M_A are quoted in Chapter 10.

The pseudoscalar form factor $g_3(q^2)$ is related to $g_1(q^2)$, assuming PCAC and the pion pole dominance, through the relation

$$g_3(q^2) = \frac{2M^2 \, g_1(q^2)}{m_\pi^2 - q^2}. \tag{12.11}$$

In order to conserve vector current at the weak vertex, the two form factors viz. $f_{PF}(q^2)$ and $f_{CT}^V(q^2)$ are expressed in terms of the isovector nucleon form factor as [481]

$$f_{PF}(q^2) = f_{CT}^V(q^2) = 2f_1^V(q^2). \tag{12.12}$$

The form factor $(f_\rho(q^2))$ is taken to be of the monopole form [481]:

$$f_\rho(q^2) = \frac{1}{1 - q^2/m_\rho^2}; \qquad \text{with} \qquad m_\rho = 0.776 \text{ GeV}. \tag{12.13}$$

In order to be consistent with the assumption of PCAC, $f_\rho(q^2)$ has also been used with the axial vector part of the contact term.

Resonance contribution

The basic (anti)neutrino induced reactions for pion production through resonance excitations are the following:

$$\nu_l(k) + N(p) \rightarrow l^-(k') + R(p_R)$$
$$\hookrightarrow N'(p') + \pi(k_\pi), \tag{12.14}$$

$$\bar{\nu}_l(k) + N(p) \rightarrow l^+(k') + R(p_R)$$
$$\hookrightarrow N'(p') + \pi(k_\pi), \tag{12.15}$$

where R stands for any resonance which contributes to the pion production. In the next section, we demonstrate by including the different positive and negative parity, spin $1/2$ and $3/2$ resonances. We have included six resonances, viz., $P_{33}(1232)$, $P_{11}(1440)$, $D_{13}(1520)$,

$S_{11}(1535)$, $S_{11}(1650)$, and $P_{13}(1720)$. However, it may be generalized to any set of resonances with the same parity and spin. The resonances which have been considered here for demonstration are summarized in Tables 11.1 and 11.2 of Chapter 11 with their properties like mass, decay width, etc. We write the matrix elements for the resonance excitations for spin $1/2$ and $3/2$ resonances using the matrix element for the purely weak excitation $\nu N \rightarrow l^- R_{\frac{1}{2}(\frac{3}{2})}$ given in Eqs. (11.2) and (11.76) and the matrix elements for the strong decay process $R_{\frac{1}{2}(\frac{3}{2})} \rightarrow N\pi$ using the Lagrangian given in Eqs. (11.34) and (11.89) for spin $1/2$ ($R_{\frac{1}{2}}$) and spin $3/2$ ($R_{\frac{3}{2}}$) resonances.

(i) Spin $\frac{1}{2}$ resonances

In Chapter 11, we have already discussed in detail resonance excitation, where the matrix element, vector and axial vector hadronic form factors, and their determination by the assumption of CVC and PCAC hypotheses are explained. Using the inputs from Chapter 11, the most general form of the hadronic currents for the s channel (direct diagram) and u channel (cross diagram) processes, shown in Figure 12.1, where a resonant state $R_{1/2}$ is produced and decays to a pion in the final state, are written as

$$j^\mu\Big|_R^{\frac{1}{2}} = i\, \mathcal{C}^\mathcal{R} \bar{u}(p') k_\pi \Gamma \frac{\not{p} + \not{q} + M_R}{(p+q)^2 - M_R^2 + iM_R\Gamma_R} \Gamma_{\frac{1}{2}}^\mu u(p), \tag{12.16}$$

$$j^\mu\Big|_{CR}^{\frac{1}{2}} = i\, \mathcal{C}^\mathcal{R} \bar{u}(p') \Gamma_{\frac{1}{2}}^\mu \frac{\not{p}' - \not{q} + M_R}{(p'-q)^2 - M_R^2 + iM_R\Gamma_R} k_\pi \Gamma u(p), \tag{12.17}$$

where M_R and Γ_R represent the mass and decay width of the resonance, respectively. $\Gamma = 1$ (γ_5) for the positive (negative) parity resonances, $\Gamma_{1/2}^\mu$ for the positive (negative) parity resonances are defined in Chapter 11 (Eqs. (11.5) and (11.6)), and $\mathcal{C}^\mathcal{R}$ is a constant which includes the coupling strength in terms of f_π, isopin factor involved in $R_{1/2} \rightarrow N\pi$ transition, etc. The constant $\mathcal{C}^\mathcal{R}$ for the various channels have been tabulated in Table 12.2. The formalism for determining the coupling strength, $f_{NR_{1/2}\pi}$ for the $R_{1/2} \rightarrow N\pi$ transitions in terms of the decay width has already been discussed in Chapter 11.

(ii) Spin $\frac{3}{2}$ resonances

The matrix element as well as the vector and axial vector transition form factors for the spin $3/2$ resonances were discussed in Chapter 11. One may write the most general form of the hadronic current for the s channel (direct diagram) and u channel (cross diagram) processes where a resonant state $R_{3/2}$ is produced and decays to a pion in the final state as

$$j^\mu\Big|_R^{\frac{3}{2}} = i\, \mathcal{C}^\mathcal{R} \bar{u}(p') k_\pi^\alpha \Gamma \frac{\not{p}_R + M_R}{p_R^2 - M_R^2 + iM_R\Gamma_R} \mathcal{P}_{\alpha\beta}(p_R) \Gamma_{\frac{3}{2}}^{\beta\mu}(p,q) u(p), \quad p_R = p + q, \tag{12.18}$$

$$j^\mu\Big|_{CR}^{\frac{3}{2}} = i\, \mathcal{C}^\mathcal{R} \bar{u}(p') \hat{\Gamma}_{\frac{3}{2}}^{\mu\alpha}(p',-q) \frac{\not{p}_R + M_R}{p_R^2 - M_R^2 + iM_R\Gamma_R} P_{\alpha\beta}^{3/2}(p_R) k_\pi^\beta \Gamma u(p), \quad p_R = p' - q. \tag{12.19}$$

In Eqs. (12.18) and (12.19), $\Gamma = 1$ (γ_5) is for the positive (negative) parity resonances, $\hat{\Gamma}_{\frac{3}{2}}^{\mu\alpha} = \gamma_0 \Gamma_{\frac{3}{2}}^{\mu\alpha\dagger} \gamma_0$. The projector operator $P_{\alpha\beta}^{3/2}(p_R)$ and $\Gamma_{\frac{3}{2}}^{\mu\nu}$ for the positive (negative) parity resonances are defined in Chapter 11. $\mathcal{C}^R$ is the coupling for the $R_{3/2} \to N\pi$ transitions and has been tabulated in Table 12.2. The coupling strength, $f_{NR_{3/2}\pi}$ for the $R_{3/2} \to N\pi$ transitions in terms of the decay width have already been discussed in Chapter 11.

Table 12.2 Coupling constant ($\mathcal{C}^R$) for spin $\frac{1}{2}$ and spin $\frac{3}{2}$ resonances in the case of charged current induced processes. Here, f^* stands for $\tilde{R}_{1/2} \to N\pi$ coupling which for $\Delta(1232)$ resonance is $f_{N\Delta\pi}$ and is $f_{NR_{1/2}\pi}(f_{NR_{3/2}\pi})$ for spin $\frac{1}{2}(\frac{3}{2})$ resonances.

Process	$\mathcal{C}^R(\text{CC } \nu)$			$\mathcal{C}^R(\text{CC } \bar{\nu})$		
	$p\pi^+$	$n\pi^+$	$p\pi^0$	$n\pi^-$	$n\pi^0$	$p\pi^-$
$P_{33}(1232)$	$\frac{\sqrt{3}f^*}{m_\pi}$	$\sqrt{\frac{1}{3}}\frac{f^*}{m_\pi}$	$-\sqrt{\frac{2}{3}}\frac{f^*}{m_\pi}$	$\frac{\sqrt{3}f^*}{m_\pi}$	$\sqrt{\frac{2}{3}}\frac{f^*}{m_\pi}$	$\sqrt{\frac{1}{3}}\frac{f^*}{m_\pi}$
$P_{11}(1440)$	0	$\sqrt{2}\frac{f^*}{m_\pi}$	$\frac{f^*}{m_\pi}$	0	$-\frac{f^*}{m_\pi}$	$\sqrt{2}\frac{f^*}{m_\pi}$
$D_{13}(1520)$	0	$\sqrt{2}\frac{f^*}{m_\pi}$	$\frac{f^*}{m_\pi}$	0	$-\frac{f^*}{m_\pi}$	$\sqrt{2}\frac{f^*}{m_\pi}$
$S_{11}(1535)$	0	$\sqrt{2}\frac{f^*}{m_\pi}$	$\frac{f^*}{m_\pi}$	0	$-\frac{f^*}{m_\pi}$	$\sqrt{2}\frac{f^*}{m_\pi}$
$S_{11}(1650)$	0	$\sqrt{2}\frac{f^*}{m_\pi}$	$\frac{f^*}{m_\pi}$	0	$-\frac{f^*}{m_\pi}$	$\sqrt{2}\frac{f^*}{m_\pi}$
$P_{13}(1720)$	0	$\sqrt{2}\frac{f^*}{m_\pi}$	$\frac{f^*}{m_\pi}$	0	$-\frac{f^*}{m_\pi}$	$\sqrt{2}\frac{f^*}{m_\pi}$

12.2.2 Neutral current

In this section, we will briefly discuss the single pion production induced by neutral currents. The differential scattering cross section for the neutral current induced process mentioned in Eq. (12.2) is given in Eq. (12.3). The transition amplitude is expressed as

$$\mathcal{M} = \frac{G_F}{\sqrt{2}} \, l_\mu j^\mu, \tag{12.20}$$

where the weak leptonic current is given in Eq. (12.5). j^μ describes the total hadronic matrix element for the neutral current induced pion production; both the non-resonant background terms and the resonance excitation followed by their decay in $N\pi$ contribute. The same analogy as we have done for the charged current induced pion production to obtain the non-resonant and resonant contributions is followed.

In the following sections, we present the formalism in brief which can be used for the non-resonant background terms and the resonant contributions to the single pion production processes in the neutral current sector.

Non-resonant background contribution

The contribution from the non-resonant background terms in the case of neutral current induced reactions $ZN \to N'\pi$ is obtained using the non-linear sigma model. The hadronic currents for the different diagrams presented in Figure 12.1 are given in Eq. (12.6) with the values of the constants $\mathcal{A}^i$ tabulated in Table 12.1.

In the absence of second class currents, the vector $(V_N^\mu(q^2))$ and the axial vector $(A_N^\mu(q^2))$ currents for the neutral current induced nucleon pole diagrams are given by,

$$V_N^\mu(q^2) = \tilde{f}_1(q^2)\gamma^\mu + \tilde{f}_2(q^2)i\sigma^{\mu\nu}\frac{q_\nu}{2M}, \tag{12.21}$$

$$A_N^\mu(q^2) = \left(\tilde{g}_1(q^2)\gamma^\mu + \tilde{g}_3(q^2)\frac{q^\mu}{M}\right)\gamma^5, \tag{12.22}$$

where $\tilde{f}_{1,2}(q^2)$ and $\tilde{g}_{1,3}(q^2)$ are the vector and axial vector neutral current form factors for the nucleons. The vector form factors are expressed as:

$$\tilde{f}_{1,2}(q^2) \xrightarrow{\text{for p}} \tilde{f}_{1,2}^p(q^2) = \left(\frac{1}{2} - 2\sin^2\theta_W\right)F_{1,2}^p(q^2) - \frac{1}{2}F_{1,2}^n(q^2), \tag{12.23}$$

$$\tilde{f}_{1,2}(q^2) \xrightarrow{\text{for n}} \tilde{f}_{1,2}^n(q^2) = \left(\frac{1}{2} - 2\sin^2\theta_W\right)F_{1,2}^n(q^2) - \frac{1}{2}F_{1,2}^p(q^2). \tag{12.24}$$

where θ_W is the Weinberg angle (Chapter 8). $F_{1,2}^p$ and $F_{1,2}^p$ are the Dirac-Pauli form factors of the nucleon (Chapter 10). On the other hand, the axial vector form factor$(\tilde{g}_1(q^2))$ is given by

$$\tilde{g}_1(q^2) = \tilde{g}_1^{p,n}(0)\left[1 - \frac{q^2}{M_A^2}\right]^{-2}, \tag{12.25}$$

with $\tilde{g}_1^{p,n}(q^2) = \pm\frac{1}{2}g_1(q^2)$, where the plus (minus) sign stands for a proton (neutron) target. The contribution of the pseudoscalar form factor $\tilde{f}_P(q^2)$ vanishes for the neutral current processes as its contribution is proportional to the lepton mass.

The form factors $f_{PF}(q^2)$ and $f_{CT}^V(q^2)$ in the neutral current interaction becomes $\tilde{f}_{PF}(q^2)$ and $\tilde{f}_{CT}^V(q^2)$ and transforms in the same way as the nucleon pole vector form factors do. Thus,

$$\tilde{f}_{PF}(q^2) = \tilde{f}_{CT}^V(q^2) = 2\tilde{f}_1(q^2), \tag{12.26}$$

with $\tilde{f}_1(q^2)$ for the neutral current induced proton and neutron processes given in Eqs. (12.23) and (12.24), respectively. However, the parameterization of f_ρ remains the same as in the case of charged current induced processes given in Eq. (12.13). It can be seen from Table 12.2 that not all the terms contribute in the case of neutral currents (NC) production of π^0 mesons. In all the cases, the contribution from the isoscalar currents arising from the strangeness content of the nucleon are neglected.

Resonant contribution

The basic (anti)neutrino induced reactions for the pion production through the resonance excitations in the neutral current sector are the following:

$$\nu_l(\bar{\nu}_l)(k) + N(p) \to \nu_l(\bar{\nu}_l)(k') + R(p_R).$$
$$\hookrightarrow N'(p') + \pi(k_\pi). \tag{12.27}$$

In the following, we will discuss spin 1/2 and 3/2 resonance excitations and their subsequent decay in $N\pi$.

(i) Spin $\frac{1}{2}$ resonances

The general form of the charged hadronic currents for the s channel (direct diagram) and u channel (cross diagram) processes when a resonance $R_{1/2}$ is produced and the particle subsequently decays to a pion in the final state, are written in Eqs. (12.16) and (12.17), respectively with the constant $\mathcal{C}_R$ given in Table 12.2. In the case of neutral currents (NC) induced pion production, the coefficients $\mathcal{C}_R$ are also given in Table 12.2 for different channels. In the case of weak vertex, the vector form factors appearing in $\Gamma_{\frac{1}{2}}$ are expressed as Eq. (11.125) (Chapter 11) in the vector sector and Eq. (11.126) (Chapter 11), in the axial vector sector. It should be noted from the structure of the vector form factors in the weak neutral current that they depend explicitly on the mixing angle θ_W unlike the weak charged current form factors. We neglect here the strangeness content of the nucleons which would contribute through the strangeness form factors both in the vector and axial vector sectors.

(ii) Spin $\frac{3}{2}$ resonances

In this case also, the most general form of the charged hadronic current for the s channel (direct diagram) and u channel (cross diagram) processes in the case of neutral current induced processes, where a resonant state $R_{3/2}$ is produced and the particle decays to a pion in the final state is given in Eqs. (12.18) and (12.19), respectively. The vector and axial vector transition form factors has already been discussed in Eqs. (11.153) and (11.154) of Chapter 11.

12.2.3 Cross sections

The cross section for the inelastic production of pions is calculated using the matrix elements for the non-resonant and resonance diagrams given in Eqs. (12.6), (12.16), (12.17), (12.18), and (12.19). However, there is ambiguity regarding the phase between various diagrams shown in Figure 12.1, when adding the matrix elements. Since the non-resonant terms are obtained from the Lagrangian derived from the chiral symmetry, the phases are predicted uniquely as written in the equations. But there could be a phase between the non-resonant and the resonance diagrams which could, in principle, be different for each resonance R considered. In the calculations based on the quark model with higher symmetry, there is no ambiguity as the nucleons and higher resonances are predicted from the same model [494] but in the case of effective Lagrangians, the phase is arbitrary, that is, the total matrix element for the hadronic current J^μ is written as

$$\langle j^\mu \rangle = \langle j^\mu \rangle_{NR} + e^{i\phi_i} \langle j^\mu \rangle_{R_i}, \tag{12.28}$$

where ϕ is the phase difference between the diagrams in the first row and the rest of the diagrams in Figure 12.1. However, we add them coherently, that is, $\phi = 0$ for the present purpose. A calculation done with $\phi_i = \pi$, $R_i = \Delta$ (neglecting other resonances) does not reproduce the experimental results [481]. A choice of $\phi_i = \pi/2$ for all i will give the incoherent sum of all the non-resonant and resonance contributions. Adding coherently $j^\mu_{R_i}$ is expressed as

$$j^\mu_{R_i} = j^\mu_{R_s} + j^\mu_{R_u},$$ (12.29)

where R can be any resonance and the net contribution of these resonances, is written as

$$j^\mu_{R_i} = j^\mu_{P_{33}(1232)} + j^\mu_{P_{11}(1440)} + j^\mu_{S_{11}(1535)} + j^\mu_{S_{11}(1650)} + j^\mu_{D_{13}(1520)} + j^\mu_{P_{13}(1720)},$$ (12.30)

with $i = s, u$. j^μ_{NR} for the charged current induced processes is given by

$$j^\mu_{NR} = j^\mu|_{NP} + j^\mu|_{CP} + j^\mu|_{CT} + j^\mu|_{PP} + j^\mu|_{PF}.$$ (12.31)

The cross section is then calculated using Eqs. (12.3), (12.4), and (12.28). The existing experimental data on the single pion production process from (almost) free nucleons are available only from the old bubble chamber experiments performed almost 40 years ago at ANL [458] and BNL [519] with deuteron and hydrogen targets. In view of this, the deuteron effects should also be taken into account.

The deuteron effect can be taken into account by writing the differential scattering cross section as

$$\left(\frac{d\sigma}{dQ^2 dW}\right)_{vd} = \int d\vec{p}^{\,d}_p |\Psi_d(\vec{p}^{\,d}_p)|^2 \frac{M}{E^d_p} \left(\frac{d\sigma}{dQ^2 dW}\right)_{vN},$$ (12.32)

where $p^\mu = (E^d_p, \vec{p}^{\,d}_p)$ is the four-momentum of the proton bound inside the deuteron with $E^d_p (= M_D - \sqrt{M^2 + \vec{p}^{\,2}_n}, \vec{p}_n = -\vec{p}_D)$ as the energy of the off-shell proton inside the deuteron, M_D is the deuteron mass, and $\psi_d(\vec{p}^{\,d}_p)$ is the wave function of the deuteron [520].

The data available from ANL and BNL on $\nu_\mu p \rightarrow \mu^- p \pi^+$ differ with each other by about $30 - 40\%$ which has been attributed in the past to the different flux normalization in these experiments. These data, when used to fix various parameters of theoretical models of reaction mechanisms for the single pion production, give rise to considerable uncertainties in the determination of these parameters, which in turn, lead to higher uncertainties in predicting the single pion production cross section from the nuclear targets. Reanalysis of the old bubble chamber data give a consistent set of data from ANL [458] and BNL [519] experiments either by minimizing the neutrino flux uncertainties [521, 522] or by reconstructing the data using the cross section ratio for the single pion production to the quasielastic processes and observed quasielastic cross sections [523]. It is hoped that the use of reanalyzed/reconstructed data on free nucleon targets will help toward a better understanding of the pion production reaction mechanism. Nevertheless, the need for $\nu(\bar{\nu})$ experiments on free proton/deuteron targets with high statistics and precision is to be emphasized.

In Figure 12.2, we show the results for the total scattering cross section for the process $\nu_\mu p \to \mu^- p \pi^+$. The results are presented in the $\Delta(1232)$ dominance model; the contributions from the non-resonant background terms are included. The results include the deuteron effect which was obtained using Eq. (12.32). It may be observed that the inclusion of the deuteron effect results in an overall reduction of $\sim 4 - 6\%$ in the total scattering cross section. The results are compared with the reanalyzed experimental data of ANL [458] and BNL [519] by Wilkinson et al. [523]. It may be noticed that due to the presence of the non-resonant background terms, there is an increase in the cross section which is about 12% at $E_{\nu_\mu} = 1$ GeV and becomes $\sim 8\%$ at $E_{\nu_\mu} = 2$ GeV. When a cut of $W \leq 1.4$ GeV or $W \leq 1.6$ GeV is applied, the increase in the total scattering cross section at $E_{\nu_\mu} = 1$ GeV is about 10% in both cases.

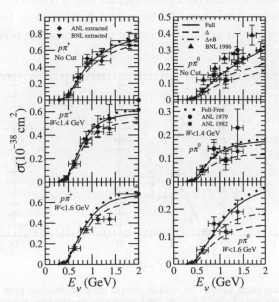

Figure 12.2 Total scattering cross section for the processes $\nu_\mu p \to \mu^- p \pi^+$ (left panel) and $\nu_\mu n \to \mu^- p \pi^0$ (right panel). The dashed line is the result calculated in the $\Delta(1232)$ dominance model, the dash-dotted line represents the result obtained when the non-resonant background terms are included. These two cases have the deuteron effect. The dotted line is the result of the full calculation without the deuteron effect while the solid line represents the results of the full model with the deuteron effect. The results in the top panel are obtained when no cut on the invariant mass is included. The middle (bottom) panel shows the results with a cut on the center of mass energy of 1.4 GeV (1.6 GeV). Diamond and down triangle represents the reanalyzed data of ANL and BNL by Wilkinson et al. [523]. Circle, star, and up triangle represent the original data from ANL [458] and BNL [519] experiments.

For the process $\nu_\mu + n \to \mu^- + p + \pi^0$, there are contributions from the non-resonant background terms as well as from the other higher resonance excitation terms besides the $\Delta(1232)$. The net contribution to the total pion production due to the presence of the non-resonant background terms in $\nu_\mu + n \to \mu^- + p + \pi^0$ reaction results in an increase in the cross section of about 26% at $E_{\nu_\mu}=1$ GeV which becomes 18% at $E_{\nu_\mu}= 2$ GeV. When other higher resonances are also taken into account, there is a further increase in the cross section by about 35% at $E_{\nu_\mu}=1$ GeV which becomes 40% at $E_{\nu_\mu}= 2$ GeV. Thus, we find that the inclusion of higher resonant terms leads to a significant increase in the cross section for the $\nu_\mu + n \rightarrow \mu^- + p + \pi^0$ process. Furthermore, it may also be concluded from these observations that with increase in neutrino energy, contributions from non-resonant background terms decrease while the total scattering cross section increases when the higher resonances are included. When a cut of $W \leq 1.4\text{GeV}$ or $W \leq 1.6\text{GeV}$ on the center of mass energy is applied, then the increase in the total scattering cross section at $E_{\nu_\mu} = 1$ GeV and 2 GeV is about 43% and 13%, respectively.

Figure 12.3 shows the total scattering cross section for the neutral current neutrino induced pion production processes. The experimental points are the data from the ANL experiment [524]. It may be observed that besides the $\Delta(1232)$ resonant term, there is significant contribution from the non-resonant background terms which results in an increase in the total scattering cross section in all the channels. Moreover, there is an $\sim 15\%$ enhancement in the cross sections in $\nu p \rightarrow \nu \pi^+ n$ and $\nu p \rightarrow \nu \pi^- p$ processes when higher resonant terms are included.

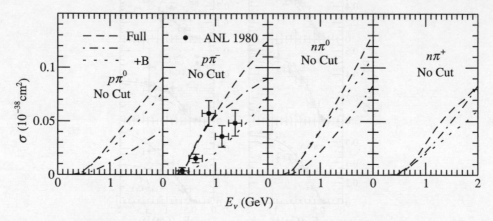

Figure 12.3   Total scattering cross section for neutral current neutrino induced pion production processes with deuteron effect and with no cut on W. The results presented from the left to the right panels are for $\nu p \rightarrow \nu p \pi^0$, $\nu n \rightarrow \nu p \pi^-$, $\nu n \rightarrow \nu n \pi^0$, and $\nu p \rightarrow \nu n \pi^+$ processes. Data points have been taken from ANL [524] experiment. Lines have the same meaning as in Fig. 12.2.

12.3   Eta Production

The (anti)neutrino induced eta production is interesting because of several reasons. Being an isoscalar particle, the η meson is an important probe that can be used to search for the strange quark content of nucleons [525]. A precise determination of the η production cross section would also help in understanding the background in the proton decay searches through the $p \rightarrow \eta e^+$ decays. Therefore, its background contribution due to the atmospheric neutrino interactions should be well estimated. Furthermore, η production is expected to be dominated by $S_{11}(1535)$ resonance excitation as this state appears near the threshold of the $N\eta$ system and has large branching ratio to η decay modes. A precise measurement of the cross section will also allow to determine the axial vector properties of this resonance.

12.3.1 Charged current

In this section, we will discuss the weak production of η mesons via the charged current interactions. The channels that may contribute to the CC-η production (Figure 12.4) are

$$\nu_\mu(k) + n(p) \longrightarrow \mu^-(k') + \eta(p_\eta) + p(p'), \tag{12.33}$$

$$\bar{\nu}_\mu(k) + p(p) \longrightarrow \mu^+(k') + \eta(p_\eta) + n(p'), \tag{12.34}$$

where the quantities in the parenthesis are the four-momenta of the particles.

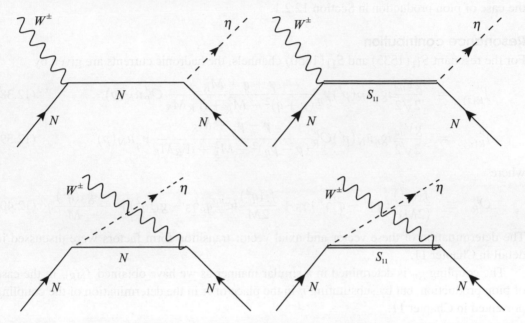

Figure 12.4 Feynman diagrams for the processes $\nu/\bar{\nu}(k) + N(p) \rightarrow \mu^\mp(k') + \eta(p_\eta) + N'(p')$. The first row from left to right: s channel nucleon pole (SC) and S_{11} resonance (SC N^*); the second row: u channel nucleon pole (UC) and S_{11} resonance (UC N^*).

The general expression of the differential scattering cross section for the reaction shown in Eqs. (12.33) and (12.34) in the laboratory frame is given in Eq. (12.3), with $\vec{p}_m = \vec{p}_\eta$ as the three-momentum of the outgoing eta meson and $E_m = E_\eta$, the energy of the eta meson. The transition matrix element, in terms of the leptonic and the hadronic currents, is given in Eq. (12.4). The leptonic current is given in Eq. (12.5); the hadronic current receives contributions from the non-resonant background terms as well as from the resonance excitations and their subsequent decay in $N\eta$.

Non-resonant background contribution

The hadronic currents for the non-resonant background terms, that is, Born diagrams (s and u channels) with nucleon poles, using the non-linear sigma model, is obtained as [488, 489],

$$J^\mu_{N(s)} = \frac{gV_{ud}}{2\sqrt{2}} \frac{D - 3F}{2\sqrt{3}f_\pi} \bar{u}_N(p') \slashed{p}_\eta \gamma^5 \frac{\slashed{p} + \slashed{q} + M}{(p+q)^2 - M^2} \mathcal{O}^\mu_N u_N(p), \tag{12.35}$$

$$J^\mu_{N(u)} = \frac{gV_{ud}}{2\sqrt{2}} \frac{D - 3F}{2\sqrt{3}f_\pi} \bar{u}_N(p') \mathcal{O}^\mu_N \frac{\not{p} - \not{p}_\eta + M}{(p - p_\eta)^2 - M^2} \not{p}_\eta \gamma^5 u_N(p), \tag{12.36}$$

where

$$\mathcal{O}^\mu_N \equiv f_1(Q^2)\gamma^\mu + f_2(Q^2)i\sigma^{\mu\rho}\frac{q_\rho}{2M_N} - g_1(Q^2)\gamma^\mu\gamma^5 - g_3(Q^2)\frac{q^\mu}{M}\gamma^5. \tag{12.37}$$

$f_{1,2}(q^2)$ are the isovector vector form factors and $g_{1,3}(q^2)$ are the axial vector and pseudoscalar form factors, respectively, for the nucleons; they are determined in the same manner as done in the case of pion production in Section 12.2.1.

Resonance contribution

For the resonant $S_{11}(1535)$ and $S_{11}(1650)$ channels, the hadronic currents are given by

$$J^\mu_{R(s)} = \frac{gV_{ud}}{2\sqrt{2}} ig_\eta \bar{u}_N(p') \not{p}_\eta \frac{\not{p} + \not{q} + M_R}{(p + q)^2 - M_R^2 + i\Gamma_R M_R} \mathcal{O}^\mu_R u_N(p), \tag{12.38}$$

$$J^\mu_{R(u)} = \frac{gV_{ud}}{2\sqrt{2}} ig_\eta \bar{u}_N(p') \mathcal{O}^\mu_R \frac{\not{p} - \not{p}_\eta + M_R}{(p - p_\eta)^2 - M_R^2 + i\Gamma_R M_R} \not{p}_\eta u_N(p), \tag{12.39}$$

where

$$\mathcal{O}^\mu_R \equiv \frac{f_1(q^2)}{(2M)^2}(\not{q}q^\mu - q^2\gamma^\mu)\gamma_5 + \frac{f_2(q^2)}{2M}i\sigma^{\mu\rho}q_\rho\gamma_5 - g_1(q^2)\gamma^\mu - \frac{g_3(q^2)}{M}q^\mu. \tag{12.40}$$

The determination of these vector and axial vector transition form factors were discussed in detail in Chapter 11.

The coupling g_η is determined in a similar manner as we have obtained $f_{NR\pi}$ in the case of pion production, but by substituting η in the place of π in the determination of the coupling presented in Chapter 11.

12.3.2 Neutral current

In this section, we will extend the formalism for the neutral current induced η production processes. The channels that contribute are given as

$$\nu(k) + N(p) \rightarrow \nu(k') + \eta(p_\eta) + N(p')$$
$$\bar{\nu}(k) + N(p) \rightarrow \bar{\nu}(k') + \eta(p_\eta) + N(p'), \qquad N = n, p \tag{12.41}$$

In the neutral current sector, the structure of current for the non-resonant as well as the resonant contributions, would be similar to that of the charged current ones; however, the form factors would be modified. The form factors for the neutral current induced processes, in the case of non-resonant background terms, are discussed in Section 12.2.2 while the neutral current form factors for the resonance excitations are presented in Chapter 11.

However in this case, there would be additional contribution from the additional isoscalar form factors, arising due to the strangeness component in the nucleon. We will not discuss this subject in this book; the reader is referred to the work of Nieves et al. and the references cited there [481].

12.3.3 Cross section

Figure 12.5 shows the results for the total scattering cross sections for the processes $\nu_\mu + n \longrightarrow \mu^- + \eta + p$ and $\bar{\nu}_\mu + p \longrightarrow \mu^+ + \eta + n$. The individual contributions from the nucleon pole (Born terms), $S_{11}(1535)$ and $S_{11}(1650)$ resonance excitations, where both direct and crossed diagrams are considered, as well as the full model (sum of all the diagrams) are shown. It may be observed from the figure that $S_{11}(1535)$ has the dominant contribution in both the channels and overestimates the full model by $\sim 45\%$. Since the individual contributions of the nucleon pole and $S_{11}(1650)$ are very small as compared to the full model, while depicting the individual contributions, we have scaled them by a factor of 5 and 10 for the process $\nu_\mu + n \longrightarrow \mu^- + \eta + p$ and $\bar{\nu}_\mu + p \longrightarrow \mu^+ + \eta + n$, respectively, in the case of nucleon pole and by a factor of 5 in the case of $S_{11}(1650)$ for both the processes.

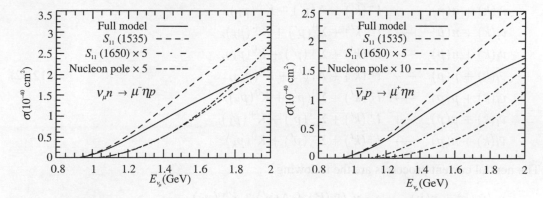

Figure 12.5 Total scattering cross section for the charged current η meson production for the processes $\nu_\mu + n \longrightarrow \mu^- + \eta + p$ and $\bar{\nu}_\mu + p \longrightarrow \mu^+ + \eta + n$. The full model consists of contributions from all the diagrams including $S_{11}(1535)$ and $S_{11}(1650)$. While showing the contribution of the Born terms, $S_{11}(1535)$ and $S_{11}(1650)$, we have taken both direct and cross diagrams. Note that, in order to present on the same scale, $S_{11}(1650)$ and nucleon Born terms are multiplied by 5 for $\nu_\mu n \rightarrow \mu^- \eta p$; for $\bar{\nu}_\mu p \rightarrow \mu^+ \eta n$, the factors are 5 and 10, respectively.

12.4 Associated Production of Strange Particles

The study of the neutrino induced $\Delta S = 0$ associated particle production processes provide an improved understanding of the basic symmetries of the standard model, the structure of the weak hadronic form factors, the strange–quark content of the nucleon, coupling constants, etc. Moreover, the kaon production through the associated production also constitutes a background in the proton decay searches, that is, $p \rightarrow K\bar{\nu}$. Therefore, an understanding and reliable estimate of the cross sections for the neutrino induced kaon production contributing as the background event is important and has been emphasized [526, 527].

The experimental observations of the neutrino induced associated particle production processes were performed earlier at BNL [528], ANL [529], and CERN[530, 453, 452]; however, these experiments have both low statistics and large systematic errors. Attempts are being made to study them in the context of the present-day neutrino experiments with high intensity $\nu(\bar{\nu})$ beams.

Theoretically, early attempts to calculate weak associated production were made by Shrock et al. [531], Mecklenburg et al. [532], Dewan et al. [533], and Amer et al. [534]. The associated particle production cross sections used, for example, in the NUANCE Monte Carlo generator [535] considers only the resonant kaon production based on the Rein and Sehgal model for the pion production [492]. Moreover, these cross sections miss the experimental data points by almost a factor of four [536]. Therefore, a better estimation of the weak interaction induced associated particle production cross section is needed.

Here, the formalism for writing the hadronic current is the same as adopted in the case of pion and eta meson production processes discussed in Sections 12.2 and 12.3, respectively. The CC induced $\Delta S = 0$ processes are the following:

$$
\begin{aligned}
\nu_l(k) + p(p) &\longrightarrow l^-(k') + \Sigma^+(p') + K^+(p_k), \\
\nu_l(k) + n(p) &\longrightarrow l^-(k') + \Lambda(p') + K^+(p_k), \\
\nu_l(k) + n(p) &\longrightarrow l^-(k') + \Sigma^0(p') + K^+(p_k), \\
\nu_l(k) + n(p) &\longrightarrow l^-(k') + \Sigma^+(p') + K^0(p_k), \\
\bar{\nu}_l(k) + p(p) &\longrightarrow l^+(k') + \Lambda(p') + K^0(p_k), \\
\bar{\nu}_l(k) + p(p) &\longrightarrow l^+(k') + \Sigma^0(p') + K^0(p_k), \\
\bar{\nu}_l(k) + p(p) &\longrightarrow l^+(k') + \Sigma^-(p') + K^+(p_k), \\
\bar{\nu}_l(k) + n(p) &\longrightarrow l^+(k') + \Sigma^-(p') + K^0(p_k).
\end{aligned}
\tag{12.42}
$$

The neutral current processes are the following

$$
\begin{aligned}
\nu_l/\bar{\nu}_l(k) + p(p) &\longrightarrow \nu_l/\bar{\nu}_l(k') + \Lambda(p') + K^+(p_k), \\
\nu_l/\bar{\nu}_l(k) + p(p) &\longrightarrow \nu_l/\bar{\nu}_l(k') + \Sigma^0(p') + K^+(p_k), \\
\nu_l/\bar{\nu}_l(k) + p(p) &\longrightarrow \nu_l/\bar{\nu}_l(k') + \Sigma^+(p') + K^0(p_k), \\
\nu_l/\bar{\nu}_l(k) + n(p) &\longrightarrow \nu_l/\bar{\nu}_l(k') + \Lambda(p') + K^0(p_K), \\
\nu_l/\bar{\nu}_l(k) + n(p) &\longrightarrow \nu_l/\bar{\nu}_l(k') + \Sigma^0(p') + K^0(p_K), \\
\nu_l/\bar{\nu}_l(k) + n(p) &\longrightarrow \nu_l/\bar{\nu}_l(k') + \Sigma^-(p') + K^+(p_K),
\end{aligned}
\tag{12.43}
$$

where $l = e, \mu$. For energies above 1.5 GeV, it is the $\Delta S = 0$ kaon production which is more dominant in comparison to the corresponding $|\Delta S| = 1$ processes as the latter is suppressed by a factor of $\tan^2 \theta_C$. However, in the case of the antikaon, the $\Delta S = 0$ mode is not possible in the three-body final state due to the change in the total strangeness quantum number by 2 units ($|\Delta S| = 2$). The only possible strangeness conserving process for $K^-, \bar{K}^0$ is through channels like,

$$
\nu_l + N \to l^- + K + \bar{K} + N', \qquad (\Delta S = 0).
\tag{12.44}
$$

However, the threshold for such type of processes is high around 1.6 GeV.

For demonstrating the results of the associated particle production, we have considered only non-resonant diagrams. However, various spin $1/2$ and $3/2$ resonances having an appreciable branching ratio for decays in the KY channel may be included, following the same procedure as in the case of pion and eta production processes. In Section 12.4.1, we describe the formalism in brief for the charged current induced associated particle production and in Section 12.4.2, the results are presented.

12.4.1 Charged current

The differential scattering cross section for the processes given in Eq. (12.42) is given in Eq. (12.3) with $E_m = E_k$ and $\vec{p}_m = \vec{p}_k$, the outgoing kaon's energy and the three-momentum, respectively. E'_p is replaced with E_Y, the energy of the outgoing hyperon. The transition matrix element for the associated particle production process is given in Eq. (12.4) with the leptonic current defined in Eq. (12.5). The contribution to the hadronic current J^μ comes from the different pieces of the Lagrangian corresponding to the Feynman diagrams shown in Figure 12.6.

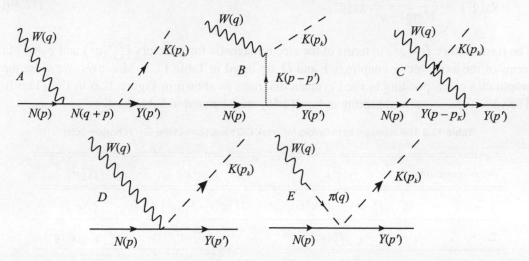

Figure 12.6 Feynman diagrams corresponding to the (anti)neutrino induced $\Delta S = 0$ associated particle production processes.

The matrix element corresponding to the nucleon–hyperon transition may be written as

$$J^\mu = \langle Y(p')|V^\mu - A^\mu|N(p)\rangle. \tag{12.45}$$

The currents V^μ and A^μ are expressed in terms of the vector $f_i(q^2)$ and axial vector $g_i(q^2)$ $N - Y$ transition form factors as [294],

$$V^\mu = \gamma^\mu f_1(q^2) + i\sigma^{\mu\nu}\frac{q^\nu}{M + M_Y}f_2(q^2) + f_3(q^2)\frac{2q^\mu}{M + M_Y}, \tag{12.46}$$

$$A^\mu = \left[\gamma^\mu g_1(q^2) + i\sigma^{\mu\nu}\frac{q^\nu}{M + M_Y}g_2(q^2) + g_3(q^2)\frac{2q^\mu}{M + M_Y}\right]\gamma^5. \tag{12.47}$$

In the vector sector, following the CVC hypothesis, $f_3(q^2) = 0$ and the form factors $f_{1,2}(q^2)$ are determined in terms of the electromagnetic form factors of protons and neutrons, viz.,

$F_{1,2}^{p,n}(q^2)$, as discussed in Chapter 10. The weak electric form factor $g_2(q^2) = 0$ due to G-invariance and SU(3) symmetry.

The axial vector form factor $g_1(q^2)$ comes from the semileptonic hyperon decays and the $\Delta S = 0$ neutrino–nucleon scattering; it is parameterized as

$$g_1(q^2) = g_A(0) \left(1 - \frac{q^2}{M_A^2}\right)^{-2}, \tag{12.48}$$

with axial dipole mass M_A taken as 1.026 GeV.

Applying PCAC, the pseudoscalar transition form factor $(g_3(q^2))$ is related to the axial vector transition form factor $(g_1(q^2))$ as (see Chapter 10 for details)

$$g_3(q^2) = \frac{(M + M_Y)^2}{2(m_K^2 - q^2)} g_1(q^2). \tag{12.49}$$

The form factors $f_{1,2}(q^2)$ in terms of the electromagnetic form factors $F_{1,2}^{p,n}(q^2)$ and $g_1(q^2)$ in terms of the axial vector couplings F and D are listed in Table 12.3. Moreover, we write the amplitudes corresponding to the Feynman diagrams as shown in Figure 12.6 in Eq. (12.50). The various parameters appearing in Eq. (12.50) are tabulated in Table 12.4.

Table 12.3 The standard form factors for weak CC transitions of the SU(3) baryon octet.

Weak transition	$f_1(q^2)$	$f_2(q^2)$	$g_1(q^2)$
p → n	$F_1^p(q^2) - F_1^n(q^2)$	$F_2^p(q^2) - F_2^n(q^2)$	$g_1(q^2)$
$\Sigma^\pm \to \Lambda$	$-\sqrt{\frac{3}{2}} F_1^n(q^2)$	$-\sqrt{\frac{3}{2}} F_2^n(q^2)$	$\sqrt{\frac{2}{3}} \frac{D}{F+D} g_1(q^2)$
$\Sigma^\pm \to \Sigma^0$	$\mp\frac{1}{\sqrt{2}}[2F_1^p(q^2) + F_1^n(q^2)]$	$\mp\frac{1}{\sqrt{2}}[2F_2^p(q^2) + F_2^n(q^2)]$	$\mp\sqrt{2}\frac{F}{F+D} g_1(q^2)$

Table 12.4 Constant factors appearing in the hadronic current. The upper sign corresponds to the processes with antineutrinos and the lower sign to that with neutrinos.

Process	A_{CT}	B_{CT}	A_{SY}	A_{UY} $Y' = \Sigma$	A_{UY} $Y' = \Lambda$	A_{TY}	A_π
$\bar{\nu}_l p \to l^+ \Sigma^- K^+$ $\nu_l n \to l^- \Sigma^+ K^0$	0	0	$D - F$	$D - F$	$\frac{1}{3}(D + 3F)$	0	0
$\bar{\nu}_l p \to l^+ \Lambda K^0$ $\nu_l n \to l^- \Lambda K^+$	$-\sqrt{\frac{3}{2}}$	$\frac{-1}{3}(D + 3F)$	$\frac{-1}{\sqrt{6}}(D + 3F)$	$-\sqrt{\frac{2}{3}}(D - F)$	0	$\frac{-1}{\sqrt{6}}(D + 3F)$	$\sqrt{\frac{3}{2}}$
$\bar{\nu}_l p \to l^+ \Sigma^0 K^0$ $\nu_l n \to l^- \Sigma^0 K^+$	$\mp\frac{1}{\sqrt{2}}$	$D - F$	$\mp\frac{1}{\sqrt{2}}(D - F)$	$\mp\sqrt{2}(D - F)$	0	$\pm\frac{1}{\sqrt{2}}(D - F)$	$\pm\frac{1}{\sqrt{2}}$
$\bar{\nu}_l n \to l^+ \Sigma^- K^0$ $\nu_l p \to l^- \Sigma^+ K^+$	-1	$D - F$	0	$F - D$	$\frac{1}{3}(D + 3F)$	$D - F$	1

The hadronic currents corresponding to the diagrams shown in Figure 12.6 now read as

$$
j^\mu|_s = iA_{SY}V_{ud}\frac{\sqrt{2}}{2f_\pi}\,\bar{u}_Y(p')\slashed{p}_k\gamma^5\frac{\slashed{p}+\slashed{q}+M}{(p+q)^2-M^2}\mathcal{H}^\mu u_N(p)
$$

$$
j^\mu|_u = iA_{UY}V_{ud}\frac{\sqrt{2}}{2f_\pi}\,\bar{u}_Y(p')\mathcal{H}^\mu\frac{\slashed{p}-\slashed{p}_k+M_Y}{(p-p_k)^2-M_Y}\slashed{p}_k\gamma^5 u_N(p)
$$

$$
j^\mu|_t = iA_{TY}V_{ud}\frac{\sqrt{2}}{2f_\pi}(M+M_Y)\,\bar{u}_Y(p')\gamma_5\,u_N(p)\,\frac{q^\mu-2p_k^\mu}{(p-p')^2-m_k^2}
$$

$$
j^\mu|_{CT} = iA_{CT}V_{ud}\frac{\sqrt{2}}{2f_\pi}\,\bar{u}_Y(p')\left(\gamma^\mu+B_{CT}\,\gamma^\mu\gamma^5\right)u_N(p)
$$

$$
j^\mu|_{\pi F} = iA_\pi V_{ud}\frac{\sqrt{2}}{4f_\pi}\,\bar{u}_Y(p')(\slashed{q}+\slashed{p}_k)u_N(p)\frac{q^\mu}{q^2-m_\pi^2} \tag{12.50}
$$

where,

$$
\mathcal{H}^\mu = f_1(q^2)\gamma^\mu+i\frac{f_2(q^2)}{2M}\sigma^{\mu\nu}q_\nu-g_1(q^2)\left(\gamma^\mu-\frac{\slashed{q}q^\mu}{q^2-m_\pi^2}\right)\gamma^5 \tag{12.51}
$$

is the transition current for $Y \leftrightarrows Y'$ with $Y = Y' \equiv$ nucleon and/or hyperon. Using the expression for hadronic current given in Eq. (12.50), the hadronic tensor $H^{\mu\nu}$ is obtained, which contracts with the leptonic tensor $L^{\mu\nu}$ to get the expression for matrix element squared.

12.4.2 Cross section

The total scattering cross sections for $\nu_\mu N \to \mu^- YK$ and $\bar{\nu}_\mu N \to \mu^+ YK$ processes are shown in Figure 12.7. It may be noted from the figure that the cross section for the reaction with a Λ in the final state are, in general, larger than that for the reactions where a Σ is produced in the final state. This can be understood by looking at the relative strengths of the couplings; for example, the ratio of the square of the couplings for the vertices $nK^0\Lambda$ and $nK^0\Sigma^0$ is $\frac{g_{NK\Lambda}^2}{g_{NK\Sigma}^2} \simeq 14$. Furthermore, the cross section for the Λ production is favoured by the available phase space due to its small mass relative to Σ. Comparing the results of the cross sections for the processes $\nu_\mu p \to \mu^- \Sigma^+ K^+$ and $\nu_\mu n \to \mu^- \Lambda K^+$, it is observed that the cross section for the $K^+\Sigma^+$ channel is $\sim 70\%$ smaller at $E_\nu = 1.5$ GeV and $\sim 25\%$ smaller at $E_\nu = 2$ GeV in comparison to the cross section obtained for the $K^+\Lambda$ channel. While the cross sections for the other two channels, viz., $\nu_\mu n \to \mu^- \Sigma^+ K^0$ and $\mu^- \Sigma^0 K^+$ are much smaller than the cross section for the $K^+\Lambda$ production.

In antineutrino induced processes, it may be noticed that unlike the neutrino induced processes, the channel with Λ in the final state is not very dominating. We find that at low energies, say $E_{\bar{\nu}_\mu} \sim 1.5$ GeV, the cross section for $\bar{\nu}_\mu p \to \mu^+ K^0\Lambda$ is around 55% larger than the cross section for $\bar{\nu}_\mu n \to \mu^+ K^0\Sigma^-$, while around 2 GeV, the production cross section for $\bar{\nu}_\mu p \to \mu^+ K^0\Lambda$ is comparable with that of $\bar{\nu}_\mu n \to \mu^+ K^0\Sigma^-$. The cross section for the reaction $\bar{\nu}_\mu p \to \mu^+ K^0\Sigma^0$ is around 40% smaller than the cross section for the $K^0\Lambda$ channel at $E_{\bar{\nu}_\mu} = 2$

GeV; whereas, $\bar{\nu}_\mu p \rightarrow \mu^+ K^+ \Sigma^-$ cross section is about 10% to the cross section of the $K^0 \Lambda$ channel.

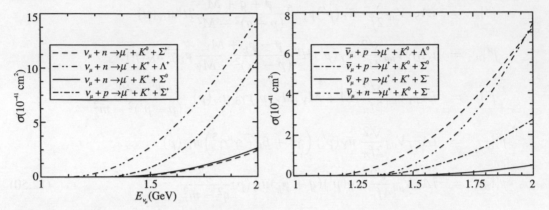

Figure 12.7 Cross section for neutrino (left) and antineutrino (right) induced $|\Delta S| = 0$ associated kaon production processes.

12.5   Kaon Production

In the few GeV energy region, single kaon production processes become important. In principle, their cross sections are smaller than the pionic processes because of the phase space and the Cabibbo suppression for the $\Delta S = 1$ reactions. Nonetheless, in the coming years of precision neutrino physics, their knowledge could be relevant for data analysis, apart from their own intrinsic value related to the role played by strange quarks in hadron physics; in fact, the process is considered a prospective candidate in proton decay searches.

In neutrino induced reactions, the first inelastic reaction creating a strange quark, without accompanying hyperons, is the single kaon production while for antineutrinos, the lowest threshold for $|\Delta S| = 1$ reactions is much lower and corresponds to the hyperon production. This CC $\Delta S = 1$ process is particularly appealing for several reasons. One of them is the better simulation of the background which is produced by atmospheric neutrino interactions in the analysis of one of the main decay channels of the proton, discussed in many SUSY GUT models, that is, $p \rightarrow \nu + K^+$ [527, 537, 538]. A second reason is its simplicity from a theoretical point of view. At low energies, it is possible to obtain model independent predictions using the non-linear sigma model and due to the absence of $S = +1$ baryonic resonances, the range of validity of the calculation could be extended to higher energies than for other channels. Furthermore, the associated production of kaons has a higher energy threshold than the single kaon production (1.10 GeV vs. 0.79 GeV). This implies that in the threshold region of associated production, the single kaon production could still be dominant in this energy region of $E_\nu \sim 0.8 - 1.2$ GeV. The basic reaction for the neutrino induced charged current kaon production is

$$\nu_l(k) + n(p) \longrightarrow l^-(k') + n(p') + K^+(p_k),$$
$$\nu_l(k) + p(p) \longrightarrow l^-(k') + p(p') + K^+(p_k),$$

$$\nu_l(k) + n(p) \quad \longrightarrow \quad l^-(k') + p(p') + K^0(p_k),$$ (12.52)

where $l = e, \mu$. The expression for the differential scattering cross section in the laboratory frame for this process is given in Eq. (12.3) with $E_m = E_k$ and $\vec{p}_m = \vec{p}_k$, respectively, the energy and three-momentum of the outgoing kaon. The transition matrix element for the single kaon production for the $\Delta S = 1$ process is given as

$$\mathcal{M} = \frac{G_F}{\sqrt{2}} \sin \theta_C \, l_\mu j^\mu,$$ (12.53)

with the leptonic current l_μ given in Eq. (12.5).

In the non-resonant sector, four different channels, viz., contact term (CT), kaon pole (KP) term, Σ or Λ hyperon exchanged in the u channel and a meson (π, η) exchange term, may contribute to the hadronic current [490] and are depicted in Figure 12.8. For the single kaon production channel, there is no s channel contribution due to the absence of $S = +1$ baryonic resonances. The current, in the case of the KP term, is proportional to q^μ, which implies that, after contraction with the leptonic tensor, the amplitude is proportional to the lepton mass and therefore, is very small.

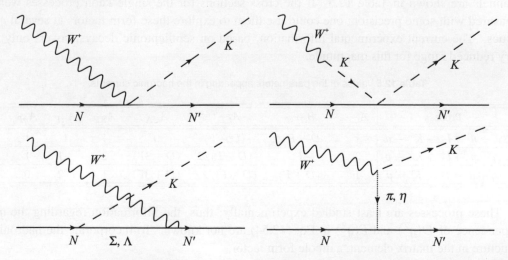

Figure 12.8 Feynman diagrams for the process $\nu N \to l N' K$. The first row from left to right: contact term (labeled CT in the text), kaon pole term (KP); second row: u channel diagram ($C\Sigma$, $C\Lambda$) and pion (eta) in flight (πP, ηP).

In analogy with the single pion production induced by neutrinos [481], as discussed earlier, the different diagrams of Figure 12.8 have the following contributions to the hadronic current:

$$j^\mu\big|_{CT} = -iA_{CT}\frac{\sqrt{2}}{2f_\pi}\bar{u}(p')(\gamma^\mu + \gamma^\mu\gamma^5 B_{CT})u(p),$$

$$j^\mu\big|_{C\Sigma} = iA_{C\Sigma}\frac{\sqrt{2}}{2f_\pi}\bar{u}(p')\left(\gamma^\mu + i\frac{\mu_p + 2\mu_n}{2M}\sigma^{\mu\nu}q_\nu + (D-F)\left(\gamma^\mu - \frac{q^\mu}{q^2 - M_k^2}\slashed{q}\right)\gamma^5\right)$$
$$\times \frac{\slashed{p} - \slashed{p}_k + M_\Sigma}{(p-p_k)^2 - M_\Sigma^2}\slashed{p}_k\gamma^5 u(p),$$

$$j^\mu\big|_{C\Lambda} = iA_{C\Lambda}\frac{\sqrt{2}}{4f_\pi}\bar{u}(p')\left(\gamma^\mu + i\frac{\mu_p}{2M}\sigma^{\mu\nu}q_\nu - \frac{D+3F}{3}\left(\gamma^\mu - \frac{q^\mu}{q^2 - M_k^2}\slashed{q}\right)\gamma^5\right)$$
$$\times \frac{\slashed{p} - \slashed{p}_k + M_\Lambda}{(p - p_k)^2 - M_\Lambda^2}\slashed{p}_k\gamma^5 u(p),$$

$$j^\mu\big|_{KP} = iA_{KP}\frac{\sqrt{2}}{4f_\pi}\bar{u}(p')(\slashed{q} + \slashed{p}_k)u(p)\frac{1}{q^2 - M_k^2}q^\mu,$$

$$j^\mu\big|_{\pi} = iA_{\pi P}(D+F)\frac{\sqrt{2}}{2f_\pi}\frac{M}{(q - p_k)^2 - M_\pi^2}\bar{u}(p')\gamma^5(q^\mu - 2p_k{}^\mu)u(p),$$

$$j^\mu\big|_{\eta} = iA_{\eta P}(D - 3F)\frac{\sqrt{2}}{2f_\pi}\frac{M}{(q - p_k)^2 - M_\eta^2}\bar{u}(p')\gamma^5(q^\mu - 2p_k{}^\mu)u(p), \tag{12.54}$$

where, $q = k - k'$ is the four-momentum transfer, $\mu_p = 1.7928\ \mu_N$, and $\mu_n = -1.9130$ μ_N are, respectively, the proton and neutron anomalous magnetic moments. The value of the various parameters appearing in the expressions of the hadronic currents of the different channels are shown in Table 12.5. If the cross sections for the single kaon processes were measured with some precision, one could use them to explore these form factors at several q^2 values. The current experimental information, based on semileptonic decays, covers only a very reduced range for this magnitude.

Table 12.5 Values of the parameters appearing in the hadronic currents.

Process	A_{CT}	B_{CT}	$A_{C\Sigma}$	$A_{C\Lambda}$	A_{KP}	$A_{\pi P}$	$A_{\eta P}$
$\nu_l + n \to l^- + K^+ + n$	1	$D - F$	$-(D - F)$	0	1	1	1
$\nu_l + p \to l^- + K^+ + p$	2	$-F$	$-(D - F)/2$	$(D + 3F)$	2	-1	1
$\nu_l + n \to l^- + K^0 + p$	1	$-(D + F)$	$(D - F)/2$	$(D + 3F)$	1	-2	0

These processes are least studied experimentally; thus, the information regarding the q^2 dependence of $D(q^2)$ and $F(q^2)$ in Eq. (12.54) are not known. To incorporate the hadronic structure in the matrix element, a dipole form factor

$$F(q^2) = 1/(1 - q^2/M_F^2)^2, \tag{12.55}$$

is used with a mass $M_F \simeq 1$ GeV. Its effect, being small at low neutrino energies, will give an idea of the uncertainties of the calculation which is explored in the next section.

However, recently the MINERvA experiment [539] has measured the differential scattering cross section for the charged current induced single kaon production as a function of the kinetic energy of the kaon produced in the final state.

12.5.1 Cross section

Figure 12.9 shows the results of the contributions of the different diagrams to the total scattering cross sections for the processes $\nu_\mu p \longrightarrow \mu^- K^+ p$ and $\nu_\mu n \to \mu^- K^0 p$. It may be observed that the contact term has a dominant contribution to the total scattering cross section in both the processes studied. The curve labeled as the full model is calculated with a dipole form factor with $M_F = 1$ GeV. The band corresponds to the variation of M_F by 10%. The process $\nu_\mu + n \rightarrow \mu^- + K^0 + p$ has a cross section of a similar size and the contact term is the largest followed by the π exchange diagram and the u channel (Λ) term. The rate of increase of the cross sections for the process $\nu_\mu n \rightarrow \mu^- p K^0$ is larger and could become more important at higher energies. We observe a destructive interference between the different terms; the cross section obtained with the full model is smaller than that produced by the contact term.

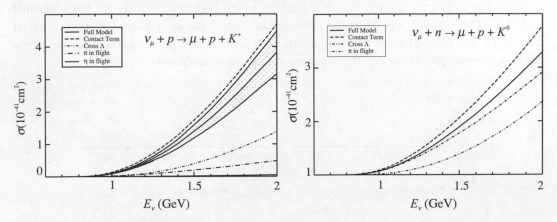

Figure 12.9 Contribution of the different terms to the total scattering cross section for the $\nu_\mu + p \longrightarrow \mu^- + K^+ + p$ (left panel) and $\nu_\mu + n \longrightarrow \mu^- + K^0 + p$ (right panel) processes.

It must be noted that due to the higher threshold of the associated kaon production, the reactions we have studied are the dominant source of kaons for a wide range of energies, and thus, their study is important for lower energy accelerator experiments as well as for atmospheric neutrino experiments.

12.6 Antikaon Production

We have seen earlier that the hyperon and kaon productions ($|\Delta S| = 1$) are important in the few GeV energy region, although they are small due to the Cabibbo suppression. In this section, we describe the weak single antikaon production off the nucleons in the charged current sector induced by antineutrinos. Other processes like the charged current induced neutrino processes as well as the neutral current induced neutrino and antineutrino processes are forbidden by the $\Delta S \neq \Delta Q$ rule. The resonant excitations, absent for the kaon case, also contribute to the single antikaon production.

The basic reaction for the antineutrino induced charged current antikaon production is

$$
\begin{aligned}
\bar{\nu}_l(k) + p(p) &\longrightarrow l^+(k') + p(p') + K^-(p_k), \\
\bar{\nu}_l(k) + n(p) &\longrightarrow l^+(k') + n(p') + K^-(p_k), \\
\bar{\nu}_l(k) + p(p) &\longrightarrow l^+(k') + n(p') + \bar{K}^0(p_k),
\end{aligned}
\tag{12.56}
$$

where $l = e, \mu$; the quantities in the parentheses represent the four-momenta of the corresponding particles.

The expression for the differential scattering cross section is given in Eq. (12.3), where $E_m = E_k$ is the energy of the outgoing antikaon and $\vec{p}_m = \vec{p}_k$ represents the three-momentum of the antikaon. The transition matrix element is defined in Eq. (12.53) with the leptonic current given in Eq. (12.5). The hadronic current receives the contribution from the non-resonant background terms as well as from the decuplet resonance terms, presented in Figure 12.10. The non-resonant as well as the decuplet resonance terms are obtained using the non-linear sigma model. The different channels which contribute to the hadronic currents are the s channel with Σ, Λ(SC), and Σ^*(SCR) as the intermediate states, the kaon pole (KP) term, contact term (CT), and meson (πP,ηP) exchange term [491]. For the single antikaon processes, there are no u channel processes with the hyperons in the intermediate state.

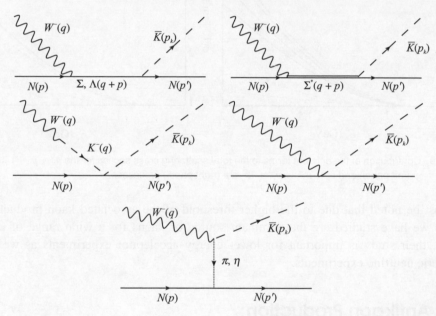

Figure 12.10 Feynman diagrams for the process $\bar{\nu}N \rightarrow lN'\bar{K}$. The first row from left to right: s channel Σ, Λ propagator (labeled SC in the text), s channel Σ^* resonance (SCR); second row: kaon pole term (KP); contact term (CT) and last row: pion(Eta) in flight (πP/ηP).

As in the case of the pion production where $P_{33}(1232)$ contributes significantly, the excitation of $\Sigma^*(1385)$ resonance and its subsequent decay in $N\bar{K}$ may be important in the single antikaon production processes. The lowest order SU(3) Lagrangian for the coupling of the pseudoscalar mesons with the decuplet–octet baryons, in the presence of the external weak current, is given by [540]

$$\mathcal{L}_{dec} = \mathcal{C} \left(\epsilon^{abc} \bar{T}^{\mu}_{ade} u^{d}_{\mu,b} B^{e}_{c} + h.c. \right), \tag{12.57}$$

where T^{μ} is the SU(3) representation of the decuplet fields,

$$
T_{ijk} = \begin{pmatrix} \Delta^{++} & \frac{1}{\sqrt{3}}\Delta^{+} & \frac{1}{\sqrt{3}}\Sigma^{*+} \\ \frac{1}{\sqrt{3}}\Delta^{+} & \frac{1}{\sqrt{3}}\Delta^{0} & \frac{1}{\sqrt{6}}\Sigma^{*0} \\ \frac{1}{\sqrt{3}}\Sigma^{*+} & \frac{1}{\sqrt{6}}\Sigma^{*0} & \frac{1}{\sqrt{3}}\Xi^{*0} \end{pmatrix} \begin{pmatrix} \frac{1}{\sqrt{3}}\Delta^{+} & \frac{1}{\sqrt{3}}\Delta^{0} & \frac{1}{\sqrt{6}}\Sigma^{*0} \\ \frac{1}{\sqrt{3}}\Delta^{0} & \Delta^{-} & \frac{1}{\sqrt{3}}\Sigma^{*-} \\ \frac{1}{\sqrt{6}}\Sigma^{*0} & \frac{1}{\sqrt{3}}\Sigma^{*-} & \frac{1}{\sqrt{3}}\Xi^{*-} \end{pmatrix}
$$

$$
\begin{pmatrix} \frac{1}{\sqrt{3}}\Sigma^{*+} & \frac{1}{\sqrt{6}}\Sigma^{*0} & \frac{1}{\sqrt{3}}\Xi^{*0} \\ \frac{1}{\sqrt{6}}\Sigma^{*0} & \frac{1}{\sqrt{3}}\Sigma^{*-} & \frac{1}{\sqrt{3}}\Xi^{*-} \\ \frac{1}{\sqrt{3}}\Xi^{*0} & \frac{1}{\sqrt{3}}\Xi^{*-} & \Omega^{-} \end{pmatrix}, \tag{12.58}
$$

$a - e$ are flavor indices, B corresponds to the baryon octet, and u_μ is the SU(3) representation of the pseudoscalar mesons interacting with weak left l_μ and right r_μ handed currents, discussed in Chapter 11. The different physical states of the decuplet corresponding to T_{ijk} are given in Table 12.6. The parameter $\mathcal{C} \simeq 1$ is fitted to the $\Delta(1232)$ decay width. The spin 3/2 propagator for Σ^* is given by

Table 12.6 The physical states of the decuplet .

T_{111}	Δ^{++}	T_{112}	$\frac{\Delta^{+}}{\sqrt{3}}$	T_{113}	$\frac{\Sigma^{*+}}{\sqrt{3}}$	T_{123}	$\frac{\Sigma^{*0}}{\sqrt{6}}$	T_{233}	$\frac{\Xi^{-}}{\sqrt{3}}$
T_{122}	$\frac{\Delta^{0}}{\sqrt{3}}$	T_{222}	Δ^{-}	T_{223}	$\frac{\Sigma^{*-}}{\sqrt{3}}$	T_{133}	$\frac{\Xi^{0}}{\sqrt{3}}$	T_{333}	Ω^{-}

$$
G^{\mu\nu}(P) = \frac{P_{RS}^{\mu\nu}(P)}{P^2 - M_{\Sigma^*}^2 + i M_{\Sigma^*}\Gamma_{\Sigma^*}}, \tag{12.59}
$$

where $P = p + q$ is the momentum carried by the resonance and $P_{RS}^{\mu\nu}$ is the projection operator for the spin 3/2 resonances, given as

$$
P_{RS}^{\mu\nu}(P) = \sum_{\text{spin}} \psi^\mu \bar{\psi}^\nu = -(\slashed{P} + M_{\Sigma^*}) \left[g^{\mu\nu} - \frac{1}{3}\gamma^\mu\gamma^\nu - \frac{2}{3}\frac{P^\mu P^\nu}{M_{\Sigma^*}^2} + \frac{1}{3}\frac{P^\mu\gamma^\nu - P^\nu\gamma^\mu}{M_{\Sigma^*}} \right], \tag{12.60}
$$

with M_{Σ^*} the resonance mass and ψ^μ, the Rarita–Schwinger spinor. The decay width of Σ^*, obtained using the Lagrangian given in Eq. (12.57), can be written as

$$
\Gamma_{\Sigma^*} = \Gamma_{\Sigma^* \to \Lambda\pi} + \Gamma_{\Sigma^* \to \Sigma\pi} + \Gamma_{\Sigma^* \to N\bar{K}}, \tag{12.61}
$$

where

$$
\Gamma_{\Sigma^* \to Y, \text{meson}} = \frac{C_Y}{192\pi} \left(\frac{\mathcal{C}}{f_\pi}\right)^2 \frac{(W + M_Y)^2 - m^2}{W^5} \lambda^{3/2}(W^2, M_Y^2, m^2)\, \Theta(W - M_Y - m). \tag{12.62}
$$

Here, m, M_Y are the masses of the emitted meson and baryon, $\lambda(x, y, z) = (x - y - z)^2 - 4yz$ and Θ is the step function. The factor C_Y is 1 for Λ and $\frac{2}{3}$ for N and Σ.

Using the symmetry arguments, we may write the most general $W^- N \to \Sigma^*$ vertex in terms of a vector and an axial vector part as

$$\langle \Sigma^*; P = p + q \,|V^\mu|\, N; p \rangle \;=\; \sin\theta_C \bar\psi_\alpha(\vec{P}) \Gamma_V^{\alpha\mu}(p, q)\, u(\vec{p}), \tag{12.63}$$

$$\langle \Sigma^*; P = p + q \,|A^\mu|\, N; p \rangle \;=\; \sin\theta_C \bar\psi_\alpha(\vec{P}) \Gamma_A^{\alpha\mu}(p, q)\, u(\vec{p}), \tag{12.64}$$

where

$$\Gamma_V^{\alpha\mu}(p, q) \;=\; \left[\frac{C_3^V}{M}(g^{\alpha\mu}\slashed{q} - q^\alpha \gamma^\mu) + \frac{C_4^V}{M^2}(g^{\alpha\mu}q \cdot P - q^\alpha P^\mu) + \frac{C_5^V}{M^2}(g^{\alpha\mu}q \cdot p - q^\alpha p^\mu) \right.$$
$$\left. + C_6^V g^{\mu\alpha} \right] \gamma_5$$

$$\Gamma_A^{\alpha\mu}(p, q) \;=\; -\left[\frac{C_3^A}{M}(g^{\alpha\mu}\slashed{q} - q^\alpha \gamma^\mu) + \frac{C_4^A}{M^2}(g^{\alpha\mu}q \cdot P - q^\alpha P^\mu) + C_5^A g^{\alpha\mu} + \frac{C_6^A}{M^2}q^\mu q^\alpha \right]. \tag{12.65}$$

Our knowledge of these form factors is quite limited. The Lagrangian given in Eq. (12.57) gives only $C_5^A(0) = -2\mathcal{C}/\sqrt{3}$ (for the $\Sigma^{*-}(1385)$ case). However, using the SU(3) symmetry, all the other form factors of $\Sigma^{*-}(1385)$ may be related to the corresponding form factors of the $\Delta(1232)$ resonance, such that $C_i^{\Sigma^{*-}}/C_i^{\Delta^+} = -1$ and $C_i^{\Sigma^{*-}}/C_i^{\Sigma^{*0}} = \sqrt{2}$. In the case of the Δ resonance, as discussed in Section 12.2, the weak vector form factors are relatively known from the electromagnetic processes and there is some information on the axial vector form factors from the study of the pion production. The vector form factors for the Δ resonances, discussed in Chapter 11, are used while the axial vector form factors $C_5^A(q^2)$ is determined from the decay width of the resonances in the pionic channels; $C_6^A(q^2)$ is determined using PCAC.

The hadronic currents for the background and resonant terms are written as

$$J^\mu|_{CT} \;=\; iA_{CT} V_{us} \frac{\sqrt{2}}{2f_\pi} \bar{u}(p')\,(\gamma^\mu + B_{CT}\,\gamma^\mu\gamma_5)\, u(p)$$

$$J^\mu|_\Sigma \;=\; iA_\Sigma (D - F) V_{us} \frac{\sqrt{2}}{2f_\pi} \bar{u}(p')\slashed{p}_k \gamma_5 \frac{\slashed{p} + \slashed{q} + M_\Sigma}{(p + q)^2 - M_\Sigma^2}\left(\gamma^\mu + i\frac{(\mu_p + 2\mu_n)}{2M}\sigma^{\mu\nu}q_\nu \right)$$
$$+\; (D - F)\left\{ \gamma^\mu - \frac{q^\mu}{q^2 - M_k^2}\slashed{q} \right\}\gamma^5 \Bigg)\, u(p)$$

$$J^\mu|_\Lambda \;=\; iA_\Lambda V_{us}(D + 3F)\frac{1}{2\sqrt{2}f_\pi} \bar{u}(p')\slashed{p}_k \gamma^5 \frac{\slashed{p} + \slashed{q} + M_\Lambda}{(p + q)^2 - M_\Lambda^2}\left(\gamma^\mu + i\frac{\mu_p}{2M}\sigma^{\mu\nu}q_\nu \right)$$
$$-\; \frac{(D + 3F)}{3}\left\{ \gamma^\mu - \frac{q^\mu}{q^2 - M_k^2}\slashed{q} \right\}\gamma^5 \Bigg)\, u(p)$$

$$J^\mu|_{KP} \;=\; iA_{KP} V_{us}\frac{\sqrt{2}}{2f_\pi} \bar{u}(p')\slashed{q}\, u(p)\frac{q^\mu}{q^2 - M_k^2}$$

$$J^\mu|_\pi = iA_\pi \frac{M\sqrt{2}}{2f_\pi} V_{us}(D+F)\frac{2p_k{}^\mu - q^\mu}{(q-p_k)^2 - m_\pi{}^2}\bar{u}(p')\gamma_5 u(p)$$

$$J^\mu|_\eta = iA_\eta \frac{M\sqrt{2}}{2f_\pi} V_{us}(D-3F)\frac{2p_k{}^\mu - q^\mu}{(q-p_k)^2 - m_\eta{}^2}\bar{u}(p')\gamma_5 u(p)$$

$$J^\mu|_{\Sigma^*} = -iA_{\Sigma^*} \frac{C}{f_\pi}\frac{1}{\sqrt{6}} V_{us} \frac{p_k^\lambda}{P^2 - M_{\Sigma^*}^2 + i\Gamma_{\Sigma^*} M_{\Sigma^*}} \bar{u}(p')P_{RS_{\lambda\rho}}(\Gamma_V^{\rho\mu} + \Gamma_A^{\rho\mu})u(p)$$

In $\Gamma_V^{\rho\mu} + \Gamma_A^{\rho\mu}$, the form factors are taken as for the Δ^+ case. The factors A_i for each diagram contributing to the hadronic current are tabulated in Table 12.7.

In analogy with the single kaon production, a global dipole form factor given in Eq. (12.55) with $M_F \simeq 1$ GeV is used in the hadronic currents, except for the resonance excitation.

Table 12.7 Constant factors appearing in the hadronic current.

Process	B_{CT}	A_{CT}	A_Σ	A_Λ	A_{KP}	A_π	A_η	A_{Σ^*}
$\bar{\nu}_l + n \to l^+ + K^- + n$	$D-F$	1	-1	0	-1	1	1	2
$\bar{\nu}_l + p \to l^+ + K^- + p$	$-F$	2	$-\frac{1}{2}$	1	-2	-1	1	1
$\bar{\nu}_l + p \to l^+ + \bar{K}^0 + n$	$-(D+F)$	1	$\frac{1}{2}$	1	-1	-2	0	-1

12.6.1 Cross section

In Figure 12.11, the different contributions of the hadronic current to the $\bar{\nu}_\mu p \to \mu^+ p K^-$ and $\bar{\nu}_\mu n \to \mu^+ n K^-$ reactions are presented. It may be observed that the cross section is dominated

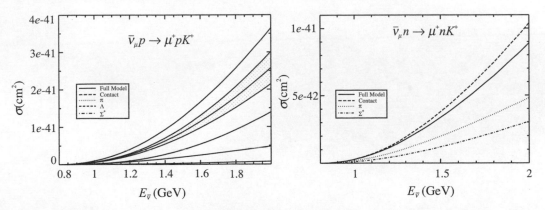

Figure 12.11 Total scattering cross section for the processes $\bar{\nu}_\mu p \to \mu^+ p K^-$ and $\bar{\nu}_\mu n \to \mu^+ n K^-$.

by the non-resonant terms, with contact term providing the largest contribution. Destructive interference leads to a total scattering cross section smaller than that predicted by the CT term. It should be noted that in the case of the $\bar{\nu}_\mu p \to \mu^+ p K^-$ process, $\Sigma^*(1385)$ has negligible contribution. The curve labeled as full model is calculated with a dipole form factor with a mass of 1 GeV. The band corresponds to a 10% variation in M_F. For the $\bar{\nu}_\mu n \to \mu^+ n K^-$ case, the contribution of Σ^* resonance is substantial due to the larger value of the couplings (see Table 12.7).

Chapter 13

Neutrino Scattering from Hadrons: Deep Inelastic Scattering

13.1 Introduction

In 1951, Lyman, Hanson, and Scott [541] were the first to observe elastic electron scattering from different nuclei using a 15.7 MeV beam obtained with the hclp of the betatron accelerator facility at Illinois. Further extensive studies using electron beams started with the development of the Mark III linear accelerator (LINAC) in 1953 at the High Energy Physics Laboratory (HEPL), Stanford. Hofstadter, Fechter, and McIntyre studied the effect of electron scattering ($E_e \sim 125 - 150$ MeV) on various nuclear targets and concluded that these nuclei have finite charge distribution. With increased energy (550 MeV) available for scattering, the first evidence of elastic scattering from the proton was observed by Chambers and Hofstadter at HEPL in 1956 [542], using a polyethylene target. Assuming that proton had an exponential density distribution, they found the r.m.s. (root mean squared) radius of the proton to be about 0.8 fm. Later, Yearian and Hofstadter performed electron scattering experiments on deuteron targets and determined the magnetic moment of the neutron [543]. At that time, the investigation of the structure of the proton and neutron was a major objective of HEPL, which was upgraded to achieve electron beams of energies up to 20 GeV; today, HEPL is known as the Stanford Linear Accelerator Center (SLAC) [544]. By the 1960s, it became possible to perform both elastic and inelastic scattering experiments at high energies and for a wide range of four-momentum transfer squared (Q^2). Thus, by the end of the 1960s, nuclear physics entered the 'deep inelastic scattering' (DIS) era, when experiments with 20–40 times higher energies were being performed at SLAC; it became possible to probe the hadron. DIS is the scattering of charged leptons/neutrinos from hadrons in the kinematic region of very high Q^2 and energy transfer ν. During the late 1960s, experiments by MIT-SLAC collaboration, led by Taylor, Kendall, and Friedman [544] confirmed the scaling phenomenon in the deep inelastic region, which was theoretically predicted by Bjorken [545, 546]. They received the 1990 Nobel

Prize in Physics for the experiments. These experimental results confirmed that, in the scaling region, the constituents of protons behave like free particles called partons. Charged partons are identified as quarks and neutral partons are identified as gluons. Later, many experiments using electron and muon beams were performed at CERN, DESY, Fermilab, etc. and their results were consistent with the earlier observations. In fact, Bjorken predicted that if neutrino beams are used in the DIS (deep inelastic scattering) region, then scaling should also be observed in the case of weak interactions [546]; this was indeed observed in the experiments performed at CERN using deuteron targets. Today, we understand nucleons to be composite systems of quarks and gluons with many internal degrees of freedom. The quantum field theory of quarks and gluons is known as quantum chromodynamics (QCD) which is a non-abelian gauge theory, with symmetry group $SU(3)_C$ in color space. These developments can be better understood

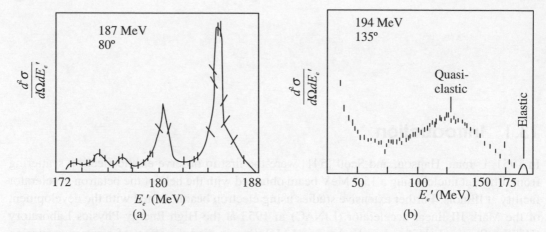

Figure 13.1 $e^- -^{12}C$ scattering cross section (a) $\theta = 80°$, $Q^2 \to 0.06$ GeV^2. The elastic peak is evident at low energy transfer; the excitation of nuclear levels can be seen with increase in energy transfer. (b) $\theta = 135°$, $Q^2 \sim 0.1$ GeV^2. The large, broad peak between 100 and 150 MeV is due to quasielastic scattering from individual neutrons and protons that make up the carbon nucleus [425].

with the help of Figures 13.1, 13.2, and 13.3. In Figure 13.1(a), the differential scattering cross section for elastic $e^- -^{12}C$ scattering at low Q^2 has been shown. The peak of the elastic scattering cross section may be seen for an electron of incident energy E_e=190 MeV, lab scattering angle $\theta = 80°$, and the outgoing electron's energy E_e'=187 MeV, which corresponds to $Q^2 \simeq 0.06$ GeV^2. With the increase in the energy transfer, that is, $E_e - E_e'$, nuclear excitations may be observed. The observation of nuclear excitations or, in general, hadron excitations imply that the object has a composite structure. At higher Q^2, corresponding to higher scattering angle, the elastic peak gets suppressed and a quasielastic peak is observed which corresponds to the scattering from individual nucleons. For example, in Figure 13.1(b), at $\theta = 135°$, $Q^2 \sim 0.1$ GeV^2, one may notice that the elastic scattering is suppressed and the broad peak between 100 and 150 MeV, which is due to the quasielastic scattering from individual neutrons and protons that make up the carbon nucleus, becomes prominent. This peak is smeared because the nucleons inside the nucleus are bound and moving with a finite momenta.

The phenomenon may also be understood in terms of the Bjorken variable x, which is defined as

$$x = -\frac{q^2}{2p \cdot q} \left(= \frac{Q^2}{2M\nu} \text{ in lab frame} \right),$$ (13.1)

where M is the nucleon mass and $\nu = E_l - E_l'$ is the energy transferred to the target particle. $Q^2 = -q^2 \simeq 4E_l E_l' \sin^2 \left(\frac{\theta}{2}\right) \geq 0$ in the limit $m_l \to 0$, x lies between 0 and 1 and is equal to 1 for an elastic scattering process as shown below.

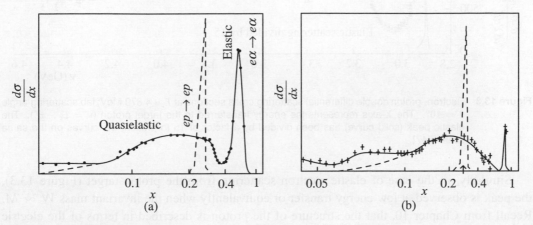

Figure 13.2 $e \ -^4He$ scattering cross section for the beam energy $E_e = 400$ MeV (a) Left: $\theta = 45°$, $Q^2 \to 0.08$ GeV^2, (b) Right: $\theta = 60°$, $Q^2 \to 0.1$ GeV^2 [425].

Let us now consider the process $e^-(k) + p(p) \to e^-(k') + N^*(p')$, where N^* may be a proton (elastic), or a Δ (resonance excitation). Using the definition of invariant mass square (W^2),

$$W^2 = p'^2 = (p+q)^2 = p^2 + q^2 + 2p \cdot q, \text{ where } q = k - k' = p' - p,$$

$$\Rightarrow W = \sqrt{s} = \sqrt{M^2 + q^2 + 2M(E_l - E_l')}.$$

For an elastic scattering process, $W = M$ and the aforementioned equation reduces to $q^2 = -2M(E_l - E_l')$. From Eq. (13.1), one may obtain $x = 1$ for the elastic scattering process and the corresponding peak in the cross section was observed. For example, in the case of elastic scattering of electrons off α particles, there is a peak at $x \sim 1$, at low Q^2 (Figure 13.2(a)); with increase in Q^2, the quasielastic peak appears which is diminished and broad. This represents the fact that there are subnuclear objects inside the helium nucleus. Had these nucleons been free particles at rest, the peak would have occurred at $x = \frac{1}{4}$ (corresponding to the two neutrons and two protons). The smeared peak implies that these nucleons are neither free nor static. With further increase in Q^2, the elastic peak gets suppressed and gradually with the decrease in x, the quasielastic peak becomes almost Q^2 independent. In general, one would expect that as $x \to 0$, the quasielastic peak should die away, whereas, in Figure 13.2(b), it may be observed that it is not the case. A finite value of the cross section represents the fact that as $x \to 0$, there are subnuclear objects other than nucleons from which e^- scattering is taking place; the scattering is understood to be the interaction of electrons with pions, rho mesons, etc. arising due to nucleon–nucleon interaction.

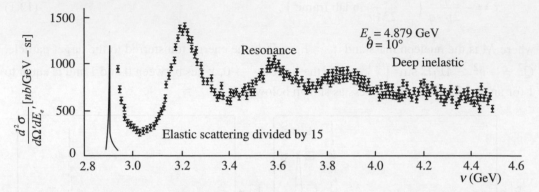

Figure 13.3 Electron–proton double differential scattering cross sections at E_e= 4.879 MeV; lab scattering angle $\theta = 10°$. The X-axis represents the energy transferred to the target proton ($\nu = E_e - E'_e$). The elastic peak (solid curve) has been divided by a factor of 15 to present the curves on the same scale [547].

Similarly, in the case of elastic electron scattering from the proton target (Figure 13.3), the peak is observed at low energy transfer or equivalently when the invariant mass $W \sim M$. Recall from Chapter 10, that the structure of the proton is described in terms of the electric and magnetic Sachs form factors; the size of the proton and its magnetic moment may also be understood with the help of these form factors. The Q^2 dependence of the form factors implies that the elastic cross section decreases with the increase in four-momentum transfer squared. In fact, with the increase in Q^2, the proton form factors decreases almost by two orders of magnitude in the Q^2 range of $1 < Q^2 < 10 \text{ GeV}^2$. With increase in energy transfer, one observes inelastic scattering which results in the production of one pion, multipions, etc. for which $W > M$. For large energy transfer, resonant states are formed, which indeed have been experimentally observed.

Now the question arises, what happens when Q^2 becomes very large? At high Q^2, the proton breaks up into a jet of hadrons (shown in Figure 13.4 (c)) and the final state is now a multiparticle state with large invariant mass. In contrast to the rapid fall of elastic proton form factors with increase in Q^2, the proton structure functions in the case of deep inelastic scattering process was observed to be independent of Q^2 and ν in the limit $Q^2 \to \infty, \nu \to \infty$ at a fixed value of x (true for a wide range of x). Thus, the cross section scales and the structure functions depend only on a single dimensionless variable, $\omega = \frac{1}{x} = \frac{2M\nu}{Q^2}$. This phenomenon is known as Bjorken scaling.

In this chapter, the general formalism for the charged lepton–nucleon DIS process is discussed in Section 13.2. In Subsection 13.2.1, a discussion on the Bjorken scaling and the Callan–Gross relation is given. Then, in Sections 13.3 and 13.4, this formalism is applied to describe the (anti)neutrino–nucleon charged and neutral current DIS processes following some discussion in Section 13.5 on non-perturbative and perturbative effects that modulate

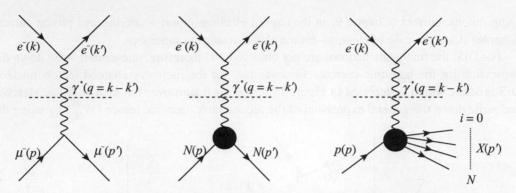

Figure 13.4 (a) Left: electron–muon scattering (b) Center: electron–proton elastic scattering, and (c) Right: electron–proton deep inelastic scattering.

the structure functions. After that, in Section 13.6, the parton model sum rules involving electromagnetic and weak structure functions have been discussed. The last section of this chapter deals with the phenomenon of quark–hadron duality.

13.2 Charged Lepton–nucleon DIS

The general reaction for the deep inelastic scattering process is given by (Figure 13.4(c))

$$l^-(k) + N(p) \longrightarrow l^-(k') + X(p'), \tag{13.2}$$

where $l = e, \mu$, $N = n, p$, and X represents the jet of hadrons in the final state.

To evaluate the differential scattering cross section for the process given in Eq. (13.2), the same method has been followed as discussed in Chapter 9 for $e^- \mu^- \to e^- \mu^-$ or in Chapter 10 for $e^- p \to e^- p$ scattering processes. We can notice that in all the three Feynman diagrams shown in Figure 13.4, the source of virtual photon is an electron (above the dashed line), which gets absorbed by a muon (scenario shown in Figure 13.4(a)), or a proton (scenario shown in Figure 13.4(b)) or it breaks the proton into a jet of hadrons (scenario shown in Figure 13.4(c)). Therefore, the leptonic current contributing to all the three processes (above the dashed line in Figure 13.4 (a), (b) and (c)) are the same. The differential scattering cross section can be expressed in terms of $\overline{\sum}\sum |\mathcal{M}|^2$ as

$$d\sigma \propto \overline{\sum}\sum |\mathcal{M}|^2, \tag{13.3}$$

which consists of the leptonic tensor,

$$L_{\mu\nu} = 4(k_\mu k'_\nu - g_{\mu\nu} k \cdot k' + k_\nu k'_\mu + m_l^2 g_{\mu\nu}), \tag{13.4}$$

and a muonic (Figure 13.4(a)) or a hadronic (Figure 13.4(b) or (c)) tensor $W_N^{\mu\nu}$. Since $\overline{\sum}\sum |\mathcal{M}|^2$ has to be a Lorentz invariant quantity, the hadronic vertex should also be a second rank tensor similar to the leptonic vertex tensor. A tensor of rank two for the lower vertex is formed by using muonic current (Chapter 9) in the case of electron–muon scattering and proton current (Chapter 10) in the case of electron–proton elastic scattering processes.

For DIS, the final state hadrons are not observed and therefore, one cannot write down the expression for the hadronic current. For example, for the inclusive charged lepton–nucleon DIS process, which is depicted in Figure 13.5, one must sum over the final hadronic states X and write down the general expression of the second rank hadronic tensor $(W_N^{\mu\nu})$ by using the

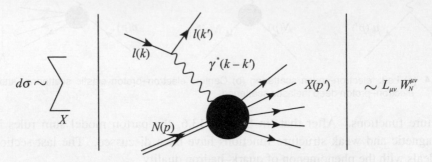

Figure 13.5 Charged lepton–nucleon inclusive scattering process.

bilinear covariants discussed in Chapter 4, along with p and q, the available four momenta. All possible combinations of tensors with four momenta p, q, and the metric tensor $g_{\mu\nu}$ at the hadronic vertex are the following

$$g^{\mu\nu}, \ p^\mu p^\nu, \ q^\mu q^\nu, \ \epsilon^{\mu\nu\lambda\sigma} p_\lambda q_\sigma, \ p^\mu q^\nu + p^\nu q^\mu, \ p^\mu q^\nu - p^\nu q^\mu,$$

using, which the most general form of the nucleon hadronic tensor is written as

$$
\begin{aligned}
W_N^{\mu\nu} = \ & -g^{\mu\nu} \, W_{1N}^{\mathrm{EM}}(\nu, Q^2) + \frac{p^\mu p^\nu}{M^2} \, W_{2N}^{\mathrm{EM}}(\nu, Q^2) \\
& - i\epsilon^{\mu\nu\lambda\sigma} \frac{p_\lambda q_\sigma}{2M^2} \, W_{3N}^{\mathrm{EM}}(\nu, Q^2) + \frac{q^\mu q^\nu}{M^2} \, W_{4N}^{\mathrm{EM}}(\nu, Q^2) \\
& + \frac{(p^\mu q^\nu + p^\nu q^\mu)}{M^2} \, W_{5N}^{\mathrm{EM}}(\nu, Q^2) + i\frac{(p^\mu q^\nu - p^\nu q^\mu)}{M^2} \, W_{6N}^{\mathrm{EM}}(\nu, Q^2),
\end{aligned}
\tag{13.5}
$$

where M is the mass of the target nucleon. The nucleon structure functions $W_{iN}^{\mathrm{EM}}(\nu, Q^2)$, $(i = 1 - 6)$ are the functions of ν and Q^2. $W_{3N}^{\mathrm{EM}}(\nu, Q^2)$ arises due to parity violation; therefore, it would not contribute in the electromagnetic interaction processes. Furthermore, the antisymmetric term that is related to $W_{6N}^{\mathrm{EM}}(\nu, Q^2)$ vanishes when contracted with the leptonic tensor, which has symmetric terms only. The conservation of vector current (CVC) at the hadronic vertex implies

$$q_\nu W_N^{\mu\nu} = q_\mu W_N^{\mu\nu} = 0,$$

which leads to the following relations

$$
\left.
\begin{aligned}
W_{4N}^{EM}(\nu, Q^2) &= \frac{M^2}{q^2} W_{1N}^{EM}(\nu, Q^2) + \left(\frac{p \cdot q}{q^2}\right)^2 W_{2N}^{EM}(\nu, Q^2), \\
W_{5N}^{EM}(\nu, Q^2) &= \frac{-p \cdot q}{q^2} W_{2N}^{EM}(\nu, Q^2).
\end{aligned}
\right\}
\tag{13.6}
$$

Thus, we are left with only two independent structure functions, which are generally chosen to be $W_{1N}^{EM}(\nu, Q^2)$ and $W_{2N}^{EM}(\nu, Q^2)$. The expression of $W_N^{\mu\nu}$ is written in terms of these two structure functions as:

$$
W_N^{\mu\nu} = \left(\frac{q^\mu q^\nu}{q^2} - g^{\mu\nu}\right) W_{1N}^{EM}(\nu, Q^2) + \left(p^\mu - \frac{p \cdot q}{q^2} q^\mu\right) \left(p^\nu - \frac{p \cdot q}{q^2} q^\nu\right) \frac{W_{2N}^{EM}(\nu, Q^2)}{M^2}.
\tag{13.7}
$$

Contraction of $L_{\mu\nu}$ with $W_N^{\mu\nu}$ leads to the following expression

$$
\begin{aligned}
L_{\mu\nu} W_N^{\mu\nu} &= 4 W_{1N}^{EM}(\nu, Q^2) \left\{-2k \cdot k' + 4(k \cdot k' - m_l^2) + 2\frac{q \cdot k q \cdot k'}{q^2} - (k \cdot k' - m_l^2)\right\} \\
&+ \frac{4 W_{2N}^{EM}(\nu, Q^2)}{M^2} \left\{2p \cdot k p \cdot k' - M^2(k \cdot k' - m_l^2) + \left(\frac{p \cdot q}{q^2}\right)^2\right. \\
&\times \left. \left\{2q \cdot k q \cdot k' - (k \cdot k' - m_l^2)q^2\right\} - \left(\frac{p \cdot q}{q^2}\right) \left\{2p \cdot k q \cdot k' + 2q \cdot k p \cdot k'\right.\right. \\
&- \left.\left. 2(k \cdot k' - m_l^2)p \cdot q\right\}\right\}.
\end{aligned}
\tag{13.8}
$$

In the limit of the massless lepton, Eq. (13.8) reduces to the following form:

$$
L_{\mu\nu} W^{\mu\nu} = 4 W_{1N}^{EM}(\nu, Q^2)\left[-q^2\right] + 4\frac{W_{2N}^{EM}(\nu, Q^2)}{M^2}\left[2p \cdot k p \cdot k' - M^2 k \cdot k'\right].
\tag{13.9}
$$

Averaging over the initial electron's and nucleon's spins and taking a sum over the final electron's and nucleon's spins, one gets the expression of $\overline{\sum}\sum |\mathcal{M}|^2$ in the Lab frame as:

$$
\overline{\sum}\sum |\mathcal{M}|^2 = \frac{e^4}{q^4} L_{\mu\nu} W^{\mu\nu} = \frac{e^4}{q^4} 2E_l E_l' \left\{2\sin^2\left(\frac{\theta}{2}\right) W_{1N}^{EM}(\nu, Q^2) + \cos^2\left(\frac{\theta}{2}\right) W_{2N}^{EM}(\nu, Q^2)\right\},
\tag{13.10}
$$

where θ is the Lab scattering angle between an incoming and an outgoing charged lepton.

To get the expression of the differential scattering cross section in the Lab frame, let us start with the general expression for the scattering cross section for a two-body scattering process; for example, $e^-(k) + p(p) \rightarrow e^-(k') + X(p')$:

$$
d\sigma = \frac{1}{4\sqrt{(p \cdot k)^2 - m_l^2 M^2}} \overline{\sum}\sum |\mathcal{M}|^2 \cdot \frac{d^3 k'}{(2\pi)^3 2E_l'} \cdot \frac{d^3 p'}{(2\pi)^3 2E_X}(2\pi)^4 \delta^4(p' + k' - p - k)
\tag{13.11}
$$

which simplifies to

$$d\sigma = \frac{1}{4ME_l}\overline{\sum}\sum|\mathcal{M}|^2\frac{1}{4\pi^2}\delta^4(p'+k'-p-k)\frac{d^3p'}{2E_X}\frac{d^3k'}{2E_l'}. \tag{13.12}$$

Performing momentum integration over d^3p' results in:

$$d\sigma = \frac{1}{4ME_l}\overline{\sum}\sum|\mathcal{M}|^2\frac{1}{4\pi^2}\delta^0(E_X+E_l'-E_l-M)\frac{1}{2E_X}\frac{|\vec{k}'|^2d|\vec{k}'|d\Omega_l'}{2E_l'} \tag{13.13}$$

$$\Rightarrow \frac{d^2\sigma}{d\Omega_l'dE_l'} = \frac{1}{4ME_l}\overline{\sum}\sum|\mathcal{M}|^2\frac{1}{4\pi^2}\delta^0(E_X+E_l'-E_l-M)\frac{E_l'}{4E_X}. \tag{13.14}$$

Using the expression of $\overline{\sum}\sum|\mathcal{M}|^2$ from Eq. (13.10), the differential scattering cross section is then given by

$$\frac{d^2\sigma}{d\Omega_l'dE_l'} = \frac{1}{4ME_l}\frac{e^4}{q^4}2E_lE_l'\left\{2\sin^2\left(\frac{\theta}{2}\right)W_{1N}^{EM}(\nu,Q^2)+\cos^2\left(\frac{\theta}{2}\right)W_{2N}^{EM}(\nu,Q^2)\right\}\frac{1}{4\pi^2}$$

$$\times\delta^0(E_X+E_l'-E_l-M)\frac{E_l'}{4}\frac{1}{M}\frac{M}{E_X}. \tag{13.15}$$

The structure functions $W_{1N}^{EM}(\nu,Q^2)$ and $W_{2N}^{EM}(\nu,Q^2)$ are redefined as

$$W_{1N}^{EM}(\nu,Q^2)\left\{\frac{1}{8M}\frac{1}{E_X}\right\}\delta^0(E_X+E_l'-E_l-M) \rightarrow W_{1N}^{EM}(\nu,Q^2), \tag{13.16}$$

$$W_{2N}^{EM}(\nu,Q^2)\left\{\frac{1}{8M}\frac{1}{E_X}\right\}\delta^0(E_X+E_l'-E_l-M) \rightarrow W_{2N}^{EM}(\nu,Q^2). \tag{13.17}$$

Keeping the same name is just a choice.

This results in the following expression for inclusive differential scattering cross section

$$\frac{d^2\sigma}{d\Omega_l'dE_l'} = \frac{\alpha^2}{4E_l^2\sin^4\left(\frac{\theta}{2}\right)}\left\{2\sin^2\left(\frac{\theta}{2}\right)W_{1N}^{EM}(\nu,Q^2)+\cos^2\left(\frac{\theta}{2}\right)W_{2N}^{EM}(\nu,Q^2)\right\}, \tag{13.18}$$

which can equivalently be also expressed in terms of Q^2 and ν distributions, that is, $\frac{d^2\sigma}{dQ^2d\nu}$. This is obtained using the following relation:

$$\frac{d^2\sigma}{dQ^2d\nu} = \frac{\pi}{E_lE_l'}\frac{d^2\sigma}{d\Omega_l'dE_l'} \tag{13.19}$$

resulting in

$$\frac{d^2\sigma}{dQ^2d\nu} = \frac{\pi\alpha^2}{4E_l^3E_l'\sin^4(\frac{\theta}{2})}\left\{2\sin^2\left(\frac{\theta}{2}\right)W_{1N}^{EM}(\nu,Q^2)+\cos^2\left(\frac{\theta}{2}\right)W_{2N}^{EM}(\nu,Q^2)\right\} \tag{13.20}$$

13.2.1 Bjorken scaling and parton model

In elastic $e^- - p$ scattering (Chapter 10), if one assumes $G_E(Q^2) = G_M(Q^2) = G(Q^2)$, where $G(Q^2) = \dfrac{1}{\left(1 + \frac{Q^2}{M_V^2}\right)^2}$ with M_V as the vector dipole mass, then the expression of the differential scattering cross section may be expressed as:

$$\frac{d^2\sigma}{d\Omega_l' dE_l'} = \frac{\alpha^2 G(Q^2)}{4E_l^2 \sin^4(\frac{\theta}{2})} \left\{ \cos^2\left(\frac{\theta}{2}\right) + \frac{Q^2}{2M^2} \sin^2\left(\frac{\theta}{2}\right) \right\} \delta\left(\nu - \frac{Q^2}{2M}\right). \quad (13.21)$$

It may be noticed that for the elastic $e^- - p$ scattering process, the form factors depend explicitly on Q^2 and an explicit mass scale (M_V) is present. However, if the proton is considered to be a point particle, that is, $G(Q^2) = 1$ and $M = m$ (mass of a point particle), then Eq. (13.21) gets modified to:

$$\frac{d^2\sigma}{d\Omega_l' dE_l'} = \frac{\alpha^2}{4E_l^2 \sin^4(\frac{\theta}{2})} \left\{ \cos^2\left(\frac{\theta}{2}\right) + \frac{Q^2}{2m^2} \sin^2\left(\frac{\theta}{2}\right) \right\} \delta\left(\nu - \frac{Q^2}{2m}\right), \quad (13.22)$$

which is the same as the differential cross section for $e\mu \to e\mu$ scattering process (Chapter 9). Comparing Eqs. (13.18) and (13.22), we may write in the limits of high ν and Q^2:

$$\left. \begin{array}{l} \nu W_2^{\text{point}}(\nu, Q^2) = \delta\left(1 - \frac{Q^2}{2m\nu}\right), \\[2mm] 2m W_1^{\text{point}}(\nu, Q^2) = \frac{Q^2}{2m\nu} \delta\left(1 - \frac{Q^2}{2m\nu}\right), \end{array} \right\} \quad (13.23)$$

where the identity $\delta\left(\nu - \frac{Q^2}{2m}\right) = \frac{1}{\nu}\delta\left(1 - \frac{Q^2}{2m\nu}\right)$ has been used. The aforementioned expressions show that for point particles (Figure 13.6), the structure functions are functions of one variable, $\frac{Q^2}{2m\nu}$, and not independent functions of Q^2 and ν (see Figure 13.7). In 1969,

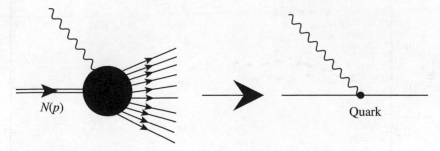

$N(p)$ Quark

Figure 13.6 Left: A virtual photon interacting with a hadron; Right: A virtual photon interacting with a point Dirac particle inside a hadron.

Bjorken had proposed that in the region of high ν and Q^2 ($\nu \to \infty$, $Q^2 \to \infty$), the structure functions $W_1(\nu, Q^2)$ and $W_2(\nu, Q^2)$ would only depend upon the ratio $\frac{Q^2}{\nu}(\sim x)$. To introduce the structure functions as a function of a single dimensionless variable $\omega = \frac{1}{x} = \frac{2M\nu}{Q^2}$, $W_1(\nu, Q^2)$ and $W_2(\nu, Q^2)$ are redefined as:

$$\left.\begin{array}{l} \nu W_2(\nu, Q^2) = F_2(x), \\[2mm] M W_1(\nu, Q^2) = F_1(x). \end{array}\right\} \tag{13.24}$$

In the case of DIS, the early results by Friedman and Kendall at SLAC [544] also showed that the results for $\frac{d^2\sigma}{d\Omega'_l dE'_l}$ in the region of high Q^2 and ν seem to depend only on one variable, that is, $\frac{Q^2}{2M\nu}$. This implies that in this kinematic limit, the cross section scales.

Now the scattering cross section given in Eq. (13.18) may be expressed in terms of the dimensionless nucleon structure functions as:

$$\frac{d^2\sigma}{d\Omega'_l dE'_l} = \frac{\alpha^2}{4M\nu E_l^2 \sin^4\left(\frac{\theta}{2}\right)} \left\{ 2\sin^2\left(\frac{\theta}{2}\right) \nu F_{1N}^{EM}(x) + \cos^2\left(\frac{\theta}{2}\right) M F_{2N}^{EM}(x) \right\}. \tag{13.25}$$

The phenomenon of scaling, that is, the dependence of $\frac{d^2\sigma}{d\Omega'_l dE'_l}$ on a single variable x suggests that in this limit of high Q^2 and ν, the deep inelastic electrons scattering from the constituents of the proton can be treated as point particles and the DIS cross section should be treated as the incoherent sum of elastic scattering from point-like constituents of the proton, as suggested by Bjorken. These constituents are called partons. The signature that the structure functions are now independent of Q^2 or ν may be realized by plotting $M W_{1N}^{EM}(\nu, Q^2)$ vs. Q^2 (or equivalently

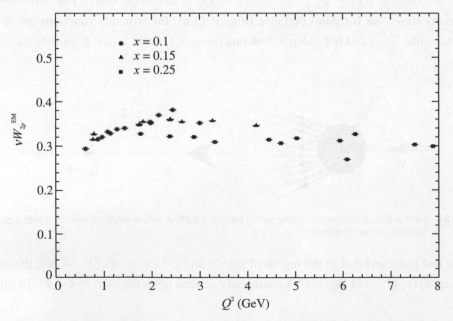

Figure 13.7 $\nu W_{2p}^{EM}(\nu, Q^2)$ vs. Q^2 at a fixed $x = \frac{1}{\omega}$. Experimental data are from SLAC [548].

ν); or $\nu W_{2N}^{EM}(\nu, Q^2)$ vs. Q^2 (or equivalently ν), at a fixed value of ω. For the first time, SLAC measured $\nu W_{2p}^{EM}(\nu, Q^2)$ vs. Q^2 at a fixed ω and found that the predictions of the parton model are correct, that is, the structure functions do scale. This can be seen from Figure 13.7.

The phenomenon of scaling observed in the SLAC experiment was explained in a simple model proposed by Feynman called the parton model which explains successfully many processes induced by electrons, muons, and neutrinos on nucleon and nuclear targets studied in the DIS region. The basic assumptions of this model are as follows:

i) A rapidly moving hadron appears as a jet of partons, all of which travel more or less in the same direction as that of the parent hadron (Figure 13.8).

Figure 13.8 A rapidly moving hadron.

ii) The basic process with free partons is calculated and then summed incoherently over the contributions of partons in the hadron (represented by the illustration on the left of Figure 13.9).

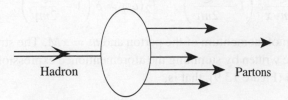

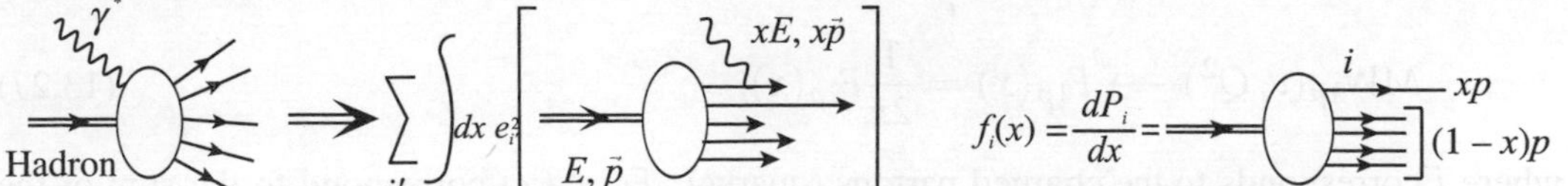

Figure 13.9 Figure on the left depicts the incoherent sum of the contributions and figure on the right represents momentum shared by the charged partons.

iii) The three-momentum of the hadron is shared among the partons and one defines the parton momentum distribution (represented on the right side of Figure 13.9) as $f_i(x) = \frac{dP_i}{dx}$, where $f_i(x)$ is the probability that the struck charged parton i carries a fraction x of the hadron's four-momentum p.

iv) These partons carry a fraction x of the hadron's momentum and energy (Table 13.1). All the fractions x add up to 1 such that

$$\sum_{i'} \int dx \, x \, f_{i'}(x) = 1,$$

where i' is the sum over the charged (quarks) as well as the neutral (gluons) partons.

Table 13.1 Kinematical quantities for hadron and parton.

Kinematic quantities	Hadron	Parton
Energy	E	xE
Momentum (longitudinal)	p_L	xp_L
Momentum (transverse)	$p_T = 0$	$xp_T = 0$
Mass	M	$m = xM$

The dimensionless structure functions for a parton struck by a charged lepton carrying a momentum fraction x and unit charge are given using Eqs. (13.23) and (13.24) as:

$$F_1(\omega) = \frac{Q^2}{4m\nu x}\delta\left(1 - \frac{Q^2}{2m\nu}\right), \quad F_2(\omega) = \delta\left(1 - \frac{Q^2}{2m\nu}\right),$$

where x is the fraction of momentum of the parton and $m = xM$. The structure functions for a proton F_{1p} and F_{2p} are written by summing the aforementioned expressions over all the partons that make up a proton (Figure 13.9), that is,

$$F_{2p}(\omega) = \sum_i \int dx\, e_i^2\, f_i(x)\, x\, \delta\left(x - \frac{1}{\omega}\right),$$

$$F_{1p}(\omega) = \frac{\omega}{2}F_{2p}(\omega),$$

leading to

$$\nu W_{2p}(\nu, Q^2) \longrightarrow F_{2p}(x) = \sum_i e_i^2\, x\, f_i(x), \tag{13.26}$$

$$M W_{1p}(\nu, Q^2) \longrightarrow F_{1p}(x) = \frac{1}{2x}F_{2p}(x), \tag{13.27}$$

where i corresponds to the charged partons (quarks). $F_{1p,2p}(x)$ correspond to the sum of the momentum fraction carried by all the quarks and antiquarks in the proton, weighted by the squares of the quark charges. Equation (13.27) leads to $F_{2p}(x) = 2xF_{1p}(x)$ which is known as the Callan–Gross relation (CGR) and is true when quarks are spin $\frac{1}{2}$ point objects.

Generally, the calculations for getting the proton ($F_{1,2}^p(x)$) or the neutron ($F_{1,2}^n(x)$) structure functions are performed in the four-flavor scheme assuming that heavy flavors (bottom and top) do not contribute as they are very massive in comparison to the nucleon's mass ($M << m_b$ or m_t). In the four-flavor scheme (considering only u, d, c, and s quarks), the proton structure function may be written in terms of the parton distribution functions as:

$$F_2^{ep}(x) = x\left[\frac{4}{9}\left(u(x) + \bar{u}(x) + c(x) + \bar{c}(x)\right) + \frac{1}{9}\left(d(x) + \bar{d}(x) + s(x) + \bar{s}(x)\right)\right]. \tag{13.28}$$

Similarly, in the case of the $e^- - n$ scattering process, where the role of up and down quarks are interchanged, one may write the structure function as:

$$F_2^{en}(x) = x\left[\frac{4}{9}\left(d(x) + \bar{d}(x) + c(x) + \bar{c}(x)\right) + \frac{1}{9}\left(u(x) + \bar{u}(x) + s(x) + \bar{s}(x)\right)\right]. \tag{13.29}$$

In these expressions, $u(x)$ is the probability distribution of the u quark in the proton or the d quark in the neutron ($u^p(x) = d^n(x) = u(x)$); similarly, $d(x)$ is the probability distribution of the d quark in the proton or the u quark in the neutron ($d^p(x) = u^n(x) = d(x)$), while the momentum distributions carried over by all the other flavors of quarks and antiquarks are taken to be the same as in the proton and neutron. $xu(x)$ represents the probability of finding an up quark with the target nucleon's momentum fraction x and so on. These probabilities are also known as parton distribution functions (PDFs). For an isoscalar nucleon target, the structure function is given by:

$$F_2^{eN}(x) = \frac{F_2^{ep}(x) + F_2^{en}(x)}{2}. \tag{13.30}$$

As partons carry the momentum fraction x of the target hadron's momentum, summing over the momentum of all the partons (charged as well as neutral ones) should result in 1, that is,

$$\int_0^1 dx \, x \, (u(x) + d(x) + s(x) + \bar{u}(x) + \bar{d}(x) + \bar{s}(x)) + \int_0^1 dx \, x \, g(x) = 1. \tag{13.31}$$

Experimentally, it was found that

$$\int_0^1 dx \, x \, (u(x) + d(x) + s(x) + \bar{u}(x) + \bar{d}(x) + \bar{s}(x)) \simeq 0.54. \tag{13.32}$$

Using this in Eq. (13.31) gives

$$\int_0^1 dx \, x \, g(x) \simeq 0.46. \tag{13.33}$$

Thus, approximately 50% of the hadron's momentum is carried by gluons. In the 'naive' parton model, one assumes that there are valence quarks (hereafter represented with a suffix v) and sea quarks (hereafter represented with a suffix s). The valence quarks contribute to the quantum numbers of hadron, while the sea quarks arise from the quark-antiquark pairs and gluons. In turn, the annihilation of the quark–antiquark pairs also gives rise to gluons. For example, in the case of proton, in the four-flavor scheme ($u, d, c,$ and s), the probability distribution of various quarks in proton are:

$$
\begin{aligned}
u(x) &= u_v(x) + u_s(x) \; ; \quad d(x) = d_v(x) + d_s(x) \; ; \quad u_v(x) = 2d_v(x), \\
s_v(x) &= \bar{u}_v(x) = \bar{d}_v(x) = \bar{s}_v(x) = c_v(x) = \bar{c}_v(x) = 0, \\
u_s(x) &= \bar{u}_s(x) = d_s(x) = \bar{d}_s(x) = s_s(x) = \bar{s}_s(x) = c_s(x) = \bar{c}_s(x) \equiv S(x).
\end{aligned}
\tag{13.34}
$$

For the nucleon structure functions, various parameterizations of parton density distribution functions are available in the literature such as GRV [549], MSTW [550], CTEQ6.6 [551], MMHT [552], etc. The dependence of the parton probability densities on the Bjorken scaling variable is shown in Figure 13.10 in the range of $2 \le Q^2 \le 10 \, \text{GeV}^2$. From the figure, one can see that valence quarks dominate in the region of mid and high x while sea quarks dominate in the low x region. Moreover, with increase in Q^2, the contribution of valence quarks get reduced, whereas the contribution of sea quarks get enhanced.

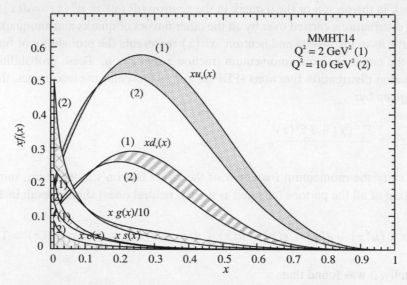

Figure 13.10 Parton density distribution functions using MMHT PDFs parameterization [552] at leading order for $2 \text{ GeV}^2 \leq Q^2 \leq 10 \text{ GeV}^2$.

The contribution of valence and sea quarks in the structure functions can be understood by taking the ratio of $\frac{F_2^{en}(x)}{F_2^{ep}(x)}$ and using Eq. (13.34) which results in:

$$\frac{F_2^{en}(x)}{F_2^{ep}(x)} = \frac{x\left[\frac{1}{9}\left(u_v(x) + 2S(x)\right) + \frac{4}{9}\left(d_v(x) + 2S(x)\right) + \frac{1}{9}\left(2S(x)\right) + \frac{4}{9}\left(2S(x)\right)\right]}{x\left[\frac{4}{9}\left(u_v(x) + 2S(x)\right) + \frac{1}{9}\left(d_v(x) + 2S(x)\right) + \frac{1}{9}\left(2S(x)\right) + \frac{4}{9}\left(2S(x)\right)\right]},$$

$$= \frac{\left[\frac{1}{9}u_v(x) + \frac{4}{9}d_v(x) + \frac{1}{9}\left(20S(x)\right)\right]}{\left[\frac{4}{9}u_v(x) + \frac{1}{9}d_v(x) + \frac{1}{9}\left(20S(x)\right)\right]}$$

which simplifies to

$$\frac{F_2^{en}(x)}{F_2^{ep}(x)} = \begin{cases} \frac{[u_v(x) + 4d_v(x)]}{[4u_v(x) + d_v(x)]} & \text{if valence quarks dominate} \\ \frac{1}{4} & \text{if valence } u \text{ quark dominates} \\ 4 & \text{if valence } d \text{ quark dominates} \\ 1 & \text{if sea-quarks dominate.} \end{cases}$$

The ratio $\frac{F_2^{en}(x)}{F_2^{ep}(x)}$ has been experimentally measured in different experiments for a wide range of x. The experimental observations are presented in Figure 13.11, from which it can be inferred that the valence u quarks in the protons (and d quarks in the neutrons) dominate at large x; the

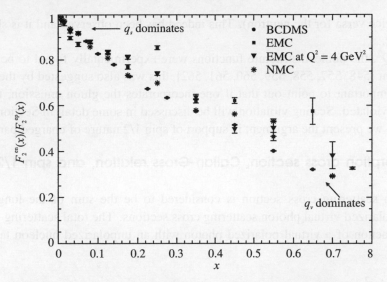

Figure 13.11 Experimental measurements for $\frac{F_2^{en}(x)}{F_2^{ep}(x)}$ vs. x [553, 554, 555, 556] for proton and neutron targets in the case of electromagnetic interaction.

sea quarks dominate at small x. However, valence d quarks (u quarks) never dominate in the case of proton (neutron). If one plots the difference of the electromagnetic structure functions in neutrons ($F_2^{en}(x)$) and protons ($F_2^{ep}(x)$) by using Eqs. (13.28), (13.29), and (13.34), that is,

$$F_2^{ep}(x) - F_2^{en}(x) = \frac{1}{3}x[u_v(x) - d_v(x)],\tag{13.35}$$

then the peak should occur at $x = 1/3$ (two valence u quarks and one valence d quark in the

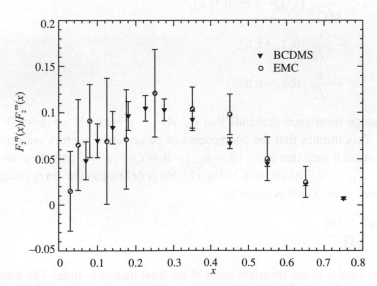

Figure 13.12 Experimental measurements for $F_2^{ep}(x) - F_2^{en}(x)$ vs. x [553, 554] for proton and neutron targets in the case of electromagnetic interaction.

proton and vice versa for the neutron). This indeed has been observed and it is shown here in Figure 13.12.

At low Q^2, the nucleon structure functions were experimentally found to be x as well as Q^2 dependent [548, 557, 558, 559, 560, 561, 562]; this was also suggested by the perturbative QCD. It is important to point out that if one incorporates the gluon emission, then Bjorken scaling gets violated. Scaling violation will be discussed in some detail in Section 13.5. In the next section, we present the argument in support of spin 1/2 nature of charged partons.

Photoabsorption cross section, Callan–Gross relation, and spin 1/2 nature of quarks

The electron scattering cross section is considered to be the sum of the longitudinal and transverse polarized virtual photon scattering cross sections. The total scattering cross section for the interaction of a virtual polarized photon with an unpolarized nucleon target is given by [238]:

$$\sigma^\lambda = \frac{4\pi^2\alpha}{K} \epsilon^\lambda_\mu \epsilon^{\lambda*}_\nu W^{\mu\nu}_N,\tag{13.36}$$

where ϵ^λ_μ is the photon polarization vector such that ϵ^0_μ describes the longitudinally polarized virtual photons for zero helicity state, that is, $\lambda = 0$, while $\epsilon^{(+1)}$ and $\epsilon^{(-1)}$ describe the transversely polarized states with helicity $\lambda = +1$ and $\lambda = -1$, respectively, corresponding to right- and left-handed virtual photons. Let us assume that the photon is traversing along the Z-axis with $q^\mu = (\sqrt{Q^2 + v^2}, 0, 0, v)$; then, the corresponding polarization states are given by:

$$\epsilon^\mu_{\lambda=0} = \frac{1}{\sqrt{Q^2}} (\sqrt{Q^2 + v^2}, 0, 0, v),\tag{13.37}$$

$$\epsilon^\mu_{\lambda=+1} = -\frac{1}{\sqrt{2}} (0, 1, +i, 0),\tag{13.38}$$

$$\epsilon^\mu_{\lambda=-1} = +\frac{1}{\sqrt{2}} (0, 1, -i, 0).\tag{13.39}$$

The Lorenz gauge invariance demands that $\partial_\mu A^\mu = 0$ with $A^\mu = \epsilon^\mu e^{-iq\cdot x}$ which leads to $q \cdot \epsilon = 0$. This implies that the components of polarization vectors are orthogonal to the momentum transfer q such that $\epsilon_{(+1)} \cdot q = \epsilon_{(-1)} \cdot q = \epsilon_{(0)} \cdot q = 0$ and obey the orthogonality relation $\epsilon_{(\pm 1)} \cdot \epsilon_{(0)} = 0$. The factor K in Eq. (13.36) is defined as the energy required to create the final hadronic state X and is given by:

$$K = \frac{W^2 - M^2}{2M},$$

with W^2 as the square of the invariant mass of the final hadronic state. The nucleon structure functions $F^{EM}_{1N}(x, Q^2)$ and $F^{EM}_{2N}(x, Q^2)$ are related to the transverse (σ_T) and longitudinal (σ_L)

photo absorption scattering cross sections and are given by using Eqs. (13.7) and (13.36) as [238]:

$$\sigma_T^{EM}(x, Q^2) = \frac{\sigma(\lambda = +1) + \sigma(\lambda = -1)}{2} = \frac{4\pi^2\alpha}{KM} F_{1N}^{EM}(x, Q^2) \tag{13.40}$$

$$\sigma_L^{EM}(x, Q^2) = \sigma(\lambda = 0) = \frac{4\pi^2\alpha}{KM} \left[\left(1 + \frac{\nu^2}{Q^2}\right) \frac{M}{\nu} F_{2N}^{EM}(x, Q^2) \right.$$
$$\left. - F_{1N}^{EM}(x, Q^2) \right]. \tag{13.41}$$

It is important to point out that in the case of electromagnetic interaction, where parity is conserved, $\sigma(\lambda = +1) = \sigma(\lambda = -1)$. Hence, the differential scattering cross section given in Eq. (13.18) may be recast as

$$\frac{d^2\sigma_N^{EM}}{d\Omega_l' dE_l'} = \left(\frac{\alpha E_l'}{4\pi^2 Q^2 M E_l(1-\epsilon)}\right) \sigma_T^{EM}(x, Q^2) \left[1 + \epsilon R_L^{EM}(x, Q^2)\right], \tag{13.42}$$

where

$$R_L^{EM}(x, Q^2) = \frac{\sigma_L^{EM}(x, Q^2)}{\sigma_T^{EM}(x, Q^2)} \tag{13.43}$$

and

$$\epsilon = \left[1 + 2\left(1 + \frac{\nu^2}{Q^2}\right)\tan^2\frac{\theta}{2}\right]^{-1} = \frac{1 - y - \frac{M^2 x^2 y^2}{Q^2}}{1 - y + \frac{y^2}{2} + \frac{M^2 x^2 y^2}{Q^2}}, \quad \text{with } y = \frac{\nu}{E}. \tag{13.44}$$

This expression for $R_L^{EM}(x, Q^2)$ defines the ratio of the probability of longitudinally polarized to transversely polarized photon absorption cross sections. It may be noticed from Eqs. (13.42) and (13.44) that for $y \rightarrow 0$, the value of ϵ will be large; therefore, the contribution to the differential scattering cross section from the longitudinal part (σ_L) will be maximum. However, for $y \rightarrow 1$, ϵ approaches zero and the transverse part (σ_T) will give the maximum contribution to the scattering cross section.

Moreover, by using Eqs. (13.40) and (13.41), one may write the nucleon structure functions $F_{1N}^{EM}(x, Q^2)$ and $F_{2N}^{EM}(x, Q^2)$ in terms of $\sigma_T^{EM}(x, Q^2)$ and $\sigma_L^{EM}(x, Q^2)$ as

$$F_{1N}^{EM}(x, Q^2) = \frac{KM}{4\pi^2\alpha}\sigma_T^{EM}(x, Q^2),$$

$$F_{2N}^{EM}(x, Q^2) = \frac{K\nu}{4\pi^2\alpha}\left(1 + \frac{\nu^2}{Q^2}\right)^{-1}[\sigma_L^{EM}(x, Q^2) + \sigma_T^{EM}(x, Q^2)]. \tag{13.45}$$

It may be noticed that $F_{1N}^{EM}(x, Q^2)$ has contribution only from transversely polarized photons while the contribution to $F_{2N}^{EM}(x, Q^2)$ comes from both the longitudinally as well as the transversely polarized photons. Therefore, in order to obtain the contribution from the longitudinal part only, the longitudinal structure function $F_{LN}^{EM}(x, Q^2)$ is defined as:

$$F_{LN}^{EM}(x, Q^2) = \frac{2x\nu(1-x)M}{4\pi^2\alpha}\sigma_L^{EM}(x, Q^2) = \gamma^2 F_{2N}^{EM}(x, Q^2) - 2x F_{1N}^{EM}(x, Q^2), \quad (13.46)$$

where $\quad \gamma^2 = \left(1 + \frac{4M^2x^2}{Q^2}\right)$.

To illustrate this further, let us discuss the ratio $R_L^{EM}(x)$ for a nucleon target defined in Eq. (13.43) by redefining $R_L^{EM}(x, Q^2)$ in terms of structure functions, that is,

$$R_L^{EM}(x, Q^2) = \frac{\sigma_L^{EM}(x, Q^2)}{\sigma_T^{EM}(x, Q^2)} = \frac{F_{LN}^{EM}(x, Q^2)}{2x F_{1N}^{EM}(x, Q^2)}$$

$$= \left(1 + \frac{4M^2x^2}{Q^2}\right)\frac{F_{2N}^{EM}(x, Q^2)}{2x F_{1N}^{EM}(x, Q^2)} - 1. \quad (13.47)$$

The ratio $\frac{2x F_{1p}^{EM}(x)}{F_{2p}^{EM}(x)}$ has been measured at SLAC [563] and found to be $\frac{2x F_{1p}^{EM}(x)}{F_{2p}^{EM}(x)} \approx 1$ (Figure 13.13: left panel). It also verifies the Callan–Gross relation (Eq.13.27). Using CGR in Eq. (13.47), we may write:

$$R_L^{EM}(x, Q^2) = \frac{\sigma_L^{EM}(x, Q^2)}{\sigma_T^{EM}(x, Q^2)} = \frac{4M^2x^2}{Q^2}, \quad (13.48)$$

which is referred to as the 'CGR limit'; it tends to zero as Q^2 goes to infinity, that is,

$$R_L^{EM}(x, Q^2) \to 0 \quad \text{as} \quad Q^2 \to \infty. \quad (13.49)$$

This is evidence for the spin 1/2 nature of quarks (Figure 13.13: right panel).

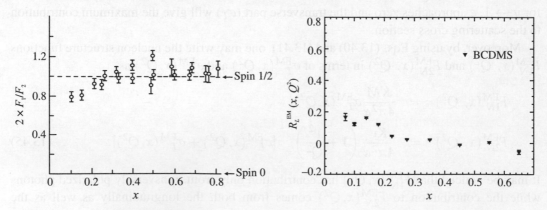

Figure 13.13   Left panel: Experimental verification of the Callan–Gross relation for the spin 1/2 nature of quarks [563]. Right panel: Experimental results for the ratio of $R_L^{EM}(x, Q^2) = \frac{F_L^{EM}(x,Q^2)}{2x F_1^{EM}(x,Q^2)}$ [557].

Furthermore, to understand the spin 1/2 nature of charged partons, let us recall the elastic electron scattering with spin 1/2 (Eq. (9.16) and Eq. (10.55)) as well as with spin zero particles (Eq. (10.24)). Notice that for the spin 1/2 particles, the expression of differential cross section have terms both with $\sin^2 \theta$ and $\cos^2 \theta$ (as shown for $e^- - p$ DIS in Eq. (13.25)), while for the spin 0 particle, terms with only $\cos^2 \theta$ contribute (Eq. (10.24)). Suppose that quarks have zero spin; then by comparing Eq. (13.25) with Eq. (10.24), it may be observed that the transverse structure function $F_{1N}^{EM}(x) \to 0$, which leads to $\sigma_T^{EM}(x, Q^2) \to 0$, implying that $R_L^{EM}(x, Q^2) \to \infty$. While, for the spin 1/2 quarks, both the structure functions $F_{1N}^{EM}(x)$ and $F_{2N}^{EM}(x)$ contribute which are related by the Callan–Gross relation (Figure 13.13).

This behavior can also be understood by looking at Figure 13.14, where the collision of a quark (constituent of proton) with an intermediate virtual photon is shown in the Breit frame. If one choose the Z-axis along the three momentum $\vec{p}$, then from the conservation of

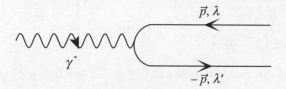

Figure 13.14 Head on collision of a quark with a virtual photon. λ and λ' denote the initial and final helicity states.

angular momentum (J_Z), a spin 0 quark cannot absorb a transverse photon as it has left- and right-handed helicity states $(\lambda = \pm 1)$ which results in

$$\sigma_T^{EM} \to 0 \quad \Rightarrow R_L^{EM} = \frac{\sigma_L^{EM}}{\sigma_T^{EM}} \to \infty.$$

For a spin half quark whose helicity is conserved in a high energy interaction process, the possible interaction is only with the transverse photons with left- and right-handed helicities. Therefore, the contribution from the longitudinal photons will be zero,

$$\sigma_L^{EM} \to 0, \quad \Rightarrow R_L^{EM} = \frac{\sigma_L^{EM}}{\sigma_T^{EM}} \to 0.$$

In the electron scattering experiments at BCDMS [557], a very small but finite (close to zero) value of the ratio $R_L^{EM}(x, Q^2)$ was measured (Figure 13.13), which is an evidence of the spin 1/2 nature of partons.

13.2.2 Differential scattering cross section in terms of dimensionless variables x and y

The differential scattering cross section $\frac{d^2\sigma}{d\Omega' dE'}$ given in Eq. (13.25) may be expressed in terms of the dimensionless variables, known as the Bjorken variable x and inelasticity $y \left(= \frac{p \cdot q}{p \cdot k} = \frac{\nu}{E_l} \right)$, using the relationship:

$$\frac{\pi}{E_l E_l'} dQ^2 d\nu = d\Omega_l' dE_l' = \frac{2ME_l}{E_l'} \pi y \, dx \, dy, \tag{13.50}$$

resulting in

$$\frac{d^2\sigma}{dxdy} = \frac{8\pi\alpha^2 ME_l}{Q^4}\left[F_{2N}^{EM}(x)\left(1-y-\frac{Mxy}{2E_l}\right) + xy^2F_{1N}^{EM}(x)\right]. \tag{13.51}$$

For the study of the nucleon structure, a complete knowledge of the structure functions over the entire range of the scaling variables is required, which for a non-zero leptonic mass, lies in the following kinematical range

$$\frac{m_l^2}{2M(E_l - m_l)} \leq x \leq 1 \tag{13.52}$$

$$(y_1 - y_2) \leq y \leq (y_1 + y_2), \tag{13.53}$$

where

$$y_1 = \frac{1 - m_l^2\sqrt{\left(\frac{1}{2ME_lx} + \frac{1}{2E_l^2}\right)}}{2\left(1 + \frac{Mx}{2E_l}\right)} \quad \text{and} \quad y_2 = \frac{\sqrt{\left(1 - \frac{m_l^2}{2ME_lx}\right)^2 - \frac{m_l^2}{E_l^2}}}{2\left(1 + \frac{Mx}{2E_l}\right)}. \tag{13.54}$$

In the massless lepton limit, this will reduce to $0 \leq x \leq 1$ and $0 \leq y \leq \frac{1}{1+\frac{Mx}{2E_l}}$. In Figure 13.15, the allowed kinematic regions for elastic, inelastic, and deep inelastic scattering processes are shown in the $Q^2 - \nu$ plane. The line of $W = M$ corresponds to the kinematical limit of elastic scattering, where $Q^2 = 2M\nu$, that is, $x = 1$. To separate the kinematical region in the (Q^2, ν)

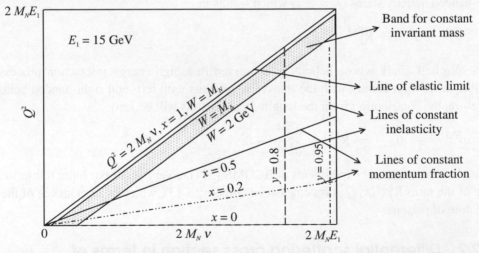

Figure 13.15 Variation of kinematic variables such as x, y, and W is shown in the $Q^2 - \nu$ plane.

plane from the resonance region, a band is shown for the constant values of invariant mass W separate from the mass of lowest lying resonance ($P_{33}(1232)$), that is, $W = M_\Delta$ to $W = 2$ GeV. It is noticeable that the lines of constant invariant mass are parallel to the line of elastic limit, that is, $W = M$ and $x = 1$. The kinematical region of DIS, corresponding to the different values of the Bjorken variable x such as 0.5, 0.2, and 0, is shown, for which the corresponding lines lie below the resonance band except for some overlapped part in the low ν region.

DIS cross section (Eq. (13.51)) may also be written in terms of the Mandelstam variables s, t, and u,

$$
\begin{aligned}
s &= (k+p)^2 = (k'+p')^2 \simeq 2p \cdot k \text{ or } 2p' \cdot k', \\
t &= (k-k')^2 = (p'-p)^2 \simeq -2k \cdot k' \text{ or } -2p \cdot p', \\
u &= (p-k')^2 = (p'-k)^2 \simeq -2p \cdot k' \text{ or } -2p' \cdot k, \text{ and using them we may write} \\
x &= \frac{Q^2}{2M\nu} = -\frac{t}{s+u} \text{ and } y = \frac{s+u}{s},
\end{aligned}
\tag{13.55}
$$

resulting in

$$
\frac{d^2\sigma^{\text{EM}}}{dtdu} = \frac{4\pi\alpha^2}{2t^2s^2(s+u)} \left\{ 2xF_{1N}^{\text{EM}}(x)(s+u)^2 - 2usF_{2N}^{\text{EM}}(x) \right\}.
\tag{13.56}
$$

13.3 Deep Inelastic Charged Current $\nu_l/\bar{\nu}_l - N$ Scattering

The charged current deep inelastic scattering for the weak interaction process (shown in Figure 13.16)

$$
\nu_l/\bar{\nu}_l(k) + N(p) \to l^-/l^+(k') + X(p'); \quad l = e, \mu
\tag{13.57}
$$

takes place via the exchange of vector bosons (W^+/W^-) when (anti)neutrino beam interact with a target nucleon (N) and gives rise to a jet of hadrons (X) in the final state. For the reaction in Eq. (13.57), the quantities within the parenthesis correspond to the four-momenta of the respective particle.

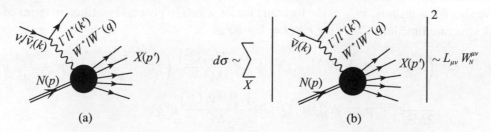

(a)   (b)

Figure 13.16   (a) Feynman diagram for the (anti)neutrino induced deep inelastic scattering process, (b) Summation over the final hadronic state X and its amplitude square corresponds to the scattering cross section.

The $V - A$ interaction allows only neutrinos with negative (left-handed) helicity and antineutrinos with positive (right-handed) helicity to participate in the process. The leptonic and hadronic vertices are shown in Figure 13.17, where the strength of the weak interaction, $\frac{-ig_W}{2\sqrt{2}}$ at the vertices is related to G_F (Chapter 8). For an inclusive $\nu_l/\bar{\nu}_l$ induced DIS process on a free nucleon target in the laboratory frame (Figure 13.16), the double differential scattering cross section is expressed as

$$\frac{d^2\sigma_N^{WI}}{d\Omega_l' dE_l'} = \frac{1}{2\pi^2} \frac{|\vec{k}'|}{|\vec{k}|} \overline{\sum}\sum |\mathcal{M}|^2 , \tag{13.58}$$

where the amplitude square is given by

$$\overline{\sum}\sum |\mathcal{M}|^2 = \frac{G_F^2}{2} \left(\frac{M_W^2}{Q^2 + M_W^2}\right)^2 L_{\mu\nu}^{WI} W_N^{\mu\nu}. \tag{13.59}$$

$L_{\mu\nu}^{WI}$ is the leptonic tensor and is expressed as

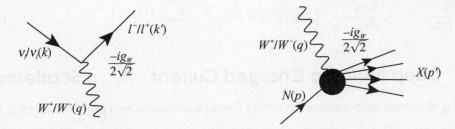

Figure 13.17 Feynman representation for leptonic and hadronic vertices (left and right columns) in the case of weak interaction. Feynman diagram showing the $\nu - N$ scattering for the summed over hadronic states X.

$$L_{\mu\nu}^{WI} = 8(\underbrace{k_\mu k_\nu' + k_\nu k_\mu' - k.k' g_{\mu\nu}}_{\text{symmetric}} \underbrace{\pm i\epsilon_{\mu\nu\rho\sigma} k^\rho k'^\sigma}_{\text{antisymmetric}}), \tag{13.60}$$

It has an antisymmetric term due to the axial vector part with a +ve sign for antineutrinos and a −ve sign for neutrinos. $W_N^{\mu\nu}$ is the hadronic tensor which in general is written in terms of the weak structure functions $W_{iN}^{WI}(\nu, Q^2)$ $(i = 1 - 6)$ as

$$W_N^{\mu\nu} = \left(\frac{q^\mu q^\nu}{q^2} - g^{\mu\nu}\right) W_{1N}^{WI}(\nu, Q^2) + \frac{W_{2N}^{WI}(\nu, Q^2)}{M^2} \left(p^\mu - \frac{p.q}{q^2} q^\mu\right)\left(p^\nu - \frac{p.q}{q^2} q^\nu\right)$$
$$- \frac{i}{2M^2}\epsilon^{\mu\nu\rho\sigma} p_\rho q_\sigma W_{3N}^{WI}(\nu, Q^2) + \frac{W_{4N}^{WI}(\nu, Q^2)}{M^2} q^\mu q^\nu$$
$$+ \frac{W_{5N}^{WI}(\nu, Q^2)}{M^2}(p^\mu q^\nu + q^\mu p^\nu) + \frac{i}{M^2}(p^\mu q^\nu - q^\mu p^\nu) W_{6N}^{WI}(\nu, Q^2). \tag{13.61}$$

In the limit of massless lepton $(m_l \to 0)$, the terms depending on $W_{4N}^{WI}(\nu, Q^2)$, $W_{5N}^{WI}(\nu, Q^2)$, and $W_{6N}^{WI}(\nu, Q^2)$ will not contribute to the scattering cross section. The additional structure function $W_{3N}^{WI}(\nu, Q^2)$ arises due to parity violation in weak interactions. Therefore, it carries opposite signs for neutrinos and antineutrinos. $W_N^{\mu\nu}$ in the limit of $m_l \to 0$ are then given by:

$$W_N^{\mu\nu} = \left(\frac{q^\mu q^\nu}{q^2} - g^{\mu\nu}\right) W_{1N}^{WI}(\nu, Q^2) + \frac{W_{2N}^{WI}(\nu, Q^2)}{M^2} \left(p^\mu - \frac{p.q}{q^2} q^\mu\right)\left(p^\nu - \frac{p.q}{q^2} q^\nu\right)$$
$$- \frac{i}{2M^2}\epsilon^{\mu\nu\rho'\sigma'} p_{\rho'} q_{\sigma'} W_{3N}^{WI}(\nu, Q^2). \tag{13.62}$$

Using Eqs. (13.59), (13.60), and (13.62), the differential scattering cross section given in Eq. (13.58) simplifies to

$$
\frac{d^2\sigma_N^{WI}}{d\Omega' dE'} = \frac{G_F^2 E_l'^2 \cos^2\left(\frac{\theta}{2}\right)}{2\pi^2} \left(\frac{M_W^2}{M_W^2 + Q^2}\right)^2 \left[2\tan^2\left(\frac{\theta}{2}\right) W_{1N}^{WI}(x,Q^2) \right.
$$
$$
\left. + W_{2N}^{WI}(x,Q^2) \pm \left(\frac{E_l + E_l'}{M}\right)\tan^2\left(\frac{\theta}{2}\right) W_{3N}^{WI}(x,Q^2)\right]. \tag{13.63}
$$

Following the same analogy as discussed in Section 13.2, the weak nucleon structure functions $W_{iN}^{WI}(\nu,Q^2)$ $(i=1,2,3)$ are written in terms of the dimensionless nucleon structure functions $F_{iN}^{WI}(x,Q^2)$ $(i=1,2,3)$ as:

$$
\left.\begin{aligned}
M W_{1N}^{WI}(\nu,Q^2) &= F_{1N}^{WI}(x,Q^2), \\
\nu W_{2N}^{WI}(\nu,Q^2) &= F_{2N}^{WI}(x,Q^2), \\
\nu W_{3N}^{WI}(\nu,Q^2) &= F_{3N}^{WI}(x,Q^2).
\end{aligned}\right\} \tag{13.64}
$$

Using Eqs. (13.64), the differential scattering cross section is then written as:

$$
\frac{d^2\sigma_N^{WI}}{d\Omega_l' dE_l'} = \frac{G_F^2 E_l'^2 \cos^2\left(\frac{\theta}{2}\right)}{2\pi^2 M\nu} \left(\frac{M_W^2}{M_W^2 + Q^2}\right)^2 \left[2\nu\tan^2\left(\frac{\theta}{2}\right) F_{1N}^{WI}(x,Q^2) \right.
$$
$$
\left. + M F_{2N}^{WI}(x,Q^2) \pm (E_l + E_l')\tan^2\left(\frac{\theta}{2}\right) F_{3N}^{WI}(x,Q^2)\right], \tag{13.65}
$$

where the term corresponding to $F_{3N}^{WI}(x,Q^2)$ will have a positive sign for neutrinos and a negative sign for antineutrinos. This scattering cross section can also be written in terms of the Bjorken scaling variables x and y as

$$
\frac{d^2\sigma_N^{WI}}{dx dy} = \frac{G_F^2 s}{2\pi} \left[xy^2 F_{1N}^{WI}(x,Q^2) + \left(1 - y - \frac{Mxy}{2E_l}\right) F_{2N}^{WI}(x,Q^2) \right.
$$
$$
\left. \pm xy\left(1 - \frac{y}{2}\right) F_{3N}^{WI}(x,Q^2)\right], \tag{13.66}
$$

where the propagator term $\left(\frac{M_W^2}{M_W^2 + Q^2}\right)^2 \approx 1$ for $M_W^2 >> Q^2$. Since the virtual bosons may have longitudinal as well as transverse polarization states, the nucleon structure functions are also written in terms of the longitudinal and transverse absorption cross section as

$$
\sigma_{(\lambda=\pm 1)}^{WI}(x,Q^2) = \frac{G\pi\sqrt{2}}{KM}\left(F_{1N}^{WI}(x,Q^2) \pm \frac{1}{2}\sqrt{1 + \frac{4M^2 x^2}{Q^2}} F_{3N}^{WI}(x,Q^2)\right), \tag{13.67}
$$

$$
\sigma_T^{WI}(x,Q^2) = \frac{\sigma_{(\lambda=+1)}^{WI}(x,Q^2) + \sigma_{(\lambda=-1)}^{WI}(x,Q^2)}{2}, \tag{13.68}
$$

$$
\sigma_{(\lambda=0)}^{WI}(x,Q^2) = \sigma_L^{WI}(x,Q^2) = \frac{G\pi\sqrt{2}}{2\,xKM} F_{LN}^{WI}(x,Q^2), \tag{13.69}
$$

where the longitudinal structure function for weak interactions is defined as

$$F_{LN}^{WI}(x, Q^2) = \left(1 + \frac{4M^2x^2}{Q^2}\right) F_{2N}^{WI}(x, Q^2) - 2xF_{1N}^{WI}(x, Q^2).$$ (13.70)

For weak interactions at low Q^2, the transverse structure function ($F_{1N}^{WI}(x, Q^2)$) vanishes while the longitudinal structure function ($F_{LN}^{WI}(x, Q^2)$) dominates. This is due to the contribution from vector (V) as well as axial vector (A) components. For the transverse structure function, both axial vector and vector components vanish at low Q^2. For the longitudinal structure function too, the vector component becomes zero by applying CVC but the axial vector component which is related to the PCAC, gives a non-zero contribution. Therefore, the ratio $R_L^{WI}(x, Q^2)$ diverges in the case of weak interaction induced processes, that is,

$$R_L^{WI}(x, Q^2) \quad = \quad \frac{F_{LN}^{WI}(x, Q^2)}{2xF_{1N}^{WI}(x, Q^2)} \text{ as } Q^2 \to 0 \quad \text{and} \quad \frac{F_{LN}^{WI,AV}(x, Q^2)}{0} \longrightarrow \infty.$$

An important feature of weak interaction processes is that through $\nu_l / \bar{\nu}_l$ scattering on nucleons, quarks and antiquarks can be directly observed which is not possible in the case of electromagnetic interactions. For the charged current weak interaction processes, there are two possibilities:

- For a neutrino beam, the mediating quanta W^+ interacts with d, s, $\bar{u}$, and $\bar{c}$ flavors on the proton target because of the charge conservation at the weak vertex (Figure 13.18 (top panel)).

- For an antineutrino beam, the mediating quanta W^- interacts with the u, c, $\bar{d}$, and $\bar{s}$ flavors on the proton target due to the conservation of charge at the weak vertex (Fig.13.18 (bottom panel)).

In the case of a neutron target, the role of the u and d quarks are interchanged. In the parton model, the partons are free inside the nucleon, then the neutrino–quark (antineutrino-antiquark) interactions will be similar to the case of neutrino–electron (antineutrino–positron) interaction processes as discussed in Chapter 9. Recall the expression of the scattering cross section for neutrino–electron (antineutrino–positron) interaction processes (Chapter 9) in terms of the Mandelstam variables, which may be written as

$$\frac{d\sigma}{dy} = \frac{G_F^2 \, s}{\pi}; \quad s = (p + k)^2 \simeq 2p \cdot k.$$ (13.71)

Similarly, for the antineutrino–electron (neutrino–positron) interaction processes, the differential cross section is given by

$$\frac{d\sigma}{dy} = \frac{G_F^2 \, s}{\pi}(1 - y)^2.$$ (13.72)

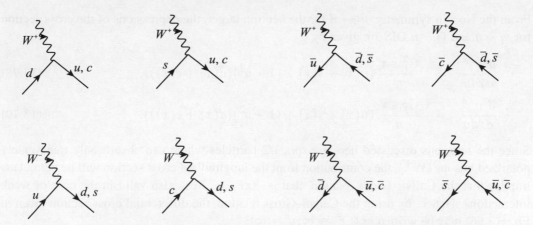

Figure 13.18 Feynman diagram for $W^\pm$ interactions with quarks and antiquarks.

The expressions of scattering cross section for $\nu_l - q$ ($\bar{\nu}_l - \bar{q}$) and $\nu_l - \bar{q}$ ($\bar{\nu}_l - q$) interaction processes would be similar to the cross section of $\nu_l - e^-$ ($\bar{\nu}_l - e^+$) given in Eq. (13.71) and $\bar{\nu}_l - e^-$ ($\nu_l - e^+$) given in Eq. (13.72), respectively. The center of mass energy (W) for $\nu_l - q$ ($\bar{\nu}_l - \bar{q}$), and $\nu_l - \bar{q}$ ($\bar{\nu}_l - q$) scattering processes is

$$W = \sqrt{s'^2}, \text{ where } s' = (xp + k)^2 \simeq 2xp \cdot k = xs. \tag{13.73}$$

Here, x is the momentum fraction carried by each quark or antiquark of the parent nucleon's momentum. Therefore, similar to Eqs. (13.71) and (13.72), the scattering cross sections are obtained as

$$\frac{d\sigma^{\nu q/\bar{\nu}\bar{q}}}{dy} = \frac{G_F^2 s x}{\pi}, \quad \frac{d\sigma^{\nu\bar{q}/\bar{\nu}q}}{dy} = \frac{G_F^2 s x}{\pi}(1 - y)^2. \tag{13.74}$$

These expressions are valid for the four flavors of quarks which are treated as massless. The integration over y in Eq. (13.74) between the limits 0 and 1 results in

$$\sigma^{\nu q/\bar{\nu}\bar{q}} = \frac{G_F^2 s x}{\pi}, \quad \sigma^{\nu\bar{q}/\bar{\nu}q} = \frac{G_F^2 s x}{3\pi} \quad \Rightarrow \quad \frac{\sigma^{\nu q/\bar{\nu}\bar{q}}}{\sigma^{\nu\bar{q}/\bar{\nu}q}} = 3. \tag{13.75}$$

By using these quark cross sections and weighing each flavor with the corresponding parton distribution functions in the nucleon, the scattering cross sections in the four-flavor scheme are obtained as:

$$\frac{d^2\sigma^{\nu p}}{dx dy} = \frac{G_F^2 s x}{\pi} \left(d(x) + s(x) + (1 - y^2)(\bar{u}(x) + \bar{c}(x)) \right). \tag{13.76}$$

$$\frac{d^2\sigma^{\bar{\nu} p}}{dx dy} = \frac{G_F^2 s x}{\pi} \left(\bar{d}(x) + \bar{s}(x) + (1 - y^2)(u(x) + c(x)) \right). \tag{13.77}$$

From the isospin symmetry, $u \leftrightarrow d$ for the neutron target, the expressions of the cross section for $\nu_l - n$ and $\bar{\nu}_l - n$ DIS are given by

$$\frac{d^2\sigma^{\nu n}}{dxdy} = \frac{G_F^2 \, s \, x}{\pi} \left(u(x) + s(x) + (1-y^2)(\bar{d}(x) + \bar{c}(x)) \right), \tag{13.78}$$

$$\frac{d^2\sigma^{\bar{\nu} n}}{dxdy} = \frac{G_F^2 \, s \, x}{\pi} \left(\bar{u}(x) + \bar{s}(x) + (1-y^2)(d(x) + c(x)) \right). \tag{13.79}$$

Since the fermions discussed here are spin 1/2 particles which can absorb only transversely polarized bosons ($W^{\pm}$), the contribution from the longitudinal cross section will be zero. This implies that the Callan–Gross relation, that is, $2xF_1 = F_2$ is also valid in the case of weak interactions. Hence, by using the Callan–Gross relation, the differential cross section given in Eq. (13.66) may be written as ($Q^2 \to \infty, \nu \to \infty$):

$$\frac{d^2\sigma^{WI}}{dxdy} = \frac{G_F^2 s}{4\pi} \left[\left(1 + (1-y)^2 \right) F_2^{WI}(x, Q^2) \pm \left(1 - (1-y)^2 \right) xF_3^{WI}(x, Q^2) \right], \tag{13.80}$$

where the terms of order M/E are neglected. On comparing Eqs. (13.76), (13.77), (13.78), and (13.79) with Eq. (13.80), one obtains

$$F_2^{\nu_l p}(x) = 2x[d(x) + s(x) + \bar{u}(x) + \bar{c}(x)] \,, \ F_2^{\bar{\nu}_l p}(x) = 2x[u(x) + c(x) + \bar{d}(x) + \bar{s}(x)],$$
$$F_2^{\nu_l n}(x) = 2x[u(x) + s(x) + \bar{d}(x) + \bar{c}(x)] \,, \ F_2^{\bar{\nu}_l n}(x) = 2x[d(x) + c(x) + \bar{u}(x) + \bar{s}(x)],$$
$$xF_3^{\nu_l p}(x) = 2x[d(x) + s(x) - \bar{u}(x) - \bar{c}(x)] \,, \ xF_3^{\bar{\nu}_l p}(x) = 2x[u(x) + c(x) - \bar{d}(x) - \bar{s}(x)],$$
$$xF_3^{\nu_l n}(x) = 2x[u(x) + s(x) - \bar{d}(x) - \bar{c}(x)] \,, \ xF_3^{\bar{\nu}_l n}(x) = 2x[d(x) + c(x) - \bar{u}(x) - \bar{s}(x)].$$

The general expression for the dimensionless nucleon structure functions in terms of parton distribution functions are given by

$$F_2^{WI}(x) = 2 \sum_{i,j} x[e_i^2 q_i(x) + e_j^2 \, \bar{q}_j(x)]; \quad xF_3^{WI}(x) = 2 \sum_{i,j} x[e_i^2 \, q_i(x) - e_j^2 \, \bar{q}_j(x)]. \tag{13.81}$$

In these expressions, i, j run for different flavors of quark/antiquark, and $q_i(x)/\bar{q}_i(x)$ represents the probability density of finding a quark/antiquark with a momentum fraction x. e_i^2 and e_j^2 represent the Cabibbo factors, that is, either $\cos^2 \theta_C$ for the favored transitions or $\sin^2 \theta_C$ for the suppressed transitions. $F_2^{WI}(x)$ and $F_3^{WI}(x)$ are the weak structure functions for neutrino and antineutrino induced processes on proton and neutron targets. These structure functions are obtained by assuming that the CKM matrix is almost unitary in its 2×2 upper left corner or equivalently, that the heavy flavors bottom and top do not mix with the lighter ones.

In the case of an isoscalar target,

$$F_2^{\nu_l(\bar{\nu}_l)N} = \frac{F_2^{\nu_l(\bar{\nu}_l)p} + F_2^{\nu_l(\bar{\nu}_l)n}}{2}, \quad F_3^{\nu_l(\bar{\nu}_l)N} = \frac{F_3^{\nu_l(\bar{\nu}_l)p} + F_3^{\nu_l(\bar{\nu}_l)n}}{2},$$

for which

$$
\left.
\begin{aligned}
F_2^{\nu_l N}(x) &= x[u(x) + \bar{u}(x) + d(x) + \bar{d}(x) + 2s(x) + 2\bar{c}(x)], \\
F_2^{\bar{\nu}_l N}(x) &= x[u(x) + \bar{u}(x) + d(x) + \bar{d}(x) + 2\bar{s}(x) + 2c(x)], \\
xF_3^{\nu_l N}(x) &= x[u(x) + d(x) - \bar{u}(x) - \bar{d}(x) + 2s(x) - 2\bar{c}(x)], \\
xF_3^{\bar{\nu}_l N}(x) &= x[u(x) + d(x) - \bar{u}(x) - \bar{d}(x) + 2c(x) - 2\bar{s}(x)].
\end{aligned}
\right\}
\tag{13.82}
$$

Assuming $s(x) = \bar{s}(x)$ and $c(x) = \bar{c}(x)$, one may write

$$
F_2^{\nu_l N}(x) = F_2^{\bar{\nu} N}(x) = x\left[u(x) + \bar{u}(x) + d(x) + \bar{d}(x) + s(x) + \bar{s}(x) + c(x) + \bar{c}(x)\right]. \tag{13.83}
$$

Under the aforesaid assumption, the average of $xF_3^{\nu_l N}(x)$ and $xF_3^{\bar{\nu}_l N}(x)$ gives us the valence quarks measurement, that is,

$$
\frac{xF_3^{\nu_l N}(x) + xF_3^{\bar{\nu}_l N}(x)}{2} = x\{u(x) - \bar{u}(x)\} + x\{d(x) - \bar{d}(x)\} = x(u_v(x) + d_v(x)). \tag{13.84}
$$

Hence, through the parity violating structure function $F_{3N}^{WI}(x, Q^2)$, the valence quark distribution inside the nucleon can be directly determined. The difference of $xF_3^{\nu_l N}(x)$ and $xF_3^{\bar{\nu}_l N}(x)$ leads us to

$$
\Delta F_{3N}^{WI}(x, Q^2) = xF_3^{\nu_l N}(x) - xF_3^{\bar{\nu}_l N}(x) = 4x\{s(x) - \bar{c}(x)\}, \tag{13.85}
$$

which provides information about the strange and charm quarks content in the nucleon.

13.3.1 Relation between electromagnetic and weak structure functions

On comparing the dimensionless nucleon structure function $F_2^{eN}(x)$ given in Eq. (13.29) with $F_2^{\nu_l N}(x)$ given in Eq. (13.82), we can write

$$
\begin{aligned}
\frac{F_2^{eN}(x)}{F_2^{\nu_l N}(x)} &= \frac{\left[\frac{5}{18}\{u(x) + \bar{u}(x) + d(x) + \bar{d}(x)\} + \frac{4}{9}\{c(x) + \bar{c}(x)\} + \frac{1}{9}\{s(x) + \bar{s}(x)\}\right]}{\left[u(x) + \bar{u}(x) + d(x) + \bar{d}(x) + s(x) + \bar{s}(x) + c(x) + \bar{c}(x)\right]} \\
&= \frac{5}{18}\left[1 - \frac{3}{5}\frac{s(x) + \bar{s}(x) - c(x) - \bar{c}(x)}{\sum_i \{q_i(x) + \bar{q}_i(x)\}}\right].
\end{aligned}
\tag{13.86}
$$

This relation is commonly known as the $\left(\frac{5}{18}\right)$th rule which confirms the fractional electric charge to the quarks. An important feature of this relation is its sensitivity to the measurement of strange sea quark distribution. In the limit, $s(x) = \bar{s}(x) = c(x) = \bar{c}(x)$:

$$
F_2^{eN}(x) \approx \frac{5}{18}F_2^{\nu_l N}(x) = \frac{5}{18}F_2^{\bar{\nu}_l N}(x). \tag{13.87}
$$

Experimental results for the ratio $\frac{5}{18}$ $\left(\frac{F_2^{\nu_l N}(x)}{F_2^{eN}(x)} \right)$ are presented in Figure 13.19, which is found to be in good agreement with the assumptions of the parton model.

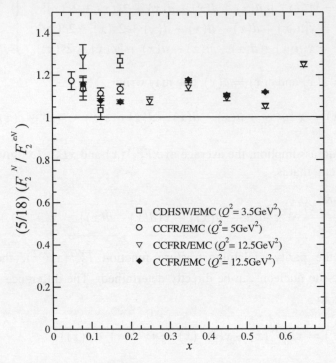

Figure 13.19 Experimental observations for the $\left(\frac{5}{18} \right)$th rule [564, 565, 566].

13.3.2 Experimental results of charged current total scattering cross section

The total scattering cross sections for the charged current DIS process has been experimentally measured by several experiments such as CCFRR [564], CCFR90 [567], CCFR96 [568], CDHS [569], BEBC-WBB [570], ANL [571], CHARM [572], etc. for neutrino and antineutrino beams, as shown in Figure 13.20. These experiments have been performed on various targets like hydrogen, deuterium, marble, iron, freon, freon–propane, etc. The world average values of total scattering cross section for the neutrino and antineutrino interaction with nucleon/nuclear targets are [573]:

$$\sigma^{\nu N}/E_\nu = 0.677 \pm 0.014 \times 10^{-38} \text{cm}^2 \text{GeV}^{-1},$$
$$\sigma^{\bar{\nu} N}/E_{\bar{\nu}} = 0.334 \pm 0.008 \times 10^{-38} \text{cm}^2 \text{GeV}^{-1}.$$

By integrating Eqs. (13.76), (13.77), (13.78) and (13.79) over x and y between the limits 0 and 1, the expressions of total scattering cross section for an isoscalar nucleon target for neutrino and antineutrino induced processes are obtained as

$$\sigma^{\nu N} = \frac{G_F^2\, s}{2\pi} \int x\left(q(x) + \frac{\bar{q}(x)}{3}\right) dx\,, \quad \sigma^{\bar{\nu} N} = \frac{G_F^2\, s}{2\pi} \int x\left(\frac{q(x)}{3} + \bar{q}(x)\right) dx. \quad (13.88)$$

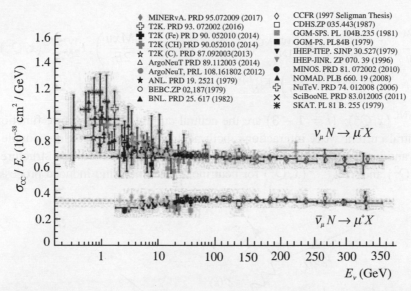

Figure 13.20 Charged current total cross section for $\nu_l - N$ and $\bar{\nu}_l - N$ processes [117].

Through the total scattering cross section, one may directly determine the total momentum carried by all the quarks and antiquarks, that is,

$$\int x(q(x) + \bar{q}(x))\, dx = \frac{3\pi}{2\, G_F^2\, s}(\sigma^{\nu N} + \sigma^{\bar{\nu} N}) \qquad (13.89)$$

and the fraction carried by the antiquarks as:

$$\frac{\int x\bar{q}(x)\, dx}{\int x(q(x) + \bar{q}(x))\, dx} = \frac{1}{2}\left(\frac{3\sigma^{\bar{\nu} N} - \sigma^{\nu N}}{\sigma^{\nu N} + \sigma^{\bar{\nu} N}}\right), \qquad (13.90)$$

which are experimentally found to be [572]:

$$\int x(q(x) + \bar{q}(x))\, dx = 0.492 \pm 0.006 \pm 0.019,$$

$$\frac{\int x\bar{q}(x)\, dx}{\int x(q(x) + \bar{q}(x))\, dx} = 0.154 \pm 0.005 \pm 0.011.$$

From these equations, it may be noticed that in the limits of high Q^2 and ν, charged partons carry only 50% of the nucleon's momentum and among them, antiquarks carry 15% of the charged partons momentum; the remaining 50% of the momentum is carried by the gluon.

13.4   Deep Inelastic Neutral Current $\nu_l/\bar{\nu}_l - N$ Scattering

The differential scattering cross section for (anti)neutrino induced neutral current (NC) deep inelastic scattering processes which are mediated by Z^0 boson (as shown in Figure 13.21) is given by

$$\frac{d^2\sigma_N^{WI}}{dxdy} = \frac{G_F^2 s}{2\pi}\left[xy^2\, F_{1N}^{WI,NC}(x,Q^2) + \left(1-y-\frac{Mxy}{2E_l}\right)F_{2N}^{WI,NC}(x,Q^2)\right.$$

$$\left.\pm xy\left(1-\frac{y}{2}\right)F_{3N}^{WI,NC}(x,Q^2)\right], \tag{13.91}$$

where $F_{iN}^{WI,NC}(x,Q^2)$; $(i=1-3)$ are the neutral current weak structure functions. In the case of neutral current weak interactions, both neutrinos and antineutrinos couple to all the quarks and antiquarks. There is no difference between the neutral current structure functions $F_{2N}^{WI,NC}(x,Q^2)$ and $xF_{3N}^{WI,NC}(x,Q^2)$ for neutrino and antineutrino induced processes. In the

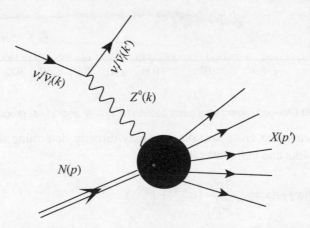

Figure 13.21 Feynman diagram for the neutral current $\nu_l(\bar{\nu}_l) - N$ DIS process.

expression of the differential cross section, the term with $xF_{3N}^{WI,NC}(x,Q^2)$ has opposite signs for the neutrino (+ve) and the antineutrino (–ve). The neutral current structure functions $F_{2N}^{WI,NC}(x,Q^2)$ and $xF_{3N}^{WI,NC}(x,Q^2)$ are written in terms of PDFs as:

$$F_2^{\nu_l(\bar{\nu}_l)p\to\nu_l(\bar{\nu}_l)X}(x) = 2x\left(\left[(e_L^{\,u})^2 + (e_R^{\,u})^2\right]\cdot[u(x)+\bar{u}(x)+c(x)+\bar{c}(x)]\right.$$

$$\left.+\left[(e_L^{\,d})^2 + (e_R^{\,d})^2\right]\cdot[d(x)+\bar{d}(x)+s(x)+\bar{s}(x)]\right) \tag{13.92}$$

$$xF_3^{\nu_l(\bar{\nu}_l)p\to\nu_l(\bar{\nu}_l)X}(x) = 2x\left(\left[(e_L^{\,u})^2 - (e_R^{\,u})^2\right]\cdot[u(x)-\bar{u}(x)+c(x)-\bar{c}(x)]\right.$$

$$\left.+\left[(e_L^{\,d})^2 - (e_R^{\,d})^2\right]\cdot[d(x)-\bar{d}(x)+s(x)-\bar{s}(x)]\right), \tag{13.93}$$

where for the neutron target, $u(x) \to d(x)$, $\bar{u}(x) \to \bar{d}(x)$ and vice versa. e_L^q and e_R^q are the left- and right-handed coupling constants. These coupling constants are determined by a combination of the weak isospin's longitudinal component (I_3) and the electric charge (Q_e) of the interacting quarks, rather than by their electric charges alone, as in charged current interactions. These coupling constants are also functions of $\sin^2 \theta_W$:

$$e_L^q = I_3^q - Q_e^q \sin^2 \theta_W , \quad e_R^q = -Q_e^q \sin^2 \theta_W. \tag{13.94}$$

Using the value of I_3 and Q_e for quarks, we get:

$$(e_L^u)^2 = \left(+\frac{1}{2} - \frac{2}{3} \sin^2 \theta_W \right)^2 , \quad (e_L^d)^2 = \left(-\frac{1}{2} + \frac{1}{3} \sin^2 \theta_W \right)^2$$

$$(e_R^u)^2 = \left(-\frac{2}{3} \sin^2 \theta_W \right)^2 , \quad (e_R^d)^2 = \left(+\frac{1}{3} \sin^2 \theta_W \right)^2.$$

By using the charged current and neutral current (anti)neutrino–nucleon scattering cross sections, one may obtain the standard model parameter $\sin^2 \theta_W$, coupling strength of weak neutral current to strange and down quarks and the ratio $R_L^{WI}(x, Q^2) = \frac{\sigma_L^{WI}(x,Q^2)}{\sigma_T^{WI}(x,Q^2)}$ which provides information about the violation of the Callan–Gross relation. The results for the differential scattering cross section by the CHARM collaboration [574] are shown in Figure 13.22 for (anti)neutrino scattering in charged and neutral current induced processes.

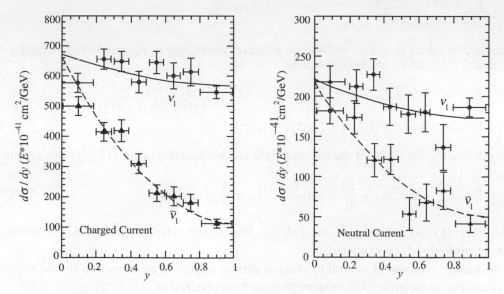

Figure 13.22 Charged and neutral current (anti)neutrino differential cross sections by CHARM collaboration [574].

Now, in order to obtain the weak mixing angle, the differential scattering cross sections for the charged and neutral current induced processes is written in terms of parton distribution functions for an isoscalar nucleon target as

$$\frac{d^2 \sigma_{\nu N}^{CC}}{dx dy} = \frac{G_F^2 sx}{2\pi} [\{q(x) + (1-y)^2 \bar{q}(x)\} + \{s(x) - c(x)\} + (1-y)^2 \{\bar{c}(x) - \bar{s}(x)\}] ,$$

$$\frac{d^2\sigma_{\bar{\nu}N}^{CC}}{dxdy} = \frac{G_F^2 sx}{2\pi}[\{\bar{q}(x) + (1-y)^2 q(x)\} + \{\bar{s}(x) - \bar{c}(x)\} + (1-y)^2\{c(x) - s(x)\}] ,$$

$$\frac{d^2\sigma_{\nu N}^{NC}}{dxdy} = \frac{G_F^2 sx}{2\pi}[\{(e_L{}^u)^2 + (e_L{}^d)^2\}\{q(x) + (1-y)^2\bar{q}(x)\}$$

$$+ \{(e_R{}^u)^2 + (e_R{}^d)^2\}\{\bar{q}(x) + q(x)(1-y)^2\}$$

$$+ \{(e_L{}^u)^2 - (e_L{}^d)^2 + (1-y)^2[(e_R{}^u)^2 - (e_R{}^d)^2]\}(c(x) - s(x))$$

$$+ \{(e_R{}^u)^2 - (e_R{}^d)^2 + (1-y)^2[(e_L{}^u)^2 - (e_L{}^d)^2]\}(\bar{c}(x) - \bar{s}(x))] , \qquad (13.95)$$

$$\frac{d^2\sigma_{\bar{\nu}N}^{NC}}{dxdy} = \frac{G_F^2 sx}{2\pi}[\{(e_R{}^u)^2 + (e_R{}^d)^2\}\{q(x) + (1-y)^2\bar{q}(x)\}$$

$$+ \{(e_L{}^u)^2 + (e_L{}^d)^2\}\{\bar{q}(x) + q(x)(1-y)^2\}$$

$$+ \{(e_R{}^u)^2 - (e_R{}^d)^2 + (1-y)^2[(e_L{}^u)^2 - (e_L{}^d)^2]\}(c(x) - s(x))$$

$$+ \{(e_L{}^u)^2 - (e_L{}^d)^2 + (1-y)^2[(e_R{}^u)^2 - (e_R{}^d)^2]\}(\bar{c}(x) - \bar{s}(x))] , \qquad (13.96)$$

where $q(x) = u(x) + d(x) + c(x) + s(x)$ and $\bar{q}(x) = \bar{u}(x) + \bar{d}(x) + \bar{c}(x) + \bar{s}(x)$. The ratio of scattering cross sections

$$R_\pm = \frac{(\sigma_{\nu N}^{NC} \pm \sigma_{\bar{\nu}N}^{NC})}{(\sigma_{\nu N}^{CC} \pm \sigma_{\bar{\nu}N}^{CC})}$$

is defined by taking into account only two flavors of quark, that is, up and down to simplify the numerical calculations and is obtained as

$$R_\pm = \frac{\{(e_L{}^u)^2 + (e_L{}^d)^2 \pm (e_R{}^u)^2 + (e_R{}^d)^2\}\{1 \pm (1-y)^2\}(\{u(x) + d(x)\} \pm \{\bar{u}(x) + \bar{d}(x)\})}{\{1 \pm (1-y)^2\}(\{u(x) + d(x)\} \pm \{\bar{u}(x) + \bar{d}(x)\})}$$

$$= \{(e_L{}^u)^2 + (e_L{}^d)^2\} \pm \{(e_R{}^u)^2 + (e_R{}^d)^2\}. \qquad (13.97)$$

By substituting the values of the left- and right-handed couplings (Eq. (13.94)), one obtains

$$R_+ = \frac{1}{2} - \sin^2\theta_W + \frac{10}{9}\sin^4\theta_W, \qquad R_- = \frac{1}{2} - \sin^2\theta_W, \qquad (13.98)$$

where Eq. (13.98) is known as the Paschos–Wolfenstein relation through which the weak mixing angle $\sin^2\theta_W$ is extracted.

Moreover, the ratio of neutral to charged current scattering cross section for the neutrino (R_ν) and the antineutrino ($R_{\bar{\nu}}$) induced processes are obtained as

$$R_\nu = \frac{\sigma_{\nu N}^{NC}}{\sigma_{\nu N}^{CC}} = \{(e_L{}^u)^2 + (e_L{}^d)^2\} + \kappa(x)\{(e_R{}^u)^2 + (e_R{}^d)^2\} , \qquad (13.99)$$

$$R_{\bar{\nu}} = \frac{\sigma_{\bar{\nu}N}^{NC}}{\sigma_{\bar{\nu}N}^{CC}} = \{(e_L{}^u)^2 + (e_L{}^d)^2\} + \kappa^{-1}(x)\{(e_R{}^u)^2 + (e_R{}^d)^2\} , \qquad (13.100)$$

where $\kappa(x) = \frac{\{\bar{q}(x) + (1-y)^2 q(x)\}}{\{q(x) + (1-y)^2\bar{q}(x)\}}.$ \qquad (13.101)

This ratio measures the relative strength of the sea and valence quarks contributions. After simplification, R_ν and $R_{\bar\nu}$ may also be written as

$$\left.\begin{array}{l} R_\nu = \frac{1}{2} - \sin^2\theta_W + \frac{5}{9}(1 + \kappa(x))\sin^4\theta_W, \\ R_{\bar\nu} = \frac{1}{2} - \sin^2\theta_W + \frac{5}{9}(1 + \kappa^{-1}(x))\sin^4\theta_W. \end{array}\right\}$$ (13.102)

These ratios have been obtained by several experimental collaborations; for example, the measured values are

$$\begin{aligned} R_\nu &= 0.3072 \pm 0.0025 \pm 0.0020 \quad \text{CDHS [575]}, \\ R_\nu &= 0.3098 \pm 0.0031 \quad \text{CHARM [576]}, \\ R_{\bar\nu} &= 0.363 \pm 0.015 \quad \text{CDHS [577]}. \end{aligned}$$

The value of R_ν has been found to be more sensitive to the predicted value of weak mixing angle by the standard model in comparison to $R_{\bar\nu}$. The experimental values of $\sin^2\theta_W$ from the different collaborations are given by

$$\begin{aligned} \sin^2\theta_W &= 0.225 \pm 0.005(\text{exp.}) \pm 0.003(\text{theor.}) \pm 0.013(m_c - 1.5\,\text{GeV}/c^2), \text{CDHS [575]} \\ &= 0.236 \pm 0.005(\text{exp.}) \pm 0.003(\text{theor.}) \pm 0.012(m_c - 1.5\,\text{GeV}/c^2), \text{CHARM [576]} \\ &= 0.2233 \pm 0.0008 \pm 0.0004, \quad \text{CDF [578]}, \end{aligned}$$

where m_c is the mass of the charm quark, while the world average value of $\sin^2\theta_W$ is [579]

$$\sin^2\theta_W = 0.23122(4) \pm 0.00017.$$

Till now, the DIS processes induced by the charged lepton and (anti)neutrino has been discussed. In the next section, the perturbative as well as the non-perturbative QCD corrections will be discussed. These corrections are important for the precise determination of the nucleon structure functions.

13.5 QCD Corrections

13.5.1 Modified parton model

In the naive parton model, the structure functions were assumed to be the function of a dimensionless variable x, that is,

$$F_{1N}(x, Q^2) \xrightarrow[Q^2 \to \infty, \nu \to \infty]{x \to \text{finite}} F_{1N}(x),$$

$$F_{2N}(x, Q^2) \xrightarrow[Q^2 \to \infty, \nu \to \infty]{x \to \text{finite}} F_{2N}(x).$$

However, QCD predicts that if one increases the four-momentum transfer squared Q^2, then each quark is found to be surrounded by a number of partons. Therefore, the effective number of resolved partons carrying a fraction of the nucleon's momentum increases with the increase

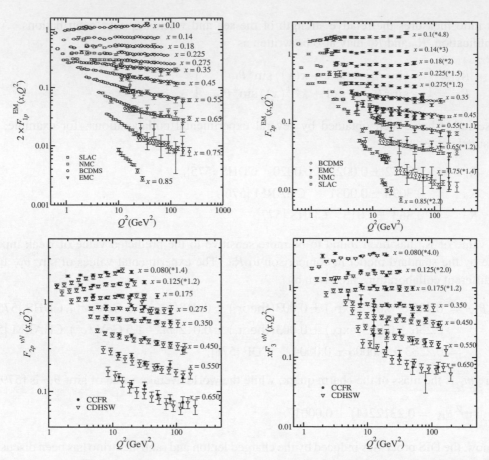

Figure 13.23 Experimental results of nucleon structure functions for (i) Top panel: electromagnetic interaction induced DIS process [548, 557, 558, 559, 560], (ii) Bottom panel: weak interaction induced DIS process [561, 562].

in Q^2 or in other words, parton distribution functions show Q^2 dependence leading to the phenomenon of scaling violation. This phenomenon was confirmed by several electron scattering experiments such as EMC, NMC, BCDMS, etc. [557, 558, 559].

In Figure 13.23 (top panel), the experimental results are shown for the electromagnetic nucleon structure functions in a wide range of x and Q^2. Later, scaling violation was also observed in (anti)neutrino scattering experiments such as CDHSW [561] and CCFR [562] for which the results are shown in Figure 13.23 (bottom panel). From these figures, it may be observed that with the increase in x and Q^2, the structure functions fall down, while for lower x and Q^2, there is a rise. This behavior of structure functions show scaling breakdown.

In perturbative QCD (pQCD), partons present inside the nucleon interact among themselves via the gluon exchange; the contribution from the gluons is responsible for the Q^2 dependence of nucleon structure functions. For example, in the case of electromagnetic interactions, $\gamma^* q \rightarrow qg$ and $\gamma^* g \rightarrow q\bar{q}$ are the possible channels, which are depicted in Figure 13.24 [400]. Generally, the Q^2 dependence of structure functions is determined by evolving the Q^2

dependent parton densities using the Dokshitzer–Gribov–Lipatov–Altarelli–Parisi (DGLAP) evolution equation [580] which is too technical to be included here. Since the structure functions are a combination of parton density distribution functions, it is important to understand their behavior in the entire kinematic region of x and Q^2.

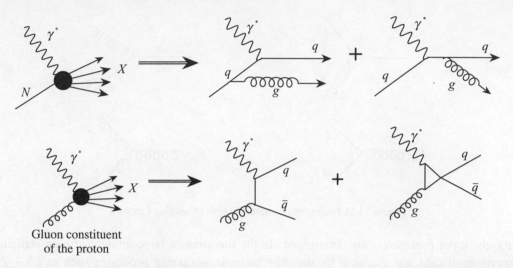

Figure 13.24   Diagrammatic representation of (i) upper panel: the process $\gamma^* q \to qg$ and (ii) lower panel: the process $\gamma^* g \to q\bar{q}$ [400].

Extraction of parton densities

Through perturbative QCD, the Q^2 evolution of parton distribution functions can be inferred but their x dependence cannot be predicted. Therefore, it is convenient to make a suitable choice of analytic parameterization consistent with the Q^2 dependence predicted by pQCD. Beyond the leading order, an exact analytic expression valid for a wide range of x and Q^2, consistent with the pQCD is very difficult to be found. Therefore, DGLAP equations are generally used to extract the parton densities beyond the leading order; they are given by

$$\frac{\partial}{\partial lnQ^2} \begin{pmatrix} q_i(x,Q^2) \\ g(x,Q^2) \end{pmatrix} = \frac{\alpha_s(Q^2)}{2\pi} \sum_j \int_x^1 \frac{dz}{z} \begin{pmatrix} P_{q_iq_j}(\frac{x}{y},\alpha_s(Q^2)) & P_{q_ig}(\frac{x}{y},\alpha_s(Q^2)) \\ P_{gq_j}(\frac{x}{y},\alpha_s(Q^2)) & P_{gg}(\frac{x}{y},\alpha_s(Q^2)) \end{pmatrix} \begin{pmatrix} q_j(y,Q^2) \\ g(y,Q^2) \end{pmatrix}, \quad (13.103)$$

where $\alpha_s(Q^2)$ is the strong coupling constant, $q(x,Q^2)$ and $g(x,Q^2)$ are the quark and gluon density distribution functions, and $P_{ij}(\frac{x}{y},\alpha_s(Q^2))(i,j = q \text{ or } g)$ are the splitting functions which are expanded in a power series of $\alpha_s(Q^2)$. The splitting function describes the probability for the splitting of a parton into two other partons having a smaller momentum fraction than the parent parton, for example, $P_{q_iq_j}$ gives the probability of splitting of a quark into a pair of quarks and a gluon. All the possibilities for the splitting of a parton at leading order are shown in Figure 13.25. The evolution equations explain that the valence quarks are surrounded by a number of virtual particles which are continuously getting absorbed and emitted. For the extraction of parton densities using the DGLAP equation, the analytic shape is assumed to be valid at an arbitrary but sufficiently large value of Q^2 (i.e., small $\alpha_s(Q^2)$) so that perturbative calculations can be applicable. Then through the DGLAP equation, the evolution of parton densities is performed up to different values of Q^2. The parton densities are convoluted over the coefficient functions (which are discussed in the next subsection) to predict the nucleon structure functions. By fitting the predicted structure functions to the available experimental

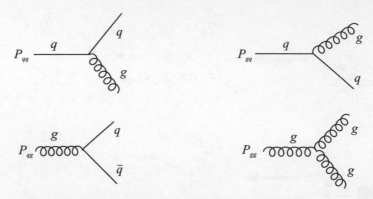

Figure 13.25 Diagrammatic representation of splitting functions.

data, the input parameters are determined. In the literature, a large amount of high statistics experimental data are available for the deep inelastic scattering processes such as $e^{\pm} - N$, $e^{\pm} - A$, $\mu^{\pm} - N$, $\mu^{\pm} - A$, $\nu_l - A$, etc., which are used to extract parton densities with high accuracy. However, besides DIS, the Drell–Yan (DY) process with hadronic probes like $p - A$, $\pi^{\pm} - A$, $d - A$, etc., are also used to obtain information about parton densities. Various groups have found the global fits of parton density distribution by using the available experimental data, for example, CTEQ [551], MSTW [550], GJR [549], MMHT [552]. Thus, using this procedure, the PDFs have been obtained in the whole range of x and at any value of Q^2 by interpolation.

The probability of the gluon emission is related to the strong coupling constant $\alpha_s(Q^2)$, which varies with the value of Q^2. For example, in the limit of $Q^2 \to \infty$, the strong coupling constant $\alpha_s(Q^2)$ becomes very small and, therefore, the higher order terms can be neglected in comparison to the leading order (LO) terms. However, for a finite value of Q^2, $\alpha_s(Q^2)$ is large and higher order terms such as next-to-leading order (NLO), and next-to-next-to-leading order (NNLO), etc., give a significant contribution. In the next section, the evolution of PDFs at NLO is discussed.

13.5.2 NLO evolution

For a precise determination of the nucleon structure functions, in the limit of finite Q^2, the next-to-leading order (NLO) followed by the next-to-next-to-leading order (NNLO) terms should be taken into account. In Figure 13.26, the parton density distribution functions are shown at LO, NLO, and NNLO for $Q^2 = 2 \text{ GeV}^2$. For example these results are obtained by using the MMHT PDFs parameterization [552]. One may observe that while we go from LO to NLO, there is a significant difference in PDFs; however, the difference is comparatively smaller between NLO and NNLO. In literature, different approaches are available for the QCD corrections at NLO and NNLO for the evolution of PDFs, which are used to obtain the nucleon structure function, for example, as discussed in Refs. [581, 582, 583, 584, 585, 586, 587]. Here, we are demonstrating in brief the basic concept of performing evolution at NLO.

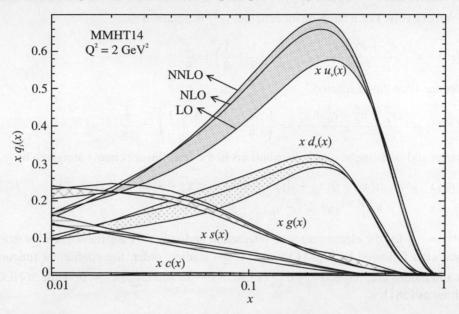

Figure 13.26 Evolution of parton density distribution functions using MMHT PDFs parameterization [552] from leading order to next-to-next-leading order at $Q^2 = 2$ GeV^2.

The nucleon structure functions may be expressed in terms of the convolution of coefficient function C_f with the density distribution of partons (f) inside the nucleon. The parton coefficient functions are generally expressed as

$$C_f(x, Q^2) = C_f^{(0)} + \frac{\alpha_s(Q^2)}{2\pi} C_f^{(1)} + \left(\frac{\alpha_s(Q^2)}{2\pi}\right)^2 C_f^{(2)} + \dots . \tag{13.104}$$

The nucleon structure function, for example, $F_{2N}(x)$ is written in terms of the coefficient function as

$$x^{-1} F_{2N}^{EM,WI}(x) = \sum_{f=q,g} C_2^{(n)}(x) \otimes f(x), \tag{13.105}$$

where superscript $n = 0, 1, 2, \dots$ for N$^{(n)}$LO and the symbol $\otimes$ is the Mellin convolution.

$$C_f(x) \otimes f(x) = \int_x^1 C_f(y) f\left(\frac{x}{y}\right) \frac{dy}{y}. \tag{13.106}$$

This Mellin convolution turns into a simple multiplication in the Mellin space. The expression for $F_{2N}(x)$ is simplified as [581]:

$$x^{-1} F_{2N}^{EM,WI}(x) = c_{2,ns}^{(n)}(x) \otimes q_{ns}(x) + \langle e^2 \rangle (c_{2,q}^{(n)}(x) \otimes q_s(x) + c_{2,g}^{(n)}(x) \otimes g(x)),$$

where $\langle e^2 \rangle$ is the average charge of partons; $q_s(x)$, $q_{ns}(x)$ are the singlet and the non-singlet quark distributions; and $g(x)$ is the gluon distribution. $c_{2,ns}(x)$ is the coefficient function for the non-singlet and $c_{2,q}^{(n)}(x)$ and $c_{2,g}^{(n)}(x)$ are the coefficient functions for the singlet quark and gluon, respectively. For a three-flavor scheme, $\langle e^2 \rangle$ is obtained as:

$$\langle e^2 \rangle = \frac{e_u^2 + e_d^2 + e_s^2}{3} = \frac{1}{3}\left(\frac{4}{9} + \frac{1}{9} + \frac{1}{9}\right) = \frac{2}{9},$$

while for the four-flavor scheme,

$$\langle e^2 \rangle = \frac{e_u^2 + e_d^2 + e_c^2 + e_s^2}{3} = \frac{1}{4}\left(\frac{4}{9} + \frac{1}{9} + \frac{4}{9} + \frac{1}{9}\right) = \frac{5}{18}.$$

The singlet and non-singlet quark distributions in the four-flavor scheme are given by

$$q_s(x) = u(x) + \bar{u}(x) + d(x) + \bar{d}(x) + s(x) + \bar{s}(x) + c(x) + \bar{c}(x)$$
$$q_{ns}(x) = F_{2N}^{EM,WI}(x) - \langle e^2 \rangle q_s,$$

with $\langle e^2 \rangle = \frac{5}{18}$ for the electromagnetic interaction and $\langle e^2 \rangle = 1$ for the weak interaction (see the discussion followed by Eq. (13.81)). At the leading order, the coefficient functions for quarks and gluons are, respectively $C_{2,q}^{(0)}(x) = \delta(1 - x)$ and $C_{2,g}^{(0)}(x) = 0$; while at NLO, they are written as [581]:

$$C_{q,2}^{(1)}(x) = c_{2,ns}^{(1)}(x) + c_{2,q}^{(1)}(x), \quad C_{g,2}^{(1)}(x) = c_{2,g}^{(1)}(x). \tag{13.107}$$

For the detailed expressions of the coefficient functions, see Ref. [581]. The expression for the weak structure function $F_{3N}^{WI}(x)$ in terms of the coefficient function and parton density distribution function is given by [588]:

$$F_{3N}^{WI}(x) = \sum_{f=q,g} C_3(x) \otimes f(x) = C_3(x) \otimes q_v(x),$$

where $q_v(x)$ is the valence quark distribution and $C_3(x)$ is the coefficient function for $F_{3N}^{WI}(x)$. It is important to point out that coefficient functions for the longitudinal structure function $F_{LN}^{EM,WI}(x, Q^2)$ are suppressed by a power of $\frac{\alpha_s(Q^2)}{2\pi}$ as compared to the case of $F_{2N}^{EM,WI}(x, Q^2)$ and $F_{3N}^{WI}(x, Q^2)$. For the detailed discussion, please see Refs. [581], [582], [588] and [589]. The parton distributions are process independent but they depend on the target. Therefore, the incorporation of the target mass gives rise to new terms that lead to additional power corrections of kinematical origin. This is known as the target mass correction (TMC) effect. In the next section, the target mass correction effect, which is taken into account for better determination of the nucleon structure functions, will be discussed.

13.5.3 TMC effect

The target mass correction (TMC) effect is a non-perturbative effect which comes into the picture at low Q^2, where perturbation theory is not valid. The TMC effect is significant at low Q^2 and high x, which is important to determine the distribution of valence quarks. Unfortunately, this kinematic region has not been explored much, unlike the region of high Q^2 and low x. The TMC effect is also known as 'kinematic higher twist effect'. In 1976, Georgi and Politzer determined the target mass corrections to electroweak structure functions, using the operator product expansion (OPE) at the leading order of QCD [590].

At a finite value of Q^2, the mass of the target nucleon and the quark masses modify the Bjorken variable x with the light cone momentum fraction. For the massless quarks, the parton light cone momentum fraction is given by the Nachtmann variable (ζ) which is defined as [591]:

$$\zeta = \frac{2x}{1 + \sqrt{1 + \frac{4M^2x^2}{Q^2}}}. \qquad (13.108)$$

Notice that ζ depends only on the hadronic mass. In the case of massive quarks, ζ gets modified to the slow rescaling variable $\bar{\zeta}$ which is given by [591]

$$\bar{\zeta} = \zeta \left(1 + \frac{m_q^2}{Q^2} \right), \qquad (13.109)$$

where m_q is the quark mass. It is noticeable that the Nachtmann variable corrects the Bjorken variable for the effects of hadronic mass while the generalized variable $\bar{\zeta}$ further corrects ζ for

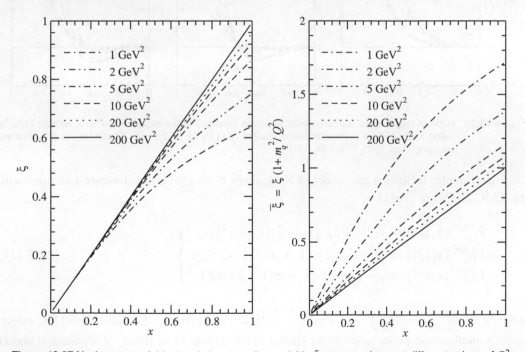

Figure 13.27 Nachtmann variable ζ and slow rescaling variable $\bar{\zeta}$ vs. x are shown at different values of Q^2.

the effects of the partonic masses [591]. The dependence of ζ and $\bar{\zeta}$ on the Bjorken variable x and four momentum transfer squared Q^2 is shown in Figure 13.27. From the figure, one can observe that for the low values of momentum transfer squared, both the Nachtman variable and the slow rescaling variable significantly deviate from the Bjorken variable, specially at high x. However, in the case of high momentum transfer squared($Q^2 \gg m_q^2$), quark masses are negligible; hence, these variables get reduced to the Bjorken variable, that is

$$\zeta \xrightarrow{Q^2 \gg m_q^2} x \quad \text{and} \quad \bar{\zeta} \xrightarrow{Q^2 \gg m_q^2} x. \tag{13.110}$$

The effect of target mass corrections obtained in the different approaches such as operator product expansion [591], the approach discussed by Ellis–Furmanski–Petronzio [592] and ζ−scaling [593, 594] are shown in Figure 13.28 at $Q^2 = 2$ GeV^2. From the figure, it can be observed that TMC effect dominates in the region of high x and low Q^2.

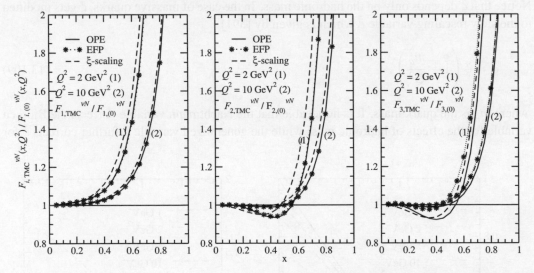

Figure 13.28 Ratio of target mass corrected weak structure functions to the structure functions without TMC by using MMHT PDFs parameterization [552] at NLO for $Q^2 = 2$ GeV^2. These results are obtained by using Refs. [591, 592, 593, 594].

For example, following the works of Schienbein et al. [591], the structure functions with the TMC effect are given by

$$\left.\begin{array}{l} F_{1N}^{TMC}(x, Q^2) \approx \dfrac{x}{\zeta\gamma} F_{1N}(\zeta) \left(1 + 2r(1 - \zeta)^2\right), \\[2mm] F_{2N}^{TMC}(x, Q^2) \approx \dfrac{x^2}{\zeta^2\gamma^3} F_{2N}(\zeta) \left(1 + 6r(1 - \zeta)^2\right), \\[2mm] F_{3N}^{TMC}(x, Q^2) \approx \dfrac{x}{\zeta\gamma^2} F_{3N}(\zeta) \left(1 - r(1 - \zeta) \ln\zeta\right), \end{array}\right\} \tag{13.111}$$

where $r = \frac{\mu x \zeta}{\gamma}$, $\mu = \left(\frac{M}{Q}\right)^2$ and $\gamma = \sqrt{1 + \frac{4M^2 x^2}{Q^2}}$, respectively. Similar to the TMC effect, there is another non-perturbative effect known as the 'higher twist effect' or 'dynamical higher twist effect', which is described in brief in the following section.

13.5.4 Higher twist effect

The possible sources of higher twist effect are QCD radiative corrections, final state interactions, finite target mass, etc., while in perturbative QCD, renormalons are responsible for this effect. Higher twist (HT) effect is a dynamical effect arising due to multiparton correlations. Therefore, this effect involves the interactions of struck quark with other quarks via the exchange of gluon; it is suppressed by the power of $\left(\frac{1}{Q^2}\right)^n$, where $n = 1, 2, \dots$. The higher twist effect is pronounced in the region of low Q^2 and high x, like the TMC effect, but is negligible for high Q^2 and low x.

For lower values of Q^2, at a few GeV or less, non-perturbative phenomena become important for a precise modeling of cross sections, in addition to higher-order QCD corrections. In the formalism of the operator product expansion (OPE), unpolarized structure functions can be expressed in terms of powers of $1/Q^2$ (power corrections):

$$F_i(x, Q^2) = F_i^{\tau=2}(x, Q^2) + \frac{H_i^{\tau=4}(x)}{Q^2} + \frac{H_i^{\tau=6}(x)}{Q^4} + \dots; \quad i = 1, 2, 3, \qquad (13.112)$$

where the first term ($\tau = 2$) is known as the twist-two or leading twist (LT) term, and corresponds to the scattering off a free quark. This term obeys the Altarelli-Parisi equation and is expressed in terms of PDFs. It is responsible for the evolution of structure functions via perturbative QCD $\alpha_s(Q^2)$ corrections. The HT terms with $\tau = 4, 6, \dots$ reflect the strength of multiparton correlations (qq and qg). The HT corrections spoil the QCD factorization; therefore, one should consider their impact on the PDFs extracted in the analysis of low Q data.

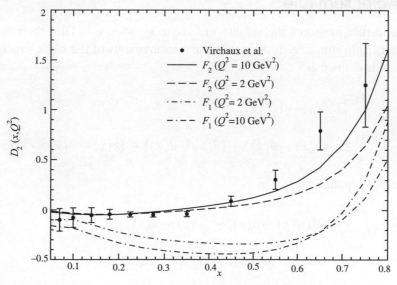

Figure 13.29 Twist-4 coefficient in the renormalon approach [595] for $F_1^{\text{EM}}(x, Q^2)$ and $F_2^{\text{EM}}(x, Q^2)$ obtained by using MMHT PDFs parameterization [552] at NLO for $Q^2 = 2$ GeV^2 and 10 GeV^2. The results are compared with the results of Virchaux et al. [596].

Due to their non-perturbative origin, current models can only provide a qualitative description for such contributions, which are usually determined via reasonable assumptions from data. In Figure 13.29, the twist-4 coefficient function obtained by following the renormalon approach [595] is shown and compared with the phenomenological coefficient function of Virchaux et al. [596]. From the figure, it may be observed that the effect of higher twist corrections are different for $F_1^{EM}(x, Q^2)$ and $F_2^{EM}(x, Q^2)$ in the entire region of x. Furthermore, to show the Q^2 dependence of the coefficient function $D_2(x, Q^2)$, the results are shown at $Q^2 = 2$ and 10 GeV^2. One may also notice that the results of $F_2^{EM}(x, Q^2)$ are consistent with the corresponding phenomenological parameterization [596].

13.6 Sum Rules in DIS

The deep inelastic cross sections for the charged lepton–nucleon scattering or (anti)neutrino–nucleon scattering are described in terms of nucleon structure functions, which depend upon the momentum distribution of these quarks and gluons. Using the explicit expressions for structure functions, one can obtain certain relations. These relations are better known as parton model sum rules. Using the theory of strong interactions (QCD), the naive parton model and its predictions for the values of the sum rules are modified. At high energies (large momentum transfer squared), the coupling strength in QCD becomes small and perturbation theory can be used. In this regime, corrections to the sum rules can be expressed as a power series expansion in the strong coupling constant $\alpha_s(Q^2)$. At lower values of Q^2, non-perturbative corrections enter and can be expressed as a power series in $1/Q^2$.

13.6.1 Adler sum rule

The Adler sum rule measures the validity of isospin symmetry in DIS, therefore, it is also known as the isospin sum rule. It deals with the measurements of the weak structure function $F_2^{WI}(x, Q^2)$. It is defined as:

$$
\begin{aligned}
S_A &= \int_0^1 \frac{dx}{x} \left[F_2^{\bar{\nu}p}(x) - F_2^{\nu p}(x) \right] \\
&= \int_0^1 \frac{dx}{x} 2x \left[(u(x) - \bar{u}(x)) - (d(x) - \bar{d}(x)) + (c(x) - \bar{c}(x)) - (s(x) - \bar{s}(x)) \right].
\end{aligned}
$$

Neglecting c and s quark PDFs,

$$
S_A = \int_0^1 \frac{dx}{x} 2x \left[(u(x) - \bar{u}(x)) - (d(x) - \bar{d}(x)) \right] = 2 = 4I_3.
$$

Alternatively, one may write this sum rule in the following forms

$$
\left.
\begin{aligned}
&\int_0^1 \frac{dx}{x} \left[F_2^{\nu n}(x) - F_2^{\nu p}(x) \right] = 2, \quad \int_0^1 \frac{dx}{x} \left[F_2^{\nu n}(x) - F_2^{\bar{\nu}n}(x) \right] = 2, \\
&\int_0^1 \frac{dx}{x} \left[F_2^{\bar{\nu}p}(x) - F_2^{\nu p}(x) \right] = 2, \quad \int_0^1 \frac{dx}{x} \left[F_2^{\bar{\nu}p}(x) - F_2^{\bar{\nu}n}(x) \right] = 2.
\end{aligned}
\right\}
\tag{13.113}
$$

The Adler sum rule [597] predicts the difference between the quark densities of the neutron and the proton integrated over x. At the leading order:

$$\int_0^1 (F_2^{vn}(x) - F_2^{vp}(x))\frac{dx}{x} = \int_0^1 2(u_v(x) - d_v(x))dx = 2.$$ (13.114)

The WA25 [598, 599] group used the neutrino beam on neutron and proton targets and instead of S_A, measured $\tilde{S}_A$ which is defined as

$$
\begin{aligned}
\tilde{S}_A &\equiv \int_0^1 dx \left[\frac{F_2^{W^+n}(x, Q^2) - F_2^{W^+p}(x, Q^2)}{2x} \right] \\
&= \int_0^1 dx \left[u_v^p(x) - d_v^p(x) - \delta\bar{u}(x) - \delta\bar{d}(x) \right] \\
&= S_A - \int_0^1 dx \left[\delta\bar{u}(x) + \delta\bar{d}(x) \right],
\end{aligned}
$$ (13.115)

where $\bar{d}^n(x) \equiv \bar{u}^p(x) - \delta\bar{u}(x)$ and $\bar{u}^n(x) \equiv \bar{d}^p(x) - \delta\bar{d}(x)$ are the charge symmetry violating parton distribution functions for antiquarks. If $\delta\bar{u}(x)$ and $\delta\bar{d}(x)$ vanish, then the charge symmetry is exact. Moreover, this sum rule is supported by the existing neutrino–nucleon DIS data, which shows no significant Q^2 variation in the range $2\,\text{GeV}^2 < Q^2 < 30\,\text{GeV}^2$ and gives $S_A^{\text{exp}} = 2.02 \pm 0.40$ [599]. However, the error bars are large and DIS $v_l - N$ experiments with better precision are required to reduce errors. The experimental results from WA25 [598, 599] for the Adler sum rule shown in Figure 13.30 are in good agreement with the theoretical predictions.

13.6.2 Gross–Llewellyn Smith sum rule

The Gross–Llewellyn Smith (GLS) sum rule is defined for an isoscalar nucleon target $'N'$ and a symmetric sea as

$$
\begin{aligned}
S_{GLS} = \int_0^1 dx\, F_3^{vN}(x) &= \int_0^1 dx \frac{1}{2} \left[F_3^{vp}(x) + F_3^{vn}(x) \right] \\
&= \int_0^1 dx \left[u(x) - \bar{u}(x) + d(x) - \bar{d}(x) + 2s(x) - 2\bar{c}(x) \right].
\end{aligned}
$$

Neglecting sea quarks,

$$\int_0^1 dx\, F_3^{vN}(x) = \int_0^1 dx \left[u(x) - \bar{u}(x) + d(x) - \bar{d}(x) \right] = 3.$$ (13.116)

Using this sum rule, one may measure the baryon number and strangeness of the nucleon. Therefore, it is sometimes called the Baryon sum rule. One may also write:

$$
\left.
\begin{aligned}
&\int_0^1 \frac{dx}{x} x \left[F_3^{vp}(x) + F_3^{\bar{v}p}(x) \right] = 6, \quad \int_0^1 \frac{dx}{x} x \left[F_3^{vp}(x) + F_3^{vn}(x) \right] = 6, \\
&\int_0^1 \frac{dx}{x} x \left[F_3^{\bar{v}p}(x) + F_3^{\bar{v}n}(x) \right] = 6, \quad \int_0^1 \frac{dx}{x} x \left[F_3^{vn}(x) - F_3^{\bar{v}n}(x) \right] = 6.
\end{aligned}
\right\}
$$

The GLS sum rule gives the QCD expectation for the integral of the valence quark densities. To the leading order in perturbative QCD, the integral $\int \frac{dx}{x} \left[xF_3(x) \right]$ is the number of valence quarks in the proton and should be equal to three [600]. QCD corrections to this integral result in a dependence on α_s [601]:

$$\int_0^1 xF_3(x, Q^2)\frac{dx}{x} = 3\left(1 - \frac{\alpha_s}{\pi} - a(n_f) \left(\frac{\alpha_s}{\pi}\right)^2 - b(n_f)\left(\frac{\alpha_s}{\pi}\right)^3\right),\tag{13.117}$$

where a and b are known functions of the number of quark flavors n_f which contribute to scattering at a given x and Q^2. This is one of the few QCD predictions that are available to order α_s^3. The world average is

$$\int_0^1 F_3(x)\, dx = 2.64 \pm 0.06,$$

which is consistent with the NNLO evaluation of Eq. (13.117) with $\Lambda = 250 \pm 50$ MeV.

13.7 Quark–hadron (QH) Duality

QCD is a theory which describes strong interactions in terms of quarks and gluons with remarkable features of asymptotic freedom at high energies (E_l) and momentum transfer squared (Q^2) and confinement at low energies and momentum transfer. At low E_l and Q^2, the effective degrees of freedom to describe strong interactions are mesons and nucleons using effective Lagrangian; while at high E_l and Q^2, the quark and gluon degrees of freedom are used to describe the strong interactions using perturbative QCD. In the case of scattering processes

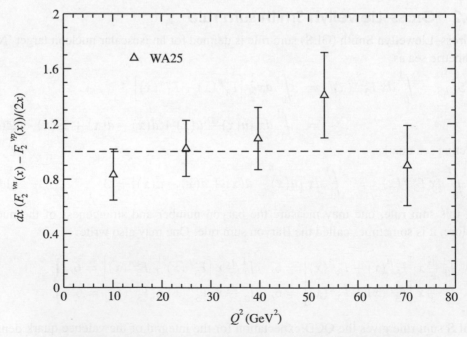

Figure 13.30 Experimental observations [598, 599] for the Adler sum rule.

induced by charged leptons and (anti)neutrino on nucleons, the inclusive cross sections at low energies are expressed in terms of structure functions corresponding to excitations of various nucleon resonances like Δ, N^*, etc. (Chapter 11, Table 11.3), lying in the first or higher resonance region. On the other hand, at high energies and Q^2, the inclusive cross sections are expressed in terms of the structure functions corresponding to the deep inelastic scattering processes from quarks and gluons. In the intermediate energy region corresponding to the transition between resonance excitations and DIS, we are yet to find a method best suited to describe the inclusive charged lepton or (anti)neutrino scattering processes. Using the kinematical cuts in the $Q^2 - \nu$ plane (Figure 13.31), one may understand the regions like elastic ($W = M$), inelastic ($M \leq W \leq 2$ GeV), DIS ($Q^2 > 1$ GeV^2, $W > 2$ GeV), soft DIS ($Q^2 < 1$ GeV^2, $W > 2$ GeV), and the shallow inelastic scattering (SIS) ($W < 2$ GeV) regions. In order to emphasize the energy dependence of different scattering processes, in Figure 13.31 the variation of $Q^2 - \nu$ plane is shown at $E_\nu = 3$ GeV (left panel) and $E_\nu = 7$ GeV (right panel). As one moves away from the higher W region, where DIS (that deals with the quarks and gluons) is the dominant process to the region of SIS (resonant+nonresonant processes having hadrons as a degree of freedom). Presently, there is no sharp kinematic boundary to distinguish these two regions. In literature, $Q^2 \geq 1$ GeV^2 has been chosen as the lower limit required for particles to be interacting with the quarks and gluonic degrees of freedom; a kinematic constraint of $W \geq 2$ GeV is also applied to safely describe the DIS region. Moreover, the kinematic region of $Q^2 \leq 1$ GeV^2, where nonperturbative QCD must be taken into serious consideration, is yet to be explored. To understand this transition region, the phenomenon of quark–hadron duality comes into play; this duality basically connects the inclusive production processes in the two regions.

Historically, the concept of what was to become 'duality' began in the 1960s with the total pion–proton cross sections being compared with the Regge fit to higher energy data. It was concluded that low-energy hadronic cross sections, on an average, could be explained in terms of high-energy behavior. In the 1970s, Poggio, Quinn, and Weinberg [602] suggested that higher energy inclusive hadronic cross sections, appropriately averaged over an energy range, should approximately coincide with the cross sections calculated using quark–gluon perturbation theory. This directly implied that the physics of quarks and gluons could describe the physics of hadrons.

In 1970, Bloom and Gilman [603] defined duality by comparing the structure functions obtained from inclusive electron–nucleon scattering with the resonance production in similar experiments and the observation that the average over resonances is approximately equal to the leading twist contribution measured in the DIS region. This seems to be valid in each resonance region individually as well as in the entire resonance region when the structure functions are summed over higher resonances. For example, in Figure 13.32, the results of proton structure function F_2^p from SLAC and the Jefferson Lab in the resonance region for $0.06 < Q^2 < 3.30$ GeV^2 are shown. The solid curve is a fit to deep inelastic scattering data at 5 GeV^2. The figure is taken from Ref. [604]. From the figure, it may be observed that the DIS scaling curve extrapolated down into the resonance region passes through the average of the 'peaks and valleys' of the resonant structure. In this picture, the resonances

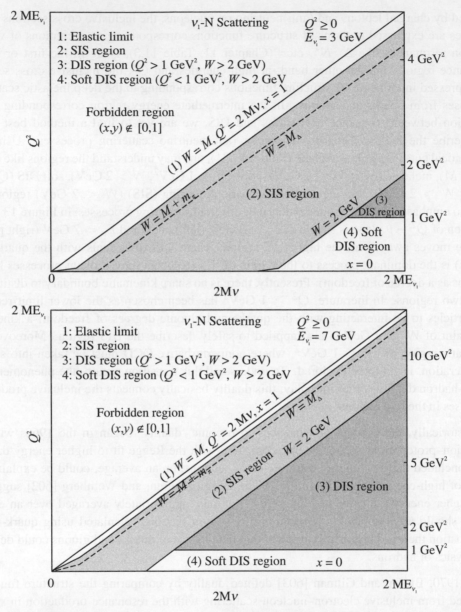

Figure 13.31 Allowed kinematical region for $\nu_l - N$ scattering in the $Q^2 - \nu$ plane for $E_\nu=3$ GeV(left panel) and $E_\nu=7$ GeV (right panel) for $Q^2 \geq 0$. Invariant mass square is defined as $W^2 = M^2 + 2M\nu - Q^2$ and the elastic limit is $x = \frac{Q^2}{2M_N\nu} = 1$. The forbidden region in terms of x and $y = \frac{\nu}{E_\nu} = \frac{(E_\nu - E_l)}{E_\nu}$ is defined as $x, y \notin [0, 1]$. The SIS region has been defined as the region for which $M + m_\pi \leq W \leq$ 2GeV and $Q^2 \geq 0$; the DIS region is defined as the region for which $Q^2 \geq 1$ GeV^2 and $W \geq 2$ GeV; and soft DIS region is defined as $Q^2 < 1$GeV^2 and $W \geq 2$ GeV (also part of SIS). Notice the unshaded band ($M < W < M + m_\pi$), where we do not expect anything from $\nu_l - N$ scattering; this region becomes important when the scattering takes place with a bound nucleon giving rise to an additional nuclear effect known as the multinucleon correlation effect.

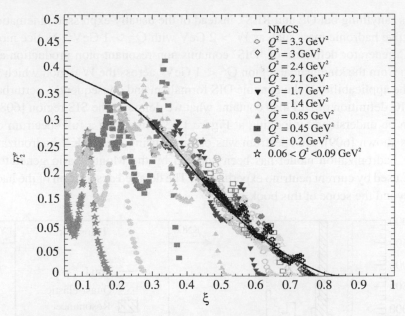

Figure 13.32 F_2^p structure function data on protons from SLAC and Jefferson Lab in the resonance region for $0.06 < Q^2 < 3.30$ GeV^2. The solid curve is a fit to deep inelastic data at 5 GeV^2. The figure is taken from Ref. [604].

can then be considered as a continuing part of the behavior observed in DIS. This would suggest that there is a connection between the behavior of resonances and QCD; perhaps even a common origin in terms of a point-like structure for both resonance and DIS interactions. Along this line, it has been conjectured that there may exist two component duality, where the resonance contribution and background contribution to the structure functions in the resonance excitation region correspond, respectively, to the valence quarks and the sea quarks contribution in structure functions in the DIS region [604]. However, these observations have to be verified by model calculations as well as by experimental data when they become available with higher precision. Currently, the observation of duality in charged lepton scattering has the following main features [605]:

- the resonance region data oscillate around the scaling DIS curve.
- the resonance data are on an average equivalent to the DIS curve.
- the resonance region data move toward the DIS curve with increase in Q^2.

As more data with better precision become available on inclusive lepton scattering from nucleons and nuclei, a verification of QH duality with sufficient accuracy will provide a way to describe lepton–nucleon and lepton–nucleus scattering over the entire SIS region. Significantly, if duality does hold for neutrino nucleon interactions, it would be possible to extrapolate the better-known neutrino DIS structure into the SIS region and give an indication of how well current event simulators, the various Monte Carlo neutrino event generator being used in the neutrino oscillation analyses (like GENIE [606], NEUT [607] etc.) model the SIS region. An initial anomaly to note is that in current Monte Carlo (MC) event simulators/generators, 'DIS' is defined as 'anything but QE and RES', instead of the usually expressed kinematic condition on the effective hadronic mass such as $W > 2$ GeV with $Q^2 > 1$ GeV^2. Notice moreover that such an MC generator definition of 'DIS' contains non-resonant pion production as well as a contribution from the kinematical region $Q^2 < 1$ GeV^2 across the W region which is certainly outside of the applicability of the genuine DIS formalism and consequently, perturbative QCD. Thus, the MC definition of DIS also contains what we define as the SIS region [608].

This can be understood by looking at Figure 13.33, in which a full spectrum of different processes is shown [609]; the spectrum was obtained using the AGKY hadronization model [610]. The hadronization model has been adopted by the Monte Carlo generators such as GENIE MC used by current neutrino experiments. The detailed description of the hadronization model is beyond the scope of this book.

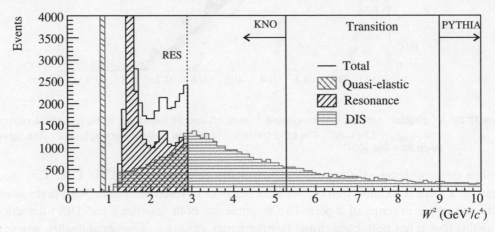

Figure 13.33 W^2 distribution of ν_μ-water target interactions in GENIE showing the quasi-elastic scattering, the resonance interactions, and the DIS region. The W distribution is further split into the three regions, KNO scaling-based model only region, PYTHIA only region, and the transition between the two regions used in the AGKY model [609].

In view of the extensive, precise experimental data on cross section and structure functions now available from JLab [611, 612, 613, 614], in a wide range of the kinematic region of electron–nucleon scattering and availability of electron DIS data from SLAC, EMC, NMC, and JLab in recent times, there is a wide international interest in studying quark–hadron duality. Phenomenological information on duality has also been accumulating for electron induced reaction channels [604]. The need for understanding (anti)neutrino-nucleon scattering cross sections in the transition region of $E_\nu \sim 1.5$–3 GeV, has also generated considerable interest in studying QH duality in the weak sector, where it has been observed in the neutral current(NC) sector through the observation of parity violating asymmetry in the scattering of polarized electron from proton and deuteron targets at JLab [614]. However, for neutrino–nucleon scattering, there is lack of experimental data and few studies are available concerning duality in the weak sector [484, 615, 605, 616, 617, 618].

The validity of quark–hadron duality in the CC and NC sectors of weak interaction will provide a way to obtain (anti)neutrino–nucleon scattering cross sections in the transition region, where either the use of effective Lagrangian or the quark–parton description is not adequate.

QH duality will facilitate a model to obtain (anti)neutrino–nucleon cross sections in the entire energy region, which can be used in various MC neutrino event generators. This will be of interest to all neutrino oscillation experiments being done around the world in the few GeV energy region [619, 620, 621].

In the next sections, some of the experimental and theoretical results, which can illustrate the physical meaning of quark–hadron duality in the electromagnetic (Section 13.8) and weak (Section 13.9) interaction processes have been discussed.

13.8 Duality in Charged Lepton–nucleon Scattering

The differential cross section for the inclusive charged lepton scattering from nucleons in the DIS region is given in terms of structure functions $W_{iN}(\nu, Q^2)$, $(i = 1, 2)$, which in the limits of high energy transfer and momentum transfer squared, are described in terms of the dimensionless structure functions $F_{1N}(x, Q^2)$ and $F_{2N}(x, Q^2)$:

$$F_{1N}(x, Q^2) = MW_{1N}(\nu, Q^2), \tag{13.118}$$
$$F_{2N}(x, Q^2) = \nu W_{2N}(\nu, Q^2). \tag{13.119}$$

In the resonance region, the inclusive cross section discussed in Chapters 11 and 12 is written as a coherent or incoherent sum of individual contributions from the resonance excitations R ($= \Delta$, N^*, etc.). This is diagrammatically shown in Figure 13.34. The expression of the cross section may be written as:

$$\frac{d^2\sigma}{d\Omega_l' dE_l'} = \frac{|\vec{k}'|\alpha^2}{2Q^4} \frac{A(p')}{\sqrt{(k \cdot p)^2 - m_l^2 M_R^2}} L_{\mu\nu} W_R^{\mu\nu}, \tag{13.120}$$

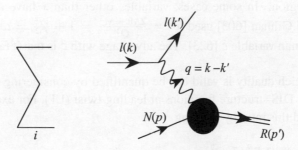

Figure 13.34 Diagrammatic representation of resonance excitations for $l + N \rightarrow l + R$.

where $L_{\mu\nu}$ is the leptonic tensor (as given in Eqs. (13.4) and (13.60)) and $W_R^{\mu\nu}$ is the hadronic tensor corresponding to the excitation of resonance R and is schematically given as

$$W_R^{\mu\nu} = \sum\sum \langle R(p')|J^\mu|N(p)\rangle^* \langle R(p')|J^\nu|N(p)\rangle, \tag{13.121}$$

$$A(p') = \frac{\sqrt{p'^2}}{\pi} \frac{\Gamma(p')}{(p'^2 - M_R^2)^2 + p'^2 \Gamma^2(p')}. \tag{13.122}$$

Here $\Gamma(p')$ is the momentum dependent width and M_R is the Breit–Wigner mass of the resonance. $\langle R(p')|J^\mu|N(p)\rangle$ corresponds to the transition matrix element for the transition $N(p) \to R(p')$ induced by the current J_μ. In the case of electromagnetic interaction, $l = e$, μ, and J^μ has only the vector contribution.

As discussed in Chapters 11 and 12, the transition matrix element of hadronic current is characterized by various transition form factors $F_i^{pR}(Q^2)$ and $F_i^{nR}(Q^2)$ for proton and neutron targets depending upon the spin of the excited resonance $R(p')$. The form factors $F_i^{pR,nR}(Q^2)$ are derived from helicity amplitudes extracted from real and virtual photon scattering experiments. These form factors then describe the structure functions in the resonance region corresponding to each resonance excitations. In the case of excitations of more than one resonance, the sum in Fig. 13.34 can be done coherently or incoherently.

Once the structure functions are defined for the DIS and resonance region, $F_{i=1-3}^R(x, Q^2)$ are theoretically calculated assuming some parameterization of quark PDFs [622]. They are also determined experimentally (in nucleon as well as nuclear targets) by analyzing the data in the two regions which are available from various experiments performed at JLab, SLAC, EMC, NMC corresponding to the various regions of Q^2 and W. For a given Q^2, x averaging is obtained over similar kinematic regions of W for each of them as follows:

$$
\begin{aligned}
\bar{F}_i^R(\Delta x, Q^2) &= \int_{x_{\min}}^{x_{\max}} dx\, F_i^R(x, Q^2)dx, \\
\bar{F}_i^{DIS}(\Delta x, Q^2) &= \int_{x_{\min}}^{x_{\max}} dx\, F_i^{DIS}(x, Q^2)dx, \\
\bar{F}_i^{exp}(\Delta x, Q^2) &= \int_{x_{\min}}^{x_{\max}} dx\, F_i^{exp}(x, Q^2)dx,
\end{aligned}
\tag{13.123}
$$

where $i = 1, 2$ and $x_{\min}$ and $x_{\max}$ would correspond to $W_{\min}$ and $W_{\max}$ relevant for each resonance or DIS region. In some cases, variables other than x have also been used. For example, Bloom and Gilman [603] used $\omega' = \frac{2M\nu + M^2}{Q^2} = 1 + \frac{W^2}{Q^2} = \omega + \frac{M^2}{Q^2}$, while others have used the Nachtman variable ζ [623]. The advantage with ζ is that it takes care of the TMC effect.

The degree to which duality is valid can be quantified by considering the ratio of integrals of the resonance and DIS structure functions at leading twist (LT). For example, Lalakulich et al. [484] have studied this ratio by defining

$$
I_i(Q^2) = \frac{\int_{\zeta_{\min}}^{\zeta_{\max}} d\zeta\, F_i^R(\zeta, Q^2)}{\int_{\zeta_{\min}}^{\zeta_{\max}} d\zeta\, F_i^{DIS}(\zeta, Q^2)},
\tag{13.124}
$$

where $i = 1, 2$, and the integration limits correspond to $\zeta_{\min} = \zeta(W = 1.6 \text{ GeV}, Q^2)$ and $\zeta_{\max} = \zeta(W = 1.1 \text{ GeV}, Q^2)$. If the results are closer to unity, it would mean that the duality holds good. The authors [615] have obtained isoscalar nucleon structure function $F_2^{eN} = (F_2^{ep} + F_2^{en})/2$, calculated as a sum of electroproduced resonances. In Figure 13.35, results for F_2^{eN} at several values of Q^2 (from 0.2 to 2 GeV^2) have been presented as a function

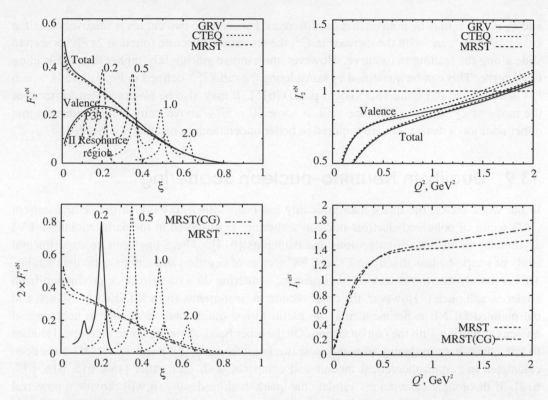

Figure 13.35 Top left panel: Duality for the isoscalar nucleon F_2^{eN} structure function. F_2^{eN} vs. ξ, for $Q^2 = 0.2, 0.5, 1$ and 2 GeV^2, compared with several leading twist parameterizations [624, 625, 626] at $Q^2 = 10$ GeV^2. Top right panel: Ratio I_2^{eN} of the integrated F_2^{eN} in the resonance region to the leading twist functions (valence and total). Bottom left panel: Duality for the isoscalar nucleon $2xF_1^{eN}$ structure function vs. ξ. The results are compared with the MRST parameterization [626] at $Q^2 = 10$ GeV^2, using F_L and F_2 (dotted) and the Callan–Gross (CG) relation, $F_2 = 2xF_1$ (dot-dashed). Bottom right panel: Ratio I_1^{eN} of the integrated $2xF_1^{eN}$ in the resonance region to the leading twist function [626]. This figure is taken from Ref. [615].

of the Nachtmann variable ξ [615]. $P_{33}(1232)$ (Δ) has the maximum contribution at the largest value of ξ that may be realized by looking at the prominent peaks, followed by peaks at smaller ξ, corresponding to the second resonance region, where the $S_{11}(1535)$ and $D_{13}(1520)$ resonances dominate. It has been found that the contribution from the $P_{11}(1440)$ resonance is small. Furthermore, it may be observed from these figures that with the increase in Q^2, the peaks in the resonance region decrease in height and move to larger ξ. In this figure (right panel), the results for the ratio I_2^{eN} has also been presented. It has been found [615] that the integrated resonance contribution is smaller at low Q^2, but increases with the increase in the value of Q^2. It may also be observed that for the results obtained with the valence-only structure function, the ratio is within $\sim 20\%$ of unity over a larger range, $Q^2 \gtrsim 0.5 \text{GeV}^2$.

Figure 13.35 (bottom panel) shows the isoscalar nucleon structure function $2xF_1^{eN}$, calculated for the four resonances ($P_{33}(1232)$, $S_{11}(1535)$, $D_{13}(1520)$, and $P_{11}(1440)$). The results are compared with the leading twist parameterization from Ref. [626] at $Q^2 = 10$ GeV^2, for the two different cases (i) using the Callan–Gross relation $F_2 = 2xF_1$, and (ii) in terms of F_L

and F_2 [615]. It may be noticed that the difference between the two curves is relatively small at $Q^2 = 10 \, \text{GeV}^2$, and with the increase in Q^2, the resonance structure function $2xF_1^{eN}$ is seen to slide along the leading twist curve. However, the average sits slightly higher than the leading twist curve. This can be quantified by considering the ratio I_1^{eN} defined in Eq. (13.124), which has been plotted in Figure 13.35 (right panel) [615]. It may also be observed that for most of the range of $Q^2 \gtrsim 0.5 \, \text{GeV}^2$, the ratio is some $30 - 50\%$ above unity, which indicates that either additional resonances are required or better understanding of duality is needed.

13.9   Duality in Neutrino–nucleon Scattering

In the weak sector, the quark-hadron duality has been shown to work in the neutral current (NC) sector of polarized electron–nucleon scattering, as observed in the parity violation (PV) asymmetry of electron from protons and deuterons [614]. There has been no experimental study of quark–hadron duality in CC and NC sectors of neutrino and antineutrino interactions. This is mainly due to the lack of (anti)neutrino scattering data on resonance production from nucleons and nuclei. However, the cross section measurements from MINERνA, NOνA, and the planned DUNE experiment, may be useful to test quark–hadron duality in nucleon and nucleus scattering with the (anti)neutrino. On the other hand, there are some theoretical studies to test quark–hadron duality in neutrino scattering, where $\bar{F}_i^R(Q^2)$ and $\bar{F}_i^{DIS}(Q^2)$ have been calculated in certain theoretical models and compared with each other [484, 615, 616, 617, 618]. If theoretical calculations validate the quark–hadron duality, it will provide a powerful method to model the (anti)neutrino–nucleus cross sections in the transition region. This is highly desirable as it will provide a method to calculate (anti)neutrino nucleus cross sections in the entire region of $\nu(\bar{\nu})$ energies relevant for ν-oscillation experiments [627, 628, 629].

In the case of weak neutral current $l = e, \mu, \nu_e, \nu_\mu, \bar{\nu}_e, \bar{\nu}_\mu$ as well as the weak charged current $l = (\nu_e, e^-), (\nu_\mu, \mu^-), (\bar{\nu}_e, e^+), (\bar{\nu}_\mu, \mu^+)$ interactions, the weak interaction vertex has both the vector and axial vector contributions. We have seen in Chapter 12 that for weak interaction induced charged current reaction $\nu_\mu p \rightarrow \mu^- \Delta^{++}$, only isospin 3/2 resonances are excited (mainly $P_{33}(1232)$). Due to isospin symmetry constraints, the neutrino–proton structure functions ($F_2^{\nu p}$, $2xF_1^{\nu p}$ and $xF_3^{\nu p}$) for these resonances are three times larger than the neutrino–neutron structure functions. Due to the dominance of Δ^{++}, the resonance structure functions are significantly larger than the DIS structure functions, $F_i^{\nu p(\text{res})} > F_i^{\nu p(\text{DIS})}$, and quark–hadron duality is violated for the interaction taking place with a proton target [615].

In the case of neutrino–neutron scattering, in addition to the dominant contribution from isospin $I = 3/2$ resonances, there is non-negligible contribution from $I = 1/2$ resonances. For example, the study made by Lalakulich et al. [484, 615], observed that $I = 1/2$ resonances, viz., $P_{11}(1440)$, $D_{13}(1520)$ and $S_{11}(1535)$, contribute to the cross section albeit small as compared to the $P_{33}(1232)$ resonance. It has been found that the leading twist curve for the νn structure functions lies above the resonance structure functions, $F_i^{\nu n(\text{res})} < F_i^{\nu n(\text{DIS})}$, and, therefore, the quark–hadron duality does not hold even in the case of neutrino–neutron scattering. However, there is need to consider the contribution from higher resonances before a conclusion is drawn (see Figure 13.36).

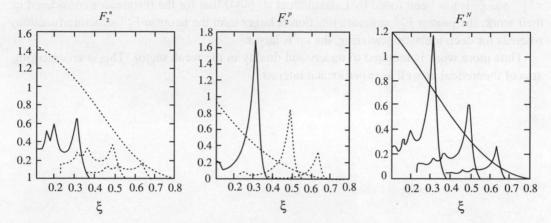

Figure 13.36 Figure from [616]: (upper) Comparison of the Rein–Sehgal F_2 structure functions vs. ζ for neutron, proton, and the isoscalar nucleon target at $Q^2 = 0.4$, 1 and 2 GeV^2 (left to right in each figure) with the appropriate DIS scaling functions at $Q^2 = 10$ GeV^2.

In Figure 13.37, the results for the weak structure function $F_2^{\nu N}$ vs. ζ are presented for the neutrino–nucleon scattering at several values of Q^2. It may be observed that $P_{33}(1232)$ resonance has the largest peak at each value of Q^2. The next peak may be seen at lower ζ (larger W) which is dominated by the $D_{13}(1520)$ and $S_{11}(1535)$ resonances. With the increase in Q^2, the contribution from $D_{13}(1520)$ and $S_{11}(1535)$ resonances become significant. These authors [484] have also obtained a theoretical curve for $F_2^{\nu N} = x(u + \bar{u} + d + \bar{d} + s + \bar{s})$, (extreme left panel) using GRV [624] and CTEQ [625] leading twist parton distributions at $Q^2 = 10$ GeV^2. It may be noticed that like in the case of electron–nucleon scattering, here also with the increase in Q^2, the resonances slide along the leading twist curve, which is required by duality. In this figure, the results for $2xF_1^{\nu N}$ vs. ζ (central panel) have also been shown. The leading twist functions correspond to the MRST parameterization [626] using the Callan–Gross

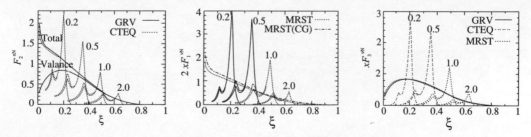

Figure 13.37 Duality for the neutrino–nucleon $F_2^{\nu N}$, $2xF_1^{\nu N}$ and $xF_3^{\nu N}$ structure functions in the resonance region at several Q^2 values, indicated against their spectra [615]. (Left) $F_2^{\nu N}$ vs. ζ. The results are compared with leading twist parameterizations [624, 625] (valence and total) at $Q^2 = 10$ GeV^2. (Center) $2xF_1^{\nu N}$ vs. ζ. The results are compared with the exact expression in Eq. (13.46) (dotted) and Callan–Gross relation (dot-dashed). (Right) $xF_3^{\nu N}$ vs. ζ. The results are compared with several leading twist parameterizations [624, 625, 626].

relation in terms of F_L. It may be observed that the resonance contributions appear to lie above the leading twist curve for a wide range of ζ. The extreme right panel of this figure shows $xF_3^{\nu N}$ vs. ζ. It has been found by Lalakulich et al. [484] that for the resonances considered in their work, the proton $F_3^{\nu p}$ structure function is larger than the neutron $F_3^{\nu n}$ structure function, whereas for deep inelastic scattering, the νn is larger.

Thus more work is required to understand duality in the weak sector. This is an emerging area of theoretical as well as experimental interest.

Chapter 14

Weak Quasielastic $\nu(\bar{\nu})$-nucleus Scattering

14.1 Introduction

In Chapters 10–13, we have discussed (anti)neutrino interactions on the free nucleon target leading to quasielastic (Chapter 10), inelastic (Chapters 11 and 12), and deep inelastic scattering (Chapters 13) depending upon the energy transferred to the target and the four-momentum transfer squared. The study of such processes is important to understand the various basic weak interaction processes induced by (anti)neutrinos from the free nucleons and determine the quasielastic form factors, transition form factors, and the structure functions of the nucleon at the $WNN(ZNN)$ vertex. In recent times, the study of (anti)neutrino reactions from the nuclear targets has been emphasized as almost all the present generation (anti)neutrino experiments use moderate to heavy nuclear targets like ^{12}C, ^{16}O, ^{40}Ar, ^{56}Fe, ^{208}Pb, where the interactions take place with the nucleons that are bound inside the nucleus. Various experiments like MINERvA, NOvA, T2K, etc., are being performed in the few GeV energy region where the contribution to the scattering cross section comes from all the possible channels, viz., quasielastic, inelastic, and deep inelastic scattering processes. A good understanding of the nuclear cross sections would be very helpful in analyzing these experiments. The precision with which the basic neutrino–nucleon cross sections in nuclear targets are known is still not better than 20–30%.

Neutrino oscillation experiments measure events that are a convolution of

(i) energy-dependent neutrino flux and

(ii) energy-dependent cross section.

Moreover, the nuclear medium effects modulating the cross sections are energy dependent. Therefore, it is highly desirable that we understand the energy dependence of nuclear medium effects in neutrino scattering processes; this understanding will help in achieving the future goals of physicists involved in studying CP violation and the mass hierarchy problem in the phenomenology of neutrino oscillations (Chapter 18).

Broadly, we may divide nuclear processes induced by neutrinos into two categories. The first one is the exclusive reactions

$$\nu_l(k) + {}^A_{Z_i}X(p) \rightarrow l^-(k') + {}^A_{Z_f}Y(p'), \tag{14.1}$$

where $Z_i(Z_f)$ is the charge of initial (final) nucleus. In this case, the final nucleus ${}^A_{Z_f}Y$ is left either in the ground state or in an excited state, which decays into some final nuclear states through electromagnetic or weak interactions by emitting photons or charged leptons which are observed. The second is the inclusive reaction in which only the leptons produced through the charged current reactions are observed and no observation is made on the hadrons or nuclei produced in the reaction. In the case of inclusive quasielastic reactions, the kinematics of the final lepton is used to ascertain the quasielastic nature of the reaction. In this chapter, we will discuss nuclear medium effects in quasielastic scattering processes; and the nuclear medium effects in inelastic and deep inelastic scattering processes will be discussed in Chapters 15 and 16 respectively.

In the region of very low energy relevant for reactor and solar neutrinos (Chapter 17), the exclusive transitions to the ground state or a few low excited states in the final nucleus are accessible. However, in the region of intermediate and high energies relevant to atmospheric and accelerator neutrinos, inclusive reactions are the most relevant processes. In this energy region, in addition to the quasielastic scattering, other neutrino processes are also very important as shown in Figure 14.1, where the total scattering cross section per nucleon per unit energy of the incoming neutrino and antineutrino in the charged current sector is presented as a function of the neutrino energy. The individual contributions to the quasielastic, inelastic, and deep inelastic scattering cross section as well as the sum of all the processes are shown and compared with the available experimental data. It is evident from the figure that in the few GeV energy region all the three processes, viz., quasielastic, inelastic, and deep inelastic scattering, have significant contributions in the neutrino and antineutrino induced processes. It may be noticed that the quasielastic scattering cross sections per unit energy peaks at $\sim$0.3 GeV while its contribution is non-negligible even at high energies. Inelastic scattering is very important in the $\sim 1 - 3$ GeV energy region as it dominates over the quasielastic and DIS processes, although the peak lies at $\sim$2 GeV. In the Deep Underground Neutrino Experiment (DUNE), it is expected that more than 30% of the events would come from the DIS region and 50% of the events would come through the nonresonant and resonant production of mesons also known as the shallow inelastic region and the DIS region. The onset for the DIS process is considered to be $\sim$2 GeV and the cross section in this region increases almost linearly with the increase in the neutrino energy up to $E_{\nu_l} \approx 100$ GeV. Moreover, it may be observed from the figure that the experimental data have large error bars. Presently, the ongoing and planned experiments are trying to obtain high statistics data with better precision.

The charged current neutrino and antineutrino induced quasielastic inclusive reactions are expressed as

$$\nu_l(k) + {}^A_ZY(p) \rightarrow l^-(k') + X(p'), \tag{14.2}$$

$$\bar{\nu}_l(k) + {}^A_ZY(p) \rightarrow l^+(k') + X(p'), \tag{14.3}$$

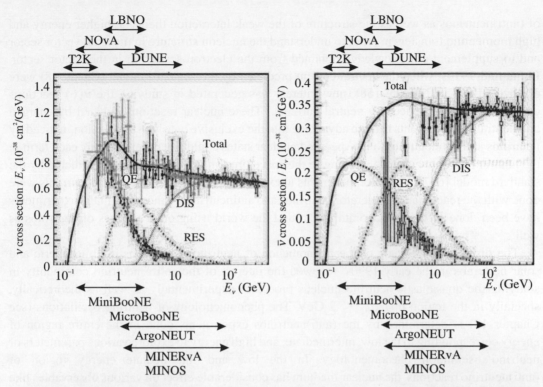

Figure 14.1 Charged current induced total scattering cross section per nucleon per unit energy of the incoming particles vs. neutrino (left panel) and antineutrino (right panel) energy. The dashed line shows the contribution from the quasielastic (QE) scattering while the dashed-dotted and dotted lines represent the contributions from the inelastic resonance (RES) and deep inelastic scattering (DIS), respectively. The sum of all the scattering cross sections (TOTAL) is shown by the solid line [630].

in which a neutrino or antineutrino scatters from a nucleon inside the nucleus. X stands for the hadronic debris produced in inclusive inelastic scattering.

The quasielastic scattering of neutrinos and antineutrinos from the nuclear targets in the entire region from low energy to very high energy has been the focus of extensive theoretical and experimental work for more than 50 years. In the very low energy region of few MeV, the QE reactions on nuclei are relevant for the astrophysical and reactor (anti)neutrinos while in the intermediate energy region of $E_{\nu(\bar{\nu})} \leq 5$ GeV, they are relevant for the accelerator and atmospheric neutrinos. While the nuclear effects are known to play an important role in the low energy region of QE scattering, it has been surprising and unexpected to see the effect of the nuclear medium even at very high energy in the region of deep inelastic scattering revealed through the EMC (European Muon Collaboration) effect (Chapter 16).

Historically, the first experiments on quasielastic $\nu(\bar{\nu})$ reactions from nuclear targets using reactor (anti)neutrinos were proposed by Pontecorvo [80]; and with the accelerator neutrinos by Pontecorvo [105], Schwartz [106], and Markov [631]. The early experiments with accelerator neutrinos were done with spark chambers (SC) on Al and Fe nuclei and with the bubble chambers (BC) on freon and propane [110, 156, 111, 385]. The early theoretical calculations for the quasielastic (anti)neutrino scattering from nuclei were done by Berman [632], Bell [633], Uberall [634], and Lovseth [635]. The primary aim of these reactions was to study the properties of (anti)neutrinos as well as the structure of the weak interaction theory at higher energy and high momentum transfers in order to understand the nucleon structure in the axial vector sector and to supplement our knowledge obtained from the electron scattering in the vector sector. In the mid-1970s, when the neutral current predicted by the standard model (Chapter 8) were discovered at CERN [636, 168], new interest was generated in studying the $\nu(\bar{\nu})$ reactions on nuclear targets induced by neutral currents. These nuclear reactions induced by charged and neutral weak currents had the advantage that the exclusive quasielastic reaction induced by neutrinos and antineutrinos to the specified nuclear states could be chosen to study each term in the neutral hadronic current according to the space time and isospin structure predicted by the standard model [637]. Various experiments to explore the structure of the neutral currents were done with the reactor and accelerator neutrinos and antineutrinos. Since then many experiments have been done in various laboratories around the world using other sources of neutrinos as well.

The discovery and study of the phenomenon of neutrino oscillations in atmospheric and solar neutrinos in the early 1990s, renewed the interest of the entire neutrino community in studying the quasielastic neutrino nucleus processes experimentally as well as theoretically, specially in the region of $E_{\nu(\bar{\nu})} \leq 2$ GeV. The phenomenology of neutrino oscillations (see Chapter 18) is determined by the (anti)neutrinos experiments done in the entire region of energy corresponding to the low, intermediate, and high energy to explore various parameters of neutrino oscillation phenomenology. In the low and intermediate energy region of (anti)neutrino reactions, the nuclear medium has considerable effect on various observables like the cross sections, angular, and energy distribution of the final lepton produced in QE reaction. Moreover, some inelastic channels in which hadrons like pions and correlated nucleons are produced could also give rise to QE-like signals through the reabsorption of pions and nucleons due to the final state interactions (FSI) which could take place in the nuclear medium. These nuclear medium effects are difficult to calculate and are model dependent. However, there are some basic nuclear medium effects which are easier to calculate and understand like the effects due to the binding energy, Fermi motion of the nucleon, and the aspects of nucleon–nucleon $(N - N)$ correlations which could be described through the wave function of nucleus and determined using well-studied nucleon–nucleon interaction potentials. In view of the various types of the nucleon–nucleon potential studied in nuclear physics, there are many approaches developed in the literature to take into account the $N - N$ interaction and the nuclear wave functions for describing the quasielastic reactions of ν and $\bar{\nu}$ from the nuclear targets. In this chapter, we focus on the QE scattering of ν and $\bar{\nu}$ from the nuclear targets in the region of low and intermediate energies.

14.2 Physics of Nuclear Medium Effects in Quasielastic Scattering

There are two major ingredients to describe the QE $\nu(\bar{\nu})$ scattering from the nuclear targets:

  (i) the theory of $\nu(\bar{\nu})$-nucleon scattering, and

  (ii) the model to describe the nucleus.

The former is described by the standard model (Chapter 10), while the latter has been studied extensively in the electron–nucleus and photon–nucleus reaction. In a simple picture, the (anti)neutrino scatters from a nucleon in which the scattering is induced by charged current (CC) interactions from a free nucleon at rest

$$
\begin{aligned}
\nu_l(k) + n(p) &\longrightarrow l^-(k') + p(p'), \\
\bar{\nu}_l(k) + p(p) &\longrightarrow l^+(k') + n(p').
\end{aligned}
\tag{14.4}
$$

The matrix element of this process can be written as

$$
\mathcal{M} = \frac{G_F}{\sqrt{2}} \cos\theta_C \, l^\mu J_\mu,
\tag{14.5}
$$

where the leptonic current l_μ and the hadronic current are defined in Chapter 10.

Using this, we can write the matrix element squared averaged over the initial and summed over the final spins of the nucleon given by:

$$
|\mathcal{M}|^2 = \frac{G_F^2}{2} \cos^2\theta_C \, L^{\mu\nu} J_{\mu\nu}.
\tag{14.6}
$$

The leptonic tensor $L^{\mu\nu}$ and the hadronic tensor $J_{\mu\nu}$ are given in Chapter 10 (section 10.5.3).

The differential scattering cross section can be expressed as

$$
\frac{d^2\sigma_{\nu l}}{d\Omega(\hat{k}')dE_l'} = \frac{1}{E_n E_p} \frac{|\vec{k}'|}{|\vec{k}|} \frac{G_F^2 \cos^2\theta_C}{128\pi^2} L^{\mu\nu} J_{\mu\nu} \delta(E_\nu - E_l + E_n - E_p).
\tag{14.7}
$$

The energy momentum conservation for the reaction given in Eq. (14.4) implies that

$$
(q+p)^2 = p'^2, \quad \text{where } q = k - k' = p' - p, \quad p^\mu = (M, \vec{0}),
\tag{14.8}
$$

leading to

$$
q_0 = \Delta E = \frac{M'^2 - M^2 - q^2}{2M} \approx \frac{-q^2}{2M}.
\tag{14.9}
$$

It shows that the energy distribution of the final leptons would have a delta function peak at the aforementioned value of energy transfer $\Delta E = \frac{-q^2}{2M}$. However, in a nucleus, the nucleons are neither at rest nor non-interacting, leading to various types of nuclear effects of kinematical as well as dynamical origin. In the following, we describe them qualitatively and then discuss some of them quantitatively in later sections.

(i) **Nucleon binding**

The nucleons in the nucleus are bound and the binding energy of the nuclei is well studied and known. Consequently, the nucleons in the nuclei are off-mass shell and do not satisfy the energy momentum relation, that is, $p^2 = M^2$. There are many theoretical models suggesting that the effective mass of nucleons is reduced in the nuclear medium and the reduction is related to the strength of the potential responsible for the nuclear binding; it is therefore model dependent. In any case, this affects the free particle kinematics used in Eq. (14.9) and the peak of the energy distribution is shifted in the energy distribution of the nucleus around the peak corresponding to $\Delta E = \frac{-q^2}{2M}$.

(ii) **Fermi motion**

The nucleons in the nucleus move with a momentum $\vec{p}$. In the Fermi gas model, this momentum is bounded by a maximum momentum p_F called Fermi momentum given in terms of the density as

$$p_F = [3\pi^2 \rho]^{\frac{1}{3}}, \tag{14.10}$$

ρ is the density of the nucleon in the nucleus.

In a shell model picture, the nucleons move in a central mean field described by a potential $V(r)$; the motion is described nonrelativistically by a Hamiltonian given by

$$H = -\frac{\vec{\nabla}^2}{2M} + V(r), \tag{14.11}$$

where $V(r)$ describes the mean field for which there are many models in the literature.

The momentum of the nucleon in a nucleus is then defined through the momentum distribution of the nucleons in the nuclei which is determined by the nucleon wave function $\psi(\vec{p})$ in momentum space obtained by solving the Schrodinger equation with H given in Eq. (14.11). This momentum distribution is called the spectral function of the nucleon $S(\vec{p}, E)$. In the simplest case of the Fermi gas model, it is given by:

$$S(\vec{p}, E) \propto \theta(p_F - p)\delta(E - \sqrt{|\vec{p}|^2 + M^2} + \epsilon), \tag{14.12}$$

where ϵ is the separation energy. In a realistic nucleus, the spectral function is related to $|\psi(\vec{p})|^2$. The cross sections from a nucleon of a given momentum $\vec{p}$ is then convoluted with the spectral function $S(\vec{p}, E)$.

Therefore, the elastic peak is not only shifted, but also smeared about the elastic peak depending upon the spectral function $S(\vec{p}, E)$. Historically, such smearing of the peak was observed in electron scattering and the process was called quasielastic elastic scattering (QEES).

(iii) **Pauli blocking**

In the conventional shell model picture of nuclei, various nuclear states are filled by neutrons and protons starting from the lowest possible state up to a certain nuclear state depending upon the number of nucleons. Similarly, in a Fermi gas picture of the nuclei, all the nuclear states in the Fermi sea are filled up to the momentum p_F (Figure 14.2), (a) and (b). This defines the ground state. In any nuclear reaction, the nucleons from a certain filled state are excited to a higher unoccupied state depending upon the energy transfer, creating a hole in the previously occupied state (Figure 14.2) (c). This is called the creation of a particle–hole $(1p - 1h)$ state in the Hilbert space of nuclei or in the Fermi sea. Since the nucleons are fermions and follow Pauli's exclusion principle, the excited particles are not allowed to occupy the already filled states. Therefore, all the nuclear states up to a certain momentum in the phase space are inaccessible for occupation after scattering. This is called Pauli blocking and leads to the reduction in the cross section which could be substantial in certain kinematical regions especially in the region of low momentum transfers.

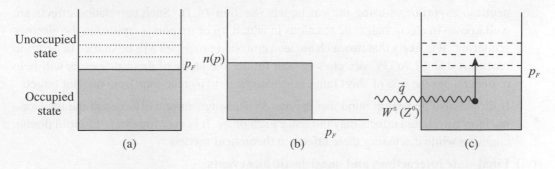

Figure 14.2 Diagrammatic representation of Pauli blocking.

(iv) Meson exchange currents

It has been established in the case of electromagnetic reactions induced by photons and electrons that in certain kinematical regions, external probes like photons and electrons interact with the mesons in flight while they are exchanged between nucleons in a nuclear medium as well as with non-nucleonic degrees of freedom like Δ in nuclei. Such effects also enter in the case of quasielastic reactions with (anti)neutrinos and give additional contribution to the vector and axial vector current matrix elements. These are called meson exchange current (MEC) effects. It has also been shown that such effects appear in the vector currents as a necessary consequence of gauge invariance of electromagnetic interactions in the presence of two nucleon potentials. Therefore, the gauge invariant part can be calculated in a model independent way. However, the contribution from the MEC in the axial vector sector and part of the contributions from the vector current are model dependent. The MEC effects are studied quite well in the low energy electroweak processes leading to the quenching of g_1 and the magnetic moment in nuclei; they are also important in quasielastic reactions.

(v) Short range and long range nucleon–nucleon correlations

The nucleons in nuclei are highly correlated with the long range correlations arising from pion exchanges as well as the short range correlations due to exchange of ρ and other heavy mesons. In simple models of the Fermi gas type, long range correlations are handled through the random phase approximation (RPA). The long range and short range correlations are taken into account in more sophisticated calculations in a model dependent way in which two-body and three-body nucleon–nucleon potentials are considered along with the central potential to calculate the nucleon wave functions in the initial and final states. Such calculations are also done in the case of low energy exclusive reactions of nuclei. In the region of high energy inclusive reactions, such multinucleon correlation effects are included in the ground state only in the context of closure approximations.

The nucleon–nucleon correlation effects due to the two particle–two hole $(2p - 2h)$ excitations calculated using the relativistic Fermi gas model in (anti)neutrino reactions have been recently shown to play a very important role in understanding present day neutrino experiments using nuclear targets (Section 14.7). Such correlation effects are well known in electromagnetic reactions in which pp or np pair is emitted in the electron scattering. It is likely that two such nucleon emission processes will be studied in neutrino reactions using LArTPC detectors in near future; the study of these processes will help to understand the role of short range correlations in $\nu(\bar{\nu})$ reactions from nuclear targets.

It should also be kept in mind that the meson exchange current effects and the nucleon–nucleon correlation effects may influence each other. It is very important to avoid double counting while discussing these effects in theoretical models.

(vi) **Final state interactions and quasielastic like events**
One of the most important nuclear medium effects is due to the final state interactions of hadrons which are produced in the reaction along with leptons. Some of these hadrons would re-scatter giving rise to inelastic reactions but some of them may be re-absorbed by the nucleus, thus mimicking the quasielastic events in which only leptons are produced. These events are called quasielastic-like events. However, such leptons will appear in a different kinematical region and model calculations show that the peak for such events are shifted by almost 150 MeV at $E_\nu = 1$ GeV. The role of such events need to be understood in the case of inclusive quasielastic reactions. Such quasielastic-like events have been a source of major concern since the early neutrino experiments with nuclear targets at CERN [110]. A theoretical understanding of such quasielastic-like events and their role in the present day neutrino oscillation experiments has been shown to be very important.

(vii) **Neutrino energy reconstruction and nuclear medium effects**
(Anti)neutrino experiments on nuclear targets have been done with a continuous beam of (anti)neutrinos having a narrow or wide band neutrino spectrum. Unlike electron beams, (anti)neutrino beams are not monoenergetic. Therefore, studying the energy dependence of (anti)neutrino cross sections is a difficult task. In some of the earlier experiments, the energy of the incident neutrino beam (and therefore, the flux) was determined from the $q^2 \rightarrow 0$ limit of the cross section which has very simple energy dependence (see Chapter 10). However, in the case of nuclear targets, even a slight deviation from $q^2 = 0$ corresponding to the spread of the experimental q^2-bin results in the nuclear effects becoming very important; the energy reconstruction also becomes model dependent. In some other experiments, the quasielastic kinematics of lepton production is inverted to obtain the incident neutrino energy in terms of the lepton variables, that is, the energy E' and the scattering angle θ. We get the reconstructed energy E_ν^{rec} and the momentum transfer Q_{rec}^2 as:

$$E_\nu^{\text{rec}} = \frac{M_p^2 + 2E_\mu M_n - M_n^2 - m_\mu^2}{2(M_n - E_\mu + \cos\theta_\mu)\sqrt{E_\mu^2 - m_\mu^2}}. \tag{14.13}$$

$$Q_{\text{rec}}^2 = 2E_\nu^{\text{rec}}(E_\mu - \cos\theta_\mu\sqrt{E_\mu^2 - m_\mu^2} - m_\mu^2). \tag{14.14}$$

In deriving this result, we have used the relation $p^2 = M_n^2$ and $p'^2 = M_p^2$. However, in the nuclear case, $M_n \to M_n - E_B$, where E_B is the binding energy and $p_n \neq 0$, that is, $p_n^2 = (M_n - E_B)^2 - \vec{p}_n^2$, where $\vec{p}_n$ has a distribution given by the spectral function. Therefore, in order to reconstruct the neutrino energy, we have to:

i) make some assumptions about the spectral function of the nucleon inside the nucleus,

ii) separate, kinematically, the quasielastic events from the quasielastic-like lepton events which have different kinematics and might overlap with the kinematical region of true quasielastic events due to the momentum distribution of target nucleons in the nucleus.

14.3 General Considerations

The general formulation for studying the quasielastic scattering of (anti)neutrinos from nuclear targets is developed in analogy with the theory of quasielastic electron scattering from nuclei using impulse approximation. In impulse approximation, the nuclear cross section is assumed to be the incoherent sum of the nucleon cross section calculated using single nucleon operators as shown in Figure 14.3.

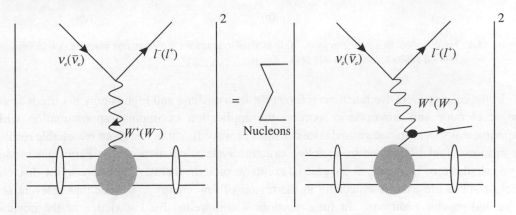

Figure 14.3 Neutrino–nucleus scattering in the impulse approximation in which W interacts with a single nucleon.

It is evident that this description is not sufficient to represent the real physical situation as the effects due to meson exchange currents (MEC) and final states interaction (and also the initial state interaction) as shown in Figure 14.4 shall always be present.

In order to study the corrections to the impulse approximation arising due to these effects, a good understanding of nuclear wave functions of the initial and final states in the presence of nucleon–nucleon correlations as well as the final state interactions is required. However, in the case of inclusive quasielastic scattering in which only leptons are observed in the final state, a good knowledge of the ground state wavefunction or ground state density would be needed in a closure approximation. Such knowledge of ground state density (and wave function) in several nuclei is available from the experiments done on quasielastic electron scattering from nuclei.

There are various theoretical approaches which have been used to calculate the $\nu(\bar{\nu})$ scattering cross section from the nuclear targets for exclusive reactions which differ from each other in their treatment of nucleons in the nucleus. The exclusive reactions are calculated using microscopic approaches in which a nuclear wavefunction, calculated in a shell model with central potential, incorporates the nucleon–nucleon correlations with a two-body potential through various approaches like the random phase approximation (RPA) [638, 639, 640, 641, 642, 643], continuous random phase approximation (CRPA) [644, 645, 646, 647, 648], quasi particle random phase approximation (QRPA) [649, 650, 651, 652, 653], projected QRPA [654], relativistic RPA [655], and relativistic nuclear energy density functional (RNEDF) [656], etc.

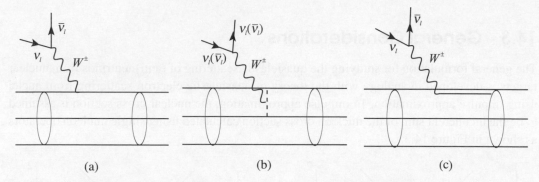

Figure 14.4   Neutrino–nucleus scattering with (a) final state interactions (b) exchange currents due to mesons and (c) exchange currents due to Δ.

In the case of inclusive reactions relevant for intermediate and high energy neutrinos from the accelerator and atmospheric sources, the application of impulse approximation with experimentally determined ground state densities or wave function has given reasonable results in various calculations done in a nuclear structure type calculation or in a Fermi gas model type calculations. However, in the case of exclusive or semi-inclusive reactions a few discrete nuclear states are excited especially in the region of low energy $\nu(\bar{\nu})$ reactions relevant for solar and reactor neutrinos. In such reactions a comprehensive knowledge of the nuclear wavefunction is required whose parameters are determined from theoretical and experimental studies of quasielastic electron nucleus scattering in nuclei like ^{12}C, ^{16}O, etc. In the case of inclusive reactions, various approaches like the nonrelativistic Fermi gas model [294], the relativistic Fermi gas model [657, 658], the local Fermi gas model [659, 660, 661, 662, 663, 664, 665, 666, 667, 668] with and without RPA, the relativistic mean field theory [669, 670, 671, 672, 673], the relativistic Green function with complex optical potential [670, 671, 672, 674] and superscaling approximation(SuSA) [637] have been used.

The theoretical structure of the basic reactions on the nucleon induced by the weak charged (CC) and neutral current(NC) is given by the standard model which has been discussed in Chapter 10. A further simplification occurs in case of nuclei, where nucleons are nonrelativistic and the nonrelativistic limit of the transition operators can be used as given in Appendix A. As discussed in Chapter 10, the strength of various form factors and their q^2 dependence may change in the presence of a nuclear medium subjected to the symmetry properties of the weak current in the CC and NC sectors but our approach is to take their values for the case of free nucleons and examine the effect of the nuclear medium on them.

In the following, we outline the basic ingredients needed to calculate the exclusive and inclusive reactions utilizing the simplest models to illustrate the methods used at low and intermediate energies.

14.4 Low Energy Quasielastic Reactions

14.4.1 Multipole expansion of the matrix elements

The low energy quasielastic reaction leads to the ground state or the excited states of the final nucleus. If the initial and final nuclear states are denoted by $|i\rangle$ and $|f\rangle$, then the transition matrix element $\mathcal{M}_{fi}$ is defined as

$$
\begin{aligned}
\mathcal{M}_{fi} &= \langle f|\mathcal{H}|i\rangle, \\
\text{where} \quad \mathcal{H} &= -\frac{G_F \cos\theta_C}{\sqrt{2}} \int d\vec{x}\, l^\mu\, J_\mu(x).
\end{aligned}
\tag{14.15}
$$

Here, l^μ is the leptonic current and J_μ is the hadronic current operator in the nucleus. Since the leptons in Eq. (14.4) are point particles, we can describe them by plane waves (neglecting the Coulomb distribution of the charged lepton in the final state), to write

$$
l_\mu(x) = l_\mu e^{-i\vec{q}\cdot\vec{x}}, \qquad q = k' - k = p - p',
\tag{14.16}
$$

such that the matrix element $\mathcal{M}_{fi}$ between the initial and final states $|i\rangle$ and $|f\rangle$ is written as

$$
\begin{aligned}
\mathcal{M}_{fi} &= -\frac{G_F \cos\theta_C}{\sqrt{2}} \langle f| \int e^{-iq\cdot x} l^\mu J_\mu(x) d\vec{x}|i\rangle \\
&= -\frac{G_F \cos\theta_C}{\sqrt{2}} \langle f| \int e^{-iq\cdot x}(l^0 J_0 - \vec{l}\cdot\vec{J}) d\vec{x}|i\rangle.
\end{aligned}
\tag{14.17}
$$

To make a multipole expansion of the matrix element [675, 676] in Eq.(14.17), we express the scalar product $\vec{l}\cdot\vec{J}e^{-iq\cdot x}$ using $\vec{l}$ in spherical basis by writing

$$
\vec{l} = \sum_\lambda l_\lambda \hat{e}_\lambda^\dagger,
\tag{14.18}
$$

where $\hat{e}_\lambda (\lambda = \pm 1, 0)$ are the components of the unit vector $(\hat{e}_x, \hat{e}_y, \hat{e}_z)$ in the spherical basis defined as

$$
e_{\pm 1} = \pm \frac{\hat{e}_x \pm i\hat{e}_y}{\sqrt{2}}, \qquad \hat{e}_0 = \hat{e}_z,
\tag{14.19}
$$

satisfying $\vec{e}_\lambda \cdot \vec{e}_{\lambda'} = \delta_{\lambda\lambda'}$, such that $l_\lambda = \vec{l} \cdot \hat{e}_\lambda$.

We use the multipole expansion for $e^{i\vec{q}\cdot\vec{x}}$, that is,

$$e^{i\vec{q}\cdot\vec{x}} = \sum_{l=0}^{\infty} \sqrt{4\pi(2l+1)}\, i^l j_l(qx) Y_{l0}(\Omega_x), \tag{14.20}$$

where $q = |\vec{q}|$ and $Y_{l0}(\Omega_x)$ are the spherical harmonics and $j_J(qx)$ are the spherical Bessel's function.

The vector spherical harmonics $\vec{\mathcal{Y}}_{Jl1}^M$ are defined as

$$\vec{\mathcal{Y}}_{Jl1}^M = \sum_{m\lambda} \langle lm1\lambda|l1JM\rangle Y_{lm}(\theta,\phi)\vec{e}_\lambda, \tag{14.21}$$

where $\langle lm1|\lambda|l1JM\rangle$ are the Clebsch–Gordan (CG) coefficients.

Using the definition of the vector spherical harmonics given in Eq. (14.21) and choosing $\hat{q} \parallel \hat{e}_z$, that is, the unit vector along the Z-axis with the orthogonality property of CG coefficients, we write

$$\vec{e}_\lambda e^{i\vec{q}\cdot\vec{x}} = \sum_l \sum_{J=0}^{\infty} \sqrt{4\pi(2l+1)}\, i^J j_l(q,x)\langle l01\lambda|l1J\lambda\rangle \vec{\mathcal{Y}}_{JL1}^\lambda, \quad x = |\vec{x}|. \tag{14.22}$$

In Eq. (14.22), we perform the expansion over l, using the values of the CG coefficients for $\lambda = \pm 1$ and $\lambda = 0$ explicitly. There would be, in general, three terms for each $\lambda(= \pm 1, 0)$, corresponding to $l = J + 1, J, J - 1$. Using the nonvanishing values of the CG coefficients in each case, the following properties of the vector spherical harmonics, that is, [676]

$$\vec{\nabla}_r \times j_J(r)\vec{\mathcal{Y}}_{JJ1}^M = -i\left(\frac{J}{2J+1}\right)^{\frac{1}{2}} j_{J+1}(r)\vec{\mathcal{Y}}_{J,J+1,1}^M + i\left(\frac{J+1}{2J+1}\right)^{\frac{1}{2}} j_{J-1}(r)\vec{\mathcal{Y}}_{J,J-1,1}^M, \tag{14.23}$$

$$\vec{\nabla}_r j_J(r)Y_{JM} = \left(\frac{J+1}{2J+1}\right)^{\frac{1}{2}} j_{J+1}(r)\vec{\mathcal{Y}}_{J,J+1,1}^M \left(\frac{J}{2J+1}\right)^{\frac{1}{2}} j_{J-1}(r)\vec{\mathcal{Y}}_{J,J-1,1}^M, \tag{14.24}$$

the expression for $\vec{e}_\lambda e^{i\vec{q}\cdot\vec{x}}$ given in Eq. (14.22) is evaluated.

After performing some basic algebraic manipulations, the following expressions are obtained:

$$\vec{e}_{\vec{q}\lambda} e^{i\vec{q}\cdot\vec{x}} = -\frac{i}{q}\sum_{J=0}^{\infty}[4\pi(2J+1)]^{\frac{1}{2}} i^J \vec{\nabla}(j_J(qx)Y_{J0}(\Omega_x)), \quad \text{for } \lambda = 0$$

$$= -\sum_{J\geq 1}[2\pi(2J+1)]^{\frac{1}{2}} i^J \left[\lambda_{j_J}(qx)\vec{\mathcal{Y}}_{JJ1}^\lambda \right.$$
$$\left. +\frac{1}{q}\vec{\nabla} \times (j_J(qx))\vec{\mathcal{Y}}_{JJ1}^\lambda\right], \quad \text{for } \lambda = \pm 1. \tag{14.25}$$

Therefore, the matrix element in Eq. (14.17) is written as

$$\langle f|\mathcal{H}|i\rangle = -\frac{G}{\sqrt{2}}\langle f|\left(-\sum_{\lambda=\pm 1} l_\lambda \sum_{J\geq 1}[2\pi(2J+1)]^{\frac{1}{2}}(-i)^J\right.$$

$$\times [\lambda \hat{T}^{\mathrm{mag}}_{J-\lambda}(q) + \hat{T}^{el}_{J-\lambda}(q)]$$

$$+ \sum_{J=0}^{\infty} [4\pi(2J+1)]^{\frac{1}{2}} (-i)^J [l_3 \hat{L}_{J0}(q) - l_0 \hat{C}_{J0}(q)] \Big) |i\rangle, \qquad (14.26)$$

where

$$\hat{C}_{JM}(q) \equiv \hat{C}^V_{JM} + \hat{C}^A_{JM} \equiv \int d\vec{x} [j_J(qx) Y_{JM}(\Omega_x)] J_0(\vec{x}), \qquad (14.27)$$

$$\hat{L}_{JM}(q) \equiv \hat{L}^V_{JM} + \hat{L}^A_{JM} \equiv \frac{i}{q} \int d\vec{x} [\vec{\nabla}(j_J(qx) Y_{JM}(\Omega_x))] \cdot \vec{J}(\vec{x}), \qquad (14.28)$$

$$\hat{T}^{el}_{JM}(q) \equiv \hat{T}^{V,el}_{JM} + \hat{T}^{A,el}_{JM} \equiv \frac{i}{q} \int d\vec{x} [\vec{\nabla} \times j_J(qx) \vec{\mathcal{Y}}^M_{JJ1}] \cdot \vec{J}(\vec{x}), \qquad (14.29)$$

$$\hat{T}^{\mathrm{mag}}_{JM}(q) \equiv \hat{T}^{V,\mathrm{mag}}_{JM} + \hat{T}^{A,\mathrm{mag}}_{JM} \equiv \int d\vec{x} [j_J(qx) \vec{\mathcal{Y}}^M_{JJ1}] \cdot \vec{J}(\vec{x}). \qquad (14.30)$$

Here, $C_{JM}, L_{JM}, T^{el}_{JM}$ and T^{mag}_{JM} are called multipoles.

In the following, we enumerate some features of the aforementioned multipoles

(i) The $C_{JM}(q), L_{JM}(q), T^{el}_{JM}(q)$, and $T^{\mathrm{mag}}_{JM}(q)$ are called, respectively, the Coulomb, longitudinal, transverse electric, and transverse magnetic multipoles.

(ii) The angular momentum sum in the expression for Coulomb and longitudinal multipoles starts from $J = 0$ as they depend upon the spherical harmonics $Y_{JM}(\Omega)$; while the transverse multipole starts from $J = 1$ as they depend upon the vector spherical harmonics $\vec{y}_{JJ1}(\Omega)$.

(iii) Since the weak current operator J^μ appearing in the definition of multipoles contains vector and axial vector currents, that is, $J^\mu = V^\mu + A^\mu$, the multipoles are classified as vector and axial vector multipoles and are written as:

$$M_{JM} \Rightarrow M^V_{JM}(q) + M^A_{JM}(q)$$

for each $M_{JM} = C_{JM}, L_{JM}, T^{el}_{JM}, T^{\mathrm{mag}}_{JM}$, where $M^V_{JM}(q)$ and $M^A_{JM}(q)$ are the multipoles corresponding to vector and axial vector currents.

(iv) The parity of the vector and axial vector multipoles $M^V_{JM}(q)$ and $M^A_{JM}(q)$ are opposite to each other. The parity of M^V_{JM} is defined in the conventional way with reference to the electromagnetic vector current. The parity of all the vector and axial vector multipoles are shown in Table 14.1.

Table 14.1 Parity of vector and axial vector multipoles.

Multipole	C^V_{JM}	L^V_{JM}	$T^{el,V}_{JM}$	$T^{\mathrm{mag},V}_{JM}$	C^A_{JM}	L^A_{JM}	$T^{el,A}_{JM}$	$T^{\mathrm{mag},A}_{JM}$
Parity	$(-1)^J$	$(-1)^J$	$(-1)^J$	$(-1)^{J+1}$	$(-1)^{J+1}$	$(-1)^{J+1}$	$(-1)^{J+1}$	$(-1)^J$

(v) In general, there are eight multipoles to be considered, four corresponding to vector currents and four corresponding to the axial vector currents. However, since the vector current is conserved,

$$q_\mu J^\mu = 0 \tag{14.31}$$

implying

$$q_0 J^0 = \vec{q} \cdot \vec{J}. \tag{14.32}$$

Taking $\vec{q} \parallel \hat{e}_z$, we get a relation between the Coulomb and longitudinal multipoles, that is,

$$q_0 \langle J_f | C_{JM}^V | J_i \rangle - \vec{q} \langle J_f | L_{JM}^V | J_i \rangle = 0. \tag{14.33}$$

Therefore, the $\nu(\bar{\nu})$ cross sections are given in terms of seven multipoles while the electron scattering is described in terms of three multipoles.

(vi) The single nucleon current operators J^μ to be used with the nuclear wave functions in the impulse approximation are derived from the definition of the matrix elements of the vector and axial vector currents between free nucleon states given in Chapter 10. In the case of a nucleus, the nucleons are treated nonrelativistically; therefore, the nonrelativistic reduction of the current operators given in Appendix A can be used. We obtain for J^μ in the lowest order of momenta, neglecting the term $O(\frac{\vec{q}^2}{M^2})$, $O(\frac{\vec{p}^2}{M^2})$, and writing J^μ as

$$J^\mu = (J^0, \vec{J}), \tag{14.34}$$

where

$$J^0 = \left(f_1(q^2) + g_1(q^2) \vec{\sigma} \cdot \frac{2\vec{p} - \vec{q}}{2M} \right) \tau^\pm \tag{14.35}$$

$$\vec{J} = \left(g_1(q^2) \vec{\sigma} - i(f_1(q^2) + 2M f_2(q^2)) \frac{\vec{\sigma} \times \vec{q}}{2M} \right) \tau^\pm + f_1(q^2) \frac{2\vec{p} - \vec{q}}{2M} \tau^\pm. \tag{14.36}$$

It should be noted that the terms involving $\frac{q_0}{2M}$ are of the order of $O(\frac{q^2}{4M^2})$ because $q_0 = -\frac{q^2}{2M}$ for elastic scattering and are, therefore, neglected in the case of nuclear transitions at low energies. The operator $\tau^{+(-)}$ corresponds to the $\nu(\bar{\nu})$ scattering processes. The nuclear operators corresponding to the nucleon operators given in Eqs. (14.35) and (14.36) are therefore given by:

$$J^0(x) = \sum_{i=1}^{A} \left[f_1(q^2) \delta(\vec{x} - \vec{x}_i) + g_1(q^2) \left(\frac{p(i)}{M} \delta(\vec{x} - \vec{x}_i) \right)_{\text{sym}} \right] \tau^\pm, \tag{14.37}$$

$$\vec{J}(x) = \sum_{i=1}^{A} \left[g_1(q^2)\vec{\sigma} + f_1(q^2)\frac{2\vec{p}\cdot\vec{q}}{2M} - i\frac{f_1(q^2) + 2M(q^2)}{2M}\vec{\sigma}(i) \times \vec{q} \right] \delta(\vec{x} - \vec{x}_i)\tau^{\pm},$$

(14.38)

where $\vec{x}_i$ is the position coordinate of the ith interacting nucleon. This shows that there are various operators in the nuclear space which enter in the current $J^0(x)$ and $\vec{J}(x)$, for the multipoles which are of the type $\tau^{\pm}(i)$, $\tau^{\pm}(i)\sigma(i)$, and $\tau^{\pm}(i)\vec{p}(i)(= -i\tau^{\pm}(i)\vec{\nabla}(i))$ multiplied by the spherical harmonics, vector spherical harmonics, and the gradient and curl operators of the vector spherical harmonics as shown in the definition of the multipoles in Eqs. (14.27)–(14.30).

14.4.2 Cross sections

The cross sections for the quasielastic scattering from nuclei using the non-relativistic kinematics for nucleons is written as:

$$\frac{d\sigma}{d\Omega} = \frac{k'E'}{(2\pi)^2} \sum_{\text{lepton spins}} \overline{\sum}_{\text{nucleon spins}} |\langle f|\mathcal{H}|i\rangle|^2.$$

(14.39)

If the initial ($|i\rangle$) and final states ($|f\rangle$) are specified by the angular momentum states $|J_iM_i\rangle$ and $|J_fM_f\rangle$, respectively, then

$$\frac{d\sigma}{d\Omega} = \frac{k'E'}{(2\pi)^2} \sum_{\text{lepton spins}} \frac{1}{2J_l + 1} \sum_{M_i}\sum_{M_f} \left| \langle J_fM_f|\mathcal{H}|J_iM_i\rangle \right|^2.$$

(14.40)

Since the operators in H_W transform as definite multipoles $M_{JM}(M_{JM} = C_{JM}, L_{JM}, T_{JM}^{el}, T_{JM}^{mag})$ with specified values of J and M, we use the Wigner–Eckart theorem to write

$$\langle J_fM_f|M_{JM}|J_iM_i\rangle = (-1)^{J_f-M_f} \begin{pmatrix} J_f & J & J_i \\ -M_f & M & M_i \end{pmatrix} \langle J_f||M_J||J_i\rangle.$$

(14.41)

For an unpolarized nucleon target and no observation made on the final nucleon, we sum over all the M_f states and average over M_i states to obtain

$$\frac{1}{2J_i + 1} \sum_{M_i}\sum_{M_f} \begin{pmatrix} J_f & J & J_i \\ -M_f & M & M_i \end{pmatrix} \begin{pmatrix} J_f & J' & J_i \\ -M_f & M' & M_i \end{pmatrix} = \delta_{JJ'}\delta_{MM'} \frac{1}{2J + 1}\frac{1}{2J_i + 1}.$$

(14.42)

We use this relation and the identity that

$$\sum_{\lambda=\pm 1} l_{\lambda}l_{\lambda}^*(a + \lambda b)^2 = (l\cdot l^* - l_3l_3^*)(|a|^2 + |b|^2) - 2i(\vec{l}\times\vec{l}^*)_3\text{Re}(a^*b),$$

(14.43)

The following result is obtained [676],

$$
\frac{1}{2J_i + 1} \sum_{M_i} \sum_{M_f} |\langle f|\hat{H}_W|i\rangle|^2 = \frac{G^2}{2} \frac{4\pi}{2J_i + 1} \Big[\sum_{J \geq 1}^{\infty} \Big(\frac{1}{2}(\vec{l} \cdot \vec{l}^* - l_3 l_3^*)(|\langle J_f||\hat{T}_J^{\mathrm{mag}}||J_i\rangle|^2
$$

$$
+ |\langle J_f||\hat{T}_J^{el}||J_i\rangle|^2)
$$

$$
- \frac{i}{2}(\vec{l} \times \vec{l}^*)_3 \, 2\mathrm{Re}\langle J_f||\hat{T}_J^{\mathrm{mag}}||J_i\rangle\langle J_f||\hat{T}_J^{el}||J_i\rangle^* \Big)
$$

$$
+ \sum_{J=0}^{\infty} \Big(l_3 l_3^* |\langle J_f||\hat{L}_J||J_i\rangle|^2 + l_0 l_0^* |\langle J_i||\hat{M}_J||J_i\rangle^*|^2
$$

$$
- 2\mathrm{Re} l_3 l_3^* \langle J_f||\hat{L}||J_i\rangle\langle ||\hat{M}_J||J_i\rangle^* \Big) \Big]. \tag{14.44}
$$

Further, using the leptonic tensor $\sum_{spin} l_\mu l_\nu = \delta(k_\mu k_\nu' + k_\nu k_\mu' - g_{\mu\nu}k \cdot k' + i\epsilon_{\mu\nu\rho\sigma}k^\rho k^\sigma)$, we obtain the following result for the quasielastic nuclear cross sections

$$
\left(\frac{d\sigma}{d\Omega} \right)_{\nu\bar{\nu}} = \left(\frac{q}{\epsilon} \right) \frac{G^2 \epsilon^2}{4\pi^2} \frac{4\pi}{2J_i + 1} \Big[\Big[\sum_{J=0}^{\infty} \{(1 + \hat{v} \cdot \vec{\beta})|\langle J_f||\hat{M}_J||J_i\rangle|^2
$$

$$
+ [1 - \hat{v} \cdot \vec{\beta} + 2(\hat{v} \cdot \vec{\beta})(\hat{q} \cdot \vec{\beta})]|\langle J_f||\hat{L}_J||J_i\rangle|^2
$$

$$
- [\hat{q} \cdot (\hat{v} + \vec{\beta})] \, 2\,\mathrm{Re}\langle J_f||\hat{L}_J||J_i\rangle\langle J_f||\hat{M}_J||J_i\rangle^* \}
$$

$$
+ \sum_{J \geq 1}^{\infty} \{[1 - (\hat{v} \cdot \hat{q})(\hat{q} \cdot \vec{\beta})][|\langle J_f||\hat{T}_J^{\mathrm{mag}}||J_i\rangle|^2 + |\langle J_f||\hat{T}_J^{el}||J_i\rangle|^2]
$$

$$
\pm [\hat{q} \cdot (\hat{v} - \vec{\beta})] \, 2\,\mathrm{Re}\langle J_f||\hat{T}_J^{\mathrm{mag}}||J_i\rangle\langle J_f||\hat{T}_J^{el}||J_i\rangle^* \} \Big]\Big], \tag{14.45}
$$

where $\hat{v} \equiv \dfrac{\vec{v}}{|\vec{v}|}$, $\hat{q} \equiv \dfrac{\vec{q}}{|\vec{q}|}$.

Since the ν_e ($\bar{\nu}_e$) used in the calculations with reactor and solar neutrinos are in the range of a few MeV, the electrons produced in these reactions can be treated relativistically. Similarly, in the case of (anti)neutrino reactions induced by the ν_μ ($\bar{\nu}_\mu$) of energies in the range of a few hundreds of MeV and GeV, the muons in the final state can also be treated relativistically. Therefore, in most calculations, Eq. (14.45) is used in the relativistic limit. In this limit, it is given by:

$$
\left(\frac{d\sigma}{d\Omega} \right)_{\nu\bar{\nu}}^{RL} = \frac{G^2 \epsilon^2}{2\pi^2} \frac{4\pi}{2J_i + 1} \Bigg[\cos^2 \frac{\theta_C}{2} \left(\sum_{J=0}^{\infty} |\left\langle J_f \left|\left| \hat{M}_J - \frac{q^0}{\vec{q}} \hat{L}_J \right|\right| J_i \right\rangle|^2 \right)
$$

$$
+ \left(\frac{q^2}{2|\vec{q}|^2} \cos^2 \frac{\theta}{2} + \sin^2 \frac{\theta}{2} \right) \left(\sum_{J \geq 1}^{\infty} \left(|\langle J_f||\hat{T}_J^{\mathrm{mag}}||J_i\rangle|^2 + |\langle J_f||\hat{T}_J^{el}||J_i\rangle|^2 \right) \right)
$$

$$
\mp \sin \frac{\theta}{2} \frac{1}{|\vec{q}|} \left(q^2 \cos^2 \frac{\theta}{2} + |\vec{q}|^2 \sin^2 \frac{\theta}{2} \right)^{1/2}
$$

$$
\left(2\,\mathrm{Re}\langle J_f||\hat{T}_J^{\mathrm{mag}}||J_i\rangle\langle J_f||\hat{T}_J^{el}||J_i\rangle^* \} \right) \Bigg]. \tag{14.46}
$$

This is the general result which is used to calculate the nuclear cross section for quasielastic ν and $\bar{\nu}$ reactions leading to discrete nuclear states. Once the initial and final states are fixed, only the multipoles which are compatible with change in the angular momentum and parity would contribute.

14.4.3 Single particle matrix element in the shell model of nuclei

The single particle wave functions in the shell model are characterized by the quantum numbers $n, l, s(= \frac{1}{2}), J$, and M, where the orbital angular momentum l is coupled to the spin s to give the total angular momentum J with $J_Z(= M)$ as the third component of the total angular momentum. The initial and final states $|i\rangle$ and $|f\rangle$ are, therefore, written as:

$$|i\rangle = |n(l\frac{1}{2})J_i M_i\rangle. \tag{14.47}$$

$$|f\rangle = |n(l'\frac{1}{2})J_f M_f\rangle. \tag{14.48}$$

where

$$|n(l\frac{1}{2})|JM\rangle = \psi_{nl\frac{1}{2}JM}(r) = N R_{nl}[Y_{lm}(\theta,\phi) \otimes \psi_{\frac{1}{2}m_j}]_{JM}. \tag{14.49}$$

Here, $R_{nl}(r)$ is a radial wave function of the nucleus obtained as a solution of nucleons moving in a potential, say a simple harmonic oscillator potential or any other potential which includes the effect of correlation in addition to a central potential.

The reduced matrix element of the multipoles are expressed as:

$$\langle n'(l'\frac{1}{2}J_f)||M_{JM}||n(l\frac{1}{2}J_i)\rangle \tag{14.50}$$

for various $M_{JM}^{V(A)}(M_{JM} = C_{JM}^{V(A)}, L_{JM}^{V(A)}, T_{JM}^{el\ V(A)}, T_{JM}^{mag\ V(A)})$ in case of vector and axial vector multipoles. The aforementioned reduced matrix elements are calculated using the expression for the weak current operator J^μ for the free nucleons given in Eq. (14.34) and the expressions in the coordinate space in the case of nuclei as given in Eqs. (14.37) and (14.38) in terms of the operators $\tau^\pm(i), \tau^\pm(i)\vec{\sigma}(i)$, and $-i\tau^\pm(i)\vec{\nabla}(i)$. The multipole operators M_{JM}, are then expressed in terms of the various nuclear operators. Introducing the notation

$$M_J^M(\vec{x}) = j_J(qx)Y_{JM}(\Omega_x), \tag{14.51}$$

$$\vec{M}_{JL}^M(\vec{x}) = j_L(qx)\vec{\mathcal{Y}}_{JL1}^M(\Omega_x), \tag{14.52}$$

and using some complicated algebraic manipulations, the various multipole operators are derived as [676]

$$M_{JM} = f_1 M_J^M(\vec{x}_i), \tag{14.53}$$

$$L_{JM} = \frac{q_0}{|\vec{q}|} f_1 M_J^M(\vec{x}_i), \tag{14.54}$$

$$T_{JM}^{el} = \frac{f_1}{M_n}\left[-\left(\frac{J}{2J+1}\right)^{\frac{1}{2}}\vec{M}_{J,J+1}^M(\vec{x}_i) + \left(\frac{J+1}{2J+1}\right)^{\frac{1}{2}}\vec{M}_{J,J-1}^M(x_i)\right]\cdot\vec{\nabla}(i)$$
$$+\frac{|\vec{q}|}{2M_n}(f_1 + 2M_nf_2)\vec{M}_{JJ}^M(\vec{x})\cdot\vec{\sigma}(i), \tag{14.55}$$

$$T_{JM}^{\text{mag}} = \frac{f_1}{iM_n}\vec{M}_{JJ}^M(\vec{x}_i)\cdot\vec{\nabla}(i) + \frac{|\vec{q}|}{2M_n}(f_1 + 2M_nf_2)\times\left[-i\left(\frac{J}{2J+1}\right)^{\frac{1}{2}}\vec{M}_{J,J+1}^M(\vec{x}_i)\right.$$
$$\left.+i\left(\frac{J+1}{2J+1}\right)^{\frac{1}{2}}\times\vec{M}_{J,J-1}^M(\vec{x}_i)\right]\cdot\vec{\sigma}(i), \tag{14.56}$$

$$T_{JM}^{\text{mag},A} = g_1\vec{M}_{JJ}^M(\vec{x}_i)\cdot\vec{\sigma}(i), \tag{14.57}$$

$$T_{JM}^{el,A} = g_1\left[-i\left(\frac{J}{2J+1}\right)^{\frac{1}{2}}\vec{M}_{J,J+1}^M(\vec{x}_i) + i\left(\frac{J+1}{2J+1}\right)^{\frac{1}{2}}\vec{M}_{J,J-1}^M(\vec{x}_i)\right]\cdot\vec{\sigma}(i), \tag{14.58}$$

$$M_{JM}^A = \frac{g_1}{iM_n}M_J^M(\vec{x}_i)[\vec{\sigma}(i)\cdot\vec{\nabla}(i)] - \frac{|\vec{q}|}{2M_ng_1}g_1L_{JM}^A, \tag{14.59}$$

$$\text{where, } L_{JM}^A \equiv g_1\left[i\left(\frac{J+1}{2J+1}\right)^{\frac{1}{2}}\vec{M}_{J,J+1}^M(\vec{x}_i) + i\left(\frac{J}{2J+1}\right)^{\frac{1}{2}}\vec{M}_{J,J-1}^M(\vec{x}_i)\right]\cdot\vec{\sigma}(i). \tag{14.60}$$

We see that all the multipoles can be evaluated using the matrix element of the operator $M_J(x_i)$, $\vec{M}_{Jl}(x_i)\cdot\vec{\nabla}(x_i)$, and $\vec{M}_J(x_i)\cdot\vec{\sigma}(i)$ taken between the states $|n(\frac{1}{2})J_fM_f\rangle$ and $|n(\frac{1}{2})J_iM_i\rangle$. The explicit expressions for these matrix elements are available in literature. Recently, Haxton and Lunardini [677] have provided a software in Mathematica to evaluate the algebraic expressions and the integrals needed to calculate these matrix elements and the cross sections in Eq. (14.46).

14.5 Quasielastic $\nu(\bar{\nu})$ Reactions at Intermediate Energies

14.5.1 Introduction

As the energy of neutrinos and antineutrinos increases, there are more nuclear states which are excited. In order to calculate the inclusive quasielastic reaction cross section, one needs to calculate the cross section of all the excited states accessible in the reaction process and sum over them. In most cases, while the ground state of the final nucleus is known reasonably well, the description of the excited states is not adequate. In many cases, while the parameters of the ground state wave functions are also determined experimentally, there is not much information about the excited states. Therefore, the microscopic calculations of the $\nu(\bar{\nu})$ scattering cross section using nuclear wave functions to calculate transitions to excited states are model dependent and suffer from large uncertainties.

The experimental studies of $\nu(\bar{\nu})$ induced reaction cross sections in nuclear targets have been made with reasonable precision in recent years with reference to the analysis of various neutrino oscillation experiments. Most of these experiments involve $\nu(\bar{\nu})$ energies in the range of 500 MeV $< E_{\nu(\bar{\nu})} < 2$ GeV with maximum flux $\nu(\bar{\nu})$ around 1 GeV. In this energy region and specially around 1 GeV, the quasielastic reactions make a dominant contribution to the cross sections and the microscopic calculations based on the summation method using multipole expansion described in the earlier section are not adequate. The subject has been reviewed by many authors in recent years [356, 447, 427, 426, 678, 679].

The simplest nuclear model to study (anti)neutrino reactions was the Fermi gas model. In this model, the free nucleon cross section $\frac{d\sigma}{dq^2}$ is multiplied by a factor $(1 - D/N)$, where N is the number of nucleon targets and D is a factor depending upon the four-momentum transfer squared q^2, N, Z, and the Fermi momentum k_F of the target nucleus [633, 635, 294, 680]. This model is extended to the higher energies, that is, the relativistic Fermi gas model (RFG), in which nucleons are treated relativistically and free [657, 658]. The nuclear effects are treated by taking into account the binding energy and the Pauli principle. Recently, the RFG model has been extended to treat the long range nucleon–nucleon correlations within Random phase approximation (RPA) and the initial state interactions by including the nucleon spectral functions, calculated with inputs from experimental information on electron nucleus scattering [356, 447, 426].

Another model which has been used is the shell model with closure approximation in which the initial state is treated in the shell model using a nuclear potential and the contributions from the higher states is summed over all the final states using closure approximation, assuming completeness of states. For this, an average excitation energy is assumed which is treated as a parameter. Since all the final state contributions are summed over, this method is expected to reproduce only the total cross section and its energy dependence.

In view of the importance of the nucleon–nucleon correlation in the initial and final states as well as the final state interactions, the relativistic mean field model (RMF) which is based on the solution of the Dirac equation in the presence of strong scalar and vector potentials, due to the mediating meson fields, has also been used [669, 670, 671, 672, 673]. Similarly, the relativistic Green function models, based on the use of a complex optical potential for nucleon–nucleus interaction to treat the final state interactions, have been used in the context of RMF as well as the conventional shell model approaches [670, 671, 672, 674].

In recent developments, the phenomenological approach based on the superscaling (SuSA) model of electron scattering has been applied to calculate the quasielastic $\nu(\bar{\nu})$ cross sections from nuclear targets [637]. In this approach, which is similar to the Bjorken scaling approach in DIS discussed in Chapter 13, the cross sections are assumed to scale and depend only upon scaling function $F(\zeta)$ which is a function of only one variable ζ instead of $|\vec{q}|$ and ω, the momentum and energy transfer. This is a good approximation in the kinematical region, where $|\vec{q}|$ and ω are transferred to a single nucleon which is treated as a free particle in the nucleus and two-body correlations or meson exchange current effects are not significant. As demonstrated in the case of electron scattering, this approximation is appropriate for momentum transfers $|\vec{q}| \geq 400$ MeV/c.

It is not possible to describe all the aforementioned models, so in the following section, we describe only the simplest model, that is, the Fermi gas model and the recent developments using this model to describe quasielastic $\nu(\bar{\nu})$ scattering in nuclei.

14.5.2 Fermi gas model

Nucleons are fermions having spin $\frac{1}{2}$. Therefore, the behavior of the neutron or the proton gas is determined by Fermi–Dirac statistics.

The assumptions of the Fermi gas model are as follows:

- This model considers the nucleus as a degenerate gas of protons and neutrons much like the free electron gas in metals. Nucleons are moving freely inside a nuclear volume (Figure 14.5).

- In such a gas at $T = 0$ K (nucleus in its ground state), all the energy levels up to a maximum, known as Fermi energy E_F, are occupied by particles. In other words, at temperature $T = 0$ K, the lowest states are filled up to a maximum momentum, called the Fermi momentum p_F.

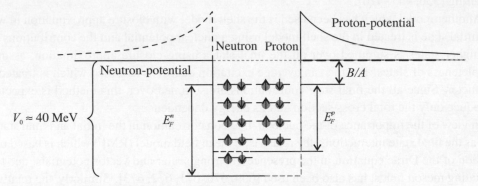

Figure 14.5 Nucleons in a square well potential states.

- Each level is occupied by two identical particles with opposite spins. Protons and neutrons are viewed as two independent systems of nucleons. Two different potential wells for protons and neutrons are considered.

- The average potential that every nucleon feels is a superposition of the potentials due to the other nucleons.

- The neutron potential well is deeper than the proton well because of the missing Coulomb repulsion. The model assumes common Fermi energy for protons and neutrons in stable nuclei; otherwise, $p \rightarrow n$ decay would happen spontaneously. This implies that there are more neutron states available and therefore, $N > Z$ for heavier nuclei.

The simplest version of the Fermi gas model, it was first applied to $\nu(\bar{\nu})$ by Berman [632], following the early work of Gatto [681] and others [294] and gives a suppression factor for the cross section as compared to the free nucleon. It should be mentioned that in these calculations, the suppression factor has a geometric origin derived from the different Fermi spheres for initial and final nucleons in the nucleus and does not depend upon the specific process of scattering.

In case of neutrino scattering on neutrons, the free neutron cross section (Eq. 14.9) is multiplied by a reduction factor $R(q)$, given by

$$
\begin{aligned}
R_N(q^2) &= 1 - \frac{D(q^2)}{N}, \text{ with } D(q^2) = Z \text{ for } \chi < u - v \\
&= \frac{A}{2}\left(1 - \frac{3}{4}\chi\left[u^2 + v^2\right] + \frac{1}{2}\chi^3 - \frac{3}{32\chi}\left[u^2 - v^2\right]^2\right) \text{ for } u - v < \chi < u + v \\
&= 0 \text{ for } \chi > u + v,
\end{aligned}
\tag{14.61}
$$

where $\chi = \frac{|\vec{q}|}{k_F}$, $u = \left(\frac{2N}{A}\right)^{\frac{1}{3}}$, and $v = \left(\frac{2Z}{A}\right)^{\frac{1}{3}}$. N, Z, and A are respectively the neutron, proton, and nucleon numbers and $\vec{q}$ is the three-momentum transfer.

This nonrelativistic global Fermi gas model was later elaborated by Loveseth [635], Yao and others [680] and was extended to relativistic nucleons by Smith and Moniz [657], Gaisser and O'Connell [658]; it is used in some of neutrino event generators for analyzing neutrino oscillation experiments.

In the Fermi gas model, it is assumed that the nucleons in a nucleus (or nuclear matter) occupy one nucleon per unit cell in phase space so that the total number of nucleons N is given by

$$
N = 2V \int_0^{p_F} \frac{d\vec{p}}{(2\pi)^3},
$$

where a factor of two is added to account for spin degree of freedom. All states up to a maximum momentum p_F ($p < p_F$) are filled. The momentum states higher than $\vec{p} > \vec{p}_F$ are unoccupied.

The occupation number $n(\vec{p})$ is defined as:

$$
\begin{aligned}
n(\vec{p}) &= 1, \vec{p} < \vec{p}_F \\
&= 0, \vec{p} > \vec{p}_F
\end{aligned}
\tag{14.62}
$$

$$
\Rightarrow \rho = \frac{N}{V} = \frac{p_F^3}{3\pi^2};
$$

$$
\text{therefore, } p_F = \left(\frac{3}{2}\pi^2\rho\right)^{\frac{1}{3}}.
\tag{14.63}
$$

Since the protons and neutrons are supposed to have different Fermi sphere,

$$
p_{Fp} = \left(3\pi^2\rho_p\right)^{\frac{1}{3}}.
$$

$$
p_{Fn} = \left(3\pi^2\rho_n\right)^{\frac{1}{3}}.
\tag{14.64}
$$

The representative values of the Fermi momentum are $p_F = 221$ MeV for carbon, $p_F = 251$ MeV for iron, etc.

Under a weak interaction induced by (anti)neutrinos, a nucleon is excited from an occupied state to an unoccupied state, that is, it creates a hole in the Fermi sea and a particle above the sea. This is known as $1p - 1h$ excitation, with the condition that the initial momentum: $|\vec{p}| < |\vec{p}_F^i|$ and the final momentum: $|\vec{p} + \vec{q}| > |\vec{p}_F^i|$. This condition is incorporated in the expression (Eq. 14.7) for the free nucleon cross section for the scattering of (anti)neutrinos from the free nucleon at rest.

Inside the nucleus

$$\frac{d^2\sigma_{vl}}{d\Omega(\hat{k}')dE_l'}\bigg|_{\text{Nucleus}} = \frac{G_F^2\cos^2\theta_C}{128\pi^2} \int \frac{1}{E_n E_p} 2d\vec{p} \frac{1}{(2\pi)^3} n_n(\vec{p})(1 - n(|\vec{p} + \vec{q}|)\frac{|\vec{k}'|}{|\vec{k}|}\delta(q_0 + E_n - E_p)L_{\mu\nu}J^{\mu\nu}.$$

$$(14.65)$$

The hadronic tensor $J_{\mu\nu}$ has to be integrated over the Fermi momentum of the initial nucleon subject to the aforementioned conditions, that is, $J_{\mu\nu}$ is replaced by

$$\frac{M^2}{E_n E_p} J_{\mu\nu}\delta(q_0 + E_n - E_p) \quad \longrightarrow \quad \int f(q,p)J_{\mu\nu}(p)\frac{d^3p}{(2\pi)^3}f(q,p)$$

$$f(q,p) = n(|\vec{p}|)(1 - n(|\vec{p} + \vec{q}|)\frac{M^2}{E_n E_p}\delta(q_0 + E_n - E_p)$$

$$n(|\vec{p}|) = \theta(p_F^i - p) \text{ and } (1 - n(|\vec{p} + \vec{q}|) = \theta(|\vec{p} + \vec{q}| - p_F^f). \quad (14.66)$$

$J_{\mu\nu}$ involves terms like $g_{\mu\nu}, q_\mu q_\nu, p_\mu p_\nu$, and $p_\mu q_\nu$.

Now, $\int f(q,p)J_{\mu\nu}(p)\frac{d^3p}{(2\pi)^3}$ can be evaluated explicitly.

These are the main features of the Smith and Moniz relativistic Fermi gas (RFG) model. Gaisser and O'Connell [658] have used the relativistic response function $R(\vec{q}, q_0)$,

in a Fermi gas model to take into account nuclear medium effects. The expression for the double differential scattering cross section is given by

$$\left(\frac{d^2\sigma}{d\Omega_l dE_l}\right)_{(v/\bar{v})} = C\left(\frac{d\sigma_{\text{free}}}{d\Omega_l}\right)_{(v/\bar{v})}R(\vec{q}, q_0),$$

$$R(\vec{q}, q_0) = \frac{1}{\frac{4}{3}\pi p_{F_N}^3}\int \frac{d^3p_N}{E_N E_{N'}}\frac{M^2}{E_N E_{N'}}\delta(E_N + q_0 - E_B - E_{N'})\theta(p_{F_N} - |\vec{p}_N|)$$

$$\times \quad \theta(|\vec{p}_N + \vec{q}| - p_{F_{N'}}),$$

$$\left(\frac{d\sigma_{\text{free}}}{d\Omega_l}\right)_{(v/\bar{v})} = \frac{G_F^2 k'E_l}{2\pi^2}\left\{(f_1^2 + g_1^2 + Q^2f_2^2)\cos^2\frac{\theta}{2} + 2\left[g_1^2\left[1 + \frac{Q^2}{4M^2}\right]\right.\right.$$

$$+ \quad \frac{Q^2}{4M^2}\mu^2f_1^2\right]\sin^2\frac{\theta}{2} \mp \frac{2g_1}{M}f_1\mu\left[Q^2 + q_0^2\sin^2\frac{\theta}{2}\right]^{\frac{1}{2}}\sin\frac{\theta}{2}\right\}, (14.67)$$

where $Q^2 = -q^2 = -(k - k')^2$ and p_{F_N} is the Fermi momentum for the initial nucleon, $N, N' = n$ or p and $C = A - Z$ for a neutrino induced process and $C = Z$ for an antineutrino induced process.

14.5.3   Local Fermi gas model

In the local Fermi gas (LFG) model, the Fermi momenta of the initial and final nucleons are not constant, but depend upon the interaction point $\vec{r}$ and are bounded by their respective Fermi momentum at r, that is, $p_{F_n}(r)$ and $p_{F_p}(r)$ for neutron and proton, respectively, where $p_{F_n}(r) = \left[3\pi^2\rho_n(r)\right]^{\frac{1}{3}}$ and $p_{F_p}(r) = \left[3\pi^2\rho_p(r)\right]^{\frac{1}{3}}$, $\rho_n(r)$ and $\rho_p(r)$ being the local neutron and proton nuclear densities, respectively. The proton density is expressed in terms of the nuclear charge density $\rho(r)$ as $\rho_p(r) = \frac{Z}{A}\rho(r)$ and the neutron density is given by $\rho_n(r) = \frac{A-Z}{A}\rho(r)$, where $\rho(r)$ is the nuclear density, determined experimentally by the electron–nucleus scattering experiments [682, 683] for the proton and neutron matter density is obtained using the Hartree–Fock calculation.

In the local density approximation, the cross section is evaluated as a function of the local Fermi momentum, $p_F(r)$ and integrated over the whole nucleus. The differential scattering cross section is given as

$$\left(\frac{d\sigma}{d\Omega_l dE_l}\right)_{\nu A} = \int \rho_n(r)d^3r \left(\frac{d\sigma}{d\Omega_l dE_l}\right)_{\nu N}, \tag{14.68}$$

where

$$\left(\frac{d\sigma}{d\Omega_l dE_l}\right)_{\nu N} = \frac{|\vec{k}'|}{64\pi^2 E_\nu M_n M_p}\overline{\sum}\sum|\mathcal{M}|^2\frac{M_n M_p}{E_n E_p}\delta(E_\nu + E_n(p) - E_l - E_p), \tag{14.69}$$

is the differential scattering cross section for the free neutrino nucleon scattering.

The modified harmonic oscillator (MHO) density

$$\rho(r) = \rho(0)\left[1 + a\left(\frac{r}{R}\right)^2 \exp\left[-\left(\frac{r}{R}\right)^2\right]\right] \tag{14.70}$$

for ^{12}C and ^{16}O and the two-parameter Fermi density (2 pF)

$$\rho(r) = \frac{\rho(0)}{\left[1 + \exp\left(\frac{r-R}{a}\right)\right]} \tag{14.71}$$

for ^{40}Ar, ^{56}Fe, and ^{208}Pb with R and a as density parameters have been used in many calculations. In Table 14.2, we show the nuclear density and other parameters needed for the numerical calculations in this chapter. In Figure 14.6, we have shown the Fermi momentum ($p_F(r)$) as a function of position (r) for various nuclei.

Fermi motion and binding energy

In a symmetric nuclear matter, each nucleon occupies a volume of $(2\pi)^3$. However, because of two possible spin orientations of nucleons, each unit cell in configuration space is occupied by two nucleons. Thus, the number of nucleons N in a certain volume V is

$$N = 2V\int^{p_F}\frac{d^3p}{(2\pi)^3}n_n(p,r)d^3r \tag{14.72}$$

$$\text{or } \rho(r) = \frac{N}{V} = 2\int\frac{d^3p}{(2\pi)^3}n_n(p,r), \tag{14.73}$$

Table 14.2   Binding energy, and Q value of the reaction for various nuclei. The last three columns are the parameters for MHO and 2pF densities. * is dimensionless for the MHO density.

Nucleus	Binding energy (MeV)	Q value (ν) (MeV)	Q value ($\bar{\nu}$) (MeV)	R_p (fm)[660]	R_n (fm)[660]	a (fm)*[660]
^{12}C	25	17.84	13.90	1.69	1.692	1.082(MHO)
^{16}O	27	19.70	14.30	1.83	1.833	1.544(MHO)
^{40}Ar	30	3.64	8.05	3.47	3.64	0.569(2pF)
^{56}Fe	36	6.52	4.35	3.97	4.05	0.593(2pF)
^{208}Pb	44	5.20	5.54	6.62	6.89	0.549(2pF)

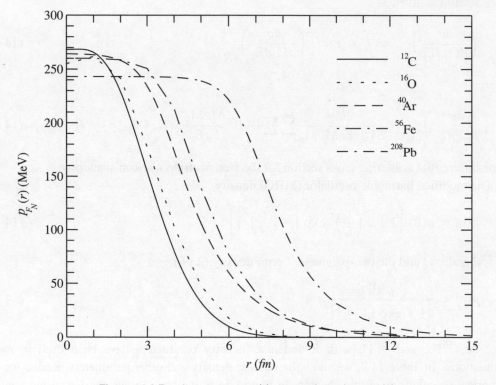

Figure 14.6 Fermi momentum $p_F(r)$ versus r for various nuclei.

where $n_N(\vec{p}, r)$ $(N = n, p)$ is the local occupation number for the nucleon. The initial nucleon has $n_i(\vec{p}, r)= 1$ for $p < p_F(r)$, where $p_F(r)$ is the maximum momentum called the Fermi momentum at position $\vec{r}$. The factor 2 in Eq. (14.72) is due to the spin degree of freedom of the nucleons.

Thus, using Eqs. (14.73) and (14.69) in Eq. (14.68), we obtain for $\nu_l n \rightarrow l^- p$ scattering process

$$\left(\frac{d\sigma}{d\Omega_l dE_l} \right)_{\nu A} = 2 \int \int d^3 r \frac{d^3 p}{(2\pi)^3} n_n(p, r) \left(\frac{d\sigma}{d\Omega_l dE_l} \right)_{\nu N}$$

$$= 2 \int \int d^3r \frac{d^3p}{(2\pi)^3} n_n(p,r) \frac{|\vec{k}'|}{64\pi^2 E_\nu M_n M_p} \overline{\sum}\sum |\mathcal{M}|^2 \frac{M_n M_p}{E_n E_p}$$
$$\times \delta(E_\nu(\vec{k}) + E_n(\vec{p}) - E_l(\vec{k}') - E_p(\vec{p}')). \tag{14.74}$$

Instead of using Eq. (14.74), we use the methods of the many-body field theory [684] where

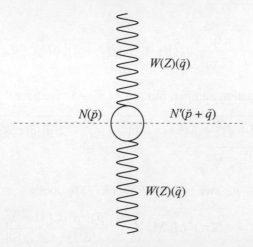

Figure 14.7 Diagrammatic representation of the particle–hole $(p - h)$ excitation induced by $W(Z)$ boson in the large mass limit of intermediate vector boson $(M_{W(Z)} \to \infty)$.

the reaction cross section for the process $\nu_l + n \to l^- + p$ in a nuclear medium is given in terms of the imaginary part of the Lindhard function Im $U_N(q_0, \vec{q})$ [639, 684] corresponding to the particle–hole $(1p - 1h)$ excitation diagram shown in Figure 14.7. This can be done by replacing the factor

$$-2\pi \int \frac{d^3p}{(2\pi)^3} n_n(p,r) \frac{M_n M_p}{E_n E_p} \delta(E_\nu(\vec{k}) + E_n(\vec{p}) - E_l(\vec{k}') - E_p(\vec{p}')) \tag{14.75}$$

by the imaginary part of the Lindhard function Im $U_N(q_0, \vec{q})$. Now the cross section is evaluated using Eq. (14.75) in Eq. (14.74). This is permissible if we assume an average value of $|\vec{p}|^2$ in $|\mathcal{M}|^2$ in the nuclear medium corresponding to the Fermi energy of the nucleons. Thus, d^3p integration is performed and the results are written in terms of the Lindhard function given in Eq. (14.76) corresponding to the particle–hole excitation given by (Figure 14.7):

$$U_N(q_0, \vec{q}) = 2 \int \frac{d^3p_n}{(2\pi)^3} \frac{M_n M_p}{E_n E_p} \frac{n_n(\vec{p}) [1 - n_p(\vec{p} + \vec{q})]}{q_0 + E_n(\vec{p}) - E_p(\vec{p} + \vec{q}) + i\epsilon}, \tag{14.76}$$

Taking the imaginary part of the Lindhard function in Eq. (14.76) corresponds to putting the intermediate particles in Figure 14.7 on shell, thereby describing the process $\nu_l + n \to l^- + p$. In the static limit for the neutron $(E_n \to M_n)$ and neglecting any Pauli blocking for the proton $(n_p \to 0)$, one recovers the result for the free nucleons.

To evaluate the imaginary part of the Lindhard function (Eq. (14.76)), we use the following relation

$$\frac{1}{\omega \pm i\eta} = \mathcal{P}\frac{1}{\omega} \mp i\pi\delta(\omega), \tag{14.77}$$

which results in

$$\text{Im}(U_N(q_0, \vec{q})) = -2\pi \int \frac{d^3 p_n}{(2\pi)^3} n_n(\vec{p}) \left[1 - n_p(\vec{p} + \vec{q})\right] \delta(q_0 + E_n - E_p) \frac{M_p M_n}{E_p E_n}, \tag{14.78}$$

where $n_n(\vec{p})$ is the occupation number. Since $\vec{q} = \vec{p}' - \vec{p}$, we have

$$
\begin{aligned}
E_p &= \sqrt{(\vec{p} + \vec{q})^2 + M^2} = \sqrt{|\vec{p}|^2 + |\vec{q}|^2 + 2|\vec{p}||\vec{q}|\cos\theta + M^2} \\
\text{and } E_n &= \sqrt{|\vec{p}|^2 + M^2}.
\end{aligned}
\tag{14.79}
$$

Using $n_n(\vec{p}) = 1$ for $p \le p_{F_n}$, we evaluate Eq. (14.78) to obtain

$$\text{Im } U_N(q_0, \vec{q}) = -(2\pi)^2 M_p M_n \int_0^{p_{F_n}} \frac{|\vec{p}_n|^2 d|\vec{p}_n|}{(2\pi)^3} \frac{[1 - n_p(\vec{p} + \vec{q})]}{\sqrt{|\vec{p}|^2 + M^2}} \int_{-1}^1 d(\cos\theta)$$

$$\left(\frac{1}{\sqrt{|\vec{p}|^2 + |\vec{q}|^2 + 2|\vec{p}||\vec{q}|\cos\theta + M^2}}\right)$$

$$\delta\left(q_0 + \sqrt{|\vec{p}|^2 + M^2} - \sqrt{|\vec{p}|^2 + |\vec{q}|^2 + 2|\vec{p}||\vec{q}|\cos\theta + M^2}\right). \tag{14.80}$$

Using the δ function property,

$$\int f(x)\delta[g(x)]\,dx = \int f(x)\frac{\delta[g(x)]}{g'(x)}dg(x) = \sum_i \frac{f(x)}{|\partial g(x_i)/\partial x|}, \tag{14.81}$$

where

$$f(x) = \frac{1}{\sqrt{|\vec{p}|^2 + |\vec{q}|^2 + 2|\vec{p}||\vec{q}|\cos\theta + M^2}}, \tag{14.82}$$

$$g(x) = q_0 + \sqrt{|\vec{p}|^2 + M^2} - \sqrt{|\vec{p}|^2 + |\vec{q}|^2 + 2|\vec{p}||\vec{q}|\cos\theta + M^2}, \tag{14.83}$$

with $x = \cos\theta$. The points x_i are the real roots of $g(x) = 0$ in the interval of integration, that is,

$$q_0 + \sqrt{|\vec{p}|^2 + M^2} = \sqrt{|\vec{p}|^2 + |\vec{q}|^2 + 2|\vec{p}||\vec{q}|\cos\theta + M^2}.$$

The $\cos\theta$ integral can be performed.

$$\cos\theta = \frac{q_0^2 - |\vec{q}|^2 + 2q_0\sqrt{|\vec{p}|^2 + M^2}}{2|\vec{p}||\vec{q}|} \le 1. \tag{14.84}$$

Further,

$$[1 - n_p(\vec{p} + \vec{q})] = \Theta(|\vec{p} + \vec{q}| - p_{F_p}) \quad \Rightarrow \quad (\vec{p} + \vec{q})^2 > p_{F_p}^2.$$

Using the expression for $\cos\theta$, this expression becomes

$$\sqrt{q_0^2 - |\vec{p}|^2 + 2q_0\sqrt{|\vec{p}|^2 + M^2}} > p_{F_p}. \tag{14.85}$$

Thus, the expression for the Lindhard function(Eq. (14.80)) is obtained as:

$$
\begin{aligned}
\text{Im}(U_N(q_0, \vec{q})) &= \frac{M_p M_n}{2\pi} \int_0^{p_{F_n}} \frac{d|\vec{p}|}{\sqrt{|\vec{p}|^2 + M^2}} \frac{|\vec{p}|}{|\vec{q}|} \Theta(1 - |\cos\theta|) \Theta(A_1 - p_{F_2}) \\
&= \frac{M_p M_n}{2\pi} \int_0^{p_{F_n}} \frac{dE}{|\vec{q}|} \Theta(1 - |\cos\theta|) \Theta(A_1 - p_{F_2}). \tag{14.86}
\end{aligned}
$$

Applying the kinematical constraint discussed earlier, we may re-write Eq. (14.86) as

$$\text{Im } U_N(q_0, \vec{q}) = -\frac{1}{2\pi} \frac{M_p M_n}{|\vec{q}|} [E_{F_1} - A] \quad \text{with} \tag{14.87}$$

$$q^2 < 0, \quad E_{F_2} - q_0 < E_{F_1} \quad \text{and} \quad \frac{-q_0 + |\vec{q}|\sqrt{1 - \frac{4M^2}{q^2}}}{2} < E_{F_1}, \tag{14.88}$$

where

$$E_{F_1} = \sqrt{p_{F_n}^2 + M_n^2}, \quad E_{F_2} = \sqrt{p_{F_p}^2 + M_p^2} \quad \text{and}$$

$$A = \text{Max} \left[M_n, E_{F_2} - q_0, \frac{-q_0 + |\vec{q}|\sqrt{1 - \frac{4M^2}{q^2}}}{2} \right]. \tag{14.89}$$

Otherwise, $\text{Im}(U_N(q_0, \vec{q})) = 0$.

The energies E_n and E_p of the neutron and proton in the Lindhard function refers to the local Fermi sea of the nucleons in the initial and final nucleus. In the Fermi sea, there is no energy gap between the occupied and unoccupied states; therefore, particle–hole($1p - 1h$) excitations can be produced with an infinitesimal energy. However, in case of finite nuclei, this is not the case. In finite nuclei, there exists a certain energy gap between the ground state of initial and final nuclei. This is the minimum excitation energy needed for transition to the ground state of the final nucleus. It is also the threshold energy (Q_{th}) of the reaction. Therefore, in nuclei, the correction due to the threshold value of the reaction Q_{th} (Table 14.2) has to be taken into account in order to get a reliable value of the cross section, specially for low energy neutrinos.

This threshold energy Q_{th} of the nuclear reactions is incorporated in these calculations by replacing the energy conserving δ function, that is, $\delta[q_0 + E_n - E_p]$ in Eq. (14.74) by $\delta[q_0 + E_n(\vec{p}) - E_p(\vec{p} + \vec{q}) - Q_{th}]$ and evaluating the Lindhard function $q_0 - Q_{th}$ instead of

q_0. Moreover, to account for the unequal Fermi sea for neutrons and protons for $N \neq Z$ nuclei, the factor $Q'_{th} = E_{Fn} - E_{Fp}$ is added to q_0 in the Lindhard function. Thus, q_0 is replaced by $q_0 - Q_{th} + Q'_{th} = E_\nu - E_l - Q_{th} + Q'_{th}$ in the Lindhard function. Because of its nature, this method only applies to inclusive processes by summing over relatively many final states. The implementation of this modification to perform numerical evaluation of cross sections requires a reasonable choice for threshold value of the nuclear reaction Q_{th}. The value of Q_{th}, for the neutrino reaction is generally taken to be the energy difference corresponding to the lowest allowed Fermi or Gamow–Teller transitions.

With the inclusion of these nuclear effects, the neutrino nuclear cross section $\sigma(E_\nu)$ is written as

$$
\sigma(E_\nu) = -\frac{G_F^2 \cos^2 \theta_C}{32\pi} \int_{r_{\min}}^{r_{\max}} r^2 dr \int_{k'\min}^{k'\max} k'^2 dk' \int_{-1}^{1} d(\cos\theta) \frac{1}{E_{\nu_e} E_e}
$$
$$
\times L^{\mu\nu} J_{\mu\nu} \operatorname{Im} U_N [E_\nu - E_l - Q_{th}, \vec{q}]. \tag{14.90}
$$

Coulomb correction

One of the important aspects of charge current neutrino interactions is the treatment of Coulomb distortion of the produced lepton in the Coulomb field of the final nucleus. At low energies of the electron relevant to β decays in the nuclei, the Coulomb distortion of electrons in the nuclear field is taken into account by multiplying the momentum distribution of the electron by a Fermi function $F(Z, E_e)$, where $F(Z, E_e)$ is given by:

$$
F(Z, E_e) = \left[1 - \frac{2}{3}(1 - \gamma_0) \right]^{-1} f(Z, E_e),
$$

with

$$
f(Z, E_e) = 2(1 + \gamma_0)) (2p_e R)^{-2(1-\gamma_0)} \frac{|\Gamma(\gamma_0 + i\eta)|^2}{(\Gamma(2\gamma_0 + 1))^2}.
$$

Here, R is the nuclear radius and $\gamma_0 = \sqrt{1 - (\alpha Z)^2}$, $\eta = \frac{\alpha Z_c}{v}$. This approximation works quite well at low energies, but it is not appropriate at higher energies, especially for high Z nuclei [685]. Therefore, at higher electron energies, a different approach is needed to describe the Coulomb distortion effect of the electron. For this purpose, we apply the methods used in electron scattering where various approximations have been used to take into account the Coulomb distortion effects of the initial and final electron. One of them is the modified effective momentum approximation (MEMA) in which the electron momentum and energy are modified by taking into account the Coulomb energy [685].

The Coulomb distortion effect on the outgoing lepton has been taken into account in MEMA in which the lepton momentum and energy are modified by replacing E_l with $E_l + V_c(r)$. The form of Coulomb potential $V_c(r)$ considered here is [686]:

$$
V_c(r) = Z_f \alpha \, 4\pi \left(\frac{1}{r} \int_0^r \frac{\rho_p(r')}{Z} r'^2 dr' + \int_r^\infty \frac{\rho_p(r')}{Z} r' dr' \right), \tag{14.91}
$$

where α is a fine structure constant($1/137.035$), Z_f is the charge of the outgoing lepton which is -1 in the case of neutrino and $+1$ in the case of antineutrino. $\rho_p(r)(\rho_n(r))$ is the proton(neutron) density of the final nucleus.

Incorporation of these considerations results in modification in the argument of the Lindhard function, that is,

$$\mathrm{Im}U_N(q_0^{\nu(\bar{\nu})}, |\vec{q}|) \longrightarrow \mathrm{Im}U_N(q_0^{\nu(\bar{\nu})} - V_c(r), \vec{q}).$$

With the inclusion of these nuclear effects, the cross section $\sigma(E_\nu)$ is written as

$$
\sigma(E_\nu) = -\frac{G_F^2 \cos^2\theta_C}{32\pi} \int_{r_{min}}^{r_{max}} r^2 dr \int_{k'_{min}}^{k'_{max}} k'^2 dk' \int_{-1}^{1} d\cos(\theta)\frac{1}{E_{\nu_l}^2 E_l} L_{\mu\nu} J^{\mu\nu}
$$
$$
\times \mathrm{Im}U_N(q_0^{\nu(\bar{\nu})} - V_c(r), \vec{q}).
\tag{14.92}
$$

Thus, in the presence of nuclear medium effects, the total cross section $\sigma(E_\nu)$, with the inclusion of Coulomb distortion effects taken into account by the Fermi function (MEMA), is written as

$$
\sigma^{FF(MEMA)}(E_\nu) = -\frac{G_F^2 \cos^2\theta_C}{32\pi} \int_{r_{min}}^{r_{max}} r^2 dr \int_{k_{min'}}^{k_{max'}} k'^2 dk' \int_{-1}^{1} d(\cos\theta)
$$
$$
\times \frac{1}{E_{\nu_e} E_e} L_{\mu\nu} J_{RPA}^{\mu\nu} \mathrm{Im}U_N^{FF(MEMA)},
\tag{14.93}
$$

where

$$
\begin{aligned}
\mathrm{Im}U_N^{FF} &= F(Z, E_e)\mathrm{Im}U_N[E_{\nu_e} - E_e - Q, \vec{q}] \quad\text{and} \\
\mathrm{Im}U_N^{MEMA} &= \mathrm{Im}U_N[E_{\nu_e} - E_e - Q - V_c(r), \vec{q}], \quad Q = Q_{th} + Q'_{th}.
\end{aligned}
\tag{14.94}
$$

Nucleon–nucleon correlations and random phase approximation

In nuclei, the strength of the electroweak couplings may not be the same as their free nucleon values due to the presence of strongly interacting nucleons. Though the conservation of vector current (CVC) forbids any change in the charge coupling, other couplings like magnetic, axial charge, and pseudoscalar couplings are likely to change from their free nucleon values. Due to PCAC, the axial current is strongly coupled to the pion field in the nuclear medium and therefore, axial couplings are more likely to change due to pionic effects modifying the nuclear response functions. To get an idea of these effects, we perform a nonrelativistic reduction of the hadronic current(J_μ given in Chapter 10). We see the occurrence of $g_1\vec{\sigma}\vec{\tau}$, $f_2\vec{\sigma}\times\vec{q}\vec{\tau}$, and $g_3\vec{\sigma}\cdot\vec{q}\vec{\tau}$ terms (Appendix A) in the weak current which are linked to the spin isospin excitation; f_2 and g_3 are coupled to the transverse and longitudinal channels. g_1 is coupled to both. There exists considerable work in understanding the quenching of magnetic moment and axial charge in nuclei due to the nucleon–nucleon correlations. In our approach, the nucleon–nucleon correlation effects are reflected in the modification of the nuclear response in longitudinal and transverse channels. We calculate this reduction in the vector–axial (VA) and axial–axial (AA) response functions due to the long range nucleon–nucleon correlations treated in the RPA, diagrammatically shown in Figure 14.8.

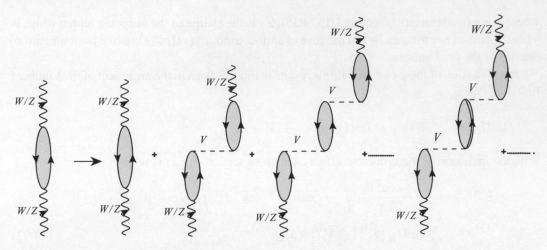

Figure 14.8 RPA effects in the $1p - 1h$ contribution to the W/Z self-energy, where particle–hole, Δ–hole, Δ–Δ, etc. excitations contribute.

The diagram shown in Figure 14.8 simulates the effects of the strongly interacting nuclear medium at the weak vertex. The $ph - ph$ interaction is shown by the dashed line and is described by the π and ρ exchanges modulated by the effect of short range correlations. For the $ph - ph$ potential, we use

$$V_N(q) = V_\pi(q) + V_\rho(q) \tag{14.95}$$

with the potential corresponding to πNN interaction which gives longitudinal part of the interaction and is defined as:

$$V_\pi(q) = \left(\frac{f_\pi^2}{m_\pi^2}\right) |\vec{q}|^2 \left[\frac{\hat{q}_i \hat{q}_j \sigma_i \sigma_j}{q_0^2 - \vec{q}^2 - m_\pi^2 + i\epsilon}\right] \vec{\tau} \cdot \vec{\tau} \tag{14.96}$$

and the ρNN interaction gives the transverse part of interaction which is given by

$$V_\rho(q) = \left(\frac{f_\rho^2}{m_\rho^2}\right) |\vec{q}|^2 \left[\frac{(\delta_{ij} - \hat{q}_i \hat{q}_j)\sigma_i \sigma_j}{q_0^2 - \vec{q}^2 - m_\rho^2 + i\epsilon}\right] \vec{\tau} \cdot \vec{\tau}, \tag{14.97}$$

where f_π and f_ρ are the coupling strengths and $\frac{f_\pi^2}{4\pi} = 0.08$, $\frac{f_\rho^2}{m_\rho^2} = 2\frac{f_\pi^2}{m_\pi^2}$. $V_\pi(q) + V_\rho(q)$ provides the spin-isospin part of the nucleon-nucleon interaction in the meson exchange model. Therefore, $V_N(q)$ may be expressed as

$$V_N(q) = \frac{f^2}{m_\pi^2} \left[V_t(q)(\delta_{ij} - \hat{q}_i \hat{q}_j) + V_l(q)\hat{q}_i \hat{q}_j\right] (\sigma_i \sigma_j)(\vec{\tau} \cdot \vec{\tau}) \tag{14.98}$$

for the *ph* case and a similar potential V_Δ is obtained in the case of $ph - \Delta h$ interaction by substituting $\vec{\sigma} \to \vec{S}$, $\vec{\tau} \to \vec{T}$, and $f \to f^* = 2.15f$. V_l is the strength of the potential in the longitudinal channel and V_t is the strength of the potential in the transverse channel. The representation into longitudinal and transverse channels is useful when one tries to sum the geometric series in Figure 14.8 where the longitudinal and transverse channels decouple and can be summed independently.

The potential $V(q)$ due to π and ρ exchange is explicitly written as:

$$V_l(q) = \frac{f^2}{m_\pi^2} \left[\frac{q^2}{-q^2 + m_\pi^2} \left(\frac{\Lambda_\pi^2 - m_\pi^2}{\Lambda_\pi^2 - q^2} \right)^2 + g' \right],$$

$$V_t(q) = \frac{f^2}{m_\pi^2} \left[\frac{q^2}{-q^2 + m_\rho^2} C_\rho \left(\frac{\Lambda_\rho^2 - m_\rho^2}{\Lambda_\rho^2 - q^2} \right)^2 + g' \right]. \tag{14.99}$$

$\frac{f^2}{4\pi} = 0.8$, $\Lambda_\pi = 1.3$ GeV, $C_\rho = 2$, $\Lambda_\rho = 2.5$ GeV, m_π and m_ρ are the pion and rho meson masses, and g' is the Landau–Migdal parameter taken to be 0.7 which has been used quite successfully to explain many electromagnetic and weak processes in nuclei [641, 687, 688].

Using the matrix elements at the weak WNN vertex and the $ph - ph$ potential, the contribution of Figure 14.8 is written as

$$U(q) = U(q) + U(q)V_N(q)U(q) + U(q)V_N(q)U(q)V_N(q)U(q) + \dots \tag{14.100}$$

Writing the potential $V_N(q)$ in terms of V_l and V_t, the series in Eq. (14.100) can be separated into the longitudinal and transverse components using the following relationship:

$$\left. \begin{aligned} (\delta_{ij} - \hat{q}_i\hat{q}_j)(\delta_{jl} - \hat{q}_j\hat{q}_l) &= \delta_{il} - \hat{q}_i\hat{q}_l \\ \hat{q}_i\hat{q}_j\hat{q}_j\hat{q}_l &= \hat{q}_i\hat{q}_l \\ (\delta_{ij} - \hat{q}_i\hat{q}_j)\hat{q}_j\hat{q}_l &= 0 \end{aligned} \right\}. \tag{14.101}$$

The longitudinal part is then written as

$$\begin{aligned} U_L(q) &= \left[U(q) + U(q)V_l\hat{q}_i\hat{q}_j\sigma_i\sigma_j U(q) + U(q)V_l\hat{q}_i\hat{q}_k\sigma_i\sigma_k U(q)V_l\hat{q}_k\hat{q}_j\sigma_k\sigma_j U(q) + \dots \right] \vec{\tau}_1 \cdot \vec{\tau}_2 \\ &= [U(q) + U(q)V_lU(q) + U(q)V_lU(q)V_lU(q) + \dots] \hat{q}_i\hat{q}_j \, \sigma_i\sigma_j \, \vec{\tau}_1 \cdot \vec{\tau}_2 \\ &= U(q)[1 + V_lU(q) + (V_lU(q))^2 + \dots] \hat{q}_i\hat{q}_j \, \sigma_i\sigma_j \, \vec{\tau}_1 \cdot \vec{\tau}_2 \\ &= \left[\frac{U(q)}{1 - U(q)V_l} \right] \hat{q}_i\hat{q}_j \, \sigma_i\sigma_j \, \vec{\tau}_1 \cdot \vec{\tau}_2. \end{aligned} \tag{14.102}$$

Similarly, the transverse part is given by

$$U_T(q) = \left[\frac{U(q)}{1 - U(q)V_t} \right] (\delta_{ij} - \hat{q}_i\hat{q}_j) \, \sigma_i\sigma_j \, \vec{\tau}_1 \cdot \vec{\tau}_2. \tag{14.103}$$

Therefore, we can write Eq. (14.100) as:

$$\bar{U}(q) = \left[\left(\frac{U(q)}{1 - U(q)V_t} \right) (\delta_{ij} - \hat{q}_i\hat{q}_j) + \left(\frac{U(q)}{1 - U(q)V_l} \right) \hat{q}_i\hat{q}_j \right] \vec{\sigma}_i \, \vec{\sigma}_j \vec{\tau}_1 \cdot \vec{\tau}_2, \tag{14.104}$$

where $U = U_N + U_\Delta$, with U_N and U_Δ as the Lindhard function for particle–hole($1p - 1h$) and Δh excitations, respectively, in the medium. The expressions for U_N and U_Δ are taken from Ref. [689].

To demonstrate how these renormalizations are done, we demonstrate it for a few cases. like the renormalization of the axial vector term of the hadronic current in Chapter 10. The nonrelativistic reduction of the axial vector term is written as

$$
\bar{u}(p')g_1(q^2)\gamma_\mu\gamma_5 u(p) = (J_0, J_i) = g_1(q^2)\left[\bar{u}(p')\gamma_0\gamma_5 u(p), \bar{u}(p')\gamma_i\gamma_5 u(p)\right]
$$
$$
= g_1(q^2)\left[\frac{\sigma_i \cdot (\vec{p} + \vec{p}')}{2E}, \left(\sigma_i + \frac{\sigma_i(\vec{\sigma} \cdot \vec{p})(\vec{\sigma} \cdot \vec{p}')}{4E^2}\right)\right] \quad (14.105)
$$

Similarly, considering the term

$$
\bar{u}(p')\, f_1(q^2)\, \gamma_\mu\, u(p),
$$

we get

$$
f_1(q^2)\, \bar{u}(p')\, \gamma_0\, u(p) = f_1(q^2)\left[1 + \frac{|\vec{p}|^2 + i\epsilon_{ijk}\, p_i p_j \sigma_k + \vec{p} \cdot \vec{q} + i\epsilon_{ijk}\, q_i p_j \sigma_k}{(2M)^2}\right] \quad (14.106)
$$

and

$$
f_1(q^2)\, \bar{u}(p')\, \gamma_i\, u(p) = f_1(q^2)\frac{1}{2M}\left[(2p_i + q_i) - i\epsilon_{ijk}\, q_j\sigma_k\right]. \quad (14.107)
$$

Now collecting J_0 and J_i terms together of the hadronic current J_μ, we get

$$
J_0 = \frac{g_1(q^2)}{2M}\, \sigma_i P_i + f_1(q^2)\left[1 + \frac{\vec{p} \cdot \vec{p}' + i\epsilon_{ijk}\, p_i' p_j \sigma_k}{(2M)^2}\right] + \frac{f_2(q^2)}{(2M)^2}\, \sigma_l q_l \sigma_m q_m \quad (14.108)
$$

and

$$
J_i = g_1(q^2)\left[\sigma_i + \frac{\sigma_i(\vec{\sigma} \cdot \vec{p})(\vec{\sigma} \cdot \vec{p}')}{(2M)^2}\right] + \frac{f_1(q^2)}{2M}\left[P_i - i\epsilon_{ijk}\, q_j\sigma_k\right] - i\frac{f_2(q^2)}{2M}\, \epsilon_{jkl}\sigma_k q_l, \quad (14.109)
$$

which leads to

$$
J_{00} = \bar{\Sigma}\Sigma\left\{\frac{g_1(q^2)}{2M}\, \sigma_i P_i + f_1(q^2)\left[1 + \frac{\vec{p} \cdot \vec{p}' + i\epsilon_{ilm}\, p_i' p_l \sigma_m}{(2M)^2}\right] + \frac{f_2(q^2)}{(2M)^2}\, \sigma_l q_l \sigma_m q_m\right\}
$$
$$
\times \left\{\frac{g_1(q^2)}{2M}\sigma_j P_j + f_1(q^2)\left[1 + \frac{\vec{p} \cdot \vec{p}' + i\epsilon_{jl'm'}\, p_j' p_{l'} \sigma_{m'}}{(2M)^2}\right] + \frac{f_2(q^2)}{(2M)^2}\sigma_{l'} q_{l'} \sigma_{m'} q_{m'}\right\}. \quad (14.110)
$$

Neglecting the terms with $1/(2M)^3$, $1/(2M)^4$ or higher powers and using the trace relations given in Appendix C, we finally get

$$
J_{00} = \left[\left(\frac{g_1(q^2)}{2M}\right)^2 \sigma_i P_i \sigma_j P_j + (f_1(q^2))^2\left\{1 + \frac{2\vec{p} \cdot \vec{p}'}{(2M)^2}\right\} + 2\frac{f_1(q^2)f_2(q^2)}{(2M)^2}\sigma_l q_l \sigma_l \sigma_m\right],
$$

$$= \left[2 \left(\frac{g_1(q^2)}{2M} \right)^2 \vec{P}^2 + 2(f_1(q^2))^2 \left\{ 1 + \frac{2\vec{p} \cdot (\vec{p} + \vec{q})}{(2M)^2} \right\} + 4 \frac{f_1(q^2)f_2(q^2)}{(2M)^2} |\vec{q}|^2 \right]. \quad (14.111)$$

Now we evaluate the expression for J_{0i}

$$
\begin{aligned}
J_{0i} &= \bar{\sum}\sum \left\{ \frac{g_1(q^2)}{2M} \sigma_i P_i + f_1(q^2) \left[1 + \frac{\vec{p} \cdot \vec{p}' + i\epsilon_{ilm} \, p'_i p_l \sigma_m}{(2M)^2} \right] \right. \\
&\quad + \frac{f_2(q^2)}{(2M)^2} \sigma_l q_l \sigma_m q_m \left\} \left\{ g_1(q^2) \left[\vec{\sigma}_j + \frac{(\vec{\sigma} \cdot \vec{p}')\sigma_j(\vec{\sigma} \cdot \vec{p})}{(2M)^2} \right] + \frac{f_1(q^2)}{2M} \right. \\
&\quad \left[P_j - i\epsilon_{jl'm'} \, q_{l'}\sigma_{m'} \right] - i\frac{f_2(q^2)}{2M} \epsilon_{jl'm'} \, \sigma_{l'}q_{m'} \right\}. \quad (14.112)
\end{aligned}
$$

Neglecting all the terms with $O(1/M^3)$ or higher powers and using trace relations, we get

$$
\begin{aligned}
J_{0i} &= 2 \left[\frac{g_1^2(q^2)}{2M} \delta_{ij}P_i + \frac{(f_1(q^2))^2}{2M} P_j \right] \\
&= 2\frac{g_1^2(q^2)}{2M} \left[(\delta_{ij} - \hat{q}_i\hat{q}_j) + \hat{q}_i\hat{q}_j \right] (2p_i + q_i) + 2\frac{(f_1(q^2))^2}{2M} (2p_i + q_i). \quad (14.113)
\end{aligned}
$$

Similarly, we find J_{ij} as follows

$$
\begin{aligned}
J_{ij} &= \bar{\sum}\sum \left\{ g_1(q^2) \left[\vec{\sigma}_i + \frac{(\vec{\sigma} \cdot \vec{p}')\sigma_i(\vec{\sigma} \cdot \vec{p})}{(2M)^2} \right] + \frac{f_1(q^2)}{2M} \left[P_i - i\epsilon_{ilm} \, q_l\sigma_m \right] \right. \\
&\quad -i\frac{f_2(q^2)}{2M} \epsilon_{ilm} \, \sigma_l q_m \right\} \times \left\{ g_1(q^2) \left[\vec{\sigma}_j + \frac{(\vec{\sigma} \cdot \vec{p}')\sigma_j(\vec{\sigma} \cdot \vec{p})}{(2M)^2} \right] + \frac{f_1(q^2)}{2M} \right. \\
&\quad \left[P_j - i\epsilon_{jl'm'} \, q_{l'}\sigma_{m'} \right] - i\frac{f_2(q^2)}{2M} \epsilon_{jl'm'} \, \sigma_{l'}q_{m'} \right\}. \quad (14.114)
\end{aligned}
$$

Neglecting the terms with $1/(2M)^3$, $1/(2M)^4$ or higher powers and using the trace relations given in Appendix C, we get

$$
\begin{aligned}
J_{ij} &= 2 \left\{ g_1^2(q^2) \left[(\delta_{ij} - \hat{q}_i\hat{q}_j) + \hat{q}_i\hat{q}_j \right] + \left(\frac{f_1(q^2)}{2M} \right)^2 P_i P_j + 2\frac{f_1(q^2)f_2(q^2)}{(2M)^2} \right. \\
&\quad |\vec{q}|^2 \left(\delta_{ij} - \hat{q}_i\hat{q}_j \right) - |\vec{q}|^2 \left(\frac{f_2(q^2)}{2M} \right)^2 (\delta_{ij} - \hat{q}_i\hat{q}_j) \right\}. \quad (14.115)
\end{aligned}
$$

Now the leading term $g_1^2(q^2)\delta_{ij}$ is split between the longitudinal and transverse components as

$$g_1^2(q^2)\delta_{ij}\text{Im}U \to g_1^2(q^2) \left[\hat{q}_i\hat{q}_j + (\delta_{ij} - \hat{q}_i\hat{q}_j) \right] \text{Im}U. \quad (14.116)$$

The RPA response of this term after summing the higher order diagrams like Figure 14.8 is modified and given by J_{ij}^{RPA}:

$$J_{ij} \to J_{ij}^{RPA} = g_1^2(q^2)\text{Im } U \left[\frac{\hat{q}_i\hat{q}_j}{|1-UV_l|^2} + \frac{\delta_{ij} - \hat{q}_i\hat{q}_j}{|1-UV_t|^2} \right], \tag{14.117}$$

Taking $\vec{q}$ along the z direction, Eq. (14.117), implies that the $g_1^2(q^2)\delta_{ij}$ contribution to the transverse (xx, yy) and longitudinal (zz) components of the hadronic tensor gets renormalized by factors $1/|1-UV_l|^2$ and $1/|1-UV_t|^2$, respectively.

The final expressions for the hadronic tensors with RPA correlations are expressed as:

$$\frac{J_{00}^{RPA}}{M^2} = \left(f_1(q^2)\right)^2 \left[\left(\frac{E(\vec{p})}{M}\right)^2 + \left(\frac{q_0 E(\vec{p}) + q^2/4}{M^2}\right) \right] - \frac{q^2}{M^2}\left(\frac{f_2(q^2)}{2}\right)^2$$

$$\left[\frac{\vec{p}^2 + q_0 E(\vec{p}) + q_0^2/4}{M^2} + \frac{q_0^2}{q^2} \right] - \frac{1}{2}\left(f_1(q^2)f_2(q^2)\right)\left(\frac{|\vec{q}|}{M}\right)^2$$

$$+ g_1^2(q^2)\left[\frac{\vec{p}^2 + q_0 E(\vec{p}) + q^2/4}{M^2} - U_L\left(\frac{q_0^2}{m_\pi^2 - q^2}\right)\left(\frac{q^2}{m_\pi^2 - q^2}\right) \right]. \tag{14.118}$$

$$\frac{J_{0z}^{RPA}}{M^2} = \frac{1}{2}\left(f_1(q^2)\right)^2 \left[\frac{E(\vec{p})}{M}\left(\frac{2p_z + |\vec{q}|}{M}\right) + \frac{q_0 p_z}{M^2} \right] - \frac{1}{2}\frac{q^2}{M^2}\left(\frac{f_2(q^2)}{2}\right)^2$$

$$\left[\frac{E(\vec{p})}{M}\left(\frac{2p_z + |\vec{q}|}{M}\right) + \frac{2q_0|\vec{q}|}{q^2} + \frac{q_0\left(2p_z + |\vec{q}|\right)}{2M^2} \right] - \frac{1}{2}\left(f_1(q^2)f_2(q^2)\right)$$

$$\left[\frac{q_0|\vec{q}|}{M^2} \right] + g_1^2(q^2)\left[U_L\frac{E(\vec{p})}{M}\left(\frac{2p_z + |\vec{q}|}{2M}\right) + \frac{q_0 p_z}{2M^2} - U_L\left(\frac{q_0|\vec{q}|}{m_\pi^2 - q^2}\right) \right.$$

$$\left. \left(\frac{q^2}{m_\pi^2 - q^2}\right) \right] \tag{14.119}$$

$$\frac{J_{zz}^{RPA}}{M^2} = \left(f_1(q^2)\right)^2 \left[\frac{p_z^2 + |\vec{q}|p_z - q^2/4}{M^2} \right] - \frac{1}{4}\frac{q^2}{M^2}\left(\frac{f_2(q^2)}{2}\right)^2 \left[\left(\frac{2p_z + |\vec{q}|}{M}\right)^2 \right.$$

$$\left. + \frac{q_0^2}{q^2} \right] - \frac{1}{2}\left(f_1(q^2)f_2(q^2)\right)\left(\frac{q_0}{M}\right)^2 + g_1^2(q^2)\left[U_L + \frac{p_z^2 + |\vec{q}|p_z - q^2/4}{M^2} \right.$$

$$\left. - U_L\left(\frac{|\vec{q}|}{m_\pi^2 - q^2}\right)\left(\frac{q^2}{m_\pi^2 - q^2}\right) \right]. \tag{14.120}$$

$$\frac{J_{RPA}^{xx}}{M^2} = \left(f_1(q^2)\right)^2 \left[\frac{p_x^2 - q^2/4}{M^2} \right] - \frac{q^2}{M^2}\left(\frac{f_2(q^2)}{2}\right)^2 \left[U_T + \frac{p_x^2}{M^2} \right]$$

$$- \frac{1}{2}\left(f_1(q^2)f_2(q^2)\right)U_T\left(\frac{q^2}{M^2}\right) + g_1^2(q^2)\left[U_L + \frac{p_x^2 - q^2/4}{M^2} \right]. \tag{14.121}$$

$$\frac{J_{xy}^{RPA}}{M^2} = ig_1(q^2)\left[f_1(q^2) + f_2(q^2)\right]\left[\frac{q_0 p_z}{M^2} - U_T\frac{|\vec{q}|E(\vec{p})}{M^2}\right]. \tag{14.122}$$

With the incorporation of these nuclear medium effects, the expression for the total scattering cross section $\sigma(E_\nu)$ is given by Eq.(14.92) with $J^{\mu\nu}$ replaced by $J_{RPA}^{\mu\nu}$ (defined in Eq. (14.117)), that is,

$$\sigma(E_\nu) = -\frac{G_F^2 a^2}{32\pi}\int_{r_{\min}}^{r_{\max}} r^2 dr \int_{k'_{\min}}^{k'_{\max}} k'^2 dk' \int_{-1}^{+1} d(\cos\theta)\frac{1}{E_{\nu_l}E_l}L_{\mu\nu}J_{RPA}^{\mu\nu}$$

$$\times \mathrm{Im}U_N(q_0^{\nu(\bar{\nu})} - V_c(r)), \tag{14.123}$$

where $J_{RPA}^{\mu\nu}$ is the hadronic tensor with its various components modified due to long range correlation effects treated in RPA. In Eq. (14.123), $a = \cos\theta_c$ for the charged current reaction. For neutral current reactions, $a = 1$ with the Lindhard function is calculated without the Coulomb potential $V_c(r)$.

14.6 Cross Sections and Effect of Nuclear Medium

We show the results for the (anti)neutrino–nucleus cross sections in ^{12}C, ^{40}Ar, ^{56}Fe, and ^{208}Pb nuclei as a function of energy in Figure 14.9, by plotting the ratio of scattering cross section per interacting nucleon to the scattering cross section on the free nucleon target. The results are obtained using the local Fermi gas (LFG) model and the local Fermi gas model with RPA effect (LFG+RPA). A large reduction in the cross section at low energies can be seen which increases with the nucleon number A. In the region of low energy, the reduction is larger in the case of $\nu_\mu(\bar{\nu}_\mu)$ induced reactions in comparison to $\nu_e(\bar{\nu}_e)$ reactions which is mainly due to the threshold effects. The reduction in the cross sections due to nuclear medium effects for the neutrino and antineutrino reactions has similar energy dependence except that it is slightly larger in the case of antineutrino reactions.

In Figure 14.10, we have shown the results of the effect of the nuclear medium in the lepton momentum distribution $\frac{d\sigma}{dp_\mu}$ for the ν_μ and $\bar{\nu}_\mu$ induced charged current quasielastic processes at $E_\nu = 1$ GeV. We find that when the nuclear medium effects are taken into account, there is a reduction as well as shift in the peak region toward the lower value of lepton momentum. This reduction in $\frac{d\sigma}{dp_\mu}$ when calculated in the local Fermi gas model without the RPA correlation effects as compared to the cross section calculated without the nuclear medium effects is around 10% in the peak region of the lepton momentum distribution, which further reduces by around 30% when RPA effects are also taken into account. In the case of antineutrinos, the reduction in $\frac{d\sigma}{dp_\mu}$ in the local Fermi gas model is around 30% which further reduces by 30% when RPA effects are also taken into account.

We show the results for the nuclear medium effects in the total cross sections σ for ^{40}Ar and compare them with the calculations done in the other variants of the Fermi gas model (Smith and Moniz [657], Llewellyn Smith [294] and Gaisser and O'Connell [658]) in Figure 14.11 by plotting fractional difference $\delta\sigma_{\text{model}}(= \frac{\sigma_{\text{free}} - \sigma_{\text{model}}}{\sigma_{\text{free}}})$. Here σ_{free} stands for the (anti)neutrino induced interaction cross section on free nucleon targets and σ_{model} stands for the (anti)neutrino

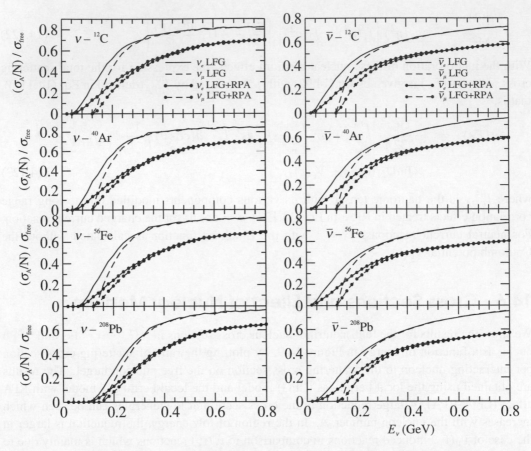

Figure 14.9 Ratio $\frac{\sigma_A/N}{\sigma_{\text{free}}}$ vs. E_ν, for neutrino (left panel) and antineutrino (right panel) induced processes in ^{12}C, ^{40}Ar, ^{56}Fe, and ^{208}Pb. The solid (dashed) line represent cross sections obtained from electron (muon) type neutrino and antineutrino beams. For neutrino induced processes, $N = A - Z$ is the neutron number and for antineutrino induced processes, $N = Z$ is the proton number. σ_A is the cross section for the nuclear target and has been evaluated using the local Fermi gas model and LFG with RPA effect (LFG+RPA). σ_{free} is the cross section for the free nucleon case.

induced interaction cross section for the nucleons bound inside the nucleus. The difference in the results for neutrino and antineutrino cross sections is mainly due to the interference terms between the vector and the axial vector contributions which come with an opposite sign for neutrino and antineutrino. In the case of LFG with RPA effects, the effect of renormalization is large and this suppresses the terms with f_2 and g_1, which results in a large change in neutrino as compared to antineutrino cross section. We find appreciable difference in the results when various nuclear models are used.

There are two types of corrections which appear when lepton mass $m_l(l = e, \mu)$ is taken into account in the cross section calculations for the reaction $\nu_l(\bar{\nu}_l) + N \rightarrow l^-(l^+) + N'$, $(N, N' = n, p)$ which can be classified as kinematical and dynamical. The kinematical effects arise due to $E_l \neq |\vec{k}'|$ in the presence of m_l and the minimum and maximum values of

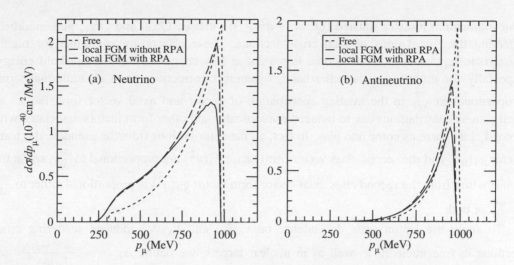

Figure 14.10 $\frac{d\sigma}{dp_\mu}$ vs. p_μ for the $\nu_\mu(\bar{\nu}_\mu)$ induced reactions on ^{12}C target at $E_\nu = 1$ GeV. The short dashed line is the result for the free case, the long dashed (solid) line is the result obtained using local Fermi gas model without (with) RPA.

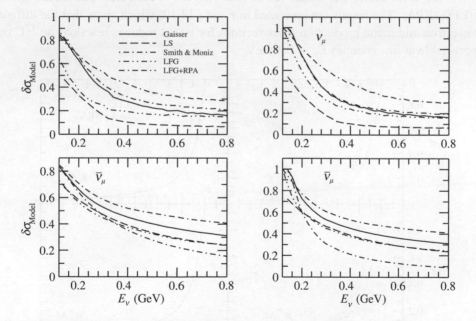

Figure 14.11 The fractional suppression in cross section $\delta\sigma_{\text{model}}(= \frac{\sigma_{\text{free}} - \sigma_{\text{model}}}{\sigma_{\text{free}}})$ vs. E_ν, where σ_{free} is the cross section obtained for free nucleons and σ_{model} is the interacting nucleon cross section in ^{40}Ar obtained by using different nuclear models. The results are presented for the cross sections obtained from different models of Fermi gas (σ_{Model}) viz. Smith and Moniz [657](dashed dotted line), Llewellyn Smith [294](dashed line), Gaisser O' Connell [658](solid line), and with (double dashed dotted line) and without RPA (dashed double dotted line) effect using the local Fermi gas model. The top panel is for neutrino and the bottom panel is for antineutrino induced processes.

four-momentum transfer squared ($Q^2 = -q^2 \geq 0$), that is, $Q^2_{\min}$ and $Q^2_{\max}$ gets modified, affecting the calculations of total cross sections. These effects are negligible for highly relativistic leptons but could become important at low energies near the threshold energy, especially for muons. On the other hand, dynamical corrections arise as additional terms proportional to $\frac{m_l^2}{M^2}$ in the existing contribution of vector and axial vector form factors as well as new contributions due to induced pseudoscalar and other form factors associated with second class currents come into play. In fact, all the contributions from the pseudoscalar form factor $g_3(q^2)$ and the second class vector form factor $f_3(q^2)$ are proportional to $\frac{m_l^2}{M^2}$, while the contribution from the second class axial vector form factor $g_3(q^2)$ is proportional either to $\frac{m_l^2}{M^2}$ or $\frac{q^2}{M^2}$ or both.

To study the lepton mass dependence on $\nu_e(\bar{\nu}_e)$ and $\nu_\mu(\bar{\nu}_\mu)$ induced scattering cross sections in free nucleons as well as in nuclear targets, we define $\Delta_I = \frac{\sigma_{\nu_e(\bar{\nu}_e)} - \sigma_{\nu_\mu(\bar{\nu}_\mu)}}{\sigma_{\nu_e(\bar{\nu}_e)}}$ for the (anti)neutrino induced reaction in ^{12}C and ^{40}Ar nuclear targets, where $I = i,\ ii,\ iii$, which respectively stands for the cross sections obtained in (i) the free (anti)neutrino–nucleon case, (ii) the local Fermi gas model (LFG) and (iii) the local Fermi gas model with RPA effect(LFG+RPA). The results are presented in Figure 14.12, which show that the differences in the electron and muon production cross sections for $\nu_l(\bar{\nu}_l)$ induced reactions on ^{12}C targets are appreciable at low energies $E_\nu < 0.4$ GeV.

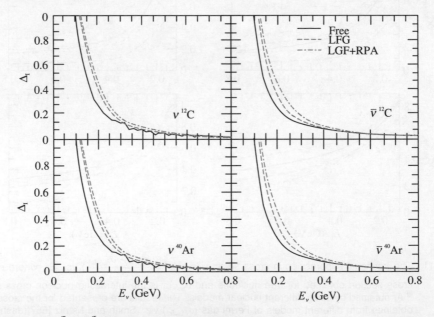

Figure 14.12 $\Delta_I = \frac{\sigma_{\nu_e(\bar{\nu}_e)} - \sigma_{\nu_\mu(\bar{\nu}_\mu)}}{\sigma_{\nu_e(\bar{\nu}_e)}}$ for neutrino (left panel) and antineutrino (right panel) induced processes in ^{12}C and ^{40}Ar targets. Here, I stands for the results of the cross sections obtained (i) for the free nucleon case (solid line), (ii) in the local Fermi gas model (dashed line) and (iii) for LFG with RPA effect (dashed dotted line).

14.7 Nuclear Medium Effects in Neutrino Oscillation Experiments

The discovery of neutrino oscillations with solar, reactor, atmospheric, and accelerator neutrinos has motivated many experiments on neutrino–nucleus cross section measurements in the entire region of low and intermediate energies. These cross sections in this energy region are significantly affected by the nuclear medium effects and the q^2 dependence of the electroweak form factors of the nucleon specially in the region of a few hundred MeV to few GeV. While the weak vector form factors $f_i(q^2)$ $(i = 1, 2)$ are well determined from the electron-nucleus scattering. The study of the nuclear medium effects on the axial vector form factor $g_i(q^2)$ was mainly confined to the study of $g_1(0)$ and $g_3(0)$ leading to the quenching of $g_1(0)$ and enhancement of $g_3(0)$ in nuclei due to various nuclear medium effects. The q^2 dependence of the axial vector form factors is parameterized in terms of the axial mass M_A in the context of dipole parameterization of $g_1(q^2)$ (Chapter 10). The q^2 dependence of $g_3(q^2)$ is related to $g_1(q^2)$ through PCAC. The effect of the nuclear medium on M_A was neglected at the beginning and was first studied by Singh and Oset [639] emphasizing the need for better experiments in the region of low q^2. The present generation of neutrino oscillation experiments done in the energy region of ~ 1 GeV in nuclei are best suited for studying the effect of the nuclear medium on M_A.

The early measurements of M_A were made from the neutrino scattering experiments on deuterium and nuclear targets. The deuterium measurements for M_A had better precision and were in agreement with the measurements made from the threshold electroproduction of pions. The nuclear measurements of M_A had large experimental uncertainties and were in general smaller than the value determined from the deuterium experiment; however, they were in agreement within the experimental error bars. A world average value of $M_A = 1.026$ GeV was generally used in many studies [443].

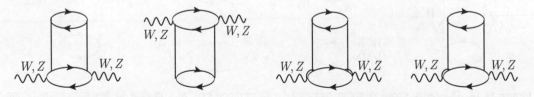

Figure 14.13 Diagrams showing some typical $2p - 2h$ contributions arising due to the $N - N$ and $N - \Delta$ correlations. Solid (dashed) lines denote nucleon (pion) propagators. Double lines represent $\Delta(1232)$ propagators. Arrows pointing to the right (left) denote particle (hole) states.

In 2009, the MiniBooNE collaboration reported new measurements on the charge changing quasielastic process using a high statistics sample of ν_μ interactions in ^{12}C with average energy $\langle E_\nu \rangle = 750$ MeV [435], which were considerably higher than the theoretical predictions. Subsequently, the results on the double differential cross sections $\frac{d^2\sigma}{d\cos\theta dE'}$ were also presented which confirmed the findings of the cross section measurements. These results could be explained with $M_A = 1.35 \pm 0.17$ GeV, which was considerably higher than the world average. This is known as the MiniBooNE puzzle.

The MiniBooNE puzzle initiated an extensive debate on the nuclear medium effects in (anti)neutrino nucleus scattering. Since the ^{12}C nucleus used in many neutrino oscillation experiments is one of the best known nucleus theoretically as well as experimentally, from the study of electron nucleus scattering regarding its wave function, it was very difficult to explain the discrepancy in the measurement of neutrino cross sections at MiniBooNE. It was attributed to other nuclear effects beyond impulse approximation due to MEC and nucleon–nucleon correlations or FSI effects leading to quasielastic-like events.

Such effects have been discussed earlier in electron scattering and applied to neutrino scattering by Martini et al. [690, 691] and later by Nieves et al. [692] and many others [693, 694]. Some of these effects beyond the impulse approximation were also included in the works of Valencia and Aligarh groups [659, 660, 661, 662, 663, 664, 665, 666, 667, 668] done on the local Fermi gas model through long range RPA correlations. However, the new results from MiniBooNE were too high to be explained by these approaches. Various calculations done earlier in the microscopic model including the initial state interactions through the nucleon spectral functions in the nucleus were also found to underestimate the cross section measurements from MiniBooNE. A summary of these theoretical calculations along with the results of MiniBooNE is given in Ref. [678].

After the experimental results of MiniBooNE, it was emphasized by Martini et al. and Nieves et al. that there are additional contributions coming from MEC and nucleon–nucleon correlation effects, which are not included in most of the earlier calculations. In earlier treatments of including such effects, the diagrams corresponding to Figure 14.13 were taken into consideration while the diagrams corresponding to Figure 14.14 and the other diagrams where W and Z bosons interact directly with the non-nucleonic degrees of freedom in the nucleus were not fully incorporated.

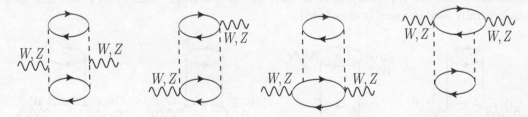

Figure 14.14 Diagrams showing some typical $2p - 2h$ contributions arising due to the meson exchange. Solid (dashed) lines denote nucleon (pion) propagators. Arrows pointing to the right (left) denote particle (hole) states.

The elaborate calculations done by Martini [691] in the microscopic model and Nieves et al. [661] in the relativistic Fermi gas model included most of the diagrams and showed that the enhancement in the cross sections in MiniBooNE comes from the contribution from the $2p - 2h$ excitations shown in Figures 14.13 and 14.14, and its interference with the $1p - 1h$ contribution. Once these contributions are included, the results from MiniBooNE can be explained using the value of $M_A = 1.049$ GeV, consistent with the world average value of M_A as shown in Figure 14.15.

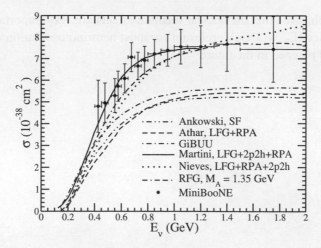

Figure 14.15 "Quasielastic-like" ν_μ-^{12}C cross sections measured by MiniBooNE [435] compared to Martini et al. calculations. The figure is taken from Ref. [356].

In Figure 14.16, the neutrino and antineutrino differential cross section $d\sigma/dQ^2$, measured in Refs. [435] and [695] and as calculated in Refs. [690] and [691] are presented. It may be observed that the inclusion of $2p-2h$ and the meson exchange currents contribution results in a better description of data in comparison to QE results without these nuclear effects.

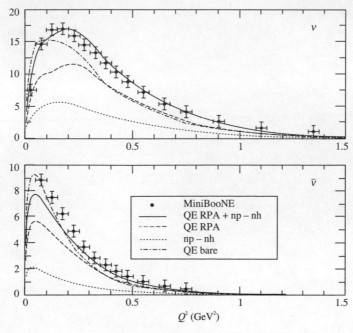

Figure 14.16 MiniBooNE flux-integrated differential cross sections $d\sigma/dQ^2$ vs. Q^2 in units of 10^{-39} cm^2/GeV^2 for neutrino (upper panels) and antineutrino (lower panels) CCQE-like scattering on carbon. The experimental MiniBooNE points are taken from [435] and [695]. The theoretical results are the ones of Refs. [690] and [691].

It has to be emphasized that the nuclear medium effects are very important in the region of intermediate energies ($E_\nu \sim 1$ GeV) relevant for most neutrino oscillation experiments being done at present and planned in the future.

Chapter 15

Inelastic Scattering of (Anti)neutrinos from Nuclei

15.1 Introduction

We have discussed inelastic processes from free nucleon targets in Chapters 11 and 12. However, most of the early experiments and also the new experiments on inelastic as well as quasielastic reactions induced by (anti)neutrinos use nuclear targets. In this chapter, we will focus on the inelastic process of producing mesons and photons from the nuclear targets. When inelastic processes like

$$\nu_l(\bar{\nu}_l) + N \rightarrow l^-(l^+) + N' + m; \quad N, N' : \text{proton or neutron and } m \text{ is a meson},$$

(discussed in Chapter 12) take place inside a nucleus, the nucleus can stay in the ground state giving almost all the transferred energy in the reaction to the outgoing meson leading to the coherent production of mesons or can be excited and/or broken up leading to the incoherent production of meson. In the subsequent sections, incoherent and coherent pion production from nuclei in the delta dominance model will be discussed with some comments on the inelastic production of kaons and photons.

The first experiments on inelastic scattering of (anti)neutrinos from nuclei were done at CERN in the early 1960s using heavy liquid bubble chambers (HLBC) filled with propane, freon and with spark chambers, and at ANL/BNL with spark and bubble chambers. The importance of nuclear medium effects in the analysis of these experiments was realized and discussed in the context of inelastic as well as quasielastic reactions. Some of the mesons, mostly pions, produced in the inelastic reactions could be absorbed in the parent nucleus giving rise to 'pionless' lepton events in charged current (CC) induced reactions enhancing the yield of quasielastic events and reducing the yield of 'pionic' events as compared to the theoretical predictions for these reactions from free nucleon targets (Chapter 12). While some theoretical calculations were made to estimate, quantitatively, the effect of nuclear medium effects in quasielastic reactions (see Chapter 14), no serious efforts were made in the case of inelastic reactions. Subsequently, many experiments were done at CERN [696, 697, 698, 699, 700], SKAT [701, 702, 703, 704], FNAL [705, 706], and CHARM [707], using propane, propane– freon, neon, marble, and aluminium targets; only a qualitative description of the nuclear medium effects was used. Most of these experiments studied the incoherent production of pions while some of them also studied the coherent production of pions [698, 699, 700, 703, 704, 706, 708, 709] from nuclei. Later, experiments done with hydrogen and deuterium targets performed at ANL [710, 711], BNL [519], CERN [470, 712, 713], and FNAL [469] were also analyzed without any consideration of deuterium effects. Since most of these experiments were done at very high energies, it was argued that nuclear medium effects may not be important in this energy region.

The theoretical and experimental interest in studying the nuclear medium effects in inelastic reactions induced by (anti)neutrinos on nuclear targets was renewed after the first evidence of neutrino oscillations was reported in the study of atmospheric neutrinos in Kamiokande and IMB experiments (Chapter 18). In these experiments, a deficit in the ratio of muon to electron yield produced in the quasielastic reactions induced by atmospheric (anti)neutrinos $\nu_\mu(\bar{\nu}_\mu)$ and $\nu_e(\bar{\nu}_e)$ when compared to the theoretical predictions was observed and attributed to neutrino oscillations of $\nu_\mu(\bar{\nu}_\mu)$ neutrinos. Since atmospheric (anti)neutrinos have quite a wide spectrum in the range of a few hundreds of MeV to hundreds of GeV, and the spectrum have higher flux at lower energies, the inelastic production of pions plays a very important role in the total event rates. A knowledge of the nuclear medium effects in the inelastic production of pions is important in the analysis of these experiments for the following reasons.

i) While identifying the quasielastic lepton events, the 'pionless' lepton events (now called the 'quasielastic like' events) discussed earlier, in the context of neutrino experiments at CERN, contribute to a major source of uncertainty in the systematics as they may overlap in energy with genuine quasielastic events due to the momentum distribution of nucleons in the nucleus. Theoretically, it is quite difficult to estimate quasielastic-like events, as it requires the knowledge of reaction mechanisms for inelastic pion production from the nucleon in the presence of nuclear medium effects as well as a mechanism for pion absorption in the nuclear medium.

ii) The charged pions $\pi^-(\pi^+)$ produced in the inelastic (anti)neutrino reactions constitute a major source to the background in identifying the genuine lepton events produced by the quasielastic reactions. This is because the charged pions produced by neutral currents (NC) in nuclei through various channels described in Chapter 12, could be misidentified with charged muons $\mu^-(\mu^+)$ in the IMB and Kamiokande detectors; these detectors use Cerenkov radiation. Even in the case of charged current (CC) induced reactions (Chapter 12), when the muon is produced with high energy and the pion is produced with low energy, the muons may escape the detector, while the pions may be misidentified as muons. This is likely to happen in the case of (anti)neutrino beams with a continuous energy spectrum of (anti)neutrinos spanning a wide energy range between $0.3 < E_{\nu(\bar{\nu})} < 3$ GeV.

iii) In the case of $\nu_e(\bar{\nu}_e)$ neutral current (NC) induced π^0 production, these pions would promptly decay into photons $(\pi^0 \to 2\gamma)$ giving electrons and positrons through pair creation $(\gamma \to e^+e^-)$. Some of these electrons (positrons) which do not escape the detector, will produce Bremsstrahlung photon signals in the detector which would mimic genuine quasielastic electron (positron) events produced by the atmospheric (anti)neutrinos.

In the aforementioned scenario, the inelastic production of pions both in the CC and NC channels contribute a major source of background in most of the present neutrino oscillation experiments.

It was therefore emphasized that the neutrino event generators NUANCE [535] and NEUT [714, 607] incorporating earlier calculations of Rein and Sehgal [492, 715] for incoherent and coherent pion production should be modified to include the nuclear medium effects. The only consideration of the nuclear medium effects included in these analyses was from the works of Smith and Moniz in the non-relativistic global Fermi gas model [657]. The other major source of nuclear medium effects in the inelastic production of pions are the well-known effects due to the final state interactions of pions with the residual nucleus. This type of nuclear medium effects are reasonably well modeled in the neutrino event generators like NUANCE [535] and NEUT [714, 607] based on the Valencia model (for discussion see Ref. [356]). These neutrino event generators were used to analyze the neutrino oscillation experiments of IMB and Kamiokande collaborations.

In view of these developments in neutrino oscillation physics and the earlier results of the LSND neutrino experiment[716] indicating the possibility of the existence of a fourth flavor, that is, sterile neutrino, new experiments like K2K, SciBooNE, MiniBooNE, T2K, NOvA, ArgoNeuT, MINERvA, etc. were planned and done in the energy region of ~ 1 GeV, using detectors with moderate nuclear targets like ^{12}C, ^{16}O, etc.(for details see Chapter 17) which measured the neutrino–nucleus cross sections for the quasielastic and inelastic production of pions.

The importance of the nuclear medium effects in the inelastic pion production in analyzing neutrino oscillation experiments was emphasized quite early by Kim et al.[717] and Singh et al. [718], in the case of incoherent production of pions and by Kim et al. [719] and Kelkar et al. [720] in the case of coherent pion production. Subsequently, many microscopic calculations employing sophisticated nuclear wave functions as well as models like relativistic Fermi gas, relativistic mean field and other approximation methods were performed to estimate the nuclear medium effects. Many of them included contributions from MEC and other correlation effects beyond impulse approximation. The details of the various methods are given in Refs. [721, 722, 723, 482, 724, 664, 725, 726, 483, 727, 668, 692, 728, 729, 730, 731, 732, 733]. For a historical development, readers are referred to excellent reviews [426, 356, 678, 731] on the subject.

In case of deuterium experiments, the early theoretical calculations for the (anti)neutrino induced disintegration of deuterons were done without nuclear medium effects, using perturbation theory in impulse approximation with sophisticated two-nucleon wave functions for the deuteron. Recently, elaborate calculations using coupled channel methods and T matrix expansion have been done showing large effects due to the final state interactions in some channels [734]. Therefore, there are two types of nuclear medium effects which are present both in the incoherent and coherent production of pions from nuclei, that is, in the production process and the final state interactions. We would describe them briefly in this chapter.

The experimental study of incoherent pion production has been recently made at K2K [735, 736], MiniBooNE [737, 738], SciBooNE [739], NOMAD [740], ArgoNeuT [741], MINERvA [742, 743, 744], T2K [745], etc. collaborations, mostly in the energy region of a few hundreds of MeV to a few GeV. Recent experimental study to measure coherent pion production cross section in this energy region has been made by K2K [746], MiniBooNE [747], SciBooNE [748, 749], NOMAD [740], ArgoNeuT [750], MINERvA [751], T2K [752], etc. collaborations.

In this chapter, we will describe some of the effects described here and their effects on the analysis of current neutrino oscillation experiments. We also briefly describe the nuclear medium effects in some other inelastic processes like the reactions in which (anti)kaons or photons are emitted in charged (neutral) current induced reactions like

$$\nu_\mu(\nu_e) + {}^A_Z X \quad \rightarrow \quad \mu^-(e^-) + {}^A_{Z+1} X, \tag{15.1}$$

$$\nu_\mu(\bar{\nu}_\mu) + {}^A_Z X \quad \rightarrow \quad \nu_\mu(\bar{\nu}_\mu) + {}^A_Z X + \gamma, \tag{15.2}$$

and reactions in which hyperons are produced, that is, $\bar{\nu}_l + {}^A_Z X \rightarrow l^+ + {}^A_\Lambda Y$. The hyperon produced in the final state is identified by its decay into pions and nucleons which is also affected by the nuclear medium effects.

15.2 Charged Current Inelastic Reactions

15.2.1 Incoherent meson production

The early experiments done on inelastic pion production emphasized the need for including nuclear medium effects but these effects were not used in the analyses of the experiments. The calculations for the final state interaction of pions in neutrino induced pion production were done by Adler et al. [753]. In recent years, these effects have been taken into account by many authors. The nuclear medium effects in the production process have been mostly calculated in the Fermi gas model [721, 722, 723, 482, 724, 664, 725, 726, 483, 727, 668, 692] with some calculations made using relativistic spectral function and relativistic plane wave approximation (RPWA) with nuclear wave functions [722, 728, 729, 730, 731, 732, 733]. The final state interactions have been calculated in various approaches and some are discussed in Section 15.4.2.

In this section, we discuss the inelastic charged current lepton production accompanied by a meson(Figure 15.1) in the relativistic Fermi gas model using local density approximation. To take into account the Fermi motion and Pauli blocking effect in the local density approximation (LDA) (Chapter 14), first a general expression for the differential scattering cross section is obtained.

The differential scattering cross section for (anti)neutrinos interacting with a bound nucleon and producing a meson m, inside a nucleus, $\nu_l(k) + N(p) \rightarrow l^-(k') + N'(p') + m(p_m)$, in LDA is written as [754]:

$$\left(\frac{d\sigma}{dE_m d\Omega_m}\right)_{\nu A} = \int d^3r \, \rho_n(r) \left(\frac{d\sigma}{dE_m d\Omega_m}\right)_{\nu N}, \tag{15.3}$$

where $\left(\frac{d\sigma}{dE_m d\Omega_m}\right)_{\nu N}$ is the free (anti)neutrino–nucleon differential scattering cross section discussed in Chapter 12, with the condition that the energy of the initial nucleon $E_N < E_F^N(r)$ and the final nucleon $E_{N'}(= E_N + q_0 - E_m) > E_F^{N'}(r)$, at the production point r.

In a symmetric nuclear matter, the differential scattering cross section is written as

$$\left(\frac{d\sigma}{dE_m d\Omega_m}\right)_{\nu A} = 2 \int d^3r \sum_{N=n,p} \int \frac{d^3 p_N}{(2m)^3} \Theta(E_F^N(r) - E_N)\Theta(E_N + q_0 - E_m - E_F^{N'}(r)) \left(\frac{d\sigma}{dE_m d\Omega_m}\right)_{\nu N},$$

where $E_N = \sqrt{|\vec{p}_N|^2 + M^2}$; $\vec{k} + \vec{p}_N = \vec{k}' + \vec{p}_N' + \vec{p}_m$.

Using energy–momentum conservation, $E_{N'}$ may be written as

$E_{N'} = \sqrt{|\vec{P}|^2 + |\vec{p}_N|^2 + 2|\vec{P}||\vec{p}_N|cos\theta_N + M^2}$, where $\vec{P} = \vec{q} - \vec{p}_m$ and $\vec{q} = \vec{k} - \vec{k}'$.

This results in

$$\left(\frac{d\sigma}{dE_m d\Omega_m}\right)_A = 2 \int_0^\infty 4\pi r^2 dr \int_\epsilon^{E_F^N(r)} dE_N \int_0^{2\pi} d\phi_N \int_{-1}^{+1} d\cos\theta_N \frac{|\vec{p}_N|E_N}{(2\pi)^3}$$

$$\times \; \Theta(E_F^N(r) - E_N) \, \Theta(E_N + q_0 - E_m - E_F^{N'}(r))$$

$$\times \; \frac{G_F^2 cos^2\theta_c}{8 M_N E_\nu} \frac{|\vec{p}_l||\vec{p}_m|}{E_N'} \frac{1}{(2\pi)^5} \delta^0(q_0 + E_N - E_F^{N'}(r) - E_m)$$

$$\times \; L^{\mu\nu} J_{\mu\nu} \, dE_l d\Omega_l. \tag{15.4}$$

In this expression, $L^{\mu\nu} = \sum \overline{\sum} l^\mu l^{\nu\dagger}$ and $J_{\mu\nu} = \sum \overline{\sum} j_\mu j_\nu^\dagger$, where l^μ is the leptonic current and j_μ is the hadronic current; both are given in Chapter 12. Furthermore, the azimuthal angle dependence have been found to be very small for pion production in electron and photon induced processes; the integration over the azimuthal angle can be replaced by 2π [754].

The delta integration in Eq. (15.4) is performed using energy conservation

$$q_0 + \sqrt{|\vec{p}_N|^2 + M^2} - \sqrt{|\vec{P}|^2 + |\vec{p}_N|^2 + 2|\vec{P}||\vec{p}_N|cos\theta_N + M^2} - E_m = 0,$$

which on simplifying gives

$$\cos\theta_N = \frac{P^2 + 2E_N P_0}{2|\vec{P}||\vec{p}_N|}; \quad P_0 = q_0 - E_m, \quad \frac{P^2 + 2E_N P_0}{2|\vec{P}||\vec{p}_N|} \leq 1 \text{ as } abs(\cos\theta_N) \leq 1,$$

and

$$E_N^2 + E_N P_0 + \frac{P^2}{4} + \frac{M^2|\vec{P}|^2}{P^2} \leq 0,$$

which results in

$$\epsilon' = \frac{-P_0 + \sqrt{P_0^2 - 4\left(\frac{P^2}{4} + \frac{M^2|\vec{P}|^2}{P^2}\right)}}{2}.$$

Thus, the lower limit in Eq. (15.4), for ϵ is constrained to

$$\epsilon = mux(M, E_F^{N'} - P_0, \epsilon').$$

The final expression to evaluate the differential scattering cross section is approximated as

$$
\begin{aligned}
\left(\frac{d\sigma}{dE_m d\Omega_m}\right)_A &= \frac{G_F^2 \cos^2\theta_c}{64\pi^5} \int_0^\infty r^2 dr \int dE_N dE_l d\Omega_l \frac{|\vec{p}_l||\vec{p}_m|}{|\vec{P}||\vec{k}|} (E_F^N(r) - \epsilon) \\
&\times \quad \Theta(E_F^N(r) - \epsilon)\Theta(-P^2)\Theta(P_0) L^{\mu\nu} J_{\mu\nu}(|\vec{p}_{N'}|, q, k_m) \\
&\times \quad \left(\frac{d\sigma}{dE_m d\cos\theta_m}\right)_N,
\end{aligned}
$$

where

$$|\vec{p}_{N'}| = \sqrt{E_{N'}^2 - M^2} \quad \text{and} \quad E_{N'} = \frac{E_F^N(r) + \epsilon}{2}.$$

This prescription is valid for any meson production, where the initial nucleon is below the Fermi sea and the final nucleon is above the Fermi sea. Moreover, the nucleus is at rest and not the nucleons.

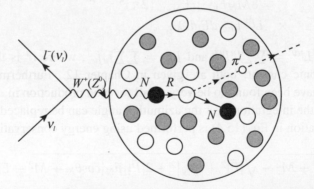

Figure 15.1   In a neutrino induced reaction on a nucleon target, when a pion is produced in the nuclear medium and comes out without FSI (final state interaction). The open (shaded circles represent protons (neutrons) inside a nucleus and the dark shaded circle represents a nucleon (p or n) with which a neutrino interacts through a CC (NC) reaction and a non-resonant or a resonant state (R) is formed, which gives rise to a p or n and a pion (π^i), where i represents the charge state. For example, in a CC reaction $\nu_l + n \rightarrow l^- + R^+$, R^+ may give a p and π^0 or a n and π^+; then the pion comes out without FSI with the residual nucleus.

The nuclear medium effects in single pion production has been widely studied; moreover, the present generation of neutrino experiments frequently analyze these events. Therefore, in the next section, we describe the nuclear medium effects in incoherent one-pion production processes. Moreover, we have seen in Chapter 12, that the pion production is mainly dominated by the formation of delta as the resonant state and its subsequent decay. Therefore, the present discussion is confined to medium modification of the delta properties.

15.2.2 Pion production in the delta dominance model

In an inelastic reaction where a Δ is produced in the intermediate state and subsequently decays to a nucleon and a pion, the various possible channels for a pion production in the (anti)neutrino interactions on a nucleon target are

$$\nu_l(k) + p(p) \rightarrow l^-(k') + \Delta^{++}(p') \qquad \nu_l(k) + n(p) \rightarrow l^-(k') + \Delta^+(p')$$
$$\searrow p + \pi^+ \qquad\qquad\qquad \searrow p + \pi^0$$
$$\searrow n + \pi^+$$

$$(15.5)$$

$$\bar{\nu}_l(k) + n(p) \rightarrow l^+(k') + \Delta^-(p') \qquad \bar{\nu}_l(k) + p(p) \rightarrow l^+(k') + \Delta^0(p')$$
$$\searrow n + \pi^- \qquad\qquad\qquad \searrow p + \pi^-$$
$$\searrow n + \pi^0$$

$$(15.6)$$

The differential cross section for the reaction $\nu(k) + p(p) \rightarrow \mu^-(k') + \Delta^{++}(p_\Delta)$ can be written as

$$d\sigma = \frac{(2\pi)^4 \delta^4(k + p - p_\Delta - k')}{4ME_\nu} \frac{d\vec{k}'}{(2\pi)^3 2E_{k'}} \frac{d\vec{p}_\Delta}{(2\pi)^3 2E_\Delta} \overline{\sum}\sum |\mathcal{M}|^2. \qquad (15.7)$$

Using the following relation,

$$d\vec{k}' = |\vec{k}'|^2 \, d|\vec{k}'| \, d\Omega_{k'} = E_{k'} |\vec{k}'| \, dE_{k'} \, d\Omega_{k'}, \qquad (15.8)$$

we get after integrating over Δ momentum,

$$\frac{d\sigma}{dE_{k'} d\Omega_{k'}} = \frac{1}{16\pi^2} \frac{1}{4ME_\nu} |\vec{k}'| \frac{1}{E_\Delta} \delta(E_p + q_0 - E_\Delta) \overline{\sum}\sum |\mathcal{M}|^2. \qquad (15.9)$$

To take into account the decay width of Δ, we replace

$$\delta(E_p + q_0 + E_\Delta) \longrightarrow -\frac{1}{\pi} \text{Im} \left[\frac{1}{E_p + q_0 - E_\Delta + i\Gamma/2} \right], \qquad (15.10)$$

where the relation

$$\frac{1}{x - x_0 \mp i\epsilon} = P\frac{1}{x - x_0} \pm i\pi\delta(x - x_0) \qquad (15.11)$$

is used.

Moreover, one may write

$$\frac{M_\Delta}{E_\Delta} \delta(p_\Delta^0 - E_\Delta) \rightarrow -\frac{1}{\pi} \text{Im} \left[\frac{1}{W - M_\Delta + i\frac{1}{2}\Gamma(W)} \right] \rightarrow \frac{1}{\pi} \left[\frac{\frac{\Gamma(W)}{2}}{(W - M_\Delta)^2 + \frac{\Gamma^2(W)}{4}} \right] \quad (15.12)$$

using

$$\delta(W - M_\Delta) = \frac{1}{p_\Delta^0/W} \delta(p_\Delta^0 - E_\Delta) \simeq \frac{M_\Delta}{E_\Delta} \delta(p_\Delta^0 - E_\Delta) \quad (15.13)$$

where $W = \sqrt{p'^2} = \sqrt{p_\Delta^{0\,2} - \vec{p}_\Delta^2}$ is the invariant mass of Δ and $\Gamma(W)$ is the rest width of Δ.

Using Eq. (15.13), Eq. (15.7) is obtained as

$$\frac{d\sigma}{dE_{k'}d\Omega_{k'}} = \frac{1}{16\pi^3} \frac{1}{4ME_\nu} \frac{|\vec{k}'|}{M_\Delta} \left[\frac{1}{W - M_\Delta + i\frac{1}{2}\Gamma(W)} \right] \overline{\sum}\sum |\mathcal{M}|^2. \quad (15.14)$$

The expression of Q^2-distribution is obtained as

$$\frac{d\sigma}{dQ^2} = \frac{\pi}{E_\nu E_{k'}} \frac{d\sigma}{d\Omega_{k'}} \quad (15.15)$$

using which, we may write

$$\frac{d\sigma}{dE_{k'}dQ^2} = \frac{1}{64\pi^2} \frac{1}{M^2 E_\nu^2} \frac{M}{M_\Delta} \frac{|\vec{k}'|}{E_{k'}} \left[\frac{\Gamma(W)}{(W - M_\Delta)^2 + \frac{\Gamma^2(W)}{4}} \right] \overline{\sum}\sum |\mathcal{M}|^2. \quad (15.16)$$

Finally, neglecting lepton mass so that $|\vec{k}'| \, d|\vec{k}'| = E_{k'} \, dE_{k'}$ and integrating over lepton energy, we get

$$\frac{d\sigma}{dQ^2} = \frac{G^2 \cos^2\theta_C}{128\pi^2} \frac{1}{M^2 E_\nu^2} \frac{M}{M_\Delta} \int_{E_{k'}^{\min}}^{E_{k'}^{\max}} dE_{k'} \, L_{\mu\nu} J^{\mu\nu} \left[\frac{\Gamma(W)}{(W - M_\Delta)^2 + \frac{\Gamma^2(W)}{4}} \right]. \quad (15.17)$$

When a neutrino interacts with a bound nucleon, the nucleon inside the nucleus is constrained to have a momentum below the Fermi momentum, while there is no such constraint on the momentum of the intermediate Δ produced in the medium. The Δ propagates in the medium and experiences all kinds of self energy (Σ_Δ) interactions. This self energy is assumed to be the function of the nuclear density $\rho(\vec{r})$. It involves the decay of Δ-isobar through $\Delta \rightarrow N\pi$ channel in the nucleus. However, their decay, is influenced by the Pauli blocking. The nucleons produced in these decay processes have to be above the Fermi momentum $k_F(r)$ of the nucleons in the nucleus thus inhibiting the decay as compared to the free decay of the Δ-isobar described by Γ in Eq. (15.13). This leads to a modification in the decay width of the Δ resonance which has been studied by many authors [755, 756, 757, 758, 759]. Further, there are additional decay channels open for the Δ resonance in the nuclear medium. In the nuclear medium, the Δ resonance decays through two- and three-body absorption processes like $\Delta N \rightarrow NN$ and $\Delta NN \rightarrow NNN$ through which Δ disappear in the nuclear medium without producing a pion, while a two-body Δ absorption process like $\Delta N \rightarrow \pi NN$ gives rise to some more pions. These nuclear medium effects on the Δ propagation are included by describing the modified mass and the decay width in terms of the Δ self energy used by the parameterization of Oset and Salcedo [755]. The real part of the delta self energy contributes to the mass of Δ and the imaginary part of the delta self energy contributes to the decay width of Δ. These modifications are parameterized by making density dependent changes in mass and decay width in a local density approximation, to the Δ-hole model.

The modification of the mass and width arises from the following sources:
(a) The intermediate nucleon state is partly blocked for the Δ decay because some of these states are occupied (Pauli blocking). The decayed nucleon must be in an unoccupied state. The Pauli correction is taken into account by assuming a local Fermi sea at each point of the nucleus of density $\rho(\vec{r})$, and forcing the nucleon to be above the Fermi sea. This leads to an energy dependent modification in the Δ decay width given as [755]

$$\Gamma \rightarrow \tilde{\Gamma} - 2\mathrm{Im}\Sigma_\Delta, \tag{15.18}$$

where $\tilde{\Gamma}$ is the Pauli blocked width of Δ in the nuclear medium and Σ_Δ is the self energy of Δ in the nuclear medium and its relativistic form is

$$\tilde{\Gamma} = \frac{1}{6\pi} \left(\frac{f_{\pi N\Delta}}{m_\pi} \right)^2 \frac{M}{\sqrt{s}} |\vec{p}'_{\mathrm{cm}}|^3 \, F(k_F, E_\Delta, p_\Delta), \tag{15.19}$$

where

$$|\vec{p}'_{\mathrm{cm}}| = \frac{\sqrt{(s - M^2 - m_\pi^2)^2 - 4M^2 m_\pi^2}}{2\sqrt{s}}, \tag{15.20}$$

and $F(k_F, E_\Delta, p_\Delta)$, the Pauli correction factor is written as [755, 760]

$$F(k_F, E_\Delta, p_\Delta) = \frac{p_\Delta |\vec{p}'_{\mathrm{cm}}| + E_\Delta E'_{p_{\mathrm{cm}}} - E_F \sqrt{s}}{2 p_\Delta |\vec{p}'_{\mathrm{cm}}|} \tag{15.21}$$

where k_F is the Fermi momentum, $E_F = \sqrt{M^2 + k_F^2}$ and $\vec{p}'_{\mathrm{cm}}$, $E'_{p_{\mathrm{cm}}}$ are the nucleon momentum and the relativistic nucleon energy in the final πN center of mass frame.

If $F(k_F, E_\Delta, p_\Delta) > 1$, it is replaced by 1. Similarly, if $F(k_F, E_\Delta, p_\Delta) < 0$, then it is replaced by 0 in Eq. (15.19).

In the aforementioned expression, $\sqrt{s}$ is the center of mass energy in the Δ rest frame averaged over the Fermi sea, $\bar{s}$ and is given as

$$\bar{s} = M^2 + m_\pi^2 + 2E_\pi \left(M + \frac{3}{5} \frac{k_F^2}{2M} \right). \tag{15.22}$$

(b) The produced nucleon in the Δ decay inside the nuclear medium feels a single particle potential due to all the other nucleons in the nucleus, known as the binding effect, which is taken care of by the real part of the Δ self energy. This effect modifies the mass of Δ in the medium as

$$M_\Delta \to \tilde{M}_\Delta = M_\Delta + \mathrm{Re}\Sigma_\Delta. \tag{15.23}$$

The Δ self energy plays a very important role in the different pion nuclear reactions. A thorough study of the Δ self energy is made, using the explicit model by Oset and Salcedo [755]. For the scalar part of the Δ self energy, the numerical results are parameterized in the approximate analytical form (excluding the Pauli corrected width), and are given as [755, 760]:

$$-\mathrm{Im}\Sigma_\Delta = C_Q \left(\frac{\rho}{\rho_0} \right)^\alpha + C_{A2} \left(\frac{\rho}{\rho_0} \right)^\beta + C_{A3} \left(\frac{\rho}{\rho_0} \right)^\gamma, \tag{15.24}$$

which is determined mainly by the one-pion interactions in the nuclear medium. This includes the two-body, three-body, and the quasielastic absorption contributions for the produced pions in the nucleus. The coefficients C_Q accounts for the quasielastic part, the term with C_{A2} for two-body absorption and the one with C_{A3} for three-body absorption, and are parameterized in the range of energies 80 MeV$< T_\pi <$320 MeV, where T_π is the pion kinetic energy, as [755, 760]:

$$C(x) = ax^2 + bx + c, \qquad x = \frac{T_\pi}{m_\pi} \tag{15.25}$$

where C stands for all the coefficients, that is, C_Q, C_{A2}, C_{A3}, α, and $\beta(\gamma = 2\beta)$. The different coefficients used here are tabulated in Table 15.1 and Table 15.2 [755].

Table 15.1 Coefficients used in Eq. (15.24) for the calculation of $\mathrm{Im}\Sigma_\Delta$ as a function of energy in the case of pion nuclear scattering.

T_π(MeV)	C_Q(MeV)	C_{A2}(MeV)	C_{A3}(MeV)	α	β	γ
85	9.7	18.9	3.7	0.79	0.72	1.44
125	11.9	17.7	8.6	0.62	0.77	1.54
165	12.0	16.3	15.8	0.42	0.80	1.60
205	13.0	15.2	18.0	0.31	0.83	1.66
245	14.3	14.1	20.2	0.36	0.85	1.70
315	9.8	13.1	14.7	0.42	0.88	1.76

Table 15.2 Coefficients used for an analytical interpolation of $C(T_\pi)$ of Eq. (15.25).

	C_Q(MeV)	C_{A2}(MeV)	C_{A3}(MeV)	α	β
a	−5.19	1.06	−13.46	0.382	−0.038
b	15.35	−6.64	46.17	−1.322	0.204
c	2.06	22.66	−20.34	1.466	0.613

The real part of the Δ self energy [760] is approximately given by

$$\mathrm{Re}\Sigma_\Delta \simeq 40.0 \left(\frac{\rho}{\rho_0}\right) \text{MeV}. \tag{15.26}$$

These considerations lead to the following modifications in the width $\tilde{\Gamma}$ and mass M_Δ of the Δ resonance:

$$\frac{\Gamma}{2} \to \frac{\tilde{\Gamma}}{2} - \mathrm{Im}\Sigma_\Delta \quad \text{and} \quad M_\Delta \to \tilde{M}_\Delta = M_\Delta + \mathrm{Re}\Sigma_\Delta. \tag{15.27}$$

With these modifications, the differential scattering cross section may be written as

$$\frac{d^2\sigma}{dE_{k'}d\Omega_{k'}} = \frac{1}{64\pi^3} \int d\vec{r} \rho_p(\vec{r}) \frac{|\vec{k}'|}{E_k} \frac{1}{MM_\Delta} \frac{\frac{\tilde{\Gamma}}{2} - \mathrm{Im}\Sigma_\Delta}{(W - \tilde{M}_\Delta)^2 + (\frac{\tilde{\Gamma}}{2} - \mathrm{Im}\Sigma_\Delta)^2} \sum\sum |\mathcal{M}|^2. \tag{15.28}$$

For the lepton production from neutron targets, $\rho_p(\vec{r})$ in this expression is replaced by $\frac{1}{3}\rho_n(\vec{r})$. Therefore, the total scattering cross section for the neutrino induced lepton production process for incoherent 1π production in the nucleus is given by

$$\sigma = \frac{1}{64\pi^3} \iint d\vec{r} \frac{d\vec{k}'}{E_k E_{k'}} \frac{1}{MM_\Delta} \frac{\frac{\tilde{\Gamma}}{2} - \mathrm{Im}\Sigma_\Delta}{(W - \tilde{M}_\Delta)^2 + (\frac{\tilde{\Gamma}}{2} - \mathrm{Im}\Sigma_\Delta)^2} \left[\rho_p(\vec{r}) + \frac{1}{3}\rho_n(\vec{r})\right] \sum\sum |\mathcal{M}|^2. \tag{15.29}$$

It should be noted that $\tilde{\Gamma}$ describes the Δ decaying into nucleons and pions. The various terms in the $\mathrm{Im}\Sigma_\Delta$ correspond to the different responses of Δ in the nuclear medium as explained earlier. The C_Q term in $\mathrm{Im}\Sigma_\Delta$ gives additional contribution to the pion production which arises solely due to nuclear medium effects. Some of the Δs are absorbed through two-body and three-body absorption processes and do not lead to pion production. These are described by C_{A2} and C_{A3} terms in the expression for $\mathrm{Im}\Sigma_\Delta$ given in Eq. (15.24) and do not contribute to the lepton production accompanied by pions. These constitute quasielastic-like events besides the pions which are physically produced but reabsorbed in the nucleus due to FSI; this will be discussed later in the text. Only the C_Q term in the expression for $\mathrm{Im}\Sigma_\Delta$(Eq. (15.24)) contributes to the lepton production accompanied by a pion.

For example, in the case of the one π^+ production process, $\tilde{\Gamma}$ and the C_Q term in $\mathrm{Im}\Sigma_\Delta$ contribute to the pion production. Further, for neutrinos giving rise to π^+ after the interaction with a nucleon target, if the nucleon happens to be a proton, then all the interactions with protons will give rise to π^+; but if the nucleon is a neutron target ($\nu_l + n \rightarrow l^- + \Delta^+$), then it may give rise to a π^+ or a π^0 which in the absence of FSI may be directly obtained using Clebsch–Gordan coefficients. Therefore, for a charged current, one π^+ production on the neutron target, $\frac{1}{3}\rho_n(\vec{r})$ in the aforementioned expression is replaced by $\frac{1}{9}\rho_n(\vec{r})$. Thus, the total scattering cross section for the neutrino induced charged current one π^+ production on a nucleon target is given by

$$\sigma = \frac{1}{64\pi^3} \int \int d\vec{r} \frac{d\vec{k'}}{E_k E_{k'}} \frac{1}{MM_\Delta} \left[\frac{\frac{\bar{\Gamma}}{2} - C_Q \left(\frac{\rho}{\rho_0}\right)^\alpha}{(W - M_\Delta - \mathrm{Re}\Sigma_\Delta)^2 + (\frac{\bar{\Gamma}}{2} - \mathrm{Im}\Sigma_\Delta)^2} \right] [a\rho_p(\vec{r}) + b\rho_n(\vec{r})] \sum\sum |\mathcal{M}|^2$$

(15.30)

Similarly, for an antineutrino reaction, the role of $\rho_p(\vec{r})$ and $\rho_n(\vec{r})$ are interchanged; thus, $[a\rho_p(\vec{r}) + b\rho_n(\vec{r})]$ in this expression is replaced by $[a\rho_n(\vec{r}) + b\rho_p(\vec{r})]$ and the antisymmetric term in the leptonic tensor $L_{\mu\nu}$ changes sign , where $a = 1$ and $b = \frac{1}{9}$ for $\pi^+(\pi^-)$ production and $a = 0$ and $b = \frac{2}{9}$ for π^0 production.

15.2.3 Quasielastic-like production of leptons

In a nuclear medium, when a neutrino interacts with a nucleon inside the nucleus, the Δ which is formed may disappear through two- and three-body absorption processes like $\Delta N \rightarrow NN$ and $\Delta NN \rightarrow NNN$ and thus, mimic the quasielastic reaction discussed in the previous section. These Δ absorption processes are described by the C_{A2} and C_{A3} terms in the expression of $\mathrm{Im}\Sigma_\Delta$ given in Eq. (15.24). To estimate the number of quasielastic-like lepton events (without a pion), we write the expression for the total scattering cross section as

$$\sigma = \frac{1}{64\pi^3} \int \int d\vec{r} \frac{d\vec{k'}}{E_k E_{k'}} \frac{1}{MM_\Delta} \frac{C_{A2} \left(\frac{\rho}{\rho_0}\right)^\beta + C_{A3} \left(\frac{\rho}{\rho_0}\right)^\gamma}{(W - M_\Delta - \mathrm{Re}\Sigma_\Delta)^2 + (\frac{\bar{\Gamma}}{2} - \mathrm{Im}\Sigma_\Delta)^2} \left[\rho_p(\vec{r}) + \frac{1}{3}\rho_n(\vec{r})\right] \sum\sum |\mathcal{M}|^2$$

(15.31)

β and γ are tabulated in Table 15.2.

15.3 Coherent Pion Production

15.3.1 Introduction

Coherent pion production has been observed experimentally at higher energies in several nuclei [703, 700, 708, 705, 706, 709] and was studied theoretically using models based on PCAC. The only nuclear medium effect considered in these calculations is the distortion of the final pion. In the CC coherent pion production induced by neutrinos ($\nu + A \rightarrow A + \mu^- + \pi^+$), the nucleus remains in its ground state. The process consists of a weak pion production followed by the strong distortion of the pion in its way out of the nucleus.

Coherent pion production processes induced by neutrinos and antineutrinos on nuclei via. charged and neutral currents have been the subject of intense studies in the last few years. The coherent pion production process

$$\nu_l \, (\bar{\nu}_l) + \mathcal{A} \to l^{\mp} \, (\nu_l/\bar{\nu}_l) + \mathcal{A} + \pi^{\pm}(\pi^0), \tag{15.32}$$

can be qualitatively interpreted as the emission of a virtual pion from the weak interaction of neutrino followed by the elastic scattering of this off-shell pion with the target nucleus till it becomes a real pion and the target nucleus remains in the ground state. In the coherent interactions on nuclei, the overall scattering amplitude is given as the sum of the constructive interference between the scattering amplitudes of the incident wave on the various nucleons in the target nucleus, which implies that all the nucleons in the nucleus must react in phase in order to have maximum constructive interferences for enhanced cross section. Thus, the momentum transferred ($|\vec{k}|$) to any nucleons in a nucleus of radius (R) must thus be small enough so that the condition

$$|\vec{k}|R < 1. \tag{15.33}$$

gets fulfilled, implying that the nucleons remain bound in the nucleus.

Coherent reactions are also characterized by the fact that the target nucleus recoils as a whole without breaking up with very little recoil energy since the effect of the incident wave is approximately the same on all the nucleons; otherwise, the coherence would disappear. In coherent interactions, enhanced cross section can occur due to the coherence effect as long as no charge, spin, isospin or any other additive quantum number is transferred to the target nucleus. If any of these are forbidden, then this would single out a specific nucleon, and destroy the coherence. For example, the total isospin (I) of the exchanged state must be zero. Indeed, the operator I_3 induced amplitude for protons and neutrons would have an opposite sign, resulting in a small effect on the nuclei with total isospin $I = 0$. If the nucleus has spin, the spin term in the coherent amplitude are suppressed in comparison to the total spin zero nuclei. Moreover, the emission of scattered particles in the forward direction, which is generally the case, implies that the coherent interactions conserve helicity.

Coherent production of pions induced by neutrinos and antineutrinos on nuclei have been reported in four possible charged as well as neutral current channels (Figure 15.2):

$$\left. \begin{aligned} \nu_\mu + \mathcal{A} &\to \mu^- + \mathcal{A} + \pi^+ \\ \bar{\nu}_\mu + \mathcal{A} &\to \mu^+ + \mathcal{A} + \pi^- \end{aligned} \right\} \quad \text{(Charged current)} \tag{15.34}$$

$$\left. \begin{aligned} \nu_\mu + \mathcal{A} &\to \nu_\mu + \mathcal{A} + \pi^0 \\ \bar{\nu}_\mu + \mathcal{A} &\to \bar{\nu}_\mu + \mathcal{A} + \pi^0 \end{aligned} \right\} \quad \text{(Neutral current)} \tag{15.35}$$

These processes can be studied in detail and with relatively large statistics and small background as the kinematical situation in coherent processes are different from other interaction processes and also due to the small pion mass and simple geometry. Experimentally, the coherent charged as well as neutral current pion production induced by neutrinos and antineutrinos have been observed on various nuclei using different techniques such as bubble chambers and spark chambers [704]–[709]. The events are characterized by the small four-momentum transfer to the nucleus and the exponential fall of the cross section with $|t|$, where $t = (q - p_\pi)^2$, $q = k - k'$. The nucleus remains undetected experimentally, but it can be estimated from the measurement of the muon and meson momenta.

In addition to $|t|$ dependence, coherent scattering also depends on the square of the four-momentum (Q^2) transferred between the leptons. Experimentally, it has also been established that the coherent scattering cross section peaks at low $Q^2 (\leq 0.1 \text{ GeV}^2)$ and the cross section rises as a function of neutrino energy which becomes logarithmic at large neutrino energies. In

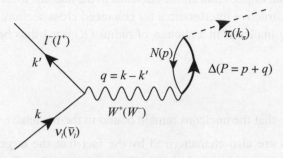

Figure 15.2 Feynman diagram considered for coherent pion production through $\Delta - h$ excitation. In the charged current reaction, k' is the four-momentum of the charged lepton, and $q (= k - k')$ is the four-momentum of the $W^\pm$ boson. In the neutral current reaction, k' is the momentum of the scattered neutrino and q is the momentum of the exchanged Z^0.

this sense, the neutrino induced coherent pion production reaction has an advantage over other existing reactions, which can enrich our understanding of the nuclear excitation mechanism; it also allows the study of the longitudinal axial vector current for very small Q^2 values, providing the most detailed test of the PCAC hypothesis at high energies. A good knowledge of coherent pion production induced by neutrinos and antineutrinos is also important in understanding the background of the quasielastic lepton production in the forward direction in the analysis of neutrino oscillation experiments with atmospheric and accelerator neutrino beams in intermediate energies. There are two approaches which have been used to calculate coherent pion production cross section:

(i) methods based on PCAC, and

(ii) microscopic models using nuclear structure.

15.3.2 PCAC based methods

Coherent pion production cross section can be calculated using Adler's theorem [753] relating the inelastic neutrino scattering cross section in the forward direction with pion nucleus scattering. In the forward direction, that is, $\theta = 0$, the initial and final leptons are produced in parallel configurations, that is, the initial and final momenta of leptons $\vec{k}$ and $\vec{k}'$ are parallel to each other. In the limit of $m_l \to 0$, the momentum transfer square $q^2 (= (k - k')^2) \approx 0$ and all the momenta k^μ, k'^μ, and q^μ are null vectors and $\vec{k}, \vec{k}'$, and $\vec{q}$ are colinear. Moreover, in the case of an inelastic reaction like $\nu + N \to \mu^- + X$, the mass of final hadron $M_X \neq M$ making $q_0 \neq 0$. A simple algebra shows that in this kinematic configuration, the momenta k^μ and k'^μ are proportional to each other and also to q^μ; thus, we can write [753]

$$k'^\mu = \frac{k' \cdot p}{q \cdot p} q^\mu \qquad k^\mu = \frac{k \cdot p}{q \cdot p} q^\mu. \tag{15.36}$$

We also know that in any neutrino reaction, the cross section $d\sigma$ is proportional to the product of leptonic and hadronic tensors, that is,

$$d\sigma \propto \sum_{\text{spin}} L^{\mu\nu} \langle X|J_\mu|N\rangle \langle X|J_\nu|N\rangle^*,$$

where $L^{\mu\nu} \propto k^\mu k'^\mu + k'^\mu k^\nu - g^{\mu\nu} k \cdot k' + i\epsilon^{\mu\nu\rho\sigma} k_\rho k'_\sigma.$

Since k^μ and k'^μ are now proportional to q^μ, $d\sigma$ becomes proportional to $q^\mu q^\nu$:

$$d\sigma \propto \langle X|q^\mu J_\mu|N\rangle \langle X|q^\nu J_\nu|N\rangle^*. \tag{15.37}$$

Using the hypothesis of CVC and PCAC, we obtain

$$q^\mu V_\mu = 0, \qquad\qquad q^\mu A_\mu = f_\pi m_\pi^2 \phi_\pi. \tag{15.38}$$

Therefore, the amplitude $q^\mu A_\mu$ is related to the matrix element of the pion field taken between $|N\rangle$ and $|X\rangle$ states, that is,

$$\langle X|q^\mu A_\mu|N\rangle = \frac{f_\pi m_\pi^2}{q^2 - m_\pi^2} T(\pi^\alpha N \to X) \tag{15.39}$$

in the pion pole approximation. We see that the matrix element for the transition $\nu + n \to \mu^- + X$ through the weak current J_μ gets nonzero contribution only from the divergence of the axial vector current (i.e., $q_\mu A^\mu$) in the $q^2 \approx 0$ limit which is related with the pion nucleon scattering corresponding to $q^2 = 0$ and $q_0 = E_\pi$ limit in the case of single pion production.

In this approximation, Adler obtained the expression of cross section for the reaction $\nu + N \to \mu^- + N + \pi^\alpha$ in the limit of $q^2 = 0$. This expression was used by Rein and Sehgal [715] to calculate the coherent production of π^0 from nuclei by extending it to $q^2 \neq 0$ values. Adler's formula is expressed as [753]

$$\frac{d\sigma}{dq^2 dy dt} = r \frac{G_F^2 f_\pi^2}{2\pi^2} \frac{(1-y)}{y} \frac{d\sigma}{dt}(\pi A \to \pi A_{gs})\Big|_{q^2=0, q_0=E_\pi} \tag{15.40}$$

with $r = 2\cos\theta_C(1)$ for the CC (NC) reaction. $t = (q - p_\pi)^2$ and $y = \frac{q_0}{E_\nu}$. $\frac{d\sigma}{dt}(\pi A \to \pi A_{gs})$ is the cross section for the pion nucleus scattering neglecting the nucleus recoil (i.e., $q_0 = E_\pi$). Rein and Sehgal [715] used a form factor $F(q^2) = \frac{1}{\left(1 - \frac{q^2}{M_A^2}\right)^2}$ to extend it to $q^2 \neq 0$ and the

following expression for the pion–nucleus scattering cross section was used [761]:

$$\frac{d\sigma}{dt}(\pi A \to \pi A_{g.s.}) = \frac{d\sigma}{dt} = |F_A(t)|^2 F_{\text{abs}} \frac{d\sigma}{dt}(\pi N \to \pi N)\Big|_{t=0'}$$

$$\text{where } F_A(t) = \int d^3 r e^{i(\vec{q} - \vec{p}_\pi) \cdot \vec{r}}(\rho_p(r) + \rho_n(r)). \tag{15.41}$$

where $\frac{d\sigma}{dt}(\pi N \rightarrow \pi N)$ is the pion nucleon cross section and F_{abs} is the factor describing the absorption of pions. This has been used in most of the early versions of neutrino events generators. Since then, many authors have extended and updated the formula by including the following corrections to the coherent pion production:

i) Lepton mass correction due to $m_l \neq 0$ relevant for charge current reactions.

ii) Improved kinematic and dynamic corrections due to $q^2 \neq 0$.

iii) Including the contribution of other mesons in addition to the pion pole as done in Adler's model and used by Rein and Sehgal [715].

iv) Improved treatment of pion nucleus cross section relevant to the experimental kinematics of present neutrino experiments.

These corrections led to reduction in the cross section as predicted by the Rein and Sehgal model [715] and are similar to those obtained with other methods using microscopic models.

15.4 Microscopic Model for Coherent Weak Pion Production

The amplitude for the charged current weak pion production from the nuclei for the delta pole term corresponding to the Feynman diagrams shown in Figure 15.3, is written in terms of l^μ, the leptonic current and hadronic current $\mathcal{J}^\mu = J_s^\mu + J_u^\mu$ as the sum of direct (s channel) and crossed (u channel) diagrams and a nuclear form factor as.

$$\mathcal{A} = \frac{G_F}{\sqrt{2}} \cos\theta_C \, l^\mu \, \mathcal{J}_\mu \, \mathcal{F}(\vec{q} - \vec{k}_\pi), \qquad (15.42)$$

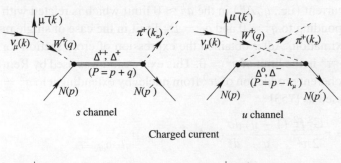

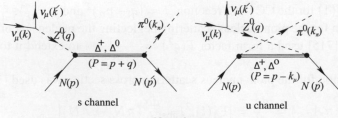

Figure 15.3 Feynman diagrams considered for neutrino induced weak coherent pion production for Δ-resonance.

where G_F is the Fermi coupling constant and $\mathcal{F}(\vec{q} - \vec{k}_\pi)$ is the nuclear form factor including the isospin factors.

The hadronic current for the s channel process using Feynman rules may be written as

$$J_s^\mu = \sqrt{3} \, \frac{f_{\pi N \Delta}}{m_\pi} \, k_\pi^\sigma \sum_r \bar{u}_r(p') \, \Delta_{\sigma\lambda} \, \mathcal{O}^{\lambda\mu} \, u_r(p). \tag{15.43}$$

The operators $\Delta_{\sigma\lambda}$ and $\mathcal{O}^{\lambda\mu}$ are discussed in Chapters 11 and 12.

Similarly, one may write the hadronic current for the u channel process.

For the coherent process, we write the spinors in the rest frame of the nucleus in the following form

$$u_r(\vec{p}) = \frac{\not{p} + M}{\sqrt{2M(E + M)}} u_r(0), \tag{15.44}$$

$$\bar{u}_r(\vec{p}') = \bar{u}_r(0) \frac{\not{p}' + M}{\sqrt{2M(E' + M)}}, \tag{15.45}$$

$$\sum_r u_r(0) \bar{u}_r(0) = \frac{1 + \gamma_0}{2}. \tag{15.46}$$

Considering only $C_5^A(Q^2)$ which is the most dominant term in $\mathcal{O}^{\lambda\mu}$, J_μ^s may be written as

$$J_\mu^s = \chi \, \mathrm{Tr} \left[(1 + \gamma_0) \, (\not{p} + M) \, (\not{p}' + M) \, k_\sigma^\pi \, (\not{P} + M_\Delta) \left(g^{\sigma\lambda} - \frac{1}{3} \gamma^\sigma \gamma^\lambda - \frac{2}{3M_\Delta^2} P^\sigma P^\lambda \right. \right.$$
$$\left. \left. + \frac{P^\sigma \gamma^\lambda - \gamma^\sigma P^\lambda}{3M_\Delta} \right) g_{\lambda\mu} \right] \tag{15.47}$$

with

$$\chi = \sqrt{3} \left(\frac{f_{\pi N \Delta}}{m_\pi} \right) \left(\frac{1}{2} \right)^2 \left(\frac{1}{2M} \right)^2 \frac{C_5^A(Q^2)}{(P^2 - M_\Delta^2) + i\Gamma M_\Delta}$$

Using trace algebra, we find the final expression as

$$J_\mu^s = 8 \, \chi \left[\alpha_s \, k_\mu^\pi + \beta_s \, P_\mu + \gamma_s \, p_\mu \right], \tag{15.48}$$

where α_s, β_s, and γ_s are the coefficients associated with k_μ, P_μ, and p_μ respectively that can be obtained after the trace calculation.

The nuclear form factor $\mathcal{F}(\vec{q} - \vec{k}_\pi)$ is given as

$$\mathcal{F}(\vec{q} - \vec{k}_\pi) = \int d^3r \, \rho(\vec{r}) \, e^{-i(\vec{q} - \vec{k}_\pi) \cdot \vec{r}}, \tag{15.49}$$

with $\rho(\vec{r})$ as the nuclear matter density which is a function of nucleon relative coordinates. It is the linear combination of proton and neutron densities incorporating the isospin factors.

15.4.1 Cross sections

The differential cross section for a pion produced in the charged current weak production process induced by neutrinos can be written as

$$\left[\frac{d^5\sigma}{dE_\pi d\Omega_\pi dQ^2}\right]_{CC} = \frac{\pi}{8}\frac{1}{(2\pi)^5}\frac{M}{E_k^2}\,|\,\vec{k}_\pi\,|\,\frac{1}{\mathcal{R}}\,\overline{\sum}\sum|\mathcal{A}|^2, \tag{15.50}$$

where

$$\mathcal{R} = \left[\left(E_{p'} + E_{k'} - E_k\cos\theta_{kk'}\right) - \frac{|\vec{k}_\pi|}{|\vec{q}|}\left(E_{k'} - E_k\cos\theta_{kk'}\right)\cos\theta_{\pi q}\right] \tag{15.51}$$

is a kinematical factor incorporating the recoil effects, which is very close to unity for low $Q^2(=-q^2)$, relevant for coherent reactions.

We find that the recoil of the nucleus gives less than $(3-4\%)$ correction in the energy region of around 1 GeV. In the charged current weak reaction, we could also measure the energy and the momentum of the leptons. It also allows an approximate separation of the coherent cross section from the non-coherent background. We can measure the differential cross section for the charged lepton production,

$$\left(\frac{d^5\sigma}{d\Omega_\pi d\Omega_{kk'}dE_k'}\right)_{CC} = \frac{1}{8}\frac{1}{(2\pi)^5}\frac{|\,\vec{k}'\,|\,|\,\vec{k}_\pi\,|}{E_k}\,\mathcal{R}\,\overline{\sum}\sum|\mathcal{A}|^2 \tag{15.52}$$

where

$$\mathcal{R} = \left[\frac{M\,|\vec{k}_\pi|}{E_{p'}|\vec{k}_\pi| + E_\pi(|\vec{k}_\pi| - |\vec{q}|\cos\theta_{\pi q})}\right] \tag{15.53}$$

15.4.2 Final state interactions

The pions which are produced in these processes while traveling inside the nucleus, can be absorbed, can change direction, energy, charge, or even produce more pions due to various processes like elastic and charge exchange scattering with the nucleons present in the nucleus through strong interactions. Therefore, the production cross sections for the pions from the nuclear targets are affected by the presence of strong interactions of the final state pions in the nuclear medium (Figure 15.4). For example, a pion produced in the nuclear medium may get absorbed by the nucleons and thus, mimic a quasielastic-like event (see Figure 15.5).

There are many approaches developed initially in the inelastic production of pions from electron–nucleus and proton–nucleus scattering which can be applied to (anti)neutrino–nucleus scattering. The various processes which affect the pions after they are produced are pion absorption, elastic and charge exchange scattering of pions. The earlier calculations of the

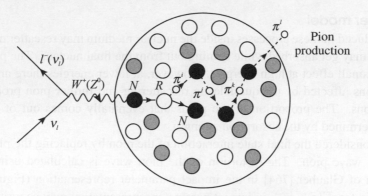

Figure 15.4 Pion production inside a nuclear target and its interaction with the nucleons in the nucleus while coming out. The pions may undergo elastic scattering, charge exchange (for example, $\pi^+ + n \rightarrow \pi^0 + p$) reaction, etc. and therefore, the charge of the pion may change along the way before it comes out of the nucleus. Circles have the same meaning as in Figure 15.1.

final state interactions (FSI) of pions applied to $\nu(\bar{\nu})$ scattering from nucleons were done by Adler et al. [753], who used a multipole scattering theory of pions. The other methods use the approach of distorted wave Born approximation (DWBA) in which the plane wave pion is replaced by a distorted pion wave obtained by solving the Klein–Gordon equation for pions in an optical potential [762]. In case of high energy and forward angles corresponding to a very low q^2 relevant for the coherent production, the Glauber model for calculating the pion distortion is most appropriate [720, 763]. In a microscopic approach, the pion trajectory is traced within the nucleus from its point of production, weighted by the probability for various final state interactions like its absorption, elastic scattering, and charge exchange scattering with the nucleon until it comes out of the nucleus. This is the method used in most of the neutrino event generators like NUANCE and NEUT; it is used even today in modern neutrino event generators. In the following, we outline the basic ingredients of the last two methods.

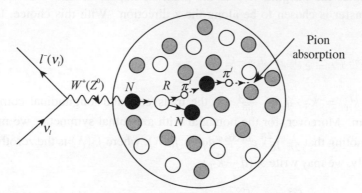

Figure 15.5 A neutrino induced reaction on a nucleon target when a pion is absorbed in the nuclear medium while coming out. Circles have the same meaning as in Figure 15.1.

FSI in Glauber model

The pions produced in these processes inside the nuclear medium may rescatter or may produce more pions or may get absorbed while coming out from the final nucleus. The pion absorption is a relatively small effect at high energies. However, at lower energies there are considerable number of pions affected by the quasielastic rescattering and further pion production on the nuclear nucleons. The proportion of the pions that eventually comes out of the nucleus is essentially determined by the absorption strength.

We have considered the final state interaction of the pion by replacing the plane wave pion by a distorted wave pion. The distortion of the pion wave is calculated using the Eikonal approximation of Glauber [764] in the impact parameter representation (Figure 15.6). The nuclear form factor $\mathcal{F}(\vec{q} - \vec{k}_\pi)$, in impact parameter representation may be written as

$$\mathcal{F}(\vec{q} - \vec{k}_\pi) = \int d^2b \, dz \, \rho(\vec{b}, z) \, e^{i(\vec{q} - \vec{k}_\pi)\cdot(\vec{b} + \hat{q}z)} \tag{15.54}$$

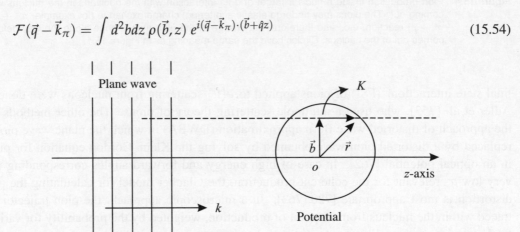

Figure 15.6 Definition of impact parameter in scattering processes at high energies. We choose the z-axis to lie in the direction of propagation of $\vec{k}$ and $\vec{K} = \vec{k} + \vec{k}'$.

where d^2b denotes the integration over the plane of impact vector. $\vec{r} = (\vec{b}, z)$, and $\vec{q}$ the momentum transfer is chosen to be along the z direction. With this choice, the exponential factor becomes

$$e^{i(\vec{q} - \vec{k}_\pi)\cdot(\vec{b} + \hat{q}z)} = e^{-i\vec{k}_\pi^t \cdot \vec{b}} \, e^{i(|\vec{q}| - k_\pi^l)z}$$

where $\vec{k}_\pi^t$ and $k_\pi^l = \vec{k}_\pi \cdot \hat{q} = \frac{\vec{k}_\pi \cdot \vec{q}}{|\vec{q}|}$ are the transverse and longitudinal components of the pion momentum. Moreover, for the potential with azimuthal symmetry, we may carry the ϕ integration by noting that $\frac{1}{2\pi} \int_0^{2\pi} e^{i\lambda \cos \phi} d\phi = J_0(\lambda)$, where $J_0(\lambda)$ is the zeroth order Bessel's function. Finally, we may write $\mathcal{F}(\vec{q} - \vec{k}_\pi)$

$$\tilde{\mathcal{F}}(\vec{q} - \vec{k}_\pi) = 2\pi \int_0^\infty b \, db \int_{-\infty}^\infty dz \, \rho(\vec{b}, z) \, J_0(k_\pi^t b) \, e^{i(|\vec{q}| - k_\pi^l)z} \, e^{-if(\vec{b}, z)} \tag{15.55}$$

where

$$f(\vec{b}, z) = \int_z^\infty \frac{1}{2|\vec{k}_\pi|} \Pi(\rho(\vec{b}, z')) dz' \qquad (15.56)$$

and the pion self energy Π related to the pion optical potential $V_{opt}(E_\pi, \vec{r})$ as $\Pi(\rho(\vec{r})) = 2E_\pi V_{opt}(E_\pi, \vec{r})$ is defined as

$$\Pi(\rho) = \frac{4}{9} \left(\frac{f_{\pi N\Delta}}{m_\pi}\right)^2 \frac{M^2}{W^2} |\vec{p}_\pi|^2 \rho \frac{1}{W - \tilde{M}_\Delta + \frac{i\tilde{\Gamma}}{2}}. \qquad (15.57)$$

Microscopic model of pion propagation

Now, we will describe in brief the prescription of Vicente Vacas [765] which we have followed while presenting the numerical results for the incoherent one pion production with FSI. In this prescription, a pion of given momentum and charge is moved along the z-direction with a random impact parameter $\vec{b}$, with $|\vec{b}| < R$, where R is the nuclear radius which is taken to be a point where nuclear density $\rho(R)$ falls to $10^{-3}\rho_0$, and ρ_0 is the central density. To start with, the pion is placed at a point $(\vec{b}, z_{in})$, where $z_{in} = -\sqrt{R^2 - |\vec{b}|^2}$ and then it is moved in small steps δl along the z-direction until it comes out of the nucleus or interacts with the nucleon. If $P(k_\pi, r, \lambda)$ is the probability per unit length at the point r of a pion of momentum $\vec{k}_\pi$ and charge λ, then $P\delta l << 1$. A random number x is generated such that $x \in [0, 1]$ and if $x > P\delta l$, then it is assumed that the pion has not interacted while traveling a distance δl. However, if $x < P\delta l$, then the pion has interacted and depending upon the weight factor of each channel given by its cross section, it is decided if the interaction is quasielastic, charge exchange reaction, pion production, or pion absorption. For example, for the quasielastic scattering

$$P_{N(\pi^\lambda, \pi^{\lambda'})N'} = \sigma_{N(\pi^\lambda, \pi^{\lambda'})N'} \times \rho_N(r),$$

where N is a nucleon, ρ_N is its density, and σ is the elementary cross section for the reaction $\pi^\lambda + N \to \pi^{\lambda'} + N'$ obtained from the phase shift analysis.

For a pion to be absorbed, P is expressed in terms of the imaginary part of the pion self energy Π, that is, $P_{abs} = -\frac{Im\Pi_{abs}(k_\pi)}{k_\pi}$, where the self energy Π is related to the pion optical potential V.

15.5 Results for Cross Sections

(i) Incoherent pion production

In Figure 15.7, the results for Q^2-distribution $\frac{d\sigma}{dQ^2}$ and momentum distribution $\frac{d\sigma}{dp_\pi}$ are shown for the charged current $\nu_\mu(\bar{\nu}_\mu)$ induced incoherent one π^+ (π^-) production cross section. These results are presented for the differential scattering cross section calculated with and without the nuclear medium effects and with nuclear medium effects including the pion absorption effects. For the Q^2-distribution shown in Figure 15.7, we see that the reduction in the cross section as compared to the cross section calculated without the nuclear medium effects is around 35% in the peak region. When pion absorption effects are also taken into

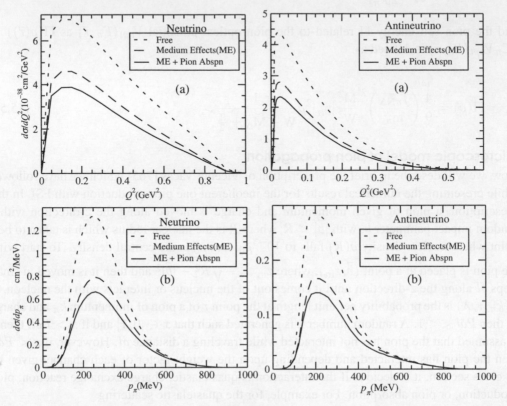

Figure 15.7 $\frac{d\sigma}{dQ^2}$ and $\frac{d\sigma}{dp_\pi}$ for the $\nu_\mu(\bar{\nu}_\mu)$ induced charged current one $\pi^+(\pi^-)$ process on ^{12}C target at $E_\nu = 1$ GeV [766].

account, there is a further reduction of around 15%. The results for the antineutrino induced one π^- production cross section are qualitatively similar in nature but quantitatively, we find that the peak shifts toward a slightly lower value of Q^2. In Figure 15.7, the results for the pion momentum distribution have been shown. In this case, the reduction in the cross section in the peak region is around 40% when nuclear medium effects are taken into account, which further reduces by about 15% when pion absorption effects are also taken into account.

In Figure 15.8, the results for the total scattering cross section σ for charged current $\nu_\mu(\bar{\nu}_\mu)$ induced one $\pi^+(\pi^-)$ production cross section are shown. We see that with the inclusion of nuclear medium effects, the reduction in the cross section from the cross section calculated without the nuclear medium effects for neutrino energies between 1–2 GeV is 30–35%, which further reduces by 15% when pion absorption effects are also taken into account. The results with antineutrinos are similar in nature.

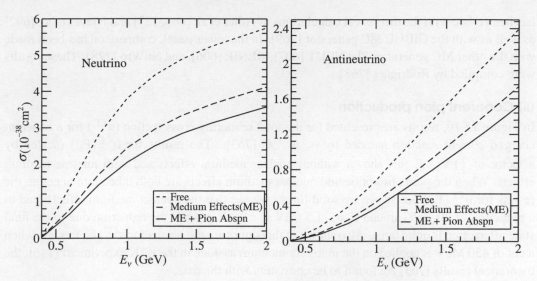

Figure 15.8 σ for ν_μ ($\bar{\nu}_\mu$) induced charged current incoherent $\pi^+(\pi^-)$ production on ^{12}C target [766].

Figure 15.9 shows the pion kinetic energy distribution $\frac{d\sigma}{dT}$ in the CC $1\pi^+$, ν_μ induced interaction on CH_2 obtained in the MiniBooNE [767] experiment. In the top panel, comparison

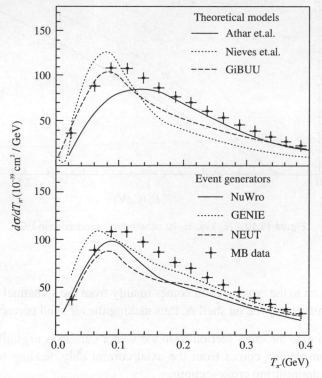

Figure 15.9 Comparison of theoretical and event generator calculations for the differential cross sections in pion kinetic energy with the MiniBooNE [767] ν_μ CH_2 CC $1\pi^+$ production data [768].

has been made with the theoretical calculations of Athar et al. [668, 727] and Nieves et al. [692] as well as with the GiBUU MC generator [769]. In the lower panel, comparison has been made with the other MC generators like NEUT [607], GENIE [606], and NuWro [728]. These results were compiled by Rodrigues [768].

(ii) Coherent pion production

In Figure 15.10, results are presented for the total scattering cross section (σ^{CC}) for a coherent charged current reaction induced by ν_μ in ^{12}C [763]. The results for $\sigma^{CC}(E_\nu)$ (scaled by a factor of $\frac{1}{2}$) vs. E_ν are shown without nuclear medium effects and with nuclear medium effects. When the pion absorption and nuclear medium effects are both taken into account, the results for $\sigma^{CC}(E_\nu)$ are shown by solid lines. We see that the nuclear medium effects lead to a reduction of 30–35% around $E_\nu=1$–1.5 GeV in $\sigma^{CC}(E_\nu)$ while the reduction due to the final state interaction is quite large. Also shown in this figure are the results for $\sigma^{CC}(E_\nu)$ vs. E_ν when a cut of 450 MeV is applied on the muon momentum as done in the K2K experiment [746]; the theoretical results [763] are found to be consistent with the data.

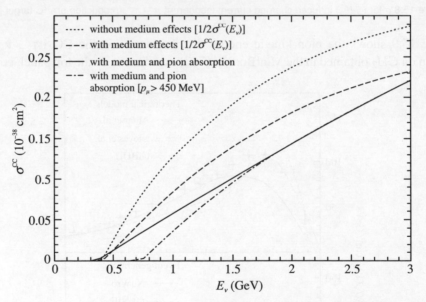

Figure 15.10 $\sigma(E_\nu)$ vs. E_ν for coherent π^+ production in ^{12}C

To conclude:

(i) the contribution to the cross section comes mainly from the s-channel diagram ($> 90\%$) which is dominated by the on shell Δ, thus making the off shell correction quite small.

(ii) the contribution to the cross section from the vector current is negligibly small (<2 %); the major contribution comes from the axial current only, leading to near equality of neutrino and antineutrino cross sections.

The differential cross sections $\frac{d\sigma}{dq^2}$ and $\frac{d\sigma}{dk_\pi}$ for charged pion production at $E_\nu = 1$ GeV are presented in Figures.15.11, where nuclear medium and final state interactions effects are shown explicitly. In the inset, we exhibit the differential cross sections $< \frac{d\sigma}{dq^2} >$ and $< \frac{d\sigma}{dk_\pi} >$ averaged over the K2K and MiniBooNE neutrino spectra without applying any cuts on the muon momentum.

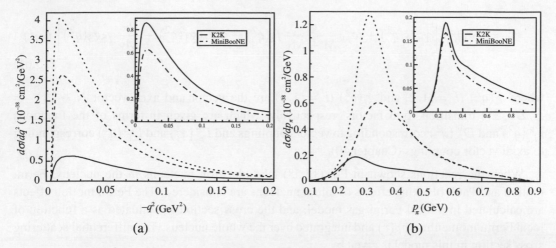

Figure 15.11 (a) $\frac{d\sigma}{dq^2}$ vs. $-q^2$ at $E_\nu=1$ GeV, for coherent π^+ production in ^{12}C nucleus without(dotted), with(dashed) nuclear medium effects and with nuclear medium and pion absorption effects (solid). In the inset, the final results for $< \frac{d\sigma}{dq^2} >$ vs. $-q^2$ averaged over the K2K and MiniBooNE spectrum are shown. (b) $\frac{d\sigma}{dk_\pi}$ vs. k_π at $E_\nu=1$ GeV.

In an improved calculation, Alvarez-Ruso et al. [770] have also included the contribution of the non-resonant diagram; it has been found that these contributions are very small at $E_\nu \sim 1$ GeV.

15.6 Pion Production through Hyperon Excitation

The $\Delta S = 1$ hyperon (Y) production processes induced by antineutrinos producing pions are given by (Chapter 10 for detail)

$$\bar{\nu}_l(k) + N(p) \rightarrow l^+(k') + Y(p')$$
$$\hookrightarrow N'(p') + \pi(k_\pi), \tag{15.58}$$

where $Y = \Lambda, \Sigma^{0,-}$. The differential scattering cross section for this process in the laboratory frame may be written as

$$d\sigma = \frac{\delta^4(k + p - k' - p')}{16\pi^2 E_\nu M_N} \frac{d^3k'}{2E_{k'}} \frac{d^3p'}{2E_{p'}} \sum\overline{\sum}|\mathcal{M}|^2, \tag{15.59}$$

where M_N is the nucleon mass and $\mathcal{M}$ is the matrix element written as

$$\mathcal{M} = \frac{G_F}{\sqrt{2}} \sin \theta_c \bar{v}_l(k') \gamma^\mu (1 - \gamma^5) v_\nu(k) J_\mu, \tag{15.60}$$

where J_μ is the hadronic current, which is written as

$$J_\mu = \bar{u}_Y(p') \left[\gamma_\mu f_1(q^2) + i\sigma_{\mu\nu} \frac{q^\nu}{M + M_Y} f_2(q^2) - \gamma_\mu \gamma_5 g_1(q^2) - \frac{q_\mu}{M_Y} \gamma_5 g_3(q^2) \right] u_N(p) \tag{15.61}$$

where $f_i(q^2)$ $(i = 1, 2)$ and $g_i(q^2)$ $(i = 1, 3)$ are the vector and axial vector $N - Y(Y = \Lambda, \Sigma^-, \Sigma^0)$ transition form factors, respectively, which are given in terms of the functions $F_i^V(q^2)$ and $D_i^V(q^2)$ corresponding to vector couplings and $F_i^A(q^2)$ and $D_i^A(q^2)$ corresponding to axial vector couplings (Chapter 10).

When the reactions shown in Eq. (15.58) take place on nucleons in the nucleus, Fermi motion and Pauli blocking effects of initial nucleons are considered. The Fermi motion effects are calculated in a local Fermi gas model, and the cross section is evaluated as a function of local Fermi momentum $p_F(r)$ and integrated over the whole nucleus. The differential scattering cross section in this model is given by

$$\frac{d\sigma}{dQ^2 dE_l} = 2 \int d^3r \int \frac{d^3p}{(2\pi)^3} n_N(p, r) \left[\frac{d\sigma}{dQ^2 dE_l} \right]_{\text{free}}, \tag{15.62}$$

where $n_N(p, r)$ is the occupation number of the nucleon.

The produced hyperons are affected by the FSI within the nucleus through the hyperon–nucleon quasielastic and charge exchange scattering processes like $\Lambda + n \to \Sigma^- + p$, $\Lambda + n \to \Sigma^0 + n$, $\Sigma^- + p \to \Lambda + n$, $\Sigma^- + p \to \Sigma^0 + n$, etc. Because of such types of interaction in the nucleus, the probability of Λ or Σ production changes; this has been taken into account by using the prescription given in Ref. [771]. In this prescription, an initial hyperon produced at a position r within the nucleus interacts with a nucleon to produce a new hyperon state within a short distance dl with a probability $P = P_Y dl$, where P_Y is the probability per unit length given by

$$P_Y = \sigma_{Y+n\to f}(E) \rho_n(r) + \sigma_{Y+p\to f}(E) \rho_p(r),$$

where f denotes a possible final hyperon–nucleon $[Y_f(\Sigma \text{ or } \Lambda) + N(n \text{ or } p)]$ state with energy E in the hyperon–nucleon center of mass system, $\rho_n(r)[\rho_p(r)]$ is the local density of the neutron(proton) in the nucleus, and σ is the total cross section for a charged current channel like $Y(\Sigma \text{ or } \Lambda) + N(n \text{ or } p) \to f$ [771]. Now a particular channel is selected, which gives rise to a hyperon Y_f in the final state with the probability P. For the selected channel, the Pauli blocking effect is taken into account by first randomly selecting a nucleon in the local Fermi sea. Then a random scattering angle is generated in the hyperon–nucleon center of mass system assuming the cross sections to be isotropic. Using this information, the hyperon and nucleon momenta are calculated and the Lorentz boosted to the lab frame. If the nucleon in the final state has momenta above the Fermi momenta, we have a new hyperon type (Y_f) and/or a new direction and energy of the initial hyperon (Y_i). This process is continued until the hyperon gets out of the nucleus. The decay modes of hyperons to pions is highly suppressed in the nuclear medium [772], making them live long enough to pass through the nucleus and decay outside the nuclear medium. Therefore, the produced pions are less affected by the strong interaction of the nuclear field; their FSI have not been taken into account.

In Figure 15.12, the results for the cross sections $\sigma(E_{\bar{\nu}})$ for one pion production obtained from the hyperon excitation and the Δ excitation are compared. These results are presented with

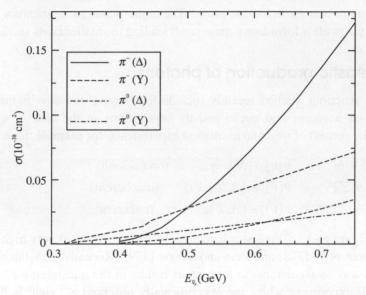

Figure 15.12 $\sigma(E_{\bar{\nu}_\mu})$ vs. $E_{\bar{\nu}_\mu}$ for π^- & π^0 production in ^{12}C, in the Δ dominance model and via intermediate hyperons.

the nuclear medium and FSI effects for the pions interacting with the residual nucleus for π^- as well as π^0 productions obtained in the delta dominance model and the pions obtained from the hyperons where the medium effects in hyperon production and FSI due to hyperon–nucleon interaction in the nuclear medium have been taken into account. When the results for the cross sections are compared in these two processes, we see that, at lower antineutrino energies $E_{\bar{\nu}} <$ 500 MeV, the contribution of π^- from the hyperon excitation is more than the pions coming from the Δ excitation. For $E_{\bar{\nu}} > 500$ MeV, the contribution from Δ excitation starts dominating and at $E_{\bar{\nu}} = 1$ GeV, the contribution of π^- from the hyperon excitation is around 20% of the total π^- production. In $\bar{\nu}_l$ induced π^0 production, the contribution of π^0 from the hyperon excitation is larger up to an antineutrino energy of 650 MeV than the contribution of π^0 from the Δ excitation. At $E_{\bar{\nu}} = 1$ GeV, the contribution of π^0 from the hyperon excitation is around 30% of the total π^0 production. For a more detailed discussion, please see Ref. [773, 774, 775].

15.6.1   Inelastic production of kaons

The inelastic production of kaons from free nucleon has been discussed in Chapter 12. These amplitudes were used to study the coherent production of kaons and antikaons from nuclei [776] which has been recently observed by the MINERvA collaboration [777]. It was found that in the theoretical calculations, the cross section for coherent (anti)kaon production is considerably smaller than the pion production apart from the $\tan^2 \theta_C$ suppression which arises due to the reduction in the values of nuclear form factor $F(k_\pi - q)$. Since the coherent production of particles is generally forward peaked, i.e. low q^2, most of the momentum is transferred to the nuclear system. Since the (anti)kaons have higher mass, their production needs higher (anti)neutrino energies. Thus, the momentum transfer to the nucleus is also relatively higher than the momentum transfer in case of pions. The higher value of momentum transfer to the nucleus makes the nuclear form factor quite small leading to smaller cross sections.

15.6.2   Inelastic production of photons

Other inelastic reactions studied recently include the photon emission induced by neutral currents (NC) on nucleons and nuclei and its implications in the case of quasielastic $e^{\mp}$ production in the context of neutrino oscillation experiments, for example the reactions

$$\nu(\bar{\nu}) + N \quad \rightarrow \quad \nu(\bar{\nu}) + N + \gamma \qquad \text{(on nucleon)} \tag{15.63}$$

$$\nu(\bar{\nu}) + (A, Z) \quad \rightarrow \quad \nu(\bar{\nu}) + (A, Z) + \gamma \quad \text{(incoherent)} \tag{15.64}$$

$$\nu(\bar{\nu}) + (A, Z) \quad \rightarrow \quad \nu(\bar{\nu}) + (A, Z)_{g.s.} + \gamma \quad \text{(coherent)} \tag{15.65}$$

and shown in Figure 15.13. The first calculation of these processes at very high energies were done by Gershtein et al. [778] and Rein and Sehgal [779]. Recently, new interest in studying these reactions was awakened due to anomalous results in the quasielastic e^- production in the MiniBooNE experiment where the experimentally observed e^- yield is higher than the theoretical prediction. It has been suggested that this higher yield may be due to additional photons which are produced in these reactions; these photons decay through pair production and add to the electron yield. In this context many calculations have been done using various

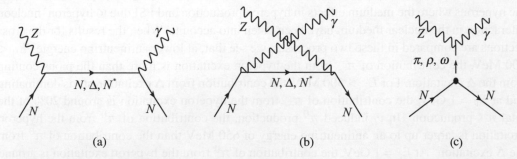

Figure 15.13 Feynman diagrams for NC photon emission. The first two diagrams are direct and crossed baryon pole terms with nucleons and resonances in the intermediate state (N,Δ(1232), N*(1440), N*(1520), etc.) The third diagram represents t channel meson (π, ρ, ω) exchange contributions.

models and they are applied to incoherent and coherent production of photons from nuclei for the photon production using effective Lagrangians [780]. This topic has been recently reviewed by Alvarez-Ruso et al. [356].

It has been found that the photon emission by $\Delta(1232)$ and $N^*(1520)$ excited by (anti) neutrinos make the dominant contribution in the basic process. Consequently, the nuclear medium effects are very important due to the renormalization of Δ in nuclei as it is the case of the pion production. When the effect due to the Pauli blocking and Fermi motion are also taken into account, the cross sections are reduced by a factor of 30% in the region of 500 MeV to 1.5 GeV. The reduction does not depend very strongly upon the mass number A and decreases very slightly with increase in A. The incoherent cross sections are larger than the coherent cross sections. However, in the forward direction $q^2 \approx 0$, the coherent cross sections are comparable to the incoherent cross section albeit small. The theoretical calculations in various models for ^{12}C agree with each other but are not sufficient to explain the excess $e^{\pm}$ events in the MiniBooNE experiment [781]. The high energy neutrino experiment NOMAD has also searched for the neutral current (NC) single photon production at $E_\nu \approx 25$ GeV and presented a limit for this process [782]:

$$\frac{\sigma(\text{single photon})}{\sigma(\nu_\mu A \to \mu^- X)} < 4 \times 10^{-4} \quad 90\% \ (C.L.). \tag{15.66}$$

The theoretically predicted cross sections are consistent with this limit in the relevant kinematic region.

Chapter 16

Deep Inelastic Scattering of (Anti)neutrinos from Nuclei

16.1 Introduction

We have discussed charged lepton and (anti)neutrino induced deep inelastic scattering (DIS) off free protons in Chapter 13. The interaction cross section and the structure functions of nucleons are modified when the scattering takes place from the nucleons bound inside a nucleus. The reactions are shown in Figure 16.1, where $l = e, \mu$. A is the target nucleus and X is the jet of hadrons. k, p and k', p' are the four momenta of the initial and the final state particles respectively. Historically, the first observations of modifications of the nuclear structure

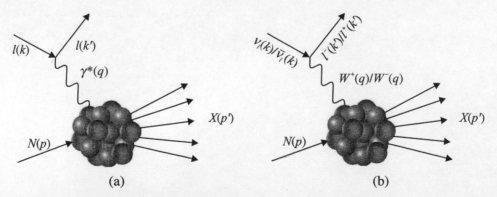

Figure 16.1 The deep inelastic charged lepton (a) and (anti)neutrino (b) scattering processes with bound nucleons for the electromagnetic and weak interactions, respectively.

functions were made by the European Muon Collaboration (EMC) at CERN in 1981–83. The EMC collaboration studied the ratio of structure function $F_2(x, Q^2)$ per nucleon for iron to deuterium targets, that is, $R(x, Q^2) = \frac{F_{2Fe}(x,Q^2)/A}{F_{2D}(x,Q^2)/2}$ in the energy region of 120-280 GeV and its deviation from unity. This effect is known as the EMC effect. Since then, the EMC effect has been confirmed and studied with improved precision in many DIS experiments using electrons, muons [783, 784, 785, 786, 787, 788], neutrinos, and antineutrinos [789, 790, 791, 792, 793, 794] from different nuclear targets as well as in the Drell–Yan processes using proton and pions [795, 796, 797]. Some of these results are presented in Figure 16.2. From the figure, it may be observed that the ratio is different from unity in almost the entire region of the Bjorken scaling variable $0 < x < 1$. From these experiments, some general features of the ratio $R(x, Q^2)$ may be inferred:

- The x dependence of $R(x, Q^2)$ has considerable structure, that is, it is different in different regions of x.

- The shape of the effect is almost independent of A.

- The functional form of $R(x, Q^2)$ is relatively independent in the region of high Q^2.

Generally, the nuclear medium effects manifested through the ratio $R(x, Q^2)$ are broadly divided into four regions of x in which the x dependence is attributed to different physical effects. These are:

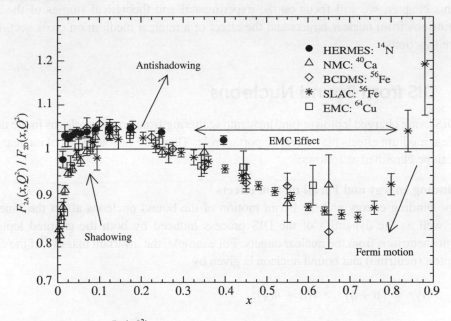

Figure 16.2 Ratio $R(x, Q^2) = \frac{F_{2A}(x,Q^2)}{F_{2D}(x,Q^2)}$; ($A$ = target nucleus) vs. x shows the nuclear medium effects in structure function. Experimental data are taken from the Refs. [783, 786, 787, 797, 784].

1. **Shadowing effect:** In the region of low $x(< 0.1)$, a suppression is found in the ratio $R(x, Q^2) = \frac{F_{2A}(x,Q^2)}{F_{2D}(x,Q^2)}$ which is known as the shadowing effect. This suppression becomes more pronounced with the increase in the mass number A.

2. **Antishadowing effect:** This is the region of $0.1 \leq x \leq 0.2$, where there is an enhancement in the ratio of structure functions ($R(x, Q^2)$); the region has been found to have almost no nuclear mass dependence.

3. **EMC effect:** The ratio $R(x, Q^2)$ shows a dip in the region of $0.2 < x < 0.7$; this is known as the EMC effect, named after the first experimental observation by the EMC collaboration [798].

4. **Fermi motion:** The nucleons bound inside the nuclear target move with some Fermi momentum which increases with the increase in the mass number. This is a kinematic effect and is responsible for the abrupt rise in the ratio of structure functions in the region of higher $x \geq 0.7$.

The discovery of the EMC effect was unexpected and quite surprising in view of the very high energy and momentum transfers involved in the DIS process which are several orders of magnitude larger than the binding energy and Fermi momentum of nucleons inside the nucleus. The possibility that the role of quarks and gluons may be modified in the nuclear environment was anticipated and discussed earlier by the nuclear physics community but the evidence of large nuclear effects as seen in the EMC experiment was surprising.

In this chapter, we will focus on the experimental and theoretical studies of the DIS of (anti)neutrinos from nuclear targets and the effect of a nuclear medium on cross sections and structure functions.

16.2 DIS from Bound Nucleons

In the DIS using charged lepton or (anti)neutrino scattering from bound nucleons in the nucleus, the nuclear medium effects play a very important role. There are various types of nuclear effects which can be classified as follows:

(i) **Binding energy and Fermi motion effects**

The binding energy and the Fermi motion of the bound nucleons affect the kinematics as well as the dynamics of the DIS process induced by both the charged lepton and (anti)neutrinos from the nuclear targets. For example, the invariant mass W of the charged lepton (neutrino) and bound nucleon is given by

$$W^2 = (p + q)^2 \geq (M + m_\pi)^2 \tag{16.1}$$

to initiate the inelastic processes. Since

$$W^2 = p^2 + q^2 + 2p \cdot q = q^2 + (M - \varepsilon)^2 + 2q_0(M - \varepsilon) - \vec{p} \cdot \vec{q}, \tag{16.2}$$

where ε is the separation energy of the bound nucleon, the kinematics of the DIS process is modified due to the nucleon target being bound. Moreover, the bound nucleons can have any momentum $\vec{p}$ lying between $0 < |\vec{p}| < p_F$, where p_F is the Fermi momentum (Chapter 14). Therefore, an integration of the cross section and structure functions over all the momentum weighted by the nucleon spectral function describing the momentum distribution is required to be performed. There are many models for the nucleon spectral function discussed in literature [799, 800, 801, 802, 803, 804, 805].

(ii) Off shell effect

The bound nucleons, that is, protons and neutrons in a nucleus are not on-shell, in other words $p^2 \neq M^2$. Therefore, the structure functions will not only depend upon the variables ν and Q^2 (or x in the scaling region) but also on the off shell virtuality (χ) of the particle measured by $\chi = \frac{p^2 - M^2}{M^2}$. Moreover, being off shell, the nucleons do not satisfy the Dirac equation. Therefore, the electromagnetic (weak) structure functions will not be described in terms of only two (three) structure functions, but more than two (three) structure functions which vanish for $p^2 = M^2$. While the corrections to the existing structure functions $F_2^{EM}(x, Q^2)$ and $F_2^{WI}(x, Q^2)$ due to $\chi \neq 0$ have been calculated by doing an expansion in the powers of χ, the additional structure functions have not been studied much in literature. However, in case of many nuclei, the virtuality is quite small; therefore, the cross section can still be described by on-shell structure functions [799, 800].

(iii) Mesonic effects

Most low energy nuclear processes are very well described in impulse approximation (IA). However, there are certain processes even in the region of low and intermediate energies, where the interaction takes place from the mesons in flight between the two bound nucleons; these are called meson exchange current (MEC) effects [806, 807, 808]. In the case of DIS, the meson cloud of nucleons are taken into account in the evaluation of the nucleon structure functions. The charged lepton and (anti)neutrinos can also scatter from mesons like π and ρ in the nuclear volume. Since these mesons are also off-shell, their meson off-shell structure functions need to be calculated in order to estimate their contributions to the nuclear structure functions. The contribution from the meson structure functions are calculated in a model similar to the model used for nucleons utilizing free meson structure functions and convoluting them with the momentum distribution of mesons [809].

(iv) Effect of short range correlations (SRC) and multiquark clusters

In the conventional picture of nuclei, the nucleons move in the central or mean field leading to the various nuclear states which describe the static properties of nuclei and nuclear transitions induced by photons, leptons, and hadrons. This also leads to an average density of the nucleus. The DIS cross section and the EMC effect seem to depend upon the average density. However, in various nuclei, especially at low A, there is evidence of the local density being higher than the average density leading to clustering. The clustering of nucleons leading to the cluster model of nuclei is well known for nuclear phenomenon at low and intermediate energies. In the region of high energy and momentum transfers, where quarks and gluons are expected to play a significant role, multiquark clustering may also be present. For example, in the DIS of electron from nuclei, there is some evidence that there exists a 20–30% probability of 6-quark clustering.

The α-cluster model description of ^{8}Be, ^{9}Be, ^{12}C, and $(\alpha - d)$ cluster model of ^{6}Li has been quite successful in describing nuclear phenomenon. This type of clustering can arise due to the short range correlations (SRC) between nucleons. Such SRC are well known leading to pairing even in intermediate and high A nuclei. Due to SRC and clustering, there would be considerable overlap of nucleons leading to quark exchange between the two nucleons. The $p - n$ correlations are known to be stronger than the $p - p$ and $n - n$ correlations, affecting the proton structure functions more than the neutron structure functions in nuclei.

Recent experiments of DIS for electrons at JLab from light nuclei like ^{4}He, ^{9}Be and ^{12}C, the observations show that the cross sections and EMC effect depend upon the local density, which is considerably higher than the average density of the nuclei.

(v) **Non-isoscalarity**

In defining the ratio $R(x, Q^2)$, the cross section per nucleon or the structure function per nucleon for nuclear targets used in the numerator corresponds to isoscalar nuclei which is obtained after applying corrections due to $N \neq Z$ in some nuclear targets like ^{56}Fe or ^{208}Pb. Similarly, in the denominator, $F_2^D = \frac{F_2^p + F_2^n}{2}$, where it is taken to be the average of structure functions for protons and neutrons. The structure function for the neutron is obtained from the DIS experiments performed on the deuterium target; deuterium corrections are then applied. In this case, the nuclear effects in the deuterium are calculated. While there are many calculations of nuclear effects in deuterium, there are no comprehensive theoretical studies of the non-isoscalarity corrections in the heavier nuclei. In the case of heavier nuclei, such corrections are calculated phenomenologically for the DIS of electrons; the same are applied in the case of neutrino scattering.

(vi) **Multiple scattering**

The DIS of charged leptons and (anti)neutrinos are understood in terms of the interaction of virtual photons and $W^\pm(Z^0)$ bosons with partons. In a nucleus, these virtual photons and $W^\pm(Z^0)$ bosons traverse the nuclear medium before scattering with the partons. During the passage, the particles may undergo hadronization, that is, fluctuate momentarily into hadrons or quark–antiquark pairs and gluons, which then undergo scattering with nuclear constituents through strong interactions leading to substantial modification in the cross sections in some kinematic regions. This is typical of a nuclear medium effect; there is considerable evidence of this type of scattering taking place in nuclei.

It is not possible to go into the details of all the nuclear medium effects described here. However, we will describe some of the effects which make dominant contribution to the structure function in the following sections, focusing on the DIS of (anti)neutrino from nuclear targets.

First, we shall discuss the phenomenological approaches which are used to extract the nuclear medium effects in the DIS of the charged lepton-, (anti)neutrino- nucleus scattering as well as the Drell–Yan processes in proton–nucleus and pion–nucleus scattering experiments.

16.3 Extraction of Structure Functions from Cross Section Measurements

The DIS scattering cross sections for the charged lepton as well as the (anti)neutrino induced processes have been measured in various experiments [810, 811, 566, 562, 561]. The structure functions are extracted from these cross section measurements, after making a suitable choice of the kinematical variables and making some reasonable assumptions about the structure functions $F_i(x, Q^2)(i = 1, 2, 3, L)$. In the case of electromagnetic interaction, the cross sections are expressed in terms of the longitudinal and transverse structure functions (Chapter 13) which are respectively described by $F_{LN}(x, Q^2)$ (Eq. (13.46)) and $F_{1N}(x, Q^2)$ (Eq. (13.45)). The separation of longitudinal and transverse cross sections is generally done by using the Rosenbluth technique [812]. This technique involves the measurements of cross sections at two (at least) or more values of photon polarization vector (ϵ) obtained at fixed (x, Q^2). A linear fit is performed as shown in Figure 16.3. In this way, one may extract $F_i(x, Q^2)(i = 1, 2, L)$

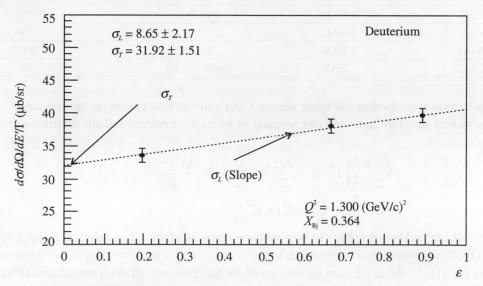

Figure 16.3 σ_L and σ_T separation using Rosenbluth technique [813].

(Eqs. (13.45) and (13.46)) and $R_L(x, Q^2)$ (Eq. (13.43)) using the cross section measurements on $\sigma_T(x, Q^2)$ and $\sigma_L(x, Q^2)$. A parameterization of $R_L(x, Q^2)$ has been given by Whitlow et al. [810] using the $e - p$ and $e - d$ scattering data from SLAC:

$$R_L(x, Q^2) = \frac{0.0635}{\ln(Q^2/0.04)}\Theta(x, Q^2) + \frac{0.5747}{Q^2} - \frac{0.3534}{Q^4 + 0.09},$$

$$\text{where} \quad \Theta(x, Q^2) = 1 + 12\left(\frac{Q^2}{1 + Q^2}\right)\left(\frac{0.125^2}{x^2 + 0.125^2}\right). \tag{16.3}$$

It is important to notice (Eq. (13.47)) that for the extraction of $F_2(x, Q^2)$, the precise measurement of $R_L(x, Q^2)$ is required which is planned to be determined in several nuclear targets in future experiments at the JLab.

In the case of (anti)neutrino scattering, the early experiments from the nuclear targets in the high energy region were done at CERN; later experiments at FNAL, BNL, ITEP, JINR, etc. were performed by using the nucleon and nuclear targets in the lower energy region (see Table 16.1). For example, in the NuTeV [566], high statistics measurements of the differential

Table 16.1 Some of the experiments which studied (anti)neutrino induced DIS cross section and structure functions from nuclear targets.

Experiment	Laboratory	Target	Energy range (GeV)
CCFR-I			30–300
CCFR-II	FNAL	Fe	30–300
CCFR-III			30–360
CDHSW	CERN	Fe	20–212
CHARM	CERN	$CaCO_3$	10–160
CHARM-II	CERN	glass	5–100
BEBC	CERN	Ne (D_2)	15–160
NuTeV	FNAL	Fe	30–360
NOMAD	CERN	C, Fe	5–200
CHORUS	CERN	Emulsion, Pb	10–200

cross section were performed using neutrino and antineutrino beams on an iron target. The extraction of structure functions can be made by using the neutrino and antineutrino scattering cross sections

$$\frac{d^2\sigma_N^{WI}}{dxdy} = \frac{G_F^2 s}{2\pi}\left[\left(1-y-\frac{Mxy}{2E_l}+\frac{y^2}{2}\frac{1+4M^2x^2/Q^2}{1+R_L}\right)F_{2N}^{WI}(x,Q^2)\right.$$
$$\left.\pm xy\left(1-\frac{y}{2}\right)F_{3N}^{WI}(x,Q^2)\right],\qquad(16.4)$$

which is obtained by using Eq. (13.66) and the modified Callan–Gross relation, Eq. (13.47). In Eq. (16.4), a positive sign corresponds to the neutrino and a negative sign to the antineutrino. Using Eq. (16.4), the expression for the sum of the neutrino and antineutrino induced DIS cross section may be written as:

$$\frac{d^2\sigma^\nu}{dx\,dy}+\frac{d^2\sigma^{\bar\nu}}{dx\,dy} = \frac{G_F^2 s}{2\pi}\left[2\left(1-y-\frac{Mxy}{2E_l}+\frac{y^2}{2}\frac{1+4M^2x^2/Q^2}{1+R_L}\right)F_{2N}^{avg}\right.$$
$$\left.+y(1-\frac{y}{2})\Delta xF_{3N}\right],\qquad(16.5)$$

where $F_{2N}^{avg}(=\frac{1}{2}[F_{2N}^\nu+F_{2N}^{\bar\nu}])$ is the average of F_{2N}^ν and $F_{2N}^{\bar\nu}$ and $\Delta xF_{3N}=[xF_{3N}^\nu-xF_{3N}^{\bar\nu}]$. At the leading order, ΔxF_{3N} is given by [566]

$$\Delta xF_{3N} = 2x(s+\bar s-c-\bar c)$$
$$= 4x(s-c)\,; \text{assuming symmetric sea, i.e. } s=\bar s,\ c=\bar c.\qquad(16.6)$$

To extract F_{2N}^{avg} from the cross section measurements (using Eq. (16.5)), ΔxF_{3N} is taken from a NLO QCD model and the input value of the ratio $R_L(x,Q^2)=\frac{\sigma_L(x,Q^2)}{\sigma_T(x,Q^2)}$ is taken from the empirical fit to the world average data for the electromagnetic induced process (using Eq. (16.3)).

By taking the difference of the neutrino and antineutrino differential scattering cross sections (Eq. (16.4)), one may write:

$$\frac{d^2\sigma^\nu}{dx\,dy} - \frac{d^2\sigma^{\bar{\nu}}}{dx\,dy} = \frac{G_F^2 s}{2\pi}\left[\left(1 - y - \frac{Mxy}{2E_l} + \frac{y^2}{2}\frac{1 + 4M^2 x^2/Q^2}{1 + R_L}\right)\Delta F_{2N}\right.$$
$$\left. + \left(y - \frac{y^2}{2}\right) x F_{3N}^{\mathrm{avg}}(x, Q^2)\right],$$

(16.7)

where $x F_{3N}^{\mathrm{avg}}(x, Q^2) = \frac{1}{2}[x F_{3N}^\nu + x F_{3N}^{\bar{\nu}}]$ and the difference ΔF_{2N} is given by [566]

$$\Delta F_{2N} = F_{2N}^\nu - F_{2N}^{\bar{\nu}} = 2(s - \bar{s} + \bar{c} - c)$$
$$= 0 \; ; \quad \text{when symmetric sea is assumed.}$$

(16.8)

ΔF_{2N} is almost negligible, therefore, Eq. (16.4) will have only the average of parity violating structure functions, that is,

$$\frac{d^2\sigma^\nu}{dx\,dy} - \frac{d^2\sigma^{\bar{\nu}}}{dx\,dy} = \frac{G_F^2 s}{2\pi}\left[y - \frac{y^2}{2}\right]\left[x F_{3N}^{\mathrm{avg}}(x, Q^2)\right]$$

(16.9)

From the cross section measurements, using Eq. (16.9), $x F_{3N}^{\mathrm{avg}}(x, Q^2)$ is obtained.

Hence, by using $l^{\pm}$ and $\nu_l/\bar{\nu}_l$ cross sections measurements, one may extract the structure functions following the methods described earlier.

16.4 Phenomenological Study

Various attempts have been made to understand the nuclear medium effects phenomenologically [814, 815, 816, 817, 818, 819, 820, 821]. Such studies have been made to obtain a nuclear correction factor by doing the analysis of the experimental data on EMC effect from the charged lepton–nucleus scattering, (anti)neutrino–nucleus scattering, and Drell–Yan processes from pion–nucleus scattering, and proton–nucleus scattering. The DIS region has been explored by using the charged lepton beam and the (anti)neutrino beam via the scattering experiments. Neutrinos have an edge over the charged lepton because of their ability to interact with particular quark flavors which would help to understand the partons distribution inside the target nucleon. Another important aspect is that nuclear medium modifications for the weak structure functions may be different from the electromagnetic structure function [822]. The early pioneers in these studies were the experiments with the bubble chambers (BC) led by the Gargamelle heavy liquid bubble chamber (normal filling of CF_3Br). This was followed by the experiments with the smaller ANL and BNL BC as well as the much larger BEBC at CERN and the 15' BC at FNAL filled with hydrogen, deuterium as well as various heavier nuclei such as Ne and propane. With the ANL, BNL, BEBC, and the 15' BC, the use of proton and deuterium targets offered an ideal tool to probe the structure of the free nucleon via the flavor separation offered by the weak charged current. However, the overall statistics was quite limited and insufficient. The first higher statistics ν and $\bar{\nu}$ nucleus measurements were performed by massive nuclear target detectors like CDHS(W)–iron and CHARM/CHARM II–marble/glass. These early experiments were followed by the CCFR and NuTeV–iron experiments and the CHORUS–lead experiments. As opposed to the high resolution of the earlier low statistics bubble chamber experiments, most of these experimental measurements from heavy nuclear targets could not resolve details of the hadronic shower and concentrated on the inclusive ν and $\bar{\nu}$ cross section measurements [608].

The goal of combining many DIS experimental results on heavy nuclei ranging from ^{12}C to ^{208}Pb was not considered to be a problem if the PDFs (parton distribution function) of the nucleons bound inside the nucleus were assumed to be the same as the free nucleon PDFs. But after the discovery of the EMC effect, it was realized that due to the nuclear environment, the structure functions of a bound nucleon are different from the free nucleon structure functions (see Figure 16.2). Currently, the analyses of both free and bound nucleons are based on the same factorization theorems [823, 824] used in the case of nucleons that do not in any way consider the nuclear environment. However, the nuclear PDFs must account for the nuclear medium effects like shadowing, antishadowing, and the EMC effect at the leading twist. The PDFs of a free proton are well studied with several global analyses being regularly updated [825]. The partonic structure of the bound nucleons must reflect the nuclear environment and, consequently, the nucleus cannot simply be considered as an ensemble of Z free protons PDFs and (A–Z) free neutron PDFs. Nuclear PDFs have been determined by several groups [817, 816, 826, 815] using global fits to experimental data that include, mainly, deep inelastic scattering and Drell–Yan lepton pair production on nuclei.

In the phenomenological analyses of structure functions, the general approach is that these nuclear PDFs are obtained using the charged lepton–nucleus scattering data. The ratios of the structure functions, for example, $\frac{F_{2A}}{F_{2A'}}$, $\frac{F_{2A}}{F_{2D}}$ are analyzed, where A, A' represent any two nuclei and D stands for the deuteron, to determine the nuclear correction factor $R_i(x, Q, A)$. The nuclear correction factor is then multiplied with free nucleon PDFs to get nuclear PDFs, that is,

$$f_i^{(p/A)}(x, Q) = R_i(x, Q, A) f_i^{\text{free proton}}(x, Q).$$

While determining the nuclear correction factor, the information regarding nuclear modification is also obtained from the Drell–Yan cross section ratio like $\frac{\sigma_{pA}^{DY}}{\sigma_{pD}^{DY}}$, $\frac{\sigma_{pA}^{DY}}{\sigma_{pA'}^{DY}}$, where p stands for proton beam. Furthermore, the information about the nuclear correction factor is also supplemented by high energy reaction data from the experiments at LHC, RHIC, etc. This approach has been used by Hirai et al. [815], Eskola et al. [816], Bodek and Yang [827], de Florian and Sassot [817], and others (Table 16.2). The same nuclear correction factor is taken for the weak DIS processes. In a recent analysis, de Florian et al. [817] have analyzed ν-A DIS data, and the data from charged lepton-nucleus scattering and Drell-Yan processes. Their [817] conclusion is that the same nuclear correction factor can describe the nuclear medium effect in $l^{\pm}$-A and ν-A DIS processes.

In another approach, nuclear PDFs are directly obtained by analyzing the experimental data, that is, without using nucleon PDFs or the nuclear correction factor. This approach has

Table 16.2 The developments in the global DGLAP analysis of nPDFs since 1998. DIS: deep inelastic scattering; DY: Drell–Yan di-lepton production; nFFs: nuclear fragmentation functions.

Phenomenological group	Data types used
EKS98 [828, 829]	$l+A$ DIS, p+A DY
HKM [830]	$l+A$ DIS
HKN04 [831]	$l+A$ DIS, p+A DY
nDS [814]	$l+A$ DIS, p-A DY
EKPS [832]	$l+A$ DIS, p+A DY
HKN07 [815]	$l+A$ DIS, p+A DY
EPS08 [833]	$l+A$ DIS, p+A DY, $h^{\pm}, \pi^0, \pi^{\pm}$ in d+Au
EPS09 [816]	$l+A$ DIS, p+A DY, π^0 in d+Au
nCTEQ [826, 834]	$l+A$ DIS, p+A DY
nCTEQ [835]	$l+A$ and $\nu+A$ DIS, p+A DY
DSSZ [817]	$l+A$ and $\nu+A$ DIS, p+A DY, $\pi^0, \pi^{\pm}$ in d+Au, computed with nFFs

been recently used by the nCTEQ [819, 818] group in getting $F_{2A}^{\text{EM}}(x, Q^2)$, $F_{2A}^{\text{WI}}(x, Q^2)$, and $F_{3A}^{\text{WI}}(x, Q^2)$, by analyzing the charged lepton-A DIS data and DY p-A data sets together, and the $\nu(\bar{\nu}) - A$ DIS data sets separately. Their observation is that the nuclear medium effects in $F_{2A}^{\text{EM}}(x, Q^2)$ in electromagnetic interactions are different from $F_{2A}^{\text{WI}}(x, Q^2)$ in weak interactions in the region of low x. Thus, in this region, there is a disagreement between the observation of de Florian and Sassot [817] and Kovarik et al. [817, 819].

The study of nuclear medium effects due to modifications of sea quark distributions, nuclear shadowing, and mesonic contributions are important in this region.

In the nCTEQ framework, the parton distributions of the nucleus are constructed as:

$$f_i^{(A,Z)}(x, Q) = \frac{Z}{A} f_i^{p/A}(x, Q) + \frac{A-Z}{A} f_i^{n/A}(x, Q), \tag{16.10}$$

where Z is the number of protons and A is the number of protons plus neutrons in the nucleus. Isospin symmetry is used to construct the PDFs of a bound neutron, $f_i^{n/A}(x, Q)$, from those of the proton by exchanging up and down quark distributions.

The parameterization of the individual parton distributions at the input scale Q_0 are similar in form to that used in the free proton CTEQ fits [836], and takes the following form:

$$x f_i^{p/A}(x, Q_0) = c_0 x^{c_1}(1 - x)^{c_2} e^{c_3 x}(1 + e^{c_4} x)^{c_5},$$
$$\text{for} \quad i = u_v, d_v, g, \bar{u} + \bar{d}, s + \bar{s}, s - \bar{s},$$
$$\frac{\bar{d}(x, Q_0)}{\bar{u}(x, Q_0)} = c_0 x^{c_1}(1 - x)^{c_2} + (1 + c_3 x)(1 - x)^{c_4}. \tag{16.11}$$

The input scale is chosen to be the same as for the free proton fits [836], namely $Q_0 = 1.3$ GeV and the DGLAP equation is used to evolve to higher Q. Figure 16.4, shows the A-dependence of the nCTEQ bound proton PDFs at the scale $Q = 10$ GeV for a range of nuclei from the free proton ($A = 1$) to lead ($A = 208$). Large variation in the quark distribution function may be observed when one goes from lighter targets to very heavy nuclear targets.

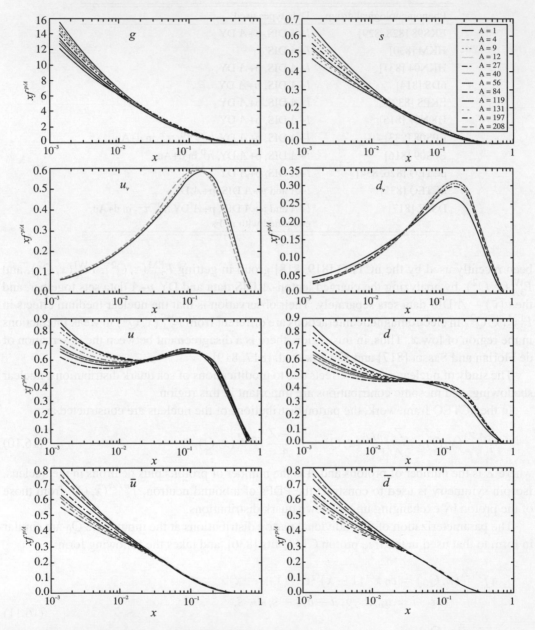

Figure 16.4 The A-dependence of the nCTEQ bound proton PDFs at the scale $Q = 10$ GeV for a range of nuclei from the free proton ($A = 1$) to lead ($A = 208$) [819].

To understand the variation in different phenomenological studies, the nuclear correction factor for the structure function $F_2(x, Q^2)$ in neutrino and antineutrino scattering from ^{56}Fe for $Q^2 = 5$ and 20 GeV^2 has been shown in Figure 16.5. The results are presented from the nCTEQ analysis of NuTeV data [566, 837], the correction factor from HKN07 [815], SLAC/NMC [837] parameterization, and the theoretical work of Kulagin and Petti [800]. The figure has been taken from Ref. [837]. Large variation in the correction factor may be observed.

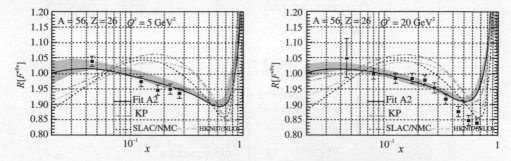

Figure 16.5 Nuclear correction factor R for the structure function F_2 in neutrino and antineutrino scattering from Fe for $Q^2 = 5$ and 20 GeV^2. The solid curve shows the result of the nCTEQ analysis of NuTeV data [566, 837], the uncertainty from the fit is represented by the shaded band. The dashed-double dotted line represents the correction factor from HKN07 [815], the dashed line represents the results for the SLAC/NMC [837] parameterization and the dashed-dotted line represents the results of Kulagin and Petti [800]. The figure has been taken from Ref. [837].

Recently, the MINERvA collaboration [627] has presented their results for the cross section ratios

$$\frac{\left(\frac{d\sigma}{dx}\right)_i}{\left(\frac{d\sigma}{dx}\right)_{CH}}, \quad i = {}^{12}C, {}^{56}Fe \text{ and } {}^{208}Pb; \text{ where CH is a hydrocarbon,}$$

as a function of x in several nuclear targets, and compared their data with different phenomenological analysis available in the literature, for example, from GENIE [606] Monte Carlo generator which uses Bodek and Yang parameterization of PDFs [827] and also with the parameterization of Cloet et al. [838]. Bodek and Yang use A-dependent parametrization of the x dependent effects based on charged lepton scattering data and Cloet et al. [838] prescription is based on the convolution of Nambu–Jona–Lasinio nuclear wave function with free nucleon valence PDFs. It may be observed from Figure 16.6 that there is a need to have a better understanding of nuclear effects in the weak interaction induced DIS processes.

In Figure 16.7, the results of differential scattering cross section are presented at neutrino energy of 65 GeV by using the recent nuclear PDFs parameterizations of nCTEQ15 [819] and nCTEQnu [839]. These nuclear PDFs, that is, nCTEQ15 and nCTEQnu are respectively obtained by analyzing the experimental data of charged lepton–nucleus and neutrino–nucleus induced processes. From the figure, it may be observed that the results of charged lepton based nPDFs and neutrino based nPDFs differ from each other in the region of low and high x albeit the difference is small.

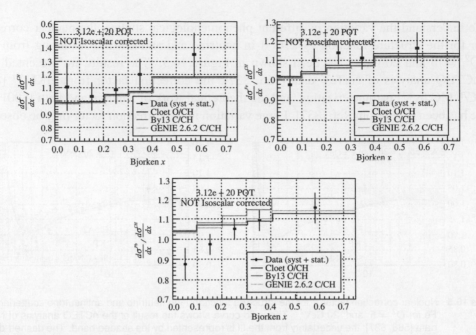

Figure 16.6 DIS cross section ratios as a function of x for MINERvA data points and various parameterizations of x-dependent nuclear effects [606, 827, 838]. The error bars on the data are the combined statistical and systematic uncertainties.

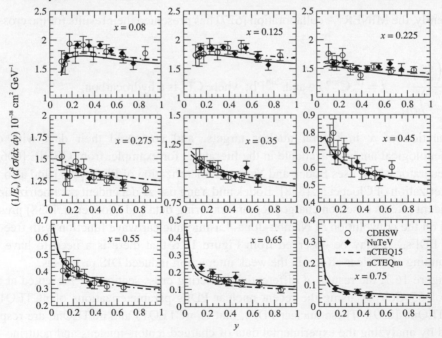

Figure 16.7 Differential scattering cross section $\frac{1}{E_\nu} \frac{d^2\sigma}{dx\,dy}$ vs. y, at different x for $\nu_\mu - {}^{56}Fe$ DIS process at $E_\nu = 65$ GeV [566, 561, 819, 839].

In the next section, different theoretical attempts made to understand the nuclear medium effects are discussed in brief; a field theoretical prescription for taking into account nuclear medium effects which is based on the model developed by the Aligarh–Valencia group is discussed in some detail.

16.5 Theoretical Study

Theoretically, many models have been proposed to study these effects on the basis of nuclear binding, nuclear medium modifications including short range correlations in nuclei [840, 841, 842, 802, 843, 844, 845, 846, 809, 847, 848, 799, 849, 800, 850, 851, 852, 853, 854, 855, 856, 857, 858, 859, 860, 861, 862, 863], pion excess in nuclei [842, 843, 809, 858, 859, 860], multi-quark clusters [861, 862, 863], dynamical rescaling [864, 865], nuclear shadowing [866, 867], etc. In spite of these efforts, no comprehensive theoretical/phenomenological understanding of the nuclear modifications of the bound nucleon structure functions across the complete range of x and Q^2 consistent with the presently available experimental data exists [844, 868, 845, 846].

16.5.1 Deep inelastic charged leptons and (anti)neutrino scattering from nuclei

We have defined the cross sections for the deep inelastic scattering of charged leptons and (anti)neutrinos from nucleon targets in Chapter 13 (see Eqs. (13.18) and (13.63)) in terms of leptonic ($L_{\mu\nu}$) and hadronic ($W_N^{\mu\nu}$) tensors. In a similar way, we define the cross sections for the DIS charged lepton and charged current (anti)neutrino scattering from nuclear targets, that is, $\frac{d^2\sigma^{l\pm}}{d\Omega_l' dE_l'}$ and $\frac{d^2\sigma_N^{\nu(\bar{\nu})}}{d\Omega_l' dE_l'}$.

$$\frac{d^2\sigma_A^j}{d\Omega_l' dE_l'} = C_j \times \frac{|\vec{k}'|}{|\vec{k}|} L_{\mu\nu}^j W_{A,j}^{\mu\nu}, \tag{16.12}$$

where j stands for either the electromagnetic or weak interaction channel and the constant factor $C_j = \frac{\alpha^2}{Q^4}$ for the electromagnetic interaction and $C_j = \frac{G_F^2}{(2\pi)^2}\left(\frac{M_W^2}{M_W^2 + Q^2}\right)^2$ for the weak interaction induced processes. $L_{\mu\nu}^j$ is the leptonic tensor as given in Eqs. (13.4) and (13.60) for the corresponding interaction channels. $W_{A,j}^{\mu\nu}$ is the nuclear hadronic tensor which is written in terms of the nuclear structure functions as:

$$W_{A,WI}^{\mu\nu} = \left(\frac{q^\mu q^\nu}{q^2} - g^{\mu\nu}\right) W_{1A}^{WI}(\nu, Q^2) + \frac{W_{2A}^{WI}(\nu, Q^2)}{M_A^2}\left(p_A^\mu - \frac{p_A \cdot q}{q^2} q^\mu\right)\left(p_A^\nu - \frac{p_A \cdot q}{q^2} q^\nu\right)$$
$$- \frac{i}{2M_A^2}\epsilon^{\mu\nu\rho\sigma} p_{A\rho} q_\sigma W_{3A}^{WI}(\nu, Q^2). \tag{16.13}$$

for the (anti)neutrino induced processes that deal with the parity violation. For the charged lepton induced processes, it is given in terms of only two nuclear structure functions

$$W_{A,EM}^{\mu\nu} = \left(\frac{q^\mu q^\nu}{q^2} - g^{\mu\nu} \right) W_{1A}^{EM}(\nu, Q^2) + \left(p_A^\mu - \frac{p_A \cdot q}{q^2} q^\mu \right) \left(p_A^\nu - \frac{p_A \cdot q}{q^2} q^\nu \right) \frac{W_{2A}^{EM}(\nu, Q^2)}{M_A^2},$$

(16.14)

because $W_{3A}(\nu, Q^2) = 0$ due to parity invariance. The contraction of leptonic and hadronic tensors simplifies the expression of differential scattering cross section to the following forms:

$$\frac{d^2\sigma_A^{WI}}{d\Omega_l' dE_l'} = \frac{G_F^2 E_l'^2}{2\pi^2} \left(\frac{M_W^2}{M_W^2 + Q^2} \right)^2 \left[2W_{1A}^{WI}(\nu, Q^2) \sin^2 \frac{\theta}{2} + W_{2A}^{WI}(\nu, Q^2) \cos^2 \frac{\theta}{2} \right.$$

$$\left. \pm W_{3A}^{WI}(\nu, Q^2) \left(\frac{E_l + E_l'}{M_A} \right) \sin^2 \frac{\theta}{2} \right], \quad (16.15)$$

and

$$\frac{d^2\sigma_A^{EM}}{d\Omega_l' dE_l'} = \frac{4\alpha^2 E_l'^2}{Q^4} \left[2W_{1A}^{EM}(\nu, Q^2) \sin^2 \frac{\theta}{2} + W_{2A}^{EM}(\nu, Q^2) \cos^2 \frac{\theta}{2} \right].$$

(16.16)

In Eq. (16.15), the positive sign is for the neutrino and the negative sign is for the antineutrino induced process.

The weak nuclear structure functions $W_{iA}^{WI}(\nu, Q^2)$ ($i = 1, 2, 3$), are defined in terms of the dimensionless nuclear structure functions $F_{iA}^{WI}(x, Q^2)$ as:

$$\left. \begin{array}{l} M_A W_{1A}^{WI}(\nu_A, Q^2) = F_{1A}^{WI}(x_A, Q^2), \\ \nu_A W_{2A}^{WI}(\nu_A, Q^2) = F_{2A}^{WI}(x_A, Q^2), \\ \nu_A W_{3A}^{WI}(\nu_A, Q^2) = F_{3A}^{WI}(x_A, Q^2), \end{array} \right\}$$

(16.17)

where the energy transfer $\nu_A = \frac{p_A \cdot q}{M_A} = q^0$.

In Subsection 16.5.1, a general idea of nuclear medium effects such as Fermi motion, Pauli blocking, and nuclear binding of nucleons is discussed in brief.

Nuclear binding, Pauli blocking, and Fermi motion of nucleons

The general approach to calculate the DIS cross section is based on an impulse approximation in which the nuclear cross section is the incoherent sum of scattering from neutrons and protons. Additional contribution due to off-shell effects and other degrees of freedom like mesons, resonances, quark clusters, or gluons are calculated as corrections. In this approach, the DIS process is considered as a two-step process, in which $W^\pm(Z^0)$ bosons or virtual photons scatter from the partons in the bound hadrons in the nucleus, that is, nucleons, mesons, etc. described by the structure function of the hadron and then convoluted with the momentum distribution of struck hadrons inside the nucleus. The effect of nuclear binding and Fermi motion and the nucleon–nucleon correlation can be included in the momentum distribution, described by the spectral function of nucleons, or the Fermi smearing function in the momentum space depending upon what nuclear effects are included in deriving the spectral functions. In this approach, the nuclear structure functions ($F_i^A(x, Q^2)$) can be formally written as:

$$F_i^A(x, Q^2) = \sum_C \int \frac{d^4 p}{(2\pi)^4} S_C(E, \vec{p}) F_i^C(x, Q^2, p),\tag{16.18}$$

where $i = 1, 2$ in the case of charged lepton and $i = 1, 2, 3$ in case of (anti)neutrino DIS from nucleons. In Eq. (16.18), the summation is performed over the cluster C which could be protons, neutrons, mesons like π and ρ or quark clusters. $S_C(E, \vec{p})$ is the spectral function of the cluster C, with energy E and momentum $\vec{p}$ in the nucleus. The integration is performed over all the possible momenta of the cluster subject to the energy momentum conservation of the cluster inside the nucleus. Explicit expressions for the nuclear structure functions $F_i^A(x, Q^2)$ are derived starting from the relation between the nuclear hadronic tensor $W_{\mu\nu}^A(x, Q^2)$ and nucleon hadronic tensor $W_{\mu\nu}^N(x, Q^2)$ for electromagnetic and weak interactions, which will be discussed later in the text.

Over the last 35 years, there have been various models proposed to calculate nuclear structure functions. It is not possible to describe all of them here and our present discussions will be limited to two models which have been recently used to study the nuclear medium effects in neutrino and antineutrino scattering from nuclear targets. First, we shall discuss the Aligarh–Valencia model [809, 850, 851, 852, 854, 855, 869, 822, 870, 871] and then, in brief, the Kulagin and Petti model [800, 799].

16.5.2 Aligarh–Valencia model

In this model, the nuclear hadronic tensor $W_{\mu\nu}^A$ is evaluated using local density approximation. To calculate the scattering cross section for a neutrino interacting with a target nucleon in the nuclear medium, we start with a flux of neutrinos hitting a collection of target nucleons over a given length of time. A majority of neutrinos will simply pass through the target without interacting while a certain fraction will interact with the target nucleons leaving the pass-through fraction and entering the fraction of neutrinos yielding final state leptons and hadrons. Here, we will introduce the concept of 'neutrino self-energy' that has a real and imaginary part. The real part modifies the lepton mass (it is similar to the delta mass or nucleon mass modified in the nuclear medium), while the imaginary part is related to this fraction of interacting neutrinos and gives the total number of neutrinos that have participated in the interactions that give rise to the charged leptons and hadrons. When a neutrino interacts with a potential provided by a nucleus (in the present scenario), then the interaction in the language of many body field theory, can be understood as the modification of the fermion two points function represented by the diagrams shown in Figure 16.8.

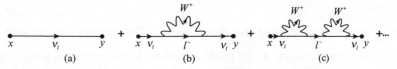

Figure 16.8 Representation of neutrino self-energy.

The first diagram (a) in Figure 16.8 is just the free field fermion propagator. The diagrams, Figure 16.8 (b) and (c) contribute to the neutrino self-energy in the lowest order and higher order (Chapter 4), respectively. Using the Feynman rules, the self energy $\Sigma(k)$ in the lowest order may be written as (Figure 16.9):

$$-i\Sigma(k) = \sum_{\text{spins}} \int \frac{d^4k'}{(2\pi)^4} \left(-\frac{ig}{2\sqrt{2}}\gamma^\mu(1-\gamma_5)\right) \frac{i(\slashed{k}'+m_l)}{k'^2 - m_l^2 + i\epsilon} \times$$

$$\left(-\frac{ig}{2\sqrt{2}}\gamma^\nu(1-\gamma_5)\right) \frac{-ig_{\mu\nu}}{(k-k')^2 - M_W^2 + i\epsilon}. \tag{16.19}$$

Notice that Σ has real and imaginary parts. The imaginary part of the neutrino self-energy accounts for the depletion of the initial neutrinos flux out of the non-interacting channel, into the quasielastic or the inelastic channels.

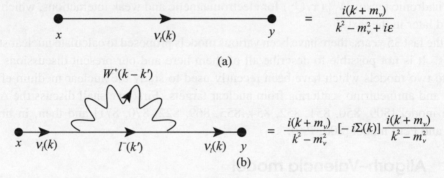

Figure 16.9 (a) Free field fermion propagator. (b) The term that contributes to the neutrino self-energy in the lowest order.

When the neutrinos interact with the nucleus, the virtual W^+ in Figure 16.9, creates particle–hole (p,h) excitations of the type $(1p, 1h)$, $(1\Delta, 1h)$, ... or (X, h) leading to quasielastic, inelastic Δ production or DIS with production of a set of particles X. The (X, h) excitations as shown in Figure 16.10 give rise to reactions like $\nu_l(\bar\nu_l) + A \to l^-(l^+) + X$. The neutrino self-energy is then evaluated corresponding to the diagram shown in Figure 16.10. The cross section for an element of volume dV in the rest frame of the nucleus is related to the probability per unit time (Γ) of the ν_l interacting with a nucleon bound inside a nucleus. The quantity $\Gamma dt dS$ provides probability times a differential of area (dS) which is nothing but the cross section ($d\sigma$) [809], that is,

$$d\sigma = \Gamma dt ds = \Gamma \frac{dt}{dl} ds dl = \Gamma \frac{1}{v} dV = \Gamma \frac{E_l}{|\vec{k}|} d^3r, \tag{16.20}$$

where $v \left(= \frac{|\vec{k}|}{E_l}\right)$ is the velocity of the incoming ν_l. The probability per unit time of the interaction of ν_l with the nucleons in the nuclear medium to give the final state is related to the imaginary part of the ν_l self-energy as [809]:

$$-\frac{\Gamma}{2} = \frac{m_\nu}{E_\nu(\vec{k})} \, \text{Im}\Sigma(k), \tag{16.21}$$

where $\Sigma(k)$ is the neutrino self-energy (shown in Figure 16.10 (a) and (b)). By using Eq. (16.21) in Eq. (16.20), we obtain:

$$d\sigma = \frac{-2m_\nu}{|\vec{k}|} \text{Im}\Sigma(k)d^3r. \tag{16.22}$$

$\Sigma(k)$ is calculated from Figure 16.10, that is,

$$\Sigma(k) = \frac{iG_F}{\sqrt{2}} \int \frac{d^4q}{(2\pi)^4} \frac{4L_{\mu\nu}^{\text{WI}}}{m_l} \frac{1}{(k'^2 - m_l^2 + i\epsilon)} \left(\frac{M_W}{q^2 - M_W^2}\right)^2 \Pi^{\mu\nu}(q), \tag{16.23}$$

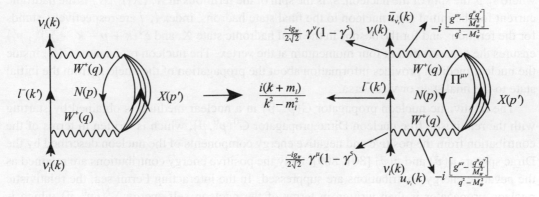

Figure 16.10 Diagrammatic representation of the neutrino self-energy.

where we have used the relation $\frac{g_W^2}{8M_W^2} = \frac{G_F}{\sqrt{2}}$ and the properties of gamma matrices. The imaginary part of the neutrino self-energy may be obtained by following the Cutkowsky rules [872] as:

$$\text{Im}\Sigma(k) = \frac{G_F}{\sqrt{2}} \frac{4}{m_\nu} \int \frac{d^4k'}{(2\pi)^4} \frac{\pi}{E(\vec{k'})} \theta(q^0) \left(\frac{M_W}{Q^2 + M_W^2}\right)^2 \text{Im}\left[L_{\mu\nu}\Pi^{\mu\nu}(q)\right]. \tag{16.24}$$

In this expression, $\Pi^{\mu\nu}(q)$ is the W boson self-energy, which is written in terms of the nucleon (G_l) and meson (D_j) propagators (depicted in Figure 16.11) following the Feynman rules and is given by:

$$\Pi^{\mu\nu}(q) = \left(\frac{G_F M_W^2}{\sqrt{2}}\right) \times \int \frac{d^4p}{(2\pi)^4} G(p) \sum_X \sum_{s_p, s_l} \prod_{i=1}^N \int \frac{d^4p_i'}{(2\pi)^4} \prod_l G_l(p_l') \prod_j D_j(p_j')$$

$$\langle X|J^\mu|N\rangle\langle X|J^\nu|N\rangle^* (2\pi)^4 \, \delta^4(k + p - k' - \sum_{i=1}^N p_i'), \tag{16.25}$$

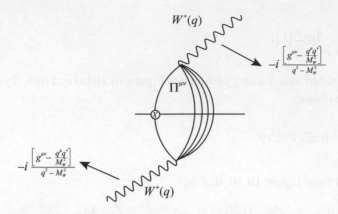

Figure 16.11 Diagrammatic representation of intermediate vector boson W self-energy.

where s_p is the spin of the nucleon, s_l is the spin of the fermions in X, $\langle X|J^\mu|N\rangle$ is the hadronic current for the initial state nucleon to the final state hadrons, index l, j are respectively, stands for the fermions and for the bosons in the final hadronic state X, and $\delta^4(k + p - k' - \sum_{i=1}^N p'_i)$ ensures the conservation of four momentum at the vertex. The nucleon propagator $G(p)$ inside the nuclear medium provides information about the propagation of the nucleon from the initial state to the final state or vice versa.

The relativistic nucleon propagator $G(p^0, \vec{p})$ in a nuclear medium is obtained by starting with the relativistic free nucleon Dirac propagator $G^0(p^0, \vec{p})$, which is written in terms of the contribution from the positive and negative energy components of the nucleon described by the Dirac spinors $u(\vec{p})$ and $v(\vec{p})$ [801, 809]. Only the positive energy contributions are retained as the negative energy contributions are suppressed. In the interacting Fermi sea, the relativistic nucleon propagator is then written in terms of the nucleon self-energy $\Sigma^N(p^0, \vec{p})$ which is shown in Figure 16.12. In the nuclear many body technique, the quantity that contains all the information on single nucleon properties in the nuclear medium is the nucleon self-energy $\Sigma^N(p^0, \vec{p})$. For an interacting Fermi sea, the relativistic nucleon propagator is written in terms of the nucleon self-energy and in nuclear matter, the interaction is taken into account through Dyson series expansion. The Dyson series expansion may be understood as the quantum field theoretical analog of the Lippmann–Schwinger equation for the dressed nucleons, which is in principle an infinite series in perturbation theory. This perturbative expansion is summed in a ladder approximation and we will now illustrate the methods.

Let us start with the expression of relativistic free nucleon Dirac propagator $G^0(p)$ given by

$$G^0(p) = \frac{1}{\not{p} - M + i\epsilon} = \frac{\not{p} + M}{(p^2 - M^2 + i\epsilon)} \tag{16.26}$$

which may be rewritten in terms of positive and negative energy states as:

$$G^0(p) = \frac{M}{E_N(\vec{p})} \left\{ \frac{\sum_r u_r(\vec{p})\bar{u}_r(\vec{p})}{p^0 - E_N(\vec{p}) + i\epsilon} + \frac{\sum_r v_r(-\vec{p})\bar{v}_r(-\vec{p})}{p^0 + E_N(\vec{p}) - i\epsilon} \right\}, \tag{16.27}$$

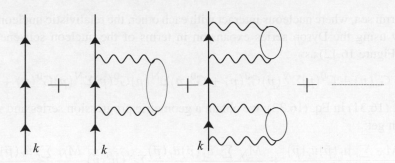

Figure 16.12 Diagrammatic representation of nucleon self-energy in the nuclear medium.

where $\sum_{r=1}^{2} u_r(\vec{p})\bar{u}_r(\vec{p}) = \frac{\not{p}+M}{2M}$, $\sum_{r=1}^{2} v_r(-\vec{p})\bar{v}_r(-\vec{p}) = \frac{\not{p}-M}{2M}$, M is the nucleon mass, $E_N(\vec{p}) = \sqrt{\vec{p}^2 + M^2}$ is the relativistic energy of an on-shell nucleon, and $\bar{u}_r(\vec{p})\,u_r(\vec{p}) = 1$. Notice that we are using a different normalization than the one discussed in Chapter 2. This has been chosen for convenience.

The Dirac propagator in a non-interacting Fermi sea may also be written as

$$G^0(p) = \frac{M}{E_N(\vec{p})} \left\{ \sum_r u_r(\vec{p})\bar{u}_r(\vec{p}) \left[\frac{1 - n(\vec{p})}{p^0 - E_N(\vec{p}) + i\epsilon} + \frac{n(\vec{p})}{p^0 - E_N(\vec{p}) - i\epsilon} \right] \right.$$
$$\left. + \frac{\sum_r v_r(-\vec{p})\bar{v}_r(-\vec{p})}{p^0 + E_N(\vec{p}) - i\epsilon} \right\}, \tag{16.28}$$

where $n(\vec{p})$ is the occupation number of a nucleon lying within the Fermi sea such that

$$n(\vec{p}) = \begin{cases} 1 & \text{for } \vec{p} \le \vec{p}_{F_N} \\ 0 & \text{for } \vec{p} > \vec{p}_{F_N} \end{cases} \tag{16.29}$$

with $\vec{p}_{F_N}$ as the Fermi momentum of the nucleon.

The nucleon propagator retains the contribution only from the positive energy components because the negative energy components are much suppressed. Hence, Eq. (16.28) reduces to

$$G^0(p) = \frac{M}{E_N(\vec{p})} \sum_r u_r(\vec{p})\bar{u}_r(\vec{p}) \left[\frac{1 - n(\vec{p})}{p^0 - E_N(\vec{p}) + i\epsilon} + \frac{n(\vec{p})}{p^0 - E_N(\vec{p}) - i\epsilon} \right]. \tag{16.30}$$

The first term of the nucleon propagator within the square bracket contributes when the momentum of the nucleon will be greater or equal to the Fermi momentum $\vec{p} \ge \vec{p}_F$, that is, for the particles above the Fermi sea while the second term within the square bracket contributes when the nucleon momentum will be less than the Fermi momentum $\vec{p} < \vec{p}_F$, that is, for the particles below the Fermi sea. This representation is known as the Lehmann's representation. With further simplification of Eq. (16.30), we may write:

$$G^0(p) = \frac{\not{p} + M}{p^2 - M^2 + i\epsilon} + 2\,i\pi\theta(p^0)\delta(p^2 - M^2)n(\vec{p})(\not{p} + M). \tag{16.31}$$

Inside the Fermi sea, where nucleons interact with each other, the relativistic nucleon propagator is written by using the Dyson series expansion in terms of the nucleon self-energy $\Sigma^N(p)$ (depicted in Figure 16.12) as:

$$G(p) \;=\; G^0(p) + G^0(p)\Sigma^N(p)G^0(p) + G^0(p)\Sigma^N(p)G^0(p)\Sigma^N(p)G^0(p) + \dots . \text{(16.32)}$$

By using Eq. (16.31) in Eq. (16.32), we obtain a geometric progression series and with further simplification get:

$$G(p) = \frac{M_N}{E(\vec{p})} \frac{\sum_r u_r(\vec{p})\bar{u}_r(\vec{p})}{p^0 - E(\vec{p})} + \frac{M_N}{E(\vec{p})} \frac{\sum_r u_r(\vec{p})\bar{u}_r(\vec{p})}{p^0 - E(\vec{p})} \Sigma^N(p^0,\vec{p}) \frac{M_N}{E(\vec{p})} \frac{\sum_r u_r(\vec{p})\bar{u}_r(\vec{p})}{p^0 - E(\vec{p})} + \dots$$

$$= \frac{M_N}{E(\vec{p})} \frac{\sum_r u_r(\vec{p})\bar{u}_r(\vec{p})}{p^0 - E(\vec{p}) - \Sigma^N(p^0,\vec{p})\frac{M_N}{E(\vec{p})}}. \tag{16.33}$$

Notice that $\Sigma^N(p^0,\vec{p})$ has a finite imaginary part, and therefore, we may write:

$$\Sigma^N(p^0,\vec{p}) = \text{Re}\{\Sigma^N(p^0,\vec{p})\} + i\text{Im}\{\Sigma^N(p^0,\vec{p})\}. \tag{16.34}$$

The real part of the self-energy is related to the effective mass while the imaginary part is related to the lifetime of the particle ($\text{Im}\Sigma^N = \frac{1}{\tau} = \Gamma$) which allows us to write Eq. (16.33) as:

$$G(p) = \frac{M}{E_N(\vec{p})} \sum_r u_r(\vec{p})\bar{u}_r(\vec{p}) \left[\frac{\{p^0 - E_N(\vec{p}) - \frac{M}{E_N(\vec{p})}\text{Re}(\Sigma^N)\} + i\{\frac{M}{E_N(\vec{p})}\text{Im}(\Sigma^N)\}}{\{p^0 - E_N(\vec{p}) - \frac{M}{E_N(\vec{p})}\text{Re}(\Sigma^N)\}^2 + \{\frac{M}{E_N(\vec{p})}\text{Im}(\Sigma^N)\}^2} \right]. \tag{16.35}$$

The nucleon self-energy $\Sigma^N(p^0,\vec{p})$ is spin diagonal, that is, $\Sigma^N_{\alpha\beta}(p^0,\vec{p}) = \Sigma^N(p^0,\vec{p})\delta_{\alpha\beta}$, where α and β are spinorial indices. The inputs required for the NN interaction are incorporated by relating them to the experimental elastic NN cross section. Furthermore, the RPA-correlation effect (described in Chapter 14) is taken into account using the spin–isospin effective interaction as the dominating part of the particle–hole (ph) interaction. Using the modified expression for the nucleon self-energy, the imaginary part of it is obtained. Due to the RPA effect, the imaginary part of the nucleon self-energy is quenched, especially at low energies and high densities. This quenching depends on the nucleon energy p^0 as well as nucleon momentum $\vec{p}$ in the interacting Fermi sea. The imaginary part of the nucleon self-energy fulfills the low-density theorem.

The dressed nucleon propagator $G(p)$ in a nuclear medium is then expressed as [809]:

$$G(p) = \frac{M}{E_N(\vec{p})} \sum_r u_r(\vec{p})\bar{u}_r(\vec{p}) \left[\int_{-\infty}^{\mu} d\omega \frac{S_h(\omega,\vec{p})}{p^0 - \omega - i\epsilon} + \int_{\mu}^{\infty} d\omega \frac{S_p(\omega,\vec{p})}{p^0 - \omega + i\epsilon} \right], \tag{16.36}$$

where chemical potential μ is given by

$$\mu = \frac{p_F^2}{2M} + \text{Re}\Sigma^N \left[\frac{p_F^2}{2M}, p_F \right].$$

ω is the removal energy, $S_h(\omega, \vec{p})$ and $S_p(\omega, \vec{p})$ are the hole and particle spectral functions, respectively, that allow us to know about the probability distribution of finding a nucleon with removal energy ω and three momentum $\vec{p}$ inside the nucleus. Thus, in the aforementioned expression, $S_h(\omega, \vec{p}) \, d\omega$ is basically the joint probability of removing a nucleon from the ground state and $S_p(\omega, \vec{p}) \, d\omega$ is the joint probability of adding a nucleon to the ground state of a nucleus. Hence, the momentum distribution of the nucleon in an interacting Fermi sea is given by

$$
\int_{-\infty}^{\mu} S_h(\omega, \vec{p}) \, d\omega = n_I(\vec{p}), \text{ and}
$$
$$
\int_{\mu}^{+\infty} S_p(\omega, \vec{p}) \, d\omega = 1 - n_I(\vec{p}), \tag{16.37}
$$

that leads to the spectral function sum rule:

$$
\int_{-\infty}^{\mu} S_h(\omega, \vec{p}) \, d\omega + \int_{\mu}^{+\infty} S_p(\omega, \vec{p}) \, d\omega = 1. \tag{16.38}
$$

Applying the Sokhatsky–Weierstrass theorem, that is,

$$
\lim_{\epsilon \to 0^+} \left(\frac{1}{x \pm i c} \right) = \mathcal{P} \left(\frac{1}{x} \right) \mp i\pi\delta(x) \tag{16.39}
$$

on Eq. (16.36) and then comparing the imaginary part of the obtained expression with Eq. (16.35), we obtain the expressions for the hole and particle spectral functions which are given by [809, 801]:

$$
S_h(p^0, \vec{p}) = \frac{1}{\pi} \frac{\frac{M}{E_N(\vec{p})} \mathrm{Im}\Sigma^N(p^0, \vec{p})}{\left(p^0 - E_N(\vec{p}) - \frac{M}{E_N(\vec{p})} \mathrm{Re}\Sigma^N(p^0, \vec{p}) \right)^2 + \left(\frac{M}{E_N(\vec{p})} \mathrm{Im}\Sigma^N(p^0, \vec{p}) \right)^2} \tag{16.40}
$$

when $p^0 \leq \mu$,

$$
S_p(p^0, \vec{p}) = -\frac{1}{\pi} \frac{\frac{M}{E_N(\vec{p})} \mathrm{Im}\Sigma^N(p^0, \vec{p})}{\left(p^0 - E_N(\vec{p}) - \frac{M}{E_N(\vec{p})} \mathrm{Re}\Sigma^N(p^0, \vec{p}) \right)^2 + \left(\frac{M}{E_N(\vec{p})} \mathrm{Im}\Sigma^N(p^0, \vec{p}) \right)^2} \tag{16.41}
$$

when $p^0 > \mu$. With these theoretical inputs, one is prepared to write the cross section by using Eqs. (16.22) and (16.24) in terms of the imaginary part of the W boson self-energy as:

$$
\frac{d^2\sigma_A^{WI}}{d\Omega' dE'_l} = -\frac{G_F^2}{(2\pi)^2} \frac{|\vec{k}'|}{|\vec{k}|} \left(\frac{M_W^2}{M_W^2 + Q^2} \right)^2 \int Im(L_{\mu\nu}\Pi^{\mu\nu}) d^3r. \tag{16.42}
$$

The hadronic tensor $W_A^{\mu\nu}$ is now given by

$$W_A^{\mu\nu} = -\int d^3r \, Im\Pi^{\mu\nu}(q) \tag{16.43}$$

Using Eq. (16.36) and the expressions for the free nucleon and meson propagators in Eq. (16.25), and finally substituting them in Eq. (16.43), we obtain the nuclear hadronic tensor $W_A^{\mu\nu}$ for a nucleus having $N \neq Z$ in terms of the nucleonic hadronic tensor $W_N^{\mu\nu}$ convoluted with the hole spectral function (S_h^τ, ($\tau = p, n$)) for a nucleon bound inside the nucleus:

$$W_A^{\mu\nu} = 2 \sum_{\tau=p,n} \int d^3r \int \frac{d^3p}{(2\pi)^3} \frac{M_N}{E_N(\vec{p})} \int_{-\infty}^{\mu_\tau} dp^0 S_h^\tau(p^0, \vec{p}, \rho^\tau(r)) \, W_N^{\mu\nu}(p, q), \tag{16.44}$$

where factor 2 is due to the two possible spin projections of protons or neutron, $\mu_p(\mu_n)$ is the chemical potential for the proton (neutron). $S_h^p(\omega, \vec{p}, \rho_p(r))$ and $S_h^n(\omega, \vec{p}, \rho_n(r))$ are the hole spectral functions for the proton and neutron, respectively, which provide information about the probability distribution of finding a proton and neutron with removal energy ω and three momentum $\vec{p}$ inside the nucleus. $\rho^\tau(r)$ is the charge density of the proton or neutron and has already been discussed in Chapter 14.

In the local density approximation, Fermi momentum is not fixed but depends upon the interaction point (r) in the nucleus and is bounded by $p_F(r)$ at the point r by

$$p_{F_{p(n)}}(r) = \left(\frac{3\pi^2 \rho_{p(n)}(r)}{2}\right)^{1/3}. \tag{16.45}$$

for the proton (neutron) target. Through the hole spectral function ($S_h(p^0, \vec{p}, \rho(r))$), the effects of Fermi motion, Pauli blocking, and nucleon correlations are incorporated. The behavior of hole spectral function vs. removal energy $\omega = p^0 - M$ is shown in Figure 16.13 for ^{12}C, ^{40}Ca, ^{56}Fe, ^{118}Sn, and ^{208}Pb. From the figure, it may be noticed that when the nucleon momentum is less than the Fermi momentum, the spectral function has a sharp and narrow distribution similar to the delta function while for $p > p_F$, the distribution has a wide range though it is very small in magnitude. This behavior of the spectral function is different as found in the case of independent particle model. It may be seen from Figure 16.13, that the hole spectral function has a smaller magnitude for heavier nuclear targets because of the enhancement in the probability of interaction among the nucleons. For an isoscalar nuclear target, the spectral function is properly normalized to reproduce the correct baryon number and binding energy for a given nucleus [869]

$$4\int \frac{d^3p}{(2\pi)^3} \int_{-\infty}^{\mu} S_h(\omega, \vec{p}, p_F(r)) d\omega = \rho(r) \tag{16.46}$$

or equivalently

$$\int d^3r \, 4\int \frac{d^3p}{(2\pi)^3} \int_{-\infty}^{\mu} S_h(\omega, \vec{p}, p_F(r)) d\omega = A. \tag{16.47}$$

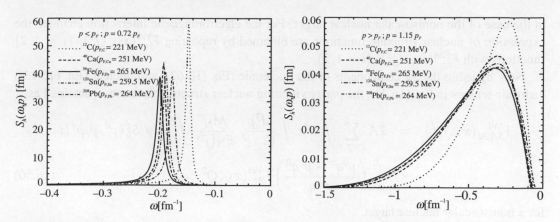

Figure 16.13 Results for $S_h(\omega, \vec{p})$ vs. ω are shown for (a) $p < p_F$ and (b) $p > p_F$ in various nuclei like ^{12}C, ^{40}Ca, ^{56}Fe, ^{120}Sn, and ^{208}Pb.

For a nonisoscalar nuclear target, these spectral functions are normalized individually for the proton (Z) and neutron ($N = A - Z$) numbers, that is,

$$2 \int d^3r \int \frac{d^3p}{(2\pi)^3} \int_{-\infty}^{\mu_p} S_h^p(\omega, \vec{p}, \rho_p(r)) \, d\omega = Z \,,$$

$$2 \int d^3r \int \frac{d^3p}{(2\pi)^3} \int_{-\infty}^{\mu_n} S_h^n(\omega, \vec{p}, \rho_n(r)) \, d\omega = N.$$

To evaluate the dimensionless nuclear structure functions by using Eq. (16.44), the appropriate components of nucleonic $W_N^{\mu\nu}$ (in Eq. (13.62)) and nuclear $W_A^{\mu\nu}$ (in Eq. (16.13)), hadronic tensors along the X, Y and Z axes are chosen.

By taking the zz component of the hadronic tensors ($W_N^{\mu\nu}$ of Eq. (13.62) and $W_A^{\mu\nu}$ of Eq. (16.13)), for a nonisoscalar nuclear target, the following expression is obtained [852]:

$$F_{2A,N}^{\mathrm{WI}}(x_A, Q^2) = 2 \sum_{\tau=p,n} \int d^3r \int \frac{d^3p}{(2\pi)^3} \frac{M_N}{E_N(\vec{p})} \int_{-\infty}^{\mu_\tau} dp^0 \, S_h^\tau(p^0, \vec{p}, \rho^\tau(r))$$

$$\times \left[\left(\frac{Q}{q^z}\right)^2 \left(\frac{|\vec{p}|^2 - (p^z)^2}{2M_N^2}\right) + \frac{(p^0 - p^z \, \gamma)^2}{M_N^2} \right.$$

$$\times \left. \left(\frac{p^z Q^2}{(p^0 - p^z \gamma)q^0 q^z} + 1\right)^2 \right] \left(\frac{M_N}{p^0 - p^z \gamma}\right) F_{2\tau}^{\mathrm{WI}}(x_N, Q^2), \quad (16.48)$$

where $\gamma = \frac{q^z}{q^0} = \sqrt{1 + \frac{4M^2 x^2}{Q^2}}$. The choice of xx components of the nucleonic (Eq. (13.62)) and nuclear (Eq. (16.13)) hadronic tensors lead to the expression of $F_{1A,N}^{\mathrm{WI}}(x, Q^2)$ as:

$$F_{1A,N}^{\mathrm{WI}}(x_A, Q^2) = 2 \sum_{\tau=p,n} A M_N \int d^3r \int \frac{d^3p}{(2\pi)^3} \frac{M_N}{E_N(\vec{p})} \int_{-\infty}^{\mu_\tau} dp^0 \, S_h^\tau(p^0, \vec{p}, \rho^\tau(r))$$

$$\times \left[\frac{F_{1\tau}^{\mathrm{WI}}(x_N, Q^2)}{M_N} + \left(\frac{p^x}{M_N}\right)^2 \frac{F_{2\tau}^{\mathrm{WI}}(x_N, Q^2)}{\nu} \right] \quad (16.49)$$

in the case of the nonisoscalar nuclear target. For the electromagnetic interaction channel, the expression of nuclear structure functions are obtained by replacing $F_{i\tau}^{WI}(x_N, Q^2)$; $(i = 1, 2)$ functions with $F_{i\tau}^{EM}(x_N, Q^2)$; $(i = 1, 2)$.

Now by using the xy components of the nucleonic (Eq. (13.62)) and nuclear (Eq. (16.13)) hadronic tensors in Eq. (16.44), the parity violating nuclear structure function is obtained as:

$$
\begin{aligned}
F_{3A,N}^{WI}(x_A, Q^2) &= 2A \sum_{\tau=p,n} \int d^3r \int \frac{d^3p}{(2\pi)^3} \frac{M_N}{E_N(\vec{p})} \int_{-\infty}^{\mu_\tau} dp^0 S_h^\tau(p^0, \vec{p}, \rho^\tau(r)) \\
&\quad \times \frac{q^0}{q^z} \left(\frac{p^0 q^z - p^z q^0}{p \cdot q} \right) F_{3\tau}^{WI}(x_N, Q^2),
\end{aligned}
\tag{16.50}
$$

for a nonisoscalar nuclear target.

For an isoscalar target, the factor of 2 in Eqs. (16.48), (16.49), and (16.50), will be replaced by 4 and the contribution will come from the nucleon's hole spectral function $S_h(p^0, \vec{p}, \rho(r))$ instead of the individual contribution from proton and neutron targets in $S_h^\tau(p^0, \vec{p}, \rho^\tau(r))$; $(\tau = p, n)$.

Furthermore, the intermediate vector bosons may interact with virtual mesons like π and ρ, being exchanged between nucleons. Therefore, the mesonic effect have also been incorporated in the Aligarh–Valencia model and are discussed in the following sub-subsection.

Mesonic cloud contribution

There are virtual mesons (mainly pion and rho mesons) associated with each nucleon bound inside the nucleus. This mesonic cloud gets strengthened by the strong attractive nature of the nucleon–nucleon interaction, which leads to a reasonably good probability of interaction of virtual bosons (IVB) with a meson instead of a nucleon [809, 799]. Although the contribution from the pion cloud is larger than the contribution from the rho meson cloud, the rho contribution is non-negligible, and both of them are positive in the entire region of x. The mesonic contribution is smaller in lighter nuclei, while it becomes more pronounced in heavier nuclear targets. The effect of including the spectral function leads to a reduction in the nuclear structure function from the free nucleon structure function, while the inclusion of the mesonic cloud contribution leads to an enhancement of the nuclear structure function, and can explain the experimental data [809, 851, 869].

To obtain the contribution from the virtual mesons, the neutrino self-energy is evaluated by considering a diagram similar to the one shown in Figure 16.10 using a meson propagator instead of a nucleon propagator. These mesons arise in the nuclear medium through particle–hole, delta–hole $(1\Delta - 1h)$, $1p1h - 1\Delta 1h$, etc. interactions as depicted in Figure 16.14.

In the case of mesons, we replace

$$
-2\pi \frac{M}{E_N(\vec{p})} S_h(p_0, \vec{p}, \rho(r)) W_N^{\mu\nu}(p, q)
$$

in Eq. (16.44) by

$$
\text{Im} D_i(p) \, \theta(p_0) \, 2W_i^{\mu\nu}(p, q)
$$

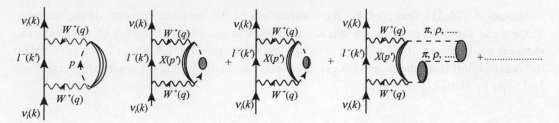

Figure 16.14 Neutrino self-energy diagram accounting for neutrino–meson DIS (a) the bound nucleon propagator is substituted with a meson (π or ρ) propagator (b) by including particle–hole ($1p$–$1h$), delta–hole (1Δ–$1h$), $1p1h - 1\Delta1h$, etc. interactions.

that leads to the following expression:

$$W_{A,i}^{\mu\nu} = 3 \int d^3r \int \frac{d^4p}{(2\pi)^4} \, \theta(p_0)(-2) \, \mathrm{Im} D_i(p) \, 2m_i W_i^{\mu\nu}(p,q), \qquad (16.51)$$

where $i = \pi, \rho$, is a factor of 3 because of the three charge states of mesons. $D_i(p)$ is the meson propagator in the nuclear medium which is given by:

$$D_i(p) = [p_0{}^2 - \vec{p}^2 - m_i^2 - \Pi_i(p_0, \vec{p})]^{-1}, \qquad (16.52)$$

with the mass of meson m_i and meson self-energy

$$\Pi_\pi(p_0, \vec{p}) = \left(\frac{f^2}{m_\pi^2}\right) \frac{F_\pi^2(p)\vec{p}^2 \Pi_\pi^*(p)}{1 - \left(\frac{f^2}{m_\pi^2}\right) V_L'(p)\Pi^*(p)}, \quad \Pi_\rho(p^0, \vec{p}) = \left(\frac{f^2}{m_\pi^2}\right) \frac{C_\rho \, \vec{p}^2 \Pi_\rho^*(p) \, F_\rho^2(p)}{1 - \frac{f^2}{m_\pi^2} V_T'(p)\Pi^*(p)}. \qquad (16.53)$$

Coupling constant $f = 1.01$, $C_\rho = 3.94$; $V_L'(p)$ ($V_T'(p)$) is the longitudinal (transverse) part of the spin–isospin interaction which is responsible for the enhancement in the meson structure function and $\Pi_i^*(p)$ is the irreducible meson self-energy that contains the contribution of particle–hole and delta–hole excitations. $F_i(p)$ is the $\pi NN/\rho NN$ form factor which is given by:

$$F_i(p) = \frac{(\Lambda_i^2 - m_i^2)}{(\Lambda_i^2 + \vec{p}^2)} \qquad (16.54)$$

with the parameter $\Lambda_i = 1$ GeV. It must be pointed out that the value of parameters Λ_π and Λ_ρ which is used in the expressions of πNN and ρNN form factors given in Eq. (16.53), respectively, is taken to be 1 GeV. This choice has been made in order to explain the experimental data from JLab [788] and other experiments performed using the charged lepton beam induced DIS processes on several nuclear targets. A detailed discussion is given in Refs.[851, 869, 822].

Equation (16.51) also contains the contribution of the mesonic contents of the nucleon. Since these mesonic contents are already contained in the sea contribution of nucleons, the mesonic contribution of the nucleon is to be subtracted from Eq. (16.51) in order to calculate the contribution from the meson excess in the nuclear medium. This is obtained by replacing $\mathrm{Im}D_i(p)$ by $\delta\mathrm{Im}D_i(p)$ [809] as

$$\mathrm{Im}D_i(p) \;\to\; \delta\mathrm{Im}D_i(p) \equiv \mathrm{Im}D_i(p) - \rho\,\left.\frac{\partial\mathrm{Im}D_i(p)}{\partial\rho}\right|_{\rho=0}. \tag{16.55}$$

Therefore, the actual mesonic contribution is given by

$$W_{A,i}^{\mu\nu} = 3\int d^3r \int \frac{d^4p}{(2\pi)^4}\,\theta(p_0)(-2)\,\delta\mathrm{Im}D_i(p)\,2m_i W_i^{\mu\nu}(p,q). \tag{16.56}$$

Following the same procedure as in the case of nucleon, one may also obtain the expressions of mesonic nuclear structure functions $F_{i'A,i}^{WI}(x,Q^2)(x,Q^2)$; $(i'=1,2)$:

$$
\begin{aligned}
F_{2A,i}^{WI}(x,Q^2) \;=\; & -6\kappa \int d^3r \int \frac{d^4p}{(2\pi)^4}\,\theta(p^0)\,\delta\mathrm{Im}D_i(p)\,2m_i\left(\frac{m_i}{p^0-p^z\,\gamma}\right)\times \\
& \left[\frac{Q^2}{(q^z)^2}\left(\frac{|\vec{p}|^2-(p^z)^2}{2m_i^2}\right)+\frac{(p^0-p^z\,\gamma)^2}{m_i^2}\right. \\
& \left.\times\left(\frac{p^z\,Q^2}{(p^0-p^z\,\gamma)q^0 q^z}+1\right)^2\right]F_{2i}^{WI}(x_i),
\end{aligned}\tag{16.57}
$$

and

$$
\begin{aligned}
F_{1A,i}^{WI}(x,Q^2) \;=\; & -2\kappa AM \int d^3r \int \frac{d^4p}{(2\pi)^4}\theta(p^0)\,\delta\mathrm{Im}D_i(p)\,2m_i\left[\frac{F_{1i}^{WI}(x_i)}{m_i}\right. \\
& \left.+\frac{|\vec{p}|^2-(p^z)^2}{2(p^0 q^0-p^z q^z)}\frac{F_{2i}^{WI}(x_i)}{m_i}\right],
\end{aligned}\tag{16.58}
$$

where $\kappa=1$ for pions and $\kappa=2$ for rho mesons, $x_i=-\frac{Q^2}{2p\cdot q}$ which carries a negative sign because meson's and nucleon's momenta are opposite in direction. Notice that the ρ-meson has an extra factor of 2 compared to pionic contribution because of the two transverse polarizations of the ρ-meson.

For pionic PDFs, different parameterizations are available in literature such as GRV [873], CTEQ5L [874], SMRS [875], Conway [876], MRST98 [877], etc., which are used while performing numerical calculations. The partonic structure of a pion has been experimentally observed in the Drell–Yan scattering process of pions from bound nucleons in experiments, like E615 [876], NA3 [878], NA10 [879], etc. The Drell–Yan mechanism provides information about the pion's partonic structure in the region of valence quarks while in the region corresponding to sea quarks, data are available from the deep inelastic scattering processes [873].

16.5.3 Kulagin–Petti model

Kulagin and Petti in a series of papers have developed a model for calculating various nuclear effects in the DIS of charged leptons and (anti)neutrinos from nuclear targets [799, 800, 880, 843]. In this model, the nuclear hadronic tensor $W_{\mu\nu}^A$ is written as:

$$W_{\mu\nu}^A(P_A, q) = \sum_{\tau=p,n} \int \frac{d^4 p}{(2\pi)^4} \, \text{Tr} \left[\widehat{\mathcal{W}}_{\mu\nu}^N(p, q) \mathcal{A}^N(p; A) \right], \tag{16.59}$$

where $W_{\mu\nu}^A(P_A, q)$ is the nuclear tensor for nucleus A with momentum P_A and $\widehat{\mathcal{W}}_{\mu\nu}^N(p, q)$ is the nucleon tensor for a bound nucleon. $\mathcal{A}^N(p; A)$ is the probability distribution for finding the nucleon with momentum p in the nucleus A given by:

$$\mathcal{A}_{\alpha\beta}^N(p; A) = \int dt \, d^3\vec{r} \, e^{ip_0 t - i\vec{p}\cdot\vec{r}} \langle A | \overline{\Psi}_\beta^N(t, \vec{r}) \, \Psi_\alpha^N(0) | A \rangle. \tag{16.60}$$

Assuming the nucleons to be nonrelativistic, Eq. (16.59) is simplified to be [799]

$$\frac{W_{\mu\nu}^A(P_A, q)}{A} = \sum_{\tau=p,n} \int \frac{d^4 p}{(2\pi)^4} \frac{M}{M+E} \mathcal{P}^\tau(E, \vec{p}) \, W_{\mu\nu}^\tau(p, q), \tag{16.61}$$

where $W_{\mu\nu}^A$ and $W_{\mu\nu}^N$ are given in Eqs. (16.59), (16.61), and (13.62). $\mathcal{P}^\tau(E, \vec{p})$, ($\tau = p, n$) is the spectral function of the nucleon determined from the $(e, e'N)$ reaction and is given by

$$\mathcal{P}(E, \vec{p}) = 2\pi \sum_n |\langle (A-1)_n, -\vec{p} \, | \psi(\vec{p}) | A \rangle|^2 \, \delta \left(E + E_n^{A-1} + E_R - E_0^A \right), \tag{16.62}$$

where $E_R = \frac{p^2}{2M_{(A-1)}}$ is the recoil energy of the residual nucleus and E is the energy of the nucleon in the $(e, e'N)$ reaction. E_n^{A-1} is the energy of the residual nucleus and E_0^A is the ground state energy of the target nucleus. $\mathcal{P}(E, \vec{p})$ is calculated using the ground state and its excitations using a mean field as well as the two nucleon short range correlations (SRC). The spectral function $\mathcal{P}(E, \vec{p})$ is expressed as:

$$\mathcal{P}_0(E, \vec{p}) = \mathcal{P}_{\text{MF}}(E, \vec{p}) + \mathcal{P}_{\text{cor}}(E, \vec{p}). \tag{16.63}$$

with

$$\mathcal{P}_{\text{MF}}(E, \vec{p}) = 2\pi \, n_{\text{MF}}(\vec{p}) \delta \left(E + E^{(1)} + E_R(\vec{p}) \right), \tag{16.64}$$

corresponding to the mean field (MF). $\mathcal{P}_{\text{cor}}(E, \vec{p})$ is the contribution from nucleon–nucleon short range correlations and described by the authors of Ref. [881]. The normalization of $\mathcal{P}(E, \vec{p})$ is given by:

$$\int \frac{d^4 p}{(2\pi)^4} \mathcal{P}^{p,n}(E, \vec{p}) = (Z, N). \tag{16.65}$$

Using the expressions discussed in this section, the following results are obtained [799]:

$$F_T^A(x, Q^2) = \sum_{\tau=p,n} \int \frac{d^4 p}{(2\pi)^4} \mathcal{P}^\tau(E, \vec{p}) \left(1 + \frac{\gamma p_z}{M}\right) \left(F_T^\tau + \frac{2x'^2 \vec{p}_\perp^2}{Q^2} F_2^\tau\right), \tag{16.66a}$$

$$F_L^A(x, Q^2) = \sum_{\tau=p,n} \int \frac{d^4 p}{(2\pi)^4} \mathcal{P}^\tau(E, \vec{p}) \left(1 + \frac{\gamma p_z}{M}\right) \left(F_L^\tau + \frac{4x'^2 \vec{p}_\perp^2}{Q^2} F_2^\tau\right), \tag{16.66b}$$

$$\gamma^2 F_2^A(x, Q^2) = \sum_{\tau=p,n} \int \frac{d^4 p}{(2\pi)^4} \mathcal{P}^\iota(E, \vec{p}) \left(1 + \frac{\gamma p_z}{M}\right) \left(\gamma'^2 + \frac{6x'^2 \vec{p}_\perp^2}{Q^2}\right) F_2^\tau. \tag{16.67}$$

$$xF_3^A(x, Q^2) = \sum_{\tau=p,n} \int \frac{d^4 p}{(2\pi)^4} \mathcal{P}^\tau(E, \vec{p}) \left(1 + \frac{p_z}{\gamma M}\right) x' F_3^\tau, \tag{16.68}$$

where in the integrand, F_i^τ ($i = T, L, 2, 3$) are the structure functions of the bound proton ($\tau = p$) and neutron ($\tau = n$) with the four momentum $p = (M + E, \vec{p})$; $\vec{p}_\perp$ is the transverse component of the nucleon momentum. The Bjorken variable for the bound nucleon is given by

$$x' = \frac{x}{[1 + \frac{(E+\gamma p_z)}{M}]} \quad \text{with} \quad \gamma' = \sqrt{1 + \frac{4x'^2 p^2}{Q^2}}.$$

In this model, the pionic contribution to the nuclear hadronic tensor is evaluated and given by:

$$W_{\mu\nu}^\pi(P_A, q) = \frac{1}{2} \int \frac{d^4 k}{2\pi^4} D_\pi(k) W_{\mu\nu}^\pi(k, q), \tag{16.69}$$

where $W_{\mu\nu}^\pi(k, q)$ is the hadronic tensor of the pion with momentum k. $D_\pi(k)$ describes the momentum distribution of pions in a nucleus [799].

Shadowing effect

The nuclear shadowing/antishadowing is generally understood as arising due to the coherent scattering of hadrons produced in the hadronization process of mediating vector bosons. The destructive interference of the amplitudes due to the multiple partons scattering gives shadowing and their constructive interference gives antishadowing. These coherent effects arise when the coherence length is larger than the average distance between the nucleons bound inside the nucleus and the expected coherence time is $\tau_c \geq 0.2\ fm$. However, the shadowing effect gets saturated if the coherence length becomes larger than the average nuclear radius, that is, in the region of low x. Due to these effects, the nuclear structure functions are modified in the low region of x [851]. The shadowing effect is different in electromagnetic and weak processes [822]. This is because the electromagnetic and weak interactions take place through the interaction of photons and $W^{\pm}/Z^0$ bosons, respectively, with the target hadrons and the hadronization processes of photons and $W^{\pm}/Z^0$ bosons are different. Moreover, in the case of weak interaction, the additional contribution of the axial current which is not present in the case of electromagnetic interaction may influence the behavior of $F_{2A}^{WI}(x, Q^2)$, especially if pions also play a role in the hadronization process through PCAC. Furthermore, in this region of low x, sea quarks also play an important role which could be different in the case of electromagnetic and weak processes. For example, the sea quark contribution, though very small, is not same for $F_2^{EM}(x, Q^2)$ and $F_2^{WI}(x, Q^2)$ even at the free nucleon level and could evolve differently in a nuclear medium. Generally, the shadowing effect is understood using the Glauber–Gribov multiple scattering theory.

For example, the nuclear structure function incorporating the shadowing effect $F_{iA,shd}^{EM/WI}$ (x, Q^2) may be defined as

$$F_{iA,shd}^{EM/WI}(x, Q^2) = \delta R_i(x, Q^2) \times F_{i,N}^{EM/WI}(x, Q^2), \qquad (16.70)$$

where $i = 1, 2, 3$ and $\delta R_i(x, Q^2)$ is the shadowing correction factor discussed in Ref. [799]. The details are beyond the scope of this book.

16.5.4   Isoscalarity corrections: Phenomenological approach

In the case of heavier nuclear targets, where neutron number $(N = A - Z)$ is larger than the proton number (Z) and their densities are also different, isoscalarity corrections become important. As most of the $\nu_l/\bar{\nu}_l$ experiments use heavy nuclear targets $(N \neq Z)$, phenomenologically, the isoscalarity correction is taken into account by multiplying the experimental results with a correction factor defined as

$$R_A^{Iso}(x) = \frac{0.5A(1 + y(x))}{N(Z/N + y(x))}, \qquad (16.71)$$

where $y(x)$ is the ratio of (free) neutron to proton cross sections at a particular value of x. As an example at $x = 0.25$ with the ratio of neutrino cross sections off neutron and protons $y(x) = \frac{\sigma^{\nu n}}{\sigma^{\nu p}} = 1.7$, $R_A^{Iso}(x)$ would be 0.98 for Fe and 0.948 for Pb. Although the non-isoscalarity of a nucleus is considered by some to be a 'nuclear effect', this neutron excess can simply hide the more subtle nuclear effects, and therefore should be properly accounted for.

There are also other phenomenological methods to take into account the isoscalarity correction, for example by taking into account the experimental results for a nonisoscalar nuclear target with a correction factor defined as

$$R_A^{Iso} = \frac{[F_2^{\nu/\bar{\nu}p} + F_2^{\nu/\bar{\nu}n}]/2}{[ZF_2^{\nu/\bar{\nu}p} + (A - Z)F_2^{\nu/\bar{\nu}n}]/A}, \qquad (16.72)$$

where Z is the atomic number of the target nucleus, $F_2^{\nu/\bar{\nu}p}$ and $F_2^{\nu/\bar{\nu}n}$ are the weak structure functions for the proton and the neutron, respectively.

The per nucleon nuclear structure function is then compared with the structure function for the isoscalar nucleon target which is obtained from the proton data or deuteron data after applying deuteron correction.

16.6   Results and Discussions

In this section, a few results on nuclear structure functions obtained by using the model of Aligarh–Valencia group (discussed in Subsections 16.5.1, 16.5.2, and 16.5.3) are presented. This model describes the nuclear structure functions $F_{iA}^{EM/WI}(x_A, Q^2)$ $(i = 1 - 3)$ in terms of the nucleon structure functions $F_{iN}^{EM/WI}(x_N, Q^2)$ convoluted with the spectral function which takes into account Fermi motion, binding energy, and nucleon correlation effects followed by the mesonic and shadowing effects. For the evaluation of $F_{iN}^{EM/WI}(x_N, Q^2)$ at the leading order (LO), free nucleon PDFs are used. The results presented here are obtained with the nucleonic PDF parameterizations of MMHT [552] as well as CTEQ6.6 in the MS-bar scheme [551]. $F_{iA,\pi}^{EM/WI}(x, Q^2)$ and $F_{iA,\rho}^{EM/WI}(x, Q^2)$ are the structure functions having pion and rho mesons contributions. In the literature, various pionic PDFs parameterizations are available like that of Gluck et al. [873], Wijesooriya et al. [874], Sutton et al. [875], Conway et al. [876], etc. The numerical results presented here are with the pionic PDF parameterization of Gluck et al. [873]. To evaluate the nucleon structure functions in the kinematic region of low and moderate Q^2, where the higher order perturbative corrections and the non-perturbative effects become important, PDF evolution is performed up to NNLO and the target mass correction effect is incorporated in the calculations. For the evolution of nucleon PDFs at the next-to-leading order (NLO) and next-to-next-to-leading order (NNLO) obtained from the leading order, the works of Vermaseren et al. [581] and Moch et al. [589, 588] have been followed. The target mass correction effect has been included following the method of Schienbein et al. [591]. The theoretical results are compared with the available experimental data. The total nuclear structure functions including all contributions is defined as:

$$F_{iA}^{EM/WI}(x, Q^2) = \underbrace{F_{iA,N}^{EM/WI}(x, Q^2)}_{\text{spectral function}} + \underbrace{F_{iA,\pi}^{EM/WI}(x, Q^2) + F_{iA,\rho}^{EM/WI}(x, Q^2)}_{\text{mesonic contributions}} + \underbrace{F_{iA,shd}^{EM/WI}(x, Q^2)}_{\text{shadowing effect}} \tag{16.73}$$

for $i = 1 - 2$. In this model, the full expression for the parity violating weak nuclear structure function is given by,

$$F_{3A}^{WI}(x, Q^2) = \underbrace{F_{3A,N}^{WI}(x, Q^2)}_{\text{spectral function}} + \underbrace{F_{3A,shd}^{WI}(x, Q^2)}_{\text{shadowing effect}}. \tag{16.74}$$

Notice that this structure function has no mesonic contribution and the contribution to the nucleon structure function comes from the valence quarks distributions.

In Figure 16.15, the results for the ratio $R(x, Q^2) = \frac{F_{iA}^{WI}(x,Q^2)}{F_{iN}^{WI}(x,Q^2)}$; $(i = 2, 3)$ of nuclear structure functions to the free nucleon structure functions are presented at different values of x in iron and lead which are treated as isoscalar. Numerical calculations have been performed at NNLO by using the MMHT nucleon PDF parameterization [552]. In the figure, theoretical

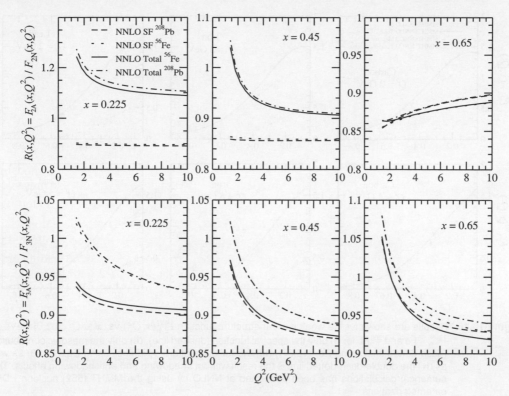

Figure 16.15 $R(x,Q^2) = \frac{F_{iA}^{WI}(x,Q^2)}{F_{iN}^{WI}(x,Q^2)}$; $(i = 2,3)$ vs. Q^2 in ^{56}Fe and ^{208}Pb using the full model. The results are obtained by using MMHT nucleon PDFs parameterization [552] at NNLO (solid line) following the works of Vermaseren et al. [581] and Moch et al. [589, 588] for the free nucleon target; for the nuclear targets, the results are obtained with only the spectral function and the full model.

results are presented by taking into account nuclear effects like Fermi motion, binding energy, and nucleon correlations through spectral function ('NNLO SF') as well as for the full calculations ('NNLO total') which incorporate the mesonic contributions and shadowing effect. One may notice that the ratio $R(x,Q^2)$ is not unity in the entire range of x and Q^2 showing the modification of nuclear structure functions in the nuclear medium. The contributions of mesons is found to be important in the region of low and intermediate x and decreases with the increase in x. Moreover, it is important to point out that as there is no mesonic contribution in $F_{3A}^{WI}(x,Q^2)$, the difference in the results of the spectral function only and the full model (which has contribution from the shadowing effect only) is comparatively smaller than $F_{2A}^{WI}(x,Q^2)$. It is also noticeable that the nuclear medium effects on $R(x,Q^2)$ are larger in lead due to the mass dependence of nuclear medium effects.

In Figure 16.16, the results are presented for $F_{2A}^{WI}(x,Q^2)$ vs. x in ^{12}C, ^{56}Fe, and ^{208}Pb, for isoscalar nuclear targets at different values of Q^2. The results are obtained using the spectral function (dashed line), mesonic effect (dash-dotted line), and the final result by including the shadowing and antishadowing effects as shown by the solid line. The mesonic contributions give an enhancement in the nuclear structure functions which is significant in the low and

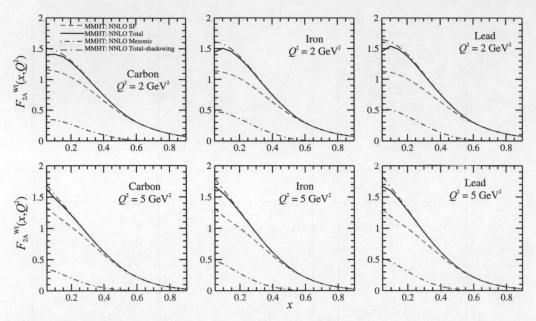

Figure 16.16   Results are shown for the weak nuclear structure function $F_{2A}^{WI}(x, Q^2)$ vs. x at $Q^2 = 2$, $5\ \text{GeV}^2$, in ^{12}C, ^{56}Fe and ^{208}Pb for **(i)** only the spectral function (dashed line), **(ii)** only the mesonic contribution (dash-dotted line) using Eq. (16.57), **(iii)** the full calculation (solid line) using Eq. (16.73) as well as **(iv)** the double-dash-dotted line is the result without shadowing and antishadowing effects. The numerical calculations has been performed at NNLO by using the MMHT [552] nucleon PDFs parameterizations.

intermediate region of x. Moreover, the effect is more pronounced at low Q^2 and becomes larger with increase in mass number A. To depict the coherent nuclear effects (shadowing and antishadowing effects) which cause a suppression in the structure functions at low x, the results are shown with the double-dash-dotted line, and it may be observed that with the increase in mass number of the nuclear target (^{56}Fe vs. ^{208}Pb), the strength of suppression becomes larger. For the (anti)neutrino scattering cross sections and structure functions, high statistics measurements have been done by CCFR [562], CDHSW [561], and NuTeV [566] experiments in iron and by CHORUS [882] collaboration in lead nuclear targets. These experiments have been performed in a wide energy range, that is, $20 \leq E_\nu \leq 350\ \text{GeV}$ and the differential scattering cross sections have been measured. The results for $F_{2A}^{\nu+\bar{\nu}}(x, Q^2)$ and $xF_{3A}^{\nu+\bar{\nu}}(x, Q^2)$ are presented in Figures 16.17 and 16.18 treating these nuclei as isoscalar targets. The experimental results are compared with the theoretical results calculated using the full model at NNLO.

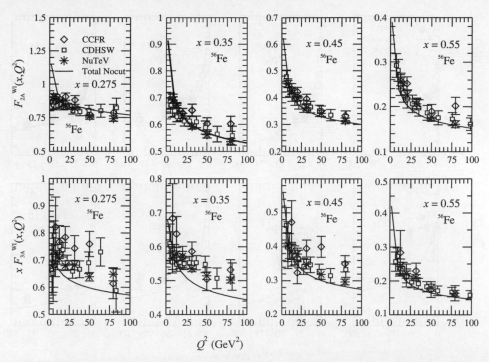

Figure 16.17 $F_{2A}^{WI}(x, Q^2)$ vs. Q^2 in ^{56}Fe using the full model. The results are obtained by using CTEQ6.6 nucleon PDFs at NLO in the MS-bar scheme (dotted line), MMHT at NLO (dashed line) and NNLO (solid line). The experimental points are the data from CDHSW [561], CCFR [562], and NuTeV [566] experiments.

The agreement among the two results is satisfactory except at very low x and Q^2. We have found that the theoretical results differ from the experimental data in the region of low x and low Q^2; however, with the increase in x and Q^2, they are found to be in reasonably good agreement with the experimental determination.

In Figure 16.19, the variation of nuclear medium effects in the electromagnetic and weak interactions has been shown by using different nuclear targets. The ratio R' deviates from unity in the region of low x even for the free nucleon case which implies the nonzero contribution from strange and charm quarks distributions. However, for $x \geq 0.4$, where the contribution of strange and charm quarks are almost negligible, the ratio approaches unity. Furthermore, if one assumes $s = \bar{s}$ and $c = \bar{c}$, then in the region of small x, this ratio would be unity for an isoscalar nucleon target following the $\left(\frac{5}{18}\right)^{\text{th}}$-sum rule. For heavier nuclear targets like ^{56}Fe and ^{208}Pb, the nuclear medium effects become more pronounced. This shows that the difference in charm and strange quark distributions could be significant in heavy nuclei.

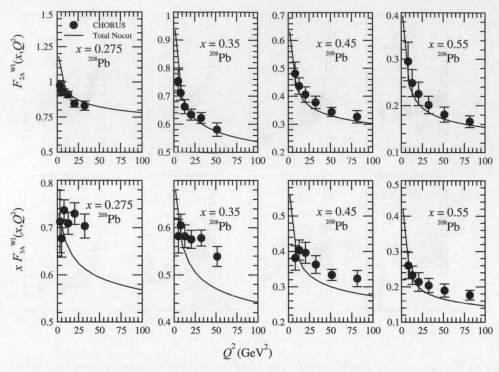

Figure 16.18 $F_{2A}^{WI}(x, Q^2)$ vs. Q^2 in ^{208}Pb using the full model. The results are obtained using CTEQ6.6 nucleon PDFs at NLO in the MS-bar scheme (dotted line), MMHT at NLO (dashed line) and NNLO (solid line). The experimental points are the data from CHORUS [562].

In Figure 16.20, the results are shown for $\frac{1}{E_\nu} \frac{d^2\sigma_A^{WI}}{dxdy}$ vs. y for $\nu_l - A$, ($A = {}^{56}$ Fe, ^{208}Pb) scattering at NNLO at different values of x for the incoming beam energy of 35 GeV. Theoretical results are presented for the spectral function only (dashed line), using the full model (solid line) and are compared with the experimental data of NuTeV [566] (top panel) and CHORUS [882] (bottom panel) collaborations. It may be noticed that the mesonic effects are important only in the low and intermediate region of x i.e. $x < 0.65$.

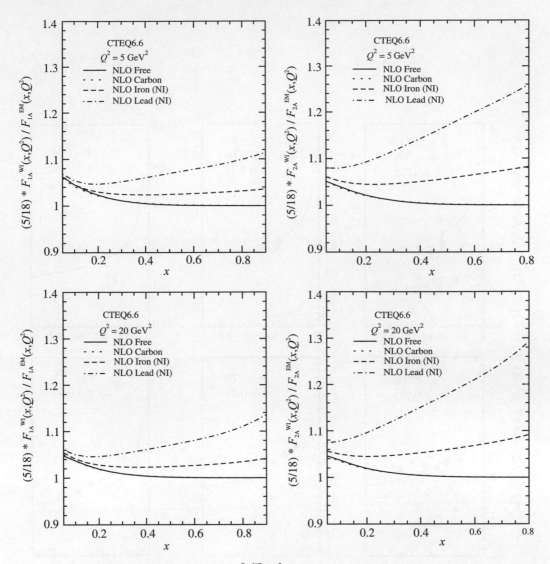

Figure 16.19 Results for the ratio $R'(x, Q^2) = \frac{5}{18} \frac{F_{iA}^{WI}(x,Q^2)}{F_{iA}^{EM}(x,Q^2)}$; $(i = 1, 2)$ are obtained by using the full model at NLO in $A = {}^{12}$C, ^{56}Fe and ^{208}Pb at $Q^2 = 5$ and 20 GeV^2. Numerical results are obtained by using the CTEQ6.6 nucleon PDFs in the $\overline{MS}$ scheme [551] (NI stands for nonisoscalar).

Thus, the nuclear medium effects are important in the DIS region and are better phenomenological as well as theoretical models which cover a wide range of x and Q^2 for intermediate and heavy mass nuclei are highly desirable.

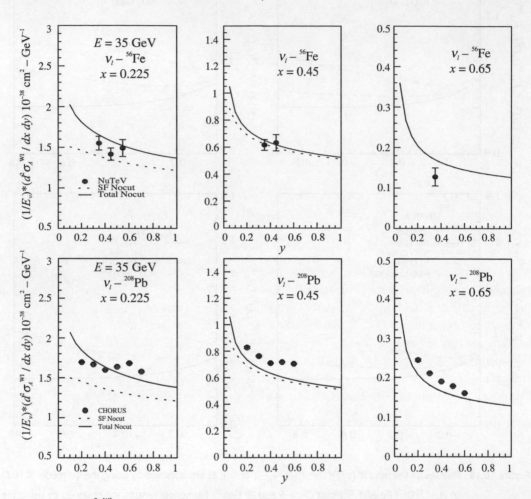

Figure 16.20 $\frac{1}{E_\nu} \frac{d^2\sigma_A^{WI}}{dxdy}$ vs. y are shown at different values of x for the incoming beam of energy $E = 35$ GeV. The numerical results are obtained with the spectral function only (dashed line) and with the full model (solid line) at NNLO, Top panel: for $\nu_l - {}^{56}$Fe and Bottom panel: for $\nu_l - {}^{208}$Pb DIS processes. The results are compared with the experimental data points of NuTeV [566] and CHORUS [882]. Nuclear targets are treated to be isoscalar.

Chapter **17**

Neutrino Sources and Detection of Neutrinos

17.1 Introduction

There are many sources of neutrinos in the universe. For example, neutrinos are produced inside the sun, the earth, and the entire atmosphere. Neutrinos are also produced during the birth, collision, and death of stars. Particularly huge flux of neutrinos is emitted during supernovae explosions. Most of the neutrinos that pass through the earth come from the sun and are produced in the nuclear fusion reactions going on inside the sun's core. In addition, (anti)neutrinos are produced in nuclear power plants. The beams of high energy protons striking a target material produce intense flux of pions and kaons, which then decay to produce neutrinos. There are neutrinos around us which were born almost 13.8 billion years ago,

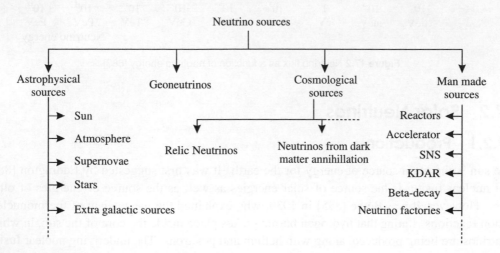

Figure 17.1 Different sources of neutrinos.

soon after the birth of the universe; they constitute cosmic neutrinos (relic of the Big Bang). Neutrinos produced inside the core of the earth are called geoneutrinos.

Figure 17.1 represents the different sources of neutrinos. Besides these sources, neutrinos are also produced by our body; they are also present in vegetables and fruits, etc. In fact, we are living in a sea of neutrinos having a wide range of energies. Simply speaking, through an area of 1 cm^2 of our body, almost a billion neutrinos pass every second. The theoretically obtained energy ranges of these neutrinos are quite broad, ranging from micro electron volts for the neutrinos left over from the Big Bang, right up to peta electron volts for the neutrinos produced in the violent gamma-ray bursts in the universe. Figure 17.2 shows the predicted neutrino flux as a function of neutrino energy from a variety of neutrino sources. Some of these neutrinos have been studied by various neutrino experiments. In this chapter, we will discuss some of the important neutrino sources and their detection techniques in various energy regions.

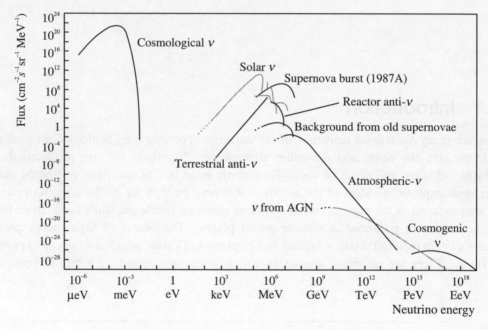

Figure 17.2 Neutrino flux as a function of neutrino energy [883].

17.2   Solar Neutrinos

17.2.1   Production

The sun is the main source of energy for the earth. It was first suggested by Eddington [884] that nuclear fusion is the source of solar energies as well as the source of energies of other stars. However, it was Bethe [885] in 1939, who explained the mechanism of thermonuclear fusion reactions, stating that hydrogen burning takes place inside the core of the sun, in which neutrinos are being produced, along with helium and positrons. The underlying nuclear fusion process is:

$$\begin{aligned}
{}^1_1\text{H} + {}^1_1\text{H} &\rightarrow {}^2_1\text{D} + e^+ + \nu_e \\
{}^2_1\text{D} + {}^1_1\text{H} &\rightarrow {}^3_2\text{He} + \gamma \\
{}^3_2\text{He} + {}^3_2\text{He} &\rightarrow {}^4_2\text{He} + 2p \quad \text{leading to,} \\
4p &\rightarrow {}^4\text{He} + 2e^+ + 2\nu_e + 24.73 \text{ MeV.}
\end{aligned}$$

Finally, two positrons annihilate with the two electrons present in the solar matter, such that the reaction may be summarized as:

$$2e^- + 4p \quad \rightarrow \quad {}^4\text{He} + 2\nu_e + 26.73 \text{ MeV.} \tag{17.1}$$

In other words, four protons are consumed in the production of 26.73 MeV of nuclear fusion energy. About 97–98% of the total energy released in this process is in the form of heat and light (photons). The remaining 2–3% is carried away by the neutrinos. The average energy of two neutrinos released in the aforementioned reaction is about 0.53 MeV. Since photons interact via electromagnetic interaction, after they are produced in the core of the sun, they are scattered and rescattered due to the interaction with solar matter during their passage through the sun, after they are produced in the core of the sun. It has been shown that it takes almost 10^4, years on an average, for the photons to come out of the solar core. In contrast, neutrinos, being weakly interacting particles, do not interact much with the sun's interior matter and travel almost with the speed of light; they reach the earth in about 8 minutes. Thus, neutrinos carry direct information about the sun's interior.

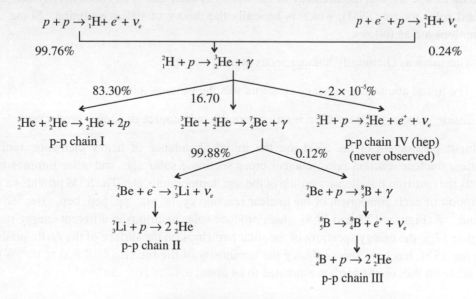

Figure 17.3 Proton–proton chain.

Solar neutrinos go through the nuclear processes driven by two different cycles of reactions, which are known as the pp (proton–proton) cycle and the CNO (carbon–nitrogen–oxygen) cycle. The pp cycle (Figure 17.3) dominates in low-mass stars and the CNO cycle (Figure 17.4) dominates in heavy stars. For example, in case of the sun, about 99% of the energy is produced

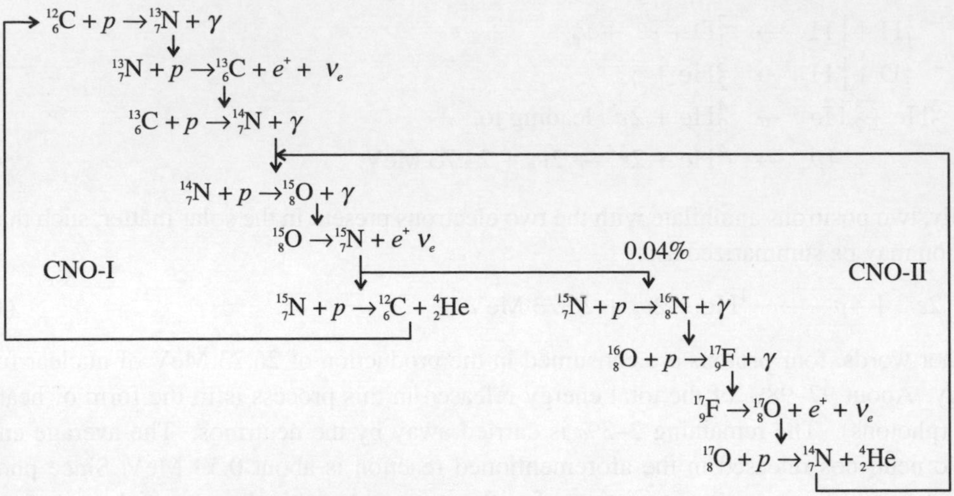

$$\longrightarrow {}^{12}_{6}C + p \rightarrow {}^{13}_{7}N + \gamma$$
$$\downarrow$$
$${}^{13}_{7}N + p \rightarrow {}^{13}_{6}C + e^+ + \nu_e$$
$$\downarrow$$
$${}^{13}_{6}C + p \rightarrow {}^{14}_{7}N + \gamma$$
$$\downarrow$$
$${}^{14}_{7}N + p \rightarrow {}^{15}_{8}O + \gamma$$
$$\downarrow$$
$${}^{15}_{8}O \rightarrow {}^{15}_{7}N + e^+ \nu_e \qquad\qquad 0.04\%$$

CNO-I $\qquad\qquad\qquad\qquad\qquad\qquad$ CNO-II

$${}^{15}_{7}N + p \rightarrow {}^{12}_{6}C + {}^{4}_{2}He \qquad\qquad {}^{15}_{7}N + p \rightarrow {}^{16}_{8}N + \gamma$$
$$\downarrow$$
$${}^{16}_{8}O + p \rightarrow {}^{17}_{9}F + \gamma$$
$$\downarrow$$
$${}^{17}_{9}F \rightarrow {}^{17}_{8}O + e^+ + \nu_e$$
$$\downarrow$$
$${}^{17}_{8}O + p \rightarrow {}^{14}_{7}N + {}^{4}_{2}He$$

Figure 17.4 Carbon–nitrogen–oxygen cycle.

by the pp cycle. The pp cycle is quite slow in comparison to the CNO cycle. The detailed computer simulations of the electron neutrino flux, created by thermonuclear reactions in the interior of the sun, were performed by Bahcall et al. [182]. The present status of these studies, the prediction level, and the accuracy of the model is such that it has been acronymed as the standard solar model (SSM), which is basically the theory of stellar evolution. Some of its assumptions are as follows:

- The sun was chemically homogeneous at time $t = 0$.

- The initial abundance of heavy elements was the same as it is now.

- Inside the sun, nuclear fusion reactions are the only source of energy production.

The input parameters of the SSM are the initial abundance of heavy elements, radiative opacities, nuclear reaction rates, nuclear cross sections, solar age, and solar luminosity. It predicts the neutrino flux, density profile of the sun, temperature, etc. The SSM provides a good description of each component of the nuclear reaction cycle, viz., pp, pep, hep, ^{7}Be, ^{8}B, ^{13}N, ^{15}O and ^{17}F (Figures 17.3 and 17.4) which produce solar neutrinos in different energy ranges. In Figure 17.5, the energy spectrum of the solar neutrinos on the surface of the earth, predicted using the SSM, has been shown. Using the luminosity of the sun ($L_\odot = 3.826 \times 10^{26}$W), the solar neutrino flux on the earth is estimated to be about 6.47×10^{10} cm^{-2} s^{-1}.

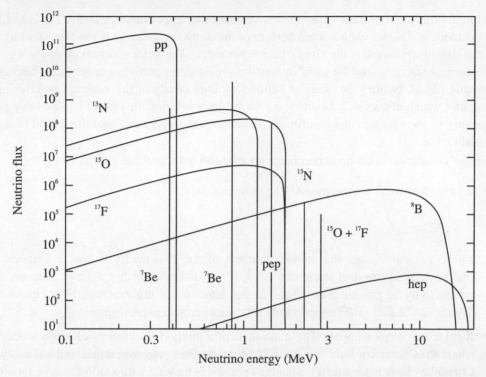

Figure 17.5 Solar neutrino spectrum in the SSM [182].

17.2.2 Detection

The $\nu_e - e^-$ cross section at 1 MeV is about 10^{-44} cm^2, and similarly, $\nu_e - N$ cross sections are small as discussed in Chapter 10. Therefore, to detect neutrinos, large detectors with high detection sensitivity are required. Furthermore, to avoid the background arising due to the presence of cosmic rays (mostly $\mu^\pm$, p, and n), and which have an average flux of about 200 particles/m^2/s, the detectors must be placed deep underground. Pontecorvo suggested performing a solar neutrino experiment with the reaction $\nu_e + Cl \rightarrow e^- + Ar$; however, it was not feasible when proposed. Later, Davis, Jr. and Bahcall in 1964, suggested the radiochemical experiment with solar neutrinos, which was performed by Davis, Jr. at the Homestake Mine located in a rock cavity, 4,850 feet below the surface in Lead, South Dakota. The setup had a tank filled with 100,000 gallons of tetrachloroethylene (C$_2$Cl$_4$) to detect solar neutrinos. In general, each molecule of C$_2$Cl$_4$ contains one atom of $^{37}_{17}$Cl, the desired isotope; the other three chlorine atoms contain two less neutrons. When a neutrino having energy more than the threshold energy (E_{th}= 0.814 MeV) of the reaction strikes $^{37}_{17}$Cl, it produces an atom of $^{37}_{18}$Ar and an electron is released, that is, $\nu_e + ^{37}_{17} Cl \rightarrow e^- + ^{37}_{18} Ar$. The atoms of $^{37}_{18}$Ar are allowed to accumulate for several months, and are then removed by purging the tank with helium gas. The argon is then analyzed for radioactivity ($^{37}_{18}Ar \rightarrow ^{37}_{17} Cl + e^- + \bar{\nu}_e$). Davis, Jr. and his collaborators reported a solar neutrino deficit which is the difference between the predicted solar neutrinos using the SSM and the observed ones. This is known as the solar neutrino anomaly. After their experiments, several other experiments like SAGE, GALLEX, Kamiokande, SNO, etc. which were performed to study the solar neutrino flux, also agreed with the deficit observed by the Homestake experiment. The solar neutrino anomaly was one of the longest standing and the most interesting problems in particle astrophysics, because it is the only object, besides the study of helioseismology (study of the structure and dynamics of the sun), which allows us a direct view into the solar interior. In Table 17.1, we have given a summary of the various solar neutrino experiments, the target used, and the period of active data taking, etc.

Broadly, the solar neutrino experiments are categorized into three groups:

- **Radiochemical experiment**: The basic reaction is

$$\nu_e + {}^A_{Z_1}X \rightarrow e^- + {}^A_{Z_2}Y,$$

where Z_1 and Z_2 are the atomic numbers of the two nuclei X and Y respectively. The nucleus in the final state, that is, ${}^A_{Z_2}Y$ is radioactive with a definite lifetime. The radioactivity of this nucleus is used in the detection of this reaction. The Homestake, SAGE, GALLEX, GNO experiments are radiochemical experiments.

- **Real time experiment**: The main detection method is neutrino–electron scattering, where the Cherenkov light originating from an outgoing electron is detected and analyzed. Cherenkov light is produced if a charged particle is moving with a velocity v, in a medium of refractive index n, such that, $\beta = \frac{v}{c} > 1$. The direction of the emitted light is also correlated with the direction of the incident neutrinos. Kamiokande, Super-Kamiokande, SNO etc. are real time experiments.

- **Organic liquid scintillator experiment**: A liquid scintillator is a mixture of an organic base solvent like pseudocumene (C_9H_{12}), linear alkyl benzene ($C_nH_{2n+1} - C_6H_5$, where $n = 10 - 13$), etc. and flour. Flour is used to achieve light emission. Neutrinos are detected in a charged current interaction of ν_e with a ${}^{12}C$ target, which is naturally contained in the scintillators. In the future, for next generation neutrino experiments, the plan is to use liquid scintillators based on water. The KamLAND, a liquid scintillator detector, although meant for detecting reactor antineutrinos, measured solar neutrino fluxes from ${}^8B$ and ${}^7Be$ sources in 2011 and 2014, respectively.

The threshold energies for radiochemical experiments are much lower than real time experiments (Table 17.1). The thresholds for real time experiments are high due to the presence of large environmental background caused by natural radioactivities (uranium, thorium, etc.) as well as the presence of spallation products (β-unstable nuclei like ${}^{16}N$, ${}^{12}N$, etc.) produced by cosmic ray muons. Therefore, to observe solar neutrinos, radiochemical experiments like GALLEX, GNO, etc., which have lower energy threshold, are preferred. However, if one has to understand the deficit in ${}^8B$ neutrino spectrum, due to the MSW (Mikheyev–Smirnov–Wolfenstein) effect (Chapter 18), then the real time experiments become important as the MSW effect not only distorts the ${}^8B$ neutrino spectrum, but also produces ν_μ and ν_τ, to which radiochemical experiments are not sensitive. This is due to the high energy threshold for charged current (CC) reactions induced by ν_μ and ν_τ, which is not possible in the case of solar neutrinos. However, in a real time experiment, through neutral current (NC) interactions, one should be able to observe the reaction:

$$\nu_i + A \rightarrow \nu_i + A',$$

where $i = \mu, \tau$, and A' is produced and detected in real time.

Table 17.1 shows the deficit of solar neutrinos observed in many experiments. However, it was the Sudbury Neutrino Observatory (SNO), using a real time Cherenkov detector, with 1000 tons of heavy water (D_2O), which clearly demonstrated the deficit of solar neutrino flux; the deficit was clearly proven by observing charged and neutral current processes simultaneously in the reactions:

$$\nu_e + {}_1^2 D \rightarrow e^- + p + p, \quad E_{th} = 1.442 \, \text{MeV(CC)}, \tag{17.2}$$

$$\nu_i + e^- \rightarrow \nu_i + e^- \text{(ES)}, \tag{17.3}$$

$$\nu_i + {}_1^2 D \rightarrow \nu_i + n + p, \quad E_{th} = 2.225 \, \text{MeV(NC)}, \tag{17.4}$$

where $i = e, \mu, \tau$. ES stands for elastic scattering. The neutrino fluxes were observed to be:

$$\Phi^{CC} = \left(1.76^{+0.06}_{-0.05}(\text{stat})^{+0.09}_{-0.09}(\text{syst}) \right) \times 10^6 \text{cm}^{-2}\text{s}^{-1}, \tag{17.5}$$

$$\Phi^{ES} = \left(2.39^{+0.24}_{-0.24}(\text{stat})^{+0.12}_{-0.12}(\text{syst}) \right) \times 10^6 \text{cm}^{-2}\text{s}^{-1}, \tag{17.6}$$

$$\Phi^{NC} = \left(5.09^{+0.44}_{-0.43}(\text{stat})^{+0.46}_{-0.43}(\text{syst}) \right) \times 10^6 \text{cm}^{-2}\text{s}^{-1}, \quad \text{giving} \tag{17.7}$$

$$\frac{\Phi^{CC}_{\nu_e}}{\Phi^{NC}_{\nu_e,\nu_\mu,\nu_\tau}} = 0.340 \pm 0.023^{+0.029}_{-0.031}, \tag{17.8}$$

$$\frac{\Phi^{CC}_{\nu_e}}{\Phi^{ES}} = 0.712 \pm 0.075^{+0.045}_{-0.044}. \tag{17.9}$$

It was found that Φ^{NC}, which was sensitive to all the three flavors of neutrinos, closely agrees with the predictions of SSM, while Φ^{CC}, which was sensitive to only ν_e, clearly showed a deficit.

Table 17.1 also summarizes the results for the ratio of the solar neutrino data to the results predicted from the Monte Carlo simulations (mainly based on SSM) (column 5). It may be clearly observed that for the charged current reactions (only ν_e induced), there is a deficit observed by many detectors like Kamiokande, SAGE, GALLEX, SNO, etc. In future, there are plans to detect solar neutrinos by using even larger detectors, as well as detectors having greater sensitivity.

Table 17.1 Solar neutrino detectors and their main characteristics. LS stands for liquid scintillator detectors.

Experiment	Reaction	Active mass (ton) & detector	E_{th} (MeV)	Data/MC	Location	Period
Homestake	$\nu_e\,{}^{37}_{17}Cl \to {}^{37}_{18}Ar\,e^-$	615 C_2Cl_4	0.814	0.32± 0.05	Homestake Gold Mine, South Dakota, USA	1967– 1994
Kamiokande II III	$\nu_e e^- \to \nu_e e^-$ $\nu_e e^- \to \nu_e e^-$	3000 H_2O	9.3/7.5 /7.0	0.44± 0.13 ± 0.08	Mozumi Mine, Japan	1986–90 1990–95
SAGE	$\nu_e\,{}^{71}_{31}Ga \to {}^{71}_{32}Ge\,e^-$	50 molten metal, Ga	0.233	0.52± 0.03	Baksan Neutrino Observatory, Russia.	1990– 2007
GALLEX	$\nu_e\,{}^{71}_{31}Ga \to {}^{71}_{32}Ge\,e^-$	30 $GaCl_3-HCl$	0.233	0.52± 0.03	Gran Sesso Laboratory, Italy.	1991– 1997
GNO	$\nu_e\,{}^{71}_{31}Ga \to {}^{71}_{32}Ge\,e^-$	30 $GaCl_3-HCl$	0.233	0.52± 0.03	Gran Sesso Laboratory, Italy.	1998– 2003
Super-K I II III IV	$\nu_e e^- \to \nu_e e^-$	50000 H_2O	4.5 6.5 4.5 3.5	0.42± 0.06	Under Mount Ikeno, Japan.	1996– 2001 2003– 2005 2006– 2008 2008– Present
SNO	$\nu_e D \to e^-\,p\,p$ $\nu_x D \to \nu_x\,p\,n$ $\nu_x\,e^- \to \nu_x\,e^-$	1000 pure D_2O	6.75/5/6	0.36±0.06 .30±.04 0.94±0 .14	Creighton mine, Sudbury, Canada	1999– 2006
Borexino I II	$\nu_x e^- \to \nu_x e^-$	300 C_9H_{12}	0.250	0.43± 0.10	Gran Sasso, Italy	2007– 2010 2011– Present
SNO+	$\nu_x e^- \to \nu_x e^-$ $\bar{\nu}_e\,p \to e^+\,n$	780 LS	¡1.0	-	Creighton Mine, Sudbury, Canada	Future
JUNO	$\bar{\nu}_e\,p \to e^+\,n$	20000 LS	1.0	-	Kaiping South China	Future
Hyper-K	$\nu_e e^- \to \nu_e e^-$	10^9 liter Water	~ 2.0	-	Kamioa, Japan	Future

17.3   Atmospheric Neutrinos

17.3.1   Introduction

Atmospheric neutrinos are typically produced in the interaction of primary cosmic ray particles with the earth's atmosphere. Cosmic rays are high energy particles, mainly protons and alpha particles, also categorized as primary cosmic rays. The observed energy and isotropy of the cosmic rays suggest that they come from outside the solar system, but within our galaxy, apart from a small component coming from solar flares. There is also a small fraction of cosmic rays that have other galactic and extra galactic sources. The exact source of cosmic rays is not yet known but may possibly be black holes, neutron stars, pulsars, quasars, or even the Big Bang. The composition of the primary cosmic rays are protons (90.6%), alpha particles (9%), and a small fraction of CNO nuclei (0.4%) above ~100 MeV/nucleus, and 95.2% protons, 4.5% alpha particles, and 0.3% of CNO nuclei above ~2 GeV/nucleus. When these primary cosmic rays interact with the earth's atmosphere, they produce secondary cosmic rays, consisting mainly of pions and kaons:

$$\left.\begin{array}{c} p + A_{\text{air}} \rightarrow n + \pi^+(K^+) + X \\ \downarrow \\ n + A_{\text{air}} \rightarrow p + \pi^-(K^-) + X. \end{array}\right\} \tag{17.10}$$

The cosmic ray spectrum starts from the sub-GeV energy region and extends roughly up to 10^{11} GeV [886]. In Figure 17.6, typical cosmic ray spectra for the proton and the alpha particles have been shown. When primary cosmic rays enter into the top of the earth's atmosphere,

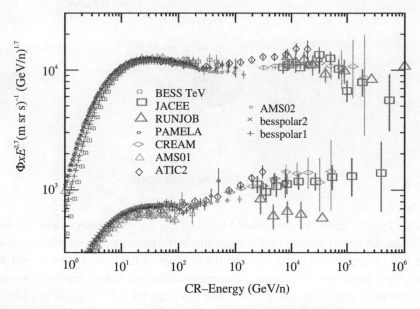

Figure 17.6   Primary cosmic ray flux for the proton (top line with experimental points) and alpha particles (bottom line with experimental points) [887, 888, 889, 890, 891, 892, 893, 117].

they propagate up to an atmospheric depth of about 15 km and produce secondaries through interaction. Subsequently, new particles are produced and gradually, the energy decreases. The energy of the charged particles is gradually absorbed by ionizing the matter present in the earth's atmosphere, as they pass through it.

The earth's geomagnetic field plays a very important role in the modulation of the intensity of cosmic rays.Suppose that the geomagnetic effect of the earth's magnetic field is switched off. Then the primary cosmic ray flux will almost be isotropic. However, the earth's geomagnetic field affects the cosmic rays, both outside, as well as inside the earth's atmosphere (Figure 17.7). Outside the atmosphere, it does not allow low energy particles to come in. Inside the atmosphere, it bends the charged secondaries. Of course, this depends upon the point of interaction, the direction and radius of curvature of the cosmic rays. Particularly, it is the strength of the horizontal component of the geomagnetic field (B_h) which plays a decisive role in the modulation of the cosmic ray flux. In literature, this is known as the effect due to rigidity cut off. The number of cosmic muons depend considerably on B_h. For example, at the INO site (Theni $9°96'7''$, $77°26'7''$ in India), this strength is $\sim$ 40000 nT, while at the Super-K (Kamioka mine $36°26''$, $137°10''$ in Japan), it is $\sim$ 30000 nT and at the South Pole ($-90°, 0°$), the strength is $\sim$ 16000 nT. Therefore, the maximum effect would be at the INO site and the minimum at the South Pole site. In fact, the east–west effect is a well-established phenomenon, where it was observed that the cosmic ray intensity is, in fact, greater in the west than in the east, showing that most of the primary cosmic ray particles are positively charged. The effect of the geomagnetic field in turn results in bending of the μ^+ beam, which enhances the east–west effect, while the μ^- bending tends to cancel east–west effect.

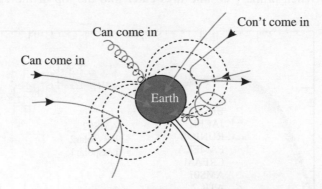

Figure 17.7 Rigidity cut off [894].

Since atmospheric neutrinos are created through the hadronic interactions of the primary cosmic rays and the nuclei present in the atmosphere, the number of produced pions and kaons which give rise to neutrinos depend more on the number of nucleons than the total number of nuclei, implying a larger contribution of atmospheric neutrinos from a heavier nucleus than the proton. For example, α particles carry $\sim$15% and CNO nuclei $\sim$3.6% above $\sim$2 GeV/nucleon, of the total nucleons in the cosmic rays despite 4.5% alpha particles and 0.3% of CNO nuclei being present in the cosmic rays.

The differential energy spectrum follows a power law of the form:

$$N(E)dE \propto E^{-\gamma}dE \tag{17.11}$$

with features known as "knee" and "ankle" at $2 - 5 \times 10^{15}$eV and $2 - 8 \times 10^{18}$eV, respectively. Hence, the spectrum steepens to $\gamma \simeq 3$ at the 'knee'. At the 'ankle', there is a flattening of spectrum, again resulting in $\gamma \simeq 2.7$. This has been shown in Figure 17.8.

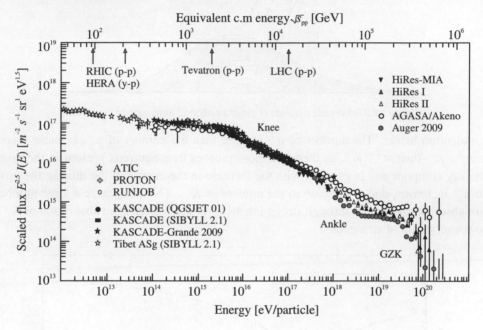

Figure 17.8 Primary cosmic ray flux for the proton showing knee, ankle, and GZK region [117].

The produced muon (μ^+) is also unstable and decays to a positron (e^+), a $\bar{\nu}_\mu$, and a ν_e (Figure 17.9). This decay chain is represented as:

$$
\begin{aligned}
&\pi^\pm \to \mu^\pm + \nu_\mu(\bar{\nu}_\mu) \quad (99.988\%) \qquad &&K^\pm \to \mu^\pm + \nu_\mu(\bar{\nu}_\mu) \quad (63.55\%) \\
&\quad\downarrow &&\quad \to \pi^\pm + \pi^o \quad (20.66\%) \\
&\quad e^\pm + \nu_e(\bar{\nu}_e) + \bar{\nu}_\mu(\nu_\mu) \quad (100\%) &&\quad \to \pi^\pm + \pi^+ + \pi^- \quad (5.6\%) \\
&K_s^o \to \pi^+ + \pi^- \quad (69.2\%) &&\quad \to \pi^o + \mu^\pm + \nu_\mu(\bar{\nu}_\mu) \quad (3.2\%) \quad (17.12) \\
&K_l^o \to \pi^+ + \pi^- + \pi^o \quad (12.54\%) &&\quad \to \pi^o + e^\pm + \nu_e(\bar{\nu}_e) \quad (4.25\%) \\
&\quad \to \pi^\mp + \mu^\pm + \nu_\mu(\bar{\nu}_\mu) \quad (27\%) &&\quad \to \pi^\pm + \pi^o + \pi^o \quad (1.76\%). \\
&\quad \to \pi^\mp + e^\pm + \nu_e(\bar{\nu}_e) \quad (40.55\%)
\end{aligned}
$$

The sources of neutrinos of energies up to around 100 GeV are mainly pion decays, while the higher energy neutrinos are produced from kaon decays, as the mean free path for the pions becomes sufficiently large at higher energies and they are able to reach the earth before decaying into muons and neutrinos (Figure 17.10). The most abundant charged particles at sea level are muons, which are produced high in the atmosphere ($\sim$15 km) and come to ground,

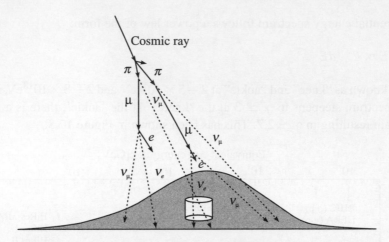

Figure 17.9 Schematic diagram of the production of atmospheric neutrinos.

after ionization losses. The number of μ^+ is more than the number of μ^-, because there are more π^+, K^+ than π^-, K^-, as there are more protons than neutrons present in the primary cosmic ray components. In general, with the increase in the energy of the muons, the ratio of μ^+ to μ^- increases, due to increase in the number of K^+. This is because a large number of K^- are absorbed by protons through strong interactions while there is no absorption of K^+ due to the conservation of strangeness.

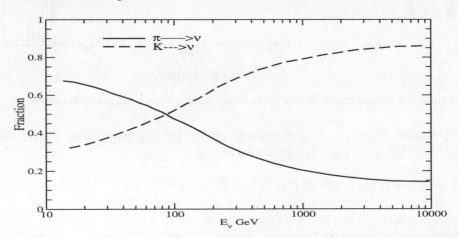

Figure 17.10 Fractional contribution of pions and kaons to the flux of neutrinos for the vertical direction.

The atmospheric neutrino flux is a convolution of the primary spectrum at the top of the atmosphere with a yield (Y) of neutrinos per primary particle. The flux of the neutrinos of type i can be represented as:

$$\Phi_{\nu_i} = \Phi_p \otimes R_p \otimes Y_{p \to \nu_i} + \sum_A \left\{ \Phi_A \otimes R_A \otimes Y_{A \to \nu_i} \right\}, \qquad (17.13)$$

where $\Phi_{p(A)}$ is the flux of the primary protons (nuclei of mass number A) outside the influence of the geomagnetic field and $R_{p(A)}$ represents the filtering effect of the geomagnetic field.

In the pion decays at rest (Chapter 6), the energy carried by the neutrinos is given by:

$$E_{\nu_\mu(\bar{\nu}_\mu)} = \frac{m_\pi^2 - m_\mu^2}{2m_\pi} \sim 30\,\text{MeV},$$

and the muons carry the rest of the energy. When a muon decays, each particle in the final state $(e^\pm + \nu_e(\bar{\nu}_e) + \bar{\nu}_\mu(\nu_\mu))$ carries almost one-third of the muon's energy, that is, approximately 36.5 MeV. When the pions are in flight, then also the energy distribution of the decay products of muons holds approximately the same distribution. Therefore, for all types of the pion spectrum, the flux ratio of the neutrinos from Eq. (17.12) for $\Phi(\nu_\mu + \bar{\nu}_\mu)$ to $\Phi(\nu_e + \bar{\nu}_e)$ is expected to be in the ratio 2:1. Moreover, the ratio of $\Phi(\nu_\mu)$ to $\Phi(\bar{\nu}_\mu)$ is expected to be ~ 1, because π^+ and π^- decay to ν_μ and $\bar{\nu}_\mu$, respectively. Moreover, inspecting Eq. (17.12) indicates that the $\Phi(\nu_e)$ to $\Phi(\bar{\nu}_e)$ ratio should also be equal to the $\Phi(\pi^+)$ to $\Phi(\pi^-)$ ratio. In the experiments, the atmospheric muon ratio $\Phi(\mu^+)$ to $\Phi(\mu^-)$ is observed with a reasonably good precision using the $\Phi(\pi^+)$ to $\Phi(\pi^-)$ ratio, which in turn gives the ratio of $\Phi(\nu_e)$ to $\Phi(\bar{\nu}_e)$ and $\Phi(\nu_\mu)$ to $\Phi(\bar{\nu}_\mu)$ with a good precision. Figure 17.11 shows the μ^+ and μ^- fluxes as a function of the muon momentum at the INO, the Super-K, and the South Pole by integrating over all the zenith and azimuthal bins for two different heights from the sea level, as well as at the sea level. It may be observed that the geomagnetic field has a strong effect which is described in terms of the rigidity cut off; the flux is maximum at the South Pole site and minimum at the INO site. Moreover, the number of muons decreases considerably at the sea level as compared to a height of say 8 km above the sea level.

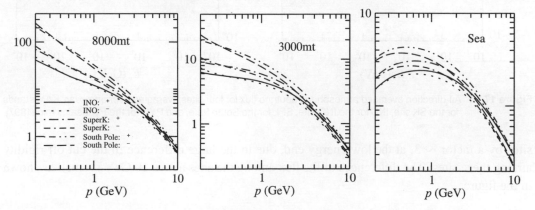

Figure 17.11 μ^+ (solid line) and μ^- (dashed line) fluxes as a function of the muon momentum at the INO, Super-K and South Pole by integrating over all the zenith and azimuthal bins for the two different heights from the sea level, as well as at the sea level.

In Figure 17.12, we show one year average of the atmospheric neutrino fluxes at the SK site $(36°26'', 137°10'')$, the INO site $(9°59'', 77°16'')$, the South Pole $(-90°00'', 0°00'')$, and the Pyhäsalmi $(63°40'', 6°41'')$ mine (Finland), averaging over all the directions [897]. The qualitative features are same at all the sites. However, we find a difference of flux among the

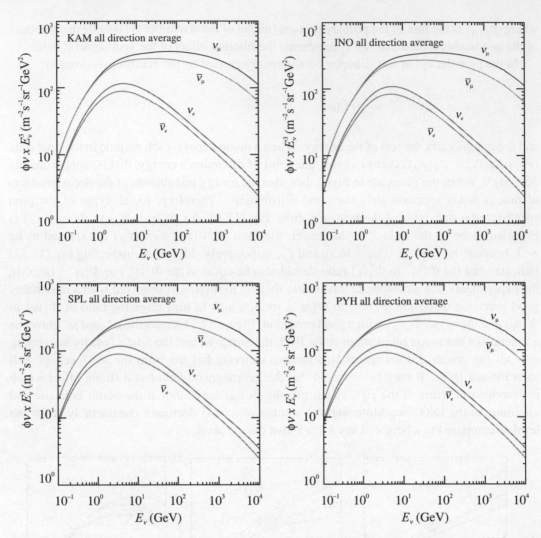

Figure 17.12 All-direction averaged atmospheric neutrino flux for four sites averaging over one year. KAM stands for the SK site, INO for the INO site, SPL for the South Pole, and PYH for the Pyhäsalmi mine [897].

sites by a factor ~ 3, at the low energy end, due to the large difference in the cut off rigidity among these sites. The differences of the flux among the sites above 10 GeV is small as shown in the figure.

17.3.2 Detection

The pioneering experiments to observe atmospheric neutrinos were started in the mid-1960s. These experiments were carried out in the Kolar Gold Field mines in India [895] and the East Rand Proprietary mine in South Africa [896]. These experiments were carried out in extremely deep underground laboratories at the depth of about 8000 meters water equivalent; the neutrino interactions occurring in the surrounding rocks were observed. At such depth, it was expected that the charged particles traversing the detectors almost horizontally would be of atmospheric neutrino origin. Most of the observed events were attributed to charged current interactions originating due to the interaction of muon type neutrinos, since it was required that the particle should penetrate through the rock as well as the detector.

The next generation of atmospheric neutrino experiments [898, 899, 900, 901, 902, 903] began in the mid-1980s in Europe (NUSEX and Frejus detectors), USA (IMB-3 detector), and Japan (Kamiokande detector). The main aim of these experiments was to look for nucleon decay candidates to verify the predictions of the grand unified theories (GUTs). In these experiments, the most serious background for proton decay was the events generated by atmospheric neutrinos. Therefore, the first requirement of these experiments was to study the atmospheric neutrino events in order to understand the proton decay background. Kamiokande measured the number of single Cherenkov ring e-like and μ-like events, which were mostly due to the charged current interactions of ν_e and ν_μ, respectively. One of the important observations made in 1988 was that the number of μ-like events had a significant deficit compared with the Monte Carlo prediction, while the number of e-like events were in agreement to the Monte Carlo predictions. This is known as the atmospheric neutrino anomaly. Kamiokande's results were later verified by the IMB water Cherenkov experiment in 1991 and in the updated analysis of the Kamiokande experiment in 1992, as well as in the Soudan-2 experiment in 1997.

Although not very conclusive, during the mid-1990s, the Kamiokande experiment found that the μ-like events show a deficit which was dependent on the zenith angle, and therefore on the neutrino flight length. The zenith angle distributions were studied for the multi-GeV fully contained events, as well as for the partially contained events. Fully contained events are those where the neutrino interacts within the volume of the inner detector, and no particle escapes the detector, while partially contained events are those where one or more particles penetrate far enough to leave the detector, and hence, their energy cannot be reliably measured. For example, higher energy charged current interactions may result in the muon exiting the detector. The vertically downward going neutrinos travel about 15 km, while the vertically upward going neutrinos travel about 12800 km (the earth's diameter) before interacting with the detector. The asymmetry in the downward going neutrinos vs. the upward going neutrinos was observed in 1996, in the Super-Kamiokande experiment, with a 50 kT water Cherenkov detector, where the data clearly showed the deficit of the upward going events in the multi-GeV energy range. Moreover, a smaller μ-like/e-like event ratio than the predictions of the Monte Carlo were model also observed. Similar observations were made by the MACRO and the Soudan-2 experiments.

Presently, there are five experiments, viz., the Super-Kamiokande, the MINOS, the SNO, the IceCube, and the ANTARES, which are sensitive to atmospheric neutrinos; many new experiments are being planned for the future. Some important features of these experiments are summarized in Table 17.2.

Table 17.2 Atmospheric neutrino detectors.

Experiment	Reaction	Active mass (ton) & detector	E_{th} (MeV)	Data/MC	Location	Period
Frejus	$p(X, Y)n\pi$ $\pi \to \mu + \nu_\mu$ $\mu \to e + \nu_\mu + \nu_e$	900 plastic flash tube	3.5×10^3	$1.06^{+0.19}_{-0.16}$ ± 0.15	Frejus highway tunnel, France	1984– 1988
IceCube	$\nu_l\, N \to \nu_l$ +Cascasde $\nu_l\, N \to l^-$ +Cascasde	0.66×10^9 Antarctic Ice	$\approx 10^3$		South Pole, Antarctica	2011– Present
IMB-3	$\nu_\mu \to \nu_\tau$	8000 Water Cherenkov	15.0	$0.54\pm$ $0.005\pm$ 0.12	Morton Salt Mine, Ohio, USA	1982– 1991
Kamiokande	$\nu_e e^- \to \nu_e e^-$	3000 H_2O	7.5	$0.44\pm$ $0.03\pm$ 0.08	Mozumi Mine, Japan	1986– 1995
MACRO	$0\nu\beta\beta$	5300 CAL			Gran Sesso, Italy	1995– 2000
MINOS	$\nu_\mu + \text{nucleus}$ $\to \mu^- + X$	5400 CAL	500		Illinois & Minnesota, USA	2005– 2012
NUSEX		150 ICAL	~ 250	$0.99+$ $0.035 - 0.25$	Mont Blanc tunnel, Europe	1982– 1988
Soudan-2		963 ICAL	< 500	$0.55^{+0.27}_{-0.19}$	Soudan Mine, USA	1989– 2001
Super-K I II III IV	$\nu_e e^- \to \nu_e e^-$ $\nu_e\, n \to e^- p$ $\nu_e\, p \to e^+ n$	50000 H_2O	SubGeV $< 1.3\text{GeV}$ MultiGeV $> 1.3\text{GeV}$	$0.42\pm$ 0.06	Under Mount Ikeno, Japan.	1996–2001 2003–2005 2006–2008 2008-present
JUNO	$\bar{\nu}_e p \to e^+ n$	20000 LS	1.0	-	Kaiping, China	Future
Hyper-K	$\nu_e e^- \to \nu_e e^-$	10^9 liter Water	~ 2.0	-	Kamioa, Japan	Future
DUNE	$\nu_e(^{40}\text{Ar},^{40}\text{K}^*)e^-$	68000 LAr-TPC	4.5	-	Homestake Mine, USA	Future
INO		50000 ICAL	< 1.0	-	Tamil Nadu, India	Future
PINGU	$\nu_l\, N \to \nu_l$ +Cascasde $\nu_l\, N \to l^-$ +Cascasde	6×10^6 Ice	$< 10^4$	-	South Pole, Antarctica	Future

17.4 Reactor Antineutrinos

The antineutrino was detected for the first time in 1956 by Reines and Cowan, at Savannah nuclear reactor power plant; since then, reactor antineutrinos have played an important role in both the discovery, as well as in the precision measurements of some of the neutrino oscillation parameters. The isotopes rich in neutrons like ^{235}U, ^{238}U, ^{239}Pu, ^{241}Pu, through the fission processes in the power reactors, undergo a series of decays which may be broadly represented by the reaction:

$$\ _{Z}^{A}X \rightarrow \ _{Z+1}^{A}Y + e^{-} + \bar{\nu}_{e}, \tag{17.14}$$

which produce approximately six antineutrinos per fission. The mixture of the isotopes produced is complex (mixture of several isotopes), which results into a continuous energy spectrum of $\bar{\nu}_e$s, with energies ranging between 0-8 MeV and the average energy around 2-3 MeV, along with the emission of $\sim$ 200MeV of fission energy. The flux of the antineutrinos (Fig. 17.13) is proportional to the thermal power of the reactors core. For example, the fission

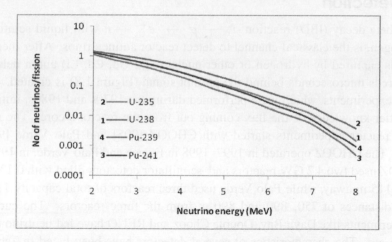

Figure 17.13 Typical antineutrino spectra [904].

reaction in ^{235}U produces fission products such as Ba, Kr, Sr, Cs, I, and Xe having atomic masses in the intermediate mass range, viz., 95 and 145:

$$^{235}U + n \rightarrow \ ^{144}Ba + ^{90}Kr + 2n + 200 \text{ MeV},$$
$$^{235}U + n \rightarrow \ ^{141}Ba + ^{92}Kr + 3n + 170 \text{ MeV},$$
$$^{235}U + n \rightarrow \ ^{139}Te + ^{94}Zr + 3n + 197 \text{ MeV}. \tag{17.15}$$

Typically, if in one fission reaction, 200 MeV of energy is released which corresponds to 3.2×10^{-11}J of energy; on an average, if six $\bar{\nu}_e$s are released, then for a Giga-Watt reactor, there would be typically 2×10^{20} $\bar{\nu}_e$s being produced in each second corresponding to a flux density given by:

$$\Phi(\bar{\nu}_e) \sim 1.6 \times 10^{19} \frac{P/GW}{L^2/m^2} m^{-2} s^{-1}, \tag{17.16}$$

where P is thermal power in Giga-Watt (GW) and L is the distance in meters from the reactor. In Figure 17.13, we have shown typical antineutrino spectra emitted by various sources.

Instead of fission rates, generally, fission fractions in the core simulation are used, which is the ratio of the fission rate of an isotope over the total rate. The reactor antineutrino spectrum is calculated as:

$$\Phi(E_\nu) = \frac{W_{th}}{\sum_i f_i e_i} \cdot \sum_i f_i \cdot S_i(E_\nu), \tag{17.17}$$

where E_ν is the antineutrino energy, and f_i, e_i, and $S_i(E_\nu)$ are the fission fraction, thermal energy released in each fission, and the antineutrino spectrum per fission for the *ith* isotope, respectively.

17.4.1 Detection

The inverse beta decay (IBD) reaction, $\bar{\nu}_e + p \rightarrow e^+ + n$ with liquid scintillators (LS) rich in hydrogen, is the classical channel to detect reactor antineutrinos. After thermalization, the neutron is captured by hydrogen or other nuclei (e.g., Gd, Cd, Li) and a delayed signal, tens or hundreds microseconds behind the prompt signal (Figure 1.8) is created. The reactor antineutrino experiments, which were performed during the 1970s and 1980s, initiated efforts to have a better knowledge of the flux coming out from the reactor's core. The modern day antineutrino reactor experiments started with CHOOZ [905] and Palo Verde [906] reactor experiments. The CHOOZ operated in 1997–1998 in France and Palo Verde, in 1998–2000, in USA. CHOOZ used two 4.2 GW reactors and scintillator detectors filled with 0.1% Gd placed 998 m and 1115 m away, while Palo Verde used three reactors of total capacity 11.6 GW and detectors at distances of 750, 890 and 890 m from the three reactors. The current reactor neutrino experiments like Daya Bay, Double Chooz and RENO have led neutrino physics into the precision era. The characteristics of several detectors have been listed in Table 17.3 and Table 17.4, for the short baseline and long baseline experiments, respectively.

A multi-purpose detector, the KamLAND, is placed at the site of the former Kamiokande experiment with a vertical overburden of 2,700 meters water equivalent (m.w.e.), with the primary goal to search for reactor antineutrino ($\bar{\nu}_e$) oscillations. It is surrounded by 55 Japanese nuclear reactor cores. The $\bar{\nu}_e$ flux weighted average baseline is 180 km. The KamLAND consists of 1 kton of ultrapure liquid scintillator. The JUNO, near Guangzhou in South China, is a next generation experiment. A total power of 26.6 GW will be available by 2020, when JUNO is scheduled to start taking data.

Table 17.3 Reactor (short baseline) neutrino detectors.

Experiment	Reaction	Active mass (ton) & detector	E_{th} (MeV)	Location	Period
CHOOZ	$\bar{\nu}_e p \to e^+ n$	5 Gd-LS	1.8	Ardennes, France	1997– 1998
Gösgen	$\bar{\nu}_e p \to e^+ n$	377 liter LS	1.8	Gösgen Switzerland	1985–
Double CHOOZ	$\bar{\nu}_e p \to e^+ n$	10.16 Gd-LS	1.8	Ardennes, France	2010– 2015
Nucifer	$\bar{\nu}_e p \to e^+ n$	850liter Gd-LS	1.8	CEA-Saclay, France	2012
NEOS	$\bar{\nu}_e p \to e^+ n$	~1 Gd-LS	1.8	Younggwang, Korea	2015– 2016
Neutrino-4	$\bar{\nu}_e p \to e^+ n$	$1.8 m^3$ volume Gd-LS	~1.8	Dimitrovgrand, Russia.	2016– Present
STEREO	$\bar{\nu}_e p \to e^+ n$	Gd-LS	~2.0	Grenoble, France	2016– Present
RENO	$\bar{\nu}_e p \to e^+ n$	16 Gd-LS	1.8	Younggwang, Korea	2011– Present

Table 17.4 Reactor (long baseline) neutrino detectors.

Experiment	Reaction	Active mass (ton) & detector	E_{th} Energy (MeV)	Location	Period
Palo Verde	$\bar{\nu}_e p \to e^+ n$	11.34 GD-LS	0.04 (low) 0.6 (high)	Arizona, USA	1998– 2000
KamLAND	$\bar{\nu}_e p \to e^+ n$	~1000 Water	~ 1.8	Kamioka, Japan	2002– Present
Daya Bay	$\bar{\nu}_e p \to e^+ n$	20 (GD-LS) +20 (LS) +37 (oil)	~ 1.8	Southern coast of China	2011– Present
RENO-50	$\bar{\nu}_e p \to e^+ n$	18000 LS	~1.8	Younggwang, Korea	Future
JUNO	$\bar{\nu}_e p \to e^+ n$	20000 LS	1.0	Kaiping South China.	Future

17.5 Supernova Neutrinos

17.5.1 Introduction

Supernova neutrinos and antineutrinos of all flavors are associated with one of the most explosive and violent astrophysical events: the supernova explosion, which occurs during the death phase of a massive star ($M_{\text{Star}} \gtrsim 8M_\odot$). Such massive stars become unstable at the end of their lives. They collapse and eject their outer mantle through a supernova explosion. These supernova explosions are the sources of the most powerful cosmic neutrinos in the MeV range. Colgate and White [907] and independently Arnett [908], were among the first who pointed out the importance of neutrinos in the context of stellar core collapse and theorized the mechanism through which massive stars die in a supernova (SN) explosion. The huge amount of energy released in a supernova explosion is due to the difference in the gravitational binding energy of a star and a neutron star:

$$E_{BE} \approx \left(-\frac{3}{5}\frac{GM_S^2}{R_S}\right) - \left(-\frac{3}{5}\frac{GM_{NS}^2}{R_{NS}}\right) \text{ and } R_S >> R_{NS}$$

$$\approx 3.6 \times 10^{47} \left(\frac{M_{NS}}{1.5\,M_\odot}\right)^2 \left(\frac{R_{NS}}{10\,\text{km}}\right)^{-1} \text{ J}, \qquad (17.18)$$

where M_S and R_S are the mass and radius of a star which can undergo supernova explosion, $M_\odot$ is mass of the sun, M_{NS} and R_{NS} are, respectively, the mass and the radius of a neutron star. During such an event, the exploding star becomes comparable in brightness to its entire host galaxy! In course, colossal amounts of energy is released, equal to the total amount of energy that the sun radiates in 10 million years. The supernovae arise from two different final stages of stars. They are either caused by the thermonuclear explosions of a white dwarf within a binary system or by the core collapse of massive stars. They are in general classified as type I and type II supernovae. The type I supernovae occur in binary systems; a binary system is a system of two stars orbiting one another, and among them, one of the stars is a white dwarf, while the other can be a giant star or an even smaller white dwarf. Type I supernovae are further divided into subclasses, for example, type Ia, Ib, Ic according to the existence of hydrogen, helium, or silicon spectral lines. We shall discuss the formation and death of stars in Chapter 19, where the supernova explosion mechanism will also be explored.

In a supernova explosion, most of the gravitational energy released in the core collapse is carried by neutrinos. Such neutrino bursts carry $\approx 3.6 \times 10^{47}$ J of energy in a very short period of time [909]. It is considered that these neutrinos could provide valuable information about the proto-neutron star core, its equation of state, core collapse, and supernova explosion mechanism; they can help to have a better understanding of supernova physics [909, 910]. After the observation of supernova neutrinos from SN1987A at Kamiokande, IMB, and BAKSAN [911, 912], the feasibility of detecting such events in the future has now been given serious consideration. For example, experiments like the Super-K [913], LVD [914], AMANDA [915], BOREXino [916], OMNIS [917], LAND [918], HALO [919], ICARUS [920], etc. are in various stages of operation while SNO+ [921], Hyper-K [922] experiments are being developed. Experiments like DUNE [629] and JUNO [923, 924] have been planned to study the physics related to supernova neutrinos in the near future. A list of the present and future experiments, having sensitivity to supernova neutrinos/antineutrinos is given in Table 17.5. These experiments are planned to use detectors with various nuclei as the target materials.

The supernova neutrino/antineutrino fluxes are determined from numerical simulations of the core collapse supernova explosion of a star and depend on the initial properties of the

Table 17.5 Supernova neutrino detectors.

Experiment	Reaction	Active mass (ton) & detector	E_{th} (MeV)	Location	Period
Baikal		Cherenkov	10^4	Lake Baikal, Russia	1980–1998
IMB	$\nu_\mu \to \nu_\tau$	8000 Water Cherenkov	15.0	Morton Salt Mine, Ohio, USA	1982–1991
Kamiokande	$\nu_e e^- \to \nu_e e^-$	3000 H_2O	7.5	Mozumi Mine, Japan.	1986–1995
LSD	$\bar{\nu}_e\, p \to e^+\, n$	90 LS	5	Mont Blanc Laboratory	1984–1999
LVD	$\bar{\nu}_e\, p \to e^+\, n$	1800 LS	0.5,4.0	Gran Sasso, Italy	1992-2016
ICARUS	$\nu_e(^{40}\mathrm{Ar},^{40}\mathrm{K})e^-$	600 LAr	~ 5	Gran Sasso, Italy	Future
KamLAND	$\bar{\nu}_e\, p \to e^+\, n$	~ 1000	~ 1.8	Kamioka, Japan	2002–Present
MiniBooNE	$\nu_\mu \to \nu_e$	800 Mineral oil		Fermilab, USA	2002–2017
HALO	$\nu_e + \mathrm{Pb} \to e^- + \mathrm{Bi}^*$ $\mathrm{Bi}^* \to \mathrm{Bi} + \gamma + n$ $\nu_x + \mathrm{Pb} \to \nu_x + \mathrm{Pb}^*$ $\mathrm{Pb}^* \to \mathrm{Pb} + \gamma + n$	1000 He & Pb	18	Creighton Mine in Sudbury, Canada	2015–Present
JUNO	$\bar{\nu}_e\, p \to e^+\, n$	20000 LS	1.0	Kaiping South China.	Future
DUNE	$\nu_e(^{40}\mathrm{Ar},^{40}\mathrm{K}^*)e^-$	68000 LAr-TPC	4.5	Homestake Mine, USA.	Future

collapsing star like its mass, density, and equation of state, as well as, on various physical processes controlling the explosion like the initial prompt burst of the neutrinos following the neutronization, accretion, and cooling in the late phases and the neutrino transport in the dense star matter [925, 926, 927, 928]. The neutrino/antineutrino fluxes are found to be sensitive to the luminosities L_ν of the various neutrino/antineutrino flavors, which are believed to be equal for all the six flavors of neutrinos/antineutrinos ν_e, $\bar{\nu}_e$, ν_x, and $\bar{\nu}_x$ ($x = \mu, \tau$), due to the assumption of equipartition of total available energy amongst various flavors. Some recent calculations have also been done assuming luminosities for ν_x different from the luminosities of $\nu_e(\bar{\nu}_e)$ and varying them in the range of $0.5L_{\nu_e} < L_{\nu_x} < 2L_{\nu_e}$, keeping the luminosities of ν_e and $\bar{\nu}_e$ to be the same [929, 930, 931, 932, 933, 934]. The simulated neutrino/antineutrino fluxes and mean energies of their various flavors are, in general, distinct from each other due to the differences in their interaction with dense star matter, which has an excess of neutrons over protons. This difference leads to ν_e losing more energy as compared to $\bar{\nu}_e$, which loses more energy than ν_x ($x = \mu$, τ, and their antineutrinos), as $\nu_x(\bar{\nu}_x)$ interacts only through neutral current interaction (due to higher threshold energy for charged current reactions induced by ν_x), while ν_e and $\bar{\nu}_e$ interact through neutral as well as charged current interactions. This gives a hierarchical structure of mean neutrino/antineutrino energies ($E_{\nu(\bar{\nu})}$) for various flavors, that is, $\langle E_{\nu_e} \rangle < \langle E_{\bar{\nu}_e} \rangle < \langle E_{\nu_x} \rangle$. The various simulations agree on this hierarchical structure of mean neutrino/antineutrino energies, but differ on the actual values which are generally taken to be in the range of $\langle E_{\nu_e} \rangle \approx 10 - 12$ MeV, $\langle E_{\bar{\nu}_e} \rangle \approx 12 - 15$ MeV, and $\langle E_{\nu_x} \rangle \approx 16 - 25$ MeV [929, 930, 931, 932, 933, 934]. However, a lower value of $\langle E_{\nu_e} \rangle$ has also been obtained in some recent studies [935], when additional medium effects were taken into account. These additional medium effects are generated by neutrino–neutrino self interactions [936, 937, 938, 939, 940] and neutrino–matter interactions when primary neutrinos of all the flavors propagate through a medium of very high neutrino and matter densities, causing flavor conversions [941, 942, 943, 944].

The simulated neutrino flavor spectra at the surface of a star are subjected to various uncertainties arising from the uncertainties in the theoretical parameters used in simulation studies of the explosion and propagation of neutrinos inside the dense star matter. This leads to large variations in the predicted spectra for various flavors of neutrino/antineutrino [929, 930] calculated by different authors. Figure 17.14 represents supernova (a) neutrino and (b) antineutrino fluxes simulated by some of the groups working on simulation for example, Totani et al. [925], Duan et al. [937] and Gava et al. [938] for electron neutrinos.

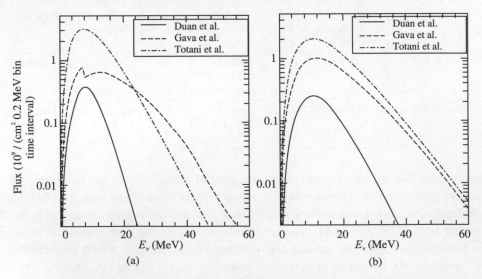

Figure 17.14 Supernova (a) neutrino and (b) antineutrino fluxes simulated by some of the groups like Totani et al. [925] (dashed-dotted line), Duan et al. [937] (solid line) and Gava et al. [938] (dashed line).

17.5.2 Neutrino emission in supernova explosions

Neutrinos are produced during the collapse of the core when electrons are captured by protons and nuclei leading to neutronization of the core; neutrinos are also produced through the process of thermal emission. Some of the several neutronization reactions are:

- Electron capture by nuclei:

$$e^- + {}^A_Z X \rightarrow \nu_e + {}^A_{Z-1} Y. \tag{17.19}$$

- Electron capture by free proton:

$$e^- + p \rightarrow \nu_e + n. \tag{17.20}$$

Some of the thermal emission processes are:

- Pair annihilation:

$$e^+ + e^- \rightarrow \nu_e + \bar{\nu}_e. \tag{17.21}$$

- Plasmon decay:

$$\gamma \rightarrow \nu + \bar{\nu}. \tag{17.22}$$

- Photoannihilation:

$$e^\pm + \gamma \rightarrow e^\pm + \nu + \bar{\nu}. \tag{17.23}$$

- Electron–nucleon Bremsstrahlung:

$$e^\pm + N \rightarrow e^\pm + N + \nu + \bar{\nu}. \tag{17.24}$$

- Nucleon–nucleon Bremsstrahlung:

$$N + N \rightarrow N + N + \nu + \bar{\nu}. \tag{17.25}$$

Each neutrino emission process has an inverse process corresponding to absorption. Both the absorption and the scattering processes delay the free escape of neutrinos from a collapsing core. The most important processes are as follows:

- Free nucleon scattering:

$$\begin{aligned} \nu + n &\rightarrow \nu + n, \\ \nu + p &\rightarrow \nu + p. \end{aligned} \tag{17.26}$$

- Coherent scattering by heavy nuclei ($A > 1$):

$$\nu + {}^A_Z X \rightarrow \nu + {}^A_Z X. \tag{17.27}$$

- Nucleon absorption:

$$\nu_e + n \rightarrow e^- + p. \tag{17.28}$$

- Electron–neutrino scattering:

$$e^- + \nu \rightarrow e^- + \nu. \tag{17.29}$$

Similar processes also occur for antineutrinos.

17.5.3 Detection

Many neutrino detectors are connected through the supernova early warning system (SNEWS), which will trigger experiments to record and save additional data if a sudden influx of neutrinos (indicating a supernova) arrives. On February 23, 1987, scientists caught neutrinos from a supernova (SN1987A) in a nearby galaxy, the large magellanic cloud (LMC), 50 kilo parsec away from the earth. The two water Cherenkov detectors, one in Japan, Kamiokande-II, and the other in the United States, the IMB, observed 19 events within 13 seconds, with timing consistent with the optical observation of the SN1987A explosion. Two smaller scintillation detectors, the BAKSAN and LSD also reported several events; however, the evidences were not very conclusive. The observation of neutrinos from SN1987A has been a major event in experimental neutrino physics which has been recognized by awarding the Nobel prize in physics to M. Koshiba, the leader of the Kamiokande experiment in 2002.

17.6 Geoneutrinos

Geoneutrinos are electron antineutrinos released in the decay of radioactive elements; they are distributed throughout the earth's interior. These geoneutrinos are, in fact, the only source of information about the geochemical composition of the earth's core. They are emitted in the decay chain of ^{40}K, ^{232}Th, and ^{238}U:

$$^{238}\text{U} \rightarrow {}^{206}\text{Pb} + 8\,{}^4\text{He} + 8e^- + 6\bar{\nu}_e + 51.7\,\text{MeV}, \tag{17.30}$$

$$^{235}\text{U} \rightarrow {}^{207}\text{Pb} + 7\,{}^4\text{He} + 4e^- + 4\bar{\nu}_e + 46.4\,\text{MeV}, \tag{17.31}$$

$$^{232}\text{Th} \rightarrow {}^{208}\text{Pb} + 6\,{}^4\text{He} + 4e^- + 4\bar{\nu}_e + 42.7\,\text{MeV}, \tag{17.32}$$

$$^{40}\text{K} \rightarrow {}^{40}\text{Ca} + e^- + \bar{\nu}_e + 1.31\,\text{MeV}\;(89.3\%), \tag{17.33}$$

$$^{40}\text{K} + e^- \rightarrow {}^{40}\text{Ar} + \nu_e + 1.505\,\text{MeV}\;\;(10.7\%). \tag{17.34}$$

The first direct measurement of the geoneutrino flux was made by the BOREXino and KamLAND collaborations. Only ^{238}U and ^{232}Th chains are measurable with the neutron inverse beta decay process.

The geoneutrino spectrum extends up to ~ 4.5 MeV (Figure 17.15) and the contributions originating from the different elements can be distinguished according to their different endpoints, that is, geoneutrinos with $E > 2.25$ MeV are produced only in the uranium chain. In the future, a few detectors are planned to further study the geoneutrinos, like SNO+, which is filled with 780 tons of liquid scintillator in a 12 m diameter acrylic sphere, shielded by ultra pure water along with 9000 photomultipliers.

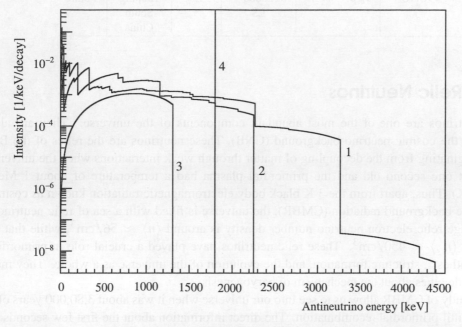

Figure 17.15 $\bar{\nu}_e$ energy distributions for (1) ^{238}U (solid), (2) ^{232}Th (dotted), and (3) ^{40}K (dash-dotted) decay. The vertical line (4) represents the $\bar{\nu}_e$ detection threshold for neutron inverse β decay. [945]

The other planned detectors aimed at measuring geoneutrinos are quite amazing. For example, the Hanohano deep-ocean transportable detector which is planned to be a mobile, sinkable 10 kton liquid scintillation detector, would be carried by a ship and deployed in the deep ocean at some 3 to 5 km depth, near Hawaii. The Daya Bay 2 experiment, is characterized by a very large mass of some 20 kton and is supposed to detect up to 400 geoneutrinos per year. However, due to the presence of nuclear power plants in the close vicinity, the detection of geoneutrinos would be quite challenging. The LENA (Low Energy Neutrino Astronomy) detector consists of a huge 50 kton multidisciplinary liquid scintillation neutrino detector; the detection of geoneutrinos is one of its main scientific goals. Table 17.6 lists some detectors which are sensitive to geoneutrinos.

Table 17.6 Geoneutrino detectors.

Experiment	Reaction	Active mass (ton) & detector	Eth (MeV)	Location	Period
KamLAND	$\bar{\nu}_e\, p \rightarrow e^+\, n$	~1000 water	~ 1.8	Kamioka, Japan	2002– Present
SNO+	$\nu_x e^- \rightarrow \nu_x e^-$ $\bar{\nu}_e\, p \rightarrow e^+\, n$	780 LS	< 1.0	Creighton Mine, Sudbury, Canada	Future
JUNO	$\bar{\nu}_e\, p \rightarrow e^+\, n$	20000 LS	1.0	Kaiping South China	Future

17.7 Relic Neutrinos

Relic neutrinos are one of the most abundant components of the universe. They are also known as the cosmic neutrino background (CNB). These neutrinos are the relics of the Big Bang, originating from the decoupling of matter through weak interactions when the universe was about one second old and the primordial plasma had a temperature of about 1 MeV ($\simeq 10^{10}$ K). Thus, apart from the 3 K black body electromagnetic radiation known as cosmic microwave background radiation (CMBR), the universe is filled with a sea of relic neutrinos. The average relic electron neutrino number density is around $\langle n \rangle \approx 56/\text{cm}^3$, while that of CMBR is $\langle n_\gamma \rangle \approx 450/\text{cm}^3$. These relic neutrinos have played a crucial role in primordial nucleosynthesis, structure formation, and the evolution of the universe as a whole. They may also provide clues about the mechanism of baryogenesis.

The study of CMBR allow us to see into our universe when it was about 3,80,000 years old, right up until primordial recombination. The direct information about the first few seconds of the universe evolution, principally, can be obtained by the detection of relic neutrinos.

These relic neutrinos were formed with a thermal equilibrium spectra, which for neutrinos (ν) and antineutrinos ($\bar{\nu}$), are given by the Fermi–Dirac distribution (with zero neutrino mass):

$$n_{\nu,\bar{\nu}}(p)dp = \frac{1}{(2\pi\hbar)^3} \frac{4\pi p^2 dp}{\exp(pc/kT)+1}. \tag{17.35}$$

The theory of primordial nucleosynthesis gives us the relation between the temperatures of relic neutrinos and photons, that is,

$$T_\nu = (4/11)^{1/3} T_\gamma,$$

arising from electron–positron annihilation. Given this relation and the current value of $T_\gamma = 2.725 \pm 0.001$ K, the temperature of relic neutrinos should be $T_\nu \approx 1.945$ K $\simeq 2 \times 10^{-4}$eV.

The detection of relic neutrinos seems to be hardly possible due to the weak interaction of neutrinos with matter and also due to their low energy ($\approx 10^{-4}$eV). Several methods have been discussed in literature to look for CNB, but there is not much progress. The most promising suggested detection methodology is through IBD in the process $\nu_e + {}^3\text{H} \rightarrow e^- + {}^3\text{He}$, where the signal will correspond to the peak in the electron spectrum. It has been proposed that such detection is possible in the KATRIN experiment [946, 947].

17.8 Accelerator Neutrinos

Pontecorvo and Schwartz, independently, proposed the idea of performing neutrino experiments with accelerators. They proposed the possibility of an experiment making use of a neutrino beam produced by pion decays at proton accelerators. The first accelerator neutrino experiments were performed by a team led by Lederman, Schwartz, and Stenberger in 1962 at the AGS accelerator, Brookhaven and very soon after this experiment, the neutrino beams at CERN were obtained. The accelerator facilities are used to accelerate the protons to very high energies. These highly energetic protons are smashed into a target; the target can be any material, although it has to be able to withstand very high temperatures. When a proton traveling near the speed of light hits a target, it slows down and the proton's energy is used to produce a jet of hadrons. There are different kinds of particles in this jet; however, the most common are pions and kaons. Neutral pions are not part of story, so we can ignore them; however, the charged pions so produced are unstable and decay essentially into muons and neutrinos. A typical sketch of neutrino beam production from hadrons at an accelerator facility is shown in Figure 17.16. A meson, carrying electric charge, can be collimated using electric and magnetic fields

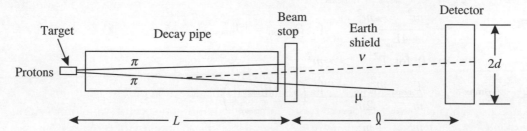

Figure 17.16 Typical sketch of neutrino beam production by accelerators.

known as magnetic horns. Thus, to get a neutrino beam in a certain direction, one points the pion in the direction of the detector. A properly designed horn system can enhance the neutrino flux. To estimate the neutrino flux with better accuracy, it is important to precisely measure the momentum and the angular spectra of the mesons. The probability P of the mesons to decay while traveling a distance L through the decay pipe is given by

$$P = 1 - e^{-\frac{L}{L_0}}, \tag{17.36}$$

where L_0 is the pion decay length and is given by:

$$L_0 = \gamma\beta\tau_{\text{meson}}. \tag{17.37}$$

Here, τ_{meson} is the decay time of the meson and

$$\beta = \frac{|\vec{p}_{\text{meson}}|}{E_{\text{meson}}} \quad \text{and} \quad \gamma = \frac{E_{\text{meson}}}{m_{\text{meson}}}.$$

For example, for a 5 GeV pion, the decay length would be $L_o \sim 280$ m. The energy spectrum of neutrino is obtained using the kinematics of the two-body mesonic decay. Let E_ν and $\cos\theta_\nu$ be, respectively, the energy and angle in the Lab frame and E_ν^{CM} and $\cos\theta_\nu^{\text{CM}}$ be the energy and angle in the CM frame. Then,

$$E_\nu = \gamma\left(1 + \beta\cos\theta_\nu^{\text{CM}}\right)E_\nu^{\text{CM}}, \text{ and}$$

$$\cos\theta = \frac{\cos\theta_\nu^{\text{CM}} + \beta}{1 + \beta\cos\theta_\nu^{\text{CM}}}, \text{ where}$$

$$E_\nu^{\text{CM}} = \frac{m_{\text{meson}}^2 - m_\mu^2}{2m_{\text{meson}}}, \text{ which is 29.8 MeV for } \pi\text{-decay.} \qquad (17.38)$$

The limits on $\cos\theta_\nu^{\text{CM}}$, that is, $-1 \leq \cos\theta_\nu^{\text{CM}} \leq +1$, correspond to the minimum and maximum energies in the Lab frame, that is,

$$
\begin{aligned}
E_\nu^{\text{min}} &= \frac{E_{\text{meson}}}{m_{\text{meson}}} \frac{m_{\text{meson}}^2 - m_\mu^2}{2m_{\text{meson}}}\left(1 - \frac{|\vec{p}_{\text{meson}}|}{E_{\text{meson}}}\right) \\
&= \frac{m_{\text{meson}}^2 - m_\mu^2}{2\left(E_{\text{meson}} + |\vec{p}_{\text{meson}}|\right)} \\
&\simeq \frac{m_{\text{meson}}^2 - m_\mu^2}{4E_{\text{meson}}} \qquad (\because E_{\text{meson}} = |\vec{p}|_{\text{meson}}) \\
&\sim 0 \text{ for } E_{\text{meson}}^2 >> m_{\text{meson}}^2. \\
E_\nu^{\text{max}} &= \frac{E_{\text{meson}}}{m_{\text{meson}}} \frac{m_{\text{meson}}^2 - m_\mu^2}{2m_{\text{meson}}}\left(1 + \frac{|\vec{p}_{\text{meson}}|}{E_{\text{meson}}}\right) \\
&\simeq \frac{m_{\text{meson}}^2 - m_\mu^2}{2m_{\text{meson}}^2}\left(E_{\text{meson}} + |\vec{p}_{\text{meson}}|\right).
\end{aligned}
$$

It may be noted that the different neutrino energies may be obtained by varying the lifetime and the Lorentz boost (γ_{meson}) of the mesons. Figure 17.17 shows the neutrino fluxes which could be obtained at various accelerators in 1973. This figure has been taken from Ref. [294]. Using modern day accelerators, two different types of neutrino beams are obtained. One is known as the wide band beam and the other is known as the narrow band beam.

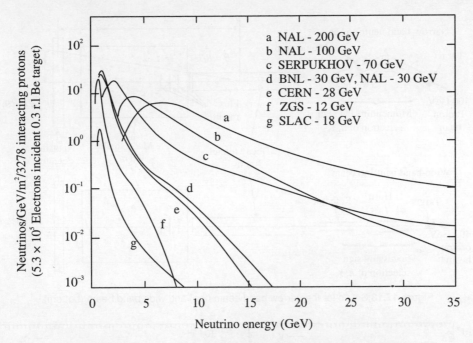

Figure 17.17 Flux of neutrinos from the older generation accelerators [294].

17.8.1 Wide and narrow band of neutrino beams

In a beam of primary protons, the intensity of the beam depends upon the accelerator's performance. To achieve maximum intensity, the produced pions and kaons are to be focused and independent of the energies which they were produced. In this case, a cylindrical target struck by the protons is aligned with the decay tunnel and magnetic horns are placed to focus the mesons (Figure 17.18). Mesons decay in the decay pipe to give neutrinos; the neutrino beam so obtained is known as the wide band beam. In such beams, the energy spectrum and the fraction of different neutrino species can be varied by

(i) tuning the focusing system by altering the shape of the horn and using different horns in cascade, and

(ii) altering the length of the decay tunnel, which affects the decay directions of pions and kaons of different energies differently.

The drawback with a wide band beam is that it is very difficult to precisely estimate the energy spectrum and concentrate (relative amount) on the different neutrino species in the beam.

Modern day accelerators providing the neutrino and antineutrino fluxes, are available at the Fermilab in USA, JPARC in Japan, CERN in Geneva, etc. In Figure (17.19), we show the ν_μ and $\bar{\nu}_\mu$ spectra at the Fermilab which have been used by the MiniBooNE collaboration and in Figure 17.20, the low energy spectra being used by the MINERvA collaboration is shown.

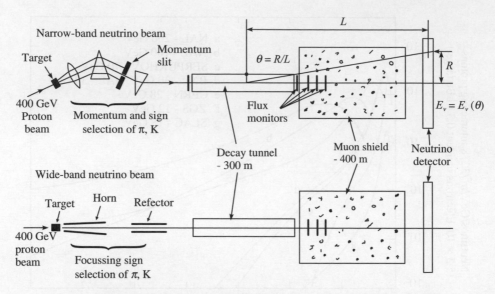

Figure 17.18 Set up for the narrow band beam (top) and wide band beam (bottom).

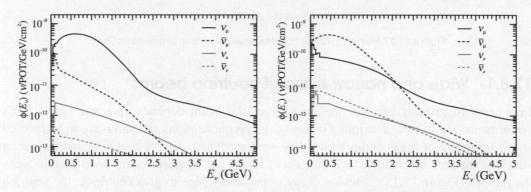

Figure 17.19 MiniBooNE spectra [948] for neutrino (ν_μ) (left) and antineutrino ($\bar{\nu}_\mu$) (right).

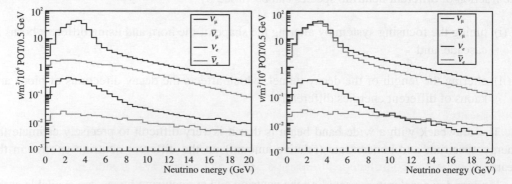

Figure 17.20 MINERvA spectra [949] for low energy neutrino (ν_μ) (left) and antineutrino ($\bar{\nu}_\mu$) (right).

The name narrow band neutrino beam refers to the selection of parent mesons in a narrow energy interval. In this case, the cylindrical target struck by the protons is not aligned with the decay tunnel and additional dipole magnets and momentum slits select the mesons of desired energy (Figure 17.18). The neutrino beam so obtained is also known as an off axis beam. Since mesons of a specific band of energies and momenta are selected, the neutrino fluxes are lesser in intensity compared to the flux obtained through an on-axis arrangement.

The angular dependence of the neutrino energy is obtained by using the four-momentum conservation rule, that is,

$$
\begin{aligned}
p_{\text{meson}} &= p_\nu + p_\mu \\
p_\nu^2 &= p_{\text{meson}}^2 + p_\mu^2 - 2p_{\text{meson}} \cdot p_\mu, \text{ leading to} \\
E_\nu &= \frac{m_{\text{meson}}^2 - m_\mu^2}{2\left(E_{\text{meson}} - |\vec{p}_{\text{meson}}|\cos\theta_\nu\right)} \\
&= \frac{m_{\text{meson}}^2 - m_\mu^2}{2|\vec{p}_{\text{meson}}|\left(1 - \cos\theta_\nu\right) + \frac{m_{\text{meson}}^2}{|\vec{p}_{\text{meson}}|}} \\
&= |\vec{p}_{\text{meson}}|\frac{m_{\text{meson}}^2 - m_\mu^2}{m_{\text{meson}}^2 + |\vec{p}_{\text{meson}}|^2\theta_\nu^2} \quad \text{(for small angle scattering)} \\
E_\nu(\theta_\nu) &= E_{\text{meson}}\frac{m_{\text{meson}}^2 - m_\mu^2}{m_{\text{meson}}^2 + E_{\text{meson}}^2\theta_\nu^2} \quad \text{(for } E_{\text{meson}} \approx |\vec{p}_{\text{meson}}|\text{)} \\
&\simeq E_\nu^{\max}\frac{1}{\gamma_{\text{meson}}^2\theta_\nu^2}
\end{aligned}
\tag{17.39}
$$

Thus, for small angles ($\theta_\nu < 1$) and relativistic mesons ($\gamma_{\text{meson}} \gg 1$), we may write:

$$
E_\nu \simeq \frac{m_{\text{meson}}^2 - m_\mu^2}{m_{\text{meson}}^2\left(1 + \gamma_{\text{meson}}^2\theta^2\right)}E_{\text{meson}},
\tag{17.40}
$$

leading to neutrino flux

$$
\Phi_\nu \simeq \frac{1}{\pi L^2}\left(\frac{E_{\text{meson}}}{m_{\text{meson}}^2}\right)^2\frac{1}{(1 + \gamma_{\text{meson}}^2\theta^2)^2},
\tag{17.41}
$$

where L is the detector distance. It may be noted that for a detector which is off-axis, the neutrino energy is no longer proportional to the pion energy but rather has a broad maximum for $\gamma_{\text{meson}}\theta = 1$. The maximum neutrino energy is $\frac{E_\nu^{\text{CM}}}{\theta_\nu}$, which therefore depends only on the value of the off-axis angle. Although the flux obtained from mesons in the off-axis beam is smaller than the flux obtained from the on-axis beam, the advantage is that all the mesons in a broad energy range contribute neutrinos in a narrow energy interval around $E_{\max} = \frac{29.8\text{MeV}}{\theta}$ for the pions and $E_{\max} = \frac{235.5\text{MeV}}{\theta}$ for the kaons. Due to this, the neutrino flux at peak is much larger than that of the on-axis beam at the same energy. Figure 17.21, shows the neutrino spectrum obtained using an on-axis arrangement and also for different off-axis angles for the T2K experiment [904].

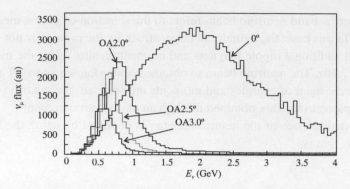

Figure 17.21 Neutrino spectrum for on-axis and different off-axis angles obtained by the T2K experiment [904].

Some of the accelerators which have been used in the neutrino experiments for both wide and narrow beams have been listed in Table 17.7. The various experiments performed using accelerator beams have been listed in Table 17.8 and Table 17.9, for the short baseline and long baseline experiments, respectively.

Table 17.7 List of accelerators.

Lab	Year	p_0 (GeV/c)	Protons Pulse (10^{12})	$\langle E_\nu \rangle$ (GeV)	Experiments
BNL	1962	15	0.3	5	Spark Ch. Observation of 2 νs
CERN	1963	20.6	0.7	1.5	HLBC, spark ch.
CERN	1969	20.6	0.63	1.5	HLBC, spark ch.
ANL	1969	12.4	1.2	0.5	Spark Chamber
ANL	1970	12.4	1.2	0.5	12′ BC
CERN	1972	26	5	1.5	GGM, Aachen-Pad.
FNAL	1974	300	10	50, 180	CITF, HPWF, 15′ BC
FNAL	1975	300, 400	10	40	HPWF
FNAL	1975	300, 400	10	50,180	CITF, HPWF
BNL	1976	28	8	1.3	7′ BC, E605, E613, E734, E776
FNAL	1976	350	13	100	HPWF, 15′ BC
CERN	1977	350	10	50,150	CDHS, CHARM, BEBC
CERN	1977	350	10	20	GGM,CDHS, CHARM, BEBC
IHEP	1977	70	10	4	SKAT, JINR
FNAL	1979	400	10	25	15′ BC
BNL	1980	28	7	3	7′ BC, E776
CERN	1983	19	5	1	CDHS, CHARM
FNAL	1991	800	10	90, 260	15′ BC, CCFRR
CERN	1995	450	11	20	NOMAD, CHORUS
FNAL	1998	800	12	70, 180	NuTeV exp't
KEK	1998	12	5	0.8	K2K long baseline osc.
FNAL	2002	8	4.5	1	MiniBooNE
FNAL	2005	120	32	4-15	MINOS, MINERνA
CERN	2006	450	50	20	OPERA, ICARUS
FNAL	2009	120	70	2	NOνA off-axis
JPARC	2009	40	300	0.8	Super K off-axis

Table 17.8 Accelerator (short baseline) neutrino detectors.

Experiment	Reaction	Active mass (ton) & detector	Location (MeV)	Period
CHARM I CHARM II	$\bar{\nu}_\mu A \to$ $\bar{\nu}_\mu \mu^+ \mu^- A$	692-glass calorimeter	CERN	1979–1984 1986-1991
CHORUS	$\nu_\tau N \to \tau^- X$	800 kg nuclear emulsion	CERN	1994–1997
LSND	$\bar{\nu}_e + p \to e^+ n$ $np \to d\gamma$	167 LS	Los Almos, USA	1993–1998
NOMAD	$\nu_\tau N \to \tau^- X$	2.7	CERN	1995–1998
KARMEN	$\bar{\nu}_e + p \to e^+ n$	65000 liter LS	ISIS, UK	1990–2001
MiniBooNE	$\nu_\mu \to \nu_e$	800 Mineral oil	Fermilab, USA	2002–2017
MicroBooNE	$\nu_e(^{40}\mathrm{Ar},^{40}\mathrm{K})e^-$	170 LAr-TPC	Fermilab, USA	2002– Present
ICARUS	$\nu_e(^{40}\mathrm{Ar},^{40}\mathrm{K})e^-$	600 LAr	Gran Sasso, Italy	Future
SBND	$\nu_e(^{40}\mathrm{Ar},^{40}\mathrm{K})e^-$	112 LAr-TPC	Fermilab, USA	Future

Table 17.9 Accelerator (long baseline) neutrino detectors.

Experiment	Reaction	Active mass (ton) & detector	Location	Period
K2K	$\nu_\mu n \to \mu^- p$	1000 (near) 50000 (far) water Cherenkov	KEK to Kamioka, Japan	1999–2004
ICARUS-T600	$\nu_e(^{40}Ar,^{40}K)e^-$	760 LArTPC	CERN to Gran Sasso, Italy	2010–2013
MINOS	$\nu_\mu + \text{nucleus}$ $\to \mu^- + X$	5400	Illinois & Minnesota, USA	2005–2012
NOνA	$K \to 2\pi/3\pi$ $\pi^+ \to \mu^+ + \nu_\mu$ $\mu^+ \to e^+ + \nu_e + \bar{\nu}_\mu$	14000 (far) 300 (near) LS	Fermilab Ash river, USA.	2011–2015
T2K	$\bar{\nu}_e\, p \to e^+ n$ $\nu_e e^- \to \nu_e e^-$	Cherenkov	Tokai to Kamioka, Japan	2013– Present
ICARUS	$\nu_e(^{40}Ar,^{40}K)e^-$	600	Gran Sasso, Italy	Future
DUNE	$\nu_e(^{40}Ar,^{40}K^*)e^-$	68000 LAr-TPC	Homestake Mine, USA	Future
T2HK		374000 Water	Tokai to Hyper-K, Japan	Future

17.9   Neutrinos from the Decay at Rest (DAR) Sources

The particles decay at rest (DAR) can provide high intensity, high purity and a well understood flux mainly in the low energy region (Figure 17.22). Various sources which have been discussed in literature to obtain neutrinos are:

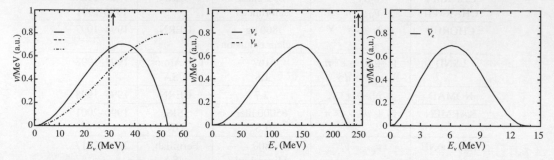

Figure 17.22 Neutrino spectrum obtained from the various sources when at rest: π and μ DAR (left), *K*DAR (center) and IsoDAR (right).

- Pion decay at rest (πDAR): $\pi^{\pm} \longrightarrow \mu^{\pm} + \nu_{\mu}(\bar{\nu}_{\mu})$.

- Muon decay at rest (μDAR): $\mu^{\pm} \longrightarrow e^{\pm} + \nu_e(\bar{\nu}_e) + \bar{\nu}_{\mu}(\nu_{\mu})$.

- Kaon decay at rest (KDAR): $K^{\pm} \longrightarrow \mu^{\pm} + \nu_{\mu}(\bar{\nu}_{\mu})$.

- Isotope decay at rest (IsoDAR): ${}_{Z}^{A}X \longrightarrow {}_{Z-1}^{A}Y + e^{-} + \bar{\nu}_e$.

The two-body decay modes of the mesons provide a beam of monoenergetic $\nu_{\mu}(\bar{\nu}_{\mu})$. The two-body leptonic decay mode of the charged kaon decay at rest (KDAR), that is, $K^{+} \rightarrow \mu^{+}\nu_{\mu}$, with a branching ratio of $63.55\pm1.1\%$, provides a unique and important source of monoenergetic muon neutrinos of energy 236 MeV. These neutrinos may be used to make high precision measurements of neutrino–nucleus cross sections for the charged current (CC) induced weak quasielastic (QE) production of muons from various nuclear targets. The high precision neutrino–nucleus cross sections measured with a well-defined monoenergetic beam of muon neutrinos may serve as a benchmark for validating many theoretical models currently being used to describe the nuclear medium effects in QE reactions, and which are relevant for the analysis of present day neutrino experiments in the low energy region of a few hundred MeV. These KDAR neutrinos are proposed to be used as a probe to study the new neutrino oscillation modes to sterile neutrinos, that is, ν_{μ} to ν_s, by performing the oscillation experiments in ν_{μ} to ν_{μ} disappearance mode and studying the CC interactions of ν_{μ} with nuclei, and/or performing the oscillation experiments in ν_{μ} to ν_e appearance mode and studying the CC interaction of ν_e with nuclei.

KPipe is an experiment which aims to study muon neutrino disappearance for a sensitive test of the $\Delta m^2 \sim 1$ eV^2 anomalies, possibly indicative of one or more sterile neutrinos. This experiment is planned to be located at the J-PARC Material and Life Science (MLS) experimental facility's spallation neutron source. Presently, this is the world's most intense source of monoenergetic (236 MeV) muon neutrinos from the charged kaon decay at rest [950]. KPipe plans to use a long, liquid scintillation based detector, which would be oriented radially with respect to an intense source of monoenergetic 236 MeV ν_μs. IsoDAR is a proposed experiment at KamLAND, mainly focussed to study the disappearance of the electron type antineutrinos and to search for the possible existence of sterile neutrinos. The target is to use the high intensity ^{8}Li β-decay at rest. DAEδALUS is a proposed decay at rest experiment for studying CP violation in the leptonic sector. The plan is to use the IsoDAR proton beam as an injector to produce higher energy neutrinos through a stopped pion beam. This source will produce both muon and electron neutrinos as well as muon antineutrinos. The muon antineutrinos will then oscillate to produce electron antineutrinos, whose appearance may be measured and compared to expectation which can be observed in appearance channels.

17.10 Neutrinos from Spallation Neutrons

The low energy neutrino beams ($E_\nu \leq 52.8$ MeV) are generally obtained from the muons decaying at rest. The neutrino energy spectrum obtained from the muons decaying at rest is given by

$$\phi(E_{\nu_e}) = \frac{12}{E_0^4} E_{\nu_e}^2 (E_0 - E_{\nu_e}), \quad E_0 = 52.8 \text{ MeV}. \tag{17.42}$$

This is known as the Michel spectrum and is shown in Fig. (17.22). The neutrino energy spectrum and its energy range is similar to the energy spectrum and energy range of neutrinos coming from the core collapse in a supernova. This similarity in the energy range and spectrum of the supernova neutrinos with the muons decaying at rest, opens up the possibility of connecting the ground based neutrino nuclear experiments with the study of neutrino nuclear cross sections in a supernova. Such a study will also be useful in understanding the r-process nucleosynthesis leading to the formation of heavy elements in the interstellar medium.

Low energy neutrino beams can also be obtained at the Spallation Neutron Source (SNS) facilities, where low energy neutrino reaction experiments can be performed on the nuclear targets to study the selective nuclear transitions. Since SNS produces pulsed neutron beams, the neutrinos are also pulsed, enabling easy separation of signals from the background. The idea behind an SNS facility is to get neutrons when a highly intense beam, for example, at SNS in ORNL, USA, 10^{14} protons of 1 GeV energy are bombarded on a liquid mercury target, in 700 ns wide bursts, with a frequency of 60 Hz. These neutrons then thermalize in hydrogenous and helium moderators surrounding the target before they are delivered to neutron scattering instruments. Besides neutrons, the interactions of the proton beam in the mercury target also produces pions which are stopped inside the dense mercury target before decaying and producing neutrinos. The flux of the neutrinos is quite high $\approx 2 \times 10^7$ neutrinos/cm^2/s for the three flavors of neutrinos at a distance of 20 m from the spallation target. The neutrino spectra are monoenergetic for ν_μ and continuous for ν_e and $\bar{\nu}_\mu$ as shown in Figure 17.22. Out of these flavors of neutrinos, ν_e produces electrons which are observed to study the neutrino–nucleus cross sections for charged current reactions. At ORNL, the COHERENT experiment has recently used neutrinos from the SNS source to study the coherent elastic neutrino–nucleus scattering (CEνNS) from a target using cesium iodide scintillator crystal doped with sodium to increase the prominence of light signals from neutrino interactions [951]. This is a pioneering experiment which has measured for the first time the coherent ν-nucleus scattering, a process of great significance in astrophysics.

17.11 Neutrinos from Muon Storage Ring (MSR)

Beams of $\bar{\nu}_e$ and $\bar{\nu}_\mu$ are produced from the decay of stored $\mu^\pm$ beams. Koshkarev [952], in 1974, first discussed the idea of neutrino production using muon storage ring. In 1980, Neuffer [953] gave a detailed description of a muon storage ring as a neutrino source for doing neutrino oscillation experiments.

Protons of 80–120 GeV energy are used to produce pions off a conventional solid target. The pions are then focused with the help of a magnetic horn and quadrupole magnets. The pions are transported to a chicane (a double bend for sign selection). The pions that decay in the first straight section of the ring can yield muons, which are captured in the ring. The circulating muons then subsequently decay into electrons and neutrinos (Figure 17.23). The nuSTORM facility has been designed to deliver beams of $\nu_e(\bar{\nu}_e)$ and $\nu_\mu(\bar{\nu}_\mu)$. A detector located at a distance of about 2 km from the end of one of the straight sections will be able to make sensitive searches for the existence of sterile neutrinos [954, 955].

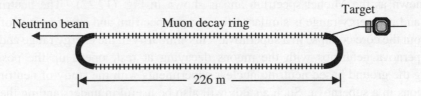

Figure 17.23 Schematic diagram of a muon storage ring.

17.12 Beta-beam Neutrinos

The pure sources of electron neutrinos or antineutrinos from β-beams were first proposed by Zucchelli in 2002 [956]. The idea is to obtain ν_e and $\bar{\nu}_e$ beams through the β-decay of accelerated radioactive ions circulating in a storage ring. These ions would be produced by intense radioactive sources. The β-beams provide a source of pure single flavor, well collimated and intense (anti)neutrino beams with a well-defined energy spectrum obtained from the β-decay of accelerated radioactive ions boosted by a suitable Lorentz factor γ. The radioactive ion and the Lorentz boost factor γ can be properly chosen to provide low [957, 958], intermediate, and high energy [959] neutrino beams according to the needs of a planned experiment.

In the feasibility study of β-beams, ^{6}He ions with a Q value of 3.5 MeV and ^{18}Ne ions with a Q value of 3.3 MeV are considered to be the most suitable candidates to produce antineutrino and neutrino beams. The possibility of accelerating these ions using the existing CERN-SPS, up to its maximum power enabling it to produce β-beams with $\gamma = 150$ (250) for ^{6}He (^{18}Ne) ions has been discussed in literature [960]; this may be used to plan a baseline neutrino experiment at $L = 130$ km to the underground Frejus laboratory with the 440 kT water Cherenkov detector [959]–[961]. The feasibility of such an experimental setup and its response to β-beam neutrinos corresponding to various values of the Lorentz boost factor γ has been studied by Autin et al. [962]. In high γ range, this provides greater sensitivity to the determination of the mixing angle θ_{13} and the CP violating phase angle δ [963]. In addition, such a facility is also expected to provide the low energy neutrino nuclear cross sections corresponding to very low γ, which may be useful in calibrating various detectors planned for the observation of supernova neutrinos [957, 964] and neutrinoless double β-decay [958].

The energy spectrum of β-beam (anti)neutrinos from an ^{18}Ne (^{6}He) ion source in the forward angle ($\theta = 0^o$) geometry corresponding to the Lorentz boost factor γ is given by [965]:

$$
\begin{aligned}
\Phi_{\text{lab}}(E_\nu, \theta = 0) &= \frac{\Phi_{cm}(E_\nu \gamma [1 - \beta])}{\gamma[1 - \beta]}, \\
\Phi_{cm}(E_\nu) &= bE_\nu^2 E_e p_e F(Z', E_e) \Theta(E_e - m_e).
\end{aligned}
\tag{17.43}
$$

In this equation, $b = \ln 2/m_e^5 f t_{1/2}$ and $E_e(= Q - E_\nu)$, p_e are the energy and momentum of the outgoing electron, Q is the Q value of the β-decay of the radioactive ion $A(Z, N) \rightarrow A(Z', N') + e^-(e^+) + \bar{\nu}_e(\nu_e)$, and $F(Z', E_e)$ is the Fermi function. In Figure 17.24, we show the representative spectra for (anti)neutrinos corresponding to the Lorentz boost factor $\gamma = 250$ (150).

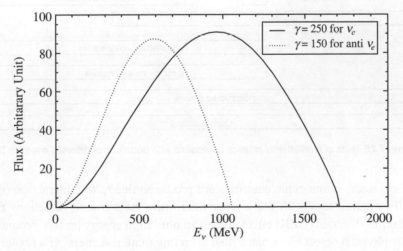

Figure 17.24 Neutrino (solid line) energy spectrum obtained with ^{18}Ne boosted at $\gamma = 250$ and antineutrino (dashed line) energy spectrum obtained with ^{6}He boosted at $\gamma = 150$.

17.13 Very High Energy Cosmic Neutrinos

Cosmic neutrinos are very important probes to learn the fundamental physics beyond the TeV scale [966]. The energy scale which can be achieved with cosmic sources at the surface of the earth, is well beyond the energy available from solar neutrinos, atmospheric neutrinos, supernova neutrinos, reactor antineutrinos, or even with the present accelerators (Figure 17.25).

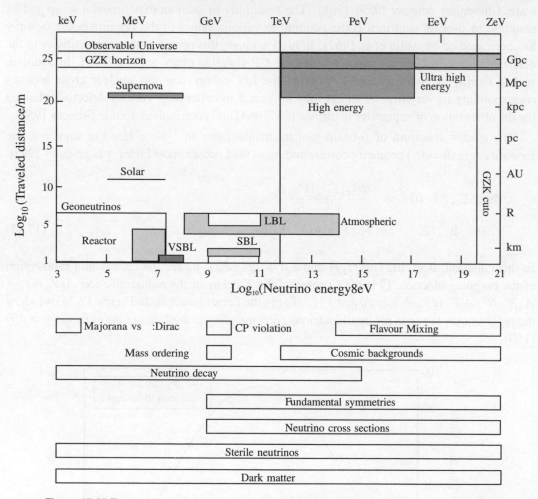

Figure 17.25 Tests of fundamental physics accessible with neutrinos of different energies [966].

At the EeV scale, cosmogenic neutrinos are produced due to the interaction of ultra high energy (UHE) cosmic rays with the cosmic microwave background radiations through the Greisen–Zatsepin–Kuzmin (GZK) effect. When an ultra high energy proton, produced in some far away astrophysical object like a black hole accreting matter, scatters off a photon present as CMBR, it produces very high energy pions, that is,

$$p + \gamma \to n + \pi^+. \tag{17.44}$$

The threshold energy required by the proton in the rest frame of the cosmic background radiation is obtained by the following considerations. The temperature of the CMBR is approximately 2.7 K, which corresponds to $235\mu eV = 2.35 \times 10^{-13}$ GeV.

Now, the four-momentum of both the sides of Eq. (17.44), must be the same, that is,

$$(p_\gamma + p_p)^2 = (p_\pi + p_n)^2,$$
$$\text{at threshold} \quad M_p^2 + 2p_p \cdot p_\gamma = (M_\pi + M_n)^2. \tag{17.45}$$

To evaluate $p_p \cdot p_\gamma$, we go to the CM frame, where extremely relativistic protons and photons coming from the opposite direction (say Z-direction), have four-momentum components $p_p = (E_p, 0, 0, E_p)$ and $p_\gamma = (E_\gamma, 0, 0, -E_\gamma)$, respectively, such that $p_p \cdot p_\gamma = 2E_p E_\gamma$, which gives

$$M_p^2 + 4E_p E_\gamma = (M_\pi + M_n)^2$$
$$\Rightarrow E_p = \frac{(M_\pi + M_n)^2 - M_p^2}{4E_\gamma}$$
$$\sim 3 \times 10^{20} eV, \quad \text{using } E_\gamma \approx 2.35 \times 10^{-13} \text{GeV}. \tag{17.46}$$

GZK limit is a theoretical upper limit on the energy of cosmic ray protons, traveling from other galaxies, through the intergalactic medium, to our galaxy. Moreover, the propagation of neutrinos from the source to the detectors is well understood and predicted by the standard model.

The AMANDA detector located beneath the Amundsen–Scott South Pole station is the largest neutrino telescope running today. It detects neutrinos through their interactions with nuclei of oxygen or hydrogen atoms contained in the ice. The interaction produces a muon and hadronic shower. There is a possibility to observe the very high energy cosmic neutrinos in the coming years. In 2013, the IceCube collaboration announced evidences for the detection of extraterrestrial high energy neutrinos. They have reported the energy range of these neutrinos to be between 30 TeV to 2 PeV [967].

Finally, we summarize in Figure 17.25, the physics goals that could be achieved by the various neutrino sources of different energies.

Chapter 18

Neutrino Mixing and Oscillations

18.1 Introduction

Neutrino oscillation is purely a quantum mechanical phenomenon, in which a neutrino of one flavor, say ν_α develops a component of a neutrino of another flavor, say ν_β, where $\alpha, \beta = e, \mu, \tau$, and $\alpha \neq \beta$. This phenomenon of neutrino oscillation implies that neutrinos have definite mass and their flavor eigenstates are different from their mass eigenstates.

The idea of neutrino masses, mixing, and oscillations was first proposed in 1957–58 [179] just after the two-component neutrino theory was confirmed by the Goldhaber et al. [153] experiment. Pontecorvo [179] conjectured the idea of neutral particle (not necessarily being elementary particle) oscillation based on the analogy with kaon regeneration (K^0–$\bar{K}^0$ oscillation), and argued that : 'If the two- component neutrino theory should turn out to be incorrect (which at present seems to be rather improbable) and if the conservation law of neutrino charge would not apply, then in principle neutrino $\leftrightarrow$ antineutrino transitions could take place in vacu'. At that time only one type of neutrino existed, therefore, for the two-component theory, there were only left-handed neutrinos ν_L and right-handed antineutrinos $\bar{\nu}_R$, implying oscillations between ν_L and $\bar{\nu}_R$. These transitions are however not possible because of the conservation of total angular momentum.

Around the same time, Davis [968] conducted an experiment using reactor antineutrino beams and looked for the electron in the reaction

$$\bar{\nu}_e + {}^{37}_{17}\mathrm{Cl} \rightarrow e^- + {}^{37}_{18}\mathrm{Ar}, \tag{18.1}$$

which showed negative results. Contrary to the negative observations, a rumor reached Pontecorvo that Davis has observed the aforementioned event before the actual publication of the work. Based on the assumed positive results, in 1958, Pontecorvo, in an another paper [969], concluded that: "The result of the Davis experiment was nonzero probability for the process under study, if confirmed, definitely shows that the (strict) law of neutrino charge conservation is not valid." He suggested that the events could be due to the transitions $\bar{\nu}_R \rightarrow \nu_R$. However, this explanation contradicted the two-component neutrino theory, where only the left-handed neutrinos (ν_L) or the right-handed antineutrinos $(\bar{\nu}_R)$ participate in the weak interaction. Therefore, the suggestion of ν_R and similarly, $\bar{\nu}_L$ by Pontecorvo implies the existence of a class of neutrinos which are noninteracting; they are called sterile neutrinos.

Based on this argument, he suggested that for non-identical ν and $\bar{\nu}$, the following processes should be observed in nuclei:

$$p \rightarrow n + \beta^+ + \nu, \quad n \rightarrow p + \beta^- + \bar{\nu}, \tag{18.2}$$

while if the strict law of neutrino charge conservation is not valid, then the following processes should be observed:

$$p \rightarrow n + \beta^+ + \bar{\nu}, \quad n \rightarrow p + \beta^- + \nu. \tag{18.3}$$

In 1962, Maki, Nakagawa, and Sakata introduced the mixing of two neutrinos [71] by contemplating that: "It should be stressed at this stage that the definition of the particle state of neutrino is quite arbitrary; we can speak of 'neutrinos' which are different from weak neutrinos but expressed by the linear combinations of the latter. We assume that there exists a representation which defines the 'true neutrinos' through some orthogonal transformation applied to the representation of weak neutrinos:

$$
\begin{aligned}
\nu_1 &= \nu_e \cos\delta + \nu_\mu \sin\delta, \\
\nu_2 &= -\nu_e \sin\delta + \nu_\mu \cos\delta,
\end{aligned}
$$

where δ is a real constant."

In 1967, when the two flavors of neutrinos ν_e and ν_μ were proved to exist, Pontecorvo [970] in a paper reiterated that: "If leptonic charge is not an exactly conserved quantum number (and in this case, the neutrino mass would be different from zero), then oscillations of the type $\bar{\nu} \leftrightarrow \nu$, $\nu_\mu \leftrightarrow \nu_e$, which are similar to oscillations in a beam of K^0 mesons become possible for neutrino beams." This paper also discussed the observation of such a possibility through solar neutrino oscillations and the following conclusions were made: "From the point of view of detection possibilities, an ideal object is the sun. If the oscillation length is much smaller than the radius of the solar region which effectively produces neutrinos which will give the main contribution to the experiments which are being planned now...." Few years later, Davis et al. [971, 972] confirmed that the detected flux of solar neutrinos was about 2–3 times smaller than the theoretically predicted flux by Bahcall [973, 182] in the standard solar model. For the solar neutrino problem, the prediction made by Pontecorvo was naturally accepted. Apart from the solar neutrinos, for the atmospheric neutrinos, the atmospheric neutrino flux was expected to be in the ratio 2 : 1 for ν_μ and ν_e type neutrinos because of the following reactions:

$$
\begin{aligned}
\pi^\pm &\longrightarrow \mu^\pm + \nu_\mu(\bar{\nu}_\mu), \\
\mu^\pm &\longrightarrow e^\pm + \nu_e(\bar{\nu}_e) + \bar{\nu}_\mu(\nu_\mu).
\end{aligned}
$$

The observations on atmospheric neutrinos later showed that there exists a discrepancy between the observed flux and the predicted flux; this discrepancy in the flux is known as the atmospheric neutrino anomaly. The ratio $\frac{(\nu_\mu + \bar{\nu}_\mu)}{(\nu_e + \bar{\nu}_e)}$ deviates from the predicted flux $2 : 1$, which was possible only if neutrinos could oscillate from one flavor to another. A similar anomaly was seen in the region of very low energy with reactor antineutrinos [974, 975] when the precise energy spectrum of antineutrinos emitted from the fission of ^{235}U, ^{239}Pu, ^{238}U, and ^{241}Pu were calculated [976, 977], and compared with the experimentally observed spectrum [978].

Experimentally, the discovery of the neutrino oscillations was confirmed through the observation of atmospheric neutrinos in Kamiokande [898, 901, 902], SOUDAN [979], Super-Kamiokande [903, 980], solar neutrinos at Super-K [981, 982, 983, 984], SAGE [985], BOREXino [986, 987], Gallex/GNO [988], SNO [989] and reactor antineutrinos at KamLAND [978], Daya Bay [990], RENO [991], and Double Chooz [992, 993]. Moreover, the accelerator experiments like K2K [994], MiniBooNE [995], NOMAD [996], CHORUS [997], MINOS [998, 999], T2K [1000, 1001], NOvA [1002, 1003, 628, 1004], OPERA [119, 1005, 1006, 1007, 1008], etc. also confirmed the phenomenon of neutrino oscillation. In fact, acknowledging the efforts made to understand neutrino oscillation physics, the Nobel prize in Physics for the year 2015 was awarded jointly to T. Kajita "for the discovery of atmospheric neutrino oscillations" and A. McDonald "for the discovery of neutrino oscillations in the solar sector".

In this chapter, we are going to study neutrino oscillations in vacuum in two- and three-flavor scenarios and obtain the expression of survival probability (chances of neutrinos of a particular flavor to keep the same identity) and transition probability (the probability of neutrinos oscillating from one flavor to another). Then, we shall discuss the effect of the interaction of neutrinos with matter on the neutrino oscillation probability. This is known as the Mikheyev–Smirnov–Wolfenstein (MSW) effect after the scientists Mikheyev, Smirnov, and Wolfenstein [1009, 1010]; it enhances the neutrino oscillation probability. Finally, we shall discuss in brief, the sensitivity of the different experiments toward the determination of oscillation parameters and conclude the chapter with a discussion on sterile neutrinos.

18.2 Neutrino Mixing

18.2.1 Two-flavor neutrino oscillations in vacuum

The oscillation of neutrino from one flavor to another is only possible when the neutrino mass eigenstates (or propagation states), say ν_1 and ν_2, are different from the flavor states, say ν_α and ν_β, where $\alpha, \beta = e, \mu$ or for any two neutrino flavors among e, μ, and τ. This implies that the quantized neutrino field of a definite flavor does not coincide with the quantized neutrino field that has definite mass. In fact, a given flavor state ν_α is expressed as an orthonormal linear combination of mass eigenstates ν_i, $i = 1, 2$ with mass m_i, such that

$$|\nu_\alpha\rangle = \sum_{j=1}^{2} U_{\alpha j}|\nu_j\rangle, \tag{18.4}$$

where $U_{\alpha j}$ is a 2×2 unitary matrix ($U^\dagger = U^{-1}$) and is generally expressed as:

$$\hat{U} = \begin{bmatrix} U_{e1} & U_{e2} \\ U_{\mu 1} & U_{\mu 2} \end{bmatrix} \quad , \quad \hat{U}^\dagger = \begin{bmatrix} U_{e1}^* & U_{\mu 1}^* \\ U_{e2}^* & U_{\mu 2}^* \end{bmatrix} , \quad \text{such that} \quad \hat{U}\hat{U}^\dagger = \begin{bmatrix} 1 & 0 \\ 0 & 1 \end{bmatrix}. \quad (18.5)$$

Similarly, the mass eigenstates in terms of flavor eigenstates may be expressed as:

$$|\nu_j\rangle = \sum_\alpha (U^\dagger)_{j\alpha} |\nu_\alpha\rangle = \sum_\alpha U_{\alpha j}^* |\nu_\alpha\rangle. \quad (18.6)$$

Since a 2×2 unitary matrix U is real, $U_{\alpha j}^* = U_{\alpha j}$ and

$$|\bar{\nu}_\alpha\rangle = U_{\alpha j}^* |\bar{\nu}_j\rangle = U_{\alpha j} |\bar{\nu}_j\rangle, \quad (18.7)$$

where $|\bar{\nu}_j\rangle$ ($|\bar{\nu}_\alpha\rangle$) are the mass (flavor) eigenstates of an antineutrino. This implies that:

$$\sum_\alpha U_{\alpha j} U_{\alpha k}^* = \delta_{jk} \quad \text{and} \quad \sum_j U_{\alpha j} U_{\beta j}^* = \delta_{\alpha \beta}. \quad (18.8)$$

Since a 2×2 matrix is parameterized in terms of one real parameter say θ, we may write:

$$\begin{bmatrix} \nu_e \\ \nu_\mu \end{bmatrix} = \begin{bmatrix} \cos\theta & \sin\theta \\ -\sin\theta & \cos\theta \end{bmatrix} \begin{bmatrix} \nu_1 \\ \nu_2 \end{bmatrix} = U \begin{bmatrix} \nu_1 \\ \nu_2 \end{bmatrix}, \quad (18.9)$$

where θ is the neutrino mixing angle. This means that the definite flavor eigenstates of neutrinos ν_e and ν_μ are composed exclusively of the mass eigenstates ν_1 and ν_2. The oscillation matrix is chosen to be unitary because in a scenario of two ν flavors in nature, where one flavor of neutrino (say ν_e) oscillates to another flavor (say ν_μ) and vice versa, there is no loss in total neutrino flux implying that the total probability of oscillation (including all possible cases) is unity, that is,

$$P(\nu_e \to \nu_e) + P(\nu_e \to \nu_\mu) = P(\nu_\mu \to \nu_\mu) + P(\nu_\mu \to \nu_e) = 1. \quad (18.10)$$

We demonstrate the oscillation probability in a two-flavor scenario, in vacuum, in Figure 18.1. Suppose we start with a π^+ decay at rest (at point A in Figure 18.1(i)), which gives rise to l_α^+ and ν_α. The detector is placed at B and we observe ν_α through charged current interactions giving rise to a charged lepton of flavor α, that is, l_α^-. Now if we observe the same flavour of charged leptons produced at the detector site B as the flavour of ν_αs that were there at the production site A, then that means no oscillation has taken place. However, if at B, we find charged lepton(s) of a different flavor, say β that means among the ν_αs, while traveling toward the detector, a few have got converted into another flavor ν_β whose interaction at the detector site B has given rise to a different lepton l_β^- (Figure 18.1(ii)). The oscillation of neutrinos of one flavor to another flavor is only possible when the propagation states (say ν_1 and ν_2) are different from the flavor states (say ν_α and ν_β), where ν_α is some linear combination of ν_1 and ν_2, while ν_β is some different linear combination of ν_1 and ν_2.

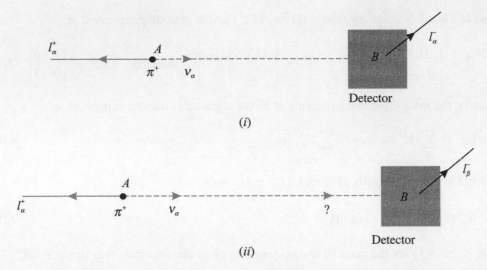

Figure 18.1 The top figure represents a no oscillation case where at "A", ν_αs are produced along with l_α^+ and travel a short distance, and at "B", the same number of l_α^- are detected through charged current interactions (ν_α, l_α^-). The bottom figure represents an oscillation case, where at "A", ν_αs are produced along with l_α^+ and travel some distance, and at "B", some of the charged leptons of flavor l_β^- are observed which is only possible when on the way, a few ν_α get converted into ν_β.

To calculate the oscillation probability in a two-flavor oscillation scenario, we start with time evolution equation for a neutrino of definite flavor α (ν_e or ν_μ) by using the Schrödinger equation

$$i\frac{\partial |\nu_\alpha(t)\rangle}{\partial t} = H\,|\nu_\alpha(t)\rangle, \tag{18.11}$$

where H, for the evolution of states in vacuum is a free Hamiltonian and its solution would be

$$|\nu_\alpha(t)\rangle = e^{-iHt}|\nu_\alpha(t=0)\rangle = e^{-iEt}|\nu_\alpha(t=0)\rangle, \tag{18.12}$$

where $|\nu_\alpha(t=0)\rangle$ or $|\nu_\alpha(0)\rangle$ is the state at the initial time $t=0$. Now to start with, if we have ν_e beam, then

$$|\nu_\alpha(0)\rangle = |\nu_\alpha(t=0)\rangle = |\nu_e\rangle. \tag{18.13}$$

Since neutrino mass eigenstates are defined by $|\nu_j\rangle$,

$$H|\nu_j\rangle = E_j|\nu_j\rangle\,;\; j=1,2, \quad \text{and} \quad E_j = \sqrt{|\vec{p}_j|^2 + m_j^2} \tag{18.14}$$

$$\Rightarrow\; |\nu_\alpha(t)\rangle = \sum_{j=1}^{2} e^{-iE_j t}\, U_{\alpha j}|\nu_j\rangle. \tag{18.15}$$

Let us say that a neutrino of generation α after a time interval t is given by the eigenstate $|\nu_\alpha, t\rangle$, such that

$$|\nu_\alpha, t\rangle = \sum_{j=1}^{2} U_{\alpha j} |\nu_j, t\rangle = \sum_{j=1}^{2} U_{\alpha j} |\nu_j, 0\rangle \, e^{-iE_j t} \tag{18.16}$$

$$= \sum_{j=1}^{2} U_{\alpha j} \, e^{-iE_j t} \left(\sum_\beta U_{\beta j}^* |\nu_\beta, 0\rangle \right). \tag{18.17}$$

Survival probability, that is, the probability that a neutrino that had started ($t = 0$) with a flavor α remains ν_α, is obtained by using amplitude defined as:

$$\mathcal{M}_{\alpha\alpha}(t) = \langle \nu_\alpha, 0 | \nu_\alpha, t \rangle = \delta_{\alpha\beta} \sum_{j=1}^{2} U_{\alpha j} \, e^{-iE_j t} U_{\beta j}^* = \sum_{j=1}^{2} U_{\alpha j} \, e^{-iE_j t} U_{\alpha j}^*, \tag{18.18}$$

where $\sum_\beta \langle \nu_\alpha(0) | \nu_\beta(0) \rangle = \delta_{\alpha\beta}$ and the survival probability

$$P_{(\nu_\alpha \to \nu_\alpha)}(t) = |\mathcal{M}_{\alpha\alpha}(t)|^2 = |\sum_{j=1}^{2} U_{\alpha j} \, e^{-iE_j t} U_{\alpha j}^*|^2. \tag{18.19}$$

Similarly, transition probability, that is, the probability that a neutrino that has started ($t = 0$) with a flavor α oscillates into another flavor ν_β, is given by:

$$P_{(\nu_\alpha \to \nu_\beta)}(t) = |\mathcal{M}_{\alpha\beta}(t)|^2 = |\sum_{j=1}^{2} U_{\alpha j} \, e^{-iE_j t} U_{\beta j}^*|^2. \tag{18.20}$$

Similarly, for antineutrinos in the two-flavor scenario, the transition probability is given by:

$$P_{(\bar{\nu}_\alpha \to \bar{\nu}_\beta)}(t) = |\mathcal{M}_{\alpha\beta}(t)|^2 = |\sum_{j=1}^{2} U_{\alpha j}^* \, e^{-iE_j t} U_{\beta j}|^2,$$

$$= |\sum_{j=1}^{2} U_{\alpha j} \, e^{-iE_j t} U_{\beta j}^*|^2 = P_{(\nu_\alpha \to \nu_\beta)}(t). \tag{18.21}$$

Suppose that a neutrino is produced with a definite energy E regardless of which ν_j it happens to be. Then for a particular mass eigenstate ν_j, it has momentum

$$p_j = \sqrt{E^2 - m_j^2} \approx E - \frac{m_j^2}{2E}. \tag{18.22}$$

We find

$$\mathcal{M}(\nu_j \text{ propagates}) \approx e^{-i\frac{m_j^2}{2E}L} \approx e^{-i\frac{m_j^2}{2p}L}. \tag{18.23}$$

Since highly relativistic neutrinos have $E \approx p$, the propagation amplitudes in both the cases are approximately equal.

Using Eq. (18.23) in Eq. (18.20), the probability of oscillation from one flavor to other flavor may be obtained as:

$$P(\nu_e \rightarrow \nu_\mu) = \left| \sum_{j=1}^{2} U_{ej} e^{-i\frac{m_j^2 t}{2E}} U_{\mu j}^* \right|^2$$

$$= |U_{e1}|^2 |U_{\mu 1}|^2 + |U_{e2}|^2 |U_{\mu 2}|^2 + U_{e1} U_{\mu 1}^* U_{e2}^* U_{\mu 2} e^{i\frac{m_2^2 t}{2E}} e^{-i\frac{m_1^2 t}{2E}}$$

$$+ U_{e1}^* U_{\mu 1} U_{e2} U_{\mu 2}^* e^{-i\frac{m_2^2 t}{2E}} e^{i\frac{m_1^2 t}{2E}}.$$

Using U from Eq. (18.9), we may write

$$P(\nu_e \rightarrow \nu_\mu) = 2\cos^2\theta \sin^2\theta - 2\cos^2\theta \sin^2\theta \left(\frac{e^{i\frac{(m_2^2 - m_1^2)t}{2E}} + e^{-i\frac{(m_2^2 - m_1^2)t}{2E}}}{2} \right)$$

$$= \sin^2(2\theta) \sin^2 \left(\frac{\Delta m^2 L}{4E} \right), \tag{18.24}$$

using $\Delta m^2 = m_2^2 - m_1^2$, $t = L/c = L$ ($c = 1$ in natural units). Therefore, from Eq. (18.24), it may be observed that in the two-flavor case, only $\sin^2 2\theta$ and Δm^2 are responsible for transitions between neutrinos of two different flavors for a fixed energy E and length L. The same expression will hold good for the antineutrinos, that is, $P(\nu_e \rightarrow \nu_\mu) = P(\bar{\nu}_e \rightarrow \bar{\nu}_\mu)$. Furthermore, the probability of transition between the two-flavor states would be the same for $\nu_\alpha \rightarrow \nu_\beta$ and $\nu_\beta \rightarrow \nu_\alpha$, that is,

$$P(\nu_\mu \rightarrow \nu_e) = P(\nu_e \rightarrow \nu_\mu). \tag{18.25}$$

Therefore,

$$P(\nu_e \rightarrow \nu_\mu) = P(\nu_\mu \rightarrow \nu_e) = P(\bar{\nu}_e \rightarrow \bar{\nu}_\mu) = P(\bar{\nu}_\mu \rightarrow \bar{\nu}_e),$$

and in fact, it will be true for any two flavors among the three different flavors of neutrinos.

For a definite L, the oscillation probability will vary with E. Therefore, a proper choice of L corresponding to the range of E ensures proper sensitivity to the oscillation probability. The quantity in the parenthesis of Eq. (18.24) is dimensionless and is evaluated numerically by inserting appropriate factors of $\hbar$ and c, that is,

$$\Delta m^2 \frac{L}{4E} = \frac{(\Delta mc^2)^2 L/c}{4E\hbar} = \frac{\Delta m^2 c^4}{eV^2} \frac{L}{km} \frac{GeV}{E} \left[\frac{eV^2 \, km}{4GeV \, (\hbar c)} \right]. \tag{18.26}$$

If Δm, L, and E are measured in eV, km, and GeV, respectively and $\hbar c = 0.197396$ GeV-fm, then this relation is expressed as:

$$1.267 \Delta m^2 (eV^2) \frac{L(km)}{E(GeV)}.$$

Equation (18.24) can also be written in terms of oscillation length as:

$$P(\nu_e \to \nu_\mu) \;=\; \tfrac{1}{2}\sin^2(2\theta)\big(1 - \cos 2\pi \tfrac{L}{L_{\mathrm{osc}}}\big), \tag{18.27}$$

where

$$L_{\mathrm{osc}} \;=\; \frac{4\pi E}{\Delta m^2} = 4\pi \hbar c \frac{E}{\Delta m^2} \simeq 2.48 \frac{\frac{E}{\mathrm{GeV}}}{\frac{\Delta m^2}{\mathrm{eV}^2}}\,\mathrm{km}. \tag{18.28}$$

Neutrino oscillation length L_{osc} should be either equal to or less than the source to detector distance otherwise the oscillations will not be developed. Thus, the quantity that determines whether oscillation has occurred or not when neutrinos of a particular flavor have traveled a distance between the source and the detector is given by:

$$\frac{\pi L}{L_{\mathrm{osc}}} = \frac{L \Delta m^2}{4E} = 1.267 \frac{\left(\frac{\Delta m^2}{\mathrm{eV}^2}\frac{L}{\mathrm{km}}\right)}{\left(\frac{E}{\mathrm{GeV}}\right)}.$$

In Figure 18.2, we have plotted the survival probability $P(\nu_\alpha \to \nu_\alpha)$ vs. $\frac{\Delta m^2}{4}\frac{L}{E}$ curve for $\sin^2 2\theta_{12} = 0.83$, where it may be observed that:

- if the detector is very close to the neutrino source, that is, $L << L_{\mathrm{osc}}$ leading to $\frac{L}{E} << \frac{1}{\Delta m^2}$, there will be no oscillation (Figure 18.2(a)).
- the most sensitive region to observe oscillation is $\frac{L}{E} \gtrsim \frac{1}{\Delta m^2}$. This region is said to be the necessary condition for the oscillation (Figure 18.2(b)).
- for $L >> L_{\mathrm{osc}}$, several oscillations will occur and the average of the oscillation probabilities would be the most acceptable measurement (Figure 18.2(c)).

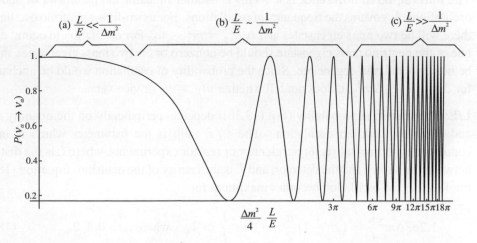

Figure 18.2 Oscillation probability vs. $\frac{\Delta m^2}{4}\frac{L}{E}$ curve for $\sin^2 2\theta = 0.83$. (a) shows no oscillation scenario in the region $\frac{1}{\Delta m^2} >> \frac{L}{E}$; (b) shows the threshold condition for oscillation at $\frac{1}{\Delta m^2} \sim \frac{L}{E}$ and (c) shows the oscillation scenario in the region $\frac{1}{\Delta m^2} << \frac{L}{E}$.

The survival probability $P(\nu_\alpha \to \nu_\alpha)$, that is, the probability of a neutrino ν_α of flavor, say $\alpha = e$ or μ retaining its original flavor, is given by:

$$
\begin{aligned}
P(\nu_\alpha \to \nu_\alpha) &= 1 - P(\nu_\alpha \to \nu_{\beta \neq \alpha}) \\
&= 1 - \sin^2(2\theta)\sin^2\left(1.267\,\Delta m^2(\text{eV}^2)\frac{L(\text{km})}{E(\text{GeV})}\right).
\end{aligned}
\tag{18.29}
$$

Thus, the transition probability $P(\nu_\alpha \to \nu_\beta)$, that is, the probability of a neutrino ν_α of flavor, say $\alpha = e$ changing into another flavor β, say ν_μ, is given by:

$$
\begin{aligned}
P(\nu_\alpha \to \nu_\beta) &= P(\nu_\alpha \to \nu_{\beta \neq \alpha}) \\
&= \sin^2(2\theta)\sin^2\left(1.267\,\Delta m^2(\text{eV}^2)\frac{L(\text{km})}{E(\text{GeV})}\right).
\end{aligned}
\tag{18.30}
$$

The survival probability is measured by the disappearance channel, that is, we measure the fraction of neutrinos of the original flavor left. On the other hand, the transition probability is measured by the appearance channel, that is, we measure the fraction of neutrinos of new flavor.

The different parameters used to describe the neutrino oscillation phenomenon are as follows:

i) **The mixing angle θ:** This is one of the two fundamental parameters of neutrino oscillation phenomenology. Notice that if $\theta = 0$, the flavor eigenstates are identical to the mass eigenstates, that is, a ν_α will propagate from the source to the detector as ν_α. If $\theta = \frac{\pi}{4}$, then the oscillations are said to be maximal and at some point along the path between the source and the detector, all of the neutrinos of one flavor (say α) will oscillate to another flavor (say β). A nonzero value of θ signifies how much the flavor eigenstates are different from the mass eigenstates.

ii) **The mass squared difference Δm^2:** This is another fundamental parameter of neutrino oscillations. It governs the frequency of oscillations. For a two-flavor neutrino oscillation, there will be two mass eigenstates and $\Delta m_{12}^2 = m_1^2 - m_2^2$. For oscillation to ocuur, either one of the neutrino mass eigenstate should be nonzero or the two mass eigenstates should be nonzero and non-degenerate. Since the probability of oscillation would be unchanged for $\Delta m_{21}^2 \to -\Delta m_{12}^2$, it does not tell whether $m_1 > m_2$ or vice versa.

iii) **L/E:** The transition probability (Eq. (18.30)) depends periodically on the quantity L/E and describes neutrino oscillation. The L/E ratio is the parameter which is in the control of experimentalists for accelerator or reactor experiments, where L is the distance between the source and the detector, and E is the energy of the neutrino. Equation (18.30) implies that the oscillation becomes maximum for

$$
1.267\Delta m^2 \frac{L}{E} = (2n+1)\frac{\pi}{2} \Rightarrow \Delta m^2 \frac{L}{E} > 1, \quad \text{where } n = 0,1,2...
\tag{18.31}
$$

or

$$
\frac{L}{E} = (2n+1)\frac{\pi}{2.534\Delta m^2}.
$$

For example, if we take $n = 0$ (corresponding to the first oscillation maxima), $\Delta m^2 = 2.5 \times 10^{-3} \text{eV}^2$ and $L = 500$ km, 1000 km, and 2000 km, the first maxima should be observed respectively at the energies $E = 1.01$ GeV, 2.02 GeV, and 4.04 GeV, which can be clearly seen from the plots shown in Figure 18.3. It may be observed that for $E < 500 \text{MeV}$ and $\frac{L}{E} \gg \frac{1}{\Delta m^2}$, the average of the oscillation probability would be the most acceptable measurement.

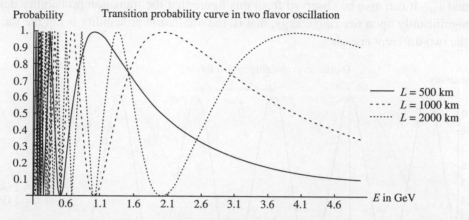

Figure 18.3 Transition probability curve for two-flavor neutrino oscillations for different values of L.

iv) The frequency of oscillation is controlled by Δm^2 and is shown in Figure 18.4, which is obtained using Eq. (18.30) for a constant L and θ. This figure is a plot of the transition probability as a function of neutrino energy (0.1 GeV to 5 GeV) at constant $L = 2000$ km and $\theta = 45°$, for the several values of Δm^2 viz., $\Delta m^2 = 2.2 \times 10^{-3} \text{eV}^2$, 2.5×10^{-3} eV^2 and 2.8×10^{-3} eV^2. It may be observed that Δm^2 is directly proportional to E, that is, with the increase in Δm^2, the probability curve gets shifted toward the higher energies.

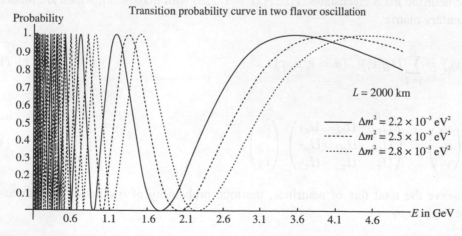

Figure 18.4 Transition probability curve for two-flavor neutrino oscillations for different values of Δm^2.

v) In Figure 18.5, we have plotted transition probability $P(\nu_\alpha \to \nu_\beta)$ for the two values of energy $E = 1$ GeV and 3 GeV, using $\Delta m^2 = 2.5 \times 10^{-3}$ and $\sin^2(2\theta)=1$. The peak occurs at the oscillation length $L^{osc} = 992$ km and 2976 km respectively which can be seen from Figure 18.5. It may be observed that at some distance from the source L, all the neutrinos of one flavor are converted into the other flavor, whereas at some other length, there is no oscillation, while at other values of L, we have a mixture of ν_e and ν_μ. It can also be observed from this figure that the transition probability depends significantly upon neutrino energy, that is, the oscillation probability is not the same for the two different energies.

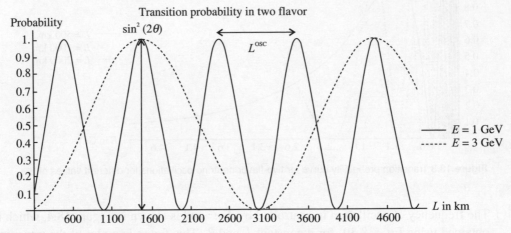

Figure 18.5 L^{osc} curves obtained using Eq. (18.30) at E_ν=1 GeV and 3 GeV, for $\Delta m^2 = 2.5 \times 10^{-3}$eV^2 and $\sin^2(2\theta)$=1.

18.2.2 Three-flavor neutrino oscillation in vacuum

The three-flavor neutrinos, viz., ν_e, ν_μ, ν_τ, while propagating in space, travel as some admixture of three neutrino mass eigenstates, viz., ν_i ($i = 1, 2, 3$) with masses m_i, which are related by a 3×3 unitary matrix

$$|\nu_\alpha\rangle = \sum_{i=1}^{3} U_{\alpha i}|\nu_i\rangle \quad (\alpha = e, \mu, \tau). \tag{18.32}$$

$$\begin{pmatrix} \nu_e \\ \nu_\mu \\ \nu_\tau \end{pmatrix} = \begin{pmatrix} U_{e1} & U_{e2} & U_{e3} \\ U_{\mu1} & U_{\mu2} & U_{\mu3} \\ U_{\tau1} & U_{\tau2} & U_{\tau3} \end{pmatrix} \begin{pmatrix} \nu_1 \\ \nu_2 \\ \nu_3 \end{pmatrix}. \tag{18.33}$$

To conserve the total flux of neutrinos, the total probability of oscillation in a three-flavor scheme is given by:

$$P(\nu_e \to \nu_e) + P(\nu_e \to \nu_\mu) + P(\nu_e \to \nu_\tau) = 1. \tag{18.34}$$

This probability is obtained for the other flavors in a similar manner. If the neutrinos are Dirac particles, then the neutrino mixing matrix is similar to the quark mixing matrix in both the number of mixing angles and CP phase as discussed in Chapter 6. Consequently, the lepton mixing matrix U is given by the Pontecorvo–Maki–Nakagawa–Sakata [179, 71] (PMNS) mixing matrix as:

$$
U = \begin{pmatrix} 1 & 0 & 0 \\ 0 & c_{23} & s_{23} \\ 0 & -s_{23} & c_{23} \end{pmatrix} \begin{pmatrix} c_{13} & 0 & s_{13}e^{-i\delta} \\ 0 & 1 & 0 \\ -s_{13}e^{i\delta} & 0 & c_{13} \end{pmatrix} \begin{pmatrix} c_{21} & s_{12} & 0 \\ -s_{12} & c_{12} & 0 \\ 0 & 0 & 1 \end{pmatrix} \tag{18.35}
$$

$$
= \begin{pmatrix} c_{12}c_{13} & s_{12}c_{13} & s_{13}e^{-i\delta} \\ -s_{12}c_{23} - c_{12}s_{13}s_{23}e^{i\delta} & c_{12}c_{23} - s_{12}s_{13}s_{23}e^{i\delta} & c_{13}s_{23} \\ s_{12}s_{23} - c_{12}s_{13}c_{23}e^{i\delta} & -c_{12}s_{23} - s_{12}s_{13}c_{23}e^{i\delta} & c_{13}c_{23} \end{pmatrix}, \tag{18.36}
$$

where $s_{ij} = \sin\theta_{ij}$ and $c_{ij} = \cos\theta_{ij} (i,j = 1,2,3)$. In this parameterization of the mixing matrix, the mixing parameters can take values in the ranges $0 \leq \theta_{ij} \leq \frac{\pi}{2}$ $(i,j = 1,3; i \neq j)$; a $\delta \neq 0, \pi$ would lead to CP violation. The parameters of the matrix are determined in oscillation experiments. The values of δ different from 0 and π imply CP violation in neutrino oscillations in vacuum [1011, 1012, 1013].

Following the same procedure as used in the case of two-flavor neutrino oscillation, we write the oscillation amplitude for the oscillation from flavor α to flavor β as:

$$
\mathcal{M}(\nu_\alpha \longrightarrow \nu_\beta) = \sum_i U_{\alpha i} e^{-i\frac{m_i^2}{2E}t} U_{\beta i}^*,
$$

which leads to the oscillation probability given by:

$$
P(\nu_\alpha \longrightarrow \nu_\beta) = |\mathcal{M}(\nu_\alpha \longrightarrow \nu_\beta)|^2 = \sum_{i=1}^{3}\sum_j U_{\alpha i} U_{\beta i}^* U_{\alpha j}^* U_{\beta j} e^{i\frac{L}{2E}(m_j^2 - m_i^2)}. \tag{18.37}
$$

The transition probability $P(\nu_\alpha \longrightarrow \nu_\beta)$ may be written as:

$$
P(\nu_\alpha \longrightarrow \nu_\beta) = \sum_{i=j} U_{\alpha i}^* U_{\beta i} U_{\alpha j} U_{\beta j}^* + \sum_{i\neq j} U_{\alpha i}^* U_{\beta i} U_{\alpha j} U_{\beta j}^* - 2\sum_{i\neq j} U_{\alpha i}^* U_{\beta i} U_{\alpha j} U_{\beta j}^*
$$

$$
\times \sin^2\left(\frac{L\Delta m_{ji}^2}{4E}\right) + i\sum_{i\neq j} U_{\alpha i}^* U_{\beta i} U_{\alpha j} U_{\beta j}^* \sin\left(\frac{L\Delta m_{ji}^2}{2E}\right) \tag{18.38}
$$

which is evaluated by first taking the first and second terms and writing:

$$
P_1 + P_2 = \sum_i \sum_j U_{\alpha i}^* U_{\beta i} U_{\alpha j} U_{\beta j}^* = \left|\sum_i U_{\alpha i} U_{\beta i}^*\right|^2 = \begin{cases} 1 & \text{for } \alpha = \beta, \\ 0 & \text{otherwise.} \end{cases} \tag{18.39}
$$

Now taking the third term of Eq. (18.38),

$$
\begin{aligned}
P_3 &= -2\sum_{i\neq j} U_{\alpha i}^* U_{\beta i} U_{\alpha j} U_{\beta j}^* \sin^2\left(\frac{L\Delta m_{ji}^2}{4E}\right) \\
&= -2\left\{\sum_{i>j} U_{\alpha i}^* U_{\beta i} U_{\alpha j} U_{\beta j}^* \sin^2\left(\frac{L\Delta m_{ij}^2}{4E}\right) + \sum_{i<j} U_{\alpha i}^* U_{\beta i} U_{\alpha j} U_{\beta j}^* \sin^2\left(\frac{L\Delta m_{ji}^2}{4E}\right)\right\} \\
&= -2\left\{\sum_{i>j} \sin^2\left(\frac{L\Delta m_{ij}^2}{4E}\right)\left[\left(U_{\alpha i}^* U_{\beta i} U_{\alpha j} U_{\beta j}^*\right) + \left(U_{\alpha i}^* U_{\beta i} U_{\alpha j} U_{\beta j}^*\right)^*\right]\right\} \\
&= -4\sum_{i>j} \mathrm{Re}\left(U_{\alpha i}^* U_{\beta i} U_{\alpha j} U_{\beta j}^*\right) \times \sin^2\left(\frac{L\Delta m_{ij}^2}{4E}\right). \qquad (18.40)
\end{aligned}
$$

Similarly, the last term of Eq. (18.38) leads to:

$$
\begin{aligned}
P_4 &= i\sum_{i\neq j} U_{\alpha i}^* U_{\beta i} U_{\alpha j} U_{\beta j}^* \sin\left(\frac{L\Delta m_{ji}^2}{2E}\right) \\
&= -i\sum_{i>j} 2i \times \mathrm{Im}\left(U_{\alpha i}^* U_{\beta i} U_{\alpha j} U_{\beta j}^*\right) \times \sin\left(\frac{L\Delta m_{ij}^2}{2E}\right).
\end{aligned}
$$

Adding all the probabilities from P_1 to P_4, we may write:

$$
\begin{aligned}
P(\nu_\alpha \longrightarrow \nu_\beta) &= \delta_{\alpha\beta} - 4\sum_{i>j} \mathrm{Re}\left(U_{\alpha i}^* U_{\beta i} U_{\alpha j} U_{\beta j}^*\right) \times \sin^2\left(\frac{\Delta_{ij}}{2}\right) \\
&\quad + 2\sum_{i>j} \mathrm{Im}\left(U_{\alpha i}^* U_{\beta i} U_{\alpha j} U_{\beta j}^*\right) \times \sin\left(\Delta_{ij}\right), \qquad (18.41)
\end{aligned}
$$

where $\quad \Delta_{ij} \equiv \dfrac{\Delta m_{ij}^2}{2E}L = \dfrac{L}{2E}\left(m_i^2 - m_j^2\right). \qquad (18.42)$

For the antineutrino oscillation $\bar\nu_\alpha \longrightarrow \bar\nu_\beta$, the oscillation probability has the same form as that of Eq. (18.41), except that the last term has a negative sign because for antineutrinos, $U_{\alpha i}$ is replaced by its complex conjugate, which is given as:

$$
\begin{aligned}
P(\bar\nu_\alpha \longrightarrow \bar\nu_\beta) &= \delta_{\alpha\beta} - 4\sum_{i>j} \mathrm{Re}\left(U_{\alpha i}^* U_{\beta i} U_{\alpha j} U_{\beta j}^*\right) \times \sin^2\left(\frac{L\Delta m_{ij}^2}{4E}\right) \\
&\quad - 2\sum_{i>j} \mathrm{Im}\left(U_{\alpha i}^* U_{\beta i} U_{\alpha j} U_{\beta j}^*\right) \times \sin\left(\frac{L\Delta m_{ij}^2}{2E}\right). \qquad (18.43)
\end{aligned}
$$

For the survival probability, that is, $\alpha = \beta$, we may write the oscillation probability for the neutrinos as:

$$
\begin{aligned}
P_{\nu_\alpha \to \nu_\alpha}(L, E) &= 1 - 4 \sum_{i>j} \mathrm{Re}\left(|U_{\alpha i}|^2 |U_{\alpha j}|^2\right) \sin^2\left(\frac{\Delta m_{ij}^2}{4E} L\right) \\
&\quad + 2 \sum_{i>j} \mathrm{Im}\left(|U_{\alpha i}|^2 |U_{\alpha j}|^2\right) \sin\left(\frac{\Delta m_{ij}^2}{2E} L\right).
\end{aligned}
\tag{18.44}
$$

Since $|U_{\alpha i}|^2 |U_{\alpha j}|^2$ is a real quantity, the imaginary part in this equation vanishes, and the general expression for the survival probability is obtained as:

$$
P_{\nu_\alpha \to \nu_\alpha}(L, E) = 1 - 4 \sum_{i>j} \left(|U_{\alpha i}|^2 |U_{\alpha j}|^2\right) \sin^2\left(\frac{\Delta m_{ij}^2}{4E} L\right).
\tag{18.45}
$$

For the three flavors of neutrinos $i, j = 1, 2, 3$, as $i > j$, the mass squared difference terms are Δm_{32}^2, Δm_{31}^2, and Δm_{21}^2.

$$
\Delta m_{32}^2 = m_3^2 - m_2^2 = (m_3^2 - m_1^2) + (m_1^2 - m_2^2) = \Delta m_{31}^2 - \Delta m_{21}^2
\tag{18.46}
$$

and only two of the three Δ_{ij}s, in Eq. (18.41) are independent.

18.2.3 Survival probability for $\nu_\alpha \to \nu_\alpha$

In the disappearance channel, we now express Eq. (18.45) for three-flavor survival probability, as:

$$
\begin{aligned}
P_{\nu_\alpha \to \nu_\alpha} &= 1 - 4\left[|U_{\alpha 3}|^2 |U_{\alpha 1}|^2 \sin^2\left(\frac{\Delta_{31}}{2}\right) + |U_{\alpha 3}|^2 |U_{\alpha 2}|^2 \sin^2\left(\frac{\Delta_{32}}{2}\right) \right. \\
&\quad \left. + |U_{\alpha 2}|^2 |U_{\alpha 1}|^2 \sin^2\left(\frac{\Delta_{21}}{2}\right)\right].
\end{aligned}
\tag{18.47}
$$

Using Eq. (18.46), we may write:

$$
\begin{aligned}
P_{\nu_\alpha \to \nu_\alpha} &= 1 - 4\left[|U_{\alpha 3}|^2 |U_{\alpha 1}|^2 \sin^2\left(\frac{\Delta_{31}}{2}\right) + |U_{\alpha 3}|^2 |U_{\alpha 2}|^2 \sin^2\left(\frac{\Delta_{31}}{2} - \frac{\Delta_{21}}{2}\right) \right. \\
&\quad \left. + |U_{\alpha 2}|^2 |U_{\alpha 1}|^2 \sin^2\left(\frac{\Delta_{21}}{2}\right)\right].
\end{aligned}
\tag{18.48}
$$

Now using the trigonometric identity

$$
\sin^2(A - B) = \sin^2 A + \sin^2 B - 2\sin^2 A \sin^2 B - \frac{1}{2}\sin 2A \sin 2B,
\tag{18.49}
$$

in the third term of Eq. (18.48), that is, $\sin^2\left(\frac{\Delta_{31}}{2} - \frac{\Delta_{21}}{2}\right)$, we may write,

$$
\begin{aligned}
P_{\nu_\alpha \to \nu_\alpha} = {} & 1 - 4\Big[|U_{\alpha3}|^2|U_{\alpha1}|^2 \sin^2\left(\frac{\Delta_{31}}{2}\right) + |U_{\alpha3}|^2|U_{\alpha2}|^2 \sin^2\left(\frac{\Delta_{31}}{2}\right) \\
& + |U_{\alpha3}|^2|U_{\alpha2}|^2 \sin^2\left(\frac{\Delta_{21}}{2}\right) + |U_{\alpha2}|^2|U_{\alpha1}|^2 \sin^2\left(\frac{\Delta_{21}}{2}\right)\Big] \\
& + 2|U_{\alpha3}|^2|U_{\alpha2}|^2\Big[4\sin^2\left(\frac{\Delta_{31}}{2}\right)\sin^2\left(\frac{\Delta_{21}}{2}\right) + \sin\Delta_{31}\sin\Delta_{21}\Big]. \quad (18.50)
\end{aligned}
$$

The unitarity condition leads to

$$
|U_{\alpha1}|^2 + |U_{\alpha2}|^2 + |U_{\alpha3}|^2 = 1
$$

and the expression of survival probability is written as:

$$
\begin{aligned}
P_{\nu_\alpha \to \nu_\alpha} = {} & 1 - 4|U_{\alpha3}|^2\left(1 - |U_{\alpha3}|^2\right)\sin^2\left(\frac{\Delta_{31}}{2}\right) - 4|U_{\alpha2}|^2\left(1 - |U_{\alpha2}|^2\right)\sin^2\left(\frac{\Delta_{21}}{2}\right) \\
& + 2|U_{\alpha3}|^2|U_{\alpha2}|^2\Big[4\sin^2\left(\frac{\Delta_{31}}{2}\right)\sin^2\left(\frac{\Delta_{21}}{2}\right) + \sin\Delta_{31}\sin\Delta_{21}\Big]. \quad (18.51)
\end{aligned}
$$

18.2.4 Transition probability for $\nu_\alpha \to \nu_\beta$

Now we will obtain the transition probability when ν_α get converted into another flavor ν_β after the neutrino has traveled a distance L with an energy E. We start with the general expression for probability, given in Eq. (18.41), and simplify it for the three flavor case. This expression takes into account the CP violation phase, therefore we will obtain an expression which can conclude about CP violation in the leptonic sector. In the appearance channel, following a similar step as followed for the disappearance channel, we start with Eq. (18.41), and write:

$$
\begin{aligned}
P_{\nu_\alpha \to \nu_\beta} = {} & \delta_{\alpha\beta} - 4\Big[\mathrm{Re}\left(U_{\alpha3}^*U_{\beta3}U_{\alpha2}U_{\beta2}^*\right)\sin^2\left(\frac{\Delta_{32}}{2}\right) + \mathrm{Re}\left(U_{\alpha3}^*U_{\beta3}U_{\alpha1}U_{\beta1}^*\right) \\
c \quad & \sin^2\left(\frac{\Delta_{31}}{2}\right) + \mathrm{Re}\left(U_{\alpha2}^*U_{\beta2}U_{\alpha1}U_{\beta1}^*\right)\sin^2\left(\frac{\Delta_{21}}{2}\right)\Big] + 2\Big[\mathrm{Im}\left(U_{\alpha3}^*U_{\beta3}U_{\alpha2}U_{\beta2}^*\right) \\
& \sin\Delta_{32} + \mathrm{Im}\left(U_{\alpha3}^*U_{\beta3}U_{\alpha1}U_{\beta1}^*\right)\sin\Delta_{31} + \mathrm{Im}\left(U_{\alpha2}^*U_{\beta2}U_{\alpha1}U_{\beta1}^*\right)\sin\Delta_{21}\Big]. \quad (18.52)
\end{aligned}
$$

Substituting Eq. (18.46) in Eq. (18.52), we obtain:

$$
\begin{aligned}
P_{\nu_\alpha \to \nu_\beta} = {} & -4\Big[\mathrm{Re}\left(U_{\alpha3}^*U_{\beta3}U_{\alpha2}U_{\beta2}^*\right)\sin^2\left(\frac{\Delta_{31}}{2} - \frac{\Delta_{21}}{2}\right) + \mathrm{Re}\left(U_{\alpha3}^*U_{\beta3}U_{\alpha1}U_{\beta1}^*\right)\sin^2\left(\frac{\Delta_{31}}{2}\right) \\
& + \mathrm{Re}\left(U_{\alpha2}^*U_{\beta2}U_{\alpha1}U_{\beta1}^*\right)\sin^2\left(\frac{\Delta_{21}}{2}\right)\Big] + 2\Big[\mathrm{Im}\left(U_{\alpha3}^*U_{\beta3}U_{\alpha2}U_{\beta2}^*\right)\sin\left(\Delta_{31} - \Delta_{21}\right) \\
& + \mathrm{Im}\left(U_{\alpha3}^*U_{\beta3}U_{\alpha1}U_{\beta1}^*\right)\sin\Delta_{31} + \mathrm{Im}\left(U_{\alpha2}^*U_{\beta2}U_{\alpha1}U_{\beta1}^*\right)\sin\Delta_{21}\Big]. \quad (18.53)
\end{aligned}
$$

Now using Eq. (18.49), we may write:

$$
\begin{aligned}
P_{\nu_\alpha \to \nu_\beta} &= -4\Big[\operatorname{Re}\left(U_{\alpha3}^* U_{\beta3} U_{\alpha2} U_{\beta2}^*\right)\left(\sin^2\left(\frac{\Delta_{31}}{2}\right) + \sin^2\left(\frac{\Delta_{21}}{2}\right) - 2\sin^2\left(\frac{\Delta_{31}}{2}\right)\sin^2\left(\frac{\Delta_{21}}{2}\right)\right. \\
&\quad \left. -\frac{1}{2}\sin\Delta_{31}\sin\Delta_{21}\right) + \operatorname{Re}\left(U_{\alpha3}^* U_{\beta3} U_{\alpha1} U_{\beta1}^*\right)\sin^2\left(\frac{\Delta_{31}}{2}\right) + \operatorname{Re}\left(U_{\alpha2}^* U_{\beta2} U_{\alpha1} U_{\beta1}^*\right) \\
&\quad \sin^2\left(\frac{\Delta_{21}}{2}\right)\Big] + 2\Big[\operatorname{Im}\left(U_{\alpha3}^* U_{\beta3} U_{\alpha2} U_{\beta2}^*\right)\left(\sin\Delta_{31}\cos\Delta_{21} - \cos\Delta_{31}\sin\Delta_{21}\right) \\
&\quad + \operatorname{Im}\left(U_{\alpha3}^* U_{\beta3} U_{\alpha1} U_{\beta1}^*\right)\sin\Delta_{31} + \operatorname{Im}\left(U_{\alpha2}^* U_{\beta2} U_{\alpha1} U_{\beta1}^*\right)\sin\Delta_{21}\Big]. \quad (18.54)
\end{aligned}
$$

Making use of the unitarity relation $U_{\alpha1} U_{\beta1}^* + U_{\alpha2} U_{\beta2}^* + U_{\alpha3} U_{\beta3}^* = 0$, Eq. (18.54) may be written as:

$$
\begin{aligned}
P_{\nu_\alpha \to \nu_\beta} &= 4|U_{\alpha3}|^2 |U_{\beta3}|^2 \sin^2\left(\frac{\Delta_{31}}{2}\right) + 4|U_{\alpha2}|^2 |U_{\beta2}|^2 \sin^2\left(\frac{\Delta_{21}}{2}\right) \\
&\quad + 2\operatorname{Re}\left(U_{\alpha3}^* U_{\beta3} U_{\alpha2} U_{\beta2}^*\right)\left[4\sin^2\left(\frac{\Delta_{31}}{2}\right)\sin^2\left(\frac{\Delta_{21}}{2}\right) + \sin\Delta_{21}\sin\Delta_{31}\right] \\
&\quad + 2\Big[\operatorname{Im}\left(U_{\alpha3}^* U_{\beta3} U_{\alpha2} U_{\beta2}^*\right)\sin\Delta_{31}\cos\Delta_{21} - \operatorname{Im}\left(U_{\alpha3}^* U_{\beta3} U_{\alpha2} U_{\beta2}^*\right)\cos\Delta_{31}\sin\Delta_{21} \\
&\quad + \operatorname{Im}\left(U_{\alpha3}^* U_{\beta3} U_{\alpha1} U_{\beta1}^*\right)\sin\Delta_{31} + \operatorname{Im}\left(U_{\alpha2}^* U_{\beta2} U_{\alpha1} U_{\beta1}^*\right)\sin\Delta_{21}\Big]. \quad (18.55)
\end{aligned}
$$

18.2.5 CP violation in the leptonic sector

The expression of the neutrino transition probability given in Eq. (18.41) may be rewritten as:

$$
P_{\nu_\alpha \to \nu_\beta}(L, E) = \delta_{\alpha\beta} + R_{\alpha\beta} + \frac{1}{2} A_{\alpha\beta}, \text{ where} \quad (18.56)
$$

$$
R_{\alpha\beta} = -2\sum_{i>j}\operatorname{Re}\left(U_{\alpha i}^* U_{\alpha j} U_{\beta i} U_{\beta j}^*\right)\left(1 - \cos\left(\frac{\Delta m_{ij}^2}{2E}L\right)\right), \quad (18.57)
$$

$$
A_{\alpha\beta} = 4\sum_{i>j}\operatorname{Im}\left(U_{\alpha i}^* U_{\alpha j} U_{\beta i} U_{\beta j}^*\right)\sin\left(\frac{\Delta m_{ij}^2}{2E}L\right) \quad (18.58)
$$

$$
\text{and} \quad P_{\bar\nu_\alpha \to \bar\nu_\beta}(L, E) = \delta_{\alpha\beta} + R_{\alpha\beta} - \frac{1}{2} A_{\alpha\beta}. \quad (18.59)
$$

Using these two expressions, we may write:

$$
R_{\alpha\beta} = \frac{1}{2}\left(P_{\nu_\alpha \to \nu_\beta} + P_{\bar\nu_\alpha \to \bar\nu_\beta}\right) - \delta_{\alpha\beta} \text{ and} \quad (18.60)
$$

$$
A_{\alpha\beta} = P_{\nu_\alpha \to \nu_\beta} - P_{\bar\nu_\alpha \to \bar\nu_\beta}, \quad (18.61)
$$

$R_{\alpha\beta}$ and $A_{\alpha\beta}$ are respectively associated with the CP even and CP odd parts of the transition probabilities. $A_{\alpha\beta}$ is a measure of CP violation and will hereafter be written as $A_{\alpha\beta}^{\mathrm{CP}}$. Assuming CPT invariance leads to $A_{\alpha\beta}^{\mathrm{CP}} = -A_{\beta\alpha}^{\mathrm{CP}}$, implying $A_{\alpha\alpha}^{\mathrm{CP}} = 0$. If the CP invariance also holds good, then $A_{\alpha\beta}^{\mathrm{CP}} = 0$ for $\alpha \neq \beta$, and therefore it may be concluded that $A_{\alpha\beta}^{\mathrm{CP}}$ is a measure of CP asymmetry. A non-zero value of $A_{\alpha\beta}^{\mathrm{CP}}$ implies CP violation in the leptonic sector. For a three-flavor neutrino oscillation scenario, $A_{e\mu}^{\mathrm{CP}}$, $A_{\tau e}^{\mathrm{CP}}$, and $A_{\mu\tau}^{\mathrm{CP}}$ are explicitly defined. Following a cyclic permutation among the three flavors of neutrinos, for example, $A_{e\mu}^{\mathrm{CP}} = -A_{\mu e}^{\mathrm{CP}}$, $A_{\tau e}^{\mathrm{CP}} = -A_{\tau\mu}^{\mathrm{CP}}$ leads to

$$\sum_{\alpha=e,\mu,\tau} A_{\alpha\beta}^{\mathrm{CP}} = 0, \text{ as} \tag{18.62}$$

$$\sum_{\alpha=e,\mu,\tau} P_{\nu_\alpha \to \nu_\beta} = 1 \text{ and } \sum_{\alpha=e,\mu,\tau} P_{\bar{\nu}_\alpha \to \bar{\nu}_\beta} = 1.$$

This relation also gives

$$\sum_{\alpha=\mu,\tau} A_{\alpha e}^{\mathrm{CP}} = 0, \quad \sum_{\alpha=e,\tau} A_{\alpha\mu}^{\mathrm{CP}} = 0 \quad \text{and} \quad \sum_{\alpha=e,\mu} A_{\alpha\tau}^{\mathrm{CP}} = 0. \tag{18.63}$$

We now introduce the Jarlskog invariant $J_{\alpha\beta}$ [1014] defined as:

$$J_{\alpha\beta}^{ij} = \mathrm{Im}\left(U_{\alpha i} U_{\alpha j}^* U_{\beta i}^* U_{\beta j} \right), \tag{18.64}$$

such that

$$J_{\alpha\beta}^{ij} = -J_{\alpha\beta}^{ji} \text{ and } J_{\beta\alpha}^{ij} = -J_{\alpha\beta}^{ij}, \quad i,j = 1,2,3. \tag{18.65}$$

Thus, we may write

$$J_{(\mu,e)} = -J_{(e,\mu)} = J_{(e,\tau)} = -J_{(\tau,e)} = J_{(\tau,\mu)} = -J_{(\mu,\tau)} = \hat{J}\sin\delta \tag{18.66}$$

$$\text{with} \quad \hat{J} = s_{12}c_{12}s_{13}c_{13}^2 s_{23}c_{23} = \sin\theta_{12}\cos\theta_{12}\sin\theta_{13}\cos^2\theta_{13}\sin\theta_{23}\cos\theta_{23}. \tag{18.67}$$

Therefore, the value of $J_{\alpha\beta}$ is the same for all flavors. Here,

$$J_{\alpha\beta} = \mathrm{Im}(U_{\alpha 3}^* U_{\beta 3} U_{\alpha 1} U_{\beta 1}^*) = -\mathrm{Im}(U_{\alpha 3}^* U_{\beta 3} U_{\alpha 2} U_{\beta 2}^*) = -\mathrm{Im}(U_{\alpha 2}^* U_{\beta 2} U_{\alpha 1} U_{\beta 1}^*).$$

Using Eqs.(18.58) and (18.65), we may write:

$$A_{\alpha\beta}^{\mathrm{CP}} = 4J_{\alpha\beta}^{21}\left(\sin\frac{\Delta m_{21}^2}{2E}L + \sin\frac{\Delta m_{32}^2}{2E}L - \sin\frac{\Delta m_{31}^2}{2E}L \right). \tag{18.68}$$

Using the relations $\Delta_{31}^2 = \Delta_{21}^2 + \Delta_{32}^2$ and $\sin\phi + \sin\psi - \sin(\phi+\psi) = 4\sin\frac{\phi}{2}\sin\frac{\psi}{2}\sin\frac{\phi+\psi}{2}$, we may write:

$$A_{e\mu}^{\mathrm{CP}} = 16\hat{J}\sin\delta\left(\sin\frac{\Delta m_{12}^2}{4E}L \; \sin\frac{\Delta m_{23}^2}{4E}L \; \sin\frac{\Delta m_{12}^2 + \Delta m_{23}^2}{4E}L \right). \tag{18.69}$$

In terms of the Jarlskog invariant, the transition probability, Eq. (18.55), may be written as:

$$P_{\nu_\alpha \to \nu_\beta} = 4|U_{\alpha 3}|^2 |U_{\beta 3}|^2 \sin^2 \left(\frac{\Delta_{31}}{2}\right) + 4|U_{\alpha 2}|^2 |U_{\beta 2}|^2 \sin^2 \left(\frac{\Delta_{21}}{2}\right) + 2\,\mathrm{Re}\left(U_{\alpha 3}^* U_{\beta 3} U_{\alpha 2} U_{\beta 2}^*\right)$$

$$\times \left[4\sin^2 \left(\frac{\Delta_{31}}{2}\right) \sin^2 \left(\frac{\Delta_{21}}{2}\right) + \sin \Delta_{21} \sin \Delta_{31}\right] - 2J_{\alpha\beta} \sin \Delta_{31} \cos \Delta_{21}$$

$$+2J_{\alpha\beta} \sin \Delta_{21} \cos \Delta_{31} + 2J_{\alpha\beta} \sin \Delta_{31} - 2J_{\alpha\beta} \sin \Delta_{21} \qquad (18.70)$$

which simplifies to

$$\begin{aligned} P_{\nu_\alpha \to \nu_\beta} =\ & 4|U_{\alpha 3}|^2 |U_{\beta 3}|^2 \sin^2 \left(\frac{\Delta_{31}}{2}\right) + 4|U_{\alpha 2}|^2 |U_{\beta 2}|^2 \sin^2 \left(\frac{\Delta_{21}}{2}\right) \\ & +2\,\mathrm{Re}\left(U_{\alpha 3}^* U_{\beta 3} U_{\alpha 2} U_{\beta 2}^*\right)\left[4\sin^2 \left(\frac{\Delta_{31}}{2}\right) \sin^2 \left(\frac{\Delta_{21}}{2}\right) + \sin \Delta_{21} \sin \Delta_{31}\right] \\ & +4J_{\alpha\beta}\left[\sin \Delta_{31} \sin^2 \left(\frac{\Delta_{21}}{2}\right) - \sin \Delta_{21} \sin^2 \left(\frac{\Delta_{31}}{2}\right)\right]. \end{aligned} \qquad (18.71)$$

It may be recalled that the oscillation probability for antineutrinos are obtained by replacing $U_{\alpha i}$ in Eq. (18.32) by its complex conjugate $U_{\alpha j}^*$, which is equivalent to flipping the sign of δ in Eq. (18.35). If CPT invariance is assumed to be an exact conservation law, then

$$P_{\nu_\alpha \to \nu_\beta} = P_{\bar\nu_\beta \to \bar\nu_\alpha} \quad \text{and} \quad P_{\bar\nu_\alpha \to \bar\nu_\beta} = P_{\nu_\beta \to \nu_\alpha}.$$

Equivalently, the survival probability as well as the transition probability for antineutrinos may be obtained from Eqs.(18.51) and (18.55) respectively by changing the sign in the Jarlskog invariant term (Eq. (18.71)).

Conditions for non-zero CP violation

- The probability expression given in Eq. (18.71) alongwith $J_{\alpha\beta}$ provides information about the CP violation phase δ_{CP}. It has two parts, one is real and the other is imaginary. The real part has no δ_{CP} term and is known as the CP conserving part. The imaginary part is written in terms of the Jarlskog invariant and its magnitude remains the same for each channel, that is, either we start with ν_μ and end up with ν_e or start with a ν_e and end up with ν_μ and so on.

- The Jarlskog invariant changes sign, that is, for channel $\nu_\mu \to \nu_e$, it is positive; for $\nu_e \to \nu_\mu$, it is negative.

- If any one of the mixing angles is 0 or $\frac{\pi}{2}$, then there is no CP violation in the neutrino sector (as seen from Eq. (18.67)).

- To observe CP violation, we must look for the appearance of a new flavor, that is, we have to measure the probability $P(\nu_\alpha \to \nu_\beta)$ because the probability $P(\nu_\alpha \to \nu_\alpha)$ is always CP conserving.

- To have CP violation, the phase δ_{CP} cannot be 0 or π.

- For the CP violation, all the mass eigenstates should be non-degenerate. If any of the frequencies, say, Δ_{31} or Δ_{21} is zero, the imaginary part of Eq. (18.71) vanishes.

To measure δ_{CP}, we should be able to experimentally measure $(P_{\nu_\alpha \to \nu_\beta} - P_{\bar\nu_\alpha \to \bar\nu_\beta})$, that is,

$$(P_{\nu_\alpha \to \nu_\beta} - P_{\bar\nu_\alpha \to \bar\nu_\beta}) = 8\hat{J} \sin\delta \left[\sin\Delta_{31} \sin^2\left(\frac{\Delta_{21}}{2}\right) - \sin\Delta_{21} \sin^2\left(\frac{\Delta_{31}}{2}\right) \right]. \quad (18.72)$$

Notice that except for $\sin\delta$, all the other terms are constant for a particular L and E in this expression. Therefore, we may write

$$(P_{\nu_\alpha \to \nu_\beta} - P_{\bar\nu_\alpha \to \bar\nu_\beta}) \quad = \quad C \sin\delta, \quad (18.73)$$

where

$$C = 8\hat{J} \left[\sin\Delta_{31} \sin^2\left(\frac{\Delta_{21}}{2}\right) - \sin\Delta_{21} \sin^2\left(\frac{\Delta_{31}}{2}\right) \right].$$

Measuring $(P_{\nu_\alpha \to \nu_\beta} - P_{\bar\nu_\alpha \to \bar\nu_\beta})$, δ_{CP} may be calculated.

The T2K and NOvA accelerator experiments have $L = 295$ km and $L = 810$ km baseline, operating at an average neutrino energy $E_\nu = 0.6$ GeV and $E_\nu = 2.0$ GeV, respectively. Both observe transition probability for the channel $\nu_\mu \to \nu_e$ and $\bar\nu_\mu \to \bar\nu_e$. They study the difference of these probabilities, that is, $A_{\alpha\beta}$, given by Eq. (18.61). From the data of T2K [1000], the best fit value for δ_{CP} has been obtained as -1.87 (-1.43) radian for normal (inverted) ordering, which is a measure of CP violation and it excludes CP conserving values 0 and π at 2σ level.

18.2.6 Series expansions for neutrino oscillation probabilities

In this section, we present the series expansion formulas for three-flavor neutrino oscillation probabilities in vacuum up to second order in α and s_{13}, where $\alpha \equiv \Delta m_{21}^2 / \Delta m_{31}^2$ is the mass hierarchy parameter and $s_{13} \equiv \sin\theta_{13}$ is the mixing parameter. We consider the $\nu_\mu \to \nu_e$ channel for the series expansion.

From Eq. (18.55), considering $\alpha = \mu$ and $\beta = e$, and using Eq. (18.66) and Eq. (18.67), we may write

$$\begin{aligned}
P(\nu_\mu \to \nu_e) \quad = \quad & 4|U_{\mu3}|^2|U_{e3}|^2 \sin^2\left(\frac{\Delta_{31}}{2}\right) + 4|U_{\mu2}|^2|U_{e2}|^2 \sin^2\left(\frac{\Delta_{21}}{2}\right) \\
& + 2\,\mathrm{Re}\left(U_{\mu3}^* U_{e3} U_{\mu2} U_{e2}^*\right)\left[4\sin^2\left(\frac{\Delta_{31}}{2}\right)\sin^2\left(\frac{\Delta_{21}}{2}\right) + \sin\Delta_{21}\sin\Delta_{31}\right] \\
& + 4s_{12}c_{12}s_{13}c_{13}^2 s_{23}c_{23}\sin\delta_{CP}\left[\sin\Delta_{31}\sin^2\left(\frac{\Delta_{21}}{2}\right)\right. \\
& \left. - \sin\Delta_{21}\sin^2\left(\frac{\Delta_{31}}{2}\right)\right].
\end{aligned} \quad (18.74)$$

Using the parameterization of PMNS matrix, we get:

$$
\begin{aligned}
P(\nu_\mu \to \nu_e) &= 4s_{12}^2 c_{13}^2 \left(c_{12}^2 c_{23}^2 + s_{12}^2 s_{13}^2 s_{23}^2 - 2c_{12}c_{23}s_{12}s_{13}s_{23} \cos\delta_{CP} \right) \sin^2\left(\frac{\Delta_{21}}{2}\right) \\
&\quad + 4c_{13}^2 s_{23}^2 s_{13}^2 \sin^2\left(\frac{\Delta_{31}}{2}\right) + 2c_{13}^2 s_{23}s_{13}s_{12} \left(c_{12}c_{23} \cos\delta_{CP} \right. \\
&\quad \left. - s_{12}s_{13}s_{23} \right) \left[4\sin^2\left(\frac{\Delta_{31}}{2}\right)\sin^2\left(\frac{\Delta_{21}}{2}\right) + \sin\Delta_{21}\sin\Delta_{31} \right] \\
&\quad + 4s_{12}c_{12}s_{13}c_{13}^2 s_{23}c_{23} \sin\delta_{CP} \\
&\quad \left[\sin\Delta_{31}\sin^2\left(\frac{\Delta_{21}}{2}\right) - \sin\Delta_{21}\sin^2\left(\frac{\Delta_{31}}{2}\right) \right].
\end{aligned}
\tag{18.75}
$$

If we take

$$
\Delta m_{21}^2 = 7.5 \times 10^{-5} \mathrm{eV}^2, \ \Delta m_{31}^2 = 2.5 \times 10^{-3} \mathrm{eV}^2,
$$

$$
\alpha = \frac{\Delta m_{21}^2}{\Delta m_{31}^2} \simeq 0.03, \ \text{and} \ \Delta = \frac{\Delta_{31}}{2}.
$$

For $\theta_{13} = 8.5°$, leading to $s_{13} = \sin\theta_{13} \simeq 0.15$ and $\alpha = 0.03$, which are much smaller than unity, we can expand Eq. (18.75) up to the second order in α and s_{13}, which results in

$$
\begin{aligned}
P(\nu_\mu \to \nu_e) &= 4s_{12}^2 c_{13}^2 \left(c_{12}^2 c_{23}^2 + s_{12}^2 s_{13}^2 s_{23}^2 - 2c_{12}c_{23}s_{12}s_{13}s_{23} \cos\delta_{CP} \right) \sin^2(\alpha\Delta) \\
&\quad + 4c_{13}^2 s_{23}^2 s_{13}^2 \sin^2(\Delta) + 2c_{13}^2 s_{23}s_{13}s_{12} \left(c_{12}c_{23} \cos\delta_{CP} - s_{12}s_{13}s_{23} \right) \\
&\quad \times \left[4\sin^2(\Delta)\sin^2(\alpha\Delta) + \sin(2\alpha\Delta)\sin(2\Delta) \right] \\
&\quad + 4s_{12}c_{12}s_{13}c_{13}^2 s_{23}c_{23} \sin\delta_{CP} \left[\sin(2\Delta)\sin^2(\alpha\Delta) - \sin(2\alpha\Delta)\sin^2(\Delta) \right].
\end{aligned}
\tag{18.76}
$$

Further, making the approximation $\sin(\alpha\Delta) \simeq \alpha\Delta$ and taking only α and s_{13} up to the second order, we obtain

$$
\begin{aligned}
P(\nu_\mu \to \nu_e) &\simeq 4s_{12}^2 c_{13}^2 c_{12}^2 c_{23}^2 \alpha^2 \Delta^2 + 4c_{13}^2 s_{23}^2 s_{13}^2 \sin^2(\Delta) \\
&\quad + 4c_{13}^2 s_{23}s_{13}s_{12}c_{12}c_{23} \cos(\delta_{CP})\alpha\Delta \sin(2\Delta) \\
&\quad - 8s_{12}c_{12}s_{13}c_{13}^2 s_{23}c_{23} \sin(\delta_{CP})\alpha\Delta \sin^2(\Delta).
\end{aligned}
\tag{18.77}
$$

For example, taking $\theta_{13} = 8.5°$ leading to $c_{13}^2 \simeq 1$, we obtain:

$$
\begin{aligned}
P(\nu_\mu \to \nu_e) &= \alpha^2 c_{23}^2 \sin^2(2\theta_{12})\Delta^2 + 2\alpha s_{13} \sin(2\theta_{23}) \sin(2\theta_{12}) \\
&\quad \times \cos(\Delta + \delta_{CP})\Delta \sin(\Delta) + 4s_{23}^2 s_{13}^2 \sin^2(\Delta).
\end{aligned}
\tag{18.78}
$$

In the T2K kinematic region, we plot a curve for the transition probability $P(\nu_\mu \to \nu_e)$ vs. E, shown in Figure 18.6, for the different values of δ_{CP}, using Eq. (18.78) with $L = 295$ km, $E = 0.05$ GeV to 1 GeV and the global fit value of $\Delta m_{21}^2 = 7.5 \times 10^{-5} \mathrm{eV}^2$, $\Delta m_{31}^2 = 2.5 \times 10^{-3} \mathrm{eV}^2$, $\theta_{12} = 34.5°$, $\theta_{23} = 45°$ and $\theta_{13} = 8.5°$.

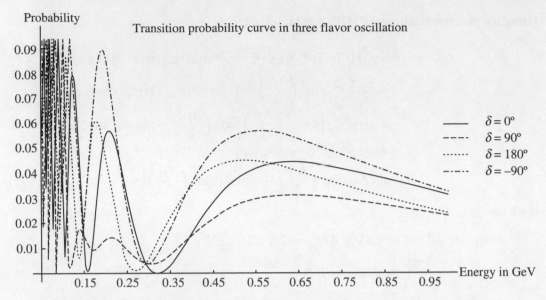

Figure 18.6 Appearance probability curve in three-flavor neutrino oscillations for different values of δ_{cp}.

A similar plot has been shown in Figure 18.7, where the transition probability curve for $P(\bar{\nu}_\mu \to \bar{\nu}_e)$ vs. E has been shown by replacing δ_{CP} with $-\delta_{CP}$ in Eq. (18.78). The observation of transition probability in the appearance experiments where both the neutrino and antineutrino channels are simultaneously studied gives information about δ_{CP}.

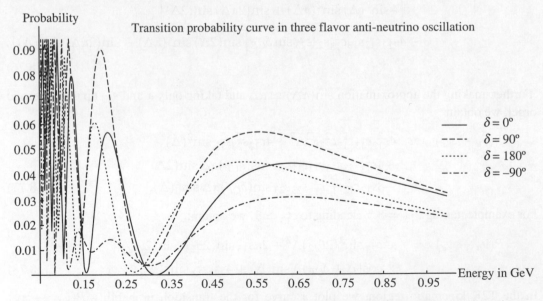

Figure 18.7 Appearance probability curve in three-flavor antineutrino oscillations for different values of δ_{CP}.

18.2.7 Neutrino mass hierarchy: Normal and inverted

When there are three neutrino flavor eigenstates and three mass eigenstates, there are two nonequivalent orderings for the neutrino masses as shown in Figure 18.8. They are called normal ordering (NO) when $m_1 < m_2 < m_3$, and inverted ordering (IO) when $m_3 < m_1 < m_2$. For such cases, two of the neutrino mass eigenstates are nearly degenerate. If these states are chosen to be m_1 and m_2, then,

$$|\Delta m_{21}^2| \equiv \Delta m_{\text{small}}^2 \quad \text{and} \quad |\Delta m_{31}^2| \cong |\Delta m_{32}^2| \equiv \Delta m_{\text{big}}^2 \ . \tag{18.79}$$

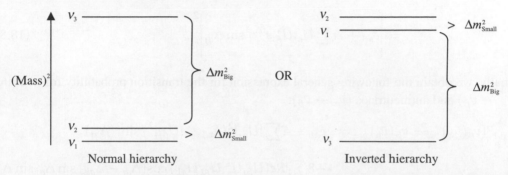

Normal hierarchy   Inverted hierarchy

Figure 18.8 A three-neutrino (mass)^2 spectrum in which the $\nu_2 - \nu_1$ splitting $\Delta m_{\text{small}}^2$ is much smaller than the splitting Δm_{big}^2 between ν_3 and the $\nu_2 - \nu_1$ pair. The later pair may be at either the bottom or the top of the spectrum.

The smaller one is always Δm_{21}^2, while the larger one is generally denoted by $\Delta m_{3\ell}^2$, with $\ell = 2$ for normal ordering and $\ell = 1$ for inverted ordering [1015]. Hence,

$$\Delta m_{3\ell}^2 = \begin{cases} \Delta m_{32}^2 = m_3^2 - m_2^2 > 0 & \text{for NO}, \\ \Delta m_{31}^2 = m_3^2 - m_1^2 < 0 & \text{for IO}. \end{cases} \tag{18.80}$$

From the analysis of neutrino oscillation data, it follows that one mass squared difference is much smaller than the other. Correspondingly, two scenarios for the three-neutrino mass spectra are possible:

1. Normal hierarchy (NH): $\Delta m_{21}^2 \ll \Delta m_{32}^2$.

2. Inverted hierarchy (IH): $\Delta m_{21}^2 \ll |\Delta m_{31}^2|$.

Let us denote two independent neutrino mass squared differences Δm_S^2 (solar) and Δm_A^2 (atmospheric). We have

$$\Delta m_{21}^2 = \Delta m_S^2, \ \Delta m_{32}^2 = \Delta m_A^2 \ (\text{NH}), \tag{18.81}$$

$$\Delta m_{21}^2 = \Delta m_S^2, \ \text{and} \ \Delta m_{31}^2 = -\Delta m_A^2 \ (\text{IH}). \tag{18.82}$$

Let us recall the expression of transition probability given in Eq. (18.37)

$$P(\nu_\alpha \longrightarrow \nu_\beta) = \left| \sum_j U_{\alpha j} e^{2i\Delta_{qj}} U_{\beta j}^* \right|^2 , \tag{18.83}$$

where q is an arbitrary fixed index and

$$\Delta_{qj} = \frac{\Delta m_{qj}^2 L}{4E}.$$

Using the unitarity of the mixing matrix, we can rewrite this expression as:

$$P(\nu_\alpha \longrightarrow \nu_\beta) = \left| \delta_{\alpha\beta} + \sum_j U_{\alpha j}(e^{2i\Delta_{qj}} - 1)U_{\beta j}^* \right|^2 \tag{18.84}$$

$$= \left| \delta_{\alpha\beta} + 2i \sum_j U_{\alpha j} U_{\beta j}^* e^{i\Delta_{qj}} \sin \Delta_{qj} \right|^2. \tag{18.85}$$

Finally, we obtain the following general expression for the transition probability for neutrinos $(\nu_\alpha \to \nu_\beta)$ and antineutrinos $(\bar{\nu}_\alpha \to \bar{\nu}_\beta)$:

$$\begin{aligned} P(\nu_\alpha(\bar{\nu}_\alpha) \longrightarrow \nu_\beta(\bar{\nu}_\beta)) = {} & \delta_{\alpha\beta} - 4\sum_j |U_{\alpha j}|^2(\delta_{\alpha\beta} - |U_{\beta j}|^2)\sin^2(\Delta_{qj}) \\ & + 8\sum_{j>k} Re(U_{\beta j}U_{\alpha j}^* U_{\beta k}^* U_{\alpha k})\cos(\Delta_{qj} - \Delta_{qk})\sin\Delta_{qj}\sin\Delta_{qk} \\ & \mp 8\sum_{j>k} Im(U_{\beta j}U_{\alpha j}^* U_{\beta k}^* U_{\alpha k})\sin(\Delta_{qj} - \Delta_{qk})\sin\Delta_{qj}\sin\Delta_{qk}. \end{aligned}$$

In the case of the NH, it is natural to choose $q = 2$. From the general expression, we have:

$$\begin{aligned} P^{NH}(\nu_\alpha(\bar{\nu}_\alpha) \longrightarrow \nu_\beta(\bar{\nu}_\beta)) = {} & \delta_{\alpha\beta} - 4|U_{\alpha 1}|^2(\delta_{\alpha\beta} - |U_{\beta 1}|^2)\sin^2\Delta_S - 4|U_{\alpha 3}|^2(\delta_{\alpha\beta} - |U_{\beta 3}|^2) \\ & \sin^2\Delta_A - 8\,Re\,[U_{\beta 3}U_{\alpha 3}^* U_{\beta 1}^* U_{\alpha 1}]\cos(\Delta_A + \Delta_S)\sin\Delta_A\sin\Delta_S \\ & \mp 8\,Im\,[U_{\beta 3}U_{\alpha 3}^* U_{\beta 1}^* U_{\alpha 1}]\sin(\Delta_A + \Delta_S)\sin\Delta_A\sin\Delta_S. \quad (18.86) \end{aligned}$$

In the case of the IH, we choose $q = 1$. For the transition probability, we obtain the following expression:

$$\begin{aligned} P^{IH}(\nu_\alpha(\bar{\nu}_\alpha) \longrightarrow \nu_\beta(\bar{\nu}_\beta)) = {} & \delta_{\alpha\beta} - 4|U_{\alpha 2}|^2(\delta_{\alpha\beta} - |U_{\beta 2}|^2)\sin^2\Delta_S - 4|U_{\alpha 3}|^2(\delta_{\alpha\beta} - |U_{\beta 3}|^2) \\ & \sin^2\Delta_A - 8\,Re\,[U_{\beta 3}U_{\alpha 3}^* U_{\beta 2}^* U_{\alpha 2}]\cos(\Delta_A + \Delta_S)\sin\Delta_A\sin\Delta_S \\ & \pm 8\,Im\,[U_{\beta 3}U_{\alpha 3}^* U_{\beta 2}^* U_{\alpha 2}]\sin(\Delta_A + \Delta_S)\sin\Delta_A\sin\Delta_S. \quad (18.87) \end{aligned}$$

Notice that the expressions given in Eq. (18.86) and Eq. (18.87) differ by the change $U_{\alpha 1} \to U_{\alpha 2}$ and by the sign of the last term. Also notice that for the CP asymmetry from Eq. (18.86) and Eq. (18.87), we may write:

$$A_{\alpha\beta}^{CP} = -16\,Im\,[U_{\beta 3}U_{\alpha 3}^* U_{\beta 1}^* U_{\alpha 1}]\sin(\Delta_A + \Delta_S)\sin\Delta_A\sin\Delta_S \tag{18.88}$$

in the case of NH and

$$A_{\alpha\beta}^{CP} = 16\,Im\,[U_{\beta 3}U_{\alpha 3}^* U_{\beta 2}^* U_{\alpha 2}]\sin(\Delta_A + \Delta_S)\sin\Delta_A\sin\Delta_S \tag{18.89}$$

in the case of IH. If we take $\Delta m_S^2 = 7.39 \times 10^{-5} \text{eV}^2$ and $\Delta m_A^2 = 2.5 \times 10^{-3} \text{eV}^2$, then

$$\frac{\Delta m_S^2}{\Delta m_A^2} \simeq 3 \cdot 10^{-2}, \quad \text{and} \quad \sin^2 \theta_{13} \simeq 2.4 \cdot 10^{-2}. \tag{18.90}$$

It may be noticed that the effect of $\frac{L}{E}$ $(\frac{\Delta m_A^2 L}{2E} \gtrsim 1)$ on the neutrino oscillations are large in the energy region of the atmospheric neutrinos; we can neglect the small contributions from Δm_S^2 and $\sin^2 \theta_{13}$ in the transition probabilities. We obtain the following expressions for the survival probabilities for $\nu_\mu (\bar{\nu}_\mu)$:

$$
\begin{aligned}
P^{NH}(\nu_\alpha \to \nu_\alpha) &\simeq P^{IH}(\nu_\alpha \to \nu_\alpha) \simeq 1 - 4|U_{\mu 3}|^2(1 - |U_{\mu 3}|^2) \sin^2 \Delta m_A^2 \frac{L}{4E} \\
&= 1 - \sin^2 2\theta_{23} \sin^2 \Delta m_A^2 \frac{L}{4E}.
\end{aligned}
\tag{18.91}
$$

Now, we have $P(\nu_\mu \to \nu_e) \simeq 0$ with this approximation:

$$P(\nu_\mu \to \nu_\tau) \simeq 1 - P(\nu_\mu \to \nu_\mu) \simeq \sin^2 2\theta_{23} \sin^2 \Delta m_A^2 \frac{L}{4E}. \tag{18.92}$$

Thus, in the energy region of the atmospheric neutrinos, predominantly $\nu_\mu \rightleftarrows \nu_\tau$ oscillations take place.

Let us now consider $\bar{\nu}_e$ oscillation in the energy region of the KamLAND reactor $(\frac{\Delta m_S^2 L}{2E} \gtrsim 1)$. Neglecting the contribution of $\sin^2 \theta_{13}$:

$$P^{NH}(\bar{\nu}_e \to \bar{\nu}_e) \simeq P^{IH}(\bar{\nu}_e \to \bar{\nu}_e) \simeq 1 - \sin^2 2\theta_{12} \sin^2 \Delta m_S^2 \frac{L}{4E}, \tag{18.93}$$

$$P^{NH}(\bar{\nu}_e \to \bar{\nu}_\mu) \simeq P^{IH}(\bar{\nu}_e \to \bar{\nu}_\mu) \simeq \sin^2 2\theta_{12} \cos^2 \theta_{23} \sin^2 \Delta m_S^2 \frac{L}{4E} \tag{18.94}$$

and

$$P^{NH}(\bar{\nu}_e \to \bar{\nu}_\tau) \simeq P^{IH}(\bar{\nu}_e \to \bar{\nu}_\tau) \simeq \sin^2 2\theta_{12} \sin^2 \theta_{23} \sin^2 \Delta m_S^2 \frac{L}{4E} \tag{18.95}$$

which gives:

$$P(\bar{\nu}_e \to \bar{\nu}_e) = 1 - P(\bar{\nu}_e \to \bar{\nu}_\mu) - P(\bar{\nu}_e \to \bar{\nu}_\tau) \tag{18.96}$$

and

$$\frac{P(\bar{\nu}_e \to \bar{\nu}_\tau)}{P(\bar{\nu}_e \to \bar{\nu}_\mu)} \simeq \tan^2 \theta_{23} \simeq 1. \tag{18.97}$$

Using Eq. (18.91) and (18.93), the Super-Kamiokande detector analyzed their atmospheric neutrino samples; K2K and MINOS used them in the analysis of accelerator neutrino experiments and KamLAND used it in the analysis of reactor antineutrino experiments.

18.3   Neutrino Oscillation in Matter

While deriving the probabilities for neutrino oscillation in vacuum in the previous section, we have not considered the fact that neutrinos may also be affected by the matter they traverse, leading to possible modifications of the transition probabilities. There are examples like the solar neutrinos which are produced deep inside the core of the sun and traverse different layers like the core, radiation zone, convection zone, photosphere, chromosphere to corona which have different matter densities (Figure 18.9). Moreover, neutrinos passing through the earth like

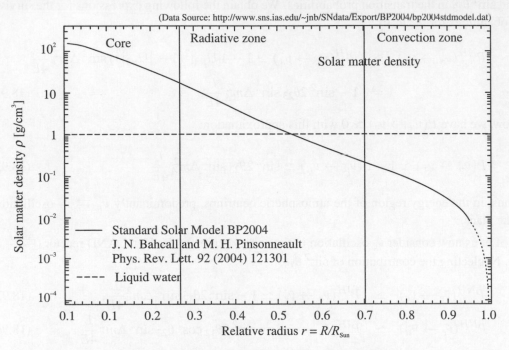

(Data Source: http://www.sns.ias.edu/~jnb/SNdata/Export/BP2004/bp2004stdmodel.dat)

Figure 18.9 Matter density variation inside the sun [1016].

atmospheric neutrinos coming from the other side of the globe traverse the core, mantle, and the crust of earth having different densities, as shown in Figure 18.10, before getting detected. During a solar eclipse, neutrinos from the sun have to pass through the moon and the earth before reaching the detection point. Moreover, neutrinos coming from different sources to the detectors like DUNE, JUNO, KamLAND, MINOS, T2K etc. get affected by the earth matter. In a pioneering work, Wolfenstein [1010] and later Mikheyev and Smirnov [1009] showed that the parameters of neutrino oscillation may be significantly modified if the neutrino travels through matter rather than vacuum. In the next section, we study the neutrino–matter interactions which affects the neutrino properties and modifies the oscillation probabilities when neutrinos travel through the matter.

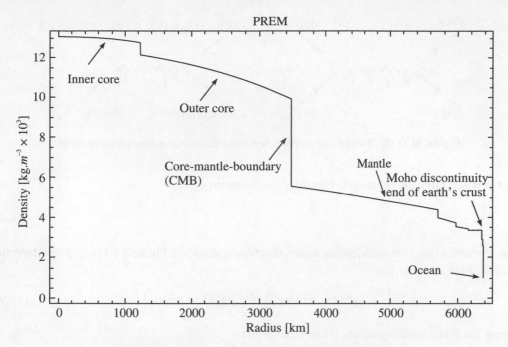

Figure 18.10 Matter density variation inside the earth [1017].

18.3.1 Effective potential for neutrino–matter interactions

When neutrinos propagate through matter, their properties may change because of their interaction with matter. Neutrinos of all flavors interact with matter via charge current (CC) and neutral current (NC) interaction; the nature of their interaction with matter depends upon their energy. For example, neutrinos (ν_e) being produced in the sun have energies less than 20 MeV, and they oscillate into ν_μ or ν_τ ($\nu_e \to \nu_\mu$) and ($\nu_e \to \nu_\tau$), through charged current reactions like:

$$\nu_\mu + n \to \mu^- + p \quad \text{and} \quad \nu_\tau + n \to \tau^- + p$$

are not possible due to the large mass of μ (m_μ=105.658 MeV) and τ (m_τ=1776.84 MeV) leptons; however, neutral current reactions like $\nu_l + n/p \to \nu_l + n/p$, $l = \mu, \tau$ are possible. Therefore, while the original matter which consists of mainly electrons, protons, and neutrons undergo neutral current interactions with all flavors of neutrinos through $\nu_l e \to \nu_l e, \nu_l p \to \nu_l p$, $\nu_l n \to \nu_l n$ ($l = e, \mu, \tau$) processes, they undergo only charged current (CC) interactions with ν_e and $\bar{\nu}_e$, that is, $\nu_e n \to e^- p$ and $\bar{\nu}_e p \to e^+ n$

In this section, we calculate the effective potentials derived from the effective CC and NC interaction through coherent neutrino scattering in matter. The Feynman diagrams for the CC and NC reactions are shown in Figure 18.11. First, we estimate the effective potential arising due to CC interactions.

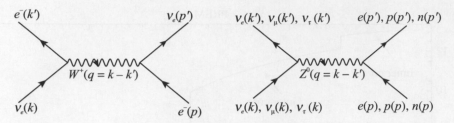

Figure 18.11 (a) Charged current interaction (left). (b) Neutral current interaction (right).

(i) Charged current reactions: Consider the reaction:

$$\nu_e + e^- \rightarrow \nu_e + e^-. \tag{18.98}$$

The Feynman diagram contributing to this reaction is shown in Figure 18.11(a). The interaction Hamiltonian is written as:

$$H_{CC} = \frac{G_F}{\sqrt{2}} \left[\left(\bar{u}_e | \gamma^\mu (1 - \gamma^5) | u_{\nu_e} \right) \left(\bar{u}_{\nu_e} \gamma_\mu (1 - \gamma^5) u_e \right) \right]. \tag{18.99}$$

Using the Fierz transformation, it can be written as:

$$H_{CC} = \frac{G_F}{\sqrt{2}} \left[\left(\bar{u}_e | \gamma^\mu (1 - \gamma^5) | u_e \right) \left(\bar{u}_{\nu_e} \gamma_\mu (1 - \gamma^5) u_{\nu_e} \right) \right], \tag{18.100}$$

where G_F is the Fermi coupling constant and u_{ν_e} and u_e represents the fields of electron type neutrino and electron, respectively.

The aforementioned interaction Hamiltonian is written for free electrons and neutrinos. It occurs when a neutrino moves in the medium of matter with solar, atmospheric, or earth matter environment in which the electrons are essentially nonrelativistic as are protons and neutrons. For example, even in the core of the sun where the temperature is fifteen million degree Celsius, the electron's energy are in the keV region corresponding to $\frac{kT}{m_e} \sim 10^{-3}$. A nonrelativistic reduction of the spinors $u_{e,p,n}$ in Eq. (18.100) can be made before taking the expectation value between the initial and final states of the nonrelativistic electrons, that is, $\vec{p} \rightarrow 0$, making the neutrino momenta to be the same in the initial as well as in the final states. We find that the energy corresponding to the Hamiltonian in Eq. (18.100) is given by:

$$\mathcal{E} = \langle H \rangle = \frac{G_F}{\sqrt{2}} \left[\left\langle \bar{u}_e(\vec{p} = 0) | \gamma^\mu (1 - \gamma^5) | u_e(\vec{p} = 0) \right\rangle \left\langle \bar{u}_\nu(\vec{k}) \gamma_\mu (1 - \gamma_5) u_\nu(\vec{k}) \right\rangle \right]. \tag{18.101}$$

Here, $\quad \langle \bar{u}_e \gamma^0 u_e \rangle = \langle u_e{}^\dagger u_e \rangle = N_e, \qquad \langle \bar{u}_e \gamma^i u_e \rangle = \langle \frac{\vec{p}}{m_e} \rangle \simeq 0,$

$\quad \langle \bar{u}_e \gamma^0 \gamma^5 u_e \rangle \simeq \langle \frac{\vec{\sigma} \cdot \vec{p}}{m_e} \rangle \simeq 0, \qquad \langle \bar{u}_e \gamma^i \gamma^5 u_e \rangle \simeq \langle \vec{\sigma} \rangle,$

where N_e is the number density of the electrons and $\langle \vec{\sigma} \rangle$ is the average spin of the electrons. Since the matter in the sun or in the atmosphere is normalized in the spin space, that is, $\langle \vec{\sigma} \rangle \simeq 0$, the energy density corresponding to the Hamiltonian is written as

$$H_{CC} = \frac{G_F}{\sqrt{2}} N_e . \langle \bar{u}_{\nu_e} \gamma_0 (1 - \gamma^5) u_{\nu_e} \rangle = \sqrt{2} G_F N_e (u_{\nu_e}^\dagger u_{\nu_e}) = \sqrt{2} G_F N_e. \qquad (18.102)$$

(ii) Neutral current reactions: Consider the reaction:

$$\nu_\alpha + X_\beta \rightarrow \nu_\alpha + X_\beta,$$

where $\alpha = e, \mu, \tau$ and $X_\beta = e, p, n$. The Feynman diagram corresponding to this equation is shown in Figure 18.11(b). All the neutrino flavors contribute equally to the effective potential in this case. The effective potential is obtained by including the electron, proton, and neutron background such that the interaction Hamiltonian for the NC is given by:

$$
\begin{aligned}
H_{NC} = \frac{G_F}{\sqrt{2}} \Bigg[& \langle \bar{u}_e \gamma^\mu (g_V^e - g_A^e \gamma^5) u_e \rangle \sum_\alpha \bar{u}_{\nu_\alpha} \gamma^\mu (g_V^\nu - g_A^\nu \gamma^5) u_{\nu_\alpha} \\
& + \langle \bar{u}_p \gamma^\mu (g_V^p - g_A^p \gamma^5) u_p \rangle \sum_\alpha \bar{u}_{\nu_\alpha} \gamma^\mu (g_V^\nu - g_A^\nu \gamma^5) u_{\nu_\alpha} \\
& + \langle \bar{u}_n \gamma^\mu (g_V^n - g_A^n \gamma^5) u_n \rangle \sum_\alpha \bar{u}_{\nu_\alpha} \gamma^\mu (g_V^\nu - g_A^\nu \gamma^5) u_{\nu_\alpha} \Bigg],
\end{aligned}
\qquad (18.103)
$$

where

$$g_V^p = \frac{1}{2} - 2 \sin^2 \theta_w = -g_V^e, \qquad g_A^p = \frac{1}{2} = -g_A^e,$$

$$g_V^n = g_A^n = -\frac{1}{2}, \quad \text{and for neutrinos } g_V^\nu = 1 \text{ and } g_A^\nu = 1.$$

Since, in matter, the number of electrons are equal to the number of protons, and g_A and g_V for electrons and protons are equal in magnitude and opposite in sign, their contributions cancel out and we are left only with the contribution from the neutron

$$
\begin{aligned}
H_{NC}^n &= \frac{G_F}{\sqrt{2}} \left[\langle \bar{u}_n \gamma^0 (-\frac{1}{2} + \frac{1}{2} \gamma^5) u_n \rangle \sum_\alpha \bar{u}_{\nu_\alpha} \gamma^0 (1 - \gamma^5) u_{\nu_\alpha} \right] \\
&\simeq -\frac{G_F}{\sqrt{2}} N_n \sum_\alpha \bar{u}_{\nu_\alpha} \gamma^0 u_{\nu_\alpha}, \\
\Rightarrow V_{NC}^n &= G_F \left(-\frac{N_n}{\sqrt{2}} \right).
\end{aligned}
\qquad (18.104)
$$

The effective potential due to CC and NC reactions is

$$V = V_{CC}^e + V_{NC}^n = \sqrt{2} G_F \left(N_e \delta_{\alpha e} - \frac{1}{2} N_n \right). \qquad (18.105)$$

We have derived the expressions for V_{CC} and V_{NC} for neutrinos; for antineutrinos, their signs will get reversed, that is,

$$V_{CC}(\bar{\nu}_e) = -\sqrt{2} G_F N_e, \qquad V_{NC}(\bar{\nu}_\alpha) = -G_F \left(-\frac{N_n}{\sqrt{2}} \right). \qquad (18.106)$$

The value of V_{CC} is given by:

$$V_{CC} \simeq 7.6 \times 10^{-14} \left(\frac{\rho}{g/cm^3} \right) Y_e \quad eV, \tag{18.107}$$

where $Y_e \sim 0.5$. The strength of V_{CC} inside the sun, the earth, and supernova is given in Table 18.1. Due to this extra potential V_{CC}, the effective mass of an electron in the matter gets modified. Using the relativistic energy–momentum relation,

$$m_{e(\text{matter})}^2 = (E + V_{CC})^2 - |\vec{p}|^2 = E^2 - |\vec{p}|^2 + V_{CC}^2 + 2V_{CC}E.$$

Table 18.1 The impact of matter potential in various medium.

Medium	Matter density(g/cm^3)	$V_{CC}(eV)$
Solar core	~ 100	$\sim 10^{-12}$
Earth core	~ 10	$\sim 10^{-13}$
Supernova	$\sim 10^{14}$	~ 1

Since $E >> V_{CC}$, V_{CC}^2 is neglected; then,

$$m_{e(\text{matter})}^2 = m_e^2 + 2V_{CC}E \quad \Rightarrow \Delta m^2 = 2EV_{CC},$$

where $\Delta m^2 = m_{e(\text{matter})}^2 - m_e^2$. Using V_{CC} from Eq. (18.106), we may write:

$$\Delta m^2 = 2\sqrt{2}EG_F N_e. \tag{18.108}$$

18.3.2 Interaction Hamiltonian in matter

Let us now consider the evolution of two neutrino flavors in ordinary matter. First, we consider the case where the density of matter ρ is constant. For this, we work out the Hamiltonian for neutrinos traveling in vacuum and then we will see how it gets modified in the case of matter.

For medium with constant density, let us consider the time-dependent Schrödinger equation in the Lab frame for a neutrino traveling through vacuum ($c = \hbar = 1$), that is,

$$i\frac{\partial}{\partial t}|\nu(t)\rangle = H|\nu(t)\rangle, \tag{18.109}$$

where for two neutrino flavors e and μ

$$|\nu(t)\rangle = \begin{pmatrix} \nu_e(t) \\ \nu_\mu(t) \end{pmatrix}, \tag{18.110}$$

where $\nu_\alpha(t)(\alpha = e, \mu)$ is the amplitude for ν being a ν_α at time t. Now writing the Hamiltonian for neutrino oscillation in vacuum

$$\langle \nu_\alpha | H_{\text{vacuum}} | \nu_\beta \rangle = \langle \sum_i U_{\alpha i} \nu_i | H_{\text{vacuum}} | \sum_j U_{\beta j} \nu_j \rangle$$

$$= \sum_i U_{\alpha i}^* \langle \nu_i | H_{\text{vacuum}} | \nu_i \rangle U_{\beta i} = \sum_i U_{\alpha i}^* E_i U_{\beta i}, \qquad (18.111)$$

where U is the same as used in Eq. (18.9) and U^* is the complex conjugate of U, such that

$$\Rightarrow \langle \nu_\alpha | H_{\text{vacuum}} | \nu_\beta \rangle = \begin{pmatrix} \cos\theta & -\sin\theta \\ \sin\theta & \cos\theta \end{pmatrix} \begin{pmatrix} E_1 & 0 \\ 0 & E_2 \end{pmatrix} \begin{pmatrix} \cos\theta & \sin\theta \\ -\sin\theta & \cos\theta \end{pmatrix}$$

$$= \begin{pmatrix} E_1 \cos^2\theta + E_2 \sin^2\theta & E_1 \sin\theta\cos\theta - E_2 \sin\theta\cos\theta \\ E_1 \sin\theta\cos\theta - E_2 \sin\theta\cos\theta & E_1 \sin^2\theta + E_2 \cos^2\theta \end{pmatrix}$$

$$= \begin{pmatrix} H_{\alpha\alpha} & H_{\alpha\beta} \\ H_{\beta\alpha} & H_{\beta\beta} \end{pmatrix}. \qquad (18.112)$$

In this expression, E_1 and E_2 are ultra-relativistic energies of the neutrino mass eigenstates. In Eq. (18.112), let us first solve $H_{\alpha\alpha}$:

$$H_{\alpha\alpha} = E_1 \cos^2\theta + E_2 \sin^2\theta = \sqrt{|\vec{p}|^2 + m_1^2} \cos^2\theta + \sqrt{|\vec{p}|^2 + m_2^2} \sin^2\theta$$

$$= -\cos(2\theta) \frac{\Delta m^2}{4|\vec{p}|} + |\vec{p}| + \frac{m_1^2 + m_2^2}{4|\vec{p}|}, \text{ where } \Delta m^2 = m_2^2 - m_1^2.$$

Similarly, we may write:

$$H_{\beta\beta} = \cos(2\theta) \frac{\Delta m^2}{4|\vec{p}|} + |\vec{p}| + \frac{m_1^2 + m_2^2}{4|\vec{p}|}$$

and $\quad H_{\alpha\beta} = H_{\beta\alpha} = \sin(2\theta) \dfrac{\Delta m^2}{4|\vec{p}|}.$

Therefore, collecting all the terms appearing in Eq. (18.112), we get:

$$H_{\text{vacuum}} = \frac{\Delta m^2}{4|\vec{p}|} \begin{pmatrix} -\cos(2\theta) & \sin(2\theta) \\ \sin(2\theta) & \cos(2\theta) \end{pmatrix} + \left(|\vec{p}| + \frac{m_1^2 + m_2^2}{4|\vec{p}|} \right) \begin{pmatrix} 1 & 0 \\ 0 & 1 \end{pmatrix}.$$

We will concentrate only on the first term of H_{vacuum} as the relative phases of the interfering terms is important and consequently, the relative energies are important. Therefore, the Hamiltonian for the highly relativistic neutrinos may be approximated as:

$$H_{\text{vacuum}} \simeq \frac{\Delta m^2}{4E} \begin{pmatrix} -\cos(2\theta) & \sin(2\theta) \\ \sin(2\theta) & \cos(2\theta) \end{pmatrix}.$$

The equation of motion becomes

$$i \begin{bmatrix} \dot{\nu}_e(t) \\ \dot{\nu}_\mu(t) \end{bmatrix} = H_{\text{vacuum}} \begin{bmatrix} \nu_e(t) \\ \nu_\mu(t) \end{bmatrix} = \frac{\Delta m^2}{4E} \begin{pmatrix} -\cos(2\theta) & \sin(2\theta) \\ \sin(2\theta) & \cos(2\theta) \end{pmatrix} \begin{bmatrix} \nu_e(t) \\ \nu_\mu(t) \end{bmatrix}. \qquad (18.113)$$

In matter, the modified Hamiltonian may be written as:

$$H_M = H_{\text{vacuum}} + V_{CC}^e \begin{pmatrix} 1 & 0 \\ 0 & 0 \end{pmatrix} + V_{NC}^n \begin{pmatrix} 1 & 0 \\ 0 & 1 \end{pmatrix}.$$

Here, V_{CC} is the extra potential faced by the electron type neutrino during its charged-current interaction and V_{NC} is the extra potential arising due to the NC interaction of all flavors of neutrino. The V_{NC} term includes contributions from both diagonal elements, so we use an identity matrix. We can re-write the aforementioned expression as:

$$H_M = H_{\text{vacuum}} + \frac{1}{2} V_{CC} \begin{pmatrix} 1 & 0 \\ 0 & -1 \end{pmatrix} + \frac{1}{2} V_{CC} \begin{pmatrix} 1 & 0 \\ 0 & 1 \end{pmatrix} + V_{NC} \begin{pmatrix} 1 & 0 \\ 0 & 1 \end{pmatrix}. \quad (18.114)$$

We ignore the last two terms appearing in the Hamiltonian due to the fact that they are diagonal and do not affect the relative energy eigenvalues. Therefore, Eq. (18.114) becomes

$$
\begin{aligned}
H_M &= \frac{\Delta m^2}{4E} \begin{bmatrix} -\cos(2\theta) & \sin(2\theta) \\ \sin(2\theta) & \cos(2\theta) \end{bmatrix} + \frac{1}{2} \sqrt{2} G_F N_e \begin{bmatrix} 1 & 0 \\ 0 & -1 \end{bmatrix} \\
&= \frac{\Delta m^2}{4E} \begin{bmatrix} -(\cos 2\theta - r) & \sin 2\theta \\ \sin 2\theta & (\cos 2\theta - r) \end{bmatrix},
\end{aligned} \quad (18.115)
$$

where $r = \frac{2\sqrt{2} G_F N_e E}{\Delta m^2}$. If the mixing angle in the presence of matter is denoted by θ_M and the mass difference squared in matter is denoted by Δm_M^2, then by diagonalizing the matrix, we may write the Hamiltonian part of the oscillation in matter, corresponding to the Hamiltonian in the case of vacuum. Let us start with

$$H_M = \frac{\Delta m_M^2}{4E} \begin{pmatrix} -\cos(2\theta_M) & \sin(2\theta_M) \\ \sin(2\theta_M) & \cos(2\theta_M) \end{pmatrix}.$$

The equation of motion in the matter is given by:

$$i \begin{bmatrix} \dot{v}_e(t) \\ \dot{v}_\mu(t) \end{bmatrix} = H_M \begin{bmatrix} v_e(t) \\ v_\mu(t) \end{bmatrix} = \frac{\Delta m_M^2}{4E} \begin{pmatrix} -\cos(2\theta_M) & \sin(2\theta_M) \\ \sin(2\theta_M) & \cos(2\theta_M) \end{pmatrix} \begin{bmatrix} v_e(t) \\ v_\mu(t) \end{bmatrix}, \quad (18.116)$$

with

$$
\begin{aligned}
\sin^2(2\theta_M) &= \frac{\sin^2(2\theta)}{\sin^2(2\theta) + (\cos(2\theta) - r)^2}, \\
\Delta m_M^2 &= \Delta m^2 \sqrt{\sin^2(2\theta) + (\cos(2\theta) - r)^2}. \quad (18.117)
\end{aligned}
$$

The calculated values of the effective masses m_{1M}^2 and m_{2M}^2 are

$$
\begin{aligned}
m_{1M}^2 &= -\frac{1}{2} \Delta m^2 \sqrt{\sin^2(2\theta) + (\cos(2\theta) - r)^2}, \\
m_{2M}^2 &= \frac{1}{2} \Delta m^2 \sqrt{\sin^2(2\theta) + (\cos(2\theta) - r)^2}. \quad (18.118)
\end{aligned}
$$

It may be noticed that at zero density $m_{1M}^2 \simeq m_1^2$, and $m_{2M}^2 \simeq m_2^2$ for neutrinos as well as for antineutrinos. It can be clearly seen from Eq. (18.118) that when the density of matter increases from zero to $\cos 2\theta$, the effective mass squared m_{1M}^2 increases for the neutrinos and decreases linearly for the antineutrinos because the effective potential (due to CC interaction) is positive for neutrinos and negative for antineutrinos. Similar observations can be made for m_{2M}^2.

18.3.3 Probability for oscillation in matter

For the two-flavor oscillation scenario in matter, replacing the mixing angle θ(in vacuum) by θ_M(in matter), we may write

$$\begin{pmatrix} \nu_e \\ \nu_\mu \end{pmatrix} = U(\theta_M) \begin{pmatrix} \nu_1^M \\ \nu_2^M \end{pmatrix} = \begin{pmatrix} \cos\theta_M & \sin\theta_M \\ -\sin\theta_M & \cos\theta_M \end{pmatrix} \begin{pmatrix} \nu_1^M \\ \nu_2^M \end{pmatrix},$$

$$\Rightarrow \quad |\nu_e(t=0)\rangle = |\nu_1^M\rangle \cos\theta_M + |\nu_2^M\rangle \sin\theta_M,$$

$$\text{and} \quad |\nu_\mu(t=0)\rangle = -|\nu_1^M\rangle \sin\theta_M + |\nu_2^M\rangle \cos\theta_M,$$

where ν_1^M and ν_2^M are the mass eigenstates in matter; these are different from the mass eigenstates in vacuum, viz., ν_1 and ν_2.

First, we take the density of matter to be constant and then the solutions will be approximated for matter with varying densities. Following the analogy of Eq. (18.9), we write:

$$|\nu_e(t)\rangle = |\nu_1^M(t)\rangle \cos\theta_M + |\nu_2^M(t)\rangle \sin\theta_M$$

$$= |\nu_1^M\rangle e^{-i\frac{\Delta m_M^2}{4E}t} \cos\theta_M + |\nu_2^M\rangle e^{+i\frac{\Delta m_M^2}{4E}t} \sin\theta_M,$$

$$\Rightarrow \langle \nu_\mu|\nu_e(t)\rangle = -\sin\theta_M e^{-i\frac{\Delta m_M^2}{4E}t} \cos\theta_M + \cos\theta_M e^{+i\frac{\Delta m_M^2}{4E}t} \sin\theta_M$$

$$= \sin\theta_M \cos\theta_M \left\{ 2i \sin(\frac{\Delta m_M^2}{4E}t) \right\}. \tag{18.119}$$

The probability that $|\nu(t)\rangle$ (either ν_e or ν_μ) is detected as ν_μ after a time t is given by:

$$P_M(\nu_e \to \nu_\mu) = |\langle \nu_\mu|\nu_e(t)\rangle|^2$$

$$= \left| \sin\theta_M \cos\theta_M \left\{ 2i \sin(\frac{\Delta m_M^2}{4E}t) \right\} \right|^2$$

$$= \sin^2(2\theta_M) \sin^2 \left(1.267 \Delta m_M^2(eV) \frac{L(\text{km})}{E(\text{GeV})} \right), \tag{18.120}$$

where we used $L = t$ (in natural units). The probability in the matter $P_M(\nu_e \to \nu_\mu)$ is similar to the probability in the vacuum except that here we have replaced θ by θ_M and Δm^2 by Δm_M^2. In the limits of zero matter density, we get back to the case of oscillation in vacuum, that is,

$$\theta_M \to \theta \quad \text{and} \quad \Delta m_M^2 \to \Delta m^2.$$

The oscillation probability curve in matter for normal and inverted mass hierarchy are shown in Figure 18.12 for $L = 810$ km, $E = 1 - 6$ GeV, $\Delta m^2 = 2.5 \times 10^{-3} \text{eV}^2$, $\theta = 8.9°$

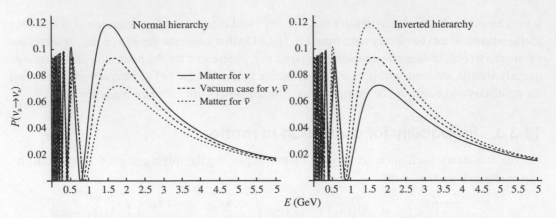

Figure 18.12 Oscillation probability for ν and $\bar{\nu}$ in matter and vacuum as illustration of matter effect. Normal and inverted hierarchy, using NOvA experiment parameters, $L = 810$ km, $E = 1$–6GeV, $\Delta m^2 = 2.5 \times 10^{-3}$eV^2, $\theta = 8.9°$.

18.3.4 Resonance condition and level crossing

In this section, we obtain a mathematical condition for the resonance in terms of the mixing angle in matter and the effective potential. Using the first equation of Eq. (18.117), we obtain:

$$\tan 2\theta_M = \frac{\sin 2\theta}{\cos 2\theta - r}. \tag{18.121}$$

We define $\tan 2\theta_M$ in terms of the vacuum oscillation length(L_v^{osc}), which is defined as the length for which the argument in $\sin^2(L_v \Delta m^2 / 4E)$ becomes π. This implies $L_v^{\text{osc}} = \frac{4\pi E}{\Delta m^2}$. For the $\nu_e - e^-$ scattering, the interaction length is given by:

$$L_{\text{int}} = \frac{4\pi}{2\sqrt{2}G_F N_e}, \text{then} \quad \frac{L_v^{\text{osc}}}{L_{\text{int}}} = \frac{2\sqrt{2}G_F N_e E}{\Delta m^2} = r. \tag{18.122}$$

Therefore, substituting $r = \frac{L_v^{\text{osc}}}{L_{\text{int}}}$ in Eq. (18.121), we get the mixing angle in terms of the oscillation and the interaction lengths as:

$$\tan 2\theta_M = \frac{L_{\text{int}} \sin 2\theta}{L_{\text{int}} \cos 2\theta - L_v^{\text{osc}}}. \tag{18.123}$$

We define $F = 2\sqrt{2}G_F N_e E$ which is positive in normal matter and $F' = \Delta m^2 \cos(2\theta)$, such that

$$\tan 2\theta_M = \frac{\tan 2\theta}{1 - F/F'}. \tag{18.124}$$

- There is maximal mixing at resonance, that is, the value of effective mixing is $\frac{\pi}{4}$ which leads to the possibility of total transitions between the two flavors, that is, the resonance region is wide enough. This is called the MSW effect.

- Resonance can exist only if $\theta_M < \frac{\pi}{4}$ because $\cos 2\theta_M < 0$, if $\theta_M > \frac{\pi}{4}$. Therefore, neutrino oscillation behaves differently in matter and vacuum, where probability is symmetric under the exchange of $\theta \rightarrow \frac{\pi}{2} - \theta$. For antineutrinos, the potential is reversed as there can be a resonance only if $\theta_M > \frac{\pi}{4}$.

In Eq. (18.124), we see that the term $\tan 2\theta_M$ reaches its maximum value for $\theta_M = \frac{\pi}{4}$ or $\tan 2\theta_M = \infty$, or we can say that for maximal mixing, the denominator of Eq. (17.124) becomes equal to zero, that is, $F = F'$ which gives

$$\Delta m^2 \cos 2\theta = 2\sqrt{2} G_F N_e E. \tag{18.125}$$

This condition is known as the MSW resonance condition. When this condition is met, neutrino mixing in matter becomes maximum.

18.4 Neutrino Oscillation: Experimental Status

Neutrino oscillation effects have been observed in several experiments using the neutrinos from following sources:

i) Solar neutrinos: Radiochemical experiments such as Homestake [1018], Gallex/GNO [988], and SAGE [985], as well as time and energy dependent rates from the four phases in Super-Kamiokande [981, 982, 983, 984], the three phases of SNO [989], and BOREXino [986, 987].

ii) Atmospheric neutrinos: Kamiokande [898, 901, 902], SOUDAN [979], MACRO [1019], Super-Kamiokande [903, 980], SNO [1020], MINOS [1021], IceCube [1022], ANTARES [1023].

iii) Accelerator neutrinos: ν_μ and $\bar{\nu}_\mu$ disappearance results from accelerator long baseline experiments MINOS [998] and T2K [1024], and ν_μ, $\bar{\nu}_\mu$ disappearance in NOνA [1025, 628, 1004].

iv) Accelerator neutrinos: Long baseline (LBL) ν_e, $\bar{\nu}_\mu$, ν_τ appearance results from MINOS [999], NOνA [1025, 628], T2K [1026], OPERA [119, 1005, 1006, 1007, 1008], etc.

v) Accelerator neutrinos: Short baseline (SBL) ν_e, $\bar{\nu}_e$ appearance experiments like at MiniBooNE [1027], LSND [1028], etc.

vi) Reactor antineutrinos: $\bar{\nu}_e$ disappearance experiments at Double Chooz [1029], Daya Bay [1030], and RENO [1031].

vii) The energy spectrum of reactor $\bar{\nu}_e$ disappearance at LBL in KamLAND [978].

In the scenario of three neutrino mixing, there are only two independent Δm_{ij}^2 say $\Delta m_{21}^2 \neq 0$ and $\Delta m_{31}^2 \neq 0$. The general convention is that θ_{12}, $\Delta m_{21}^2 (> 0)$, and θ_{23}, Δm_{31}^2, respectively represent the parameters which drive the solar (ν_e) and the dominant atmospheric neutrino (ν_μ and $\bar{\nu}_\mu$) oscillations. The parameter θ_{13} is associated with the smallest mixing angle in the PMNS mixing matrix. With the available experimental data, the following parameters are determined, viz. θ_{12}, Δm_{21}^2, θ_{23}, $|\Delta m_{31(32)}^2|$, and θ_{13}.

The value of Δm_{ij}^2 are sensitive to different oscillation lengths and neutrino's energy and have been tabulated in Table 18.2. In Table 18.3, we have listed the various neutrino sources like accelerator, reactor, solar, and atmospheric which are sensitive to the different values of oscillation parameters, viz., Δm_{ij}^2 and θ_{ij}. The observation of neutrino oscillation in the various experiments imply that neutrinos have mass and there is physics beyond the standard model. The globally best fitted values of oscillation parameters are listed in Table 18.4 and have been taken from PDG [117].

Table 18.2 Δm^2 reach of experiments using various neutrino sources. These experiments have different L and E range.

Neutrinos (baseline)	L(km)	E(GeV)	$\dfrac{L\text{(km)}}{E\text{(GeV)}}$	$\Delta m_{ij}^2(\text{eV}^2)$ Reach
Accelerator (short baseline)	1	1	1	1
Reactor (medium baseline)	1	10^{-3}	10^3	10^{-3}
Accelerator (long baseline)	10^3	10	10^2	10^{-2}
Atmospheric	10^4	1	10^4	10^{-4}
Solar	10^8	10^{-3}	10^{11}	10^{-11}

Table 18.3 Neutrino oscillation experiments sensitive to different θ_{ij} and Δm_{ij}^2.

Experiment	Dominant	Important
Solar experiments	θ_{12}	$\Delta m_{21}^2, \theta_{13}$
Reactor LBL (KamLAND)	Δm_{21}^2	θ_{12}, θ_{13}
Reactor MBL (Daya-Bay, Reno, Double Chooz)	θ_{13}	$|\Delta m_{3\ell}^2|$
Atmospheric Experiments	θ_{23}	$|\Delta m_{3\ell}^2|, \theta_{13}, \delta_{CP}$
Accelerator LBL ν_μ Disapp (MINOS, NOvA, T2K)	$|\Delta m_{3\ell}^2|, \theta_{23}$	
Accelerator LBL ν_e App (MINOS, NOvA, T2K)	δ_{CP}	$\theta_{13}, \theta_{23}, \text{sign}(\Delta m_{3\ell}^2)$

The determination of the mixing angles yields at present a maximum allowed CP violation

$$|J_{CP}^{max}| = 0.0329 \pm 0.0009 \, (\pm 0.0027) \tag{18.126}$$

at 1σ (3σ) for both orderings. The preference of the present data for non-zero δ_{CP} implies a best fit $J_{CP}^{best} = -0.032$, which is favored over CP conservation at the $\sim 1.2\sigma$ level.

The oscillation angles are measured to be $\theta_{12} \simeq 34°$, $\theta_{23} \simeq 45°$ and $\theta_{13} \simeq 8°$ [117]. The CP violating phase is not yet measured but recent observations indicate that it may be $\delta^{CP} \simeq -90°$ (maximal CP violation), for example, T2K and NOvA have measured $(P(\nu_\mu \to \nu_e) - P(\bar{\nu}_\mu \to \bar{\nu}_e))$. The T2K experiment hints that δ_{CP} is neither zero nor π [1000]. Recent measurements from T2K [1000] and NOvA [1004] also suggest normal neutrino mass hierarchy.

For details on neutrino oscillation physics, readers are referred to Refs.[1032, 1033].

Table 18.4 The best fit values of the three neutrino oscillation parameters from a global fit of the current neutrino oscillation data. The values (in brackets) correspond to $m_1 < m_2 < m_3$ ($m_3 < m_1 < m_2$). The definition of Δm^2 used is: $\Delta m^2 = m_3^2 - \frac{m_2^2 + m_1^2}{2}$. Thus, $\Delta m^2 = \Delta m_{31}^2 - \frac{\Delta m_{21}^2}{2} > 0$, if $m_1 < m_2 < m_3$ and $\Delta m^2 = \Delta m_{32}^2 + \frac{\Delta m_{21}^2}{2} < 0$, for $m_3 < m_1 < m_2$.

Parameter	best fit ($\pm 1\sigma$)
$\Delta m_{21}^2 \, 10^{-5} \mathrm{eV}^2$	$7.54^{+0.26}_{-0.22}$
$\lvert \Delta m^2 \rvert \, 10^{-3} \mathrm{eV}^2$	$2.43 \pm 0.06 \, (2.38 \pm 0.06)$
$\sin^2 \theta_{12}$	0.308 ± 0.017
$\sin^2 \theta_{23}, \; \Delta m^2 > 0$	$0.437^{0.033}_{0.023}$
$\sin^2 \theta_{23}, \; \Delta m^2 < 0$	$0.455^{0.039}_{0.031}$
$\sin^2 \theta_{13}, \; \Delta m^2 > 0$	$0.0234^{0.0020}_{0.0019}$
$\sin^2 \theta_{13}, \; \Delta m^2 < 0$	$0.0240^{0.0019}_{0.0022}$
δ / π (2σ range quoted)	$1.39^{0.38}_{0.27} \, (1.31^{0.29}_{0.33})$

18.5 Sterile Neutrinos

The standard model (SM) describes the phenomenology of the weak interactions of the three flavors of neutrinos ν_e, ν_μ, and ν_τ, which are considered to be massless, left-handed fermions interacting with matter through the exchange of $W^\pm$ and Z^0 bosons. Moreover, the SM also allows for the existence of massless right-handed neutrinos ν_R which have no interactions with matter and can be called sterile neutrinos. In fact, the idea of non-interacting neutrinos was first conceived by Pontecorvo in 1957 in the context of neutrino reactions with Argon. Subsequently, the discovery of the phenomenon of neutrino oscillations in solar, atmospheric, reactor, and accelerator neutrinos imply that the left-handed neutrinos which interact with matter are not massless. In fact, neutrinos which are the eigenstates of weak interaction Hamiltonian are a coherent mixture of the mass eigenstates of three neutrinos ν_1, ν_2, and ν_3 with very small masses $m_i (i = 1 - 3)$. The phenomenology of the neutrino oscillations is described satisfactorily in terms of the mass squared differences Δm_{ij}^2 ($i, j = 1, 2, 3$) and the three mixing angles θ_{ij} ($i, j = 1, 2, 3$). The latest values of Δm_{ij}^2 and θ_{ij} are given in Table 18.4. The three flavors of neutrino are also favored by the LEPP data on Z^0 decays and cosmological observations.

However, there are some neutrino oscillation experiments in the last 20 years which have reported anomalous results; the results are not consistent with the three flavor neutrino phenomenology and point towards the existence of additional flavors of neutrinos which are sterile. These sterile neutrinos do not interact with matter in order to be consistent with the LEPP observations but affect the phenomenology of neutrino interactions and neutrino oscillations. These experiments could beare either disappearance experiments with $\nu_\mu(\bar{\nu}_\mu) \to \nu_\mu(\bar{\nu}_\mu)$ and $\nu_e(\bar{\nu}_e) \to \nu_e(\bar{\nu}_e)$ or appearance experiments like: $\nu_\mu(\bar{\nu}_\mu) \to \nu_e(\bar{\nu}_e)$. There exists definite evidence of anomalous results in $\nu_\mu(\bar{\nu}_\mu) \to \nu_e(\bar{\nu}_e)$ and $\bar{\nu}_e \to \bar{\nu}_e$ experiments. In the following, we give a brief account of the experiments which reported anomalous results.

- **LSND experiment [716]:** The liquid scintillator neutrino detector (LSND) was a beam dump experiment at the Los Alamos Meson Physics Facility (LAMPF). It was a very short baseline neutrino (SBN) oscillation experiment: the length was 20 m corresponding to $\frac{E}{L} \sim 1\text{eV}^2$ in which the low energy muon antineutrinos from π-decay and subsequently muon decay were used. The LSND experiment observed an unexpected excess of $\bar{\nu}_e$ events through the inverse β-decay, that is, $\bar{\nu}_e + p \rightarrow n + e^+$, that are interpreted as $\bar{\nu}_\mu \rightarrow \bar{\nu}_e$ oscillations. However, the $\Delta m_{12}^2 = 1 - 10\text{eV}^2$ implied by this experiment was too large to be compatible with the other experimental results on Δm_{12}^2 and also with our present results on Δm_{12}^2 based on the three-flavor analysis of neutrino oscillation experiments. Attempts were made to explain these results by proposing the existence of a neutrino with a fourth flavor as sterile neutrino ν_s and oscillation of the active neutrinos to sterile neutrinos. The experiment also provided limits on the additional mixing parameters Δm_{14}^2 and θ_{14}.

- **MiniBooNE experiment [1027]:** The MiniBooNE was an accelerator experiment designed to have the same order of $\frac{L}{E}$ as the LSND experiment to test the anomalous results with the conventional beam of $\nu_\mu(\bar{\nu}_\mu)$ with a broad energy spectrum peaking at 1.25 GeV. The signature of $\nu_e(\bar{\nu}_e)$ respectively in this oscillation experiment was observed by the quasielastic reactions induced by $\nu_e(\bar{\nu}_e)$ which produced charged leptons. An excess of electron-like events was observed in the low energy region around the threshold region of the analysis corresponding to 200–350 MeV. The later experiments with better statistics confirmed the excess of $\nu_e/\bar{\nu}_e$ like events which also suggested the existence of a fourth flavor of sterile neutrinos with Δm_{14}^2 in the range of $1 - 5\text{eV}^2$ depending upon a reasonable range of the value of the mixing angle θ_{14}. The $\frac{L}{E}$ corresponding to the MiniBooNE experiment corresponds to a slightly smaller value of $\frac{L}{E}$ compared to the $\frac{L}{E}$ in the LSND experiment but is marginally consistent with it.

- **Gallium anomaly:** A deficit of electron antineutrinos as compared to the predictions was observed in the radiochemical experiments using Gallium (^{71}Ga) targets producing ^{70}Ge in the GALLEX [988] and SAGE [985] experiments. The experiments were being calibrated for solar neutrino experiments and the deficit is known as Gallium anomaly. In the GALLEX experiment, a ^{51}Cr source was used, while in the SAGE experiment, ^{51}Cr as well as ^{37}Ar sources for neutrinos were used. Initially, the deficit was found to be around 15%. These experiments involve nuclei like ^{71}Ge and ^{70}Ge in the initial and the final states leading to some uncertainties in the calculation of weak nuclear cross sections uncertainties from the excited nuclear states. The deficit has been confirmed in later analysis to the 3σ level.

- **Reactor anomaly:** The neutrino oscillation experiments with the reactor antineutrinos at KamLAND [978], Double Chooz [992], Daya Bay [1030], and RENO [1031] have played a very important role in the determination of the neutrino oscillation parameters, specially Δm_{21}^2 and θ_{12} and more recently, θ_{13} and Δm_{13}^2. The reactor antineutrino anomaly (RAA) was reported in the context of θ_{13} experiments once the predictions of the antineutrino flux from nuclear reactors were re-evaluated and updated. It refers to a deficit of antineutrino flux as compared to the predicted flux by about 6% in the antineutrino oscillation experiments operating with 10–100 m from the nuclear reactor and corresponds to a 2.8σ effect. Since the main isotopes contributing to the antineutrino spectra are ^{235}U, ^{238}U, ^{239}Pu, and ^{241}Pu, flux calculations are subject to the theoretical uncertainties arising due to the nuclear structure of these nuclei. The deficit is considered to be a real effect and attempts have been made to explain it in terms of the neutrino oscillations to sterile neutrinos ν_s.

The anomalies in the short baseline experiments at LSND and MiniBooNE and suggestions for the sterile neutrinos have initiated a search for sterile neutrinos in the cosmological observations. The limits on the mixing parameters of the fourth neutrino, that is, Δm_{i4}^2 and $|U_{i4}|$ $(i = \mu, \tau)$ are provided by NOMAD [1034], MINOS [1035], SK [1036], and IceCube [1037] experiments. Similarly, limits on the mixing parameters $|U_{e3}|^2 + |U_{e4}|^2$ exist from the SNO experiment [1038]. In the case of astrophysical neutrinos of very high energy from galactic and extra galactic sources, the anomalous events observed by ANITA [1039] collaboration could point to the existence of a fourth flavor of neutrinos. In cosmological observations, quantities like the total number of neutrino species and sum of their masses can be inferred from the study of the cosmic microwave background (CMB), the abundance of light elements in the Big Bang nucleosynthesis (BBN) and large scale structures (LSS) in the universe. While those observations are subject to considerable uncertainties, no definite conclusion can be drawn regarding the existence or nonexistence of additional flavors of neutrinos. If they exist, the mixing parameters like Δm_{i4}^2 and θ_{i4} are not consistent with the SBN results [1040].

18.6 Phenomenology of Sterile Neutrinos

The most economical model for sterile neutrinos is to add a single sterile neutrino as a fourth flavor of neutrino ν_s leading to a $3 + 1$ flavor mixing scenario for neutrinos as shown in Figure 18.13. This is called $3 + 1$ flavor instead of four flavor neutrino oscillation, because the fourth flavor neutrino is sterile, that is, it is a non-interacting neutrino while the other three flavors are active and interact with matter. This imposes certain restrictions on the mixing parameters of the fourth flavor ν_s, that is, Δm_{i4}^2 and θ_{i4} $(i = e, \mu, \tau, s)$, which are defined through a 4×4 mixing matrix U as:

$$\begin{pmatrix} \nu_e \\ \nu_\mu \\ \nu_\tau \\ \nu_s \end{pmatrix} = \begin{pmatrix} U_{e1} & U_{e2} & U_{e3} & U_{e4} \\ U_{\mu 1} & U_{\mu 2} & U_{\mu 3} & U_{\mu 4} \\ U_{\tau 1} & U_{\tau 2} & U_{\tau 3} & U_{\tau 4} \\ U_{s1} & U_{s2} & U_{s3} & U_{s4} \end{pmatrix} \begin{pmatrix} \nu_1 \\ \nu_2 \\ \nu_3 \\ \nu_4 \end{pmatrix}, \tag{18.127}$$

where ν_i $(i = e, \mu, \tau, s)$ are the flavor states and ν_i $(i = 1, 2, 3, 4)$ are the mass states with mass m_i. This implies that sterile neutrinos are almost the same as the ν_4 mass state. The sterile neutrino flavor state is a mixture of the four mass states. However, three of the mass states must have a very small mixture of sterile neutrinos in order to explain the data in the standard model. This implies the existence of additional splitting Δm_{i4}^2 $(i = 1, 2, 3)$ and more parameters to describe the mixing matrix. For example, for a 4×4 matrix, there would be a

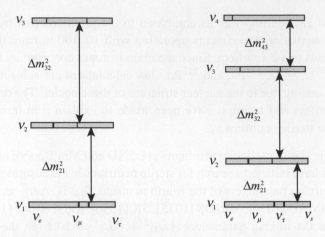

Figure 18.13 Pictorial representation of normal neutrino mass ordering and mixing for the three (left) and four (right) neutrino picture.

total of nine parameters with rotation angles θ_{ij} $(i \neq j)$ and 3 phases. However, in view of the success of the three-flavor oscillation phenomenology, most of the additional parameters will be small and not effective. Moreover, the anomalies in the short baseline experiments indicate that the mass splitting Δm_{41}^2 is large, that is, almost ten times the other mass splittings like Δm_{21}^2 and Δm_{32}^2. We can therefore work in an approximation where $\Delta m_{21}^2 \simeq \Delta m_{32}^2 \simeq 0$, the short baseline approximation, in which only one splitting Δm_{41}^2, that is, splitting between the most active and sterile neutrino is important. We also assume that all the neutrinos are Dirac neutrinos. A similar analysis can be done for Majorana neutrinos [1041]. In this approximation:

$$P_{\nu_e \to \nu_e} = 1 - 4(1 - |U_{e4}|^2)|U_{e4}|^2 \sin^2(\Delta_{41}^2), \tag{18.128}$$

$$P_{\nu_\mu \to \nu_\mu} = 1 - 4(1 - |U_{\mu 4}|^2)|U_{\mu 4}|^2 \sin^2(\Delta_{41}^2), \tag{18.129}$$

$$P_{\nu_\mu \to \nu_e} = 4|U_{\mu 4}|^2|U_{e4}|^2 \sin^2(\Delta_{41}^2), \tag{18.130}$$

where L and E are given in kilometers and GeV, or meters and MeV, respectively. Additionally, there are expressions for oscillations probabilities to the τ channel:

$$P_{\nu_\tau \to \nu_\tau} = 1 - 4(1 - |U_{\tau 4}|^2)|U_{\tau 4}|^2 \sin^2(\Delta_{41}^2), \tag{18.131}$$

$$P_{\nu_\tau \to \nu_\mu} = 4|U_{\tau 4}|^2|U_{\mu 4}|^2 \sin^2(\Delta_{41}^2), \tag{18.132}$$

$$P_{\nu_\tau \to \nu_e} = 4|U_{\tau 4}|^2|U_{e4}|^2 \sin^2(\Delta_{41}^2), \tag{18.133}$$

with $\Delta_{41}^2 = \frac{1.27 \Delta m_{41}^2 (eV^2)\ L(Km)}{E(GeV)}$.

18.7   Present and Future Experiments

The present experimental results on the parameters describing sterile neutrino mixing in the $3 + 1$ flavor model derived from short baseline experiments in the appearance and disappearance mode suggest the presence of sterile neutrinos. However, they do not show the expected overlapping regions in the parameter space which shows the inadequacy of the 3+1 flavor mixing model. This has led to the consideration of other theoretical models like 3+2 flavor mixing with two sterile neutrinos or 3+1 flavor decay model in which the sterile neutrino decays to invisible particles. On the experimental side, more neutrino oscillation experiments have been proposed to get data with better statistics leading to the determination of mixing parameters with improved precision.

There are a number of experiments proposed to search for sterile neutrinos which have been planned with (anti)neutrinos from radioactive sources, reactors, accelerators and atmospheric neutrinos. In both category of disappearance and appearance experiments [191]. In addition, the experiments planned to directly measure the mass of ν_e ($\bar{\nu}_e$) neutrinos can also be used to search for the existence of sterile neutrinos.

- Disappearance experiments: Most (anti)neutrino oscillation experiments with radioactive sources and reactors are disappearance type experiments in which radioactive isotopes are artificially produced; these isotopes decay by emitting ν_e or $\bar{\nu}_e$ which are used to perform $\nu_e \to \nu_e$ or $\bar{\nu}_e \to \bar{\nu}_e$ type of disappearance experiments. The neutrinos and antineutrinos could have mono energetic or continuous energy spectrum depending upon the decay process. The earlier GALLEX and SAGE were of this type of experiments. Further experiments like SOX at LNGS (Italy), BEST in Baksan, JUNO in China, and ISODAR@KAMLAND in Japan, and ISODAR@JUNO in China are planned with radioactive sources [189].

All the reactor experiments are disappearance experiments with antineutrino beams. In addition to the earlier reactor experiments, Double CHOOZE, Daya Bay, and RENO which continue with their search for sterile neutrinos, there are new reactor experiments [189]. The present reactor experiments DANSS in Russia has already reported results for $\Delta m_{14}^2 = 1.4$ eV^2 and $\sin^2(2\theta_{14}) = 0.05$ [1042]. On the other hand, the preliminary results from STEREO at ILL in France [1043] have excluded the parameter space implied by the original reactor antineutrino anomaly (RAA) and Neutrino-4 in Russia [1044] which implies $\Delta m_{14}^2 = 7.3$ eV^2 and $\sin^2(2\theta_{14}) = 0.39$. Future experiments include NEOS in Korea and SOLID in Belgium.

The disappearance experiments with ν_μ and $\bar{\nu}_\mu$ are done with accelerator and atmospheric neutrinos in different energy regions. The present experiments of MINOS and NOvA at Fermilab in USA with accelerator neutrinos are upgrading for the sterile neutrino search while the Super-Kamiokande in Japan and IceCube in Antarctica are continuing with their search for sterile neutrinos with atmospheric neutrinos. The short baseline neutrino (SBN) experiment at the Fermilab with three detectors of LArTPC type in the booster neutrino beam (BNB) baseline has an elaborate program for sterile neutrino search in the disappearance mode. The new proposal in the very low energy region of CCM at LANSC [1045], OscSNS at SNS in USA [1046], and JSNS at JPARC [950] are also proposed. All these experiments are designed for the $\nu_\mu \to \nu_\mu$ and $\bar{\nu}_\mu \to \bar{\nu}_\mu$ type of disappearance experiments.

- Appearance experiments: The short baseline neutrino (SBN) program at the Fermilab is a dedicated three detector setup designed to search for the sterile neutrino in the ν_e appearance mode in addition to the ν_μ disappearance mode. All the three detectors are liquid argon time projection chambers (LArTPC) called SBND, MicroBooNE, and ICARUS, placed in the BNB baseline in that order with SBND being nearest to the source. The expected signal from a sterile neutrino increases with the baseline and is largest in the ICARUS detector. This is the most ambitious program as it will perform neutrino oscillation experiments in the ν_e appearance as well as the ν_μ disappearance with the same setup and reduce systematics. Other appearance experiments in the kinematic region of LSND have been proposed at the OscSNS facility in USA [1046] and JSNS at JPARC in Japan [950] using neutrino beams from π and K decays at rest.

- Neutrino mass measurement experiments can also be used to search for the presence of sterile neutrinos. These experiments are based on the $\bar{\nu}_e$ mass determination from the energy spectrum from β-decay and electron capture on nuclei. These are (anti)neutrinos of $\nu_e(\bar{\nu}_e)$ type and are a superposition of mass eigen states ν_i $(i = 1, 2, 3, 4)$ weighted by the mixing matrix elements U_{ei}. Therefore, a mixing of the fourth matrix with $m_4 \rightarrow m_i$ $(i = 1, 2, 3)$ would manifest itself as a kink like signature at the end point of the β decay spectrum. In case of electron capture based on experiments, a detailed analysis of the mass determination of ν_e is required. Experiments like KATRIN and Prospect-8 based on measurements of electrons and experiments like ECHO and Holmes based on ν_e mass determination from electron capture on ^{163}Ho nucleus, using very low energy temperature measurements, are proposed to search for sterile neutrinos.

Chapter **19**

Neutrino Astrophysics and the Synthesis of Elements

19.1 Introduction

Neutrinos play a very important role in astrophysics. Due to their weakly interacting nature, they give information about the interior of stars, supernova explosions, the distant galaxies, the possible origin of the cosmic rays, etc. In Chapter 17, we have observed that astrophysical sites are the major contributors to low energy neutrinos. In this chapter, we discuss the importance of neutrinos in the creation of the chemical elements (see Figure 19.1) in the universe. We know that the universe started around 13.8 billion years ago with a singularity, that is, Big Bang and since then, it has been expanding and becoming cooler. The very early universe consisted only of radiation, which during the expansion and cooling phase gave rise to quark–antiquark and lepton–antilepton pairs. With further fall in temperature, the quarks combined to form

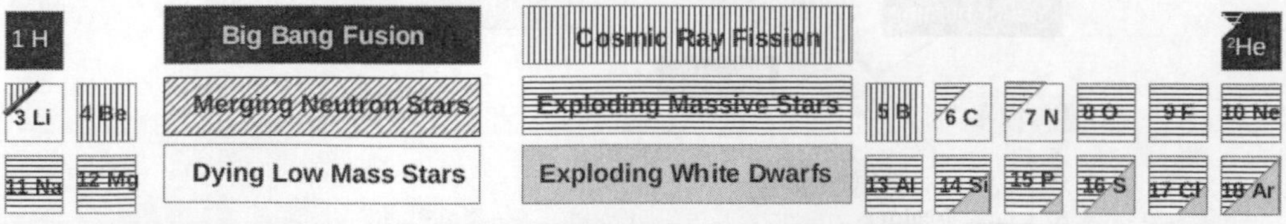

Figure 19.1 Origin of the elements through different processes [1047]. The figure has been taken from Ref. [1048].

nucleons, which in turn fused to form the lighter elements like hydrogen, helium, and lithium, the first ever nuclei created in the universe. Thus, the early universe consisted of about 75% hydrogen nuclei, 25% helium nuclei, and traces of lithium nuclei. It should be noted that hydrogen is the only element that was solely created during the Big Bang nucleosynthesis; all the other elements including helium and lithium are synthesized by several processes as shown in Figure 19.1. Therefore, all the hydrogen in water molecules were produced during the first few minutes of the Big Bang.

The process of creation of new nuclei from pre-existing nucleons is known as nucleosynthesis. The nucleosynthesis of the lighter elements does not require the emission or absorption of neutrinos while all the elements heavier than lithium require neutrinos directly or indirectly in their synthesis. The nucleosynthesis of intermediate and heavy elements require a very high temperature and pressure environment. Elements up to iron were/are synthesized in the core of stars through the nuclear fusion reaction and it is believed that the heavier elements were synthesized outside the newly formed neutron star in a core collapse supernova. Without neutrinos, we cannot think of energy from the stars. Moreover, neutrino properties figure prominently in many astrophysical environment. Neutrinos are involved in different types of nucleosynthesis processes like the ν-process, νp-process, etc., in the creation of proton-rich nuclei as well as in the synthesis of neutron-rich nuclei through the r-process and s-process. Figure 19.2 shows some of the processes responsible for the synthesis of elements which we shall discuss later in the chapter. During the death phase of massive stars, in a supernova

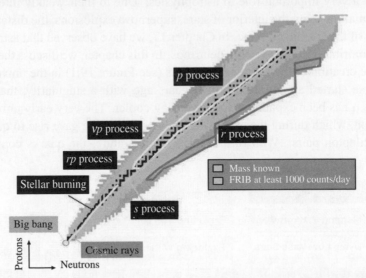

Figure 19.2 Formation of the elements in the universe through the Big Bang nucleosynthesis, stellar nucleosynthesis, and the supernova nucleosynthesis by the s- and r-processes. This figure has been taken from Ref. [1049].

explosion, almost 10^{58} neutrinos are released in a very short span of time of the order of a few seconds and these neutrinos move outwards carrying almost 10^{59} MeV of energies. These neutrinos give rise to a neutrino driven wind where matter from the neutron star is pushed outward through the interactions of neutrinos with matter as shown in Figure 19.3.

The neutrinos share energy with matter and the temperature of the matter rises resulting in large mass outflow within the neutron star through various reactions. Table 19.1 illustrates some of the neutrino processes which have played a very important role in supernova and proto-neutron star matter. As these neutrinos leave the neutron star, they take away energy and therefore, cool the neutron star which also gives rise to the possibilities of νp and weak r-processes [1050, 910] (discussed later in Section 19.6.2). This results in the formation of heavier nuclei through proton and neutron capture. In this chapter, we will learn about the formation of different elements through various processes, and the important role of neutrinos in the formation of elements.

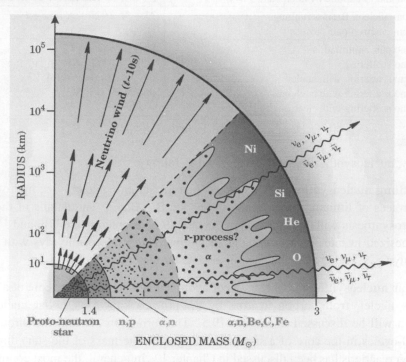

Figure 19.3 Neutrino driven wind [1052].

This chapter also throws some light on the interesting question that every inquisitive mind, at least once, must have pondered on: "how the universe came into existence?"; the quest to understand this is still going on. The idea that the universe started with a Big Bang and is expanding was first given by Lemaitre in 1927, on purely theoretical grounds by assuming that the universe started with a "primordial atom". It was Hubble, who in 1929 [1051] confirmed the expansion of the universe by observing the galactic redshifts. This expansion when extrapolated back in time takes us to the point, known as "singularity", from where space and time started. Today, the most accepted theory describing the origin and expansion of the universe is known as the "Big Bang theory" (Figure 19.4). What existed before the Big Bang and how it started is still a mystery and a topic of research. What we know today as the Big Bang is assigned a time $t = 0$, when the universe was extremely hot($T > 10^{40}$ K) with infinite density and consisted of radiation only. As time evolved, the universe started expanding and due to this expansion, the temperature decreased. Through this expansion, the universe has passed through various phases (known in literature as eras) and built up matter from radiation through the process of nucleosynthesis.

Table 19.1 Different neutrino processes in supernova and proto-neutron star matter.

Process	Reaction[a]
Electron and ν_e absorption by nuclei	$e^- + (A, Z) \rightleftharpoons (A, Z - 1) + \nu_e$
Electron and ν_e capture by nucleons	$e^- + p \rightleftharpoons n + \nu_e$
Positron and $\bar{\nu}_e$ capture by nucleons	$e^+ + n \rightleftharpoons p + \bar{\nu}_e$
Nucleon–nucleon Bremsstrahlung	$N + N \rightleftharpoons N + N + \nu + \bar{\nu}$
Electron–positron pair process	$e^- + e^+ \rightleftharpoons \nu + \bar{\nu}$
Neutrino pair annihilation	$\nu_e + \bar{\nu}_e \rightleftharpoons \nu_x + \bar{\nu}_x$
Neutrino scattering	$\nu_x + \{\nu_e, \bar{\nu}_e\} \rightleftharpoons \nu_x + \{\nu_e, \bar{\nu}_e\}$
Neutrino scattering with nuclei	$\nu + (A, Z) \rightleftharpoons \nu + (A, Z)$
Neutrino scattering with nucleons	$\nu + N \rightleftharpoons \nu + N$
Neutrino scattering with electrons and positrons	$\nu + e^\pm \rightleftharpoons \nu + e^\pm$

$$^a \; \nu \in \{\nu_e, \bar{\nu}_e, \nu_\mu, \bar{\nu}_\mu, \nu_\tau, \bar{\nu}_\tau\}, \nu_x \in \{\nu_\mu, \bar{\nu}_\mu, \nu_\tau, \bar{\nu}_\tau\}$$

Broadly, nucleosynthesis can be categorized as follows:

- **Big Bang nucleosynthesis:** This created the first atomic nuclei, viz., hydrogen, helium and traces of lithium, a few minutes after the Big Bang. The details of the Big Bang nucleosynthesis will be discussed in Section 19.2. Elements like ^{9}Be and ^{11}B were/are synthesized in interstellar space due to the interaction of cosmic rays with gas clouds, mainly C, N, and O atoms present in interstellar matter.

- **Stellar nucleosynthesis:** The fusion of hydrogen and helium in the core of a star created many nuclei, from carbon to iron, by the process known as stellar nucleosynthesis, which will be discussed in Section 19.5. The formation of elements through pp- and CNO-cycles in the core of a star (depending upon the mass of the star) during nuclear fusion reactions has been discussed in Chapter 17. In general, the most common process through which different elements are formed in the universe is the fusion of two nuclei to give rise to a heavier nucleus. The nuclear processes are responsible for the production of energy and synthesis of elements in the various astrophysical sites.

- **Supernova nucleosynthesis:** Heavy mass stars end their life with a supernova explosion where huge amount of energy is released in a few seconds. Supernova nucleosynthesis creates isotopes of lighter mass nuclei as well as nuclei heavier than iron. In the case of heavier elements, different processes contribute to the synthesis of proton- and neutron-rich elements. The synthesis of neutron-rich elements is well studied both theoretically as well as experimentally; however, the synthesis of proton-rich nuclei is not well known. Two processes, in neutron-rich environments like during the supernova explosion and the neutron star merger, known as the r-process (rapid capture of neutrons) and the s-process (slow capture of neutrons) are assumed to be responsible for the creation of a majority of the elements. In the s-process, the neutron capture happens in time scale τ_n

much longer than the mean time for the β-decay τ_β, that is, $\tau_n \gg \tau_\beta$. In the r-process, the neutron capture happens in time scale τ_n much shorter than the mean time for the β-decay τ_β, that is, $\tau_n \ll \tau_\beta$.

The synthesis of the proton-rich elements proceeds through different processes like p-process, ν-process, γ-process, rp-process, np-process and νp-process (Section 19.6.2). In all these processes, there is a capture of proton from the surroundings.

Let us first understand the formation of primordial atoms during the course of journey from the Big Bang (Figure 19.4) and thereafter.

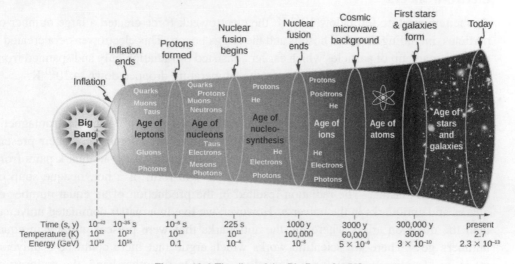

Figure 19.4 Timeline of the Big Bang [1053].

19.2 The Big Bang and the Nucleosynthesis of the Lighter Elements

- **Planck and Grand Unification eras**

The timeline of the Big Bang started with the Planck era which extended from 0 to 10^{-43} s, known as Planck's time. Presently, limited information is available about this era; it is believed that during this time, all the fundamental forces of nature, viz., gravitational, electromagnetic, weak, and strong, had the same strength or in other words, they were unified into a single force called "superforce". At the end of the Planck era, the temperature of the universe was about 10^{40} K and its size as small as 10^{-35} m known as the Planck's length. After the Planck era came the Grand Unification era which spanned the time from 10^{-43} to 10^{-36} s. During this era, the gravitational force separated out from the other three forces.

- **Inflation era**

  In this era, the temperature of the universe dropped to 10^{36} K. During this time, the strong force got separated out from the electroweak force. The separation of the strong force released huge amounts of energy which caused the universe to expand exponentially. The size of the universe increased by a factor of about 10^{26} in a fraction of a second which spanned the time from 10^{-36} to 10^{-32} s. At the end of this era, the universe cooled down to a temperature of about 10^{33} K, to a size of about 10 cm. Up to the inflation era, the universe entirely consisted of radiation.

- **Electro-weak era**

  As the temperature fell below 10^{33} K, the electroweak force created a large number of particles, including W and Z bosons and the Higgs boson. This electroweak era created a universe made up of particles which earlier consisted of radiation only and spanned from 10^{-36} to 10^{-12} s. By the end of this era, the temperature dropped down to 10^{20} K.

- **Quark era**

  As the temperature further fell below 10^{20}K and almost up to 10^{16}K, the electromagnetic and weak forces got disentangled and the four fundamental forces took their present form. This era is called the quark era as a large number of quark–antiquark pairs from the radiation were created which were present in the form of a hot, opaque soup of quark–gluon plasma. The radiation resulted in the production of an equal number of quarks and antiquarks in the universe. However, we live in a matter dominated universe, then the question arises where did the antiquarks that were present initially in equal numbers go? There are scientific works which argue that there could be a universe made up of antimatter running backward in time. On the other hand, the majority of the scientists believe that a process known as baryogenesis created the asymmetry in the number of quarks and antiquarks; there is no experimental evidence that such a process has occurred. The quark era spanned from 10^{-12} to 10^{-6} s.

- **Hadron era**

  After the quark era, came the hadron era spanning from 10^{-6} to 1 s. The temperature of the universe fell to about 10^{16}K to 10^{12}K and allowed the quark–gluon plasma to form primordial nucleons.

- **Lepton era**

  As the temperature further fell from 10^{12}K to 10^{10}K, the highly energetic photons gave rise to lepton–antilepton pairs (electron–positron and neutrino–antineutrino pairs). The lepton era spanned from 1 second to 3 minutes during which the mass of the universe was dominated by leptons. The electrons and the positrons got annihilated to give energy in the form of photons. Also, the colliding photons gave electron–positron pairs. However, a process known as leptogenesis, similar to the baryogenesis in the case of quarks, resulted in the dominance of leptons over the antileptons. During this era, neutrinos decoupled from the quark–gluon plasma and traveled freely in the universe. These neutrinos today constitute the "cosmic neutrino background" and are also known as relic neutrinos which have been discussed in Chapter 17. At the end of the lepton era, the universe consisted of highly energetic photons, hadrons (protons and neutrons), and leptons. The temperature fell to about 10^{10} K.

- **Era of nucleosynthesis**

This is the time during which the first ever nuclei in the universe were created. The era spanned from 3 to 20 minutes and the temperature fell below 10^{10} K. During this time, the collisions between the protons and the neutrons became very effective and the protons and the neutrons combined to form atomic nuclei through the following processes:

$$p + n \longrightarrow {}^2\text{H} + \gamma, \qquad p + {}^2\text{H} \longrightarrow {}^3\text{He} + \gamma,$$
$$ {}^2\text{H} + {}^2\text{H} \longrightarrow {}^3\text{He} + n, \qquad {}^2\text{H} + {}^2\text{H} \longrightarrow {}^3\text{H} + p,$$
$$ {}^3\text{He} + {}^2\text{H} \longrightarrow {}^4\text{He} + p.$$

The formation of ^{7}Li and ^{7}Be took place through the processes:

$$n + {}^3\text{He} \longrightarrow {}^3\text{H} + p, \qquad {}^3\text{He} + {}^4\text{He} \longrightarrow {}^7\text{Be} + \gamma,$$
$$ {}^3\text{H} + {}^2\text{H} \longrightarrow {}^4\text{He} + n, \qquad {}^3\text{H} + {}^4\text{He} \longrightarrow {}^7\text{Li} + \gamma,$$
$$n + {}^7\text{Be} \longrightarrow {}^7\text{Li} + p, \qquad p + {}^7\text{Li} \longrightarrow {}^4\text{He} + {}^4\text{He}.$$

Thus, the atomic nuclei of the three lightest elements were created when the universe was only a few minutes old. As the universe became about 20 minutes old, the temperature fell below 10^9 K and the nuclear fusion process stopped. At the end of this era, the baryonic matter dominated the universe with a composition of about 75% hydrogen nuclei, 25% helium nuclei, and traces of lithium nuclei.

- **Photon era**

Next came the photon era which spanned for an extremely long time, from 20 minutes to 2.4×10^5 years and the universe was again dominated by radiation. During this era, the universe was filled with a hot opaque soup of atomic nuclei and electrons. However, the temperature was still very high, that is, from 10^9 K to 10^6 K, for the atomic nuclei to combine with electrons to form atoms.

- **Era of recombination and decoupling**

Next came the era of recombination and decoupling which spanned from 2.4×10^5 to 3×10^5 years and the temperature dropped from 10^6 K to around 3000 K. The atomic nuclei of hydrogen and helium captured electrons to form atoms by the process known as "recombination". With the formation of atoms, the photons were able to move freely throughout the universe, and the universe finally got transparent to light. This is the earliest era which is accessible today. The photons which were interacting with the atomic nuclei and electrons till the recombination era were now released in the universe; they are still present today as cosmic microwave background radiation (CMBR). These photons decoupled because they did not have sufficient energy to interact and participate in particle creation. The cosmic microwave background radiation were first observed by

Penzias and Wilson in 1965 [1054], and their observation is a strong evidence in support of the Big Bang theory. The density (400–500 photons/cm^3) of these radiations were almost isotropic throughout the universe. At the end of this era, that is, around 3,00,000 years after the Big Bang, the universe consisted of a fog of about 75% hydrogen, 25% helium, and traces of lithium atoms. This fog, also known as the interstellar matter, acted (and they still act) as the progenitor of the stars. It may be recalled that the nuclei of these atoms were synthesized when the universe was only a few minutes old, whereas, the atoms were created much later.

- **Dark era**

From 3,00,000 to 150 million years, the universe is said to have been in the dark era as during this time, neither were there stars to give light nor did the cosmic microwave radiation interact with the atoms to give any light.

- **Era of stars and galaxies and the present universe**

From 150 million years to 1 billion years, the universe expanded and its temperature fell down to about 20 to 30 K which was sufficient for interstellar matter to form stars. Therefore, this is the era which marks the beginning of the formation of stars and galaxies. A star is formed when a disturbance, due to a shock wave, compresses the interstellar matter into a confined region. After millions of years, this region becomes gradually hotter and hotter, hot enough for nuclear fusion to take place. The origin of the shock wave responsible for the formation of the very first star is not known and is well beyond the understanding of the present scientific theories. Thus, it must be noted that the formation of the first star in the universe is still not known. From 1 billion years to 13.8 billion years, the universe appeared to be similar to what we find today. At present, the average temperature of the universe is about 2.73 K and its average density is about 9.9×10^{-30} gm/cm^3.

In the following, we will discuss how the stars are formed and what is their fate. Also, we will discuss how the chemical elements heavier than lithium were created in the universe.

19.3 Interstellar Matter

Before going into the details of star formation, we first focus on interstellar matter (ISM). The space between two stars is not empty, but filled with interstellar matter which consists of gas and dust clouds. Modern observations of the ISM indicate that in terms of mass, it consists of 71% hydrogen, 27% helium, and 2% heavier elements while initially, at the time when the stars and galaxies were not formed, the ISM consisted of 75% hydrogen, 25% helium, and traces of lithium without the presence of any heavier element. It should be emphasized that the difference in terms of the mass have been used in the formation of stars. The density of ISM is about 10^{-24} gm/cm^3(or 1 atom/cm^3). Moreover, this distribution of ISM is not uniform in the universe. There are regions in space where the density of ISM is significantly higher by about tens of thousand times; and these regions are known as interstellar cloud or nebula. The nebulae are found in the galactic plane and are enormous in size, typically, in the range 1

– 10,000 AU (astronomical unit), with 1 AU = 1.5×10^{11} m. The mass of these clouds vary between 10^4 to 10^6 $M_\odot$, where $M_\odot$ is the mass of the sun (2×10^{30}kg); the temperature of these clouds is extremely low, that is, between 10 to 30 K. These clouds are known to be the progenitors of stars.

19.4 Formation of Stars

The matter particles in the interstellar cloud, where the density is comparatively high, experiences gravitational attraction. However, their random motion, known as gas pressure, prevents them from collapsing. Stars are formed when the matter comprising the interstellar cloud collapses; what triggers a collapse is not well understood. However, there are evidences which suggest that the collapse is triggered by the perturbations created by shock waves in interstellar matter. The possible sources of these shock waves are as follows:

i) intergalactic waves, the origin of which is not yet known. These waves are considered to be the predominant trigger for the collapse of matter to form stars, and may be responsible for the formation of the very first star in the universe.

ii) pressure waves caused by the supersonic bursts during the formation of very massive stars. The star undergoes a period where it releases enormous quantities of matter in space traveling at great speeds. These cause pressure waves which may trigger disturbances when they pass through any nebula.

iii) shock waves, released when a massive star at its end phase explodes and ejects lots of matter which move out in space almost at the rate of about 10000 km/s.

When a shock wave hits the interstellar cloud, it affects a particular portion of the cloud and compresses it (Figure 19.5). If the gas pressure of the cloud fights back against the perturbation

Distribution of matter in interstellar cloud

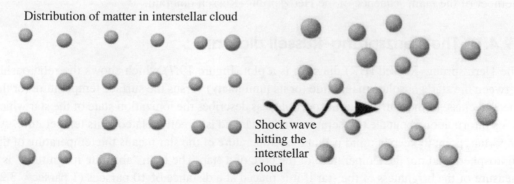

Shock wave hitting the interstellar cloud

Figure 19.5 Interstellar cloud and the shock wave hitting the cloud.

created by the shock waves, then the perturbation dies out. However, if the perturbation dominates the gas pressure, then the disturbance compresses the gas to a confined region as shown in Figure 19.6(a) which, in turn, increases the gravitational force between the particles within that vicinity. Due to this increase in the gravitational force, the particles get closer and closer resulting in further increase of the gravitational force, which results in the accumulation of matter in the confined space. The gravitational energy of these infalling particles is converted into kinetic energy and particle collision transfers the kinetic energy into thermal energy.

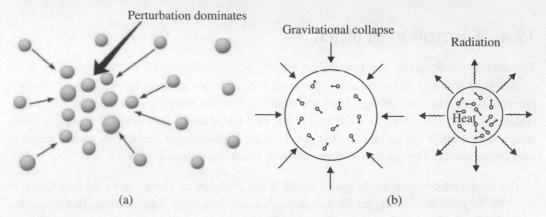

Figure 19.6 (a) Accumulation of the matter of the interstellar cloud in a region. (b) Hydrostatic equilibrium.

As more and more particles collapse, the temperature and density of the gas increase which, in turn, raises the opacity of the gas. When this contracting gas becomes opaque to its own radiation, a photosphere is formed which defines the boundary between the inside and outside of this collapsing cloud. When hydrostatic equilibrium is achieved, that is, the gravitational force is balanced by the thermal pressure, this collapsing gas becomes a protostar (Figure 19.6(b)). The temperature of the protostar is hotter than its surroundings and its formation takes about millions of years. The protostar further shrinks and heats. Eventually, the temperature of the core reaches the threshold of the hydrogen burning ($T \sim 10^9$ K). This takes more than millions of years. When hydrogen in the core starts to fuse, a star becomes a member of the main sequence of the Hertzsprung–Russell diagram.

19.4.1 The Hertzsprung–Russell diagram

The Hertzsprung–Russell (HR) diagram is a plot (Figure 19.7) which shows the relationship between the star's absolute magnitude (or its luminosity) versus the surface temperature (or the spectral class) of the star. The term spectral class describes the ionization state of the star which gives information about the temperature of the star and is directly related to its temperature. An important point to keep in mind is that the temperature of the star means the temperature of the photosphere and not the temperature of the core of a star. The term "absolute magnitude" is a measure of the brightness of the star if it is placed at a distance of 10 parsecs (1 parsec = 3.26 light years).

Most of the stars fall in the diagonal group starting from the top left to the bottom right of the plot (Figure 19.7). They are called the main sequence stars in which hydrogen fusion is taking place in the core. The hottest stars ($T > 30,000$ K) appear blue in color and belong to the spectral class "O" while the coldest stars ($T < 4,000$ K) appear red and belong to the spectral class "M". Our sun, having a surface temperature of about 5780 K, belongs to the spectral class "G" and appears orange in color. The luminosity of a star depends on the temperature (T) and radius (R) of the star, that is, $L \propto R^2T^4$. Therefore, a hot star of the same size is more luminous than another star of the same size having a lower surface temperature. The period of time a star lives in the main sequence depends upon its mass. The mass and the lifetime of a star are inversely proportional and hence, as the mass increases, the main sequence lifetime decreases. This relationship can be understood as follows:

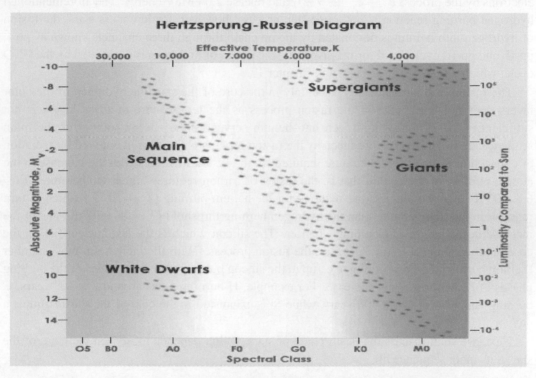

Figure 19.7 The Hertzsprung–Russell diagram: This classifies stars according to their luminosity, spectral class, surface temperature and evolutionary stage [1055].

With increase in the mass of a star, its gravitational collapse and density of the core increases, which results in an increased fusion rate in order to balance the collapse. This causes the star to fuse hydrogen in the core within a shorter time.

For example, our sun's life, as a main sequence star is about 9×10^9 years; however, for a star of mass around $5M_\odot$, the main sequence life is about 1.6×10^8 years and for a star of mass around $25M_\odot$, the main sequence life is about 2.88×10^6 years.

The stars on the lower left side of the HR diagram are white dwarfs. In the name "white dwarf", white color signifies the higher temperature of the star and dwarf signifies their small size as compared to the main sequence stars. In contrast, the stars on the upper right side of the HR diagram are called red giants and red supergiants signifying their temperature (red color means comparatively cooler objects) and their huge size.

19.5   Stellar Nucleosynthesis

In this section, we give a general description for the hydrogen fusion process and in Section 19.5.1, we discuss the fate of stars having mass in the range $0.08M_\odot$ to $50M_\odot$. When a star is in the main sequence, the fusion of hydrogen creates new chemical elements depending upon the mass of the star. The dominant reaction that takes place during the hydrogen fusion process is $4p \rightarrow^4 He + 2e^+ + 2\nu_e + 24.73$ MeV (Chapter 17). These two positrons annihilate with two electrons by the process $e^- + e^+ \rightarrow \gamma + \gamma$ and release 2.04 MeV energy. The aforementioned hydrogen burning reaction is not a one-step process; however, for low mass stars, the fusion of hydrogen into helium is dominated by the pp chain through three channels known as pp-I, pp-II, and pp-III cycles, and for massive stars, the fusion of hydrogen is dominated by the CNO cycle, which were already discussed in Chapter 17.

The fusion of hydrogen takes place only in the core of the star; the hydrogen of the outer layers does not participate in the fusion process as the temperature of these layers is not sufficient. The fusion of hydrogen into helium serves as the energy source for the main sequence stars. The energy produced by the fusion process is transferred outward by radiation or convection. Since the temperature required for the fusion increases with the increase in the atomic number of the nuclei, that is, the fusion of carbon requires higher temperature ($T \sim 6 \times 10^8$ K) than the temperature required for the helium burning ($T \sim 10^8$ K) which in turn requires higher temperature than the hydrogen burning threshold ($T \sim 4 \times 10^6$ K) due to the increase in the height of Coulomb's barrier. The silicon which is the last nuclear fuel in the core requires about 3×10^9 K to start the fusion process. With the increase in mass number (helium burning, carbon burning, etc., up to the silicon burning), the time scales of the burning phases of heavier elements decrease. For example, H burning takes around 7×10^6 years, C burning happens for about 600 years, while Si is exhausted in the core of the star in almost a day.

Next, we shall discuss the death phase of stars, which depends mostly on the mass of the star in its main sequence life.

19.5.1   Death of a star

Brown dwarf

Stars having mass $< 0.08 \ M_\odot$ do not even ignite the hydrogen for fusion as the temperature of their core is less than the threshold for hydrogen burning. Although such stars become hot due to gravitational collapse, it is not hot enough to start the fusion of hydrogen into helium. After reaching the maximum possible temperature, depending upon the mass, the star starts to cool. Such objects are called brown dwarfs.

Black dwarf

For stars having mass in the range $0.08 \ M_\odot < M < 0.4M_\odot$, the temperature of the core becomes high enough for the hydrogen fusion to take place. However, after the exhaustion of hydrogen fuel, the stars again collapse and heat up but the thermal energy produced by the gravitational collapse is not sufficient enough to ignite helium burning. Due to this contraction and rise in temperature due to convection, the star moves toward the lower left portion of the HR diagram and becomes a white dwarf. Once a white dwarf shrinks to its minimum size, it radiates energy. As there exists no fresh source of energy, it starts to cool gradually and ultimately becomes a black dwarf. In such type of stars, the energy is transmitted outward by convection only.

White dwarf

Now we consider the evolution of stars having mass lying in the range 0.4 $M_\odot$ and 4 $M_\odot$. This category is very exciting as our sun belongs to this mass range. During the main sequence, the energy produced by the hydrogen burning prevents the core from collapse. When the core runs out of hydrogen fuel, most of the it consists of helium nuclei and there is no source of energy to balance the gravitational force. The core starts shrinking due to gravitational collapse, as in the case of protostar, and this contraction converts the gravitational energy into thermal energy which in turn increases the temperature of the core. This energy is more than the energy produced by the fusion process which results in the increase in the temperature of the shell surrounding the core. Hence, hydrogen burning starts in this surrounding shell.

Now, there are two sources of energy at the center of the star, viz., the gravitational energy due to the collapse of the outer core and the fusion energy in the shell surrounding it. These two sources result in an increase in the radiation output arising from the center toward the outer envelope of the star and cause the outer portion of the star to expand. Due to this expansion, the outer envelope cools, but the core remains at a very high temperature, and the star becomes a red giant and moves towards the upper right side of the HR diagram. In Figure 19.8, we have shown the expansion of the outer envelope of the sun during its red giant phase. Although the

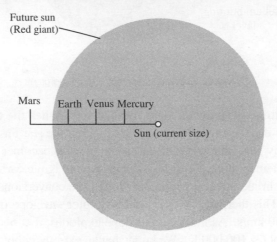

Figure 19.8 The future of sun (a red giant).

outer envelope expands, the core continues to collapse due to gravitation thereby increasing the temperature of the core until it becomes sufficient for helium burning. Helium burning proceeds through two cycles, viz., α cycle and triple α cycle:

$$3\alpha \text{ cycle}$$

$$^4\text{He} + {}^4\text{He} \longrightarrow {}^8\text{Be} + \gamma,$$

$$^4\text{He} + {}^8\text{Be} \longrightarrow {}^{12}\text{C} + \gamma,$$

$$^4\text{He} + {}^{12}\text{C} \longrightarrow {}^{16}\text{O} + \gamma,$$

$$\alpha \text{ cycle}$$

$$^4\text{He} + {}^{12}\text{C} \longrightarrow {}^{16}\text{O} + \gamma$$

$$^4\text{He} + {}^{16}\text{O} \longrightarrow {}^{20}\text{Ne} + \gamma$$

$$^4\text{He} + {}^{20}\text{Ne} \longrightarrow {}^{24}\text{Mg} + \gamma$$

$$^4\text{He} + {}^{24}\text{Mg} \longrightarrow {}^{28}\text{Si} + \gamma$$

$$^4\text{He} + {}^{28}\text{Si} \longrightarrow {}^{32}\text{S} + \gamma$$

As helium burning takes place at the center of the core, there is a carbon shell which is surrounded by a helium burning shell which, in turn, is surrounded by a hydrogen burning shell as shown in Figure 19.9. A red giant uses helium as a fuel for only about 10 to 20% of

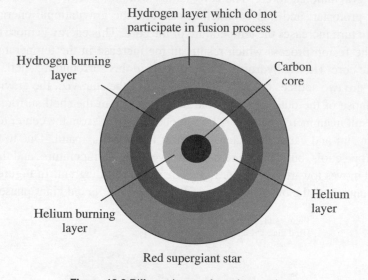

Figure 19.9 Different layers of a red supergiant.

the time as it uses hydrogen as a fuel in the main sequence. When the core runs out of helium fuel, it again contracts due to gravitational collapse making the star unstable. With increase in the gravitational energy, the luminosity increases and the star becomes a red super giant. The core collapses but the temperature is not sufficient enough to ignite carbon burning. Therefore, the core continues to shrink and heat up quickly due to the conversion of gravitational energy into thermal energy. This thermal energy causes the outer envelope of the star to blow with a speed of about 20–30 km/s. As the outer envelope explodes, it cools down but the core left behind is extremely hot ($\sim 100,000$ K). We know that an extremely hot object emits ultraviolet radiation; the ultraviolet radiation emitted by the core makes the outer envelope of the star which was blown outward glow. Such a glowing object is called a planetary nebula (Figure 19.10). The word "planetary nebula" is a misnomer as the glowing object is neither associated with planets nor with nebula, but is a remnant of the outer envelope of a star.

The material blown out with the outer envelope gets dispersed in the ISM while the hot and bright core left behind remains very luminous for a short time and quickly moves toward the bottom of the HR diagram becoming a white dwarf. White dwarfs have approximately the same size as that of the earth but they have a density of $\sim 10^6$ gm/cm^3 while the average density of earth is 5.5 gm/cm^3. White dwarfs must have mass in the range 0.8 M$_\odot$ < M_{WD} < 1.4 M$_\odot$. This upper limit of 1.4M$_\odot$ is known as the "Chandrasekhar limit" [1056].

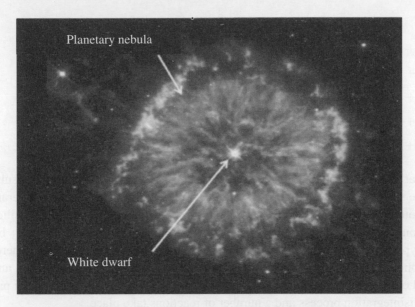

Figure 19.10 Planetary nebula and the white dwarf [1057].

19.5.2 Chandrasekhar limit

In the case of a main sequence star, the core is a gas (consisting mainly of hydrogen and helium). With increase in the gravitational energy, the temperature of the core increases. As more pressure is exerted on the core, the matter becomes so compressed that the electrons no longer move freely in the core but become degenerate electron gas, as the compression of the core causes the energy levels to be completely filled. The size of the core, now, is not proportional to its temperature, but is determined by the electron degeneracy pressure. It was Chandrasekhar, who in 1930, calculated this limit for the case of a white dwarf. It must be pointed out that in the case of low mass stars (M_{core} < 1.4 M$_\odot$), it is the electron degeneracy pressure that halts the collapsing core.

19.5.3 Death of middle mass stars

Next, we consider stars having mass more than 4M$_\odot$. Specifically, there are two categories, viz., the middle mass stars (having mass 4M$_\odot$ < M < 8M$_\odot$) and the heavy mass stars (having mass > 8M$_\odot$). Although the final products of these two categories of stars are different, before the outer envelope explodes, the two types of stars spend their lives more or less in a similar way. Therefore, we will discuss them together for the intermediate steps. It may be recalled that in the case of low mass stars, the core has carbon at its center surrounded by a helium shell which is surrounded by the hydrogen shell, but the core does not have sufficient temperature to ignite carbon burning. In massive stars, the temperature and the pressure are high enough to ignite the fusion of carbon and oxygen into heavier elements by the processes:

Carbon burning

$$^{12}C + {}^{12}C \longrightarrow {}^{24}Mg + \gamma,$$
$$^{12}C + {}^{12}C \longrightarrow {}^{23}Mg + n,$$
$$^{12}C + {}^{12}C \longrightarrow {}^{23}Na + p,$$
$$^{12}C + {}^{12}C \longrightarrow {}^{20}Ne + {}^{4}He,$$
$$^{12}C + {}^{12}C \longrightarrow {}^{16}O + {}^{4}He + {}^{4}He,$$

Oxygen burning

$$^{16}O + {}^{16}O \longrightarrow {}^{32}S + \gamma,$$
$$^{16}O + {}^{16}O \longrightarrow {}^{31}S + n,$$
$$^{16}O + {}^{16}O \longrightarrow {}^{31}P + p,$$
$$^{16}O + {}^{16}O \longrightarrow {}^{28}Si + {}^{4}He,$$
$$^{16}O + {}^{16}O \longrightarrow {}^{24}Mg + {}^{4}He + {}^{4}He.$$

When a star runs out of its oxygen fuel, its core consists mainly of silicon and sulfur. If the mass of the star is sufficiently high such that the compression increases the temperature of the core to about 3×10^9 K, the fusion of silicon takes place. The fusion of two silicon nuclei to form iron and nickel is forbidden because of the very high Coulomb barrier; however, a different type of nuclear reaction, known as photodisintegration takes place, where a highly energetic photon interacts with a heavy nucleus and disintegrates it into a helium nucleus or a proton and other lighter nuclei. The silicon and sulfur capture these helium nuclei produced in the photodisintegration process and a number of reactions take place

$$^{28}Si + {}^{4}He \longrightarrow {}^{32}S + \gamma, \qquad {}^{32}S + {}^{4}He \longrightarrow {}^{36}Ar + \gamma,$$
$$^{36}Ar + {}^{4}He \longrightarrow {}^{40}Ca + \gamma, \qquad {}^{40}Ca + {}^{4}He \longrightarrow {}^{44}Ti + \gamma,$$
$$^{44}Ti + {}^{4}He \longrightarrow {}^{48}Cr + \gamma, \qquad {}^{48}Cr + {}^{4}He \longrightarrow {}^{52}Fe + \gamma,$$
$$^{52}Fe + {}^{4}He \longrightarrow {}^{56}Ni + \gamma, \qquad {}^{56}Ni + {}^{4}He \longrightarrow {}^{60}Zn + \gamma.$$

In general, these reactions are reversible (the direct and reverse reactions occur at the same rate), although the nuclear equilibrium is not perfect. For example, the reaction produces neon at around 10^9 K but the reaction reverses direction above 1.5×10^9 K. As the temperature reaches $\sim 7 \times 10^9$ K, elements like iron, cobalt, and nickel resist the photodisintegration process and are produced in the core. The burning of silicon as a fuel lasts for only about one day. As both nickel and cobalt are radioactive nuclei with half-lives of 6.02 days and 77.3 days, respectively, ^{56}Ni decays to ^{56}Co via β^+ decay which in turn decays to ^{56}Fe. Hence, at the end, the core consists of mostly iron and iron like elements. The fusion process stops at iron as it is the most stable element. Therefore, in order to create nuclei heavier than iron, the fusion of iron must consume energy so as to overcome the Coulomb barrier. The lighter elements release energy by the fusion process while the heavier elements release energy by the fission process; however, iron and iron like elements require energy for both fusion as well as fission processes.

Now, the core resembles a multilayered onion structure as shown in Figure 19.11 and fusion takes place at the boundary between the two layers. This iron core is a degenerate gas consisting of iron nuclei and electrons. The iron core grows until it reaches the critical mass, known as the Chandrasekhar mass limit ($M_{Ch} \sim 1.4 M_\odot$), and then collapses. From here, the star begins its

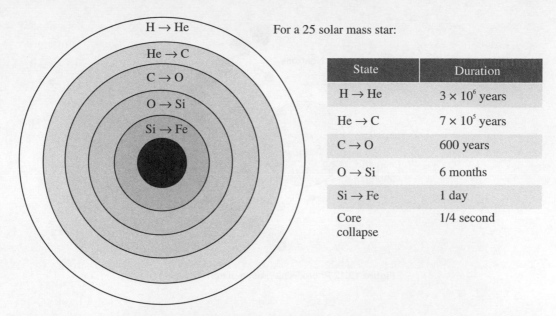

For a 25 solar mass star:

State	Duration
H $\rightarrow$ He	3×10^6 years
He $\rightarrow$ C	7×10^5 years
C $\rightarrow$ O	600 years
O $\rightarrow$ Si	6 months
Si $\rightarrow$ Fe	1 day
Core collapse	1/4 second

Figure 19.11 Onion like layers in the core of a massive star ($M_{\text{star}} = 25 M_\odot$). The table shows the duration of fusion taking place during the fusion of various elements.

journey toward death. As the mass of this core exceeds the Chandrasekhar limit, the pressure of the degenerate electron gas becomes unable to balance the force of gravity. Due to this, the core contracts more rapidly, which in turn increases the temperature of the core, releasing very high energy gamma rays. As the temperature of the core reaches to about 8×10^9 K and the density becomes $\sim 10^9$ gm/cm^3, two processes take place that consume energy:

i) The photodisintegration of iron by high energy photons into helium nuclei and neutrons, as shown in Figure 19.12, via the reaction

$$\gamma + {}^{56}\text{Fe} \longrightarrow 13\,{}^4\text{He} + 4n - 124.4\,\text{MeV}.$$

This reaction takes place by absorbing ~ 2 MeV per nucleon, which is contrary to the fusion process where ~ 2 MeV per nucleon is released. This loss in energy enhances the gravitational collapse, which almost becomes a free fall of the matter from the outer layers of the core there by further increasing the temperature and density of the inner core.

ii) With further increase in temperature, the photons become so energetic that the disintegration of helium into proton and neutron takes place; this disintegration absorbs ~ 6 MeV per nucleon in the reaction

$$\gamma + {}^4\text{He} \longrightarrow 2p + 2n - \text{energy}.$$

Since the core still continues to contract, a stage is reached where the density becomes sufficient for the protons and electrons to combine and form neutrons giving rise to the proto neutron star.

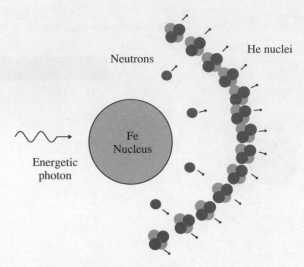

Figure 19.12 Photodisintegration of iron nucleus.

The process of electron capture by protons (Table 19.1) releases neutrinos in large numbers; neutrinos also carry substantial amounts of energy. Neutrinos released due to the capture of electrons in the stellar interior, initially come out of the core as their mean free path is sufficiently large due to the comparatively low density of the core. As the density rises to about 10^{11} gm/cm^3, the mean free path for the neutrinos reduce in such a way that they get trapped inside the core and are unable to come out. During the electron capture by heavy nuclei, several new nuclei having masses in the range $A \sim 80 - 100$, with almost 50 neutrons, are produced. This process absorbs energy; it also reduces the number of particles which in turn reduces the pressure, but the core collapse continues. When the collapsing core reaches a density of $\sim 10^{14}$ gm/cm^3, which is approximately the same as the density of the atomic nuclei, neutrons become degenerate due to the repulsive nature of the nuclear forces at short distances. This neutron degeneracy pressure halts the collapse like the electron degeneracy in the case of low mass stars. The inner core fights with the the gravitational force of the infalling matter, and due to this the infalling matter rebounds, which results in the formation of shock waves (like the sea waves at the shore when they strike rocks and produce sound waves which can be heard even from far distances). These outgoing shock waves moving almost at a speed of about 30,000 km/s, reverse the infalling motion of the material in the star and accelerate it outward. Thus, the outer envelope explodes, and constitutes the core collapse supernovae, which may give rise to the birth of a neutron star or a black hole. With this explosion, almost 10^{58} neutrinos of all flavors (notice from Table 19.1 how neutrinos of all flavors may be produced) are released which were earlier trapped due to the high density of the core. These neutrinos act as energy carriers and carry about 10^{59} MeV of energy deposited in the core of the star.

Neutron stars and pulsars

Stars having mass in the range $4M_\odot < M < 8M_\odot$ end their life as neutron stars. Neutron stars must have mass greater than 1.4 $M_\odot$ and lesser than 2.17 $M_\odot$; this limit is known as the

Tolman–Oppenheimer–Volkoff limit [1058]. A typical neutron star has a diameter of about 20 km which implies the enormous density of the neutron star (about 10^{14} gm/cm^3). Due to such high density, the neutron star has immense gravity which, in turn, makes the escape velocity enormously large varying between 100,000 km/s to 150,000 km/s. Neutron stars have very strong magnetic fields varying in the range 10^4 to 10^{11}T at the surface. During the core collapse, the size of the star shrinks; thus, in order to conserve the angular momentum, the rotation rate of the star increases. Some neutron stars may rotate several hundred times per second in order to conserve the angular momentum. These rotating neutron stars emit electromagnetic radiation originating from their magnetic poles. If the magnetic pole and the rotation axis do not coincide, then for a distant observer, these radiations appear as pulses emitting from a fixed source at a fixed interval of time. These electromagnetic radiations are known as pulsars, discovered in 1967 by Burnell and Hewish.

Black hole

The heavier mass stars having mass more than $8M_\odot$ spend their life in the same manner as the middle mass stars but their fate is different. In the case of heavier mass stars, the mass of the core exceeds $2.17M_\odot$, and the neutron degeneracy becomes unable to counter the core collapse. Matter from the outer shells of the star falls freely toward the core and the core becomes a black hole. A black hole is a celestial object on which the escape velocity exceeds the speed of light which means that whatever light or matter falls on the black hole, it can never come back. Since no light can escape from a black hole, there is no possibility of seeing it directly. The details of a black hole are beyond the scope of this book.

Next, we will discuss the supernova nucleosynthesis which is responsible for the creation of heavier nuclei.

19.6 The Supernova Nucleosynthesis: Formation of Heavy Elements

Before discussing the supernova nucleosynthesis, we first point out the difference between the stellar and supernova nucleosynthesis. In the case of stellar nucleosynthesis, the elements are created in the core of the star and the process takes place for a longer time (approximately billions of years) depending upon the mass of the star whereas in the latter case, the chemical elements are created during the supernova explosion which takes place in a very short time (a few seconds).

In the formation of the neutron star, we have seen that a supernova explosion takes place when a massive star runs out of its nuclear fuel and the explosion of the outer envelope of the star takes place. Such kinds of supernova are known as Type II, Ib, and Ic. The classification of supernova into different types are characterized by the presence of various chemical elements in their spectrum. However, another category of supernova exists, known as type Ia supernova (see Figure 19.13), which occurs in a system of binary stars. One of the stars in the binary system is a white dwarf and the other star can be a main sequence star or a red giant or a white dwarf. In type Ia supernova, the white dwarf, which is composed mainly of carbon and oxygen accretes mass from the other star until it reaches the temperature for carbon burning. The white dwarf is degenerate so it does not expand with the increase in temperature. Thus, due to the fusion process, the temperature of the core increases rapidly which, in turn, increases the fusion rate. The entire star participates in the fusion process and exhausts its nuclear fuel in a shorter time. This generates huge energy that overcomes electron degeneracy and the star expands in a supernova explosion. This supernova explosion disperses chemical elements in space.

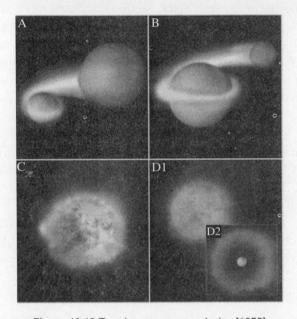

Figure 19.13 Type Ia supernova explosion [1059].

19.6.1 Nucleosynthesis of neutron-rich elements

s-process

The slow neutron capture process (s-process), depicted diagrammatically in Figure 19.14, is responsible for the creation of isotopes of lighter elements like carbon, nitrogen, oxygen, as well as the creation of heavier elements up to bismuth. In the s-process, an atomic nucleus captures a neutron at a time. If the resulting nucleus is stable, it may or may not capture another neutron but if the resulting nucleus is unstable, β-decay occurs before the capture of the next neutron. The neutron density required for the s-process to take place lies between 10^6 to 10^{11} neutrons/cm^3. The s-process is supposed to take place in thousands of years. If neutron capture rates are slow compared to β-decay rates, only isotopes near the stability line are synthesized. The s-process occurs in massive stars where the burning of carbon and helium takes place in consecutive layers called the weak s-process and in the low mass stars when they become red giants called the main s-process.

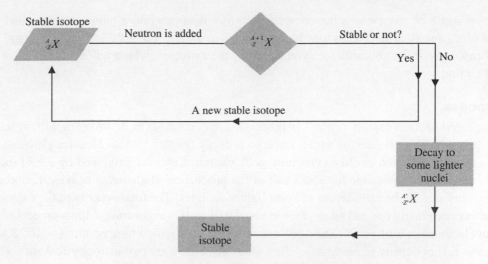

Figure 19.14 Slow neutron capture process.

The weak s-process creates isotopes of iron like elements up to strontium and yttrium with atomic number $Z = 38, 39$ and neutron number $N = 42$ to 52. The required density of neutrons in such a scenario is provided by a process like:

$$^{22}\text{Ne} + {}^4\text{He} \longrightarrow {}^{25}\text{Mg} + n.$$

In a massive star, where the fusion of carbon, helium, and hydrogen takes place in consecutive layers, mixing of elements between the layers occur. The carbon produced in the helium burning shell mixes with the hydrogen burning shell and produces ^{14}N through the following processes:

$$^{12}\text{C} + p \longrightarrow {}^{13}\text{N} + \gamma, \qquad {}^{13}\text{N} \longrightarrow {}^{13}\text{C} + e^{+} + \nu_e, \qquad {}^{13}\text{C} + p \longrightarrow {}^{14}\text{N} + \gamma.$$

This ^{14}N is again mixed with the helium burning shell and produces isotopes of nuclei up to neon. At the exhaustion of helium fuel, the capture of helium by ^{22}Ne produces large amounts of neutrons which leads to the production of various isotopes of carbon, nitrogen, and oxygen.

$$^{14}\text{N} + {}^4\text{He} \longrightarrow {}^{18}\text{F} + n, \qquad {}^{18}\text{F} + e^{-} \longrightarrow {}^{18}\text{O} + \nu_e, \qquad {}^{18}\text{O} + {}^4\text{He} \longrightarrow {}^{22}\text{Ne} + \gamma.$$

The s-process also takes place in low mass stars during their red giant phase and requires comparatively low neutron density of the order of 10^7 neutrons/cm^3.

The production of bismuth, polonium, and lead are cyclic in the s-process:

$$^{209}\text{Bi} + n \longrightarrow {}^{210}\text{Bi} + \gamma, \qquad\qquad {}^{210}\text{Bi} \longrightarrow {}^{210}\text{Po} + e^{-} + \bar{\nu}_e$$

$$^{210}\text{Po} \longrightarrow {}^{206}\text{Pb} + {}^4\text{He}, \qquad\qquad {}^{206}\text{Pb} + 3n \longrightarrow {}^{209}\text{Pb}$$

$$^{209}\text{Pb} \longrightarrow {}^{209}\text{Bi} + e^{-} + \bar{\nu}_e.$$

The net result of this cycle is the conversion of four neutrons into a helium nucleus and two pairs of e^- and $\bar{\nu}_e$. The process may be represented as $4n \rightarrow^4 He + 2e^- + 2\bar{\nu}_e +$ energy. The elements heavier than bismuth are synthesized by the r-process, which will be discussed in the next section.

r-process

In the rapid neutron capture process (r-process), a nucleus absorbs neutrons rapidly in such a way that it happens before the nuclei undergo β-decay (Figure 19.15). Heavier elements and more neutron-rich isotopes like europium, gold, platinum, etc. are produced by the r-process. The r-process is responsible for about half of the production of elements heavier than iron; it also contributes to the abundances of some lighter nuclides. The total process of the capture of neutron is extremely fast and takes place in about 0.01 to 10 s in contrast to the s-process which occurs in thousands of years. The r-process occurs at a very high temperature ($\sim 10^9$ K) and requires higher density of neutrons $\sim 10^{20}$ gm/cm^3. There are two astrophysical sites where such density of neutrons is found: the core collapse supernova explosion (supernova type II) and the neutron star merger.

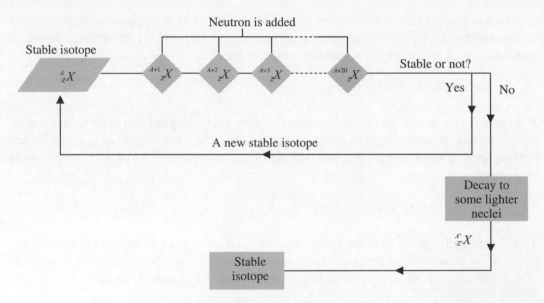

Figure 19.15 Rapid neutron capture process.

Now, we will explain type II supernova or the supernova explosion that takes place during the core collapse of a massive star. As we have seen earlier in the case of middle and heavy mass stars, the core collapse is stopped by neutron degeneracy pressure. The infalling matter from the outer envelope of the star produces a shock wave that blows up the outer envelope of the star at very high speeds in a few seconds resulting in a supernova explosion. The synthesis of elements during the supernova explosion is known as supernova nucleosynthesis. During the propagation of the shock wave from the core, the r-process creates heavier elements. The neutron density during a supernova explosion is 10^{14} gm/cm^3 while the density required for the r-process to take place is 10^{20} gm/cm^3. Now, the question arises from where would the neutrons come so as to reach such density to initiate the r-process? The answer is that during the core collapses, the core compresses the electrons in such a way that β-decay($n \longrightarrow p + e^- + \bar{\nu}_e$) is stopped. However, the capture of electrons by the atomic nuclei takes place which further enhances the density of the neutrons up to what is required for the r-process to start.

Neutrinos play a very important role in the understanding of the supernova explosion because neutrinos are the only source which provides information about the interior of the core. The nuclear fusion reactions taking place in the core produce neutrinos and antineutrinos of electron type only. However, the other flavors of neutrinos, viz., muon and tauon type neutrinos and antineutrinos are also produced by the following reactions of pair production and annihilation:

$$N + N \to N + N + \nu_l + \bar{\nu}_l, \; e^- + e^+ \to \nu_l + \bar{\nu}_l, \; \gamma + \gamma \to \nu_l + \bar{\nu}_l, \; \nu_e + \bar{\nu}_e \to \nu_l + \bar{\nu}_l.$$

Therefore, all the flavors of neutrinos and antineutrinos are produced in the core. These neutrinos are emitted from the core of a neutron star at very high speed ($\sim$ 30,000 km/s) when the supernova explosion takes place. As the high energy neutrinos travel outward, they interact with the matter of the outer layers of the star. This process continues for more than 10 s and it is assumed that during this outflow of matter, the r-process takes place. The outflow of matter from the surface of the neutron star to the outer shells at such high speed is known as the neutrino driven wind, shown in Figure 19.3, which is a very important topic from the point of view of the creation of heavier elements as it is supposed to be an astrophysical site, responsible for the creation of almost half of the heavy elements through the r-process.

Now, we will discuss the r-process taking place in the neutron star merger. Two neutron stars in a binary system orbit each other. As time passes, the gravitational radiation (energy carried by the gravitational waves) between the two increases which causes them to approach each other and orbit in the form of spiral geometry which increases the gravitational pull between the two. When these neutron stars meet each other, a massive neutron star or a black hole (depending upon the mass of the merged star) is formed.

In 2017, two gravitational wave observatories LIGO and VIRGO [1060] detected gravitational waves which were assumed to be emitted by the merger of two neutron stars. Merging neutron stars have a high density of neutronized matter and during their spiral path, they throw away heavy elements created by the r-process. Hence, the merging of two neutron stars gives information about the formation of heavier elements due to the r-process (Figure 19.16). The bulk of the matter thrown by the neutron star merger seems to be of two types:

(a) highly radioactive r-process matter of heavier nuclei with comparative lower mass ($A <$ 140) which appears blue in color due to their higher temperature,

(b) higher mass number r-process nuclei ($A > 140$) rich in actinides (such as uranium, thorium, californium, etc.) and appear red in color due to their relatively lower temperature.

Figure 19.16 Two merging neutron stars [1061].

19.6.2   Nucleosynthesis of proton-rich elements

The nucleosynthesis of proton-rich nuclei in the universe is not very well understood. There is not a single process which is responsible for the synthesis of all the proton-rich nuclei; however, there are different processes, in various astrophysical sites responsible for their creation, which are discussed in the following sections.

p-process

In a p-process, an atomic nucleus absorbs one or more proton (depending upon the stability of the synthesized nuclei) from the surroundings before the β-decay takes place. Generally, the proton capture reactions are represented by the type (p, γ). The possible synthesis site for the p-process is still not known.

ν-process

The ν-process takes place in the outer layers of the star during the supernova explosion, where almost 10^{58} neutrinos of all flavors are released, and it is possible for the neutrinos to interact directly with the nuclei. Since all the flavors of neutrinos are released, both charged as well as neutral current induced neutrino interactions contribute to this process. The supernova neutrinos have average energy of the order of few tens of MeV. Thus, the charged current reactions are induced only by ν_e and $\bar{\nu}_e$; the other flavors participate only through the neutral current interactions. This process creates a few isotopes of ^{7}Li, ^{11}B, ^{15}N, ^{19}F, ^{138}La and ^{180}Ta as well as the long-lived radioactive nuclei like ^{22}Na and ^{26}Al.

The light nuclei ^{7}Li and ^{11}B are synthesized by the interactions of neutrinos, induced both by charged as well as neutral currents, with helium and carbon as:

$$
\begin{aligned}
\nu_e + {}^4\text{He} &\longrightarrow e^- + p + {}^3\text{H}, & \nu_e + {}^{12}\text{C} &\longrightarrow e^- + p + {}^{11}\text{B}, \\
\nu_x + {}^4\text{He} &\longrightarrow \nu_x + n + {}^3\text{He}, & \bar{\nu}_e + {}^{12}\text{C} &\longrightarrow e^+ + n + {}^{11}\text{C}, \\
{}^3\text{H} + {}^4\text{He} &\longrightarrow {}^7\text{Li} + \gamma, & \nu_x + {}^{12}\text{C} &\longrightarrow \nu_x + n + {}^{11}\text{C}, \\
{}^3\text{He} + {}^4\text{He} &\longrightarrow {}^7\text{Be} + \gamma, & & \\
{}^7\text{Be} + e^- &\longrightarrow {}^7\text{Li} + \nu_e, & &
\end{aligned}
$$

where ν_x can be $\nu_{e,\mu,\tau}$ or $\bar{\nu}_{e,\mu,\tau}$. ^{11}C has a half-life of ~ 20 min, which then decays to ^{11}B. Similarly, the synthesis of ^{19}F is also induced both by the charged as well as the neutral current reactions:

$$
\begin{aligned}
{}^{18}\text{O} + p &\longrightarrow {}^4\text{He} + {}^{15}\text{N}, & \nu_x + {}^{20}\text{Ne} &\longrightarrow \nu_x + {}^{20}\text{Ne}^*, \\
{}^{15}\text{N} + {}^4\text{He} &\longrightarrow {}^{19}\text{F} + \gamma, & {}^{20}\text{Ne}^* &\longrightarrow {}^{19}\text{Ne} + n, \\
& & {}^{19}\text{Ne} &\longrightarrow {}^{19}\text{F} + e^+ + \nu_e.
\end{aligned}
$$

The heavier isotopes, that is, ^{138}La and ^{180}Ta are produced only by the charged current interaction of ν_e and $\bar{\nu}_e$, through the reaction

$$
\nu_e + {}^{138}\text{Ba} \longrightarrow {}^{138}\text{La} + e^- \qquad \text{and} \qquad \nu_e + {}^{180}\text{Hf} \longrightarrow {}^{180}\text{Ta} + e^-.
$$

The radioactive nuclei ^{22}Na and ^{26}Al are also synthesized in a similar manner as discussed earlier.

γ-process

The photodisintegration of the elements produced through the s- and r-processes also yields proton-rich elements by the process known as γ-process. This process takes place at sufficiently high temperature $T \sim 2 - 3 \times 10^9$ K during the core collapse supernova as well as in type Ia supernova explosions.

rp-process

In the rapid proton capture process, the atomic nuclei capture more protons before the β-decay takes place. The rp-process requires very high proton densities of the order of 10^{28} protons/cm^3 as well as high temperature $\sim 2 \times 10^9$ K. This process is dominant in the binary system of neutron stars and creates a proton-rich nuclei having mass number $A \leq 104$.

pn-process

The neutron-rich rapid proton (np) capture process is much faster than the p-process and proceed by the (n, p) type of reactions. This process requires a considerable neutron density which is provided by the type Ia supernova explosion. It is assumed that pn-processes take place in type Ia supernova explosions.

νp-process

The νp-process is also a process of rapid capture of protons, where the proton captures an antineutrino to produce a neutron in the final state. This neutron then interacts with an atomic nucleus by (n, p) type of reaction and creates a proton-rich nuclei. A high flux of antineutrinos is required for this process which is available from the core collapse supernova explosion.

Thus, neutrinos have been involved in the formation of elements beyond hydrogen; they are still playing an important role in the production of elements in the astrophysical environment.

Chapter 20

Neutrino Interactions Beyond the Standard Model

20.1 Introduction

The phenomenon of neutrino oscillation discussed in Chapter 18 implies that neutrinos have finite mass. The phenomenology of neutrino oscillation determines only the mass square differences $\Delta m_{ij}^2 (i \neq j)$ but not the absolute masses of the neutrinos. The scalar and fermionic structure of the standard model does not allow neutrinos to have non-zero mass. There is also the possibility that neutrinos may have magnetic moments (intrinsic or transition) considerably larger than the prediction of the standard model. Moreover, there are anomalous results in the measurements of neutrino and antineutrino oscillation parameters in various regions of energies which suggest that the standard model description in terms of three weak flavor doublets of leptons and quarks is not adequate to describe weak interactions and there may exist additional flavor of neutrinos which are non-interacting i.e. sterile. All these processes suggest that although the standard model has been a spectacular success in describing most electroweak processes, there is need for physics beyond the standard model.

Moreover, there are many other rare physical processes driven by weak interactions which have been studied for a long time theoretically as well as experimentally and are not explained by the standard model. The experimental observations of these processes would establish the physics beyond the standard model. The subject of physics beyond the standard model is too vast to be described in space of a chapter but we discuss here, some of the neutrino processes to introduce the subject.

  (i) Neutrinoless double beta decay (NDBD) and Majorana neutrinos

 (ii) Lepton flavor violating (LFV) decays of elementary particles

(iii) Flavor changing neutral current (FCNC)

(iv) Existence of non-standard interaction in high precision weak processes.

20.2   Netrinoless Double-beta Decay

20.2.1   General considerations

The problem of double-beta decays (DBD) involving two-neutrino double-beta decay $(2\nu\beta\beta)$ and neutrinoless double-beta decay $(0\nu\beta\beta)$ has been with us for more than 80 years after it was first discussed by Goeppert-Mayer in the case of $(2\nu\beta\beta)$ in 1935 [1062] soon after the Fermi theory of β-decay was formulated [23] and the process of $0\nu\beta\beta$ was discussed by Furry in 1939[1063] after a new theory of neutrino was given by Majorana [121].

There are many nuclei in which ordinary β-decay, that is, $(A, Z) \to (A, Z+1) + e^- + \bar{\nu}_e$ is not allowed but the process of $2\nu\beta\beta$ decay $(A, Z) \to (A, Z+2) + e^- + e^- + \bar{\nu}_e + \bar{\nu}_e$ is kinematically allowed. This process occurs in the second order of weak interactions in which two nucleons decay simultaneously with the nucleus giving rise to two antineutrinos and two electrons, thus, involving five particle phase space in the final state as shown in Figure 20.1(a). The process is allowed in the standard model and the transition matrix element is proportional to $(G_F \cos\theta_C)^2$. Although the process is very rare, it has been observed in many nuclei with lifetimes in the range exceeding $10^{18} - 10^{19}$ years (see Table 20.1)[1064].

Table 20.1 Some of the measured values of $T^{2\nu}_{1/2}$ in various nuclei.

Isotope	$T^{2\nu}_{1/2}$ years	Isotope	$T^{2\nu}_{1/2}$ years
^{48}Ca	$4.4^{+0.6}_{-0.5} \times 10^{19}$	^{116}Cd	$(2.8 \pm 0.2) \times 10^{19}$
^{76}Ge	$(1.5 \pm 0.1) \times 10^{21}$	^{128}Te	$(1.9 \pm 0.4) \times 10^{24}$
^{82}Se	$(0.92 \pm 0.07) \times 10^{20}$	^{130}Te	$6.8^{+1.2}_{-1.1} \times 10^{20}$
^{96}Zr	$(2.3 \pm 0.2) \times 10^{19}$	^{150}Nd	$(8.2 \pm 0.9) \times 10^{18}$
^{100}Mo	$(7.1 \pm 0.4) \times 10^{18}$	^{150}Nd–^{150}Sm(0^+_1)	$1.33^{+0.45}_{-0.26} \times 10^{20}$
^{100}Mo–^{100}Ru(0^+_1)	$5.9^{+0.8}_{-0.6} \times 10^{20}$	^{238}U	$(2.0 \pm 0.6) \times 10^{21}$

Soon after the new theory of neutrino was proposed by Majorana in which the neutrino was considered to be its own antiparticle, Furry analyzed the $0\nu\beta\beta$ decay mode of nuclei in which the Racah sequence [1065], that is,

$$
\begin{aligned}
(A, Z) &\to (A, Z+1) + e^- + \bar{\nu}_e \\
\nu_e + A(Z+1) &\to A(Z+2) + e^-
\end{aligned}
$$

$$\text{would lead to } (A, Z) \to (A, Z+2) + e^- + e^- \tag{20.1}$$

in case of virtual neutrinos as shown in Figure 20.1(b), if the neutrino is its own antiparticle, that is, $\bar{\nu} = \nu$ assuming that it is a Majorana neutrino. The decay rates for both types of double-beta decays were calculated in the context of the Fermi theory. Since the $0\nu\beta\beta$ decays are favored by the phase space consideration, the lifetimes of these decays were found to be in the range of $10^{15} - 10^{16}$ years for $0\nu\beta\beta$ and $10^{18} - 10^{22}$ years for $2\nu\beta\beta$ decays based on the phase space consideration. Clearly, $0\nu\beta\beta$ decay should have been discovered first but this did not happen [1066]. This implies that the physics of $2\nu\beta\beta$ and $0\nu\beta\beta$ decays are completely different from each other. We now know that it does not happen in the standard model which will be explained now.

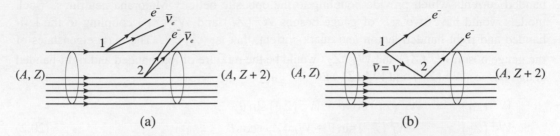

Figure 20.1 (a)$(A, Z) \to (A, Z+2) + e^- + e^- + \bar{v}_e + \bar{v}_e$ and (b)$(A, Z) \to (A, Z+2) + e^- + e^-$

At the first vertex in Figure 20.1(b), where an electron (e^-) is emitted, the electron is accompanied by an antineutrino which is right-handed, while the neutrino which is absorbed at the second vertex to produce the second electron, is left-handed. Thus, the process involves not only the condition that $\bar{v} = v$ but also that neutrino's helicity should flip. This process of $0v\beta\beta$ decay violates lepton number conservation by two units, that is, $\Delta L_e = 2$. It cannot happen for the massless neutrinos which conserve helicity. Therefore, in order that the Feynman diagram corresponding to Figure 20.1(b) gives a non-zero value of the transition matrix element, the neutrinos must have mass and the weak interaction must be lepton flavor violating (LFV) or both. The other alternative is to have additional interactions of Majorana neutrinos responsible for the flipping of the neutrino helicity. Both the possibilities may be present simultaneously. In both cases, we need to go beyond the standard model to describe the phenomenon of neutrinoless double-beta decay.

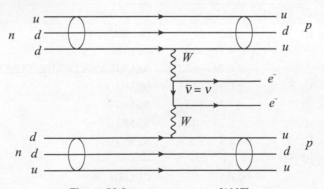

Figure 20.2 $nn \to pp + e^- + e^-$[1067].

There exist many theoretical models which have been proposed to generate the mass of Majorana neutrinos as well as theoretical models which predict new interactions for neutrinos by enlarging the $SU(2) \times U(1)$ group structure of the standard models like $SU(5), SO(10)$ or models based on super symmetry and have been applied to calculate the decay rates of neutrinoless double-beta decay of various nuclei. The subject has been presented in the classic review of Haxton et al. [1068] and Doi et al. [1069] as well as in many recent articles [1066, 1070, 1071, 1072, 1073] in view of the current interest in the phenomenology of neutrino oscillation with massive neutrinos. One of the simplest models is the left–right symmetric model having a minimal group structure of $SU(2)_L \times SU(2)_R \times U(1)$ allowing for right-handed currents which provide coupling to the opposite helicity Majorana neutrinos. Such models would have two sets of gauge bosons $W_L^{\pm}(Z_L)$ and $W_R^{\pm}(Z_R)$ coupling to the left-handed and right-handed lepton and quark current, that is, (j_μ^L, j_μ^R). The mass eigenstates of the gauge boson $W_1^{\pm}(Z_1)$ and $W_2^{\pm}(Z_2)$ would be the mixture of left-handed and right-handed gauge bosons $W_L^{\pm}(Z_L)$ and $W_R^{\pm}(Z_R)$ as

$$
\begin{aligned}
W_1^{\pm}(Z_1) &= W_L^{+}(Z_L)\cos\theta + W_R^{\pm}(Z_R)\sin\theta \\
W_2^{\pm}(Z_2) &= -W_L^{\pm}(Z_L)\sin\theta + W_R^{\pm}(Z_R)\cos\theta
\end{aligned}
\tag{20.2}
$$

with $\theta < 1$ and $M_1 << M_2$ to reproduce the phenomenology of the standard model at the lower energy scale. There are many more sophisticated models of describing the physics beyond the standard model with more Higgs scalars or more gauge fields than in the SM gauge fields or heavy leptons, proposed in the context of the Grand Unified theories and supersymmetry (SUSY) which predict new interactions allowing the $0\nu\beta\beta$ decays.

Experimentally, there have been many attempts to search for $0\nu\beta\beta$ decays in various nuclei, which provide lower limits for their lifetime. The average mass of the electron neutrino generated by the neutrino mixing matrix parameters is estimated from these limits on the $0\nu\beta\beta$ decay lifetime of nuclei. A summary of such experiments [1074, 1075, 1076, 1077, 1078, 1079, 1080, 1081, 1082, 1083, 1084] is given in Table 20.2. Some of these experiments are being updated while new experiments are also planned.

Table 20.2 $T_{1/2}^{0\nu}$ and $\langle m_{\beta\beta} \rangle$ limits (90% C.L.) from the most recent experiments on $0\nu\beta\beta$ in various nuclei.

Isotope	$T_{1/2}^{0\nu}$ ($\times 10^{25}$ years)	$\langle m_{\beta\beta} \rangle$ (eV)	Experiment	Reference
^{48}Ca	$> 5.8 \times 10^{-3}$	$< 3.5 - 22$	ELEGANT-IV	[1074]
^{76}Ge	> 8.0	$< 0.12 - 0.26$	GERDA	[1075]
	> 1.9	$< 0.24 - 0.52$	MAJORANA DEMONSTRATOR	[1076]
^{82}Se	$> 3.6 \times 10^{-2}$	$< 0.89 - 2.43$	NEMO-3	[1066]
^{96}Zr	$> 9.2 \times 10^{-4}$	$< 7.2 - 19.5$	NEMO-3	[1077]
^{100}Mo	$> 1.1 \times 10^{-1}$	$< 0.33 - 0.62$	NEMO-3	[1078]
^{116}Cd	$> 1.0 \times 10^{-2}$	$< 1.4 - 2.5$	NEMO-3	[1079]
^{128}Te	$> 1.1 \times 10^{-2}$	—	—	[1080]
^{130}Te	> 1.5	$< 0.11 - 0.52$	CUORE	[1081]
^{136}Xe	> 10.7	$< 0.061 - 0.165$	KamLAND-Zen	[1082]
	> 1.8	$< 0.15 - 0.40$	EXO-200	[1083]
^{150}Nd	$> 2.0 \times 10^{-3}$	$< 1.6 - 5.3$	NEMO-3	[1084]

There are other processes equivalent to neutrinoless double beta-decays discussed earlier, which are also possible, depending on the nuclear structure of various nuclei and kinematical considerations $\beta^+\beta^+$ decays,

$$
(A, Z) \rightarrow (A, Z - 2) + e^+ + e^+ + \nu_e + \nu_e,
$$

the electron capture (EC) processes:

$$e^- + (A, Z) \rightarrow (A, Z - 2) + e^+$$

and double electron capture (EC/EC) processes:

$$2e^- + (A, Z) \rightarrow (A, Z - 2).$$

It should be noted that a possible $0\nu\beta\beta$ decay of the EC/EC type needs an additional particle in the final state to conserve energy and momentum which shold be emitted from the final nuclear state in order to be physically possible. These processes are difficult to observe but provide a strong motivation for studying the physics beyond the standard model.

20.2.2 Decay rates of $0\nu\beta\beta$

In a general description of weak interaction the $0\nu\beta\beta$ decay involving two neutrinos in the nucleus is depicted at the quark level as shown in Figure 20.2. In the simplest extension of the standard model based on the left–right symmetric models, the general Hamiltonian for $0\nu\beta\beta$ decay at the quark level (Figure 20.2) can be written as [1067]

$$\mathcal{H} = \frac{G_F \cos\theta_c}{\sqrt{2}} (l_L^\mu J_{L\mu}^\dagger + \kappa l_L^\mu J_{R\mu}^\dagger + \eta l_R^\mu J_{L\mu}^\dagger + \lambda l_{\mu R}^\mu J_{\mu R}^\dagger) \tag{20.3}$$

$$\text{where } l_{L,R}^\mu = \bar{e}\gamma^\mu(1 - \gamma_5)v_{e_L}, \quad l_R^\mu = \bar{e}\gamma^\mu(1 + \gamma_5)v_{e_R} \tag{20.4}$$

$$J_L^\mu = \bar{u}_e\gamma^\mu(1 - \gamma_5)d; \quad J_R^\mu = \bar{u}\gamma^\mu(1 + \gamma_5)d. \tag{20.5}$$

In the standard model $k = \eta = \lambda = 0$, while in the left–right symmetric models,

$$\eta = k \approx \tan\theta, \quad \lambda \approx \left(\frac{M_1}{M_2}\right)^2 + \tan^2\theta$$

$$\text{leading to } \mathcal{H}^\beta = G_F \cos\theta\sqrt{2}\left[(\bar{e}_L\gamma_\mu v_{eL})\left(J_{L\mu}^\dagger + \eta J_{R\mu}^\dagger\right) \right.$$

$$\left. + (\bar{e}_R\gamma_\mu v_{eR})\left(\eta J_{L\mu}^\dagger + \lambda J_{R\mu}^\dagger\right) + h.c.\right], \tag{20.6}$$

and

$$W_L = \cos\theta W_1 - \sin\theta W_2, \quad W_R = \sin\theta W_1 + \cos\theta W_2, \tag{20.7}$$

with $\eta = \tan\theta$ describing the mixing of $(W_L(Z))$ and $(W_R(Z))$ bosons as given in Eq. (20.2). The matrix elements of the nucleonic current $J_{\mu,L}^\dagger$ and $J_{\mu,R}^\dagger$, corresponding to FIg 20.1(b) are defined as (Chapter 10), where the sum over i is performed over the number of nucleons in the nucleus

$$J_L^{\mu\dagger} = \sum_i \bar{u}_p(i)\left[f_1(q^2)\gamma^\mu + if_2(q^2)\frac{\sigma^{\mu\nu}}{2M_p}q_\nu - g_1(q^2)\gamma^\mu\gamma_5 - g_3(q^2)q^\mu\gamma_5\right]u_n(i), \tag{20.8}$$

for the left-handed nucleon currents, in which the form factors $f_j(q^2)$ $(j = 1, 2)$ and $g_j(q^2)$ $(j = 1, 3)$, are known and

$$J_R^{\mu\dagger} = \sum_i \bar{u}_p(i)\left[f_1'(q^2)\gamma^\mu + if_2'(q^2)\frac{\sigma^{\mu\nu}}{2M_p}q_\nu + g_1'(q^2)\gamma^\mu\gamma_5 + g_3'(q^2)q^\mu\gamma_5\right]u_n(i), \tag{20.9}$$

for the right-handed currents. The form factors $f_j(q^2)$ and $g_j(q^2)$ appearing in the matrix elements of the left- handed currents in Eq. (20.8) are discussed in Chapter 10.

It should be realized that while the bare couplings of the right-handed leptonic and hadronic currents in Eq. (20.9) are determined by the coefficients k, η, and λ, the form factors $f'_j(q^2)$ and $g'_j(q^2)$, appearing in the matrix elements of the right-handed currents $f'_j(g'_j)$ are related with $f_j(q^2)$ and $g_j(q^2)$ defined for the left-handed currents as they are derived from the bare couplings due to the renormalization in the presence of strong interactions which conserve parity and therefore would be same as the form factors in the left handed current.

The process of $(0\nu\beta\beta)$ occurs in the second order as shown in Figure 20.1(b) and the decay rate is given by

$$dW \approx 2\pi \int |\mathcal{M}^{fi}|^2 \delta(E_f - E_i - E_{e_1} - E_{e_2}) F(Z, E_{e1}) F(Z, E_{e2}) d^3p_1 d^3p_2, \quad (20.10)$$

where $F(Z, E_{e1})$ and $F(Z, E_{e2})$ are the Fermi functions describing the Coulomb distortion of electrons in the field of final nucleus given in Chapter 5. $\mathcal{M}^{fi}$ is the transition matrix element calculated in the second order of perturbation theory leading to

$$\mathcal{M}^{fi} = L^{\mu\nu} J^{fi}_{\mu\nu}, \quad (20.11)$$

where $L^{\mu\nu}$ and $J_{\mu\nu}$ are the leptonic and hadronic tensors calculated by using the leptonic and hadronic currents l^μ and J_μ defining the interaction Hamiltonian given in Eq. (20.3). Without going into the details of calculation, we demonstrate the main features of particles physics and nuclear physics which enter into these calculations [1033].

Leptonic component

In calculating the second order of the matrix elements, the leptonic part depends on the product of two neutrino fields at space–time points x_1 and x_2 in Figure 20.1(b) which are contracted, leading to the lepton tensor given by reference of Engel and Menendez [1071] and Bilenky [1033]

$$L^{\mu\nu} \propto \int \int dx_2 dx_1 \sum_i \bar{e}_L(x_1) \gamma^\mu (1 - \gamma_5) U_{ei} \underline{\nu_{iL}(x_1) \bar{e}_L(x_2) \gamma^\nu (1 - \gamma_5) U_{ei} \nu_{iL}(x_2)}$$

$$= -\int \int dx_2 dx_1 \sum_i \bar{e}(x_1) \gamma^\mu (1 - \gamma_5) U_{ei} \underline{\nu_{iL}(x_1) \bar{\nu}^c_{iL}(x_2)} \gamma_\nu (1 + \gamma_5) U_{ei} e^C_L(x_2), \quad (20.12)$$

where ν_i are the Majorana mass eigenstates with mass m_i and U_{ei} are the matrix elements of the mixing matrix. $\bar{\nu}_{iL}$ is the charge conjugate of the neutrino field which is equal to the neutrino field in the case of Majorana neutrinos $\nu^c_i = \nu_i$, where for a fermion field $\psi^C(x) = i\gamma^2 \psi^*(x)$. The contraction of neutrino fields $\nu_{eL}(x_1)$ and $\bar{\nu}_{eL}(x_2)$ therefore gives the neutrino propagator leading to the leptonic tensor

$$L^{\mu\nu} \propto \int \int dx_1 dx_2 \Big[\sum_i \frac{d^4q}{(2\pi)^4} e^{-iq\cdot(x_1-x_2)} \bar{u}_e(p_1) \gamma_\mu (1 - \gamma_5) e^{+i(p_1\cdot x_1 + p_2\cdot x_2)}$$

$$\frac{\slashed{q} + m_i}{q^2 - m_i^2} \gamma_\nu (1 + \gamma_5) u^c_e(p_2) U^2_{ei} \Big], \quad (20.13)$$

where q is the four momentum of the initial neutrino and p_1 and p_2 are the momenta of the two electrons. Now, we consider the following case of the weak interaction models:

(a) Models in which there are Majorana neutrinos with no right-handed currents; here, $\not{q}$ terms operating on the leptonic part, vanish. The propagator term then reduces to

$$\frac{m_i}{q^2 - m_i^2} \quad \propto \quad \frac{m_i}{q^2} \quad \text{if } m_i^2 << q^2$$

$$\propto \quad -\frac{1}{m_i} \quad \text{if } m_i^2 >> q^2. \tag{20.14}$$

implying that in the case of light neutrinos, the contribution to the transition matrix element for the leptonic part has different mass (m_i) and q^2 dependence than the case of heavy neutrinos, depending upon the mass (m_i) of the virtual neutrino exchanged between the two nucleons in the nuclei. In the case of the light neutrino scenario, it depends upon the effective mass $m_{\beta\beta}$ of Majorana neutrinos, given by:

$$m_{\beta\beta} = \sum_i U_{ei}^2 m_i. \tag{20.15}$$

(b) Models with $V + A$ currents in which the leptonic currents have opposite chirality; here, one gets from the neutrino propagator, a term like

$$\frac{\not{q}}{q^2 - m_i^2} \quad \rightarrow \quad \frac{\not{q}}{q^2}, \; m_i^2 << q^2 \tag{20.16}$$

$$\rightarrow \quad -\frac{\not{q}}{m_i^2}, \; m_i^2 > q^2, \tag{20.17}$$

In the light neutrino scenario there is no ν-mass dependence in the transition matrix element.

It is worth emphasizing that in all the scenarios discussed till now, the kinematics becomes different and the neutrino mass plays a very important role. Since the mass m_i of Majorana neutrinos are yet to be determined experimentally, theoretical modeling of Majorana masses have been a very active field of physics beyond the standard model in recent years, in the context of $0\nu\beta\beta$ decays.

Hadronic component

The hadronic part of the matrix element $J_{\mu\nu}^{if}$ is calculated by taking the matrix element of the T-product $T(J_\mu(x_1)J_\nu(x_2))$ between the nuclear states $|i\rangle$ and $|f\rangle$, that is, [1033]

$$J_{\mu\nu}^{fi} = \langle f|TJ_\mu(x_1)J_\nu(x_2)|i\rangle,$$

$$J_{\mu\nu}^{fi} = \langle f|J_\mu(x_1)J_\nu(x_2)\theta(x_{1_0} - x_{2_0})|i\rangle + \langle f|J_\nu(x_2)J_\mu(x_1)\theta(x_{2_0} - x_{1_0})|i\rangle. \tag{20.18}$$

Considering the first term, where $x_{1_0} > x_{2_0}$, after saturating it with a set of intermediate states $|n\rangle$, we write

$$J_{\mu\nu}^{if} = \sum_n \langle f|J_{\mu L}(x_1)|n\rangle\langle n|J_{\nu L}(x_2)|i\rangle + (\mu \rightarrow \nu, x_{1_0} \rightarrow x_{2_0}), \tag{20.19}$$

where $|n\rangle$ is a complete set of intermediate states, that is, $\sum_n |n\rangle\langle n| = 1$.

$$J_\mu(x) = e^{iHx_0} J_{\mu L}(\vec{x}) e^{-iHx_0} \tag{20.20}$$

$$J_{\mu\nu}^{fi} = \sum_n \langle f|J_{\mu L}(\vec{x}_1)|n\rangle \langle n|J_{\nu L}(\vec{x}_2)|i\rangle e^{-i(E_n - E_f)x_{10}} e^{-i(E_n - E_i)x_{20}} + (\mu \to \nu, x_{10} \to x_{20}). \tag{20.21}$$

Now the matrix element $M^{fi} = L^{\mu\nu} J_{\mu\nu}^{fi}$ is evaluated using Eq. (20.11) and combining the various factors in Eq. (20.13) and (20.21); we integrate first over x_{10} and x_{20} and then integrate over q_0 to obtain the factor $2\pi\delta(E_f + E_{e_1} + E_{e_2} - E_i)$ and we get the following expression:

$$
\begin{aligned}
M_{0\nu}^{fi} &\propto \left(\frac{G_F^2 \cos^2\theta_C}{\sqrt{2}} \bar{u}(p)\gamma^\mu\gamma^\nu(1+\gamma^5)u^c(p_2) \right) \int d^3x_1 \int d^3x_2 e^{i(\vec{p}_1\cdot\vec{x}_1 + \vec{p}_2\cdot\vec{x}_2)} \\
&\quad \sum_i U_{ei}^2 m_i \int \frac{e^{i\vec{q}\cdot(\vec{x}_1 - \vec{x}_2)}}{(2\pi)^3 q_i^0} d^3q \, 2\pi\delta(E_f + p_1^0 + p_2^0 - E_i) \\
&\quad \times \sum_n \left[\frac{\langle f|J_{\mu L}(\vec{x}_1)|n\rangle \langle n|J_{\nu L}(\vec{x}_2)|i\rangle}{E_n + p_2^0 + q_i^0 - E_i} + \frac{\langle f|J_{\nu L}(\vec{x}_2)|n\rangle \langle n|J_{\mu L}(\vec{x}_1)|i\rangle}{E_n + p_1^0 + q_1^0 - E_i} \right]. \tag{20.22}
\end{aligned}
$$

The matrix element in Eq. (20.22) is further evaluated, making the following approximations for the nucleons in nuclei and the kinematics of the virtual neutrinos and nucleons

1. The mass of the neutrinos are small, therefore, the energy $q_i^0 = \sqrt{|q_i|^2 + m_i^2} \simeq |\vec{q}_i|$. Since the virtual neutrinos are confined within the volume of the nuclei, with radius $\approx$ few fm, $|\vec{q}_i|$ is of the order of $\frac{1}{r}(= 100 \text{ MeV})$.

2. The order of the magnitude of the momenta of the electrons emitted in the double beta-decay is a few MeV and the nuclear radius is about a few fm; therefore, the magnitude of $\vec{p}_1\cdot\vec{x}_1 \approx \vec{p}_2\cdot\vec{x}_2 \approx p\cdot R$ is very small of the order of 10^{-2} and the factor $e^{i(\vec{p}_1\cdot\vec{x}_1 + \vec{p}_2\cdot\vec{x}_2)} \approx 1$.

3. The evaluation of the nuclear matrix element M_{fi} requires the knowledge of the nuclear ground state of the initial nucleus as well as the excited states of the final nucleus. Since the energy of virtual neutrinos could be of the order of 100 MeV almost all the excited states $|n\rangle$ need to be considered. This is not easy. Therefore, nuclear matrix elements are evaluated in closure approximation. For this, we assume an average excitation energy $\bar{E}$ for E_n, so that the summation over $|n\rangle$ states could be performed assuming the completeness relation. Therefore, we write:

$$\sum_n \frac{\langle f|J_\mu(\vec{x}_1)|n\rangle \langle n|J_\nu(\vec{x}_2)|i\rangle}{E_n + p_2^0 + q_i^0 - E_i} \simeq \frac{\langle f|J_\mu(x_1)J_\nu(x_2)|i\rangle}{\bar{E} + p_2^0 + |\vec{q}| - E_i}, \tag{20.23}$$

$$\sum_n \frac{\langle f|J_\nu(\vec{x}_2)|n\rangle \langle n|J_\mu(\vec{x}_1)|i\rangle}{E_n + p_1^0 + q_i^0 - E_i} \simeq \frac{\langle f|J_\nu(x_2)J_\mu(x_1)|i\rangle}{\bar{E} + p_1^0 + |\vec{q}| - E_i}. \tag{20.24}$$

4. Neglecting nuclear recoil, we can write:

$$E_i = E_f + p_1^0 + p_2^0 = \frac{E_i + E_f}{2} + \frac{p_1^0 + p_2^0}{2}$$

implying

$$|\vec{q}| + \bar{E} + p_{1,2}^0 - E_i = |\vec{q}| + \bar{E} - \frac{E_i + E_f}{2} \pm \frac{(p_1^0 - p_2^0)}{2}. \tag{20.25}$$

Since $\frac{p_1^0 - p_2^0}{2}$ is much smaller than other terms, the denominator in both terms is the same and depends only upon $\vec{q}$, making it easier to perform d^3q integration in Eq. (20.22), because $J_\mu(x) J_\nu(y)$ is independent of q in the leading order (as given here in Eq. (20.29)). Therefore, performing the $\vec{q}$ integration leads to

$$\frac{1}{(2\pi)^3} \int \frac{e^{iq|\vec{x}_1 - \vec{x}_2|} d^3q}{q(q + \bar{E} - \frac{1}{2}(E_i + E_f))} = \frac{1}{2\pi^2 r} \int \frac{\sin qr \, dq}{q + \bar{E} - \frac{E_i + E_f}{2}} = \frac{1}{4\pi R} H(r, \bar{E}), \tag{20.26}$$

with $\vec{r} = |\vec{x}_1 - \vec{x}_2|$ and R being the nuclei radius.

such that $H(r, \bar{E}) = \frac{2R}{\pi r} \int_0^\infty \frac{\sin qr dq}{q + \bar{E} - \frac{1}{2}(E_i + E_f)}$ which is called the neutrino potential.

5. The weak interaction current operator $J^\mu(x)$ in the nuclei is calculated in the impulse approximation using nonrelativistic nucleons (Chapter 14). In this approximation,

$$J_{0L}(\vec{x}) \simeq \sum_i \delta(\vec{x} - \vec{x}_i) f_1(0) \tau_i^+ \tag{20.27}$$

$$\vec{J}_L(\vec{x}) \simeq \sum_i \delta(\vec{x} - \vec{x}_i) g_1(0) \vec{\sigma}_i \tau_i^+ \tag{20.28}$$

in the lower order, where $\sum_i$ is over nucleons such that

$$J_{\mu L}(\vec{x}_1) J_{\nu L}(\vec{x}_2) = \sum_{i,j} \tau_i^+ \tau_j^+ \delta(\vec{x}_1 - \vec{x}_i) \delta(\vec{x}_2 - \vec{x}_j) [f_1^2(0) - g_1^2(0) \vec{\sigma}_i \cdot \vec{\sigma}_j]. \tag{20.29}$$

Note that because of Eqs. (20.27) and (20.28), only $\mu = 0, \nu = 0$ and $\mu = i, \nu = j$ components contribute in the leptonic part $L^{\mu\nu}$. The matrix element in Eq. (20.22) is therefore written as:

$$M_{0\nu}^{fi} \propto 2\pi\delta(E_f - E_i - E_1 - E_2) \left(\frac{G_F \cos\theta_C}{\sqrt{2}}\right)^2 m_{\beta\beta}$$

$$\times \bar{u}(p_1)(1 + \gamma^5) u^c(p_2) M_{0\nu}, \tag{20.30}$$

where $M_{0\nu} = f_1^2(0) M_{0\nu}^F - g_1^2(0) M_{0\nu}^{GT}$

$$M_{0\nu}^F = \langle f| \sum_{i,j} H(r_{ij}, \bar{E}) \tau_i^+ \tau_j^+ |i\rangle \tag{20.31}$$

$$M_{0\nu}^{GT} = \langle f| \sum_{i,j} H(r_{ij}, \bar{E}) \vec{\sigma}_i \cdot \vec{\sigma}_j \tau_i^+ \tau_j^+ |i\rangle \tag{20.32}$$

and $m_{\beta\beta}$ is given by Eq. (20.15). Using this value of $M_{0\nu}$ in Eq. (20.30), the square of the matrix element $\sum_{\text{leptons}} |M_{0\nu}^{fi}|^2$ is calculated as:

$$\sum_{\text{leptons}} |M_{0\nu}^{fi}|^2 \propto \left(\frac{G_F \cos\theta_C}{\sqrt{2}} \right)^4 m_{\beta\beta}^2 \, 8p_1 \cdot p_2 |M_{0\nu}|^2, \qquad (20.33)$$

using $\sum_{\text{spins}} |\bar{u}(p_1)(1 + \gamma^5)u(p_2)|^2 = 8p_1 \cdot p_2$.

Nuclear matrix elements $M_{0\nu}$

All the nuclei which are likely candidates for neutrinoless β-decay are heavy nuclei (far from the closed shell); some of them are also deformed nuclei. Therefore, the theoretical calculation of the nuclear matrix element is a formidable nuclear problem. The different approaches used for calculating the nuclear matrix elements (NME) are the nuclear shell model (NSM) and the quasi particle random phase approximation (QRPA) [1068, 1069]. Recently, calculations based on the projected Hartree–Fock–Bogoliobov (PHFB) model, the interacting boson model (IBM), and the energy density functional (EDF) methods have also been used [1070, 1071, 1072, 1073, 1068, 1069].

In the closure approximation, the NME depends upon the nuclear wave functions and on the neutrino potential $H(r)$ given in Eq. (20.26) arising due to the virtual neutrino propagator which is of long range. The calculation of $M_{0\nu}$ requires a very good knowledge of nuclear wave function in the entire range of r, and the various nuclear structure models give different results. Moreover, in the case of calculations using multipole expansion (Chapter 14), the sum over various multipoles has a very slow convergence. Since the energy of the virtual neutrinos could be of the order of 100 MeV, there are many multipoles which contribute. This necessitates a good knowledge of almost all the excited state wave functions of the final nucleus, making the calculations of NME very difficult. It is because of this reason that closure approximation is used (see Chapter 14) Moreover, in the case of particle physics models based on $0\nu\beta\beta$ decays mediated by heavy neutrinos or other heavy particles, the short range correlation effects in the nuclear wave functions make the calculations of NME even more difficult. This leads to considerable uncertainty in the results for NME [1071, 1067, 1085], Various calculations could give results differing from each other by a factor of 2-3. For details, see Engel and Menendez [1071].

There is some experimental information available on the nuclear wave functions in the case of a few nuclei in which $2\nu\beta\beta$ decays have been observed. The experimental strength of GT transitions for $0^+ \rightarrow 1^+$ transitions is also available in some nuclei from the $(^2\text{He}, d)$ and $(^3\text{He}, t)$ reactions. In the case of transitions involving ground states, limited information is also available from electron capture and muon capture experiments. However, these are limited in scope and describe very few excitations corresponding to the low lying excited states in the final nucleus. In summary, the NME are the major source of uncertainty in the study of $0\nu\beta\beta$ decays.

Phase space and decay rates

The decay rate is obtained using Eqs. (20.11), (20.30) and (20.33) after integrating the $\sum |M^{fi}|^2$, over the phase space of two electrons. The phase space is considerably modified in the presence of the Coulomb field of the nucleus $(A, Z+2)$. Therefore, the decay rate is given by [1067]:

$$
\begin{aligned}
\Gamma_{ov} &= m_{\beta\beta}^2 |M_{ov}|^2 \int_{m_e} F(E_1, Z+2) F(E_2, Z+2) \frac{d^3 p_1}{2E_1 (2\pi)^3} \frac{d^3 p_2}{2E_2 (2\pi)^3} \\
&\times \; 8(E_1 E_2 - |\vec{p}_1||\vec{p}_2| \cos\theta) 2\pi \delta(E_f - E_i - E_1 - E_2) \\
&= \; m_{\beta\beta}^2 |M_{ov}|^2 G(Q, Z),
\end{aligned}
\tag{20.34}
$$

where

$$
\begin{aligned}
G(Q, Z) &\propto \int \int F(E_1, Z+2) F(E_2, Z+2) |\vec{p}_1||\vec{p}_2| (E_1 E_2 - |\vec{p}_1||\vec{p}_2| \cos\theta) dE_1 dE_2 \\
&\delta(Q - E_1 - E_2),
\end{aligned}
\tag{20.35}
$$

and $Q \approx M_i - M_f - 2m_e$, neglecting the nuclear recoil. $m_{\beta\beta}$ and $|M_{ov}|^2$ are given in Eqs. (20.15) and (20.33) and $|\vec{p}_1|(|\vec{p}_2|)$ are the magnitude of 3-momenta $\vec{p}_1 (\vec{p}_2)$.

In the case of other particle physics models, the decay rate has many more terms; for example, in the minimal extension of the standard model having the right-handed currents based on $SU(2)_L \times SU(2)_R \times U(1)$, the decay rate in Eq. (20.34) is now given by [1067]:

$$
\begin{aligned}
\Gamma^{ov} &= C_{mm} \frac{m_{\beta\beta}^2}{m_e^2} + C_{\eta\eta} \langle \eta \rangle^2 + C_{\lambda\lambda} \langle \lambda \rangle^2 \\
&+ C_{m\eta} \left(\frac{m_{\beta\beta}}{m_e} \right) \langle \eta \rangle + C_{m\lambda} \left(\frac{m_{\beta\beta}}{m_e} \right) \langle \lambda \rangle + C_{\eta\lambda} \langle \eta \rangle \langle \lambda \rangle,
\end{aligned}
\tag{20.36}
$$

where $\langle \eta \rangle = \eta \sum_i U_{ei} V_{ei}$, $\langle \lambda \rangle = \lambda \sum_i U_{ei} V_{ei}$ with V_{ej} as the mixing matrix elements among the right-handed neutrino states. Equation (20.36) reduces to Eq. (20.34) when $\langle \eta \rangle$, $\langle \lambda \rangle = 0$ in the standard model. For example, the element C_{mm} becomes

$$
C_{mm} = |M_{0v}|^2 G(Q, Z) m_e^2.
\tag{20.37}
$$

20.2.3 Experiments

The experimental observation of the neutrinoless double beta-decay is very challenging as it is perhaps the smallest theoretically predicted process accessible to observation after the proton decay. The typical energy for a double beta-decay is of the order of a few MeV, which is shared by four leptons in the case of $2\nu\beta\beta$ and two electrons in the case of $0\nu\beta\beta$ decay. The decay spectrum of the two electrons as a function of $E_e = E_{e_1} + E_{e_2}$ is a delta function in the case of $0\nu\beta\beta$; the decay rate has Q^5 dependence. Therefore, amongst the heavy nuclei which are likely candidates of $0\nu\beta\beta$ decay, only those with larger Q value and larger nuclear matrix element for the transitions are favored for experimental studies. The list of several such nuclei which have been studied is given in Table 20.2. Since this is a very rare process, a significant amount of source material should be made available either naturally or by isotopical enrichment Moreover, a low-level counting technique for detecting charged particles should be used along with sufficient arrangements made for shielding from background events. The major sources of background are charged particles produced by cosmic rays and by the natural and artificial radioactivity in the surrounding material leading to charged particles through primary or secondary interactions. In this scenario, the deep underground experiments planned for the study of proton decays and atmospheric neutrinos are the most sutaible laboratories for performing DBD (double beta-decay) experiments in both modes of $2\nu\beta\beta$ and $0\nu\beta\beta$ decays. Since $0\nu\beta\beta$ decay is a rarer process compared to $2\nu\beta\beta$ decays by at least 5–6 orders of magnitude, the charged particles from $2\nu\beta\beta$ are themselves a serious background. However, their energy spectra are quite different from the electrons from $0\nu\beta\beta$ decays.

There is an extensive experimental program being pursued for the study of DBD which started in 1948 and also focussed on $0\nu\beta\beta$ decay in early years using Geiger, proportional, and scintillation counters using few grams of source material. These days, sophisticated detectors of various types with considerable high energy resolution are being used with hundreds of kilograms of source material. These are categorized according to various techniques used for the detection of signals of charged particles. Excellent reviews of the experimental developments are given by many reviewers [1070, 1068, 1069, 1086, 1087, 1088]. In the following, we present a summary of various experiments without describing them individually. The details of each experiment mentioned below can be found in the above references and Giuliani et al. [1089].

Semiconductor detectors

High purity germanium (HPGe) detectors enriched with ^{76}Ge are the most suitable detectors in the source as detector configuration for studying $0\nu\beta\beta$ decays and have been used in the Hydelberg-Moscow and IGEX experiments as well as in the current generation of GERDA and Majorana DEMONSTRATOR experiments in underground laboratories at Gran Sasso, Aquila Italy and Sanford, Lead USA. The results for the upper limits for $0\nu\beta\beta$ lifetimes for various nuclei from these experiments are already available and are shown in Table 20.2. The upgrades of GERDA have shown the technical feasibility of doing large scale $0\nu\beta\beta$ decay experiments and the experiments proposed at LEGEND-200 and LEGEND-1000 with 200 and 1000 kg of enriched ^{76}Ge.

Cryogenic detectors

Bolometers are cryogenic calorimeters that operate at very low temperatures of ≈ 10 mK. A heat absorber is connected via a weak thermal link to a low temperature thermal bath and the increase in temperature due to the energy deposited by the charge particles produced in $0\nu\beta\beta$ is measured by a sensitive thermometer. A variety of materials with $\beta\beta$ emitting nuclei like ^{130}Te, ^{116}Cd, ^{82}S, ^{100}Mo can be used as bolometric absorbers. Experiments like CUORE-O, Coricino, and CUORE use enriched ^{130}Te source from which experimental limits are already available (Table 20.2). The upgrade of CUORE like CUPID with ^{100}Mo and ^{82}Se $\beta\beta$ emitters are planned at LNGS, Italy in the near future. Other advanced detectors of this type are AMORE-I and AMORE-II planned to be operated in South Korea with 5 kg and 200 kg of ^{100}Mo crystals as source material with sensitivity up to $10^{25} - 10^{26}$ years for the β lifetime of ^{100}Mo, which is an improvement by a factor of 10–100 from the present limits.

Scintillator detectors

Another type of detectors which have been used in the source–detector configuration are the scintillators detector. The advantage of these detectors is that they can be used with $\beta\beta$ emitters with higher values of Q like ^{48}Ca. The latest detector in this category is the ELEGANT-IV setup in Japan providing limits on $0\nu\beta\beta$ decays lifetime for ^{48}Ca(Table 20.2). The series of CANDLES detectors using ^{48}Ca have been used to carry out experiments at the Kamioka Observatory. The latest in the series of CANDLES-III detectors have already reported early results for their experiment. In another experiment, ^{116}Cd was used in the Solotvino salt mine in Ukraine to obtain an upper limit on the lifetime of ^{116}Cd (1.7×10^{23} years).

In addition to these detector experiments based on inorganic scintillators, KamLAND-ZEN and SNO+ detectors use ^{136}Xe and ^{130}Te $\beta\beta$ emitters in large liquid scintillators. The KamLAND-ZEN loaded with 320 kg and 400 kg enriched ^{136}Xe has already reported results (Table 20.2) and an update to KamLAND-II with 1 ton ^{136}Xe is already planned. Another experiment, ZICOS loaded with 45 kg of liquid scintillator enriched ^{96}Zr is also being planned.

Ionization detectors

Experiments based on these detectors are called passive experiments and use the principle of ionizations and scintillations and make use of time projection chambers (TPC). The $\beta\beta$ emitter is used in the form of gas or thin foils; the emitted electrons are tracked in energy and angle between electrons. In most of the experiments, ^{136}Xe has been used or is planned to be used. EXO-200 with 110 kg of enriched ^{136}Xe has already produced limits on ^{136}Xe lifetime against $0\nu\beta\beta$ and updates like nEXO with 500 kg of ^{136}Xe is already planned. Other experiments with ^{136}Xe like NEXT-100 with 100 kg of ^{136}Xe in Spain and PandaX with 200 kg of ^{136}Xe in China, are already planned for future operations.

Tracking calorimeter detectors

Tracking calorimeter detectors like NEMO-3 situated in the Frejus laboratory in France use a thin foil of source material surrounded by a gas tracking layer in which two electrons are tracked in energy and the angle between them. It has been used to report limits on the $0\nu\beta\beta$ lifetime of ^{100}Mo,^{116}Co,^{82}Se, and ^{96}Zr. The SuperNEMO experiment, an update of the NEMO-3 plans to use 20 modules containing 6.3 kg of ^{82}Se in each module to be surrounded by scintillator blocks to track the electrons and is projected to reach a half life sensitivity of $\approx 10^{26}$ years, that is, an improvement by a factor 10^{2} on the present limits.

The observation of NDBD will establish the mass of the neutrino and its Majorana nature in the context of theoretical developments and confirm the existence of a physics beyond the standard model (BSM). However, there are many theoretical models formulated for describing the physics beyond the standard model like Grand Unified Theories (GUT), string theory, supersymmetry (SUSY) and others in which the neutrino masses and new interactions are predicted depending upon the parameters of the model. We would, therefore, need more experiments with greater sensitivity to BSM physics in order to find the correct theoretical description of the physics beyond the standard model.

20.3 Lepton Flavor Violating Processes

There are many other leptonic, hadronic, and nuclear processes, which are allowed if the lepton flavor number is not conserved as in the case of the neutrinoless double beta decay (NDBD). The predictions for the rates of these processes are based on various theories using beyond the standard model (BSM) physics. In NDBD, the electron lepton number L_e is violated by two units, that is, $\Delta L_e = 2$, but there could be processes in which the muon lepton number L_μ is violated with one or more units as well as the processes in which L_e and L_μ, both are violated. All these processes have the capability of discriminating among the various models of the BSM physics. Such processes are predicted in $\Delta S = 0$ as well as in $\Delta S = 1$ sectors involving muons and electrons.

The leptonic processes involving heavy leptons $\mu^\pm(\tau^\pm)$ in which these particles decay into lighter leptons $e^\pm(\mu^\pm)$ without any neutrinos, violate L_e, L_μ or both and are not allowed in the standard model. The observation of such processes is a signature of BSM physics. The present limits on many leptonic processes with lepton flavor violation (LFV) [1090] are given in Table 20.3.

Table 20.3 Table for rare leptonic processes with lepton flavour violation (LFV).

Decay mode	Present limit	Decay mode	Present limit
$\mu^- \to e^- \gamma$	$< 1.2 \times 10^{-11}$	$\mu^- \text{Ti} \to e^- \text{Ti}$	$< 6.1 \times 10^{-13}$ (PSI)
$\mu^+ \to e^+ e^+ e^-$	$< 1.0 \times 10^{-12}$	$\mu^- \text{Ti} \to e^- \text{Ti}$	$< 4.6 \times 10^{-12}$ (TRIUMF)
$\mu^+ e^- \to \mu^- e^+$	$< 8.3 \times 10^{-11}$	$\mu^- + \text{Cu} \to e^+ + \text{Co}$	$< 2.6 \times 10^{-8}$
$\tau \to e\gamma$	$< 2.7 \times 10^{-6}$	$\mu^- + \text{S} \to e^+ + \text{Si}$	$< 9 \times 10^{-10}$
$\tau \to \mu\gamma$	$< 3 \times 10^{-6}$	$\mu^- \text{Au} \to e^- \text{Au}$	$< 7 \times 10^{-13}$ (SINDRUM)
$\tau \to \mu\mu\mu$	$< 1.9 \times 10^{-16}$		
$\tau \to eee$	$< 2.9 \times 10^{-6}$		

These processes can be calculated using the two-body or three-body kinematics for decay and scattering processes described elsewhere in the book provided that the interaction Lagrangians are known. There are various calculations done for these processes with phenomenological Lagrangians based on theoretical models to describe the physics beyond the standard model. Among the theoretical models, supersymmetry (SUSY) models have received much attention. Minimal supersymmetry models (MSSM), SUSY with explicit symmetry breaking and R-parity violation or SUSY models proposed in context of Grand Unified Theories (GUT) have been used extensively in literature to calculate these processes. Theoretical models based on the right-handed neutrinos, charged Higgs bosons, more gauge fields than SM gauge fields and leptoquarks have also been used.

There are also LFV decay processes involving pions, kaons, or Z bosons which have been searched for the presence of BSM physics. Some of these processes with present limits are listed in Table 20.4. They can also be calculated in the models proposed for BSM physics and would be able to discriminate between various models if they are observed experimentally.

Table 20.4 Rare hadronic processes with lepton flavor violation (LFV).

Decay mode	Limit on branching ratio	Decay mode	Limit on branching ratio
$\pi^0 \to \mu e$	$< 8.6 \times 10^{-9}$	$D^+ \to \pi^- e^+ e^+$	$< 3.6 \times 10^{-6}$
$K_L^0 \to \mu e$	$< 4.7 \times 10^{-12}$	$D^+ \to \pi^- \mu^+ \mu^+$	$< 4.8 \times 10^{-6}$
$K^+ \to \mu^+ \mu^- e^-$	$< 2.1 \times 10^{-10}$	$D^+ \to \pi^- e^+ \mu^+$	$< 5.0 \times 10^{-5}$
$K_L^0 \to \pi^0 \mu^+ e^-$	$< 3.1 \times 10^{-9}$	$D_s^+ \to \pi^- e^+ e^+$	$< 6.9 \times 10^{-4}$
$Z^0 \to \mu e$	$< 1.7 \times 10^{-6}$	$D_s^+ \to \pi^- \mu^+ \mu^+$	$< 2.9 \times 10^{-5}$
$Z^0 \to \tau e$	$< 9.8 \times 10^{-6}$	$D_s^+ \to \pi^- e^+ \mu^+$	$< 7.3 \times 10^{-4}$
$Z^0 \to \tau \mu$	$< 1.2 \times 10^{-5}$	$B^+ \to \pi^- e^+ e^+$	$< 1.6 \times 10^{-6}$
$K^+ \to \pi^- e^+ e^+$	$< 6.4 \times 10^{-10}$	$B^+ \to \pi^- \mu^+ \mu^+$	$< 1.4 \times 10^{-6}$
$K^+ \to \pi^- \mu^+ \mu^+$	$< 3.0 \times 10^{-9}$	$B^+ \to \pi^- e^+ \mu^+$	$< 1.3 \times 10^{-6}$
$K^+ \to \pi^- e^+ \mu^+$	$< 5.0 \times 10^{-10}$		

20.4 Flavor Changing Neutral Currents

We have seen in earlier chapters that the suppression of FCNC induced by $s \to d$ and $b \to s$ transitions at the quark level was a special feature of the standard model (SM). These processes could occur through higher order loop diagrams in the standard model with fine tuning of heavy flavor quark masses (Figure 20.3). These processes are also predicted in many extensions of the SM (Figure 20.4) like the Z, Z' models, leptoquark models, MSSM models with various extensions. An experimental observation of such decays would play an important role in studying the BSM physics.

20.4.1 Particle decay processes

FCNC processes are driven by quark level transitions like $s \to d\nu\bar{\nu}$ and $b \to s\nu\bar{\nu}$. Some of the FCNC processes in the strangeness sector that involve neutrinos are $K^\pm \to \pi^\pm\nu\bar{\nu}$, $K_L^0 \to \pi^0\nu\bar{\nu}$. Processes like $K^\pm \to \pi^+ e^\pm \mu^\mp$ and $K^- \to \pi^- e^\pm \mu^\mp$ involve FCNC as well as the lepton flavor violation (LFV) with $|\Delta L_e| = |\Delta L_\mu| = 1$. Processes like $K^\pm \to \pi^\pm\nu\bar{\nu}$ and $K_L^0 \to \pi^0\nu\bar{\nu}$ can occur in the standard model through the higher order loop diagrams involving $W(Z)$ bosons and/or u, c, t quarks in the intermediate state (Figure 20.3). Preliminary results from the NA62 experiment from CERN (Table 20.5) has generated great excitement in the study of these reactions; forthcoming experiments at JPARC would play a decisive role in studying the FCNC processes [1091].

The new contributions from BSM physics would add to the standard model contribution either coherently or incoherently, depending upon the structure of the model which could have

 i) Additional quarks or gauge bosons or Higgs (Figure 20.4(a)).

 ii) Additional light neutrinos ν_R or sterile neutrinos ν_S (Figure 20.4(b)).

 iii) New interaction with lepton flavor violation (LFV) (Figure 20.4(c)).

 iv) Leptoquark (Figure 20.4(d)).

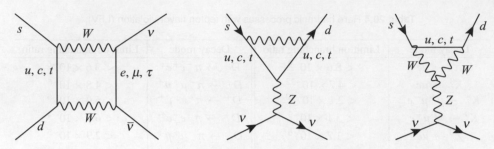

Figure 20.3 Flavor changing neutral current processes in higher order loop diagrams in SM.

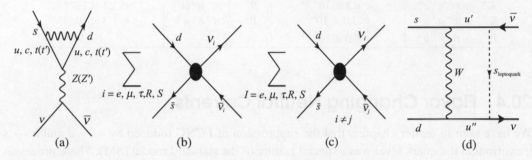

Figure 20.4 Flavor changing neutral current processes in some models beyond the standard model (BSM) physics.

Since the two neutrinos in the final state are not observed, these contributions would modify the decay rates. The existence of new particles and new interactions envisaged in the BSM physics models are constrained to satisfy well-established observations on the number of light neutrinos from LEP, that is, $n = 2.9840 \pm 0.0082$ or the limits on the strength of right-handed

Table 20.5 Recent experimental limits of FCNC kaon decays.

Decay mode	Present limit	Experiment
$K^+ \to \pi^+ \nu\bar{\nu}$	$< 1.85 \times 10^{-10}$	NA62 (2019)
	$< 2.88 \times 10^{-10}$	BNL787
$K^+ \to \pi^+ e^- \mu^+$	$< 1.3 \times 10^{-11}$	BNL 865

current couplings or the coupling of new Higgs or Z bosons. These BSM physics models are also applied to predict the rates for other FCNC decays like $K^\pm \to \pi^\pm e^+ \mu^-$ and $K^\pm \to \pi^\pm e^- \mu^+$.

20.4.2 Production of strange and heavy flavored hadrons in scattering experiments

The presence of FCNC can be experimentally tested in various scattering experiments induced by neutrinos, electrons, or in $e^- e^+$ collisions at LEP. In the low energy region, where MINERvA operates, the observation of processes like

$$\nu + p \quad \rightarrow \quad \nu + \Sigma^+$$
$$\nu + n \quad \rightarrow \quad \nu + \Lambda^0 \, (\Sigma^0)$$

or at the electron accelerators at JLAB and MAINZ [1092, 1093], processes like

$$e^- + p \quad \rightarrow \quad e^- \Sigma^+$$
$$e^- + D \quad \rightarrow \quad e^- \Lambda^0 p$$
$$e^- + D \quad \rightarrow \quad e^- n \Sigma^+ \tag{20.38}$$

could be searched for. In the very high energy region of LEP, the search for single top quark productions or leptoquark production induced by FCNC have been made. Any observation of such events would give information about the BSM physics through the study of FCNC.

20.5 Nonstandard Interaction (NSI) in High Precision Low Energy Weak Processes

In recent years, there has been remarkable progress in high precision experiments at very low energies involving nuclear and neutron β decays. This has renewed interest in examining the low energy effects of BSM in high precision weak processes at low energies. These efforts involve probing for the presence of exotic current couplings of scalar, tensor, pseudoscalar, and right-handed currents in β decays as well as the high precision study of symmetry properties of weak currents under discrete symmetries like parity (P), time reversal (T), charge conjugation (C), and CP symmetry [1094, 1095, 226, 1096]. Various beta decay parameters can be measured with high precision and the parameters which were earlier inaccessible or were measured with poor precision can be determined with high precision using polarized neutron and nuclear sources; these parameters can test various models proposed for physics beyond the standard model (BSM). In this low energy region, the weak interaction is described in terms of (u, d) quarks and (e, ν_e) and (μ, ν_μ) leptons. In the quark picture, the interaction Lagrangian $\mathcal{L}^{\text{eff}}$ for $d \rightarrow u e \nu$ including the nonstandard interactions can be written as [1094, 1095, 226]:

$$
\begin{aligned}
\mathcal{L}_{\text{eff}} = -\frac{G_F V_{ud}}{\sqrt{2}} \Big[\; & (1 + \epsilon_L) \; \bar{e}\gamma_\mu (1 - \gamma_5)\nu_e \cdot \bar{u}\gamma^\mu (1 - \gamma_5)d \; + \; \tilde{\epsilon}_L \, \bar{e}\gamma_\mu (1 + \gamma_5)\nu_e \cdot \bar{u}\gamma^\mu (1 - \gamma_5)d \\
& + \epsilon_R \, \bar{e}\gamma_\mu (1 - \gamma_5)\nu_e \cdot \bar{u}\gamma^\mu (1 + \gamma_5)d \; + \; \tilde{\epsilon}_R \, \bar{e}\gamma_\mu (1 + \gamma_5)\nu_e \cdot \bar{u}\gamma^\mu (1 + \gamma_5)d \\
& + \epsilon_T \, \bar{e}\sigma_{\mu\nu} (1 - \gamma_5)\nu_e \cdot \bar{u}\sigma^{\mu\nu} (1 - \gamma_5)d \; + \; \tilde{\epsilon}_T \, \bar{e}\sigma_{\mu\nu} (1 + \gamma_5)\nu_e \cdot \bar{u}\sigma^{\mu\nu} (1 + \gamma_5)d \\
& + \epsilon_S \, \bar{e}(1 - \gamma_5)\nu_e \cdot \bar{u}d \; + \; \tilde{\epsilon}_S \, \bar{e}(1 + \gamma_5)\nu_e \cdot \bar{u}d \\
& - \epsilon_P \, \bar{e}(1 - \gamma_5)\nu_e \cdot \bar{u}\gamma_5 d \; - \; \tilde{\epsilon}_P \, \bar{e}(1 + \gamma_5)\nu_e \cdot \bar{u}\gamma_5 d \; + \ldots \Big] + \text{h.c.} ,
\end{aligned}
\tag{20.39}
$$

The aforementioned Lagrangian depicts the SM $(V - A)$ Fermi interaction, with

$$V_{ud} = \cos\theta_c. \tag{20.40}$$

ϵ_i and $\tilde{\epsilon}_i$ are complex coefficients which are functions of the masses and couplings of the new particles and depend upon the detailed predictions of the models proposed for BSM. In general, these coefficients are expressed as:

$$\epsilon_i, \tilde{\epsilon}_i \propto \left(\frac{M_W}{\Lambda}\right)^2, \tag{20.41}$$

where Λ is the energy scale of new physics (NP) and is of the order of $\Lambda \sim$ TeV. This means that $\epsilon_i, \tilde{\epsilon}_i \approx 10^{-3}$ [226].

The Fermi constant, obtained from muon decay, phenomenologically, is $G_F = 1.1663787$ $(6) \times 10^{-5}$ GeV^{-2}. The Fermi coupling G_F is also determined from the nuclear β decays using mainly the Fermi decays induced by the vector currents which would get additional contributions due to ϵ_L and ϵ_R terms in Eq. (20.39). Neglecting the ϵ_i' terms as they involve right-handed neutrinos, we, therefore, write

$$\tilde{V}_{ud} \equiv V_{ud}\left(1 + \epsilon_L + \epsilon_R\right)\left(1 - \frac{\delta G_F}{G_F}\right), \tag{20.42}$$

where δG_F is the change in G_F due to other corrections in SM and new physics (NP) due to BSM. The effective Lagrangian is rewritten to first order in ϵ_i as:

$$\begin{aligned}
\mathcal{L}_{\text{eff}} = & -\frac{G_F \tilde{V}_{ud}}{\sqrt{2}} \left\{ \bar{e}\gamma_\mu(1-\gamma_5)\nu_e \cdot \bar{u}\gamma^\mu \left[1 - (1-2\epsilon_R)\gamma_5\right] d \right. \\
& + \epsilon_S\, \bar{e}(1-\gamma_5)\nu_e \cdot \bar{u}d - \epsilon_P\, \bar{e}(1-\gamma_5)\nu_e \cdot \bar{u}\gamma_5 d + \epsilon_T\, \bar{e}\sigma_{\mu\nu}(1-\gamma_5)\nu_e \cdot \bar{u} \\
& \left. \sigma^{\mu\nu}(1-\gamma_5)d \right\} + \text{h.c.}
\end{aligned} \tag{20.43}$$

The matrix elements of the hadronic current in this equation between the nucleon states are defined as (Chapter 10)

$$\langle p(p_p)| \, \bar{u}\gamma_\mu d \, |n(p_n)\rangle = \bar{u}_p(p_p)\left[f_1(q^2)\,\gamma_\mu + \frac{f_2(q^2)}{2M_N}\sigma_{\mu\nu}q^\nu + \frac{f_3(q^2)}{2M_N}q_\mu\right]u_n(p_n), \tag{20.44a}$$

$$\langle p(p_p)| \, \bar{u}\gamma_\mu\gamma_5 d \, |n(p_n)\rangle = \bar{u}_p(p_p)\left[g_1(q^2)\gamma_\mu + \frac{g_2(q^2)}{2M_N}\sigma_{\mu\nu}q^\nu + \frac{g_3(q^2)}{2M_N}q_\mu\right]\gamma_5 u_n(p_n), \tag{20.44b}$$

$$\langle p(p_p)| \, \bar{u} d \, |n(p_n)\rangle = g_S(0)\,\bar{u}_p(p_p)\,u_n(p_n) + \mathcal{O}(q^2/M_N^2), \tag{20.44c}$$

$$\langle p(p_p)| \, \bar{u}\gamma_5 d \, |n(p_n)\rangle = g_P(0)\,\bar{u}_p(p_p)\,\gamma_5\,u_n(p_n) + \mathcal{O}(q^2/M_N^2), \tag{20.44d}$$

$$\langle p(p_p)| \, \bar{u}\sigma_{\mu\nu} d \, |n(p_n)\rangle = g_T(0)\,\bar{u}_p(p_p)\,\sigma_{\mu\nu}u_n(p_n) + \mathcal{O}(q/M_N), \tag{20.44e}$$

where $f_i(q^2), g_i(q^2)$ $(i = 1, 2, 3, S, P, T)$ are the form factors. It should be noted that $f_2(q^2)$, $f_3(q^2), g_2(q^2), g_3(q^2)$ are additional couplings induced by the strong interactions of nucleons which may get additional contributions from $g_S, g_P,$ and g_T, which are new contributions due to NP only.

In the limit $q^2 \to 0$, corresponding to β decays, in which only S,V,T,A,P form factors at $q^2 = 0$ make significant contributions (see Chapters 4 and 10), we can write

$$-\mathcal{L}_{n \to p e^- \bar{\nu}_e} = \bar{p}\, n \, \left(C_S \bar{e} \nu_e - C_S' \bar{e} \gamma_5 \nu_e \right) + \bar{p} \gamma^\mu n \left(C_V \bar{e} \gamma_\mu \nu_e - C_V' \bar{e} \gamma_\mu \gamma_5 \nu_e \right)$$

$$+ \frac{1}{2} \bar{p} \sigma^{\mu\nu} n \left(C_T \bar{e} \sigma_{\mu\nu} \nu_e - C_T' \bar{e} \sigma_{\mu\nu} \gamma_5 \nu_e \right) - \bar{p} \gamma^\mu \gamma_5 n \left(C_A \bar{e} \gamma_\mu \gamma_5 \nu_e - C_A' \bar{e} \gamma_\mu \nu_e \right)$$

$$+ \bar{p} \gamma_5 n \, \left(C_P \bar{e} \gamma_5 \nu_e - C_P' \bar{e} \nu_e \right) + \text{h.c.} \tag{20.45}$$

where C_i and $C_i' (i = S, V, T, A, P)$ are expressed in terms of ϵ_i and $\tilde{\epsilon}_i (i = S, P, T, A, P)$. That is,

$$C_i(C_i') = \frac{G_F \cos \theta_C}{\sqrt{2}} \bar{C}_i(\bar{C}_i'), \tag{20.46}$$

and $\bar{C}_i(\bar{C}_i')$ are expressed in terms of ϵ_i and $\tilde{\epsilon}_i$ as [226]

$$\begin{aligned}
\overline{C}_V + \overline{C}_V' &= 2 f_1 (1 + \epsilon_L + \epsilon_R) & \overline{C}_V - \overline{C}_V' &= 2 f_1 (\tilde{\epsilon}_L + \tilde{\epsilon}_R) \\
\overline{C}_A + \overline{C}_A' &= -2 g_1 (1 + \epsilon_L - \epsilon_R) & \overline{C}_A - \overline{C}_A' &= 2 g_1 (\tilde{\epsilon}_L - \tilde{\epsilon}_R) \\
\overline{C}_S + \overline{C}_S' &= 2 g_S \epsilon_S & \overline{C}_S - \overline{C}_S' &= 2 g_S \tilde{\epsilon}_S \\
\overline{C}_P + \overline{C}_P' &= 2 g_P \epsilon_P & \overline{C}_P - \overline{C}_P' &= -2 g_P \tilde{\epsilon}_P \\
\overline{C}_T + \overline{C}_T' &= 8 g_T \epsilon_T & \overline{C}_T - \overline{C}_T' &= 8 g_T \tilde{\epsilon}_T,
\end{aligned} \tag{20.47}$$

The various observables studied in the β decays of polarized neutrons and polarized nuclei like the decay rates, angular and energy distributions of electrons (protons), various spin correlations between the spins of the decaying neutron and outgoing electrons and neutrinos, helicity of (anti)neutrino and polarization of electrons (positrons) etc. as discussed in Chapter 4, are used to determine the coefficients C_i and C_i' which would give the values of ϵ_L and ϵ_R. The SM without right handed currents ($\tilde{\epsilon}_i = 0$) predicts that

$$C_V = C_V', \quad C_A = C_A', C_i(C_i') = 0 \quad (i \neq V, A). \tag{20.48}$$

The most recent values of these coupling constants assuming no right-handed currents are given in Table 20.6, with $|C_V| = 0.98554 \frac{G_F}{\sqrt{2}}$: The nuclear β decays do not give any limits

Table 20.6 Recent values of various couplings C_i ($i = V, A, S, T$) determined from β decays which taken from [226].

$\text{Re}\left(\frac{C_A}{C_V}\right) = -1.27290(17)$	$\text{Im}\left(\frac{C_A}{C_V}\right) = -0.00034(59)$
$\text{Re}\left(\frac{C_S}{C_V}\right) = 0.0014(12)$	$\text{Im}\left(\frac{C_S}{C_V}\right) = -0.007(30)$
$\text{Re}\left(\frac{C_T}{C_V}\right) = 0.0020(22)$	$\text{Im}\left(\frac{C_T}{C_V}\right) = 0.0004(33)$

on C_P which is obtained from other probes using particle decays and production of hadrons in electroweak interactions, which also provide limits on C_T and C_S. From these values of C_i and

C_i', limits on ϵ_i and $\tilde{\epsilon}_i$ are derived. Experiments in the high energy scattering experiments at LEP and LHC also provide limits on ϵ_i and $\tilde{\epsilon}_i$. Analysis with nonzero values for the right-handed current couplings are also performed in most of these β decay experiments; they are described in detail by Dolinski et al. [1070] and Cirigliano et al. [1094, 1095]. The bounds on scalar and tensor couplings for the β decays are $|\tilde{\epsilon}_S| < 0.0063$ and $0.006 < \tilde{\epsilon}_T < 0.024$. The limits from the other experiments at LEP and LHC on new interactions are:

$$|\epsilon_{S,P}|, |\tilde{\epsilon}_{S,P}| < 5.8 \times 10^{-3}; \qquad |\epsilon_T|, |\tilde{\epsilon}_T| < 1.3 \times 10^{-3}; \qquad |\epsilon_R| < 2.2 \times 10^{-3} \tag{20.49}$$

A discussion of these results is beyond the scope of this book.

20.6 Summary

The present experimental efforts in observing neutrinoless double beta decay in many nuclei leading to the sucessful observation of these decays will go a long way in determining the properties of neutrinos regarding their mass and nature of being Dirac or Majorana type. Various experimental efforts at LEP of producing tau particles and gauge bosons and their rare decays as well as the production of kaons and other hadrons with heavy flavor contents at CERN and their decay governed by FCNC processes will also provide valuable information in distinguishing among the different models of BSM physics. In the field of lepton physics, intense beams of muons to be made available at various laboratories may also help in observing many rare decays of muons and muon capture processes in nuclei with LFV. In view of this, we expect very exciting times in the study of BSM physics in the near future.

Appendix A

Lorentz Transformation and Covariance of the Dirac Equation

A.1 Lorentz Transformations

Lorentz transformation (L.T.) is the rotation in Minkowski space. Minkowski space is the mathematical representation of the space time in Einstein's theory of relativity. Unlike the Euclidean space, Minkowski space treats space and time on a different footing and the space time interval between the two events would be the same in all frames of reference.

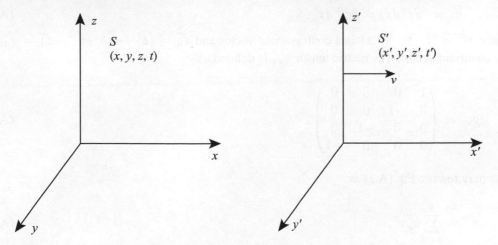

Figure A.1 Lorentz transformation.

If an event is observed in an inertial frame S at the coordinates (x, y, z, t) (Fig.A.1) and in another inertial frame S' at the coordinates (x', y', z', t') and the frame S' is moving with a constant velocity v with respect to S in the direction of x, then the measurements in S and S' are related by

$$x' = \gamma(x - vt), \qquad\qquad\qquad x = \gamma(x' + vt'),$$
$$y' = y, \qquad\qquad\qquad\qquad\quad y = y',$$

$$z' = z, \qquad\qquad\qquad\qquad\quad z = z',$$
$$t' = \gamma(t - vx), \qquad\qquad\qquad t = \gamma(t' + vx'), \qquad\qquad \text{(A.1)}$$

where $\gamma = \dfrac{1}{\sqrt{1 - \frac{v^2}{c^2}}}$ and $c = 1$.

The L.H.S. of these equations describe the transformation of the coordinates from S to S' and the R.H.S. of the equations describe the transfer of coordinates from S' to S. Equation (A.1) may also be written as

$$\begin{bmatrix} t' \\ x' \\ y' \\ z' \end{bmatrix} = \begin{bmatrix} \gamma & -\gamma v & 0 & 0 \\ -\gamma v & \gamma & 0 & 0 \\ 0 & 0 & 1 & 0 \\ 0 & 0 & 0 & 1 \end{bmatrix} \begin{bmatrix} t \\ x \\ y \\ z \end{bmatrix}. \qquad\qquad \text{(A.2)}$$

The invariant length element squared ds^2 is defined as

$$\begin{aligned} ds^2 &= c^2 dt^2 - dx^2 - dy^2 - dz^2, \\ &= c^2 dt'^2 - dx'^2 - dy'^2 - dz'^2, \\ &= g_{\mu\nu} dx^\mu dx^\nu = g_{\mu\nu} dx'^\mu dx'^\nu, \\ &= dx^\mu dx_\mu = dx'^\mu dx'_\mu, \end{aligned} \qquad\qquad \text{(A.3)}$$

where $x^\mu = \begin{pmatrix} t & x & y & z \end{pmatrix}$ is a contravariant vector and $x_\mu = \begin{pmatrix} t & -x & -y & -z \end{pmatrix} = g_{\mu\nu} x^\nu$ is a covariant vector. The metric tensor $g_{\mu\nu}$ is defined as

$$g_{\mu\nu} = \begin{pmatrix} 1 & 0 & 0 & 0 \\ 0 & -1 & 0 & 0 \\ 0 & 0 & -1 & 0 \\ 0 & 0 & 0 & -1 \end{pmatrix}. \qquad\qquad \text{(A.4)}$$

One may rewrite Eq. (A.2) as

$$x'^\mu = \sum_{\nu=0}^{3} a^\mu_\nu x^\nu, \qquad\qquad \text{(A.5)}$$

where a^μ_ν is the Lorentz transformation matrix. Any set of quantities which have four components and transform like Eq. (A.5) under L.T. form a four vector, that is,

$$A'^\mu = \sum_{\nu=0}^{3} \Lambda^\mu_\nu A^\nu, \qquad\qquad \text{(A.6)}$$

with the properties

$$\Lambda_\nu^\mu \Lambda_\sigma^\nu = \delta_\sigma^\mu \quad \text{and} \quad \det\left(\Lambda_\nu^\mu\right) = \pm 1.$$

For an infinitesimal L.T., Λ_ν^μ may also be written as $\Lambda_\nu^\mu = \delta_\nu^\mu + \omega_\nu^\mu$. The tensor ω_ν^μ generates the infinitesimal L.T. and it is considered to be a quantity that is smaller than unity. The major contributor is δ_ν^μ (unit matrix) and ω_ν^μ generates infinitesimal boost or infinitesimal rotation or both. This can be understood as demonstrated in the next section.

Rotation of coordinates around the z-axis in 3-D

If we consider the rotation around the z-axis by an angle θ, then

$$
\begin{aligned}
x_1' &= \cos\theta\, x_1 - \sin\theta\, x_2, \\
x_2' &= \sin\theta\, x_1 + \cos\theta\, x_2, \\
x_3' &= x_3.
\end{aligned}
\tag{A.7}
$$

In the matrix form, Eq. (A.7) can be written as

$$
\begin{bmatrix} x_1' \\ x_2' \\ x_3' \end{bmatrix} = \begin{bmatrix} \cos\theta & -\sin\theta & 0 \\ \sin\theta & \cos\theta & 0 \\ 0 & 0 & 1 \end{bmatrix} \begin{bmatrix} x_1 \\ x_2 \\ x_3 \end{bmatrix}.
$$

For an infinitesimal rotation, that is, $(\theta \to 0), \cos\theta \to 1, \sin\theta \to \theta$

$$
\begin{aligned}
\begin{bmatrix} x_1' \\ x_2' \\ x_3' \end{bmatrix} &= \begin{bmatrix} 1 & -\theta & 0 \\ \theta & 1 & 0 \\ 0 & 0 & 1 \end{bmatrix} \begin{bmatrix} x_1 \\ x_2 \\ x_3 \end{bmatrix}, \\
&= \left[\begin{pmatrix} 1 & 0 & 0 \\ 0 & 1 & 0 \\ 0 & 0 & 1 \end{pmatrix} + \begin{pmatrix} 0 & -\theta & 0 \\ \theta & 0 & 0 \\ 0 & 0 & 0 \end{pmatrix} \right] \begin{bmatrix} x_1 \\ x_2 \\ x_3 \end{bmatrix}, \\
x_i' &= (\delta_j^i + \epsilon_j^i) x_j,
\end{aligned}
\tag{A.8}
$$

where ϵ_j^i represents an infinitesimal rotation if $\theta = \epsilon$. Equation (A.7) can be rewritten as

$$
\begin{aligned}
x_1' &= x_1 - \epsilon x_2, \\
x_2' &= \epsilon x_1 + x_2, \\
x_3' &= x_3.
\end{aligned}
\tag{A.9}
$$

The matrix representing the infinitesimal change is antisymmetric. Reconsidering Eq. (A.2),

$$
\begin{bmatrix} x'^0 \\ x'^1 \\ x'^2 \\ x'^3 \end{bmatrix} = \begin{bmatrix} \gamma & -\gamma\beta & 0 & 0 \\ -\gamma\beta & \gamma & 0 & 0 \\ 0 & 0 & 1 & 0 \\ 0 & 0 & 0 & 1 \end{bmatrix} \begin{bmatrix} x^0 \\ x^1 \\ x^2 \\ x^3 \end{bmatrix},
$$

$$
= \begin{bmatrix} \cosh\omega & -\sinh\omega & 0 & 0 \\ -\sinh\omega & \cosh\omega & 0 & 0 \\ 0 & 0 & 1 & 0 \\ 0 & 0 & 0 & 1 \end{bmatrix} \begin{bmatrix} x^0 \\ x^1 \\ x^2 \\ x^3 \end{bmatrix},
$$

where $\cosh\omega = \gamma = \frac{1}{\sqrt{1-\beta^2}}$ and $\sinh\omega = \gamma\beta = \frac{\beta}{\sqrt{1-\beta^2}}$. Now $x'^\mu = \Lambda^\mu_\nu x^\nu$ and Λ^μ_ν is given by

$$
\Lambda^\mu_\nu = \begin{bmatrix} \cosh\omega & -\sinh\omega & 0 & 0 \\ -\sinh\omega & \cosh\omega & 0 & 0 \\ 0 & 0 & 1 & 0 \\ 0 & 0 & 0 & 1 \end{bmatrix}. \tag{A.10}
$$

For an infinitesimal boost or rotation, that is,

$$
\Lambda^\mu_\nu = \begin{bmatrix} 1 & -\epsilon & 0 & 0 \\ -\epsilon & 1 & 0 & 0 \\ 0 & 0 & 1 & 0 \\ 0 & 0 & 0 & 1 \end{bmatrix} = \delta^\mu_\nu + \epsilon^\mu_\nu. \tag{A.11}
$$

Therefore, Eq. (A.11) implies that

$$
\epsilon^{\mu\nu} = g^{\nu\lambda}\epsilon^\mu_\lambda = \begin{bmatrix} 0 & +\epsilon & 0 & 0 \\ -\epsilon & 0 & 0 & 0 \\ 0 & 0 & 0 & 0 \\ 0 & 0 & 0 & 0 \end{bmatrix} = -\epsilon^{\nu\mu}
$$

$$
x'^\mu = \Lambda^\mu_\nu x^\nu = (\delta^\mu_\nu + \epsilon^\mu_\nu)x^\nu,
$$

$$
\Rightarrow \qquad g_{\mu\nu}x'^\mu x'^\nu = g_{\mu\nu}x^\mu x^\nu,
$$

$$
\Rightarrow \qquad g_{\mu\nu}\Lambda^\mu_\rho\Lambda^\nu_\sigma x^\rho x^\sigma = g_{\rho\sigma}x^\rho x^\sigma,
$$

$$
\Rightarrow \qquad \Lambda^\rho_\mu\Lambda^\mu_\sigma = \delta^\rho_\sigma.
$$

Now,

$$
\begin{aligned}
x'_\mu x'^\mu &= \Lambda^\alpha_\mu x_\alpha \Lambda^\mu_\beta x^\beta, \\
&= (\delta^\alpha_\mu + \epsilon^\alpha_\mu)x_\alpha(\delta^\mu_\beta + \epsilon^\mu_\beta)x^\beta, \\
&= (\delta^\alpha_\mu\delta^\mu_\beta + \epsilon^\alpha_\mu\epsilon^\mu_\beta + \delta^\alpha_\mu\epsilon^\mu_\beta + \epsilon^\alpha_\mu\delta^\mu_\beta)x_\alpha x^\beta. \tag{A.13}
\end{aligned}
$$

The second term on the R.H.S. of Eq. (A.13) can be neglected as it is the product of two infinitesimal numbers, that is,

$$x'_\mu x'^\mu = x_\alpha x^\alpha + (\epsilon_\alpha^\beta + \epsilon_\beta^\alpha)x_\alpha x^\beta.$$

Since $x'_\mu x'^\mu$ should be an invariant quantity,

$$(\epsilon_\alpha^\beta + \epsilon_\beta^\alpha) = 0,$$
$$\epsilon_\beta^\alpha = -\epsilon_\alpha^\beta, \tag{A.14}$$

that is, Eq. (A.14) shows antisymmetry. Therefore, the generator of an L.T. must be an antisymmetric matrix.

A.2 Covariance of Dirac Equation

Physical observables in all the inertial systems are the same. Therefore, it is important that the Dirac equation, upon which our physical interpretation is based, must be covariant under L.T. To prove that the Dirac equation is covariant under L.T., consider a spin $\frac{1}{2}$ particle (fermion) moving in space and an observer in the rest frame of reference (S), making measurements and determining that the properties of this fermion are described by the Dirac equation

$$(i\hbar\gamma^\mu\partial_\mu - m_0 c)\psi(x) = 0. \tag{A.15}$$

Covariance of Dirac equation means the following:

1. There must be an explicit rule to enable the observer in S' to calculate $\psi'(x')$, if $\psi(x)$ of the observer in S is given. Hence, $\psi'(x')$ of the S' frame describes a physical state of the fermion, as $\psi(x)$ describes the physical state in the S frame.

2. $\psi'(x')$ must be a solution of the Dirac equation in S', having the form

   $$(i\hbar\gamma'^\mu\partial'_\mu - m_0 c)\psi'(x') = 0, \tag{A.16}$$

   with γ'^μ satisfying the relations $\gamma'^\mu\gamma'^\nu + \gamma'^\nu\gamma'^\mu = 2g^{\mu\nu}$; $\gamma'^{0\dagger} = \gamma'^0$; $\gamma'^{02} = 1$ and $\gamma'^{i\dagger} = -\gamma'^i$, where $i = 1, 2, 3$. Rewriting Eq. (A.16), we obtain

   $$i\gamma'^0\frac{\partial\psi'(x')}{\partial t'} = \left(-i\gamma'^k\frac{\partial}{\partial x'^k} + m_0\right)\psi'(x'). \tag{A.17}$$

   Multiplying by γ'^0 from the L.H.S., we obtain

   $$i\frac{\partial\psi'(x')}{\partial t'} = \left(-i\gamma'^0\gamma'^k\frac{\partial}{\partial x'^k} + m_0\gamma'^0\right)\psi'(x').$$

This is an equation similar to the Schrodinger equation. Therefore

$$i\frac{\partial \psi'(x')}{\partial t'} = \left(-i\gamma'^0\gamma'^k\frac{\partial}{\partial x'^k} + m_0\gamma'^0\right)\psi'(x') = \hat{H}'\psi', \tag{A.18}$$

where

$$\hat{H}' = -i\gamma'^0\gamma'^k\frac{\partial}{\partial x'^k} + m_0\gamma'^0. \tag{A.19}$$

This should be a Hermitian to get real eigenvalues, that is,

$$(\hat{H}')^\dagger = \hat{H}'. \tag{A.20}$$

It can be shown that all 4×4 γ'^μ matrices which satisfy all the properties obeyed by γ^μ are identical up to a unitary transformation, that is,

$$\gamma'^\mu = \hat{U}^\dagger \gamma^\mu \hat{U}, \qquad \hat{U}^\dagger = \hat{U}^{-1}. \tag{A.21}$$

Since the observables do not change under unitary transformation, for all purposes, we may replace γ'^μ by γ^μ. Now we construct the transformation between $\psi(x)$ and $\psi'(x')$. This transformation is required to be linear, since both the Dirac equation and the L.T. are linear in the space time coordinates. This requires that

$$\psi'(x') = \psi'(\hat{\Lambda}x) = \hat{S}(\hat{\Lambda})\psi(x) = \hat{S}(\hat{\Lambda})\psi(\hat{\Lambda}^{-1}x'), \tag{A.22}$$

where we have used $x = \hat{\Lambda}^{-1}x'$ as $x' = \hat{\Lambda}x$, since

$$x'^\mu = \Lambda^\mu_\nu x^\nu, \qquad \Lambda^\mu_\nu = \frac{\partial x'^\mu}{\partial x^\nu}. \tag{A.23}$$

Moreover, $\psi(x)$ may be written as

$$\psi(x) = \hat{S}^{-1}(\hat{\Lambda})\psi'(x') = \hat{S}^{-1}(\hat{\Lambda})\psi'(\hat{\Lambda}x) = \hat{S}(\hat{\Lambda}^{-1})\psi'(\hat{\Lambda}x). \tag{A.24}$$

Thus, the Dirac equation in frame S, that is,

$$\left(i\gamma^\mu\frac{\partial}{\partial x^\mu} - m_0\right)\psi(x) = 0,$$

can also be written as

$$\left(i\gamma^\mu\hat{S}^{-1}(\hat{\Lambda})\frac{\partial}{\partial x^\mu} - m_0 S^{-1}(\hat{\Lambda})\right)\psi'(x') = 0. \tag{A.25}$$

Multiplying L.H.S. of Eq. (A.25) by $\hat{S}(\hat{\Lambda})$ and using $\hat{S}(\hat{\Lambda})\hat{S}^{-1}(\hat{\Lambda}) = 1$, results in

$$\left(i\hat{S}(\hat{\Lambda})\gamma^\mu\hat{S}^{-1}(\hat{\Lambda})\frac{\partial}{\partial x^\mu} - m_0\right)\psi'(x') = 0, \tag{A.26}$$

and

$$\frac{\partial}{\partial x^\mu} = \frac{\partial x'^\nu}{\partial x^\mu}\frac{\partial}{\partial x'^\nu} = \Lambda^\nu_\mu \frac{\partial}{\partial x'^\nu}.$$

We may write Eq. (A.26) as

$$\left[i\left(\hat{S}(\hat{\Lambda})\gamma^\mu \hat{S}^{-1}(\hat{\Lambda})\Lambda^\nu_\mu\right)\frac{\partial}{\partial x'^\nu} - m_0 \right]\psi'(x') = 0. \tag{A.27}$$

Now if the Dirac equation is to be invariant under L.T., then we must identify

$$\hat{S}(\hat{\Lambda})\gamma^\mu \hat{S}^{-1}(\hat{\Lambda})\Lambda^\nu_\mu = \gamma^\nu, \tag{A.28}$$

$$\hat{S}(\hat{\Lambda})\gamma^\mu \hat{S}^{-1}(\hat{\Lambda}) = \Lambda^\mu_\nu \gamma^\nu, \tag{A.29}$$

$$S^{-1}(\hat{\Lambda})\gamma^\nu \hat{S}(\hat{\Lambda}) = \Lambda^\nu_\mu \gamma^\mu. \tag{A.30}$$

There must exist a matrix connecting the two representations γ^μ and γ'^μ. It turns out that this matrix does exist and one of the possible forms could be

$$S(\Lambda) = e^{-\frac{i}{4}\sigma^{\mu\nu}\epsilon_{\mu\nu}},$$

where

$$\Lambda^\nu_\mu = \delta^\nu_\mu + \epsilon^\nu_\mu, \qquad \text{and} \qquad \sigma^{\mu\nu} = \frac{i}{2}[\gamma^\mu, \gamma^\nu]. \tag{A.31}$$

Moreover,

$$\hat{S}^{-1} = \gamma^0 \hat{S}^\dagger \gamma^0. \tag{A.32}$$

Another choice of S could be a 4×4 matrix given by

$$S = \begin{pmatrix} a_+ & 0 & 0 & a_- \\ 0 & a_+ & a_- & 0 \\ 0 & a_- & a_+ & 0 \\ a_- & 0 & 0 & a_+ \end{pmatrix} = \begin{pmatrix} a_+ I_{2\times2} & a_- \sigma_1 \\ a_- \sigma_1 & a_+ I_{2\times2} \end{pmatrix}, \tag{A.33}$$

where

$$a_+ = \sqrt{\frac{1}{2}(\gamma+1)}, \qquad a_- = -\sqrt{\frac{1}{2}(\gamma-1)},$$

$$I_{2\times2} = \begin{pmatrix} 1 & 0 \\ 0 & 1 \end{pmatrix}, \qquad \sigma_1 = \begin{pmatrix} 0 & 1 \\ 1 & 0 \end{pmatrix}, \qquad \text{and} \quad \gamma = \frac{1}{\sqrt{1-\frac{v^2}{c^2}}}.$$

The components of the Dirac spinors do not transform as a four vector when one goes from one inertial system to another

$$\psi'(x') = \hat{S}(\hat{\Lambda})\psi(x).$$

$\psi^\dagger \psi$ is not L.I. as

$$
\begin{aligned}
\psi^\dagger \psi &= |\psi_1|^2 + |\psi_2|^2 + |\psi_3|^2 + |\psi_4|^2, \\
(\psi^\dagger \psi)' &= (\psi'^\dagger)\psi', \\
&= (S\psi)^\dagger(S\psi) = \psi^\dagger S^\dagger S\psi, \\
&\neq \psi^\dagger \psi \qquad \text{since }, \ S^\dagger S \neq 1.
\end{aligned}
\tag{A.34}
$$

$$
\Rightarrow \qquad
\begin{aligned}
S^\dagger &\neq S^{-1}, \\
\gamma^0 S^\dagger \gamma^0 &= S^{-1}, \\
\gamma^0 &= \begin{pmatrix} I & 0 \\ 0 & -I \end{pmatrix}.
\end{aligned}
\tag{A.35}
$$

It turns out that in the case of spinors, we need minus signs for the third and fourth components, for Lorentz invariance and for this, we introduce the adjoint of the Dirac spinor

$$
\overline{\psi} = \psi^\dagger \gamma^0 = \begin{pmatrix} \psi_1^* & \psi_2^* & -\psi_3^* & -\psi_4^* \end{pmatrix},
$$

where

$$
\begin{aligned}
\overline{\psi}\psi &= |\psi_1|^2 + |\psi_2|^2 - |\psi_3|^2 - |\psi_4|^2, \\
(\overline{\psi}\psi)' &= (\psi^\dagger \gamma^0 \psi)' = \psi'^\dagger \gamma^0 \psi', \\
&= (S\psi)^\dagger \gamma^0 (S\psi) = \psi^\dagger S^\dagger \gamma^0 S\psi,
\end{aligned}
\tag{A.36}
$$

$$
\begin{aligned}
&= \psi^\dagger \gamma^0 \gamma^0 S^\dagger \gamma^0 S\psi, \\
&= \overline{\psi} S^{-1} S\psi, \\
&= \overline{\psi}\psi.
\end{aligned}
\tag{A.37}
$$

Therefore, $\overline{\psi}\psi$ is invariant under Lorentz transformation.

Now, we look for the four current density under Lorentz transformation

$$
\begin{aligned}
j'^\mu(x') &= \overline{\psi}'(x')\gamma^\mu \psi'(x'), \\
&= \psi'^\dagger(x')\gamma^0 \gamma^\mu \psi'(x'), \\
&= \left[\hat{S}(\hat{\Lambda})\psi(x)\right]^\dagger \gamma^0 \gamma^\mu \left[\hat{S}(\hat{\Lambda})\psi(x)\right], \\
&= \psi^\dagger(x)\gamma^0 \gamma^0 \hat{S}^\dagger(\hat{\Lambda})\gamma^0 \gamma^\mu \hat{S}(\hat{\Lambda})\psi(x), \\
&= \overline{\psi}(x)\hat{S}^{-1}(\hat{\Lambda})\gamma^\mu \hat{S}(\hat{\Lambda})\psi(x), \\
&= \overline{\psi}(x)\Lambda^\mu_\nu \gamma^\nu \psi(x), \\
&= \Lambda^\mu_\nu \overline{\psi}(x)\gamma^\nu \psi(x), \\
&= \Lambda^\mu_\nu j^\nu(x),
\end{aligned}
\tag{A.38}
$$

which transforms as a four vector.

$$
\begin{aligned}
\overline{\psi}'(x')\gamma^5\psi'(x') &= \psi'^\dagger(x')\gamma^0\gamma^5\psi'(x'), \\
&= \left[\hat{S}\psi(x)\right]^\dagger \gamma^0\gamma^5 \left[\hat{S}\psi(x)\right], \\
&= \psi^\dagger(x)\gamma^0\gamma^0\hat{S}^\dagger\gamma^0\gamma^5\hat{S}\psi(x), \\
&= \overline{\psi}(x)\hat{S}^{-1}\gamma^5\hat{S}\psi(x), \\
&= \overline{\psi}(x)\hat{S}^{-1}\hat{S}\gamma^5\psi(x), \qquad (\gamma^5\hat{S} = \hat{S}\gamma^5) \\
&= \overline{\psi}(x)\gamma^5\psi(x), \tag{A.39}
\end{aligned}
$$

which is invariant under L.T.

   Similarly, one can show that

1. $\overline{\psi}'(x')\gamma_5\gamma^\mu\psi'(x') = \det(\Lambda)\Lambda^\mu_\nu\overline{\psi}(x)\gamma_5\gamma^\mu\psi(x)$ transforms as a pseudovector.

2. $\overline{\psi}'(x')\hat{\sigma}^{\mu\nu}\psi'(x') = \Lambda^\mu_\alpha\Lambda^\nu_\beta\overline{\psi}(x)\hat{\sigma}^{\alpha\beta}\psi(x)$ is a tensor of rank two, with

$$
\hat{\sigma}_{\mu\nu} = \frac{i}{2}[\gamma_\mu, \gamma_\nu] = \frac{i}{2}(\gamma_\mu\gamma_\nu - \gamma_\nu\gamma_\mu)
$$

A.3   Bilinear Covariants

The transition matrix element for any given physical process is a scalar quantity. The matrix element is written using currents involved in the leptonic and hadronic vertices. These currents are obtained using the Lagrangian which involves nature and strength of the interaction. Therefore, in order to construct these currents, one needs bilinear covariants such that the current takes the following form

$$
j_\mu = \overline{\psi}(p') \, (\text{bilinear covariants}) \, \psi(p).
$$

Since $\overline{\psi}$ is a 1×4 matrix and ψ is a 4×1 matrix, in order to make j_μ a constant, these bilinear covariants must be some 4×4 matrices. These bilinear covariants are nothing but different combinations of γ matrices, which gives five types of bilinear covariants depending upon their behavior under Lorentz (described here earlier) and parity (Chapter 2) transformations. The quantities which are invariant under L.T. as well as under parity transformation (P.T.) are called scalars or pure scalars while the quantities which are invariant under L.T. but change sign under P.T. are called pseudoscalars. The quantities which transform like a four vector under L.T. and also change sign under P.T. are called vectors; the quantities which transform like a four vector under L.T. but do not change sign under P.T. are called pseudovectors or axial vectors. We also have quantities which transform like a tensor.

$$
\begin{aligned}
\overline{\psi}(\vec{x}, t) \, \sigma^{oi} \, \psi(\vec{x}, t) &\xrightarrow{\hat{P}} -\overline{\psi}(-\vec{x}, t) \, \sigma^{oi} \, \psi(-\vec{x}, t). \\
\overline{\psi}(\vec{x}, t) \, \sigma^{ij} \, \psi(\vec{x}, t) &\xrightarrow{\hat{P}} \overline{\psi}(-\vec{x}, t) \, \sigma^{ij} \, \psi(-\vec{x}, t).
\end{aligned}
$$

In total, there are sixteen components; the tensor component is

$$\sigma^{\mu\nu} = \frac{i}{2}(\gamma^\mu \gamma^\nu - \gamma^\nu \gamma^\mu).$$

$$\overline{\psi}(x)\, 1\, \psi(x) \quad \longrightarrow \quad \text{Scalar (one component)} \tag{A.40}$$

$$\overline{\psi}(x)\, \gamma^5\, \psi(x) \quad \longrightarrow \quad \text{Pseudoscalar (one component)} \tag{A.41}$$

$$\overline{\psi}(x)\, \gamma^\mu\, \psi(x) \quad \longrightarrow \quad \text{Vector (four components)} \tag{A.42}$$

$$\overline{\psi}(x)\, \gamma^\mu \gamma^5\, \psi(x) \quad \longrightarrow \quad \text{Axial vector (four components)} \tag{A.43}$$

$$\overline{\psi}(x)\, \sigma^{\mu\nu}\, \psi(x) \quad \longrightarrow \quad \text{Antisymmetric tensor (six components).} \tag{A.44}$$

A.4   Nonrelativistic Reduction

Nonrelativistic reduction is applied to the process where kinetic energy of the particles is very small as compared to the rest of the mass ($F_k << M$). For example, in the β-decay processes

$$n \quad \rightarrow \quad p + e^- + \overline{\nu}_e,$$
$$X \quad \rightarrow \quad Y + e^- + \overline{\nu}_e,$$

$E_n \approx M_n$ and $E_p \approx M_p$ or $E_X \approx M_X$ and $E_Y \approx M_Y$ and if four momenta for n or X is p and for proton or Y is p', then in the nonrelativistic limit

$$E + M \quad \simeq \quad 2E \text{ or } 2M,$$
$$E' + M \quad \simeq \quad 2E' \text{ or } 2M,$$

$$u(p) \quad = \quad \sqrt{\frac{E+M}{2M}} \begin{pmatrix} \chi^s \\ \frac{\vec{\sigma}\cdot\vec{p}}{2M}\chi^s \end{pmatrix},$$

$$\overline{u}(p') \quad = \quad \sqrt{\frac{E'+M}{2M}} \left(\chi^{s\dagger} \quad -\frac{\vec{\sigma}\cdot\vec{p}'}{2M}\chi^{s\dagger} \right).$$

1. Scalar interaction (I_4):

$$\overline{u}(p')u(p) \quad = \quad \sqrt{\frac{(E+M)(E'+M)}{2M.2M}} \left(\chi^{s\dagger} \quad -\frac{\vec{\sigma}\cdot\vec{p}'}{2M}\chi^{s\dagger} \right) \begin{pmatrix} \chi^s \\ \frac{\vec{\sigma}\cdot\vec{p}}{2M}\chi^s \end{pmatrix},$$

$$= \quad \left(\chi^{s\dagger}\chi^s - \frac{\vec{\sigma}\cdot\vec{p}'\vec{\sigma}\cdot\vec{p}}{4M_N^2}\chi^{s\dagger}\chi^s \right),$$

$$= \quad \left(\chi^{s\dagger}\chi^s - \frac{\vec{p}\cdot\vec{p}' + i\vec{\sigma}\cdot\vec{p}\times\vec{p}'}{4M_N^2}\chi^{s\dagger}\chi^s \right).$$

However, $\vec{p} \approx \vec{p}\,'$ (in the nonrelativistic limit), as kinetic energy of the outgoing hadron is almost zero which implies negligible momentum transfer. Therefore,

$$\overline{u}(p')u(p) \quad = \quad \left(1 - \frac{|\vec{p}|^2}{4M^2} \right) = \left(1 - \frac{E^2 - M_N^2}{4M^2} \right) \simeq 1. \tag{A.45}$$

2. Pseudoscalar interaction (γ_5):

$$\bar{u}(p')\gamma^5 u(p) = \left(\chi^{s\dagger} \quad -\frac{\vec{\sigma}\cdot\vec{p}'}{2M}\chi^{s\dagger}\right)\begin{pmatrix} 0 & I \\ I & 0 \end{pmatrix}\begin{pmatrix} \chi^s \\ \frac{\vec{\sigma}\cdot\vec{p}}{2M}\chi^s \end{pmatrix},$$

$$= \frac{\vec{\sigma}\cdot\vec{p}}{2M}\chi^{s\dagger}\chi^s - \frac{\vec{\sigma}\cdot\vec{p}}{2M}\chi^{s\dagger}\chi^s,$$

$$= 0. \tag{A.46}$$

Therefore, nonrelativistic reduction shows that the pseudoscalar term vanishes in β-decay, since the energy involved is very small.

3. Vector interaction (γ_μ): For the zeroth component, we substitute $\mu = 0$ in $\bar{u}(p)\gamma^\mu u(p)$,

$$\bar{u}(p')\gamma^0 u(p) = \left(\chi^{s\dagger} \quad -\frac{\vec{\sigma}\cdot\vec{p}'}{2M}\chi^{s\dagger}\right)\begin{pmatrix} I & 0 \\ 0 & -I \end{pmatrix}\begin{pmatrix} \chi^s \\ \frac{\vec{\sigma}\cdot\vec{p}}{2M}\chi^s \end{pmatrix},$$

$$= \left(\chi^{s\dagger}\chi^s + \frac{\vec{p}\cdot\vec{p}' + i\vec{\sigma}\cdot\vec{p}\times\vec{p}'}{4M_N^2}\chi^{s\dagger}\chi^s\right), = 1,$$

because $\vec{p} \approx \vec{p}'$ and $T_N' = 0$. Substituting $\mu = i$ in $\bar{u}(p)\gamma^\mu u(p)$, we have

$$\bar{u}(p')\gamma^i u(p) = \left(\chi^{s\dagger} \quad -\frac{\vec{\sigma}\cdot\vec{p}'}{2M}\chi^{s\dagger}\right)\begin{pmatrix} 0 & \sigma^i \\ -\sigma^i & 0 \end{pmatrix}\begin{pmatrix} \chi^s \\ \frac{\vec{\sigma}\cdot\vec{p}}{2M}\chi^s \end{pmatrix},$$

$$= \left(\chi^{s\dagger} \quad -\frac{\sigma_j p_j'}{2M}\chi^{s\dagger}\right)\begin{pmatrix} \frac{\sigma_i\sigma_l p_l}{2M}\chi^s \\ -\sigma_i\chi^s \end{pmatrix},$$

$$= \frac{\sigma_i\sigma_l p_l}{2M} + \frac{\sigma_j\sigma_i p_j'}{2M},$$

$$= \frac{\delta_{il} p_l + i\epsilon_{ilm}\sigma_m p_l}{2M} + \frac{\delta_{ji} p_j' + i\epsilon_{jik} p_j'\sigma_k}{2M},$$

$$= \frac{p_i + p_i'}{2M} + \frac{i(\epsilon_{ilm}\sigma_m p_l - \epsilon_{ijk} p_j'\sigma_k)}{2M}.$$

4. Axial vector interaction ($\gamma_\mu\gamma_5$): The zeroth component for the axial vector interaction given by $\bar{u}(p')\gamma^\mu\gamma_5 u(p)$, can be obtained as

$$\bar{u}(p')\gamma^0\gamma_5 u(p) = \left(\chi^{s\dagger} - \frac{\vec{\sigma}\cdot\vec{p}'}{2M}\chi^{s\dagger}\right)\begin{pmatrix} I & 0 \\ 0 & -I \end{pmatrix}\begin{pmatrix} 0 & I \\ I & 0 \end{pmatrix}\begin{pmatrix} \chi^s \\ \frac{\vec{\sigma}\cdot\vec{p}}{2M}\chi^s \end{pmatrix},$$

$$= \left(\chi^{s\dagger} - \frac{\vec{\sigma}\cdot\vec{p}'}{2M}\chi^{s\dagger}\right)\begin{pmatrix} \frac{\vec{\sigma}\cdot\vec{p}}{2M}\chi^s \\ -\chi^s \end{pmatrix},$$

$$= \frac{\vec{\sigma} \cdot (\vec{p} + \vec{p}')}{2E},$$

$$= \frac{\vec{\sigma} \cdot \vec{P}}{2M} \quad \text{(for } E = E' = M \text{ and } \vec{P} = \vec{p} + \vec{p}'.\text{)} \tag{A.47}$$

Similarly, for the ith component, we find

$$\bar{u}(p')\gamma_i\gamma_5 u(p) = \left(\chi^{s\dagger} - \frac{\vec{\sigma}\cdot\vec{p}'}{2M}\chi^{s\dagger} \right) \begin{pmatrix} 0 & \sigma_i \\ -\sigma_i & 0 \end{pmatrix} \begin{pmatrix} 0 & I \\ I & 0 \end{pmatrix} \begin{pmatrix} \chi^s \\ \frac{\vec{\sigma}\cdot\vec{p}}{2M}\chi^s \end{pmatrix},$$

$$= \left(\chi^{s\dagger} - \frac{\vec{\sigma}\cdot\vec{p}'}{2M}\chi^{s\dagger} \right) \begin{pmatrix} \sigma_i \chi^s \\ -\sigma_i \frac{\vec{\sigma}\cdot\vec{p}}{2M}\chi^s \end{pmatrix},$$

$$= \left(\vec{\sigma} + \frac{\vec{\sigma}(\vec{\sigma}\cdot\vec{p})(\vec{\sigma}\cdot\vec{p}')}{4E^2} \right) \quad \text{(for } E = E' = M\text{).} \tag{A.48}$$

5. Tensor interaction ($\sigma_{\mu\nu}q^\nu$): We expand $\sigma_{\mu\nu}q^\nu$ as

$$\sigma_{\mu\nu}\, q^\nu = \sigma_{\mu 0}\, q^0 - \sigma_{\mu i}\, q^i.$$

Since $q^0 = E - E' = 0$ for $E = E'$, only $\nu = i$ will contribute, that is,

$$\sigma_{\mu\nu}\, q^\nu = -\sigma_{\mu i}\, q^i = -\left[\sigma_{0i}\, q^i,\ \sigma_{ji}\, q^i \right]$$

$$= -i\left[\begin{pmatrix} 0 & \vec{\sigma}\cdot\vec{q} \\ \vec{\sigma}\cdot\vec{q} & 0 \end{pmatrix},\ \begin{pmatrix} \vec{\sigma}\times\vec{q} & 0 \\ 0 & \vec{\sigma}\times\vec{q} \end{pmatrix} \right]. \tag{A.49}$$

Hence,

$$i\,\bar{u}(p')\,\sigma_{\mu\nu}\, q^\nu u(p) = -i\left[\bar{u}(p')\,\sigma_{0i}\, q^i\, u(p),\ \bar{u}(p')\,\sigma_{ji}\, q^i\, u(p) \right]. \tag{A.50}$$

We have

$$-i\,\bar{u}(p')\,\sigma_{0i}\, q^i u(p) = -\left[\left(\chi^{s\dagger} - \frac{\vec{\sigma}\cdot\vec{p}'}{2M}\chi^{s\dagger} \right) \begin{pmatrix} 0 & \vec{\sigma}\cdot\vec{q} \\ \vec{\sigma}\cdot\vec{q} & 0 \end{pmatrix} \begin{pmatrix} \chi^s \\ \frac{\vec{\sigma}\cdot\vec{p}}{2M}\chi^s \end{pmatrix} \right]$$

$$= -\left[\frac{(\vec{\sigma}\cdot\vec{q})(\vec{\sigma}\cdot\vec{q})}{2M} \right] \tag{A.51}$$

and

$$-i\,\bar{u}(p')\,\sigma_{ji}\, q^j u(p) = -\left[\left(\chi^{s\dagger} - \frac{\vec{\sigma}\cdot\vec{p}'}{2M}\chi^{s\dagger} \right) \begin{pmatrix} \vec{\sigma}\times\vec{q} & 0 \\ 0 & \vec{\sigma}\times\vec{q} \end{pmatrix} \begin{pmatrix} \chi^s \\ \frac{\vec{\sigma}\cdot\vec{p}}{2M}\chi^s \end{pmatrix} \right]$$

$$= i\left[\left(\vec{\sigma}\times\vec{q} - \frac{(\vec{\sigma}\times\vec{q})(\vec{\sigma}\cdot\vec{p})(\vec{\sigma}\cdot\vec{p}')}{4M^2} \right) \right]. \tag{A.52}$$

Appendix B

Cabibbo Theory

B.1 Cabibbo Theory, SU(3) Symmetry, and Weak N–Y Transition Form Factors

For the $\Delta S = 0$ processes,

$$\nu_l(k) + n(p) \longrightarrow l^-(k') + p(p');$$ (B.1)

$$\bar{\nu}_l(k) + p(p) \longrightarrow l^+(k') + n(p'); \qquad l = e, \mu, \tau,$$ (B.2)

and for the $|\Delta S| = 1$ processes,

$$\bar{\nu}_l(k) + p(p) \longrightarrow l^+(k') + \Lambda(p'),$$ (B.3)

$$\bar{\nu}_l(k) + p(p) \longrightarrow l^+(k') + \Sigma^0(p'),$$ (B.4)

$$\bar{\nu}_l(k) + n(p) \longrightarrow l^+(k') + \Sigma^-(p'); \qquad l = e, \mu, \tau,$$ (B.5)

the matrix elements of the vector (V_μ) and the axial vector (A_μ) currents between a nucleon $N'(= p, n)$ or a hyperon $Y(= \Lambda, \Sigma^0$ and $\Sigma^-)$ and a nucleon $N = n, p$ are written as:

$$\langle Y(p')|V_\mu|N(p)\rangle = \bar{u}(p') \left[\gamma_\mu f_1^{NY}(q^2) + i\sigma_{\mu\nu} \frac{q^\nu}{M + M'} f_2^{NY}(q^2) \right.$$

$$\left. + \frac{2\,q_\mu}{M + M'} f_3^{NY}(q^2) \right] u(p),$$ (B.6)

and

$$\langle Y(p')|A_\mu|N(p)\rangle = \bar{u}(p') \left[\gamma_\mu \gamma_5 g_1^{NY}(q^2) + i\sigma_{\mu\nu} \frac{q^\nu}{M + M'} \gamma_5 g_2^{NY}(q^2) \right.$$

$$\left. + \frac{2\,q_\mu}{M + M'} \gamma_5 g_3^{NY}(q^2) \right] u(p),$$ (B.7)

where M and M' are the masses of the nucleon and hyperon, respectively. $f_1^{NY}(q^2)$, $f_2^{NY}(q^2)$, and $f_3^{NY}(q^2)$ are the vector, weak magnetic and induced scalar $N - Y$ transition form factors and $g_1^{NY}(q^2)$, $g_2^{NY}(q^2)$ and $g_3^{NY}(q^2)$ are the axial vector, induced tensor (or weak electric), and induced pseudoscalar form factors, respectively.

In the Cabibbo theory, the weak vector (V_μ) and the axial vector (A_μ) currents corresponding to the $\Delta S = 0$ and $\Delta S = 1$ hadronic currents whose matrix elements are defined between the states $|N\rangle$ and $|N'\rangle$ or $|Y\rangle$ are assumed to belong to the octet representation of SU(3).

Accordingly, they are defined as:

$$
\begin{aligned}
V_i^\mu &= \bar{q}F_i\gamma^\mu q, \\
A_i^\mu &= \bar{q}F_i\gamma^\mu\gamma^5 q,
\end{aligned}
\tag{B.8}
$$

where $F_i = \frac{\lambda_i}{2}(i = 1 - 8)$ are the generators of flavor SU(3) and λ_is are the well-known Gell–Mann matrices written as

$$
\lambda_1 = \begin{pmatrix} 0 & 1 & 0 \\ 1 & 0 & 0 \\ 0 & 0 & 0 \end{pmatrix}, \qquad
\lambda_2 = \begin{pmatrix} 0 & -i & 0 \\ i & 0 & 0 \\ 0 & 0 & 0 \end{pmatrix},
$$

$$
\lambda_3 = \begin{pmatrix} 1 & 0 & 0 \\ 0 & -1 & 0 \\ 0 & 0 & 0 \end{pmatrix}, \qquad
\lambda_4 = \begin{pmatrix} 0 & 0 & 1 \\ 0 & 0 & 0 \\ 1 & 0 & 0 \end{pmatrix},
$$

$$
\lambda_5 = \begin{pmatrix} 0 & 0 & -i \\ 0 & 0 & 0 \\ i & 0 & 0 \end{pmatrix}, \qquad
\lambda_6 = \begin{pmatrix} 0 & 0 & 0 \\ 0 & 0 & 1 \\ 0 & 1 & 0 \end{pmatrix},
$$

$$
\lambda_7 = \begin{pmatrix} 0 & 0 & 0 \\ 0 & 0 & -i \\ 0 & i & 0 \end{pmatrix}, \qquad
\lambda_8 = \frac{1}{\sqrt{3}} \begin{pmatrix} 1 & 0 & 0 \\ 0 & 1 & 0 \\ 0 & 0 & -2 \end{pmatrix}.
\tag{B.9}
$$

The generators obey the following algebra of SU(3) generators

$$
\begin{aligned}
[F_i, F_j] &= if_{ijk}F_k, \\
\{F_i, F_j\} &= \frac{1}{3}\delta_{ij} + d_{ijk}F_k, \quad i, j, k = 1 - 8.
\end{aligned}
\tag{B.10}
$$

$F_{i,j,k} = \frac{1}{2}\lambda_{i,j,k}$, f_{ijk} and d_{ijk} are the structure constants, and are antisymmetric and symmetric, respectively, under the interchange of any two indices. These are obtained using the λ_i given in Eq. (B.9) and have been tabulated in Table B.1.

From the property of the SU(3) group, it follows that there are three corresponding SU(2) subgroups of SU(3) which must be invariant under the interchange of quark pairs ud, ds, and us respectively, if the group is invariant under the interchange of u, d, and s quarks. Each of these SU(2) subgroups has raising and lowering operators. One of them is SU(2)$_I$, generated by the generators ($\lambda_1, \lambda_2, \lambda_3$) to be identified with the isospin operators (I_1, I_2, I_3) in the isospin space. For example, $I_\pm$ of isospin space is given by

$$
I_\pm = I_1 \pm iI_2 = F_1 \pm iF_2 = \frac{1}{2}(\lambda_1 \pm i\lambda_2).
\tag{B.11}
$$

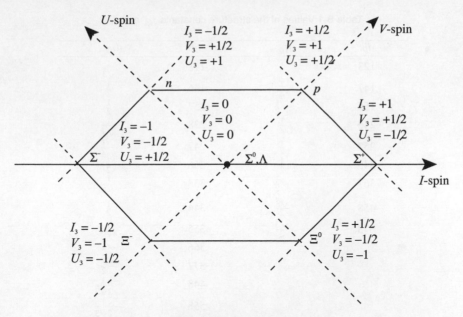

Figure B.1 Baryon octet

The other two are defined as $SU(2)_U$ and $SU(2)_V$ generated by the generators $(\lambda_6, \lambda_7, \frac{1}{2}(\sqrt{3}\lambda_8 - \lambda_3))$ and $(\lambda_4, \lambda_5, \frac{1}{2}(\sqrt{3}\lambda_8 + \lambda_3))$, respectively, in the U-spin and V-spin space with $(d\ s)$ and $(u\ s)$ forming the basic doublet representation of $SU(2)_U$ and $SU(2)_V$.

The $(I\ I_3)$, $(U\ U_3)$, and $(V\ V_3)$ quantum numbers of $(u\ d\ s)$ quarks are assigned as:

$$u \text{ quark:} \quad (I, I_3) = \left(\frac{1}{2}, +\frac{1}{2}\right), \quad (V, V_3) = \left(\frac{1}{2}, +\frac{1}{2}\right),$$

$$d \text{ quark:} \quad (I, I_3) = \left(\frac{1}{2}, -\frac{1}{2}\right), \quad (U, U_3) = \left(\frac{1}{2}, +\frac{1}{2}\right),$$

$$s \text{ quark:} \quad (U, U_3) = \left(\frac{1}{2}, -\frac{1}{2}\right), \quad (V, V_3) = \left(\frac{1}{2}, -\frac{1}{2}\right).$$

From the Gell–Mann matrices λ_i, one may obtain the raising and lowering operators with U-spin and V-spin in analogy with I-spin as:

$$U_\pm = U_1 \pm iU_2 = F_6 \pm iF_7,$$
$$V_\pm = V_1 \pm iV_2 = F_4 \pm iF_5.$$

Accordingly, the I-spin, U-spin, and V-spin for the baryon octet are represented diagrammatically in Figure B.1; for the pseudoscalar $(J^P = 0^-)$ and vector $(J^P = 1^-)$ meson nonet are shown in Figure B.2.

Table B.1 Values of the structure constants f_{ijk} and d_{ijk}.

ijk	f_{ijk}	ijk	d_{ijk}
123	1	118	$\frac{1}{\sqrt{3}}$
147	$\frac{1}{2}$	146	$\frac{1}{2}$
156	$-\frac{1}{2}$	157	$\frac{1}{2}$
246	$\frac{1}{2}$	228	$\frac{1}{\sqrt{3}}$
257	$\frac{1}{2}$	247	$-\frac{1}{2}$
345	$\frac{1}{2}$	256	$\frac{1}{2}$
367	$-\frac{1}{2}$	338	$\frac{1}{\sqrt{3}}$
458	$\frac{\sqrt{3}}{2}$	344	$\frac{1}{2}$
678	$\frac{\sqrt{3}}{2}$	355	$\frac{1}{2}$
		366	$-\frac{1}{2}$
		377	$-\frac{1}{2}$
		448	$-\frac{1}{2\sqrt{3}}$
		558	$-\frac{1}{2\sqrt{3}}$
		668	$-\frac{1}{2\sqrt{3}}$
		778	$-\frac{1}{2\sqrt{3}}$
		888	$-\frac{1}{\sqrt{3}}$

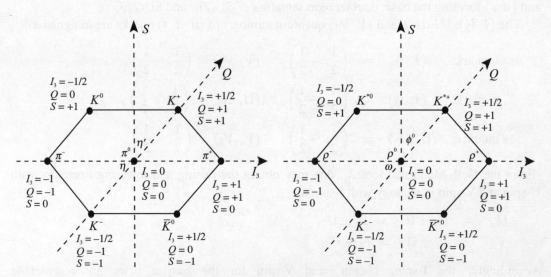

Figure B.2 Pseudoscalar ($J^P = 0^-$) and vector ($J^P = 1^-$) meson nonet.

In neutron β-decay, a d quark is transformed into a u quark, the vector and the axial vector currents for this transition can be written as

$$\bar{\psi}_u \gamma_\mu \psi_d = \bar{q}\gamma_\mu \begin{pmatrix} 0 & 1 & 0 \\ 0 & 0 & 0 \\ 0 & 0 & 0 \end{pmatrix} q = \bar{q}\gamma_\mu \left(\frac{\lambda_1 + i\lambda_2}{2}\right) q = V_\mu^{1+i2},$$

$$\bar{\psi}_u \gamma_\mu \gamma_5 \psi_d = \bar{q}\gamma_\mu \gamma_5 \begin{pmatrix} 0 & 1 & 0 \\ 0 & 0 & 0 \\ 0 & 0 & 0 \end{pmatrix} q = \bar{q}\gamma_\mu \gamma_5 \left(\frac{\lambda_1 + i\lambda_2}{2}\right) q = A_\mu^{1+i2}.$$

Similarly, the vector and axial vector currents for the $u \rightarrow d$ transformation can be written as:

$$\bar{\psi}_d \gamma_\mu \psi_u = V_\mu^{1-i2},$$

$$\bar{\psi}_d \gamma_\mu \gamma_5 \psi_u = A_\mu^{1-i2},$$

and the $s \rightarrow u$ and $u \rightarrow s$ transformations are written as

$$\bar{\psi}_u \gamma_\mu \psi_s = \bar{q}\gamma_\mu \left(\frac{\lambda_4 + i\lambda_5}{2}\right) q = V_\mu^{4+i5},$$

$$\bar{\psi}_u \gamma_\mu \gamma_5 \psi_s = \bar{q}\gamma_\mu \gamma_5 \left(\frac{\lambda_4 + i\lambda_5}{2}\right) q = A_\mu^{4+i5},$$

$$\bar{\psi}_s \gamma_\mu \psi_u = \bar{q}\gamma_\mu \left(\frac{\lambda_4 - i\lambda_5}{2}\right) q = V_\mu^{4-i5}, \tag{B.12}$$

$$\bar{\psi}_s \gamma_\mu \gamma_5 \psi_u = \bar{q}\gamma_\mu \gamma_5 \left(\frac{\lambda_4 - i\lambda_5}{2}\right) q = A_\mu^{4-i5}. \tag{B.13}$$

The electromagnetic current which is a vector current is written using the charge operator $e = I_3 + \frac{Y}{2} = \lambda_3 + \frac{1}{2\sqrt{3}}\lambda_8$ as:

$$J_\mu^{\text{em}} = V_\mu^3 + \frac{1}{\sqrt{3}} V_\mu^8. \tag{B.14}$$

Therefore, the charge changing weak vector and axial vector currents are written, in Cabibbo theory, as:

$$V_\mu^\pm = \left[V_\mu^1 \pm iV_\mu^2\right] \cos\theta_c + \left[V_\mu^4 \pm iV_\mu^5\right] \sin\theta_c,$$

$$A_\mu^\pm = \left[A_\mu^1 \pm iA_\mu^2\right] \cos\theta_c + \left[A_\mu^4 \pm iA_\mu^5\right] \sin\theta_c. \tag{B.15}$$

In the Cabibbo theory, the isovector part of the electromagnetic current J_{em}^μ, that is, V_μ^3 along with the weak vector currents $V_\pm^\mu$ are assumed to transform as a triplet under SU(2) of isospin. Similarly, the axial vector currents are also assumed to transform as an octet under SU(3). The weak transition form factors $f_i(q^2)$ and $g_i(q^2)$; $i = 1 - 3$ are determined using the Cabibbo theory of $V-A$ interaction extended to the strange sector.

The form factors defined in the matrix element of an octet of the vector (axial vector) currents taken between the octets of the initial and the final baryon states as defined in Eqs. (B.6) and (B.7) can, therefore, be expressed in terms of the two couplings of the vector (axial vector) currents corresponding to the symmetric and antisymmetric octets according to the decomposition:

$$8 \times 8 = 1 + 8^S + 8^A + 10 + \overline{10} + 27, \tag{B.16}$$

and the corresponding SU(3) Clebsch–Gordan coefficients. In general, the expression for the matrix element of the transition between the two states of baryons (say B_i and B_k), through the SU(3) octet (V_j or A_j) of currents can be written as:

$$< B_i|V_j|B_k > = if_{ijk}F^V + d_{ijk}D^V, \tag{B.17}$$

$$< B_i|A_j|B_k > = if_{ijk}F^A + d_{ijk}D^A. \tag{B.18}$$

F^V and D^V are determined from the experimental data on the electromagnetic form factors, and F^A and D^A are determined from the experimental data on semileptonic decays of neutron and hyperons. The physical baryon octet states are written in terms of their octet state B_i as:

$$p = \frac{1}{\sqrt{2}}(B_4 + iB_5), \qquad n = \frac{1}{\sqrt{2}}(B_6 + iB_7),$$

$$\Sigma^{\pm} = \frac{1}{\sqrt{2}}(B_1 \pm iB_2), \qquad \Xi^- = \frac{1}{\sqrt{2}}(B_4 - iB_5),$$

$$\Xi^0 = \frac{1}{\sqrt{2}}(B_6 - iB_7), \qquad \Sigma^0 = B_3, \qquad \Lambda^0 = B_8. \tag{B.19}$$

The matrix element for the electromagnetic transition in Eq. (A.14) between two octet states B_i and B_k is defined as:

$$\langle B_i|V_3 + \frac{1}{\sqrt{3}}V_8|B_k\rangle = i\left[f_{i3k} + \frac{1}{\sqrt{3}}f_{i8k}\right]F^V + \left[d_{i3k} + \frac{1}{\sqrt{3}}d_{i8k}\right]D^V. \tag{B.20}$$

Applying Eq. (B.20) to the matrix element of electromagnetic current between proton states, we get

$$
\begin{aligned}
\langle p|J^{em}|p\rangle &= \frac{1}{2}\langle B_4 + iB_5|J^{em}|B_4 + iB_5\rangle \\
&= \frac{1}{2}\{\langle B_4|J^{em}|B_4\rangle + i\langle B_4|J^{em}|B_5\rangle - i\langle B_5|J^{em}|B_4\rangle + \langle B_5|J^{em}|B_5\rangle\} \\
&= \frac{1}{2}\left[\left\{i(f_{434} + \frac{1}{\sqrt{3}}f_{484})F^V + (d_{434} + \frac{1}{\sqrt{3}}d_{484})D^V\right\}\right. \\
&\quad + i\left\{i(f_{435} + \frac{1}{\sqrt{3}}f_{485})F^V + (d_{435} + \frac{1}{\sqrt{3}}d_{485})D^V\right\} \\
&\quad - i\left\{i(f_{534} + \frac{1}{\sqrt{3}}f_{584})F + (d_{534} + \frac{1}{\sqrt{3}}d_{584})D\right\} \\
&\quad + \left.\left\{i(f_{535} + \frac{1}{\sqrt{3}}f_{585})F^V + (d_{535} + \frac{1}{\sqrt{3}}d_{585})D^V\right\}\right] \\
&= \frac{1}{2}\left[i(0+0)F^V + \left(\frac{1}{2} + \frac{1}{\sqrt{3}}\cdot\frac{-1}{2\sqrt{3}}\right)D^V + i\left\{i\left(-\frac{1}{2} - \frac{1}{\sqrt{3}}\cdot\frac{\sqrt{3}}{2}\right)F^V + (0+0)D^V\right\}\right.
\end{aligned}
$$

$$- \quad i\left\{i\left(\frac{1}{2} + \frac{1}{\sqrt{3}}\cdot\frac{\sqrt{3}}{2}\right)F^V + (0+0)D^V\right\} + \left\{i(0+0)F^V + \left(\frac{1}{2} + \frac{1}{\sqrt{3}}\cdot\frac{-1}{2\sqrt{3}}\right)D^V\right\}\right]$$

$$= \quad \frac{1}{2}\left[\frac{1}{3}D^V + F^V + \frac{1}{3}D^V + F^V\right]$$

$$= \quad \frac{D^V}{3} + F^V. \tag{B.21}$$

Similarly, for the $n \to n$ transition, the electromagnetic current is calculated as

$$\langle n|J_{em}|n\rangle \quad = \quad \frac{1}{2}\langle B_6 + iB_7|J_{em}|B_6 + iB_7\rangle$$

$$= \quad \frac{1}{2}\left\{\langle B_6|J_{em}|B_6\rangle + i\langle B_6|J_{em}|B_7\rangle - i\langle B_7|J_{em}|B_6\rangle + \langle B_7|J_{em}|B_7\rangle\right\}$$

$$= \quad \frac{1}{2}\left[\left\{i(f_{636} + \frac{1}{\sqrt{3}}f_{686})F^V + (d_{636} + \frac{1}{\sqrt{3}}d_{686})D^V\right\}\right.$$

$$+ \quad i\left\{i(f_{637} + \frac{1}{\sqrt{3}}f_{687})F^V + (d_{637} + \frac{1}{\sqrt{3}}d_{687})D^V\right\}$$

$$- \quad i\times\left\{i(f_{736} + \frac{1}{\sqrt{3}}f_{786})F^V + (d_{736} + \frac{1}{\sqrt{3}}d_{786})D^V\right\}$$

$$+ \quad \left.\left\{i(f_{737} + \frac{1}{\sqrt{3}}f_{787})F^V + (d_{737} + \frac{1}{\sqrt{3}}d_{787})D^V\right\}\right]$$

$$= \quad \frac{1}{2}\left[i(0+0)F^V + \left(-\frac{1}{2} + \frac{1}{\sqrt{3}}\cdot\frac{-1}{2\sqrt{3}}\right)D^V + i\left\{i\left(\frac{1}{2} - \frac{1}{\sqrt{3}}\cdot\frac{\sqrt{3}}{2}\right)F^V + (0+0)D^V\right\}\right.$$

$$- \quad i\left\{i\left(-\frac{1}{2} + \frac{1}{\sqrt{3}}\cdot\frac{\sqrt{3}}{2}\right)F^V + (0+0)D^V\right\} + \left.\left\{i(0+0)F^V + \left(-\frac{1}{2} + \frac{1}{\sqrt{3}}\cdot\frac{-1}{2\sqrt{3}}\right)D^V\right\}\right]$$

$$= \quad \frac{1}{2}\left[-\frac{2}{3}D^V - \frac{2}{3}D^V\right] = -\frac{2D^V}{3}. \tag{B.22}$$

The matrix elements of the electromagnetic currents for the protons and neutrons are described in terms of the electromagnetic form factors as:

$$\langle p(p')|J_\mu^{em}|p(p)\rangle \quad = \quad \bar{u}(p')\left[\gamma_\mu F_1^p(q^2) + i\sigma_{\mu\nu}\frac{q^\nu}{2M}F_2^p(q^2)\right]u(p),$$

$$\langle n(p')|J_\mu^{em}|n(p)\rangle \quad = \quad \bar{u}(p')\left[\gamma_\mu F_1^n(q^2) + i\sigma_{\mu\nu}\frac{q^\nu}{2M}F_2^n(q^2)\right]u(p).$$

Therefore, each of the form factors defined earlier, that is, $F_i^p(q^2)$ and $F_i^n(q^2)$; $(i = 1, 2)$ can be written in terms of their SU(3) coupling constants as

$$f_i(q^2) \quad = \quad aF_i^V(q^2) + bD_i^V(q^2) \qquad i = 1, 2, 3 \tag{B.23}$$

$$g_i(q^2) \quad = \quad aF_i^A(q^2) + bD_i^A(q^2) \qquad i = 1, 2, 3. \tag{B.24}$$

Comparing Eqs. (B.21) and (B.22) with Eq. (B.23), we obtain the Clebsch–Gordan coefficients for the electromagnetic $p \to p$ and $n \to n$ transitions as

$$p \to p, \qquad a \quad = \quad 1 \qquad b = 1/3, \tag{B.25}$$

$$n \to n, \qquad a \quad = \quad 0 \qquad b = -2/3. \tag{B.26}$$

Using Eqs. (B.25) and (B.26) in Eq. (B.23), we obtain

$$F_i^{p \to p}(q^2) = F_i^p(q^2) = F_i^V(q^2) + \frac{1}{3}D_i^V(q^2), \tag{B.27}$$

$$F_i^{n \to n}(q^2) = F_i^n(q^2) = -\frac{2}{3}D_i^V(q^2). \tag{B.28}$$

Solving Eqs. (B.27) and (B.28), we get

$$F_i^V(q^2) = F_i^p(q^2) - F_i^n(q^2), \tag{B.29}$$

$$D_i^V(q^2) = -\frac{3}{2}F_i^n(q^2). \tag{B.30}$$

Once $F_i^V(q^2)$ and $D_i^V(q^2)$ are determined in terms of the electromagnetic form factors of the nucleon, they can be used to determine all the form factors in the case of the matrix element of the weak vector current for the various $\Delta S = 0, 1$ transitions.

The Clebsch–Gordan coefficients (a and b) for the various transitions can be obtained as follows. For example, let us obtain the values of a and b for $p(uud) \to \Lambda(uds)$ transition. Using Eq. (B.12), the current for the $u \to s$ transition can be written as

$$
\begin{aligned}
\langle \Lambda | j_4 - ij_5 | p \rangle &= \langle \Lambda | j_4 | p \rangle - i \langle \Lambda | j_5 | p \rangle \\
&= \frac{1}{\sqrt{2}} \left[\langle B_8 | j_4 | B_4 \rangle + i \langle B_8 | j_4 | B_5 \rangle \right] - \frac{i}{\sqrt{2}} \left[\langle B_8 | j_5 | B_4 \rangle + i \langle B_8 | j_5 | B_5 \rangle \right] \\
&= \frac{1}{\sqrt{2}} \left[if_{844}F + d_{844}D + i(if_{845}F + d_{845}D) \right] \\
&\quad - \frac{i}{\sqrt{2}} \left[if_{854}F + d_{854}D + i(if_{855}F + d_{855}D) \right] \\
&= -\sqrt{\frac{3}{2}}F - \frac{1}{\sqrt{6}}D. \tag{B.31}
\end{aligned}
$$

Similarly, the coefficients for the various transitions can be obtained and are presented in Table B.2.

B.2   Octet Representation of Mesons

In the SU(3) symmetry, the quark states can be represented as

$$q = \begin{pmatrix} u \\ d \\ s \end{pmatrix} \tag{B.32}$$

and the corresponding antiquark states can be represented as

$$\bar{q} = \begin{pmatrix} \bar{u} & \bar{d} & \bar{s} \end{pmatrix}. \tag{B.33}$$

Table B.2 Values of the coefficients a and b given in Eqs. (B.23) and (B.24).

Interaction	Transition	a	b
Electromagnetic	$p \to p$	1	$\frac{1}{3}$
interaction	$n \to n$	0	$-\frac{2}{3}$
Weak vector	$n \to p$	1	1
and axial vector	$\Lambda \to p$	$-\sqrt{\frac{3}{2}}$	$-\frac{1}{\sqrt{6}}$
	$\Sigma^0 \to p$	$-\frac{1}{\sqrt{2}}$	$\frac{1}{\sqrt{2}}$
	$\Sigma^- \to n$	-1	1
	$\Sigma^\pm \to \Lambda$	0	$\sqrt{\frac{2}{3}}$
	$\Sigma^- \to \Sigma^0$	$\sqrt{2}$	0
	$\Xi^- \to \Lambda$	$\sqrt{\frac{3}{2}}$	$-\frac{1}{\sqrt{6}}$
	$\Xi^- \to \Sigma^0$	$\frac{1}{\sqrt{2}}$	$\frac{1}{\sqrt{2}}$
	$\Xi^0 \to \Sigma^+$	1	1
	$\Xi^- \to \Xi^0$	1	-1

Mesons are the bound state of $q\bar{q}$, which can be represented as

$$q_i\bar{q}_j = \underbrace{q_i\bar{q}_j - \frac{1}{3}\delta_{ij}\sum_k q_k\bar{q}_k}_{\text{octet of the pseudoscalar mesons}} + \underbrace{\frac{1}{3}\delta_{ij}\sum_k q_k\bar{q}_k}_{\text{singlet of the pseudoscalar mesons}}, \tag{B.34}$$

$i, j, k = 1 - 3$. In the notation of group theory, Eq. (B.34) can be written as

$$3 \otimes \bar{3} = 8 \oplus 1. \tag{B.35}$$

The operator for the octet of the pseudoscalar mesons can be defined as

$$P_{ji}|0> = |P_{ji}> = q_i\bar{q}_j - \frac{1}{3}\delta_{ij}\sum_k q_k\bar{q}_k. \tag{B.36}$$

Using Eq. (B.36), the elements of P_{ji} can be obtained as follows

$$P_{21} = q_1\bar{q}_2 - \frac{1}{3}\delta_{12}\sum_k q_k\bar{q}_k$$
$$= u\bar{d} = \pi^+, \tag{B.37}$$
$$P_{31} = q_1\bar{q}_3 - \frac{1}{3}\delta_{13}\sum_k q_k\bar{q}_k$$
$$= u\bar{s} = K^+, \tag{B.38}$$

$$
\begin{aligned}
P_{33} &= q_3 \bar{q}_3 - \frac{1}{3} \delta_{33} \sum_k q_k \bar{q}_k \\
&= s\bar{s} - \frac{1}{3}(u\bar{u} + d\bar{d} + s\bar{s}) \\
&= -\frac{1}{3}(u\bar{u} + d\bar{d} - 2s\bar{s}) = -\frac{\sqrt{6}}{3}\eta.
\end{aligned}
\tag{B.39}
$$

In the matrix form, P_{ji} can be written as

$$
P = \begin{pmatrix} \frac{1}{\sqrt{6}}\eta + \frac{1}{\sqrt{2}}\pi^0 & \pi^+ & K^+ \\ \pi^- & \frac{1}{\sqrt{6}}\eta - \frac{1}{\sqrt{2}}\pi^0 & K^0 \\ K^- & \bar{K}^0 & -\frac{2}{\sqrt{6}}\eta \end{pmatrix}.
\tag{B.40}
$$

This matrix can be rewritten in terms of the generators of the SU(3) symmetry as

$$
\begin{aligned}
P_{ji} &= \frac{1}{\sqrt{2}} \sum_{A=1}^{8} (\lambda_A)_{ji}\, \pi_A, \\
P &= \frac{1}{\sqrt{2}} \begin{pmatrix} \pi_3 + \frac{\pi_8}{\sqrt{3}} & \pi_1 + i\pi_2 & \pi_4 + i\pi_5 \\ \pi_1 - i\pi_2 & -\pi_3 + \frac{\pi_8}{\sqrt{3}} & \pi_6 + i\pi_7 \\ \pi_4 - i\pi_5 & \pi_6 - i\pi_7 & -\frac{2}{\sqrt{3}}\pi_8 \end{pmatrix}.
\end{aligned}
\tag{B.41}
$$

B.3 Octet Representation of Baryons

Baryons are the bound state of three quarks. In the notation of group theory, the bound state of baryons can be represented as follows:

First, we write for the two bound quarks

$$
3 \otimes 3 = 6 \oplus \bar{3},
\tag{B.42}
$$

where 6 represents the symmetric part with 6 components and $\bar{3}$ represents the antisymmetric part with 3 components. Adding the third quark, we have

$$
\begin{aligned}
3 \otimes 3 \otimes 3 &= (6 \otimes 3) \quad \oplus \quad (\bar{3} \otimes 3) \\
&= \underbrace{10 \oplus 8'}_{\text{symmetric}} \quad \oplus \quad \underbrace{8 \oplus 1}_{\text{antisymmetric}}.
\end{aligned}
\tag{B.43}
$$

The wavefunction of the lowest lying baryons ($J^P = \frac{1}{2}^+$) is antisymmetric in nature. Therefore, to obtain the antisymmetric part in Eq. (B.43) for the octet representation of baryons, we proceed in the following way. The bound state for two quarks can be written as

$$
q_i q_j = \frac{1}{2} \underbrace{(q_i q_j + q_j q_i)}_{\text{symmetric part}} + \frac{1}{2} \underbrace{(q_i q_j - q_j q_i)}_{\text{antisymmetric part}},
\tag{B.44}
$$

or

$$q_i q_j = \frac{1}{\sqrt{2}} S_{ij} + \frac{1}{\sqrt{2}} A_{ij}, \tag{B.45}$$

where the symmetric and the antisymmetric tensors with 6 and 3 components respectively, are defined as

$$S_{ij} = \frac{1}{\sqrt{2}} (q_i q_j + q_j q_i), \tag{B.46}$$

$$A_{ij} = \frac{1}{\sqrt{2}} (q_i q_j - q_j q_i). \tag{B.47}$$

We define T_k which can be written in terms of A_{ij} as

$$T_k = \epsilon_{klm} A^{lm}, \tag{B.48}$$

or

$$A^{ij} = \frac{1}{2} \epsilon_{ijk} T_k. \tag{B.49}$$

Now, considering the antisymmetric tensor, we have

$$T^k q_j = \underbrace{\left(T^k q_j - \frac{1}{3} \delta_j^k \sum_i T^i q_i \right)}_{\text{octet of the baryons}} + \underbrace{\frac{1}{3} \delta_j^k \sum_i T^i q_i}_{\text{singlet state}}. \tag{B.50}$$

The operator for the octet of baryons can be written as

$$\bar{B}_{jk} = T^k q_j - \frac{1}{3} \delta_j^k \sum_i T^i q_i. \tag{B.51}$$

Using Eqs. (B.49) and (B.47) in Eq. (B.51), we get

$$\bar{B}_{jk} |0> = |B_{jk}> = \frac{1}{2\sqrt{2}} \left[\epsilon_{klm} (q_l q_m - q_m q_l) q_j - \frac{1}{3} \delta_j^k \epsilon_{ilm} ((q_l q_m - q_m q_l) q_i \right] |0>. \tag{B.52}$$

The elements of B_{jk} can be obtained, using Eq. (B.52), as:

$$\begin{aligned} B_{13} &= \frac{1}{2\sqrt{2}} \left[\epsilon_{3lm} (q_l q_m - q_m q_l) q_1 - \frac{1}{3} \delta_1^3 \epsilon_{ilm} ((q_l q_m - q_m q_l) q_i \right], \\ &= \frac{1}{2\sqrt{2}} \left[\epsilon_{312} (q_1 q_2 - q_2 q_1) q_1 + \epsilon_{321} (q_2 q_1 - q_1 q_2) q_1 \right], \\ &= \frac{1}{\sqrt{2}} \left[q_1 q_2 q_1 - q_2 q_1 q_1 \right], \end{aligned}$$

$$= \frac{1}{\sqrt{2}} [udu - duu] = p. \tag{B.53}$$

Similarly, the element B_{32} is obtained as

$$
\begin{aligned}
B_{32} &= \frac{1}{2\sqrt{2}} \left[\epsilon_{2lm}(q_l q_m - q_m q_l)q_3 - \frac{1}{3}\delta_3^2 \epsilon_{ilm}((q_l q_m - q_m q_l)q_i \right], \\
&= \frac{1}{2\sqrt{2}} [\epsilon_{213}(q_1 q_3 - q_3 q_1)q_3 + \epsilon_{231}(q_3 q_1 - q_1 q_3)q_3], \\
&= \frac{1}{\sqrt{2}} [q_3 q_1 q_3 - q_1 q_3 q_3], \\
&= \frac{1}{\sqrt{2}} [sus - uss] = \Xi^0. \tag{B.54}
\end{aligned}
$$

The other elements of B_{jk} can be determined in a similar manner and can be written in matrix form as:

$$
B = \begin{pmatrix}
\frac{1}{\sqrt{6}}\Lambda + \frac{1}{\sqrt{2}}\Sigma^0 & \Sigma^+ & p \\
\Sigma^- & \frac{1}{\sqrt{6}}\Lambda - \frac{1}{\sqrt{2}}\Sigma^0 & n \\
\Xi^- & \Xi^0 & -\frac{2}{\sqrt{6}}\Lambda
\end{pmatrix}. \tag{B.55}
$$

In terms of SU(3) generators, the matrix can be written, in a similar manner as written for the mesons in Eq. (B.41) as

$$
B = \frac{1}{\sqrt{2}} \begin{pmatrix}
B_3 + \frac{B_8}{\sqrt{3}} & B_1 + iB_2 & B_4 + iB_5 \\
B_1 - iB_2 & -B_3 + \frac{B_8}{\sqrt{3}} & B_6 + iB_7 \\
B_4 - iB_5 & B_6 - iB_7 & -\frac{2}{\sqrt{3}}B_8
\end{pmatrix}. \tag{B.56}
$$

Some Properties of Pauli and Dirac Matrices and Spin Density Matrices

C.1 Trace Properties of Pauli and Dirac Matrices

$$
\sigma_x = \sigma_1 = \begin{pmatrix} 0 & 1 \\ 1 & 0 \end{pmatrix} \qquad \sigma_y = \sigma_2 = \begin{pmatrix} 0 & -i \\ i & 0 \end{pmatrix}
$$

$$
\sigma_z = \sigma_3 = \begin{pmatrix} 1 & 0 \\ 0 & -1 \end{pmatrix}
$$

$$
\sigma_1^2 = \sigma_2^2 = \sigma_3^2 = I_{2\times 2}
$$

$$
\mathrm{Tr}(\sigma_i) = 0
$$

$$
\mathrm{Tr}(\sigma_i \sigma_j) = 2\delta_{ij}
$$

$$
\mathrm{Tr}(\sigma_i \sigma_j \sigma_k) = 2i\epsilon_{ijk}
$$

$$
\mathrm{Tr}(\sigma_i \sigma_j \sigma_k \sigma_l) = 2(\delta_{ij}\delta_{kl} - \delta_{ik}\delta_{jl} + \delta_{il}\delta_{jk})
$$

$$
\gamma^\mu = (\gamma^0, \gamma^i)
$$

$$
\gamma_\mu = g_{\mu\nu}\gamma^\nu
$$

$$
\gamma_\mu = (\gamma^0, -\gamma^i)
$$

$$
\gamma^\mu \gamma^\nu + \gamma^\nu \gamma^\mu = 2g^{\mu\nu} I_{4\times 4}, \ \text{(where } I_{4\times 4} \text{ is the } 4 \times 4 \text{ identity matrix),}
$$

$$
\gamma_\mu \not{p} \gamma^\mu = \gamma_\mu p_\nu \gamma^\nu \gamma^\mu = \gamma_\mu p_\nu (2g^{\mu\nu} - \gamma^\mu \gamma^\nu) = 2\gamma_\mu p^\mu - \gamma_\mu \gamma^\mu \not{p} = -2\not{p}
$$

$$
\not{p}\not{q} = p_\mu \gamma^\mu q_\nu \gamma^\nu = p_\mu q_\nu (2g^{\mu\nu} - \gamma^\nu \gamma^\mu) = 2p \cdot q - \not{q}\not{p}
$$

$$
\mathrm{Tr}(\gamma_\mu \gamma_\nu) = \frac{1}{2}\mathrm{Tr}\left(\gamma_\mu \gamma_\nu + \gamma_\nu \gamma_\mu\right) = \frac{1}{2}\mathrm{Tr}\left(2g_{\mu\nu}\right) = g_{\mu\nu}\mathrm{Tr}(I_{4\times 4}) = 4g_{\mu\nu}
$$

$$
\mathrm{Tr}(\gamma^\mu \gamma^\nu \gamma^\rho \gamma^\sigma) = \mathrm{Tr}(\gamma^\mu \gamma^\nu (2g^{\sigma\rho} - \gamma^\sigma \gamma^\rho))
$$

$$
\begin{aligned}
&= 2g^{\sigma\rho}\mathrm{Tr}(\gamma^\mu\gamma^\nu) - \mathrm{Tr}(\gamma^\mu\gamma^\nu\gamma^\sigma\gamma^\rho) \\
&= 2g^{\sigma\rho}4g^{\mu\nu} - \mathrm{Tr}(\gamma^\mu(2g^{\nu\sigma} - \gamma^\sigma\gamma^\nu)\gamma^\rho) \\
&= 2g^{\sigma\rho}4g^{\mu\nu} - 2g^{\nu\sigma}4g^{\mu\rho} + \mathrm{Tr}(\gamma^\mu\gamma^\sigma\gamma^\nu\gamma^\rho) \\
&= 2g^{\sigma\rho}4g^{\mu\nu} - 2g^{\nu\sigma}4g^{\mu\rho} + 2g^{\mu\sigma}4g^{\nu\rho} - \mathrm{Tr}(\gamma^\sigma\gamma^\mu\gamma^\nu\gamma^\rho) \\
&= 4(g^{\mu\nu}g^{\sigma\rho} - g^{\mu\rho}g^{\nu\sigma} + g^{\mu\sigma}g^{\nu\rho}). \quad (\because \mathrm{Tr}(ABCD) = \mathrm{Tr}(BCDA))
\end{aligned}
$$

Here, we have used the properties of cyclic permutation of gamma matrices.

$$
\begin{aligned}
\mathrm{Tr}(\not{k}\not{p}) &= \frac{1}{2}\mathrm{Tr}(\not{k}\not{p} + \not{p}\not{k}) = \frac{1}{2}\mathrm{Tr}\left(k^\mu p^\nu \gamma_\mu\gamma_\nu + p^\nu k^\mu \gamma_\nu\gamma_\mu\right) \\
&= \frac{1}{2}k^\mu p^\nu \mathrm{Tr}\left(2g_{\mu\nu}I_{4\times4}\right) = 4k\cdot p \\
\mathrm{Tr}(\not{k}\not{p}\not{k}'\not{p}') &= k_\mu p_\nu k'_\theta p'_\phi \mathrm{Tr}\left(\gamma^\mu\gamma^\nu\gamma^\theta\gamma^\phi\right) \\
&= k_\mu p_\nu k'_\theta p'_\phi \mathrm{Tr}\left(g^{\mu\nu}g^{\theta\phi} - g^{\mu\theta}g^{\nu\phi} + g^{\mu\phi}g^{\nu\theta}\right) \\
&= 4\Big((k\cdot p)(k'\cdot p') - (k\cdot k')(p\cdot p') + (k\cdot p')(p\cdot k')\Big) \\
\mathrm{Tr}(\not{p}_1\not{p}_2\not{p}_3\cdots\not{p}_{2n}) &= p_1\cdot p_2 \mathrm{Tr}(\not{p}_3\cdots\not{p}_{2n}) - p_1\cdot p_3 \mathrm{Tr}(\not{p}_2\not{p}_4\cdots\not{p}_{2n}) \\
&\quad + \ldots + p_1\cdot p_{2n}\mathrm{Tr}(\not{p}_2\not{p}_3\cdots\not{p}_{2n-1})
\end{aligned}
$$

$$
\begin{aligned}
\mathrm{Tr}(\gamma_5) &= 0 \\
\mathrm{Tr}(\gamma_5\not{p}_1\not{p}_2) &= 0 \\
\mathrm{Tr}(\gamma_\mu\gamma_5) &= 0 \\
\mathrm{Tr}(\gamma_5 \times \text{odd no. of } \gamma \text{ matrices}) &= 0 \\
\mathrm{Tr}(\text{odd no. of } \gamma \text{ matrices}) &= 0 \\
\mathrm{Tr}(\gamma_5\not{p}_1\not{p}_2\not{p}_3\not{p}_4) &= 4i\epsilon_{\mu\nu\alpha\beta}p_1{}^\mu p_2{}^\nu p_3{}^\alpha p_4{}^\beta
\end{aligned}
$$

$$
\epsilon^{\mu\nu\alpha\beta}\epsilon^{\mu'\nu'\alpha'\beta'} = -\begin{vmatrix}
g^{\mu\mu'} & g^{\mu\nu'} & g^{\mu\alpha'} & g^{\mu\beta'} \\
g^{\nu\mu'} & g^{\nu\nu'} & g^{\nu\alpha'} & g^{\nu\beta'} \\
g^{\alpha\mu'} & g^{\alpha\nu'} & g^{\alpha\alpha'} & g^{\alpha\beta'} \\
g^{\beta\mu'} & g^{\beta\nu'} & g^{\beta\alpha'} & g^{\beta\beta'}
\end{vmatrix}
$$

$$
\epsilon^{\mu\nu\alpha\beta}\epsilon_\mu{}^{\nu'\alpha'\beta'} = -\begin{vmatrix}
g^{\nu\nu'} & g^{\nu\alpha'} & g^{\nu\beta'} \\
g^{\alpha\nu'} & g^{\alpha\alpha'} & g^{\alpha\beta'} \\
g^{\beta\nu'} & g^{\beta\alpha'} & g^{\beta\beta'}
\end{vmatrix}
$$

$$
\begin{aligned}
\epsilon^{\mu\nu\alpha\beta}\epsilon_{\mu\nu}{}^{\alpha'\beta'} &= -2(g^{\alpha\alpha'}g^{\beta\beta'} - g^{\alpha\beta'}g^{\alpha'\beta}) \\
\epsilon^{\mu\nu\alpha\beta}\epsilon_{\mu\nu\alpha}{}^{\beta'} &= -6g^{\beta\beta'} \\
\epsilon^{\mu\nu\alpha\beta}\epsilon_{\mu\nu\alpha\beta} &= -24.
\end{aligned}
$$

C.2 Spin Density Matrix

Consider that a beam of spin $\frac{1}{2}$ particles is produced from the interaction with another system. When the interaction is over, the wavefunction of the whole system can be written as a sum of the products of wavefunctions of the other systems in free state:

$$\Psi'_\sigma(\vec{x}, \xi, t) = \frac{1}{(\sqrt{2\pi})^3} e^{-ip.x} \phi_\sigma(\xi, t),$$ (C.1)

where

$$\phi_\sigma(\xi, t) = \sum_{r,n} u^r_\sigma(p) \phi_n(\xi, t) c_{rn}$$ (C.2)

and $u^r(p)$ represents the wavefunction of the states with momentum p; helicity r, $\phi_n(\xi, t)$ represents the wavefunction of the other system and c_{rn} are constants. The wavefunctions $u^r(p)$ and $\phi_n(\xi, t)$ are normalized by the conditions:

$$\bar{u}^{r'}(p) u^r(p) = 2m\,\delta_{r'r}, \quad (m \text{ being the mass})$$ (C.3)

$$\phi^\dagger_{n'}(\xi, t) \phi_n(\xi, t) = \delta_{n'n}.$$ (C.4)

Operator $\hat{O}$ acts on the spin variable σ. The mean value of the operator $\hat{O}$ is obtained as

$$\hat{O} = \frac{\int \bar{\phi}(\xi) \hat{O} \phi(\xi) d\xi}{\int \bar{\phi}(\xi) \phi(\xi) d\xi}.$$ (C.5)

Using Eq. (C.2), $\bar{\phi}(\xi, t)$ can be obtained as

$$\bar{\phi}_\sigma(\xi, t) = \sum_{r,n} \phi^\dagger_n(\xi, t) \bar{u}^r_\sigma(p) c^*_{rn}.$$ (C.6)

Using Eq. (C.2) and (C.6) and the normalization condition for $\phi_n(\xi, t)$ from Eq. (C.4), the denominator of Eq. (C.5) can be written as:

$$\int \bar{\phi}(\xi) \phi(\xi) d\xi = \sum_{\sigma,\sigma'} \bar{\phi}_{\sigma'}(\xi) \phi_\sigma(\xi)$$

$$= \sum_{r,r'} \bar{u}^{r'}(p) u^r(p) \sum_{n,n'} \phi^\dagger_{n'} \phi_n c^*_{r'n'} c_{rn}$$

$$= \sum_{r,r'} \bar{u}^{r'}(p) u^r(p) \rho_{rr'}$$ (C.7)

with

$$\rho_{rr'} = \sum_n c^*_{r'n} c_{rn}.$$ (C.8)

Similarly, the numerator of Eq. (C.5) can be obtained as

$$\int \bar{\phi}(\xi)\hat{O}\phi(\xi)d\xi = \sum_{r,r'} \bar{u}^{r'}(p)\hat{O}u^r(p)\rho_{rr'}. \tag{C.9}$$

Using Eqs. (C.9) and (C.7) in Eq. (C.5), we have

$$
\begin{aligned}
<\hat{O}> \; &= \; \frac{\sum_{r,r'} \bar{u}^{r'}(p)\hat{O}u^r(p)\rho_{rr'}}{\sum_{r,r'} \bar{u}^{r'}(p)u^r(p)\rho_{rr'}}, \\
&= \; \frac{\mathrm{Tr}[\hat{O}\rho(p)]}{\mathrm{Tr}[\rho(p)]},
\end{aligned} \tag{C.10}
$$

with

$$\rho(p) = \rho_{\sigma\sigma'}(p) = \sum_{r,r'} \bar{u}_{\sigma'}^{r'}(p)u_{\sigma}^r(p)\rho_{rr'}. \tag{C.11}$$

This matrix $\rho(p)$ is called the spin density matrix.

Some properties of the spin density matrix

1. Hermiticity

$$
\begin{aligned}
\rho_{rr'}^* &= \rho_{rr'}, \\
\bar{\rho}(p) &= \rho(p).
\end{aligned}
$$

$$\tag{C.12}$$
$$\tag{C.13}$$

2. Normalization

$$\sum_r \rho_{rr} = 1, \tag{C.14}$$

$$\mathrm{Tr}[\rho(p)] = 2m. \tag{C.15}$$

3.

$$\mathrm{Tr}[\rho^2(p)] = (2m)^2 = (\mathrm{Tr}[\rho(p)])^2. \tag{C.16}$$

4.

$$\Lambda(p)\rho(p) = \rho(p)\Lambda(p) = 2m\rho(p), \tag{C.17}$$

where $\Lambda(p) = \not{p} + m$ is the projection operation. Equation (C.17) can be rewritten as

$$\rho(p) = \frac{1}{(2m)^2}\Lambda(p)\rho(p)\Lambda(p). \tag{C.18}$$

Now, in order to obtain the expression for the spin density matrix, we expand $\rho(p)$ in terms of bilinear covariants as

$$\rho(p) = a + b_\mu \gamma^\mu + c_{\mu\nu} \sigma^{\mu\nu} + d_\mu \gamma^\mu \gamma_5 + e\gamma_5. \tag{C.19}$$

Using the properties of the spin density matrix, we can evaluate the constants a to e in Eq. (C.19).

To evaluate constant a, we take the trace of Eq. (C.19)

$$\text{Tr}[\rho(p)] = a\text{Tr}[I_4] + b_\mu \text{Tr}[\gamma^\mu] + c_{\mu\nu} \text{Tr}[\sigma^{\mu\nu}] + d_\mu \text{Tr}[\gamma^\mu \gamma_5] + e\text{Tr}[\gamma_5]. \tag{C.20}$$

Using the trace properties of gamma matrices as given in Section C.1, we get

$$\text{Tr}[\rho(p)] = 4a + \frac{i}{2}c_{\mu\nu}(4g^{\mu\nu} - 4g^{\nu\mu}),$$
$$\implies a = \frac{1}{4}\text{Tr}[\rho(p)]. \tag{C.21}$$

Using the normalization condition for $\rho(p)$ matrices as given in Eq. (C.15), the constant a can be evaluated as

$$a = \frac{m}{2}. \tag{C.22}$$

To calculate b_μ, we multiply Eq. (C.19) by γ^α from the right side and then the trace of the resulting equation can be calculated as:

$$\text{Tr}[\rho(p)\gamma^\alpha] = a\text{Tr}[\gamma^\alpha] + b_\mu \text{Tr}[\gamma^\mu \gamma^\alpha] + c_{\mu\nu} \text{Tr}[\sigma^{\mu\nu}\gamma^\alpha] + d_\mu \text{Tr}[\gamma^\mu \gamma_5 \gamma^\alpha] + e\text{Tr}[\gamma_5 \gamma_\alpha],$$
$$\implies b^\alpha = \frac{1}{4}\text{Tr}[\rho(p)\gamma^\alpha].$$

Using Eq. (C.18) in the aforementioned expression, we get

$$b^\alpha = \frac{1}{4(2m)^2}\text{Tr}[\Lambda(p)\rho(p)\Lambda(p)\gamma^\alpha], \tag{C.23}$$

where

$$\Lambda(p)\gamma^\alpha = (\not{p} + m)\gamma^\alpha,$$
$$= (p_\mu \gamma^\mu \gamma^\alpha + m\gamma^\alpha),$$
$$= (2p^\alpha + \gamma^\alpha \Lambda(-p)),$$

and the anti- commutation relation of γ matrices have been used. Using this expression in Eq. (C.23), we obtain

$$b^\alpha = \frac{1}{4(2m)^2}\left(\mathrm{Tr}[\Lambda(p)\rho(p)2p^\alpha] + \mathrm{Tr}[\Lambda(p)\gamma^\alpha\Lambda(-p)]\right)$$

$$b^\alpha = \frac{1}{4(2m)^2}\left(2p^\alpha\,\mathrm{Tr}[\Lambda(p)\rho(p)]\right) \qquad (\Lambda(p)\Lambda(-p) = 0)$$

$$\implies \qquad b^\alpha = \frac{p^\alpha}{2}. \tag{C.24}$$

Next, we evaluate coefficient e, for which we multiply Eq. (C.19) by γ_5 and then take the trace of the resulting expression as

$$\mathrm{Tr}[\rho(p)\gamma_5] = a\mathrm{Tr}[\gamma_5] + b_\mu\mathrm{Tr}[\gamma^\mu\gamma_5] + c_{\mu\nu}\mathrm{Tr}[\sigma^{\mu\nu}\gamma_5] + d_\mu\mathrm{Tr}[\gamma^\mu] + e\mathrm{Tr}[I_4],$$

$$\implies \qquad e = \frac{1}{4}\mathrm{Tr}[\rho(p)\gamma_5] = \frac{1}{4(2m)^2}\mathrm{Tr}[\Lambda(p)\rho(p)\Lambda(p)\gamma_5]. \tag{C.25}$$

Evaluating $\Lambda(p)\gamma_5$ using the commutation relation of γ matrices, we obtain

$$\Lambda(p)\gamma_5 = (\not{p} + m)\gamma_5 = (p^\alpha\gamma_\alpha + m)\gamma_5,$$
$$= \gamma_5\Lambda(-p).$$

Using this expression in Eq. (C.25), we have

$$e = 0. \qquad (\text{using } \Lambda(p)\Lambda(-p) = 0) \tag{C.26}$$

To evaluate constant d, we multiply Eq. (C.19) by $\gamma^\alpha\gamma_5$ from the right side as

$$\rho(p)\gamma^\alpha\gamma_5 = a\gamma^\alpha\gamma_5 + b_\mu\gamma^\mu\gamma^\alpha\gamma_5 + c_{\mu\nu}\sigma^{\mu\nu}\gamma^\alpha\gamma_5 - d_\mu\gamma^\mu\gamma^\alpha - e\gamma^\alpha.$$

Taking the trace of this equation, we have

$$\mathrm{Tr}[\rho(p)\gamma^\alpha\gamma_5] = a\mathrm{Tr}[\gamma^\alpha\gamma_5] + b_\mu\mathrm{Tr}[\gamma^\mu\gamma^\alpha\gamma_5] + c_{\mu\nu}\mathrm{Tr}[\sigma^{\mu\nu}\gamma^\alpha\gamma_5] - d_\mu\mathrm{Tr}[\gamma^\mu\gamma^\alpha] - e\mathrm{Tr}[\gamma^\alpha],$$

$$\implies \qquad d^\alpha = -\frac{1}{4}\mathrm{Tr}[\rho(p)\gamma^\alpha\gamma_5]. \tag{C.27}$$

The spin operator is defined as

$$s_\alpha = \frac{-1}{2m}\gamma_5(\gamma_\alpha\not{p} - \not{p}\gamma_\alpha). \tag{C.28}$$

Let ξ_α be the mean value of the spin operator; it can be expressed as

$$\xi_\alpha = \frac{\mathrm{Tr}[\rho(p)s_\alpha]}{\mathrm{Tr}[\rho(p)]} = \frac{\mathrm{Tr}[\rho(p)s_\alpha]}{2m}, \tag{C.29}$$

where the normalization condition has been used. Using Eq. (C.28) in Eq. (C.29) expression, we have

$$
\begin{aligned}
\zeta_\alpha &= \frac{-1}{(2m)^2}\mathrm{Tr}[\rho(p)\gamma_5(\gamma_\alpha \not{p} - \not{p}\gamma_\alpha], \\
&= \frac{-1}{(2m)^2}\mathrm{Tr}[2\gamma_5\gamma_\alpha\rho(p)\not{p}], \\
&= \frac{1}{2m}\mathrm{Tr}[\rho(p)\gamma_\alpha\gamma_5].
\end{aligned}
\tag{C.30}
$$

Comparing Eqs. (C.27) and (C.30), we have

$$
d_\alpha = -\frac{m}{2}\zeta_\alpha.
\tag{C.31}
$$

Finally, we obtain $c_{\mu\nu}$ in Eq. (C.19). To evaluate $c_{\mu\nu}$, the following relation for $\sigma^{\mu\nu}$ have been used

$$
\sigma^{\mu\nu} = \frac{i}{2}\epsilon^{\mu\nu\rho\tau}\sigma_{\rho\tau}\gamma_5.
\tag{C.32}
$$

Using Eq. (C.32), we have

$$
c_{\mu\nu}\sigma^{\mu\nu} = c_{\mu\nu}\left[\frac{i}{2}\epsilon^{\mu\nu\rho\tau}\sigma_{\rho\tau}\gamma_5\right] = c'^{\rho\tau}\sigma_{\rho\tau}\gamma_5.
$$

Using this relation in Eq. (C.19), we have

$$
\rho(p) = a + b_\mu\gamma^\mu + c'^{\rho\tau}\sigma_{\rho\tau}\gamma_5 + d_\mu\gamma^\mu\gamma_5 + e\gamma_5.
$$

Taking the trace of this equation after multiplying it with $\sigma_{\rho\tau}\gamma_5$, we have

$$
\begin{aligned}
\mathrm{Tr}[\rho(p)\sigma_{\rho\tau}\gamma_5] &= a\mathrm{Tr}[\sigma_{\rho\tau}\gamma_5] + b_\mu tr[\gamma^\mu\sigma_{\rho\tau}\gamma_5] + c'^{\rho\tau}\mathrm{Tr}[\sigma_{\rho\tau}\sigma_{\rho\tau}] + d_\mu\mathrm{Tr}[\gamma^\mu\sigma_{\rho\tau}] + e\mathrm{Tr}[\sigma_{\rho\tau}], \\
\implies c'^{\rho\tau} &= \frac{1}{8}\mathrm{Tr}[\rho(p)\sigma^{\rho\tau}\gamma_5] = \frac{1}{8(2m)^2}\mathrm{Tr}[\Lambda(p)\rho(p)\Lambda(p)\sigma^{\rho\tau}\gamma_5].
\end{aligned}
$$

Using the properties of the spin density matrix and Eq. (C.30) in the aforementioned expression, we have

$$
c'^{\rho\tau} = \frac{i}{4}[p^\tau\zeta^\rho - p^\rho\zeta^\tau].
$$

Using this equation, $c_{\mu\nu}\sigma^{\mu\nu}$ can be written as

$$
c_{\mu\nu}\sigma^{\mu\nu} = \frac{1}{2}\not{\zeta}\,\not{p}\gamma_5.
\tag{C.33}
$$

Substituting the values of coefficients a to e from Eqs. (C.22), (C.24),(C.26), (C.31), and (C.33) in Eq. (C.19), we obtain

$$
\begin{aligned}
\rho(p) &= \frac{m}{2} + \frac{p_\mu}{2}\gamma^\mu + (\frac{1}{2}\ \not{\xi}\ \not{p}\gamma_5) + \frac{m}{2}\xi_\mu\gamma^\mu\gamma_5 \\
&= \frac{1}{2}(\not{p}+m)(1+\gamma_5\ \not{\xi}).
\end{aligned}
\tag{C.34}
$$

This is the expression for the spin density matrix of the initial state particle.

Spin density matrix of the final state particle

Consider a process in which both the initial and the final spin states are taken into account and are described by the density matrix. In this section, we obtain the expression for the density matrix of the final state particles. Consider the scattering of a spin 0 particle from a spin $\frac{1}{2}$ particle, the matrix element for the process has the following form:

$$
< f|(S-1)|i >= N\bar{u}^{r'}(p')\mathcal{M}(p',q';p,q)u^r(p)(2\pi)^4\delta(p'+q'-p-q),
\tag{C.35}
$$

where $q, p(q', p')$ are the momenta of initial (final) spin 0 and spin $\frac{1}{2}$ particles, respectively, and $\mathcal{M}$ is the matrix acting on the spin variables.

The spin wave function of the final particle is given as

$$
\chi^r(p') = \sum_{r_1} u^{r_1}(p')(\bar{u}^{r_1}(p')\mathcal{M}u^r(p)).
\tag{C.36}
$$

The initial particles are described by the density matrix

$$
\rho(p) = \sum_{r,r'} u^r(p)\bar{u}^{r'}(p)\rho_{rr'}.
\tag{C.37}
$$

The mean value of operator O', which acts on the final particle spin variables, is obtained as

$$
< \hat{O}' >= \frac{\sum_{r,r'}\bar{\chi}^{r'}(p')O'\chi^r(p')\rho_{rr'}}{\sum_{r,r'}\bar{\chi}^{r'}(p')\chi^r(p')\rho_{rr'}} = \frac{\mathrm{Tr}[O'\rho(p')]}{\mathrm{Tr}[\rho(p')]},
\tag{C.38}
$$

where, the spin density matrix for the final particle $\rho(p')$ is expressed as

$$
\rho(p') = \Lambda(p')\mathcal{M}\rho(p)\bar{\mathcal{M}}\Lambda(p').
\tag{C.39}
$$

Trace of $\rho(p')$ is obtained as

$$
\mathrm{Tr}[\rho(p')] = 2m'\mathrm{Tr}[\mathcal{M}\rho(p)\bar{\mathcal{M}}\Lambda(p')],
\tag{C.40}
$$

where m' is the mass of the final state particle. The polarization four-vector for the final state particle is calculated as

$$
\zeta'^\alpha = -\frac{\mathrm{Tr}[\rho(p')\gamma_5\gamma^\alpha]}{\mathrm{Tr}[\rho(p')]}.
\tag{C.41}
$$

Using Eqs. (C.39) and (C.40) in this equation, we get

$$\zeta^\alpha = -\frac{\mathrm{Tr}[\Lambda(p')\mathcal{M}\rho(p)\bar{\mathcal{M}}\Lambda(p')\gamma_5\gamma^\alpha]}{2m'\mathrm{Tr}[\mathcal{M}\rho(p)\bar{\mathcal{M}}\Lambda(p')]}. \tag{C.42}$$

Using the following relations in Eq. (C.42):

$$\Lambda(p')\gamma_5\gamma^\alpha\Lambda(p') = -2m'\Lambda(p')\left(g^{\alpha\beta} - \frac{p'^\alpha p'^\beta}{m'^2}\right)\gamma_5\gamma_\beta \tag{C.43}$$

and

$$\Lambda(p')\Lambda(p') = 2M'\Lambda(p'), \tag{C.44}$$

the expression for the polarization four-vector of the final particle can be obtained as

$$\zeta^\alpha = -\left(g^{\alpha\beta} - \frac{p'^\alpha p'^\beta}{m'^2}\right)\frac{\mathrm{Tr}\left[\mathcal{M}\rho(p)\bar{\mathcal{M}}\Lambda(p')\gamma_5\gamma_\beta\right]}{\mathrm{Tr}\left[\mathcal{M}\rho(p)\bar{\mathcal{M}}\Lambda(p')\right]}. \tag{C.45}$$

If the scattering of spin $\frac{1}{2}$ particles take place with another spin $\frac{1}{2}$ particle, where the spin density matrix of the final particle is taken into account, then the polarization four-vector of the final particle becomes

$$\zeta^\alpha = \left(g^{\alpha\beta} - \frac{p'^\alpha p'^\beta}{M'^2}\right)\frac{\mathcal{L}^{\mu\nu}\mathrm{Tr}\left[\gamma_\beta\gamma_5\Lambda(p')\Gamma_\mu\Lambda(p)\tilde{\Gamma}_\nu\right]}{\mathcal{L}^{\mu\nu}\mathrm{Tr}\left[\Lambda(p')\Gamma_\mu\Lambda(p)\tilde{\Gamma}_\nu\right]}. \tag{C.46}$$

Appendix **D**

Leptonic and Hadronic Tensors

D.1 Contraction of Leptonic Tensors in Electromagnetic Interactions

For the $e^-(k) + \mu^-(p) \to e^-(k') + \mu^-(p')$ scattering discussed in Chapter 9, the transition matrix element squared is written as

$$\overline{\sum}\sum | \mathcal{M} |^2 = \frac{1}{2}\frac{1}{2}\frac{e^4}{q^4} L_{\mu\nu}^{\text{electron}} L_{\text{muon}}^{\mu\nu}, \tag{D.1}$$

where $q = k - k' = p' - p$ is the momentum transfer and the factor of $\frac{1}{2}$ is for the averaging over the initial electron and muon spins.

The leptonic current is given by

$$l_\mu = \left[\bar{u}(k') \gamma_\mu u(k) \right]. \tag{D.2}$$

In Eq. (D.2),

- Adjoint Dirac spinor$(\bar{u})$ is a 1×4 matrix,

- Dirac spinor (u) is a 4×1 matrix,

- γ_μ is a 4×4 matrix,

- Ultimately, we have $(1 \times 4)(4 \times 4)(4 \times 1) = A$, a number,

- For any number A, the complex conjugate and the Hermitian conjugate are the same thing.

Therefore, instead of $L_{\mu\nu} = l_\mu l_\nu{}^*$, we may write $L_{\mu\nu} = l_\mu l_\nu{}^\dagger$

$$
\begin{aligned}
l_\mu l_\nu{}^\dagger &= \left[\bar{u}(k')\gamma_\mu u(k)\right] \times \left[\bar{u}(k')\gamma_\nu u(k)\right]^\dagger, \\
&= \left[\bar{u}(k')\gamma_\mu u(k)\right] \times \left[u^\dagger(k)\gamma_\nu^\dagger \gamma_0^\dagger (u^\dagger(k'))^\dagger\right]; \qquad \bar{u} = u^\dagger \gamma_0 \\
&= \left[\bar{u}(k')\gamma_\mu u(k)\right] \times \left[u^\dagger(k)\gamma_0 \gamma_0 \gamma_\nu^\dagger \gamma_0 u(k')\right]; \qquad \gamma_0^\dagger = \gamma_0, \ \gamma_0^2 = 1_{4\times 4}, \\
&= \left[\bar{u}(k')\gamma_\mu u(k)\right] \times \left[\bar{u}(k)\gamma_\nu u(k')\right]; \qquad \gamma_0 \gamma_\nu^\dagger \gamma_0 = \gamma_\nu,
\end{aligned}
$$

we can rewrite the aforementioned expression in the component form for an electronic tensor as:

$$
\begin{aligned}
L_{\mu\nu}^{\text{electron}} &= \sum_{\alpha,\beta,\delta,\sigma=1}^{4} \sum_{\mu,\nu=0}^{3} \left[\bar{u}_\alpha(k')\{\gamma_\mu\}_{\alpha\beta} u_\beta(k)\right]\left[\bar{u}_\delta(k)\{\gamma_\nu\}_{\delta\sigma} u_\sigma(k')\right], \\
&= \sum_{\alpha,\beta,\delta,\sigma=1}^{4} \sum_{\mu,\nu=0}^{3} \left[\bar{u}_\alpha(k')\{\gamma_\mu\}_{\alpha\beta}(\slashed{k}+m_e)_{\beta\delta}\{\gamma_\nu\}_{\delta\sigma} u_\sigma(k')\right], \\
&= \sum_{\alpha,\beta,\delta,\sigma=1}^{4} \sum_{\mu,\nu=0}^{3} \left[(\slashed{k}'+m_e)_{\sigma\alpha}\{\gamma_\mu\}_{\alpha\beta}(\slashed{k}+m_e)_{\beta\delta}\{\gamma_\nu\}_{\delta\sigma}\right], \\
&= \sum_{\sigma,\delta=1}^{4} \sum_{\mu,\nu=0}^{3} \left[\{(\slashed{k}'+m_e)\gamma_\mu(\slashed{k}+m_e)\}_{\sigma\delta}\{\gamma_\nu\}_{\delta\sigma}\right], \\
&= \sum_{\sigma=1}^{4} \sum_{\mu,\nu=0}^{3} \left[(\slashed{k}'+m_e)\{\gamma_\mu\}(\slashed{k}+m_e)\{\gamma_\nu\}\right]_{\sigma\sigma}, \\
&= \sum_{\mu,\nu=0}^{3} Tr\left[(\slashed{k}'+m_e)\gamma_\mu(\slashed{k}+m_e)\gamma_\nu\right], \\
&= 4\left[k_\mu' k_\nu + k_\mu k_\nu' - k\cdot k' g_{\mu\nu} + m_e^2 g_{\mu\nu}\right],
\end{aligned}
\tag{D.3}
$$

where we have used the trace properties,

$$
\begin{aligned}
Tr\left[\gamma_\mu \gamma_\nu\right] &= 4g_{\mu\nu} \\
Tr\left[\gamma_\mu \gamma_\nu \gamma_\theta \gamma_\phi\right] &= 4\left[g_{\mu\nu}g_{\theta\phi} - g_{\mu\theta}g_{\nu\phi} + g_{\mu\phi}g_{\nu\theta}\right].
\end{aligned}
$$

The trace of an odd number of gamma matrices is zero. Similarly,

$$
\begin{aligned}
L_{\text{muon}}^{\mu\nu} &= \sum_{\alpha,\beta,\delta,\sigma=1}^{4} \sum_{\mu,\nu=0}^{3} \left[\bar{u}_\alpha(p')\{\gamma^\mu\}_{\alpha\beta} u_\beta(p)\right]\left[\bar{u}_\delta(p)\{\gamma^\nu\}_{\delta\sigma} u_\sigma(p')\right] \\
&= \sum_{\mu,\nu=0}^{3} Tr\left[(\slashed{p}'+m_\mu)\gamma^\mu(\slashed{p}+m_\mu)\gamma^\nu\right] \\
&= 4\left[p'^\mu p^\nu + p^\mu p'^\nu - p\cdot p' g^{\mu\nu} + m_\mu^2 g^{\mu\nu}\right].
\end{aligned}
\tag{D.4}
$$

Using Eqs. (D.3) and (D.4), we get

$$
\begin{aligned}
L_{\mu\nu}^{\text{electron}} L_{\text{muon}}^{\mu\nu} &= 16 \left[k_\mu' k_\nu + k_\mu k_\nu' - k \cdot k' g_{\mu\nu} + m_e^2 g_{\mu\nu} \right] \times \left[p'^\mu p^\nu + p^\mu p'^\nu - p \cdot p' g^{\mu\nu} + m_\mu^2 g^{\mu\nu} \right] \\
&= 16 \left[2k \cdot p k' \cdot p' + 2k \cdot p' k' \cdot p - 2(p \cdot p' - m_\mu^2) k \cdot k' \right. \\
&\quad \left. - 2(k \cdot k' - m_e^2) p \cdot p' + 4(k \cdot k' - m_e^2)(p \cdot p' - m_\mu^2) \right].
\end{aligned}
\tag{D.5}
$$

$$
\overline{\sum} \sum | \mathcal{M} |^2 = \frac{8e^4}{q^4} \left[k \cdot p k' \cdot p' + k \cdot p' k' \cdot p - m_e^2 p \cdot p' - m_\mu^2 k \cdot k' + 2m_e^2 m_\mu^2 \right].
\tag{D.6}
$$

D.2　Contraction of Leptonic Tensors in the Case of Weak Interactions

For the $\nu_e + \mu^- \to e^- + \nu_\mu$ scattering discussed in Chapter 9, where the interaction is mediated by a W boson, the transition matrix element squared is expressed as

$$
\overline{\sum} \sum | \mathcal{M} |^2 = \frac{G_F^2}{2} \frac{1}{2} L_{\mu\nu}^{\text{electron}} L_{\text{muon}}^{\mu\nu},
\tag{D.7}
$$

where the factor of $\frac{1}{2}$ is for the averaging over the initial muon spin.

Leptonic tensor, $L_{\mu\nu} = l_\mu l_\nu^\dagger$ with

$$
l_\mu = \bar{u}(k') \gamma_\mu (1 - \gamma_5) u(k).
$$

Therefore,

$$
L_{\mu\nu} = \left[\bar{u}(k') \gamma_\mu (1 - \gamma_5) u(k) \right] \cdot \left[\bar{u}(k) \gamma_\nu (1 - \gamma_5) u(k') \right].
$$

We can rewrite this expression in the component form as:

$$
\begin{aligned}
L_{\mu\nu}^{\text{electron}} &= \sum_{\alpha,\beta,\delta,\sigma=1}^{4} \sum_{\mu,\nu=0}^{3} \left[\bar{u}_\alpha(k') \{\gamma_\mu(1-\gamma_5)\}_{\alpha\beta} u_\beta(k) \right] \left[\bar{u}_\delta(k) \{\gamma_\nu(1-\gamma_5)\}_{\delta\sigma} u_\sigma(k') \right] \\
&= \sum_{\sigma=1}^{4} \sum_{\mu,\nu=0}^{3} \left[(\not{k}' + m_e) \{\gamma_\mu(1-\gamma_5)\} (\not{k} + m_{\nu_e}) \{\gamma_\nu(1-\gamma_5)\} \right]_{\sigma\sigma} \\
&= 8 \left[k_\mu k_\nu' + k_\mu' k_\nu - k \cdot k' g^{\mu\nu} + i\epsilon_{\mu\nu\alpha\beta} k^\alpha k'^\beta \right]; \quad \text{with } m_{\nu_e} = 0.
\end{aligned}
\tag{D.8}
$$

Similarly, for the muonic tensor,

$$
L_{\text{muon}}^{\mu\nu} = 8 \left[p_\mu p_\nu' + p_\mu' p_\nu - g^{\mu\nu} p \cdot p' - i\epsilon^{\mu\nu\theta\phi} p_\theta p_\phi' \right]; \quad \text{with } m_{\nu_\mu} = 0.
\tag{D.9}
$$

Using Eqs (D.8) and (D.9), the transition matrix element squared is obtained as

$$
\begin{aligned}
\overline{\sum}\sum|\mathcal{M}|^2 &= 16\, G_F^2 \left[k_\mu k'_\nu + k'_\mu k_\nu - g^{\mu\nu} k\cdot k' + i\epsilon_{\mu\nu\alpha\beta} k^\alpha k'^\beta \right], \\
&\times \left[p_\mu p'_\nu + p'_\mu p_\nu - g^{\mu\nu} p\cdot p' - i\epsilon^{\mu\nu\theta\phi} p_\theta p'_\phi \right], \\
&= 16 G_F^2 \left[2k\cdot p k'\cdot p' + 2k\cdot p' k'\cdot p + 2\left(\delta^\alpha_\theta \delta^\beta_\phi - \delta^\alpha_\phi \delta^\beta_\theta\right) k^\alpha k'^\beta p_\theta p'_\phi \right], \\
&= 32\, G_F^2 \left[k\cdot p k'\cdot p' + k\cdot p' k'\cdot p + k\cdot p k'\cdot p' - k\cdot p' k'\cdot p \right], \\
&= 64\, G_F^2 \left[k\cdot p k'\cdot p' \right].
\end{aligned}
\tag{D.10}
$$

D.3 Contraction of Weak Leptonic Tensor with Hadronic Tensor

Contracting the various terms of hadronic tensor $J^{\mu\nu}$ with the leptonic tensor $L_{\mu\nu}$, we get

$$
\begin{aligned}
\overline{\sum}\sum|\mathcal{M}|^2 &= \frac{G_F^2 \cos\theta_C}{2}\frac{1}{2}\left(\frac{1}{4M_p M_n}\right)\left(\frac{1}{4m_l m_\nu}\right) \\
&\times \Big\{ (f_1(q^2))^2 \left[(p_1.p_2)(p_3.p_4) + (p_1.p_4)(p_2.p_3) - M_p M_n (p_1.p_3) \right] \\
&+ \left(g_1^2(q^2)\right)\left[(p_1.p_2)(p_3.p_4) + (p_1.p_4)(p_2.p_3) + M_p M_n (p_1.p_3) \right] \\
&+ \left(2f_1(q^2)g_1(q^2)\right)\left[(p_1.p_4)(p_2.p_3) - (p_1.p_2)(p_3.p_4) \right] \\
&+ \left(\frac{f_2(q^2)}{2M}\right)^2 \left[(p_1.p_2)(p_3.q)(p_4.q) - (p_1.p_2)(p_3.p_4)q^2 + (p_1.p_4)(p_2.q)(p_3.q) \right. \\
&- (p_1.p_4)(p_2.p_3)q^2 + \frac{1}{2}(p_1.p_3)(p_2.p_4)q^2 - \frac{1}{2}(p_1.p_3)M_p M_n q^2 \\
&- (p_1.q)(p_3.q)(p_2.p_4) - (p_1.q)(p_3.q)M_n M_p + (p_1.q)(p_2.q)(p_3.p_4) \\
&+ (p_1.q)(p_4.q)(p_2.p_3)] + \left(\frac{f_1(q^2)f_2(q^2)}{2M}\right)\left[-M_n(p_1.p_3)(p_4.q) - M_n(p_1.q)(p_3.p_4) \right. \\
&- M_n(p_3.q)(p_1.p_4) + M_p(p_1.p_3)(p_2.q) + M_p(p_1.q)(p_3.p_2) + M_p(p_3.q) \\
&(p_1.p_2)] + \left(\frac{2g_1(q^2)f_2(q^2)}{2M}\right)\left[-M_p(p_3.q)(p_1.p_2) + M_p(p_2.p_3)(p_1.q) - M_n(p_3.q) \right. \\
&(p_1.p_4) + M_n(p_3.p_4)(p_1.q)] + \left(g_3^2(q^2)\right)\left[(p_1.q)(p_3.q) - \frac{1}{2}q^2(p_1.p_3) \right][(p_2.p_4) \\
&- M_p M_n] + \left(g_1(q^2)g_3(q^2)\right)\left[-M_n(p_1.p_4)(p_3.q) - M_n(p_1.q)(p_3.p_4) + M_n(p_4.q) \right. \\
&(p_1.p_3) + M_p(p_1.p_2)(p_3.q) + M_p(p_1.q)(p_2.p_3) - M_p(p_2.q)(p_1.p_3)] \\
&+ \left(\frac{f_2(q^2)g_3(q^2)}{2M}\right)[-(p_4.q)(p_1.p_2)(p_3.q) + (p_4.q)(p_1.q)(p_3.p_2) + q^2(p_3.p_4)(p_1.p_2) \\
&- q^2(p_1.p_4)(p_2.p_3) - (p_2.q)(p_1.q)(p_3.p_4) + (p_2.q)(p_3.q)(p_1.p_4) \Big\},
\end{aligned}
\tag{D.11}
$$

where

$$
\begin{aligned}
p_1 &= (E_\nu, \vec{p}_\nu), \quad p_2 = (E_n, \vec{p}_n), \\
p_3 &= (E_l, \vec{p}_l), \quad p_4 = (E_p, \vec{p}_p), \\
E_n &= \sqrt{\vec{p}_n^2 + M_n^2}, \quad E_p = \sqrt{\vec{p}_p^2 + M_p^2}, \\
q_\mu &= (E_\nu - E_l, \vec{p}_\nu - \vec{p}_l).
\end{aligned}
\tag{D.12}
$$

D.5 Contraction of Weak Leptonic Tensor with Hadronic Tensor

General Expression for the Total Scattering Cross Section and Decay Rates

E.1 Cross Section

Consider a two body scattering process with four momenta $p_i \equiv (E_i, \vec{p}_i)$, $i = a, b$. There are N particles in the final state with four momenta $p_f \equiv (E_f, \vec{p}_f)$, $f = 1, 2, \ldots, N$.

The general expression for the cross section is given by

$$d\sigma = \frac{1}{4E_a E_b |\vec{v}_a - \vec{v}_b|} (2\pi)^4 \delta^4 \left(\sum_{f=1}^{N} p_f - \sum_{i=a,b} p_i \right) \prod_{f=1}^{N} \frac{d^3 p_f}{(2\pi)^3 2E_f} \sum \sum |\mathcal{M}|^2, \quad \text{(E.1)}$$

where

$$|\vec{v}_a - \vec{v}_b| = v_{\text{rel}} = |\frac{\vec{p}_a}{E_a} - \frac{\vec{p}_b}{E_b}| = \frac{1}{E_a E_b} |\vec{p}_a E_b - \vec{p}_b E_a|,$$

$$v_{\text{rel}} = \frac{1}{E_a E_b} \sqrt{|\vec{p}_a|^2 E_b^2 + |\vec{p}_b|^2 E_a^2 - 2E_a E_b |\vec{p}_a||\vec{p}_b| \cos\theta_{ab}}.$$

In the laboratory frame, where, say, particle "b" is at rest (i.e. $v_b = 0$, $E_b = m_b$)

$$E_a E_b v_{\text{rel}} = m_b E_a v_a, \quad \text{using} \quad \vec{p} = E\vec{v}.$$

$v_a = c = 1$, if the incident particle is relativistic, that is, $E_a \gg m_a$.

In the center of mass frame, where particles "a" and "b" approach each other from exactly opposite direction, that is, $\theta_{ab} = 180^o$, with the same magnitude of three momenta, that is, $|\vec{p}_a| = |\vec{p}_b| = |\vec{p}_i|$ such that

$$E_a E_b v_{\text{rel}} = |\vec{p}_i| [E_a + E_b] = |\vec{p}_i| \sqrt{s},$$

where $\sqrt{s}$ is the center of mass energy.

More conveniently, this is also written as

$$E_a E_b v_{rel} = \sqrt{(p_a \cdot p_b)^2 - m_a^2 m_b^2}.$$

Two body scattering

For a reaction, where "a" and "b" are particles in the initial state and "1" and "2" are particles in the final state, that is,

$$a + b \rightarrow 1 + 2, \tag{E.2}$$

the general expression for the differential scattering cross section is given by

$$d\sigma = \frac{1}{|\vec{v}_a - \vec{v}_b|} \frac{1}{2E_a} \frac{1}{2E_b} (2\pi)^4 \delta^4 \left(\sum_{f=1,2} p_f - \sum_{i=a,b} p_i \right) \frac{d^3 p_1}{(2\pi)^3 2E_1} \frac{d^3 p_2}{(2\pi)^3 2E_2} \overline{\sum} \sum |\mathcal{M}|^2. \tag{E.3}$$

In any experiment, one observes either particle "1" or particle "2"; therefore, the kinematical quantities of the particle which is not to be observed are fixed by doing phase space integration. For example, if particle "2" is not to be observed, then

$$\int \delta^3 (\vec{p}_a + \vec{p}_b - \vec{p}_1 - \vec{p}_2) d^3 p_2 = 1, \tag{E.4}$$

which gives the constraint on $\vec{p}_2 = \vec{p}_a + \vec{p}_b - \vec{p}_1 = \vec{q} + \vec{p}_b$, where $\vec{q}$ is the three momentum transfer and $E_2 = \sqrt{|\vec{q} + \vec{p}_b|^2 + m_2^2}$, which results in

$$d\sigma = \frac{1}{4\sqrt{(p_a \cdot p_b)^2 - m_a^2 m_b^2}} \frac{1}{(2\pi)^2} \delta^0 (E_a + E_b - E_1 - E_2) \overline{\sum} \sum |\mathcal{M}|^2 \frac{d^3 p_1}{2E_1} \frac{1}{2E_2}, \tag{E.5}$$

$$d^3 p_1 = |\vec{p}_1|^2 d|\vec{p}_1| d\Omega_1 = E_1 |\vec{p}_1| dE_1 d\Omega_1,$$

$$\frac{d\sigma}{dE_1 d\Omega_1} = \frac{1}{64\pi^2 \sqrt{(p_a \cdot p_b)^2 - m_a^2 m_b^2}} \delta^0 (E_a + E_b - E_1 - E_2) \overline{\sum} \sum |\mathcal{M}|^2 \frac{|\vec{p}_1|}{E_2}. \tag{E.6}$$

Integrating over the energy of particle "1", using the delta function integration property, we get

$$\int \delta f(x) dx = \left| \frac{\partial f(x)}{\partial x} \right|_{x=x_0}^{-1}. \tag{E.7}$$

Now, we obtain the angular distribution for the outgoing particle "1".

For this, we evaluate

$$\int \delta^0 \left(E_a + E_b - E_1 - \sqrt{|\vec{p}_a + \vec{p}_b - \vec{p}_1|^2 + m_2^2} \right) dE_1. \tag{E.8}$$

Using $\vec{p}_i = \vec{p}_a + \vec{p}_b$, $E_i = E_a + E_b$, $f(E_1) = E_i - E_1 - \sqrt{|\vec{p}_i - \vec{p}_1|^2 + m_2^2}$,

$$\frac{\partial f(E_1)}{\partial E_1} = -1 - \frac{\partial \left(\sqrt{|\vec{p}_i - \vec{p}_1|^2 + m_2^2}\right)}{\partial E_1},$$

$$= -1 - \frac{\frac{\partial}{\partial E_1}\left(|\vec{p}_i|^2 + |\vec{p}_1|^2 - 2\vec{p}_i \cdot \vec{p}_1 + m_2^2\right)}{2\left(\sqrt{|\vec{p}_i - \vec{p}_1|^2 + m_2^2}\right)},$$

$$= \frac{-(E_1 + E_2)|\vec{p}_1|^2 + \vec{p}_i \cdot \vec{p}_1 E_1}{|\vec{p}_1|^2 E_2}, \quad \text{where } E_2 = \sqrt{|\vec{p}_i - \vec{p}_1|^2 + m_2^2},$$

$$= -\left(\frac{E_i|\vec{p}_1|^2 - \vec{p}_i \cdot \vec{p}_1 E_1}{|\vec{p}_1|^2 E_2}\right),$$

$$\left|\frac{\partial f(E_1)}{\partial E_1}\right| = \left(\frac{E_i|\vec{p}_1|^2 - \vec{p}_i \cdot \vec{p}_1 E_1}{|\vec{p}_1|^2 E_2}\right). \tag{E.9}$$

Thus,

$$\int \delta^0\left(E_a + E_b - E_1 - \sqrt{|\vec{p}_a + \vec{p}_b - \vec{p}_1|^2 + m_2^2}\right) dE_1 = \frac{|\vec{p}_1|^2 E_2}{E_i|\vec{p}_1|^2 - \vec{p}_i \cdot \vec{p}_1 E_1}, \tag{E.10}$$

which results in,

$$\frac{d\sigma}{d\Omega_1} = \frac{1}{64\pi^2 \sqrt{(p_a \cdot p_b)^2 - m_a^2 m_b^2}} \overline{\sum}\sum |\mathcal{M}|^2 \frac{|\vec{p}_1|^3}{(E_i|\vec{p}_1|^2 - \vec{p}_i \cdot \vec{p}_1 E_1)}. \tag{E.11}$$

If the scattering $a + b \rightarrow 1 + 2$ takes place in the lab frame, $\vec{p}_b = 0$, $E_b = m_b$ and $E_i = E_a + m_b$, it results in

$$\frac{d\sigma}{d\Omega_1}\bigg|_{\text{Lab}} = \frac{1}{64\pi^2 m_b E_a} \overline{\sum}\sum |\mathcal{M}|^2 \frac{|\vec{p}_1|^3}{((E_a + m_b)|\vec{p}_1|^2 - \vec{p}_a \cdot \vec{p}_1 E_1)}. \tag{E.12}$$

If the scattering $a + b \rightarrow 1 + 2$ takes place in the center of mass frame, where, $|\vec{p}_a| = |\vec{p}_b| = |\vec{p}|$, $|\vec{p}_1| = |\vec{p}_2| = |\vec{p}\,'|$ and $E_a + E_b = \sqrt{s} = E_1 + E_2$, is the center of mass energy, we write

$$\frac{d\sigma}{d\Omega_1}\bigg|_{\text{CM}} = \frac{1}{64\pi^2 s} \frac{|\vec{p}\,'|}{|\vec{p}|} \overline{\sum}\sum |\mathcal{M}|^2. \tag{E.13}$$

Energy distribution of the outgoing particle "1"

Here, we evaluate energy distribution in the lab frame. The delta function given in Eq. (E.8), in the lab frame may be written as

$$\int \delta^0\left(E_a + m_b - E_1 - \sqrt{|\vec{p}_a - \vec{p}_1|^2 + m_2^2}\right) d\cos\theta_{a1}. \tag{E.14}$$

Here,

$$f(\cos\theta_{a1}) = E_a + m_b - E_1 - \sqrt{|\vec{p}_a|^2 + |\vec{p}_1|^2 - 2|\vec{p}_a||\vec{p}_1|\cos\theta_{a1} + m_2^2},$$

$$\frac{\partial f(\cos\theta_{a1})}{\partial\cos\theta_{a1}} = -\frac{\partial\left(\sqrt{|\vec{p}_a|^2 + |\vec{p}_1|^2 - 2|\vec{p}_a||\vec{p}_1|\cos\theta_{a1} + m_2^2}\right)}{\partial\cos\theta_{a1}},$$

$$= \frac{|\vec{p}_a||\vec{p}_1|}{E_2}, \tag{E.15}$$

$$\frac{d\sigma}{dE_1}\bigg|_{\text{Lab}} = \frac{1}{32\pi m_b E_a |\vec{p}_a|}\overline{\sum}\sum|\mathcal{M}|^2. \tag{E.16}$$

Three body scattering

For a general reaction, where "a" and "b" are particles in the initial state and "1", "2", and "3" are particles in the final state, that is,

$$a + b \rightarrow 1 + 2 + 3, \tag{E.17}$$

the expression of differential scattering cross section is given by

$$d\sigma = \frac{1}{|\vec{v}_a - \vec{v}_b|}\frac{1}{2E_a}\frac{1}{2E_b}(2\pi)^4\delta^4\left(\sum_{f=1,2,3}p_f - \sum_{i=a,b}p_i\right)\overline{\sum}\sum|\mathcal{M}|^2\frac{d^3p_1}{(2\pi)^32E_1}$$

$$\times\frac{d^3p_2}{(2\pi)^32E_2}\frac{d^3p_3}{(2\pi)^32E_3}. \tag{E.18}$$

Suppose, now the particle in state "2" is not to be observed, then

$$\int\delta^3(\vec{p}_a + \vec{p}_b - \vec{p}_1 - \vec{p}_2 - \vec{p}_3)d^3p_2 = 1, \tag{E.19}$$

gives the constraint on $\vec{p}_2 = \vec{p}_a + \vec{p}_b - \vec{p}_1 - \vec{p}_3$ and $E_2 = \sqrt{|\vec{p}_a + \vec{p}_b - \vec{p}_1 - \vec{p}_3|^2 + m_2^2}$, which results in

$$d\sigma = \frac{1}{4\sqrt{(p_a\cdot p_b)^2 - m_a^2 m_b^2}}\frac{1}{(2\pi)^5}\delta^0\left(E_a + E_b - E_1 - E_2 - E_3\right)$$

$$\times\overline{\sum}\sum|\mathcal{M}|^2\frac{d^3p_1}{2E_1}\frac{1}{2E_2}\frac{d^3p_3}{2E_3} \tag{E.20}$$

$$d^3p_3 = |\vec{p}_3|^2 d|\vec{p}_3|d\Omega_3 = E_3|\vec{p}_3|dE_3d\Omega_3$$

$$\frac{d\sigma}{dE_3d\Omega_3} = \frac{1}{4\sqrt{(p_a\cdot p_b)^2 - m_a^2 m_b^2}}\frac{1}{(2\pi)^5}\delta^0\left(E_a + E_b - E_1 - E_2 - E_3\right)$$

$$\times\overline{\sum}\sum|\mathcal{M}|^2\frac{d^3p_1}{2E_1}\frac{1}{2E_2}\frac{E_3|\vec{p}_3|}{2E_3}. \tag{E.21}$$

Integrating over the energy of particle "3", using the delta function integration property given in Eq. (E.8),

$$\int \delta^0 \left(E_a + E_b - E_1 - E_3 - \sqrt{|\vec{p}_a + \vec{p}_b - \vec{p}_1 - \vec{p}_3|^2 + m_2^2} \right) dE_3, \tag{E.22}$$

and $\vec{p}_i = \vec{p}_a + \vec{p}_b$, $E_i = E_a + E_b$, $f(E_3) = E_i - E_1 - E_3 - \sqrt{|\vec{p}_i - \vec{p}_1 - \vec{p}_3|^2 + m_2^2}$,

$$\frac{\partial f(E_3)}{\partial E_3} = -1 - \frac{\partial \left(\sqrt{|\vec{p}_i - \vec{p}_1 - \vec{p}_3|^2 + m_2^2} \right)}{\partial E_3},$$

$$= -1 - \frac{\frac{\partial}{\partial E_3} \left(|\vec{p}_i - \vec{p}_1|^2 + |\vec{p}_3|^2 - 2(\vec{p}_i - \vec{p}_1) \cdot \vec{p}_3 + m_2^2 \right)}{2 \left(\sqrt{|\vec{p}_i - \vec{p}_1 - \vec{p}_3|^2 + m_2^2} \right)},$$

$$= -1 - \frac{E_3 |\vec{p}_3|^2 - (\vec{p}_i - \vec{p}_1) \cdot \vec{p}_3 E_3}{|\vec{p}_3|^2 E_2}, \quad \text{where} \ E_2 = \sqrt{|\vec{p}_i - \vec{p}_1 - \vec{p}_3|^2 + m_2^2},$$

$$= -\left(\frac{(E_i - E_1)|\vec{p}_3|^2 - (\vec{p}_i - \vec{p}_1) \cdot \vec{p}_3 E_3}{|\vec{p}_3|^2 E_2} \right),$$

$$\left| \frac{\partial f(E_3)}{\partial E_3} \right| = \left(\frac{(E_i - E_1)|\vec{p}_3|^2 - (\vec{p}_i - \vec{p}_1) \cdot \vec{p}_3 E_3}{|\vec{p}_3|^2 E_2} \right), \tag{E.23}$$

$$\int \delta^0 \left(E_a + E_b - E_1 - E_3 - \sqrt{|\vec{p}_a + \vec{p}_b - \vec{p}_1 - \vec{p}_3|^2 + m_2^2} \right) dE_3 = \frac{|\vec{p}_3|^2 E_2}{(E_i - E_1)|\vec{p}_3|^2 - (\vec{p}_i - \vec{p}_1) \cdot \vec{p}_3 E_3}.$$

In the lab frame $\vec{p}_b = 0$, $E_b = m_b$, $E_i = E_a + m_b$

$$\frac{d\sigma}{d\Omega_3} = \frac{1}{2^{10}\pi^5 m_b E_a} \overline{\sum} \sum |\mathcal{M}|^2 \frac{d^3 p_1}{E_1} \frac{|\vec{p}_3|^3}{(E_a + m_b - E_1)|\vec{p}_3|^2 - (\vec{p}_a - \vec{p}_1) \cdot \vec{p}_3 E_3},$$

$$\frac{d\sigma}{dE_1 d\Omega_1 d\Omega_3} = \frac{1}{2^{10}\pi^5 m_b E_a} \overline{\sum} \sum |\mathcal{M}|^2 \frac{|\vec{p}_1||\vec{p}_3|^3}{(E_a + m_b - E_1)|\vec{p}_3|^2 - (\vec{p}_a - \vec{p}_1) \cdot \vec{p}_3 E_3}. \tag{E.24}$$

E.2 Decay Rate

Two body decay

The decay of pions is an example of two body decay, given by

$$\pi^- \longrightarrow \mu^- + \bar{\nu}_\mu$$
$$\pi^+ \longrightarrow \mu^+ + \nu_\mu$$
$$\pi^0 \longrightarrow \gamma + \gamma.$$

For the decay of a particle "a" into "1" and "2" in the final state

$$a \rightarrow 1 + 2, \tag{E.25}$$

the general expression of the decay rate is given by

$$d\Gamma = \frac{1}{2E_a}(2\pi)^4\delta^4\left(p_a - \sum_{f=1,2} p_f\right)\overline{\sum}\sum|\mathcal{M}|^2\frac{d^3p_1}{(2\pi)^32E_1}\frac{d^3p_2}{(2\pi)^32E_2}. \tag{E.26}$$

For the decay at rest $E_a = m_a$,

$$d\Gamma = \frac{1}{32\pi^2m_a}\delta^0\left(m_a - E_1 - E_2\right)\overline{\sum}\sum|\mathcal{M}|^2\frac{d^3p_1}{E_1E_2}, \tag{E.27}$$

where $\vec{p}_2 = -\vec{p}_1$ and $E_2 = \sqrt{|\vec{p}_1|^2 + m_2^2}$.

$$\frac{d\Gamma}{d\Omega_1 dE_1} = \frac{1}{32\pi^2m_a}\delta^0\left(m_a - E_1 - \sqrt{|\vec{p}_1|^2 + m_2^2}\right)\overline{\sum}\sum|\mathcal{M}|^2\frac{|\vec{p}_1|}{\sqrt{|\vec{p}_1|^2 + m_2^2}}, \tag{E.28}$$

$$\int\delta^0\left(m_a - E_1 - \sqrt{|\vec{p}_1|^2 + m_2^2}\right)dE_1 = \frac{\sqrt{|\vec{p}_1|^2 + m_2^2}}{(E_1 + \sqrt{|\vec{p}_1|^2 + m_2^2})} = \frac{E_2}{m_a}, \quad (m_a = E_1 + E_2) \tag{E.29}$$

$$\frac{d\Gamma}{d\Omega_1} = \frac{|\vec{p}_1|}{32\pi^2m_a^2}\overline{\sum}\sum|\mathcal{M}|^2. \tag{E.30}$$

Three body decay

The decay of muons, neutron, lambda, etc. are examples of three body decay, given by

$$\begin{aligned}
\mu^- &\longrightarrow \bar{\nu}_e + \nu_\mu + e^- \\
\mu^+ &\longrightarrow \bar{\nu}_\mu + \nu_e + e^+ \\
n &\longrightarrow \bar{\nu}_e + e^- + p \\
\Lambda &\longrightarrow \bar{\nu}_e + e^- + p.
\end{aligned} \tag{E.31}$$

For the decay of a particle "a" into "1", "2" and "3" in the final state

$$a \rightarrow 1 + 2 + 3, \tag{E.32}$$

the general expression of the decay rate is given by

$$d\Gamma = \frac{1}{2E_a}(2\pi)^4\delta^4\left(p_a - \sum_{f=1,2,3} p_f\right)\overline{\sum}\sum|\mathcal{M}|^2\frac{d^3p_1}{(2\pi)^32E_1}\frac{d^3p_2}{(2\pi)^32E_2}\frac{d^3p_3}{(2\pi)^32E_3}. \tag{E.33}$$

In all the examples cited in Eq. (E.31), at least one particle (say "1") is massless, and the other particle (say "2") is either massless or has negligible mass in comparision to the third particle in the final state. However, in processes like beta decay, where the kinetic energies of the outgoing particles are low, one cannot ignore the mass of e^-. Therefore, particle "2" is kept with the mass term alongwith the mass of particle "3".

$$\delta^4 \left(p_a - \sum_{f=1,2,3} p_f \right) = \delta^0 \left(m_a - \sum_{f=1,2,3} E_f \right) \delta^3 \left(\sum_{f=1,2,3} \vec{p}_f \right)$$

$$\int \delta^3(\vec{p}_1 + \vec{p}_2 + \vec{p}_3)d^3p_3 = 1; \quad \text{if } \vec{p}_3 = -(\vec{p}_1 + \vec{p}_2),$$

where δ^3 integration is performed over particle "3".

$$E_3^2 = |\vec{p}_1 + \vec{p}_2|^2 + m_3^2 = |\vec{p}_1|^2 + |\vec{p}_2|^2 + 2|\vec{p}_1||\vec{p}_2| \cos\theta + m_3^2,$$

$$E_3^2 = E_1^2 + E_2^2 - m_2^2 + 2E_1 \sqrt{E_2^2 - m_2^2} \cos\theta + m_3^2,$$

$$E_3 = \sqrt{E_1^2 + E_2^2 - m_2^2 + 2E_1 \sqrt{E_2^2 - m_2^2} \cos\theta + m_3^2}, \quad (E.34)$$

$$d^3p_1 = |\vec{p}_1|^2 d|\vec{p}_1|d\Omega_1 = E_1^2 dE_1 d\Omega_1 = E_1^2 dE_1 d\cos\theta d\phi,$$

$$d^3p_1 = 2\pi E_1^2 dE_1 d\cos\theta. \quad \text{(if there is no azimuthal dependence)}$$

For the zenith angle integration,

$$\int_{-1}^{+1} d(\cos\theta)\delta^0 \left(m_a - \sum_{f=1,2,3} E_f \right) = \int_{-1}^{+1} d(\cos\theta)\delta^0 \left(m_a - E_1 - E_2 \right.$$
$$\left. - \sqrt{E_1^2 + E_2^2 - m_2^2 + 2E_1 \sqrt{E_2^2 - m_2^2} \cos\theta + m_3^2} \right).$$

Assume

$$f(\cos\theta) = \left(m_a - E_1 - E_2 - \sqrt{E_1^2 + E_2^2 - m_2^2 + 2E_1 \sqrt{E_2^2 - m_2^2} \cos\theta + m_3^2} \right)$$

$$\implies \frac{\partial f(\cos\theta)}{\partial \cos\theta} = \frac{-E_1 \sqrt{E_2^2 - m_2^2}}{E_3}$$

$$\implies \left| \frac{\partial f(\cos\theta)}{\partial \cos\theta} \right|^{-1} = \frac{E_3}{E_1 \sqrt{E_2^2 - m_2^2}} \quad (E.35)$$

which results in

$$\mathbf{I}(E_1, E2) = \int_{-1}^{+1} d(\cos\theta) \frac{E_1 \sqrt{E_2^2 - m_2^2}}{E_3} \delta^0 (m_a - E_1 - E_2 - E_3) \qquad = \qquad 1, \qquad \text{(E.36)}$$

$$\text{if, } E_3^- < (m_a - E_1 - E_2) \qquad < \qquad E_3^+,$$

where

$$E_3^+ = \sqrt{E_1^2 + E_2^2 - m_2^2 + 2E_1 \sqrt{E_2^2 - m_2^2} + m_3^2} \qquad \text{for} \qquad \cos\theta = +1$$

$$E_3^- = \sqrt{E_1^2 + E_2^2 - m_2^2 - 2E_1 \sqrt{E_2^2 - m_2^2} + m_3^2} \qquad \text{for} \qquad \cos\theta = -1$$

$$\text{otherwise} \quad 0$$

After performing the zenith angle integration and assuming that the decay takes place at rest ($E_a = m_a$),

$$d\Gamma = \frac{1}{m_a(4\pi)^4} \overline{\sum}\sum |\mathcal{M}|^2 \frac{d^3 p_2 dE_1}{E_2 |\vec{p}_2|} \times \mathbf{I}(E_1, E2). \qquad \text{(E.37)}$$

Now to integrate over E_1, first the limits on the integration are obtained by equating

$$M_a - E_1 - E_2 = \sqrt{E_1^2 + E_2^2 - m_2^2 \pm 2E_1 \sqrt{E_2^2 - m_2^2} + m_3^2}$$

which results in

$$E_1^\pm = \frac{\frac{1}{2}\left(m_a^2 + m_2^2 - m_3^2\right) - m_a E_2}{m_a \mp |\vec{p}_2| - E_2}$$

$$\frac{d\Gamma}{dE_2} = \frac{1}{m_a(4\pi)^3} \int_{E_1^-}^{E_1^+} \overline{\sum}\sum |\mathcal{M}|^2 \times \mathbf{I}(E_1, E2) dE_1$$

$$\Gamma = \frac{1}{m_a(4\pi)^3} \int dE_2 \int_{E_1^-}^{E_1^+} \overline{\sum}\sum |\mathcal{M}|^2 \times \mathbf{I}(E_1, E2) dE_1. \qquad \text{(E.38)}$$

Expressions of $N(q^2)$, the Coefficients of the Polarization Observables

F.1 Expression of $N(q^2)$ in Terms of Mandelstam Variables

The expression of $N(q^2)$ is expressed in terms of the Mandelstam variables and the form factors as:

$$
\begin{aligned}
N(q^2) &= f_1^2(q^2)\left(\frac{1}{2}\left(2\left(M^2-s\right)\left(M'^2-s\right)-t\left(\Delta^2-2s\right)+t^2+m_l^2\left(\Delta^2-2s-t\right)\right)\right) \\
&+ \frac{f_2^2(q^2)}{(M+M')^2}\left(\frac{1}{4}\left(-2t\left(M^4-2s\left(M^2+M'^2\right)+M'^4+2s^2\right)+2t^2\left((M+M')^2\right.\right.\right. \\
&\quad\left.-\ 2s\right)+m_l^2\left(2\Delta(M+M')\left(M^2+M'^2-2s\right)+t\left((M-3M')(M+M')+4s\right)+t^2\right) \\
&\quad\left.\left.+\ m_l^4(-((3M-M')(M+M')+t)))\right)\right) \\
&+ g_1^2(q^2)\left(\frac{1}{2}\left(2(M^2-s)(M'^2-s)-t\left((M+M')^2-2s\right)+t^2+m_l^2\left((M+M')^2\right.\right.\right. \\
&\quad\left.\left.-\ 2s-t\right)\right)\right)+\frac{|g_2(q^2)|^2}{(M+M')^2}\left(\frac{1}{4}\left(4\left(\Delta^2-t\right)\left(\left(M^2-s\right)\left(M'^2-s\right)+st\right)\right.\right. \\
&\quad+m_l^2\left(4\Delta\left(M^3+M^2M'-M(3s+t)+M's\right)+2\Delta^2\left((M+M')^2-2s-t\right)\right. \\
&\quad\left.-\ (4s+t)\left(\Delta^2-t\right)\right)+2\Delta^2\left(-2\left(M^2-s\right)\left(M'^2-s\right)-t\left((M+M')^2+2s\right)+t^2\right) \\
&\quad\left.\left.+\ m_l^4\left(\Delta^2+4M\Delta-t\right)\right)\right)+\frac{g_3^2(q^2)}{(M+M')^2}\left(m_l^2\left(m_l^2-t\right)\left(\Delta^2-t\right)\right) \\
&+ \frac{f_1(q^2)f_2(q^2)}{(M+M')}\left(-\left(t(M+M')\left(\Delta^2-t\right)+m_l^2\left(-\Delta\left(M'^2-s\right)+M't\right)+m_l^4M\right)\right) \\
&\pm f_1(q^2)g_1(q^2)\left(-\left(t\left(M^2+M'^2-2s-t\right)+m_l^2\left(M^2-M'^2+t\right)\right)\right)
\end{aligned}
$$

$$
\pm \; \frac{Re[f_1(q^2)g_2(q^2)]}{(M+M')} \left(-\Delta \left(t \left(M^2 + M'^2 - 2s - t \right) + m_l^2 \left(M^2 - M'^2 + t \right) \right) \right)
$$

$$
\pm \; \frac{f_2(q^2)g_1(q^2)}{(M+M')} \left(-(M+M') \left(t \left(M^2 + M'^2 - 2s - t \right) + m_l^2 \left(M^2 - M'^2 + t \right) \right) \right)
$$

$$
\pm \; \frac{Re[f_2(q^2)g_2(q^2)]}{(M+M')^2} \left(\Delta(-(M+M')) \left(t \left(M^2 + M'^2 - 2s - t \right) + m_l^2 \left(M^2 - M'^2 + t \right) \right) \right)
$$

$$
+ \; \frac{Re[g_1(q^2)g_2(q^2)]}{(M+M')} \left(\left(\Delta \left(-t(M+M')^2 + t^2 \right) + m_l^2 \left(M^3 + M^2 M' + \Delta \left((M+M')^2 \right. \right. \right. \right.
$$

$$
- \; 2s - t) - 3Ms - Mt + M's) + m_l^4 M \right) \right)
$$

$$
+ \; \frac{g_1(q^2)g_3(q^2)}{(M+M')} \left(-2m_l^2 \left(m_l^2 M + M^3 - M^2 M' - M(s+t) + M's \right) \right)
$$

$$
+ \; \frac{Re[g_2(q^2)g_3(q^2)]}{(M+M')^2} \left(m_l^2 \left(-2\Delta \left(m_l^2 M + M^3 - M^2 M' - M(s+t) + M's \right) \right. \right.
$$

$$
- \; \left(\Delta^2 - t \right) \left(m_l^2 + 2M^2 - 2s - t \right) \right) \right)
\tag{F.1}
$$

where $(+)-$ sign represents the (anti)neutrino induced scattering and the Mandelstam variables are defined as

$$
s = M^2 + 2ME,
\tag{F.2}
$$

$$
t = M^2 + M'^2 - 2ME',
\tag{F.3}
$$

with

$$
\Delta = M' - M.
\tag{F.4}
$$

F.2   Expressions of $A^h(q^2)$, $B^h(q^2)$, and $C^h(q^2)$

The expressions $A^h(q^2)$, $B^h(q^2)$, and $C^h(q^2)$ are expressed in terms of the Mandelstam variables and the form factors as:

$$
A^h(q^2) = -2 \left[f_1^2(q^2) \left(\pm \frac{1}{2}(M+M') \left(\Delta^2 - t \right) \right) \pm \frac{f_2^2(q^2)}{(M+M')^2} \left(\frac{1}{2} t(M+M') \left(\Delta^2 - t \right) \right) \right.
$$

$$
\pm \; g_1^2(q^2) \left(\frac{1}{2}\Delta \left((M+M')^2 - t \right) \right) \pm \frac{|g_2(q^2)|^2}{(M+M')^2} \left(\frac{1}{2} t\Delta \left((M+M')^2 - t \right) \right)
$$

$$
\pm \; \frac{f_1(q^2)f_2(q^2)}{(M+M')} \left(\frac{1}{2} \left(4MM't + t^2 - \Delta^2 (M+M')^2 \right) \right)
$$

$$
+ \; f_1(Q^2)g_1(Q^2) \left(-M' \left(M^2 + M'^2 - 2s - t \right) \right)
$$

$$
+ \; \frac{Re[f_1(q^2)g_2(q^2)]}{(M+M')} \left(\frac{1}{2} \left(\left(M^2 + M'^2 - 2s - t \right) \left(-t - 2M'\Delta + \Delta^2 \right) + m_l^2 \left(\Delta^2 - t \right) \right) \right)
$$

$$
+ \; \frac{f_1(q^2)g_3(q^2)}{(M+M')} \left(-m_l^2 \left(\Delta^2 - t \right) \right)
$$

$$+ \quad \frac{f_2(q^2)g_1(q^2)}{(M+M')} \left(\frac{1}{2} \left(\left(M^2 - M'^2 - t\right) \left(M^2 + M'^2 - 2s - t\right) + m_l^2 \left((M+M')^2 - t \right) \right) \right)$$

$$+ \quad \frac{Re[f_2(q^2)g_2(q^2)]}{(M+M')^2} \left(\frac{1}{2} \left((M+M') \left(\Delta^2 - t \right) \left(m_l^2 + M^2 + M'^2 - 2s - t \right) \right. \right.$$

$$+ \quad \Delta \left(m_l^2 \left((M+M')^2 - t \right) + \left(M^2 - M'^2 - t \right) \left(M^2 + M'^2 - 2s - t \right) \right) \Big) \Big)$$

$$+ \quad \frac{f_2(q^2)g_3(q^2)}{(M+M')^2} \left(-m_l^2 (M+M') \left(\Delta^2 - t \right) \right)$$

$$\pm \quad \left. \frac{Re[g_1(q^2)g_2(q^2)]}{(M+M')} \left(\frac{1}{2} \left((M+M')^2 - t \right) \left(-\Delta^2 - t \right) \right) \right] \qquad\qquad (\text{F.5})$$

$$B^h(q^2) \quad = \quad \frac{2}{M'} \left[f_1^2(q^2) \left(\pm \frac{1}{4} \left(t \left(\Delta^2 - 2s \right) - t^2 - 2M'\Delta \left(M^2 - s \right) + m_l^2 (M^2 + 2MM' - M'^2 + t) \right) \right) \right.$$

$$\pm \quad \frac{f_2^2(q^2)}{(M+M')^2} \left(\frac{1}{4} \left(t(M+M') \left(M^3 + M^2 M' - M(M'^2 + 2s + t) + M'^3 - M't \right) \right. \right.$$

$$+ \quad m_l^2 \left(M^4 + t(M+M')^2 - M'^4 \right) \Big) \Big)$$

$$\pm \quad g_1^2(q^2) \left(\frac{1}{4} \left(\left(-2M'(M+M')(M^2 - s) + t \left((M+M')^2 - 2s \right) - t^2 \right. \right. \right.$$

$$+ \quad m_l^2 (M^2 - 2MM' - M'^2 + t) \Big) \Big) \Big)$$

$$\pm \quad \frac{|g_2(q^2)|^2}{(M+M')^2} \left(\frac{1}{4} \Delta \left(2M' \left(-2m_l^2 M^2 - M^4 + M^2(M'^2 + s + t) + s \left(t - M'^2 \right) \right) \right. \right.$$

$$+ \quad \Delta \left(-2M'(M+M')(M^2 - s) + t \left((M+M')^2 - 2s \right) - t^2 \right)$$

$$+ \quad m_l^2 (M^2 - 2MM' - M'^2 + t) \Big) \Big)$$

$$\pm \quad \frac{f_1(q^2)f_2(q^2)}{(M+M')} \left(\frac{1}{2} \left(M^4 M' + M^3 t - M^2 M'(M'^2 + s) - Mt(M'^2 + 2s + t) \right. \right.$$

$$+ \quad M'(M'^2 - t)(s+t) + m_l^2 \left(M^3 + M^2 M' + M(M'^2 + t) - M'^3 + M't \right) \Big) \Big)$$

$$+ \quad f_1(q^2)g_1(q^2) \left(\frac{1}{2} \left(t \left(M^2 + M'^2 - 2s \right) - 2s \left(s - M^2 \right) - t^2 - m_l^2(M^2 + M'^2 - 2s - t) \right) \right)$$

$$+ \quad \frac{Re[f_1(q^2)g_2(q^2)]}{(M+M')} \left(-\frac{1}{2} \left(M^4 M' - 2M^3 s + M^2 \left(M'^3 - M'(s+t) - \Delta(2s+t) \right) \right. \right.$$

$$+ \quad 2Ms(s+t) - M'^3 s - M'^2 \Delta t + M'st + 2\Delta s^2 + 2\Delta st + \Delta t^2 + m_l^2 \left(M^3 + M^2 \Delta \right.$$

$$- \quad M(3s+t) + \Delta(M'^2 - 2s - t) \Big) + m_l^4 M \Big) \Big) + \frac{f_1(q^2)g_3(q^2)}{(M+M')} \left(m_l^2 M \left(m_l^2 + M^2 - s - t \right) \right)$$

$$+ \quad \frac{f_2(q^2)g_1(q^2)}{(M+M')} \left(\frac{1}{2} \left(-M' \left(M^2 - s \right) (M^2 + M'^2 - 2s) + t \left(M^3 + 2M^2 M' + MM'^2 + M'^3 \right. \right. \right.$$

$$- \quad 3M's) - t^2(M+M') - m_l^2 \left(M^2 M' + M \left(M'^2 + s \right) + M' \left(M'^2 - 2s - t \right) \right) + m_l^4 M \Big) \Big)$$

$$+ \quad \frac{Re[f_2(q^2)g_2(q^2)]}{(M+M')^2} \left(\frac{1}{4} \left(2 \left(M^3 \left(-M'^3 + M'(3s+t) + \Delta t \right) + M^2 \left(M'^4 - M'^3 \Delta \right. \right. \right. \right.$$

$$
\begin{aligned}
- \quad & M'^2(3s+t) + M'\Delta(3s+2t) + 2st\Big) + MM's(M'^2 - 2s - 3t) + s(M'^2 - t) \\
\times \quad & \Big(2(s+t) - M'^2\Big) + M\Delta t(M'^2 - t) + M'\Delta(s+t)\Big(M'^2 - 2s - t\Big)\Big) \\
+ \quad & m_l^2\Big(-2M^4 - 2M^3 M' + M^2(-2M'\Delta + 2s + t) - 2M\Delta\Big(M'^2 + s\Big) + 2MM'(3s+t) \\
+ \quad & 2M'\Delta\Big(-M'^2 + 2s + t\Big) - (M'^2 - t)(4s+t)\Big) - m_l^4\Big(3M^2 - M'^2 + t\Big)\Big)\Big) \\
+ \quad & \frac{f_2(q^2)g_3(q^2)}{(M+M')^2}\Big(\frac{1}{2}m_l^2\Big(m_l^2\Big(M^2 + 2MM' - M'^2 + t\Big) - 2M'\Delta(M^2 - s) \\
+ \quad & t\Big(\Delta^2 - 2s\Big) - t^2\Big)\Big) \\
\pm \quad & \frac{Re[g_1(q^2)g_2(q^2)]}{(M+M')}\Big(\frac{1}{2}\Big(M'\Big(-2m_l^2 M^2 - M^4 + M^2\Big(M'^2 + s + t\Big) + s(t - M'^2)\Big) \\
+ \quad & \Delta\Big(m_l^2\Big(M^2 - 2MM' - M'^2 + t\Big) - 2M'(M+M')\Big(M^2 - s\Big) \\
+ \quad & t\Big((M+M')^2 - 2s\Big) - t^2\Big)\Big)\Big)\Big] \tag{F.6}
\end{aligned}
$$

$$
\begin{aligned}
C^h(q^2) \;=\; & 2\Bigg[\pm\frac{Im[f_1(q^2)g_2(q^2)]}{(M+M')}\Big(-t + 2M\Delta + \Delta^2\Big) \pm \frac{Im[f_2(q^2)g_2(q^2)]}{(M+M')}\Big(-t \\
+ \quad & \frac{\Delta\Big(M^2 - M'^2 + t\Big)}{M+M'} + \Delta^2\Big) + \frac{Im[g_1(q^2)g_2(q^2)]}{(M+M')}\Big(M^2 + M'^2 - 2s - t + m_l^2\Big) \\
+ \quad & \frac{Im[g_3(q^2)g_2(q^2)]}{(M+M')^2}\Big(2m_l^2\Delta\Big)\Bigg] \tag{F.7}
\end{aligned}
$$

F.3   Expressions of $A^l(q^2)$, $B^l(q^2)$ and $C^l(q^2)$

The expressions $A^l(q^2)$, $B^l(q^2)$, and $C^l(q^2)$ are expressed in terms of the Mandelstam variables and the form factors as:

$$
\begin{aligned}
A^l(q^2) \;=\; & 2\Bigg[f_1^2(q^2)\Big(\frac{1}{2}m_l\Big(M^2 + 2MM' - s\Big)\Big) \\
+ \quad & \frac{f_2^2(q^2)}{(M+M')^2}\Big(\frac{1}{4}m_l\Big(-m_l^2((3M - M')(M+M') + t) - 2M^4 + 2M^2\Big(M'^2 + s\Big) + 4MM't \\
- \quad & 2\Big(M'^2 - t\Big)(s+t)\Big)\Big) + g_1^2(q^2)\Big(\frac{1}{2}m_l\Big(M^2 - 2MM' - s\Big)\Big) \\
+ \quad & \frac{|g_2(q^2)|^2}{(M+M')^2}\Big(\frac{1}{4}m_l\Big(\Big(\Delta^2 - t\Big)\Big(m_l^2 + 4M^2 + 2M'^2 - 2s - 2t\Big) + 2\Delta\Big(2m_l^2 M - 4M^2\Delta \\
+ \quad & 3M\Big(M'^2 - s - t\Big) + M'\Big(-M'^2 + s + t\Big)\Big) + 2\Delta^2\Big(M^2 - 2MM' - s\Big)\Big)\Big) \\
+ \quad & \frac{g_3^2(q^2)}{(M+M')^2}\Big(m_l^3\Big(\Delta^2 - t\Big)\Big) \\
+ \quad & \frac{f_1(q^2)f_2(q^2)}{(M+M')}\Big(-\frac{1}{2}m_l\Big(2m_l^2 M - \Delta\Big(2M^2 - M'^2 - s\Big) - t(3M + M')\Big)\Big)
\end{aligned}
$$

$$+ f_1(q^2)g_1(q^2)\left(-m_l\left(M^2-s\right)\right) + \frac{Re[f_1(q^2)g_2(q^2)]}{(M+M')}\left(-m_l\Delta\left(M^2-s\right)\right)$$

$$+ \frac{f_2(q^2)g_1(q^2)}{(M+M')}\left(-m_l(M+M')\left(M^2-s\right)\right) + \frac{Re[f_2(q^2)g_2(q^2)]}{(M+M')^2}\left(-m_l\Delta(M+M')\left(M^2-s\right)\right)$$

$$+ \frac{Re[g_1(q^2)g_2(q^2)]}{(M+M')}\left(\frac{1}{2}m_l\left(2m_l^2M+4M^3+2M^2(\Delta-2M')+MM'(3M'-4\Delta)\right.\right.$$

$$-\ 3M(s+t)-M'^3+M'(s+t)-2\Delta s\Big)\Big)$$

$$+ \frac{g_1(q^2)g_3(q^2)}{(M+M')}\left(-\left(2m_l^3M-m_l\Delta\left(M'^2-s-t\right)\right)\right)$$

$$+ \frac{Re[g_3(q^2)g_2(q^2)]}{(M+M')^2}\left(-m_l\left(m_l^2\left(\Delta^2+2M\Delta-t\right)-t(M'^2-s-t)\right)\right)\Bigg]$$ (F.8)

$$B^l(q^2) = \frac{2}{M'}\left[f_1^2(q^2)\left(\frac{1}{4m_l}M'\left(m_l^2\left(-\left(M^2+2MM'-M'^2+t\right)\right)\right)+2\left(M^2-s\right)(M'^2-s)\right.\right.$$

$$-\ t\Delta^2+2st+t^2\Big)\Big) + \frac{f_2^2(q^2)}{(M+M')^2}\left(\frac{1}{8m_l}M'\left(m_l^4((3M-M')(M+M')+t)\right.\right.$$

$$+\ m_l^2\left(\Delta^2-t\right)\left(2(M+M')^2+t\right)-2t\left(M^4-2s\left(M^2+M'^2\right)+M'^4+2s^2\right)$$

$$+\ 2t^2\left((M+M')^2-2s\right)\Big)\Big) + g_1^2(q^2)\left(\frac{1}{4m_l}M'\left(m_l^2\left(-\left(M^2-2MM'-M'^2+t\right)\right)\right.\right.$$

$$+\ 2\left(M^2-s\right)(M'^2-s)-t(M+M')^2+2st+t^2\Big)\Big)$$

$$+ \frac{|g_2(q^2)|^2}{(M+M')^2}\left(\frac{1}{8m_l}M'\left(m_l^4\left(-\left(\Delta^2+4M\Delta-t\right)\right)+m_l^2\left(-2\Delta^2\left(M^2-2MM'-M'^2+t\right)\right.\right.$$

$$-\ (4M^2-t)\left(\Delta^2-t\right)+4M\Delta(2M\Delta+t)\Big)+2\Delta^2\left(2\left(M^2-s\right)(M'^2-s)\right.$$

$$-\ t(M+M')^2+2st+t^2\Big)-8\Delta^2\left(\left(M^2-s\right)(M'^2-s)+st\right)$$

$$+\ 4\left(\Delta^2-t\right)\left(\left(M^2-s\right)(M'^2-s)+st\right)\Big)\Big) + \frac{g_3^2(q^2)}{(M+M')^2}\left(-\frac{1}{2}m_lM'\left(m_l^2-t\right)\left(\Delta^2-t\right)\right)$$

$$+ \frac{f_1(q^2)f_2(q^2)}{(M+M')}\left(\frac{1}{2m_l}M'\left(m_l^2-t\right)\left(m_l^2M+(M+M')\left((M-M')^2-t\right)\right)\right)$$

$$+ f_1(q^2)g_1(q^2)\left(\frac{1}{2m_l}M'\left(m_l^2-t\right)\left(M^2+M'^2-2s-t\right)\right)$$

$$+ \frac{Re[f_1(q^2)g_2(q^2)]}{(M+M')}\left(\frac{1}{2m_l}M'\Delta\left(m_l^2-t\right)\left(M^2+M'^2-2s-t\right)\right)$$

$$+ \frac{f_2(q^2)g_1(q^2)}{(M+M')}\left(\frac{1}{2m_l}M'\left(m_l^2-t\right)(M+M')\left(M^2+M'^2-2s-t\right)\right)$$

$$+ \frac{Re[f_2(q^2)g_2(q^2)]}{(M+M')^2}\left(\frac{1}{2m_l}M'\Delta\left(m_l^2-t\right)(M+M')\left(M^2+M'^2-2s-t\right)\right)$$

$$+ \frac{Re[g_1(q^2)g_2(q^2)]}{(M+M')}\left(\frac{1}{2m_l}M'\left(-m_l^4M+m_l^2\left(-2M^3+M^2(2M'-\Delta)+M(2M'\Delta+t)\right.\right.\right.$$

$$+\ \Delta\left(M'^2-t\right)\Big)+\Delta\left(2\left(M^2-s\right)\left(M'^2-s\right)-t\left((M+M')^2-2s\right)+t^2\right)$$

$$- 2\Delta \left(\left(M^2 - s \right) \left(M'^2 - s \right) + st \right) \right) \right) + \frac{g_1(q^2) g_3(q^2)}{(M + M')} \left(m_l M M' \left(m_l^2 - t \right) \right)$$

$$+ \frac{Re[g_2(q^2) g_3(q^2)]}{(M + M')^2} \left(\frac{1}{2} m_l M' \left(m_l^2 - t \right) \left(\Delta^2 + 2M\Delta - t \right) \right) \right] \tag{F.9}$$

$$C^l(q^2) = 2 \left[\frac{Im[g_1(q^2) g_2(q^2)]}{(M + M')} \left(m_l (M + M') \right) + \frac{Im[g_3(q^2) g_2(q^2)]}{(M + M')^2} \left(2 m_l t \right) \right] \tag{F.10}$$

References

[1]   W. Pauli, Letter to L. Meitner and her colleagues dated 4 December 1930 (letter open to the participants of the conference in Tübingen) (1930), recorded in W. Pauli, *Wissenschaftlicher Briefwechsel mit Bohr, Einstein, Heisenberg u.a.,* Band 11 (Springer, Berlin, 1985) p. 39. A reference to 'neutrino' is seen in a letter from Heisenberg to Pauli on 1 December 1930: 'Zu Deinen Neutronen möchte ich noch bemerken: ...'.

[2]   N. Bohr, in *Convegno di Fisica Nucleare* (R. Accad. d'Italia, Roma 1932), p. 119; J. Chem. Soc. 1932, 349 (Faraday Lectures); earlier reference is seen in a letter from Pauli to Bohr dated 5 March 1929, '... Willst Du den armen Energiesatz noch weiter maltraitieren?' (W . Pauli, Wissenschaftlicher Briefwechs el, op. cit. p. 493).

[3]   Pauli, W. 1931. *Physical Review* 38: 579.

[4]   Chadwick, J. 1914. 'The Intensity Distribution in the Magnetic Spectrum of Beta Particles from Radium (B+ C).' *Verh. Phys. Gesell.* 16: 383–391.

[5]   Kronig, R. de L. 1928. 'Der Drehimpuls des Stickstoffkerns.' *Naturwissenschaften* 16 (19): 335–335.

[6]   Heitler, W., and G. Herzberg. 1929. *Nature* 17: 673.

[7]   Becquerel, Henri. 1896. 'On the Rays Emitted by Phosphorescence.' *Compt. Rend. Hebd. Seances Acad. Sci.* 122: 420–421.

[8]   Curie, Marie, and Gabriel Lippmann. 1898. *Rayons émis par les composés de l'uranium et du thorium.* Gauthier-Villars.

Ward, F. A. B. 1963. *Atomic Physics: Descriptive Catalogue (Science Museum).* London: Her Majesty's Stationery Office.

Chadwick, James. 1931. 'Radioactivity and Radioactive Substances.'

Giesel, F. 1899. *Ann. Phys.* 69: 834.

Curie, Pierre, and Marie Curie. 1900. 'Sur la Charge Électrique des Rayons Déviables du Radium.' *CR Acad. Sci. Paris* 130: 647–650.

Becquerel, H. 1900. *Comprend.* 130: 809.

Wien, W. 1903. *Phys. Zeit.* 4: 624.

Rutherford, E., and F. Soddy. 1902. *Phil. Msg.* 4: 387.

Kaufmann, W. 1902. *Phys. Zeit.* 4: 54.

Kaufmann, Walter. 1906. 'Über die Konstitution des Elektrons.' *Annalen der Physik* 324 (3): 487–553.

Bestelmeyer, Adolf. 1907. 'Spezifische Ladung und Geschwindigkeit der durch Röntgenstrahlen erzeugten Kathodenstrahlen.' *Annalen der Physik* 327 (3): 429–447.

Rutherford, E. 1934. *Nature* 134: 90.

[9]   Rutherford, Ernest. 1914. 'XXXVII. The Connexion between the β and γ ray Spectra.' *The London, Edinburgh, and Dublin Philosophical Magazine and Journal of Science* 28 (165): 305–319.

[10]   Villard, M. M. 1900. 'Sur le Rayonnement du Radium.' *CR Acad. Sci. Paris* 130: 1178.

[11]   Falconer, Isobel. 1987. 'Corpuscles, Electrons and Cathode rays: JJ Thomson and the 'Discovery of the Electron'.' *The British journal for the history of science* 20 (3): 241–276.

[12]   Rutherford, Ernest. 1911. 'LXXIX. The Scattering of α and β Particles by Matter and the Structure of the Atom.' *The London, Edinburgh, and Dublin Philosophical Magazine and Journal of Science* 21 (125): 669–688.

[13]   Bohr, Niels. 1913. 'XXXVII. On the Constitution of Atoms and Molecules.' *The London, Edinburgh, and Dublin Philosophical Magazine and Journal of Science* 26 (153): 476-502.

[14]   Meitner, L. 1922. 'Über den Zusammenhang zwischen β-und γ-Strahlen.' *Zeitschrift für Physik* 9 (1): 145–152.

Meitner, L. 1922. 'Über dieβ-Strahl-Spektra und ihren Zusammenhang mit der γ-Strahlung.' *Zeitschrift für Physik* 11 (1): 35–54.

[15]   Ellis, Charles Drummond. 1922. 'β-ray Spectra and their Meaning.' *Proceedings of the Royal Society of London. Series A, Containing Papers of a Mathematical and Physical Character* 101 (708): 1–17.

Chadwick, James, and Charles D. Ellis. 1922. 'A Preliminary Investigation of the Intensity Distribution in the β-Ray Spectra of Radium B and C.' In *Proceedings of the Cambridge Philosophical Society* 21: 274–280.

Ellis, Charles Drummond, and William A. Wooster. 1927. 'The Average Energy of Disintegration of Radium E.' *Proceedings of the Royal Society of London. Series A, Containing Papers of a Mathematical and Physical Character* 117 (776): 109–123.

[16]   Meitner, Lise, and Wilhelm Orthmann. 1930. 'Über eine Absolute Bestimmung der Energie der Primáren β-Strahlen von Radium E.' *Zeitschrift für Physik* 60 (3-4): 143–155.

[17]   Pauli, Wolfgang. 1961. *Aufsatze und Vortrage uber Physik und Erkenntnistheorie.* Vieweg.

[18]   Private communication of Goudsmith to L. M. Brown quoted in *Phys. Today,* Sept. 1978, p. 23.

[19]   Chadwick, James. 1932. 'The Existence of a Neutron.' *Proceedings of the Royal Society of London. Series A, Containing Papers of a Mathematical and Physical Character* 136 (830): 692–708.

[20]   F. Rasetti, in E. Fermi, *Note e Memorie,* ed. by E . Amaldi et al. vol. I. (Academia Naz .dei Lincei, Roma 1962), p. 538.

[21]   Heisenberg, Werner. 1932. 'On the Structure of Atomic Nuclei.' *Z. Phys.* 77: 1–11.

[22]   Iwanenko, D. 1932. 'The Neutron Hypothesis.' *Nature* 129 (3265): 798–798.

[23]   Fermi, Enrico. 1934. 'E. Fermi.' *Nuovo cimento* 11: 157.

[24]   F. Perrin, F. 1933. 'Structure et Proprietes des Noyaux Atomiques, Rapports et Discussions du Septiem e Conseil de Physique, idem.' *Compt. Rendus* 197 (1625): 327.

[25]   W. Pauli, in *Discussion du Rapport de M. Heisenberg 'La Structure du Noyau' in Structure et Propri etes des Noyaux Atomiqu es, Rapports et Dis- cussi ons du Septseme Conseil de Physique,* Brussels, October 1933, ed . by Institut International de Physique Solvay (Gaut hier-Villars, Pari s 1934), p.324.

[26]   Gamow, George, and Edward Teller. 1936. 'Selection Rules for the β-Disintegration.' *Physical Review* 49 (12): 895.

[27]   Bethe, Hans Albrecht, and Robert Fox Bacher. 1936. 'Nuclear Physics A. Stationary States of Nuclei.' *Reviews of Modern Physics* 8 (2): 82.

[28]   Dirac, Paul Adrien Maurice. 1928. 'The Quantum Theory of the Electron.' *Proceedings of the Royal Society of London. Series A, Containing Papers of a Mathematical and Physical Character* 117 (778): 610–624.

Dirac, Paul Adrien Maurice. 1928. 'The Quantum Theory of the Electron. Part II.' *Proceedings of the Royal Society of London. Series A, Containing Papers of a Mathematical and Physical Character* 118 (779): 351–361.

Dirac, P. A. M. 1930. *Proceedings of Royal Society A* 126 (360): 692.

[29]   Cook, C. Sharp, and Lawrence M. Langer. 1948 'The Beta-Spectra of Cu^{64} as a Test of the Fermi Theory.' *Physical Review* 73 (6): 601.

Reitz, John R. 1950. 'The Effect of Screening on Beta-ray Spectra and Internal Conversion.' *Physical Review* 77 (1): 10.

[30]   Orear, J., Harris, G. and Taylor, S., 1956. Spin and Parity Analysis of bevatron τ Mesons. *Physical Review,* 102(6): 1676.

[31]   Lee, T. D., and C. N. Yang. 1956. 'Mass Degeneracy of the Heavy Mesons.' *Physical Review* 102 (1): 290.

[32]   Lee, Tsung-Dao, and Chen-Ning Yang. 1956. 'Question of Parity Conservation in Weak Interactions.' *Physical Review* 104 (1): 254.

[33]   Wu, Chien-Shiung, Ernest Ambler, R. W. Hayward, D. D. Hoppes, and Ralph Percy Hudson. 1957. 'Experimental Test of Parity Conservation in Beta Decay.' *Physical Review* 105 (4): 1413.

[34]   Frauenfelder, H., R. Bobone, E. Von Goeler, N. Levine, H. R. Lewis, R. N. Peacock, A. Rossi, and G. De Pasquali. 1957. 'Parity and the Polarization of Electrons from Co^{60}.' *Physical Review* 106 (2): 386.

[35]   Page, Lorne A., and Milton Heinberg. 1957. 'Measurement of the Longitudinal Polarization of Positrons Emitted by Sodium-22.' *Physical Review* 106 (6): 1220.

[36]   Osipowicz, A. et al. [KATRIN Collaboration]. 2001. 'KATRIN: A next generation tritium beta decay experiment with sub-eV sensitivity for the electron neutrino mass.' *arXiv preprint hep-ex/0109033*.

[37]   A. Salam, Il Nuovo Cim. X299 (1957); Conf. Proc. C 680519, 367 (1968).

[38]   Landau, Lev. 1957. 'On the Conservation laws for Weak Interactions.' *Nuclear Physics* 3 (1): 127–131.

[39]   Lee, Tsung D., and Chen Ning Yang. 1957. 'Parity Nonconservation and a Two-component Theory of the Neutrino.' *Physical Review* 105 (5): 1671.

[40]   Weyl, Hermann. 1929. 'Elektron Und Gravitation. I.' *Zeitschrift für Physik* 56 (5-6): 330–352.

[41]   Sudarshan, Eo CG, and R. E. Marshak. 1958. 'Chirality Invariance and the Universal Fermi Interaction.' *Physical Review* 109 (5): 1860.

[42]   Feynman, Richard P., and Murray Gell-Mann. 1958. 'Theory of the Fermi Interaction.' *Physical Review* 109 (1): 193.

[43]   Sakurai, Jun John. 1958. 'Mass Reversal and Weak Interactions.' *Il Nuovo Cimento (1955-1965)* 7 (5): 649–660.

[44]   Gershtein, S. S., and Ya B. Zeldovich. 1955. 'Meson Corrections in the Theory of Beta Decay.' *Zh. Eksp. Teor. Fiz.* 2: 698–699. [Sov. Phys. JETP 2, 576 (1956)].

[45]   Goldberger, M. L., and S. B. Treiman. 1958. 'Decay of the π Meson.' *Physical Review* 110 (5): 1178.

[46]   Nambu, Yoichiro. 1960. 'Axial Vector Current Conservation in Weak Interactions.' *Physical Review Letters* 4 (7): 380.

[47]   Adler, Stephen L. 1965. 'Calculation of the Axial-vector Coupling Constant Renormalization in β Decay.' *Physical Review Letters* 14 (25): 1051.

[48]   Weisberger, William I. 1965. 'Renormalization of the Weak Axial-vector Coupling Constant.' *Physical Review Letters* 14 (25): 1047.

[49]   Schwinger, Julian. 1957. 'A Theory of the Fundamental Interactions.' *Ann. Phys* 2 (407): 34.

[50]   Bludman, Sidney A. 1958. 'On the Universal Fermi Interaction.' *Il Nuovo Cimento (1955-1965)* 9 (3): 433–445.

[51]   Lopes, J. Leite. 1958. 'A model of the Universal Fermi Interaction.' *Nuclear Physics* 8: 234–236.

[52]   Heisenberg, W. 1938. 'Über die in der Theorie der Elementarteilchen Auftretende Universelle Lánge.' *Annalen der Physik* 424 (1-2): 20–33.

[53]   Tomonaga, S., H. Tamaki. 1937. *Sei. Papers Inst. Phys.-Chem. Res. (Tokyo)* 33: 288.

[54]   Leipunski, A. I. 1936. *Proc. Cam. Phil. Soc.* 32: 301.

[55]   Fierz, M. 1936. *Helv. Phys. Aeta* 9: 245.

[56]   Nishijima, Kazuhiko. 1955. 'Charge Independence Theory of V Particles.' *Progress of Theoretical Physics* 13 (3): 285–304.

[57]   Gell-Mann, M. 1956. *Nuovo Cim.* 4 (S2): 848.

[58]   Gell-Mann, Murray, and Maurice Lévy. 1960. 'The Axial Vector Current in Beta Decay.' *Il Nuovo Cimento (1955–1965)* 16 (4): 705–726.

[59]   Cabibbo, Nicola. 1963. 'Unitary Symmetry and Leptonic Decays.' *Physical Review Letters* 10 (12): 531.

[60]   Sakata, Shoichi. 1956. 'On a Composite Model for the New Particles.' *Progress of theoretical physics* 16 (6): 686–688.

[61]   Gell-Mann, M. and A. Pais. 1955. *Physical Review* 97 (5): 1387.

[62]   Zweig, G. 1964. CERN Geneva TH-401.

[63]   Zweig, G., and S. U. An. 1980. 'Model for strong interaction symmetry and its breaking.' In *Developments in the Quark Theory of Hadrons, Volume 1.* Edited by D. Lichtenberg and S. Rosen. pp 22–101.

[64]   Glashow, Sheldon L., Jean Iliopoulos, and Luciano Maiani. 1970. 'Weak Interactions with Lepton-hadron Symmetry.' *Physical review D* 2 (7): 1285.

[65]   Bjørken, B. J., and Sheldon L. Glashow. 1964. 'Elementary Particles and SU (4).' *Physics Letters* 11: 255–257.

[66]   Aubert, Jean-Jacques, U. Becker, P. J. Biggs, J. Burger, M. Chen, G. Everhart, P. Goldhagen et al. 1974. 'Experimental Observation of a Heavy Particle J.' *Physical Review Letters* 33 (23): 1404.

[67]   Augustin, J-E., Adam M. Boyarski, Martin Breidenbach, F. Bulos, J. T. Dakin, G. J. Feldman, G. E. Fischer et al. 1974. 'Discovery of a Narrow Resonance in e^+e^- Annihilation.' *Physical Review Letters* 33 (23): 1406. [*Adv. Exp. Phys.* 5, 141 (1976)].

[68]   Herb, S. W., D. C. Hom, L. M. Lederman, J. C. Sens, H. D. Snyder, J. K. Yoh, J. A. Appel et al. 1977. 'Observation of a Dimuon Resonance at 9.5 GeV in 400-GeV Proton-nucleus Collisions.' *Physical Review Letters* 39 (5): 252.

[69]   Behrends, S., K. Chadwick, J. Chauveau, P. Ganci, T. Gentile, Jan M. Guida, Joan A. Guida et al. 1983. 'Observation of Exclusive Decay Modes of b-flavored Mesons.' *Physical Review Letters* 50 (12): 881.

[70]   Kobayashi, M., and T. Maskawa. 1973. *Prog. Theor. Phys.* 49: 652.

[71]   Maki, Ziro, Masami Nakagawa, and Shoichi Sakata. 1962. 'Remarks on the Unified Model of Elementary Particles.' *Progress of Theoretical Physics* 28 (5): 870–880.

[72]   Budde, R., M. Chretien, J. Leitner, N. P. Samios, M. Schwartz, and J. Steinberger. 1956. 'Properties of Heavy Unstable Particles Produced by 1.3-Bev π^- Mesons.' *Physical Review* 103 (6): 1827.

[73]   Christenson, James H., James W. Cronin, Val L. Fitch, and René Turlay. 1964. 'Evidence for the 2 π Decay of the K 2 0 Meson.' *Physical Review Letters* 13 (4): 138.

[74]   Bethe, Hans, and Rudolph Peierls. 1934. 'The "Neutrino".' *Nature* 133 (3362): 532–532.

[75]   Rodeback, George W., and James S. Allen. 1952. 'Neutrino Recoils Following the Capture of Orbital Electrons in A 37.' *Physical Review* 86 (4): 446.

[76]   Snell, Arthur H., and Frances Pleasonton. 1955. 'Spectrometry of Recoils from Neutrino Emission in Argon-37.' *Physical Review* 97 (1): 246.

[77]   Jaeobsen, J. C., O. Kofoed-Hansen, Det. Kgl. Danske Viedensk. 1945. *Selskab. Mat . Fys. Medd.* 23 (12): 1.

[78]   Sherwin, Chalmers W. 1948. 'Momentum Conservation in the Beta-decay of P 32 and the Angular Correlation of Neutrinos with Electrons.' *Physical Review* 73 (3): 216.

[79]   Crane, H. R., and J. Halpern. 1938. 'New Experimental Evidence for the Existence of a Neutrino.' *Physical Review* 53 (10): 789.

[80]   Pontecorvo, B. 1946. *Report PD-205*. CRL, Canada: Chalk River Laboratory.

[81]   Alvarez, L. W. 1949. 'University of California Radiation Lab.' *Report No UCRL-328*.

[82]   Fermi, Enrico. 1950 *Nuclear physics: A Course given by Enrico Fermi at the University of Chicago*. University of Chicago Press.

[83]   Reines, F. and C. L. Cowan. 1953. *Physical Review* 92: 830.

[84]   Cowan Jr, C. L., F. Reines, F. B. Harrison, E. C. Anderson, and F. N. Hayes. 1953. 'Large Liquid Scintillation Detectors.' *Physical Review* 90 (3): 493.

[85]   Reines, Frederick, and Clyde L. Cowan. 1956. 'The Neutrino.' *Nature* 178 (4531): 446–449.

[86]   Reines, Frederick, and Clyde L. Cowan Jr. 1959. 'Free Antineutrino Absorption Cross Section. I. Measurement of the Free Antineutrino Absorption Cross Section by Protons.' *Physical Review* 113 (1): 273.

[87]   Davis Jr, Raymond. 1952. 'Nuclear Recoil Following Neutrino Emission from Beryllium 7.' *Physical Review* 86 (6): 976.

[88]   Davis Jr, Raymond. 1955. "Attempt to Detect the Antineutrinos from a Nuclear Reactor by the $Cl^{37}(\nu, e^-)$ A^{37} Reaction." *Physical Review* 97 (3): 766.

[89]   Neddermeyer, Seth H., and Carl D. Anderson. 1937. 'Note on the Nature of Cosmic-ray Particles.' *Physical Review* 51 (10): 884.

[90] Street, J. C., and E. C. Stevenson. 1937. 'New Evidence for the Existence of a Particle of Mass Intermediate between the Proton and Electron.' *Physical Review* 52 (9): 1003.

[91] Nishina, Yoshio, Masa Takeuchi, and Torao Ichimiya. 1937. 'On the Nature of Cosmic-ray Particles.' *Physical Review* 52 (11): 1198.

[92] Yukawa, Hideki. 1935. 'On the Interaction of Elementary Particles. I.' *Proceedings of the Physico-Mathematical Society of Japan. 3rd Series* 17: 48–57.

[93] Lattes, Cesare Mansueto Giulio, Hugh Muirhead, Giuseppe P. S. Occhialini, and Cecil Frank Powell. 1947. 'Processes involving charged mesons.' *Nature* 159 (4047): 694–697.

[94] Perkins, D. H. 1947. 'Nuclear Disintegration by Meson Capture.' *Nature* 159 (4030): 126–127.

[95] Occhialini, G. P. S., and C. F. Powell. 1947. 'Nuclear Disintegrations Produced by Slow Charged Particles of Small Mass.' *Nature* 159 (4032): 186–190.

[96] Tanikawa, Y. 1947. 'On the Cosmics-Ray Meson and the Nuclear Meson.' *Progress of Theoretical Physics* 2 (4): 220–221.

[97] Marshak, R. E., and Hans A. Bethe. 1947. 'On the Two-meson Hypothesis.' *Physical Review* 72 (6): 506.

[98] Conversi, Marcello, Ettore Pancini, and Oreste Piccioni. 1947. 'On the Disintegration of Negative Mesons.' *Physical Review* 71 (3): 209.

[99] Sakata, S., and T. Inoue. 1946. *Progress of Theoretical Physics* 1: 143.

[100] Hincks, E. P., and B. Pontecorvo. 1949. 'The Penetration of μ-Meson Decay Electrons and Their Bremsstrahlung Radiation.' *Physical Review* 75 (4): 698.

[101] Steinberger, J. 1948. 'On the Range of the Electrons in Meson Decay.' *Physical Review* 74 (4): 500.

[102] Leighton, Robert B., Carl D. Anderson, and Aaron J. Seriff. 1949. 'The Energy Spectrum of the Decay Particles and the Mass and Spin of the Mesotron.' *Physical Review* 75 (9): 1432.

[103] Feinberg, G. 1958. 'Decays of the μ Meson in the Intermediate-Meson Theory.' *Physical Review* 110 (6): 1482.

[104] Oneda, S., J. C. Pati, and B. Sakita. 1960. '$|\Delta I| = \frac{1}{2}$ Rule and the Weak Four-Fermion Interaction.' *Physical Review* 119 (1): 482.

[105] Pontecorvo, B. 1960. *Soviet Physics JETP* 10: 1236–1240. [*Zh. Eksp. Teor. Fiz.* 37, 1751 (1959)].

[106] Schwartz, Melvin. 1960. 'Feasibility of using High-energy Neutrinos to Study the Weak Interactions.' *Physical Review Letters* 4 (6): 306.

[107] Lee, T. D., and Chen-Ning Yang. 1960. 'Theoretical Discussions on Possible High-energy Neutrino Experiments.' *Physical Review Letters* 4 (6): 307.

[108] Cabibbo, Nicola, and Raul Gatto. 1961. 'Consequences of Unitary Symmetry for Weak and Electromagnetic Transitions.' *Il Nuovo Cimento (1955–1965)* 21 (5): 872–877.

[109]   Yamaguchi, Yoshio. 1960. 'Interactions Induced by High Energy Neutrinos.' *Progress of Theoretical Physics* 23 (6): 1117–1137.

[110]   Danby, Gaillard, Jean Maurice Gaillard, Konstantin Goulianos, Leon M. Lederman, N. Mistry, M. Schwartz, and J. Steinberger. 1962. 'Observation of High-energy Neutrino Reactions and the Existence of Two Kinds of Neutrinos.' *Physical Review Letters* 9 (1): 36.

[111]   Bienlein, J. K. et al. 1964. *Physics Letters* 13: 80.

[112]   Pontecorvo, Bruno. 1947. 'Nuclear Capture of Mesons and the Meson Decay.' *Physical Review* 72 (3): 246.

[113]   Puppi, G. 1948. 'Sui Mesoni dei Raggi Cosmici.' *Il Nuovo Cimento (1943–1954)* 5 (6): 587–588.

[114]   Klein, O. 1948. 'Mesons and Nucleons.' 897.

[115]   Tiomno, Jayme, and John A. 1949. Wheeler. 'Energy Spectrum of Electrons from Meson Decay.' *Reviews of Modern Physics* 21 (1): 144.

[116]   Perl, Martin L., G. S. Abrams, A. M. Boyarski, Martin Breidenbach, D. D. Briggs, F. Bulos, William Chinowsky et al. 1975. 'Evidence for Anomalous Lepton Production in $e^+ - e^-$ Annihilation.' *Physical Review Letters* 35 (22): 1489.

        Perl, Martin L., G. J. Feldman, G. S. Abrams, M. S. Alam, A. M. Boyarski, Martin Breidenbach, F. Bulos et al. 1976. 'Properties of anomalous $e\mu$ events produced in e^+e^- annihilation.' *Physics Letters B* 63 (4): 466–470.

[117]   Tanabashi, Masaharu, K. Hagiwara, K. Hikasa, K. Nakamura, Y. Sumino, F. Takahashi, J. Tanaka et al. 2018. 'Review of Particle Physics.' *Physical Review D* 98 (3): 030001.

[118]   Kodama, K., N. Ushida, C. Andreopoulos, N. Saoulidou, G. Tzanakos, P. Yager, B. Baller et al. 2001. 'Observation of Tau Neutrino Interactions.' *Physics Letters B* 504 (3): 218–224.

[119]   Agafonova, N., A. Aleksandrov, O. Altinok, M. Ambrosio, A. Anokhina, S. Aoki, A. Ariga et al. 2010. 'Observation of a First $\nu\tau$ Candidate Event in the OPERA Experiment in the CNGS Beam.' *Physics Letters B* 691 (3): 138–145.

[120]   Li, Z., K. Abe, C. Bronner, Y. Hayato, M. Ikeda, K. Iyogi, J. Kameda et al. 2018. 'Measurement of the Tau Neutrino Cross Section in Atmospheric Neutrino Oscillations with Super-Kamiokande.' *Physical Review D* 98 (5): 052006.

[121]   Majorana, E. 1937. *Il Nuovo Cimento* 14: 171–184.

[122]   Bergkvist, Karl-Erik. 1972. 'A High-luminosity, High-resolution Study of the End-point Behaviour of the Tritium β-Spectrum (I). Basic Experimental Procedure and Analysis with Regard to Neutrino Mass and Neutrino Degeneracy.' *Nuclear Physics B* 39: 317–370.

[123]   Tretyakov, E. F. 1975. *Izv. Akad. Nauk. USSR, Ser. Fiz.* 39: 583. *Proc. Intern. Nuetrino Conf.* (Aachen, 1976) pp. 663–670.

[124] Lubimov, V. A., E. G. Novikov, V. Z. Nozik, E. F. Tretyakov, and V. S. Kosik. 1980. 'An Estimate of the νe Mass from the β-spectrum of Tritium in the Valine Molecule.' *Physics Letters B* 94 (2): 266–268.

[125] Stueckelberg, E. C. G. 1936. *Nature* 131: 1070.

Helv. Phys. Acta. 9: 533 (1936).

[126] Moller, C. 1937. *Phys. Zeits d. Sowjetunion* 11: 9.

[127] Bambynek, W., H. Behrens, M. H. Chen, B. Crasemann, M. L. Fitzpatrick, K. W. D. Ledingham, H. Genz, M. Mutterer, and R. L. Intemann. 1977. 'Orbital Electron Capture by the Nucleus.' *Reviews of Modern Physics* 49 (1): 77.

Bambynek, W., H. Behrens, M. H. Chen, B. Crasemann, M. L. Fitzpatrick, K. W. D. Ledingham, H. Genz, M. Mutterer, and R. L. Intemann. 1977. 'Erratum: Orbital Electron Capture by the Nucleus.' *Reviews of Modern Physics* 49 (4): 961.

[128] De Rújula, Alvaro. 1981. 'A New Way to Measure Neutrino Masses.' *Nuclear Physics B* 188 (3): 414–458.

[129] Aker, M., K. Altenmüller, M. Arenz, et al., [KATRIN Collaboration]. 2019. 'Improved Upper Limit on the Neutrino Mass from a Direct Kinematic Method by KATRIN.' *Physical Review Letters* 123 (22): 221802. [arXiv:1909.06048].

[130] Assamagan, K., Ch Brönnimann, M. Daum, H. Forrer, R. Frosch, P. Gheno, R. Horisberger et al. 1996. 'Upper Limit of the Muon-neutrino Mass and Charged-pion Mass from Momentum Analysis of a Surface Muon Beam.' *Physical Review D* 53 (11): 6065.

[131] Anderhub, H. B., J. Boecklin, H. Hofer, F. Kottmann, P. Le Coultre, D. Makowiecki, H. W. Reist, B. Sapp, and P. G. Seiler. 1982. 'Determination of an Upper Limit of the Mass of the Muonic Neutrino from the Pion Decay in Flight.' *Physics Letters B* 114 (1): 76–80.

[132] Barate, R., and Aleph Collaboration. 1998. 'An Upper Limit on the τ Neutrino Mass from Three-and Five-prong Tau Decays.' *The European Physical Journal C-Particles and Fields* 2 (3): 395–406.

[133] Albrecht, H., U. Binder, P. Böckmann, R. Gláser, G. Harder, A. Krüger, A. Nippe et al. 1988. 'An Improved Upper Limit on the $\nu\tau$-mass from the Decay $\pi^- \to \pi^-\pi^-\pi^-\pi^+$ $\pi^+\nu\tau$.' *Physics Letters B* 202 (1): 149–153.

[134] Cinabro, D., S. Henderson, K. Kinoshita, T. Liu, M. Saulnier, R. Wilson, H. Yamamoto et al. 1993. 'Limit on the Tau Neutrino Mass.' *Physical review letters* 70 (24): 3700.

[135] Alexander, Gideon, J. Allison, N. Altekamp, K. Ametewee, K. J. Anderson, S. Anderson, S. Arcelli et al. 1996. 'Upper Limit on the $\nu\tau$ mass from $\tau \to 3h\nu\tau$ Decays.' *Zeitschrift für Physik C: Particles and Fields* 72 (2): 231.

[136] Dylla, H. Frederick, and John G. King. 1973. 'Neutrality of Molecules by a New Method.' *Physical Review A* 7 (4): 1224.

[137] Zorn, Jens C., George E. Chamberlain, and Vernon W. Hughes. 1963. 'Experimental Limits for the Electron-Proton Charge Difference and for the Charge of the Neutron.' *Physical Review*129 (6): 2566.

[138] Barbiellini, Guido, and Giuseppe Cocconi. 1987. 'Electric Charge of the Neutrinos from SN1987A.' *Nature* 329 (6134): 21–22.

[139] Degrassi, G., A. Sirlin, and W. J. Marciano. 1989. 'Effective Electromagnetic form Factor of the Neutrino.' *Physical Review D* 39 (1): 287.

[140] Auerbach, L. B., R. L. Burman, D. O. Caldwell, E. D. Church, J. B. Donahue, A. Fazely, G. T. Garvey et al. 2001. 'Measurement of Electron-neutrino Electron Elastic Scattering.' *Physical Review D* 63 (11): 112001.

[141] Deniz, M., S. T. Lin, V. Singh, J. Li, H. T. Wong, S. E. L. C. U. K. Bilmis, C. Y. Chang et al. 2010. 'Measurement of ν^- e-electron Scattering Cross Section with a CsI (Tl) Scintillating Crystal Array at the Kuo-Sheng Nuclear Power Reactor.' *Physical Review D* 81 (7): 072001.

[142] Hirsch, Martin, Enrico Nardi, and Diego Restrepo. 2003. 'Bounds on the Tau and Muon Neutrino Vector and Axial Vector Charge Radius.' *Physical Review D* 67 (3): 033005.

[143] Krakauer, D. A., R. L. Talaga, R. C. Allen, H. H. Chen, R. Hausammann, W. P. Lee, X-Q. Lu et al. 1990. 'Limits on the Neutrino Magnetic Moment from a Measurement of Neutrino-electron Elastic Scattering.' *Physics Letters B* 252 (1): 177–180.

[144] Dorenbosch et al. [CHARM Collaboration], Z. Phys. C 41, 567 (1989) Erratum: [Z.Phys. C 51, 142 (1991)].

[145] Ahrens, L. A., S. H. Aronson, P. L. Connolly, B. G. Gibbard, M. J. Murtagh, S. J. Murtagh, S. Terada et al. 1990. 'Determination of Electroweak Parameters from the Elastic Scattering of Muon Neutrinos and Antineutrinos on Electrons.' *Physical Review D* 41 (11): 3297.

[146] Derbin, A. V. 1994. 'Restriction on the Magnetic Dipole Moment of Reactor Neutrinos.' *Physics of Atomic Nuclei* 57 (2): 222–225.

[147] Derbin, A. V., L. A. Popeko, A. V. Chernyi, and G. A. Shishkina. 1986. 'New Experiment on Elastic Scattering of Reactor Neutrinos by Electrons.' *JETP Lett* 43(4).

Vidyakin, G. S., V. N. Vyrodov, I. I. Gurevich, Yu V. Kozlov, V. P. Martemyanov, S. V. Sukhotin, V. G. Tarasenkov, E. V. Turbin, and S. Kh Khakhimov. 1992. 'Limitations on the Magnetic Moment and Charge Radius of the Electron-anti-neutrino.' *JETP Lett* 55 (4).

Popeko et al. 1993. *Pisma Zh. Eksp. Teor. Fiz.* 57: 755.

[148] Kyuldjiev, Assen V. 1984. 'Searching for Effects of Neutrino Magnetic Moments at Reactors and Accelerators.' *Nuclear Physics B* 243 (3): 387–397.

[149] Grotch, H., and Richard Wallace Robinett. 1988. 'Limits on the τ Neutrino Electromagnetic Properties from Single Photon Searches at e^+e^- Colliders.' *Zeitschrift für Physik C Particles and Fields* 39 (4): 553–556.

[150] Abreu, Paulo, W. Adam, T. Adye, E. Agasi, I. Ajinenko, R. Aleksan, G. D. Alekseev et al. 1997. 'Search for New Phenomena Using Single Photon Events at LEP1.' *Zeitschrift für Physik C Particles and Fields* 74: 577–586.

[151] Acciarri, M., M. Aguilar-Benitez, S. Ahlen, J. Alcaraz, G. Alemanni, J. Allaby, A. Aloisio et al. 1997. 'Search for New Physics in Energetic Single Photon Production in e^+e^- Annihilation at the Z Resonance.' *Physics Letters B* 412 (1-2): 201–209.

[152] Cooper-Sarkar, Amanda M., Subir Sarkar, J. Guy, W. Venus, P. O. Hulth, and K. Hultqvist. 1992. 'Bound on the Tau Neutrino Magnetic Moment from the BEBC Beam Dump Experiment.' *Physics Letters B* 280 (1-2): 153–158.

[153] Goldhaber, Maurice, L. Grodzins, and A. W. Sunyar. 1958. 'Helicity of Neutrinos.' *Physical review* 109 (3): 1015.

[154] Klein, O. 1939. In *Proc. Con. 'Les Nouvelles Theorische la Physique', Warsaw (Paris)*, p. 81.

[155] Burns, R., K. Goulianos, E. Hyman, L. Lederman, W. Lee, N. Mistry, J. Rettberg, M. Schwartz, J. Sunderland, and G. Danby. 1965. 'Search for Intermediate Bosons in High-energy Neutrino Interactions.' *Physical Review Letters* 15 (1): 42.

[156] Block, M. M. 1964. 'Neutrino Interactions and a Unitary Universal Model.' *Physical Review Letters* 12 (10): 262.

[157] Weinberg, Steven. 1967. 'A Model of Leptons.' *Physical review letters* 19 (21): 1264.

[158] Higgs, Peter W. 1964. 'Broken Symmetries and the Masses of Gauge Bosons.' *Physical Review Letters* 13 (16): 508.

[159] Englert, François, and Robert Brout. 1964. 'Broken Symmetry and the Mass of Gauge Vector Mesons.' *Physical Review Letters* 13 (9): 321.

[160] Guralnik, Gerald S., Carl R. Hagen, and Thomas WB Kibble. 1964. 'Global Conservation Laws and Massless Particles.' *Physical Review Letters* 13 (20): 585.

[161] Kibble, Tom WB. 1967. 'Symmetry Breaking in Non-Abelian Gauge Theories.' *Physical Review* 155 (5): 1554.

[162] Goldstone, Jeffrey. 1961. 'Field theories with Superconductor Solutions.' *Il Nuovo Cimento (1955–1965)* 19 (1): 154–164.

[163] Nambu, Yoichiro, and Giovanni Jona-Lasinio. 1961. 'Dynamical Model of Elementary Particles Based on an Analogy with Superconductivity. II.' *Physical Review* 124 (1): 246.

[164] Hasert, F. J., S. Kabe, W. Krenz, J. Von Krogh, D. Lanske, J. Morfin, K. Schultze et al. 1974. 'Observation of Neutrino-like Interactions without Muon or Electron in the Gargamelle Neutrino Experiment.' *Nuclear Physics B* 73 (1): 1–22.

[165] Cnops, A. M., P. L. Connolly, S. A. Kahn, H. G. Kirk, M. J. Murtagh, R. B. Palmer, N. P. Samios et al. 1978. 'Measurement of the Cross Section for the Process $\nu_\mu + e^- \rightarrow \nu_\mu + e^-$ at High Energies.' *Physical Review Letters* 41 (6): 357.

Baker, N. J., P. L. Connolly, S. A. Kahn, M. J. Murtagh, R. B. Palmer, N. P. Samios, M. Tanaka et al. 1989. 'Measurement of Muon-neutrino—electron Elastic Scattering in the Fermilab 15-foot Bubble Chamber.' *Physical Review D* 40 (9): 2753.

[166] Heisterberg, R. H., L. W. Mo, T. A. Nunamaker, K. A. Lefler, A. Skuja, A. Abashian, N. E. Booth, C. C. Chang, C. Li, and C. H. Wang. 1980. 'Measurement of the Cross Section for $\nu_\mu + e^- \rightarrow \nu_\mu + e^-$.' *Physical Review Letters* 44 (10): 635.

[167] Faissner, H., H. G. Fasold, E. Frenzel, T. Hansl, D. Hoffmann, K. Maull, E. Radermacher et al. 1978. 'Measurement of Muon-neutrino and-antineutrino Scattering off Electrons.' *Physical Review Letters* 41 (4): 213.

[168] Hasert, F. J., et al. [Gargamelle Neutrino Collaboration]. 1973. *Physics Letters B* 46: 121 [138].

[169] Prescott, C. Y. et al. 1978. *Physics Letters B* 77: 347.

[170] Arnison, G., A. Astbury, B. Aubert, C. Bacci, G. Bauer, A. Bezaguet, R. Bóck et al. 1983. 1983. 'Experimental Observation of Isolated Large Transverse Energy Electrons with associated missing energy at $s = 540$ GeV.' *Physics Letters B* 122 (1): 103–116.

[171] Banner, M., B. Madsen, J. C. Chollet, J. L. Siegrist, H. Hänni, L. Di Lella, H. M. Steiner et al. 1983. 'Observation of Single Isolated Electrons of High Transverse Momentum in Events with Missing Transverse Energy at the CERN $\bar{p}$ p collider.' *Phys. Lett. B* 122, no. CERN-EP-83-25: 476–485.

[172] Abreu, Paulo, and DELPHI Collaboration. 2000. 'Measurements of the Z Partial Decay Width into $c\bar{c}$ and Multiplicity of Charm Quarks per β-decay.' *The European Physical Journal C-Particles and Fields* 12 (2): 225–241.

[173] Barate, R., and ALEPH collaboration. 2000. 'Measurement of the Z Resonance Parameters at LEP.' *The European Physical Journal C-Particles and Fields* 14 (1): 1–50.

[174] Acciarri, M., and L3 Collaboration. 2000. 'Measurements of Cross Sections and Forward-backward Asymmetries at the Z Resonance and Determination of Electroweak Parameters.' *The European Physical Journal C-Particles and Fields* 16 (1): 1–40.

[175] Tevatron New Physics Higgs Working Group [CDF and D0 Collaborations]. 2012. *FERMILAB-CONF-12-318-E; CDF Note 10884; D0 Note 6348.* arXiv:1207. 0449 [hep-ex].

[176] Chatrchyan, S. [CMS collaboration]. 2012. *Physics Letters B* 716 (30).

[177] ATLAS Collaboration, and G. Collaboration. 'Aad et al. 2012. '*Observation of a New Particle in the Search for the Standard Model Higgs Boson with the ATLAS detector at the LHC, Phys. Lett. B* 716: 1–29.

[178] Aad, G., et al. [ATLAS and CMS Collaborations]. 2016. *Journal of High Energy Physics* 1608: 45.

[179] B. Pontecorvo, B. 1957. *Sov. Phys. JETP* 6: 429. [*Zh. Eksp. Teor. Fiz.* 33: 549 (1957)]

[180] Gribov, V., and B. Pontecorvo. 1969. 'Neutrino Astronomy and Lepton Charge.' *Physics Letters B* 28 (7): 493–496.

[181] Bahcall, J. N., and S. C. Frautschi. 1969. *Physics Letters. B* 29: 623.

[182] Bahcall, John N. 1989. *Neutrino astrophysics*. Cambridge University Press.

[183] Nakamura K., T. Kajita, M. Nakahata, A. Suzuki. 1994. 'Kamiokande.' In Fukugita M. and Suzuki A., eds. *Physics and Astrophysics of Neutrinos*. Tokyo: Springer.

[184] Bionta, R. M., G. Blewitt, C. B. Bratton, B. G. Cortez, S. Errede, G. W. Forster, W. Gajewski et al. 1983. 'Search for Proton Decay into $e^+\pi^0$.' *Physical Review Letters* 51 (1): 27.

[185] Ahmad, Q. Retal, R. C. Allen, T. C. Andersen, J. D. Anglin, G. Bühler, J. C. Barton, E. W. Beier et al. 2001. 'Measurement of the Rate of $\nu_e + d \rightarrow p + p + e^-$ Interactions Produced by B 8 Solar Neutrinos at the Sudbury Neutrino Observatory.' *Physical Review Letters* 87 (7): 071301.

[186] Wolfenstein, Lincoln. 1979. 'Neutrino Oscillations and Stellar Collapse.' *Physical Review D* 20 (10): 2634.

[187] Mikheev, S. P., and A. Yu Smirnov. 1985. 'Resonance Enhancement of Oscillations in Matter and Solar Neutrino Spectroscopy.' *Soviet Journal of Nuclear Physics* 42 (6): 913–917.

Mikheev, S. P., and A. Yu Smirnov. 1985. 'Resonance Amplification of Oscillations in Matter and Spectroscopy of Solar Neutrinos.' *Yadernaya Fizika* 42 (6): 1441–1448.

[188] Bethe, Hans Albrecht. 1986. 'Possible Explanation of the Solar-neutrino Puzzle.' *Physical Review Letters* 56 (12): 1305.

[189] Böser, S., et al. 2019. arXiv:1906.01739 [hep-ex].

[190] Giunti, Carlo, and Thierry Lasserre. 2019. 'eV-scale Sterile Neutrinos.' *Annual Review of Nuclear and Particle Science* 69: 163–190. doi:10.1146/annurev-nucl-101918-023755, arXiv:1901.08330 [hep-ph].

[191] Diaz, A., C. A. Argüelles, G. H. Collin, J. M. Conrad, and M. H. Shaevitz. 2019. 'Where Are We With Light Sterile Neutrinos?.' *arXiv preprint arXiv:1906.00045 [hep-ex]*.

[192] Klein, O. 1926. 'Theory of Relativity, Z.' *Phys* 37: 895. [*Surveys High Energ. Phys.* 5, 241 (1986)].

[193] Gordon, W. 1926. 'The Compton Effect According to Schrödinger's Theory.' *Z. Phys* (40): 117Á–133.

[194] Pauli, W., and V. F. Weisskopf. 1934. *Helv. Phys. Acta* 7: 709.

[195] Lorenz, L. 1867. 'Ueber die Identität der Schwingungen des Lichts mit den elektrischen Strömen.' *Annalen der Physik* 207 (6): 243–263. [XXXVIII. On the identity of the vibrations of light with electrical currents, The London, Edinburgh, and Dublin Philosophical Magazine and Journal of Science, 34: 230, 287–301] doi: 10.1080/14786446708639882

[196] Dirac, Paul Adrien Maurice. 1936. 'Relativistic Wave Equations.' *Proceedings of the Royal Society of London. Series A-Mathematical and Physical Sciences* 155 (886): 447–459.

[197] Fierz, Markus, and Wolfgang Ernst Pauli. 1939. 'On Relativistic Wave Equations for Particles of Arbitrary Spin in an Electromagnetic Field.' *Proceedings of the Royal Society of London. Series A. Mathematical and Physical Sciences* 173 (953): 211–232.

[198] Bargmann, V., and E. P. Wigner. 1948. *Proc. Nat. Acad. Sci.* 34: 211.

[199] Duffin, R. J. 1938. 'On the Characteristic Matrices of Covariant Systems.' *Physical Review* 54 (12): 1114.

[200] Kemmer, Nicholas. 1939. 'The Particle Aspect of Meson Theory.' *Proceedings of the Royal Society of London. Series A. Mathematical and Physical Sciences* 173 (952): 91–116.

[201] Proca, A. 1936. *Journal de Physique Archives, Radium* 7 (8): 347. doi: 10.1051/jphysrad: 0193600708034700

[202] Rarita, William, and Julian Schwinger. 1941. 'On a Theory of Particles with Half-integral Spin.' *Physical Review* 60, no. 1: 61.

[203] Weinberg, S. 1965. In S. Deser and K. W. Ford, eds. *Lectures on Fields and Particles, Vol. 2*, p. 405. [Brandeis University Summer Institute in Theoretical Physics]. New Jersey: Prentice-Hall.

[204] Ryder, L. H. 1987. *Quantum Field Theory*. Cambridge: Cambridge University Press.

[205] Bhabha, H. J. 1945. 'Relativistic Wave Equations for the Elementary Particles.' *Reviews of Modern Physics* 17 (2-3): 200.

[206] Harish-Chandra, H. C. 1948. 'Relativistic equations for elementary particles.' *Proceedings of the Royal Society of London. Series A. Mathematical and Physical Sciences* 192 (1029): 195–218.

[207] Fermi, Enrico. 1934. 'An Attempt of a Theory of Beta Radiation. 1.' *Z. Phys.* 88, no. UCRL-TRANS-726: 161–177.

[208] Yukawa, Hideki. 1950. 'Quantum Theory of Non-local Fields. Part I. Free Fields.' *Physical Review* 77 (2): 219.

[209] Yukawa, Hideki. 1950. 'Quantum Theory of Non-local Fields. Part II. Irreducible Fields and their Interaction.' *Physical Review* 80 (6): 1047.

[210] Mandl, F., and G. Shaw. 2010. *Quantum Field Theory*, 2nd ed. Hoboken, New Jersey: Wiley-Blackwell.

[211] Bjorken, James D., and Sidney D. Drell. 1965. Relativistic Quantum Mechanics. New York: McGraw-Hill.

[212] Noether, Emmy. 1918. 'Invariante Variationsprobleme. Nachrichten der Königlichen Gesellschaft der Wissenschaften zu Göttingen, Mathematisch-Physikalische Klasse pp. 235–257 (1918). Translated as 'Invariant Variation Problems' by M. A. Tavel.' *Transp. Theor. Stat. Phys* 1: 183–207.

[213] Gupta, Suraj N. 1950. 'Theory of Longitudinal Photons in Quantum Electrodynamics.' *Proceedings of the Physical Society. Section A* 63 (7): 681.

[214] Bleuler, K. 1950. 'A New Method of Treatment of the Longitudinal and Scalar Photons.' *Helv. Phys. Acta* 23: 567–586.

[215] Dyson, Freeman J. 1949. 'The Radiation Theories of Tomonaga, Schwinger, and Feynman.' *Physical Review* 75 (3): 486.

[216] Wick, Gian-Carlo. 1950. 'The Evaluation of the Collision Matrix.' *Physical Review* 80 (2): 268.

[217] Bogolyubov, N. N., and D. V. Shirkov. 1984. Introduction to the Theory of Quantized Fields [in Russian]. Moscow: Nauka.

[218] Greiner, W., and J. Reinhardt. 2009. *Quantum Electrodynamics*. Berlin: Springer-Verlag.

[219] Fermi, E. 1933. *Ricerca Seient.* 4 (2): 491; *Z. Phys.* 88: 161; *Nuovo Cimento* 11: 1.

[220] Glashow, Sheldon L. 1961. 'Partial-symmetries of Weak Interactions.' *Nuclear Physics* 22 (4): 579–588.

[221] Fukugita, M., and T. Yanagida. 2003. Physics of Neutrinos: and Applications to Astrophysics. Berlin: Springer-Verlag.

[222] Winter, K. 2000. *Neutrino Physics* (Cambridge Monographs on Particle Physics, Nuclear Physics and Cosmology) 2nd ed. Cambridge: Cambridge University Press.

[223] Marshak, R. E., M. Riazuddin, and C. P. Ryan. 1969. *Theory of Weak Interactions in Particle Physics*. New York: Wiley Interscience.

[224] Behrens, Heinrich, and Wolfgang Búhring. 1982. *Electron Radial Wave Functions and Nuclear Betadecay*. No. 67. Oxford University Press, USA.

[225] Hardy, J. C., and I. S. Towner. 2015. 'Superallowed $0^+ \rightarrow 0^+$ Nuclear β Decays: 2014 Critical Survey, with Precise Results for V_{ud} and CKM Unitarity.' *Physical Review C* 91 (2): 025501.

[226] Gonzalez-Alonso, M., O. Naviliat-Cuncic, and N. Severijns. 2019. 'New Physics Searches in Nuclear and Neutron β Decay.' *Progress in Particle and Nuclear Physics* 104: 165–223.

[227] Johnson, C. H., Frances Pleasonton, and T. A. Carlson. 1963. 'Precision Measurement of the Recoil Energy Spectrum from the Decay of He^6.' *Physical Review* 132 (3): 1149.

[228] Adelberger, E. G., C. Ortiz, A. García, H. E. Swanson, M. Beck, O. Tengblad, M. J. G. Borge, I. Martel, and H. Bichsel. 1999. 'the "ISOLDE Collaboration".' *Physical Review Letters* 83: 1299.

[229] Gorelov, A., D. Melconian, W. P. Alford, D. Ashery, G. Ball, J. A. Behr, P. G. Bricault et al. 2005. 'Scalar Interaction Limits from the $\beta - \nu$ Correlation of Trapped Radioactive Atoms.' *Physical Review Letters* 94 (14): 142501.

[230] Wauters, Frederik, I. Kraev, D. Zákoucký, M. Beck, M. Breitenfeldt, V. De Leebeeck, V. V. Golovko et al. 2010. 'Precision Measurements of the Co 60 β-asymmetry Parameter in Search for Tensor Currents in Weak Interactions.' *Physical Review C* 82 (5): 055502.

[231] Soti, Gergely, F. Wauters, M. Breitenfeldt, P. Finlay, P. Herzog, A. Knecht, U. Köster et al. 2014. 'Measurement of the β-asymmetry Parameter of ^{67}Cu in Search for Tensor-type Currents in the Weak Interaction.' *Physical Review C* 90 (3): 035502.

[232] Wauters, F., V. De Leebeeck, I. Kraev, M. Tandecki, E. Traykov, S. Van Gorp, Natalis Severijns, and D. Zákoucký. 2009. 'β Asymmetry Parameter in the Decay of In 114.' *Physical Review C* 80 (6): 062501.

[233] Carnoy, A. S., J. Deutsch, T. A. Girard, and René Prieels. 1991. 'Limits on Nonstandard Weak Currents from the Polarization of ^{14}O and ^{10}C Decay Positrons.' *Physical Review C* 43 (6): 2825.

[234] Wichers, V. A., T. R. Hageman, J. Van Klinken, H. W. Wilschut, and D. Atkinson. 1987. 'Bounds on Right-handed Currents from Nuclear Beta Decay.' *Physical review letters* 58 (18): 1821.

[235] Patrignani, C. P. D. G., D. H. Weinberg, C. L. Woody et al. 2016. *Chin. Phys.* 40: 100001.

[236] Garwin, Richard L., Leon M. Lederman, and Marcel Weinrich. 1957. 'Observations of the Failure of Conservation of Parity and Charge Conjugation in Meson Decays: The Magnetic Moment of the Free Muon.' *Physical Review* 105 (4): 1415.

[237] Aron, Walter, and A. J. Zuchelli. 1957. 'Contribution to Lamb Shift Due to Finite Proton Size.' *Physical Review* 105 (5): 1681.

[238] Renton, Peter. 1990. *Electroweak Interactions: An Introduction to the Physics of Quarks and Leptons*. Cambridge University Press.

[239] Commins, Eugene D., and Philip H. Bucksbaum. 1983. *Weak Interactions of Leptons and Quarks*. England: Cambridge University Press.

[240] Jackson, J. D., S. B. Treiman, and H. W. Wyld Jr. 1957. 'Coulomb Corrections in Allowed Beta Transitions.' *Nuclear Physics* 4: 206–212.

[241] Jackson, J. D., S. B. Treiman, and H. W. Wyld Jr. 1957. 'Possible Tests of Time Reversal Invariance in Beta Decay.' *Physical Review* 106 (3): 517.

[242] Schopper, H. F. 1966. *Weak Interactions and Nuclear Beta Decay*. Amsterdam: North-Holland.

[243] Daniel, H. 1968. 'Shapes of Beta-ray Spectra.' *Reviews of Modern Physics* 40 (3): 659.

[244] Paul, H. 1970. 'Least-squares Adjustment of the Coupling Constants in β-decay.' *Nuclear Physics A* 154 (1): 160–176.

[245] Brosi, A. R., B. H. Ketelle, H. C. Thomas, and R. J. Kerr. 1959. 'Decay Schemes of Sm^{145} and Pm^{145}.' *Physical Review* 113 (1): 239.

[246] Gerber, G., D. Newman, A. Rich, and E. Sweetman. 1977. 'Precision Measurement of Positron Polarization in ^{68}Ga Decay Based on the use of a New Positron Polarimeter.' *Physical Review D* 15 (5): 1189.

[247] Koks, F. W. J., and J. Van Klinken. 1976. 'Investigation on β-polarization at Low Velocities with β-particles from the Decay of Tritium.' *Nuclear Physics A* 272 (1): 61–81.

[248] Roesch, L. Ph, V. L. Telegdi, P. Truttmann, A. Zehnder, L. Grenacs, and L. Palffy. 1982. 'Direct Measurement of the Helicity of the Muonic Neutrino.' *American Journal of Physics* 50 (10): 931–935.

[249] Bopp, Peter, D. Dubbers, L. Hornig, E. Klemt, J. Last, H. Schütze, S. J. Freedman, and O. Schärpf. 1986. 'Beta-Decay Asymmetry of the Neutron and $\frac{gA}{gV}$.' *Physical Review Letters* 56 (9): 919.

[250] Liaud, P., K. Schreckenbach, R. Kossakowski, H. Nastoll, A. Bussiere, J. P. Guillaud, and L. Beck. 1997. 'The Measurement of the Beta Asymmetry in the Decay of Polarized Neutrons.' *Nuclear Physics A* 612 (1): 53–81.

[251] Yerozolimsky, B., I. Kuznetsov, Yu Mostovoy, and I. Stepanenko. 1997. 'Corrigendum: Corrected Value of the Beta-emission Asymmetry in the Decay of Polarized Neutrons Measured in 1990.' *Physics Letters B* 412 (3-4): 240–241.

[252] Mund, D., B. Márkisch, M. Deissenroth, J. Krempel, Marc Schumann, H. Abele, A. Petoukhov, and T. Soldner. 2013. 'Determination of the Weak Axial Vector Coupling $\lambda = g_A/g_V$ from a Measurement of the β-Asymmetry Parameter A in Neutron Veta Decay.' *Physical Review Letters* 110 (17): 172502.

[253] Valverde, A. A., M. Brodeur, T. Ahn, J. Allen, D. W. Bardayan, F. D. Becchetti, D. Blankstein et al. 2018. 'Precision half-life Measurement of ^{11}C: The Most Precise Mirror Transition $\mathcal{F}_\sqcup$ Value.' *Physical Review C* 97 (3): 035503.

[254] Kuznetsov, I. A., A. P. Serebrov, I. V. Stepanenko, A. V. Alduschenkov, M. S. Lasakov, A. A. Kokin, Yu A. Mostovoi, B. G. Yerozolimsky, and Maynard S. Dewey. 1995. 'Measurements of the Antineutrino Spin Asymmetry in Beta Decay of the Neutron and Restrictions on the Mass of a Right-handed Gauge Boson.' *Physical review letters* 75 (5): 794.

[255] Serebrov, A. P., I. A. Kuznetsov, I. V. Stepanenko, A. V. Aldushchenkov, M. S. Lasakov, Yu A. Mostovoi, B. G. Erozolimskii et al. 1998. 'Measurement of the Antineutrino Escape Asymmetry with Respect to the Spin of the Decaying Neutron.' *Journal of Experimental and Theoretical Physics* 86 (6): 1074–1082.

[256] Kreuz, Michael, T. Soldner, S. Baeßler, B. Brand, F. Glúck, U. Mayer, D. Mund et al. 2005. 'A Measurement of the Antineutrino Asymmetry B in Free Neutron Decay.' *Physics Letters B* 619 (3-4): 263–270.

[257] Schumann, Marc, T. Soldner, M. Deissenroth, F. Glúck, J. Krempel, M. Kreuz, B. Márkisch, D. Mund, A. Petoukhov, and H. Abele. 2007. 'Measurement of the Neutrino Asymmetry Parameter B in Neutron Decay.' *Physical Review Letters* 99 (19): 191803.

[258] Fierz, Markus. 1937. 'Zur Fermischen Theorie des β-zerfalls.' *Zeitschrift Fúr Physik* 104 (7-8): 553–565.

Fierz, M. 1939. 'Force-free particles with any spin.' *Helv. Phys. Acta* 12: 3–37.

[259] Greiner, W., and B. Múller. 1993. Gauge Theory of Weak Interactions. Berlin: Springer- Verlag.

[260]   Behrends, R. E., R. J. Finkelstein, and A. Sirlin. 1956. 'Radiative Corrections to Decay Processes.' *Physical Review* 101 (2): 866.

[261]   Kinoshita, Toichiro, and Alberto Sirlin. 1959. 'Radiative Corrections to Fermi Interactions.' *Physical Review* 113 (6): 1652.

[262]   Berman, Sam M., and A. Sirlin. 1962. 'Some Considerations on the Radiative Corrections to Muon and Neutron Decay.' *Annals of Physics* 20 (1): 20–43.

[263]   Berman, Sam M. 1958. 'Radiative Corrections to Muon and Neutron Decay.' *Physical Review* 112 (1): 267.

[264]   Sirlin, A. 1978. 'Current Algebra Formulation of Radiative Corrections in Gauge Theories and the Universality of the Weak Interactions.' *Reviews of Modern Physics* 50 (3): 573.

[265]   Jonker, M., J. Panman, F. Udo, J. V. Allaby, U. Amaldi, G. Barbiellini, A. Baroncelli et al. 1980. 'Experimental Study of Inverse Muon Decay.' *Physics Letters B* 93 (1-2): 203–209.

[266]   Mishra, S. R., K. T. Bachmann, R. E. Blair, C. Foudas, B. J. King, W. C. Lefmann, W. C. Leung et al. 1990. 'Inverse Muon Decay, $\nu_\mu + e \to \mu^- + \nu_e$, at the Fermilab Tevatron.' *Physics Letters B* 252 (1): 170–176.

[267]   Vilain, Pierre, Gaston Wilquet, R. Beyer, W. Flegel, H. Grote, T. Mouthuy, H. Øveras et al. 1995. 'A Precise Measurement of the Cross Section of the Inverse Muon Decay $\nu_\mu + e^- \to \mu^- + \nu_e$.' *Physics Letters B* 364 (2): 121–126.

[268]   Heisenberg, W. 1957. *Rev. Mod. Phys.* 29: 269.

[269]   Conversi, M., E. Pancini and O. Piccioni. 1945. *Physical Review* 68 (9-10): 232.

[270]   Hildebrand, Roger H. 1962. 'Observation of μ^- Capture in Liquid Hydrogen.' *Physical Review Letters* 8 (1): 34.

[271]   Fermi, Enrico, Edward Teller, and Victor Weisskopf. 1947. 'The Decay of Negative Mesotrons in Matter.' *Physical Review* 71 (5): 314.

[272]   Yukawa, Hideki, and Shoichi Sakata. 1935. 'On the Theory of the β-Disintegration and the Allied Phenomenon.' *Proceedings of the Physico-Mathematical Society of Japan. 3rd Series* 17: 467–479.

[273]   Alvarez, Luis W. 1938. 'The Capture of Orbital Electrons by Nuclei.' *Physical Review* 54 (7): 486.

[274]   Lee, T. D., Marshall Rosenbluth, and Chen Ning Yang. 1949. 'Interaction of Mesons with Nucleons and Light Particles.' *Physical Review* 75 (5): 905.

[275]   Goldberger, M. L., and S. B. Treiman. 1958. 'Form Factors in β Decay and μ Capture.' *Physical Review* 111 (1): 354.

[276]   Lopes, J. Leite. 1958. 'Capture of Negative Muons by Light Nuclei.' *Physical Review* 109 (2): 509.

[277]   Wolfenstein, L. 1958. *Nuovo Cim.* 8: 382.

[278] Leader, Elliot, and Enrico Predazzi. 1996. *An Introduction to Gauge Theories and Modern Particle Physics*. Cambridge University Press.

[279] Aitchison, I. J. R., and A. J. G. Hey. 2012. *Gauge Theories in Particle Physics: A Practical Introduction, Volume 1: From Relativistic Quantum Mechanics to QED,* 4th ed., and *Volume 2: Non-Abelian Gauge Theories: QCD and The Electroweak Theory,* 4th ed. Boca Raton: CRC Press.

[280] Bég, M. A. B., and Alberto Sirlin. 'Gauge Theories of Weak Interactions (Circa 1973–74 CE).' *Annual Review of Nuclear Science* 24 (1): 379–450.

[281] Tsai, Yung-Su. 1976. 'Erratum: Decay correlations of heavy leptons in $e^+ + e^- \to l^+ + l^-$.' *Physical Review D* 13 (3): 771.

[282] Thacker, H. B., and J. J. Sakurai. *Physics Letters B* 36: 103.

[283] Burmester, J., L. Criegee, H. C. Dehne, K. Derikum, R. Devenish, G. Flúgge, J. D. Fox et al. 1977. 'Anomalous Muon Production in e^+e^- Annihilations as Evidence for Heavy Leptons.' *Physics Letters B* 68 (3): 297–300.

[284] Gentile, Simonetta, and Martin Pohl. 1996. 'Physics of Tau Leptons.' *Physics Reports* 274 (5-6): 287–374.

[285] Bacino, W., T. Ferguson, L. Nodulman, W. Slater, H. K. Ticho, A. Diamant-Berger, M. Faessler et al. 1978. 'Measurement of the Threshold Behavior of $\tau^+\tau^-$ Production in e^+e^- Annihilation.' *Physical Review Letters* 41 (1): 13.

[286] Albrecht, H., H. Ehrlichmann, G. Harder, A. Krüger, A. Nau, A. W. Nilsson, A. Nippe et al. 1990. 'Determination of the Michel Parameter in Tau Decay.' *Physics Letters B* 246 (1-2): 278–284.

[287] Shimizu N. et al. [Belle Collaboration]. 2018. *Progress of Theoretical and Experimental Physics* 2018 (2): ARTN-023C01.

[288] Olive, Keith A. 2014. 'Review of Particle Physics.' *Chinese Physics C* 38 (9): 090001–090001.

[289] Heister, A., and ALEPH Collaboration. 2001. 'Measurement of the Michel Parameters and the ν_τ Helicity in τ Lepton Decays.' *The European Physical Journal C-Particles and Fields* 22 (2): 217–230.

[290] Alexander, J. P., C. Bebek, B. E. Berger, K. Berkelman, K. Bloom, D. G. Cassel, H. A. Cho et al. 1997. 'Determination of the Michel Parameters and the τ Neutrino Helicity in τ Decay.' *Physical Review D* 56 (9): 5320.

[291] Lattes, Cesare Mansueto Giulio, Hugh Muirhead, Giuseppe PS Occhialini, and Cecil Frank Powell. 1947. 'Processes Involving Charged Mesons.' *Nature* 159 (4047): 694–697.

Occhialini, G. P. S., and C. F. Powell. 1947. 'Nuclear Disintegrations Produced by Slow Charged Particles of Small Mass.' *Nature* 159 (4032): 186–190.

C. M. G. Lattes, G. P. S. Occhialini and C. F. Powell, Nature 160, 453 (1947).

[292] Britton, D. I., S. Ahmad, D. A. Bryman, R. A. Burnham, E. T. H. Clifford, P. Kitching, Y. Kuno et al. 1992. 'Measurement of the $\pi^+ \to e^+\nu$ Branching Ratio.' *Physical Review Letters* 68 (20): 3000.

Czapek, Gerhard, Andrea Federspiel, Andreas Flükiger, Daniel Frei, Beat Hahn, Carl Hug, Edwin Hugentobler et al. 1993. 'Branching Ratio for the Rare Pion Decay into Positron and Neutrino.' *Physical Review Letters* 70 (1): 17.

[293] Ruderman, M., and R. Finkelstein. 1949. 'Note on the Decay of the π-meson.' *Physical Review* 76 (10): 1458.

[294] Llewellyn Smith, C. H. 1972 *Phys. Rept.* 3: 261.

[295] Wick, G. C., A. S. Wightman and E. P. Wigner. 1952. *Physical Review* 88: 101 .

Feinberg, G., and Steven Weinberg. 1959. 'On the Phase Factors in Inversions.' *Il Nuovo Cimento (1955–1965)* 14 (3): 571–592.

[296] Gell-Mann, Murray. 1958. 'Test of the Nature of the Vector Interaction in β Decay.' *Physical Review* 111 (1): 362.

[297] Goldberger, M. L., and S. B. Treiman. 1958. 'Conserved Currents in the Theory of Fermi Interactions.' *Physical Review* 110 (6): 1478.

[298] Adler, Stephen L., and Roger F. Dashen. 1968. *Current Algebras and Applications to Particle Physics*. Vol. 30. Benjamin.

[299] Fukuda, H., and Y. Miyamoto. 1949. 'Selection Rule for Meson Problem.' *Progress of Theoretical Physics* 4 (3): 389–391.

van Wyk, C. B. 1950. 'Selection Rules for Closed Loop Processes.' *Physical Review* 80 (3): 487.

Nishijima, K. 1951. 'Generalized Furry's Theorem for Closed Loops.' *Progress of Theoretical Physics* 6 (4): 614–615.

Michel, L. 1952. *Program Cosmic Ray Physics - 3*. New York: Interscience.

Pais, A., and R. Jost. 1952. 'Selection Rules Imposed by Charge Conjugation and Charge Symmetry.' *Physical Review* 87 (5): 871.

Lee, T. D., and C. N. Yang. 1956. *Nuovo Cim.* 3: 79.

Goebel, Charles. 1956. 'Selection Rules for $N\bar{N}$ Annihilation.' *Physical Review* 103 (1): 258.

[300] Weinberg, Steven. 1958. 'Charge Symmetry of Weak Interactions.' *Physical Review* 112 (4): 1375.

[301] Moulson, M. 2007. *arXiv:1301.3046 [hep-ex]*.

[302] Chounet, L-M., J-M. Gaillard, and Mary K. Gaillard. 1972. 'Leptonic Decays of Hadrons.' *Physics Reports* 4 (5): 199–323.

[303] García, A., P. Kielanowski, H. Araki, and A. Bohm (ed.). 1985. *Lecture Notes in Physics, Volume 222*. New York: Springer.

[304] Leutwyler, H., and M. Roos. 1984. 'Determination of the Elements V_{us} and V_{ud} of the Kobayashi-Maskawa Matrix.' *Zeitschrift für Physik C Particles and Fields* 25 (1): 91–101.

[305] Gaillard, Mary Katherin, and Benjamin W. Lee. 1974. 'Rare Decay Modes of the K Mesons in Gauge Theories.' *Physical Review D* 10 (3): 897.

[306] Cabibbo, Nicola, Earl C. Swallow, and Roland Winston. 2003. 'Semileptonic Hyperon Decays.' *Annual Review of Nuclear and Particle Science* 53 (1): 39–75.

[307] Hom, D. C., L. M. Lederman, H. P. Paar, H. D. Snyder, J. M. Weiss, J. K. Yoh, J. A. Appel et al. 1976. 'Observation of High-mass Dilepton Pairs in Hadron Collisions at 400 GeV.' *Physical Review Letters* 36 (21): 1236.

[308] Basile, M., G. Bonvicini, G. Cara Romeo, L. Cifarelli, A. Contin, G. D'Alí, P. Di Cesare et al. 1981. 'A Comparison between Beauty and Charm Production in pp Interactions.' *Il Nuovo Cimento A (1965-1970)* 65 (3): 391–399.

[309] Bari, G., M. Basile, G. Bruni, G. Cara Romeo, R. Casaccia, L. Cifarelli, F. Cindolo et al. 1991. 'The Λ_b^0 Beauty Baryon Production in Proton-proton Interactions at $\sqrt{S} = 62$ GeV: A Second Observation.' *Il Nuovo Cimento A (1965–1970)* 104 (12): 1787–1800.

[310] Abe, F., H. Akimoto, A. Akopian, M. G. Albrow, S. R. Amendolia, D. Amidei, J. Antos et al. 1995. 'Observation of Top Quark Production in $\bar{p}p$ collisions with the Collider Detector at Fermilab.' *Physical review letters* 74 (14): 2626.

[311] Abachi, S., B. Abbott, M. Abolins, B. S. Acharya, I. Adam, D. L. Adams, M. Adams et al. 1995. 'Search for High Mass Top Quark Production in $\bar{p}p$ Collisions at $\sqrt{s} = 1.8$ TeV.' *Physical Review Letters* 74 (13): 2422.

[312] Wolfenstein, Lincoln. 1983. 'Parametrization of the Kobayashi-Maskawa Matrix.' *Physical Review Letters* 51 (21): 1945.

[313] Goldhaber, G., F. M. Pierre, G. S. Abrams, M. S. Alam, A. M. Boyarski, Martin Breidenbach, W. C. Carithers et al. 1976. 'Observation in e^+e^- Annihilation of a Narrow State at 1865 MeV/c^2 Decaying to $K\pi$ and $K\pi\pi\pi$.' *Physical Review Letters* 37 (5): 255.

[314] Peruzzi, I., M. Piccolo, G. J. Feldman, H. K. Nguyen, James E. Wiss, G. S. Abrams, M. S. Alam et al. 1976. 'Observation of a Narrow Charged State at 1876 MeV/c^2 Decaying to an Exotic Combination of $K\pi\pi$.' *Physical Review Letters* 37 (10): 569.

[315] Cazzoli, E. Go, A. M. Cnops, P. L. Connolly, R. I. Louttit, M. J. Murtagh, R. B. Palmer, N. P. Samios, T. T. Tso, and H. H. Williams. 1975. 'Evidence for $\Delta S = -\Delta Q$ Currents or Charmed-Baryon Production by Neutrinos.' *Physical Review Letters* 34 (17): 1125.

[316] Aaltonen, T., V. M. Abazov, B. Abbott, B. S. Acharya, M. Adams, T. Adams, J. P. Agnew et al. [CDF and D0 Collaborations]. 2014. *Physical Review D* 89 (7): 072001

[317] Gamow, G., and E. Teller. 1937. 'Some Generalizations of the β Transformation Theory.' *Physical Review* 51 (4): 289.

[318] Kemmer, N. 1937. 'Field Theory of Nuclear Interaction.' *Physical Review* 52 (9): 906.

[319] Gardner, Susan, W. C. Haxton, and Barry R. Holstein. 2017. 'A New Paradigm for Hadronic Parity Nonconservation and Its Experimental Implications.' *Annual Review of Nuclear and Particle Science* 67: 69–95.

[320] Haxton, Wick C., and Barry R. Holstein. 2013. 'Hadronic Parity Violation.' *Progress in Particle and Nuclear Physics* 71: 185–203.

[321] YndurÃ ain, F. J., and Å. 1983. Quantum Chromodynamics. 'An Introduction to the Theory of Quarks and Gluons.'

[322] Weyl, Hermann. 1917. 'The Theory of Gravitation.' *Annalen Phys* 54: 117.

[323] Einstein, Albert. 1948. 'A Generalized Theory of Gravitation.' *Reviews of Modern Physics* 20 (1): 35.

[324] Klein, Oskar. 1926. 'The Atomicity of Electricity as a Quantum Theory Law.' *Nature* 118 (2971): 516–516.

[325] Wentzel, G. 1937. *Helv. Phys. Acta* 10: 108.

[326] Yang, Chen-Ning, and Robert L. Mills. 1954. 'Conservation of Isotopic Spin and Isotopic Gauge Invariance.' *Physical Review* 96 (1): 191.

[327] Utiyama, Ryoyu. 1956. 'Invariant Theoretical Interpretation of Interaction.' *Physical Review* 101 (5): 1597.

[328] Shaw, R. 1955. Unpublished Ph.D. thesis, Cambridge University.

[329] Fermi, Enrico. 1933. 'Tentativo di Una Teoria Dell'emissione dei Raggi Beta.' *Ric. Sci.* 4: 491–495.

[330] Nambu, Yoichiro. 1960. 'Quasi-particles and Gauge Invariance in the Theory of Superconductivity.' *Physical Review* 117 (3): 648.

[331] Goldstone, Jeffrey, Abdus Salam, and Steven Weinberg. 1962. 'Broken Symmetries.' *Physical Review* 127 (3): 965.

[332] Gilbert, Walter. 1964. 'Broken Symmetries and Massless Particles.' *Physical Review Letters* 12 (25): 713.

[333] Schwinger, Julian. 1962. 'Gauge Invariance and Mass.' *Physical Review* 125 (1): 397.

[334] Anderson, Philip W. 1963. 'Plasmons, Gauge Invariance, and Mass.' *Physical Review* 130 (1): 439.

[335] Guralnik, G. S. 1964. 'Photon as a Symmetry-breaking Solution to Field Theory. II.' *Physical Review* 136 (5B): B1417.

[336] Guralnik, Gerald Stanford. 1964. 'Photon as a Symmetry-breaking Solution to Field Theory. I.' *Physical Review* 136 (5B): B1404.

[337] Guralnik, G. S., and C. R. Hagen. 1965. *Nuovo Cimento* 43 (1): 1.

[338] 't Hooft, Gerardus, and Veltman, Martinus. 1972. 'Regularization and Renormalization of Gauge Fields.' *Nuclear Physics B* 44 (1): 189–213.

[339] Lee, Benjamin W., and Jean Zinn-Justin. 1972. 'Spontaneously Broken Gauge Symmetries. I. Preliminaries.' *Physical Review D* 5 (12): 3121.

[340] Glashow, Sheldon L. 1959. 'The Renormalizability of Vector Meson Interactions.' *Nuclear Physics* 10: 107–117.

[341] Salam, A., and J. C. Ward. 1959. *Nuovo Cimento* 11: 568.

[342] Salam, A., and J. C. Ward. 1964. *Physics Letters* 13: 168.

[343] Fabri, Elio, and Luigi E. Picasso. 1966. 'Quantum Field Theory and Approximate Symmetries.' *Physical Review Letters* 16 (10): 408.

[344] Bardeen, J., L. N. Cooper, and J. R. Schrieffer. 1957. *Physical Review* 108: 1175.

[345] Landau, L. D. 1937. *Zh. Eksp. Teor. Fiz.* 7: 19 [*Phys. Z. Sowjetunion* 11: 26 (1937)] [*Ukr. J. Phys.* 53: 25 (2008)]

Ginzburg, V. L., and L. D. Landau. 1960. *Zh. Eksp. Teor. Fiz.* 20: 1064.

[346] Higgs, Peter Ware. 1964. 'Broken Symmetries, Massless Particles and Gauge Fields.' *Phys. Lett.* 12: 132–133.

[347] Prescott, C. Yi, W. B. Atwood, R. L. A. Cottrell, H. DeStaebler, Edward L. Garwin, A. Gonidec, Roger H. Miller et al. 1979. 'Further Measurements of Parity Non-conservation in Inelastic Electron Scattering.' *Physics Letters B* 84 (4): 524–528.

[348] Abe, K., L. A. Ahrens, K. Amako, S. H. Aronson, E. W. Beier, J. L. Callas, D. Cutts et al. 1987. 'Measurement of the Weak-neutral-current Coupling Constants of the Electron and Limits on the Electromagnetic Properties of the Muon Neutrino.' *Physical review letters* 58 (7): 636.

[349] Abe, K., L. A. Ahrens, K. Amako, S. H. Aronson, E. W. Beier, J. L. Callas, D. Cutts et al. 1989. 'Determination of $\sin^2 \theta W$ from Measurements of Differential Cross Sections for Muon-neutrino and -antineutrino Scattering by Electrons.' *Physical review letters* 62 (15): 1709.

[350] Baker, N. J., P. L. Connolly, S. A. Kahn, M. J. Murtagh, R. B. Palmer, N. P. Samios, M. Tanaka et al. 1989. 'Measurement of Muon-neutrino—Electron Elastic Scattering in the Fermilab 15-foot Bubble Chamber.' *Physical Review D* 40 (9): 2753.

[351] Vilain, Pierre, A. Capone, H. E. Roloff, A. Ereditato, T. Mouthuy, V. D. Khovanskii, G. Wilquet et al. 1994. 'Precision Measurement of Electroweak Parameters from the Scattering of Muon-neutrinos on Electrons.' *Phys. Lett. B* 335, no. CERN-PPE-94-124: 246–252.

[352] Abe, K., L. A. Ahrens, K. Amako, S. H. Aronson, E. W. Beier, J. L. Callas, P. L. Connolly et al. 1986. 'Erratum: Precise determination of $\sin^2 \theta_W$ from measurements of the differential cross sections for $\nu_\mu p \to \nu_\mu p$ and $\bar{\nu}_\mu p \to \bar{\nu}_\mu p$ [Phys. Rev. Lett. 56, 1107 (1986)].' *Physical Review Letters* 56: 1883.

[353] Aguilar-Arevalo, A. A., C. E. Anderson, A. O. Bazarko, S. J. Brice, B. C. Brown, L. Bugel, J. Cao et al. 2010. 'Measurement of the Neutrino Neutral-current Elastic Differential Cross Section on Mineral Oil at $E_\nu \backsim 1$ GeV.' *Physical Review D* 82 (9): 092005.

[354]   Reines, F., H. S. Gurr, and H. W. Sobel. 1976. 'Detection of $\bar{\nu}_e - e$ Scattering.' *Physical Review Letters* 37 (6): 315.

[355]   Ahmad, Q. R., R. C. Allen, T. C. Andersen, J. D. Anglin, J. C. Barton, E. W. Beier, M. Bercovitch et al. 2002. 'Measurement of Day and Night Neutrino Energy Spectra at SNO and Constraints on Neutrino Mixing Parameters.' *Physical Review Letters* 89 (1): 011302.

[356]   Alvarez-Ruso, Luis, Y. Hayato, and J. Nieves. 2014. 'Progress and Open Questions in the Physics of Neutrino Cross Sections at Intermediate Energies.' *New Journal of Physics* 16 (7): 075015.

[357]   Sakumoto, W. K., P. De Barbaro, A. Bodek, H. S. Budd, B. J. Kim, F. S. Merritt, M. J. Oreglia et al. 1990. 'Calibration of the CCFR Target Calorimeter.' *Nuclear Instruments and Methods in Physics Research Section A: Accelerators, Spectrometers, Detectors and Associated Equipment* 294 (1-2): 179–192.

[358]   Allasia, D., Carlo Angelini, A. Baldini, M. Baldo-Ceolin, S. Barlag, L. Bertanza, A. Bigi et al. 1983. 'Measurement of the Neutral Current Coupling Constants in Neutrino and Antineutrino Interactions with Deuterium.' *Physics Letters B* 133 (1-2): 129–134.

[359]   Paschos, E. A., and L. Wolfenstein. 1973. 'Tests for Neutral Currents in Neutrino Reactions.' *Physical Review D* 7 (1): 91.

[360]   Souder, Paul A., R. Holmes, D-H. Kim, K. S. Kumar, M. E. Schulze, K. Isakovich, G. W. Dodson et al. 1990. 'Measurement of Parity Violation in the Elastic Scattering of Polarized Electrons from C 12.' *Physical Review Letters* 65 (6): 694.

[361]   Spayde, D. T., T. Averett, D. Barkhuff, D. H. Beck, E. J. Beise, C. Benson, H. Breuer et al. 2000. 'Parity Violation in Elastic Electron-proton Scattering and the Proton's Strange Magnetic form Factor.' *Physical Review Letters* 84 (6): 1106.

[362]   Heil, W., J. Ahrens, H. G. Andresen, A. Bornheimer, D. Conrath, K-J. Dietz, W. Gasteyer et al. 1989. 'Improved Limits on the Weak, Neutral, Hadronic Axial Vector Coupling Constants from Quasielastic Scattering of Polarized Electrons.' *Nuclear Physics B* 327 (1): 1–31.

[363]   Baunack, S., K. Aulenbacher, D. Balaguer Rios, L. Capozza, J. Diefenbach, B. Gläser, D. Von Harrach et al. 2009. 'Measurement of Strange Quark Contributions to the Vector form Factors of the Proton at $Q^2 = 0.22$ (GeV/c)^2.' *Physical Review Letters* 102 (15): 151803.

[364]   Aniol, K. A., D. S. Armstrong, Todd Averett, Maud Baylac, Etienne Burtin, John Calarco, G. D. Cates et al. 2004. 'Parity-violating Electroweak Asymmetry in (e) Over-right- arrowp Scattering.' *Physical Review C* 69 (6): 065501_1.

[365]   Armstrong, D. S., J. Arvieux, Razmik Asaturyan, Todd Averett, S. L. Bailey, Guillaume Batigne, Douglas H. Beck et al. 2005. 'Strange-quark Contributions to Parity-violating Asymmetries in the Forward G0 Electron-proton Scattering Experiment.' *Physical Review Letters* 95 (9): 092001.

[366] Acha, A., K. A. Aniol, D. S. Armstrong, John Arrington, Todd Averett, S. L. Bailey, James Barber et al. 2007. 'Precision Measurements of the Nucleon Strange Form Factors at $Q^2 \sim 0.1$ GeV^2.' *Physical Review Letters* 98 (3): 032301.

[367] Erler, J., and M. Schott. 2019. *Prog. Part. Nucl. Phys.* 106: 68.

[368] Abrahamyan, Sea, Z. Ahmed, H. Albataineh, K. Aniol, D. S. Armstrong, W. Armstrong, T. Averett et al. 2012. 'Measurement of the Neutron Radius of ^{208}Pb through Parity Violation in Electron Scattering.' *Physical Review Letters* 108 (11): 112502.

[369] Anthony, P. L., R. G. Arnold, C. Arroyo, K. Bega, J. Biesiada, P. E. Bosted, G. Bower et al. 2005. 'Precision Measurement of the Weak Mixing Angle in Moeller Scattering.' *Physical Review Letters* 95 (8): 081601.

[370] Androic, D., D. S. Armstrong, J. Arvieux, S. L. Bailey, D. H. Beck, E. J. Beise, J. Benesch et al. 2012. 'First Measurement of the Neutral Current Excitation of the Delta Resonance on a Proton Target.' *arXiv preprint arXiv:1212.1637*.

[371] Musolf, M. J., T. W. Donnelly, J. Dubach, S. J. Pollock, S. Kowalski, and E. J. Beise. 1994. 'Intermediate-energy Semileptonic Probes of the Hadronic Neutral Current.' *Physics Reports* 239 (1-2): 1–178.

[372] Nath, L. M., K. Schilcher, and M. Kretzschmar. 1982. 'Parity-violating Effects in Electroproduction of the Δ (1232) by Polarized Electrons.' *Physical Review D* 25 (9): 2300.

[373] Zhu, Shi-Lin, Claudio Masumi Maekawa, G. Sacco, Barry R. Holstein, and M. J. Ramsey- Musolf. 2001. 'Electroweak Radiative Corrections to Parity-violating Electroexcitation of the Δ.' *Physical Review D* 65 (3): 033001.

[374] Wang, D., K. Pan, R. Subedi, X. Deng, Z. Ahmed, K. Allada, K. A. Aniol et al. 2014. 'Measurement of Parity Violation in Electron–quark Scattering.' *Nature* 506 (7486): 67–70.

[375] Matsui, K., T. Sato, and T-SH Lee. 2005. 'Quark-hadron Duality and Parity Violating Asymmetry of Electroweak Reactions in the Δ Region.' *Physical Review C* 72 (2): 025204.

[376] Gorchtein, Mikhail, C. J. Horowitz, and Michael J. Ramsey-Musolf. 2011. 'Model Dependence of the γZ Dispersion Correction to the Parity-violating Asymmetry in Elastic *ep* Scattering.' *Physical Review C* 84 (1): 015502.

[377] Hall, Nathan Luke, Peter Gwithian Blunden, Wally Melnitchouk, Anthony W. Thomas, and Ross D. Young. 2013. 'Constrained γZ Interference Corrections to Parity-violating Electron Scattering.' *Physical Review D* 88 (1): 013011.

[378] Owens, J. F., Alberto Accardi, and Wally Melnitchouk. 2013. 'Global Parton Distributions with Nuclear and Finite-Q^2 Corrections.' *Physical Review D* 87 (9): 094012.

[379] Bouchiat, M. A., and C. C. Bouchiat. 1974. 'I. Parity Violation Induced by Weak Neutral Currents in Atomic Physics.' *Journal de Physique* 35 (12): 899–927.

[380] Bouchiat, Marie-Anne, and Claude Bouchiat. 1997. 'Parity Violation in Atoms.'
       Reports on Progress in Physics 60 (11): 1351.

[381] Barkov, L. M., and M. S. Zolotorev. 1979. 'Parity Violation in Atomic Bismuth.'
       Physics Letters B 85 (2-3): 308–313.

[382] Safronova, M. S., D. Budker, D. DeMille, Derek F. Jackson Kimball, A. Derevianko,
       and Charles W. Clark. 2018. 'Search for New Physics with Atoms and Molecules.'
       Reviews of Modern Physics 90 (2): 025008.

[383] Ginges, J. S. M., and Victor V. Flambaum. 2004. 'Violations of Fundamental
       Symmetries in Atoms and Tests of Unification Theories of Elementary Particles.'
       Physics Reports 397 (2): 63–154.

[384] Roberts, B. M., V. A. Dzuba, and V. V. Flambaum. 2015. 'Parity and Time-reversal
       Violation in Atomic Systems.' *Annual Review of Nuclear and Particle Science* 65:
       63–86.

[385] Bernardini, G. et al. 1964. *Physics Letters* 13: 86.

[386] Arnison, G., A. Astbury, B. Aubert, C. Bacci, G. Bauer, A. Bezaguet, R. Böck et al.
       1983. 'Experimental Observation of Lepton Pairs of Invariant Mass Around 95 GeV/c^2
       at the CERN SPS Collider.' *Physics Letters B* 126 (5): 398–410.

[387] Bagnaia, Po, B. Madsen, J. C. Chollet, J. L. Siegrist, H. Hänni, L. Di Lella, H. M.
       Steiner et al. 1983. 'Evidence for Z0 $\rightarrow e^+ e^-$ at the CERN $\bar{p}$ p Collider.' *Phys. Lett.
       B* 129, no. CERN-EP-83-112: 130–140.

[388] Schael, S. et al. [ALEPH and DELPHI and L3 and OPAL and SLD Collaborations
       and LEP Electroweak Working Group and SLD Electroweak Group and SLD Heavy
       Flavour Group]. 2006. *Phys. Rept.* 427: 257.

[389] Aaltonen, Terhi, B. Álvarez González, Silvia Amerio, D. Amidei, A. Anastassov, A.
       Annovi, J. Antos et al. 2012. 'Precise Measurement of the *W*-boson Mass with the
       CDF II Detector.' *Physical Review Letters* 108 (15): 151803.

[390] Straessner, Arno. 2004. 'Measurement of the *W* Boson Mass at LEP.' *arXiv preprint
       hep-ex/0405005.*

[391] Pich, Antonio. 2011. 'Flavour Physics and CP Violation.' *arXiv preprint arXiv:1112.
       4094.* doi:10.5170/CERN-2013-003.119.

[392] Aubert, Bernard, Y. Karyotakis, J. P. Lees, V. Poireau, E. Prencipe, X. Prudent,
       V. Tisserand et al. 2010. 'Measurements of Charged Current Lepton Universality and
       $|V_{us}|$ Using Tau Lepton Decays to $e^- \bar{\nu}_e \nu_\tau, \mu^- \bar{\nu}_\mu \nu_\tau, \pi^- \nu_\tau$, and $K^- \nu_\tau$.' *Physical Review
       Letters* 105 (5): 051602.

[393] Lazzeroni, Cristina, Angela Romano, A. Ceccucci, H. Danielsson, V. Falaleev,
       L. Gatignon, S. Goy Lopez et al. 2011. 'Test of Lepton Flavour Universality in
       $K^+ \rightarrow \ell^+ \nu$ Decays.' *Physics Letters B* 698 (2): 105–114.

[394] Antonelli, M., V. Cirigliano, G. Isidori, F. Mescia, M. Moulson, H. Neufeld, E. Passemar
       et al. 2010. 'An Evaluation of $|V_{us}|$ and Precise Tests of the Standard Model from World

Data on Leptonic and Semileptonic Kaon Decays.' *The European Physical Journal C* 69 (3-4): 399–424.

[395] Barate, R., ALEPH collaboration, DELPHI Collaboration, L3 Collaboration, and OPAL Collaboration. 2003. 'LEP Working Group for Higgs Boson Searches.' *Phys. Lett. B* 565: 61.

[396] Quigg, C. 1983. *Gauge Theories of the Strong, Weak, and Electromagnetic Interactions.* Menlo Park, California: The Benjamin/Cummings Publishing Company, Inc.

[397] Pich, Antonio. 2012. 'The Standard Model of Electroweak Interactions.' *arXiv preprint arXiv:1201.0537.*

[398] Aaboud, Morad, Georges Aad, Brad Abbott, O. Abdinov, Baptiste Abeloos, Syed Haider Abidi, O. S. AbouZeid et al. 2018. 'Measurement of the Higgs Boson Mass in the $H \to ZZ \to 4\ell$ and $H \to \gamma\gamma$ Channels with $s = 13$ TeV pp Collisions using the ATLAS Detector.' *Physics Letters B* 784: 345–366.

[399] Aad, Georges, B. Abbott, J. Abdallah, O. Abdinov, Rosemarie Aben, Maris Abolins, O. S. AbouZeid et al. 2015. 'Combined Measurement of the Higgs Boson Mass in pp Collisions at $\sqrt{s} = 7$ and 8 TeV with the ATLAS and CMS Experiments.' *Physical Review Letters* 114 (19): 191803.

[400] Halzen, F., and A. D. Martin. 1984. *Quarks And Leptons: An Introductory Course In Modern Particle Physics.* New York: Wiley.

[401] Geiregat, D., Gaston Wilquet, U. Binder, H. Burkard, U. Dore, W. Flegel, H. Grote et al. 1990. 'A New Measurement of the Cross Section of the Inverse Muon Decay Reaction $\nu\mu^+e^- \to \mu^- + \nu e$.' *Physics Letters B* 247 (1): 131–136.

[402] Blietschau, J., H. Deden, F. J. Hasert, W. Krenz, J. Morfin, K. Schultze, L. Welch et al. 1978. 'Upper Limit to the Cross-section for the Process $\nu\mu^+e^- \to \nu\mu^+e^-$.' *Physics Letters B* 73 (2): 232–234.

[403] Vilain, Pierre, G. Wilguet, W. Flegel, H. Grote, T. Mouthuy, H. Øveras, J. Panman et al. 1992. 'Neutral Current Coupling Constants from Neutrino-and Antineutrino-electron Scattering.' *Physics Letters B* 281 (1-2): 159–166.

[404] Blietschau, J. et al. [Gargamelle Collaboration]. 1976. *Nucl. Phys. B* 114: 189.

Alibran, P., N. Armenise, E. Bellotti, A. Blondel, D. Blum, S. Bonetti, G. Bonneaud et al. 1978. 'Observation and Study of the $\nu\mu e^- \to \nu\mu e^-$ Reaction in Gargamelle at High Energy.' *Physics Letters B* 74 (4-5): 422–428.

Armenise, N., O. Erriquez, M. T. Fogli-Muciaccia, S. Natali, S. Nuzzo, F. Romano, G. Bonneaud et al. 1979. 'High Energy Elastic $\nu\mu$ Scattering off Electrons in Gargamelle.' *Physics Letters B* 86 (2): 225–228.

Bertrand, D. et al. 1979. *Physics Letters B* 84: 354.

[405] Armenise, Nicola, O. Erriquez, M. T. Fogli-Muciaccia, S. Natali, S. Nuzzo, F. Romano, F. Ruggieri et al. 1979. 'Upper Limit to the Cross Section for $\bar{\nu}\nu^+e^- \to \bar{\nu}\nu^+e^-$ at High Energy.' *Physics Letters B* 81 (3-4): 385–388.

[406] Berge, J. P., D. Bogert, R. Hanft, D. Hamilton, G. Harigel, J. A. Malko, G. I. Moffatt et al. 1979. 'A Search at High Energies for Antineutrino-electron Elastic Scattering.' *Physics Letters B* 84 (3): 357–359.

[407] Allen, R. C., H. H. Chen, P. J. Doe, R. Hausammann, W. P. Lee, X. Q. Lu, H. J. Mahler et al. 1993. 'Study of Electron-neutrino—Electron Elastic Scattering at LAMPF.' *Physical Review D* 47 (1): 11.

[408] Avignone III, F. T. 1970. '$V - A$ Elastic Scattering of Electrons by Fission Antineutrinos.' *Physical Review D* 2 (11): 2609.

[409] Gurevich, GS Vidyakin VN VyrodovJ I., Yu V. Kozlov, V. P. Martem'yanov, S. V. Sukhotin, V. G. Tarasenkov, E. V. Turbin, and S. Kh Khakimov. 1989. 'Study of the Scattering of Fission Antineutrinos by Electrons with an Organofluoric-scintillator Detector.' *JETP Lett* 49 (12).

[410] Fujikawa, Kazuo, and Robert E. Shrock. 1980. 'Magnetic Moment of a Massive Neutrino and Neutrino-spin Rotation.' *Physical Review Letters* 45 (12): 963.

[411] Barber, D. P., U. Becker, H. Benda, A. Boehm, J. G. Branson, J. Bron, D. Buikman et al. 1979. 'Study of Electron-Positron Collisions at Center-of-mass Energies of 27.4 and 27.7 GeV at Petra.' *Physical Review Letters* 43 (13): 901.

[412] Marshall, R. 1985. In J. Grunhaus, ed. *Proc. XVI Int. Symp. on Multiparticle Dynamics, Kiryat Anavim, Israel.* Singapore: Kim Hup Lee.

[413] Amaldi, Ugo, Albrecht Bóhm, L. S. Durkin, Paul Langacker, Alfred K. Mann, William J. Marciano, Alberto Sirlin, and H. H. Williams. 1987. 'Comprehensive Analysis of Data Pertaining to the Weak Neutral Current and the Intermediate-vector-boson Masses.' *Physical Review D* 36 (5): 1385.

[414] Galster, S., H. Klein, J. Moritz, K. H. Schmidt, D. Wegener, and J. Bleckwenn. 1971. 'Elastic Electron-deuteron Scattering and the Electric Neutron Form Factor at Four-momentum Transfers $5fm - 2 < q^2 < 14fm - 2$.' *Nuclear physics B* 32 (1): 221–237.

[415] Budd, Howard, A. Bodek, and J. Arrington. 2005. 'Vector and Axial Form Factors Applied to Neutrino Quasielastic Scattering.' *Nuclear Physics B-Proceedings Supplements* 139: 90–95.

[416] Bradford, R., A. Bodek, H. S. Budd, and J. Arrington. 2006. 'NuInt05, Proceedings of the 4th International Workshop on Neutrino-Nucleus Interactions in the Few-GeV Region, Okayama, Japan, 26–29 September 2005.' *Nucl. Phys. Proc. Suppl* 159: 127.

[417] Kelly, J. J. 2004. 'Simple Parametrization of Nucleon Form Factors.' *Physical Review C* 70 (6): 068202.

[418] Punjabi, V., C. F. Perdrisat, M. K. Jones, E. J. Brash, and C. E. Carlson. 2015. 'The Structure of the Nucleon: Elastic Electromagnetic Form Factors.' *The European Physical Journal A* 51 (7): 79.

[419] Collaboration, Precision Neutron Decay Matrix Elements PNDME, Rajan Gupta, Yong-Chull Jang, Huey-Wen Lin, Boram Yoon, and Tanmoy Bhattacharya. 2017. 'Axial-

vector Form Factors of the Nucleon from Lattice QCD.' *Physical Review D*96 (11): 114503.

[420] Meyer, Aaron S., Minerba Betancourt, Richard Gran, and Richard J. Hill. 2016. 'Deuterium Target Data for Precision Neutrino-nucleus Cross Sections.' *Physical Review D* 93 (11): 113015.

[421] Green, Jeremy, Nesreen Hasan, Stefan Meinel, Michael Engelhardt, Stefan Krieg, Jesse Laeuchli, John Negele, Kostas Orginos, Andrew Pochinsky, and Sergey Syritsyn. 2017. 'Up, Down, and Strange Nucleon Axial Form Factors from Lattice QCD.' *Physical Review D* 95 (11): 114502.

[422] Alexandrou, Constantia, Martha Constantinou, Kyriakos Hadjiyiannakou, Karl Jansen, Christos Kallidonis, Giannis Koutsou, and A. Vaquero Aviles-Casco. 2017. 'Nucleon Axial Form Factors using $N_f = 2$ Twisted Mass Fermions with a Physical Value of the Pion Mass.' *Physical Review D* 96 (5): 054507.

[423] Yao, De-Liang, Luis Alvarez-Ruso, and Manuel J. Vicente-Vacas. 2017. 'Extraction of Nucleon Axial Charge and Radius from Lattice QCD Results using Baryon Chiral Perturbation Theory.' *Physical Review D* 96 (11): 116022.

[424] Capitani, Stefano, Michele Della Morte, Dalibor Djukanovic, Georg M. von Hippel, Jiayu Hua, Benjamin Jäger, Parikshit M. Junnarkar, Harvey B. Meyer, Thomas D. Rae, and Hartmut Wittig. 2019. 'Isovector Axial form Factors of the Nucleon in Two-flavor Lattice QCD.' *International Journal of Modern Physics A* 34 (02): 1950009.

[425] Close, F. E. 1980. *An Introduction to Quarks and Partons.* London: Academic Press.

[426] Morfin, J. G., J. Nieves, and J. T. Sobczyk. 2012. *Adv. High Energy Phy.* 2012: 934597.

[427] Gallagher, H., G. Garvey, and G. P. Zeller. 2011. 'Neutrino-nucleus Interactions.' *Annual Review of Nuclear and Particle Science* 61: 355–378.

[428] Miller, K. L., S. J. Barish, A. Engler, R. W. Kraemer, B. J. Stacey, M. Derrick, E. Fernandez et al. 1982. 'Study of the Reaction $\nu_\mu d \to \mu^- pps$.' *Physical Review D* 26 (3): 537.

[429] Baker, N. J., A. M. Cnops, P. L. Connolly, S. A. Kahn, H. G. Kirk, M. J. Murtagh, R. B. Palmer, N. P. Samios, and M. Tanaka. 1981. 'Quasielastic Neutrino Scattering: A Measurement of the Weak Nucleon Axial-vector Form Factor.' *Physical Review D* 23 (11): 2499.

[430] Kitagaki, T., S. Tanaka, H. Yuta, K. Abe, K. Hasegawa, A. Yamaguchi, K. Tamai et al. 1983. 'High-energy Quasielastic $\nu_\mu n \to \mu^- p$ Scattering in Deuterium.' *Physical Review D* 28 (3): 436.

[431] Bodek, A., S. Avvakumov, R. Bradford, and H. Budd. 2008. 'Vector and Axial Nucleon form Factors: A Duality Constrained Parameterization.' *The European Physical Journal C* 53 (3): 349–354.

[432] Lyubushkin, V., B. Popov, J. J. Kim, L. Camilleri, J-M. Levy, Mauro Mezzetto, Dmitry Naumov et al. 2009. 'A Study of Quasi-elastic Muon Neutrino and Antineutrino

Scattering in the NOMAD Experiment.' *The European Physical Journal C* 63 (3): 355–381.

[433] Fields, L., J. Chvojka, L. Aliaga, O. Altinok, B. Baldin, A. Baumbaugh, A. Bodek et al. 2013. 'Measurement of Muon Antineutrino Quasielastic Scattering on a Hydrocarbon Target at $E_\nu \backsim 3.5$ GeV.' *Physical Review Letters* 111 (2): 022501.

[434] Aguilar-Arevalo, A. A., A. O. Bazarko, S. J. Brice, B. C. Brown, L. Bugel, J. Cao, L. Coney et al. 2008. 'Measurement of Muon Neutrino Quasielastic Scattering on Carbon.' *Physical Review Letters* 100 (3): 032301.

[435] Aguilar-Arevalo, A. A., C. E. Anderson, A. O. Bazarko, S. J. Brice, B. C. Brown, L. Bugel, J. Cao et al. 2010. 'First Measurement of the Muon Neutrino Charged Current Quasielastic Double Differential Cross Section.' *Physical Review D* 81 (9): 092005.

[436] Dorman, M., and MINOS collaboration. 2009. 'Preliminary Results for CCQE Scattering with the MINOS near Detector.' In *AIP Conference Proceedings* 1189 (1): 133–138. American Institute of Physics.

[437] Adamson, P., I. Anghel, A. Aurisano, G. Barr, M. Bishai, A. Blake, G. J. Bock et al. 2015. 'Study of Quasielastic Scattering using Charged-current ν_μ-iron Interactions in the MINOS near Detector.' *Physical Review D* 91 (1): 012005.

[438] Gran, Richard, E. J. Jeon, E. Aliu, S. Andringa, S. Aoki, J. Argyriades, K. Asakura et al. 2006. 'Measurement of the Quasielastic Axial Vector Mass in Neutrino Interactions on Oxygen.' *Physical Review D* 74 (5): 052002.

[439] Abe, Ko, C. Andreopoulos, M. Antonova, S. Aoki, A. Ariga, S. Assylbekov, D. Autiero et al. 2016. 'Measurement of Muon Antineutrino Oscillations with an Accelerator-produced off-axis Beam.' *Physical review letters* 116 (18): 181801.

Abe, Kou, J. Adam, Hiroaki Aihara, T. Akiri, C. Andreopoulos, Shigeki Aoki, Akitaka Ariga et al. 2015. 'Measurements of Neutrino Oscillation in Appearance and Disappearance Channels by the $t2k$ Experiment with 6.6×10^{20} Protons on Target.' *Physical Review D* 91 (7): 072010.

Abe, Kou, J. Adam, H. Aihara, T. Akiri, C. Andreopoulos, S. Aoki, A. Ariga et al. 2015. 'Measurement of the ν_μ Charged-current Quasielastic Cross Section on Carbon with the ND280 Detector at T2K.' *Physical Review D* 92 (11): 112003.

[440] Nakajima, Y., J. L. Alcaraz-Aunion, S. J. Brice, L. Bugel, J. Catala-Perez, G. Cheng, J. M. Conrad et al. 2011. 'Measurement of Inclusive Charged Current Interactions on Carbon in a Few-GeV Neutrino Beam.' *Physical Review D* 83 (1): 012005.

[441] Cheng, G., W. Huelsnitz, A. A. Aguilar-Arevalo, J. L. Alcaraz-Aunion, S. J. Brice, B. C. Brown, L. Bugel et al. 2012. 'Dual Baseline Search for Muon Antineutrino Disappearance at 0.1 eV$^2 < \Delta m^2 < 100$ eV^2.' *Physical Review D* 86 (5): 052009.

[442] Fiorentini, G. A., D. W. Schmitz, P. A. Rodrigues, L. Aliaga, O. Altinok, B. Baldin, A. Baumbaugh et al. 2013. 'Measurement of Muon Neutrino Quasielastic Scattering on a Hydrocarbon Target at $E_\nu \backsim 3.5$ GeV.' *Physical Review Letters* 111 (2): 022502.

[443] Bernard, Veronique, Latifa Elouadrhiri, and Ulf-G. Meissner. 2001. *Journal of Physics G: Nuclear and Particle Physics* 28 (1): R1.

[444] Schindler, M. R., T. Fuchs, J. Gegelia, and S. Scherer. 2007. 'Axial, Induced Pseudoscalar, and Pion-nucleon Form Factors in Manifestly Lorentz-invariant Chiral Perturbation Theory.' *Physical Review C* 75 (2): 025202.

[445] Tsapalis, A. 2007. 'Nucleon and Pion-Nucleon Form-Factors From Lattice QCD.' *eConf* 70910 (MENU-2007-168): 168.

[446] Pate, S. F., and J. P. Schaub. 2011. 'Strange Quark Contribution to the Nucleon Spin from Electroweak Elastic Scattering Data.' In *Journal of Physics: Conference Series* 295 (1): 012037. IOP Publishing.

[447] Formaggio, Joseph A., and G. P. Zeller. 2012. 'From eV to EeV: Neutrino Cross Sections Across Energy Scales.' *Reviews of Modern Physics* 84 (3): 1307.

[448] Fearing, Harold W., P. C. McNamee, and R. J. Oakes. 1969. 'Weak Electron Scattering: $e^- + p \to \Lambda + v$.' *Il Nuovo Cimento A (1965–1970)* 60 (1): 10–24.

[449] Veltman, M. J. G., and S. M. Berman. 1964. 'Transverse Muon Polarization in Neutrino Induced Interactions as a Test for Time Reversal Violation.' *Physics letters* 12 (3): 275–278.

[450] Fujii, A., and Y. Yamaguchi. 1966. 'Nucleon Polarization in Elastic Lepton or Antilepton Production by High-energy Neutrino or Antineutrino.' *Il Nuovo Cimento A (1971–1996)* 43 (2): 325–333.

[451] Eichten, T., H. Faissner, S. Kabe, W. Frenz, J. Von Krogh, J. Morfin, K. Schultze et al. 1972. 'Observation of 'Elastic' Hyperon Production by Antineutrinos.' *Physics Letters B* 40 (5): 593–596.

[452] Erriquez, O., MT Fogli Muciaccia, S. Natali, S. Nuzzo, A. Halsteinslid, C. Jarlskog, K. Myklebost et al. 1978. 'Production of Strange Particles in Antineutrino Interactions at the CERN PS.' *Nuclear Physics B* 140 (1): 123–140.

[453] Enriquez, O. et al. 1977. *Physics Letters B* 70: 383.

[454] Brunner, J., H-J. Grabosch, H. H. Kaufmann, R. Nahnhauer, S. Nowak, S. Schlenstedt, V. V. Ammosov et al. 1990. 'Quasielastic Nucleon and Hyperon Production by Neutrinos and Antineutrinos with Energies below 30 GeV.' *Zeitschrift für Physik C Particles and Fields* 45 (4): 551–555.

[455] Fanourakis, G., L. K. Resvanis, G. Grammatikakis, P. Tsilimigras, A. Vayaki, U. Camerini, W. F. Fry, R. J. Loveless, J. H. Mapp, and D. D. Reeder. 1980. 'Study of Low-energy Antineutrino Interactions on Protons.' *Physical Review D* 21 (3): 562.

[456] Bilenky, S. M., and Samoil M. Bilen'kij. *Basics of introduction to Feynman diagrams and electroweak interactions physics*. Atlantica Séguier Fronti 'eres, 1994.

[457] E. C. M. Young, CERN-67-12.

[458] Radecky, G. M., V. E. Barnes, D. D. Carmony, A. F. Garfinkel, M. Derrick, E. Fernandez, L. Hyman et al. 1982. 'Study of Single-pion Production by Weak Charged Currents in Low-energy νd Interactions.' *Physical Review D* 25 (5): 1161.

Barish, S. J., J. Campbell, G. Charlton, Y. Cho, M. Derrick, R. Engelmann, L. G. Hyman et al. 1977. 'Study of Neutrino Interactions in Hydrogen and Deuterium: Description of the Experiment and Study of the Reaction $\nu^+ d \to \mu^- + p + p_s$.' *Physical Review D* 16 (11): 3103.

Barish, S. J., M. Derrick, T. Dombeck, L. G. Hyman, K. Jaeger, B. Musgrave, P. Schreiner et al. 1979. 'Study of Neutrino Interactions in Hydrogen and Deuterium. II. Inelastic Charged-current Reactions.' *Physical Review D* 19 (9): 2521.

[459] Kitagaki, T., H. Yuta, S. Tanaka, A. Yamaguchi, K. Abe, K. Hasegawa, K. Tamai et al. 1990. 'Study of $\nu d \to \mu^- p p_s$ and $\nu d \to \mu^- \Delta^{++}(1232) n_s$ using the BNL 7-foot Deuterium-filled Bubble Chamber.' *Physical Review D* 42 (5): 1331.

[460] Dennery, Philippe. 1962. 'Pion Production in Neutrino-Nucleon Collisions.' *Physical Review* 127 (2): 664.

[461] Dombey, Norman. 1962. 'Weak Pion Production.' *Physical Review* 127 (2): 653.

[462] Bell, J. S., and S. M. Berman. 1962. *Nuovo Cimento C* 25: 404.

[463] Bell, John Stewart, and CH Llewellyn Smith. 1970. 'Near-forward Neutrino Reactions on Nuclear Targets.' *Nuclear Physics B* 24 (2): 285–304.

[464] Adler, Stephen L. 1968. 'Photo-, Electro-, and Weak Single-pion Production in the (3, 3) Resonance Region.' *Annals of Physics* 50 (2): 189–311.

[465] Albright, C. H., and L. S. Liu. 1965. 'Baryon Resonance Production by Neutrinos and the Relativistic Generalizations of SU (6).' *Physical Review* 140 (6B): B1611.

[466] Albright, C. H., and Lu Sun Liu. 1965. 'Weak N^* Production and SU (6).' *Physical Review* 140 (3B): B748.

[467] Kim, C. W. 1965. 'Production of N^* in Neutrino Reactionsin Neutrino Reactions.' *Il Nuovo Cimento (1955–1965)* 37 (1): 142–148.

[468] Lee, W., E. Maddry, P. Sokolsky, L. Teig, A. Bross, T. Chapin, L. Holloway et al. 1977. 'Single-pion Production in Neutrino and Antineutrino Reactions.' *Physical Review Letters* 38 (5): 202.

[469] Bell, J., J. P. Berge, D. V. Bogert, R. J. Cence, C. T. Coffin, R. N. Diamond, F. A. DiBianca et al. 1978. 'Cross-Section Measurements for the Reactions $\nu p \to \nu^- \pi^+ p$ and $\nu p \to \nu^- K^+ p$ at High Energies.' *Physical Review Letters* 41 (15): 1008.

[470] Allen, P., Vanna T. Cocconi, Gottfried Kellner, P. O. Hulth, Bianca Conforto, L. Pape, D. Lanske et al. 1980. 'Single π^+ Production in Charged Current Neutrino-hydrogen Interactions.' *Nucl. Phys. B* 176, no. CERN-EP-80-69: 269–284.

[471] Schreiner, Philip A., and Frank Von Hippel. 1973. 'Neutrino Production of the Δ (1236).' *Nuclear Physics B* 58 (2): 333–362.

[472] Fogli, Gian Luigi, and G. Nardulli. 1979. 'A New Approach to the Charged Current Induced Weak One-pion Production.' *Nuclear Physics B* 160 (1): 116–150.

[473] Sato, Toru, D. Uno, and T-SH Lee. 2003. 'Dynamical Model of Weak Pion Production Reactions.' *Physical Review C* 67 (6): 065201.

[474] Fogli, Gian Luigi, and G. Nardulli. 1980. 'Neutral Current Induced One-pion Production: A New Model and Its Comparison with Experiment.' *Nuclear Physics B* 165 (1): 162–184.

[475] Kamano, H., S. X. Nakamura, T-SH Lee, and T. Sato. 2012. 'Neutrino-induced Forward Meson-production Reactions in Nucleon Resonance Region.' *Physical Review D* 86 (9): 097503.

[476] Nakamura, S. X., H. Kamano, T-SH Lee, and T. Sato. 2015. 'Neutrino-induced Meson Productions off Nucleon at Forward Limit in Nucleon Resonance Region.' In *AIP Conference Proceedings* 1663 (1): 070005. AIP Publishing LLC.

[477] Paschos, Emmanuel A., Ji–Young Yu, and Makoto Sakuda. 2004. 'Neutrino Production of Resonances.' *Physical Review D* 69 (1): 014013.

[478] Lalakulich, Olga, and Emmanuel A. Paschos. 2005. 'Resonance Production by Neutrinos: $J = 3/2$ Resonances.' *Physical Review D* 71 (7): 074003.

[479] Barbero, C., G. Lopez Castro, and A. Mariano. 2008. 'Single Pion Production in $CC\nu_\mu N$ Scattering within a Consistent Effective Born Approximation.' *Physics Letters B* 664 (1-2): 70–77.

[480] Barbero, C., G. López Castro, and A. Mariano. 2014. 'One Pion Production in Neutrino–nucleon Scattering and the Different Parameterizations of the Weak $N \to \Delta$ Vertex.' *Physics Letters B* 728: 282–287.

[481] Hernandez, E., J. Nieves, and M. Valverde. 2007. 'Weak Pion Production off the Nucleon.' *Physical Review D* 76 (3): 033005.

[482] Leitner, Tina, L. Alvarez-Ruso, and U. Mosel. 2006. 'Charged Current Neutrino-nucleus Interactions at Intermediate Energies.' *Physical Review C* 73 (6): 065502.

[483] Leitner, T., O. Buss, L. Alvarez-Ruso, and U. Mosel. 2009. 'Electron-and Neutrino-nucleus Scattering from the Quasielastic to the Resonance Region.' *Physical Review C* 79 (3): 034601.

[484] Lalakulich, Olga, Emmanuel A. Paschos, and Giorgi Piranishvili. 2006. 'Resonance Production by Neutrinos: The Second Resonance Region.' *Physical Review D* 74 (1): 014009.

[485] Lalakulich, Olga, Tina Leitner, Oliver Buss, and Ulrich Mosel. 2010. 'One Pion Production in Neutrino Reactions: Including Nonresonant Background.' *Physical Review D* 82 (9): 093001.

[486] González-Jiménez, Raúl, Natalie Jachowicz, Kajetan Niewczas, Jannes Nys, V. Pandey, Tom Van Cuyck, and Nils Van Dessel. 2017. 'Electroweak Single-pion Production off the Nucleon: From Threshold to High Invariant Masses.' *Physical Review D* 95 (11): 113007.

[487] Rafi Alam, M., M. Sajjad Athar, S. Chauhan, and S. K. Singh. 2016. 'Weak Charged and Neutral Current Induced One Pion Production off the Nucleon.' *International Journal of Modern Physics E* 25 (2): 1650010.

[488] Alam, M. Rafi, I. Ruiz Simo, L. Alvarez-Ruso, M. Sajjad Athar, and MJ Vicente Vacas. 2015. 'Weak Production of Strange Particles and η Mesons off the Nucleon.' In *AIP Conference Proceedings*, vol. 1680 (1): 020001. AIP Publishing LLC.

[489] Alam, M. Rafi, L. Alvarez-Ruso, M. Sajjad Athar, and MJ Vicente Vacas. 2015. 'Weak η Production off the Nucleon.' In *AIP Conference Proceedings* 1663 (1): 120014. AIP Publishing LLC.

[490] Alam, M. Rafi, I. Ruiz Simo, M. Sajjad Athar, and MJ Vicente Vacas. 2010. 'Weak Kaon Production off the Nucleon.' *Physical Review D* 82 (3): 033001.

[491] Alam, M. Rafi, I. Ruiz Simo, M. Sajjad Athar, and MJ Vicente Vacas. 2012. 'ν Induced K Production off the Nucleon.' *Physical Review D* 85 (1): 013014.

[492] Rein, Dieter, and Lalit M. Sehgal. 1981. 'Neutrino-excitation of Baryon Resonances and Single Pion Production.' *Annals of Physics* 133 (1): 79–153.

[493] Tiator, L., D. Drechsel, S. S. Kamalov, and M. Vanderhaeghen. 2011. 'Electromagnetic Excitation of Nucleon Resonances.' *The European Physical Journal Special Topics* 198 (1): 141.

[494] Liu, Jun, Nimai C. Mukhopadhyay, and Lisheng Zhang. 1995. 'Nucleon to Δ Weak Excitation Amplitudes in the Nonrelativistic Quark Model.' *Physical Review C* 52 (3): 1630.

[495] Hemmert, Thomas R., Barry R. Holstein, and Nimai C. Mukhopadhyay. 1995. 'NN, $N\Delta$ Couplings and the Quark Model.' *Physical Review D* 51 (1): 158.

[496] Serot, Brian D., and Xilin Zhang. 2012. 'Neutrinoproduction of Photons and Pions from Nucleons in a Chiral Effective Field Theory for Nuclei.' *Physical Review C* 86 (1): 015501.

[497] Leitner, Tina, O. Buss, U. Mosel, and L. Alvarez-Ruso. 2009. 'Neutrino-induced Pion Production at Energies Relevant for the MiniBooNE and K2K Experiments.' *Physical Review C* 79 (3): 038501.

[498] Sobczyk, Jan T., and Jakub Żmuda. 2013. 'Impact of Nuclear Effects on Weak Pion Production at Energies Below 1 GeV.' *Physical Review C* 87 (6): 065503.

[499] Graczyk, Krzysztof M., and Jan T. Sobczyk. 2008. 'Lepton Mass Effects in Weak Charged Current Single Pion Production.' *Physical Review D* 77 (5): 053003.

[500] Nambu, Yoichiro, and Giovanni Jona-Lasinio. 1961. 'Dynamical Model of Elementary Particles Based on an Analogy with Superconductivity. II.' *Physical Review* 124 (1): 246.

[501] Koch, Volker. 1997. 'Aspects of Chiral Symmetry. ' *International Journal of Modern Physics E* 6 (2): 203–249.

[502] Scherer, Stefan, and Matthias R. Schindler. 2012. 'A Primer for Chiral Perturbative Theory.'

[503] Georgi, Howard, and Sheldon L. Glashow. 1974. 'Unity of all Elementary-particle Forces.' *Physical Review Letters* 32 (8): 438.

[504] Nath, Pran, and R. Arnowitt. 2000. 'Grand Unification and B & L Conservation.' *Physics of Atomic Nuclei* 63 (7): 1151–1157. [arXiv:hep-ph/9808465]

[505] McGrew, C., R. Becker-Szendy, C. B. Bratton, J. L. Breault, D. R. Cady, D. Casper, S. T. Dye et al. 1999. 'Search for Nucleon Decay using the IMB-3 Detector.' *Physical Review D* 59 (5): 052004.

[506] Hirata, K. S., T. Kajita, T. Kifune, K. Kihara, M. Nakahata, K. Nakamura, S. Ohara et al. 1989. 'Experimental Limits on nucleon Lifetime for Lepton$^+$ Meson Decay Modes.' *Physics Letters B* 220 (1-2): 308–316.

[507] Abe, K., Y. Haga, Y. Hayato, M. Ikeda, K. Iyogi, J. Kameda, Y. Kishimoto et al. 2017. 'Search for Proton Decay via $p \to e^+ \pi^0$ and $p \to \mu^+ \pi^0$ in 0.31 Megaton• Years Exposure of the Super-Kamiokande Water Cherenkov Detector.' *Physical Review D* 95 (1): 012004.

[508] Shiozawa, Masato, B. Viren, Y. Fukuda, T. Hayakawa, E. Ichihara, K. Inoue, K. Ishihara et al. 1998. 'Search for Proton Decay via $p \to e^+ \pi^0$ in a Large Water Cherenkov Detector.' *Physical Review Letters* 81 (16): 3319.

Nishino, Haruki, S. Clark, Kou Abe, Yoshinari Hayato, Takashi Iida, Motoyasu Ikeda, Jun Kameda et al. 2009. 'Search for Proton Decay via $p \to e^+ \pi^0$ and $p \to \mu^+ \pi^0$ in a Large Water Cherenkov Detector.' *Physical Review Letters* 102 (14): 141801.

[509] Abe, K., Y. Hayato, K. Iyogi, J. Kameda, M. Miura, S. Moriyama, M. Nakahata et al. 2014. 'Search for Proton Decay via $p \to \nu k^+$ Using 260 Kiloton• Year Data of Super-kamiokande.' *Physical Review D* 90 (7): 072005.

[510] Wall, D., W. W. M. Allison, G. J. Alner, D. S. Ayres, W. L. Barrett, C. Bode, P. M. Border et al. 2000. 'Search for Nucleon Decay with Final States $l^+ \eta^0, \bar{\nu} \eta^0$, and $\bar{\nu} \pi^{+,0}$ using Soudan 2.' *Physical Review D* 62 (9): 092003.

[511] Berman, Sam M., and M. Veltman. 1965. 'Baryon-resonance Production by Neutrinos.' *Il Nuovo Cimento (1955–1965)* 38 (2): 993–1005.

[512] Altarelli, Guido, Richard A. Brandt, and Giuliano Preparata. 1971. 'Light-Cone Analysis of Massive μ-Pair Production.' *Physical Review Letters* 26 (1): 42.

[513] Ravndal, F. 1972. *Lett. Nuovo Cim.* 3S2: 631. [*Lett. Nuovo Cim.* 3: 631 (1972)].

[514] K. S. Kuzmin, V. V. Lyubushkin and V. A. Naumov, Mod. Phys. Lett. A 19, 2815 (2004) [Phys. Part. Nucl. 35, S133 (2004)].

[515] Kuzmin, Konstantin S., Vladimir V. Lyubushkin, and Vadim A. Naumov. 2005. 'Extended Rein–Sehgal Model for Tau Lepton Production.' *Nuclear Physics B-Proceedings Supplements* 139: 158–161.

[516] Wu, Jia-Jun, and Bing-Song Zou. 2015. 'Hyperon Production from Neutrino–nucleon Reaction.' *Few-Body Systems* 56 (4-5): 165–183.

[517] Wu, Jia-Jun, T. Sato, and T-SH Lee. 2015. 'Incoherent Pion Production in Neutrino-deuteron Interactions.' *Physical Review C* 91 (3): 035203.

[518] Kamano, H., S. X. Nakamura, T-SH Lee, and T. Sato. 2013. 'Nucleon Resonances within a Dynamical Coupled-channels Model of πN and γN Reactions.' *Physical Review C* 88 (3): 035209.

[519] Kitagaki, T., H. Yuta, S. Tanaka, A. Yamaguchi, K. Abe, K. Hasegawa, K. Tamai et al. 1990. 'Study of $\nu d \to \mu^- p p_s$ and $\nu d \to \mu^- \Delta^{++}(1232)n_s$ using the BNL 7-foot Deuterium-filled Bubble Chamber.' *Physical Review D* 42 (5): 1331.

[520] Lacombe, M. et al. 1981. *Physics Letters B* 101: 139.

[521] Graczyk, K. M., D. Kielczewska, P. Przewlocki, and J. T. Sobczyk. 2009. 'C_5^A Axial form Factor from Bubble Chamber Experiments.' *Physical Review D* 80 (9): 093001.

[522] Graczyk, Krzysztof M., Jakub Żmuda, and Jan T. Sobczyk. 2014. 'Electroweak form Factors of the Δ (1232) Resonance.' *Physical Review D* 90 (9): 093001.

[523] Wilkinson, Callum, Philip Rodrigues, Susan Cartwright, Lee Thompson, and Kevin McFarland. 2014. 'Reanalysis of Bubble Chamber Measurements of Muon-neutrino Induced Single Pion Production.' *Physical Review D* 90 (11): 112017.

[524] Derrick, M., E. Fernandez, L. Hyman, G. Levman, D. Koetke, B. Musgrave, P. Schreiner et al. 1980. 'Study of the Reaction $\nu n \to \nu p \pi^-$.' *Physics Letters B* 92 (3-4): 363–366.

[525] Dover, Carl B., and Paul M. Fishbane. 1990. 'η and η'scattering: A Probe of the Strange-quark ($s s^-$) Content of the Nucleon?.' *Physical Review Letters* 64 (26): 3115.

[526] Solomey, Nickolas. 2005. 'A Proposed Study of Neutrino-induced Strange-particle Production Reactions at Minerva.' *Nuclear Physics B-Proceedings Supplements* 142: 74–78.

[527] Mann, W. A., T. Kafka, M. Derrick, B. Musgrave, R. Ammar, D. Day, and J. Gress. 1986. 'K-meson Production by ν_μ-deuterium Reactions Near Threshold: Implications for Nucleon-decay Searches.' *Physical Review D* 34 (9): 2545.

[528] Baker, N. J., P. L. Connolly, S. A. Kahn, H. G. Kirk, M. J. Murtagh, R. B. Palmer, N. P. Samios, and M. Tanaka. 1981. 'Strange-particle Production from Neutrino Interactions in the BNL 7-foot Bubble Chamber.' *Physical Review D* 24 (11): 2779.

[529] Barish, S. J., M. Derrick, L. G. Hyman, P. Schreiner, R. Singer, R. P. Smith, H. Yuta et al. 1974. 'Strange-particle Production in Neutrino Interactions.' *Physical Review Letters* 33 (24): 1446.

[530] Deden, H., F. J. Hasert, W. Krenz, J. Von Krogh, D. Lanske, J. Morfin, M. Pohl et al. 1975. 'Strange Particle Production and Charmed Particle Search in the Gargamelle Neutrino Experiment.' *Physics Letters B* 58 (3): 361–366.

[531] Shrock, Robert E. 1975. 'Associated Production by Weak Charged and Neutral Currents.' *Physical Review D* 12 (7): 2049.

[532] Mecklenburg, W. 1978. 'Neutrino-induced Associated Production.' *Acta Physica Austriaca* 48 (4): 293–316.

[533] Dewan, H. K. 1981. 'Strange-particle Production in Neutrino Scattering.' *Physical Review D* 24 (9): 2369.

[534] Amer, A. A. 1978. 'Production of Strange Particles by Neutrinos and Antineutrinos.' *Physical Review D* 18 (7): 2290.

[535] Casper, D. 2002. *Nucl. Phys. Proc. Suppl.* 112: 161.

[536] Datchev, K. 2002. *APS Meeting Abstracts* p. 1037P.

[537] Undagoitia, T. Marrodan, F. von Feilitzsch, M. Göger-Ne, C. Grieb, K. A. Hochmuth, L. Oberauer, W. Potzel, and M. Wurm. 2006. 'Proton Decay in the Large Liquid Scintillator Detector LENA: Study of the Background.' In *Journal of Physics: Conference Series*, vol. 39 (1): 269. IOP Publishing.

[538] Kobayashi, K., M. Earl, Y. Ashie, J. Hosaka, K. Ishihara, Y. Itow, J. Kameda et al. 2005. 'Search for Nucleon Decay via Modes Favored by Supersymmetric Grand Unification Models in Super-Kamiokande-I.' *Physical Review D* 72 (5): 052007.

[539] Marshall, C. M., L. Aliaga, O. Altinok, L. Bellantoni, A. Bercellie, M. Betancourt, A. Bodek et al. 2016. 'Measurement of K^+ Production in Charged-current ν_μ Interactions.' *Physical Review D* 94 (1): 012002.

[540] Butler, Malcolm N., Martin J. Savage, and Roxanne P. Springer. 1993. 'Strong and Electromagnetic Decays of the Baryon Decuplet.' *Nuclear Physics B* 399 (1): 69–85.

[541] Lyman, E. M. 1951. 'A. 0. Hanson, and MB Scott.' *Phys. Rev* 84 (626): 1.

[542] Chambers, E. E., and R. Hofstadter. 1956. 'Structure of the Proton.' *Physical Review* 103 (5): 1454.

[543] Yearian, M. R., and R. Hofstadter. 1958. *Physical Review* 110 (2): 552.

[544] Friedman, J. I., H. W. Kendall, and R. E. Taylor. 1990. *SLAC-REPRINT-1991-019*.

[545] Bjorken, J. D. 1970. *Conf. Proc. C* 700612V1: 1. [*Acta Phys. Polon. B* 2: 5 (1971)].

[546] Bjorken, James D. 1969. 'Asymptotic Sum Rules at Infinite Momentum.' *Physical Review* 179 (5): 1547.

[547] Bartel, W., B. Dudelzak, H. Krehbiel, J. McElroy, U. Meyer-Berkhout, W. Schmidt, V. Walther, and G. Weber. 'Electroproduction of Pions near the Δ (1236) Isobar and the Form Factor $G * M$ (q^2) of the $(\gamma N\Delta)$-vertex.' *Physics Letters B* 28 (2): 148–151.

[548] Whitlow, L. W., E. M. Riordan, S. Dasu, Stephen Rock, and Arie Bodek. 1990. 'A Precise Extraction of $R =$ sigma-L/sigma-T from a Global Analysis of the SLAC Deep Inelastic ep and ed Scattering Cross-sections.' *Phys. Lett.* 250, no. SLAC-PUB-5284: 193–198.

Whitlow, L. W., E. M. Riordan, S. Dasu, Stephen Rock, and A. Bodek. 1991. 'Precise Measurements of the Proton and Deuteron Structure Functions from a Global Analysis of the SLAC Deep Inelastic Electron Scattering Cross-sections.' *Phys. Lett.* 282, no. SLAC-PUB-5442: 475–482.

[549] Glück, M., P. Jimenez-Delgado, and E. Reya. 2008. 'Dynamical Parton Distributions of the Nucleon and Very Small-x Physics.' *The European Physical Journal C* 53 (3): 355–366.

[550] Martin, Alan D., W. James Stirling, Robert S. Thorne, and G. Watt. 2009. 'Parton Distributions for the LHC.' *The European Physical Journal C* 63 (2): 189–285.

[551] Nadolsky, Pavel M., Hung-Liang Lai, Qing-Hong Cao, Joey Huston, Jon Pumplin, Daniel Stump, Wu-Ki Tung, and C-P. Yuan. 2008. 'Implications of CTEQ Global Analysis for Collider Observables.' *Physical Review D* 78 (1): 013004.

[552] Harland-Lang, Lucian A., A. D. Martin, P. Motylinski, and R. S. Thorne. 2015. 'Parton Distributions in the LHC Era: MMHT 2014 PDFs.' *The European Physical Journal C* 75 (5): 204.

[553] Aubert, Jean-Jacques, C. Peroni, Gabriel Bassompierre, K. Moser, Terence Sloan, V. Korbel, V. Gibson et al. 1987. 'Measurements of the Nucleon Structure Functions F_2^N in Deep Inelastic Muon Scattering from Deuterium and Comparison with those from Hydrogen and Iron.' *Nucl. Phys.* B293, no. CERN-EP-87-66: 740–786.

[554] Benvenuti, Alberto C., D. Bollini, G. Bruni, F. L. Navarria, W. Lohmann, R. Voss, V. I. Genchev et al. 1990. 'A Comparison of the Structure Functions F2 of the Proton and the Neutron from Deep Inelastic Muon Scattering at High Q2.' *Physics Letters B* 237 (3-4): 599–604.

[555] Allasia, D., P. Amaudruz, M. Arneodo, A. Arvidson, B. Badelek, Guenter Baum, J. Beaufays et al. 1990. 'Measurement of the Neutron and the Proton F2 Structure Function Ratio.' *Physics Letters B* 249 (2): 366–372.

[556] Arneodo, M., A. Arvidson, B. Badelek, M. Ballintijn, Guenter Baum, J. Beaufays, I. G. Bird et al. 1994. 'Reevaluation of the Gottfried Sum.' *Physical Review D* 50 (1): R1.

[557] Benvenuti, Alberto C., D. Bollini, G. Bruni, L. Monari, F. L. Navarria, A. Argento, J. Cvach et al. 1989. 'Test of QCD and a Measurement of Λ from Scaling Violations in the Proton Structure Function $F^2(x, Q^2)$ at High Q^2.' *Physics Letters B* 223 (3-4): 490–496.

Benvenutti, A. C., D. Bollini, G. Bruni, T. Camporesi, G. Heiman, L. Monari, F. L. Navarria et al. 1987. 'A high statistics Measurement of the Nucleon Structure Function $F^2(x, Q^2)$ from Deep Inelastic Muon-carbon Scattering at High Q2.' *Physics Letters B* 195 (1): 91–96.

[558] Arneodo, M., A. Arvidson, B. Badelek, M. Ballintijn, Guenter Baum, J. Beaufays, I. G. Bird et al. 1996. 'The A Dependence of the Nuclear Structure Function Ratios.' *Nuclear Physics B* 481 (1-2): 3–22.

Arneodo, M., A. Arvidson, B. Badelek, M. Ballintijn, Guenter Baum, J. Beaufays, I. G. Bird et al. 1997. 'Measurement of the Proton and Deuteron Structure Functions, F2p and F2d, and of the Ratio $\sigma L \sigma T$.' *Nuclear Physics B* 483 (1-2): 3–43.

[559] Aubert, Jean-Jacques, C. Peroni, Gabriel Bassompierre, K. Moser, Terence Sloan, A. W. Edwards, V. Korbel et al. 1985. 'A detailed Study of the Proton Structure Functions in Deep Inelastic Muon-proton Scattering.' *Nucl. Phys. B* 259, no. CERN-EP-85-34: 189–265.

[560] Arneodo, M., A. Arvidson, B. Bade'ek, M. Ballintijn, Guenter Baum, J. Beaufays, I. G. Bird et al. 1995. 'Measurement of the Proton and the Deuteron Structure Functions, F2p and F2d.' *Physics Letters B* 364 (2): 107–115.

[561] Berge, P., H. Burkhardt, F. Dydak, R. Hagelberg, M. W. Krasny, H. J. Meyer, P. Palazzi et al. 1991. 'A Measurement of Differential Cross-sections and Nucleon Structure Functions in Charged-current Neutrino Interactions on Iron.' *Zeitschrift für Physik C Particles and Fields* 49 (2): 187–223.

[562] Oltman, E., P. Auchincloss, R. E. Blair, C. Haber, S. R. Mishra, M. Ruiz, F. J. Sciulli et al. 1992. 'Nucleon Structure Functions from High Energy Neutrino Interactions.' *Zeitschrift für Physik C Particles and Fields* 53 (1): 51–71.

[563] Bodek, A., Martin Breidenbach, D. L. Dubin, J. E. Elias, Jerome I. Friedman, Henry W. Kendall, J. S. Poucher et al. 1979. 'Experimental Studies of the Neutron and Proton Electromagnetic Structure Functions.' *Physical Review D* 20 (7): 1471.

[564] MacFarlane, D. B., M. V. Purohit, R. L. Messner, D. B. Novikoff, R. E. Blair, F. J. Sciulli, M. H. Shaevitz et al. 1984. 'Nucleon Structure Functions from High Energy Neutrino Interactions with Iron and QCD Results.' *Zeitschrift fřr Physik C Particles and Fields* 26 (1): 1–12.

[565] Aubert, Jean-Jacques, G. Bassompierre, K. H. Becks, C. Best, E. Boehm, X. De Bouard, F. W. Brasse et al. 1986. 'A Detailed Study of the Nucleon Structure Functions in Deep Inelastic Muon Scattering in Iron.' *Nuclear Physics B* 272 (1): 158–192.

[566] Tzanov, M., D. Naples, S. Boyd, J. McDonald, V. Radescu, R. A. Johnson, N. Suwonjandee et al. 2006. 'Precise Measurement of Neutrino and Antineutrino Differential Cross Sections.' *Physical Review D* 74 (1): 012008.

[567] Auchincloss, Priscilla S., R. Blair, C. Haber, E. Oltman, W. C. Leung, M. Ruiz, S. R. Mishra et al. 1990. 'Measurement of the Inclusive Charged-current Cross Section for Neutrino and Antineutrino Scattering on Isoscalar Nucleons.' *Zeitschrift für Physik C Particles and Fields* 48 (3): 411–431.

[568] Seligman, W. G. 1997. 'A Next-to-Leading Order QCD Analysis of Neutrino - Iron Structure Functions at the Tevatron.' PhD thesis, Nevis Labs, Columbia University.

[569] Berge, P., A. Blondel, P. Böckmann, H. Burkhardt, F. Dydak, J. G. H. De Groot, A. L. Grant et al. 1987. 'Total Neutrino and Antineutrino Charged Current Cross Section Measurements in 100, 160, and 200 GeV Narrow Band Beams.' *Zeitschrift für Physik C Particles and Fields* 35 (4): 443–452.

[570] Colley, D. C., G. T. Jones, S. O'Neale, S. J. Sewell, G. Bertrand-Coremans, H. Mulkens, J. Sacton et al. 1979. 'Cross Sections for Charged Current ν and $\bar{\nu}$ Interactions in the Energy Range 10 to 50 GeV.' *Zeitschrift für Physik C Particles and Fields* 2 (3): 187–190.

[571] Barish, S. J., M. Derrick, T. Dombeck, L. G. Hyman, K. Jaeger, B. Musgrave, P. Schreiner et al. 1979. 'Study of Neutrino Interactions in Hydrogen and Deuterium. II. Inelastic Charged-current Reactions.' *Physical Review D* 19 (9): 2521.

[572] Allaby, James V., Ugo Amaldi, Guido Barbiellini, M. Baubillier, F. Bergsma, A. Capone, W. Flegel et al. 1988. 'Total Cross Sections of Charged-current Neutrino and Antineutrino Interactions on Isoscalar Nuclei.' *Zeitschrift für Physik C Particles and Fields* 38 (3): 403–410.

[573] Nakamura, S. X., T. Sato, T-SH Lee, B. Szczerbinska, and K. Kubodera. 2010. 'Dynamical Model of Coherent Pion Production in Neutrino-nucleus Scattering.' *Physical Review C* 81 (3): 035502.

[574] Collab, C. H. A. R. M., M. Jonker, J. Panman, F. Udo, J. V. Allaby, U. Amaldi, G. Barbiellini et al. 1981. 'Experimental Study of Differential Cross Sections $d\sigma dy$ in Neutral Current Neutrino and Antineutrino Interactions.' *Physics Letters B* 102 (1): 67–72.

[575] Abramowicz, H., R. Belusevic, A. Blondel, H. Blümer, P. Böckmann, H. D. Brummel, P. Buchholz et al. 1986. 'Precision Measurement of $\sin^2 \theta_W$ from Semileptonic Neutrino Scattering.' *Physical Review Letters* 57 (3): 298.

[576] Allaby, James V., Ugo Amaldi, Guido Barbiellini, M. Baubillier, F. Bergsma, A. Capone, W. Flegel et al. 1986. 'A Precise Determination of the Electroweak Mixing Angle from Semi-leptonic Neutrino Scattering.' *Physics Letters B* 177 (3-4): 446–452.

[577] Abramowicz, H., J. G. H. de Groot, T. Hansl-Kozanecka, J. Knobloch, J. May, E. L. Navarria, P. Palazzi et al. 1985. 'Measurement of the Neutral to Charged Current Cross Section Ratios in Neutrino and Antineutrino Nucleon Interactions and Determination of the Weinberg Angle.' *Zeitschrift für Physik C Particles and Fields* 28 (1): 51–56.

[578] Aaltonen, T., S. Amerio, D. Amidei, A. Anastassov, A. Annovi, J. Antos, G. Apollinari et al. [CDF Collaboration]. 2014. *Physical Review D* 89 (7): 072005.

[579] Tanabashi, M. et al. (Particle Data Group). 2018. 'Particle Physics Booklet.' *Physical Review D* 98: 030001. http://pdg.lbl.gov/2019/download/db2018.pdf.

[580] Altarelli, Guido, and Giorgio Parisi. 1977. 'Asymptotic Freedom in Parton Language.' *Nuclear Physics B* 126 (2): 298–318.

Gribov, V. N., and L. N. Lipatov. 1972. *Sov. J. Nucl. Phys.* 15: 438. [*Yad. Fiz.* 15: 781 (1972)].

Lipatov, L. N. 1975. *Sov. J. Nucl. Phys.* 20: 94. [*Yad. Fiz.* 20: 181 (1974)].

Dokshitzer, Y. L. 1977. *Sov. Phys. JETP* 46: 641. [*Zh. Eksp. Teor. Fiz.* 73: 1216 (1977)].

[581] Vermaseren, Jos AM, Andreas Vogt, and S. Moch. 2005. 'The Third-order QCD Corrections to Deep-inelastic Scattering by Photon Exchange.' *Nuclear Physics B* 724 (1-2): 3–182.

[582] Van Neerven, W. L., and A. Vogt. 2000. 'NNLO Evolution of Deep-inelastic Structure Functions: The Non-singlet Case.' *Nuclear Physics B* 568 (1-2): 263–286.

[583] Furmanski, W., and Roberto Petronzio. 1982. 'Lepton-hadron Processes Beyond Leading Order in Quantum Chromodynamics.' *Zeitschrift für Physik C Particles and Fields* 11 (4): 293–314.

[584] Hirai, M., S. Kumano, and M. Miyama. 1998. 'Numerical Solution of Q2 Evolution Equations for Polarized Structure Functions.' *Computer Physics Communications* 108 (1): 38–55.

[585] Kumano, S., and J. T. Londergan. 1992. 'A FORTRAN Program for Numerical Solution of the Altarelli-Parisi Equations by the Laguerre method.' *Computer physics communications* 69 (2-3): 373–396.

[586] Coriano, Claudio, and Çetin Şavkli. 1999. 'QCD Evolution Equations: Numerical Algorithms from the Laguerre Expansion.' *Computer physics communications* 118 (2-3): 236–258.

[587] Ratcliffe, Philip G. 2001. 'Matrix Approach to a Numerical Solution of the Dokshitzer-Gribov-Lipatov-Altarelli-Parisi Evolution Equations.' *Physical Review D* 63 (11): 116004.

[588] Moch, S., Jos AM Vermaseren, and Andreas Vogt. 2009. 'Third-order QCD Corrections to the Charged-current Structure Function F3.' *Nuclear Physics B* 813 (1-2): 220–258.

[589] Moch, S., Jos AM Vermaseren, and Andreas Vogt. 2005. 'The Longitudinal Structure Function at the Third Order.' *Physics Letters B* 606 (1-2): 123–129.

[590] Georgi, Howard, and H. David Politzer. 1976. 'Freedom at Moderate Energies: Masses in Color Dynamics.' *Physical Review D* 14 (7): 1829.

[591] Schienbein, Ingo, Voica A. Radescu, G. P. Zeller, M. Eric Christy, C. E. Keppel, Kevin S. McFarland, W. Melnitchouk et al. 2008. 'Target Mass Corrections.' *Journal of Physics G: Nuclear and Particle Physics* 35 (5): 053101.

[592] Ellis, R. Keith, W. Furmanski, and Roberto Petronzio. 1983. 'Unravelling Higher Twists.' *Nuclear Physics B* 212 (1): 29–98.

[593] Aivazis, M. A. G., Fredrick I. Olness, and Wu-Ki Tung. 1994. 'Leptoproduction of Heavy Quarks. I. General Formalism and Kinematics of Charged Current and Neutral Current Production Processes.' *Physical Review D* 50 (5): 3085.

[594] Kretzer, S., and M. H. Reno. 2002. 'Tau Neutrino Deep Inelastic Charged Current Interactions.' *Physical Review D* 66 (11): 113007.

[595] Dasgupta, M., and B. R. Webber. 1996. 'Power Corrections and Renormalons in Deep Inelastic Structure Functions.' *Physics Letters B* 382 (3): 273–281.

[596] Virchaux, Marc, and Alain Milsztajn. 1992. 'A Measurement of αs and of Higher Twists from a QCD Analysis of High Statistics F2 Data on Hydrogen and Deuterium Targets.' *Physics Letters B*274 (2): 221–229.

[597] Adler, Stephen L. 1966. 'Sum Rules Giving Tests of Local Current Commutation Relations in High-energy Neutrino Reactions.' *Physical Review* 143 (4): 1144.

[598] Allasia, D., Carlo Angelini, A. Baldini, S. Barlag, L. Bertanza, A. Bigi, V. Bisi et al. 1984. 'Measurement of the Neutron and Proton Structure Functions from Neutrino and Antineutrino Scattering in Deuterium.' *Physics Letters B* 135 (1-3): 231–236.

[599] Allasia, D., Carlo Angelini, A. Baldini, L. Bertanza, A. Bigi, V. Bisi, F. Bobisut et al. 1985. 'Q^2 Dependence of the proton and Neutron Structure Functions from Neutrino and Antineutrino Scattering in Deuterium.' *Zeitschrift für Physik C Particles and Fields* 28 (3): 321–333.

[600] Gross, David J., and CH Llewellyn Smith. 1969. 'High-energy Neutrino-nucleon Scattering, Current Algebra and Partons.' *Nuclear Physics b* 14 (2): 337–347.

[601] Larin, S. A., and Jos AM Vermaseren. 1991. 'The $\alpha s3$ corrections to the Bjorken Sum Rule for Polarized Electroproduction and to the Gross-Llewellyn Smith sum rule.' *Physics Letters B* 259 (3): 345–352.

[602] Poggio, E. C., Helen R. Quinn, and Steven Weinberg. 1976. 'Smearing Method in the Quark Model.' *Physical Review D* 13 (7): 1958.

[603] Bloom, Elliott D., and Frederick J. Gilman. 1970. 'Scaling, Duality, and the Behavior of Resonances in Inelastic Electron-proton Scattering.' *Physical Review Letters* 25 (16): 1140.

[604] Melnitchouk, Wolodymyr, Rolf Ent, and C. E. Keppel. 2005. 'Quark–hadron Duality in Electron Scattering.' *Physics Reports* 406 (3-4): 127–301.

[605] Lalakulich, Olga, Ch Praet, Natalie Jachowicz, Jan Ryckebusch, T. Leitner, O. Buss, and U. Mosel. 2009. 'Neutrinos and Duality.' In *AIP Conference Proceedings* 1189 (1): 276–282. American Institute of Physics.

[606] Andreopoulos, Costas, A. Bell, D. Bhattacharya, F. Cavanna, J. Dobson, S. Dytman, H. Gallagher et al. 2010. 'The GENIE Neutrino Monte Carlo Generator.' *Nuclear Instruments and Methods in Physics Research Section A: Accelerators, Spectrometers, Detectors and Associated Equipment* 614 (1): 87–104.

[607] Hayato, Yoshinari. 2009. 'A Neutrino Interaction Simulation Program Library NEUT.' *Acta Phys. Polon.* 40: 2477–2489.

[608] Athar, M. S. and J. G. Morfin. *Jour. Phys. G.* (forthcoming).

[609] T. Katori, P. Lasorak, S. Mandalia and R. Terri, JPS Conf. Proc. 12, 010033 (2016).

[610] Yang, T., C. Andreopoulos, H. Gallagher, K. Hofmann, and P. Kehayias. 2009. 'A Hadronization Model for Few-GeV Neutrino Interactions.' *The European Physical Journal C* 63 (1): 1–10.

[611] Mamyan, Vahe. 2012. 'Measurements of F_2 and $R = \sigma_L/\sigma_T$ on Nuclear Targets in the Nucleon Resonance Region.' *arXiv preprint arXiv:1202.1457*.

[612] Malace, S. P. 2009. 'Jefferson Lab E00-115 Collab.' *Phys. Rev. C* 80: 035207.

Malace, S. P., W. Melnitchouk, and A. Psaker. 2011. 'Evidence for quark-hadron duality in $\gamma^* p$ helicity cross sections.' *Physical Review C* 83 (3): 035203.

[613] https://www.jlab.org.

[614] Wang, Diancheng, Kai Pan, R. Subedi, Xiaoyan Deng, Z. Ahmed, K. Allada, K. A. Aniol et al. 2013. 'Measurements of Parity-Violating Asymmetries in Electron-

Deuteron Scattering in the Nucleon Resonance Region.' *Physical Review Letters* 111 (8): 082501.

[615] Lalakulich, O., W. Melnitchouk, and E. A. Paschos. 2007. 'Quark-hadron Duality in Neutrino Scattering.' *Physical Review C* 75 (1): 015202.

Lalakulich, Olga, W. Melnitchouk, E. A. Paschos, Christophe Praet, Natalie Jachowicz, and Jan Ryckebusch. 2007. 'Duality in Neutrino Reactions.' In *AIP Conference Proceedings*, vol. 967 (1): 243–248. American Institute of Physics.

Lalakulich, Olga, Natalie Jachowicz, Christophe Praet, and Jan Ryckebusch. 2009. 'Quark-hadron Duality in Lepton Scattering off Nuclei.' *Physical Review C* 79 (1): 015206.

[616] Graczyk, Krzysztof M., Cezary Juszczak, and Jan T. Sobczyk. 2007. 'Quark–hadron Duality in the Rein–Sehgal Model.' *Nuclear Physics A* 781 (1-2): 227–246.

[617] Paschos, E. A., and D. Schalla. 2013. *Adv. High Energy Phys.* 2013: 270792.

[618] Paschos, E. A. 1996. 'A Non-perturbative Effect in Deep Inelastic Scattering.' *Physics Letters B* 389 (2): 383–387.

[619] https://j-parc.jp

[620] http://www.fnal.gov/

[621] http://ep-news.web.cern.ch/content/cerns-strategy-neutrino-physics-0

[622] http://hepdata.cedar.ac.uk/

[623] Nachtmann, Otto. 1973. 'Positivity Constraints for Anomalous Dimensions.' *Nuclear Physics B* 63: 237–247.

[624] Glück, M., E. Reya, and A. Vogt. 1998. 'Dynamical Parton Distributions Revisited.' *The European Physical Journal C-Particles and Fields* 5 (3): 461–470.

[625] Pumplin, Jon, Alexander Belyaev, Joey Huston, Daniel Stump, and Wu-Ki Tung. 2006. 'Parton Distributions and the Strong Coupling: CTEQ6AB PDFs.' *Journal of High Energy Physics* 2006 (02): 032.

[626] Martin, A. D., R. G. Roberts, William James Stirling, and R. S. Thorne. 2004. 'Physical Gluons and High-ET jets.' *Physics Letters B* 604 (1-2): 61–68.

[627] Mousseau, J., M. Wospakrik, L. Aliaga, O. Altinok, L. Bellantoni, A. Bercellie, M. Betancourt et al. 2016. 'Measurement of Partonic Nuclear Effects in Deep-inelastic Neutrino Scattering using MINERvA.' *Physical Review D* 93 (7): 071101.

[628] Adamson, P., C. Ader, M. Andrews, N. Anfimov, I. Anghel, K. Arms, E. Arrieta-Diaz et al. 2016. 'First Measurement of Electron Neutrino Appearance in NOvA.' *Physical Review Letters* 116 (15): 151806.

Adamson, Ph, C. Ader, M. Andrews, N. Anfimov, I. Anghel, K. Arms, E. Arrieta-Diaz et al. 2016. 'First measurement of muon-neutrino disappearance in NOvA.' *Physical Review D* 93 (5): 051104.

[629] Strait, J. et al. [DUNE Collaboration]. 2015. *arXiv:1601.05823 [physics.ins-det]*.

[630] Lipari, Paolo, Maurizio Lusignoli, and Francesca Sartogo. 1995. 'The Neutrino Cross Section and Upward Going Muons.' *Physical Review Letters* 74 (22): 4384.

[631] Markov, M. A. 1964. *Neutrino preprint* JINR-D577 (1960). Moscow: Nauka.

[632] Berman, S. M. 1961. *Lectures on Weak Interactions*. CERN-62–20

[633] Bell, John Stewart, and Martinus JG Veltman. 1963. 'Polarisation of Vector Bosons Produced by Neutrinos.' *Phys. Lett.* 5, no. CERN-TH-348: 151–152.

Veltman, M. J. G., and J. S. Bell. 1963. 'Intermediate Boson Production by Neutrinos.' *Physics Letters: A* 5 (1): 94–96.

[634] Úberall, H. 1964. 'Polarization Effects in the Production of Intermediate Bosons.' *Physical Review* 133 (2B): B444.

[635] Løvseth, J. 1963. 'On the Angular Distribution of the Brookhaven 1962 Neutrino Experiment.' *Phys. Letters* 5.

[636] Hasert, F. J., Helmut Faissner, Wulf Dieter Krenz, J. Von Krogh, D. Lanske, J. Morfin, K. Schultze et al. 1973. 'Search for Elastic Muon-neutrino Electron Scattering.' *Physics letters. Section B* 46 (1): 121–124.

[637] Donnelly, Thomas Wallace, and Roberto D. Peccei. 1979. 'Neutral Current Effects in Nuclei.' *Physics Reports* 50 (1): 1–85.

[638] Auerbach, N., Nguyen Van Giai, and O. K. Vorov. 1997. 'Neutrino Scattering from 12 C and 16 O.' *Physical Review C* 56 (5): R2368.

[639] Singh, S. K., and E. Oset. 1993. 'Inclusive Quasielastic Neutrino Reactions in ^{12}C and ^{16}O at Intermediate Energies.' *Physical Review C* 48 (3): 1246.

[640] Kosmas, T. S., and E. Oset. 1996. 'Charged Current Neutrino-nucleus Reaction Cross Sections at Intermediate Energies.' *Physical Review C* 53 (3): 1409.

[641] Singh, Shri Krishna, Nimai C. Mukhopadhyay, and E. Oset. 1998. 'Inclusive Neutrino Scattering in ^{12}C: Implications for v_μ to v_e oscillations.' *Physical Review C* 57 (5): 2687.

[642] Volpe, C., N. Auerbach, G. Colo, T. Suzuki, and Nguyen Van Giai. 2000. 'Microscopic Theories of Neutrino-^{12}C Reactions.' *Physical Review C* 62 (1): 015501.

[643] Volpe, C., N. Auerbach, G. Colo, and N. Van Giai. 2002. 'Charged-current Neutrino-^{208}Pb Reactions.' *Physical Review C* 65 (4): 044603.

[644] Kolbe, E., K. Langanke, S. Krewald, and F-K. Thielemann. 1992. 'Inelastic Neutrino Scattering on ^{12}C and ^{16}O Above the Particle Emission Threshold.' *Nuclear Physics A* 540 (3-4): 599–620.

[645] Kolbe, E., K. Langanke, F-K. Thielemann, and P. Vogel. 1995. 'Inclusive ^{12}C (v_μ, μ) ^{12}N Reaction in the Continuum Random Phase Approximation.' *Physical Review C* 52 (6): 3437.

[646] Jachowicz, Natalie, Stefan Rombouts, Kristiaan Heyde, and Jan Ryckebusch. 1999. 'Cross Sections for Neutral-Current Neutrino-nucleus Interactions: Applications for ^{12}C and ^{16}O.' *Physical Review C* 59 (6): 3246.

[647] Jachowicz, Natalie, Kristiaan Heyde, Jan Ryckebusch, and Stefan Rombouts. 2002. 'Continuum Random Phase Approximation Approach to Charged-current Neutrino-nucleus Scattering.' *Physical Review C* 65 (2): 025501.

[648] Botrugno, Antonio. 2005. 'Excitation of Nuclear Giant Resonances in Neutrino Scattering off Nuclei.' *Nuclear Physics A* 761 (3-4): 200–231.

[649] Lazauskas, R., and C. Volpe. 2007. 'Neutrino Beams as a Probe of the Nuclear Isospin and Spin–isospin Excitations.' *Nuclear Physics A* 792 (3-4): 219–228.

[650] Cheoun, Myung-Ki, Eunja Ha, K. S. Kim, and Toshitaka Kajino. 2010. 'Neutrino–nucleus Reactions via Neutral and Charged Currents by the Quasi-particle Random Phase Approximation (QRPA).' *Journal of Physics G: Nuclear and Particle Physics* 37 (5): 055101.

[651] Chasioti, V. Ch, and T. S. Kosmas. 2009. 'A Unified Formalism for the Basic Nuclear Matrix Elements in Semi-leptonic Processes.' *Nuclear Physics A* 829 (3-4): 234–252.

[652] Tsakstara, V., and T. S. Kosmas. 2011. 'Analyzing Astrophysical Neutrino Signals using Realistic Nuclear Structure Calculations and the Convolution Procedure.' *Physical Review C* 84 (6): 064620.

[653] Tsakstara, V., and T. S. Kosmas. 2012. 'Nuclear Responses of ^{64,66}Zn Isotopes to Supernova Neutrinos.' *Physical Review C* 86 (4): 044618.

[654] Samana, A. R., Francesco Krmpotić, Nils Paar, and C. A. Bertulani. 2011. 'Neutrino and Antineutrino Charge-Exchange Reactions on ^{12}C.' *Physical Review C* 83 (2): 024303.

[655] Paar, Nils, Dario Vretenar, Tomislav Markctin, and Peter Ring. 2008. 'Inclusive Charged-current Neutrino-nucleus Reactions Calculated with the Relativistic Quasiparticle Random-phase Approximation.' *Physical Review C* 77 (2): 024608.

[656] Paar, Nils, Hrvoje Tutman, Tomislav Marketin, and Tobias Fischer. 2013. 'Large-scale Calculations of Supernova Neutrino-induced Reactions in $Z = 8 - 82$ Target Nuclei.' *Physical Review C* 87 (2): 025801.

[657] Smith, R. A., and Ernest J. Moniz. 1972. 'Neutrino Reactions on Nuclear Targets.' *Nuclear Physics B* 43: 605–622.

[658] Gaisser, T. K., and J. S. O'Connell. 1986. 'Interactions of Atmospheric Neutrinos in Nuclei at Low Energy.' *Physical Review D* 34 (3): 822.

[659] Singh, S. K., and E. Oset. 1992. 'Quasielastic Neutrino (antineutrino) Reactions in Nuclei and the Axial-vector form Factor of the Nucleon.' *Nuclear Physics A* 542 (4): 587–615.

[660] Nieves, J., Jose Enrique Amaro, and M. Valverde. 2004. 'Inclusive Quasielastic Charged-current Neutrino-nucleus Reactions.' *Physical Review C* 70 (5): 055503.

[661] Nieves, J., I. Ruiz Simo, and MJ Vicente Vacas. 2013. 'Two Particle–hole Excitations in Charged Current Quasielastic Antineutrino-nucleus Scattering.' *Physics Letters B* 721 (1-3): 90–93.

[662] Athar, M. Sajjad, Shakeb Ahmad, and S. K. Singh. 2005. 'Supernova Neutrino Induced Inclusive Reactions on Fe 56 in Terrestrial Detectors.' *Physical Review C* 71 (4): 045501.

[663] Athar, M. Sajjad, S. Ahmad, and S. K. Singh. 2005. 'Charged Lepton Production from iron Induced by Atmospheric Neutrinos.' *The European Physical Journal A-Hadrons and Nuclei* 24 (3): 459–474.

[664] Athar, M. Sajjad, Shakeb Ahmad, and S. K. Singh. 2007. 'Charged Current Antineutrino Reactions from ^{12}C at MiniBooNE energies.' *Physical Review D* 75 (9): 093003.

[665] Akbar, F., M. Rafi Alam, M. Sajjad Athar, S. Chauhan, S. K. Singh, and F. Zaidi. 2015. 'Electron and Muon Production Cross-sections in Quasielastic $\nu(\bar{\nu})$-Nucleus Scattering for $E_\nu < 1$ GeV.' *International Journal of Modern Physics E* 24 (11): 1550079.

[666] Athar, M. Sajjad, and S. K. Singh. 2004. '$\nu e(\bar{\nu}e)$–40Ar Absorption Cross Sections for Supernova Neutrinos.' *Physics Letters B*591 (1-2): 69-75.

[667] Athar, M. Sajjad, Shakeb Ahmad, and S. K. Singh. 2006. 'Neutrino Nucleus Cross Sections for Low Energy Neutrinos at SNS Facilities.' *Nuclear Physics A* 764: 551–568.

[668] Athar, M. Sajjad, S. Chauhan, and S. K. Singh. 2010. 'Theoretical Study of Lepton Events in the Atmospheric Neutrino Experiments at SuperK.' *The European Physical Journal A* 43 (2): 209–227.

[669] Kim, Hungchong, J. Piekarewicz, and C. J. Horowitz. 1995. 'Relativistic Nuclear Structure Effects in Quasielastic Neutrino Scattering.' *Physical Review C* 51 (5): 2739.

[670] Meucci, Andrea, Carlotta Giusti, and Franco Davide Pacati. 2004. 'Neutral-current Neutrino-nucleus Quasielastic Scattering.' *arXiv preprint nucl-th/0405004.*

[671] Meucci, Andrea, Maria Benedetta Barbaro, J. A. Caballero, Carlotta Giusti, and J. M. Udias. 2011. 'Relativistic Descriptions of Final-state Interactions in Charged-current Quasielastic Neutrino-nucleus Scattering at MiniBooNE Kinematics.' *Physical Review Letters* 107 17): 172501.

[672] González-Jiménez, Raúl, J. A. Caballero, Andrea Meucci, Carlotta Giusti, Maria Benedetta Barbaro, M. V. Ivanov, and J. M. Udías. 2013. 'Relativistic Description of Final-state Interactions in Neutral-current Neutrino and Antineutrino Cross Sections.' *Physical Review C* 88 (2): 025502.

[673] Caballero, J. A., Jose Enrique Amaro, Maria Benedetta Barbaro, T. W. Donnelly, and J. M. Udias. 2007. 'Scaling and Isospin Effects in Quasielastic Lepton–nucleus Scattering in the Relativistic Mean Field Approach.' *Physics Letters B* 653 (2-4): 366–372.

[674] Meucci, Andrea, and Carlotta Giusti. 2014. 'Final-state Interaction Effects in Neutral-current Neutrino and Antineutrino Cross Sections at MiniBooNE kinematics.' *Physical Review D* 89 (5): 057302.

[675] Donnelly, T. W., and J. D. Walecka. 1976. 'Semi-leptonic Weak and Electromagnetic Interactions with Nuclei: Isoelastic Processes.' *Nuclear Physics A* 274 (3-4): 368–412.

[676] Walecka, J. D., V. M. Hughes, and C. S. Wu. 1975. 'Muon Physics.' *Academis, New York USA.*

[677] Haxton, Wick, and Cecilia Lunardini. 2008. 'SevenOperators, a Mathematica Script for Harmonic Oscillator Nuclear Matrix Elements Arising in Semileptonic Electroweak Interactions.' *Computer Physics Communications* 179 (5): 345–358.

[678] Katori, T., and M. Martini. 2018. *J. Phys. G* 45: 013001.

[679] Garvey, G. T., D. A. Harris, H. A. Tanaka, R. Tayloe, and G. P. Zeller. 2015. 'Recent Advances and Open Questions in Neutrino-induced Quasi-elastic Scattering and Single Photon Production.' *Physics Reports* 580: 1–45.

[680] Yao, York-Peng. 1968. 'Nuclear Effects on the Quasi-Elastic Neutrino Scattering $\nu^+ n \to \mu^- + p$.' *Physical Review* 176 (5): 1680.

[681] Gatto, R. 1953. 'On the Scattering of π-mesons by Nuclei.' *Il Nuovo Cimento (1943–1954)* 10 (11): 1559–1581.

Nuovo Cimento 2: 670 (1955).

[682] De Jager, C. W., H. De Vries, and C. De Vries. 1974. 'Nuclear Charge-and Magnetization-density-distribution Parameters from Elastic Electron Scattering.' *Atomic data and nuclear data tables* 14 (5-6): 479–508.

[683] De Vries, H., C. W. De Jager, and C. De Vries. 1987. 'Nuclear charge-density-distribution Parameters from Elastic Electron Scattering.' *Atomic data and nuclear data tables* 36 (3): 495–536.

[684] Fetter, A. L., and J. D. Walecka. 1971. 'Quantum Theory of Many Particle Systems New York.'

[685] Engel, Jonathan. 1998. 'Approximate Treatment of Lepton Distortion in Charged-current Neutrino Scattering from Nuclei.' *Physical Review C* 57 (4): 2004.

[686] Preston, Melvin Alexander. 1962. 'Physics of the Nucleus.'.

[687] Gil, A., J. Nieves, and E. Oset. 1997. 'Many Body Approach to the Inclusive (e, e') Reaction from the Quasielastic to the Δ Excitation Region.' *arXiv preprint nucl-th/9711009.*

[688] Carrasco, R. C., and E. Oset. 1992. 'Interaction of Real Photons with Nuclei from 100 to 500 MeV.' *Nuclear Physics A* 536 (3-4): 445–508.

[689] Oset, E., D. Strottman, H. Toki, and J. Navarro. 1993. 'Core Polarization Phenomena in Pion-nucleus Charge-exchange Reactions above the Delta Resonance.' *Physical Review C* 48 (5): 2395.

Oset, E., P. Fernandez de Cordoba, L. L. Salcedo, and R. Brockmann. 1990. 'Decay Modes of Sigma and Lambda Hypernuclei.' *Physics Reports* 188 (2): 79–145..

[690] Martini, M., M. Ericson, and G. Chanfray. 2011. 'Neutrino Quasielastic Interaction and Nuclear Dynamics.' *Physical Review C* 84 (5): 055502.

[691] Martini, Marco, and Magda Ericson. 2013. 'Quasielastic and Multinucleon Excitations in Antineutrino-nucleus Interactions.' *Physical Review C* 87 (6): 065501.

[692] Nieves, J., I. Ruiz Simo, and MJ Vicente Vacas. 2011. 'Inclusive Charged-current Neutrino-nucleus Reactions.' *Physical Review C* 83 (4): 045501.

[693] Simo, I. Ruiz, C. Albertus, J. E. Amaro, Maria Benedetta Barbaro, J. A. Caballero, and T. W. Donnelly. 2014. 'Relativistic Effects in Two-particle Emission for Electron and Neutrino Reactions.' *Physical Review D* 90 (3): 033012.

[694] Lovato, Alessandro, Stefano Gandolfi, Joseph Carlson, Steven C. Pieper, and Rocco Schiavilla. 2014. 'Neutral Weak Current Two-Body Contributions in Inclusive Scattering from ^{12}C.' *Physical Review Letters* 112 (18): 182502.

[695] Aguilar-Arevalo, A. A., B. C. Brown, L. Bugel, G. Cheng, E. D. Church, J. M. Conrad, R. Dharmapalan et al. 2013. 'First Measurement of the Muon Antineutrino Double-differential Charged-current Quasielastic Cross Section.' *Physical Review D* 88 (3): 032001.

[696] Lerche, Wolfgang, Martin Pohl, M. Dewitt, C. Vander Velde-Wilquet, P. Vilain, D. Haidt, C. Matteuzzi et al. 1978. 'Experimental Study of the Reaction $vp \rightarrow \mu^- p\pi^+$: Gargamelle Neutrino Propane Experiment.' *Physics Letters B* 78 (4): 510–514.

[697] Bologneşe, T., J. P. Engel, J. L. Guyonnet, and J. L. Riester. 1979. 'Single Pion Production in Antineutrino Induced Charged Current Interactions.' *Physics Letters B* 81 (3-4): 393–396.

[698] Isiksal, Engin, Dieter Rein, and Jorge G. Morfín. 1984. 'Evidence for Neutrino-and Antineutrino-induced Coherent π0 Production.' *Physical Review Letters* 52 (13): 1096.

[699] Marage, Pierre, M. Aderholz, N. Armenise, T. Azemoon, K. W. J. Barnham, J. H. Bartley, J. P. Baton et al. 1984. 'Observation of Coherent Diffractive Charged Current Interactions of Antineutrinos on Neon Nuclei.' *Physics Letters B* 140 (1-2): 137–141.

[700] Marage, Pierre, M. Aderholz, P. Allport, N. Armenise, J. P. Baton, M. Berggren, D. Bertrand et al. 1986. 'Coherent Single Pion Production by Antineutrino Charged Current Interactions and Test of PCAC.' *Zeitschrift für Physik C Particles and Fields* 31 (2): 191–197.

[701] V. V. Ammosov, V. V. et al. 1989. *Sov. J. Nucl. Phys.* 50: 67. [*Yad. Fiz.* 50: 106 (1989)].

[702] Grabosch, H. J., H. H. Kaufmann, R. Nahnhauer, S. Nowak, S. Schlenstedt, V. V. Ammosov, D. S. Baranov et al. 1989. 'Cross-section Measurements of Single Pion Production in Charged Current Neutrino and Antineutrino Interactions.' *Zeitschrift für Physik C Particles and Fields* 41 (4): 527–531.

[703] Grabosch, H-J., H. H. Kaufmann, U. Krecker, R. Nahnhauer, S. Nowak, S. Schlenstedt, H. Vogt et al. 1986. 'Coherent Pion Production in Neutrino and Antineutrino Interactions on Nuclei of Heavy Freon Molecules.' *Zeitschrift fŘr Physik C Particles and Fields* 31 (2): 203–211.

[704] Faissner, Helmut, E. Frenzel, M. Grimm, T. Hansl-Kozanecka, D. Hoffmann, E. Radermacher, D. Rein et al. 1983. 'Observation of Neutrino and Antineutrino Induced Coherent Neutral Pion Production off Al127.' *Physics Letters B* 125 (2-3): 230–236.

[705] Aderholz, Michael, M. M. Aggarwal, Homaira Akbari, P. P. Allport, P. V. K. S. Baba, S. K. Badyal, Marie Barth et al. 1989. 'Coherent Production of π^+ and π^- Mesons

by Charged-current Interactions of Neutrinos and Antineutrinos on Neon Nuclei at the Fermilab Tevatron.' *Physical Review Letters* 63 (21): 2349.

[706] Willocq, Stephane, M. Aderholz, H. Akbari, P. P. Allport, S. K. Badyal, H. C. Ballagh, M. Barth et al. 1993. 'Coherent Production of Single Pions and ρ Mesons in Charged-current Interactions of Neutrinos and Antineutrinos on Neon Nuclei at the Fermilab Tevatron.' *Physical Review D* 47 (7): 2661.

[707] Bergsma, F., J. Dorenbosch, J. V. Allaby, U. Amaldi, G. Barbiellini, W. Flegel, Livio Lanceri et al. 1985. 'Measurement of the Cross Section of Coherent π^0 Production by Muon-neutrino and Antineutrino Neutral-current Interactions on Nuclei.' *Physics Letters B* 157 (5-6): 469–474.

[708] Marage, Pierre, P. P. Allport, N. Armenise, J. P. Baton, M. Berggren, W. Burkot, M. Calicchio et al. 1989. 'Coherent Production of π^+ Mesons in ν-neon Interactions.' *Zeitschrift fŘr Physik C Particles and Fields* 43 (4): 523–526.

[709] Vilain, Pierre, Gaston Wilquet, R. Beyer, W. Flegel, H. Grote, T. Mouthuy, H. Øveras et al. 1993. 'Coherent Single Charged Pion Production by Neutrinos.' *Physics Letters B* 313 (1-2): 267–275.

[710] Campbell, J., G. Charlton, Y. Cho, M. Derrick, R. Engelmann, J. Fetkovich, L. Hymah et al. 1973. 'Study of the Reaction $\nu p \to \mu^- \pi^+ p$.' *Physical Review Letters* 30 (8): 335.

[711] Barish, S. J., J. Campbell, G. Charlton, Y. Cho, M. Derrick, R. Engelmann, L. G. Hyman et al. 1977. 'Study of Neutrino Interactions in Hydrogen and Deuterium: Description of the Experiment and Study of the Reaction $\nu + d \to \mu^- + p + p_s$.' *Physical Review D* 16 (11): 3103.

[712] Allen, P., H. Grässler, R. Schulte, G. T. Jones, B. W. Kennedy, S. W. O'Neale, W. Gebel et al. 1986. 'A Study of Single-meson Production in Neutrino and Antineutrino Charged-current Interactions on Protons.' *Nuclear Physics B* 264: 221–242.

[713] Allasia, D., C. Angelini, G. W. van Apeldoorn, A. Baldini, S. M. Barlag, L. Bertanza, F. Bobisut et al. 1990. 'Investigation of Exclusive Channels in ν/ν-deuteron Charged Current Interactions.' *Nuclear Physics B* 343 (2): 285–309.

[714] Hayato, Y. 2002. *Nucl. Phys. Proc. Suppl.* 112: 171.

[715] Rein, Dieter, and Lalit M. Sehgal. 1983. 'Coherent π^0 Production in Neutrino Reactions.' *Nuclear Physics B* 223 (1): 29–44.

[716] A. Aguilar-Arevalo, *et al.*, Phys. Rev. D64, 112007 (2011).

[717] Kim, Hungchong, S. Schramm, and C. J. Horowitz. 1996. 'Delta Excitations in Neutrino-nucleus Scattering.' *Physical Review C* 53 (5): 2468.

[718] Singh, S. K., M. J. Vicente-Vacas, and E. Oset. 1998. 'Nuclear Effects in Neutrino Production of Δ at Intermediate Energies.' *Physics Letters B* 416 (1-2): 23–28.

[719] Kim, Hungchong, S. Schramm, and C. J. Horowitz. 1996. 'Detection of Atmospheric Neutrinos and Relativistic Nuclear Structure Effects.' *Physical Review C* 53 (6): 3131.

[720] Kelkar, N. G., E. Oset, and P. Fernández De Córdoba. 1997. 'Coherent Pion Production in Neutrino Nucleus Collision in the 1 GeV Region.' *Physical Review C* 55 (4): 1964.

[721] Benhar, Omar, Nicola Farina, Hiroki Nakamura, Makoto Sakuda, and Ryoichi Seki. 2005. 'Electron-and Neutrino-nucleus Scattering in the Impulse Approximation Regime.' *Physical Review D* 72 (5): 053005.

[722] Benhar, Omar, and Davide Meloni. 2007. 'Total Neutrino and Antineutrino Nuclear Cross Sections Around 1 GeV.' *Nuclear Physics A* 789 (1-4): 379–402.

[723] Ahmad, Shakeb, M. Sajjad Athar, and S. K. Singh. 2006. 'Neutrino Induced Charged Current 1 π^+ Production at Intermediate Energies.' *Physical Review D* 74 (7): 073008.

[724] Singh, S. K., M. Sajjad Athar, and Shakeb Ahmad. 2006. 'Nuclear Cross Sections in ^{16}O for β Beam Neutrinos at Intermediate Energies.' *Physics Letters B* 641 (2): 159–163.

[725] Praet, Christophe, Olga Lalakulich, Natalie Jachowicz, and Jan Ryckebusch. 2009. 'Δ-mediated Pion Production in Nuclei.' *Physical Review C* 79 (4): 044603.

[726] Leitner, Tina, Ulrich Mosel, and Stefan Winkelmann. 2009. 'Neutrino-induced Coherent Pion Production off Nuclei Reexamined.' *Physical Review C* 79 (5): 057601.

[727] Athar, M. S., S. Chauhan and S. K. Singh. 2010. *J. Phys. G* 37: 015005.

[728] Golan, Tomasz, Cezary Juszczak, and Jan T. Sobczyk. 2012. 'Effects of Final-state Interactions in Neutrino-nucleus Interactions.' *Physical Review C* 86 (1): 015505.

[729] Hernandez, E., J. Nieves, and M. Valverde. 2010. 'Coherent Pion Production off Nuclei at T2K and MiniBooNE Energies Revisited.' *Physical Review D* 82 (7): 077303.

[730] Ivanov, Martin V., Guillermo D. Megias, Raúl González-Jiménez, Oscar Moreno, Maria Benedetta Barbaro, Juan A. Caballero, and T. William Donnelly. 2016. 'Charged-current Inclusive Neutrino Cross Sections in the SuperScaling Model Including Quasielastic, Pion Production and Meson-exchange Contributions.' *Journal of Physics G: Nuclear and Particle Physics* 43 (4): 045101.

[731] Nakamura, S. X., H. Kamano, Y. Hayato, M. Hirai, W. Horiuchi, S. Kumano, T. Murata et al. 2017. 'Towards a Unified Model of Neutrino-nucleus Reactions for Neutrino Oscillation Experiments.' *Reports on Progress in Physics* 80 (5): 056301.

[732] Nikolakopoulos, Alexis, Raúl González-Jiménez, Kajetan Niewczas, Jan Sobczyk, and Natalie Jachowicz. 2018. 'Modeling Neutrino-induced Charged Pion Production on Water at T2K Kinematics.' *Physical Review D* 97 (9): 093008.

[733] González-Jiménez, R., A. Nikolakopoulos, N. Jachowicz, and J. M. Udías. 2019. 'Nuclear Effects in Electron-Nucleus and Neutrino-nucleus Scattering within a Relativistic Quantum Mechanical Framework.' *Physical Review C* 100 (4): 045501.

[734] Nakamura, S. X., H. Kamano, and T. Sato. 2019. 'Impact of Final State Interactions on Neutrino-nucleon Pion Production Cross Sections Extracted from Neutrino-deuteron Reaction Data.' *Physical Review D* 99 (3): 031301.

[735] Mariani, C., and K2K collaboration. 2007. 'Neutral Pion Cross Section Measurement at K2K.' In *AIP Conference Proceedings*, vol. 967 (1): 174–178. American Institute of Physics.

[736] Rodriguez, A., L. Whitehead, J. L. Alcaraz, S. Andringa, S. Aoki, J. Argyriades, K. Asakura et al. 2008. 'Measurement of Single Charged Pion Production in the Charged-current Interactions of Neutrinos in a 1.3 GeV Wide Band Beam.' *Physical Review D*78 (3): 032003.

[737] Wascko, M. O. [MiniBooNE Collaboration]. 2006. *Nucl. Phys. Proc. Suppl.* 159: 50.

[738] Aguilar-Arevalo, A. A., C. E. Anderson, A. O. Bazarko, S. J. Brice, B. C. Brown, L. Bugel, J. Cao et al. 'Measurement of ν_μ-induced Charged-current Neutral Pion Production Cross Sections on Mineral Oil at $E_\nu \in 0.5 - 2.0$ GeV.' *Physical Review D* 83 (5): 052009.

[739] Hiraide, K. 2008. 'Measurement of Charged Current Charged Single Pion Production in SciBooNE.' *arXiv preprint arXiv:0810.3903*.

[740] Kullenberg, Christopher Thomas, S. R. Mishra, M. B. Seaton, J. J. Kim, X. C. Tian, A. M. Scott, M. Kirsanov et al. 2009. 'A Measurement of Coherent Neutral Pion Production in Neutrino Neutral Current Interactions in the NOMAD Experiment.' *Physics Letters B* 682 (2): 177–184.

[741] Acciarri, R., C. Adams, J. Asaadi, B. Baller, T. Bolton, C. Bromberg, F. Cavanna et al. 2018. 'First Measurement of the Cross Section for ν_μ and $\bar{\nu}_\mu$ Induced Single Charged Pion Production on Argon using ArgoNeuT.' *Physical Review D* 98 (5): 052002.

[742] Eberly, B., and MINERνA collaboration. 2015. 'Muon Neutrino Charged Current Inclusive Charged Pion ($CC\pi^\pm$) Production in MINERνA.' In *AIP Conference Proceedings* 1663 (1): 070006. AIP Publishing LLC.

[743] Eberly, B., L. Aliaga, O. Altinok, MG Barrios Sazo, L. Bellantoni, M. Betancourt, A. Bodek et al. 2015. 'Charged Pion Production in ν_μ Interactions on Hydrocarbon at $\langle E_\nu \rangle = 4.0$ GeV.' *Physical Review D* 92 (9): 092008.

[744] Le, T., J. L. Palomino, L. Aliaga, O. Altinok, A. Bercellie, A. Bodek, A. Bravar et al. 2015. 'Single Neutral Pion Production by Charged-current $\bar{\nu}_\mu$ Interactions on Hydrocarbon at $\langle E_\nu \rangle 3.6$ GeV.' *Physics Letters B* 749: 130–136.

[745] Abe, Ko, C. Andreopoulos, M. Antonova, S. Aoki, A. Ariga, S. Assylbekov, D. Autiero et al. 2017. 'First Measurement of the Muon Neutrino Charged Current Single Pion Production Cross Section on Water with the T2K Near Detector.' *Physical Review D* 95 (1): 012010.

[746] Hasegawa, M., E. Aliu, S. Andringa, S. Aoki, J. Argyriades, K. Asakura, R. Ashie et al. 2005. 'Search for Coherent Charged Pion Production in Neutrino-carbon Interactions.' *Physical Review Letters* 95 (25): 252301.

[747] Aguilar-Arevalo, A. A., C. E. Anderson, A. O. Bazarko, S. J. Brice, B. C. Brown, L. Bugel, J. Cao et al. 2008. 'First Observation of Coherent π^0 Production in Neutrino–nucleus Interactions with $E_\nu < 2$ GeV.' *Physics Letters B* 664 (1-2): 41–46.

[748] Hiraide, K., J. L. Alcaraz-Aunion, S. J. Brice, L. Bugel, J. Catala-Perez, G. Cheng, J. M. Conrad et al. 2008. 'Search for Charged Current Coherent Pion Production on Carbon in a Few-GeV Neutrino Beam.' *Physical Review D* 78 (11): 112004.

[749] Kurimoto, Y., J. L. Alcaraz-Aunion, S. J. Brice, L. Bugel, J. Catala-Perez, G. Cheng, J. M. Conrad et al. 2010. 'Improved Measurement of Neutral Current Coherent π^0 Production on Carbon in a Few-GeV Neutrino Beam.' *Physical Review D* 81 (11): 111102.

[750] Acciarri, R., C. Adams, J. Asaadi, B. Baller, T. Bolton, C. Bromberg, F. Cavanna et al. 2014. 'First Measurement of Neutrino and Antineutrino Coherent Charged Pion Production on Argon.' *Physical Review Letters* 113 (26): 261801.

[751] Higuera, A., A. Mislivec, L. Aliaga, O. Altinok, A. Bercellie, M. Betancourt, A. Bodek et al. 2014. 'Measurement of Coherent Production of $\pi^{\pm}$ in Neutrino and Antineutrino Beams on Carbon from E_ν of 1.5 to 20 GeV.' *Physical Review Letters* 113 (26): 261802.

[752] Abe, K., C. Andreopoulos, M. Antonova, S. Aoki, A. Ariga, S. Assylbekov, D. Autiero et al. 2016. 'Measurement of Coherent π^+ Production in Low Energy Neutrino-Carbon Scattering.' *Physical Review Letters* 117 (19): 192501.

[753] Adler, Stephen L. 1964. 'Tests of the Conserved Vector Current and Partially Conserved Axial-vector Current Hypotheses in High-energy Neutrino Reactions.' *Physical Review* 135 (4B): B963.

[754] Hernández, E., J. Nieves, and MJ Vicente Vacas. 2013. 'Single π Production in Neutrino- nucleus Scattering.' *Physical Review D*87 (11): 113009.

[755] Oset, E., and L. L. Salcedo. 1987. 'Delta Self-energy in Nuclear Matter.' *Nuclear Physics A* 468 (3-4): 631–652.

[756] Hirenzaki, S., J. Nieves, E. Oset, and M. J. Vicente-Vacas. 1993. 'Coherent π^0 Electroproduction.' *Physics Letters B* 304 (3-4): 198–202.

[757] Carrasco, R. C., J. Nieves, and E. Oset. 1993. 'Coherent (γ, π^0) Photoproduction in a Local Approximation to the Delta-hole Model.' *Nuclear Physics A* 565 (4): 797–817.

[758] Oset, E., and W. Weise. 1978. 'Microscopic Calculation of Medium Corrections to Pion-nuclear Elastic Scattering.' *Physics Letters B* 77 (2): 159–164.

[759] Hofmann, Hartmut M. 1979. 'Effects of Pion-absorption in Pion-helium Scattering.' *Zeitschrift für Physik A Atoms and Nuclei* 289 (3): 273–282.

[760] Garcia-Recio, C. et al. 1991. *Nucl. Phys. A* 526: 685.

[761] Alvarez-Ruso, L. 2011. 'Review of Weak Coherent Pion Production.' In *AIP Conference Proceedings* 1405 (1): 140–145. American Institute of Physics.

[762] Alvarez-Ruso, Luis, Li Sheng Geng, Satoru Hirenzaki, and MJ Vicente Vacas. 2007. 'Charged Current Neutrino-induced Coherent Pion Production.' *Physical Review C* 75 (5): 055501.

[763] Singh, S. K., M. Sajjad Athar, and Shakeb Ahmad. 2006. 'Nuclear Effects in Neutrino Induced Coherent Pion Production at K2K and MiniBooNE.' *Physical Review Letters* 96 (24): 241801.

[764] Glauber, R. J. 1959. 'Lectures in Theoretical Physics, ed. WE Brittin and LG Dunham.' *Interscience, New York* 1: 315.

[765] Vicente-Vacas, M. J., M. Kh Khankhasayev, and S. G. Mashnik. 1994. 'Inclusive Pion Double Charge Exchange above 0.5 GeV.' *arXiv preprint nucl-th/9412023.*

[766] Sajjad Athar, Mohammad, Shikha Chauhan, Shri Krishna Singh, and Manuel José Vicente Vacas. 2009. 'Neutrino Nucleus Cross-sections.' *International Journal of Modern Physics E* 18 (07): 1469–1481.

[767] Aguilar-Arevalo, A. A., C. E. Anderson, A. O. Bazarko, S. J. Brice, B. C. Brown, L. Bugel, J. Cao et al. 2011. 'Measurement of Neutrino-induced Charged-current Charged Pion Production Cross Sections on Mineral Oil at $E_\nu \backsim 1$ GeV.' *Physical Review D* 83 (5): 052007.

[768] Rodrigues, P. A. 2015. 'Comparing Pion Production Models to MiniBooNE Data.' In *AIP Conference Proceedings* 1663 (1): 030006. AIP Publishing LLC.

[769] Buss, O., T. Gaitanos, K. Gallmeister, H. Van Hees, M. Kaskulov, O. Lalakulich, A. B. Larionov, T. Leitner, J. Weil, and U. Mosel. 2012. 'Transport-theoretical Description of Nuclear Reactions.' *Physics Reports* 512 (1-2): 1–124.

[770] Alvarez-Ruso, L., L. S. Geng, S. Hirenzaki, and M. J. Vacas. 2007. 'Coherent Pion Production in Neutrino-nucleus Collisions.' *arXiv preprint arXiv:0709.0728.*

[771] Singh, Shri Krishna, and MJ Vicente Vacas. 2006. 'Weak Quasielastic Production of Hyperons.' *Physical Review D* 74 (5): 053009.

[772] Oset, E., P. Fernandez de Cordoba, L. L. Salcedo, and R. Brockmann. *Phys. Rep.* 188: 79.

[773] Alam, M. Rafi, S. Chauhan, M. Sajjad Athar, and S. K. Singh. 2013 '$\bar{\nu}l$ Induced Pion Production from Nuclei at $\backsim 1$ GeV.' *Physical Review D* 88 (7): 077301.

[774] Alam, M. Rafi, M. Sajjad Athar, S. Chauhan, and S. K. Singh. 2015. 'Quasielastic Hyperon Production In-nucleus Interactions.' *Journal of Physics G: Nuclear and Particle Physics* 42 (5): 055107.

[775] Fatima, Atika, Mohammad Sajjad Athar, and S. K. Singh. 2019. 'Weak Quasielastic Hyperon Production Leading to Pions in the Antineutrino-nucleus Reactions.' *Frontiers in Physics* 7: 13.

[776] Alvarez-Ruso, Luis, J. Nieves, I. Ruiz Simo, M. Valverde, and MJ Vicente Vacas. 2013. 'Charged Kaon Production by Coherent Scattering of Neutrinos and Antineutrinos on Nuclei.' *Physical Review C* 87 (1): 015503.

[777] Wang, Z., C. M. Marshall, L. Aliaga, O. Altinok, L. Bellantoni, A. Bercellie, M. Betancourt et al. 2016. 'Evidence of Coherent K^+ Meson Production in Neutrino-Nucleus Scattering.' *Physical review letters* 117 (6): 061802.

[778] Gershtein, S. S. 1980. 'Yu. Y. Komachenko and MY Khlopov.' *Sov. J. Nucl. Phys* 32: 861.

[779] Rein, Dieter, and Lalit M. Sehgal. 1981. 'Coherent Production of Photons by Neutrinos.' *Physics Letters B* 104 (5): 394–398.

[780] Wang, En, Luis Alvarez-Ruso, and J. Nieves. 2014. 'Photon Emission in Neutral-current Interactions at Intermediate Energies.' *Physical Review C* 89 (1): 015503.

[781] Aguilar-Arevalo, A. A., C. E. Anderson, A. O. Bazarko, S. J. Brice, B. C. Brown, L. Bugel, J. Cao et al. 2009. 'Unexplained Excess of Electronlike Events from a 1-GeV Neutrino Beam.' *Physical Review Letters* 102 (10): 101802.

[782] Kullenberg, C. T., Sanjib R. Mishra, D. Dimmery, X. C. Tian, Dario Autiero, S. Gninenko, André Rubbia et al. 2012. 'A Search for Single Photon Events in Neutrino Interactions.' *Physics Letters B* 706 (4-5): 268–275.

[783] Gomez, J., R. G. Arnold, Peter E. Bosted, C. C. Chang, A. T. Katramatou, G. G. Petratos, A. A. Rahbar et al. 1994. 'Measurement of the A Dependence of Deep-inelastic Electron Scattering.' *Physical Review D* 49(9): 4348.

[784] Ackerstaff, K., A. Airapetian, N. Akopov, I. Akushevich, M. Amarian, E. C. Aschenauer, H. Avakian et al. 2000. 'Nuclear Effects on $R = \sigma L / \sigma T$ in Deep-inelastic Scattering.' *Physics Letters B* 475 (3-4): 386–394. [*Phys. Lett. B* 567: 339 (2003)].

[785] Bari, G., A. C. Benvenuti, D. Bollini, G. Bruni, T. Camporesi, G. Heiman, L. Monari et al. 1985. 'A measurement of Nuclear Effects in Deep Inelastic Muon Scattering on Deuterium, Nitrogen and Iron Targets.' *Physics Letters B* 163 (1-4): 282–286.

[786] Benvenuti, Alberto C., D. Bollini, G. Bruni, F. L. Navarria, A. Argento, J. Cvach, K. Dieters et al. 1987. 'Nuclear Effects in Deep Inelastic Muon Scattering on Deuterium and Iron Targets.' *Physics Letters B* 189 (4): 483–487.

[787] Amaudruz, P., M. Arneodo, A. Arvidson, B. Badelek, M. Ballintijn, Guenter Baum, J. Beaufays et al. 1995. 'A Re-evaluation of the Nuclear Structure Function Ratios for D, He, 6Li, C and Ca.' *Nuclear Physics B* 441 (1-2): 3–11.

[788] Seely, J., A. Daniel, D. Gaskell, J. Arrington, N. Fomin, P. Solvignon, R. Asaturyan et al. 2009. 'New Measurements of the European Muon Collaboration Effect in Very Light Nuclei.' *Physical review letters* 103 (20): 202301.

[789] Ammosov, V. V., and A. E. Asratyan. 1984. 'Observation of the EMC Effect in Nu-barNe Interactions.' *JETP Letters* 39 (7): 393–397.

[790] Cooper, A. M., J. Derkaoui, M. L. Faccini-Turluer, Michael Andrew Parker, A. Petridis, R. A. Sansum, C. Vallee et al. 1984. 'An Investigation of the EMC Effect using Antineutrino Interactions in Deuterium and Neon.' *Physics Letters B* 141 (1-2): 133–139.

[791] Asratian, A. E. et al. 1986. *Sov. J. Nucl. Phys.* 43: 380.

[792] Guy, J., Biagio Saitta, G. Van Apeldoorn, P. Allport, Carlo Angelini, N. Armenise, A. Baldini et al. 1987. 'A Study of the EMC Effect using Neutrino and Antineutrino Interactions in Neon and Deuterium.' *Zeitschrift für Physik C Particles and Fields* 36(3): 337–348.

[793] Lassila, K. E., and U. P. Sukhatme. 1991. 'Analog of the "Emc Effect" in Neutrino-nucleus Interactions.' *International Journal of Modern Physics A* 6 (04): 613–623.

[794] Mousseau, J.A., 2016. *First Search for the EMC Effect and Nuclear Shadowing in Neutrino Nuclear Deep Inelastic Scattering at MINERvA*. Springer. doi:10.1007/978-3-319-44841-1.

[795] Stirling, William James, and M. R. Whalley. 1993. 'A compilation of Drell-Yan cross sections.' *Journal of Physics G: Nuclear and Particle Physics* 19 (D): D1.

[796] http://durpdg.dur.ac.uk/review/dy/

[797] Ashman, J., B. Badelek, Guenter Baum, J. Beaufays, C. P. Bee, C. Benchouk, I. G. Bird et al. 1993. 'A Measurement of the Ratio of the Nucleon Structure Function in Copper and Deuterium.' *Zeitschrift für Physik C Particles and Fields* 57 (2): 211–218.

[798] Aubert, Jean-Jacques, G. Bassompierre, K. H. Becks, C. Best, E. Bóhm, X. de Bouard, F. W. Brasse et al. 1983. 'The Ratio of the Nucleon Structure Functions $F2n$ for Iron and Deuterium.' *Physics Letters B* 123 (3-4): 275–278.

[799] Kulagin, Sergey A., and R. Petti. 2006. 'Global study of Nuclear Structure Functions.' *Nuclear Physics A* 765 (1-2): 126–187.

[800] Kulagin, Sergey A., and R. Petti. 2007. 'Neutrino Inclusive Inelastic Scattering off Nuclei.' *Physical Review D* 76 (9): 094023.

[801] de Cordoba, P. Fernandez, and E. Oset. 1992. 'Semiphenomenological Approach to Nucleon Properties in Nuclear Matter.' *Physical Review C* 46 (5): 1697.

[802] Degli Atti, C. Ciofi, and S. Liuti. 1989. 'On the Effects of Nucleon Binding and Correlations in Deep Inelastic Electron Scattering by Nuclei.' *Physics Letters B* 225 (3): 215–221.

[803] Vagnoni, Erica, Omar Benhar, and Davide Meloni. 2017. 'Inelastic Neutrino-Nucleus Interactions within the Spectral Function Formalism.' *Physical review letters* 118 (14): 142502.

[804] Ankowski, Artur M., and Jan T. Sobczyk. 2008. 'Construction of Spectral Functions for Medium-mass Nuclei.' *Physical Review C* 77 (4): 044311.

[805] Sargsian, M. M., S. Simula, and M. I. Strikman. 2002. 'Neutron Structure Function and inclusive Deep Inelastic Scattering from ^{3}H and ^{3}He at large Bjorken x.' *Physical Review C* 66 (2): 024001.

[806] Katori, Teppei. 2015. 'Meson Exchange Current (MEC) Models in Neutrino Interaction Generators.' In *AIP Conference Proceedings* 1663(1): 030001. AIP Publishing LLC.

[807] Mathiot, J-F. 1989. 'Electromagnetic Meson-exchange Currents at the Nucleon Mass Scale.' *Physics Reports* 173 (2-3): 63–172.

[808] Barbaro, M. B. et al. 2016. 'The Role of Meson Exchange Currents in Charged Current (Anti)neutrino-nucleus Scattering.' *Nuclear Theory* 35: 60–71. arXiv:1610.02924 [nucl-th].

[809] Marco, E., E. Oset, and P. Fernández De Córdoba. 1996. 'Mesonic and Binding Contributions to the EMC Effect in a Relativistic Many-body Approach.' *Nuclear Physics A* 611 (4): 484–513.

[810] Whitlow, L. W., E. M. Riordan, S. Dasu, Stephen Rock, and Arie Bodek. 1990. 'A Precise Extraction of R= sigma-L/sigma-T from a Global Analysis of the SLAC Deep Inelastic Ep and Ed Scattering Cross-sections.' *Phys. Lett.* 250, no. SLAC-PUB-5284: 193-198.

[811] Dasu, S., P. De Barbaro, A. Bodek, H. Harada, M. W. Krasny, Karol Lang, E. M. Riordan et al. 1994. 'Measurement of Kinematic and Nuclear Dependence of $R = \frac{\sigma_L}{\sigma_T}$ in Deep Inelastic Electron Scattering.' *Physical Review D* 49 (11): 5641.

[812] Rosenbluth, M. N. 1950. 'High Energy Elastic Scattering of Electrons on Protons.' *Physical Review* 79 (4): 615.

[813] https : //www.jlab.org/expprog/proposals/14/PR12 − 14 − 002.pdf

[814] De Florian, D., and R. Sassot. 2004. 'Nuclear Parton Distributions at Next to Leading Order.' *Physical Review D* 69 (7): 074028.

[815] Hirai, M., S. Kumano, and T-H. Nagai. 2007. 'Determination of Nuclear Parton Distribution Functions and their Uncertainties at Next-to-leading Order.' *Physical Review C* 76 (6): 065207.

[816] Eskola, K. J., Hannu Paukkunen, and C. A. Salgado. 2009. 'EPS09—a New Generation of NLO and LO Nuclear Parton Distribution Functions.' *Journal of High Energy Physics* 2009 (04): 065.

[817] de Florian, Daniel, Rodolfo Sassot, Pia Zurita, and Marco Stratmann. 2012. 'Global Analysis of Nuclear Parton Distributions.' *Physical Review D* 85 (7): 074028.

[818] Kovařík, K., I. Schienbein, F. I. Olness, J. Y. Yu, C. Keppel, J. G. Morfín, J. F. Owens, and T. Stavreva. 2012. 'Nuclear Corrections in ν A DIS and Their Compatibility with Global NPDF Analyses.' *Few-Body Systems* 52 (3-4): 271–277.

[819] Kovaříí k, K., A. Kusina, T. Ježo, D. B. Clark, C. Keppel, F. Lyonnet, J. G. Morfín et al. 2016. 'nCTEQ15: Global Analysis of Nuclear Parton Distributions with Uncertainties in the CTEQ framework.' *Physical Review D* 93 (8): 085037.

[820] Eskola, Kari J., Petja Paakkinen, Hannu Paukkunen, and Carlos A. Salgado. 2017. 'EPPS16: Nuclear Parton Distributions with LHC Data.' *The European Physical Journal C* 77 (3): 163.

[821] Khanpour, Hamzeh, and S. Atashbar Tehrani. 2016. 'Global Analysis of Nuclear Parton Distribution Functions and their Uncertainties at Next-to-next-to-leading Order.' *Physical Review D* 93 (1): 014026.

[822] Haider, H., F. Zaidi, M. Sajjad Athar, S. K. Singh, and I. Ruiz Simo. 2016. 'Nuclear Medium Effects in F2AEM $(x, Q2)$ and F2AWeak $(x, Q2)$ Structure Functions.' *Nuclear Physics A* 955: 58–78.

[823] Collins, John C., Davison E. Soper, and George Sterman. 1985. 'Factorization for Short Distance Hadron-hadron Scattering.' *Nuclear Physics B* 261: 104–142.

[824] Bodwin, Geoffrey T. 1985. 'Factorization of the Drell-Yan Cross Section in Perturbation Theory.' *Physical Review D* 31 (10): 2616.

[825] Jimenez-Delgado, Pedro, and Ewald Reya. 2014. 'Delineating Parton Distributions and the Strong Coupling.' *Physical Review D* 89 (7): 074049.

[826] Schienbein, I., J. Y. Yu, K. Kovařík, C. Keppel, J. G. Morfin, F. I. Olness, and J. F. Owens. 2009. 'Parton Distribution Function Nuclear Corrections for Charged Lepton and Neutrino Deep Inelastic Scattering Processes.' *Physical Review D* 80 (9): 094004.

[827] Bodek, Arie, and Un-ki Yang. 2010. 'Axial and Vector Structure Functions for Electron- and Neutrino-Nucleon Scattering Cross Sections at all Q^2 using Effective Leading order Parton Distribution Functions.' *arXiv preprint arXiv:1011.6592*.

[828] Eskola, Kari J., V. J. Kolhinen, and C. A. Salgado. 1999. 'The Scale Dependent Nuclear Effects in Parton Distributions for Practical Applications.' *The European Physical Journal C-Particles and Fields* 9 (1): 61–68.

[829] Eskola, Kari J., V. J. Kolhinen, and P. V. Ruuskanen. 1998. 'Scale Evolution of Nuclear Parton Distributions.' *Nuclear Physics B* 535 (1-2): 351–371.

[830] Hirai, M., S. Kumano, and M. Miyama. 2001. 'Determination of Nuclear Parton Distributions.' *Physical Review D* 64 (3): 034003.

[831] Hirai, M., S. Kumano, and T-H. Nagai. 2004. 'Nuclear Parton Distribution Functions and their Uncertainties.' *Physical Review C* 70 (4): 044905.

[832] Eskola, Kari J., Vesa J. Kolhinen, Hannu Paukkunen, and Carlos A. Salgado. 2007. 'A Global Reanalysis of Nuclear Parton Distribution Functions.' *Journal of High Energy Physics* 2007 (05): 002.

[833] Eskola, Kari J., Hannu Paukkunen, and Carlos A. Salgado. 2008. 'An Improved Global Analysis of Nuclear Parton Distribution Functions Including RHIC data.' *Journal of High Energy Physics* 2008 (07): 102.

[834] Stavreva, T., I. Schienbein, F. Arleo, K. Kovařík, F. Olness, J. Y. Yu, and J. F. Owens. 2011. 'Probing Gluon and Heavy-Quark Nuclear PDFs with $\gamma + Q$ Production in pA Collisions.' *Journal of High Energy Physics* 2011 (1): 152.

[835] Kovařík, K., I. Schienbein, F. I. Olness, J. Y. Yu, C. Keppel, J. G. Morfín, J. F. Owens, and T. Stavreva. 2011. 'Nuclear Corrections in Neutrino-Nucleus Deep Inelastic Scattering and their Compatibility with Global Nuclear Parton-Distribution-Function Analyses.' *Physical Review Letters* 106 (12): 122301.

[836] Owens, J. F., J. Huston, C. E. Keppel, S. Kuhlmann, J. G. Morfin, F. Olness, J. Pumplin, and D. Stump. 2007. 'Impact of New Neutrino Deep Inelastic Scattering and Drell-Yan Data on Large-x Parton Distributions.' *Physical Review D* 75 (5): 054030.

[837] Kopeliovich, B. Z., J. G. Morfin, and I. Schmidt. 2013. 'Nuclear Shadowing in Electro-Weak Interactions.' *Prog. Part. Nucl. Phys.* 68: 314–372. doi: 10.1016/j.ppnp.2012.09.004.

[838] Cloet, I. C., Wolfgang Bentz, and Anthony William Thomas. 2006. 'EMC and Polarized EMC Effects in Nuclei.' *Physics Letters B* 642 (3): 210–217.

[839] H. Haider, private communication.

[840] Akulinichev, S. V., S. A. Kulagin, and G. M. Vagradov. 1985. 'The Role of Nuclear Binding in Deep Inelastic Lepton-Nucleon Scattering.' *Physics Letters B* 158 (6): 485–488.

Akulinichev, S. V., S. Shlomo, S. A. Kulagin, and G. M. Vagradov. 1985. 'Lepton-nucleus Deep-inelastic Scattering.' *Physical Review Letters* 55 (21): 2239.

Akulinichev, S. V., and S. Shlomo. 1990. 'Nuclear Binding Effect in Deep-inelastic Lepton Scattering.' *Physics Letters B* 234 (1-2): 170–174.

[841] Dunne, Gerald V., and Anthony W. Thomas. 1986. 'On the Interpretation of the European Muon Collaboration effect.' *Physical Review D* 33 (7): 2061.

[842] Bickerstaff, R. P., and Anthony William Thomas. 1989. 'The EMC Effect-with Emphasis on Conventional Nuclear Corrections.' *Journal of Physics G: Nuclear and Particle Physics* 15 (10): 1523.

[843] Kulagin, Sergey A. 1989. 'Deep-inelastic Scattering on Nuclei: Impulse Approximation and Mesonic Corrections.' *Nuclear Physics A* 500 (3): 653–668.

[844] Arneodo, M. 1994. 'Nuclear Effects in Structure Functions.' *Phys. Rept.* 240: 301–393. doi: 10.1016/0370-1573(94)90048-5.

[845] Hen, Or, Douglas W. Higinbotham, Gerald A. Miller, Eli Piasetzky, and Lawrence B. Weinstein. 2013. 'The EMC Effect and High Momentum Nucleons in Nuclei.' *International Journal of Modern Physics E* 22 (07): 1330017.

[846] Piller, Gunther, and Wolfram Weise. 2000. 'Nuclear Deep-inelastic Lepton Scattering and Coherence Phenomena.' *Physics Reports* 330 (1): 1–94.

[847] Benhar, O., V. R. Pandharipande, and I. Sick. 1999. 'Density Dependence of the EMC Effect.' *Physics Letters B* 469 (1-4): 19–24.

Benhar, O., V. R. Pandharipande, and I. Sick. 1997. 'Nuclear Binding and Deep Inelastic Scattering.' *Physics Letters B* 410 (2-4): 79–85.

[848] Smith, Jason R., and Gerald A. Miller. 2003. 'Chiral Solitons in Nuclei: Saturation, EMC Effect, and Drell-Yan Experiments.' *Physical Review Letters* 91 (21): 212301.

Smith, Jason R., and Gerald A. Miller. 2007. 'Erratum: Chiral Solitons in Nuclei: Saturation, EMC Effect, and Drell-Yan Experiments.' *Physical Review Letters* 98 (9): 099902.

[849] degli Atti, C. Ciofi, L. L. Frankfurt, L. P. Kaptari, and M. I. Strikman. 2007. 'Dependence of the Wave Function of a Bound Nucleon on its Momentum and the EMC Effect.' *Physical Review C* 76 (5): 055206.

[850] Athar, M. Sajjad, Shri Krishna Singh, and MJ Vicente Vacas. 2008. 'Nuclear effects in $F3$ structure function of nucleon.' *Physics Letters B* 668 (2): 133–142.

[851] Athar, M. Sajjad, I. Ruiz Simo, and MJ Vicente Vacas. 2011. 'Nuclear Medium Modification of the $F2(x, Q2)$ Structure Function.' *Nuclear Physics A* 857 (1): 29–41.

[852] Haider, Huma, I. Ruiz Simo, M. Sajjad Athar, and MJ Vicente Vacas. 2011. 'Nuclear Medium Effects in $\nu(\nu)$-nucleus Deep Inelastic Scattering.' *Physical Review C* 84 (5): 054610.

[853] Frankfurt, Leonid, and Mark Strikman. 2012. 'QCD and QED Dynamics in the EMC Effect.' *International Journal of Modern Physics E* 21 (04): 1230002.

[854] Haider, H., I. Ruiz Simo, and M. Sajjad Athar. 2012. '$\nu(\bar{\nu})^{-}$ ^{208}Pb Deep-inelastic Scattering.' *Physical Review C* 85 (5): 055201.

[855] Haider, H., I. Ruiz Simo, and M. Sajjad Athar. 2013. 'Effects of the Nuclear Medium and Non-isoscalarity in Extracting $\sin^2 \theta_W$ using the Paschos-Wolfenstein Relation.' *Physical Review C* 87 (3): 035502.

[856] Haider, H., M. Sajjad Athar, S. K. Singh, and I. Ruiz Simo. 2015. 'Parity Violating aSymmetry with Nuclear Medium Effects in Deep Inelastic $e \to$ scattering.' *Nuclear Physics A* 940: 138–157.

[857] Malace, Simona, David Gaskell, Douglas W. Higinbotham, and Ian C. Cloét. 2014. 'The Challenge of the EMC Effect: Existing Data and Future Directions.' *International Journal of Modern Physics E* 23 (08): 1430013.

[858] Ericson, Magda, and Anthony W. Thomas. 1983. 'Pionic Corrections and the EMC Enhancement of the Sea in Iron.' *Physics Letters B* 128 (1-2): 112–116.

[859] Bickerstaff, R. P., and G. A. Miller. 1986. 'Origins of the EMC Effect.' *Physics Letters B* 168 (4): 409–414.

[860] Berger, Edmond L., and F. Coester. 1987. 'Nuclear Effects in Deep Inelastic Lepton Scattering.' *Annual Review of Nuclear and Particle Science* 37 (1): 463–491.

[861] Jaffe, Robert L. 1983. 'Quark Distributions in Nuclei.' *Physical Review Letters* 50 (4): 228.

[862] Mineo, H., Wolfgang Bentz, N. Ishii, Anthony William Thomas, and K. Yazaki. 2004. 'Quark Distributions in Nuclear Matter and the EMC Effect.' *Nuclear Physics A* 735 (3-4): 482–514.

[863] Cloet, I. C., Wolfgang Bentz, and Anthony William Thomas. 2005. 'Spin-dependent Structure Functions in Nuclear Matter and the Polarized EMC Effect.' *Physical Review Letters* 95 (5): 052302.

[864] Nachtmann, O., and H. J. Pirner. 1984. 'Color-conductivity in Nuclei and the EMC-effect.' *Zeitschrift für Physik C Particles and Fields* 21 (3): 277–280.

[865] Close, Francis Edwin, Robert Gwilym Roberts, and Graham G. Ross. 1983. 'The Effect of Confinement Size on Nuclear Structure Functions.' *Physics Letters B* 129 (5): 346–350.

[866] Frankfurt, Leonid, and Mark Strikman. 1988. 'Hard Nuclear Processes and Microscopic Nuclear Structure.' *Physics Reports* 160 (5-6): 235–427.

[867] Armesto, Nestor. 2006. 'Nuclear Shadowing.' *Journal of Physics G: Nuclear and Particle Physics* 32 (11): R367.

[868] Geesaman, Donald F., Koichi Saito, and Anthony W. Thomas. 1995. 'The Nuclear EMC Effect.' *Annual Review of Nuclear and Particle Science* 45 (1): 337–390.

[869] Haider, H., F. Zaidi, M. Sajjad Athar, S. K. Singh, and I. Ruiz Simo. 2015. 'Nuclear Medium Effects in Structure Functions of Nucleon at Moderate $Q2$.' *Nuclear Physics A* 943: 58–82.

[870] Zaidi, F., H. Haider, M. Sajjad Athar, S. K. Singh, and I. Ruiz Simo. 2019. 'Nucleon and Nuclear Structure Functions with Nonperturbative and Higher Order Perturbative QCD Effects.' *Physical Review D* 99 (9): 093011.

[871] Zaidi, F., H. Haider, M. Sajjad Athar, S. K. Singh, and I. Ruiz Simo. 2020. *Phys. Rev. D* 101: 033001.

[872] Itzykson, C., and J. B. Zuber. 1986. *Quantum Field Theory.* New York: McGRAW-HILL Publication.

[873] Glúck, M., E. Reya, and A. Vogt. 1992. 'Pionic Parton Distributions.' *Zeitschrift für Physik C Particles and Fields* 53 (4): 651–655.

[874] Wijesooriya, K., P. E. Reimer, and R. J. Holt. 2005. 'Pion Parton Distribution Function in the Valence Region.' *Physical Review C* 72 (6): 065203.

[875] Sutton, P. J., Alan D. Martin, R. G. Roberts, and WJames Stirling. 1992. 'Parton Distributions for the Pion Extracted from Drell-Yan and Prompt Photon Experiments.' *Physical Review D* 45 (7): 2349.

[876] Conway, J. S., C. E. Adolphsen, J. P. Alexander, K. J. Anderson, J. G. Heinrich, J. E. Pilcher, A. Possoz et al. 1989. 'Experimental Study of Muon Pairs Produced by 252-GeV Pions on Tungsten.' *Physical Review D* 39 (1): 92.

[877] Martin, Alan D., R. G. Roberts, WJames Stirling, and R. S. Thorne. 1998. 'Parton Distributions: A New Global Analysis.' *The European Physical Journal C-Particles and Fields* 4 (3): 463–496.

[878] Badier, J., J. Boucrot, J. Bourotte, G. Burgun, O. Callot, Ph Charpentier, M. Crozon et al. 1983. 'Experimental Determination of the π Meson Structure Functions by the Drell-Yan Mechanism.' *Zeitschrift für Physik C Particles and Fields* 18 (4): 281–287.

[879] Betev, B., J. J. Blaising, P. Bordalo, A. Boumediene, L. Cerrito, A. Degre, A. Ereditato et al. 1985. 'Differential Cross-section of High-mass Muon Pairs Produced by a 194 GeV/c π^- Beam on a Tungsten Target.' *Zeitschrift für Physik C Particles and Fields* 28 (1): 9–14.

[880] Kulagin, S. A., and R. Petti. 2010. 'Structure.Functions for Light Nuclei.' *Physical Review C* 82 (5): 054614.

[881] Múther, H., and A. Polls. 2000. 'Two-body Correlations in Nuclear Systems.' *Progress in Particle and Nuclear Physics* 45 (1): 243–334.

[882] Ónengút, G., R. Van Dantzig, M. De Jong, RUDOLF GERHARD CHRISTIAAN Oldeman, M. Güler, S. Kama, U. Köse et al. 2006. 'Measurement of Nucleon Structure Functions in Neutrino Scattering.' *Physics Letters B* 632 (1): 65–75.

[883] Katz, Ulrich F., and Ch Spiering. 2012. 'High-energy Neutrino Astrophysics: Status and Perspectives.' *Progress in Particle and Nuclear Physics* 67 (3): 651–704.

[884] Eddington, A. S. 1920. 'The Internal Constitution of the Stars.' *The Scientific Monthly* 11 (4): 297–303.

[885] Bethe, Hans Albrecht. 1939. 'Energy Production in Stars.' *Physical Review* 55 (5): 434.

[886] Maurin, D., F. Melot, and Richard Taillet. 2014. 'A Database of Charged Cosmic Rays.' *Astronomy & Astrophysics* 569: A32. https://lpsc.in2p3.fr/cosmic-rays-db/.

[887] Abe, Koh, Hideyuki Fuke, Sadakazu Haino, T. Hams, M. Hasegawa, A. Horikoshi, A. Itazaki et al. 2016. 'Measurements of Cosmic-ray Proton and Helium Spectra from the BESS-Polar Long-duration Balloon Flights Over Antarctica.' *The Astrophysical Journal* 822 (2): 65.

[888] Abe, K., H. Fuke, S. Haino, T. Hams, M. Hasegawa, A. Horikoshi, A. Itazaki et al. 2014. 'Time Variations of Cosmic-ray Helium Isotopes with BESS-Polar I.' *Advances in Space Research* 53 (10): 1426–1431.

[889] Incagli, Marco. 2015. 'Results from the AMS02 Experiment on the International Space Station.' In *EPJ Web of Conferences* 95: 03017. EDP Sciences.

Aguilar, M., D. Aisa, B. Alpat, A. Alvino, G. Ambrosi, K. Andeen, L. Arruda et al. 2015. 'Precision Measurement of the Proton Flux in Primary Cosmic Rays from Rigidity 1 GV to 1.8 TV with the Alpha Magnetic Spectrometer on the International Space Station.' *Physical Review Letters* 114 (17): 171103.

Haino, S. 2015. *The Helium Spectrum from AMS.* AMS Days at CERN: The Future of Cosmic Ray Physics and Latest Results. (Geneva, Switzerland, 2015).

[890] Haino, Sadakazu, T. Sanuki, K. Abe, K. Anraku, Y. Asaoka, H. Fuke, M. Imori et al. 2004. 'Measurements of Primary and Atmospheric Cosmic-ray Spectra with the BESS-TeV Spectrometer.' *Physics Letters B* 594 (1-2): 35–46.

[891] Derbina, V. A., V. I. Galkin, M. Hareyama, Y. Hirakawa, Y. Horiuchi, M. Ichimura, N. Inoue et al. 2005. 'Cosmic-ray Spectra and Composition in the Energy Range of 10-1000 TeV per Particle Obtained by the RUNJOB Experiment.' *The Astrophysical Journal Letters* 628 (1): L41.

[892] Wilczynski, H. [JACEE Collaboration]. 1997. 'JACEE Results on Very High Energy Interactions.' *Nucl. Phys. Proc. Suppl. B* 52: 81. doi: 6/S0920-5632(96)00851-1.

[893] Yamamoto for the BESS Collaboration, Akira. 2008. 'BESS-Polar: Search for Cosmic-ray Antiparticles of Primary Origins.' *Journal of the Physical Society of Japan* 77, no. Suppl. B: 45–48.

[894] Gaisser, T.K., and M. Honda. 2002. 'Flux of Atmospheric Neutrinos.' *Annual Review of Nuclear and Particle Science* 52 (1): 153–199.

[895] Achar, C. V., M. G. K. Menon, V. S. Narasimham, P. V. Murthy, B. V. Sreekantan, K. Hinotani, S. Miyake et al. 1965. 'Detection of Muons Produced by Cosmic Ray Neutrinos Deep Underground.' *Physics Letters* 18: 196–199.

[896] Reines, F., M. F. Crouch, T. L. Jenkins, W. R. Kropp, H. S. Gurr, G. R. Smith, J. P. F. Sellschop, and B. Meyer. 1965. 'Evidence for High-energy Cosmic-ray Neutrino Interactions.' *Physical Review Letters* 15 (9): 429.

[897] Athar, M. Sajjad, M. Honda, T. Kajita, K. Kasahara, and S. Midorikawa. 2013. 'Atmospheric Neutrino Flux at INO, South Pole and Pyhäsalmi.' *Physics Letters B* 718 (4-5): 1375–1380.

[898] Hirata, K. S., A. K. Mann, M. Takita, W. Frati, K. Takahashi, T. Kajita, A. Suzuki et al. 1988. 'Experimental Study of the Atmospheric Neutrino Flux.' *Phys. Lett.* 205, no. UPR-0149E: 416.

[899] Casper, D., R. Becker-Szendy, C. B. Bratton, D. R. Cady, R. Claus, S. T. Dye, W. Gajewski et al. 1991. 'Measurement of Atmospheric Neutrino Composition with the IMB-3 Detector.' *Physical Review Letters* 66 (20): 2561.

[900] Allison, W. W. M., G. J. Alner, D. S. Ayres, W. L. Barrett, C. Bode, P. M. Border, C. B. Brooks et al. 1997. 'Measurement of the Atmospheric Neutrino Flavour Composition in Soudan 2.' *Physics Letters B* 391 (3-4): 491–500.

[901] Hirata, K. S., K. Inoue, T. Ishida, T. Kajita, K. Kihara, M. Nakahata, K. Nakamura et al. 1992. 'Observation of a Small Atmospheric $v\mu/ve$ Ratio in Kamiokande.' *Physics Letters B* 280 (1-2): 146–152.

[902] Fukuda, Y., K. Nishikawa, M. Koga, J. Suzuki, Y. Suzuki, A. K. Mann, H. Miyata et al. 1994. 'Atmospheric v_{μ}/v_e Ratio in the Multi-GeV Energy Range.' *Phys. Lett. B* 335, no. NGTHEP-94-1: 237–245.

[903] Fukuda, Y., T. Hayakawa, E. Ichihara, K. Inoue, K. Ishihara, Hirokazu Ishino, Y. Itow et al. 1998. 'Evidence for Oscillation of Atmospheric Neutrinos.' *Physical Review Letters* 81 (8): 1562.

[904] Mueller, Th A., D. Lhuillier, Muriel Fallot, A. Letourneau, S. Cormon, M. Fechner, Lydie Giot et al. 2011. 'Improved Predictions of Reactor Antineutrino Spectra.' *Physical Review C* 83 (5): 054615.

Le, Trung. 2009. 'Overview of the T2K Long Baseline Neutrino Oscillation Experiment.' *arXiv preprint arXiv:0910.4211.*

[905] Apollonio, Marco, A. Baldini, C. Bemporad, E. Caffau, Fabrizio Cei, Y. Declais, H. De Kerret et al. 1999. 'Limits on Neutrino Oscillations from the CHOOZ Experiment.' *Physics Letters B* 466 (2-4): 415–430.

[906] Boehm, F. 2000. 'Palo Verde Experiment.' *Physical Review Letters* 84: 3764.

[907] Colgate, Stirling A., and Richard H. White. 1966. 'The Hydrodynamic Behavior of Supernovae Explosions.' *The Astrophysical Journal* 143: 626.

[908] Arnett, W. David. 1966. 'Gravitational Collapse and Weak Interactions.' *Canadian Journal of Physics* 44 (11): 2553–2594.

[909] Raffelt, Georg G. 1996. *Stars as Laboratories for Fundamental Physics: The Astrophysics of Neutrinos, Axions, and other Weakly Interacting Particles.* Chicago: University of Chicago Press.

[910] Haxton, W. C. 2012. 'Neutrino Astrophysics.' *arXiv:nucl-th/1209.3743*.

[911] Hirata, K. et al. [KAMIOKANDE-II Collaboration]. 1987. *Physical Review Letters* 58: 1490.

[912] Bionta, R. M. et al. 1987. *Physical Review Letters* 58: 1494.

[913] Bays, Kirk, T. Iida, K. Abe, Y. Hayato, K. Iyogi, J. Kameda, Yusuke Koshio et al. 2012. 'Supernova Relic Neutrino Search at Super-kamiokande.' *Physical Review D* 85 (5): 052007.

[914] Agafonova, N. Yu, M. Aglietta, P. Antonioli, G. Bari, A. Bonardi, V. V. Boyarkin, G. Bruno et al. 2008. 'On-line Recognition of Supernova Neutrino Bursts in the LVD.' *Astroparticle Physics*28 (6): 516–522.

[915] Ahrens, Jens, E. Andrés, X. Bai, G. Barouch, S. W. Barwick, R. C. Bay, T. Becka et al. 2002. 'Observation of High Energy Atmospheric Neutrinos with the Antarctic Muon and Neutrino Detector Array.' *Physical Review D* 66 (1): 012005.

[916] Cadonati, L., F. P. Calaprice, and M. C. Chen. 2002. 'Supernova Neutrino Detection in Borexino.' *Astroparticle Physics* 16 (4): 361–372.

[917] Boyd, R. N., and A. S. J. Murphy. 2001. 'The Observatory for Multiflavor NeutrInos from Supernovae.' *Nuclear Physics A* 688: 386–389. doi: 10.1016/S0375-9474(01) 00732-1.

Smith, P. F. 2001. 'Astroparticle Phys., 8, 27 (1997).' *Astroparticle Phys* 16: 75.

[918] Hargrove, C. K., I. Batkin, M. K. Sundaresan, and J. Dubeau. 1996. 'A Lead Astronomical Neutrino Detector: LAND.' *Astroparticle Physics* 5 (2): 183–196.

[919] Duba, C. A., F. Duncan, J. Farine, A. Habig, A. Hime, R. G. H. Robertson, K. Scholberg et al. 2008. 'HALO–the Helium and Lead Observatory for Supernova Neutrinos.' In *Journal of Physics: Conference Series*, vol. 136, no. 4, p. 042077. IOP Publishing.

[920] Bueno, A. 2005. 'The ICARUS project.' *Nuclear Physics B-Proceedings Supplements* 143: 262–265.

[921] M.A. Schumaker [SNO+ Collaboration], Nucl. Phys. B – Proc. Supp. 00, 1 (2010).

[922] Abe, K., T. Abe, H. Aihara, Y. Fukuda, Y. Hayato, K. Huang, A. K. Ichikawa et al. 2011. 'Letter of Intent: The Hyper-Kamiokande Experiment—Detector Design and Physics Potential—.' *arXiv preprint arXiv:1109.3262*.

[923] Djurcic, Z. et al. [JUNO Collaboration]. 2015. 'JUNO Conceptual Design Report.' *arXiv:1508.07166 [physics.ins-det]*.

[924] An, Fengpeng, Guangpeng An, Qi An, Vito Antonelli, Eric Baussan, John Beacom, Leonid Bezrukov et al. 2016. 'Neutrino Physics with JUNO.' *Journal of Physics G: Nuclear and Particle Physics* 43 (3): 030401.

[925] Totani, T., K. Sato, H. E. Dalhed, and J. R. Wilson. 1998. 'Future Detection of Supernova Neutrino Burst and Explosion Mechanism.' *The Astrophysical Journal* 496 (1): 216.

[926] Mezzacappa, Anthony. 2005. 'Ascertaining the Core Collapse Supernova Mechanism: The State of the Art and the Road Ahead.' *Annu. Rev. Nucl. Part. Sci.* (55): 467–515.

[927] Janka, H-Th, K. Langanke, Andreas Marek, G. Martínez-Pinedo, and B. Múller. 2007. 'Theory of Core-collapse Supernovae.' *Physics Reports* 442 (1-6): 38–74.

[928] Dasgupta, B., and A. Dighe, J. 2008. *J. Phys. Conf. Ser.* 136: 042072. doi: 10.1088/1742-6596/136/4/042072.

[929] Keil, Mathias Th. 2003. 'Supernova Neutrino Spectra and Applications to Flavor Oscillations.' *ArXiv Preprint Astro-Ph/0308228.*

[930] Fischer, T., S. C. Whitehouse, Anthony Mezzacappa, F-K. Thielemann, and Matthias Liebendoerfer. 2009. 'The Neutrino Signal from Protoneutron Star Accretion and Black Hole Formation.' *Astronomy & Astrophysics* 499 (1): 1–15.

[931] Keil, Mathias Th, Georg G. Raffelt, and Hans-Thomas Janka. 2003. 'Monte Carlo Study of Supernova Neutrino Spectra Formation.' *The Astrophysical Journal* 590 (2): 971.

[932] Fogli, G. et al. 2009. *J. Cosmol. Astropart. Phys.* 0910: 002. doi: 10.1088/1475-7516/2009/10/002.

[933] Vaananen, D., and C. Volpe. 2011. 'The Neutrino Signal at HALO: Learning about the Primary Supernova Neutrino Fluxes and Neutrino Properties.' *J. Cosmol. Astropart. Phys.* 1110: 019. doi: 10.1088/1475-7516/2011/10/019.

[934] Choubey, Sandhya, Basudeb Dasgupta, Amol Dighe, and Alessandro Mirizzi. 2010. 'Signatures of Collective and Matter Effects on Supernova Neutrinos at Large Detectors.' *arXiv preprint arXiv:1008.0308.*

[935] Hüdepohl, Lorenz, B. Müller, H-T. Janka, Andreas Marek, and Georg G. Raffelt. 2010. 'Neutrino Signal of Electron-Capture Supernovae from Core Collapse to Cooling.' *Physical Review Letters* 104 (25): 251101. [Erratum 105: 249901 (2010)].

[936] Pantaleone, James. 1992. 'Neutrino Oscillations at High Densities.' *Physics Letters B* 287 (1-3): 128–132.

[937] Duan, Huaiyu, and Alexander Friedland. 2011. 'Self-induced Suppression of Collective Neutrino Oscillations in a Supernova.' *Physical Review Letters* 106 (9): 091101.

[938] Gava, Jerome, James Kneller, Cristina Volpe, and G. C. McLaughlin. 2009. 'Dynamical Collective Calculation of Supernova Neutrino Signals.' *Physical review letters* 103 (7): 071101.

[939] Duan, H., G. M. Fuller, and Y-Z. Qian. 2010. 'Collective Neutrino Oscillations.' *Ann. Rev. Nucl. Part. Sci.* 60: 569. doi: 10.1146/annurev.nucl.012809.104524.

[940] Duan, Huaiyu, George M. Fuller, J. Carlson, and Yong-Zhong Qian. 2007. 'Analysis of Collective Neutrino Flavor Transformation in Supernovae.' *Physical Review D* 75 (12): 125005.

[941] Mirizzi, A., G. G. Raffelt, and P. D. Serpico. 2006. 'Earth Matter Effects in Supernova Neutrinos: Optimal Detector Locations.' *J. Cosm. Astropart. Phys.* 0605: 012. doi: 10.1088/1475-7516/2006/05/012.

[942] Lunardini, Cecilia, and A. Yu Smirnov. 2001. 'Supernova Neutrinos: Earth Matter Effects and Neutrino Mass Spectrum.' *Nuclear Physics B* 616 (1-2): 307–348.

[943] Takahashi, Keitaro, and Katsuhiko Sato. 2002. 'Earth Effects on Supernova Neutrinos and their Implications for Neutrino Parameters.' *Physical Review D* 66 (3): 033006.

[944] Takahashi, Keitaro, and Katsuhiko Sato. 2003. 'Effects of Neutrino Oscillation on Supernova Neutrino: Inverted Mass Hierarchy.' *Progress of theoretical physics* 109 (6): 919–931.

[945] Araki, T., S. Enomoto, K. Furuno, Y. Gando, K. Ichimura, H. Ikeda, K. Inoue et al. 2005. 'Experimental Investigation of Geologically Produced Antineutrinos with KamLAND.' *Nature* 436 (7050): 499–503.

[946] Faessler, A., R. Hodak, S. Kovalenko, and F. Simkovic. 2013. 'Search for the Cosmic Neutrino Background and KATRIN.' *Rom. J. Phys.* 58 (9–10): 1221. arXiv:1304.5632 [nucl-th].

[947] Mertens, Susanne, Antonio Alborini, Konrad Altenmüller, Tobias Bode, Luca Bombelli, Tim Brunst, Marco Carminati et al. 2019. 'A Novel Detector System for KATRIN to Search for keV-scale Sterile Neutrinos.' *Journal of Physics G: Nuclear and Particle Physics* 46 (6): 065203.

[948] https://www-boone.fnal.gov/.

[949] https://minerva.fnal.gov/.

[950] Axani, S., G. Collin, J. M. Conrad, M. H. Shaevitz, J. Spitz, and T. Wongjirad. 2015. 'Decisive Disappearance Search at High Δ_m^2 with Monoenergetic Muon Neutrinos.' *Physical Review D* 92 (9): 092010.

[951] Akimov, D., J. B. Albert, P. An, C. Awe, P. S. Barbeau, B. Becker, V. Belov et al. 2018. 'COHERENT 2018 at the Spallation Neutron Source.' *arXiv preprint arXiv:1803.09183*.

[952] Koshkarev, D. G. 1974. *CERN Internal Report*. CERN/ISR-DI/74-62.

[953] Neuffer, D. 'Design Considerations for a Muon Storage Ring (1980).' In *Telmark Conference on Neutrino Mass, Barger and Cline eds., Telmark, Wisconsin*.

[954] Adey, David, Ryan Bayes, Alan D. Bross, and Pavel Snopok. 2015. 'nuSTORM and A Path to a Muon Collider.' *Annual Review of Nuclear and Particle Science* 65: 145–175.

[955] Adey, D., S. K. Agarwalla, C. M. Ankenbrandt, R. Asfandiyarov, J. J. Back, G. Barker, E. Baussan et al. 2013. 'nuSTORM-neutrinos from STORed Muons: Proposal to the Fermilab PAC.' *arXiv preprint arXiv:1308.6822*.

[956] Zucchelli, P. 2002. 'A Novel Concept for a $\nu e/\nu e$ Neutrino Factory: the Beta-beam.' *Physics Letters B* 532 (3-4): 166–172

[957] Volpe, Cristina. 2004. 'What About a Beta-beam Facility for Low-energy Neutrinos?.' *Journal of Physics G: Nuclear and Particle Physics* 30 (7): L1.

[958] Volpe, Cristina. 2005. 'Neutrino–nucleus Interactions as a Probe to Constrain Double-beta Decay Predictions.' *Journal of Physics G: Nuclear and Particle Physics* 31 (8): 903.

[959] Mezzetto, Mauro. 2003. 'Physics Potential of the SPL Super Beam.' *Journal of Physics G: Nuclear and Particle Physics* 29 (8): 1781.

[960] Mezzetto, Mauro. 2005. 'SPL and Beta Beams to the Frejus.' *Nuclear Physics B-Proceedings Supplements* 149: 179–181.

[961] Mezzetto, M. 2006. *Nucl. Phys. Proc. Suppl.* 155: 214–217. doi: 0.1016/j.nuclphysbps. 2006.02.112.

[962] Autin, Bruno, M. Benedikt, M. Grieser, S. Hancock, H. Haseroth, A. Jansson, U. Kóster et al. 2003. 'The Acceleration and Storage of Radioactive ions for a Neutrino Factory.' *Journal of Physics G: Nuclear and Particle Physics* 29 (8): 1785.

[963] Burguet-Castell, Jordi, D. Casper, E. Couce, Juan José Gómez-Cadenas, and P. Hernandez. 2005. 'Optimal β-beam at the CERN-SPS.' *Nuclear Physics B* 725 (1-2): 306–326.

[964] Jachowicz, Natalie, and G. C. McLaughlin. 2006. 'Reconstructing Supernova-neutrino Spectra Using Low-energy Beta Beams.' *Physical Review Letters* 96 (17): 172301.

Jachowicz, N., and G. C. McLaughlin. 2006. *Eur. Phys. J.* A27 (S1): 41–48. doi: 10.1140/epja/i2006-08-005-x.

[965] Serreau, J. and C. Volpe. 2004. 'Neutrino-nucleus interaction rates at a low-energy β-beam facility.' *Physical Review C* 70 (5): 055502.

Volpe, C. 2006. 'Physics with a very first low-energy beta-beam.' *Nuclear Physics B-Proceedings Supplements* 155 (1): 97-101.

[966] Ackermann, Markus, Markus Ahlers, Luis Anchordoqui, Mauricio Bustamante, Amy Connolly, Cosmin Deaconu, Darren Grant et al. 2019. 'Astrophysics Uniquely Enabled by Observations of High-energy Cosmic Neutrinos.' *arXiv preprint arXiv:1903.04334*.

[967] Stecker, Floyd W. 'PeV Neutrinos Observed by IceCube from Cores of Active Galactic Nuclei.' *Physical Review D* 88, no. 4 (2013): 047301.

[968] Davis Jr, Raymond, and Don S. Harmer. 1959. 'Attempt to Observe the Cl $^{37}(\bar{\nu}e^-)$ Ar^{37} Reaction Induced by Reactor Antineutrinos.' *Bull. Am. Phys. Soc.* 4: 217.

[969] Pontecorvo, B. 1958. *Sov. Phys. JETP* 7: 172. [*Zh. Eksp. Teor. Fiz.* 34 (1957) 247].

[970] Pontecorvo, B. 1968. *Sov. Phys. JETP* 26: 984. [*Zh. Eksp. Teor. Fiz.* 53 (1967) 1717–1725].

[971] Davis Jr, Raymond. 1964. 'Solar Neutrinos. II. Experimental.' *Physical Review Letters* 12 (11): 303.

[972] Davis, R., K. Lande, C. K. Lee, B. T. Cleveland, and J. Ullman. 1990. *In Proceedings, 21st International Cosmic Ray Conference, Adelaide, Australia, January 6–19, 1990, Vol. 7*, pp. 155–158.

[973] Bahcall, John N. 1964. 'Solar Neutrinos. i. Theoretical.' *Physical Review Letters* 12 (11): 300.

[974] Araki, T., K. Eguchi, S. Enomoto, K. Furuno, K. Ichimura, H. Ikeda, K. Inoue et al. 2005. 'Measurement of Neutrino Oscillation with KamLAND: Evidence of Spectral Distortion.' *Physical Review Letters* 94 (8): 081801.

[975] Mention, G., M. Fechner, Th Lasserre, Th A. Mueller, D. Lhuillier, M. Cribier, and A. Letourneau. 2011. 'Reactor Antineutrino Anomaly.' *Physical Review D* 83 (7): 073006.

[976] Huber, Patrick. 2011. 'Determination of Antineutrino Spectra from Nuclear Reactors.' *Physical Review C* 84 (2): 024617. [Erratum: Physical Review C 85: 029901(2012).].

[977] Mueller, Th A., D. Lhuillier, Muriel Fallot, A. Letourneau, S. Cormon, M. Fechner, Lydie Giot et al. 2011. 'Improved Predictions of Reactor Antineutrino Spectra.' *Physical Review C* 83 (5): 054615.

[978] Gando, A., Y. Gando, K. Ichimura, H. Ikeda, K. Inoue, Y. Kibe, Y. Kishimoto et al. 2011. 'Constraints on $\theta_1 3$ from a Three-flavor Oscillation Analysis of Reactor Antineutrinos at KamLAND.' *Physical Review D* 83 (5): 052002.

[979] Allison, W. W. M., G. J. Alner, D. S. Ayres, W. L. Barrett, C. Bode, P. M. Border, C. B. Brooks et al. 1997. 'Measurement of the Atmospheric Neutrino Flavour Composition in Soudan 2.' *Physics Letters B* 391 (3-4): 491–500.

Kafka, T. 1999. In 'Proceedings of 5th Int. Workshop on Topics in Astroparticle and Underground Physics, Gran Sasso, Italy, Sep. 1997.' *Nuclear Physics B - Proceedings Supplements* 70 (1–3): 204–206.

[980] Wendell, R. 2014. 'Atmospheric Results from Super-Kamiokande. - 2014.' Talk given at the *XXVI International Conference on Neutrino Physics and Astrophysics, Boston, USA, June 2–7*, 2014.

[981] Hosaka, J., K. Ishihara, J. Kameda, Yusuke Koshio, A. Minamino, C. Mitsuda, M. Miura et al. 2006. 'Solar Neutrino Measurements in Super-Kamiokande-I.' *Physical Review D* 73 (11): 112001.

[982] Cravens, J. P., K. Abe, T. Iida, K. Ishihara, J. Kameda, Yusuke Koshio, A. Minamino et al. 2008. 'Solar Neutrino Measurements in Super-Kamiokande-II.' *Physical Review D* 78 (3): 032002.

[983] Abe, K., Y. Hayato, T. Iida, M. Ikeda, C. Ishihara, K. Iyogi, J. Kameda et al. 2011. 'Solar neutrino results in Super-Kamiokande-III.' *Physical Review D* 83 (5): 052010.

[984] Koshio, Y. 2014. 'Solar Results from Super-Kamiokande'. Talk given at the *XXVI International Conference on Neutrino Physics and Astrophysics, Boston, USA, June 2–7*, 2014.

[985] Abdurashitov, J. N., V. N. Gavrin, V. V. Gorbachev, P. P. Gurkina, T. V. Ibragimova, A. V. Kalikhov, N. G. Khairnasov et al. 2009. 'Measurement of the Solar Neutrino Capture Rate with Gallium Metal. III. Results for the 2002–2007 Data-taking Period.' *Physical Review C* 80 (1): 015807.

[986] Bellini, Gianpaolo, J. Benziger, D. Bick, S. Bonetti, G. Bonfini, M. Buizza Avanzini, B. Caccianiga et al. 2011. 'Precision M of the Be 7 Solar Neutrino Interaction Rate in Borexino.' *Physical Review Letters* 107 (14): 141302.

[987] Bellini, G., J. Benziger, S. Bonetti, M. Buizza Avanzini, B. Caccianiga, L. Cadonati, Frank Calaprice et al. 2010. 'Measurement of the Solar B 8 Neutrino Rate with a Liquid Scintillator Target and 3 MeV Energy Threshold in the Borexino Detector.' *Physical Review D* 82 (3): 033006.

[988] Kaether, Florian, Wolfgang Hampel, Gerd Heusser, Juergen Kiko, and Till Kirsten. 2010. 'Reanalysis of the GALLEX Solar Neutrino Flux and Source Experiments.' *Physics Letters B* 685 (1): 47–54.

[989] Aharmim, B., S. N. Ahmed, A. E. Anthony, N. Barros, E. W. Beier, Alain Bellerive, B. Beltran et al. 2013. 'Combined Analysis of all Three Phases of Solar Neutrino Data from the Sudbury Neutrino Observatory.' *Physical Review C* 88 (2): 025501.

[990] An, Feng Peng, A. B. Balantekin, H. R. Band, M. Bishai, S. Blyth, D. Cao, G. F. Cao et al. 2017. 'Measurement of Electron Antineutrino Oscillation Based on 1230 Days of Operation of the Daya Bay Experiment.' *Physical Review D* 95 (7): 072006.

[991] Bak, G., J. H. Choi, H. I. Jang, J. S. Jang, S. H. Jeon, K. K. Joo, Kiwon Ju et al. 2018. 'Measurement of Reactor Antineutrino Oscillation Amplitude and Frequency at RENO.' *Physical Review Letters* 121 (20): 201801.

[992] Abe, Y., S. Appel, T. Abrahao, Helena Almazan, C. Alt, J. C. dos Anjos, J. C. Barriere et al. 2016. 'Measurement of θ_{13} in Double Chooz using Neutron Captures on Hydrogen with Novel Background Rejection Techniques.' *Journal of High Energy Physics* 2016 (1): 163.

[993] Abe, Y., J. C. Dos Anjos, J. C. Barriere, E. Baussan, I. Bekman, M. Bergevin, T. J. C. Bezerra et al. 2014. 'Improved Measurements of the Neutrino Mixing Angle θ_{13} with the Double Chooz Detector.' *Journal of High Energy Physics* 2014 (10): 86.

[994] Ahn, M. H., E. Aliu, S. Andringa, S. Aoki, Y. Aoyama, J. Argyriades, K. Asakura et al. 2006. 'Measurement of Neutrino Oscillation by the K2K Experiment.' *Physical Review D* 74 (7): 072003.

[995] Aguilar-Arevalo, A. A., C. E. Anderson, S. J. Brice, B. C. Brown, L. Bugel, J. M. Conrad, R. Dharmapalan et al. 2010. 'Event Excess in the MiniBooNE Search for $\nu_\mu \rightarrow \nu_e$ Oscillations.' *Physical Review Letters* 105 (18): 181801.

Sorel, M. 2008. 'MiniBooNE: First Results on the Muon-to-electron Neutrino Oscillation Search.' In *Journal of Physics: Conference Series* 110 (8): 082020. IOP Publishing.

[996] Popov, B. A. 2004. 'Final Results on the Search for $\nu_\mu \rightarrow \nu_e$ Oscillations in the NOMAD Experiment.' *Physics of Atomic Nuclei* 67 (11): 1942–1947.

[997] Eskut, E., A. Kayis-Topaksu, G. Onengüt, R. Van Dantzig, M. De Jong, J. Konijn, O. Melzer et al. 2001. 'New Results from a Search for $\nu\mu \rightarrow \nu\tau$ and $\nu\upsilon \rightarrow \nu\tau$ oscillation.' *Physics Letters B* 497 (1-2): 8–22.

[998] Adamson, P., I. Anghel, C. Backhouse, G. Barr, M. Bishai, A. Blake, G. J. Bock et al. 2013. 'Measurement of Neutrino and Antineutrino Oscillations using Beam and Atmospheric Data in MINOS.' *Physical Review Letters* 110 (25): 251801.

[999] Adamson, P., I. Anghel, C. Backhouse, G. Barr, M. Bishai, A. Blake, G. J. Bock et al. 2013. 'Electron Neutrino and Antineutrino Appearance in the Full MINOS Data Sample.' *Physical Review Letters* 110 (17): 171801.

[1000] Abe, Katsushige, R. Akutsu, A. Ali, J. Amey, C. Andreopoulos, L. Anthony, M. Antonova et al. 2018. 'Search for C P Violation in Neutrino and Antineutrino Oscillations by the T2K Experiment with 2.2×10^{21} Protons on Target.' *Physical Review Letters* 121 (17): 171802.

[1001] Abe, K., J. Amey, C. Andreopoulos, M. Antonova, S. Aoki, A. Ariga, Y. Ashida et al. 2017. 'Measurement of Neutrino and Antineutrino Oscillations by the T2K Experiment Including a New Additional Sample of ν_e Interactions at the Far Detector.' *Physical Review D* 96 (9): 092006.

[1002] Adamson, P., L. Aliaga, D. Ambrose, Nikolay Anfimov, A. Antoshkin, E. Arrieta-Diaz, K. Augsten et al. 2017. 'Constraints on Oscillation Parameters from ν_e Appearance and ν_μ Disappearance in NOvA.' *Physical Review Letters* 118 (23): 231801.

[1003] Acero, M. A., P. Adamson, L. Aliaga, T. Alion, V. Allakhverdian, N. Anfimov, A. Antoshkin et al. 2018. 'New Constraints on Oscillation Parameters from ν_e Appearance and ν_μ Disappearance in the NOvA Experiment.' *Physical Review D* 98 (3): 032012.

[1004] Acero, M. A. et al. [NOvA Collaboration]. 2019. 'First Measurement of Neutrino Oscillation Parameters using Neutrinos and Antineutrinos by NOvA.' *Phys. Rev. Lett.* 123 (15): 151803. arXiv:1906.04907.

[1005] OPERA collaboration, N. Agafonova, A. Aleksandrov, A. Anokhina, S. Aoki, A. Ariga, T. Ariga et al. 2014. 'Observation of Tau Neutrino Appearance in the CNGS Beam with the OPERA Experiment.' *Progress of Theoretical and Experimental Physics* 2014 (10): 101C01.

[1006] Agafonova, N., A. Aleksandrov, A. Anokhina, S. Aoki, A. Ariga, T. Ariga, D. Bender et al. 2015. 'Discovery of τ Neutrino Appearance in the CNGS Neutrino Beam with the OPERA Experiment.' *Physical Review Letters* 115 (12): 121802.

[1007] Agafonova, N., A. Aleksandrov, A. Anokhina, S. Aoki, Akitaka Ariga, Tomoko Ariga, T. Asada et al. 2013. 'New Results on $\nu_\mu \rightarrow \nu_\tau$ Appearance with the OPERA Experiment in the CNGS beam.' *Journal of High Energy Physics* 2013 (11): 36.

[1008] Agafonova, N. et al. [OPERA Collaboration]. 2019. *SciPost Phys. Proc.* 1: 028. doi: 10.21468/SciPostPhysProc.1.028.

[1009] Mikheyev, S. P., and A. Yu Smirnov. 1986. 'Resonant Amplification of ν Oscillations in Matter and Solar-neutrino Spectroscopy.' *Il Nuovo Cimento C* 9 (1): 17–26.

[1010] Wolfenstein, Lincoln. 1978. 'Neutrino Oscillations in Matter.' *Physical Review D* 17 (9): 2369.

[1011] Cabibbo, Nicola. 1978. 'Time Reversal Violation in Neutrino Oscillation.' *Physics Letters B* 72 (3): 333–335.

[1012] Bilenky, S. M., J. Hosek, and S. T. Petcov. 1980. 'On Oscillations of Neutrinos with Dirac and Majorana Masses.' *Phys. Lett. B* 94: 495–498. doi: 10.1016/0370-2693(80)90927-2.

[1013] Barger, V., K. Whisnant, and R. J. N. Phillips. 1980. 'CP Nonconservation in Three-neutrino Oscillations.' *Physical Review Letters* 45 (26): 2084.

[1014] Jarlskog, Cecilia. 1985. 'Commutator of the Quark Mass Matrices in the Standard Electroweak Model and a Measure of Maximal CP Nonconservation.' *Physical Review Letters* 55 (10): 1039.

[1015] Gonzalez-Garcia, M. C., Michele Maltoni, and Thomas Schwetz. 2014. 'Updated fit to three neutrino mixing: status of leptonic CP violation.' *Journal of High Energy Physics* 2014 (11): 52.

[1016] http://backreaction.blogspot.co.uk/2009/09/light-bulbs-and-solar-energy-production.html.

[1017] https://commons.wikimedia.org/wiki/File:RadialDensityPREM.jpg.

[1018] Cleveland, Bruce T., Timothy Daily, Raymond Davis Jr, James R. Distel, Kenneth Lande, C. K. Lee, Paul S. Wildenhain, and Jack Ullman. 1998. 'Measurement of the Solar Electron Neutrino Flux with the Homestake Chlorine Detector.' *The Astrophysical Journal* 496 (1): 505.

[1019] MACRO collaboration. 2004. 'Measurements of Atmospheric Muon Neutrino Oscillations, Global Analysis of the Data Collected with MACRO Detector.' *The European Physical Journal C-Particles and Fields* 36 (3): 323–339.

[1020] Aharmim, B., S. N. Ahmed, T. C. Andersen, A. E. Anthony, N. Barros, E. W. Beier, Alain Bellerive et al. 2009. 'Measurement of the Cosmic Ray and Neutrino-induced Muon Flux at the Sudbury Neutrino Observatory.' *Physical Review D* 80 (1): 012001.

[1021] Adamson, P., I. Anghel, A. Aurisano, G. Barr, M. Bishai, A. Blake, G. J. Bock et al. 2014. 'Combined Analysis of ν_μ Disappearance and $\nu_\mu \to \nu_e$ Appearance in MINOS using Accelerator and Atmospheric Neutrinos.' *Physical Review Letters* 112 (19): 191801.

[1022] Terliuk, Andrii. 2019. 'Atmospheric Neutrino Oscillations with IceCube.' *PoS*: 007.

[1023] Albert, A., Michel André, Marco Anghinolfi, Gisela Anton, M. Ardid, J-J. Aubert, J. Aublin et al. 2019. 'Measuring the Atmospheric Neutrino Oscillation Parameters and Constraining the 3 + 1 Neutrino Model with Ten Years of ANTARES Data.' *Journal of High Energy Physics* 2019 (6): 113.

[1024] Abe, K., J. Adam, H. Aihara, T. Akiri, C. Andreopoulos, S. Aoki, A. Ariga et al. 2014. 'Precise Measurement of the Neutrino Mixing Parameter θ_{23} from Muon Neutrino Disappearance in an off-axis Beam.' *Physical Review Letters* 112 (18): 181801.

[1025] Sanchez, M. 2015. 'Results and Prosects from the NO*nu*A Experiment.' Talk given at the *XVII International Workshop on Neutrino Factories and Future Neutrino Facilities, Rio de Janeiro, Brazil, August 10–15, 2015.*

[1026] Abe, K., J. Adam, H. Aihara, T. Akiri, C. Andreopoulos, S. Aoki, Akitaka Ariga et al. 2014. 'Observation of Electron Neutrino Appearance in a Muon Neutrino Beam.' *Physical Review Letters* 112 (6): 061802.

[1027] Aguilar-Arevalo, A. A., B. C. Brown, L. Bugel, G. Cheng, E. D. Church, J. M. Conrad, R. Dharmapalan et al. 2012. 'A Combined $\nu_\mu \to \nu_e$ and $\bar{\nu}_\mu \to \bar{\nu}_e$ Oscillation Analysis of the MiniBooNE Excesses.' *arXiv preprint arXiv:1207.4809.*

[1028] Athanassopoulos, C., L. B. Auerbach, D. A. Bauer, R. D. Bolton, B. Boyd, R. L. Burman, D. O. Caldwell et al. 1995. 'Candidate Events in a Search for $\bar{\nu}_\mu \to \bar{\nu}_e$ Oscillations.' *Physical Review Letters* 75 (14): 2650.

Athanassopoulos, C., L. B. Auerbach, R. L. Burman, I. Cohen, D. O. Caldwell, B. D. Dieterle, J. B. Donahue et al. 1996. 'Evidence for $\nu_\mu \to \nu_e$ Oscillations from the LSND Experiment at the Los Alamos Meson Physics Facility.' *Physical Review Letters* 77 (15): 3082.

Athanassopoulos, C., L. B. Auerbach, R. L. Burman, D. O. Caldwell, E. D. Church, I. Cohen, J. B. Donahue et al. 1998. 'Results on $\nu_\mu \to \nu_e$ Neutrino Oscillations from the LSND Experiment.' *Physical Review Letters* 81 (9): 1774.

Athanassopoulos, C., L. B. Auerbach, R. L. Burman, D. O. Caldwell, E. D. Church, I. Cohen, J. B. Donahue et al. 1998. 'Results on $\nu_\mu \to \nu_e$ Oscillations from Pion Decay in Flight Neutrinos.' *Physical Review C* 58 (4): 2489.

Aguilar, A., L. B. Auerbach, R. L. Burman, D. O. Caldwell, E. D. Church, A. K. Cochran, J. B. Donahue et al. 2001. 'Evidence for Neutrino Oscillations from the Observation of ν_e Appearance in a ν_μ Beam.' *Physical Review D* 64 (11): 112007.

[1029] Abe, Y., Christoph Aberle, J. C. Dos Anjos, J. C. Barriere, M. Bergevin, A. Bernstein, T. J. C. Bezerra et al. 2012. 'Reactor ν_e disappearance in the Double Chooz experiment.' *Physical Review D* 86 (5): 052008.

[1030] Zhang, C. 2014. 'Recent Results from Daya Bay, Talk given at the XXVI International Conference on Neutrino Physics and Astrophysics.' *Boston, USA*: 2–7.

[1031] Choi, J. H., W. Q. Choi, Y. Choi, H. I. Jang, J. S. Jang, E. J. Jeon, K. K. Joo et al. 2016. 'Observation of Energy and Baseline Dependent Reactor Antineutrino Disappearance in the RENO Experiment.' *Physical review letters* 116 (21): 211801.

[1032] Giunti, Carlo, and Chung W. Kim. 2007. *Fundamentals of Neutrino Physics and Astrophysics*. Oxford University Press.

[1033] Bilenky, Samoil. 2010. *Introduction to the Physics of Massive and Mixed Neutrinos*. Vol. 817. Springer.

[1034] Astier, P., D. Autiero, A. Baldisseri, M. Baldo-Ceolin, M. Banner, G. Bassompierre, K. Benslama et al. 2001. 'Final NOMAD Results on $\nu_\mu \to \nu_\mu$ and $\nu_e \to \nu_\tau$ Oscillations

Including a New Search for ν_τ Appearance using Hadronic τ Decays.' *Nuclear Physics B* 611 (1-3): 3–39.

[1035] Adamson, P., D. J. Auty, D. S. Ayres, C. Backhouse, G. Barr, M. Bishai, A. Blake et al. 2011. 'Active to Sterile Neutrino Mixing Limits from Neutral-current Interactions in MINOS.' *Physical Review Letters* 107 (1): 011802.

[1036] Abe, K., Y. Haga, Y. Hayato, M. Ikeda, K. Iyogi, J. Kameda, Y. Kishimoto et al. 2015. 'Limits on Sterile Neutrino Mixing using Atmospheric Neutrinos in Super-Kamiokande.' *Physical Review D* 91 (5): 052019.

[1037] Aartsen, M. G., M. Ackermann, J. Adams, J. A. Aguilar, M. Ahlers, M. Ahrens, I. Al Samarai et al. 2017. 'Search for Sterile Neutrino Mixing using Three Years of IceCube DeepCore Data.' *Physical Review D* 95 (11): 112002.

[1038] Bellini, G. et al. [Borexino Collaboration]. 2013. *Physical Review D* 88 (7): 072010.

[1039] Cherry, John F., and Ian M. Shoemaker. 2019. 'Sterile Neutrino Origin for the Upward Directed Cosmic Ray Showers Detected by ANITA.' *Physical Review D* 99 (6): 063016.

[1040] Hamann, Jan, Steen Hannestad, Georg G. Raffelt, and Yvonne YY Wong. 2011. 'Sterile Neutrinos with eV Masses in Cosmology—How Disfavoured Exactly?.' *Journal of Cosmology and Astroparticle Physics* 2011 (9): 034.

[1041] Abazajian, Kevork N., M. A. Acero, S. K. Agarwalla, A. A. Aguilar-Arevalo, C. H. Albright, S. Antusch, C. A. Arguelles et al. 2012. 'Light Sterile Neutrinos: A White Paper.' *arXiv preprint arXiv:1204.5379.*

[1042] Alekseev, I., V. Belov, V. Brudanin, M. Danilov, V. Egorov, D. Filosofov, M. Fomina et al. 2018. 'Search for Sterile Neutrinos at the DANSS Experiment.' *Physics Letters B* 787: 56–63.

[1043] Almazán, H. et al. [STEREO Collaboration]. 2018. *Physical Review Letters* 121: 161801.

[1044] Serebrov, A. P. et al. [NEUTRINO-4 Collaboration]. 2019. *Pisma Zh. Eksp. Teor. Fiz.* 109 (4): 209–218. [*JETP Lett.* 109 (4) 213–221].

[1045] http://meetings.aps.org/Meeting/APR19/Session/Z14.9.

[1046] Allen, R., F. T. Avignone, J. Boissevain, Y. Efremenko, M. Elnimr, T. Gabriel, F. G. Garcia et al. 2013. 'The OscSNS White Paper.' *arXiv preprint arXiv:1307.7097.*

[1047] Wallerstein, George, Icko Iben, Peter Parker, Ann Merchant Boesgaard, Gerald M. Hale, Arthur E. Champagne, Charles A. Barnes et al. 1997. 'Synthesis of the Elements in Stars: Forty Years of Progress.' *Reviews of Modern Physics* 69 (4): 995.

[1048] https://en.wikipedia.org/wiki/Nucleosynthesis/.

[1049] http://www.int.washington.edu/PROGRAMS/14-2b/.

[1050] Langanke, K., G. Martinez-Pinedo, and A. Sieverding. 2018. 'Neutrino nucleosynthesis: An overview.' *AAPPS Bulletin* 28 (6): 41–48. doi: 10.22661/AAPPSBL.2018.28.6.41.

[1051] Hubble, Edwin P. 1926. 'Extragalactic Nebulae.' *The Astrophysical Journal* 64.

Hubble, Edwin P. 1926. *Astrophys. J.* 63: 236.

Hubble, Edwin P. 1929. 'A Spiral Nebula as a Stellar System, Messier 31.' *The Astrophysical Journal* 69.

[1052] Cowan, John J., and Friedrich-Karl Thielemann. 2004. 'R-process Nucleosynthesis in Supernovae.' *Physics Today* 57 (10): 47–54.

[1053] https://cnx.org/contents/bIMtPPGL@7/Evolution-of-the-Early-Universe.

[1054] Penzias, Arno A., and Robert Woodrow Wilson. 1965. 'A Measurement of Excess antenna Temperature at 4080 Mc/s.' *The Astrophysical Journal* 142: 419–421.

[1055] http://homepages.spa.umn.edu/ llrw/a1001 s02/HRdiag.html.

[1056] Chandrasekhar, Subrahmanyan. 1931. 'The Maximum mass of Ideal White Dwarfs.' *The Astrophysical Journal* 74: 81.

[1057] http://cse.ssl.berkeley.edu/bmendez/ay10/2000/cycle/planetarynebula.html.

[1058] Margalit, B., and B. D. Metzger. 2017. *Astrophys. J.* 850: L19.

[1059] https://phys.org/news/2010-05-supernova-universal-mysteries.html.

[1060] Abbott, Benjamin P., Rich Abbott, T. D. Abbott, Fausto Acernese, Kendall Ackley, Carl Adams, Thomas Adams et al. 2017. 'GW170817: Observation of Gravitational Waves from a Binary Neutron Star Inspiral.' *Physical Review Letters* 119 (16): 161101.

[1061] https://www.insidescience.org/news/gravitational-waves-throw-light-neutron-star-mergers

[1062] Goeppert-Mayer, Maria. 1935. 'Double Beta-disintegration.' *Physical Review* 48 (6): 512.

[1063] Furry, W. H. 1939. 'On Transition Probabilities in Double Beta-disintegration.' *Physical Review* 56 (12): 1184.

[1064] Barabash, A. S. 2010. 'Precise Half-life Values for Two-neutrino Double-β Decay.' *Physical Review C* 81 (3): 035501.

[1065] Racah, Giulio. 1937. 'Symmetry Between Particles and Anti-particles.' *Nuovo cimento* 14: 322–328.

[1066] Barabash, A. S. 2011. 'Experiment Double Beta Decay: Historical Review of 75 Years of Research.' *Physics of Atomic Nuclei* 74 (4): 603–613.

[1067] Kai Zuber, Neutrino Physics (II-edition), CRC Press (2012).

[1068] Stephenson Jr, WC Haxtonand GJ. 1984. 'Double Beta Decay.' *Prog. Part. Nucl. Phys* 12: 409–479.

[1069] Kotani, Tsuneyuki, and Eiichi Takasugi. 1985. 'Double Beta Decay and Majorana Neutrino.' *Progress of Theoretical Physics Supplement* 83: 1–175.

[1070] Dolinski, M. J., A. W. P. Poon, and W. Rodejohann. 2019. 'Neutrinoless Double-Beta Decay: Status and Prospects.' *Ann. Rev. Nucl. Part. Sci.* 69: 219–251. arXiv:1902.04097 [nucl-ex].

[1071] Engel, Jonathan, and Javier Menéndez. 2017. 'Status and Future of Nuclear Matrix Elements for Neutrinoless Double-beta Decay: A Review.' *Reports on Progress in Physics* 80 (4): 046301.

[1072] Vergados, John D., Hiroyasu Ejiri, and F. Simkovic. 2016. 'Neutrinoless Double Beta Decay and Neutrino Mass.' *International Journal of Modern Physics E* 25 (11): 1630007.

[1073] Deppisch, Frank F., Martin Hirsch, and Heinrich Päs. 2012. 'Neutrinoless Double-beta Decay and Physics Beyond the Standard Model.' *Journal of Physics G: Nuclear and Particle Physics* 39 (12): 124007.

[1074] Umehara, S., T. Kishimoto, I. Ogawa, R. Hazama, H. Miyawaki, S. Yoshida, K. Matsuoka et al. 2008. 'Neutrino-less Double-β Decay of ^{48}Ca Studied by CaF$_2$ (Eu) Scintillators.' *Physical Review C* 78 (5): 058501.

[1075] Agostini, M., A. M. Bakalyarov, M. Balata, I. Barabanov, L. Baudis, C. Bauer, E. Bellotti et al. 2018. 'Improved Limit on Neutrinoless Double-β Decay of ^{76}Ge from GERDA Phase II.' *Physical Review Letters* 120 (13): 132503.

[1076] Aalseth, C. E., N. Abgrall, E. Aguayo, S. I. Alvis, M. Amman, I. J. Arnquist, F. T. Avignone III et al. 2018. 'Search for Neutrinoless Double-β Decay in Ge 76 with the Majorana Demonstrator.' *Physical Review Letters* 120 (13): 132502.

[1077] Argyriades, J., R. Arnold, C. Augier, J. Baker, A. S. Barabash, A. Basharina-Freshville, M. Bongrand et al. 2010. 'Measurement of the Two Neutrino Double Beta Decay Half-life of Zr-96 with the NEMO-3 detector.' *Nuclear Physics A* 847 (3-4): 168-179.

[1078] Arnold, R., C. Augier, J. D. Baker, A. S. Barabash, A. Basharina-Freshville, S. Blondel, S. Blot et al. 2015. 'Results of the Search for Neutrinoless Double-β Decay in ^{100}Mo with the NEMO-3 Experiment.' *Physical Review D* 92 (7): 072011.

[1079] Arnold, R., C. Augier, J. D. Baker, A. S. Barabash, A. Basharina-Freshville, S. Blondel, S. Blot et al. 2017. 'Measurement of the $2\nu\beta\beta$ Decay Half-life and Search for the $0\nu\beta\beta$ Decay of ^{116}Cd with the NEMO-3 Detector.' *Physical Review D* 95 (1): 012007.

[1080] Arnaboldi, C., C. Brofferio, C. Bucci, S. Capelli, O. Cremonesi, E. Fiorini, A. Giuliani et al. 2003. 'A Calorimetric Search on Double Beta Decay of ^{130}Te.' *Physics Letters B* 557 (3-4): 167–175.

[1081] Alduino, C., F. Alessandria, K. Alfonso, E. Andreotti, C. Arnaboldi, F. T. Avignone III, O. Azzolini et al. 2018. 'First Results from CUORE: A Search for Lepton Number Violation via $0\nu\beta\beta$ Decay of ^{130}Te.' *Physical review letters* 120 (13): 132501.

[1082] Gando, A., Y. Gando, T. Hachiya, A. Hayashi, S. Hayashida, H. Ikeda, K. Inoue et al. 2016. 'Publisher's Note: Search for Majorana Neutrinos Near the Inverted Mass Hierarchy Region with KamLAND-Zen [*Physical Review Letters* 117: 082503].' [Addendum: *Physical Review Letters* 117 (10): 109903.]

[1083] Albert, J. B., G. Anton, I. Badhrees, P. S. Barbeau, R. Bayerlein, D. Beck, V. Belov et al. 2018. 'Search for Neutrinoless Double-beta Decay with the Upgraded EXO-200 Detector.' *Physical Review Letters* 120 (7): 072701.

[1084] Arnold, R., C. Augier, J. D. Baker, A. S. Barabash, A. Basharina-Freshville, S. Blondel, S. Blot et al. 2016. 'Measurement of the $2\nu\beta\beta$ Decay Half-life of ^{150}Nd and a search for $0\nu\beta\beta$ Decay Processes with the Full Exposure from the NEMO-3 Detector.' *Physical Review D* 94 (7): 072003.

[1085] Vergados, J. D., H. Ejiri, and F. Šimkovic. 2012. 'Theory of Neutrinoless Double-beta Decay.' *Reports on Progress in Physics* 75 (10): 106301.

[1086] Elliott, Steven R., and Petr Vogel. 2002. 'Double Beta Decay.' *Annual Review of Nuclear and Particle Science* 52 (1): 115–151.

[1087] Avignone III, Frank T., Steven R. Elliott, and Jonathan Engel. 2008. 'Double Beta Decay, Majorana Neutrinos, and Neutrino Mass.' *Reviews of Modern Physics* 80 (2): 481.

[1088] Ejiri, Hiroyasu. 2005. 'Double Beta Decays and Neutrino Masses.' *Journal of the Physical Society of Japan* 74 (8): 2101—2127.

[1089] Giuliani, A., J. J. Cadenas, S. Pascoli, E. Previtali, R. Saakyan, K. Schaeffner, and S. Schoenert. 2019. 'Double Beta Decay APPEC Committee Report.' *arXiv preprint arXiv:1910.04688.*

[1090] Kuno, Yoshitaka, and Yasuhiro Okada. 2001. 'Muon Decay and Physics Beyond the Standard Model.' *Reviews of Modern Physics* 73 (1): 151.

[1091] Ruggier, C., and G. Valencia. 2019. Contributions at 2019 Conference on KAONS.

[1092] Drechsel, D., and M. M. Giannini. 1997. 'Electroproduction of Hyperons.' *Physics Letters B* 397 (3-4): 311–316.

[1093] Jin, Xuemin, and Robert L. Jaffe. 1997. 'Weak Hyperon Production in ep Scattering.' *Physical Review D* 55 (9): 5636.

[1094] Cirigliano, Vincenzo, Martín González-Alonso, and Michael L. Graesser. 2013. 'Non-standard Charged Current Interactions: Beta Decays versus the LHC.' *Journal of High Energy Physics* 2013 (2): 46.

[1095] Cirigliano, Vincenzo, Susan Gardner, and Barry R. Holstein. 2013. 'Beta decays and non-standard interactions in the LHC era.' *Progress in Particle and Nuclear Physics* 71: 93–118.

[1096] Vos, K. K., H. W. Wilschut, and R. G. E. Timmermans. 2015. 'Symmetry Violations in Nuclear and Neutron β Decay.' *Reviews of Modern Physics* 87 (4): 1483.

Index